Lecture Notes in Mechanical Engineering

Lecture Notes in Mechanical Engineering (LNME) publishes the latest developments in Mechanical Engineering—quickly, informally and with high quality. Original research reported in proceedings and post-proceedings represents the core of LNME. Volumes published in LNME embrace all aspects, subfields and new challenges of mechanical engineering. Topics in the series include:

- Engineering Design
- Machinery and Machine Elements
- Mechanical Structures and Stress Analysis
- Automotive Engineering
- Engine Technology
- Aerospace Technology and Astronautics
- Nanotechnology and Microengineering
- Control, Robotics, Mechatronics
- MEMS
- Theoretical and Applied Mechanics
- Dynamical Systems, Control
- Fluid Mechanics
- Engineering Thermodynamics, Heat and Mass Transfer
- Manufacturing
- Precision Engineering, Instrumentation, Measurement
- Materials Engineering
- Tribology and Surface Technology

To submit a proposal or request further information, please contact the Springer Editor of your location:

China: Dr. Mengchu Huang at mengchu.huang@springer.com
India: Priya Vyas at priya.vyas@springer.com
Rest of Asia, Australia, New Zealand: Swati Meherishi at swati.meherishi@springer.com
All other countries: Dr. Leontina Di Cecco at Leontina.dicecco@springer.com

To submit a proposal for a monograph, please check our Springer Tracts in Mechanical Engineering at http://www.springer.com/series/11693 or contact Leontina.dicecco@springer.com

Indexed by SCOPUS. The books of the series are submitted for indexing to Web of Science.

More information about this series at http://www.springer.com/series/11236

S. Vijayan · Nachiappan Subramanian ·
K. Sankaranarayanasamy
Editors

Trends in Manufacturing and Engineering Management

Select Proceedings of ICMechD 2019

Springer

Editors
S. Vijayan
Sri Sivasubramaniya Nadar College of Engineering
Chennai, Tamil Nadu, India

Nachiappan Subramanian
University of Sussex
Brighton, UK

K. Sankaranarayanasamy
National Institute of Technology Puducherry
Puducherry, India

ISSN 2195-4356 ISSN 2195-4364 (electronic)
Lecture Notes in Mechanical Engineering
ISBN 978-981-15-4744-7 ISBN 978-981-15-4745-4 (eBook)
https://doi.org/10.1007/978-981-15-4745-4

This Springer imprint is published by the registered company Springer Nature Singapore Pte Ltd.
The registered company address is: 152 Beach Road, #21-01/04 Gateway East, Singapore 189721, Singapore

Preface

We are delighted to publish the selected papers presented in the Second International Conference on Mechanical Engineering Design 2019 (ICMechD 2019) held at Sri Sivasubramaniya Nadar College of Engineering, Chennai, India, during April 25–26, 2019.

In this book, around 100 quality papers in the areas of advanced manufacturing technology, development and characterization of metal matrix composite, structures and processing of polymer matrix composite, material processing technology, material joining technology, production and operation management, optimization techniques and energy engineering are included for publication.

We are happy to put together this collection of thoughtful papers on the theme of "Trends in Manufacturing and Engineering Management." We are sure that this book will help to nurture knowledge among the research society.

We are grateful to the reviewers for their valuable suggestions for improving the quality of the papers. Also, we thank the session chairs and organizing committee members for their steadfast support and suggestions.

Chennai, India — S. Vijayan
Brighton, UK — Nachiappan Subramanian
Puducherry, India — K. Sankaranarayanasamy

Contents

Development and Characterization of Metal Matrix Composite

About the Editors

Dr. S. Vijayan, is currently working as Associate Professor in Department of Mechanical Engineering at Sri Sivasubramaniya Nadar College of Engineering has 20 years of teaching experience. He has received his Bachelor of Engineering (B.E.) in Mechanical Engineering from the Madurai Kamarajar University in 1997, completed his post graduate Masters of Engineering (M.E.) in Industrial Engineering from Thigarajar College of Engineering in 1998 and Doctorate Ph.D. from the Anna University in the year 2010. To his credit he has completed two research sponsored projects to the tune of 30 lakh funded by Naval research Board, DRDO and SSN Trust. He has published 25 International journals papers to his credit. He has published two books for VDM-Verlag and Lambert, respectively. He served in administrative responsibilities such as Chairman Valuation in Anna University and Additional Controller of Examination in SSN Institutions.

Dr. Nachiappan Subramanian is currently working as Professor in Operations and Logistics Management & Supply Chains, University of Business School, University of Sussex, United Kingdom. Formerly he served as Associate Professor in Nottingham University Business School China. In total, he has 20 years of experience including 18 years in academia and 2 years in a consulting company. He has received his Bachelor of Engineering (B.E.) in Mechanical Engineering from the Madurai Kamarajar University in 1995, completed his post graduate Masters of Engineering (M.E.) in Production Engineering from Thigarajar College of Engineering in 1998 and Doctorate Ph.D. from the Madurai Kamarajar University in the year 2006. He has completed his postdoctoral research from Nottingham University and Royal Melbourne Institute of Technology University. To his credit, he has completed several sponsored projects to the tune of £ 438,370. To date, he has published 89 peer-reviewed refereed research papers. His research interest areas are in sustainable supply chains (environmental and social/humanitarian issues), risk and resilience, technology enabled operations and marketing interface, and in performance measurement.

Dr. K. Sankaranarayanasamy is serving currently as Director of NIT Pondicherry, formerly a Professor of Mechanical Engineering at National Institute of Technology Tiruchirappalli, received his B.E. Mechanical Engineering (Hons) in 1981 from PSG College of Technology, India. He received his M.Tech Production Engineering in 1983 and Ph.D. in 1989 from Indian Institute of Technology Madras, India. His research targets finite element simulation of welding processes, Ergonomics Study and Gear Design. He has authored more than 75 papers in international and national journals and has attended conferences in Slovak Republic, China, America and Austria. To his credit, he has completed several sponsored projects (total worth of Rs. 60 lakhs) funded by BHEL, Trichy, NLC, Neyveli and DST. He served in administrative responsibilities such as Head of the Department, Dean and Director of NIT.

Advanced Manufacturing Technology

Effect of Process Parameters in Electric Discharge Machining of AZ31 Magnesium Alloy

M. Somasundaram and J. Pradeep Kumar

Abstract In modern-day scenario, AZ31 magnesium alloy is considered as a suitable alternative to iron and aluminum in a wide variety of medical, aerospace and automobile applications due to high strength-to-weight ratio and other excellent physical properties. In all the applications, the geometrical accuracy of the products plays a vital role. Electrical discharge machining (EDM) has emerged as a prominent manufacturing process to achieve geometrical accuracy as per the requirements and standards. A hole of diameter 10 mm in a flat plate made of AZ31 alloy is identified as the part feature for investigation in this study. The aim of this work is to optimize and investigate the effect of both qualitative and quantitative process parameters such as discharge current (I), pulse-on time (T_{ON}), pulse-off time (T_{OFF}), electrode material (M) on critical-to-quality geometric features such as overcut (OC), tapercut (TC), circularity (CIR) and cylindricity (CYL). The experiments are carried out based on Taguchi's L16 orthogonal array. Pulse-on time and discharge current are found to have a significant effect on the geometrical accuracy when compared with other process parameters. Regression models are developed for prediction of response parameters, and the results predicted by the models are found to correlate significantly with the results from experiments.

Keywords Electrical discharge machining (EDM) · AZ31 magnesium alloy · Discharge current (I) · Pulse-on time (T_{ON}) · Pulse-off time (T_{OFF}) · Overcut (OC) · Tapercut (TC) · Circularity (CIR) · Cylindricity (CYL)

M. Somasundaram (✉) · J. Pradeep Kumar
Department of Production Engineering, PSG College of Technology, Coimbatore, India
e-mail: somaattitude@gmail.com

J. Pradeep Kumar
e-mail: jpk.prod@psgtech.ac.in

S. Vijayan et al. (eds.), *Trends in Manufacturing and Engineering Management*, Lecture Notes in Mechanical Engineering,
https://doi.org/10.1007/978-981-15-4745-4_1

1 Introduction

The magnesium and its alloys have been established as the best alternative to iron and aluminum in a wide variety of applications due to high strength-to-weight ratio and other excellent physical properties [1]. However, machining magnesium alloys using conventional methods cause a built-up edge and chatter. The most important precaution needed to bear in mind, while machining magnesium alloy is that the formation of fine chips and dust is highly flammable [2]. Due to the development of new difficulties to machine materials (carbide, composite materials, etc.) and complex geometrical shapes of engineering components, the available traditional finishing processes alone are incapable of producing required surface finish and other geometrical features. Even if these processes can be used, they require expensive equipments and large labor, hence making them economically incompetent. Developments in advanced machining processes in the last few decades have attributed to the relaxation of limitations of tool hardness requirement or/and restricted predefined relative motion of cutting edges with respect to the workpiece surface. Use of alternative source of energy like electrical, chemical, mechanical or thermal to assist or perform machining relaxed the constraint of using cutting tool harder than the workpiece. Electric discharge machining (EDM), electrochemical machining (ECM), ultrasonic machining (USM), abrasive jet machining (AJM), laser beam machining (LBM), etc. are a few instances of such techniques. Being the most versatile of all the non-conventional machining, electric discharge machining (EDM) process becomes viable for the machining of these magnesium alloys materials [3–5]. In all applications, geometrical features of machined parts are more important. This work is carried out to study the effect of process parameters while machining a hole feature of diameter 10 mm. Existing research works carried out using AZ31 magnesium alloy focus on the measurement of surface roughness [6]. But the present work focuses on geometrical features of the holes which is essential to make a proper hole on AZ31 magnesium alloy while machining using EDM.

2 Details of Experiments

In this work, magnesium AZ31 alloy is chosen as a workpiece material. The workpiece is in the form of a flat plate with length (50 mm), width (30 mm) and thickness (6 mm). Tool materials such as copper, brass and graphite are chosen as an electrode. The electrode in the form of a rod with a diameter of 10 mm and a length of 25 mm is machined and used to carry out experiments. The workpiece material (anode) used is AZ31 magnesium alloy. The tool (cathode) materials such as copper, brass and graphite are used. The experimental trials are conducted using ALTRA ZNC ORB 5530 EDM machine, based on Taguchi's L16 orthogonal array. Figure 1 shows the machined workpiece and their corresponding electrode. The process parameters such as discharge current (I_P), pulse-on time (T_{ON}), pulse-off time (T_{OFF}) and tool

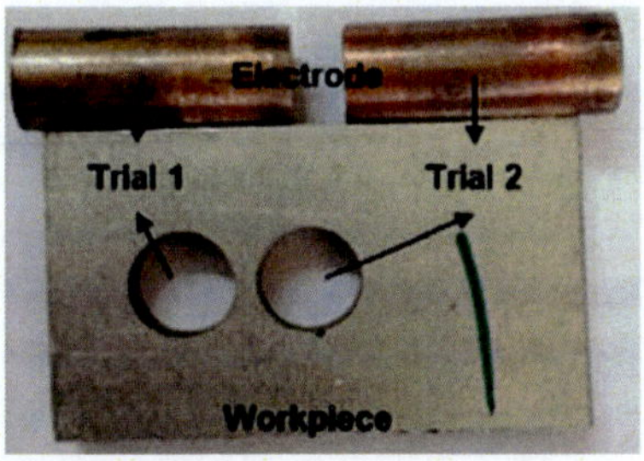

Fig. 1 Machined workpiece

Table 1 Control factors and their levels

Factors	Level 1 (−1)	Level 2 (0)	Level 3 (1)	Level 4 (2)
I_P (Amps)	3	6	9	12
T_{ON} (μs)	10	20	30	40
T_{OFF} (μs)	5	6	7	8
M	Copper (Cu)	Brass (Br)	Graphite (Gr)	Copper (Cu)

material (M) are selected as control factors, and overcut (OC), tapercut (TC), circularity (CIR) and cylindricity (CYL) are considered as response variables. Table 1 shows the control factors and their levels. Overcut is equal to half the difference of the diameter of the hole to the diameter of the tool. The diameter of the hole and tool is measured using a digital vernier caliper. Tapercut is the difference between hole diameter of top and bottom surface of the plate. It was measured with the help of a profile projector. Circularity describes how close an object should be to a true circle. Cylindricity is used when the hole part features must have good circularity and straightness. Circularity applies only to cross sections, where the cylindricity applies simultaneously to the entire surface. Both circularity and cylindricity were measured using Mitutoyo Coordinate Measuring Machine (CMM).

3 Result and Discussion

Two trials are carried out for each combination of process parameters, and the average of two values is presented in Table 2.

3.1 Effect of Process Parameters on Response Parameters

Analysis of variance (ANOVA) is used to study the effect of process parameters on response variables. ANOVA is carried out at a confidence level of 95%. Since it is always desirable to achieve the lowest overcut, tapercut, circularity and cylindricity,

Table 2 Design matrix and their output response parameters

S. No.	I_P (Amps)	T_{ON} (μs)	T_{OFF} (μs)	*M*	OC (mm)	TC (mm)	CIR (mm)	CYL (mm)
1	3	10	5	Cu	0.025	0.057	0.0178	0.0428
2	3	20	6	Br	0.055	0.069	0.0105	0.0783
3	3	30	7	Gr	0.14	0.079	0.0092	0.1269
4	3	40	8	Cu	0.085	0.047	0.0068	0.1074
5	6	10	6	Gr	0.07	0.08	0.0416	0.0683
6	6	20	5	Cu	0.035	0.059	0.0063	0.0316
7	6	30	8	Cu	0.065	0.033	0.234	0.0865
8	6	40	7	Br	0.095	0.039	0.0051	0.0801
9	9	10	7	Cu	0.03	0.02	0.0311	0.0635
10	9	20	8	Gr	0.165	0.023	0.0058	0.0501
11	9	30	5	Br	0.15	0.028	0.0451	0.1029
12	9	40	6	Cu	0.205	0.015	0.0068	0.0849
13	12	10	8	Br	0.005	0.039	0.008	0.0766
14	12	20	7	Cu	0.13	0.06	0.025	0.0515
15	12	30	6	Cu	0.165	0.03	0.0321	0.0791
16	12	40	5	Gr	0.27	0.05	0.0647	0.1375

smaller the better characteristic is selected. It is found that pulse-on time shows a maximum contribution of 48% in the assessment of overcut as shown in Fig. 2. The increase in pulse-on time resulted in increasing of overcut as shown in Fig. 3, and this is mainly due to increases in spark energy as pulse-on time increases.

From Fig. 4, it is observed that the discharge current shows a maximum contribution of 28% on tapercut. As discharge current increases, tapercut decreases up to a certain saturated point as shown in Fig. 5. This phenomenon occurs mainly due to an increase in spark energy as discharge current increases.

From Fig. 6, it is observed that interaction between discharge current and pulse-off time shows a maximum contribution of 31% on circularity. Circularity increases

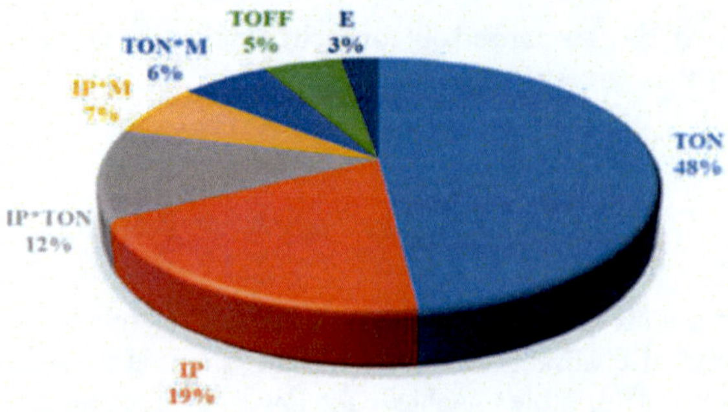

Fig. 2 Contribution of process parameters on OC

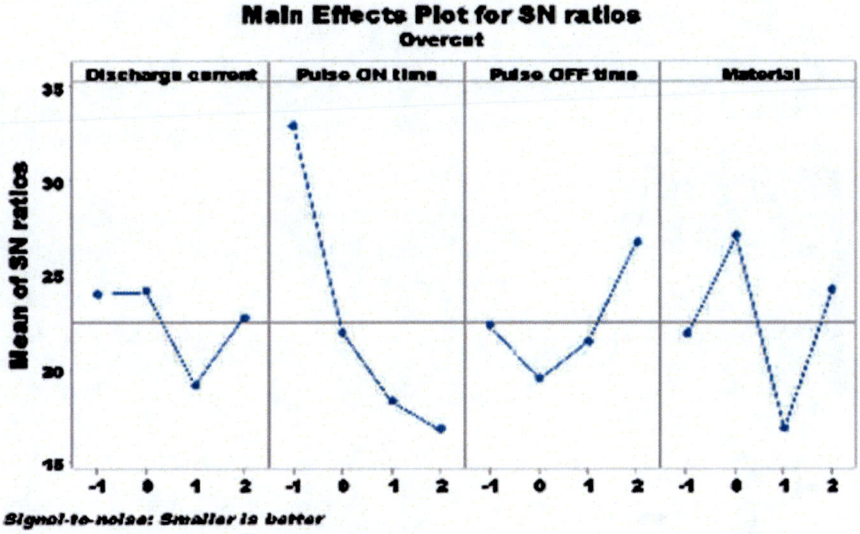

Fig. 3 Main effects plot for SN ratio OC

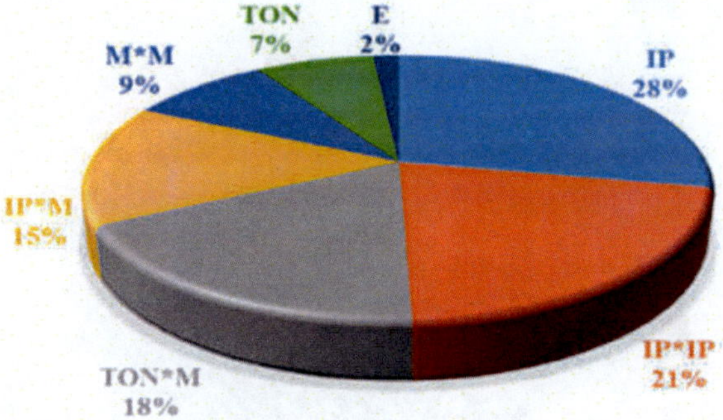

Fig. 4 Contribution of process parameters on TC

as an increase in discharge current. Increasing in pulse-off time leads to a decrease in circularity, and it is mainly due to proper flushing time as an increase in pulse-off time as shown in Fig. 7.

From Fig. 8, it is observed that pulse-on time shows a maximum contribution of 45% on cylindricity. Cylindricity decreases initially with an increment in pulse-on time but increased again at a higher value of pulse-on time as shown in Fig. 9. This is because of the wear of electrode and deposition of the black carbon layer on the tool surface.

Copper electrode helps to get minimum tapercut and cylindricity as shown in Figs. 5 and 9. Since it has less tool wear rate and good thermal conductivity, minimum circularity and overcut are achieved with the help of the brass electrode as shown in

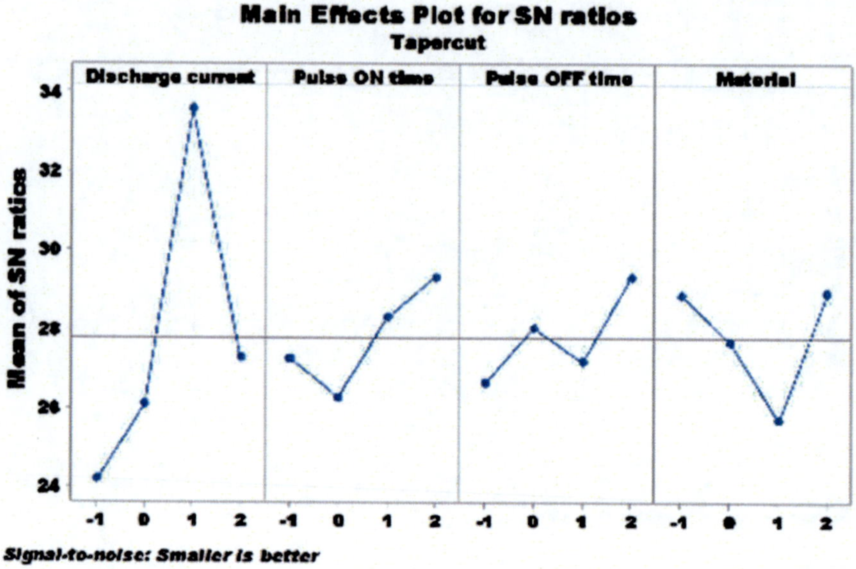

Fig. 5 Main effects plot for SN ratio TC

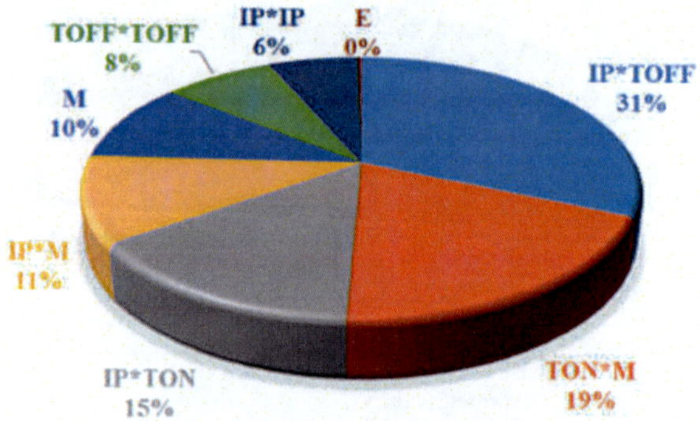

Fig. 6 Contribution of process parameters on CIR

Figs. 3 and 7. Since brass has less thermal conductivity and greater tool wear rate which made brass suitable for features that can be machined at a particular plane. In case of geometrical features when compared with copper and brass, graphite found to be very less significant.

Regression model for prediction of overcut (OC) is developed using the results from experiments as shown in Eq. (1). Similar models are developed for prediction of other response variables. The experimental response variables show significant correlation with predicted response variables as shown in Figs. 10, 11, 12 and 13.

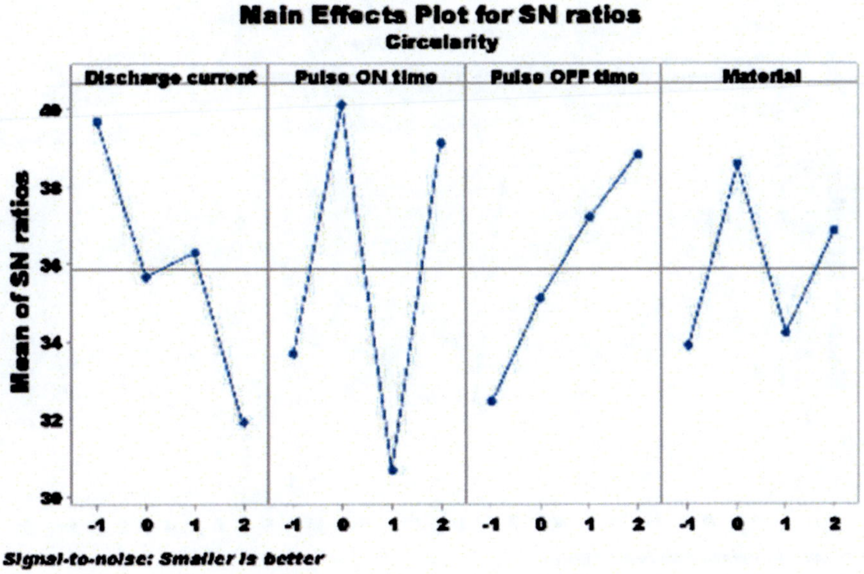

Fig. 7 Main effects plot for SN ratio CIR

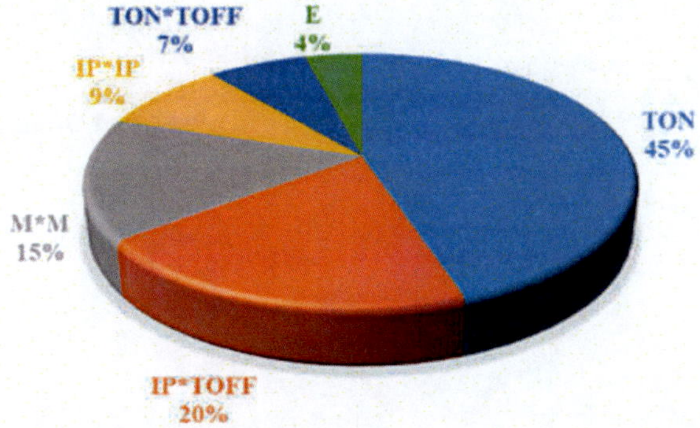

Fig. 8 Contribution of process parameters on CYL

$$
\begin{aligned}
\text{OC (mm)} = {} & 0.314 + 0.0659 * I_P + 0.0105 * T_{ON} + 0.1972 * T_{OFF} \\
& - 0.1273M - 0.1638 * I_P * I_P - 0.0075 * T_{ON} * T_{ON} \\
& - 0.0894 * T_{OFF} * T_{OFF} + 0.0656 * M * M + 0.0573 * I_P * T_{ON} \\
& - 0.0019 * I_P * T_{OFF} - 0.0223 * I_P * M \\
& - 0.1575T_{ON} * T_{OFF} + 0.1675 * T_{ON} * M
\end{aligned} \tag{1}
$$

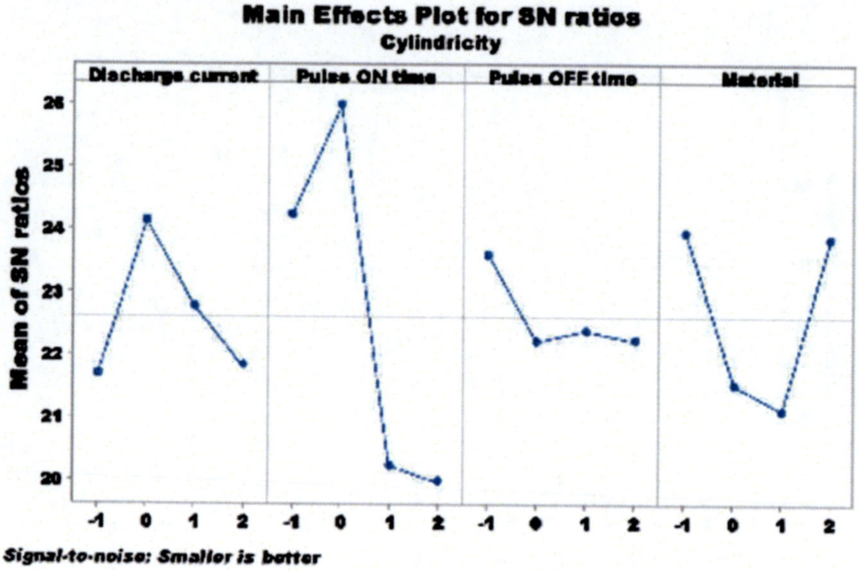

Fig. 9 Main effects plot for SN ratio CYL

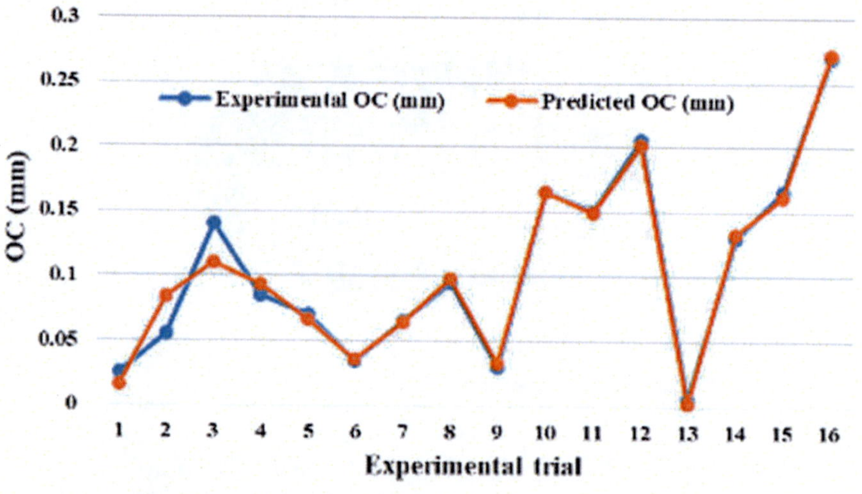

Fig. 10 Experimental OC versus predicted OC

4 Conclusion

An experimental setup for significant manufacturing processes, electric discharge machining, is used, in order to study the effect of discharge current, pulse-on time,

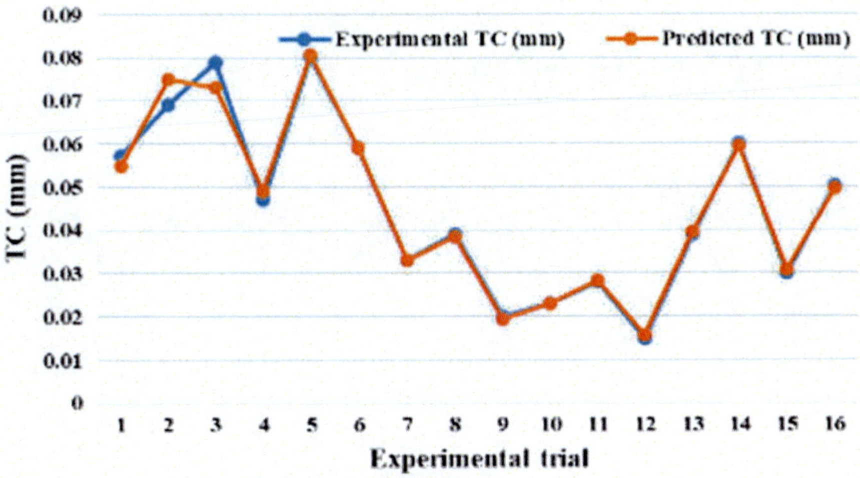

Fig. 11 Experimental TC versus predicted TC

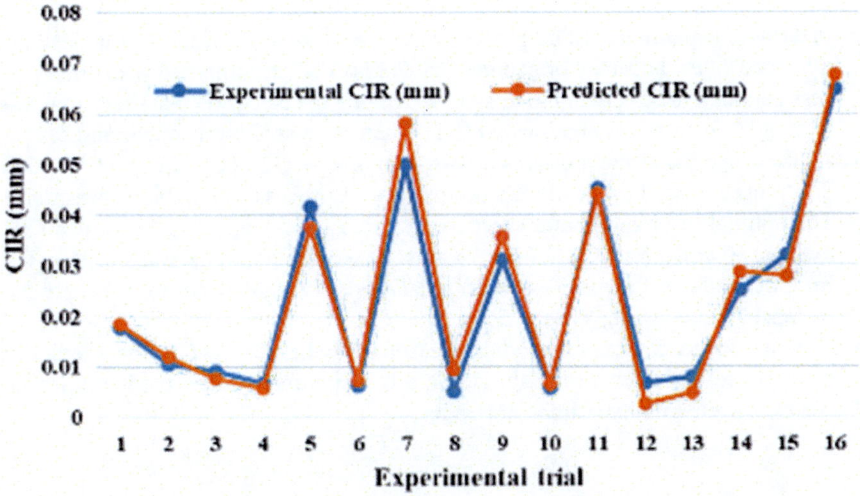

Fig. 12 Experimental CIR versus predicted CIR

pulse-off time and tool material on the overcut, tapercut, circularity, and cylindricity of a through-hole made in AZ31 magnesium alloy. Based on the results from experiments, the following conclusions are made.

- Among the mentioned quantitative EDM process parameters, pulse-on time and discharge current show the significant effect on geometrical features. It was observed that pulse-off time shows less significant in case of geometrical features.

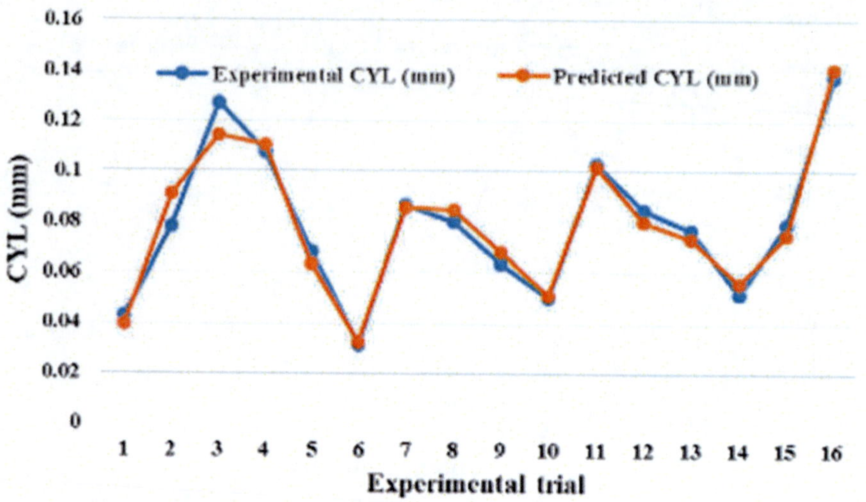

Fig. 13 Experimental CYL versus predicted CYL

- In terms of a qualitative EDM process parameter, tool material also shows less significance on the accuracy of geometrical features when compared with other process parameters considered in this work. Among the electrode materials, graphite is found to have a less significant effect when compared with copper and brass.
- As the objective of the present study is to minimize OC, TC, CIR and CYL. The S/N plots are used to decide the optimal set of process to obtain best response. The optimal machining parameters for OC are $I_p = 6$ amps; $T_{ON} = 10$ µs; $T_{OFF} = 8$ µs; $M =$ Br, for TC are $I_P = 9$ amps; $T_{ON} = 40$ µs; $T_{OFF} = 8$ µs; $M =$ Cu, for CIR are $I_P = 3$ amps; $T_{ON} = 20$ µs; $T_{OFF} = 7$ µs; $M =$ Br, for CYL are $I = 6$ amps; $T_{ON} = 20$ µs; $T_{OFF} = 5$ µs; $M =$ Cu.
- The developed regression models are found to be accurate in the prediction of response variables within the domain of experiments. The predicted results correlate well with experimental results.

References

1. Lee ES, Won JK, Shin TH, Kim SH (2012) Investigation of machining characteristics for electrochemical micro-deburring of the AZ31 lightweight magnesium alloy. Int J Precis Eng Manuf 13(3):339–345
2. Abdul-Rani AM, Razak MA, Littlefair G, Gibson I, Nanimina AM (2017) Improving EDM process on AZ31 magnesium alloy towards sustainable biodegradable implant manufacturing. Procedia Manuf 7:504–509
3. Kumar P, Parkash R (2016) Experimental investigation and optimization of EDM process parameters for machining of aluminum boron carbide (Al–B4C) composite. Mach Sci Technol 20(2):330–348

4. Kathiravan S, Pradeep Kumar J (2017) Experimental investigation on influence of process parameter in EDM of P20 mold steel. J Adv Nat Appl Sci 8:161–167
5. Balaji, R.L., Kumar, J.P., (2016) Experimental study on electrical discharge machining (EDM) of monel metal. J Eng Sci Res Appl 10–13
6. Razak MA, Abdul-Rani AM, Rao TVVLN, Pedapati SR, Kamal S (2016) Electrical discharge machining on biodegradable AZ31 magnesium alloy using Taguchi method. Procedia Eng 148:916–922

Desirability Approach Machining Study on Aluminum Composite Through Wire-Cut Electric Discharge Technique

K. Rajkumar, C. Balasubramaniyan, K. Ramraji, A. Gnanavelbabu, and P. Sabarinathan

Abstract Aluminum matrix composites (AMCs) are exceptionally designed for automobile components like brake pads, cylinder body, and pistons owing to high temperature and wear resistance. Machining is a paramount issue to achieve required dimensions and tolerance for those products. The present investigation includes fabrication of composite and wire-cut electric discharge machining with various parameters. AMC designed by B_4C and CNT particles of hybrid composite was fabricated by liquid metallurgy process. Carbon nanotubes were added to tailor the high temperature withstanding and reduce friction during the sliding. The role of B_4C is to improve the hardness and wear resistance of the composite. The combined actions of these reinforcements enhance overall mechanical properties of the composite. The objective of the present work is to study the machinability aspects of the composite by wire-cut EDM. Input machining parameters considered are pulse on time, pulse off time, wire feed, gap in voltage, and servo speed. Kerf width and material removal rate were studied with response surface method (RSM). Pulse on time and voltage increase the material removal rate; however, voltage has a more relatively significant effect on the MRR.

Keywords Electric discharge · Metal composite · Spark erosion · CNT

K. Rajkumar (✉) · K. Ramraji · P. Sabarinathan
Department of Mechanical Engineering, Sri Sivasubramaniya Nadar College of Engineering, Chennai, Tamilnadu, India
e-mail: rajkumark@ssn.edu.in

C. Balasubramaniyan
Department of Mechanical Engineering, Sri Venkateswaraa College of Technology, Sriperumbudur, Tamilnadu, India

A. Gnanavelbabu
Department of Industrial Engineering, College of Engineering, Chennai, Tamilnadu, India

S. Vijayan et al. (eds.), *Trends in Manufacturing and Engineering Management*, Lecture Notes in Mechanical Engineering,
https://doi.org/10.1007/978-981-15-4745-4_3

1 Introduction

Metal matrix composites (MMCs) are now replacing metals in various engineering applications. Specifically, aluminum matrix composites (AMCs) occupy a position of great significance in industries, because of their good thermo-mechanical properties, good damage resistance, and wear resistance. Generally, machining of MMCs is a laborious work owing to the presence of hard and brittle reinforcement particle and non-homogenous structure [1]. It was also realized that their machining behavior is unlike any of the other conventional processes observed so far. It can be machined through some of the conventional methods. But, severe damage to the workpiece and the cutting tool are observed. Workpiece damages such as particle cracking, particle pulling out and particle debonding were occurred [2, 3]. Surface roughness was also observed to be high, and in order to overcome that, self-lubricating particles are added to the composites. This reduces wear between the tool and workpiece. Otherwise, it decreases strength, hardness, and low fracture toughness which are undesirable [4]. To counteract unsatisfactory properties, some novel approach has to be followed, i.e., adding concurrent lubricants and strength improvement particles. Tribological investigations are evident that the relationship between work material grain size and reinforcement particle size has a major impact on the wear behavior [5].

In recent times, carbon nanotubes are considered as reinforcement in aluminum composites. There are major reasons for selection of CNT as reinforcement that increases stiffness, strength and vibration absorption properties of the composite. This also makes the machining process as smoother than ever and facilitates easy chip breaking [6, 7].

Nowadays, machining of MMCs is viable through unconventional processes [8]. The major advantages of these processes are having a good surface finish and easy disposal of removed material. Also, regardless of the material properties, it can machine the component with less tool wear. Shubhra paliwal et al. [9] studied the EDM process by analyzing pulse on and off time on the machining behavior of the material. Hewidy et al. [10] conducted EDM experiments using a RSM for the Inconel 601 material. They concluded that RSM method has an advantage of describing the effect between input variables and response parameters.

Tilekar et al. [11] determined that high electrical conductivity and low melting point of aluminum are directly affecting machining. As an increase in current, more melting and vaporization of material are observed, thereby increasing MRR. Bobbili et al. [12] experimented EDM with hot-pressed boron carbide and observed deep craters and microcracks at high Ton and peak current which deteriorated surface finish. Moreover, higher input energy significantly affected the wear rate of the wire and caused breakage of the wire frequently. Udaya Prakash et al. [13] conducted machinability studies on the Al alloy with fly ash and B_4C composite using a wire-cut EDM. They determined that gap voltage is the more significant parameter than pulse on time, pulse off time, wire feed rate, and percentage of reinforcement. However, fabricated Al/boron carbide/CNT hybrid composite is heterogeneous in nature. Therefore, it required a non-traditional wire-cut EDM machining approach to maintain the surface integrity. The aim of this study was to investigate the material removal rate and kerf dimension of the hybrid composite by varying WEDM processes parameters.

2 Experimental Procedure

2.1 Material Selection

Aluminum (Al 6061) was used as matrix element in the composite. B_4C (40 μm) was used as one of the reinforcing agents and has exotic properties like high hardness, low specific gravity, and neutron absorption. CNT (74 nm) was used as another reinforcing fiber and also solid lubricant.

2.2 Composite Fabrication: Al-B_4C-CNT

Fabrication of composite was done using a stir casting apparatus. Aluminum 6061 ingots were melted in the furnace around 800 °C. Flux was used to increase the wettability between melt matrix and reinforcements and prevents oxidation of melt. A steel stirrer was used to mix the reinforcements in the melt uniformly and runs at 150 rpm with the help of a motor. Simultaneously, B_4C and CNT were added as the reinforcements. CNT was functionalized using nitric acid solution before mixing it with the aluminum melt. The composition of the composite was Al-80 wt% B_4C-18.5 wt% and CNT-1.5 wt%. After uniform dispersion of the reinforcements in the aluminum matrix melt, the mixture was poured into a rectangular die of dimension 12 * 100 *10 mm.

2.3 Wire-Cut EDM Process

The machining parameters that influence the metal removal rate and kerf width were T_{on}, T_{off}, voltage, wire feed (in machine code values), and its levels are presented in Table 1. A face-centered composite design (FCD) with total of 30 experimental runs (six center points) has been selected, and experimentation was conducted on the 4-axis CNC Eco-cut WEDM. The results were analyzed using Design–Expert software.

The study focuses on the effects of these inputs on the responses, namely MRR, kerf width, and surface roughness. A wire-cut EDM machine uses bronze-coated copper wire of 0.25 mm diameter. 15 straight slit cuts were made on the composite. During the machining process, a EDM oil dielectric fluid was used. The dielectric

Table 1 Process parameters and levels

S. No.	Process parameter		Levels		
		Units	1	2	3
1	Pulse on time (T_{on})	(μs)	126	128	130
2	Pulse off time (T_{off})	(μs)	50	55	60
3	Spark voltage (SV)	(V)	20	30	40
4	Wire feed (WF)	(mm/s)	2	3	4

fluid flow is perpendicular to the direction of the wire feed. The typical cut job is shown in Fig. 1.

The kerf width was measured using an optical microscope. The typical cuts on the workpiece are shown in Fig. 2. Kerf width was measured by using ImageJ software. Material removal rate was calculated using a mathematical Eq. (1)

$$\text{MRR} = V_c * b * t \tag{1}$$

where b—kerf width (mm), t—thickness of the workpiece in mm, and surface roughness was calculated in R_a using a tally surf roughness meter with cutoff length 0.25 mm.

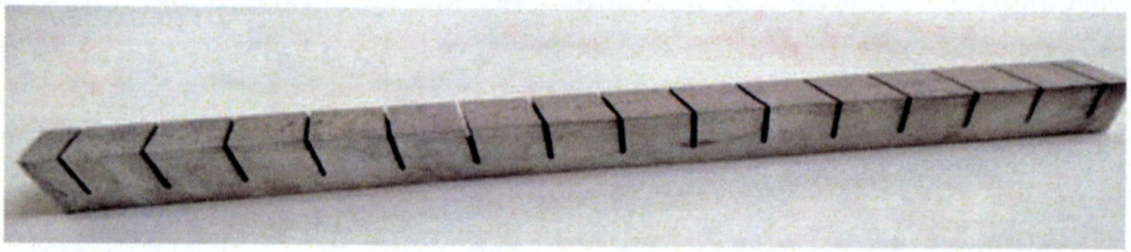

Fig. 1 Typical wire EDM processed workpiece

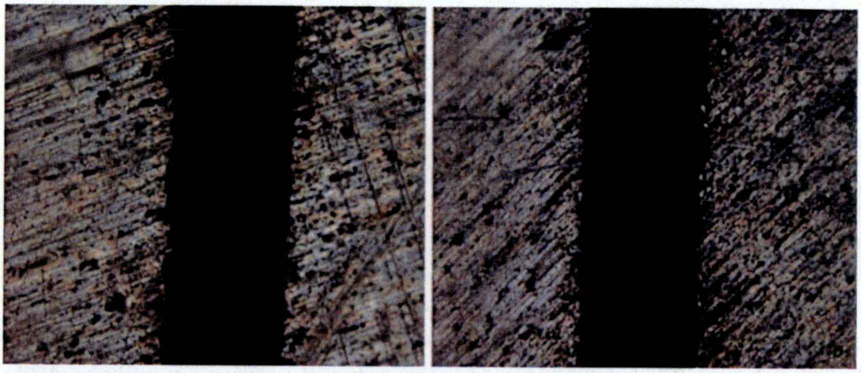

Fig. 2 Typical wire EDM cut on the composite

3 Results and Discussion

The effects of the process parameters on the MRR, surface roughness, and Kerf width are shown in Table 2.

Table 2 Experiments results and responses

Std	Run	SV (V)	T_{on} (μs)	T_{off} (μs)	WF (m/min)	MRR (mm^3/min)	Kerf (mm)	Surface roughness (Ra)
1	1	20.00	126.00	50.00	2.00	0.297	0.364	1.489
9	2	20.00	126.00	50.00	4.00	0.332	0.359	1.412
16	3	40.00	130.00	60.00	4.00	0.451	0.369	1.676
3	4	20.00	130.00	50.00	2.00	0.322	0.382	1.521
13	5	20.00	126.00	60.00	4.00	0.288	0.345	1.405
30	6	30.00	128.00	55.00	3.00	0.341	0.369	1.565
19	7	30.00	126.00	55.00	3.00	0.311	0.374	1.592
27	8	30.00	128.00	55.00	3.00	0.305	0.373	1.543
8	9	40.00	130.00	60.00	2.00	0.42	0.391	1.745
29	10	30.00	128.00	55.00	3.00	0.311	0.369	1.645
23	11	30.00	128.00	55.00	2.00	0.353	0.371	1.622
22	12	30.00	128.00	60.00	3.00	0.313	0.372	1.601
4	13	40.00	130.00	50.00	2.00	0.424	0.382	1.673
28	14	30.00	128.00	55.00	3.00	0.342	0.372	1.556
14	15	40.00	126.00	60.00	4.00	0.389	0.359	1.645
5	16	20.00	126.00	60.00	2.00	0.298	0.353	1.539
15	17	20.00	130.00	60.00	4.00	0.279	0.372	1.422
21	18	30.00	128.00	50.00	3.00	0.358	0.374	1.572
7	19	20.00	130.00	60.00	2.00	0.31	0.378	1.51
12	20	40.00	130.00	50.00	4.00	0.426	0.379	1.562
11	21	20.00	130.00	50.00	4.00	0.342	0.391	1.465
25	22	30.00	128.00	55.00	3.00	0.35	0.372	1.573
24	23	30.00	128.00	55.00	4.00	0.326	0.348	1.532
17	24	20.00	128.00	55.00	3.00	0.322	0.354	1.495
2	25	40.00	126.00	50.00	2.00	0.423	0.349	1.675
6	26	40.00	126.00	60.00	2.00	0.412	0.381	1.645
20	27	30.00	130.00	55.00	3.00	0.373	0.392	1.567
18	28	40.00	128.00	55.00	3.00	0.422	0.364	1.632
26	29	30.00	128.00	55.00	3.00	0.324	0.369	1.541
10	30	40.00	126.00	50.00	4.00	0.382	0.342	1.622

In order to understand the effects of process parameters on MRR, contour plots are shown in Figs. 3 and 4. The effect of variation in Ton and voltage on the MRR is shown in Fig. 3. It is seen that the MRR increases with increase in voltage. Besides, it also increases current. As a result of high intensity sparks, more material was melted and evaporated. It is found that 38 V is sufficient for material removal. T_{on} also has a significant effect on MRR. Only after a certain value of T_{on}, higher values of MRR can be achieved. It is found that 128 μs is sufficient for higher material removal. Higher Ton values mean that longer duration of the spark. This leads to more material erosion. Fig. 4 also reveals that voltage was also significant, and it increases MRR. But considering effect of T_{off} on the MRR is a little effect, as longer T_{off} only reduces the MRR. The low T_{off} (50 μs) is producing higher MRR.

Figures 5 and 6 show the kerf width with effect of voltage, current on and off time. It is seen that the kerf dimension decreases gradually up to 30 V. Then, it increases with further increase in voltage. When the cut is initiated, there was no debris and balance between material erosion and spark intensity. With increase in voltage, the kerf width increases due to higher order of material removal. This is attributed to the high intensity sparks. The localized melting results in resolidified layer on both sides of the cut surfaces. It is also revealed that the kerf dimension increases with increase in T_{on} value due to longer time spark eroded side materials. Moreover, Fig. 6 shows the kerf narrow region at pulse off time 55 and voltage 30.

Figure 7 shows the effect of wire speed and voltage on the surface roughness of the composite. From the plot, it is clear that when the wire speed increases, the surface roughness decreases. Due to wire electrode moving fast on the workpiece,

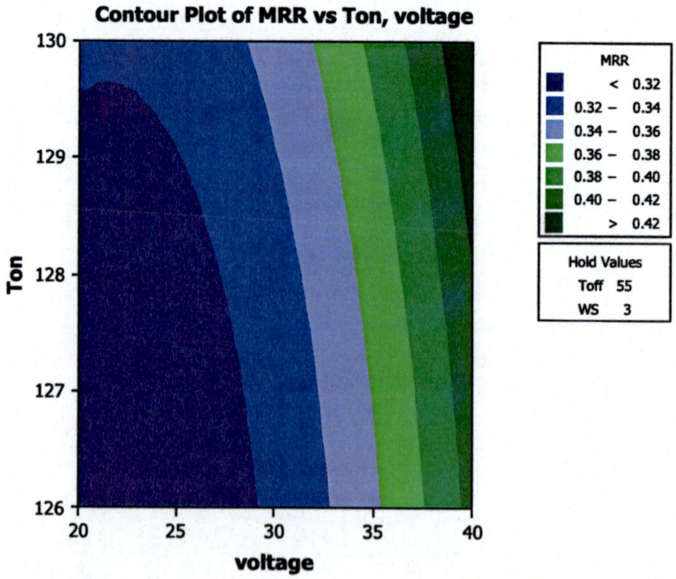

Fig. 3 Contour plot of MRR versus T_{on}, voltage

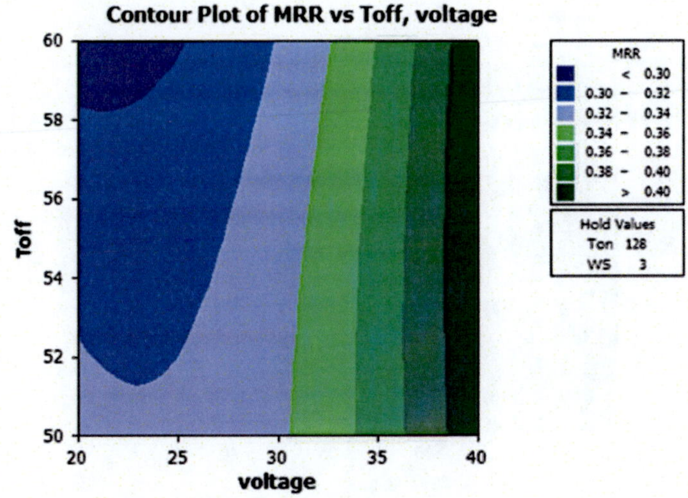

Fig. 4 Contour plot of MRR versus T_{off}, voltage

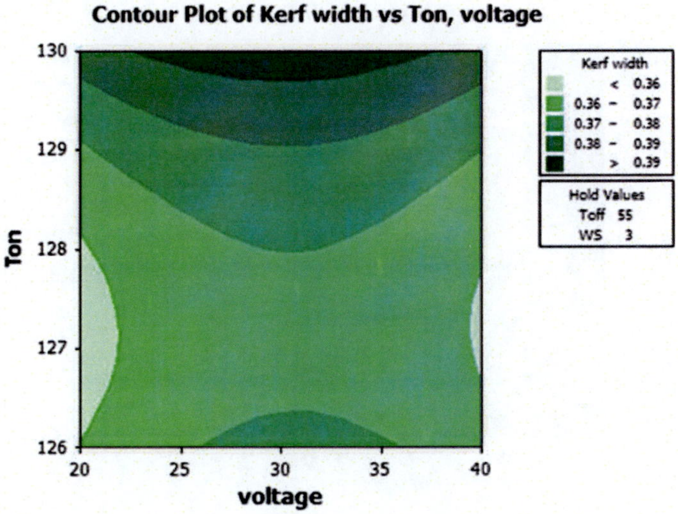

Fig. 5 Contour plot of kerf versus T_{on}, voltage

the availability of spark time on the local spot is low, resulting in little erosion of material. This results in a smoother surface. MRR and surface roughness are inversely proportional, and hence, as voltage increases, the surface roughness also increases. In Fig. 8, it can be concluded that similar observation is evident. The combination of high wire speed and low T_{on} is evident of smoother surface of cut. Moreover, it is

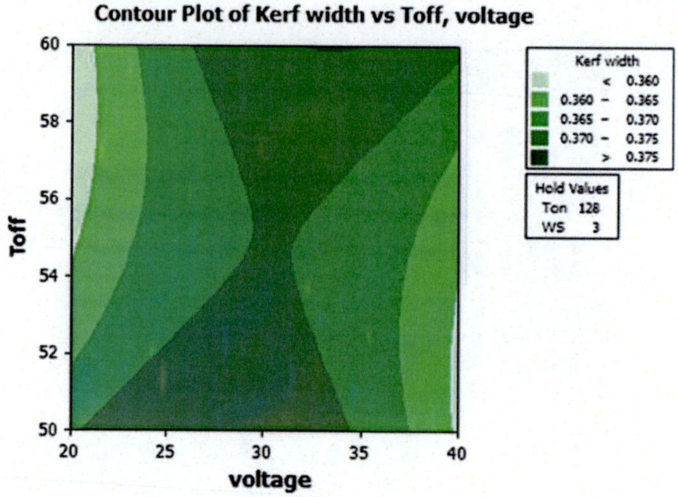

Fig. 6 Contour plot of kerf versus T_{off}, voltage

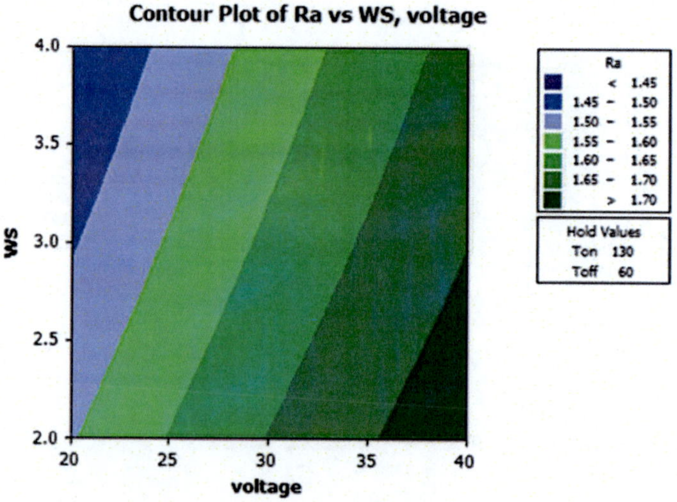

Fig. 7 Contour plot of R_{a} versus WS, voltage

observed that Ra with slope is steeper than the T_{on} which indicates that voltage has more significant effect on surface roughness.

Figure 9 shows the desirability values of optimized multiple responses of wire-cut EDM on the given composite material. The desirability of the cut is having a high MRR, a low kerf width, and a low surface roughness, and in order to achieve that, optimization was done, to identify the best balance between the three requirements.

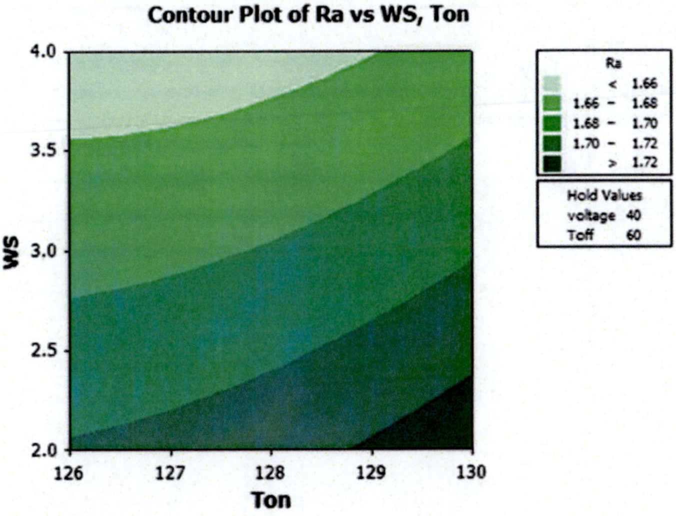

Fig. 8 Contour plot of R_a versus WS, T_{on}

The corresponding input parameters were found to be exhibiting the best results; the desirability of the cut was relatively higher when the combination of input parameters was of 39.99 V, 128.94 μs-T_{on}, 50 μs-T_{off}, and 4 mm/s-WS, than any other combinations.

4 Conclusions

AA6061-B_4C-CNT hybrid composite was successfully fabricated, and wire-cut EDM process was carried out with face-centered composite experiment design. From the experiment results, the following conclusions are drawn:

1. MRR increases with increase in voltage. It is found that 38 V is sufficient for material removal. T_{on} also has a significant effect on MRR. It is found that 128 μs shows off higher material removal. Higher T_{on} leads to more material erosion. It is observed that T_{off} on the MRR has a little effect, but longer T_{off} reduces the MRR. By comparing results, voltage has a more significant effect on the MRR than T_{on}.
2. WEDM experimentation reveals that kerf dimension decreases gradually up to 30 V. But again, it increases with voltage. This could be due to the high order of material removal. The localized melting generates a resolidified layer on both sides of the cut surface, affecting the kerf dimension. It is shown that the kerf dimension increases with increase in T_{on}, but kerf narrow region is obtained at pulse off time 55 μs and voltage 30.

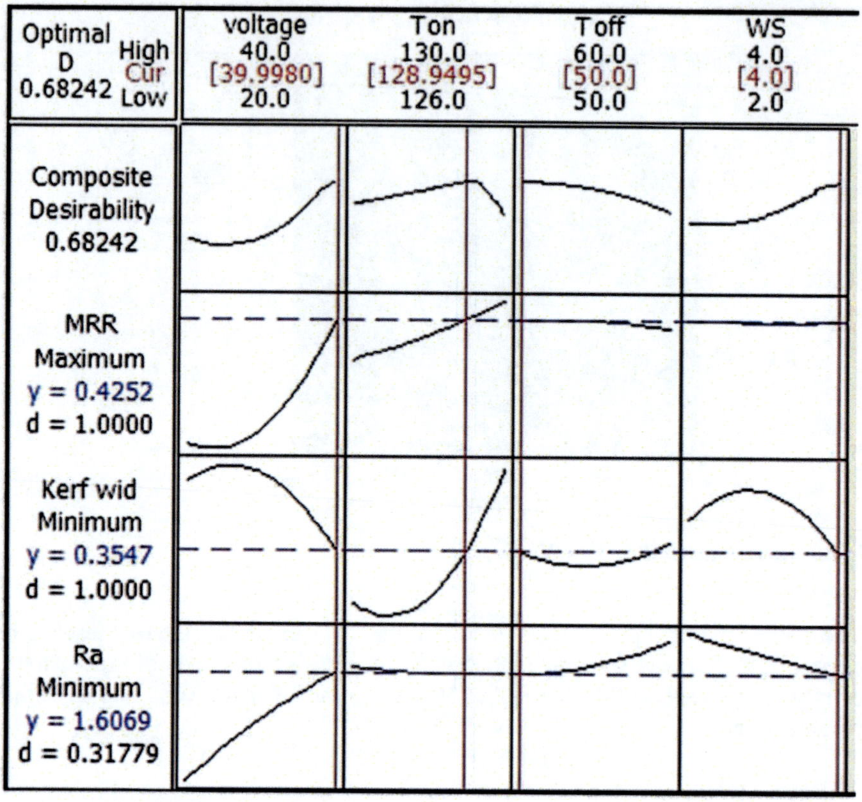

Fig. 9 Desirability values of process parameters

3. It is evident that the smoothness of cut surface is obtained with increasing in wire speed. The reason is that a fast-moving wire electrode has less time for on a local spot leading to a reduction of erosion of workpiece material. The combination of high wire speed and low T_{on} offered smoother surface of cut.
4. The combination of 39.99 V, 128.94 μs-T_{on}, 50 μs-T_{off}, and 4 mm/s-wire speed provided a relatively more desirable cut than any other combinations of input parameters.

References

1. Bharath V, Nagaral M, Auradi V, Kori SA (2014) Preparation of 6061Al-Al2O3 MMC's by stir casting and evaluation of mechanical and wear properties. Procedia Mater. Sci. 6:1658–1667
2. Sankar M, Gnanavelbabu A, Rajkumar K (2014) Effect of reinforcement particles on the abrasive assisted electrochemical machining of Aluminium-Boron carbide-Graphite composite. Procedia Eng 97:381–389

3. Yadav RK, Abhishek K, Mahapatra SS (2015) A simulation approach for estimating flank wear and material removal rate in turning of Inconel 718. Simul Model Pract Theory 52:1–14
4. Kong L, Zhu S, Bi Q, Qiao Z, Yang J, Liu W (2014) Friction and wear behavior of self-lubricating ZrO2 (Y2O3)–CaF2–Mo–graphite composite from 20 to 1000 °C. Ceram Int 40(7):10787–10792
5. Vangla P, Gali ML (2016) Effect of particle size of sand and surface asperities of reinforcement on their interface shear behaviour. Geotext Geomembr 44(3):254–268
6. Banks-Sills L, Shiber DG, Fourman V, Eliasi R, Shlayer A (2016) Experimental determination of mechanical properties of PMMA reinforced with functionalized CNTs. Compos B Eng 95:335–345
7. Moghadam AD, Omrani E, Menezes PL, Rohatgi PK (2015) Mechanical and tribological properties of self-lubricating metal matrix nanocomposites reinforced by carbon nanotubes (CNTs) and graphene–a review. Compos B Eng 77:402–420
8. Rajkumar K, Santosh S, Ibrahim SJS, Gnanavelbabu A (2014) Effect of Electrical discharge machining parameters on microwave heat treated Aluminium-Boron carbide-Graphite composites. Procedia Eng 97:1543–1550
9. Paliwal S, Solanki P (2014) Parameter optimization of wire electrical discharge machining for minimum surface roughness and kerf width using taguchi method. In: 4th SARC international conference
10. Hewidy MS, El-Taweel TA, El-Safty MF (2005) Modelling the machining parameters of wire electrical discharge machining of Inconel 601 using RSM. J Mater Process Technol 169(2):328–336
11. Tilekar S, Das SS, Patowari PK (2014) Process parameter optimization of wire EDM on Aluminum and mild steel by using taguchi method. Procedia Mater Sci 5:2577–2584
12. Bobbili R, Madhu V, Gogia AK (2015) An experimental investigation of wire electrical discharge machining of hot-pressed boron carbide. Defence Technol 11(4):344–349
13. Prakash JU, Moorthy TV, Peter JM (2013) Experimental investigations on machinability of aluminium alloy (A413)/Flyash/B4C hybrid composites using wire EDM. Procedia Eng 64:1344–1353

A Comparative Study on Abrasive Water Jet Machining Characteristics of Entry and Exit Layers of Glass and Basalt Woven Polymer Composites

K. Ramraji, K. Rajkumar, M. Rajesh, and A. Gnanavelbabu

Abstract Abrasive water jet machining (AWJM) induced kerf and kerf tapper angle variation is much affected by the properties of the entry and exit layer of fiber reinforced polymer of composite. In this work, experimental investigation on the machining characteristics of vinyl ester composite fabricated by basalt and glass fiber as top and bottom layers separately interlayered with flax fiber. Hybridization of basalt/glass composite was made by the additional layers of flax woven. Two different stacked composites were fabricated by using hand layup process followed by static compression loading. The mechanical properties such as tensile and flexural strength were studied. It was found that top and bottom basalt layers interlayered flax fiber composite exhibited higher tensile and flexural properties. The tensile and flexural strength improvements were 7.15% and flexural 13.3% over the top and bottom glass layers interlayered flax fiber composite. The fabricated composites were machined by abrasive water jet machining (AWJM) using constant cutting parameters with pressure, nozzle speed, and standoff distance. Experimental results reveal that minimum kerf taper and higher cutting surface quality obtained with entry and exit layers of glass fiber composite.

Keywords Polymer composite · AWJM · Flax fiber · Kerf · Interlayered fiber

1 Introduction

Nowadays, FRP composite is losing its merits by environmental concerns and government regulation [1]. Natural fibers most widely accepted have many advantage properties as reinforcement to composite. Owing to relatively low density and high strength, the natural fiber is popular now. The natural fibers are extracted from the

K. Ramraji · K. Rajkumar (✉) · M. Rajesh
Department of Mechanical Engineering, Sri Sivasubramaniya Nadar College of Engineering, Chennai, Tamil Nadu, India
e-mail: rajkumark@ssn.edu.in

A. Gnanavelbabu
Department of Industrial Engineering, College of Engineering, Chennai, Tamil Nadu, India

S. Vijayan et al. (eds.), *Trends in Manufacturing and Engineering Management*, Lecture Notes in Mechanical Engineering,
https://doi.org/10.1007/978-981-15-4745-4_4

plants, nutshells, palm shells, wood flour, and animals [2]. Moreover, they have enormous merits over synthetic fiber such as renewable in nature and less abrasive [3, 4]. However, its mechanical properties are much lower than synthetic fiber [5]. This limits to use in the semi-structural application.

Natural fiber composite exhibits poor resistance to moisture absorption because of its hydrophilic behavior. Thus, it makes less attractive but rather using suitable chemical treatments rectifying hydrophilic problem [6]. The reinforcing of natural and synthetic fibers in a polymer matrix provides the best properties. The hybridization of fibers is remarking superiority and reduces relay on the synthetic fibers. Hybrid composite is being now enough strength, stiffness, and rigidity to satisfy the needs of semi-structural applications [7, 8]. Apart from that, composite required machining process, to shape, size, and tolerance for an application.

Conventional machining processes namely drilling, turning, and milling will tend to pull out fibers from the composite. Additionally, it leads to delamination and damage to composite. Almost 60% of the composites are rejected due to damages like fiber pullout and delamination [9]. In order to improve the machining quality of composite, these works made an attempt with abrasive water jet machining (AWJM) process. This process is environmentally safe and free from heat-induced detects. Researchers are more interested in AWJM of the composite to evaluate kerf ratio, kerf angle, and surface quality parameters [10, 11].

Moreover, the selection of proper machining parameters is more concern to the researcher and machinist for quality machining. A few researchers revealed that standoff distance is the most significant factor to reduce the kerf ratio followed by the traverse speed. The minimum traverse speed and maximum pressure were smoothly removed material from the FRP composite and also minimize delamination fracture [12]. The combination of traverse rate and standoff distance increases kerf ratio but decreases standoff distance resulting in a smooth surface and lowers the kerf ratio as a result of the increased kinetic energy of jet. Increase in water pressure and abrasive flow rate improved surface quality of composite [13, 14]. Selvam et al. [15] investigated AWJM cutting parameters on the surface and kerf taper by response surface model (RSM). They concluded that kerf and surface roughness are a function of traverse speed and abrasive flow rate.

A comparative study on the kerf and surface quality by changing entry and exit layer of glass and basalt woven fiber polymer composite by AWJM process has been made in this work.

Table 1 Properties of flax, basalt, and glass fibers [2, 16]

Properties	Unit	Flax	Basalt	Glass
Density	g/cm^3	2.80	1.40	2.55
Tensile module	GPa	87	51	73
Tensile strength	MPa	4700	1034	2400
Elongation at break	mm	3.15	1.5	3

2 Materials and Method

2.1 Materials

The interplay hybrid FRP composite was made by woven fabrics of basalt, flax, E-glass, and vinyl ester resin of grade VBR 4508 was used as matrix. Regular accelerator, promotor, and catalyst were used for curing agent. The properties of flax, basalt, and glass fabric are summarized in Table 1.

2.2 Fabrication of Composite Laminates

The flax fabric was treated in sodium hydroxide (NaOH) solution with the concentration of 5% in one liter of water for the duration of two hours at room temperature. NaOH-treated fibers were washed by tap water followed with distilled water. FRP composite was fabricated by a hand layup process with dead weight. The stacking sequences of FRP composites were G2F7 and B2F7. G2F7 has two layers of E-glass woven fabric (top and bottom) and seven layers of flax woven fabric. B2F7 has two layers of basalt woven fabric (top and bottom) and seven layers of flax woven fabrics. The ratio of reinforcement weight fraction to the matrix weight fraction is maintained at 40:60. The size of the laminate was a square of 300 × 300 mm. The stacking sequence of laminates was shown in Fig. 1. The actual thickness of the laminate was approximately 7.2 mm for both B2F7 and for G2F7.

2.3 Tensile and Flexural Strength

Universal testing machine (UTM) was used to evaluate the tensile and flexural properties of the composite. To measure tensile property, the test sample was made in accordance with ASTM-D 638 and tested at a uniform crosshead speed of 5 mm/min. Three-point bending tester was performed to the specimen as per ASTM D790 the standard for the flexural property evaluation. The sample was tested with a strain rate of 5 mm/min. Typical tensile and flexural fracture specimens are shown in Fig. 2a–d.

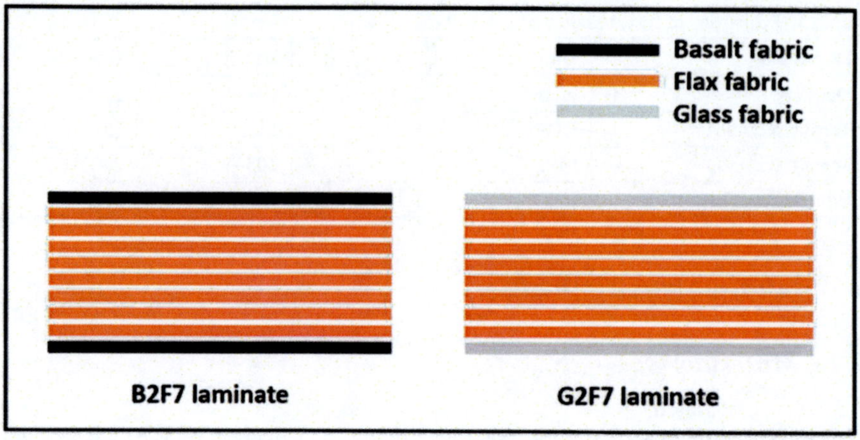

Fig. 1 Stacking layers of B2F7 and G2F7 composite laminates

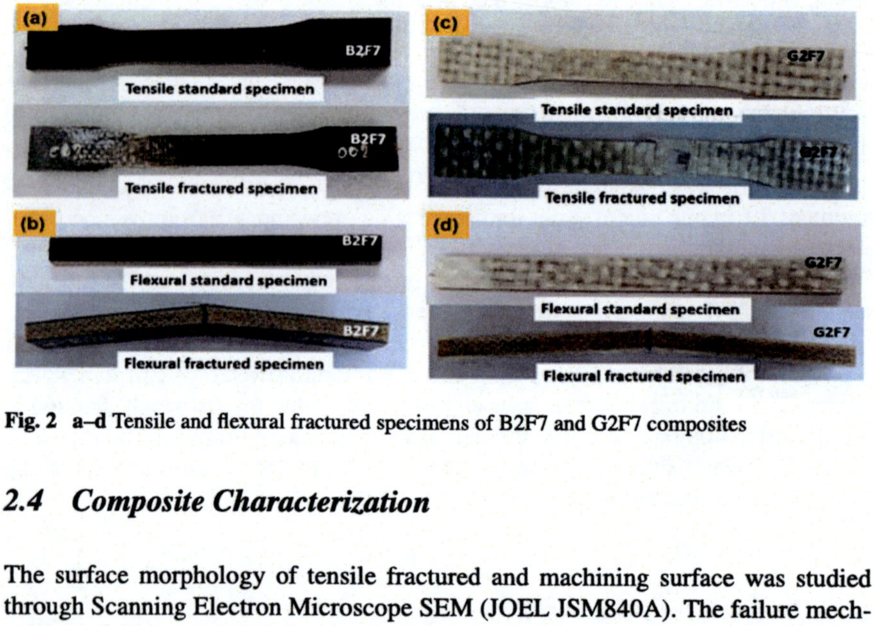

Fig. 2 **a–d** Tensile and flexural fractured specimens of B2F7 and G2F7 composites

2.4 Composite Characterization

The surface morphology of tensile fractured and machining surface was studied through Scanning Electron Microscope SEM (JOEL JSM840A). The failure mechanisms of the composite were analyzed through fiber and matrix intact by SEM images.

2.5 AWJM Process

A two-axis OMAX-1530(CNC) water jet cutting machine was used for experimentation. AWJM nozzle assembly consists of an orifice of (0.35 mm diameter) with a sapphire jewel and carbide focusing tube of 0.35 mm internal diameter and length of 75 mm. The impinging angle of the jet was maintained at 90°. The mixing tube diameter was 0.762 mm, and the abrasive size was 80 mesh and tool offset of 0.3 mm. These standard set of AWJM geometry are used widely in most industrial applications. The photocopy of AWM machining of composite setup is shown in Fig. 3. The selected parameters and their ranges listed in Table 2.

In this study, surface roughness was measured over the cut by SURFTEST SJ-210. Three different locations of readings for Ra were measured, and then, the average value was calculated. The kerf taper was measured using a video measuring system (VMS), as shown in Fig. 4. Kerf taper was calculated by measuring top, and bottom of kerf average value of three readings reported in order to minimize the error. The kerf taper is calculated as given in Eq. (1)

$$\text{Kerf Taper} = \frac{L_t - L_b}{2T} \tag{1}$$

Where L_t—top kerf width (mm), L_b—bottom kerf width (mm), and T—thickness of the laminate (mm).

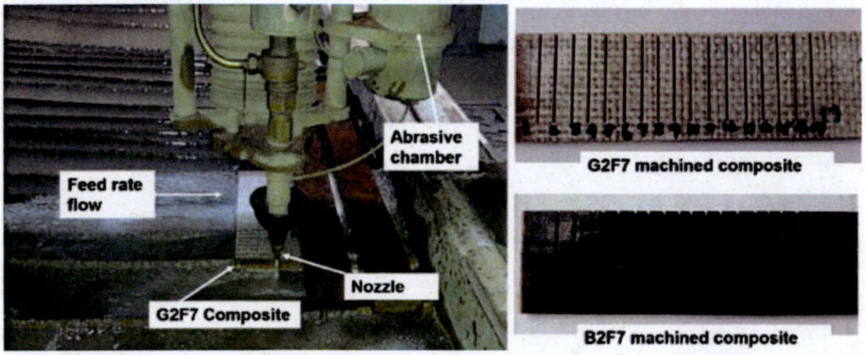

Fig. 3 Photocopy of AWJ machining setup and typical machined G2F7 and B2F7 composites

Table 2 AWJM parameters and their ranges

Process parameters	Units	Levels		
Hydraulic pressure (HP)	MPa	160	240	320
Standoff distance (SOD)	mm	1.5	2.5	3.5
Traverse rate (TR)	mm/min	150	225	300

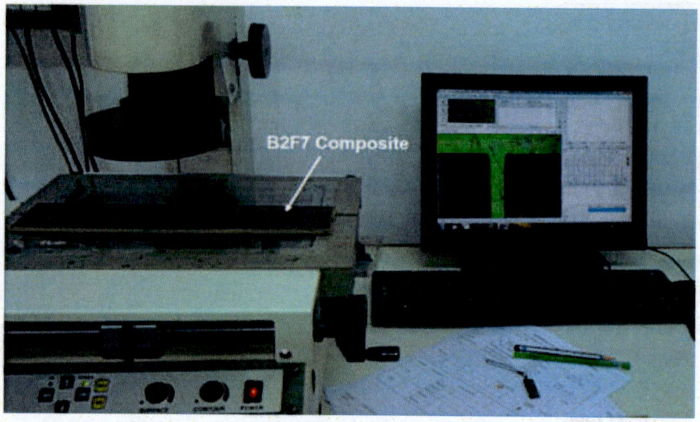

Fig. 4 Photocopy of VMS for kerf width measurement

3 Results and Discussion

3.1 Tensile and Flexural Strength

Tensile and flexural strength of B2F7 and G2F7 composites are represented in Fig. 5a, b. From this figure, it clearly shows that the maximum tensile and flexural strength were obtained with B2F7 composite. The tensile strength and flexural strength improvement were 7.15% and 13.3%, respectively, over the G2F7 composite. This could be a typical reaction between vinyl ester resin and basalt fiber reacting a good interfacial strength. The basalt surface layers provide the load-bearing strength to the composite. This is certainly delayed fracture and pullout of basalt fibers. As

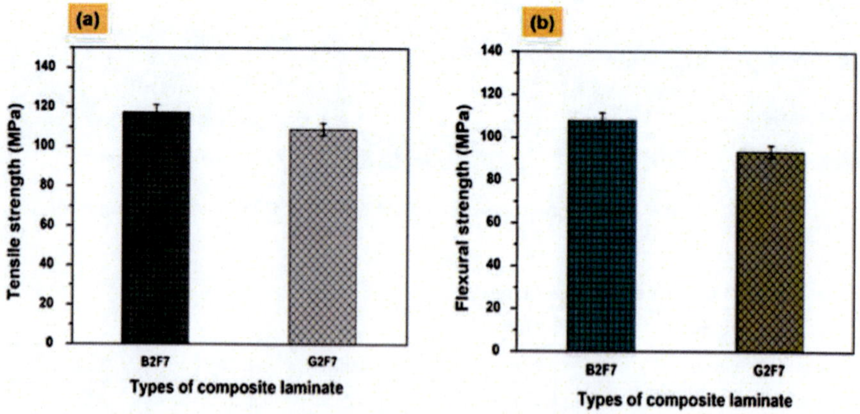

Fig. 5 **a, b** Tensile and flexural strength of B2F7 and G2F7 composites

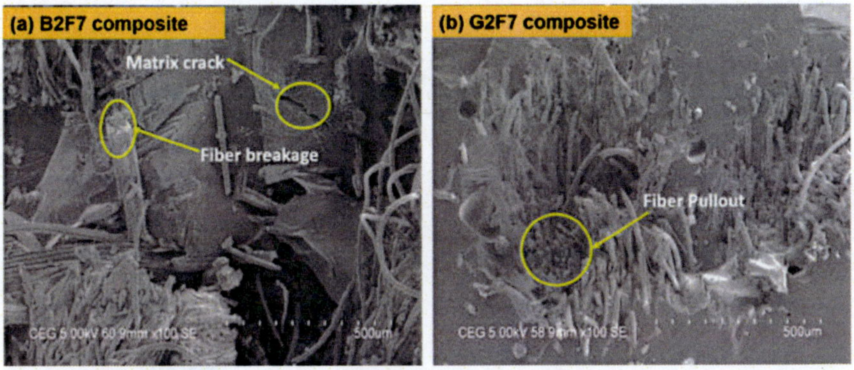

Fig. 6 Typical SEM images of tensile fracture surface

in flexural strength, the top and bottom layers are so important to redistribute the bending stress to next interlayers. The sufficient flexibility provided by the basalt fibers increases the bending strength of the composite.

The typical SEM images of the tensile fracture surface are shown in Fig. 6a, b. It is observed that good adhesion was seen between basalt fiber and matrix, as in the case of B2F7composite. G2F7 composite reveals matrix crack, fiber pullout, and breakage, as shown in Fig. 6b This is due to that changes in fiber surface morphology during loading which affecting the fiber/matrix interfacial adhesion strength.

3.2 Machining Characteristics of Interplied Composites

3.2.1 Effect of Water Pressure on Kerf Taper and Surface Roughness by Water Pressure

The effect of water pressure on the kerf taper of B2F7 and G2F7 composites is shown in Fig. 7a. This figure shows that the kerf taper ratio decreases with an increase in water pressure for both the B2F7 and G2F7 composites. Moreover, G2F7 composite exhibited minimum kerf taper. The reduction of kerf taper observed to be 14% at a pressure of 300 MPa. B2F7 composite is rather produced higher taper angle due to more brittle nature of basalt fiber. The brittle fracture of the top layer of basalt fiber increases with increase in water jet energy. This results in larger access by the water jet next layer of flax fabrics loading to higher kerf taper. As water jet moving in a transverse direction, the energy of water jet gets lowered due to friction with cutting surface. It reduces the jet ability to the removal of material at bottom layers. The effect of surface roughness by water pressure for B2F7 and G2F7 composites is shown in Fig. 7b. It can be seen that surface quality increases when an increase in water pressure for both the B2F7 and G2F7 composites. Moreover, G2F7 composite shows

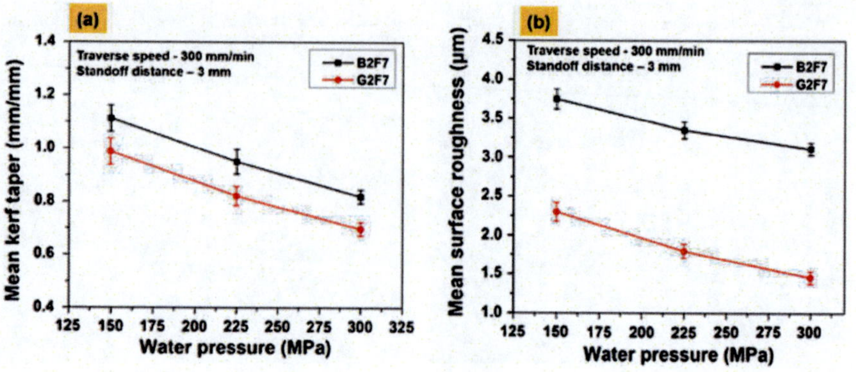

Fig. 7 **a, b** Kerf taper and surface roughness with water pressure for the B2F2 and G2F2 composites

a minimum surface roughness with 53.05% lower than B2F7 composite. Owing to uneven machining in B2F7 composite, the surface roughness was higher.

3.2.2 Effect of Mean Kerf Taper and Surface Roughness by Nozzle Traverse Speed

The effect of nozzle traverse speed on the kerf taper of B2F7 and G2F7 composites is shown in Fig. 8a. This figure clearly reveals that kerf taper increases with increasing the nozzle traverse speed for both the B2F7 and G2F7 composites. Moreover, G2F7 composite shows a minimum kerf taper with 6.2% at a traverse speed of 125 mm/min, B2F7 composite. This could be due to the energy unbalance between energy requirement for cutting and energy supplied to the point of cutting. This unbalanced cutting

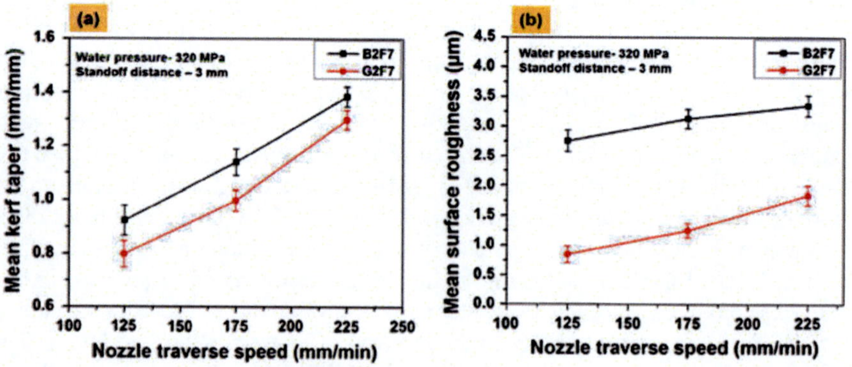

Fig. 8 **a, b** Kerf taper and surface roughness with nozzle traverse speed for B2F2 and G2F2 composites

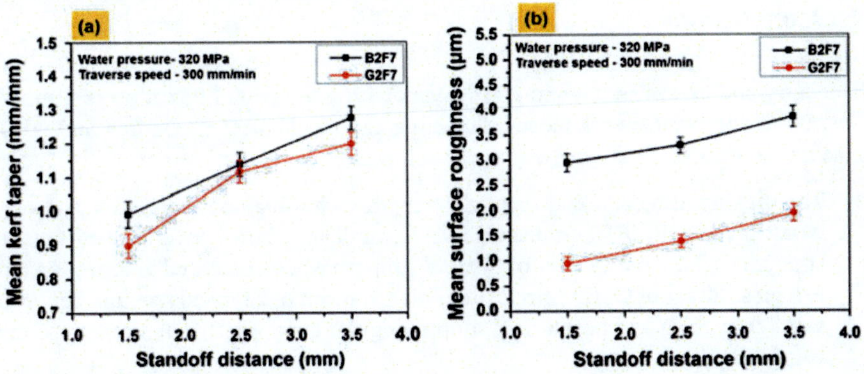

Fig. 9 **a, b** Kerf taper and surface roughness with standoff distance for B2F7 and G2F7 composites

energy increased with traverse speed left larger kerf taper. The effect of nozzle traverse speed on the surface roughness of B2F7 and G2F7 composites is shown in Fig. 8b. Surface quality decreases with increasing the nozzle traverse speed for both the B2F7 and G2F7 composites. This is attributed to the un even cutting with varied water jet energy in the traverse direction. Moreover, G2F7 composite shows a minimum kerf taper with 45.1% lower than B2F7 composite.

3.2.3 Effect of Mean Kerf Taper and Surface Roughness by Standoff Distance

The effect of standoff distance on the kerf taper of B2F7 and G2F7 composites is shown in Fig. 9a. Kerf taper ratio increases with every step increment of standoff distance for both B2F7 and G2F7 composites. This could be to water jet flaring with distance and loss of water jet energy. Moreover, G2F7 composite shows a minimum kerf taper with a reduction of 5.9% at a standoff distance of 1.5 mm as compared with B2F7 composite. The effect of standoff distance on the surface roughness of B2F7 and G2F7 composites is shown in Fig. 9b. The surface quality of the machined surface decreases with increasing the standoff distance for both B2F7 and G2F7 composites. This is a reduction in localized material erosion by flared water jet which results in non-uniform cutting over the surface. Interestingly, G2F7 composite reveals minimum surface roughness by 49.2% at standoff distance 1.5 mm than the B2F7 composite.

4 Conclusions

The entry and exit of basalt and glass layers with interlayered flax fiber reinforced polymeric composite were most influencing the mechanical properties and AWJ machining process. From the results, the following conclusion were made,

1. The maximum tensile and flexural strength were obtained with entry and exit basalt layer of B2F7 composite. B2F7 composite shows an improved tensile strength by 7.15 and 13.3% for flexural strength when compared to the entry and exit glass layers of G2F7 composite. This is a result of a typical reaction between vinyl ester resin and basalt fiber, improving the good interfacial strength of the composite.
2. Water jet pressure was considered as the most significant control factor in influencing surface quality (Ra) and kerf taper. Increase in pressure results in a better machining performance for both B2F7 and G2F7 composites. G2F7 composite shows a minimum kerf taper by 14% at a pressure of 300 MPa, as compared to B2F7 composite. As a result of more brittle nature of basalt fiber (B2F7) increases taper angle. Whilst, G2F7 composite shows a minimum surface roughness with 53.05% at a pressure of 300 MPa compared with G2F7. This could be uneven machining in B2F7 composite leads to poor surface quality.
3. Interestingly, another machining parameters, traverse rate, and standoff distance at low level improved the machining performance of both the composites.
4. These experiments confirm that greater kinetic energy and pressure energy required for abrasive water jet machining of interlayered composites.

References

1. Frollini E, Silva CG, Ramires EC (2013) Advanced fibre-reinforced polymer (FRP) composites for structural applications: 2. Phenolic resins as a matrix material in advanced fiber-reinforced polymer (FRP) composites. Elsevier Inc. Chapters
2. Westman MP, Fifield LS, Simmons KL, Laddha S, Kafentzis TA (2010) Natural fiber composites: a review (No. PNNL-19220). Pacific Northwest National Lab (PNNL), Richland, WA (United States)
3. Monteiro SN, Lopes FPD, Ferreira AS, Nascimento DCO (2009) Natural-fiber polymer-matrix composites: cheaper, tougher, and environmentally friendly. JOM 61(1):17–22
4. Kalia S, Kaith BS, Kaur I (2009) Pretreatments of natural fibers and their application as reinforcing material in polymer composites—a review. Polym Eng Sci 49(7):1253–1272
5. Joshi SV, Drzal LT, Mohanty AK, Arora S (2004) Are natural fiber composites environmentally superior to glass fiber reinforced composites? Compos A Appl Sci Manuf 35(3):371–376
6. Ramraji K, Rajkumar K, Sabarinathan P (2019) Tailoring of tensile and dynamic thermo-mechanical properties of interleaved chemical-treated fine almond shell particulate flax fiber stacked vinyl ester polymeric composites. In: Proceedings of the Institution of Mechanical Engineers, Part L: Journal of Materials: Design and Applications, p 1464420719849616
7. Gujjala R, Ojha S, Acharya SK, Pal SK (2014) Mechanical properties of woven jute–glass hybrid-reinforced epoxy composite. J Compos Mater 48(28):3445–3455

8. Sanjay MR, Arpitha GR, Yogesha B (2015) Study on mechanical properties of natural-glass fibre reinforced polymer hybrid composites: a review. Mater Today: Proc 2(4–5):2959–2967
9. Ramkumar J, Malhotra SK, Krishnamurthy R (2004) Effect of workpiece vibration on drilling of GFRP laminates. J Mater Process Technol 152(3):329–332
10. Wang J, Wong WCK (1999) A study of waterjet cutting of metallic coated sheet steels. Int J Mach Tools Manuf 39:855–870 (1999)
11. Gnanavelbabu A, Saravanan P, Rajkumar K, Karthikeyan S (2018) Experimental investigations on multiple responses in abrasive waterjet machining of Ti-6Al-4 V alloy. Mater Today: Proc 5(5):13413–13421
12. Irina Wong MM, Azmi AI, Lee CC, Mansor AF (2018) Kerf taper and delamination damage minimization of FRP hybrid composites under abrasive water-jet machining. Int J Adv Manufact Technol 94(5–8):1727–1744
13. Azmir MA, Ahsan AK (2009) A study of abrasive water jet machining process on glass/epoxy composite laminate. J Mater Process Technol 209(20):6168–6173
14. El-Hofy M, Helmy MO, Escobar-Palafox G, Kerrigan KM, Scaife R, El-Hofy H (2018) Abrasive water jet machining of multidirectional CFRP laminates. Procedia CIRP 68:535–540 (Elsevier)
15. Selvam R, Karunamoorthy L, Arunkumar N (2017) Investigation on performance of abrasive water jet in machining hybrid composites. Mater Manuf Processes 32(6):700–706
16. Singha K (2012) A short review on basalt fiber. Int J Text Sci 1(4):19–28

Electrolyte and Machining Parameters Optimization of Wire Electrochemical Cutting of Aluminum/Titanium Diboride Composite

K. Rajkumar, M. Rajesh, K. Ramraji, P. Sabarinathan, and A. Gnanavelbabu

Abstract A hard to machining characteristics of ceramic reinforced aluminum composite requires unconventional machining methods. Wire electrochemical machining (WECM) is one of the superior machining processes that produced a smoother and brighter surface with complex profile. In this investigation, optimization of process parameters is mainly focused for slit profile cutting on the aluminum 6061/TiB_2 composite by a WECM method. Liquid metallurgy route is employed to fabricate the composite with varied TiB_2 weight percentages (5 and 15%). The parameters of wire electrochemical cutting process are applied voltage (10–14 V), electrolyte flow rate (2–8 l/min), and electrolyte concentration (12–18%). These were optimized for achieving optimal material removal rate and surface roughness. The optimization of process parameters was carried out based on the response surface methodology. Electrolyte flow rate significantly increases MRR, surface roughness, and slit width. When considering the optimal MRR and slit width condition, the experiment results reveal a better surface finish at low-level electrolyte concentration. Inter-electrode voltage increases the strength of electrochemical ionization resulting in higher MRR, and on the other hand, it is considerably affected surface roughness.

Keywords Wire electrochemical machining · Surface roughness · TiB_2 · Electrolyte · Composite machining

1 Introduction

Aluminum metal matrix composites have opened a new window to various industrial sectors like automobile components—drive shaft, brakes, cylinder liners, engine

K. Rajkumar (✉) · M. Rajesh · K. Ramraji
Department of Mechanical Engineering, Sri Sivasubramaniya Nadar College of Engineering, Chennai, Tamil Nadu, India
e-mail: rajkumark@ssn.edu.in

P. Sabarinathan · A. Gnanavelbabu
Department of Industrial Engineering, College of Engineering, Chennai, Tamil Nadu, India

S. Vijayan et al. (eds.), *Trends in Manufacturing and Engineering Management*, Lecture Notes in Mechanical Engineering,
https://doi.org/10.1007/978-981-15-4745-4_5

locks, and motor casing [1]. Generally, aluminum metal matrix composites are reinforced with mono or combination of reinforcements such as oxides, nitride, carbide, and borides [2] to increase its mechanical properties. The addition of the reinforcement leads to complexity in machining of composite. Normally, the conventional machining process cannot help in the context of hard materials like a metal matrix composites and ceramic materials [3]. A selected machining technology requires to produce good surface integrity for components made from particulate reinforced composites. The past and present scenarios show an electrochemical machining that fulfills the industry requirements without much disturbing the materials being machined. Another variant of an electrochemical process namely wire electrochemical machining process is a hybrid unconventional metal cutting process for very hard materials [4]. The reasons for selecting a wire electrochemical machining process, specifically for metal matrix composites, are produced complex shapes, profiles, and slits without residual stress.

Normally, material removal rate (MRR), surface roughness, and slit width are the essential characteristics for the WEDM machining operations [5]. These characteristic parameters depend on the machining gap, concentration of electrolyte, type of electrolyte, wire diameter, and current source characteristics. Senthilkumar et al. [6] obtained higher MRR and lower surface roughness by performing electrochemical machining of MMC with different process parameters. It is reported that increases in voltage and feed rate resulting in lower surface roughness and higher MRR. Furthermore, increasing in concentration and flow rate of electrolyte showed the same result. Taweel et al. [7] were optimized wire electrochemical turning experimental results using a response surface method. The input parameters such as feed rate, rotational speed, and overlap distance were optimized to obtain higher material removal rate and lower surface roughness. In case of ECM optimization, Bhattacharyya et al. [8] developed a mathematical model to correlate machining response to process parameters.

In the present study, the WEDM on the Al-5 & 15wt% TiB_2 composites has been done. Process parameters like inter-gap voltage, concentration, and flow rate of electrolyte were optimized in order to obtain better machining responses such as surface roughness, material removal rate, and slit width.

2 Materials and Method

Composites made from Al6061 matrix reinforced with TiB_2 ceramic particles (size of 10 μm) with two different proportions of (5% and 15wt%) using a stir casting method. The fabricated composites were heat treated by T-6 heat treatment method. Figure 1 shows the experimental machine setup of the wire electrochemical machine.

In a wire electrochemical machining process, copper wire of 0.5 mm diameter can be used as a cathode and the Al6061/TiB_2 as an anode. Both electrodes were immersed in a sodium chloride electrolyte solution with wire feed rate of 1 mm/min. experiments were conducted by varying the input parameters such as

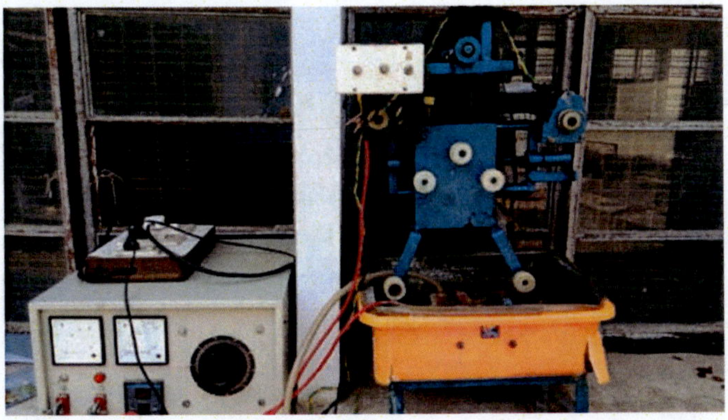

Fig. 1 Experimental machine setup WECM

Table 1 WECM process parameters

Parameters	Value
Tool (mm)	Copper wire Ø 0.5
Electrolyte concentration (%)	12, 15, 18
Voltage (V)	10, 12, 14
Feed rate (mm/min)	1
Electrolyte flow rate (l/min)	2, 5, 8
Electrolyte	Sodium chloride

DC voltage of (10–14 V), electrolyte concentration (12–18%), and flow rate (2–8 l/min). The machining output parameters like MRR, slit width, and surface roughness are reported. MRR was calculated using a mass loss method. Slit width was observed by optical microscope. The surface roughness of machined surface was measured by using a talysurf meter with cut-off length 0.8 mm. Table 1 shows complete experimental parameters for WECM.

2.1 Response Surface Methodology

Material removal rate, surface roughness, and slit width were classified as response functions to the voltage (A), electrolyte concentration (B), and electrolyte flow rate (C) and were considered as control variables. The experiments were conducted using a design of experiment technique with central composite design. The experiments were run as sequence of L19 orthogonal array. Table 2 shows the process parameters levels as per the central composite design. Table 3 presents an experimental design matrix and 5 and 15% reinforced composite experimental results.

Table 2 Experimental parameters and their levels

Process parameters	Levels		
	−1	0	1
Voltage (V)	10	12	14
Electrolyte flow rate (l/min)	2	5	8
Electrolyte concentration (%)	12	15	18

Table 3 Experimental design matrix and results

Composition			5% reinforcement			15% reinforcement		
(A) Voltage (V)	(B) Electrolyte concentration (%)	(C) Electrolyte flow rate (l/min)	MRR (gm/min)	Surface roughness (μm)	Slit width (mm)	MRR (gm/min)	Surface roughness (μm)	Slit width (mm)
12	15	5	120	4.311	0.802	96	4.744	0.761
10	15	5	108	3.822	0.772	76	4.322	0.731
10	18	8	129	4.434	0.783	99	4.985	0.745
12	15	8	130	4.115	0.784	106	4.398	0.748
14	15	5	124	4.641	0.819	113	5.349	0.775
10	12	2	79	3.308	0.760	56	3.621	0.734
10	12	8	106	2.724	0.715	84	3.346	0.709
12	15	5	122	4.311	0.802	94	4.823	0.789
12	15	5	117	4.421	0.794	99	4.912	0.795
12	12	5	104	3.311	0.771	80	3.712	0.744
12	15	5	124	4.484	0.814	95	4.658	0.779
12	18	5	126	4.985	0.831	105	5.373	0.785
14	18	2	125	5.581	0.869	105	6.387	0.809
12	15	2	108	4.511	0.819	84	4.856	0.772
14	12	8	125	2.986	0.781	105	3.975	0.745
14	12	2	90	3.689	0.811	80	4.324	0.771
12	15	5	123	4.254	0.811	100	5.01	0.804
10	18	2	107	5.092	0.828	81	5.468	0.761
14	18	8	146	5.225	0.841	127	5.985	0.788

3 Results and Discussion

3.1 Mathematical Modeling

The mathematical model for the material removal rate when considering with various process parameters for 5 and 15% reinforcement composites is shown in Eqs. 1 and 2,

$$\begin{aligned} \mathrm{MRR}(5\%) = &+ 120.25 + 8.10 * A + 12.90 * B + 12.70 * C + 0.62 * A * B \\ &+ 0.88 * A * C - 2.38 * B * C - 3.06 * A2 - 4.06 * B2 - 0.057 * C2 \end{aligned} \quad (1)$$

$$MRR(15\%) = +93.95 + 13.40 * A + 11.20 * B + 11.50 * C \quad (2)$$

The mathematical model for the surface roughness of machined surface of 5 and 15% reinforcement composites is shown in Eqs. 3 and 4,

$$\text{Surface Roughness } (5\%) = +4.22 + 0.27 * A + 0.93 * B - 0.27 * C \quad (3)$$

$$\text{Surface Roughness } (15\%) = +4.75 + 0.43 * A + 0.92 * B - 0.20 * C \quad (4)$$

The mathematical model for the obtained slit width while cutting with wire as considering various process parameters for 5 and 15% reinforcement composites is shown in Eqs. 5 and 6,

$$\text{Slit width } (5\%) = +0.80 + 0.026 * A + 0.031 * B - 0.018 * C \quad (5)$$

$$\text{Slit width } (15\%) = +0.77 + 0.021 * A + 0.018 * B - 0.011 * C \quad (6)$$

3.2 *Parametric Influence on Material Removal Rate*

Analyzing experimental work by DOE method and regression model for MRR, and the effect caused by control variables is given in ANOVA Table 4. Composite 5 and 15% show that the developed MRR model was well-suited for machining of composite material, as seen that model is highly significance. Likewise, each control variable was affected material removal rate within confidence limit of 95%, as clearly shown in Table 4.

Figure 2a, b shows a 3D model graph of MRR of composite machining with electrolyte flow rate, concentration, and voltage. Electrolyte flow rate and concentration were directly affected by MRR. When increased electrolyte flow rate hydrogen bubbles are removed which were generated at cathodic surface, this increases the ionic strength of electrolyte in the slit cutting zone. Finally, it leads to high material removal rate at anode surface. Besides, high flow rate of electrolyte can speed up ionization process between inter-electrode gaps. This results in more anodic dissolution of material and easy flushing of formed debris. Further, electrolyte high flow rate generates a fresh surface over and over again so that fresh un-machined surface available for further electrochemical machining. This also resulted in increased MRR with the flow rate. Similarly, the electrical conductivity of the electrolyte between the wire and composite is increased by electrolyte concentration. Thus, electrolyte conductivity between electrodes increased which speed up the electrochemical reaction. This makes a process to steadily improve its cutting ability on the composite. It is also observed that increases in the voltage and produces high machining current

Table 4 ANOVA table for MRR

Source	5% reinforcement					15% reinforcement				
	Sum of squares	df	Mean square	F value	*p*-value Prob > *F* value	Sum of squares	df	Mean square	F value	*p*-value Prob > *F* value
Model	4179.78	9	464.42	58.91	<0.0001	4372.50	3	1457.5	106.93	<0.0001
A-voltage	656.10	1	656.10	83.22	<0.0001	1795.60	1	1795.6	131.74	<0.0001
B-electrolyte concentration	1664.10	1	1664.1	211.0	<0.0001	1254.40	1	1254.4	92.03	<0.0001
C-electrolyte flow rate	1612.90	1	1612.9	204.5	<0.0001	1322.50	1	1322.5	97.03	<0.0001
Residual	70.95	9	7.88			204.45	15	13.63		
Pure error	30.80	4	7.70			26.80	4	6.70		
Pred R-squared	0.8911					0.9258				
Adj R-squared	0.9666					0.9464				

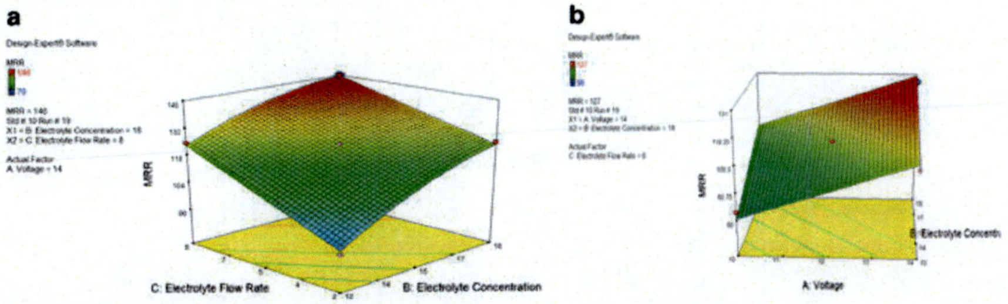

Fig. 2 Effect of MRR **a** electrolyte flow rate and concentration **b** voltage and electrolyte concentration

in the inter-electrode gap. It might result in increased current density between the electrodes that increased material removal rate at anode surface [9].

3.3 Parametric Influence on Surface Roughness

Table 5 shows an ANOVA table for surface roughness of machined surface of composites. This table indicates developed surface roughness model which perfectly reflects the control variables behavior to the experimental results. The control variables were adequately fit into the developed mathematical model.

Figure 3a, b shows the 3D model graph of surface roughness with the electrolyte flow rate, concentration, and voltage. It is observed from the plot that surface roughness slightly increases with increase in flow rate. Owing to flow streaks on the surface increases unsteady rotating eddies on the surface. This results in high turbulence over the surface. Surface roughness increases with increasing electrolyte concentration. At low concentration, the ion depletion in electrolyte between inter-electrode gap results in reduced surface roughness [10]. The current density increases with increase in electrolyte concentration which leads to pitting on the cutting surface. It is observed that surface roughness affected with inter-electrode potential difference increases. When voltage increased the metal, dissolution was also increased. This leads to an uneven dissolution around the cutting surface resulting in poor surface finish.

3.4 Parametric Influence on Slit Width

Table 6 shows an ANOVA table for slit width of composites. This table clearly indicates that developed slit width model with consideration of all control variables is well-fitted by seeing an evidence of less than 5% P-value. All the control variables were significantly affected the process

Table 5 ANOVA table for surface roughness

Source	5%					15%				
	Sum of squares	df	Mean square	*F* value	*p*-value prob > *F* value	Sum of squares	df	Mean square	*F* value	*p*-value prob > *F* value
Model	10.13	3	3.38	130.74	<0.0001	10.72	3	3.57	159.4	<0.0001
A-voltage	0.75	1	0.75	29.12	<0.0001	1.83	1	1.83	81.65	<0.0001
B-electrolyte concentration	8.65	1	8.65	334.93	<0.0001	8.50	1	8.50	379.28	<0.0001
C-electrolyte flow rate	0.73	1	0.73	28.17	<0.0001	0.39	1	0.39	17.26	0.0008
Residual	0.39	15	0.026			0.34	15	0.022		
Pure error	0.035	4	8.766E−003			0.076	4	0.019		
Pred R-squared	0.9360					0.9546				
Adj R-squared	0.9558					0.9635				

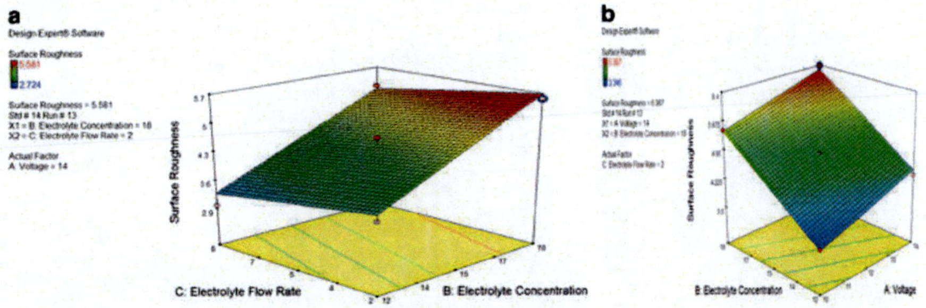

Fig. 3 Surface roughness plot **a** electrolyte flow rate and concentration **b** voltage and electrolyte concentration

The slit width of the composites as function of different process variables are shown in Fig. 4a, b. It is observed that high voltage increased order of MRR. Increase in metal dissolution increases the slit width. Similarly, at high flow rate the burrs were removed and enlarge the slit width. Hence, the medium flow rate was optimal for machining of composites.

Slit width was observed using optical microscopy images. Figure 5a, b shows the slit profiles of machined composites. 5% reinforcement composite shows a line straight geometric profile (Fig. 5a). This is the effect of lesser reinforcement particle's presence in composite. It is understood that these particles slightly interfere the machining. The presence of larger amount of nonconductive ceramic particles, in case of 15% reinforcement composite, leads to inhomogeneous dissolution and uneven distribution of current intensity resulted in an entry side enlarge of slit (Fig. 5b).

4 Conclusions

In this present study, the various compositions of aluminum/titanium diboride composites were machined using a wire electrochemical machining process. The machining parameters were optimized by using response surface methodology. Mathematical model for the responses such as MRR, slit width, and surface quality was developed, and the optimal conditions are presented.

1. Electrolyte flow rate and concentration were directly affected MRR of composite machining. An electrolyte flow rate reduces hydrogen bubbles attachment over the cathodic surface. This improves ionic reaction of electrolyte in cutting zone; therefore, high material removal rate is observed at anode surface.
2. The electrolyte electrical conductivity between electrodes increases with electrolyte concentration. This also increases electrochemical reaction resulting in higher MRR.
3. An increase in electrolyte flow rate substantially increases the surface roughness. Owing to flow streaks over the cutting surface an unsteady rotating eddy

Table 6 ANOVA table for slit width

Source	5%					15%				
	Sum of squares	df	Mean square	*F* value	*p*-value prob >*F* value	Sum of squares	df	Mean square	F value	*p*-value prob >*F* value
Model	0.020	3	6.708E−003	162.56	<0.0001	9.003E−003	3	3.001E−003	11.12	0.0004
A-Voltage	6.917E−003	1	6.917E−003	167.61	<0.0001	4.326E−003	1	4.326E−003	16.03	0.0012
B-Electrolyte concentration	9.860E−003	1	9.860E−003	238.92	<0.0001	3.422E−003	1	3.422E−003	12.68	0.0028
C-Electrolyte flow rate	3.349E−003	1	3.349E−003	81.15	<0.0001	1.254E−003	1	1.254E−003	4.65	0.0478
Residual	6.190E−004	15	4.127E−005			4.049E−003	15	2.700E−004		
Pure error	2.552E−004	4	6.380E−005			2.552E−004	4	6.380E−005		
Pred R-squared	0.9534					0.5916				
Adj R-squared	0.9642					0.6277				

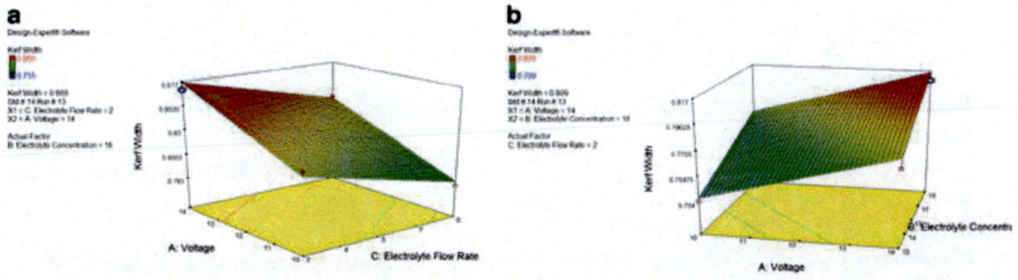

Fig. 4 Slit width plot **a** electrolyte flow rate and concentration **b** voltage and electrolyte concentration

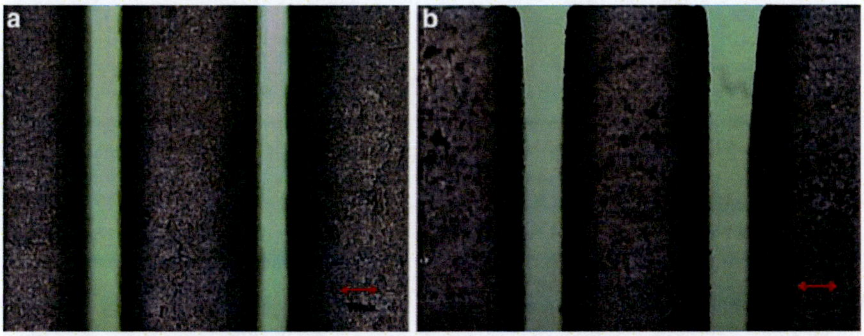

Fig. 5 Slit profiles of machined composites. **a** 5% composite and **b** 10% composite

increases which results in an uneven machining. Surface roughness increases with increasing electrolyte concentration. The current density increases with increasing electrolyte concentration that leads to pitting on the cutting surface.

4. It is observed that high voltage increased slit width dimension. Because of metal dissolution increases the slit width. Similarly, high flow rate removes burrs but enlarges the slit width.
5. The optimum process parameters for efficient machining of 5% reinforced composite were voltage (10.2 V), electrolyte concentration (12.8%), and electrolyte flow rate (8 l/min). The optimum process parameters for efficient machining of 15% reinforced composite were 14 V, 12% electrolyte concentration, and 8 l/min flow rate.

References

1. Hemanth J (2009) Quartz (SiO_2p) reinforced chilled metal matrix composite (CMMC) for automotive applications. Mater Des 30(2):323–329
2. Shorowordi KM, Laoui T, Haseeb ASMA, Celis JP, Froyen L (2003) Microstructure and interface characteristics of B4C, SiC and Al_2O3 reinforced Al matrix composites: a comparative study. J Mater Process Technol 142(3):738–743
3. McGeough JA (1988) Advanced methods of machining. Springer Science & Business Media (1988)
4. Sen M, Shan HS (2005) A review of electrochemical macro-to micro-hole drilling processes. Int J Mach Tools Manuf 45(2):137–152
5. Sankar M, Gnanavelbabu A, Rajkumar K (2014) Effect of reinforcement particles on the abrasive assisted electrochemical machining of Aluminium-Boron carbide-Graphite composite. Procedia Eng 97:381–389
6. Senthilkumar C, Ganesan G, Karthikeyan R (2009) Study of electrochemical machining characteristics of Al/SiC p composites. Int J Adv Manufact Technol 43(3–4):256–263
7. El-Taweel TA, Gouda SA (2011) Performance analysis of wire electrochemical turning process—RSM approach. Int J Adv Manufact Technol 53(1–4):181–190
8. Bhattacharyya B, Sorkhel SK (1999) Investigation for controlled electrochemical machining through response surface methodology-based approach. J Mater Process Technol 86(1–3):200–207
9. Bhattacharya B, Sorkhel SK (1997) Response surface methodology based analysis for achieving controlled electrochemical machining. In: Proceedings of the 17th all India machine tool design and research conference, pp 307–311
10. Sankar M, Gnanavelbabu A, Rajkumar K, Thushal NA (2017) Electrolytic concentration effect on the abrasive assisted-electrochemical machining of an aluminum–boron carbide composite. Mater Manuf Processes 32(6):687–692

Modeling and Parametric Optimization of Process Parameters of Wire Electric Discharge Machining on EN-31 by Response Surface Methodology

Sushant B. Patil, Swarup S. Deshmukh, Vijay S. Jadhav, and Ramakant Shrivastava

Abstract Nowadays, the majority of industry uses nonconventional machines; wire electric discharge machining is one of them. In this experimentation, optimize process parameters of wire electric discharge machining with help of response surface methodology. Central composite design is used for the design of experiments. The process parameters considered for this study are a pulse on time, wire feed rate, pulse off time, and servo voltage. For this experimentation work, EN-31 used as workpiece material. The high percentage of carbon present in the material due to this is used for manufacturing punches and dies. To find out significant factors, ANOVA is calculated. Analysis of variance for MRR clearly shows that pulse on time and servo voltages are the most significant parameters. From result analysis, the high value of MRR is obtained at high value of pulse on time and low value of servo voltage. In the case of surface roughness also pulse on time and servo voltage are the most significant factors as compared to others; the low value of surface roughness is obtained at a low value of pulse on time and high value of servo voltage.

Keywords WEDM · EN-31 · MRR · SR · ANOVA · RSM

S. B. Patil (✉) · S. S. Deshmukh · V. S. Jadhav · R. Shrivastava
Department of Mechanical Engineering, Government College of Engineering, Karad, Maharashtra 415124, India
e-mail: patilsush96@gmail.com

S. S. Deshmukh
e-mail: deshmukh.swarup@gmail.com

V. S. Jadhav
e-mail: vijay.jadhav@gcekarad.ac.in

R. Shrivastava
e-mail: ramakant.shrivastava@gmail.com

S. Vijayan et al. (eds.), *Trends in Manufacturing and Engineering Management*, Lecture Notes in Mechanical Engineering,
https://doi.org/10.1007/978-981-15-4745-4_6

1 Introduction

In wire electric discharge machining electro-thermal mechanism used for cutting workpiece as same as electric discharge machining. The wire of various materials like brass, zinc-coated brass, diffused brass, molybdenum is used as an electrode in case of wire electric discharge machine. The wire electrode continuously travels along the cutting path generated by the numerical control program. In the wire electric discharge machining workpiece acts as anode and wire electrode as cathode and the proper gap is provided between them. Due to this, high-intensity spark is generated between them; this spark is responsible for melting workpiece. In cutting region, the temperature goes too high due to this high-intensity spark energy so deionized water helps to reduce the temperature in that region and helps to remove debris particle which is produced after cutting workpiece material. This is the non-contact type of cutting process, and hence no stress is present in the workpiece after machining. The complex, contour, and irregular shape profiles easily cut from the workpiece this is one of the major advantages. Kansal et al. [1] optimized process parameters of PMEDM using response surface methodology. In this experimentation, they used silicon powder in a dielectric fluid. Central composite design with face-centered used for the design of experiments. Finally, they concluded that after the addition of silicon powder in dielectric fluid improvement in value of response variables takes place. Kung and Chiang et al. [2] carried out parametric optimization of wire electric discharge machining on aluminum oxide-based ceramic with help of response surface methodology. After increasing the values of T_{on} and duty factor, MRR and SR increase up to a certain limit then after it decreases. Patel et al. [3] performed parametric optimization of process parameters of EDM on Al_2O_3/Sic/Tic ceramic composite. They finally concluded that pulse on time is a major significant factor as compared to other factors for surface roughness. After increasing the value of the duty cycle, surface roughness first increases then decreases. Patil et al. [4] carried out an analysis of wire electric discharge machining on Al/SiC_p MMC using response surface methodology. The conclusion of that experimentation is that (T_{on}), (T_{off}), and volume fraction (V%) of ceramic reinforcement are the most significant parameters for cutting rate. For surface roughness (V%) of ceramic, reinforcement and (T_{on}) are the most significant parameters. Ojha et al. [5] studied material removal rate and tool wear rate as response variables after the addition of chromium powder in a dielectric fluid. They get a high value of material removal rate at higher powder concentration and tool wear rate is decreases after increasing value of tool diameter. Shandilya et al. [6] carried out the optimization of process parameters of WEDM on MMC by response surface methodology. Good surface finish and small size of the crater is getting at a low level of process parameters. Khanna et al. [7] performed an analysis of cryogenically treated workpiece on wire electric discharge machining by response surface methodology. They performed single and multi-objective optimization to achieve maximum value of MRR and the minimum value of surface roughness. The error between actual and predicted values for MRR lies between −2 and +4% and for SR lies between −7 and +7%. Gopalakannan et al. [8] used central composite design and desirability approach for the optimization of process factors of EDM on MMC. The low value of EWR is observed at high value of T_{off} (pulse off time) and SR is influenced by current and T_{on} (pulse on time). Sharma et al. [9] performed single and multi-objective optimization of process variables of WEDM using response surface methodology (RSM). For this research work, they used HSLA (High Strength Low Alloy) steel used as workpiece material and brass wire as an electrode. The result of ANOVA shows that (CS), T_{on}, and T_{off} are the most significant parameters and for (DD), (SV), and (T_{on}) are the most significant parameters. Singh et al. [10] investigated surface characteristics after powder mixed electric discharge machining and compare results of surface roughness and recast layer in between powder mixed and without powder mixed electric discharge machining. The workpiece material used for this study is aluminum 6061. Finally, they concluded that surface roughness and recast layer is reduced in the case of powder electric discharge machining as compared to normal electric discharge machining. Dubey et al. [11] studied the material removal rate in powder mixed electric discharge machining on AA7075/B_4C composite. In this experimentation for improving material removal rate chromium powder is added in a dielectric fluid. From SEM images of the recast, layer shows that recast layer thickness is more at high value of process variables as compared to initial values of process variables.

2 Experimental Setup

This experimentation is performed on the ELECTRONICA-ECOCUT WEDM which is shown in Fig. 1. This is five-axis machine, for normal cutting of workpiece nozzle move in *x*- and *y*-direction and for taper cutting nozzle moves in *U* and *V* axis. The R_a is measured with the help of MITUTOYO SJ-210 surface roughness tester which is shown in Fig. 2. The EN-31 material is selected as workpiece material for this study. Due to the high % of carbon majority application is in the punches and dies industry. The chemical composition of EN-31 is shown in Table 1. Process

Fig. 1 Machine setup (Electronica-Ecocut)

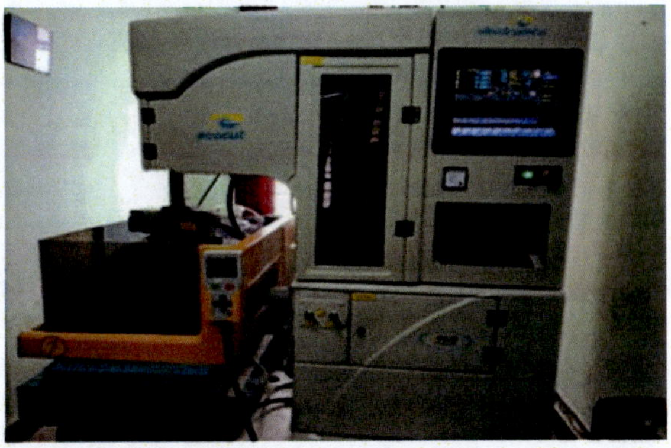

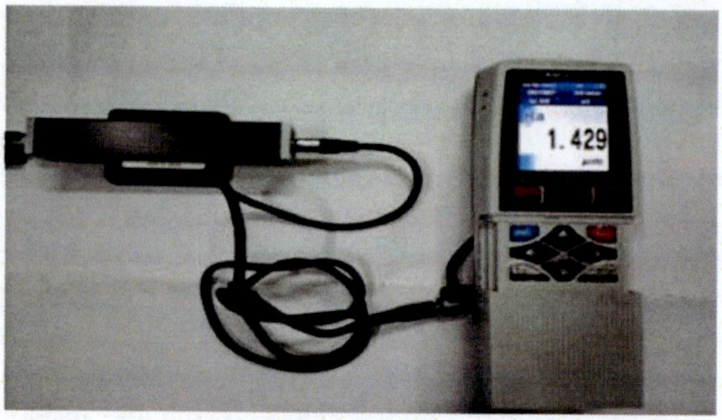

Fig. 2 Surface roughness tester (Mitutoyo SJ-210)

Table 1 Chemical composition of EN-31

Elements	C%	Mn%	S%	P%	Cr%
Contents	0.90–1.20	0.30–0.75	0.050 Max	0.050 Max	1.00–1.60

parameters and their levels in terms of actual and coded value is shown in Table 2. The zinc (Zn) coated Brass wire of 0.25 mm which is used as an electrode in wire electric discharge machining. So the weight (Wt.) of the workpiece is measured with the help of PESCO weight balance machine having least count of 0.001 gm. The PESCO weight balance machine is shown in Fig. 3. The equation used for calculation of MRR which is given in Eqs. 1.

Table 2 Process Parameters and their levels in terms of coded and actual value

Factor symbol	Parameter	Unit	Levels	
			Low (−1)	High (+1)
A	Pulse on time (T_{on})	μs	110 (−1)	120 (+1)
B	Pulse off time (T_{off})	μs	45 (−1)	55 (+1)
C	Servo Voltage (SV)	V	20 (−1)	40 (+1)
D	Wire Feed (WF)	m/min	2 (−1)	6 (+1)

Fig. 3 PESCO weight balance machine

Fig. 4 Machined specimen

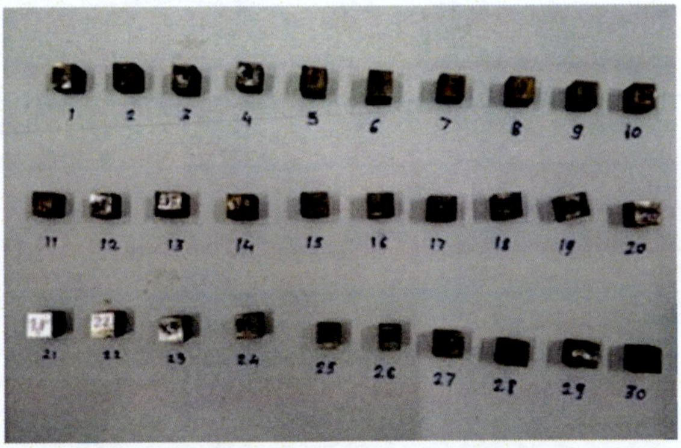

$$MRR = \frac{\text{Weight of workpiece before machining} - \text{Weight of workpiece after machining}}{\text{Time}} \frac{\text{gm}}{\text{min}} \quad (1)$$

Central composite design is used for design of experiments. Central composite design is an experimental design in which factors value less or greater than factors level are tested in a systematic way in order to better relate the factor to the response in a precise way. In this experimentation full factorial design with all possible combinations of process parameters considered at a low level and high level, in this design eight axial (star) points at the face of cube and centre point is equal to six. The face-centered central composite design is selected in which alpha value is equal to one. In this study, total 30 experiments are conducted according to design matrix given by central composite design. The 30 specimens cut according to design matrix which is shown in Fig. 4. The design matrix is developed with the help of "Design Expert 11.0" which is shown in Table 3. The graphical analysis, i.e., (surface plots, contour plots), analysis of variance (ANOVA), regression analysis is done with the help of "Design Expert 11.0" software.

3 Results and Discussion

To find out significant factors from model and to check the lack of fit of model, analysis of variance is calculated. Result of analysis of variance for MRR is shown in Table 4.

3.1 Analysis of Material Removal Rate

According to the fit summary linear model is significant for analysis of material removal rate. The value of R^2 is 96.07% and the value of adjusted R^2 is 95.44% this clearly indicates that the regression model gives a good relationship between process factors and the response variable. The factors having P-value less than 5% (i.e., 0.05)

Table 3 Design matrix and output response

Exp. Run	Process parameters				Response variable	
	A: T_{on}	B: T_{off}	C: SV	D: WF	MRR	SR
	μs	μs	Volt	m/min	gm/min	Micrometer
1	110	45	20	2	0.220637	1.405
2	120	45	20	2	0.448101	2.870
3	110	55	20	2	0.227062	1.391
4	120	55	20	2	0.407225	3.345
5	110	45	40	2	0.157805	1.390
6	120	45	40	2	0.371077	1.578
7	110	55	40	2	0.14984	1.401
8	120	55	40	2	0.306302	1.490
9	110	45	20	6	0.22588	1.341
10	120	45	20	6	0.45933	3.488
11	110	55	20	6	0.210858	1.296
12	120	55	20	6	0.427424	3.417
13	110	45	40	6	0.16485	1.368
14	120	45	40	6	0.331006	1.620
15	110	55	40	6	0.159317	1.443
16	120	55	40	6	0.31628	1.599
17	110	50	30	4	0.202971	1.404
18	120	50	30	4	0.38827	2.367
19	115	45	30	4	0.286261	1.494
20	115	55	30	4	0.283548	1.519
21	115	50	20	4	0.331577	2.276
22	115	50	40	4	0.215709	1.387
23	115	50	30	2	0.266409	1.469
24	115	50	30	6	0.266356	1.535
25	115	50	30	4	0.254277	1.516
26	115	50	30	4	0.245908	1.492
27	115	50	30	4	0.294183	1.564
28	115	50	30	4	0.285822	1.475
29	115	50	30	4	0.283582	1.555
30	115	50	30	4	0.279677	1.399

Table 4 ANOVA for MRR

Source	Sum of squares	DF	Mean square	*F*-value	*p*-value	Check significance
Model	0.2034	4	0.0509	152.83	<0.0001	Significant
A-T_{on}	0.1674	1	0.1674	502.97	<0.0001	Significant
B-T_{off}	0.0017	1	0.0017	5.24	0.0309	
C-SV	0.0343	1	0.0343	103.11	<0.0001	Significant
D-Wire feed	2.601E−06	1	2.601E−06	0.0078	0.9303	
Residual	0.0083	25	0.0003			
Lack of fit	0.0065	20	0.0003	0.8748	0.6294	Not significant
Pure error	0.0018	5	0.0004			
Cor. Total	0.2128	29				
Std. Dev.	0.0182		R^2	0.9607		
Mean	0.2823		Adjusted R^2	0.9544		
C.V.%	6.46		Predicted R^2	0.9415		

these factors are most significant. The factor A-Pulse on time, B-Pulse off time, C-servo voltage has a significant effect. Among them, A-pulse on time, C-servo voltage these parameters are most significant for MRR. The value of lack of fit is also non-significant this is desirable. Figure 5 shows that the normal probability plot for MRR is clearly indicated that residuals are present on a straight line, it means that error follows a normal distribution. Figure 6 shows the plot for predicted versus actual

Fig. 5 Normal probability plot for MRR

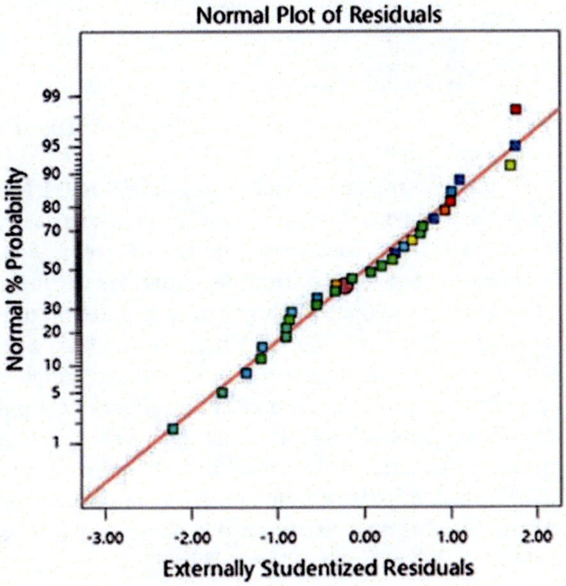

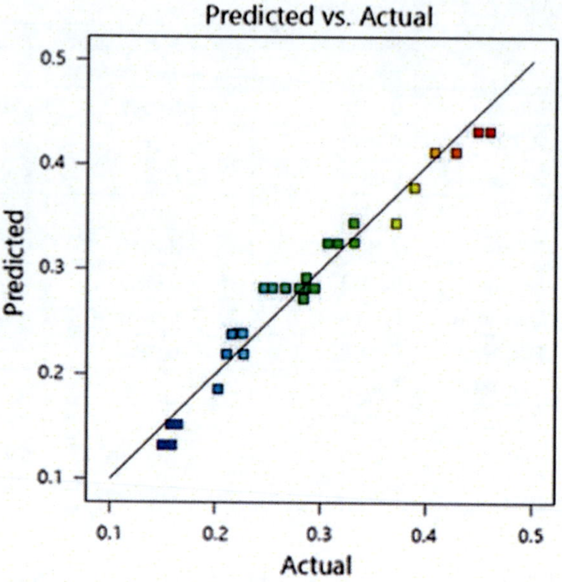

Fig. 6 Plot of actual versus predicted response of MRR

value it shows that model is fitted well. After backward elimination, i.e., eliminates non-significant terms equation of MRR in coded terms and actual terms is given in Eqs. 2 and 3 respectively.

$$\text{(In Coded terms)} - \text{MRR} = 0.2823 + 0.0964A - 0.0098B - 0.0437C \quad (2)$$

$$\text{(In actual factors)} - \text{MRR} = -1.70634 + 0.019287 * T_{\text{on}} - 0.001968T_{\text{off}} - 0.004366 * \text{SV} \quad (3)$$

Figure 7 shows the surface plot for MRR in relation to the process parameters of the pulse on time and servo voltage. From the surface plot, it is clearly observed that as a pulse on time increases material removal rate increases, and as servo voltage increases material removal rate decreases. Hence, the maximum value of the material removal rate is obtained at high value of pulse on time (120 μs) and a low value of servo voltage (20 V). The minimum value of the material removal rate is obtained at a low value of pulse on time (110 μs) and a high value of servo voltage (40 V). From Fig. 8, it is clearly indicated the effect of the pulse on time and pulse off time on material removal rate. The material removal rate increases as the value of pulse off time decreases. In this case, the high value of material removal rate is obtained at high value of pulse on time (120 μs) and low value of pulse off time (45 μs) and minimum value material removal is obtained at low value of pulse on time (110 μs) and high value of pulse off time (55 μs).

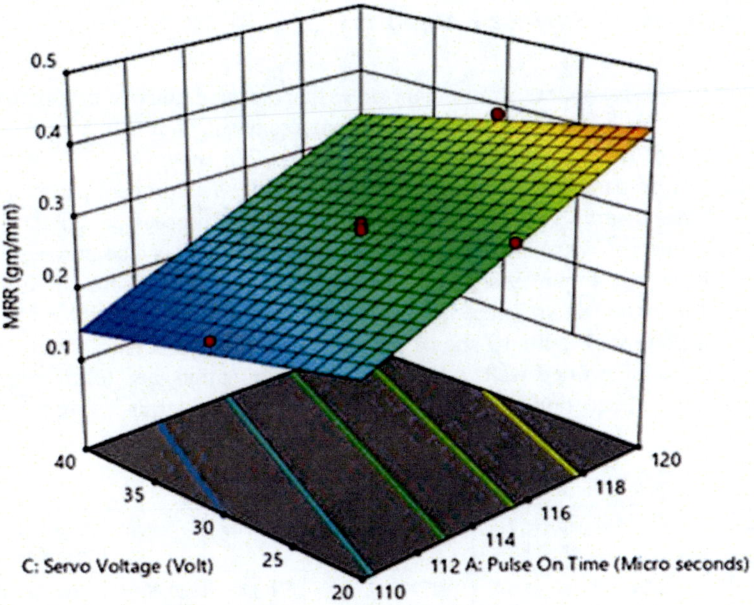

Fig. 7 Effect of T_{on} and SV on MRR

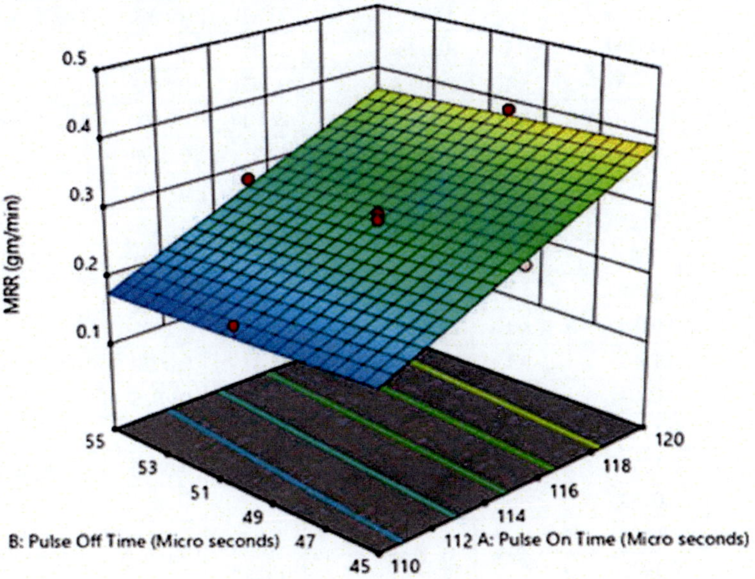

Fig. 8 Effect of T_{on} and T_{off} on MRR

3.2 Analysis of Surface Roughness

In case of surface roughness, fit summary states that the quadratic model is significant. Result of analysis of variance for surface roughness is shown in Table 5.

The value of R^2 is 0.9858 and adjusted R^2 is 0.9725 shows a good relationship between process parameters and the response variable. The process parameters A-pulse on time, C-servo voltage, and interaction effect of process parameter A with process parameter C have a significant effect. The lack of fit is also non-significant which is desirable. Figure 9 shows that normal probability plot for SR it clearly indicated that residuals are present on a straight line it means that error follows a normal distribution. Figure 10 shows the plot for predicted versus actual value it shows that model is fitted well. After backward elimination, i.e., (eliminates non-significant terms) equation of SR in coded terms and actual terms is given in Eqs. 4

Table 5 ANOVA for SR

Source	Sum of squares	DF	Mean square	*F*-value	*p*-value	Check significance
Model	12.21	14	0.8723	74.34	<0.0001	Significant
A-T_{on}	4.84	1	4.84	412.58	<0.0001	Significant
B-T_{off}	0.0067	1	0.0067	0.5701	0.4619	
C-SV	3.17	1	3.17	270.10	<0.0001	Significant
D-WF	0.0328	1	0.0328	2.79	0.1154	
AB	0.0045	1	0.0045	0.3826	0.5455	
AC	3.06	1	3.06	261.14	<0.0001	Significant
AD	0.0600	1	0.0600	5.12	0.0390	
BC	0.0085	1	0.0085	0.7213	0.4091	
BD	0.0124	1	0.0124	1.06	0.3196	
CD	0.0081	1	0.0081	0.6903	0.4191	
A^2	0.2252	1	0.2252	19.19	0.0005	
B^2	0.0184	1	0.0184	1.56	0.2303	
C^2	0.1503	1	0.1503	12.81	0.0027	
D^2	0.0204	1	0.0204	1.74	0.2075	
Residual	0.1760	15	0.0117			
Lack of fit	0.1577	10	0.0158	4.32	0.0600	Not significant
Pure error	0.0183	5	0.0037			
Cor. Total	12.39	29				
Std. Dev.	0.1083		R^2	0.9858		
Mean	1.76		Adjusted R^2	0.9725		
C.V.%	6.14		Predicted R^2	0.9042		
			Adeq. precision	27.7738		

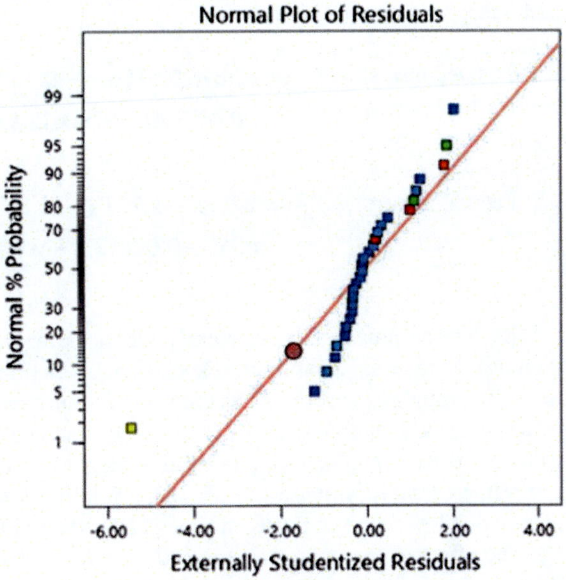

Fig. 9 Normal probability plot for SR

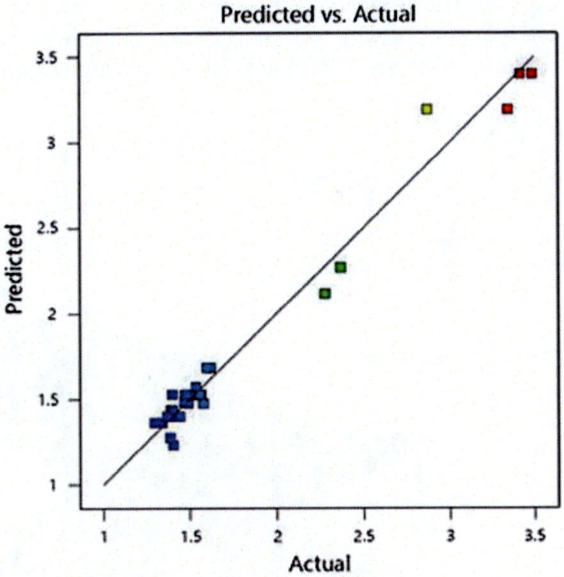

Fig. 10 Plot of actual versus Predicted response

and 5 respectively.

$$\text{(In Coded terms)} - \text{SR} = 1.53 + 0.5186A - 0.4196C + 0.0427D - 0.4376AC + 0.0612AD + 0.2224A^2 + 0.1684C^2 \quad (4)$$

$$\text{(In actual factors)} - \text{SR} = 82.54 - 1.70T_{\text{on}} + 0.86\text{SV} - 0.68\text{WF} - 0.0087T_{\text{on}} * \text{SV} + 0.0061T_{\text{on}} * \text{WF} + 0.0088T_{\text{on}}^2 + 0.0016\text{SV}^2 \quad (5)$$

The surface plot in Fig. 11 clearly indicates the variation of surface roughness with respect to process parameters such as pulse on time and servo voltage. The value of surface roughness increases as the value of pulse on time increases similarly high value of surface roughness is obtained at a low value of servo voltage. The maximum value of surface roughness is obtained at high value of pulse on time (120 μs) and low value of servo voltage (20 V) and a minimum value of surface roughness is obtained at low value of pulse on time (110 μs) and high value of servo voltage (40 V). The effect of the pulse on time and wire feed on surface roughness is shown in Fig. 12 with the help of the surface plot. Surface plot in Fig. 12 clearly indicates that high value of surface roughness is obtained at a high value of pulse on time (120 μs) and low value of wire feed rate (2 m/min) and low value of surface roughness is get at low value of pulse on time (110 μs) and high value of feed rate (6 m/min).

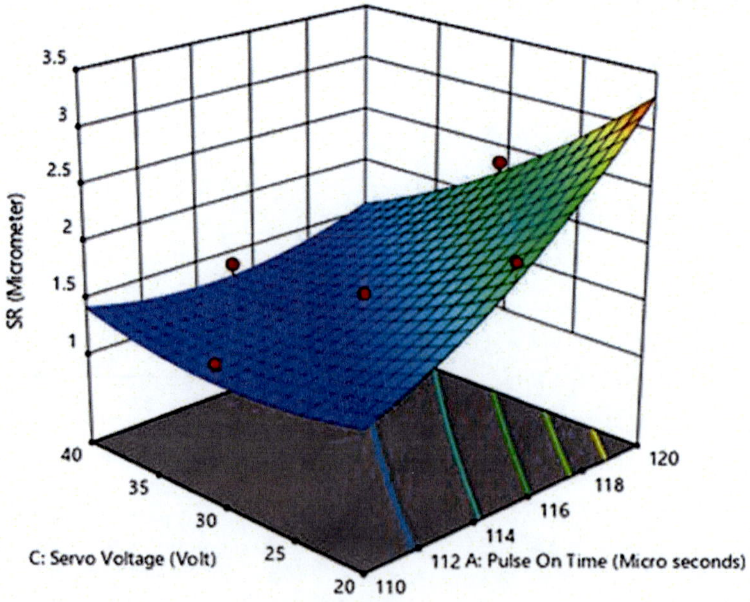

Fig. 11 Effect of T_{on} and SV on SR

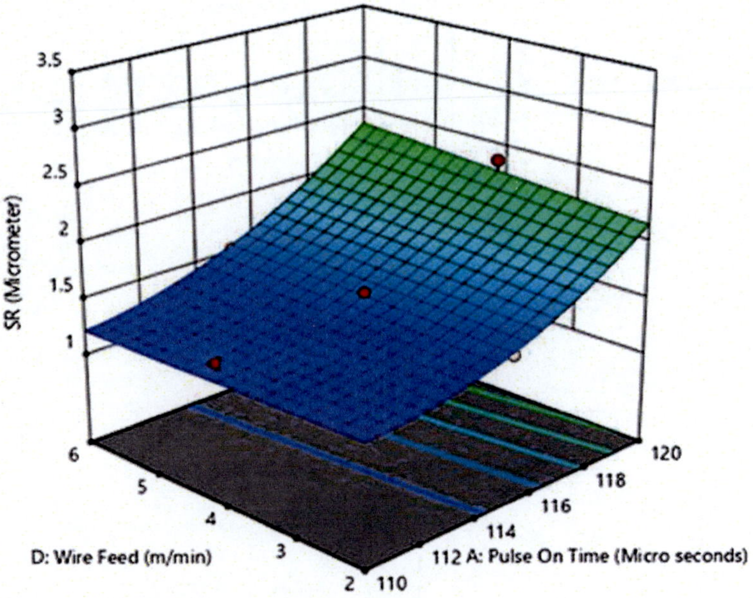

Fig. 12 Effect of T_{on} and WF on SR

3.3 Confirmation Tests and Their Comparison with Results

The confirmation test table clearly shows that error is very small lies between ± 4%. Result of confirmation test is shown in Table 6. Finally, confirmation test confirms the good reproducibility of experimentation results.

4 Conclusions

Modeling and parametric optimization of process parameters of WEDM using response surface methodology reveals the following conclusions by variation of process parameters within the specified range. The pulse on time and servo voltage affects both material removal rate and surface roughness and pulse off time and wires feed rate are least significant factors as compared to the pulse on time and servo voltage. The material removal rate increases with increases in pulse on time and decreases in the value of servo voltage. Hence, maximum value of material removal rate is secured at pulse on time is equal to 120 μs and servo voltage is equal to 20 V and minimum value of surface roughness is got at pulse on time is equal to 110 μs and servo voltage is equal to 40 V. The ANOVA of surface roughness after backward elimination clearly shows that pulse on time and servo voltages are the most significant parameters for surface roughness. The surface plot for surface

Table 6 Confirmation test

Exp. No.	Process parameters				MRR			SR		
	T_{on}	T_{off}	SV	WF	Exp.	Predicted	Error	Exp.	Predicted	Error
1	110	55	20	2	0.227012	0.219042	3.51	1.391	1.401	−0.71
2	120	50	30	4	0.388270	0.378685	2.46	2.367	2.269	4.14
3	115	50	30	4	0.245908	0.288225	−0.17	1.492	1.528	−2.41

roughness clearly shows that surface roughness increases as the value of pulse on time increases and the value of servo voltage decreases. The minimum value of surface roughness is obtained at a low value of pulse on time and high value of servo voltage, i.e., (T_{on} = 110 μs and SV = 40 V). From the confirmation test the error between actual and predicted values for material removal rate and surface roughness lies within ± 4% range. This error is very small it confirms the good reproducibility of experimental results.

References

1. Kansal HK, Singh S, Kumar P (2005) Parametric optimization of powder mixed electrical discharge machining by response surface methodology. J Mater Proc Technol 169:427–436
2. Kung K-Y, Chiang K-T (2008) Modelling and analysis of machinability evaluation in the wire electrical discharge machining (WEDM) process of aluminium oxide-based ceramic. Mater Manufact Proc 23:241–250
3. Patel KM, Pandey PM, Venkateswara Rao P (2009) Determination of an parametric combination using a surface roughness prediction model for EDM of $Al_2O_3/SiC_w/TiC$ ceramic composite. Mater Manufact Proc 24:675–682
4. Patil NG, Brahmankar PK (2010) On the response surface modelling of wire electrical discharge machining of Al/SiCp metal matrix composites (MMCs). J Mach Form Technol 2(1/2):47–70
5. Ojha K, Garg RK, Singh KK (2011) Experimental investigation and modeling of PMEDM process with chromium powder suspended dielectric. Int J Appl Sci Eng 2:65–81
6. Shandilya P, Jain PK, Jain NK (2012) Parametric optimization during wire electrical discharge machining using response surface methodology. Proced Eng 38:2371–2377
7. Khanna R, Singh H (2013) Performance analysis for D-3 material using response surface methodology on WEDM. Int. J Mach Mach Mater 14(1):45–65
8. Gopalakannan S, Senthilvelan T (2014) Optimization of machining parameters for EDM operations based on central composite design and desirability approach. J Mechan Sci Technol 28(3):1045–1053
9. Sharma N, Khanna R, Gupta RD, Sharma R (2013) Modeling and multiresponse optimization on WEDM for HSLA by RSM. Int J Adv Manuf Technol 67:2269–2281
10. Singh B, Kumar J, Kumar S (2014) Experimental investigation on surface characteristics in powder-mixed electrodischarge machining of AA6061/10%SiC composite. Mater Manufact Proc 29:287–297
11. Dubey V, Singh B (2018) Study of material removal rate in powder mixed EDM of AA7075/B4C composite in IMME17. Mater Today: Proc 5:7466–7475

Wire-Cut Electric Discharge Machining on Nickel–Aluminium–Bronze Using Brass Wire Electrode

Earnest Beni, Poovazhagan Lakshmanan, and S. C. Amith

Abstract Wire-cut electric discharge machining (WEDM) is a widely used non-traditional subtractive machining process for removing materials from conductive workpiece to produce parts with intricate shapes/complex profiles. Nickel–aluminium–bronze (NAB alloy) is a high-strength, difficult-to-machine nickel-based alloy. In this work, a trial was conducted to machine NAB alloy using WEDM technique. The machining voltage, pulse-on time, pulse-off time, and wire feed were varied for the experiments. The 10 × 10 mm square-shaped samples were cut on NAB alloy with 0.25-mm-diameter brass wire electrode. Based on the analysis, it is found that pulse-on time, machining voltage, and wire tension are the significant parameters that affect the kerf width, material removal rate, and surface roughness. Scanning electron microscopy (SEM) is used to identify the microstructure and the surface morphology of the machined work piece. A higher pulse-on time setting leads to thicker recast layer. At lower value of pulse-on time and higher value of pulse-off time, the wire deposition on the machined surface is low.

Keywords Brass wire · NAB alloy · Wire-cut EDM · Taguchi's method · DOE

1 Introduction

Nickel–aluminium–bronze (NAB) alloy is the most widely used group of nickel–aluminium alloy. NAB alloys can be obtained in both cast or wrought forms. They possess the following unique combination of properties: high strength, high corrosion resistance, good wear resistance, good cryogenic properties, and excellent damping capacity. The commercial uses of NAB alloy range from propellers for ships, landing gear bushings, bearings for aircraft and sea water pumps, and non-sparking tools in oil and gas industries [1]. Generally, nickel-based alloys are difficult to machine by conventional machining processes [2].

E. Beni · P. Lakshmanan (✉) · S. C. Amith
Department of Mechanical Engineering, Sri Sivasubramaniya Nadar College of Engineering, Chennai 603110, Tamil Nadu, India
e-mail: poovazhaganl@ssn.edu.in

S. Vijayan et al. (eds.), *Trends in Manufacturing and Engineering Management*, Lecture Notes in Mechanical Engineering,
https://doi.org/10.1007/978-981-15-4745-4_7

Electrical discharge machining (EDM) is one of the well-established non-traditional machining processes intended to machine geometrically complex shapes and extremely hard conductive materials. EDM is the non-contact-type machining process, which leads to the following benefits: eliminating vibration, chatter, and mechanical stresses [3]. EDM is the indispensable machining technology to various sectors of industries such as automobile, aeronautical, and medical due to its ability to cut very hard materials [4]. Wire-cut EDM (WEDM) is introduced in the late 1960s and has revolutionized the metal working, tool and die, and mould industries. Mechanism of material removal in WEDM is based on the thermo-electric principles. Melting point of the work material is the only property to be considered during WEDM [5].

Brajesh Kumar Lodhi et al. (2014) used Taguchi L9 orthogonal array to optimize the WEDM parameters for better surface roughness during the machining of AISI D3 steel. It was concluded that discharge current was the most influencing machining parameter for surface roughness [6]. Bijaya Bijeta Nayak et al. (2016) optimized WEDM process parameters while machining deep cryo-treated inconel 718 using hybrid approach [7]. Ashish Goyal (2017) investigated material removal rate (MRR) and surface roughness of inconel 625 super alloy using cryogenic-treated zinc-coated wire. ANOVA optimization revealed that current intensity and tool electrodes are dominant parameters that affect the MRR and surface roughness [8]. Amitesh Goswami et al. (2017) optimized the trim-cut WEDM machining parameters while machining nimonic alloy 80A. Results show that trim-cut WEDM produces very good surface finish as compared to the rough-cut WEDM [9]. Coral reef microstructures with few voids were observed while machining inconel 718 in WEDM using uncoated brass wire electrode of diameter 200 μm [10]. Inconel 625 materials were machined on a wire-cute EDM by using brass as an electrode; influence of MRR, SR, discharge current, voltage, and pulse-on time were optimized by using Taguchi method [11]. The copper electrode with zinc coating and uncoated brass electrode wire were used for wire EDM machining of IN 718; coated wire showed better performance like reduction in wire consumption when compared to uncoated wire electrode [12].

It is apparent from the detailed literature review that no experimental work has been performed on the machinability, kerf width (KW), and surface roughness (SR) study of NAB alloy using brass wire in WEDM process. Hence, in this current research work attempt has been made to examine the machining characteristics (KW, MRR, and SR) of NAB alloy using untreated brass wire in WEDM.

Table 1 Properties of workpiece material

Material	Chemical composition	Hardness (HV)	Thickness (mm)	Thermal conductivity (W/m^0K)
NAB alloy	Cu-balance, Al-10%, Ni-5%, Fe-4%, Mn-1.5%	200	5.5 mm	42

2 Experimentation

2.1 Materials and Methods

Material employed for experimentation is nickel–aluminium–bronze (NAB) alloy with brass wire as electrode for WEDM process. NAB is a high-strength Cu-Al alloy with the applications involving heavy loads, abrasive wear resistance, friction, and corrosion. The addition of nickel increases strength of the alloy without deteriorating its outstanding ductility, toughness, and corrosion resistance. The chemical composition and properties of NAB alloy are given in Table 1.

The wire-cut EDM machine used in this study was Sodick model A280 with the wire electrode made from brass (Cu—35 wt%Zn). The wire was 0.25 mm in diameter and can tolerate a tension up to 95 kgf/mm^2. Elektra Maxicut-made CNC-controlled wire-cut electrical discharge machine is used to conduct wire-cut EDM experiments. It consists of CNC motion control unit, fixture for holding the workpiece and wire spool. NAB base plate was of 250 × 150 × 6 mm dimensions that was used for this study purpose.

2.2 Design of Experiment Based on Taguchi Method

A total of 9 experiments were conducted using Taguchi L9 experimental design methodology. In our study, the machining parameters were taken as pulse-on time, pulse-off time, peak current, and wire tension. There are 3 levels of factors in each process parameters as shown in Table 2. Using these factors and levels, L9 experiments were selected which are shown in Table 3. The responses that are taken into account are the kerf width, material removal rate, and the surface roughness.

The sample was cut by using brass wire in the wire-cut EDM machine of 10 × 10 × 6 mm dimension for each set of parameters based on the L9 orthogonal array. In each experiment, the time taken for machining was noted. After machining the dimension of the sample and the inner dimension of the base plate from which the sample was cut are measured for calculating the kerf width and material removal rate. The EDMed samples were then cleaned using ethanol before SEM analysis.

Table 2 Levels of various process parameters

Parameters	Designation	Levels		
		L1	L2	L3
Pulse ON (μs)—T_{on}	A	1	3	6
Pulse OFF (μs)—T_{off}	B	4	7	10
Current (A)—I	C	2	3	5
Wire tension (N)—WT	D	0.98	1.47	2.45

Table 3 Experimental design—L9 orthogonal array (tool: brass wire, workpiece: NAB alloy)

Experiment number	A	B	C	D	Kerf width (mm)	MRR (mm^3/min)	SR (μm)
1	1	4	2	0.98	0.215	2.95	2.245
2	1	7	3	1.47	0.195	3.29	1.958
3	1	10	5	2.45	0.31	6.82	2.256
4	3	4	3	2.45	0.31	5.67	2.289
5	3	7	5	0.98	0.27	6.59	2.494
6	3	10	2	1.47	0.295	4.98	2.175
7	6	4	5	1.47	0.28	6.83	2.277
8	6	7	2	2.45	0.29	4.54	2.587
9	6	10	3	0.98	0.325	5.40	2.556

3 Results and Discussion

3.1 SEM Analysis of WEDM Samples

The SEM microstructures reveal the surface quality for each sample under different parameters. A smooth surface is obtained in Fig. 1a for the parameters as pulse-on time: 1 μs, pulse-off time: 4 μs, current: 2 A, and wire tension: 0.98 N. Figure 1c shows the formation of a rough surface obtained at higher setting parameters as pulse-on time: 6 μs, pulse-off time: 10 μs, current: 5 A, and wire tension 2.45 N. The smoother surface as seen in Fig. 1a may be due to the lower intensity discharges that took place during machining; thus, there is very small debris that can be observed in this surface. Figure 1c shows the lumps and globules of debris with some deep craters and microcracks on the surface which may have caused due to higher values machining parameters. These deeper craters may be the places where the individual discharge was able to penetrate far into the workpiece.

The recast layers can be clearly seen thicker in Fig. 1d compared to Fig. 1b, as Experiment No. 1 (pulse-on time: 1 μs, pulse-off time: 4μs, current: 2A, and wire tension: 0.98 N) was used to machine the sample shown in Fig. 1b and Experiment No. 9 (pulse-on time: 6 μs, pulse-off time: 10 μs, current: 5 A, and wire tension 2.45 N)

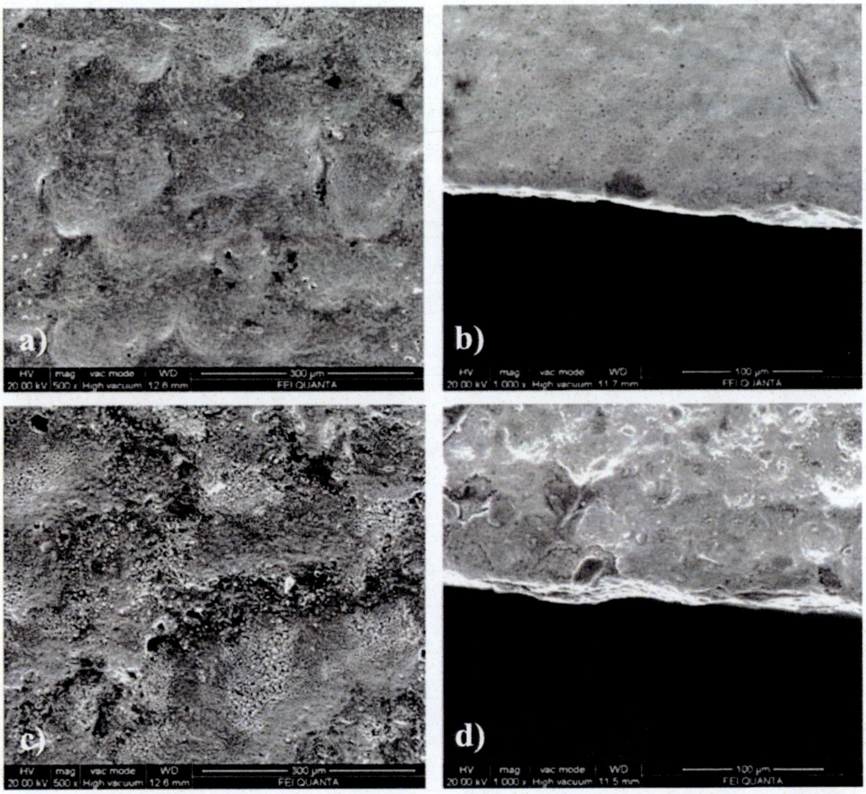

Fig. 1 SEM micrographs of **a** sample 1 WEDM surface, **b** sample 1 cross section, **c** sample 9 WEDM surface, and **d** sample 9 cross section

was used to machine the sample shown in Fig. 1d. This difference in the recast layer thickness is due to the resolidified materials formed on the surface layer after the spark melts a small portion of workpiece and wire electrode during machining.

3.2 *Effect of Process Parameters on MRR, Kerf Width, and Surface Roughness*

The higher the MRR, better the machining. Figure 2 shows the effect of MRR for the various machining parameters. When the pulse-on time increases the MRR also increases; it is similar to the peak current and wire tension. High current generates large number of electrons. These electrons striking the work surface in a single discharge, which increases the erosion process. Therefore kerf width and MRR also increased. This can be confirmed by Figs. 3 and 2 respectively. Similarly, the surface

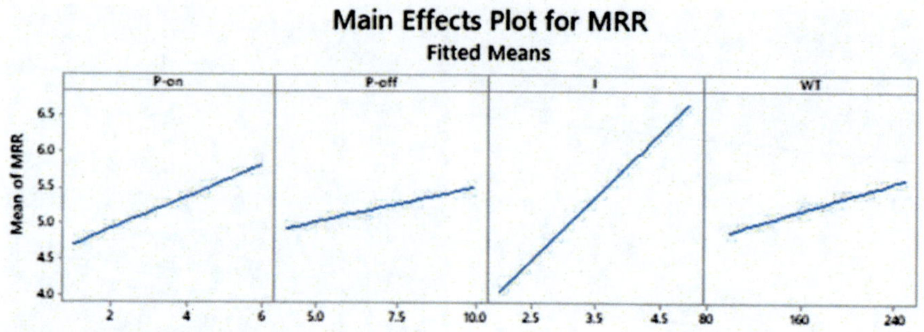

Fig. 2 Effect plot for material removal rate

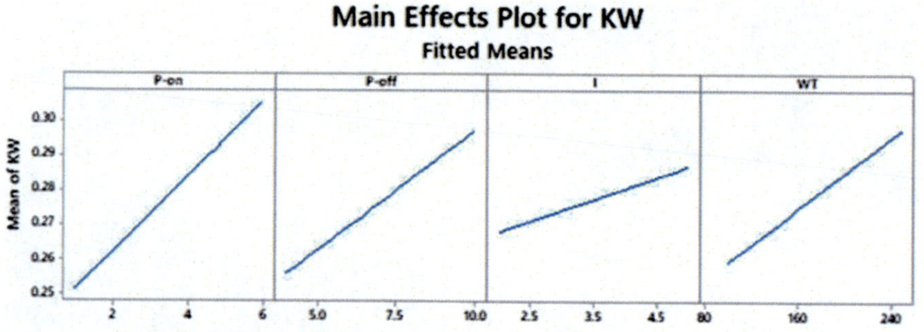

Fig. 3 Effect plot for kerf width

roughness also increases when the peak current is high which results in high discharge frequency, leading to the formation of deeper and larger craters in the surface. Table 4 shows the interaction between various process parameters and their significance based on the rankings, with percentage contribution for the peak current to be the highest at 87.2% and the wire tension and the pulse-off time percentage contribution is the least at 31% and 32.3%, respectively. The pulse-on time is also a significant factor with percentage contribution of 53.8%.

Table 4 Ranking means for brass wire

Level	Pulse on	Pulse off	Current	WT
1	2.249	2.563	2.253	2.561
2	2.786	2.468	2.444	2.476
3	2.787	2.791	3.125	2.786
Delta	0.538	0.323	0.872	0.310
Rank	2	3	1	4

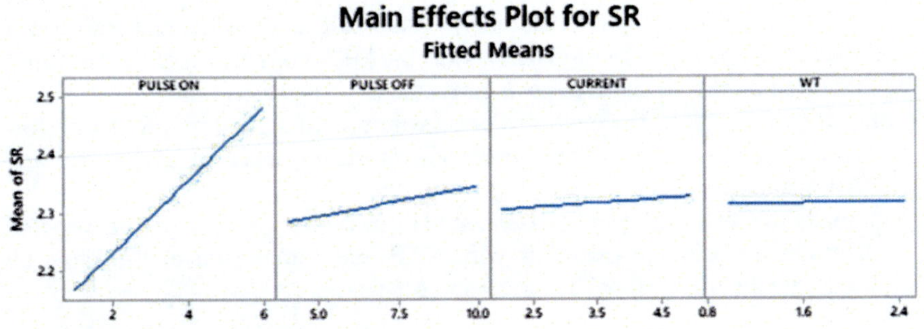

Fig. 4 Effect plot for surface roughness

In all the graphs, the pulse-off time also showed some increase in MRR and kerf width. This particular phenomenon requires supporting reasons to clarify; thus, the pulse-off effect is not considered as a major significance causing parameter.

The effect of various parameters like pulse-on time, pulse-off time, peak current, and the wire tension has been studied through machining of NAB with using brass electrode wire. The surface roughness was unchanged by using different parameters as shown in Fig. 3. The SR had been only influenced by the pulse-on time. Wire tension has the least effect on the surface roughness. This is because the number of electrons hitting the surface for elapsed time causes more erosion of material, and thus, surface roughness is increased (Fig. 4).

4 Conclusions

In this study, the effect of machining parameters on the kerf width, MRR, and the surface roughness of nickel–aluminium–bronze was investigated experimentally by using brass wire-cut EDM.

1. Peak current and pulse-on time have been found as the most significant factors, with percentage contribution of 87.2% and 53,8%, respectively. The other two process parameter interactions were also significant; however, their contribution in variance is relatively low.
2. The higher values of discharge energy and the elongated period of pulse-on time increase the machining rate and thus result in large craters and microdamages that cause the surface to be rough. This problem can be optimized by Taguchi's parameter design for WEDM.
3. The effect of various parameters like pulse-on time, pulse-off time, peak current and the wire tension has been studied through machining of NAB with using brass electrode wire. The surface roughness was unchanged by using different parameters. SR had been only influenced by the pulse-on time.

4. A linear regression graph is obtained for the MRR, kerf width, and the surface roughness responses. The result shows that the MRR and kerf width can be highly influenced by the variation in peak current and the pulse-on time.
5. Based on the results, the untreated brass wire provides good machining performance (maximum MRR and better surface roughness) when used for machining NAB material.
6. In the future, this work can be extended to discuss effect of the cryogenic cooled electrodes of different materials with NAB and comparison of the same by discussing the dimensional accuracy and MRR.

References

1. Lia L, Guob YB, Weia XT, Lib W (2013) Surface integrity characteristics in wire-EDM of inconel 718 at different discharge energy. In: Seventeenth international symposium on electro machining, Procedia CIRP, vol 6, pp 220–225
2. Amitesh G, Jatinder K (2014) Investigation of surface integrity, material removal rate and wire wear ratio for WEDM of Nimonic 80A alloy using GRA and Taguchi method. Eng Sci Technol Int J 17:173–184
3. Kanlayasiri K, Boonmung S (2007) An investigation on effects of wire-EDM machining parameters on surface roughness of newly developed DC53 die steel. J Mater Process Technol 187–188:26–29
4. Ivan R, Carol P (2016) Guide to nickel aluminium bronze for engineers. Copper Development Association Publication, No. 22
5. Ho KH, Newman ST (2003) State of the art electrical discharge machining (EDM). Int J Mach Tools Manufact 43:1287–1300
6. Brajesh KL, Sanjay A (2014) Optimization of machining parameters in WEDM of AISI D3 Steel using Taguchi Technique. In: 6th CIRP international conference on high performance cutting, HPC
7. Bijaya BN, Siba SM (2016) Optimization of WEDM process parameters using deep cryo-treated Inconel 718 as work material. Eng Sci Technol Int J 19:161–170
8. Ashish G (2009) Investigation of material removal rate and surface roughness during wire electrical discharge machining (WEDM) of Inconel 625 super alloy by cryogenic treated tool electrode. J King Saud Univ Sci 29:528–535
9. Amitesh G, Jatinder K (2017) Trim cut machining and surface integrity analysis of Nimonic 80A alloy using wire cut EDM. Eng Sci Technol Int J 20:175–186
10. Norliana MA, Darius GS, Fuad BMd (2007) A review on current research trends in electrical discharge machining (EDM). Int J Mach Tools Manuf 47:1214–1228
11. Subrahmanyam M, Nancharaiah T (2019) Optimization of process parameters in wire-cut EDM of Inconel 625 using Taguchi's approach. Mater Today: Proc
12. Reolon LW, Henning Laurindo CA, Torres RD, Amorim FL (2019) WEDM performance and surface integrity of Inconel alloy IN718 with coated and uncoated wires. Int J Adv Manufact Technol 100:1981–1991

Development and Characterization of Metal Matrix Composite

Manufacture, Mechanical Properties and Microstructural Characterization of Aluminium and Iron Metal Matrix Composite Manufactured

D. Arthur Jebastine Sunderraj, D. Ananthapadmanaban, K. Arun Vasantha Geethan, and A. John Rajan

Abstract The objective of the present work is to investigate the formation of intermetallic compounds such as iron aluminides and oxides during the formation of an alloy by melting commercially pure aluminium (Al) billets and adding iron powder (Fe) at 750 °C in the molten aluminium. The exothermic reaction during the addition of 'Fe' in the molten 'Al' resulted in the formation of more amounts of intermetallic compounds because of the rise in temperature. The composition of the 'Al' and 'Fe' powder was varied to obtain various alloys. The optical microscopy examination revealed the presence of various intermetallic compounds and oxides in the matrix. The Rockwell hardness survey performed on various samples exhibited a maximum hardness value of 79 in 'B' scale for the alloy composition of 60% Al and 40% Fe. A maximum shear strength of 288 Mpa was obtained for the samples containing 65% Al and 35% Fe.

Keywords Intermetallics · Iron aluminide · Microstructure

1 Introduction

Aluminium is a chemical element in the boron group with symbol 'Al' and atomic number 13. It is silvery-white, and it is not soluble in water under normal circumstances. Pure aluminium possesses low strength and toughness, and its strength can be increased with the addition of various elements. Aluminium is alloyed with elements

---

D. Arthur Jebastine Sunderraj (✉) · K. Arun Vasantha Geethan
Department of Mechanical Engineering College, St. Joseph's Institute of Technology, Chennai 600119, India
e-mail: arthurjebastine@gmail.com

D. Ananthapadmanaban
Department of Mechanical Engineering, Sri Sivasubramaniya Nadar College of Engineering, Chennai, Tamil Nadu 603110, India

A. John Rajan
Department of Manufacturing Engineering, School of Mechanical Engineering VIT, Vellore, Tamil Nadu, India

S. Vijayan et al. (eds.), *Trends in Manufacturing and Engineering Management*, Lecture Notes in Mechanical Engineering,
https://doi.org/10.1007/978-981-15-4745-4_8

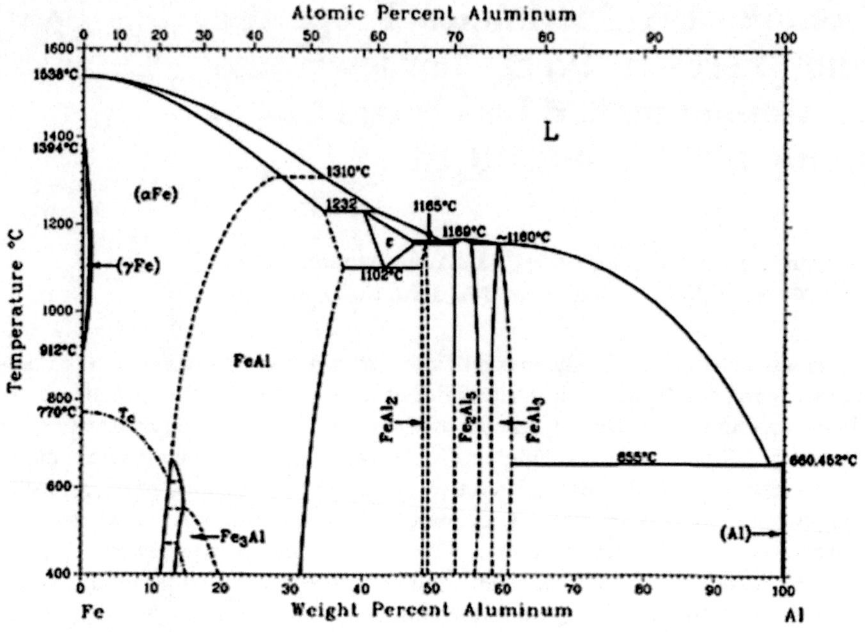

Fig. 1 Fe-Al phase diagram (*Source* ASM Hand Book)

like ‘Cu’, ‘Si’, ‘Mg’, ‘Zn’, ‘Fe’ and other elements. The solubility of aluminium in iron at 400 °C is 11.5% which is as shown in Fig. 1. Addition of excess ‘Al’ in ‘Fe’ results in the formation of intermetallic compounds such as FeAl, Fe_3Al, $FeAl_2$ and $FeAl_3$. The presence of intermetallic compounds and oxides reinforces the softer matrix aluminium and increases the mechanical properties. These intermetallic compounds are formed as a result of the reaction between two or more metals which are mixed together in certain proportions. These intermetallic compounds exhibit neither metallic nor ceramic properties. Iron aluminides are compounds formed between ‘Fe’ and ‘Al’. The phase diagram drawn between Al–Fe reveals the formation of intermetallic compounds like Fe_3Al, FeAl, $FeAl_2$, Fe_2 Al_5 and $FeAl_3$. The properties of iron aluminide are dependent on number of aluminium atoms present in the compound. The presence of more aluminium atoms in the intermetallic compounds such as $FeAl_2$, Fe_2Al_5 and $FeAl_3$ is detrimental to strength properties and induces brittleness. On the other hand, Fe_3Al and FeAl, which have a high iron composition, are used as structural materials because of their good wear resistance, corrosion resistance and specific strength properties.

In pure elements and solid solutions, the atoms are bound together with metallic bonds. In the case of intermetallic compounds covalent bond is formed which alters the crystal structure, chemical, mechanical and electrical properties. Intermetallic compounds such as Fe_3Al and FeAl are among a variety of different intermetallic compounds that can be used in heterogeneous metal (Al)–intermetallic materials and intermetallics that can be used for corrosion resistance applications at higher temperature. This alloy system containing Al–intermetallic compounds having nearly homogeneous composition exhibits properties varying from intermetallics to aluminium alloys. The literature revealed that the formation of Fe–Al intermetallic compounds took place during melting of these elements and during powder compacting and sintering of these powders. Wangyu Hu et al. [1] revealed the formation of nanocrystalline iron aluminide intermetallic compounds by mechanical alloying, sintering and high-velocity oxy-fuel (HVOF) thermal spraying from the elemental powders. The strength, ductility and toughness of the samples were increased by maintaining grain size to the nanocrystalline level. Jiqiang et al. [2] produced an ultra-fine grained Fe-40Al intermetallic compound. The X-ray diffraction of the Fe-40Al intermetallic compound revealed the presence of FeAl phase. The Fe-40Al intermetallic compound presented ultra-fine grains size in the range of 100–600 nm because of high nucleation rate and the low growth rate.

The fabrication process of the metal–intermetallic materials is one of the most important areas of research. Several research works presented the capabilities of the formation of intermetallic compounds, metal–intermetallic compound and metal–ceramic combination. These include solidification processing, powder metallurgy (PM), coating technology and layered manufacturing. Among these methods, PM is the most commonly used because of its ability to produce a wide range of compositions, its phase and property control as well as shape-forming capability. In the present research work, the Al-FeAl intermetallic compound's fabrication route consists of melting the solid aluminium billets and keeping 'Fe' powder in solid state. In the first method, 'Al' solid billets and 'Fe' powder are mixed together and then heated to 750 °C. At this temperature, the molten aluminium reacted with 'Fe' powder to form Al-FeAl rich phase. Solubility of Fe in Al is 10–12%. Since we are adding 20–40% Fe, it can be seen from the phase diagram that there is a very good possibility of formation of intermetallics.

This alloy is then heat-treated for 10 h in a muffle furnace. In the second method, 'Fe' powder is added with the molten aluminium followed by annealing the alloy for 10 h at 600 °C to form Al-FeAl system. During the solidification process, oxides such as Al_2O_3 and FeO_2 are formed which may have strengthened the Al-FeAl alloy.

Prichard [3] conducted experiments by adding silicon to Fe-15%A1 alloy which produced faceting on powder particle surfaces during vacuum sintering above 1200 °C. The faceting planes occurred on the low index planes with high symmetry. The faceting was due to a change in the solid–vapour surface energy as a function of crystallographic orientation. The extrusion of Fe–Al–Si alloys produced a wire texture, which nearly doubled the yield strength of the material in the extrusion direction compared to the transverse direction. Although the extruded alloy was fully dense, the tensile properties in the extruded direction do not represent the mechanical properties of a sintered microstructure with randomly oriented grains. The yield strength of sintered powders was comparable to the extruded powders in the same heat treatment condition. The yield strength and fracture energy of the fine grain material with approximately 10% porosity exceeded the yield strength and fracture energy of coarse grain material with approximately 3% porosity. Therefore, the processing parameters are broad enough to develop desirable mechanical properties with a wide range of microstructures. The next phase of development would be to incorporate these processing parameters into an MIM process to address the interaction of the powder with binder systems. The uniform porosity and fine grain size of the sintered Fe-15A1-2.8Si alloy, which was heat-treated at 600 °C, developed the best combination of strength and toughness. This indicates the silicon atoms are more beneficial in a solid solution rather than in an ordered precipitate.

The objective of this work is to produce Al–Fe alloys possessing optimum mechanical properties. Addition of 'Fe' in 'Al' to very high level decreases strength because more amounts of brittle phase is formed in the alloy. Therefore, an optimum level of 'Fe' percentage should be fixed for obtaining maximum strength. In the present work 'Fe' content is varied from 20–40% and 'Al' content 60–80% to produce various alloys. In the molten aluminium, 'Fe' powder is added which produces an exothermic reaction. This reaction generates heat, and this heat partially melts the 'Fe' powder to form intermetallic compounds with 'Al'.

2 Experimental Procedure

2.1 Material

The iron powder 4–8 μm particle size having purity 99.9% is used in this study. Wrought aluminium having 99.5% purity with 0.2% Si, 0.1% Fe, 0.04% Mg, 0.1% Mn and 0.05% O is used. The purchased rod is of 20 mm in diameter which is cut to billets of 20 mm in length.

2.2 Experimental Set-up

The muffle furnace having a heating chamber of 380 mm length and 180 mm width is used. The heating rate of this furnace can be varied to the required level. The maximum temperature of the furnace is 1000 °C. Small crucibles made with clay are used for melting the aluminium metal. The experimental set-up is shown in Fig. 2.

2.3 Experimental Procedure

The billets were cleaned mechanically by turning them in a lathe and finally cleaned with chemicals like HCl and NaOH. The 'Al' billets were weighed and placed inside the crucibles and heated inside the furnace. The crucibles were heated up to 750 °C and kept inside the furnace for 1 h. In the molten 'Al', 'Fe' powder is added to initiate

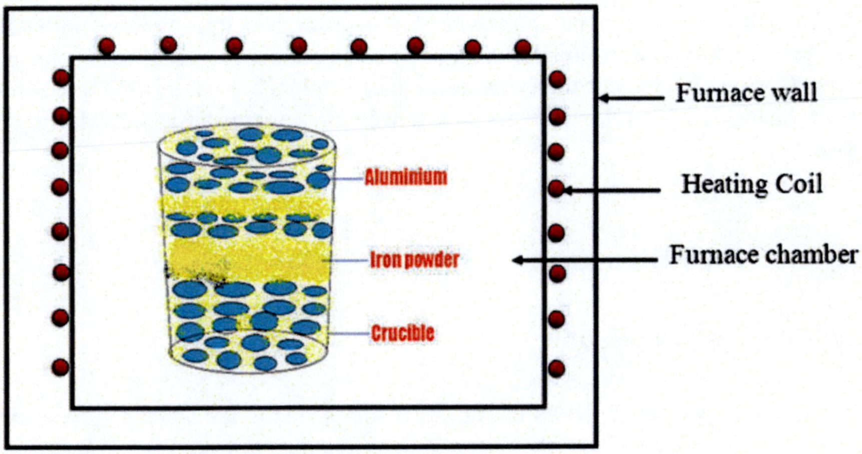

Fig. 2 Experimental set-up of a muffle furnace

exothermic reaction. Because of this exothermic reaction, the temperature inside the crucible rises to 1600 °C. Further, this alloy is kept at 750 °C for two hours and given an annealing heat treatment for 16 h at 550 °C.

The specimens obtained from the furnace were machined in a lathe to obtain circular discs as shown in Fig. 3a, b. Square and circular specimens of 6 × 6 mm and 4 mm in diameter (for metallography and shear test) were cut using wire-cut electric discharge machining technique. The WEDM specimens are shown in Fig. 3c. The metallographic samples were polished using various grades (up to 1200 grade) of

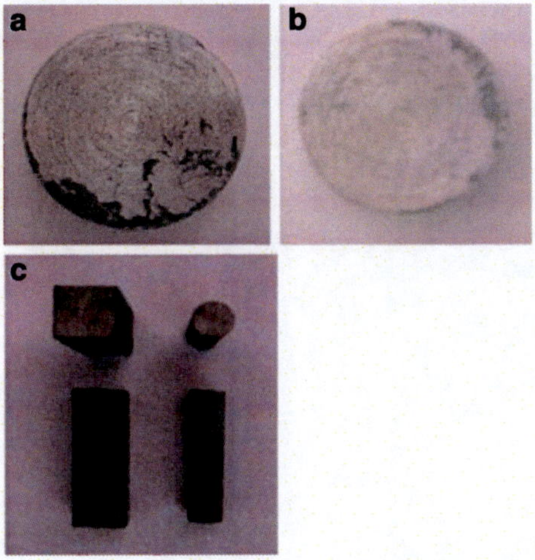

Fig. 3 **a** and **b** Al–Fe partial melt turned samples and **c** wire-cut EDM samples

silicon carbide paper to obtain mirror finish. These samples were finished with Al_2O_3 slurry to obtain mirror finish without any scratch marks. The polished specimens were etched with a chemical mixture containing 5 ml HCl + 10 ml HNO_3 + 3 ml HF + 50 ml H_2O for 1 min. The specimens were rinsed in water and dried using a drier.

3 Result and Discussion

3.1 *Optical Microscopy*

The optical micrographs shown in Fig. 4a, b correspond to the sample1 processed at 750 °C and annealed at 550 °C for 19 h inside the furnace. Figure 4a shows three different regions marked as region 'A', region 'B' and region 'C'. These three regions are rich with intermetallic compounds such as $FeAl_2$, Fe_2Al_5 and $FeAl_3$ and oxides like Al_2O_3 and FeO_2. Further, away from these dark regions a yellowish region with fine grains of 'Al' is noted. This region is rich with pure aluminium.

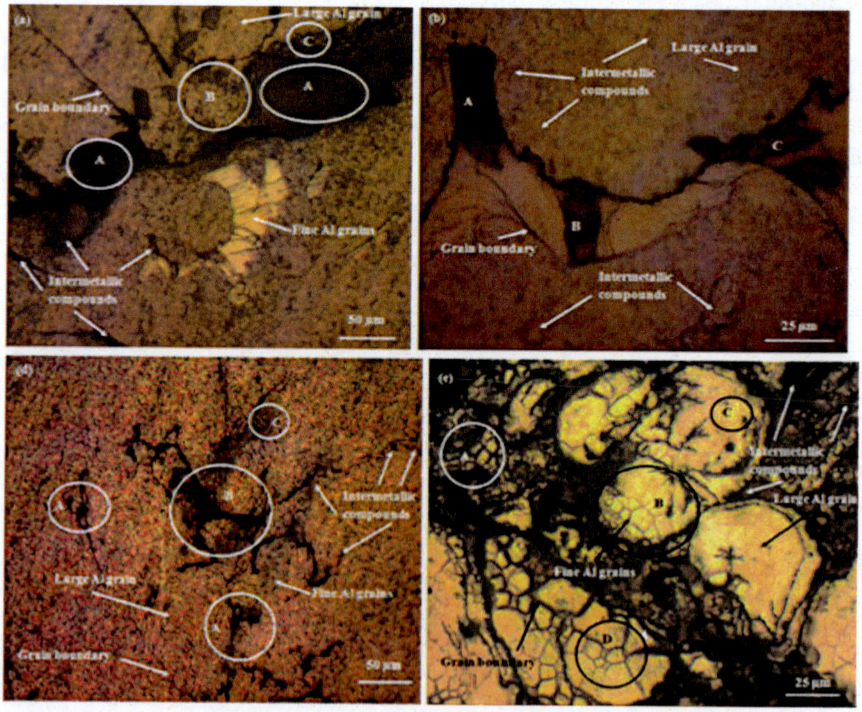

Fig. 4 **a** and **b** Optical micrographs of sample 1, **d** and **e** optical micrographs of sample 2

Table 1 Annealing temperature for various compositions

S. No	Composition in wt. percentage	Melting temperature	Annealing temperature 600 °C and time in h
1	80% Al and 20% Fe	750 °C and soaking at this temperature for 2 h	19
2	80% Al and 20% Fe	750 °C and soaking at this temperature for 2 h and mechanically agitated	16
3	75% Al and 25% Fe		16
4	65% Al and 35% Fe		16
5	70% Al and 30% Fe		16
6	60% Al and 40% Fe		16

Adjacent to this region on both sides, fine intermetallic compounds which impart dispersion strengthening effect to this soft region are seen. The micrograph shown in Fig. 4b reveals three distinct regions containing recrystallized regions rich with alternate grains of aluminium and intermetallic compounds. The grain boundaries are precipitated with 'Fe'. The recrystallization is because of pinning of these 'Fe' atoms along the grain boundaries and migration of 'Fe' atoms into 'Al' matrix and subsequent formation of newer grain boundaries. In Fig. 4c, a dark region which resembles like a large spear at the centre of the micrograph with serrations at the sides is observed and indicates the formation of intermetallic compounds and its growth into the 'Al' grains because of long annealing time and diffusion of 'Fe' atoms into the 'Al' matrix.

The micrographs shown in Fig. 4a, b correspond to the sample 2 as given in Table 1. This sample is rich with 20% 'Fe' and annealed for 16 h at 550 °C which is mechanically agitated to obtain homogeneous alloy. The microstructures show dark regions, which may possibly be intermetallics.

The optical micrographs shown in Fig. 5a, b correspond to the sample 3 processed at 750 °C and annealed at 550 °C for 16 h inside the furnace. Figure 5a shows two different regions marked as region A and region B. These three regions are rich with intermetallic compounds such as $FeAl_2$, Fe_2Al_5 and $FeAl_3$ and oxides like Al_2O_3 and FeO_2. Further, proof of intermetallics formation can be obtained by viewing the XRD patterns as shown in Figs. 9, 10 and 11. The micrographs shown in Fig. 5d–f correspond to the sample 4 as given in Table 1. This sample is rich with 35% 'Fe' and annealed for 16 h at 550 °C which is mechanically agitated to obtain homogeneous alloy.

Because of this, the micrographs shown in Fig. 5a, b reveal the presence of fragmented intermetallic compounds such as $FeAl_2$, Fe_2Al_5 and $FeAl_3$.

In Fig. 5d, the thick grain boundaries are visible and adjacent to the grain boundaries fingerprint-like intermetallic compound formation is noted which resembles like laves phase (circle mark A). The lengthy intermetallic compounds rich with 'Fe' are observed in the published literature by Chakraborty et al. [4]. The circle marks containing a sphere are possibly intermetallic compounds. The micrographs shown

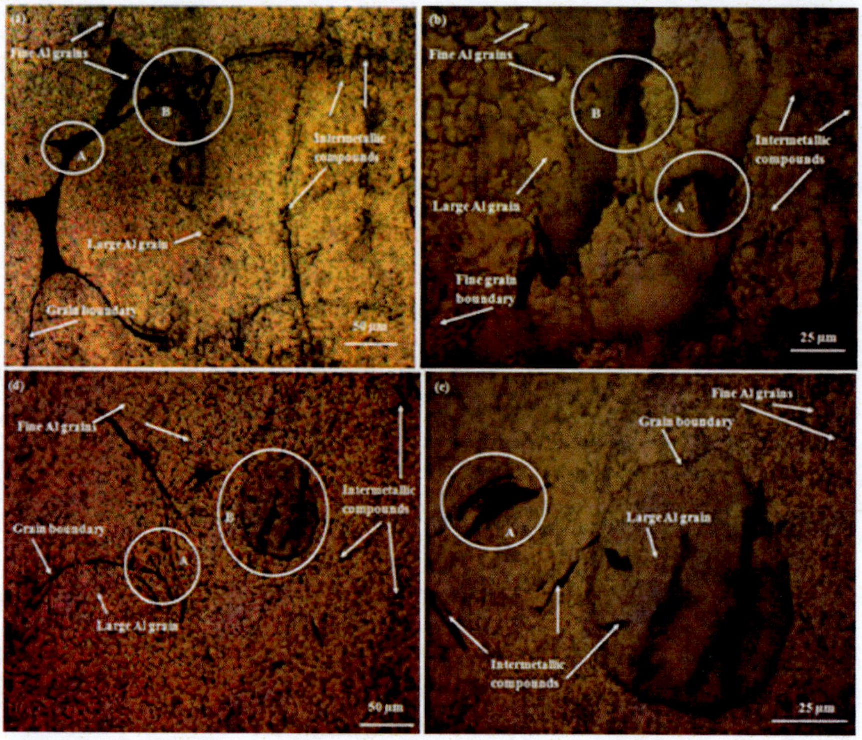

Fig. 5 **a** and **b** Optical micrographs of sample 3, **d** and **e** optical micrographs of sample 4

in Fig. 5a, b reveal the formation of long-thick-intermetallic compounds rich with $FeAl_3$ and $FeAl_6$ which is in accordance with the published work by Taylor [5].

The optical micrographs shown in Fig. 6a, b correspond to the sample5 processed at 750 °C and annealed at 550 °C for 16 h inside the furnace. Figure 6a shows two different regions marked as region 'A' and region 'B'. These two regions are rich with intermetallic compounds such as $FeAl_2$, Fe_2Al_5 and $FeAl_3$ and oxides like Al_2O_3 and FeO_2. Further, away from these dark regions yellowish bands run across the micrograph rich with fine grains of 'Al' which is noted. This region is rich with pure aluminium.

Kraishat et al. [6] observed fragmented and lengthy intermetallic compounds formed in the published literature.

3.2 *Rockwell Hardness*

The Rockwell hardness values measured across the sample at twelve points were recorded. The graphical plot of the hardness values is shown in Fig. 7. The hardness values were on the rise as the composition of 'Fe' was on the rise. A maximum

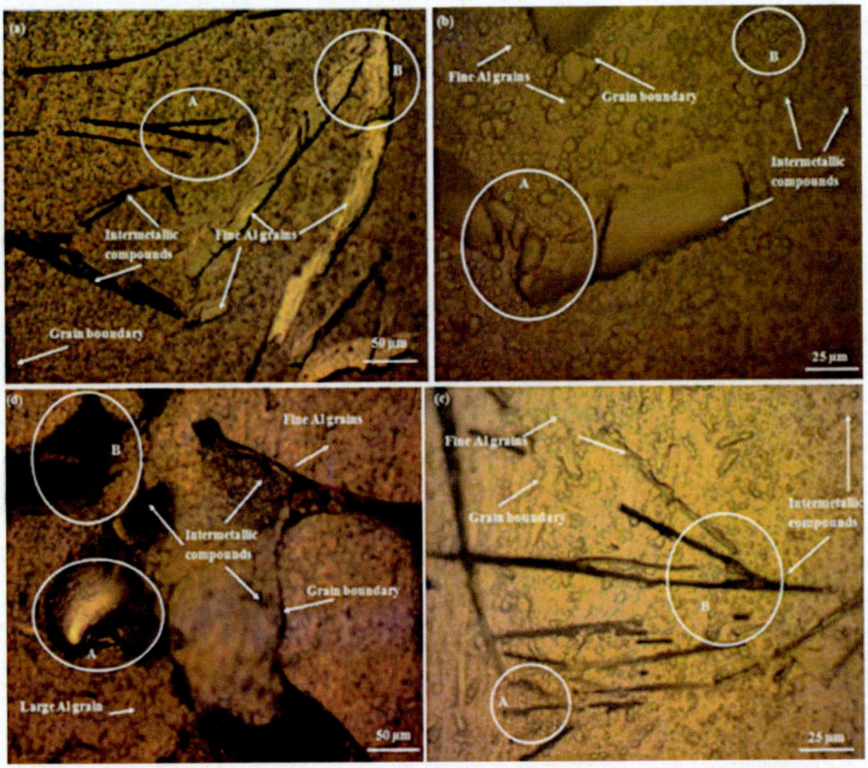

Fig. 6 **a** and **b** Optical micrographs of sample 5, **d** and **e** optical micrographs of sample 6

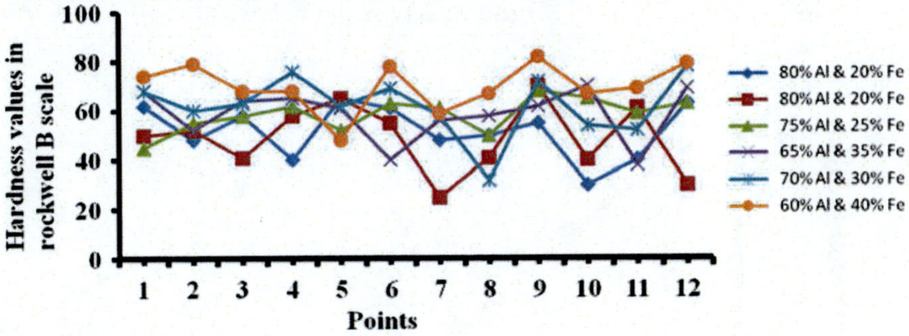

Fig. 7 Rockwell hardness values of various samples

hardness of 79 in ‘B’ scale was noted for the sample 6. The hardness value was in the range of 52–78 for the sample 5. A minimum value of 41 in B scale was recorded for the sample 1. This is due to less amount of ‘Fe’ present in the alloy.

3.3 Shear Strength

The average shear strength values are shown in Table 2. The average shear strengths of the sample 1 processed at 750 °C and annealed for 19 h was 278 MPa. The average shear strength of the sample 2 was only 265 MPa. The lowest shear strength was recorded for the sample 3 (Table 2) which was 250 MPa. The shear strength of the sample 4 was 280 MPa. For the sample 5, the shear strength was 272 MPa. The sample 6 exhibited a shear strength of 279 MPa. The shear strength of various samples is presented as a graph in Fig. 8. The variation in shear strength was attributed to variation in the composition of ‘Fe’ and difference in annealing time and stirring of the alloy during the formation of alloy.

Table 2 Shear strength values of various composition

S. No	Composition in wt. percentage	Shear strength values in N/mm^2
1	80% Al and 20% Fe	278
2	80% Al and 20% Fe	265
3	75% Al and 25% Fe	250
4	65% Al and 35% Fe	280
5	70% Al and 30% Fe	272
6	60% Al and 40% Fe	279

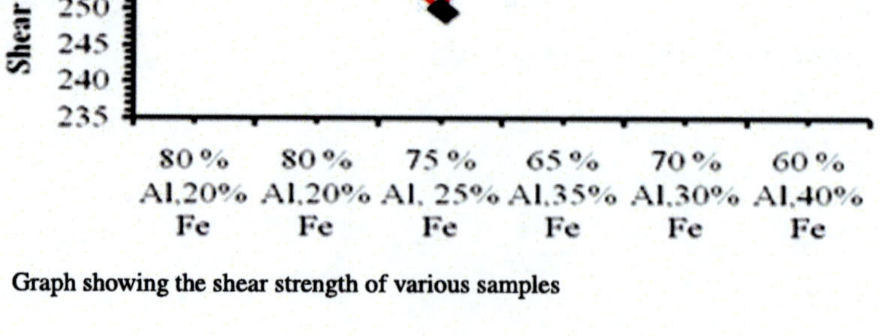

Fig. 8 Graph showing the shear strength of various samples

3.4 X-Ray Diffraction

The XRD report of the sample 2 as given in Table 1 was shown in Fig. 9 and has indicated the presence of several intermetallic compounds in the FeAl matrix. The XRD pattern for the sample 2 is shown in Fig. 9. These compounds are hard and brittle, which increase the shear strength of the annealed samples. The intermetallic compounds such as Fe_2Al, Fe_3Al, FeAl, $FeAl_5$ and Fe_2Al_3 are present in the alloy. The secondary phase particles like Al_2O_3 and FeO_2 are present. These intermetallic compounds and secondary phase particles strengthened the soft aluminium matrix. The XRD pattern reveals the presence of pure Al and Fe along with other compounds.

This XRD report of the sample 4 as given in Table 1 is shown in Fig. 10. The intermetallic compounds such as Fe_2Al, Fe_3Al, FeAl, $FeAl_5$ and Fe_2Al_3 are present in the alloy and indicated the presence of several intermetallic compounds in the FeAl matrix. It increases the shear strength of the annealed samples. The secondary phase particles like Al_2O_3 and FeO_2 were present. These intermetallic compounds and secondary phase particles strengthened the soft aluminium matrix and make the increase in strength of the original aluminium.

The XRD report reveals the composition of the various intermetallic compounds that were present in Fig. 11 of sample 6 and has indicated the presence of several intermetallic compounds in the FeAl matrix; this intermetallic compounds increase the shear strength of the annealed samples. The various compositions of alloy are Fe_2Al, Fe_3Al, FeAl, $FeAl_5$ and Fe_2Al_3 which are present in the alloy, and also, the particles like Al_2O_3 and FeO_2 are present in the secondary phase particle. These intermetallic compounds and secondary phase particles strengthened the soft aluminium matrix and give more strength to the pure aluminium.

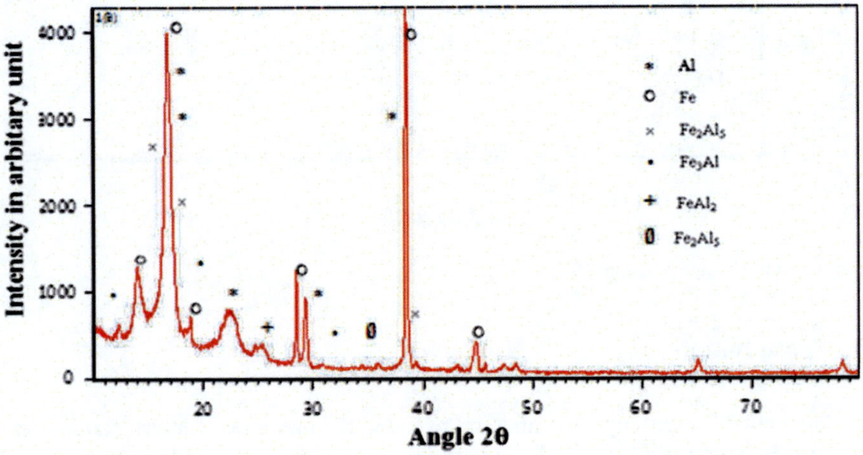

Fig. 9 XRD image of sample 2

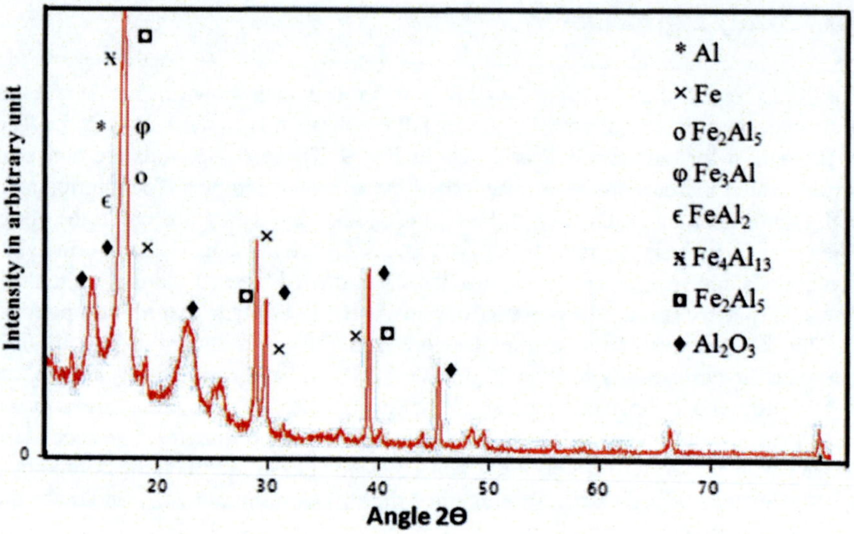

Fig. 10 XRD image of sample 4

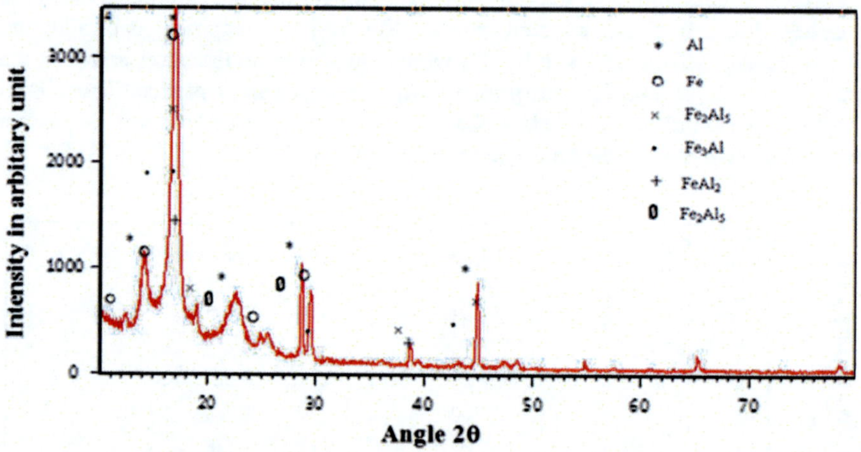

Fig. 11 XRD image of sample 6

4 Conclusions

- The optical microscopy examination revealed the presence of intermetallic compounds in all the alloys. The amount of intermetallic compounds formed was on the rise for the samples containing more amounts of 'Fe'.
- The hardness survey is carried out on the samples revealed hardness variation across the length because of the presence of intermetallic compounds. A maximum

hardness value of 79 in B scale was recorded for the sample 6. The concentration of Fe was at maximum level for this sample.
- Maximum shear strength of 280 MPa was observed for the sample 4. The minimum shear strength was observed for the sample 3.
- The formation of intermetallic compounds such as $FeAl_2$, Fe_2Al_5, $FeAl_6$ and $FeAl_3$ was observed in all the samples in varying amounts. The oxides such as FeO_2 and $Al2O_3$ were noted in few samples.

References

1. Hu W, Kato T, Fukumoto M (2003) Synthesis and nanocrystalline iron aluminide intermetallic compounds. Mater Trans 44(12):2678–2687
2. Jiqiang MA, Jun YANG, Qinling BI, Weimin LIU (2010) Preparation of an ultrane-grained Fe-40Al intermetallic Compound. Acta Metall Sin (Engl Lett) 23(1):50–56
3. Prichard PD (1998) The structure-property relationships of powder processed Fe-Al-Si alloys. PhD Thesis submitted to Iowa State University
4. Chakraborthy SP, Sharma IG (1998) A study on the preparation of iron aluminium based intermetallic alloy by aluminothermic smelting technique. J Alloy Comp 280:255–261
5. Taylor DGM (2006) Refinement of second phase dispersion in iron aluminide intermetallics by high temperature severe plastic deformation. Mater Trans 2092–9098
6. Khraisat W, Abu Jadayil W (2010) Strengthening aluminium scrap by alloying with iron. Jordan J Mechan Ind Eng 4:372–377

Influence of Graphite Particles on Microhardness and Microstructural Behavior of AA7068 Metal Matrix Composites Processed by Powder Metallurgy

K. John Joshua, S. J. Vijay, P. Ramkumar, and D. Philip Selvaraj

Abstract Automotive industries require components such as automotive shock absorbers that should be lightweight and has to withstand high wear resistance. Monolithic aluminum alloys is not enough to meet the demands. Hence aluminum alloy (AA7068) reinforced with graphite particles were processed by powder metallurgy technique with different weight percentages of graphite (3, 6, and 9%) along with base matrix. A low pressure of 318 MPa was applied for compaction and sintered at a temperature of 560 °C for one hour. Vickers microhardness test was carried out for measuring the microhardness of the composites. Using pin-on-disc wear-tester, wear experiments were conducted with a velocity of 1.2 m/s over a sliding distance of 2.5 km of load 5 N. Field Emission Scanning Electron Microscopy (FESEM) analysis was carried out to investigate the worn surface. Experimental results showed that the Vickers microhardness number has been reduced to 19 VHN by the addition of 9% graphite particles. Wear experiments revealed improved wear resistance by the addition of graphite particles. FESEM analysis exposed that abrasion, delamination, and oxidation were the dominant wear mechanisms for the composites. The automotive shock absorber produced using such a composite combination would be lightweight with improved wear resistance.

Keywords Metal matrix composites · Powder metallurgy · Microhardness · Sintering

K. John Joshua (✉)
Department of Marine Engineering, PSN College of Engineering and Technology, Tirunelveli 627152, Tamil Nadu, India
e-mail: kjohnjoshua@gmail.com

S. J. Vijay · P. Ramkumar · D. Philip Selvaraj
Department of Mechanical Engineering, Karunya Institute of Technology and Sciences, Coimbatore 641114, Tamil Nadu, India

S. Vijayan et al. (eds.), *Trends in Manufacturing and Engineering Management*, Lecture Notes in Mechanical Engineering,
https://doi.org/10.1007/978-981-15-4745-4_9

1 Introduction

Nowadays industries have focused its attention towards aluminum alloys due to its improved properties such as lightweight, corrosion resistance, improved thermal conductivity, recycling properties, etc. and are widely used in aviation, automobile and structural industries [1, 2]. AA7068 aluminum alloy provides high mechanical strength among the alloying elements and provides improved mechanical properties [3]. Graphite is a soft reinforcement and due to its self-lubricating property, it provides a better wear resistance than conventional aluminum alloys for Al-Gr composites, which prevents metal to metal contact by forming a thin layer of graphite particle [4]. Graphite serves as a solid lubricating layer and reduces the composite wear without the need for solid and liquid lubrication [5]. The addition of graphite in aluminum matrix leads to a reduction in hardness and flexural strength [6]. Graphite powder acts as a friction-reducing element, helps to get stronger bond with metals and widely used in power metallurgy [7]. Graphite has extrinsic lubricating property and hence reduces friction and wear. Large amount of graphite added to copper reduces the hardness and fracture strength of the composites [8]. Addition of graphite in Al7075/Al_2O_3/Gr composites exhibited superior wear-resistance properties. Smaller amount of aluminum graphite composite exhibits superior wear properties over the base alloys [9]. Gheorghe et al. [10] fabricated Al/Al_2O_3/Gr hybrid composites through powder metallurgy route and showed that Vickers microhardness was increased by increasing the duration of mechanical milling and also by increasing the reinforcement content of hard Al_2O_3 particles. The presence of graphite particles act as a solid lubricant and reduced the wear rate. Kaushik et al. [11] produced Al 6082 alloy-SiC-Gr hybrid composites using stir casting technique and reported that the hybrid composites showed improved wear resistance compared to single SiC reinforced composite and base alloy. The presence of graphite forms a graphitic film which acts as a self-lubricant in the hybrid composites. Akhlaghi and Pelaseyyed [12] developed A365 alloy- graphite composites using in situ powder metallurgy technique and showed that wear rate reduced as a result of increased graphite content due to the homogeneous distribution of graphite particles with reduced porosity and fine-grained base alloy. Akhlaghi and Zare-Bidaki [13] produced Al 2024/5-20 wt% flake graphite composites by in situ powder metallurgy technique and reported that the hardness of these composites decreased with increasing graphite content. Up to 5 wt% graphite addition the composites showed improved the wear resistance and beyond 10 wt% or more graphite particles addition, the wear rate increased. Riahi and Alpas [14] fabricated A356 Al–10% SiC–4% Gr and A356 Al–5% Al_2O_3–3% Gr composites using molten metal mixing technique. He reported that the composites have undergone severe wear regime and the hardness of the composites has been reduced. Graphitic composites demonstrated mild to severe wear at various load and sliding speed combinations, which were higher than A356 base alloy and A356–20% SiC composites. Babu et al. [15] studied the wear behavior of aluminum/graphite nanofiber/alumina short fiber hybrid composites with a different volume fraction of fibers (10, 15, and 20%) under dry sliding conditions. The composites with 10 and

15 vol.% of fiber showed the lowest friction and wear loss, and increased friction and wear loss for 20 vol.% of fiber. Ted and Tsao [16] fabricated aluminum/SiC/graphite hybrid composites with different volume percentages (2, 5, 8%) of graphite using a semi-solid powder densification technique and studied the tribological behavior. As the amount of graphite content increased, the hardness of the composites decreased. The wear rate of the composites increased up to 5% graphite addition and then reduced for 8% graphite addition. Rajaram et al. [17] fabricated Al-SiC/graphite hybrid composites using stir casting technique and studied the wear behavior. The wear properties showed that by the addition of graphite particles, the wear resistance of the composites increased compared to Al–Si base alloy for all load ranges.

The produced AA7068/graphite metal matrix composites were tested and their results were reported.

2 Experimental Procedure

2.1 Composite Preparation

FESEM image of AA7068 and graphite powders used in this research work was demonstrated in Fig. 1. AA7068 was used as a matrix material and graphite was used as the reinforcement material to produce metal matrix composites using powder metallurgy technique. AA7068 (Zn-8%, Mg-3%, Cu-2%, Si-0.12%, Fe-0.15%, Mn-0.1%, Cr-0.05%, Zr-0.1%, Ti-0.01%, Al-bal) powders reinforced with different weight percentages of graphite (3, 6, 9%) particles were milled in a ball mill for 40 h separately. The average particle size of AA7068 powders and graphite powders were 1.785 μm and 2.07 μm respectively. Stearic acid is mixed up to 2% to improve the bonding strength between the metal powders [18]. The ball-milled powders were compacted using a Universal Testing Machine (UTM) at a low pressure of 318 MPa and sintered at a temperature of 560 °C for an hour to produce composites of size

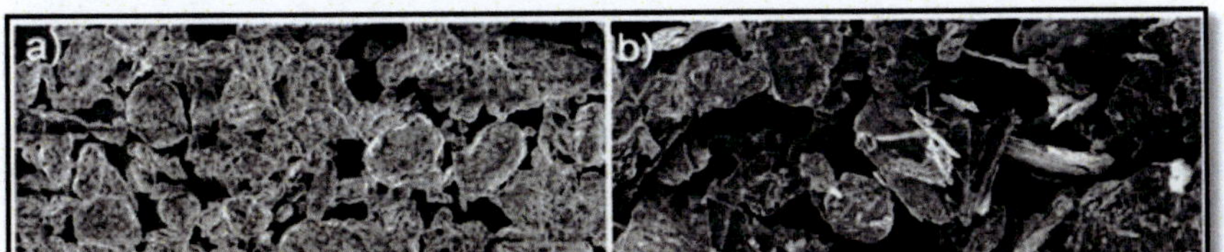

Fig. 1 FESEM image of powers **a** AA7068 after milling **b** Graphite as received

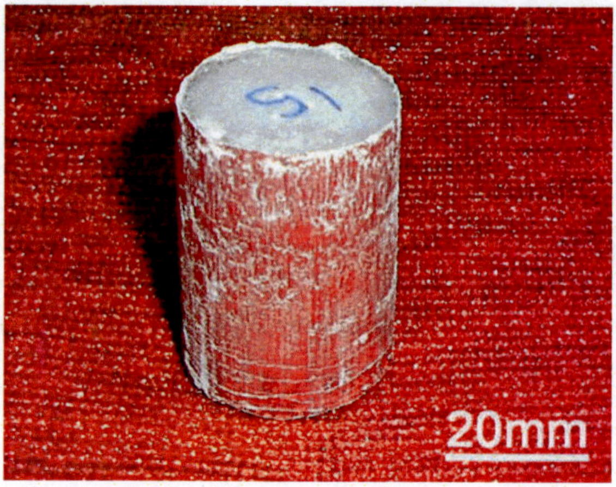

Fig. 2 Typical composite specimen

20 mm diameter and 32 ± 2 mm height. A typical composite specimen of a produced green compact is as shown in Fig. 2. The composites of AA7068 reinforced with different weight percentages of graphite (3, 6, 9%) along with plain AA7068 were produced. The produced composites were polished using 600, 800 and 1200 abrasive grit paper and with 2 μm and 0.5 μm diamond paste in twin disk polisher and etched in Keller's reagent for 10 s for cleaning [19]. Three spots were spotted and the average values were taken for measuring the Vickers microhardness. Using pin-on-disc wear-tester, experiments were conducted with a velocity of 1.2 m/s over a sliding distance of 2.5 km of load 5 N. FESEM analysis was carried out to investigate the worn surface for the produced composites.

3 Results and Discussion

3.1 Microhardness Test

The hardness values of AA7068/graphite composites were presented in Fig. 3. As the weight fraction of graphite increases, the hardness decreases. The microhardness was found to be a minimum of 19 VHN for AA7068 reinforced with a weight fraction of 9% graphite particles. Incorporation of graphite particles into aluminum alloys normally reduces the strength, ductility, and hardness of the composites. The decrease in hardness can be attributed due to the increased amount of the brittleness of graphite particles with the increased affinity of crack instigation and transmission at the graphite/base alloy boundary [13]. This result was supported by the results of other researchers [4, 9].

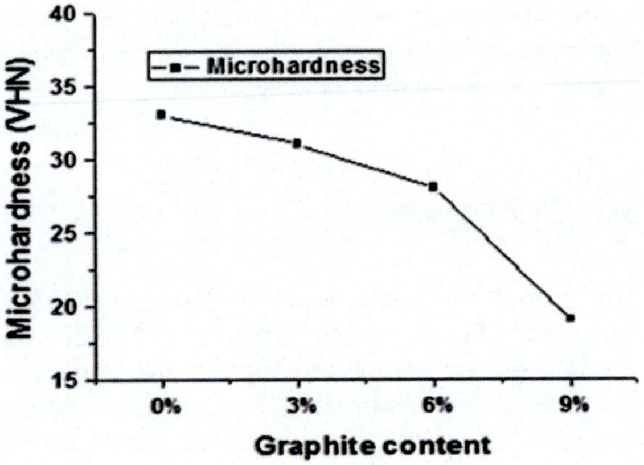

Fig. 3 Vickers microhardness of AA7068/Graphite composites

3.2 Wear Behavior

The graph of wear rate vs sliding distance and volume loss vs. sliding distance for AA7068/graphite composites were presented in Fig. 4a, b respectively. It was noticed that the wear rate and the volume loss of all the composites were lower compared to that of the matrix material. As the graphite content increased, the wear rate and the volume loss decreased for all the composites. As the sliding distance increased, wear rate and the volume loss increased for both the base alloy and the composites. The graphite particles incorporated in the base alloy act as a solid lubricant, thereby increasing the wear resistance of the composites. The increase in wear resistance was due to the direct metal contact which was prevented due to the presence of graphite

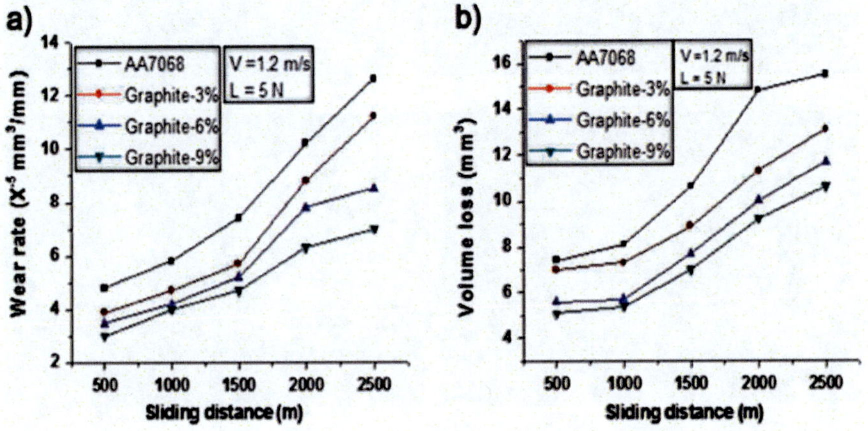

Fig. 4 **a** Wear rate vs sliding distance and **b** volume loss vs sliding distance of AA7068/Graphite composites

particles which formed a rich thin tribofilm between sliding surfaces [9]. Composites reinforced with 9% graphite showed reduced wear rate and volume loss compared to other composite combinations.

3.3 SEM Analysis

Figure 5a–d demonstrates the FESEM images of AA7068/graphite metal matrix composites. Figure 5a demonstrates the FESEM image of AA7068 base matrix. It was noticed that abrasion, delamination, and oxidation were the dominant wear modes for the matrix material. Deep, shallow and narrow grooves were noticed on the surface. Grooves present confirms the abrasive wear mechanism paralleling to the sliding direction. The evidence of grooves and plastic flow rises the shear stress which promotes delamination wear. The frictional heat generated and shear stress produces such deep grooves [20, 21]. Figure 5b demonstrates the FESEM image of AA7068 with the addition of 3% graphite particles. Shallow and narrow grooves were found on the worn surfaces paralleling to the sliding direction which shows the occurrence of plastic deformation. This behavior confirms delamination and abrasion

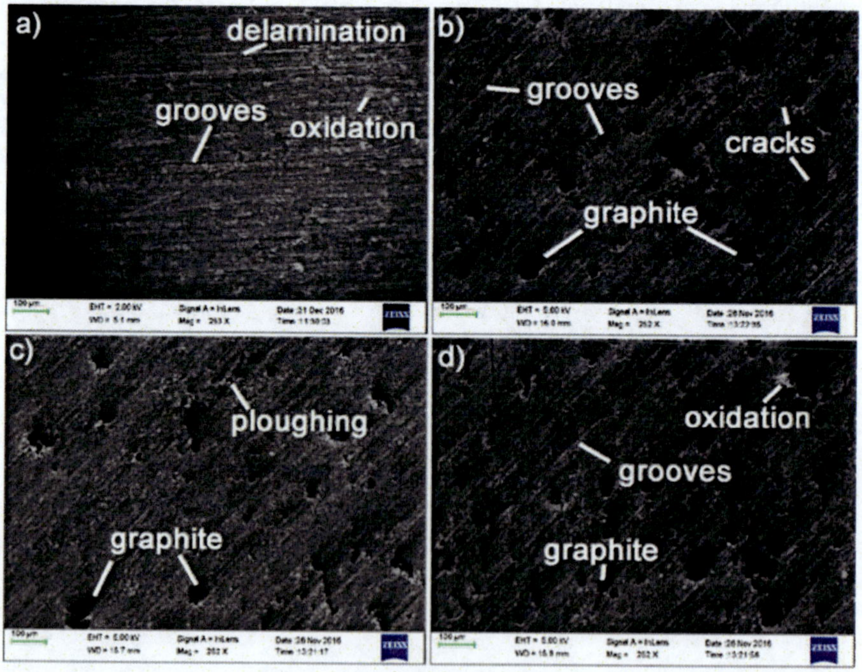

Fig. 5 FESEM analysis of **a** AA7068 **b** AA7068-3% Graphite **c** AA7068-6% Graphite and **d** AA7068-9% Graphite

wear mechanisms [9, 22, 23]. Mixture of rough, smooth regions and large cavities were found. Cracks and cavities were present on the worn surface due to delamination. Subsurface deformation, nucleation of cracks, propagation of cracks were involved in delamination. The nucleation of cracks occurs on the surface of the material at weak points. Figure 5c, d demonstrate the FESEM images of AA7068 with the addition of 6% graphite and 9% graphite respectively. Shallow, narrow grooves were present on the worn surfaces along the sliding direction of the pin. Severely damaged patches and few rough regions were found on the worn surface. Cracks, cavities, and small pits were visible on the worn surface. Direct metal contact was prevented due to the presence of graphite particles which form a rich thin tribofilm between sliding surfaces [9]. Ploughing and exposure of graphite particles were clearly visible. Ploughing of materials results in hard wear debris entrapped in between the contact surfaces due to micro-cutting.

4 Conclusions

- Using the powder metallurgy route, AA7068/graphite composites can be processed successfully.
- The microhardness of the composites decreased with an increase of graphite content and AA7068-9% graphite composite exhibited a minimum of 19 VHN. The decrease in hardness can be attributed due to the increased amount of the brittleness of graphite particles with the increased affinity of crack instigation and transmission at the graphite/base alloy boundary.
- The wear experiments revealed an increased wear resistance by increasing the graphite content. AA7068-9% graphite composite exhibited a maximum resistance which was due to the direct metal contact which was prevented due to the presence of graphite particles which formed a rich thin tribofilm between sliding surfaces.
- FESEM analysis demonstrated that dominant wear modes such as abrasion, delamination, ploughing, and oxidation were predominant for the produced composites.
- The automotive shock absorber produced using AA7068-9% graphite composite combination would be lightweight with improved wear resistance which would meet the industry demands.

Acknowledgements The authors thank Mr. J. Devamanoharan of Centre for Research in Metallurgy (CRM) Laboratory of Karunya Institute of Technology and Sciences for his assistance to carry out this experimental work.

References

1. Sandar D, Chatterjee R, Bharathi RJ, Sreehari P (2019) Investigation on the microstructure and mechanical properties of AA6082/TiO_2 surface composites produced by friction stir processing. In: Lakshminarayanan AK, Idapalapati S, Vasudevan M (eds) Conference 2018, ICEMMM. Springer, Singapore, pp 19–28
2. Kaku SMY, Khanra AK (2019) Hot deformation studies of Al–Cu–Mg powder metallurgy alloy composite. In: Lakshminarayanan AK, Idapalapati S, Vasudevan M (eds) Conference 2018, ICEMMM. Springer, Singapore, pp 75–81
3. Amin A, Ali S, Omid N, Hamid F, Saeid (2015) Optimizing consolidation behavior of Al 7068–TiC nanocomposites using Taguchi statistical analysis. Trans Nonferrous Met Soc China 25:2499–2508
4. Pardeep S, Dinesh K, Satpal S (2016) Dry sliding wear investigation of Al6082/Gr metal matrix composite by response surface methodology. J Mater Res Technol 5(1):29–36
5. Kenneth KA, Kazeem OS (2015) Microstructural characteristics, mechanical and wear behaviour of aluminium matrix hybrid composites reinforced with alumina, rice husk ash and graphite. Eng Sci Tech 18:416–422
6. Baradeswaran A, Vettivel SC, Elaya P, Selvakumar N, Franklin IR (2014) Experimental investigation on mechanical behaviour, modelling and optimization of wear parameters of B4C and graphite reinforced aluminium hybrid composites. J Mater Des 63:620–632
7. Gan Y, Yih P, Chen CQ (2011) Electroplating graphite powder with metals and its use in microwave absorbing composite. Mater Sci Eng Compos Mater 3(2):129–133
8. Jing FL, Lei Z, Jin KX, Ke CZ (2015) Sliding wear behavior of copper-based composites reinforced with graphenenanosheets and graphite. Trans Nonferrous Met Soc China 25:3354–3362
9. Baradeswaran A, Elaya PA (2014) Study on mechanical and wear properties of Al 7075/Al_2O_3/graphite hybrid composites. Compos Part B 56:464–471
10. Gheorghe I, Valeriu GG, Mihai B, Traian B, Mircea IP (2015) Studies on wear rate and microhardness of the Al/Al_2O_3/Gr hybrid composites produced via powder metallurgy. Compos Part B 69:603–611
11. Kaushik NC, Rao RN (2016) The effect of wear parameters and heat treatment on two body abrasive wear of Al–SiC–Gr hybrid composites. Tribol Intl 96:184–190
12. Akhlaghi F, Pelaseyyed SA (2004) Characterization of aluminum/graphite particulate composites synthesized using a novel method termed in-situ powder metallurgy. Mater Sci Eng, A 385:258–266
13. Akhlaghi F, Zare-Bidaki A (2009) Influence of graphite content on the dry sliding and oil impregnated sliding wear behavior of Al 2024–graphite composites produced by in situ powder metallurgy method. Wear 266:37–45
14. Riahi AR, Alpas AT (2001) The role of tribo-layers on the sliding wear behavior of graphitic aluminum matrix composites. Wear 251:1396–1407
15. Babu JSS, Kang CG, Kim HH (2011) Dry sliding wear behavior of aluminum based hybrid composites with graphite nanofiber–alumina fiber. Mater Des 32:3920–3925
16. edGuo ML, Tsao CYA (2000) Tribological behavior of self-lubricating aluminium/SiC/graphite hybrid composites synthesized by the semi-solid powder-densification method. Compos Sci Technol 60:65–74
17. Rajaram G, Kumaran S, Srinivasa Rao T (2010) Sliding wear behavior of Al-Si/graphite composite. Tribol Trans 54:1(115–121)
18. Angelo PC, Subramanian R (2008) Powder metallurgy: science, technology and application, 1st edn. Prentice Hall, India
19. John Joshua K, Vijay SJ, Philip Selvaraj D (2018) Effect of nano-TiO_2 particles on microhardness and microstructural behavior of AA7068 metal matrix composites. Ceram Int 44:20774–20781
20. Ashok KB, Murugan N, Dinaharan I (2014) Dry sliding wear behavior of stir cast AA6061-T6/AlNp composite. Trans Nonferrous Met Soc China 24:2785–2795

21. Yan TY, Lan J, Gao FF, Li QC (2015) Wear behavior and mechanism of B4C reinforced Mg-matrix composites fabricated by metal-assisted pressureless infiltration technique. Trans Nonferrous Met Soc China 25:2543–2548
22. Baradeswaran A, Elaya PA (2013) Wear and mechanical characteristics of Al 7075/graphite composites. Compos Part B 56:472–476
23. Maa W, Lua J, Wanga B (2009) Sliding friction and wear of Cu–graphite against 2024, AZ91D and Ti6Al4V at different speeds. Wear 266:1072–1081

Microhardness and Microstructural Behavior of AA7068/SiC Metal Matrix Composites Synthesized by Powder Metallurgy

K. John Joshua, P. Ramkumar, S. J. Vijay, and S. Mohanasundaram

Abstract If anybody is missing any arm or leg, a prosthetic limb can be replaced to take care of their activities by themselves. Manufacturing of prosthetic limbs for such persons should be lightweight and requires improved hardness. Hence, silicon carbide (SiC) was incorporated into aluminum alloy (AA7068) to produce composites by powder metallurgy technique with different weight percentages of SiC (3, 6 and 9%) along with base alloy. A compaction pressure of 318 MPa was applied and sintered at a temperature of 560 °C for an hour. Vickers microhardness test was carried out to find out the microhardness values. Wear experiments were conducted with a velocity of 1.2 m/s over a sliding distance of 2.5 km of load 5 N. Field emission scanning electron microscopy (FESEM) analysis was carried out to investigate the worn surface. Experimental results showed that the Vickers microhardness number has been increased to a maximum of 55 VHN by the addition of 9% SiC particles which were 66.7% more than the base alloy. Wear experiments exposed an improved wear resistance by the addition of SiC particles. AA7068-9% SiC composite exhibited maximum wear resistance. FESEM analysis revealed that abrasion and oxidation were the predominant wear modes for the produced AA7068/SiC composites. The prosthetic limb produced using such composite combination would be lightweight with the improved hardness which would serve the persons better to perform their daily activities.

Keywords Metal matrix composites · Wear · FESEM · Microhardness

K. John Joshua (✉)
Department of Marine Engineering, PSN College of Engineering and Technology, Tirunelveli 627152, Tamil Nadu, India
e-mail: kjohnjoshua@gmail.com

P. Ramkumar · S. J. Vijay · S. Mohanasundaram
Department of Mechanical Engineering, Karunya Institute of Technology and Sciences, Coimbatore 641114, Tamil Nadu, India

S. Vijayan et al. (eds.), *Trends in Manufacturing and Engineering Management*, Lecture Notes in Mechanical Engineering,
https://doi.org/10.1007/978-981-15-4745-4_10

1 Introduction

Aluminum matrix composites when reinforced with hard ceramic particles replace monolithic materials and finds a wide range of application in automobile, transportation, defense and aerospace industries due to its improved properties such as high strength, improved wear resistance, high corrosion resistance, low density, heat treatment capability, high elastic modulus, high strength-to-weight ratio and high thermal conductivity [1–4]. Among the 7xxx series aluminum alloys, AA7068 is one of the most famous one which provides high strength and improved mechanical properties [5]. Silicon carbide (SiC) is one of the most used ceramic particulates when introduced into aluminum-based matrix that exhibits various improved advantages such as high corrosion resistance, improved thermal conductivity, high strength and stiffness, improved wear resistance, good machinability and low cost [6–9]. Mohammad et al. [10] reported that the hardness and wear resistance improved for Al 5252 alloy reinforced with micro- and nano-sized SiC particles fabricated using in situ powder metallurgy method. Ali et al. [11] stated that the AA6061 reinforced with SiC synthesized through semi-solid processing exhibits improved hardness and superior wear resistance compared to the matrix alloy. Ali et al. [12] concluded that A356 alloy reinforced with nano-SiC composites fabricated through stir casting showed the improved hardness, yield strength and ultimate strength. Devaraju et al. [13] reported that aluminum alloy 6061-T6 reinforced with (SiC and Al2O3) hybrid composites fabricated using stir casting showed the improved hardness and wear resistance due to the presence of hard SiC and Al_2O_3 particles. Nidhi et al. [14] stated that Al–10 wt%SiC composites fabricated using stir casting technique showed reduced wear rate and coefficient of friction. Dolatkhah et al. [15] concluded that the Al5052/SiC metal matrix composite fabricated through friction stir processing showed that the microhardness has been improved up to 55% and wear rate was decreased up to 9.7 times compared with the base matrix. Powder metallurgy is an economical manufacturing method to produce components and offers various advantages such as better dimensional precision, improved surface finish, ability to produce complex shapes and cost-effectiveness [16].

The mechanical and microstructural properties are characterized for the fabricated AA7068/SiC composites, and their results are reported.

2 Experimental Procedure

2.1 *Composite Preparation*

Figure 1 demonstrates the FESEM image of AA7068 and SiC powders used in this research work. AA7068 was used as a matrix material, and SiC was used as the reinforcement material to produce metal matrix composites using a powder metallurgy technique. Power metallurgy provides better bonding of matrix and particles,

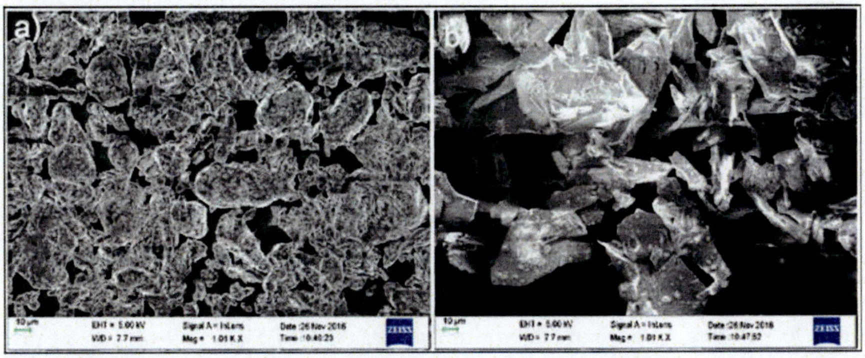

Fig. 1 FESEM image of powers **a** AA7068 after milling, **b** SiC as received

better distribution and easier control of matrix structure. AA7068 (Zn-8%, Mg-3%, Cu-2%, Si-0.12%, Fe-0.15%, Mn-0.1%, Cr-0.05%, Zr-0.1%, Ti-0.01%, Al-bal) powders reinforced with different weight percentages of SiC (3%, 6%, 9%) particles were milled in a ball mill for 40 h separately. The average particle size of AA7068 powders and SiC powders were 1.785 μm and 2.253 μm, respectively. Stearic acid is mixed up to 2% to improve the bonding strength between the metal powders. The ball-milled powders were compacted using a Universal Testing Machine (UTM) at a low pressure of 318 MPa and sintered at a temperature of 560 °C for one hour to produce composites of size 20 mm diameter and 32 ± 2 mm height. Figure 2 demonstrates

Fig. 2 Typical composite specimen

the typical composite specimen of a produced green compact. The composites of AA7068 reinforced with different weight percentages of SiC (3, 6, 9%) along with plain AA7068 were produced. The produced composites were polished using 600, 800 and 1200 abrasive grit paper and with 2 and 0.5 μm diamond paste in twin disk polisher and etched in Keller's reagent for 10 s for cleaning.

Three spots were spotted, and the average values were taken for measuring the Vickers microhardness using Vickers microhardness tester. Wear experiments were conducted on DUCOM TR 20-LE pin-on-disk apparatus. The experiments were conducted at a load of 5 N with a sliding velocity of 1.2 m/s at five sliding distances at an interval of 500 m for a total distance of 2.5 km. The exact mass was measured after every sliding distance of 500 m after the wear test. FESEM analysis was carried out to investigate the worn surface for the produced composites to study the dominant wear behavior.

3 Results and Discussion

3.1 Microhardness Test

The microhardness values of AA7068/SiC composites are presented in Fig. 3. It is noted from the graph that the microhardness increases as the SiC content increases. The microhardness rises from 33 VHN (base alloy) to 55 VHN (9 wt% SiC), showing an increase of 66.7%. It can be attributed due to the presence of hard ceramic SiC particles in the base alloy, reduced grain size and the SiC particles act as obstacles to the motion of dislocation [11, 12]. The increase in microhardness is due to the presence, pinning effect of the SiC particles and Orowan strengthening. The incorporation of SiC particles forms an excellent grain structure due to the grain boundary restrain and improvement in induced strain [13, 17].

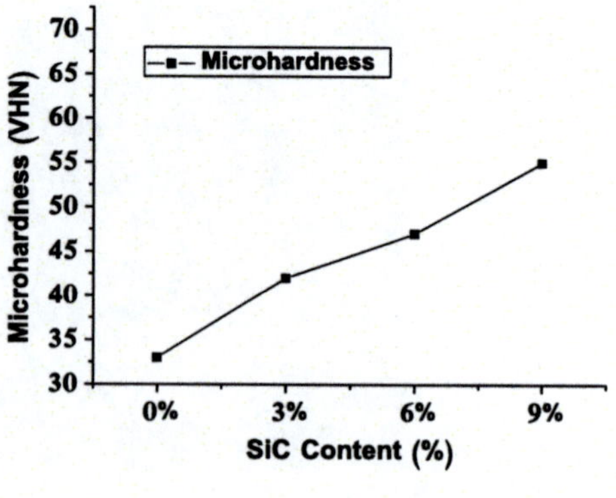

Fig. 3 Vickers microhardness of AA7068/SiC composites

3.2 Wear Behavior

The effect of sliding distance on the wear rate is presented in Fig. 4a. As the SiC content increases from 0 to 9%, the wear rate decreases for all the composites. As the sliding distance increases, wear rate increases for both base alloy and composites. According to Archard's equation, as hardness increases the wear rate decreases. The increase in wear resistance is due to an increase in the thermal stability of the base alloy because of the addition of SiC particles [10]. Composites reinforced with 9% SiC demonstrate superior wear resistance than other composites. The composites reinforced with SiC exhibit improved wear resistance compared to that of the base alloy and also increase with the increase in the SiC content. The wear rate increases with increasing sliding distance, while as the SiC weight fraction increases, wear rate decreases.

The effect of sliding distance on the volume loss is presented in Fig. 4b. As the sliding distance increases, the volume loss also increases for both base alloy and composites. The volume loss decreases as the SiC content increases from 0 to 9%. Increase in the SiC content weight percentage improves the hardness which improves the wear resistance of the composites [18]. Composites reinforced with 9% SiC show reduced volume loss than other composite combinations. The volume loss of the base alloy is higher compared to other reinforced composites. The hardness of SiC is more compared to the base alloy which naturally increases the wear resistance of all the composites. Hard SiC reinforcement particles provide protection, strength to the base alloy and also improve the thermal stability and load-bearing capacity to the composites [19].

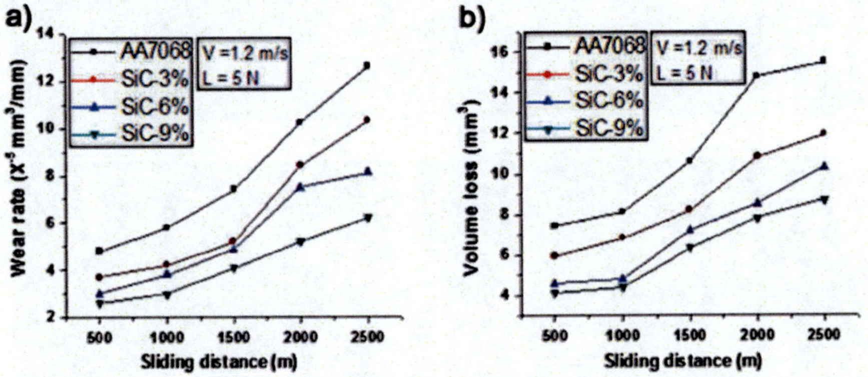

Fig. 4 a Wear rate versus sliding distance and b volume loss versus sliding distance of AA7068/SiC composites

3.3 SEM Analysis

Figure 5a–d demonstrates the FESEM micrographs of AA7068/SiC composites. Figure 5a shows the FESEM micrograph of unreinforced AA7068 matrix. It is noticed that abrasion, oxidation and delamination are the dominant wear modes present in the matrix. Large number of wider, deeper, shallow and narrow grooves paralleling to the sliding direction are noticed which confirms abrasion wear behavior. Few pits, scratches, microcuts and rough regions are visible on the worn surface. Wear debris is caused due to the hard asperities or pulled out hard particles among the sliding surfaces. This behavior ensures a severe abrasive wear mechanism [9, 15]. Delamination wear is also noticed from the micrograph which can be attributed to the quality of the interfacial bond between the particles present in the base alloy [10].

Figure 5b, c demonstrates the FESEM micrographs of AA7068-3%SiC and AA7068-6% SiC composites, respectively. Abrasion and oxidation are the predominant mechanisms for both the samples. Both worn surfaces exhibit a similar type of wear pattern. By adding SiC reinforcement particles, the hardness of the composite increases. Shallow, narrow grooves are present parallel to the sliding direction which confirms abrasion wear mode. The presence of hard ceramic particles resists

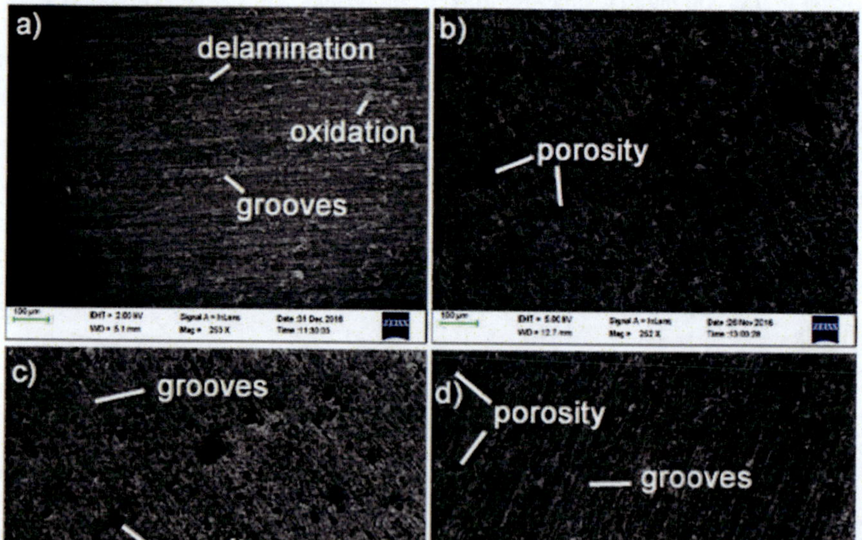

Fig. 5 FESEM analysis of **a** AA7068, **b** AA7068-3% SiC, **c** AA7068-6% SiC and **d** AA7068-9% SiC composites

the delamination process which improves the wear resistance of the composites. The incorporation of hard SiC content improves the strength and reduces the plastic deformation of the composites. During deformation, SiC particles initiate microcracks and develop a strong interfacial bonding between the reinforcement and the base alloy. By high-stress concentration, the local strain and dislocation density reach the critical values, as the SiC particles nucleate microcracks [6, 8]. It is also observed that less cracks and scratches compared to the base alloy. Addition of SiC particles protects the base alloy and hence improves wear resistance, hardness and prevents severe wear [11, 13]. Reduced grain size, improved hardness and uniform dispersion of SiC particles exhibit improved wear properties. Abrasion wear is predominant at lower loads. Addition of SiC particles reduces the grain size which reduces plastic deformation and improves the wear resistance [15]. Spherical shaped pores are evident from the worn surfaces. Pores present are likely to occur in powder metallurgy compared to the liquid melting route which may be due to reduced compaction pressure applied or different sintering parameters [20].

Figure 5d demonstrates the FESEM micrograph of AA7068-9% SiC composite. Mild abrasion and oxidation are the predominant wear modes found in the composite. Few narrow, shallow microgrooves are noticed paralleling to the sliding direction. The depth and width of the grooves present are very less and the surface is finer and smoother compared to other composite combinations. This can be attributed to the microploughing tendency, which is less in the composite. During the eventual process, first the crack initiation layer is removed and then hard SiC particles are involved in further wear reduction process [21]. Delamination is eliminated in the composites reinforced with SiC due to the incorporation of hard SiC particles which has a strong interfacial bond between the SiC and base alloy particles [10]. It is evident from the micrograph that the oxide layer is present in all the samples. The oxide layer serves as a protection layer, thereby minimizing the wear loss by reducing the contact area between the sliding surfaces [22].

4 Conclusions

- AA7068/SiC metal matrix composites have been produced successfully using the powder metallurgy route.
- The microhardness has been increased as SiC content increased. AA7068-9% SiC composite exhibited a maximum hardness of 55 VHN which was 66.7% more than the base alloy.
- The wear resistance has been improved by increasing the SiC content. Addition of 9% SiC exhibited superior wear resistance compared to other composite combinations.
- The FESEM analysis disclosed that abrasion and oxidation were the predominant wear modes found in the produced AA7068/SiC composites.

- The prosthetic limb produced using AA7068-9% SiC composite combination would be lightweight with an improved hardness which would serve the persons better to perform their daily activities.
- AA7068 was the strongest aluminum alloy having zinc and magnesium as major elements. Further, other new reinforcements such as silicon oxide, titanium carbide, titanium boride and zirconium boride can be incorporated with AA7068 base alloy using different fabrication methods such as powder metallurgy, friction stir welding and friction stir processing to produce composites. The composites produced with such combinations would meet industry demands with improved hardness, high stiffness, improved wear resistance, better thermal stability, etc., which can be widely used for various applications.

Acknowledgements The authors thank Mr. J. DevaManoharan, Centre for Research in Metallurgy Laboratory of Karunya Institute of Technology and Sciences, for his assistance to carry out the experimental work.

References

1. Shyam Kumar CN, Ranjit B, Devinder Y (2016) Wear properties of 5083Al–W surface composite fabricated by friction stir processing. Tribol Int 101:284–290
2. Ridvan G, HakanAtapek S, Ahmet K (2017) Influence of preform preheating on dry sliding wear behavior of 304 stainless steel reinforced A356 aluminum matrix composite produced by melt infiltration casting. Tribol Int 115:608–618
3. Hayrettin A, Tolga K, Ercan C, Huseyin C (2006) Wear behaviour of Al/(Al_2O_3pCSiCp) hybrid composites, Tribol Int 39:213–220
4. Ramesh CS, Keshavamurthy R, Channabasappa BH, Pramod S (2010) Friction and wear behavior of Ni–P coated Si_3N_4 reinforced Al6061 composites. Tribol Int 43:623–634
5. Amin A, Ali S, Omid N, Hamid F, Saeid S (2015) Optimizing consolidation behavior of Al 7068 − TiC nanocomposites using Taguchi statistical analysis. Trans Nonferrous Met Soc China 25:2499–2508
6. Umanath K, Palanikumar K, Selvamani ST (2013) Analysis of dry sliding wear behaviour of Al6061/SiC/Al_2O_3 hybrid metal matrix composites. Compos. Part B 53:159–168
7. Taherzadeh MR, Azari RK, Yazdani S, Brabazon D, Boostani AF (2016) Fabrication of aluminum matrix composites reinforced with nano- to micrometer-sized SiC particles. Mat Des 89:58–70
8. Min S, Yue-hui H (2010) Effects of die-pressing pressure and extrusion on the microstructures and mechanical properties of SiC reinforced pure aluminum composites. Mat Des 31:985–989
9. Uthayakumar MS, Aravindan S, Rajkumar K (2013) Wear performance of Al–SiC–B_4C hybrid composites under dry sliding conditions. Mater Des 47:456–464
10. Mohammad MG, Farshad A (2016) Wear behaviour of al 5252 alloy reinforced with micrometric and nanometricSiC particles. Tribol Int 102:28–37
11. Ali M, Mohsen Ostad S (2013) Microstructural and abrasive wear properties of SiC reinforced aluminum-based composite produced by compocasting. Trans Nonferrous Met Soc China 23:1905–1914
12. Ali M, Mohsen Ostad S (2012) Characterization of cast A356 alloy reinforced with nanoSiC composites. Trans Nonferrous Met Soc China 22:275–280

13. Devaraju A, Kumar A, Kumaraswamy A, Kotiveerachari B (2013) Influence of reinforcements (SiC&Al_2O_3) and rotational speed on wear and mechanical properties of aluminum alloy 6061-T6 based surface hybrid composites produced via friction stir processing. Mater Des 51:331–341
14. Nidhi J, Anshul B, Mondal DP, Das S, Singh M (2011) Sliding wear behavior of aluminum syntactic foam: a comparison with Al–10 wt%SiC composites. Tribol Int 44:220–231
15. Dolatkhah A, Golbabaei P, BesharatiGivi MK, Molaiekiya F (2012) Investigating effects of process parameters on microstructural and mechanical properties of Al5052/SiC metal matrix composite fabricated via friction stir processing. Mater Des 37:458–464
16. Martin F, Garcı C, Blanco Y, Aparicio L (2013) Tribo corrosion behaviour of powder metallurgy duplex stainless steels sintered in nitrogen. Tribol Int 57:76–85
17. Devaraju A, Kumar A, Kotiveerachari B (2013) Influence of addition of Gr_p/Al_2O_3p with SiCp on wear properties of aluminum alloy 6061-T6 hybrid composites via friction stir processing. Trans Nonferrous Met Soc China 23:1275–1280
18. Jayakumar L, Balamurugan K (2014) Reciprocating wear behavior of 7075Al/SiC in comparison with 6061Al/Al_2O_3 composites. Int J Refrac Met Hard Mat 46:137–144
19. Adel MH, Abdalla A, Mohammed TH, Ahmad TM (2009) Wear behavior of Al–Mg–Cu–based composites containing SiC particles. Tribol Int 42:1230–1238
20. John Joshua K, Vijay SJ, Philip Selvaraj D (2018) Effect of nano-TiO_2 particles on microhardness and microstructural behavior of AA7068 metal matrix composites. Ceram Int 44:20774–20781
21. Kaushik NC, Rao RN (2016) The effect of wear parameters and heat treatment on two body abrasive wear of Al–SiC–Gr hybrid composites. Tribol Int 96:184–190
22. Antony VKC, Selwinrajadurai J (2016) Influence of rutile (TiO_2) content on wear and microhardness characteristics of aluminium-based hybrid composites synthesized by powder metallurgy. Trans Nonferrous Metals Soc China 26:63–73

Optimization of CO_2 Laser Cutting Parameters for AA6061/B_4C/hBN Hybrid Composites using Taguchi-based Response Surface Methodology

A. Gnanavelbabu, V. Arunachalam, K. T. Sunu Surendran, K. Rajkumar, and E. Anandhababu

Abstract Aluminium hybrid composites due to its tailored properties are being used in many industrial applications. Machining of metal matrix hybrid composites with desired quality through conventional methods is a great challenge. As advancement, non-conventional cutting method such as laser cutting is effectively employed to cut hybrid composites. The present work investigates the cutting quality characteristics of AA6061/B_4C/hBN hybrid composites using pulsed CO_2 laser cutting method. Optimization of machining parameters plays a vital role in improving the quality characteristics. The parameters involved in this non-contact-type cutting process such as laser power (2000–2200 W), gas pressure (12–14 bar), speed of spot movement (600–1000 mm/s), pulsing frequency (60–100 Hz) and reinforcement (5–15 Vol.%) were varied according to Taguchi L_{27} orthogonal array, and the quality characteristics such as surface roughness, kerf width and edge slope were noted. Response surface methodology (RSM) was applied to generate quadratic models and 3D surface plots. Desirability analysis was carried out to obtain the optimal conditions. The optimal parameters obtained are: laser power—2000W, gas pressure—12.05 bar, speed of spot movement—998.07 mm/s, pulsing frequency—60 Hz and reinforcement—6.81 Vol.%.

Keywords Aluminium hybrid composites · CO_2 laser cutting · Surface response methodology (RSM) · Desirability analysis

1 Introduction

The needs of structural materials with high specific strength and stiffness have catalysed the emergence of metal matrix composite (MMC). MMCs are intrinsically

A. Gnanavelbabu (✉) · V. Arunachalam · K. T. Sunu Surendran · E. Anandhababu
Department of Industrial Engineering, CEG, Anna University, Chennai 600025, India
e-mail: agbabu@annauniv.edu

K. Rajkumar
Department of Mechanical Engineering, Sri Sivasubramaniya Nadar College of Engineering, Chennai 603110, India

S. Vijayan et al. (eds.), *Trends in Manufacturing and Engineering Management*, Lecture Notes in Mechanical Engineering,
https://doi.org/10.1007/978-981-15-4745-4_11

smart, high performing and lightweight materials that are being utilized for various automobile, aircraft and industrial applications [1]. To act as a solution over aluminium alloys, the attention shifts to aluminium metal matrix composites (AMMCs) because of the beneficial properties like lightweight, good castability and machinability, durability, good mechanical properties, etc. [2]. Further, properties of AMMCs can be enhanced by the addition of ceramic (SiC, B_4C, Al_2O_3, etc.) and solid lubricant (CNT, graphite, hBN, etc.) reinforcements [3]. Nowadays, boron carbide (B_4C) is used as ceramic reinforcement because of its superior properties like high strength, low density, high hardness and good chemical stability. Hexagonal boron nitride (hBN) solid lubricant improves the friction and wear resistance of the composite material, and moreover, it reduces the cutting zone temperature [4–7].

Laser cutting is a technology that uses a laser to cut materials and is typically used for industrial manufacturing applications. Laser beam machining (LBM) is a non-traditional subtractive manufacturing machining process, in which a laser is directed towards the work piece for machining. Chinmaya R. et al. investigated the experimental evaluation of laser-assisted machining of silicon carbide reinforced aluminium matrix composites. It is found that laser-assisted machining minimizes the specific cutting energy, roughness, tool wear and surface damage as compared to traditional machining process [8]. Adalarasan et al. confirmed the possibility of obtaining good quality characteristics such as surface roughness, material removal rate and kerf width using pulsed CO_2 laser cutting of Al6061/Al_2O_3 composites [9]. Response surface methodology (RSM) is an easy way to calculate a first-degree polynomial model, and it uses a fractional factorial design. Adalarasan et al. elucidated that the application of Taguchi-RSM was to optimize the drilling parameters for Al6061/SiC/Al_2O_3 hybrid composite and the experimental trials were designed using Taguchi's L_{18}OA [10].

The literature survey reveals that a less number of research works has been reported particularly for the optimization of CO_2 laser machining parameters on aluminium metal matrix hybrid composites. Therefore, the main objective of this research work is to optimize the effects of CO_2 laser cutting parameters over surface roughness, kerf width and edge slope on AA6061-B_4C-hBN hybrid composites using Taguchi-RSM.

2 Materials and Methodology

AA6061 was chosen as the matrix material as it is the most commonly used precipitation-hardened aluminium alloy for the development of aluminium composites due to its favourable properties like good mechanical properties and wettability. B_4C is a hard ceramic commonly used as particulate reinforcements in aluminium to improve the wear resistance and strength. Hexagonal boron nitride (hBN) is well known for its excellent lubricating properties, chemical inertness, high-temperature stability, etc. hBN in composites improve the machinability by acting as a self-lubricant and reducing the cutting zone temperature due to its good thermal conductivity. In this experiment, B_4C and hBN particles of size 20 μm and 5 μm,

Table 1 Parameters and levels

Symbol	Parameter	Low level	Medium level	High level
A	Laser power (W)	2000	2100	2200
B	Gas pressure (bar)	12	13	14
C	Speed of spot movement (mm/s)	600	800	1000
D	Pulsing frequency (Hz)	60	80	100
E	Reinforcement (Vol. %)	5	10	15

respectively, were used as the reinforcements. Three different aluminium hybrid composites were fabricated by varying volume percentages of B_4C (5–15%) and hBN (10%) by stir casting technique. CO_2 laser cutting was performed on the hybrid composites using Mitsubishi 3015-eX series (max. power: 3.8 KW) CO_2 laser cutting machine. Taguchi's L_{27} orthogonal array was utilized to design the experiment trials. The parameters and its levels are shown in Table 1. Five different controlling parameters with three levels denoted by low, medium and high were chosen for this study. Parameters such as stand-off distance and flow rate of assist gas were kept constant throughout the cutting. The optimization of the laser cutting parameters was carried out by applying RSM. Desirability graph for all responses was generated to assess the contribution of input parameters.

3 Results and Discussions

The aluminium metal matrix hybrid composites were successfully fabricated using stir casting process. The CO_2 laser cutting was performed on the composites as per L_{27} orthogonal array, and the corresponding responses such as surface roughness (Ra), kerf width (mm), edge slope (°) are shown in Table 2.

The cutting parameters and their responses were analysed using Design–Expert software. Polynomial equations of the second order were formed statistically with the laser cutting parameters and quality characteristics joined with RSM. The mathematical model is created with the help of back selection technique and elimination of higher order to maintain the model hierarchy. This model explains the individual and interaction of cutting parameters on surface roughness, kerf width and edge slope. The analysis on the influence of input parameters on responses was carried out by ANOVA, predicted versus actual graph, response surface plot and quadratic equation.

3.1 Analysis on Surface Roughness

The ANOVA for the surface roughness is shown in Table 3. SS denotes Sum of

Table 2 Parameters, levels and responses of CO_2 laser cutting

Trials	A (W)	B (Bar)	C (mm/s)	D (Hz)	E (Vol.%)	Surface roughness (Ra)	Kerf width (mm)	Edge slope (°)
1	2000	12	600	60	5	4.154	0.332	3.16
2	2000	12	600	60	10	5.098	0.227	2.16
3	2000	12	600	60	15	6.473	0.262	2.50
4	2000	13	800	80	5	4.054	0.298	2.84
5	2000	13	800	80	10	4.986	0.262	2.50
6	2000	13	800	80	15	6.331	0.270	2.57
7	2000	14	1000	100	5	3.966	0.321	3.06
8	2000	14	1000	100	10	4.965	0.294	2.80
9	2000	14	1000	100	15	6.295	0.290	2.76
10	2100	12	800	100	5	4.228	0.266	2.54
11	2100	12	800	100	10	5.116	0.252	2.40
12	2100	12	800	100	15	6.654	0.284	2.70
13	2100	13	1000	60	5	4.334	0.282	2.69
14	2100	13	1000	60	10	5.214	0.276	2.63
15	2100	13	1000	60	15	6.659	0.285	2.71
16	2100	14	600	80	5	4.490	0.300	2.86
17	2100	14	600	80	10	5.499	0.266	2.53
18	2100	14	600	80	15	6.681	0.304	2.90
19	2200	12	1000	80	5	4.797	0.253	2.41
20	2200	12	1000	80	10	5.587	0.201	1.91
21	2200	12	1000	80	15	6.836	0.319	3.04
22	2200	13	600	100	5	4.946	0.225	2.14
23	2200	13	600	100	10	5.559	0.223	2.12
24	2200	13	600	100	15	6.908	0.206	1.96
25	2200	14	800	60	5	5.078	0.240	2.29
26	2200	14	800	60	10	5.739	0.240	2.29
27	2200	14	800	60	15	6.968	0.356	3.39

Squares, "D" denotes the Degree of Freedom, M.S. denotes Mean Square and "S" denotes Significant.

The model*F*-value of 269.53 implies the model is significant. Values of "Prob >*F*" less than 0.0001 indicates that model terms are significantly contributed. In this case, A—Laser power, E—Vol.% of reinforcement, AE, E^2 are the significant model terms.

The closeness of the fit determines the actual and predicted values into the subjected desired level. From Fig. 1, it is evident that the actual and predicted values are

Table 3 ANOVA for surface roughness

Source	SS	D	M.S.	F-value	*P*-value Prob. >*F*	Remarks
Model	24.40	14	1.74	269.53	<0.0001	S
A	0.70	1	0.70	116.65	<0.0001	S
B	0.026	1	0.026	4.39	0.0579	
C	0.024	1	0.024	3.95	0.0703	
D	2.772E−003	1	2.772E−003	0.46	0.5103	
E	21.69	1	21.69	3602.55	<0.0001	S
AB	4.335E−004	1	4.335E−004	0.072	0.7930	
AC	7.633E−003	1	7.633E−003	1.27	0.2822	
AD	0.000	0	0.000			
AE	0.089	1	0.089	14.80	0.0023	
BC	2.427E−003	1	2.427E−003	0.40	0.5374	
BD	0.000	0	0.000			
BE	0.012	1	0.012	1.94	0.1893	
CD	0.000	0	0.000			
CE	4.070E−003	1	4.070E−003	0.68	0.4270	
DE	2.791E−003	1	2.791E−003	0.46	0.5089	
A	4.843E−003	1	4.843E−003	0.80	0.3874	
B^2	0.000	0	0.000			
C^2	0.000	0	0.000			
D^2	0.000	0	0.000			
E^2	0.35	1	0.35	57.57	<0.0001	S
Residual	0.072	12	6.020E−003			
Corr. total	24.47	26				

R-Squared = 0.9970; Adj. R-squared = 0.9936; Mean = 5.47

on the same line without any appreciable spreading. It ensures the closeness of the fit as well as the validity of the fitted model.

Figure 2 illustrates the 3D response surface plot for surface roughness. *X*-axis represents laser power, *Y*-axis represents Vol.% of reinforcement and *Z*-axis represents surface roughness. Increase in reinforcement percentage and laser power increases the surface roughness of the composite. The best surface roughness 4.0197 is achieved with the combination of the lowest laser power and the lowest reinforcement. Even though the minimal interaction was found between the reinforcement and laser power, laser power emerged to be a dominant parameter in affecting the surface roughness by means of its linear as well as interaction conditions. The regression equation for the surface roughness is given in Eq. 1.

$$\textbf{Quadratic equation for surface roughness} = +5.28 + 0.36 * A + 0.077 * B$$

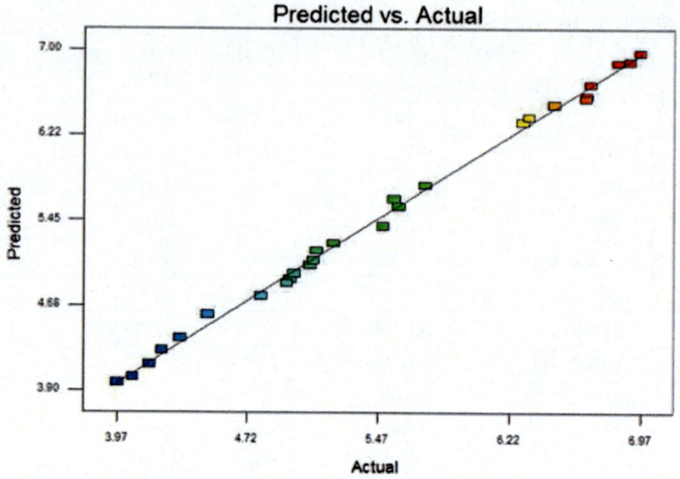

Fig. 1 Actual versus predicted values for surface roughness

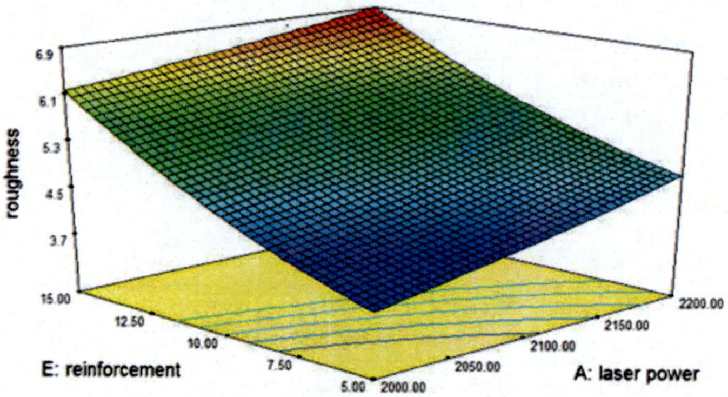

Fig. 2 Three-dimensional response surface plot for surface roughness

$$
\begin{aligned}
& -0.073 * C - 0.033 * D + 1.10 * E - 0.017 * A * B \\
& + 0.071 * A * C - 0.086 * A * E + 0.035 * B * C \\
& - 0.031 * B * E\, E + 0.018 * C * + 0.015 * D * E \\
& + 0.038 * A^2 + 0.24 * E^2 \qquad (1)
\end{aligned}
$$

3.2 *Analysis on kerf width*

The ANOVA for the kerf width is shown in Table 4.

Table 4 ANOVA for kerf width

Source	SS	D	M.S.	*F*-value	*P*-value Prob. >*F*	Remarks
Model	0.030	14	2.141E−003	3.16	0.0263	S
A	6.434E−004	1	6.434E−004	0.95	0.3489	
B	4.481E−003	1	4.481E−003	6.62	0.0244	S
C	8.405E−004	1	8.405E−004	1.24	0.2870	
D	6.401E−004	1	6.401E−004	.95	0.3500	
E	1.934E−004	1	1.934E−004	.29	0.6028	
AB	9.074E−005	1	9.074E−005	.13	0.7207	
AC	2.308E−003	1	2.308E−003	3.41	0.0897	
AD	0.000	0	0.000			
AE	7.105E−003	1	7.105E−003	10.50	0.0071	S
BC	2.420E−004	1	2.420E−004	0.36	0.5610	
BD	0.000	0	0.000			
BE	4.88E−004	1	4.88E−004	0.69	0.4216	
CD	0.000	0	0.000			
CE	1.261E−003	1	1.261E−003	1.86	0.1974	
DE	5.468E−004	1	5.468E−004	0.81	0.3865	
A^2	1.017E−003	1	1.017E−003	1.50	0.2438	
B^2	0.000	0	0.000			
C^2	0.000	0	0.000			
D^2	0.000	0	0.000			
E^2	6.913E−003	1	6.913E−003			
Residual	8.124E−003	12	6.770E−004	10.21	0.007	
Corr. total	0.038	26				

R-Squared = 0.7860; Adj. R-squared = 0.5386; Mean = 0.27

The model*F*-value of 3.16 implies the model is significant. Values of "Prob >*F*" less than 0.0263 indicates that model terms are significantly contributed. In this case, B, AE are the significant model terms.

Figure 3 represents the predicted and actual values for kerf width. Higher R^2 value indicates the closeness of all the points to the regression line. The obtained R^2 value 78.6 indicates a good fitness for kerf width, and the model is validated.

Figure 4 represents the response surface plot for kerf width. Laser power, reinforcement percentage and kerf width are on *X*, *Y*, *Z* axes, respectively. Minimum kerf width 0.200121 was achieved at minimum laser power and nominal reinforcement %. Increase in reinforcement percentage and laser power increases the kerf width of the composite. Equation 2 is the corresponding quadratic equation for kerf width.

$$\textbf{Quadratic equation for kerf width} = +0.26 - 0.011 * A + 0.032 * B + 0.014 * C$$

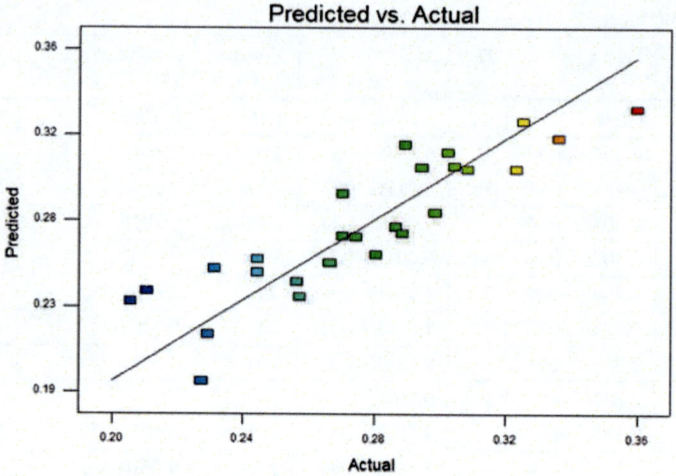

Fig. 3 Actual versus predicted values for kerf width

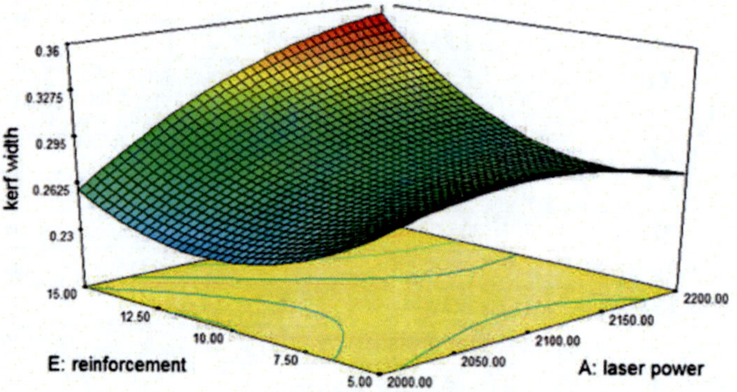

Fig. 4 Three-dimensional response surface plot for kerf width

$$
\begin{aligned}
&+0.016 * D + 3.278E - 003 * E + 7.778E - 003 * A * B \\
&+0.039 * A * C + 0.024 * A * E + 0.011 * B * C \\
&+6.250E003 * B * E + 0.010 * C * E - 6.750E \\
&-003 * D * E - 0.017 * A^2 + 0.034 * E^{2-}
\end{aligned} \quad (2)
$$

3.3 *Analysis on Edge Slope*

The ANOVA for the edge slope is shown in Table 5.

Table 5 ANOVA for edge slope

Source	SS	D	M.S.	*F*-value	*P*-value Prob. >*F*	Remarks
Model	2.73	14	0.20	3.17	0.0260	S
A	0.060	1	0.060	0.97	0.3437	
B	0.41	1	0.41	6.68	0.0239	S
C	0.079	1	0.079	1.28	0.2802	
D	0.060	1	0.060	0.97	0.3445	
E	0.016	1	0.016	0.26	0.6172	
AB	9.074E−003	1	9.074E−003	0.15	0.7077	
AC	0.21	1	0.21	3.44	0.0884	
AD	0.000	0	0.000			
AE	0.64	1	0.64	10.47	0.0071	S
BC	0.021	1	0.021	0.35	0.5667	
BD	0.000	0	0.000			
BE	0.042	1	0.042	0.68	0.4247	
CD	0.000	0	0.000			
CE	0.11	1	0.11	1.79	0.2056	
DE	0.051	1	0.051	0.82	0.3819	
A^2	0.092	1	0.092	1.50	0.2446	
B^2	0.000	0	0.000			
C^2	0.000	0	0.000			
D^2	0.000	0	0.000			
E^2	0.63	1	0.63	10.27	0.0076	S
Residual	0.74	12	0.062			
Corr. Total	3.47	26		3.17		

R-Squared = 0.7887; Adj. R-squared = 0.5423; Mean = 2.59

The model*F*-value of 3.17 implies the model is significant. Values of "Prob >*F*" less than 0.0260 indicates that model terms are significantly contributed. In this case, *B*, *AE*, E^2 are significant model terms.

Figure 5 shows the relationship between actual and predicted values of edge slope. R^2 value for edge slope is 78.87. It indicates the closeness of points to the regression line, and the model is validated.

Figure 6 represents the response surface plot for edge slope. Laser power, reinforcement percentage and edge slope are on *X*, *Y*, *Z* axes, respectively. Minimum edge slope 1.9107 achieved at minimum laser power and nominal reinforcement %. Increase in reinforcement percentage and laser power increased the edge slope of the composites. Equation 3 is the quadratic equation for edge slope.

Quadratic equation for edge slope $= +2.480 - 10 * A + 0.30 * B + 0.13 * C$

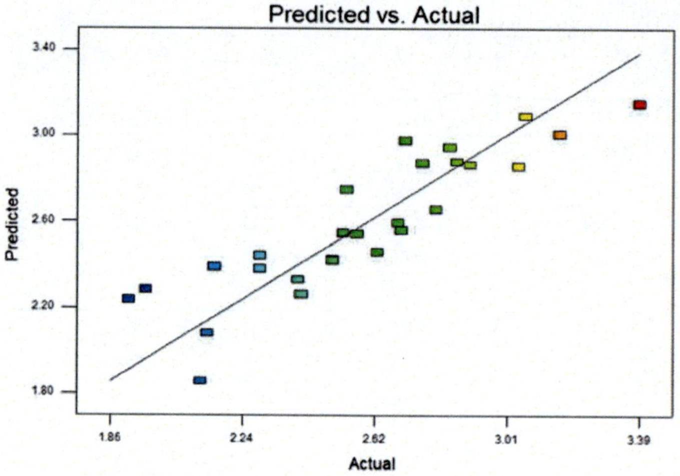

Fig. 5 Actual versus predicted values for edge slope

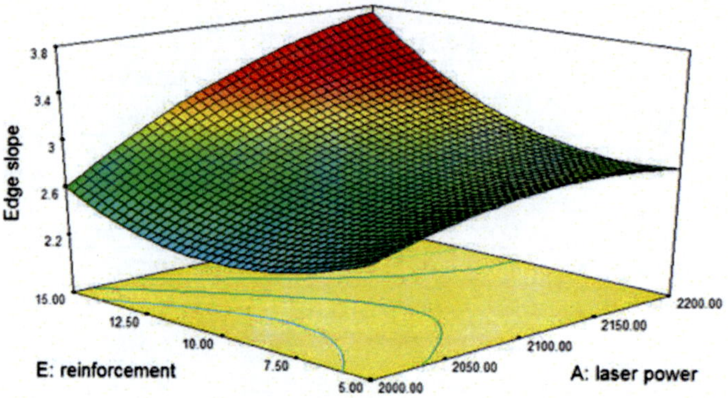

Fig. 6 Response surface plot for edge slope

$$+0.15*D+0.030*E+0.078*A*B+0.38*A*C$$
$$+0.23*A*E+0.10*B*C+0.059*B*E+0.096*C*E$$
$$-0.065*D*E-0.168A^2+0.32*E^2 \quad (3)$$

3.4 Desirability Analysis

Design of expert software was employed to perform the desirability analysis. The obtained values for parameters and responses are shown in Fig. 7. The desirability

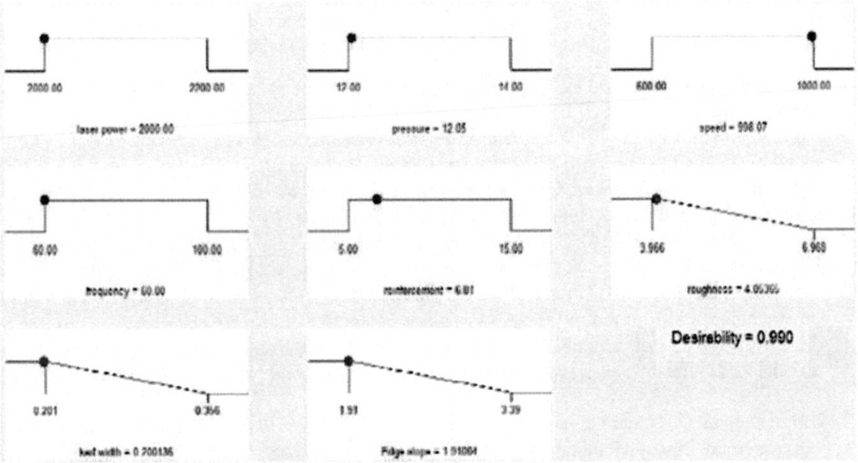

Fig.7 Desirability graph for surface roughness, kerf width and edge slope

value 0.990 proves higher proximity of outputs towards the target.

4 Conclusions

CO_2 laser cutting on AA6061/B_4C/hBN hybrid composites was successfully done as per the structured L_{27} orthogonal array. Optimization of process parameters was carried out by applying RSM, and the following conclusions were drawn.

1. Minimum surface roughness of 4.0197 Ra was obtained by maintaining the laser power at 2000 W, pressure 12.03 bar, speed 1000 mm/s, frequency 60.01 Hz and reinforcement 6.65 Vol.%.
2. Minimum kerf width 0.200121 mm was obtained at laser power 2000 W, pressure 12 bar, speed 999.04 mm/s, frequency 60.16 Hz and reinforcement 6.68 Vol.%.
3. At laser power 2000 W, pressure 12 bar, speed 999.04 mm/s, frequency 60.16 Hz and reinforcement 6.68 Vol.%, a minimum edge slope 1.9107° was achieved.
4. From the desirability analysis of the parameters, a desirability value of 0.990 was obtained. It showed that the optimal condition for pulsed CO_2 laser cutting on AA6061/B_4C/hBN composite was laser power = 2000 W, pressure = 12.05 bar, speed of spot movement = 998.07 mm/s and pulsing frequency = 60 Hz and Vol.% of % reinforcement = 6.81.

Acknowledgement The financial support for consumables in this research work was provided by "DST PURSE Phase II".

References

1. Bains PS, Sidhu SS, Payal HS (2016) Fabrication and machining of metal matrix composites. Mater Manufact Proc 31:573
2. Jojith R, Radhika N (2018) Mechanical and tribological properties of LM13/TiO_2/MoS2 hybrid metal matrix composite synthesized by stir casting. Part Sci Technol 1–13
3. Hung NP, Zhong ZW Zhong CH (1997) Grinding of metal matrix composites reinforced with silicon-carbide particles. Mater Manufact Proc 12:1075–1091
4. Yao YT, Chen LQ (2016) B_4C/Al Composites Processed by Metal assisted Pressure less Infiltration Technique and its Characterization. Mater Manufact Proc 31:1286–1291
5. Santhosh S, Rajkumar K, Gnanavelbabu A (2015) Effect of hBN solid lubricant concentration on machinability of titanium (Ti-6Al-4V) alloy. Mater Sci Forum 830:87–90
6. Gnanavelbabu A, Rajkumar K, Saravanan P (2018) Investigation on the cutting quality characteristics of abrasive water jet machining of AA6061-B4C-hBN hybrid MMC. Mater Manufact Process 1313–1323
7. Muthazhagan C, Gnanavelbabu A, Bhaskar GB, Rajkumar K (2014) Influence of graphite reinforcement on mechanical properties of Al-boron carbide composites. Adv Mater Res 845:398–402
8. Dandekar CR, Shin YC (2012) Experimental evaluation of laser-assisted machining of siliconcarbide particle-reinforced aluminium matrix composites. Int J Adv Manufact Technol 66:1603–1610
9. Adalarasan R, Santhanakumar M, Thileepan S (2016) Selection of optimal machining parameters in pulsed CO_2lasercutting of Al6061/Al_2O_3 composite using Taguchi-based response surface methodology (T-RSM). Int J Adv Manufact 93:305–317
10. Adalarasan R, Santhanakumar M (2015) Application of taguchi based response surface method (TRSM) for optimization of multi responses in drilling Al/SiC/Al_2O_3 hybrid composites. J Inst Eng (India) 96:65–71

Analysis of Corrosion Resistance in Domestic Water Geysers by Coating Nano-Film Using Thermal Spray Coating

Kabilan Sankar, Karthick Selvam, K. Joy Ashwin, and S. P. Sathya Prasanth

Abstract The lifespan of domestic water geysers drastically reduces over a period of time of usage due to environmental challenges such as corrosion and scale development. The objective to be achieved here is to increase the property of corrosion resistance of internal components of the geyser and overcoming the challenges posed by such. A steel plate is to be coated with a nano-film of specified thickness using thermal spray coating consisting of nano-alumina, nano-zinc oxide and nano-molybdenum. These elements were chosen for their high corrosion resistance and formation of a superior alloy coating. The coating elements are mixed in three different ratio combinations, and each is coated on three stainless steel sample plates. These sample plates are tested for corrosion and scale development by performing a salt spray test for a period of time till visual examination shows the development of corrosion. A comparison of the results of the three samples is analysed with industry standards to determine the suitable alternative for geyser coatings. The results have convincingly proved that there is an increase in the life expectancy of the geyser which can significantly reduce maintenance expenditure.

Keywords Domestic water geyser. corrosion resistance · Nano-Film coating

1 Introduction

Corrosion is the slow destruction of a material by chemical or electrochemical reactions that occur due to interaction with the environment. They are generally known to deform and degrade the material, therefore making it non-usable for any purpose. One such area impacted by corrosion is the domestic water geyser. Geysers used for this purpose have two types of tanks, a stainless steel tank with no coating and a mild steel tank coated with porcelain enamel. The most predominant type in India is the stainless steel tank as it is less expensive to manufacture. However, stainless steel tanks and porcelain-coated tanks also have a very limited lifetime averaging at

K. Sankar (✉) · K. Selvam · K. Joy Ashwin · S. P. Sathya Prasanth
Department of Mechanical Engineering, KCG College of Technology, Chennai, India
e-mail: sankar.kabilan@gmail.com

S. Vijayan et al. (eds.), *Trends in Manufacturing and Engineering Management*, Lecture Notes in Mechanical Engineering,
https://doi.org/10.1007/978-981-15-4745-4_12

12–18 months. Many geysers undergo corrosion at a rate with respect to the hardness of the water and frequency of usage. The high temperature and hardness of water combined create a favourable environment for the degradation of the tank at a gradual pace. McAdam et al. [1] performed experiments to determine the rate of calcium carbonate formation and how its growth can be controlled. Shi et al. [2] did a study on the impact of nano-aluminium articles in corrosion resistance coatings. This experiment provided strong backing for the selection of nano-alumina to be included in the coating. Kourtidou et al. [3] performed experiments on Ni-Al coatings in high temperature and marine environments. This project aims at developing a better alternative using nano-technology by providing a nano-film coating of commercially available materials that possess high corrosion resistance properties [4]. A study was conducted to analyze the types of geysers available in the commercial market before coming to a conclusion on material selection [5].The materials that fit this category were alumina, zinc and molybdenum. These materials were mixed in three different ratios to determine the most suitable and effective coating mixture [6–8] that can withstand the corrosion attacks [9, 10] inside the geyser. Salt spray test would then be performed to determine the most suitable coating.

2 Methodology

The nano-materials selected for the coating are alumina, zinc and molybdenum. The three elements were mixed in three different specified ratios. The nano-materials were in powder form and mixed using a ball grinding machine. After ensuring the thorough mixing of the materials, each sample ratio was coated onto a stainless steel plate of dimension 75 mm × 75 mm × 1 mm. The plates had to be pre-treated before applying the coating. The coating method used here is the plasma thermal spray coating where the materials were coated onto the plate by blowing at a temperature of 4000 °C. The plate is then cooled to allow the material to be coated well onto the plate. The three stainless steel plates coated with different ratios of nano-coating are tested using a salt spray test for a period of 96 h to determine the impact of corrosion resistance. Visual examination is performed to determine the impact of corrosion on the three sample plates, and the results are compared to determine the effect of corrosion (Fig. 1).

Fig. 1 Experimental set-up

3 Experimental Details

Alumina and zinc are commercially feasible and widely used elements. Both elements possess high corrosion resistance properties. Molybdenum is a rare earth element that is also used in the steel industry for its ability to impart high corrosion resistance properties to the alloy it is alloyed to. Hence, they are the reason for the selection of these nano-materials. The three nano-materials consist of a particle size of 80 nm. The three nano-materials are mixed in three different ratios. Three test samples are used here, and the best ratio of material coating is selected from the three. The materials are weighed as per the ratio required and put into the ball grinding machine. The ball grinding machine is used to mix the elements thoroughly so as to prevent any large concentrations of same elements when coated onto the stainless steel plate.

3.1 Reference Samples

Reference Sample 1

Reference sample 1 is one of the two benchmark samples that are considered among the industry best standards. The salt spray test results of this reference sample will be compared with the salt spray test results of test samples. Figure 2 shows an image of the plate sample used. Stainless steel's resistance to corrosion and staining make it an ideal material for many applications where corrosion resistance is required. This reference sample will be tested by salt spray testing for the duration of 96 h. The stainless steel plate used is of the grade SAE 304 with a dimension of 75 mm in length, 75 mm in width and 1 mm thickness.

Fig. 2 Reference sample 1 before salt spray testing

Fig. 3 Reference sample 2 before salt spray testing

Reference Sample 2

Reference sample 2 is one of the two benchmark samples that are considered best according to industry standards whose salt spray test results will be compared with those of the test samples. The mild steel grade used for this sample is of Indian Standard 2062 and its dimensions are 75 mm in length, 75 mm in width and a thickness of 2 mm. The enamel coating is done on top of the steel plate's surface. It is a mild steel plate consisting of a coating of enamel coating with a coating thickness of 125 μm. Figure 3 shows an image of the plate after coating.

3.2 *Thermal Spray Coating of Nano-Mixture*

The nano-materials after mixing are coated onto the stainless steel plate using thermal spray coating for a thickness of 125 μm. Initially, the surface of the stainless steel plate undergoes pre-treatment before coating can be done. The surface is wiped clean to remove off any dust or foreign particles. The surface undergoes bit blasting where it is sprayed with aluminium oxide. The surface is then treated with a nickel chromium bond coating to ensure proper adhesion of the nano-material coating. A spray gun is used to eject the nano-mixture onto the plate at a speed of 80–90 m/s with argon gas to prevent external reaction with foreign elements. The amount of material coated per minute is 30 gm/min and the spray gun is at a distance of 127 mm from the surface of the stainless steel plate. The details of the three test samples are described in detail along with an SEM image.

Test Sample 1

Test sample 1 contains nano-alumina, nano-zinc oxide and nano-molybdenum that are mixed together with the help of ball grinding, and the mixture ratio is as shown in Table 1. After that process, they are fed to the thermal spray coating machine where

Table 1 Ratio of nano-material for test sample 1

Sample number	Base metal	% of alumina	% of zinc	% of mum	Coating thickness
Test sample 1	Stainless steel	50	35	15	125 μm

they will be coated onto the stainless steel plate. The stainless steel plate used here is of the grade SAE 304, and its dimensions are 75 mm in length, 75 mm in width and 1 mm in thickness. The stainless steel undergoes pre-treatment of its surface for better adhesion of coating of the nano-material mixture. The nano-material mixture is coated on top of the stainless steel plate with a coating thickness of 125 μm (Fig. 4; Table 2).

Test Sample 2

Test sample 2 contains nano-alumina, nano-zinc oxide and nano-molybdenum that are mixed together with the help of ball grinding, and the ratio is as shown in Table 3.

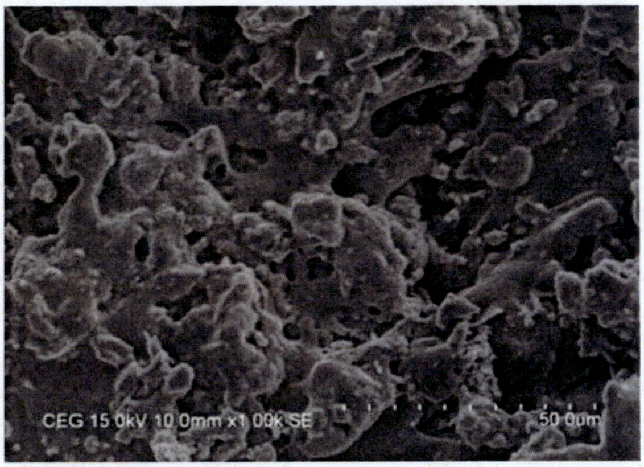

Fig. 4 SEM image of test sample 1

Table 2 Test samples before coating and after coating

Test sample	Before coating	After coating
Test sample 1		

Table 3 Ratio of nano-material for test sample 2

Sample number	Base metal	% of alumina	% of zinc	% of molybdenum	Coating thickness
Test sample 1	Stainless steel	65	25	10	125 μm

After that process, they are fed to the thermal spray coating machine where they will be coated onto the stainless steel plate. The stainless steel plate used here is of the grade SAE 304, and its dimensions are 75 mm in length, 75 mm in width and 1 mm in thickness. The nano-material mixture is coated on top of the stainless steel plate with a coating thickness of 125 μm (Fig. 5; Table 4).

Test Sample 3

The three elements are mixed together with the help of ball grinding, and in the mixture, the ratio is as shown in Table 5. After that, they are fed to the thermal spray coating machine where they will be coated onto the stainless steel plate. The stainless steel plate used here is of the grade SAE 304, and its dimensions are 75 mm in length,

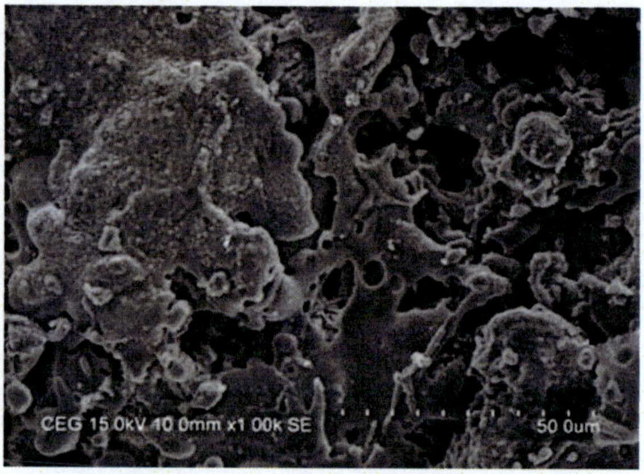

Fig. 5 SEM image of test sample 2

Table 4 Test samples before coating and after coating

Test sample	Before coating	After coating
Test sample 2		

Table 5 Ratio of nano-material for test sample 3

Sample number	Base metal	% of alumina	% of zinc	% of molybdenum	Coating thickness
Test sample 1	Stainless steel	80	15	5	125 μm

75 mm in width and 1 mm in thickness. The stainless steel undergoes pre-treatment of its surface for better adhesion of coating of the nano-material mixture. The nano-material mixture is coated on top of the stainless steel plate with a coating thickness of 125 μm (Fig. 6; Table 6).

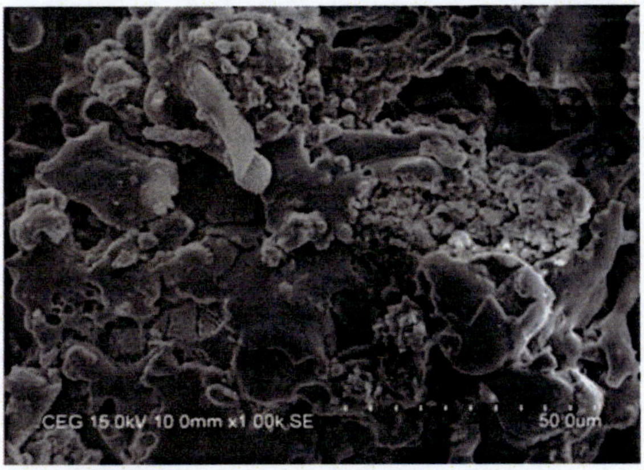

Fig. 6 SEM image of test sample 3

Table 6 Test samples before coating and after coating

Test sample	Before coating	After coating
Test sample 3		

3.3 *Salt Spray Testing of Test Samples and Reference Samples*

The salt spray test is a standardized corrosion test method used to test for corrosion resistance of materials and surface coatings. It is an accelerated corrosion test that produces a corrosive attack to coated samples in order to evaluate the suitability of the coating for use as a protective finish. The apparatus consists of a closed testing chamber where a saltwater (5% NaCl) solution is atomized by means of spray nozzles using pressurized air. This produces a corrosive environment of dense saltwater fog in the chamber so that the test samples are subjected to severely corrosive conditions. The chamber temperature is maintained at 70.5–71.5 °C with a pH value of 6.65–6.85. The components are loaded into the chamber at an angle of 30° with air pressure at 14–18 psi.

4 Examination and Analysis of Results

4.1 *Reference Sample 1*

Reference sample 2 is a commonly used and commercially feasible stainless steel plate. It is the most common type of tank available in India due to its reduced selling price. However, the sample did not last longer than 24 h in the salt spray test. This can be justified by the possibility of no additional coating on the surface and moderate corrosion resistance properties. It can be observed through visual examination that the plate has undergone extensive damage during the salt spray test. Hence, this type of coating cannot increase the life expectancy of a geyser (Table 7).

Table 7 Test samples before coating and after coating

Before testing	After testing

Table 8 Test samples before coating and after coating

Before testing	After testing

4.2 Reference Sample 2

This sample is considered to be the best in industry standard that is commercially available. However, the sample did not last long in the salt spray test. It underwent extensive corrosion damage and did not last longer than 48 h. Though enamel coating does have corrosion resistance properties, it is not enough to provide a geyser with the increased life expectancy that needs to be achieved (Table 8).

4.3 Test Sample 1

Test sample 1 image shows it turned the colour of the mixture into a dark silver. Thus, the appearance of the coating on the test plate is dark in colour compared to the other three samples. After 96 h of salt spray testing, test sample showed the endurance and good resistance to corrosion attack from the salt spray. This can be due to the possibility of an additional percentage of molybdenum and zinc oxide, both having high corrosion resistant properties. Visual examinations do show slight traces of corrosion which could have occurred due to the mixture not containing optimum ratio (Table 9).

4.4 Test Sample 2

The darkness in the colour of the coating has reduced to a minor extent which can be justified by the reduction in the percentage of molybdenum added in this mixture. After 96 h of salt spray testing, the test sample showed good endurance and very good resistance to corrosion attack from the salt spray. The percentage of alumina in this sample has been increased by 15% with a reduction of 10% in zinc oxide and

Table 9 Test samples before coating and after coating

Before testing	After testing

Table 10 Test samples before coating and after coating

Before coating	After coating

a 5% reduction in molybdenum. Visual examinations show no traces of corrosion or erosion, suggesting high corrosion resistance for this coating. This can be because alumina has bonded well with zinc oxide and molybdenum suggesting this is the optimum ratio for a stronger anti-corrosion coating (Table 10).

4.5 *Test Sample 3*

The darkness in colour is the least in test sample 3 among the three test samples. This can be justified by the very low presence of molybdenum, and a higher percentage of alumina which is white in colour. Visual examinations showed that the sample underwent quite extensive corrosion damage but not as severe as the reference test samples. This can be justified by the increased percentage of alumina and very few percentages of zinc with traces of molybdenum. Alumina's corrosion resistance

Table 11 Test samples before coating and after coating

Before testing	After testing

properties are lesser compared to that of zinc and molybdenum and are the reason why the corrosion damage is greater among the three samples (Table 11).

5 Conclusion

The three test samples and two reference samples were tested by salt spray testing for a period of 96 h. The salt spray test provided the simulation of a geyser except that the saltwater is constantly being sprayed without any intervals. This test provided us with an insight on how the five samples would survive in what is considered an extreme environment. Test sample 2 proved to be the optimum combination of nano-materials that can withstand the extreme environment of the geyser without falling victim to corrosion. Test results showed that it was expected to last well beyond the 96 h of testing after performing a visual examination of its surface. Test sample 1 ranked second in its ability to withstand corrosion. This could be due to the almost correct ratio of elements but apparently not the optimal ratio to withstand for long. Test sample 3 ranked third due to a large percentage of alumina and reduced percentages of zinc oxide and molybdenum. Both the reference samples did not last longer than 48 h. Reference sample 2 ranked fourth lasting for 48 h before giving way to attacks of corrosion. Reference sample 1 ranked fifth and the last among the five samples not lasting longer than 24 h providing evidence of why many commercially available geysers in India have a reduced life expectancy and undergo frequent maintenance activities. Therefore, it can be concluded that test sample 2 consists of the optimum ratio of nano-materials to resist attacks of corrosion faced in a domestic water geyser. This is supported by evidence from the salt spray test.

References

1. McAdam J, Parsons SA (2004) Calcium carbonate scale formation and control. Publ Views Env Sci Technol 3(2)
2. Shi X, Nguyen TA, Suo Z, Liu Y, Avci R (2009) Influence of Nano-Al concentrates on the corrosion resistance of epoxy coatings. Elsevier-Surface Coat Technol 204(3):237–245
3. Kourtidou D, Chaliampalias D, Tarani E, Pavlidou E (2019) Deposition of Ni-Al coatings by pack cementation and corrosion resistance in high temperature and marine environments. Elsevier Corrosion Sci 148:12–23
4. Gal J-Y, Bollinger J-C, Tolosa H, Gache N (1996) Calcium: a reappraisal of scale formation and inhibition. Published in Talanta 43(9):1497–1509
5. Venus Home Appliances webpage (2019) https://www.venushomeappliances.com/tank-comparision.html. Last accessed 12 May 2019
6. Khodair ZT, Khadom AA, Jasim HA (20185) Corrosion protection of mild steel in different aqueous media via epoxy/nanomaterial coating: preparation, characterization and mathematical views. J Mater Res Technol
7. Liang Y, Liu F-C, Nie M, Zhao S, Lin J, Han E-H (2013) Influence of Nano-Al concentrates on the corrosion resistance of epoxy coatings. J Mater Sci Technol 29(4):353–358
8. Koutsomichalis A, Vaxevanidis NM, Psyllaki P (2016) Wear and corrosion behaviour of zinc-alumina electrodeposited coatings on mild steel. Tribological J Bultrib 6(06)
9. Kadhim MJ, Sukkar KA, Abbas AS, Obaeed NH (2017) Investigation of nano coating for corrosion protection of petroleum pipeline steel type A106 grade B: theoretical and practical study in Iraqi Petroleum Sector. Eng Technol J 35(Part A, No. 10)
10. Bhakre VS, Nayak DK (2007) Nanotechnology in corrosion prevention. Published at international conference CORCON 2007 held at Mumbai focusing on cost of corrosion to the nation

Effect of Multi-walled Carbon Nanotubes Additions on Its Dispersion Characteristics in Titanium Oxides

Senzeni Sipho Lephuthing, Avwerosuoghene Moses Okoro, Masego Mohlala, Noxolo Malgas, Oladeji Ige, and Peter Apata Olubambi

Abstract The transition metal oxides are inorganic nanomaterials with exclusive properties. However, metal oxides cannot exceptionally accomplish all requirements to develop new technologies as they are brittle and have low fracture toughness, which has motivated researchers to adopt novel techniques in enhancing their properties. Carbon nanotubes (CNTs) have drawn attention in engineering fields due to their unique structural, electrical and mechanical properties. This work aims to achieve a homogeneous dispersion of multi-walled carbon nanotubes (MWCNTs) by investigating the effect of increased MWCNTs concentration in titanium oxide. The MWCNTs (0.5, 1, 1.5 wt.%) were dispersed in titanium oxide using high energy ball milling (HEBM) technique at a lower milling speed of 100 rpm for 6 h. The composite powders were characterized using scanning electron microscopy (SEM), transmission electron microscopy (TEM), Raman spectroscopy and X-ray diffraction (XRD) to understand the uniformity of dispersion, interfacial reaction and structural integrity of the MWCNTs after dispersion. The results showed that the MWCNTs detangles at a lower concentration of 0.5 wt%, with an increase in MWCNTs there is the formation of entanglement. In addition, further analysis on structural integrity showed minimal structure damage due to lower speed adopted for HEBM technique, and there were no interfacial reactions favoured.

Keywords Titanium oxide · Multi-walled carbon nanotubes (MWCNTs) · Homogeneous · Dispersed

---

S. S. Lephuthing (✉) · A. M. Okoro · M. Mohlala · N. Malgas · O. Ige · P. Apata Olubambi
Department of Metallurgy, School of Mining, Metallurgy and Chemical Engineering, Center for Nanoengineering and Tribocorrosion, University of Johannesburg, 2006 Gauteng, Republic of South Africa
e-mail: senzenil@uj.ac.za

S. Vijayan et al. (eds.), *Trends in Manufacturing and Engineering Management*, Lecture Notes in Mechanical Engineering,
https://doi.org/10.1007/978-981-15-4745-4_13

1 Introduction

The utilization of CNTs for diverse structural and engineering applications have spurred more research attention in the recent time. This is due to their attractive, outstanding and superlative properties [1–4]; high Young's modulus in the tera-pascal (TPa) range and excellent strength about 100 GPa to synthesize less dense materials for diverse applications. Additionally, the exclusive properties of CNTs are ascribed to their unique morphology and the existence of strong sp2 carbon–carbon (C–C) bonds in their outermost shells. These sp2 C–C bonds form a fastened honeycomb lattice and high bond strength in the structure of the material [4].

The development of durable, inert and insoluble composites has been of interest in scientific research communities. This composite materials demand have encouraged the integration of CNTs into titanium-based materials to enhance their properties for composites production. However, during the processing of these composites, CNTs tend to agglomerate into clusters and this is due to their nanostructural dimension, high aspect ratio and the strong van der Waal forces that exist between their individual tubes [5–7]. In many cases, the increase in the content of MWCNTs as reinforcements improves the properties of the matrix material but poses the tendency of agglomeration and deteriorates the dispersion characteristics of the MWCNTs within the matrix materials [7, 8] which is ascribed to the corresponding of increase in the strong van der Waal forces among individual nanotubes.

In a bid to improve the properties of the titanium oxides and achieve uniform dispersion of MWCNTs in the metal oxides, optimized dispersion techniques are highly desirable to overcome the strong van der Waal forces and achieved optimal dispersion characteristics of the MWCNTs in titanium oxides [7]. In this study, high energy ball milling (HEBM) method with regulated milling parameters was adopted [9] to 1.0 and 1.5 wt.% of MWCNTs in titanium oxide, and various characterization techniques were used to investigate the effect of an increase in MWCNTs on its dispersion.

2 Materials and Methods

2.1 Starting Materials

The starting powder materials used for this study were MWCNTs (NX7100-A-Ultra purified CNT) with an average diameter of 9.5 nm and an average length of 1.5 μm, supplied by Nanocyl, Belgium. Titanium (IV) oxide anatase powder with percentage purity (99.6%) and −325 mesh particle size was supplied by industrial analytical in Johannesburg, South Africa.

2.2 *High Energy Ball Milling (HEBM)*

High energy ball mill (Retch 400, Germany) was used to 1 and 1.5 wt.% MWCNTs into titanium oxide and this was conducted by charging the titanium oxide powder in an alumina vial (250 ml, inner diameter (ID) = 100 mm) with alumina balls of 5 mm diameter. The milling was carried out using a ball to powder ratio (BPR) was 10:1 at 100 rpm milling time and speed of 6 h and 100 rpm, respectively. A relaxation time of 10 min was observed during the milling operation to prevent the samples from overheating. The admixed powders were collected and characterized after milling.

2.3 *Characterization of As-Received and Admixed Powders*

Scanning electron microscopy (FESEM, Carl Zeiss Sigma) equipped with energy-dispersive X-ray spectroscopy (EDX) and transmission electron microscopy (JEOL-Jem 2100) were used for morphological characterization of the starting and admixed powders. X-ray diffractometer (XRD, Rigaku D/max-rB) was used for phase identification using Cu-Kα($\lambda = 0.154$ nm) radiation at a scanning rate of 1°/min over the angular range of 20–90°. Raman spectroscopy was used to ascertain the structural integrity of the MWCNTs before and after milling

3 Results and Discussion

3.1 *Morphology of As-Received and Admixed Powders*

Figure 1(a)–(d) shows the morphology characteristics of as-received powders which are SEM and TEM analysis. Figure 1(a) shows the SEM image anatase-type titanium oxide powder (TiO_2) that has a spherical shape of similar particle sizes. Figure 1(b) is SEM image of pristine MWCNTs agglomerated together which is attributed by their large aspect ratio, high surface area strong van der Waal forces between individual MWCNTs [7]. This is confirmed by TEM image in Fig. 1(c) where MWCNTs are bundled into clusters and the high-resolution transmission microscopy image (HRTEM) in Fig. 1(d) shows the walls of non-destructive individual pristine MWCNTs.

In order to attain unique properties of MWCNTs in TiO_2, an effective homogenous dispersion of MWCNTs with good interfacial bonding alongside with minimal destruction of their tubular structure has to be achieved [10]. Past works have emphasized that the addition of high concentration of MWCNTs helps to improve the properties of the TiO_2 [7, 8, 10]. However, the increase in concentration of MWCNTs during composite production poses the tendency of increase in agglomeration of the nanotubes which is as a result of the nanostructural dimension, high aspect

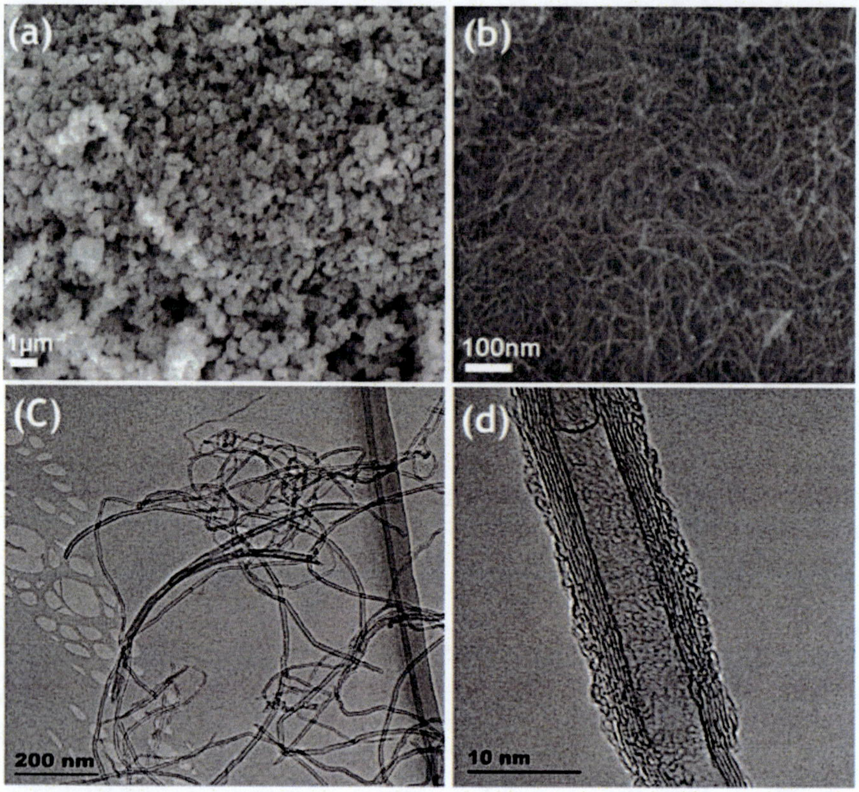

Fig. 1 SEM images of the as-received powders **(a)** Titanium oxide, **(b)** MWCNTs, **(c)** TEM image of as-received MWCNTs and **(d)** HRTEM image showing the walls of MWCNTs

ratio and the strong van der Waal forces that exist between individual MWCNTs [7] This features of MWCNTs deteriorate the dispersion characteristics in the TiO_2.

The strategies that usually adopted in achieving the detangling of MWCNTs during the powder processing are to further increase the milling parameter such as milling time, speed and BPR which could results on defects on the graphitic structure of the MWCNTs [11] which affect their properties to be attained on TiO_2 powder. Since, it is essential to efficiently transfer these unique properties to TiO_2. This can be achieved by homogeneous dispersion of MWCNTs with good interfacial bonding without damaging the walls of the nanotubes. Hence, this study adopted the powder processing using HEBM technique to disperse MWCNTs in TiO_2 powder at lower regulated milling parameters.

Figure 2(a)–(c) shows the SEM images of (0.5, 1.0, 1.5) wt% MWCNTs dispersed into TiO_2. Figure 2(a) shows the SEM image where 0.5 wt% MWCNTs dispersed in TiO_2, and it is observed that the MWCNTs are relatively detangled which could be attributed by less effectiveness of van der Waal forces between individuals. However,

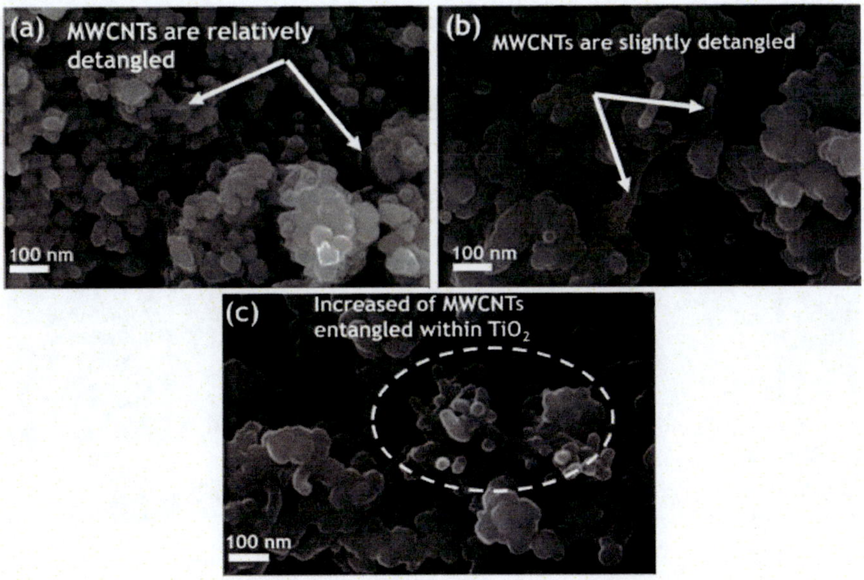

Fig. 2 SEM images of dispersed MWCNTs in titanium oxide **(a)** 0.5 wt%, **(b)** 1.0 wt% and **(c)** 1.5 wt

Fig. 2(b) shows an increase concentration in MWCNTs to 1.0 wt% dispersed in TiO_2 powder and slight entanglement of MWCNTs within TiO_2 powder begin to be noticed. Further increase in MWCNTs concentration to 1.5 wt% dispersed in TiO_2 powder is shown in Fig. 2(c). It is observed that MWCNTs are prominently entangled within MWCNTs with increase in MWCNTs concentration. This could be attributed by increase in van der Waal forces. In addition, due to lower milling parameter that were adopted also played a role in dispersion behaviour of the MWCNTs dispersed in TiO_2 powder.

Hence, it was observed that a degree of agglomeration became visible with increase in MWCNTs wt% concentration due to insufficient impact energy exerted on the composite during powder processing. Meanwhile, increase in impact energy has its concerns as it may damage the outer layer of the MWCNT by introducing open edges and vacancies. This will lead to the destruction of the basal planes of MWCNTs exposing the prismatic planes, where interfacial products are favoured [9].

Figure 3(a)–(f) shows the TEM images of (0.5, 1.0 and 1.5) wt% of MWCNTs dispersed in TiO_2 powder, and this was conducted to further ascertain the dispersion behaviour of MWCNTs as well how the walls were evolved during powder production. Figure 3(a)–(c) confirms on what has been observed on SEM images that MWCNTs are relatively detangled at lower concentration; however, as the wt% concentration increases, the entanglement begins until it becomes obvious when 1.5 wt% is dispersed in TiO_2 powder. There is bonding effect of MWCNTs on TiO_2 particles,

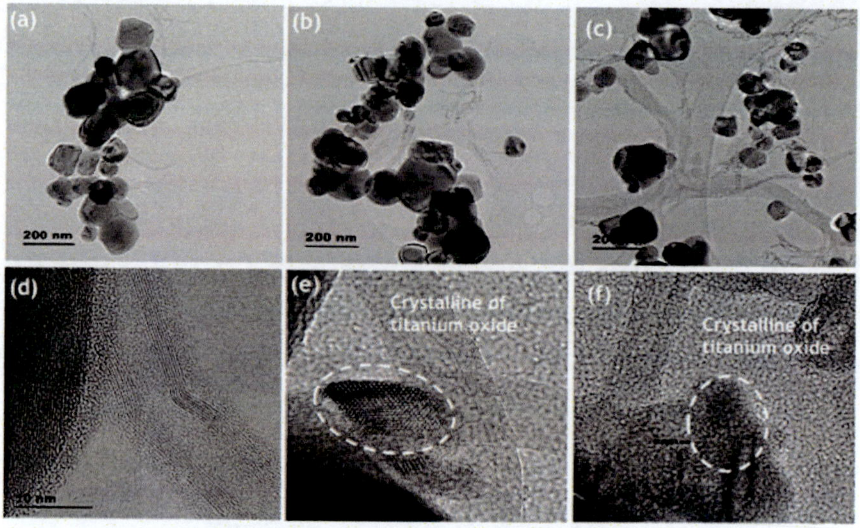

Fig. 3 TEM images of dispersed MWCNTs in titanium oxide **(a)** 0.5 wt% **(b)** 1 wt% and **(c)** 1.5 wt% and HRTEM for tubular characteristics **(d)** 0.5 wt% **(e)** 1 wt% and **(f)** 1.5 wt%

and it is observed by wrapping of MWCNTs around that TiO_2 particles which may lead to transfer of unique properties of MWCNTs in TiO_2.

Moreover, HRTEM analysis shown in Fig. 3(d)–(f) for dispersion of MWCNTs in TiO_2 powder. Figure 3(d) displays a slightly evolution on the walls compared to the pristine MWCNTs. However, when the concentration of MWCNTs dispersed in TiO_2 powder increases to 1.0 and 1.5 wt% in Fig. 3(e)–(f), there is obvious crystalline structure observed and this confirms that MWCNTs were embedded within titanium oxide and this could be possibility of good interfacial bonding that leads to the transfer of the unique properties of MWCNTs to titanium oxide [12].

3.2 *Structural Integrity and Phase Analysis of Dispersed MWCNTs in TiO_2*

Raman spectroscopy analysis was used to evaluate the structural integrity of MWCNTs after dispersion in titanium oxide. This analysis was carried out specifically to ascertain the D-band which is associated with defects in form of sp3 disorders in the C–C bonds and the G-band which is related to the plane stretching mode of the C–C bond to determine the degree of graphitization, orderliness of carbon, metallicity and crystalline nature of the MWCNTs. Meanwhile, the I_D/I_G ratio is used to assess the crystallinity and structural disorder of the MWCNTs during powder processing [13, 14].

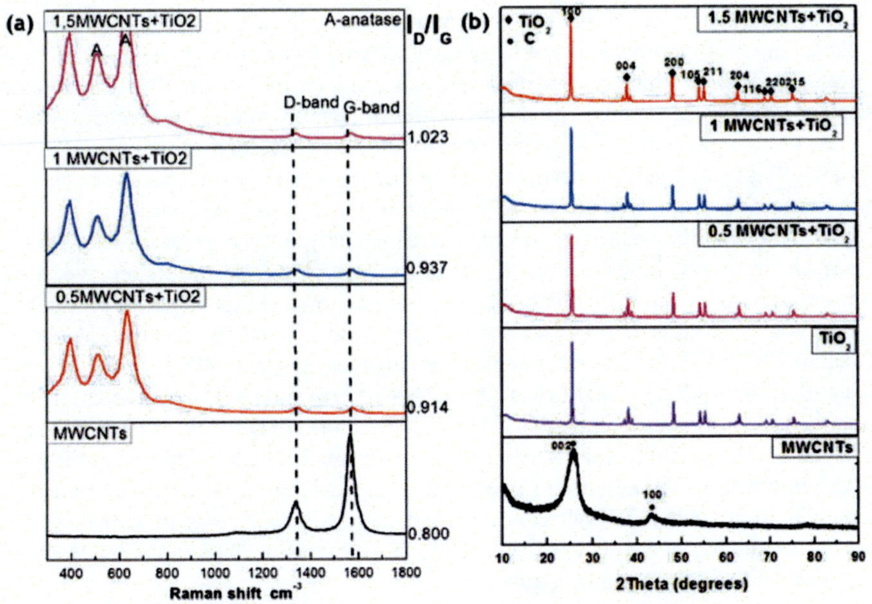

Fig. 4 (a) Raman shift spectrum and (b) XRD patterns Raman shift of dispersed 0.5, 1 and 1.5 wt% MWCNTs in titanium oxide

Figure 4a shows the Raman spectra of the pristine MWCNTs and the admixed composite powders. The graphite peaks of the pristine MWCNTs are observed correspondingly, D (1340 cm^{-1}) and G (1570 cm^{-1}) bands. This was followed by pristine TiO_2 powder which showed anatase structures at B1g (399 cm^{-1}), A1g (518 cm^{-1}) and Eg (641 cm^{-1}), respectively [15]. The dispersion of MWCNTs (0.5, 1.0 and 1.5) wt% shifted the D- and G-band from the pristine of MWCNTs D (1340 cm^{-1}) and G (1570 cm^{-1}) to (D)1341.3 cm^{-1} and (G) 1575.6 cm^{-1}, (D)1345.2 cm^{-1} and (G)1575.3 cm^{-1} and (D)1347.6 cm^{-1} and (G) 1577.7 cm^{-1}, respectively. This could have been attributed by the energy extended to the walls of MWCNTs during power processing. However, there were no interfacial reactions products which are carbide phases observed on the Raman spectra. Application of lower milling parameters adopted for MWCNTs dispersion played a role in preserving the walls [16].

The I_D/I_G ratio of the pristine MWCNTs was calculated to be 0.800, after the dispersion of MWCNTs at (0.5, 1.0 and 1.5 wt%) in titanium oxide. There was increase in I_D/I_G ratio form 0.800 to 0.914, 0.937 and 1.023, respectively. This characteristic is an indication of radical shift that broadens the intensity peaks of the D-band and G-band which is ascribed to higher stress on the walls of MWCNTs during milling. The stress on the nanotubes may lead to the formation of weak sp^3 C–C amorphous phase from broken sp^2 bonds due to harsh milling conditions which favours formation of carbides [9, 12]. Carbide phases are usually formed when there are severe damages to the walls of the nanotubes during dispersion.

The destruction in the walls of the nanotubes usually occurs when high milling speed, long milling time and extremely high ball to powder ratios are adopted during milling operation which may further affect the unique properties of the nanotubes that are transferrable to augment the properties of titanium oxides. From past studies, HEBM technique is still adjudged as the most effective route for dispersing MWCNTs in TiO_2 [9, 11] especially when regulated milling parameters are adopted to minimize the destruction to the walls of the MWCNTs during milling operation.

X-ray diffraction technique was conducted to analyse the different phases, structural changes and the evolution of the crystalline phases in the as-received and admixed composite powders. The XRD results are shown in Fig. 4b where prominent peaks are observed at 25.6 and 43.3 which corresponds to (002) and (100) planes of the MWCNTs. While, in the admixed composite powders, the following planes were identified (101), (004), (200), (105), (211), (204), (215) and (303) which ascertain the anatase type of titanium oxide powder. In addition, the XRD results further revealed that the MWCNT (002) plane was overlaps by the anatase-type titanium oxide (101) plane in the MWCNTs/titanium oxide composite. Moreover, it was also observed that with increase in MWCNT concentration, there was a decrease in peak intensity on corresponding plane (101) of MWCNTs this may be ascribed to the interlock of the MWCNTs in the titanium oxide.

4 Conclusion

It has been successfully demonstrated that MWCNTs can be dispersed using high energy ball milling HEBM technique. Though, this dispersion technique has drawbacks of destruction the MWCNTs wall due to harsh milling conditions during powder processing which poses a challenge on losing their ability to improve TiO_2. However, milling conditions were controlled to minimize the destruction as shown in this study. Therefore, the following was drawn from this study.

- The SEM results showed that the lower concentration of 0.5 wt% MWCNTs was relatively de-agglomeration of MWCNTs. However, when further increases in the MWCNTs the prominent agglomerated MWCNTs were shown.
- The TEM results confirmed the SEM results, however, at HRTEM analysis. It was observed that there was good bonding which showed by crystal structures and this my attribute to good interfacial bonding the leads to transfer of the their unique properties.
- The structural integrity of MWCNTs showed that indeed, there was good interfacial bonding achieved during powder processing. Though, the D- and G-band were stretched which could defeat of MWCNTs and there was an increase in the I_D/I_G ratio that affects the crystallinity. The destruction was minimal as there were no interfacial reactions observed of the Raman spectra.

Acknowledgements National Research Foundation of South Africa is acknowledged for funding this research study.

References

1. Mallakpour S, Khadem E (2016) Carbon nanotube–metal oxide nanocomposites: Fabrication, properties and applications. Chem Eng J 302:344–367
2. Soldano C (2015) Hybrid metal-based carbon nanotubes: novel platform for multifunctional applications. Prog Mater Sci 69:183–212
3. Mohammadi M (2016) No Title. Dev Pb-MnO_2 Compos anodes electrowinning Appl Electrochem Corros Eval. LNCS Homepage, http://www.springer.com/lncs. Last accessed 21 Nov 2016
4. Liu Y, Chae HG, Ho Choi Y, Kumar S (2014) Effect of carbon nanotubes on sintering behavior of alumina prepared by sol-gel method. Ceram Int 40(5):6579–6587
5. Zapata-Solvas E, Gómez-García D, Domínguez-Rodríguez A (2012) Towards physical properties tailoring of carbon nanotubes-reinforced ceramic matrix composites. J Eur Ceram Soc 32:3001–3020
6. Lephuthing SS, Okoro AM, Lesufi M et al (2018) Effect of milling parameters on the dispersion characteristics of multi-walled carbon nanotubes in transition metal oxides. IOP Conf Ser Mater Sci Eng 430:12002. https://doi.org/10.1088/1757-899X/430/1/012002
7. Munir KS, Li Y, Liang D, Qian M, Xu W, Wen C (2015) Effect of dispersion method on the deterioration, interfacial interactions and re-agglomeration of carbon nanotubes in titanium metal matrix composites. JMADE 88:138–148
8. Munir KS, Zheng Y, Zhang D, Lin J, Li Y, Wen C (2017) Improving the strengthening efficiency of carbon nanotubes in titanium metal matrix composites. Mater Sci Eng, A 696(March):10–25
9. Munir KS, Kingshott P, Wen C (2015) Carbon nanotube reinforced titanium metal matrix composites prepared by powder metallurgy—a review. Crit Rev Solid State Mater Sci 40(1):38–55
10. Thostenson ET, Ren Z, Chou T-W (2001) Advances in the science and technology of carbon nanotubes and their composites: a review. Compos Sci Technol 61(13):1899–1912
11. Hassan MTZ, Esawi AMK, Metwalli S (2014) Effect of carbon nanotube damage on the mechanical properties of aluminium–carbon nanotube composites. J Alloys Compd 607:215–222
12. Munir KS, Wen C (2016) Deterioration of the strong sp2 carbon network in carbon nanotubes during the mechanical dispersion processing—a review. Crit Rev Solid State Mater Sci 41(5):347–366
13. Adegbenjo AO, Olubambi PA, Potgieter JH, Shongwe MB, Ramakokovhu M (2017) Spark plasma sintering of graphitized multi-walled carbon nanotube reinforced Ti6Al4V. Mater Des 128(May):119–129
14. Ahmad I, Cao H, Chen H et al (2010) Carbon nanotube toughened aluminium oxide nanocomposite. J Eur Ceram Soc 30:865–873
15. Inam F, Yan H, Reece MJ, Peijs T (2008) Dimethylformamide : an effective dispersant for making ceramic—carbon nanotube composites. 195710. https://doi.org/10.1088/0957-4484/19/19/195710
16. Klein MV, Holy JA, Williams WS (1978) Raman scattering induced by carbon vacancies in Ti C x. Phys Rev B 17:1546

Evaluation of Process Parameters Influence on the Mechanical Properties of RF Magnetron Sputtered TiC Thin-Film Coating

Olayinka Oluwatosin Abegunde, Esther Titilayo Akinlabi, and Oluseyi Philip Oladijo

Abstract The mechanical properties of a thin-film coating are very paramount to the development and applications for both research and industrial usages. Enhanced mechanical properties give vast options of applications, curb cost and mitigate the unprecedented failure of materials in services. Optimizing the process parameter is very crucial in developing thin-film coatings mechanical properties. Taguchi was used to develop the experimental matrix using three levels and three factors. Three different sputtering parameters were varied during the experimental process from low to high level. Mechanical characterizations using nanoindentation, nanoscratch and wear tests were performed on the samples to determine the optimal process parameter that yields the best mechanical properties. From Taguchi analysis, the RF power proves to be the most effective control factor in optimizing the hardness and Young's modulus properties with a contribution rate of 70.4% and 84.4%, respectively.

Keywords ANOVA · Taguchi · RF magnetron sputtering

1 Introduction

Mechanical properties have a correlation with other properties like corrosion, optical properties and surface morphology. Mechanical properties can be improved via different manufacturing techniques such as composite, heat treatment and surface modification [1–3]. RF magnetron sputtering is a useful tool technique for improving these properties by coating the surface with materials of superior qualities. A detailed analysis of thin-film coating by these processes can be found in Reference

O. O. Abegunde (✉) · E. T. Akinlabi · O. P. Oladijo
University of Johannesburg, Johannesburg 2006, South Africa
e-mail: oabegunde@uj.ac.za

E. T. Akinlabi
Covenant University, Ota, Nigeria

O. P. Oladijo
Botswana International University of Science and Technology, Palapye, Botswana

S. Vijayan et al. (eds.), *Trends in Manufacturing and Engineering Management*, Lecture Notes in Mechanical Engineering,
https://doi.org/10.1007/978-981-15-4745-4_14

[4–7]. Although studies have been done on the applications of sputtering for surface modification [8–10], further understanding of the contribution of each process parameters of RF magnetron sputtering on the response properties still needs more research study. Understanding the influence of RF power, sputtering and temperature on the mechanical properties is important for both industrial application and research advancement. In this study, Taguchi was used to developing the experimental matrix. The control factors considered are RF power, sputtering time and temperature. An orthogonal three levels and three factors method (l_9 3^3) were used. The percentage contribution and influence of the independent process parameters' factors on the hardness and Young's modulus were analyzed using ANOVA analysis and signal-to-noise ratio.

2 Experimental Method

2.1 RF Magnetron Sputtering

TiC thin films were deposited in an RF magnetron sputtering coater which was specifically designed for the deposition process. The substrate (CpTi) was ground using silica carbide paper and then polished to remove any surface impurities. The surface of the substrate was further cleaned in acetone and stored in a desiccator to prevent contamination and surface oxidation. The purity of the TiC target used is 99.99% pure TiC. The target–substrate distance was kept fixed throughout the experiments. Evacuation of the sputtering chamber to a base pressure of 1.13×10^{-5} m bar was done for all the coating process before sputtering takes place. The reason for the low pressure is to provide a long mean free path for a collision between the target and substrate and allows for the control and minimization of contaminants. After the evacuation, the system is refilled with argon to a partial pressure of 2.5×10^{-3} m bar. Argon flow rate of 12 sccm was used, and constant working pressure was maintained. Maintaining constant pressure helps to place many argon atoms in the path of the ions and ejected coatings. After the plasma formation, the sputtering of the target is triggered, and the shutter is removed. Pre-sputtering was done for all the samples for 5 min before the main sputtering deposition. The substrate was located directly above the target on a rotating disk to ensure uniform distribution and condensation of the target on the surface of the substrate. The target material was cleaned up after each cycle to prevent build up charge, thereby reducing the arcing effect.

2.2 Process Parameters and Optimization

The sputtering experiments were conducted by varying three crucial parameters of magnetron sputtering, namely RF power, sputtering time and temperature. The experimental matrix optimization was developed using the Taguchi method. Taguchi method is a simple, efficient and powerful design tool for statistical analysis and optimization. It lessens the number of experimental runs to be conducted and curbs the effects of external factors that cannot be controlled on the system. The flowchart of the process is shown in Fig. 1, and Table 1 represents the magnetron sputtering parameters and their levels. The full orthogonal array of Taguchi L9 (33) with three factors at three levels of low, mid and high and a total of nine deposition runs is presented in Table 2. The time ranges from 2 to 3 h, power from 150 to 250 W and temperature from 80 to 1000 °C.

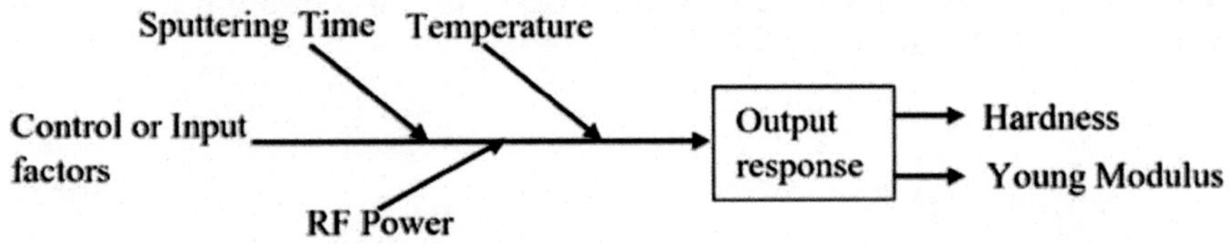

Fig. 1 Flowchart of experimental design

Table 1 Control factors level for experimental setup

Sputtering parameters	Symbol	Level 1	Level 2	Level 3
RF Power (W)	A	150	200	250
Sputtering time (Hrs)	B	2.0	2.5	3.0
Temperature (°C)	C	80	90	100

Table 2 Orthogonal array of Taguchi L_9 (3^3)

Experiment no.	RF power	Sputtering time	Temperature
L1	1	1	1
L2	1	2	2
L3	1	3	3
L4	2	1	2
L5	2	2	3
L6	2	3	1
L7	3	1	3
L8	3	2	1
L9	3	3	2

2.3 Nanoindentation

Hysitron Triboindenter (T1950, USA) was used to investigate the nanomechanical properties (Young's modulus and hardness) of thin-film coating under ambient working condition. Load controlled testing which was modeled to have a trapezoidal flow pattern of loading-dwelling-unloading was used to generate the indentation profile of the thin-film coating with a holding time of 15 s at maximum peak load. The loading and unloading rate of 10 μN s^{-1} and a maximum load of 300 μN were used. The diamond indenter of the Triboindenter is made up of Berkovich tip of 100 nm radius of curvature. Young's modulus and nanohardness of measured thin films were calculated using Oliver and Pharr analysis [11–13] from the load-displacement curves. The indentations for each thin film were repeated for 10 times, and the measurement recorded corresponds to the average of the 10 measurements. To ensure that the measurement taken is actually for the thin film and eliminate the influence of the substrate on the results, the indenter depth of penetration was maintained at 10% of the total coating thickness [14, 15].

2.4 Microscratch

The peeling behavior of the coatings was evaluated using a linear displacement microscratch tester (Anton Paar Micro Combo Tester, Austria). A Rockwell diamond ball with a radius of 100 μm was used for the nanoscratch test to determine the film adhesion toughness and failure mode. For each measurement, the applied load was progressive from 50 mN to 2 N at a sliding speed of 2 mm/min and sliding length of 1 mm. All these experiments were carried out in an ambient atmospheric condition.

2.5 Wear

The tribological properties which represent the macroscratch of the thin-film coatings were achieved on a Rtec Universal ball-on-disk tribometer at ambient condition. (temperature, pressure and humidity). An E521000 alloy steel grade 25 ball (6.350 mm in diameter) was used as the counter body while the surface of the thin-film coating serves as the disk. The movement of the motion is translational with maximum sliding displacement of 1 mm. The sliding velocity was kept constant at 1 mm/s for 2 min at normal load of 10 N for all the experiments. The friction coefficient was monitored throughout the experiment by a linear variable displacement transducer and recorded on a data acquisition computer attached to the tribometer. After completion of the wear, the wear tracks were examined on the microscope for further analysis of the surface deformation and failure. All process parameters were repeated three times and the average values were reported.

3 Result and Discussion

3.1 Analysis of the Signal-to-Noise (S/N) Ratio for Young's Modulus and Hardness

Young's modulus and hardness of the thin-film coating were characterized using an experimental approach and the optimization of measured response was analyzed using signal-to-noise ratio. The signal-to-noise (S/N) ratio is a loss function used for determining the deviation between the experimental values and desired. S/N ratio is divided into three distinctive quality characteristics for analysis namely lower the better, nominal the better and higher the better. The S/N ratio result of the hardness and young modulus is shown in Table 3. The higher the better equation was used for the S/N ratio of the hardness. Higher hardness value is a sign of improvement in the surface quality and improves the resistance to wear and corrosion, thereby reducing the loss of material due to contact with other surfaces and enhance the overall quality. Evaluation of young modulus gives in-depth behavior of specific properties of the coating such as flexibility, bond strength, fatigue life, wear, fracture toughness and residual stress [16, 17]. Low Young's modulus yields a more flexible coating behavior and improves its thermal stress resistance [17]. Higher Young's modulus results in high wear rate and short fatigue life. For this reason, the lower the better equation was used for the Young modulus S/N ratio. The average values of the hardness and Young's modulus were calculated after the sputtering deposition and were found to be 4.54 GPa and 121.32 GPa, respectively. Similar calculation was done for the S/N ratio of the hardness and Young's modulus and found to be 12.404 dB and

Table 3 S/N ratio analysis of the hardness and Young's modulus

Experiment no.	Control factors			Hardness (GPa)	S/N ratio for hardness	Young's modulus E (GPa)	S/N ratio for Young's modulus
	A RF power (W)	B sputtering time (Hrs)	C temperature (°C)				
L1	150	2.0	80	4.41	12.889	148.54	−43.4369
L2	150	2.5	90	5.00	13.970	134.19	−42.5544
L3	150	3.0	100	6.51	16.272	135.54	−42.6413
L4	200	2.0	90	2.38	7.532	96.67	−39.7058
L5	200	2.5	100	2.57	8.199	90.79	−39.1608
L6	200	3.0	80	2.37	17.605	89.13	−39.0005
L7	250	2.0	100	7.59	7.495	144.39	−43.1907
L8	250	2.5	80	4.09	12.234	113.21	−41.0777
L9	250	3.0	90	5.91	15.432	139.46	−42.8890

Table 4 Response table for signal-to-noise ratios for hardness and Young's modulus

Levels	Control factors					
	Hardness (H)			Young's modulus (E)		
	A	B	C	A	B	C
Level 1	14.380	12.675	10.873	−42.88	−42.11	**−41.17**
Level 2	7.742	11.471	12.314	**−39.29**	**−40.93**	−41.72
Level 3	**15.090**	**13.066**	**14.025**	−42.39	−41.51	−41.66
Delta	7.349	1.595	3.152	3.59	1.18	0.54

The bold values are the output response with optimal performance

−41.52bB, respectively. The response table for the signal-to-noise ratio and level values of the input or control factor is illustrated in Table 4. The significant effect of the input or control factors (RF power, sputtering time, temperature) on the coating hardness and Young's modulus obtained from Taguchi analysis was presented. The optimal sputtering process parameter levels that yield high hardness and low Young's modulus values can be easily obtained from the table. The highest S/N ratio of the control factor at different levels was utilized to determine the optimal setting. From Table 4, the levels and S/N ratios for the factors yielding optimal hardness values are factor A (level 3, S/N = 15.090), factor B (level 3, S/N = 13.060) and factor C (level 3, S/N = 14.025). On the other hand, the levels and S/N ratios for the factors yielding optimal young modulus values are factor A (level 2, S/N = −39.29), factor B (level 2, S/N = −40.93) and factor C (level 1, S/N = −41.17), respectively. This implies that an optimum value for hardness was obtained at RF power of 250 W, sputtering time 3 h and temperature 1000 C which is graphically illustrated in Fig. 2. The optimum value for the young modulus was obtained at RF power of 200 W, sputtering time of 2.5 h and temperature of 800 C graphically presented in Fig. 3.

3.2 *Analysis of ANOVA for Hardness and Young Modulus*

ANOVA is a statistical technique utilized to study the interaction and contribution of each control parameter in the design matrix. ANOVA was used to analyze the effect of RF power, sputtering time and temperature on the behavior of the hardness and Young's modulus. Table 5 shows the ANOVA analysis result, and it was calculated at a 5% significance level and 95% confidence level. The level of significance of the control or input factors was determined by comparing the F-ratio values and the percentage contribution rate which depicts the degree of influence each control parameter on the response is also calculated. From the table, the contribution rate of A, B and C factors on the hardness are 70.4%, 6.7%, and 19.7%, respectively. This shows that factor A (RF power) has a major predominant effect on the outcome of the hardness properties of the coating. A similar trend was also observed for the young modulus of the coating. The contribution rate for factor A is 84.4%, while

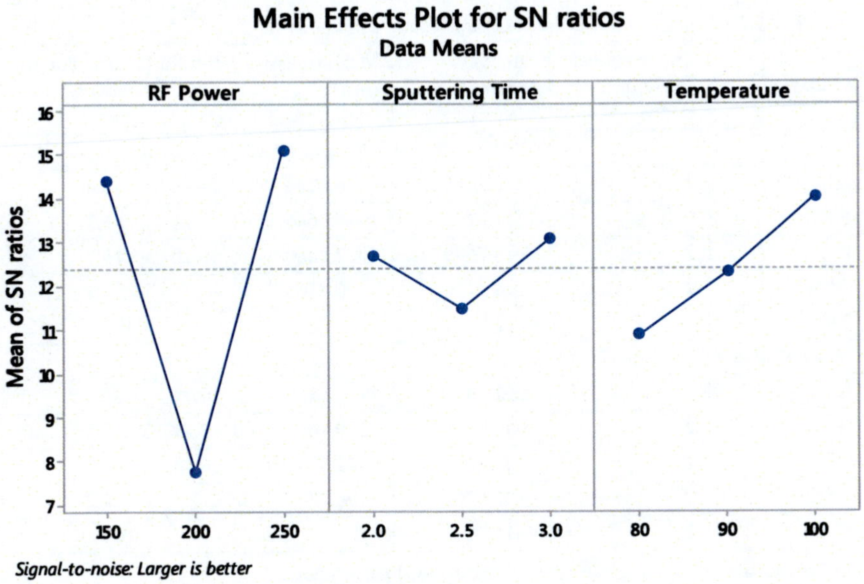

Fig. 2 S/N ratio graph of the hardness response

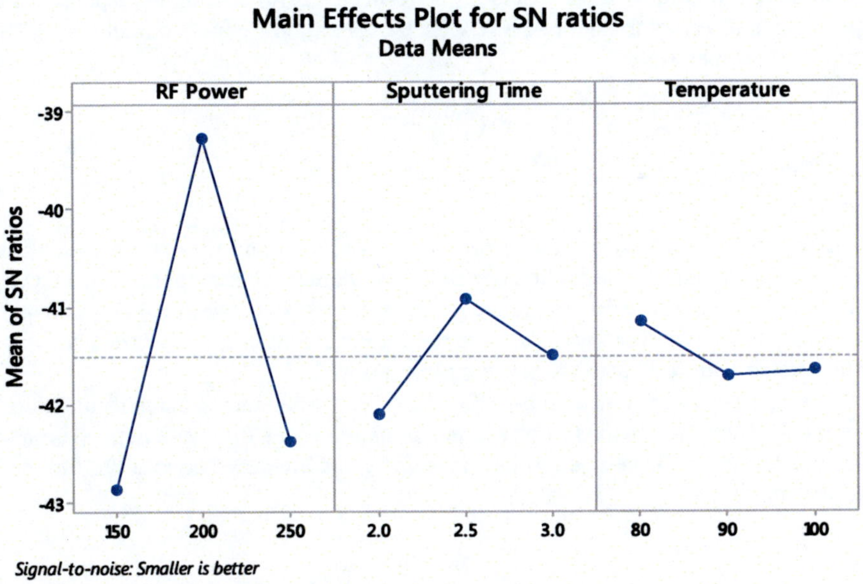

Fig. 3 S/N ratio graph of the Young's modulus response

Table 5 ANOVA analysis of the hardness and Young's modulus

Variance source	The degree of freedom (DoF)	Sum of squares (SS)	Mean square (MS)	F-ratio	Contribution rate (%)
Hardness					
A	**2**	**20.2469**	**10.1234**	**23.19**	**70.4**
B	2	1.9293	0.9646	2.11	6.7
C	2	5.6579	2.8289	6.20	19.7
Error	2	0.9126	0.4563	–	3.2
Total	8	28.7466	–	–	100
Young modulus					
A	**2**	**3892.90**	**1946.45**	**20.22**	**84.4**
B	2	440.51	220.26	2.29	9.5
C	2	85.74	42.87	0.45	1.9
Error	2	192.57	96.29	–	4.2
Total	8	4611.73	–	–	100

The bold values are the output response with optimal performance

factor B and C contribute 9.5% and 1.9%, respectively, to the output response. The percentage of errors for both ANOVA for hardness and Young's modulus is 3.2% and 4.2%, respectively.

4 Tribology

The coefficients of friction (COF) for the microscratch and wear are presented in Table 6. The range for the friction coefficient of scratch is between 0.031 and 0.075, while the coefficient of friction range for the wear test is between 0.447 and 0.496. It is noticeable that there was a substantial improvement in the COF for both wear and scratch with respect to the process parameters.

Figure 4 shows some selection of the wear scar for the microscratch result micrograph obtained. No obvious crack and delamination of the film were noticed around the crack region. These micrographs revealed good bonding between the film and the substrate.

5 Conclusion

From the Taguchi analysis of the results using signal-to-noise analysis and ANOVA, it can be concluded that the RF power is the most effective control factor on both the

Table 6 Friction coefficient for wear and scratch tests

Experimental no		L1	L2	L3	L4	L5	L6	L7	L8	L9
Coefficient of friction	Scratch	0.031	0.040	0.048	0.071	0.048	0.075	0.036	0.047	0.069
	Wear	0.447	0.462	0.484	0.471	0.465	0.496	0.469	0.480	0.462

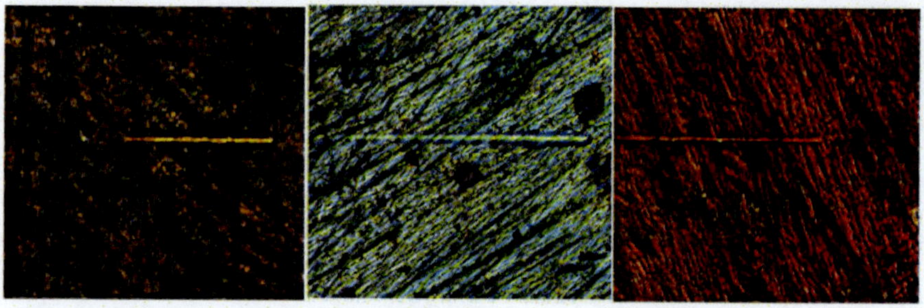

Fig. 4 Macrograph of some selected scratch surface for sample L1, L4 and L7

hardness and Young's modulus. The coefficients of friction for both microscratch and wear show considerable improvement, and the process parameters prove to be major determinant factors in the response outcome.

Acknowledgements The authors will like to acknowledge the Council for Scientific and Industrial Research (CSIR) for bursary awarded to the first Author.

References

1. Halling J (1986) The tribology of surface coatings, particularly ceramics. Proc Inst Mech Eng Part C 200(1):31–40
2. Holmberg K, Mathews A (1994) Coatings tribology: a concept, critical aspects and future directions. Thin Solid Films 253(1–2):173–178
3. Abegunde OO et al (2019) Overview of thin film deposition techniques. AIMS Mater Sci (-2):174
4. Martin P (2010) Deposition technologies: an overview. In: Handbook of deposition technologies for films and coatings, 3rd Edn. Elsevier, Oxford
5. Martin PM (2009) Handbook of deposition technologies for films and coatings: science, applications and technology, William Andrew
6. Wasa K, Hayakawa S (1992) Handbook of sputter deposition technology
7. Seshan K (2012) Handbook of thin film deposition. William Andrew
8. Alexeeva OK, Fateev VN (2016) Application of the magnetron sputtering for nanostructured electrocatalysts synthesis. Int J Hydrogen Energy, 41(5):3373–3386, 2016
9. Pethe SA et al (2012) Effect of sputtering process parameters on film properties of molybdenum back contact. Solar Energy Mater Solar Cells 100:1–5
10. Dibyendu R, Shampa A (2013) Effect of sputtering process parameters on the magnetron sputtered Ni-Ti-Cu thin films. Int J Curr Res 5(01):075–079
11. Oliver WC, Pharr GM (1992) An improved technique for determining hardness and elastic modulus using load and displacement sensing indentation experiments. J Mater Res 7(6):1564–1583
12. Oliver WC, Pharr GM (2004) Measurement of hardness and elastic modulus by instrumented indentation: advances in understanding and refinements to methodology. J Mater Res 19(1):3–20
13. Fang T, Jian S, Chuu D (2004) Nanomechanical properties of TiC, TiN and TiCN thin films using scanning probe microscopy and nanoindentation. Appl Surf Sci 228(1–4):365–372

14. Bec S et al (1996) Improvements in the indentation method with a surface force apparatus. Philos Mag A 74(5):1061–1072
15. Mani A et al (2005) Effects of residual stress on the mechanical and structural properties of TiC thin films grown by RF sputtering. Surf Coat Technol 194(2–3):190–195
16. Janos B, Lugscheider E, Remer P (1999) Effect of thermal aging on the erosion resistance of air plasma sprayed zirconia thermal barrier coating. Surf Coat Technol 113(3):278–285
17. Karthikeyan S, Balasubramanian V, Rajendran R (2014) Developing empirical relationships to estimate porosity and microhardness of plasma-sprayed YSZ coatings. Ceram Int 40(2):3171–3183

Effect of Compaction Loads in Machining of Short Carbon Fiber-Reinforced Aluminum Composite

Check for updates

S. Mohanasundaram, S. J. Vijay, Rajakumar S. Rai, I. Kantharaj, A. S. Melwyn, and S. Theophilus

Abstract Carbon fibers are preferred as reinforcements in metal matrix composites for their mechanical and metallurgical properties. In the present study, short carbon fiber-reinforced aluminum composites are fabricated with 2% reinforcement in pure aluminum powders using a powder metallurgy process. The blended carbon fibers and aluminum powders are compacted for different loading conditions such as 250, 300 and 350 kN followed by sintering at 580 °C. Abrasive jet machining trials are conducted on all combinations, by varying the machining parameters. Carbon fibers are found to be well dispersed in the aluminum matrix when examined through SEM. The bonding between the carbon fibers and aluminum is also found to be without any flaw. The mechanical properties showed improvements in wear resistance and hardness of composites when compared to pure aluminum. The machining parameters are further studied to obtain improved surface finish and minimize delamination of carbon fibers from the metal matrix.

Keywords Carbon fiber · Aluminum · Sintering · Metal matrix composites · Abrasive jet machining

1 Introduction

Aluminum matrix composite materials exhibit great potential in several manufacturing industries like aircraft, aerospace, automobiles and various other fields. The reinforcements in operating temperature should be both stable and non-reactive [1]. Reinforcements can be in the form of short fibers, continuous fibers and particles.

S. Mohanasundaram (✉) · S. J. Vijay · R. S. Rai · I. Kantharaj · A. S. Melwyn · S. Theophilus
Department of Mechanical Engineering, Karunya Institute of Technology and Sciences, Coimbatore 641 114, India
e-mail: johnmohana@yahoo.com

S. J. Vijay
e-mail: vijayjoseph.2001@gmail.com

S. Vijayan et al. (eds.), *Trends in Manufacturing and Engineering Management*, Lecture Notes in Mechanical Engineering,
https://doi.org/10.1007/978-981-15-4745-4_15

Aluminum properties are found to be reduced significantly at relatively low temperatures, which subsequently limits its functionality in high–precision mechanical components [2]. Fibers play an important role as reinforcements, fulfilling the requirements and distributing strength to the constituent matrix, influence their properties as desired and improve them [1]. Carbon fibers are used to produce materials that enhance performance in applications such as nanoelectronics, structural and medical applications. Because of its lightweight, mechanical properties and thermal properties, reinforced carbon fiber (CF) composites were extensively used as industrial materials [3]. Integrating CF into metal matrices considerably enhances the hardness, tensile strength, modulus of elasticity and other mechanical properties such as friction coefficient, wear resistance, corrosion and fatigue strength. They can also be customized to the corresponding application requirements in metal matrix composites [4, 5]. CF/Al composites were used in the waveguide mast of the Hubble Space Telescope antenna [6]. The properties of MMCs are related to matrix properties, especially at low fiber content or when short fibers or particles are used as reinforcements instead of continuous fibers [7].

Solid-state processes like powder metallurgy are used to fabricate short fiber-reinforced aluminum alloys [8] and liquid-state processes such as compo casting [9] and pressurized liquid-metal infiltration [10]. The liquid processes include melting and casting, are unable to provide good matrix reinforced homogeneity due to the high interface of energy between molten metal and reinforcements. Homogeneity solid-state or mechanical alloy route reinforcements can be produced, clustering of particle reinforcements. But it is a relatively expensive process [11].

Among the non-traditional processes, abrasive water jet (AWJ) machining has provided exceptional machining ability for the most difficult materials in which traditional methods are technically or economically impossible. AWJ finds application for a wide range of metals and non-metals, concrete, ceramics, metal matrix, laminates and fiber-reinforced composites [12]. The advantages of abrasive water jet cutting include simple fixtures required, high machining versatility, negligible heat affected zone, small forces for cutting composite materials, etc. [12, 13].

In the present study, the machining qualities of short carbon fiber-reinforced Al matrix materials prepared using cold pressing and powder metallurgy by abrasive water jet machining for two process parameters (standoff distance and transverse speed) were investigated.

2 Experimental Procedure

2.1 Materials

Aluminum is light, non-magnetic, malleable and ductile. It is widely used in several industrial applications where light and robust material is needed. Pure metal has inadequate strength and soft in nature, nevertheless once alloyed with a small amount of copper, magnesium, silicon, manganese or other elements, it results in a variety of useful properties. Pure aluminum powder (99.5%) of 16 microns was used for the fabrication of the composite.

Carbon fibers (CF) have the highest specific rigidity compared to ceramic reinforcements. They are also commercially available at less cost and are not hazard to health. These properties make them to be used as reinforcement for light metals like aluminum and its alloys. 1 mm chopped carbon fibers were used for the fabrication of the composite.

2.2 Fabrication of SCF Reinforced al Composite

Ball milling is one of the techniques used to homogeneously blend the reinforcement particles in the matrix phase [11]. Pure aluminum powder and short carbon fiber (SCF) of 2 weight percentage were mixed using ball milling for an hour. The Al matrix powder and SCF mixer are then compacted in the mild steel die, and it is cold pressed with 250, 300 and 350 kN compressive loads using universal testing machine. After cold pressing, sintering was carried out in an electric muffle furnace with an inert atmosphere. The cold-pressed Al/SCF sample is kept inside the furnace at 580 °C for 30 min and allowed to cool in the furnace. The sintered samples were of 50 mm in diameter and 25 mm in height.

2.3 SCF/Al Composite Machining

Machining was carried out in a water jet machine at SBB Water Jet (STREAMLINE SL-V E50V2), Coimbatore, with a 1.1 mm orifice. Experiments were performed at a constant pressure of 3500 bar and constant mass flow rate of 700 g/min using abrasive garnet (silicon sand of 100 mesh). The two-process parameters used to perform the machining are standoff distance and transverse speed. They were considered in abrasive water jet machining of SCF/Al composite specimens. The machining process parameters and the compacting load variation are shown in Table 1.

Table 1 Process parameters and the levels for the experiment

Process parameter	Unit	Level 1	Level 2	Level 3
Compacting load	kN	250	300	350
Standoff distance	mm	1	1.5	2
Transverse speed	mm/min	180	190	200

Table 2 Process parameter wear test

Process parameter	Unit	Value
Sliding velocity	m/sec	2
Load	N	2
Sliding distance	m	1500
Track diameter	mm	120
Speed	rpm	318.3
Time	sec	750

2.4 Wear Test

Wear is the process of friction between two interacting surfaces in a confined environment. Wear test was conducted in pin-on-disk test apparatus as per the ASTM standard G99. The specimens were machined to a size of 10 mm×10 mm×25 mm. The wear rate was calculated directly as the volume loss of the specimen for the distance traveled. During the conduct of wear test, the specimen was forced against the rotating hardened steel disk by applying the load. The test was carried out at room temperature with the process parameters shown in Table 2.

$$\text{Sliding velocity} = (\pi \text{DN})/60 \quad (1)$$

$$\text{Sliding distance} = \text{Sliding Velocity} \times \text{Time} \quad (2)$$

$$\text{Wear rate} = \text{Volume loss/Sliding distance} \quad (3)$$

2.5 Design of Experiments

Design of experiments (DOE) is the process of carrying out experiments considering the process parameters at various levels. To evaluate the performance, quality and cost of experimental design, Taguchi's method provides a simple, efficient and systematic approach for an optimal DOE [14]. Based on Taguchi's method, DOE with three factors and three levels L9 orthogonal arrays table was selected for the experimentation as shown in Table 3.

Table 3 Orthogonal array for experiments

S. No	Compacting load (kN)	Standoff distance (mm)	Transverse speed (mm/min)
1	250	1	180
2	250	1.5	190
3	250	2	200
4	300	1	190
5	300	1.5	200
6	300	2	180
7	350	1	200
8	350	1.5	180
9	350	2	190

2.6 *Surface Roughness*

Surface roughness is the degree of irregularities on the surface which is of three components, i.e., roughness, waviness and form [15]. Surface roughness is measured by Mitutoyo make surface tester. In order to reduce the measurement error, three sets of readings were observed for each sample, and the mean value is taken for SR.

3 Results and Discussion

The 9-run machining was carried out in abrasive water jet machine as per combination of factors based on the DOE, and the composite's surface roughness, hardness and wear rate are measured as in Table 4.

Table 4 Experimental results

S. No	Compacting load (kN)	Standoff distance (mm)	Transverse speed (mm/min)	Surface roughness Ra (μm)	Hardness (HV)	Wear rate mm^3/m
1	250	1	180	1.95	20.6	0.080
2	250	1.5	190	2.383		
3	250	2	200	2.755		
4	300	1	190	2.559	24.6	0.033
5	300	1.5	200	2.712		
6	300	2	180	2.327		
7	350	1	200	3.32	24.8	0.020
8	350	1.5	180	3.179		
9	350	2	190	3.256		

The powder mixer was compacted and sintered, and then, it was kept in the furnace to maintain the temperature. The sintered composites with various compacting load have shown an increase in hardness (Fig. 1). The hardness value for the specimen compacted with 300 kN and specimen compacted with 350 kN are almost similar. Hence, it may be concluded that increasing the compacting load over 300 kN is insignificant with respect to the hardness values.

Surface morphology of the sintered composite for various compacting loads is shown in Fig. 2. The microstructure for the specimen with 250 kN is found to be more porous than 300 and 350 kN. Hence, the specimen compacted with 250 kN is not as dense as the other specimens. As the compacting loads are gradually applied to the powdered mixer, the same leads to the densification of the specimen which reduces the porosity and increases its hardness. The surface of the specimen compacted with 300 kN (Fig. 2b) and 350 kN (Fig. 2c) is found to have a similar textured surface. Hence for the higher loads, the change in hardness value is negligible.

Wear tests were conducted with a load of 2 N, by rubbing the sintered composite specimens. Sliding velocity was kept as constant of 2 m/sec. Wear test results show the effect of the compacting load on the wear rate. The specimen compacted for the load of 250 kN displayed a higher wear rate than 300 and 350 kN are shown in Fig. 3. It shows that the increase in compacting load reduces the wear rate.

The hardness of the composites influenced the resistance to friction and erosion. Cavities are formed along the composites' surface due to carbon fiber particles pulling out from the composites, which in turn increases the surface roughness. The surface roughness value of the composite increases with an increase in transverse speed from 180 mm/min to 200 mm/min as shown in Fig. 4.

It is observed that material erodes at a higher rate as the transverse speed increases are shown in Fig. 5. Surface roughness of composite specimens for various standoff distances. At a load of 300 kN, as the standoff distance increases there is a gradual reduction in the surface roughness value owing to the porosity of the specimen. Since the compaction id good at a load of 350 kN the surface roughness value varies linearly as the specimen is less porous.

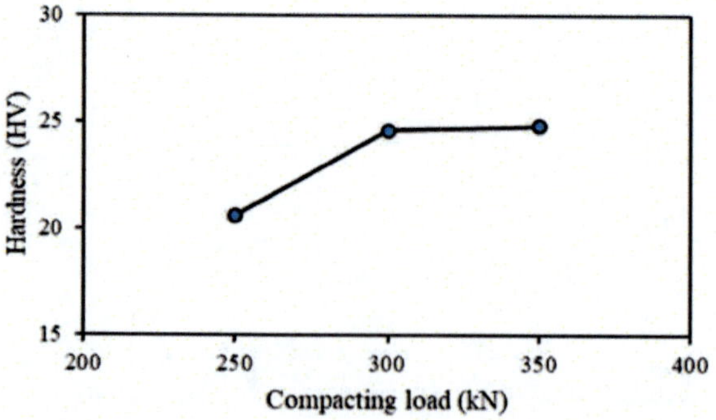

Fig. 1 Hardness of composite specimens for different compacting load

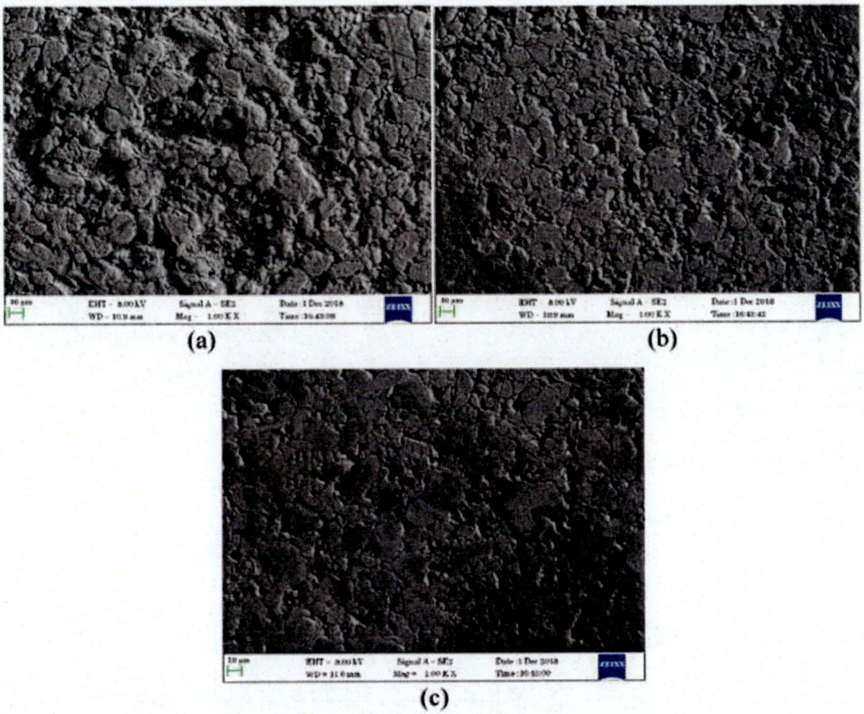

Fig. 2 Surface morphology of sintered composite (**a**) 250 kN compacting load, (**b**) 300 kN compacting load and (**c**) 350 kN compacting load

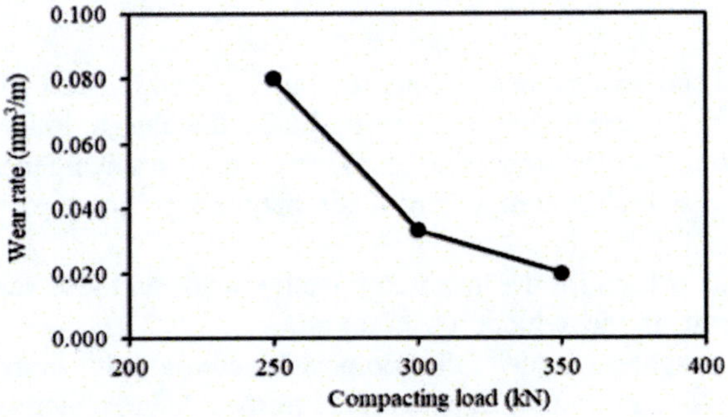

Fig. 3 Wear rate of specimens for various compacting loads

4 Conclusion

The Al/SCF powder and fiber mixer were compacted for different loads and sintered at 580 °C. The sintered specimens were machined in abrasive water jet machine for

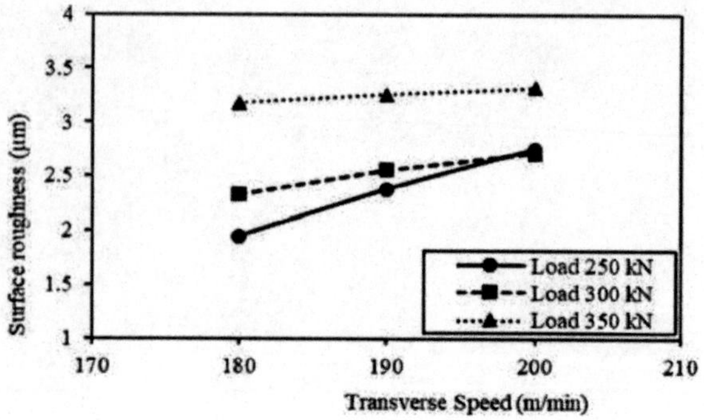

Fig. 4 Surface roughness of composite specimens for various transverse speeds

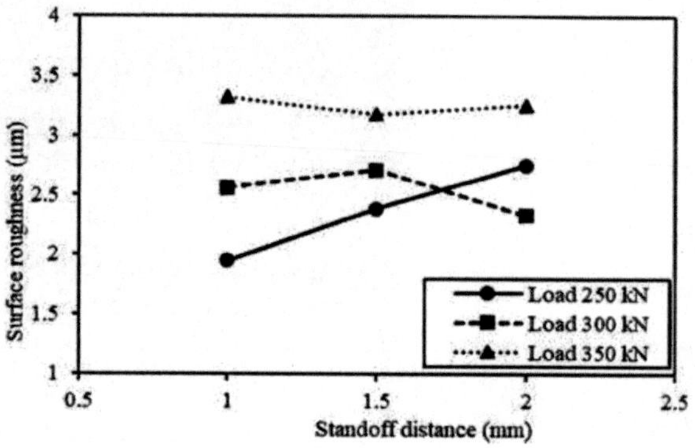

Fig. 5 Surface roughness of composite specimens for various standoff distances

the various transverse speeds and standoff distances. The following conclusions were made.

- The sintered composites with various compacting load increased the hardness of the specimen. For the higher compacting loads, the change in hardness value is very minimum. As the compacting loads increased, the specimen became denser. It is found that higher compacting loads play a significant role in enhancing hardness.
- The wear rate is high for the lower compacting loads and vice versa. Increase in hardness results in the reduction of wear rate.
- The surface roughness value of the composite increases with an increase in transverse speed. Along the surface, cavities are formed due to pull out of carbon fiber particles from the composites due to which the surface roughness is increased.

References

1. Vijaya Ramnath B, Elanchezhian C, Annamalai RM, Aravind S, Sri Ananda Atreya T, Vignesh V, Subramanian C (2014) Aluminium metal matrix composites—a review. RAMS 38(2014):55–60
2. Shirvanimoghaddam K et al (2017) Carbon fiber reinforced metal matrix composites: fabrication processes and properties. Compos Part Appl Sci Manuf 92:70–96
3. Surappa MK (2003) Aluminium matrix composites: Challenges and opportunities. Sadhana 28(1):319–334
4. Kelly A (2006) Composite materials after seventy years. J Mater Sci 41(3):905–912
5. Ramesh CS, Adarsha H, Pramod S, Khan Z (2013) Tribological characteristics of innovative Al6061–carbon fiber rod metal matrix composites. Mater Des 50:597–605
6. Miracle DB (2005) Metal matrix composites—from science to technological significance. Compos Sci Technol 65(15):2526–2540
7. Kavecký Š, Šebo P (1996) Short aluminosilicate fibre reinforced aluminium. J Mater Sci 31(3):757–764
8. Toth IJ, Brentnall WD, Menke GD (1974) Survey of aluminum matrix composites. Metallurgical Society of the American Institute of Mining, Metallurgical and Petroleum Engineers, Inc, United States
9. Das S, Dan TK, Prasad SV, Rohatgi PK (1986) Aluminium alloy—rice husk ash particle composites. J Mater Sci Lett 5(5):562–564
10. Oh S-Y, Cornie JA, Russell KC (1987) Particulate wetting and metal: ceramic interface phenomena. Ceram Eng Sci Proc 8:912–936
11. Mendoza-Duarte JM, Estrada-Guel I, Carreño-Gallardo C, Martínez-Sánchez R (2015) Study of Al composites prepared by high-energy ball milling; effect of processing conditions. J Alloys Compd 643:S172–S177
12. Kovacevic R, Hashish M, Mohan R, Ramulu M, Kim TJ, Geskin ES (1997) State of the art of research and development in abrasive waterjet machining. J Manuf Sci Eng 119(4B):776–785
13. Vikram G, Ramesh Babu N (2002) Modelling and analysis of abrasive water jet cut surface topography. Int J Mach Tools Manuf 42(12):1345–1354
14. Davim JP (2003) Design of optimisation of cutting parameters for turning metal matrix composites based on the orthogonal arrays. J Mater Process Technol 132(1):340–344
15. Choudhury SK, Chinchanikar S (2017) 1.3 finish machining of hardened steel. In: Hashmi M (ed) Comprehensive materials finishing. Elsevier, Oxford, pp 47–92

Synthesis and Characterization of Al 7072-Al_2O_3 Metal Matrix Composites

G. Mallesh, R. Pavankumar, V. G. Pradeep Kumar, and L. Laxman Naik

Abstract Composite materials are the combinations of two or more materials which are different in form and chemical composition. These composite are gradually gaining more importance as a structural material in the present engineering design and development activities because they offer very attractive mechanical properties such as high strength-to-weight ratio higher thermal and corrosive resistance. A wide range of metals and their alloys such as aluminium, titanium and magnesium are extensively used in combination with different reinforcements such as SiC, Al_2O_3, TiC, TiO_2, TiB, ZrO_2 and ZrB. Aluminium and its alloys are widely used to develop lightweight material for construction, transportation, aerospace industries, marine, automobile and similar engineering applications. In the present study, attempts are made to synthesize Al 7072-Al_2O_3 MMCs using stir casting process by varying the weight percentage of reinforcement to investigate the mechanical behaviour as per ASTM standards. EDX and XRD results confirm the presence of Al_2O_3 particles in the composite, and it is observed from the SEM images that alumina particles are homogeneously distributed in the matrix. Further, results reveal that the mechanical properties of Al 7072-Al_2O_3 were enhanced with the addition of the reinforcement.

Keywords Aluminium Matrix Composites (AMC's) · Stir Casting · Al_2O_3 · Mechanical Properties · EDX and XRD

G. Mallesh (✉) · V. G. P. Kumar
Department of Mechanical Engineering, Sri Jayachamarajendra College of Engineering, Mysuru, India
e-mail: mallesh@sjce.ac.in

V. G. P. Kumar
e-mail: mepradeepvg@gmail.com

G. Mallesh · R. Pavankumar
JSS Science and Technology University, Mysuru, Karnataka, India
e-mail: pavanjce2016@gmail.com

L. L. Naik
Department of Mechanical Engineering, Malnad College of Engineering, Hassan, Karnataka, India
e-mail: llnmce@gmail.com

S. Vijayan et al. (eds.), *Trends in Manufacturing and Engineering Management*, Lecture Notes in Mechanical Engineering,
https://doi.org/10.1007/978-981-15-4745-4_16

1 Introduction

Nowadays, AMC's are accepted as a pivotal member in many important industries. Over the years, AMC's have been accepted material for different engineering sectors. Many researchers have reported the use of aluminium in automobile bodybuilding [1], fabrication industries, aeronautical industries [2], shipbuilding and other structural applications [3–5].

AMC's reveal unique properties when reinforcements are added. Determining the proper reinforcement is an essential task to get the required composite. Commonly used reinforcements in different grades of aluminium are SiC, Al_2O_3, B_4C, TiC TiO_2, TiB_2, etc. [6]. Application of AMCs are increasing due to stringent environmental protocols, and hence, it is inevitable to develop new AMCs.

Auradi et al. [7] developed Al 6061-B_4C composites using pre-heated B_4C particles and K_2TiF_6 flux to overcome the problem of agglomeration and clustering of the particles for different wt% of the reinforcement particles. XRD, EDX and SEM analyses were conducted to know the presence and distribution of the reinforcement in the composites. It is evident from the results that B_4C particles were present and uniformly distributed in the composites developed. Further, the addition of the B_4C particles to the melt leads to a decrease in the grain size and hence improves the mechanical properties.

Stalin et al. [8] synthesized LM4-B_4C (400 mesh) composites for 5, 10 and 15 wt% of reinforcement by stir casting process to evaluate mechanical behaviour of MMCs. SEM and XRD tests confirm the presence and uniform distribution of B_4C in the matrix alloy. It is noticed that mechanical properties were improved with the addition of 10 wt% of the reinforcement. It is also evident that absorption of energy found to be increased by increasing percentage of reinforcement.

Kumar et al. [9] fabricated Al6061-SiC and Al7075-Al_2O_3 composites using stir casting route by maintaining a temperature of 720 °C with a stirring speed of 400 rpm about 10 min to study the microstructure, mechanical and wear properties. It is observed that the addition of the reinforcement in both the alloy increase mechanical, tribological properties, but Al7075-Al_2O_3 has better properties compared to Al 6061-SiC.

Kaliselvan et al. [10] synthesized Al 6061-B_4C composites by adding K_2TiF_6 flux using stir casting process with argon gas environment to study the mechanical properties of the composites. It is noticed that the addition of K_2TiF_6 results in good wettability between the particles and matrix. Further, mechanical properties like hardness increased from 51.3 to 80.8 HV and tensile strength enhanced from 185 to 215 MPa. Gopal Krishna et al. [11] developed AMCs using B_4C using stir casting process to study the mechanical characteristics of the composites for different wt% of the reinforcement as per ASTM standards. It is observed from the results that B_4C particles homogeneously distributed in the matrix and perceived the presence of particles in the matrix. Further, it is noticed that tensile strength and hardness of the composites were enhanced with the addition of 12 and 8 wt. of the reinforcement, respectively.

Ravi et al. [12] developed aluminium matrix composites using boron carbide using centrifugal casting technique by varying the wt% of the reinforcement to study the mechanical properties. It is evident from the results that tensile strength was enhanced from 117 to 145 MPa, and hardness value increased from 62 to 68 HV.

Zhao et al. [13] developed AA 6061-B_4C composites using a friction stir process to study the microstructure and mechanical properties. Results revealed that with an increasing number of FSP passes, the dispersion of B_4C particles becomes more uniform, further by way of four FSP passes, the expected AA6061/B_4C surface composite layer is successfully fabricated with homogeneously distributed B_4C particles.

Vijaya Ramnath et al. [14] proposed that Al6061 B_4C-SiC composite was produced with stir cast route for different particle size (viz. 37, 44, 63, 105, 250 μ) by maintaining melt temperature at 800 °C during the process preheated B_4C particles and degassing agents C_2Cl_6 were added into the vortex with a stirring speed of 300 rpm. It was observed from the XRD analysis the B_4C particles are homogeneously dispersed in the matrix. Further, it is noticed that hardness of the composite was maximum at 250 μ at 12 wt% and an increment in tensile strength at 105 μ for 8 wt% of B_4C.

From the available literature, it is absorbed that many researches are worked towards the development of lightweight materials for engineering applications using different reinforcements and found that the addition of the reinforcement enhances the mechanical and tribological properties. Further noticed that limited work has been done to study the mechanical characteristics of Al 7072-Al_2O_3 AMCs. Hence in this research work, attempts are made to develop Al 7072-Al_2O_3 MMCs to study the mechanical properties.

1.1 Reasons for Selecting the Matrix and Reinforcement Material

Al 7072 alloy is an extensively used aluminium alloy which offers a range of good properties [15]. Al 7072 (Fig. 1) based composites are high strength, good machinability, low thermal and electrical resistivity and easily extractable in nature. Being stiff and strong material, it finds many applications in different engineering fields [16, 17].

Among various reinforcements, Al_2O_3 (Fig. 2) is selected because of its staggering properties like chemical and thermal stability, availability, relatively good strength, high hardness strength, high electrical insulation, good corrosion resistance and biocompatibility.

Fig. 1 Wrought Aluminium 7072

Fig. 2 Al_2O_3 Powder

2 Synthesis of Al 7072-Al_2O_3 Composites

Several manufacturing techniques are used to synthesize metal matrix composites. The most commonly used techniques are liquid, solid and gaseous state [18], and these processes include squeeze casting, powder metallurgy, spray forming, diffusion bonding, sinter forging, stir casting and in situ process.

Among these manufacturing processes, stir casting is generally accepted as a promising route to manufacture metal matrix composites in large volumes with minimum cost. Hence in this research work, stir casting technique is used to develop Al 7072-Al_2O_3 MMC's.

2.1 Stir Casting

Stir casting process is an extensively used method to fabricate MMCs in liquid state in which molten metal is mixed in the furnace by means of mechanical stirring [19]. It is noticed that almost all the composites were manufactured by this process using up to 30% volume fraction of reinforcement [20, 21]. It is evident from the literature that use of stir casting process enables better mixing of matrix and reinforcement in the MMCs. Different components of stir casting process are shown in Figs. 3 and 4.

Al 7072 rods and 3, 6, 9 and 12 wt% of Al_2O_3 are weighed as per the wt% of the reinforcement and fed into the graphite crucible and heated using an electrical furnace up to 850 °C and coating was applied inside the crucible to avoid contamination.

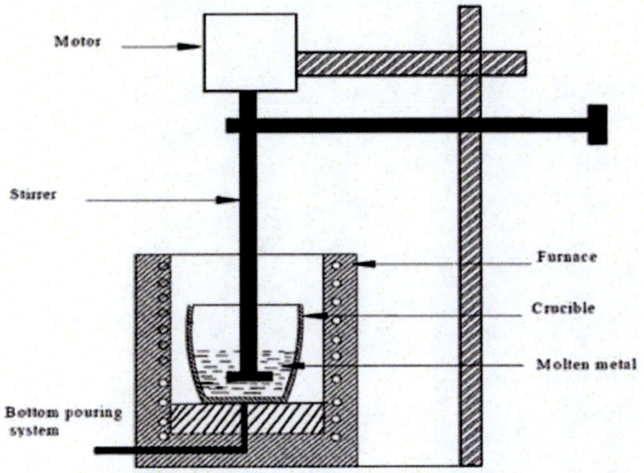

Fig. 3 Components of Stir Casting Process

Fig. 4 Stir Casting Process Set-up

Halide salt, hexaflurotitanate is added to improve wettability between Al_2O_3 particles and Al 7072. Similarly, hexachloroethane (C_2Cl_6) tablets were fed into the matrix to remove hydrogen gas in the melt. During stirring, preheated Al_2O_3 is gradually poured into the molten metal and stirring was continued about 10–12 min at the rate of 250 rpm. Temperature of the molten metal is monitored and regulated by digital temperature controller installed in the machine. Slag formed with the addition of hexaflurotitanate was removed from the crucible shown in Fig. 5. The melt was poured into a preheated (200 °C) cast iron moulds to get the rectangular plates and circular rods shown in Fig. 6.

After solidification process castings were obtained as shown in Fig. 7 and machin-

Fig. 5 Slag Removal Process

Fig. 6 Pouring of Molten Metal into the Cast iron Moulds

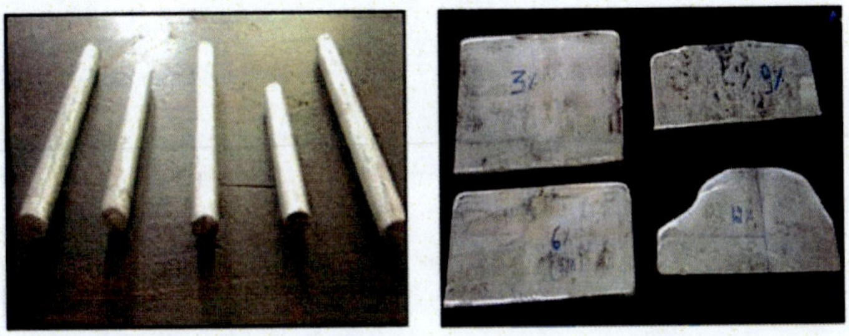

Fig. 7 Al 7072-different wt.% of Al_2O_3 MMC's

ing of different wt% of Al_2O_3 composites using wire cut EDM, CNC turning and milling was carried out as per different ASTM standards to conduct mechanical tests.

2.2 *EDX, XRD and Microstructure Characteristics of Al 7072-Al_2O_3 Composites*

EDS is an analytical tool used to know the presence of different elements in the specimen using X-ray spectrum. To stimulate the emission characteristic of X-rays from a specimen, a high-energy beam of electrons or protons or a beam of X-rays was focused on the sample being studied.

In the present research, EDS analyses of Al 7072-Al_2O_3 composites by varying the wt% of Al_2O_3 were conducted to know the presence of elements in the composites. The EDS patterns of Al 7072-Al_2O_3 with different wt% of Al_2O_3 are shown in Fig. 8(a–e). It is evident from the plots that aluminium exhibits high intensity peaks which indicate that aluminium is predominant because it is chosen as base alloy. Further, it is found that low intensity peaks of oxygen were identified which shows the presence of Al_2O_3 reinforcement [22].

The X-ray diffraction patterns (XRD) of Al 7072-Al_2O_3 composites were obtained using Panalytical X-ray diffractometer having 0.15–0.40 nm inter-atomic distances with photon energies between 3 and 8 keV. Peak values were collected over the 2θ range of 10–800 with a step size of 0.0170° and step time of 20.3142 s. The crystalline size measurements were carried out using Debye–Scherrer equation $D_{hkl} = \frac{0.98\lambda}{\beta \cos\theta}$ [23]. It is observed from the XRD patterns (Fig. 9a–e) that all the composites developed confirm the formation of Al_2O_3 particles.

The SEM micrographs of Al 7072-Al_2O_3 composites are presented in Fig. 10a–e. It is noticed from the micrographs that all the castings are free from defects such as porosity, shrinkages and slag inclusions which highlights the quality of castings and Al_2O_3 particles are distributed homogeneously in the composites which is an essential requirement to achieve better mechanical properties.

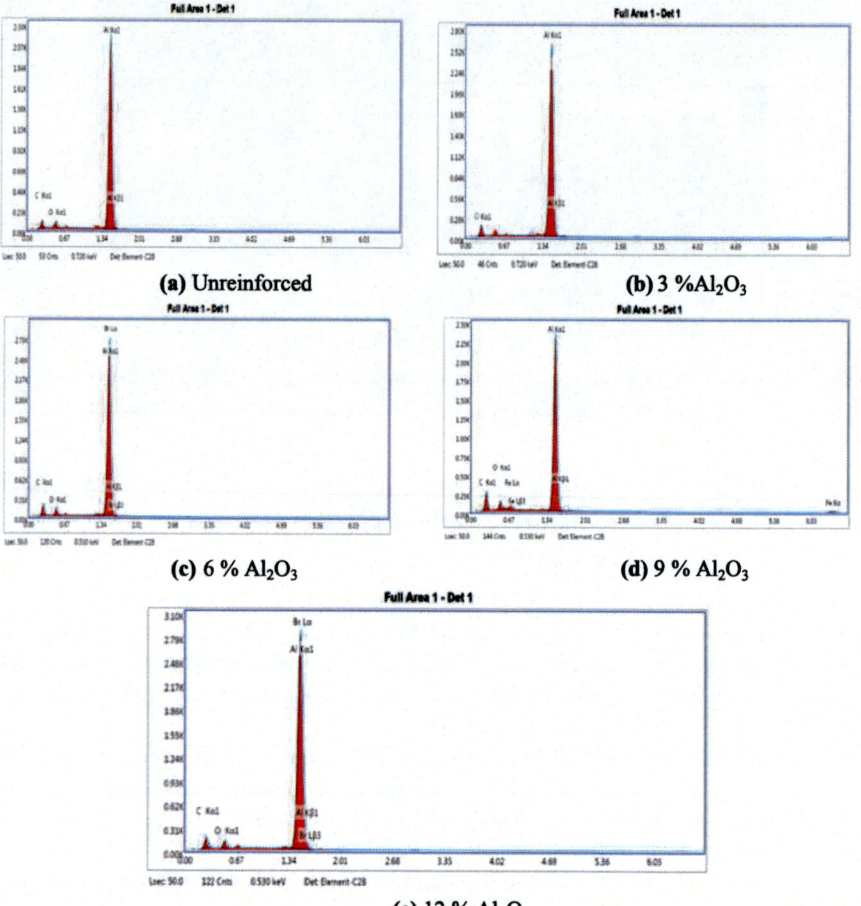

(a) Unreinforced **(b)** 3 %Al_2O_3 **(c)** 6 % Al_2O_3 **(d)** 9 % Al_2O_3 **(e)** 12 % Al_2O_3

Fig. 8 EDS Patterns of Al 7072 and different wt% of Al_2O_3 Reinforcement

3 ASTM Standard Mechanical Testing Methods

3.1 Tensile Strength

Tensile strength is the ability or capacity of the material to expand prior to failure. The tensile tests were conducted on computerized universal testing machine model TUE400(C) as per ASTM E8M-13a standard shown in Fig. 11.

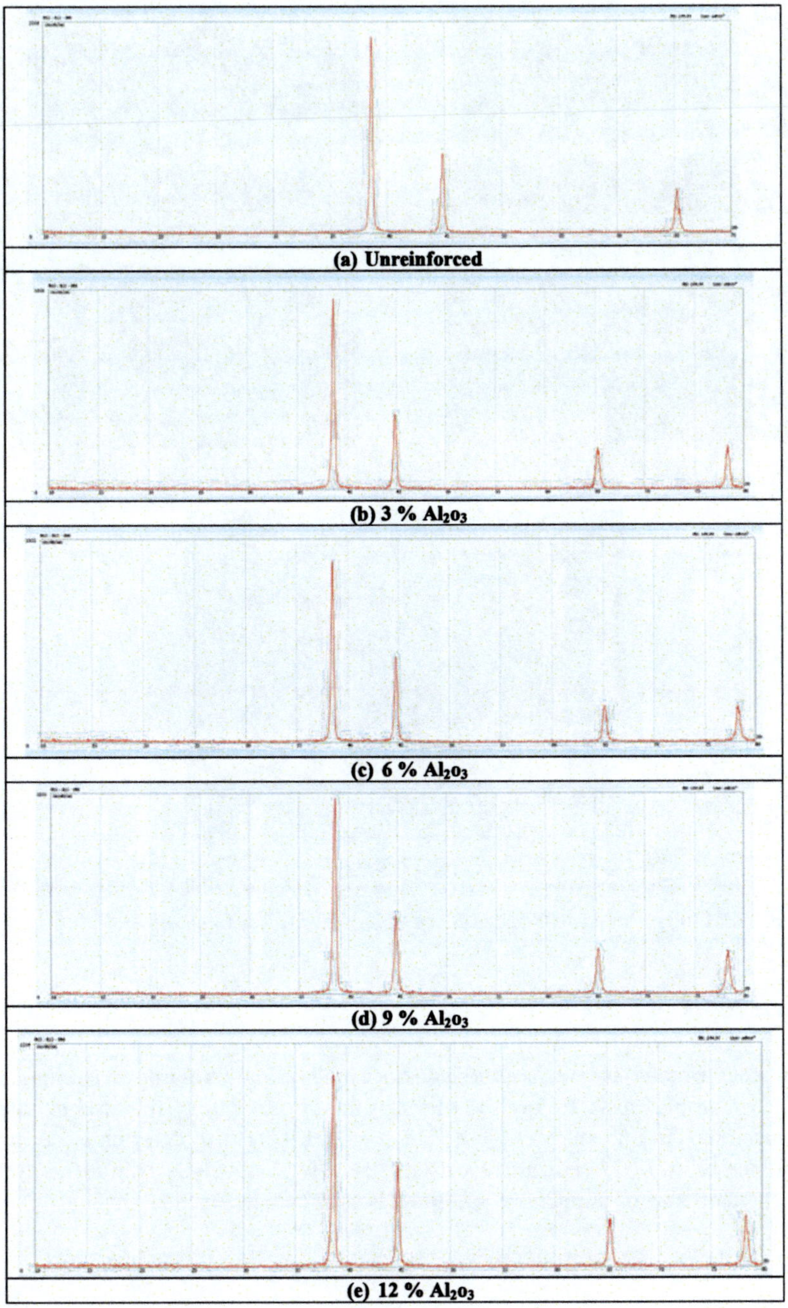

Fig. 9 XRD Plots of Al 7075 and different wt% of Al_2O_3 Reinforcement

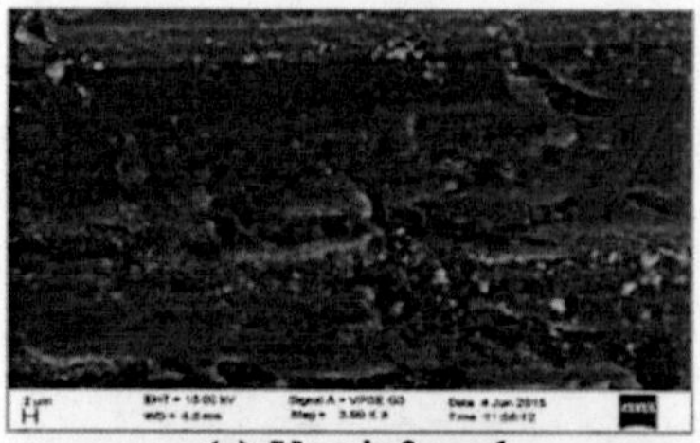

(a) Unreinforced

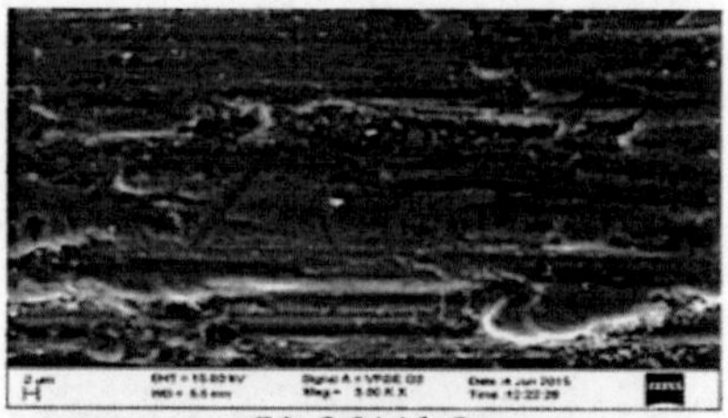

(b) 3 %Al_2O_3

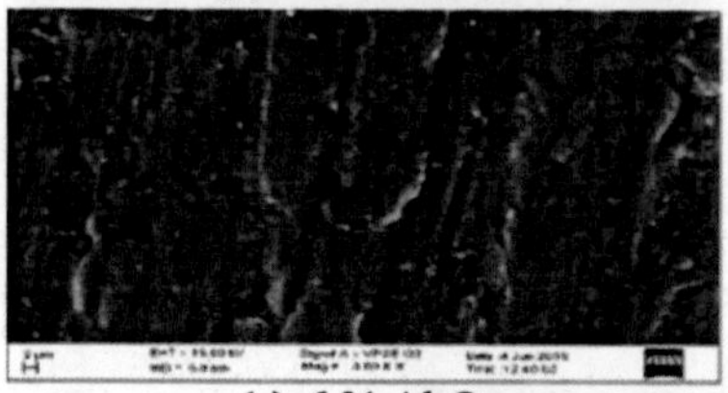

(c) 6 % Al_2O_3

(d) 9 % Al_2O_3

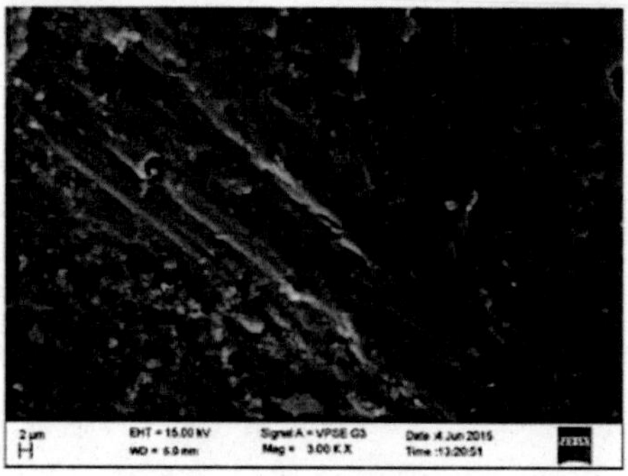

(e) 12 % Al_2O_3

Fig. 10 SEM Images of Al 7075 and different wt% of Al_2O_3 Reinforcement

3.2 Impact Strength

Ability of the material to resist a sudden/shock loads is known as impact strength. Izod or Charpy impact tests are used to measure the impact energy required to fracture the specimen. In this study, ASTM E23 standard is used to determine the impact resistance of Al 7072 reinforced with different wt% of Al_2O_3 composite system. The standard size of impact test specimen is as shown in Fig. 12.

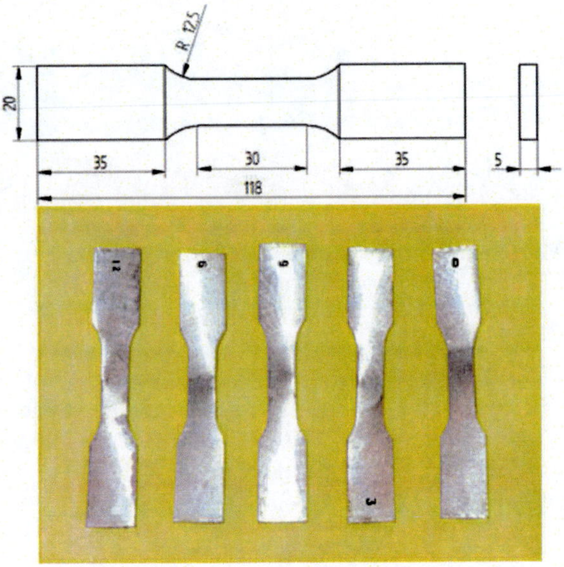

Fig. 11 ASTM E8M-13a Tensile Test Specimens

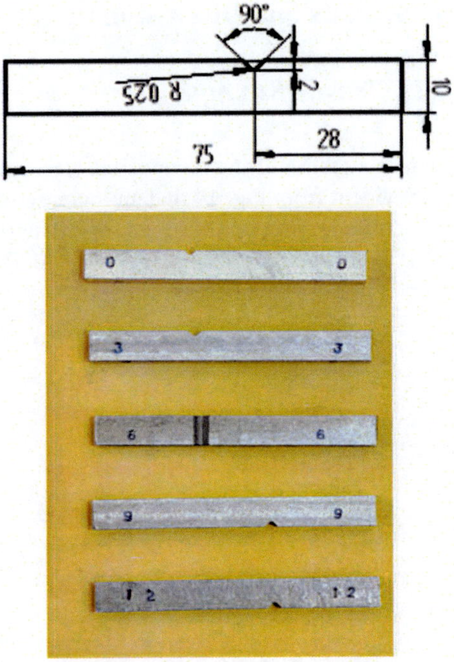

Fig. 12 ASTM E-23 Impact Test Specimens

Fig. 13 ASTM E-384 Hardness Test Specimens

3.3 *Hardness*

Hardness is the property of a material that resist plastic deformation by indentation. Vickers hardness test (VHN) is the most versatile since only one indenter is used for a range of loads for different materials. The tests were conducted for different wt% of Al_2O_3 as per ASTM E-384 standard shown in Fig. 13.

4 Results and Discussions

4.1 *Tensile Strength*

Tensile strength of the composites was tested using universal testing machine, and it is evident from Fig. 14 that an increase in the wt% of the reinforcement increases the tensile strength of the specimen. Unreinforced alloy possesses tensile strength of 51.28 MPa and addition of 3 wt% Al_2O_3 tensile strength improved by 10%, i.e. 56.64 MPa. Further increase in the 12 wt% of Al_2O_3, tensile strength reaches its peak

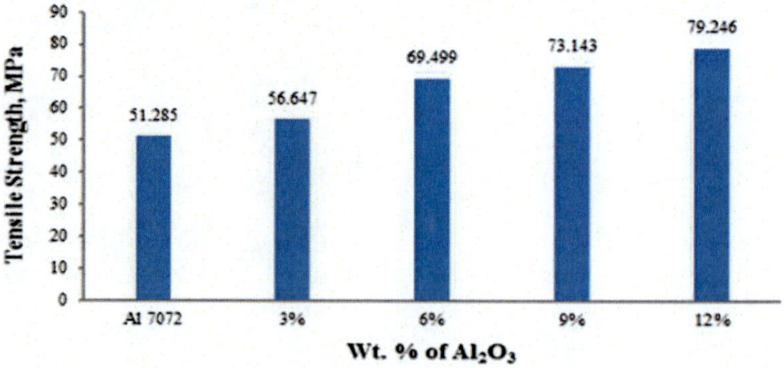

Fig. 14 Tensile Strength of Al 7072-Al_2O_3 Composite System

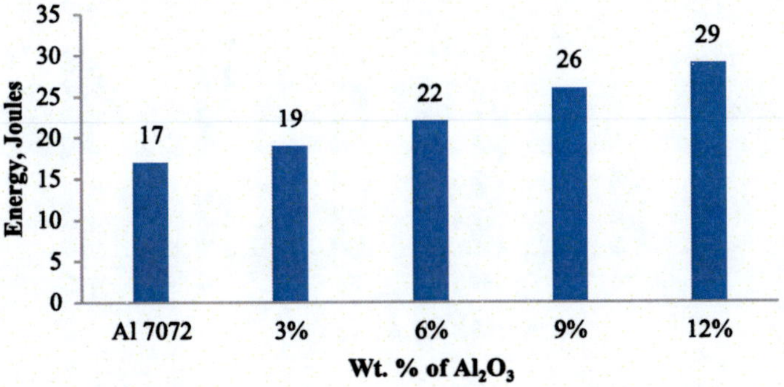

Fig. 15 Impact Strength of Al 7072-Al_2O_3 Composite System

value of 79.28 MPa, which is approx. 70% improvement [14]. Reason for this exponential improvement in tensile strength is because of grain boundary strengthening where alumina acts as a grain refiner and decreases the size of the alloy, uniform distribution of the particle in the matrix, reduction in grain size leads to enhance the tensile strength of Al 7072-Al_2O_3 composite.

It is also evident that increase in the wt% of the reinforcement there was noticeable decrease in the percentage of elongation due to the addition of hard particles alumina which leads to the formation of brittle phase in the composite in turn enhances the load carrying capacity.

4.2 Impact Strength

Izod impact tests were conducted as per ASTM standard to know the impact strength of the composites. It is observed from the results that a gradual improvement in the impact strength of the composite as percentage of reinforcement increases. The improvement in the impact strength is due to the resistance in flowing energy of the reinforcing particles at the molecule level. It is observed from Fig. 15 that impact strength of the composite developed is enhanced by 41% compared to base alloy.

4.3 Hardness

Vickers hardness tests were conducted as per ASTM standard to know the hardness number of Al 7072-Al_2O_3 composites. Figure 16 represents hardness number of 3,6,9 and 12 wt% of Al_2O_3 reinforcement. It is observed from the results that the addition of alumina enhances the hardness of the composite. Presence of Al_2O_3 particles in the composite helps to resist the movement of dislocation within the

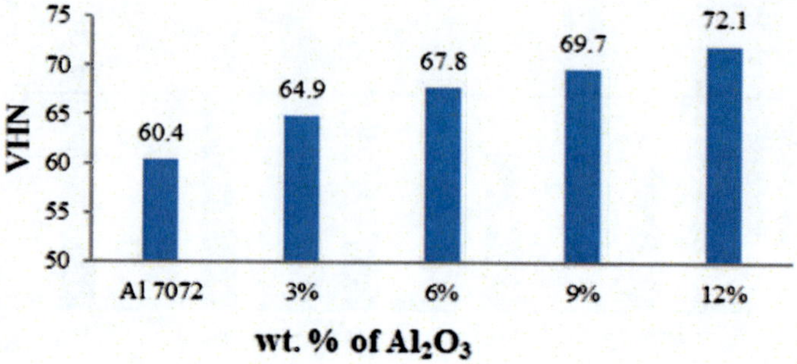

Fig. 16 Hardness Test Results

matrix, minimization of inter-atomic distance and better bonding between the matrix and the reinforcement.

5 Conclusions

- Al 7072 reinforced with different wt% of Al_2O_3 AMC's were synthesized using stir casting process.
- EDX and XRD results reveals the presence of the reinforcement in the composite developed and SEM images shows the clear and uniform distribution of Al_2O_3 in the matrix alloy.
- Mechanical tests on Al 7072-Al_2O_3 AMC's were carried out using different ASTM standards to evaluate mechanical behaviour.
- Mechanical properties such as tensile strength, impact strength and hardness of Al 7072-Al_2O_3 composites were enhanced with the addition of reinforcement.

References

1. Miller WS, Zhuang L, Bottema J, Wittebrood AJ, De Smet P, Haszler A, Vieregge A (2000) Recent development in aluminium alloys for the automotive industry. Mater Sci Eng A 280(1):37–49
2. Heinz A, Haszler A, Keidel C, Moldenhauer S, Benedictus R, Miller WS (2000) Recent development in aluminium alloys for aerospace applications. Mater Sci Eng A 280(1):102–107
3. Vijaya RB, Elanchezhian C, Annamalai RM, Aravind S, Sri Ananda Atreya T, Vignesh V, Subramanian C (2014) Aluminium metal matrix composites-a review. Rev Adv Mater Sci 38(2014):55–60
4. Shirvani MK, Abdizadehb H, Baharvandic HR. Mechanical and structural behavior of aluminium matrix composite reinforced by B_4C under different processing temperature and volume fraction of reinforcement

5. Bindumadhavan PN, Chia TK, Chandrasekaran M, Wah HK, Loh NL, Prabhakar O (2001) Effect of particle-porosity clusters on tribological behavior of cast aluminium alloy A356-SiCp metal matrix composites. Mater Sci Eng A 315(1–2):217–226
6. Hong SJ, Kim HM, Chun BS (2003) Effect of clustering on the mechanical properties of SiC particulate-reinforced aluminium alloy 2024 metal matrix composites. Mater Sci Eng A 347(1–2):198–204
7. Auradi V, Rajesh GL, Kori SA (2014) Processing of B 4 C particulate reinforced 6061 aluminum matrix composites by melt stirring involving two-step addition. Procedia Mater Sci 6:1068–1076
8. Stalin B, Murugan C (2016) Evaluation of mechanical behavior of aluminium alloy boron carbide MMC. In: International conference on emerging engineering trends and science (ICEETS–2016), pp 2348–8360
9. Kumar GBV, Rao CSP, Selvaraj N, Bhagyashekar MS (2010) Studies on Al6061-SiC and Al7075-Al2O3 metal matrix composites. J Miner Mater Charact Eng 9(01):43
10. Kalaiselvan K, Dinaharan I, Murugan N (2014) Characterization of friction stir welded boron carbide particulate reinforced AA6061 aluminum alloy stir cast composite. Mater Des 55:176–182
11. Gopal Krishna UB, Sreenivas Rao KV, Vasudeva B (2013) Effect of boron carbide reinforcement on aluminium matrix composites. Int J Metall Mater Sci Eng (IJMMSE):41–48. ISSN
12. Ravi B, Balu Naik B, Udaya Prakash J (2015) Characterization of aluminium matrix composites (AA6061/B4C) fabricated by stir casting technique. Mater Today: Proc 2(4–5): 2984–2990
13. Zhao Y, Huang X, Li Q, Huang J, Yan K (2015) Effect of friction stir processing with B 4 C particles on the microstructure and mechanical properties of 6061 aluminum alloy. Int J Adv Manufact Technol 78(9–12):1437–1443
14. Subramanya RP, Kesavan R, Vijaya Ramnath B (2018) Investigation of mechanical properties of aluminium 6061-silicon carbide, boron carbide metal matrix composite. Silicon 10(2):495–502
15. Eliasson J, Sandström R (1995) Applications of aluminium matrix composites. In: Key engineering materials. vol. 104, Trans Tech Publications, pp 3–36
16. Viala JC, Bouix J, Gonzalez G, Esnouf C (1997) Chemical reactivity of aluminium with boron carbide. J Mater Sci 32(17):4559–4573
17. Shorowordi KM, Laoui T, Haseeb ASMA, Celis JP, Froyen L (2003) Microstructure and interface characteristics of B4C, SiC and Al_2O_3 reinforced Al matrix composites: a comparative study. J Mater Process Technol 142(3):738–743
18. Kok M (2005) Production and mechanical properties of Al_2O_3 particle-reinforced 2024 aluminium alloy composites. J Mater Process Technol 161(3):381–387
19. Gopalakrishnan S, Murugan N (2012) Production and wear characterisation of AA 6061 matrix titanium carbide particulate reinforced composite by enhanced stir casting method. Compos B Eng 43(2):302–308
20. Balasivanandha PS, Karunamoorthy L, Kathiresan S, Mohan B (2006) Influence of stirring speed and stirring time on distribution of particles in cast metal matrix composite. J Mater Process Technol 171(2):268–273
21. Reddy AC (2009) Mechanical properties and fracture behavior of 6061/SiCp metal matrix composites fabricated by low pressure die casting process. J Manufact Technol Res 1(3/4):273–286
22. Suresh S, Shenbaga Vinayaga Moorthi N, Vettivel SC, Selvakumar N (2014) Mechanical behavior and wear prediction of stir cast Al–TiB_2 composites using response surface methodology. Mater Des 59:383–396
23. Kaviyarasu K, Anand P (2012) One pot synthesis and characterization of caesium doped S_nO_2 Nano crystals via a hydrothermal process. J Mater Sci Technol 28:15–20

Erosion-wear Behaviour of 304 Stainless Steel Reinforced with TiN at Elevated Temperatures

Ramokone Marcia Mafafo, Babatunde Abiodun Obadele, Bruno Pilotti, Walter Roberto Tuckart, and Peter Apata Olubambi

Abstract The erosion-wear behaviour of spark plasma-sintered austenitic stainless steels matrix composite was investigated under high temperature conditions. Erosion-wear behaviour of the composites was conducted at a constant velocity of 18 m/s and impinging angle of 90° at different temperatures of 25, 400 and 600 °C. Hard abrasive alumina particles with particle size of 40 μm were used as erodent. After SPS, relative density of spark plasma-sintered samples decreased with increasing TiN content. Furthermore, the results show a reduction in material loss with increased hardness. However, there was a significant increase in material wastage as temperature increases. Microstructures of the eroded samples indicate that the erosion damage occurs mainly by plastic deformation and brittle fracture. Large fragments in the form of wear debris, pitting, ploughing with slip on the side, misplaced material and a few grooves are observed as well on the SEM images.

Keywords Erosion–corrosion · Spark plasma sintering · 304 stainless steel · Microstructure · Wear · Debris

1 Introduction

Austenitic stainless steels (ASSs) are extensively used in nuclear reactors, biomedical implants, as well as in components for chemical and food industries. They are the most popular corrosion-resistant materials used in various applications due to

R. M. Mafafo · B. A. Obadele (✉) · P. A. Olubambi
Centre for Nanoengineering and Tribocorrosion, School of Mining, Metallurgy and Chemical Engineering, University of Johannesburg, Doornfontein Campus, Johannesburg, South Africa
e-mail: obadele4@gmail.com

B. A. Obadele
Department of Chemical, Materials and Metallurgical Engineering, Botswana International University of Science and Technology, Palapye, Botswana

B. Pilotti · W. R. Tuckart
Tribology Group, Universidad Nacional del Sur/CONICET, Av. Alem 1253, 8000 Bahía Blanca, Argentina

S. Vijayan et al. (eds.), *Trends in Manufacturing and Engineering Management*, Lecture Notes in Mechanical Engineering,
https://doi.org/10.1007/978-981-15-4745-4_17

their excellent corrosion resistance and good mechanical properties [1]. Apart from applications in corrosive environment, high impact velocity of particles is frequently encountered in components used in these industries. Examples are coal-fired boilers used in the power generation industries [2]. Although austenitic stainless steels are well known for their good corrosion properties, they are famous for poor mechanical properties such as friction and wear. According to Sun and Bell [3], when ASSs are in contact with each other or other materials, they experience severe metallic wear as a result of adhesion junction and severe plastic deformation occurring between the surfaces of the materials in contact with each other. Furthermore, ASS fails mainly because of wear rather than corrosion-related material loss problems.

Wear can be defined as the loss of materials due to the mechanical action occurring between two or more bodies in contact with each other [4]. Wear phenomena include several mechanisms such as abrasive, adhesive, corrosive and erosion wear. Erosion-wear mechanism plays an important role in determining the life span of components used in several applications such as in mining, solids transportation in pipelines, power generation and marine [5]. Erosion wear is a material deterioration process due to impacting particles on solid surfaces.

Although there are literatures available on 304 steel reinforcement with ceramics and nitrides, there is still a need to study the effect of these reinforcements at elevated temperatures. This is important as there are very few literatures available on erosion-wear or erosion-corrosion performance of spark plasma-sintered 304 stainless steel composites at high temperatures. Therefore, this study focuses on the erosion-wear behaviour of sintered 304 steel under high temperature conditions.

2 Experimental Procedures

Spark plasma-sintered compacts of 304 austenitic stainless steel (ASS) without and with 2, 4 and 6% TiN were supplied by Centre for Nanoengineering and Tribocorrosion. The samples were sintered under high vacuum conditions using graphite die and punch at an applied pressure of 50 MPa, sintering temperature of 1000 °C, heating rate of 100 °C/min and holding time of 10 min. After sintering, the samples were sand-blasted to remove any excess graphite material on them. Thereafter, the substrates were machined using the robofil 240-SL electrical discharge machine (EDM) wire cut. The sintered samples were ground and polished to 3 μm diamond finish. The samples were further polished with fumed silica using a polishing suspension of 0.2 μm to achieve mirror-like surface. The morphology of the sintered specimen before and after erosion-wear test was investigated using scanning electron microscope model TESCAN Vega 3 which is equipped with energy-dispersive spectroscope. The Rigaku Ultima IV X-ray diffractometer XRD equipped with a graphite-monochromatic Cu Kα radiation source at 40 kV and 30 mA was used to perform the XRD analysis. Collection of diffractograms took place in the 2θ range between 3° and 90° with a step size of 0.01 [6].

Table 1 Chemical compositions of admixed powders

Erosion conditions	
Erodent material	Alumina particles
Particle size (μm)	45
Particle velocity (m/s)	18
Impact angle (°)	90
Sample temperature (°C)	25, 400, 600
Duration of impact (mins)	10

Erosion-wear testing was carried out using a high temperature erosion tester. The erosion test rig used in the study was designed at the Universidad Nacional Del Sur (UNS), Argentina. The equipment works similar with that of the ASTM G76-04 standards. Prior to erosion tests, specimens were cut into approximately 35 mm × 1.5 mm × 1.5 mm dimension, ground and polished using sand papers 180, 220, 400 and 600 and cleaned in acetone for 2 min in an ultrasonic cleaner. The erosion tester consists of particle feed inlet, air compressor and furnace. The alumina particles were used as erodent material and were accelerated through the feed inlet by compressed air at a constant speed of 18 m/s and this cause the abrasive particles to impact on the sample. Erosion-wear tests were carried out at room temperature, 400 and 600 °C, and the specimens were eroded for 10 min. Table 1 shows the operating solid particle erosion-wear parameters used in the study.

3 Results and Discussion

The effects of temperatures (25, 400 and 600 °C) on the rate of erosion of 304 austenitic stainless steel without and with TiN additions at an impact angle of 90° and an impact velocity of 18 m/s are presented in Fig. 1.

From the graphs, it could be observed that the erosion-wear behaviour of these materials shows a similar behaviour. The graphs show that material mass loss increases with increasing temperature and as a result, higher material loss is experienced at 600 °C while the least material loss is observed at room temperature. Material loss increases as the temperature increases. This is observed up to a maximum temperature where a point of maximum material loss is achieved. The total material loss between room temperature and 400 °C in all the samples is small in the same order (in the range of 10–15 mg) while the total material loss between 400 and 600 °C in all the samples rises sharply (in the range of 90–40 mg). In this case, not only erosion-wear occurs but also the combined effect of erosion and oxidation takes place at elevated temperatures [7].

The combined effect of temperatures on material loss as the amount of TiN hard phase is increased in the substrate is presented in Fig. 2. From the graph, it could be observed that at 25 °C, the materials show almost similar weight loss between 28.9 and 31.4 mg. As the temperature increases, more material loss is experienced

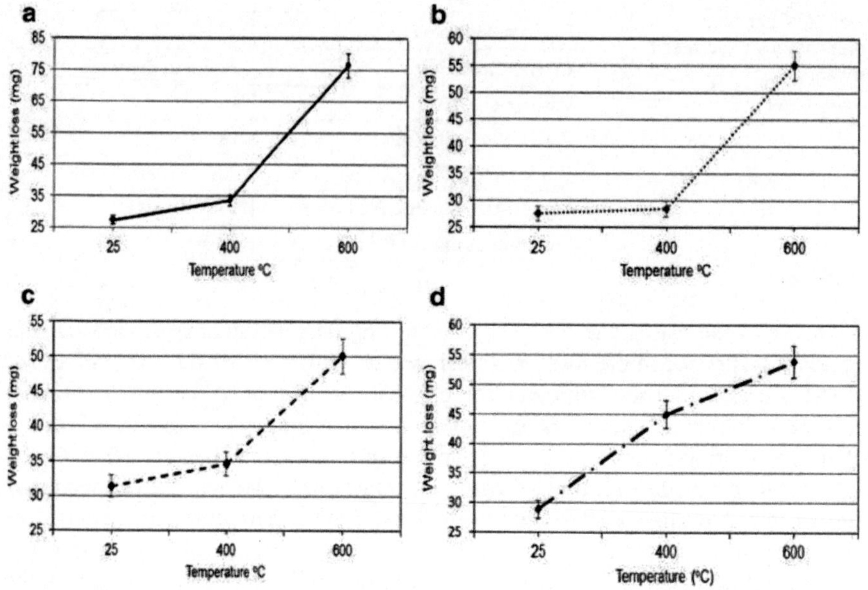

Fig. 1 Temperature versus weight loss of **a** 304 ASS **b** 304/2TiN **c** 304/4TiN and **d** 304/6TiN

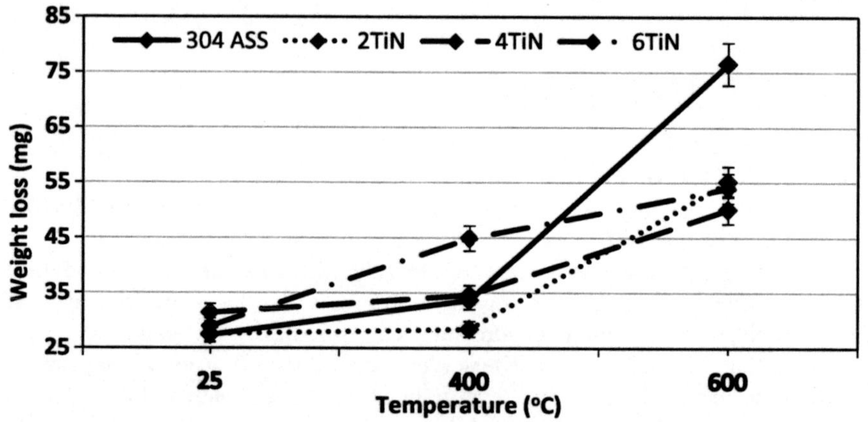

Fig. 2 Combined effect of temperature on material loss

on 304/6TiN. This could be due to the formation of debris that could contain TiN hard phase which can act as a third body. Since there is a high content of TiN in 304/6TiN, it could translate to higher TiN content as a third body. Therefore, this may promote higher weight loss. On the other hand, there could be particle loosening along the grain boundaries at high temperature and as a result, an increase in wear phenomena on the eroded surface. At 600 °C, it could be observed that 304 ASS with TiN additions display similar trend and closely related weight loss between 50 and 55.15 mg. However, for 304 ASS without TiN at 600 °C, it was observed that weight loss was the highest (76 mg) compared to 304 ASS with TiN. This could be due to sintered 304 ASS ductility behaviour as the temperature increase, thereby a loss in hardness properties [7, 8].

Figure 3 further compares the effect of temperature on the hardness of the sintered composites under erosion-wear test. Higher microhardness values are recorded at room temperature. At higher temperatures (400 and 600 °C), all the samples recorded a significant drop in microhardness values. According to studies reported by Huttunen-Saarivirta et al. [7], when a material is subjected to high temperature effect, it becomes soft and loses its hardness properties. These findings are in full agreement with the results indicated in Fig. 3.

Figure 4 compares the effect of temperature (25, 400 and 600 °C) on the erosion-wear behaviour of the composites. The rate of erosion wear indicates that there is a proportional relationship between drop in microhardness values (shown in Fig. 3) and the weight loss, as the percentage of TiN in the material increases.

This indicates that at low temperature, the sintered composites are less prone erosion-wear. On the other hand, an increase in temperature raises the materials loss. However, addition of TiN to 304 matrix reduced material loss at 600 °C from 76 mg (for 304 ASS without TiN) to 55 mg. This shows a correlation between TiN content in the metal matrix hardness values and weight loss.

Visual aid image of sintered composite surfaces before and after erosion-wear are presented in Fig. 5. Figure 5a shows the material prior to erosion-wear tests while Fig. 5b–d shows the materials after being subjected to erosion-wear at 25, 400 and 600 °C, respectively. In every case, erosion-wear region shows a circular shape at the centre and spread towards the edge of the samples. Furthermore, three layers could be identified: the first layer is at the centre, where most of the material is removed; second layer where less material is removed and lastly, the third layer,

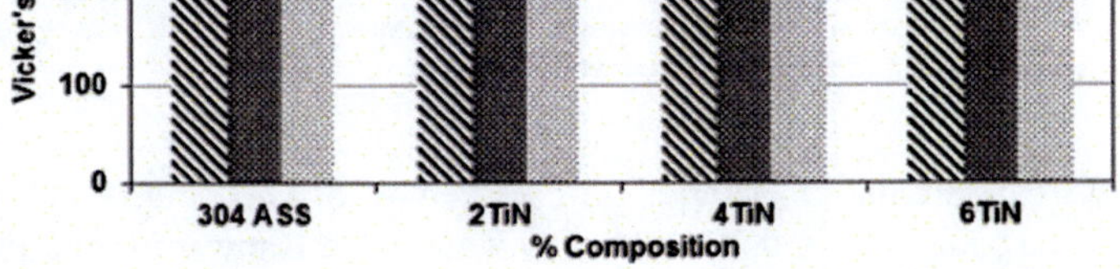

Fig. 3 Microhardness values of eroded 304 SS without and with TiN addition at different temperatures

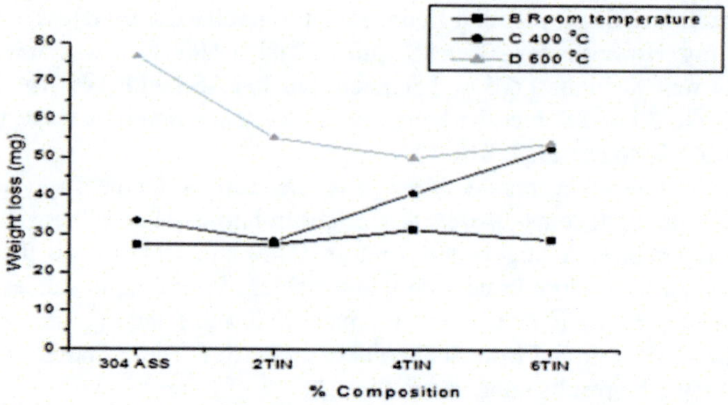

Fig. 4 Comparison of material loss in spark plasma-sintered 304 stainless steel without and with TiN

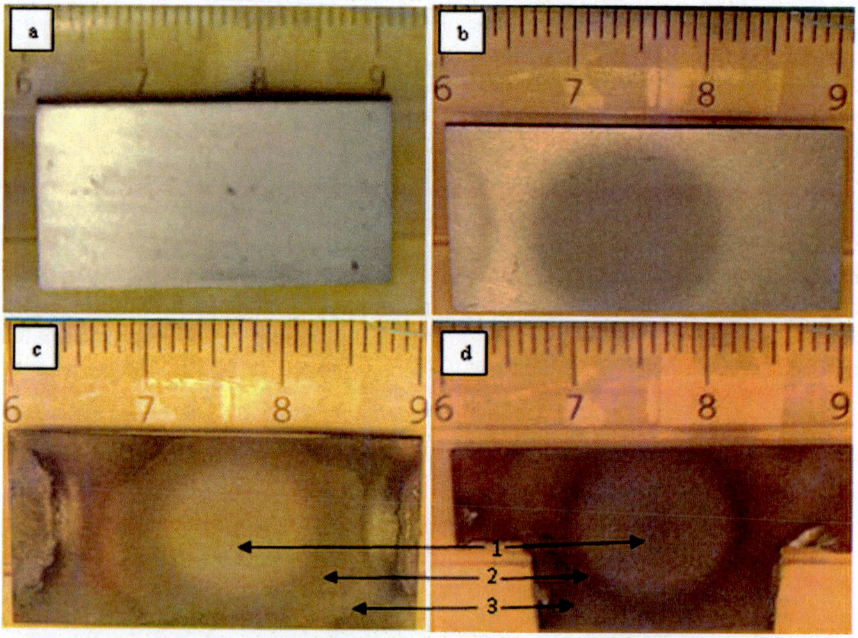

Fig. 5 Visual inspection of sintered specimen; **a** Before erosion-wear **b** Erosion-wear at 25 °C **c** Erosion-wear at 400 °C and **d** Erosion-wear at 600 °C

where very small amount of material has been removed. This could be clearly seen in Fig. 5d. This behaviour is also reported by Kumar and Kanwar [9] and Patel et al. [10]. Similarly, Kumar and Kanwar [9] found that an erosion study carried out at an impact angle of 90° results in circular shape on the region where erosion has taken place while at an impact angle of 30°, an elliptical shape is observable on the surface where erosion has occurred.

SEM micrographs of the eroded composites are displayed in Fig. 6 at 600 °C condition. The SEM of eroded samples provides useful information about the mechanism of erosion. According to Javaheri et al. [11], the mechanism at which erosion occurs by solid particle is broadly divided into two mechanisms, namely "cutting" and "deformation". On the damaged surfaces, pitting action is clearly observed. Materials were predominantly damaged by brittle fracture and plastic deformation. Large fragments of wear debris could be observed in all the samples and plastic deformation in the form of ploughing, misplaced material, micro-cut and ploughing is present. The results found in this study are in line with those conducted by Laguna-Camacho et al. [12] and Qin et al. [13]. Sintered 304 (Fig. 6a) shows an unconventional wear mechanism which was characterized by large fragments on the surface of the eroded surface, and this is an indication that the removal mechanism is through brittle fracture. Also, pitting behaviour was observed. Elsewhere, Bahri et al. [14] revealed that materials from eroded surface of 304 L SS had been removed due to ploughing and pitting. Patel et al. [10] observed that in a steady-state condition, a simultaneous action of three phases occurs on the surface of 304 steel at different locations.

During the initial phase, a crater is formed by the impacting particles and the material is displaced from the specimen. The second phase involves the material being displaced due to subsequent impact and this result in lateral material displacement

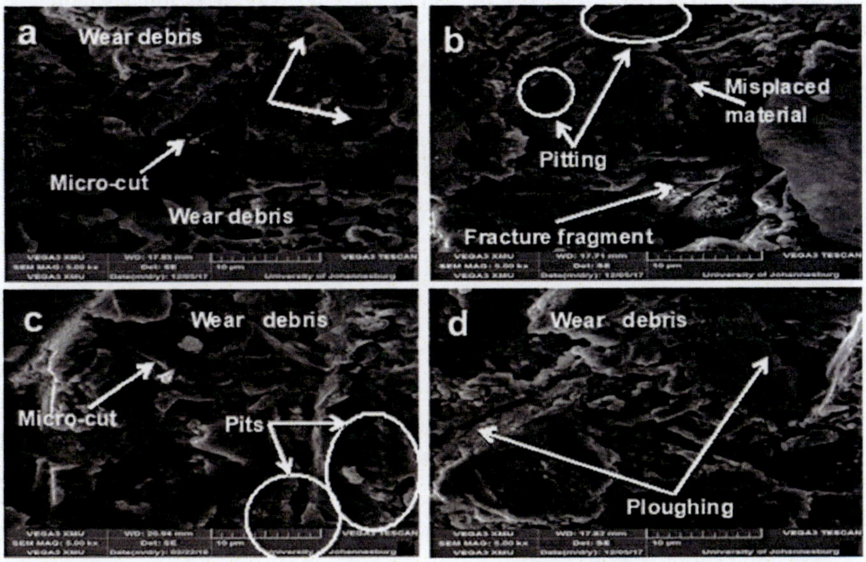

Fig. 6 SEM micrographs of **a** 304 ASS **b** 304 ASS/2TiN **c** 304 ASS/4TiN and **d** 304 ASS/6TiN at 600 °C

accompanied by fracture. The third stage is experienced after few impacts where the materials become strained and therefore removed from the surface by fracture. Qin et al. [13] conducted a study on the erosion-wear and intergranular corrosion resistance properties of AISI 304 L austenitic stainless steel after low-temperature plasma nitriding. It was reported that the most erosion damage taking place on the surface of the materials includes ploughing action, embedded particles, cutting grooves, large fragments, pitting action and wear debris on the surface of the sample that was nitride at 673 K.

4 Conclusion

In this study, solid particle erosion-wear behaviours of spark plasma-sintered 304 ASS, 304+2TiN, 304+4TiN and 304+6TiN were conducted at different temperature (25, 400 and 600 °C) conditions at an impact angle of 90° and a constant impact velocity of 18 m/s. We examined the effects of materials and temperatures on erosion-wear. An increase in temperature resulted in decrease in hardness properties of the composites. The erosion-wear rate indicated that there is a proportional relationship between decreasing percentage of hardness and material loss. Macrographs of the eroded specimens showed three layers in which the first (central) layer represents high material loss, the second layer showed lesser amount of material wastage and the third (outer) layer showed least amount of material. The SEM images indicated that the erosion damage occurs mainly by plastic deformation and brittle fracture. Large fragments in the form of wear debris, pitting, ploughing with slip on the side, misplaced material and a few grooves are observed.

Acknowledgements The authors are grateful to Tribology Group, Universidad Nacional del Sur, Bahía Blanca, Argentina for the use of equipment and facilities. The Institute of Nano Engineering Research (INER), Tshwane University of Technology, Pretoria, for the use of sintering and characterization facilities.

References

1. Obadele BA, Andrews A, Shongwe MB, Olubambi PA (2016) Tribocorrosion behaviours of AISI 310 and AISI 316 austenitic stainless steels in 3.5% NaCl solution. Mater Chem Phys 171:239–246
2. Samandi M, Shedden B, Smith D, Collins G, Hutchings R, Tendys J (1993) Microstructure, corrosion and tribological behaviour of plasma immersion ion-implanted austenitic stainless steel. Surf Coat Technol 59:261–266
3. Sun Y, Bell T (2002) Dry sliding wear resistance of low temperature plasma carburised austenitic stainless steel. Wear 253:689–693
4. Bayer RG (2002) Tribological and wear testing. characterization of materials 1–16
5. Chawla K, Singh K, Saini N, Singh J (2013) Erosion wear behavior of chromium coated steel 304 and grey cast iron. Int J Innovative Res Sci Eng Technol 2:4567–4575

6. Dada OR (2015) Characterization and beneficiation of foundry greensand waste from ferrous foundries. University of Johannesburg
7. Huttunen-Saarivirta E, Antonov M, Veinthal R, Tuiremo J, Mäkelä K, Siitonen P (2011) Influence of particle impact conditions and temperature on erosion–oxidation of steels at elevated temperatures. Wear 272:159–175
8. dos Santos JF, Garzón CM, Tschiptschin AP (2004) Improvement of the cavitation erosion resistance of an AISI 304 L austenitic stainless steel by high temperature gas nitriding. Mater Sci Eng A 382:378–386
9. Kumar N, Kanwar R (2012) To study erosion behavior of Cr_2O_3 Coating on SS-304 boiler steel tubes in simulated coal fired boiler conditions
10. Patel M, Patel D, Sekar S, Tailor P, Ramana P (2016) Study of solid particle erosion behaviour of SS 304 at room temperature. Procedia Technol 23:288–295
11. Javaheri V, Porter D, Kuokkala V-T (2018) Slurry erosion of steel–review of tests, mechanisms and materials. Wear 408:248–273
12. Laguna-Camacho JR, Vite-Torres M, Gallardo-Hernández E, Vera-Cárdenas E (2013) Solid particle erosion on different metallic materials. Tribology in Engineering. InTech
13. Qin X, Guo X, Lu J, Chen L, Qin J, Lu W (2017) Erosion-wear and intergranular corrosion resistance properties of AISI 304 L austenitic stainless steel after low-temperature plasma nitriding. J Alloy Compd 698:1094–1101
14. Bahri A, Guermazi N, Elleuch K, Ürgen M (2016) On the erosive wear of 304 L stainless steel caused by olive seed particles impact: Modeling and experiments. Tribol Int 102:608–619

A Concise Review of Nano-enhanced Phase Change Materials for Passive Cooling Applications in Buildings

Chukwumaobi K. Oluah, Esther Titilayo Akinlabi, and Howard O. Njoku

Abstract Building energy consumption has been on a steady rise over the past decade. This is due to the ever-growing human population and the effect of climate change. Space conditioning accounts for over 40% of the final global energy production. Thermal energy storage (TES) systems have shown to be quite effective in reducing buildings energy demand while maintaining the desired comfort. It finds a useful application in solar thermal systems as and heat recovery systems. Phase change materials are used in building components for storing excess heat energy for use when the need arises. The ability of PCMs to store great quantity of heat in comparatively small volume makes them suitable for this application. Some of the set backs encountered by using PCMs is its poor thermal conductivity and sub-cooling characteristics. Nanoparticles have been found very useful in ameliorating the thermal conductivity of PCMs thereby improving its general performance. This review work brings to bear recent developments on the use of nanoenhanced PCMs for passive cooling applications in buildings, and there are very little articles in literature that emphasizes the application of nanoPCMs in passive cooling systems. We further discussed the physics of operation and phase transition behaviours with mathematical representations. Some gaps currently encountered in the research of nanoPCMs was also highlighted for further research purposes.

Keywords Phase change materials (PCMs) · Thermal energy storage (TES) · Nanomaterials · Passive cooling

1 Introduction

Space conditioning in buildings have been identified as the major consumer of the final global energy production with respect to buildings. This pattern is the same for both residential and commercial buildings. This figure was reported to be about

C. K. Oluah (✉) · E. T. Akinlabi · H. O. Njoku
Department of Mechanical Engineering Sciences, University of Johannesburg, Johannesburg 2006, South Africa
e-mail: chukwumaobi.oluah@unn.edu.ng

S. Vijayan et al. (eds.), *Trends in Manufacturing and Engineering Management*, Lecture Notes in Mechanical Engineering,
https://doi.org/10.1007/978-981-15-4745-4_18

40% of annual global energy production by the international energy agency [1]. A greater majority of the global energy resource is fossil-based and comes with increased CO_2 production. The rapidly growing population of the world and the effect of climate change account for the rise in energy demand. Some of the adverse effects of global warmingas listed by the IPCC include but not restricted to the following: a continuous rise in temperature, increased hurricane, more draught, and heat waves etc. [2]. It is due to this cause that scientists have sought to reduce global CO_2 emission while maintaining reasonable thermal comfort. Over the past decades, researches have studied various novel renewable approaches to meet the energy requirements of residential or commercial buildings while keeping the environment safe. Thermal energy storage (TES) is one way, that many researchers have proposed, that can sufficiently reduce building's energy demand. It finds a useful application in solar thermal systems as well as heat recovery systems [3]. Thermal energy storage (TES) mediums are green systems which results in zero-emission of carbon into the atmosphere. Thermal energy storage (TES) systems absorbs heat, stores it and releases it when required for use; this helps in load shedding in buildings. TES is further classified based on their mode of operation as: chemical heat storage (CHTES) sensible heat storage (SHTES) or latent heat storage (LHTES) [4].

2 Phase Change Materials: Definition, Physics of Operation and Classification

Phase change materials change their physical form upon absorption or release of energy. This energy could be in the form of heat of fusion or heat of vaporization; their ability to store substantial amounts of thermal energy in proportionately small volume makes them fit for a broad range of applications. When the PCM absorbs energy from the ambient, the chemical bonds of the PCM breaks up and this leads to a change of phase. The process of the phase change is an endothermic process; therefore, heat is absorbed from the ambient causing a drop in the bulk temperature of the surrounding. When there is a drop in the ambient temperature, the PCM gives up its energy and returns to the solid state, thereby increasing the ambient temperature of the room [5]. PCMs are generally classified by considering their chemical composition as organic, inorganic or eutectic compounds [6]. Summarized below are some advantages and disadvantages of different classes of PCMs also some desired properties of PCMs (Table 1).

Some examples of the three classes PCMs applicable to buildings are outlined in the Table 2.

Table 1 Advantages and disadvantages of PCMS [6]

Types of PCM	Advantages	Disadvantages
Organic	Organic PCMs have the ability to freeze with very little supercooling or none as the case may be	The thermal conductivity of organic PCMs in their solid state are very low
	Ability to be incorporated directly	Some organic PCMs are insoluble in water
	Good thermal properties	The heat of fusion is low
	Uniformity in melting	Less dense
Inorganic	High thermal conductivity	High degree of supercooling
	Great heat storage capacity	Reduced thermal stability
	Low volume expansion upon a change of phase	Phase segregation
Eutectics	Definite melting temperature	Thermophysical properties of eutectic PCMs are not well established yet
	High specific heat capacity	Low enthalpy

Table 2 PCMs for cooling applications in buildings [7]

Compounds		Temp (°C) of fusion	Latent heat (KJ/Kg)	Ref
Organic	Dimethyl sabacate	20.8	120	[10]
	Polyglycol (E 600)	22	127	[13]
	Paraffin (C_{13}–C_{24})	22–24	189	[10]
Inorganic	$CaCl_2 \cdot 6H_2O$	29	190.8	[13]
	$Mn(No_3)2 \cdot 6H_2O$	25.8	126	[10]
	$LiNO_3 \cdot 3H_2O$	30	296	[10]
Eutectic	33.3% $MgCl_2$–$6H_2O$ + 66.6% $CaCl_2 \cdot 6H_2O$	25	127	[13]
	48% $CaCl_2$ + 4.3% NaCl + 0.4% KCl + 47.3% H_2O	27	188	[13]
	60% $Na(CH_3COO) \cdot 3H_2O$ + 40% $CO(NH_2)_2$	30	200	[13]

3 NanoPhase Change Materials

The foregoing discussions have revealed how phase change materials play a very crucial role in latent heat energy storage and dispensing systems. There exist various solar-deriventhermal technologies that employ the use of PCMs in both active and passive capacities. Enhancing the heat storage and dispensing capabilities of PCMs can be achieved by doping pure PCM with nanoparticles [8]. Metals, metallic oxides

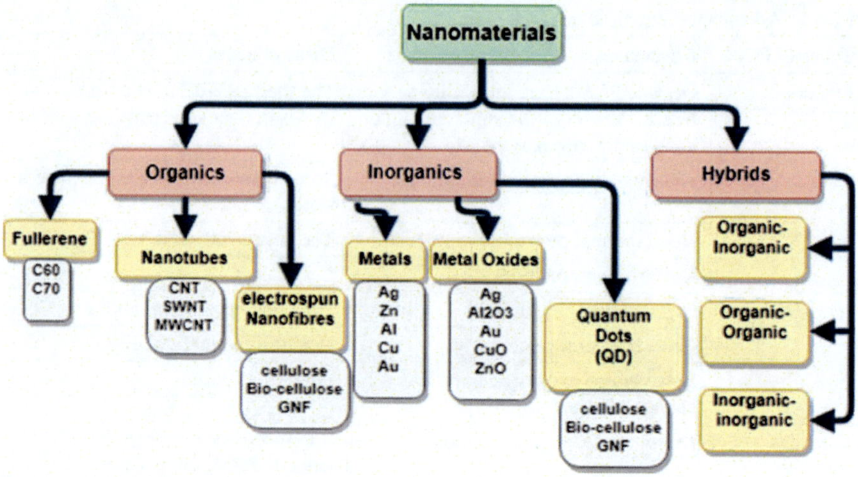

Fig. 1 Classification of nanomaterials

and carbonic nanoparticles have all been introduced in various quantities to boost the heat conduction abilities of PCMs]. Babapoor et al. further established that nanoparticles improved the heat transfer abilities of PCMs, and there seem to be a dilemma in the discourse of the outcome of nanoparticles on the heat of fusion of phase change materials, while a few authors presented results of positive improvement on the latent heat of their materials, and some others had shown a gradual decline of the latent heat value of the nanoPCM with increased quantities of nanoparticles. The duration of phase change, density and the viscosity have also been improved with nanoparticles. [9].

3.1 Classification of Nanomaterials

Nanomaterials are broadly classified as organic, inorganic and hybrids just like phase change materials [8]. Further sub-classifications are identified in the chart below (Fig. 1).

3.2 Applications of Nanomaterials

Nanomaterials find great applications in basically all spheres of human endeavor, some of the major applications are outlined in the chart below [8] (Fig. 2).

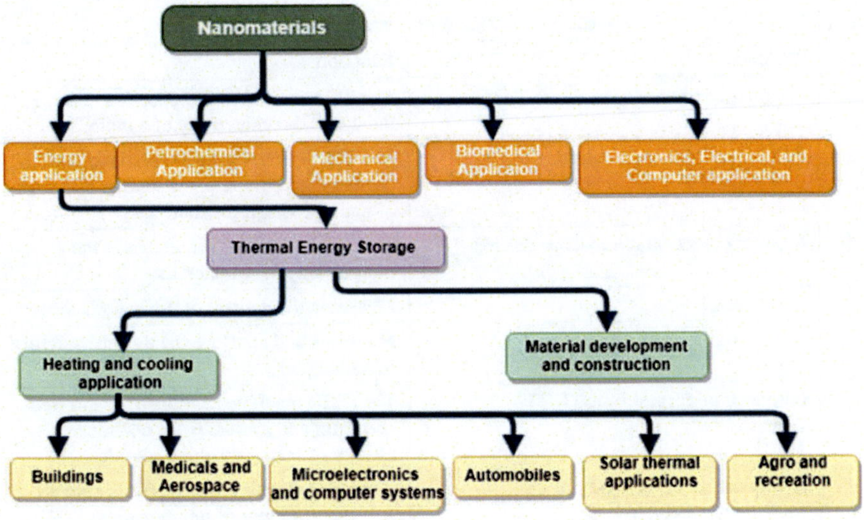

Fig. 2 Applications of nanomaterials

3.3 Preparation of NanoPCMs

There are two general methods scientists use for the preparation of nanoenhanced phase change materials; these methods are the one-step method and the two-step method. The first method being the one-step method entails the concurrent production of the nanoparticles into the base PCM while in the two-step approach, the nanomaterials are first produced before introduction into the base material [10].

3.4 Desired Properties of Nanoenhanced PCM

The essence of doping phase change materials with nanoparticles is to improve its properties leading to a better performance upon application. Below are some desirable characteristics of nanoPCMs.

Suitable Phase Change Temperature
The building needs to be maintained at a temperature interval within the human comfort limits; this means that phase change materials in use should have a phase change temperature within the interval of application required. [11].

Good Thermal Conductivity
One of the major challenges associated with PCMs are low thermal conductivity, doping with nanoparticles greatly improves the heat transfer potentials of PCMs [12].

Table 3 Summary of characterization techniques of nanoenhanced PCM [13]

s/n	Technique	Function
1	Thermal conductivity analyser (TCA)	The TCA is used to measure and analyse the thermal conductivity of nanoPCMs
2	Differential scanning calorimeter (DSC)	The DSC is used to ascertain the phase change temperature and enthalpy of fusion of nanoPCM/sample materials
4	Scanning electron microscope (SEM)	The SEM examines the image of the morphology nanomaterials
6	Hydrometer	Measures the density of the nanomaterial
7	Thermal gravimetric analyser (TGA)	Measures and analyses the thermal stability of nanomaterials
8	Fourier transform infrared (FTIR) spectrometer	The FTIR machine examines the infrared spectrum and evaluates the chemical affinity of materials being sampled
9	X-Ray diffraction (XRD)	The XRD machine measures the crystal size and structure of *nanoparticles*

Higher Latent Heat

The potential of PCMs to retain a great quantity of heat in a relatively small volume is quite desirable. It makes them suitable for quite a number of applications.

3.5 *Characterisation of NanoPCM*

Nanophase changes materials after preparation are characterized to ascertain their properties. Various techniques are used to characterize nanophase change materials some of which are summarised in the Table 3.

3.6 *Heat Transfer Models of Nanoparticles in Base PCM*

The two-dimensional energy balance equation of heat flow in the PCM containing nanoparticles is given below

$$\rho c_p \frac{\partial T}{\partial t} + (\rho c_p)\left[u\frac{\partial T}{\partial x} + v\frac{\partial T}{\partial y}\right] = k\left[\frac{\partial^2 T}{\partial x^2} + \frac{\partial^2 T}{\partial y^2}\right] \tag{1}$$

where u, v, t, ρ, k, C_p, T, are the velocity components of the momentum equation in the cartesian coordinate system, time, density, thermal conductivity and specific heat at constant pressure and temperature. The phase change in PCMs occurs over a range of temperature denoted as ΔT_m. Approximations for the thermophysical properties like density (ρ), heat capacity (ρC_p) and thermal conductivity (k) due to change in phase presented in Eqs. (2)–(4) below.

$$\rho = \begin{cases} \rho_{ns} & T < T_m \\ \frac{\rho_{ns}+\rho_{nl}}{2} & T \leq T \leq T_m + \Delta T_m \\ \rho_{nl} & T \geq T_m + \Delta T_m \end{cases} \quad (2)$$

$$k = \begin{cases} k_{ns} & T < T_m \\ \frac{k_{ns}+k_{nl}}{2} & T_m \leq T \leq T_m + \Delta T_m \\ k_{nl} & T \geq T_m + \Delta T_m \end{cases} \quad (3)$$

$$\rho C_p = \begin{cases} (\rho C_p)_{ns} & T < T_m \\ \frac{(\rho C_p)_{ns}+(\rho C_p)_{nl}}{2} + \frac{\rho_{ns}+\rho_{nl}}{2}\left(\frac{L_{hf,n}}{\Delta T_m}\right) & T \leq T \leq T_m + \Delta T_m \\ (\rho C_p)_{nf} & T \geq T_m + \Delta T_m \end{cases} \quad (4)$$

Subscripts "*ns*" and "*nl*" refers to the state properties of the nanoPCM in solid and liquid forms. $L_{hf,n}$ in Eq. (4) above represents the latent heat of fusion of the nanoPCM and expressed mathematically as $L_{hf,n} = (1 - \varphi)h_{sf}$ and φ is the weight percentage of nanoparticles contained in the base materials [14].

4 Nanoenhanced PCMs for Passive Cooling Applications

Introducing small quantities of nanoparticles in base PCMs to improve its thermophysical properties has gained momentum in the recent past. Nanomaterials are currently being used in virtually all fields of science and technology. The essence of this review is to highlight works done on various nanomaterials for passive cooling applications in building. For the want of space, these works have been summarized in a tabular form below (Table 4).

4.1 Challenges and Scope for Future Research

The use of nanoparticles have greatly improved the thermal conductivity of PCMs, however, some researchers reported a decline in latent heat as the concentration of nanodopant increased while some others reported a linear relationship with the increased concentration of nanomaterials; it is therefore important that a validation study is carried out. A research on optimum particle loading for best thermal performance of nanoPCMs should also be investigated. Health risks involved in integration on nanoparticles in building envelopes should be studied and safety measures are advised accordingly.

Table 4 Summary of recent studies on nanoenhanced PCM for passive cooling applications

Nanoparticle	Base material	Application	Major findings	References
TiO_2	Mortar bricks	Space cooling in buildings	Resistance to thermal conductivity and thermal transmissivity coefficient for TiO_2-coated mortar wall showed a reduction by 10% and 13%, respectively	[15]
TiO_2	Paraffin wax	Ceiling roofs for space cooling	An optimum value of heat capacity was attained at 0.7% loading of TiO_2. Thermal conductivity showed a liner increment as the concentration of nanoparticle increased	[16]
Aluminiumnanoparticles	PEG + Epoxy resins	Building material	Improved thermal conductivity due to nanoparticles addition	[17]
Al_2O_3 + Alumina	n-tetradecane +n hexadecane	Space cooling in buildings	Increasing the amount of nanoparticles in base PCM resulted to an improved solidification and melting rate	[18]
Graphite nanosheet	Fatty acid	Gypsum wallboards	A reduction in the energy consumption by 79% was recorded for the interior temperature regulation	[19]
AgO and TiO_2	Ethyl trans-cinnamate	CTES in building	52% improvement in thermal conductivity, improved energy efficiency and an overall freezing rate reduction of 23.9%	[20]
AgO	ICE block	VAV-NTES air conditioning system	Achieved energy savings of 36–58% and 24–51%, during peak load and everyday basis, respectively	[21]

(continued)

Table 4 (continued)

Nanoparticle	Base material	Application	Major findings	References
Nanoencapsulation	PMMA/capric–stearic acid	CTES in buildings	The TGA results showed great thermal conductivity, chemically stable and excellent thermal reliability after thermal cycling	[22]
Graphene	Beeswax	CTES in buildings	An optimum thermal conductivity of 2.8 W/mK was achieved at 0.3 wt% of graphene and a 22.5% increase in heat capacity were recorded	[23]
SiO_2	Stearic acid, capric acid, palmitic acid, and (SA-PA -CA)	CTES in Buildings	The eutectic PCM has a temperature of fusion between the range of 17.16–26 °C. The presence of silicon dioxide greatly improved its thermal conductivity and shape stabilized	[24]
Activated carbon (AC)	Oleic and capric acid	CTES in buildings	The presented showed that, upon addition of 0.1% of activated carbon, a corresponding 55% improvement in the thermal conductivity of the pure PCM	[24]
Expanded graphite	Capric-lauric acid (CA-LA)	CTES in buildings	A 114.2% improvement in thermal conductivity upon addition of 10 wt% of EG	[25]

5 Conclusion

A review of phase change materials doped with varying quantities of nanoparticles was discussed in this article; research on the use of nanomaterials to improve the performance of PCMs is relatively not a new concept, but recent advances in the manufacturing and doping of substances with nanomaterials have made this field of research gain momentum in the recent past. One major application of nanoenhanced PCMs is in cool thermal energy storage (CTES) in buildings, this review was geared towards evaluating studies on nanoPCMs for passive cooling applications in buildings; Futhermore, provided an insight into the current challenges of nanoPCMs in buildings and research gaps.

References

1. IEA Roadmap targets. https://www.iea.org/publications/freepublications/publication/IEASolar_HC_Roadmap_FoldOutPrint.pdf
2. IPCC (2007) Summary for policymakers. In: Climate change 2007: impacts, adaptation and vulnerability. contribution of working group II to the fourth assessment report of the intergovernmental panel on climate change. Cambridge University Press, Cambridge, UK, p 17
3. Oysterman E et al (2012) Review of PCM based cooling technologies for buildings energy and buildings
4. Hassan N, Mariah B, OSORIO FJB, Marllory I, Xinhai Y, Vignarooban K, Patrick P, Kannan IAM (2019) Recent developments in phase change materials for energy storage applications: a review. Int J heat mass transf 129:491–523
5. Hasnain SM (1998) Review on sustainable thermal energy storage technologies, Part I: heat storage materials and techniques. Energy Convers Manag 39(11):1127–1138. https://doi.org/10.10106/S0196-8904(98)00025-9
6. Feldman D, Shapiro M, Banu D (1986) Organic phase change materials for thermal energy storage. Sol Energy Mater 13:1–10
7. Lane GA (1980) Low temperature heat storage with phase change materials. Int J Ambient Energy 1:155–168
8. Hussain H, Saw C (2013) Afolabi Lreview on nanomaterials for thermal energy storage technologies. Nanosci Nanotechnol-Asia 3:60–71
9. Babapoor A, Karimi G, Sabbaghi S (2016) Thermal characteristic of nanocomposite phase change materials during solidification process. J Energy Storage 7:74–81
10. Farid MM, Khudhair AM, Razack SAK, Al-Hallaj S (2004) A review on phase changeenergy storage: materials and applications. Energy Conv Manag 45(9–10):1597–1615
11. Zhu Y, Qin Y, Wei C, Liang S, Luo X, Wang J, Zhang L (2018) Nanoencapsulated phasechange materials with polymer- SiO_2 hybrid shell materilas: compositions, morphologies and properties. Energy Conv Manag 164:83–92
12. Yu W, Xie H (2012) A review on nanofluids: preparation, stability mechanisms, and application. J Nanomaterials. Article ID 435873
13. Sari A, Alkan C, Doguscu DK, Biçer A (2014) Micro/nano-encapsulated n-heptadecane with polystyrene shell for latent heat thermal energy storage. Solar Energy Mater Solar Cells 126:42e50
14. Harikrishnan MD, Kalaiselvam S (2014) Thermal and rheological properties of hybrid nanocomposite phase change material for thermal. Energy Convers Manage 86:864–872

15. Carneiro JO, Vasconcelos G, Azevedo S, Jesus C, Palha C, Gomes N, Teixeira V (2014) Energy Build 81:1–8
16. Wang J, Xie H, Guo Z, Guan L, Li Y (2014) Appl Thermal Eng 73(2):1541–1547
17. Constantinescu M, Dumitrache L, Constantinescu D, Anghel E, Popa V, Stoica A, Olteanu M (2010) Latent heat nano composite building materials. Eur Polym J 46(12):2247–2254
18. Kalaiselvam S, Parameshwaran R, Harikrishnan S (2012) Analytical andexperimental investigations of nanoparticles embedded phase change materials for coolingapplication in modern buildings. Renew Energy 39(1):375–387
19. Sayyar M, Weerasiri RR, Soroushian P, Lu J (2014) Experimental and numericalstudy of shape-stable phase-change nanocomposite toward energy-efficient buildingconstructions. Energy Build 75:249–255
20. Parameshwaran R, Deepak K, Saravanan R, Kalaiselvam S. Preparation
21. Parameshwaran R, Kalaiselvam S (2014) Energy conservative air conditioning system using silver nano-based PCM thermal storage for modern buildings. Energy Energy Storage. Appl Energy 115:320–330. Build 69:202–212
22. Sarı A, Alkan C, Özcan AN (2015) Synthesis and characterization of micro/nano capsules of PMMA/capric–stearic acid eutectic mixture for low temperature-thermal energy storage in buildings. Energy Build 90:106–113
23. Amin M, Putra N, Kosasih EA, Prawiro E, Luanto RA, Mahlia T (2017) Thermal properties of beeswax/graphene phase change material as energy storage for buildingapplications. Appl Therm Eng 112:273–280
24. Luo Z, Zhang H, Gao X, Xu T, Fang Y, Zhang Z (2017) Fabrication andcharacterization of form-stable capric-palmitic-stearic acid ternary eutectic mixture/nano-SiO_2 composite phase change material. Energy Build
25. Hussain SI, Dinesh R, Roseline AA, Dhivya S, Kalaiselvam S (2017) Enhancedthermal performance and study the influence of sub cooling on activated carbon dispersedeutectic PCM for cold storage applications. Energy Build 143:17–24

Effect of Silicon Carbide in Yttria-stabilized Zirconia for Thermal Protective Structural Materials

S. Subha, Poojari Surendra, Ch. Gowtham Chowdary, V. Naga Venkata Sai Ram, and Palem Srikanth

Abstract Modern coating technology development is essentially determined by systems, which perform extreme and adverse operating conditions. The ceramic coating technology (CCT) has found applications in space vehicles, missiles, gas turbine engines, and nuclear reactors for the thermal protective structural material. In present study, theyttria-stabilized zirconia (YSZ) and with various weight percentages of 0.1 and 0.25% silica carbide nanoparticles were synthesized by a wet chemical process. The morphological and thermal properties of the prepared hybrid ceramic coating particles were studied using scanning electron microscope (SEM) and thermogravimetric analysis (TGA). The ablation performances of the hybrid coating samples were tested using oxyacetylene flame test. The grain size and crystal structure of samples were confirmed by SEM analysis and showed that the synthesize hybrid particles are composed of agglomerated nanometer sub particles. The results of TGA showed the major weight mass loss observed around 604 °C for pure YSZ and in the range of 650 °C for hybrid coating cured composites. The linear and mass ablation rates of the prepared samples were significantly decreased by 43% and 18.4%, respectively, on adding the 0.25 wt% of SiC hybrid filler.

Keywords Thermal protective coating · Silicon nanocarbide · Thermal stability · Ablation rate

1 Introduction

Zirconia-based thermal protective coatings are widely used for the thermal, oxidation, and hot corrosion protection of high temperature components in gas turbine engines. These coatings provide insulation to metallic structures in the hot section of turbine engines and offer increased operating temperature of the engine-enhanced efficiency, extended durability, and extended life of metallic components subjected to high

S. Subha (✉) · P. Surendra · Ch. Gowtham Chowdary · V. Naga Venkata Sai Ram · P. Srikanth
School of Aeronautical Sciences, Hindustan Institute of Technology and Sciences, Chennai, India
e-mail: subhasvkmr@gmail.com

S. Vijayan et al. (eds.), *Trends in Manufacturing and Engineering Management*, Lecture Notes in Mechanical Engineering,
https://doi.org/10.1007/978-981-15-4745-4_19

temperatures [1–3]. Several impurities common in fuels have been identified and associated with corrosion in the coatings. These impurities include sodium, sulfur, phosphorous, and especially vanadium. These impurities react with convectional YSZ turbine blade coating, severely limit the coating lifetime. Therefore, it is of great interest to develop alternative materials that increase the operating lifetime of the coating [4].

New zirconia-based ceramic compositions have been under investigation for TBC applications since many years. Previous work focuses on studying stabilizers such as Y_2O_3 and MgO were susceptible to chemical attacks. Indium-stabilized zirconia shows relative inertness in vandia containing atmosphere but it volatilizes at a lower temperature than zirconia [5]. Graphene-based materials are single- or few-layer platelets that can be produced in bulk quantities by chemical methods. Thermal and barrier properties of these composites are dependent upon the intrinsic properties of graphene-based materials and their state of dispersion in the matrix [6]. Nanostructurally stabilized zirconium oxide (NSZ) hard transparent films were produced without chemical stabilizers by the ion beam-assisted deposition technique (IBAD). These NSZ films have better thermal stability, probably because of the absence of porosity due to the nature of the IBAD process [7]. In another study, sol–gel method has been used to coat multiwalled carbon nanotubes (MWCNTs) with zirconia at room temperature and the coated tubes were sintered at high temperature. Sintering has improved the adhesion of the coating with CNTs and thereby improved its high temperature resistance. It is also observed that the selecting proper sintering temperature is the key to ensure a uniform coating layer on the CNTs [8].

The aim of this study is to synthesis the yttria-stabilized zirconia with silicon nanocarbide material through wet chemical method. The morphology and thermal properties of the prepared hybrid ceramic coating particles were studied by using scanning electron microscope (SEM) and thermogravimetric analysis (TGA), respectively. The ablation performances of the hybrid coating samples were tested using oxyacetylene flame test.

2 Experimental

2.1 Materials

SiC of average particle size 50 nm used for composite preparation was obtained from Sisco Research Chemicals Co. Ltd. Yttria-stabilized zirconia (YSZ) of average particle size 3 μm was procured from Sigma Aldrich Co Ltd.

2.2 *Preparation of Hybrid Coating*

The SiC was coated on yttria-stabilized zirconia (YSZ) using sol–gel method. The solution of YSZ is mixed with distilled water and ethanol in the molar ratio 0.1:3.17:1.75 [9]. The solution was magnetically stirred for 30 min to ensure the homogenous distribution of particles. Thereafter, the SiC has added to the YSZ solution and the mixture was pre-reacted for 15 min with ultrasonicator. Finally, the mixture was dried at room temperature for 24 h. The same process was repeated for other samples with SiC content (0.1, 0.25 wt%). To improve the interfacial adhesion of SiC on YSZ, the YSZ/SiC powder was sintered at 1000 °C for 1 h under nitrogen atmosphere. The pre-determined amount of yttria-stabilized zirconia (YSZ) powder was added to 150 ml of deionized water in a beaker maintain stirring at 600 rpm. To ensure the better solid suspension, the mixture was stirred for 5 h. The resultant mixture was dried in an muffle furnace at 200 °C for 2 h and dried solid mixture was grinded to form fine powders and poured into a metallic mold (30 mm in diameter and 10 mm thickness), Then the powder was dry-pressed into a disk-shaped pellet, under a pressure of 70 MPa [10]. The sintering temperature was set at 1400 °C for 2 h with the heating rate of 1 °C/min as shown in Fig. 1.

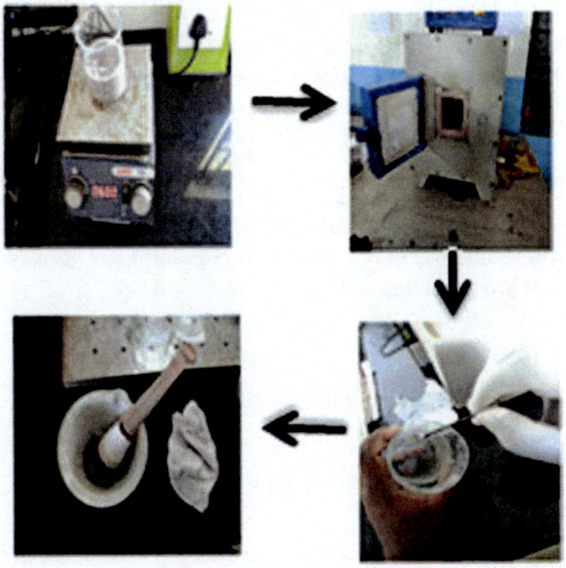

Fig. 1 Preparation of hybrid coating by sol–gel method

2.3 Thermogravimetric Analysis

The thermal stability of the prepared samples was investigated using TGA instrument (SEIKO model TG/DTA 6200). The 5 mg sample was placed in an alumina crucible and temperature was raised to 900 °C at a heating rate of 10 °C/min.

2.4 Oxyacetylene Flame Test

Ablation performance of the prepared sample was evaluated using oxygen–acetylene flame test. Each sample was fixed in a sample holder and a distance of 50 mm was maintained between the nozzle tip and the specimen. The test samples were prepared with 30 × 10 mm dimension. The oxyacetylene flame temperature was estimated to be approximately 1400 °C during the test. The linear and mass ablation rates were calculated using following Eqs. (1) and (2).

$$L_R = \frac{\Delta d}{t_a} = \frac{(d_i - d_f)}{t_a} \quad (1)$$

where *LR* is linear ablation rate; Δd is a total change in thickness of specimen before and after ablation; d_i and d_f are the initial and final thickness of the composite sample before and after ablation test; ta is ablation through time.

$$M_R = \frac{\Delta M}{t_a} = \frac{(m_i - m_f)}{t_a} \quad (2)$$

where *MR* is mass ablation rate; Δm is a total change in mass of specimen before and after ablation; m_i and m_f are initial and final mass of the composite sample before and after ablation test.

3 Results and Discussion

3.1 Microstructure of Coated YSZ-SiC

Figure 2a shows the surface morphology that YSZ grains blend and become incredibly quick at 1400 °C and the break is about all trans granular because of the grain development it was seen in Fig. 2b that the microstructure turns out to be progressively heterogeneous dispersion. The state of micron pores is increased by using higher slurries [11]. When the silicon nanocarbide is added to yttria-stabilized zirconia surface, strong stacking was diminshed (Fig. 2c) . The heterogeneous dispersion of silicon nanocarbide was scribed to various sedimentation speeds as shown in Fig. 2d. The

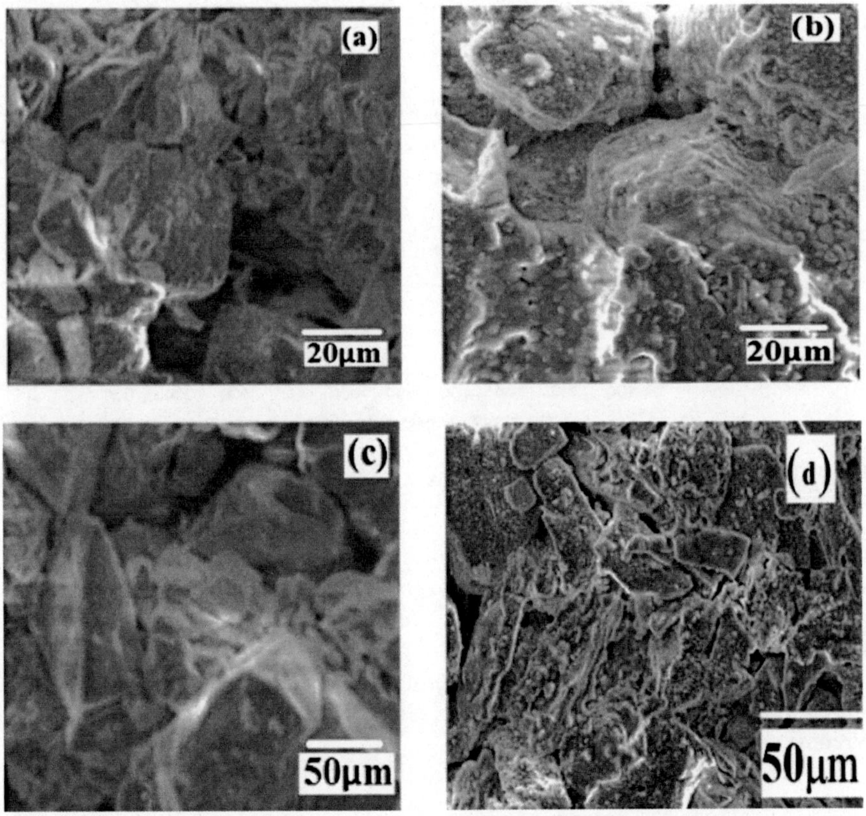

Fig. 2 High magnification of **a** Surface morphology of YSZ **b** YSZ with 0.25% silicon nanocarbide **c** Surface morphology of YSZ **d** YSZ with 0.25% silicon nanocarbide

homogeneous dispersion and progressively standard formed pore certainly contribute to a compressibility of the coating material.

3.2 *Thermal Stability*

The effect of fillers on thermal stability of YSZ-SiC hybrid coating was studied using TGA and DTA techniques. Figure 3 shows the weight loss of samples at the low-temperature stage, below 150 °C, corresponds to the evolution of organic solvents or gases trapped during the process of synthesizing. This small amount of weight loss (2–4%) includes formaldehyde, water, cross-linked molecules, methanol and phenol [12]. However, the evolution of these products was less evident for the YSZ1 and YSZ2 hybrid coatings. The major weight mass loss was observed around 604 °C for pure YSZ and in the range of 650 °C for hybrid coating cured composites [13]. The

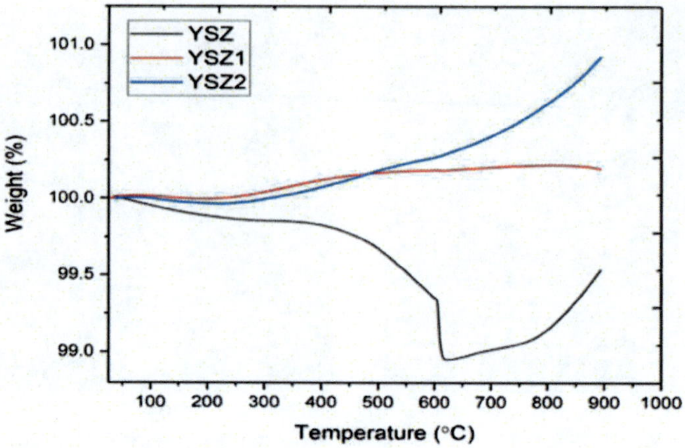

Fig. 3 TGA analysis for YSZ-SiC hybrid coating

thermal decomposition of prepared samples occurs temperature-dependent stages The decomposition is mainly due to the degradation and oxidation of zirconia [14]. Several researchers have been reported a similar observation on the decomposition temperatures [15].

In Fig. 4, it can be noticed that the maximum decomposition temperatures of both stages were increased with the addition of silicon nanocarbide into yttria-stabilized zirconia.

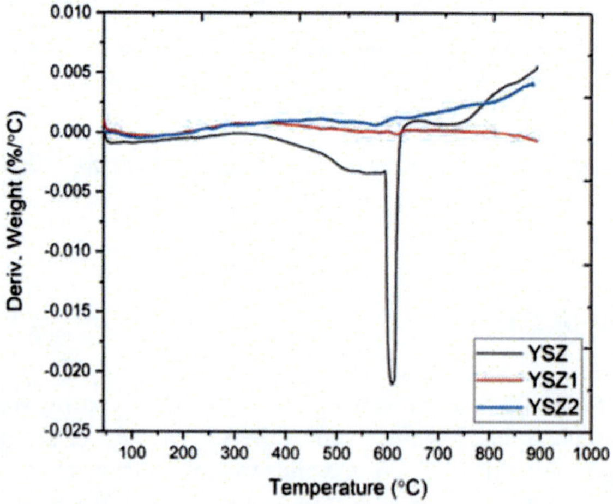

Fig. 4 DTA analysis for YSZ-SiC hybrid coating

3.3 Ablation Performance

The ablation performance of prepared samples was evaluated using an oxyacetylene flame test. The linear ablation and mass ablation rates of composites were calculated by exposing the surface of the sample to oxyacetylene flame at 1200 °C for 60 s. The images of prepared samples before and after ablation test are shown in Fig. 5a, b. A comparison of mass and linear ablation rates of the prepared samples are presented in Figs. 6 and 7, respectively. It was observed that the mass and linear ablation rates of the pure YSZ were 0.0038 g/s and 0.0028 mm/s, respectively. Whereas, the mass and linear ablation rates of the ysz1 sample were 0.0036 g/s and 0.0018 mm/s, respectively, and 0.0031 g/s and 0.0016 mm/s for ysz2 hybrid coating sample .

The addition of 0.1 wt% SiC hybrid filler into YSZ reduced the mass and linear ablation rates by 5.3% and 35.7%, respectively. The mass and linear ablation rates of the prepared samples were significantly decreased by 18.4% and 43%, respectively,

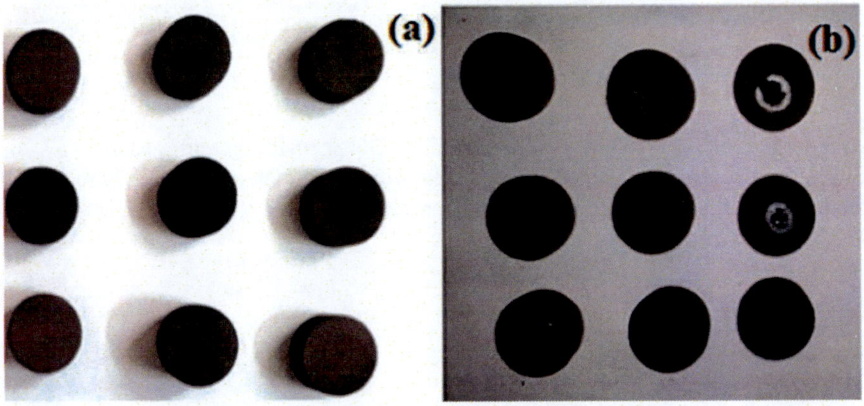

Fig. 5 Images of prepared samples: **a** before ablation test **b** after ablation test

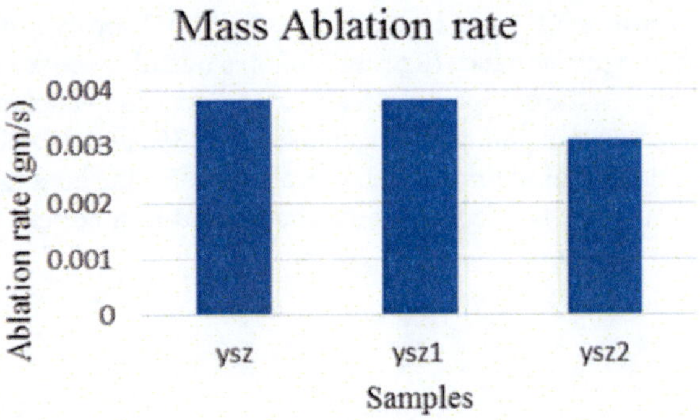

Fig. 6 Mass ablation performance of prepared sample after oxyacetylene flame test

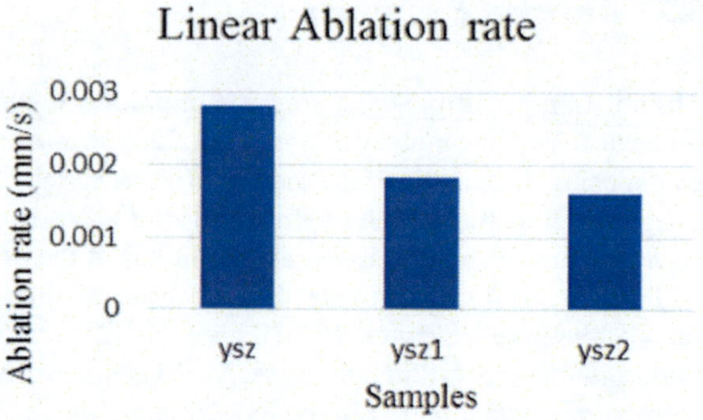

Fig. 7 Linear ablation performance of prepared sample after oxyacetylene flame test

Table 1 Ablative properties of prepared samples

Samples	Porosity (%)	Mass ablation rate (g/s)	Linear ablation rate (mm/s)
YSZ	0.338	0.0038	0.0028
YSZ1	0.419	0.0036	0.0018
YSZ2	0.433	0.0031	0.0016

after adding the 0.25 wt% of SiC hybrid filler. The porosity increases as the YSZ-to-SiC ratio increases, and because the permeability of the material, it directly increases the thermal properties. The ablative properties of prepared samples are presented in Table 1.

4 Conclusion

In this work, the effect of silicon nano-carbide coated into yttria-stabilized zirconia was studied. Thermal studies results show that the YSZ2 sample attained higher thermal stability compared to all other samples. The major weight mass loss was observed around 604 °C for pure YSZ and in the range of 650 °C for hybrid coating cured composites. Ablation performance of tested samples linearly reduced as % wt of SiC nanofillers increased in the composite. The YSZ2 sample showed that the linear and mass ablation rates were significantly decreased by 43% and 18.4%, respectively. The surface morphology of hybrid-coated sample revealed the homogeneous dispersion and porus structure formed certainly contributed to a compressibility of the coating material.

References

1. Miller RA (1987) Surface coating technology 30(1)
2. Brindley WJ, Miller RA (1989) Av. Material processes 29(8)
3. Jones RL (1996) Thermal barrier coating: metallurgical and protective coating. London
4. Michael D, Phelphs (2002) Corrosion resistant thermal barrier coating material for industrial application. Prog Therm Barrier Coating 3–11
5. S Xu (1993) Thermal barrier coating material for industrial gas turbine application. J Ceram Soc 76(8):2027–2032
6. Jun M, Qingshi M, Izzuddin ZS, Andrew M, Nobuyuki K, Wang CH, Hsu-Chiang K (2014) Development of polymer composites using modified, high-structural integrity graphene platelets. Compos Sci Technol 91:82–90
7. Namavar F, Gonghua W (2007) Thermal stability of nano structurally stabilized zirconium oxide. Nano Technol 18
8. Mo CB, Cha SI, Kim KT, Lee KH, Hong SH (2005) Fabrication of carbon nanotube reinforced alumina matrix nanocomposite by sol–gel process. Mater Sci Eng A 395:124–128
9. Montanaro L, Jorand Y, Fantozzi G, Negro A (1998) Ceramic foams by powder processing. J Eur Ceram Soc 18:1339–1350
10. Corbin SF, Apté PS (1999) Engineered porosity via tape casting, lamination and the percolation of pyrolyzable particulates. J Am Ceram Soc 82:1693–1701
11. Kim J, Lin YS (1999) Porous ion-conducting ceramic membranes. J Am Soc 46:2641–2646
12. Gu Y, Liu X, Meng G, Peng D (1999) Porous YSZ ceramics by water-based gel casting. Ceram Int 25:705–709
13. Zhang Q, Shen J, Wang J, Wu G, Chen L (2000) Sol-gel derived ZrO 2–SiO_2 highly reflective coatings. Int J Inorg Mater 2:319–323
14. Subha S, Dalbir S, Pappakudi SV, Kavitha N (2018) Experimental investigation of mechanical, thermal, and ablation performance of ZrO_2/SiC modified C-Ph composites. Int J Appl Ceram Technol 00:1–11
15. Koo JH, Miller MJ, Weispfenning J, Blackmon C (2010) Siliconepolymer composites for thermal protection system: fiber reinforcements and microstructures. J Compos Mater 45(13):1363–1380

Structures and Processing of Polymer Matrix Composite

Optimization of Process Parameter in Abrasive Water Jet Machining of Blue-Fired Grain-Reinforced Glass Fiber Polymer Composite

P. Sabarinathan, V. E. Annamalai, and K. Rajkumar

Abstract Abrasive water jet machining (AWJM) is the non-traditional machining process which is mostly employed for cutting non-conductive materials. The present work concentrates on the AWJ machining study of various compositions (5, 15 and 25%) of blue-fired alumina-reinforced polymer composite. The addition of reinforcement improved mechanical properties which may affect the machining characteristics. The experiment was conducted by varying the experimental process parameters of pressure, abrasive flow rate, and standoff distance. Kerf angle and surface roughness were affecting process parameters which are required to optimize for quality cutting. The results show that the pressure and standoff distance have a direct correlation with the surface roughness and kerf taper angle. Higher pressure and shortest standoff distance show a smooth surface roughness due to no appreciable loss of kinetic energy in cutting wear zone. Similarly, the increase in the reinforcement percentage increases surface roughness and kerf angle. Fiber delamination and grain embedment in the matrix were studied using an optical microscope.

Keywords Machining · Abrasive · Hardness · Surface roughness first section

1 Introduction

Polymer composites have evolved in modern world applications, in various sectors of automobile, aerospace, construction, and household things [1]. To improve strength and load bearing capability of polymer composite, various types of fibers and fillers are added to the matrix. Glass, carbon, and Kevlar are the commercially available synthetic fibers. Glass fiber is the most used synthetic fiber in polymer composite; this is due to its low cost, high strength-to-weight ratio, and corrosive resistance property [2]. Nowadays, researchers are showing interest toward developing new materials

P. Sabarinathan (✉) · V. E. Annamalai · K. Rajkumar
Department of Mechanical Engineering, Sri Sivasubramaniya Nadar College of Engineering, Chennai, India
e-mail: sabarinathan14010@mech.ssn.edu.in

S. Vijayan et al. (eds.), *Trends in Manufacturing and Engineering Management*, Lecture Notes in Mechanical Engineering,
https://doi.org/10.1007/978-981-15-4745-4_20

from the recycled or recovered spent materials. Recycling of spent materials from various sectors like agricultural [3], industrial [4], and electrical [5] are explored as reinforcement substitute to the polymer composites. Addition of this kind of recycled materials shows good improvement in mechanical properties.

Abrasive water jet machining (AWJM) is one of the unconventional machining processes and is most suitable to machining of versatile materials [6]. This process is not evolving with thermal heat leading to no heat-affected zone, and more flexible to cut in all direction [7]. Besides, AWJM process bestows, as exact, required dimensions with limited tolerance with complex profile.

Conventional machining like drilling of fiber-reinforced polymer composite exhibited various damages such as fiber pull out, fiber scrolling, and delamination. These are due effect of variation in cutting forces during drilling operation. AWJM process produces a minor damage like side wall damage but it is controllable by judicious selection of process parameters. The various process parameters controllable in AWJM process are pressure, standoff distance, abrasive flow rate and size, and traverse rate [8]. Cut surface quality was mostly influenced by the above-mentioned machining parameters. Besides economy of the cut, environment consciousness is most important issue for cutting operation.

Machining of polymer composite is entirely different form homogenous material cutting because of discontinuous phases of constituent materials. Azmir et al., have studied the surface roughness and kerf taper ratio on the glass fiber/epoxy composite with varying process parameters. Machining quality in terms of surface roughness was low at higher pressure of 275 MPa, lower standoff distance of 1.5 mm, and lower traverse rate of 1.5 mm/s [9]. Unde et al., have investigated the abrasive water jet machining process on carbon fiber-reinforced polymer composite [10]. The results report that standoff distance and abrasive flow rate were influencing material removal rate and delamination of fiber. Likewise, hydraulic pressure has majorly influenced the cut quality. Researchers have studied the machinability studies of with and without filler-added polymer composites. The results show that the filler added composite produces a good surface quality. Moreover, the traverse rate and jet pressure were influential parameters for the machining process [11, 12].

The present work deals influence of cutting quality on recovered blue-fired alumina-reinforced GFRP composite using an abrasive water jet machining process. Various process parameters such as abrasive flow rate, pressure, and standoff distance were used. Analysis on the response parameters such as kerf taper angle and surface roughness was done to determine the optimal and influential parameters on the surface cut quality.

2 Materials and Methods

2.1 Preparation of Various Percentage of Filler-Reinforced Polymer Composite

Vitrified grinding wheel reject of blue-fired alumina with various percentages (5, 15 and 25%) was used as secondary reinforcement for polymer composite production. Composites were produced using hand layup technique. Epoxy LY556 and hardener HY951 were raw materials for the matrix. Blue-fired alumina of 120 grit particles was mixed with resin using magnetic stirring. Mixed resin and glass mat with various weight proportions were placed inside wooden mold with cross-section of 310 × 310 × 5 mm. Five layers of resin in-between four layers of fiber was stacked consecutively. Composite samples were cured at room temperature for 24 h. After 24 h, the samples were taken out from mold and used for machining operations. Figure 1 shows the fabricated composites. These composites filler loading of 5, 15, and 25% are referred as RBA1, RBA2, and RBA3, respectively. The hardness of fabricated composites was measured using ASTM-D2240 by D-Shore hardness tester.

2.2 Machining of Polymer Composite

Typical layout of the abrasive water jet machining process is shown in Fig. 2. Two-axis abrasive water jet CNC machine (OMAX Corporation, Model: 1530) was used for the machining process. The equipment which is designed to have a nozzle diameter of 0.762 mm, abrasive mesh size of 80 grit, tool offset of 0.3 mm, and impinging angle of 90° is maintained constant throughout the experiment.

The machine is capable to work with various process parameters such as pressure, standoff distance, abrasive flow rate, abrasive mesh size, and traverse rate. From that effective parameters of water jet pressure (MPa), standoff distance and traverse rate were chosen. The machining parameters for abrasive water jet cutting were referred

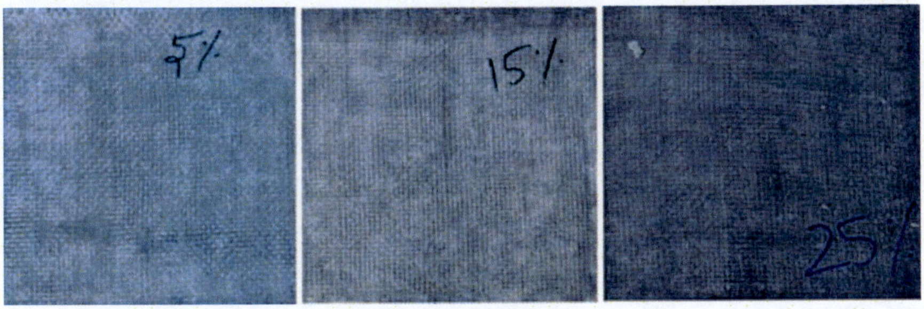

Fig. 1 Fabricated blue-fired alumina-reinforced composites

Fig. 2 Typical abrasive water jet machine

Table 1 Parameters and levels of abrasive water jet machining process

Sl. No.	Parameters	Level		
1	Pressure (MPa)	200	250	300
2	Standoff distance (mm)	1	3	5
3	Abrasive flow rate (g/min)	270	300	330

from the literature [13]. Table 1 shows various process parameters and its level used for the machining process.

2.3 *Characterization of Machined Surface*

Kerf width was measured using profile projector (Make: Metzer) with a coordinate measuring table size of 200 × 200 mm with micrometer. Typical kerf width and taper angle measurement are depicted Fig. 3. Measurement was done at bottom and top kerf, and the kerf taper angle was calculated using an Eq. (1),

$$\text{Kerf taper angle } \emptyset = \tan^{-1}\frac{(K_t - K_b)}{2t} \quad (1)$$

where t is the thickness of the polymer sample and K_t is kerf top width and K_b is kerf bottom width.

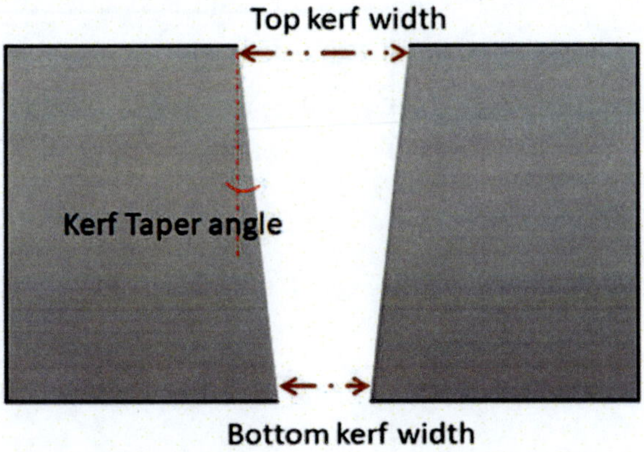

Fig. 3 Typical geometry of cut down cross-section using abrasive water jet machining process

Surface roughness of the machined surface was analyzed using a Mitutoyo Surf test SJ-301 stylus-type profilometer. The average surface roughness (Ra) was measured for all the three composites at different locations. Optical images of the machined zone were taken using a Leica S6 D stereo microscope.

3 Results and Discussion

3.1 Mechanical Properties of Produced Composites

The hardness of the blue-fired alumina-reinforced polymer composites is shown in Fig. 4. Incorporation of hard reinforcement in polymer composites shows a linear increase in hardness. This is due to higher hardness of the added filler and termination of linear polymer chains. Similar results were reported by Rajesh et al. [14].

3.2 Machining Studies on Polymer Composites

Abrasive water jet machining process was carried by varying the selected process parameter on the RBA1, RBA2, and RBA3 samples which are shown in Fig. 5.

Effect of Water Pressure

Figure 6 shows the effect of water jet pressure on kerf taper angle and surface roughness of various compositions of blue-fired alumina-reinforced polymer composites. The experiment was conducted by varying jet pressures of 200, 250, and 300 MPa and by maintaining the other parameters such as standoff distance (1 mm) and abrasive flow rate (330 g/min). The results clearly show that increase in pressure reduces the kerf taper angle. This could be higher kinetic energy of jet increasing the top kerf

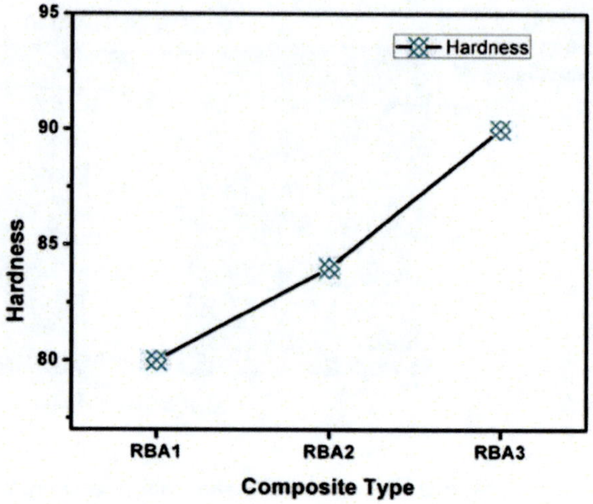

Fig. 4 Hardness value of fabricated composite

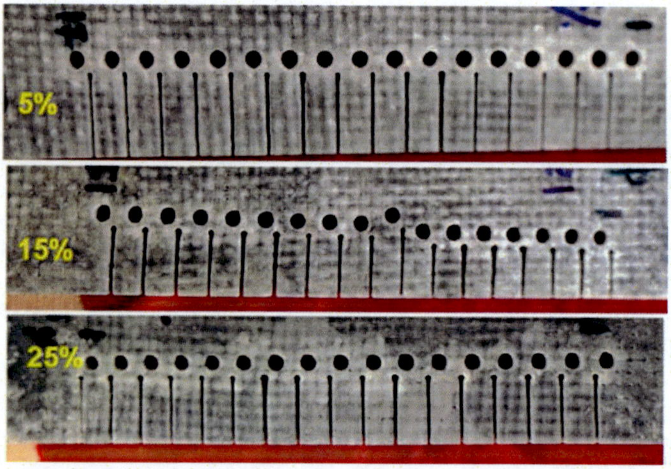

Fig. 5 Abrasive water jet-machined polymer composites

width. On the other hand, there is only slight energy loss during cutting at bottom, resulting in reduction of bottom kerf taper, as shown in Fig. 6a. The higher percentage of reinforcement (blue-fired alumina) shows the highest kerf taper angle. This is due to high pressure garnet particle striking on the composite with higher energy at entrance. Further jet energy absorbed by reinforcement particles causes them to dislocate or dislodge from its own sites. This reduces jet energy at exit side. Due to the interaction of garnet with high pressure, the reinforcement particles in matrix could be dislodged [15]. As a result, a higher top kerf width is produced.

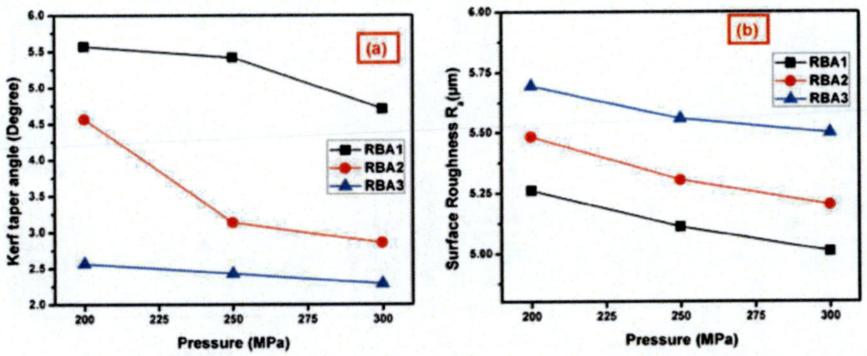

Fig. 6 Effect of water jet pressure on machined surfaces **a** kerf taper angle **b** surface roughness

Maximum surface roughness of the machined surfaces is observed at lower pressure range. Increase in pressure decreases the surface roughness value. This is due to smooth erosion of matrix at higher pressure. In other words, there is no much loss in jet energy along the jet direction resulting in a uniform cut in both cutting wear zone and deformation wear zone [16]. Composite RBA3 shows a higher surface roughness value in all the pressures. This is due to particle pulled out from the matrix at high pressure. The dislodgement of hard reinforcement can be seen in Fig. 7.

Effect of Standoff Distance

Figure 8a–b shows the kerf taper angle and surface roughness of fabricated polymer composites by varying standoff distance. The experiment was conducted by maintaining the constant pressure (300 MPa), abrasive flow rate (330 g/min), and varying

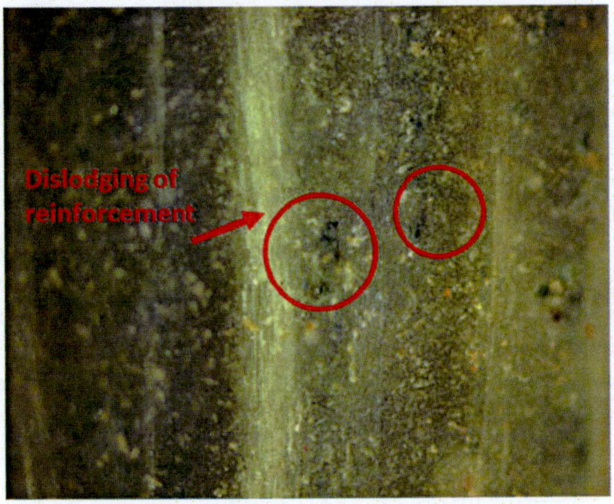

Fig. 7 Machined surface of RBA3 composite

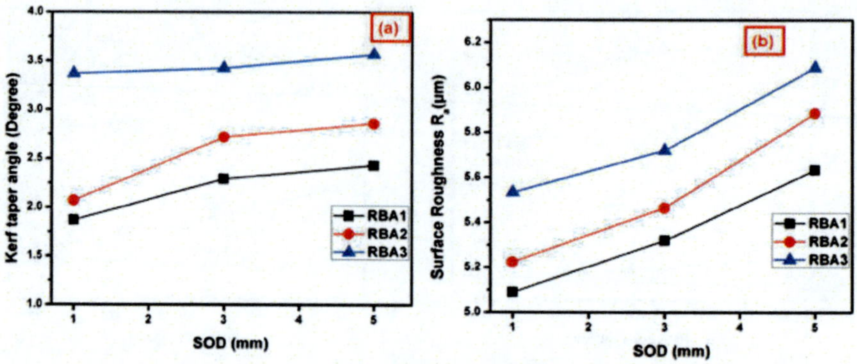

Fig. 8 Effect of standoff distance on machined surfaces **a** kerf taper angle **b** surface roughness

the standoff distance (1, 3 and 5 mm). Maximum kerf taper angle was observed at higher standoff distance and also for the 25% filler composite. An increase in standoff distance causes an expansion of water jet leading to more interaction with atmospheric air. It leads to loss in the jet energy, ultimately, increasing kerf taper angle.

Standoff distance increases dragging of water jet in atmospheric air and impairs the cutting ability of jet at top and is worst in bottom portion. Owing to uneven erosion of matrix, the reinforcement particles were protruded from the matrix surface, increasing the surface roughness. In the case of 25% filler loaded composite, reduced water jet energy at high standoff distance results in poor interaction of jet with reinforcement particle. This may lead to deflection of water jet at reinforcement sites, reducing the machining rate and uneven machining. This adversely affects surface roughness at the cutting zone.

Effect of abrasive flow rate

Figure 9a–b shows the kerf taper angle and surface roughness of fabricated polymer composite with respect to various abrasive flow rates. The experiment was conducted by maintaining the constant pressure (300 MPa), standoff distance (1 mm), and varying the abrasive flow rates (270, 300 and 330 g/min). The results show that increase in abrasive flow rate decreases kerf taper angle. At lower abrasive flow rate, the number of garnet particles interacting with the polymer composite is less. At higher abrasive flow rate, the penetration depth was higher and uniform machining was observed at both the top and bottom portions. It reduces kerf taper angle and also surface roughness of the composite.

Composite with 25% reinforcement shows a higher taper angle and surface roughness in all abrasive flow rates. Limited erosion of polymer matrix could be possibly due to the presence of more filler particles in the matrix. This increased the taper angle and surface roughness of the composite machining. Surface of the cut-down zone was characterized; it shows that the minimum filler added composite (RBA5) exhibits more uniform cutting as compared to RBA25, as shown in Fig. 10a–b.

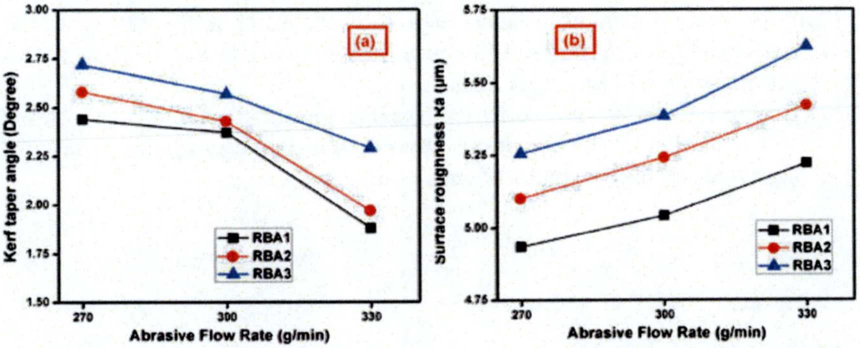

Fig. 9 Effect of abrasive flow rate on machined surfaces **a** kerf taper angle **b** surface roughness

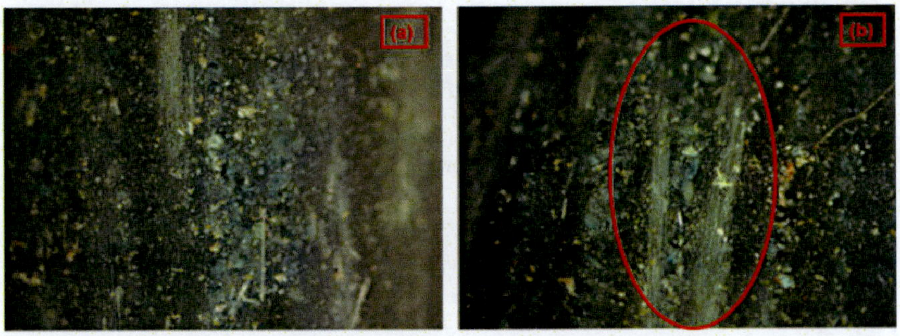

Fig. 10 Machined surface of higher abrasive flow rate **a** RBA5 **b** RBA25

4 Conclusion

From the present research work, the following conclusions are arrived.

1. The addition of blue-fired alumina abrasive grain in the epoxy matrix shows a linear increase in the hardness.
2. Machining on different percentages of (5, 15 and 25%) blue-fired alumina-reinforced polymer composite was experimentally done and uniform machining was observed at 5% composite whereas uneven machining was observed for the 25% composite.
3. Better quality cutting of polymer composites was achieved at high water pressure, low standoff distance, and high abrasive flow rate.
4. Jet pressure has direct co-relation with kerf taper and surface roughness. At higher pressure, the reinforcement is uniform and deeper machining is observed. At low pressure, jet capability was reduced to low level, increasing surface roughness and kerf taper angle.

5. With increase in standoff distance, there is an expansion in the jet size, due to interaction of jet with the atmospheric air and there is a loss in jet energy. This results in increase of kerf taper angle.
6. Optical image of machined surface shows that uniform machining has occurred at minimum filler loaded composite. Maximum filler loaded composite resulted in increased surface roughness of cut surface.

References

1. Singh N, Hui D, Singh R, Ahuja IPS, Feo L, Fraternali F (2017) Recycling of plastic solid waste: A state of art review and future applications. Compos B Eng 115:409–422
2. Gnanavelbabu A, Saravanan P, Rajkumar K, Sabarinathan P, Karthikeyan S (2018) Mechanical strengthening effect by various forms and orientation of glass fibre reinforced isopthalic polyester polymer composite. Mater Today Proc 5(13):26850–26859
3. Sabarinathan P, Rajkumar K, Gnanavelbabu A (2016) Mechanical properties of almond shell-sugarcane leaves hybrid epoxy polymer composite. Appl Mech Mater 852:43–48
4. Sabarinathan P, Annamalai VE, Kumar SS, Kennedy AX (2019) A study on recovery of alumina grains from spent vitrified grinding wheel. J Mater Cycles Waste Manag 21(1):156–165
5. Senophiyah-Mary J, Loganath R, Meenambal T (2018) A novel method for the removal of epoxy coating from waste printed circuit board. Waste Manag Res 36(7):645–652
6. Wang J (1999) Abrasive waterjet machining of polymer matrix composites–cutting performance, erosive process and predictive models. Int J Adv Manufact Technol 15(10):757–768
7. Hutyrova Z, Scucka J, Hloch S, Hlavacek P, Zelenak M (2016) Turning of wood plastic composites by water jet and abrasive water jet. Int J Adv Manufact Technol 84(5–8):1615–1623
8. Gnanavelbabu A, Rajkumar K, Saravanan P (2018) Investigation on the cutting quality characteristics of abrasive water jet machining of AA6061-B4C-hBN hybrid metal matrix composites. Mater Manuf Processes 33(12):1313–1323
9. Azmir MA, Ahsan AK (2008) Investigation on glass/epoxy composite surfaces machined by abrasive water jet machining. J Mater Process Technol 198(1–3):122–128
10. Unde PD, Gayakwad MD, Patil NG, Pawade RS, Thakur DG, Brahmankar PK (2015) Experimental investigations into abrasive waterjet machining of carbon fiber reinforced plastic. J Compos
11. Ramesha N, Akhtar S (2016) Abrasive water jet machining and mechanical behavior of banyan tree saw dust powder loaded polypropylene green composites. Polym Compos 37(6):1754–1764
12. Jani SP, Kumar AS, Khan MA, Kumar MU (2016) Machinablity of hybrid natural fiber composite with and without filler as reinforcement. Mater Manuf Processes 31(10):1393–1399
13. Dhanawade A, Kumar S (2017) Experimental study of delamination and kerf geometry of carbon epoxy composite machined by abrasive water jet. J Compos Mater 51(24):3373–3390
14. Rajesh S, VijayaRamnath B, Elanchezhian C, Aravind N, Rahul VV, Sathish S (2014) Analysis of mechanical behavior of glass fibre/Al_2O_3–SiC reinforced polymer composites. Procedia Eng 97:598–606
15. Doreswamy D, Shivamurthy B, Anjaiah D, Sharma NY (2015) An investigation of abrasive water jet machining on graphite/glass/epoxy composite. Int J Manufact Eng
16. MM IW, Azmi AI, Lee CC, Mansor AF (2018) Kerf taper and delamination damage minimization of FRP hybrid composites under abrasive water-jet machining. Int J Adv Manufact Technol 94(5–8):1727–1744

Mechanical Behavior of Raavi and Pineapple Fiber-reinforced Hybrid Polyester Composites

J. Varun Siddharth, I. Daniel Lawrence, and S. Jayabal

Abstract The raising requirement of plastic materials and their production in advanced engineering were having significant impact on environment. This investigation aims to observe the potential of agro waste residues, such as Raavi and Pineapple fiber, as reinforcement in Polyester GP51 matrix as an alternative to plastic-based components for structural applications. The composites were developed by using compression molding technique by varying the fiber content in percentage. Single fiber test and Fourier transform infrared spectroscopy were used to find the ability of the fiber. The composites were fabricated and tested as per ASTM standards. The tensile strength, flexural test, and impact strength were studied and morphological study was carried out by using scanning electron microscopy and energy dispersive X-ray analysis. The result shows that the Raavi and Pineapple fiber polymer composites have achieved considerable increasing value of mechanical properties of the composites. This investigation suggested the possible way of utilizing hybrid fiber from the agro-based waste in polymer matrix composites.

Keywords Raavi fiber · Pineapple fiber · Polymer composites · Mechanical behaviors

J. Varun Siddharth (✉)
Department of Mechanical Engineering, Anna University, Regional Campus, Madurai, India
e-mail: varunmechanical1993@gmail.com

I. Daniel Lawrence
Department of Mechanical Engineering, Loyola Institute of Technology, Chennai, India

S. Jayabal
Department of Mechanical Engineering, Government College of Engineering, Thanjavur, India

S. Vijayan et al. (eds.), *Trends in Manufacturing and Engineering Management*, Lecture Notes in Mechanical Engineering,
https://doi.org/10.1007/978-981-15-4745-4_21

1 Introduction

No the global consumer sectors accounts for major part of the plastic materials, but commonly these materials are degradable due to a resistance to microbial decay. The natural fiber-based composite material can be produced by partly replacing polyester resin with some natural fibers of different compositions with different percentage, owing to use some conventional resources in our day today life. Van Vuure et al. [1] have made comparative analysis on three different natural fiber composites based on bamboo, flax, and coir fiber and he observed the percentage of variations of tensile properties of natural fibers. Chegdani et al. [2] performed investigation on multiscale tribological behavior when drilling natural fiber composites and hence the cause by which the tribo mechanical behavior of drilling operation is affected was studied. Vitale et al. [3] studied on natural and synthetic fiber-reinforced composite sandwich panels for attaining a report on failure mode in sandwich panels of the composite. Torres et al. [4] performed statistical study on comprehensive set of long natural fiber composite systems to check the variability's in long natural fiber laminate properties. Corona et al. [5] observed on natural fiber composite selection without causing an impact over the environment. Song et al. [6] studied the dynamic tensile properties for recycled office paper fiber and old corrugated cardboard fiber at different proportions. Fan et al. [7] determined the thermal decomposition of natural fiber composites and the behavior and damage profile of natural fiber composite at microscopic and macroscopic levels. Pornwannachai et al. [8] determined the flammability and mechanical performance of each fiber about the fire resistance on natural fiber-reinforced composites. Mejri et al. [9] prepared birch fiber and high density polyethylene composite for replacing nylon which is used in spur gear manufacturing in industries. Chegdani et al. [10] studied on machinability of natural fiber-reinforced composite material by investigating the mechanics of chip formation and the multiscale cut surface by orthogonal cutting experiment. Vignesh G et al. [11] prepared hybrid composite materials with kenaf, human hair as reinforcements, and epoxy resin as matrix for investigating its mechanical properties such tensile strength, flexural strength, and impact strength. Choudhry et al. [12] conducted experimental investigation on polypropylene-based composite material with human hair as reinforcement by varying the reinforcement and matrix percentage. From the results, they found the mechanical properties of composite material are good with lower percentage of reinforcements then the higher. Pothana et al. [13] analyzed the dynamic mechanical properties of banana fiber-reinforced polyester composites with the effect of fiber loading, frequency, and temperature. Rana et al. [14] fabricated natural fibers such sisal, kenaf, hemp, jute, and coir-reinforced polypropylene composites processed by compression molding using a film stacking method and analyzed the mechanical properties of composites. Idicula et al. [15] fabricated randomly oriented intimately mixed short banana and sisal hybrid fiber-reinforced polyester composites and analyzed the dynamic and static mechanical properties. The major objective of this work is to bring new classes of fibers under process and to develop an alternative method toward the use of unconventional and recycling products. In this work, we are going to study and compare the mechanical properties, characteristics, and behavior of natural fiber composites where Pineapple and Raavi serves as reinforcements and polyester resin serves as matrix, thereby framing applications toward the use of these composite materials.

2 Materials and Fabrication Procedure

2.1 Materials

In this study, two natural fibers, namely Raavi fiber (ficusreligiosa) is extracted from the sticks of the branches of peepal tree or bodhi tree by pealing the outer layer of the sticks by mechanical means and Pineapple fiber (ananascomosus) is extracted from the leaves of the Pineapple plant by mechanical means and polyester GP51 matrix was used to fabricate natural fiber-reinforced composites to evaluate the fiber performance under different percentages of composition. The selected fibers are from different region and they exhibit variability in their physical and mechanical properties.

The natural fiber composite with 15% Raavi [Ficusreligiosa], 15% Pineapple [Ananas Comosus], 69.5% polyester GP51 resin, 0.25% of MERP catalyst, and 0.25% of octane solution is kept as composition 1 and the natural fiber composite with 10% Pineapple [Ananas Comosus], 20% Raavi [Ficusreligiosa], 65.5% polyester GP51 resin, 0.25% of MERP catalyst, 0.25% of octane solution is kept as composition 2 (Fig. 1).

Fig. 1 Pineapple 15%, Raavi 15% as composition 1 and Pineappl e10%, Raavi 20% as composition 2

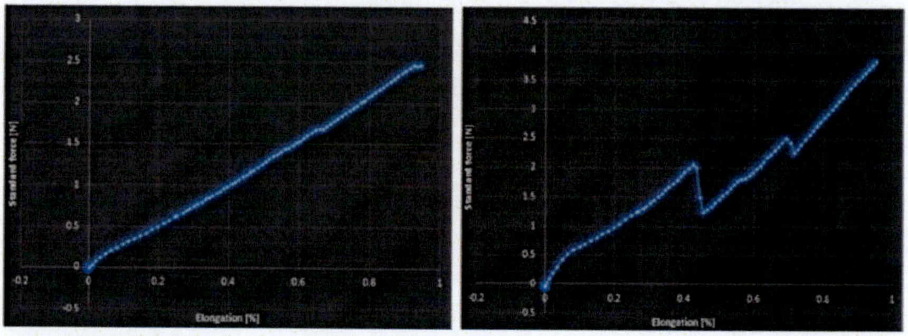

Fig. 2 Single fiber test for Pineapple and Raavi fibers

2.2 *Single Fiber Strength for Preprocessing*

In particular, the preprocessing test is conducted to check the mechanical ability of the fiber by conducting the single fiber test with respect to ASTM standard D3822 by 100 series modular universal testing machine.

Single Fiber Test for Pineapple Fiber Based on the observations which are obtained by conducting single fiber test over a Pineapple fiber, we observe that the stress of the fiber is found to be around 10,100 MPa, with the corresponding maximum force and elongation of 2.467766 N and 0.945653% by applying gradual tension of 10 mm/min, respectively. The maximum force and elongation are obtained just by a step on the point or on the region of the breaking zone (Fig. 2).

Single Fiber Test for Raavi Fiber From the observations which are obtained by conducting single fiber test over a Raavi fiber, we observe that the stress of the fiber is found to be around 16,200 MPa, with the corresponding maximum force and elongation of 3.825889 N and 0.951% by applying gradual tension of 10 mm/min, respectively. The maximum force and elongation are obtained by just a step on the point or on the region of the breaking zone.

2.3 *FTIR Test*

Fourier transform infrared spectroscopy was used to determine the functional groups in a molecule. The functional group exhibits its vibration by bending, stretching, and waging at different frequencies.

FTIR Test with Raavi Fiber The functional groups present in the molecules of Raavi fiber are observed by passing light at different frequencies through the specimen. By observing the graph, it is found that the transmittance of IR light is less in functional group region then in finger print region. The stretching vibration with the

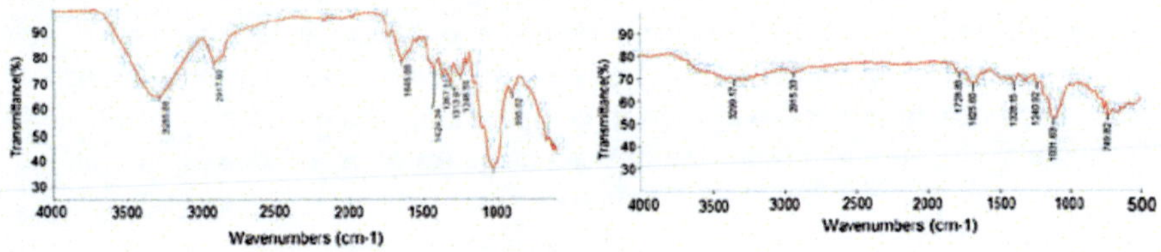

Fig. 3 FTIR graph for Raavi and Pineapple fibers

transmittance of 88% is obtained for the wave number of 2917.90; the obtained wave number shows presence of strong alkane (C–H) group. The stretching vibration with the transmittance of 65% is obtained for the wave number of 3285.88, the obtained wave number shows the presence of strong and broad alcohol (O–H) group. The values from 0 to 1800 are neglected since they come under finger print region. It is observed that Raavi fiber has formaldehyde as its chemical compound. The results are shown as graphical representation (Fig. 3).

FTIR Test with Pineapple Fiber The results are obtained by observing the graph; it is found that the transmittance of IR light is less in functional group region then in finger print region. The stretching vibration with the transmittance of 88% is obtained for the wave number of 2915.33, and the obtained wave number shows presence of strong alkane (C–H) group. The stretching vibration with the transmittance of 65% is obtained for the wave number of 3299.17, and the obtained wave number shows the presence of strong and broad alcohol (O–H) group. The values from 0 to 1800 are neglected since they come under finger print region. It is observed that Raavi fiber has formaldehyde as its chemical compound.

2.4 Fabrication Procedure

The fabrication work was executed with the help of compression molding machine by setting up the composite lamina. The compression molding machine is set up with 5–15 amps as current with the upper and lower die temperatures as 348 K and the pressure of the hydraulic actuation system is maintained at 2 bar. The lamina plate is set up with the dimensional specification of 3 mm as its thickness and the 300 mm as its length and width is taken as a boundary parameter for the experiment, whose total gross weight is said to be 450 g.

3 Result and Discussion

The prepared composite laminas are introduced to various test method in order to attain the property and characteristics of the above said compositions. The properties and the characteristics of the above said two compositions are compared with each other in order to observe the changes in the composite behaviour, characteristics, and properties with respect to change in composition, which is having similar pair of materials with varying proportion of percentage for each and every material.

3.1 Tensile Test

The tensile test is performed with the help of universal testing machine where the alternate ends of the testing specimen are held to unidirectional moving supports. The UTM with the composite specimen is integrated with the computer in order to attain the results which are obtained through the universal testing machine. The specimen for the tensile test is grooved using wood grooving machine from the lamina based on the dimensional specification which is suggested as per ASTM D638 standard.

Tensile Test Characterization of Composition 1 From the visual observation of the stress–strain graph of the composite specimen were observed that the composite specimen is capable of withstanding up to the maximum extent of plastic zone, and the elastic zone is too small where the fracture occurs immediately. The observed composite specimen is brittle in nature. Based on the visual observation of the graph represented that the ultimate stress is attained at 32 MPa whose corresponding strain percentage is 1.99 (Fig.4).

Based on the visual observation of force–length graph, the composite specimen has elongated up to the length of 1.20 mm. Since the composite specimen is brittle in nature, the reduction of size along its cross-section is too small. The specimen attains a maximum force of 1200 N at a length of 1.20 mm and the specimen breaks at 1.22 mm with a small drop in force of 1998 N for the corresponding thickness of 3 mm.

Tensile Test Characterization of Composition 2 From the visual observation of the stress–strain graph of the composite specimen were found that the composite specimen is capable of withstanding up to the maximum extent of plastic zone, and the elastic zone is too small where the fracture occurs immediately. The observed

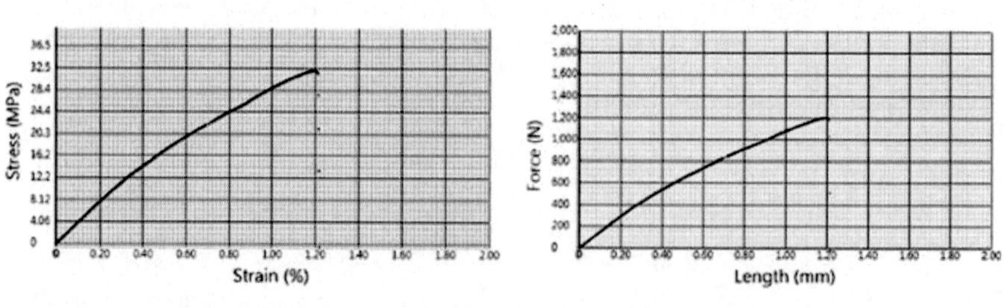

Fig. 4 Stress–strain and force–length graphs for composition 1

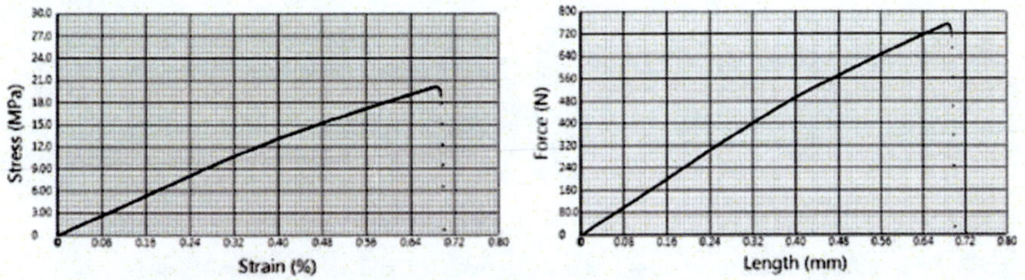

Fig. 5 Stress strain and Force length graph for composition 2

composite specimen is brittle in nature. Based on the visual observation of the graph, we can note that the ultimate stress is attained at 20.1 MPa whose corresponding strain percentage is 0.67 (Fig. 5).

Based on the visual observation of force to length graph, the composite specimen has elongated up to the length of 0.68 mm. Since the composite specimen is brittle in nature, the reduction of size along its cross-section is too small. The specimen attains a maximum force of 753 N at a length of 0.68 mm and the specimen breaks at 0.70 mm with a small drop in force of 725 N for the corresponding thickness of 3 mm.

Comparative Analysis of Tensile Test Characterstics Based on the results obtained through the experimental analysis, the mechanical characteristics of natural fiber composites for composition 1 and composition 2 are studied. The observations show that both the composites are exhibiting brittleness and the plastic behavior of the product is comparatively more than that of elastic behavior. In order to obtain better strength and withstanding ability, comparative analysis is conducted with composition 1and composition 2 (Table 1).

The above given bar graph and table are obtained by conducting tensile test between two different compositions. The results obtained through comparative analysis show that the strength and withstanding ability of the composite composition 1 are greater than that of composition 2 (Fig. 6; Table 2).

Table 1 Comparative table of tensile test

Mechanical characteristics	Composition 1	Composition 2
Stress [MPa]	32	20.1
Strain [%]	1.99	0.67
Maxload [N]	1200	753
Elongation [mm]	1.2	0.68
Break [mm]	1.22	0.7
Break Load [N]	1998	725

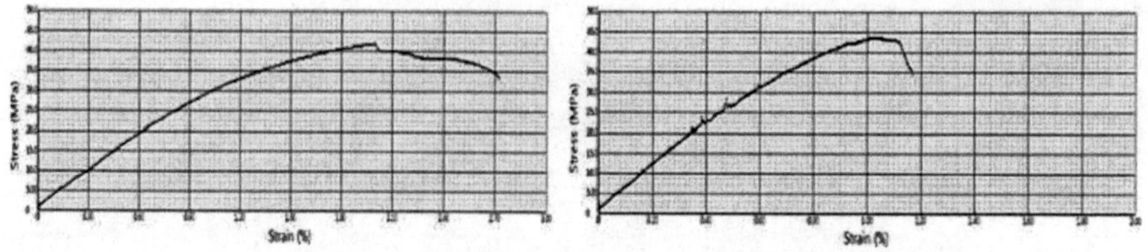

Fig. 6 Stress strain graph for flexural test in composition 1 and composition 2

Table 2 Comparative table of flexural test

Mechanical characteristics	Composition 1	Composition 2
Max force [N]	31.67	32.67
Strength [MPa]	42.22	43.56
Breaking stress [MPa]	33	34.95
Breaking strain [%]	28	18.5

3.2 *Flexural Test*

The flexural test was conducted with the help universal testing machine by using three-point bending method by replacing the default setting for tensile test.

Flexural Test Characterization of Composition 1 The results converged through stress–strain graph show that the maximum force required to break the specimen is 31.67 N and the flexural strength attained by the specimen is 42.22 MPa, respectively. The stress and strain at which the specimen breaks is 33 MPa and 2.80%, respectively.

Flexural Test Characterization of Composition 2 The results obtained through stress-strain graph show that the maximum force required to break the specimen is 32.67 N and the flexural strength attained by the specimen is 43.56 MPa, respectively. The stress and strain at which the specimen breaks is 34.95 MPa and 1.85%, respectively.

Comparative Analysis of Flexural Test Characteristics The table which is represented below is obtained by conducting flexural test between two different compositions. The results obtained through comparative analysis show that the maximum force, flexural strength, and breaking stress are greater in composition 1 and the breaking strain is greater in composition 2.

3.3 *Impact Test*

The impact test is done with the help of impact testing machine as per the ASTM standard D256. The stress–strain curve of the composite specimen pointed that the specimen is exhibiting an brittle nature.

Table 3 Comparative table of impact test

Mechanical characteristics	Composition 1	Composition 2
Internal energy [J]	22.449	22.449
External energy [J]	2.797	5.675
Energy absorbed [J]	19.651	21.774
Impact strength [J/M]	22.379	45.400

Impact Test Characterization of Composition 1 Based on the Izod test results with the test specimen, the internal energy absorbed by the specimen is 22.449 J and the remaining energy after the specimen stroked is 2.7974 J. Therefore, the total energy of the fractured specimen is 19.6516 J which is the difference between the internal energy and the remaining energy. The impact strength which is obtained in the composite specimen in terms of KJ/m^2 is 74.591 and the impact strength in composite specimen in terms of J/m is 223.797, respectively.

Impact test characterization of Composition 2 Based on the Izod test conducted with the test specimen, the internal energy absorbed by the specimen is found to be 22.449 J and the remaining energy after the specimen stroke is 5.6750 J. Therefore, the total energy which is delivered to the composite specimen is 21.774 J which is the difference between the internal energy and the remaining energy and it is the impact value of energy in the fracturing specimen. The impact strength which is obtained in the composite specimen in terms of KJ/m^2 is 151.335 and the impact strength in composite specimen is in terms of J/m is 454.007, respectively.

Comparative Analysis of Impact Test Characteristics The results obtained through comparative analysis show that the toughness of the composite composition 2 is greater than that of composition 1 (Table 3).

3.4 Analysis of SEM Images

The fractured surface of the Raavi and Pineapple fiber-reinforced composite was analyzed using scanning electron micro graphs. Observed SEM images show the fiber and resin compatibility in natural fiber polymer composites. The SEM image of tensile fractured specimen showed uniform crack propagation due to better mixing of the fiber and matrix. The fiber segregation was found to be good in the fractured region due to the improvement in compatibility between the fiber and resin matrix.

3.5 Energy Dispersive X-ray Analysis

Chemical composition (chemical characterization) of the specimens was obtained by conducting the EDX test by scanning electron microscope instrument.

Energy Dispersive X-ray Analysis of Composition 1 The obtained results of the energy dispersive X-ray analysis, carbon element contains the maximum level of un-normalized concentration and normalized concentration in weight percent of the element, atomic weight percentage and the error in the weight percent concentration at the 1 sigma level and also the observed results showed that, Phosphorus element contains the minimum level of un-normalized concentration in weight percent of the element, normalized concentration in weight percent of the element, atomic weight percentage and the error in the weight percent concentration at the 1 sigma level.

Energy Dispersive X-ray Analysis of Composition 2 The obtained results of the energy dispersive X-ray analysis, carbon element contains the maximum level, and potassium element contains the minimum level of un-normalized concentration and normalized concentration in weight percent of the element, atomic weight percentage and the error in the weight percent concentration at the 1 sigma level (Figs. 7, 8, 9; Table 4).

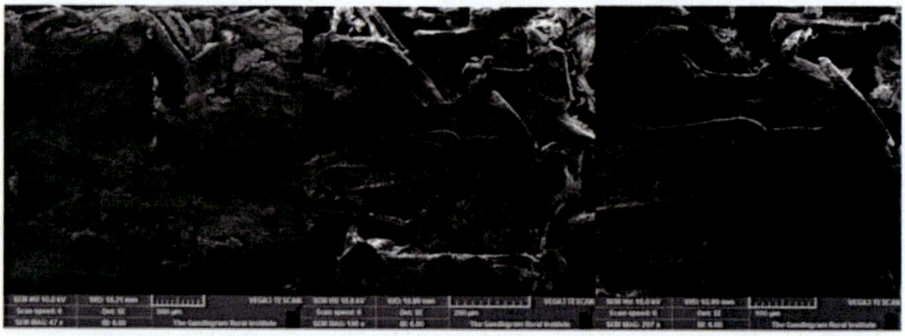

Fig. 7 SEM image of composition 1 for 500, 200, and 100 μm

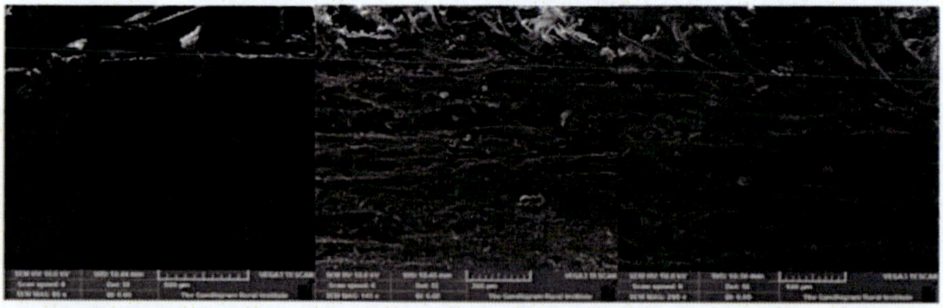

Fig. 8 SEM image of composition 2 for 500, 200, and 100 μm

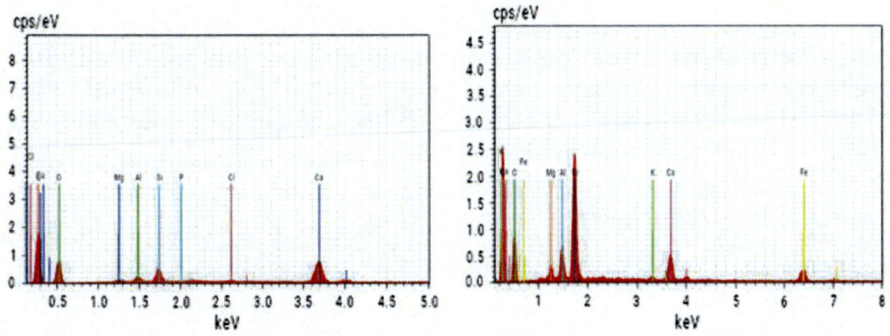

Fig. 9 EDAX image of composition 1 and composition 2

Table 4 EDX chemical composition of composition 1 and composition 2 as K-series

El	AN	Norm.C (wt%)	Atom.C (at%)	Error (1sigma)
C	6	51.87	60.91	7.67
O	8	40.86	36.02	6.78
Ca	20	4.16	1.46	0.16
Si	14	1.66	0.83	0.11
Al	13	0.86	0.45	0.08
Mg	12	0.45	0.26	0.07
Cl	17	0.14	0.05	0.04
P	15	0.01	0.01	0.03
C	6	57.57	67.78	11.84
O	8	30.06	26.57	7.28
Si	14	6.05	3.05	0.41
Al	13	1.78	0.94	0.17
Ca	20	1.76	0.62	0.11
Mg	12	1.01	0.58	0.13
Fe	26	1.64	0.42	0.11
K	19	0.13	0.05	0.04

4 Conclusion

In this investigation, a systematic procedure to evaluate and analyze the behavior of Raavi and Pineapple fiber hybrid polymer composites was done under two different percentages of compositions. The better tensile, flexural, and impact behaviors of 32, 42.22 MPa, and 22.379 J/m were obtained for Raavi (15%)- and Pineapple (15%)-reinforced polymer composites (Composition 1). The maximum tensile, flexural, and impact behaviors of 20.1, 43.56 MPa, and 45.4 J/m were obtained for Raavi (10%)- and Pineapple (20%)-reinforced polymer composites (Composition 2). The microsturcture and chemical composition were studied by scanning electron microscope. The mechanical behaviors studied through real-time experiments witnessed that both composition of Raavi- and Pineapple-reinforced polymer composites can be utilized as the replacement of plastics such low load applications because of its ease of decomposability and low cost.

References

1. Van Vuure AW, Baets J, Wouters K, Hendrickx K (2015) Compressive properties of natural fibre composites. Mater Lett 149:138–140
2. Chegdani F, El Mansori M (2018) Friction scale effect in drilling natural fiber composites. Tribol Int 119:622–630
3. Vitale JP, Francucci G, Xiong J, Stocchi A (2017) Failure mode maps of natural and synthetic fiber reinforced composite sandwich panels. Compos A Appl Sci Manuf 94:217–225
4. Torres JP, Vandi LJ, Veidt M, Heitzmann MT (2017) The mechanical properties of natural fibre composite laminates: a statistical study. Compos A Appl Sci Manuf 98:99–104
5. Corona A, Madsen B, Hauschild MZ, Birkved M (2016) Natural fibre selection for composite eco-design. CIRP Ann Manuf Technol 65(1):13–16
6. Song Z, Li S, Lei J, Noble RL, Wang Z (2018) Dynamic tensile properties of ROP/OCC natural hybrid fibers reinforced composites. Compos Struct 185:600–606
7. Fan M, Naughton A (2016) Mechanisms of thermal decomposition of natural fibre composites. Compos B Eng 88:1–10
8. Pornwannachai W, Ebdon JR, Kandola BK (2018) Fire-resistant natural fibre-reinforced composites from flame retarded textiles. Polym Degradation Stab
9. Mejri M, Toubal L, Cuillière JC, François V (2018) Hygrothermal aging effects on mechanical and fatigue behaviors of a short-natural-fiber-reinforced composite. Int J Fatigue 108:96–108
10. Chegdani F, El Mansori M (2018) Mechanics of material removal when cutting natural fiber reinforced thermoplastic composites. Polym Testing 67:275–283
11. Vignesh G. Mechanical characterization of kenaf-hair reinforced hybrid composite
12. Choudhry S, Pandey B (2012) Mechanical behaviour of polypropylene and human hair fibres and polypropylene reinforced polymeric composites. Int J Mech Ind Eng 2(1):118–121
13. Pothan LA, Oommen Z, Thomas S (2003) Dynamic mechanical analysis of banana fiber reinforced polyester composites. Compos Sci Technol 63(2):283–293
14. Rana AK, Mandal A, Bandyopadhyay S (2003) Short jute fiber reinforced polypropylene composites: effect of compatibiliser, impact modifier and fiber loading. Compos Sci Technol 63(6):801–806
15. Idicula M, Malhotra SK, Joseph K, Thomas S (2005) Dynamic mechanical analysis of randomly oriented intimately mixed short banana/sisal hybrid fibre reinforced polyester composites. Compos Sci Technol 65(7–8):1077–1087

Carbon Fiber Surface Treatment for Enhanced Interfacial Properties: A Review

Kwame Anane-Fenin, Esther Titilayo Akinlabi, and Nicolas Perry

Abstract Pristine carbon fibers are intrinsically brittle, fuzzy, non-polar and possess graphite-based planes which are highly crystallized. These characteristics are limiting to effective interfacial bonding between fiber and matrix. Therefore, the ability to modify the surface of carbon fibers present an avenue for enhancing surface functionalities and increasing surface energy which results in improved interfacial properties (wettability) and mechanical properties. In this paper, the current novel techniques for carbon fiber surface treatment are presented. The current direction of the field is reviewed, and emerging technologies are discussed. The comparative study conducted provides insight for optimal selection of treatment approaches depending on the requirement.

Keywords Interfacial bonding · Surface treatment · Wettability

1 Introduction

Carbon fiber (CF) reinforced composites (CFRPs) are known to be very reliable and find applications in varying multidisciplinary fields such as the aerospace, automotive and sports equipment manufacturing industries as a result of the excellent and superior mechanical strength and physical (lightweight) properties [1]. Unmodified or virgin carbon fiber, however, has some limitations which include a very moderately active specific surface area, non-polar properties and chemically inert characteristic, low lipophobic surface, low energy surface and the inherently low weak interfacial adhesion characteristic of the fibers which limits the efficient transfer of stresses

K. Anane-Fenin (✉) · E. T. Akinlabi
Department of Mechanical Engineering Science, University of Johannesburg, Johannesburg, South Africa
e-mail: lncs@springer.com

N. Perry
Arts et Métiers ParisTech, CNRS, 12M Bordeaux, Esplanade des Arts et Métiers, Talence, France

E. T. Akinlabi
Department of Mechanical Engineering, Coventry University Ota, Ota, Nigeria

S. Vijayan et al. (eds.), *Trends in Manufacturing and Engineering Management*, Lecture Notes in Mechanical Engineering,
https://doi.org/10.1007/978-981-15-4745-4_22

from the resin to the carbon fiber that consequently affects the mechanical integrity of the fabricated composite [2]. Pristine carbon fibers are intrinsically brittle, fuzzy, non-polar and possess graphite-based planes which are highly crystallized. These characteristics are limiting to effective interfacial bonding between fiber and matrix [3]. As a result, many CFRPs tend to demonstrate a propensity for failure within the fiber-matrix interface due to the relatively weak bonding and eventually results in the pullout effect which exposes the fiber from the matrix (pullout effect). This phenomenon compromises the mechanical integrity of the composite material like the interlaminar shear stress (ILSS).

Over a couple of decades, several techniques for modifying the surface of carbon fibers to improve the interfacial properties of fabricated composites, include plasma treatment [4], sizing [1], oxidation treatment [5], graphene oxide modification [6] coating [7], high energy irradiation [8], chemical grafting [9], electrophoretic deposition (EPD) [10], etc. All the above-stated methods have their own inherent insufficiencies. An optimal approach to fiber surface treatment is very critical to ensuring that the right amount of fiber-matrix interface bonding is obtained to prevent issues of extreme bonding which can result in brittleness [11]. Therefore, a carefully controlled interface is vital to ensuring that the requisite amount is transferred from the matrix to fiber reinforcements which ameliorate the induced internal stress concentrations and also enhance the mechanical properties [12].

The ability to modify the surface characteristics of carbon fiber provides a powerful optimization tool for enhancing the fiber-matrix adhesion properties of carbon fiber reinforced polymer composites. Therefore, in this study, a thorough review of the approach currently adopted for the modification of carbon fiber surface is presented. Comparison of the different methods is discussed, and the advantages and disadvantages also reported.

1.1 Surface Sizing

In industry, the manufacture of carbon fiber is characterized by the activation of carbon fiber to achieve electrolytic surfaces before stabilization and carbonization. The activation of carbon fiber by anodic oxidation is an important step critical for improving its surface reactivity due to the development of oxygen-rich functional groups [13]. This significantly enhances wettability and promotes the implementation of polymeric sizing, which is required for additional fiber processing. Furthermore, an overall improvement in bonding between the matrix, polymeric sizing and fiber surface is achieved [13]. Other advantages of anodic oxidation include the simplicity of the processing methodology and cost-effectiveness. It is an oxidation technique which inherently limits it to the creation of only oxygen-containing surface functional groups. The initial intent of applying sizing agents (polymer-based) during commercial carbon fibers manufacture was to provide a 100-nm-thick protective coating against fiber damage and improve handling [14]. In recent times, the dimension of exploring and developing sizing agents that also ensure enhanced interfacial bonding between carbon fiber reinforcement and the matrix is a primary focus for manufacturers [15]. Therefore, sizing agents are designed to promote increased fiber-matrix interlocking, enhanced wettability and improve the density of oxygen-containing functional groups on the surface of the fiber. To enhance the interfacial adhesion properties of fiber and matrix, nanoparticle inclusions in sizing agents have been the most preferred approach; the nanoparticles explored include carbon nanotubes [16], titanium oxide [17], graphene oxide [18] and zirconium dioxide [15].

Polyurethanes (PU) have also emerged as highly suitable sizing agents because of advantages such as excellent mechanical properties (strength and toughness), low viscosity, excellent abrasion resistance, outstanding interlocking adhesion with several matrices, cost-efficient and environmentally sustainable [19, 20]. The application of PUs for sizing has not been thoroughly explored as only a few publications are available.

The most industrially preferred approach for surface modification is sizing treatment methods and the reasons include simplicity and ease of process operation, moderate environmental impact, flexibility in tunability and controllability process is continuous and seamlessly incorporates into the overall manufacture of carbon fibers [21].

1.2 Plasma Treatment

In contradistinction, conceptually, carbon fiber surface modification using plasma permits for a wider spectrum of surface functionalization options [13]. During plasma treatment an interaction between the carbon fiber surface, neutral and excited free electrons, radicals, ions, atoms or molecules takes place. As a result, four forms of surface modifications can occur, namely activation, layer deposition, functionalization or etching. The chemical property and constitution of the precursor gas for plasma treatment coupled with selected plasma processing parameters determine the topographic, morphological and chemical characteristics of the carbon fiber. To improve the fiber-matrix (epoxy) bonding, oxygen or air plasma functionalization is usually conducted [22]. The resulting fiber surface oxidation is characterized by the development of oxygen-carrying functional groups equivalent to those observed after anodic oxidation.

Another promising approach used for enhancing the adhesion between fiber and matrix is via the modification of the fiber surface through the formation of functional groups with nitrogen carrying elements. A high chemical interaction between the amine functional groups on the fiber surface and the epoxy resin groups is envisaged [23]. The carboxylic and alcohol groups are particularly of great interest as they are known to react chemically with the epoxy groups within the matrix. In cases where amine hardeners are used, the fiber surface carbonyl groups are particularly very reactive to amine groups in general [23] Several studies [24] have shown that the carbon fiber-epoxy matrix bonding was significantly improved after oxygen or air plasma functionalization. The drawback of using air or oxygen plasma is the deterioration in mechanical strength of the fiber associated as a result of the damage inflicted on the fiber during the process [25].

Ammonia [26] and nitrogen [27] plasma treatment are often used to introduce nitrogen carrying functionalities onto the surface of carbon fibers. Ammonia particularly is an attractive precursor gas alternative for achieving nitrogen functionalization due to its capacity to form very high ratios of amine and imine functional groups on the surface of the carbon fiber [28] while avoiding the drawback of tensile strength degradation common with oxygen plasma treatment [27]. An improvement of interfacial bonding of fibers to bismaleimide resin, initially untreated and then treated with ammonia plasma is reported [25, 29]. Moosburger-Will et al. [30], functionalized the surface of unmodified carbon fibers, extracted immediately after carbonization using ammonia plasma in a low pressure environment straight. In comparison with untreated carbon fibers, an improvement in surface energy and nitrogen functionality concentration at the surface of treated fibers were reported. The enhanced wettability is comparable to activated carbon fibers processed using anodic oxidation.

1.3 Electro-grafting/polymerization

Electro-grafting or electro-polymerization is a technique that has been widely used for depositing functional coatings of polymer homogenously on to the surface of the carbon fiber. Electro-polymerization has several advantages, these include the ability to optimize the desired coating through controlling and tuning critical parameters and conditions such as polymerization time, monomer concentration, homogeneity and thickness of coating [31]. Subramanian et al. [32] were able to clearly establish the influence of electro-polymerization parameters like polymerization time, monomer functionality, selected solvent, post-coating treatment and fiber content on the mechanical properties of the coated CF. Some studies have observed that electro-polymerization establishes an interlayer that lies between the matrix and fiber capable of energy absorption from internal cracks within the composite which consequently results in a reduction or elimination of the total induced stress concentration [33].

Polymethacrylic acid (PMAA) polymer, when used for electro-grafting of monomers onto the carbon fiber, is reported to have exhibited the best tensile strength of fabricated composites [34]. These superior characteristics can be attributed to hydrogen bonds that form between the hydroxyl groups of the matrix and the carboxyl groups of PMAA, thereby forming an improved interfacial adhesion without the drawback of brittleness. In a recent study by Semitekolos et al. [35], the improvement of interfacial strength between the fiber-matrix bond within carbon fiber reinforced composites was accomplished through the reaction of carbon fiber and polymethacrylic acid by a process known as electro-polymerization.

1.4 Grafting

The inclusion of nanoscale materials (e.g., nanoparticles, carbon nanotubes, nanoclay, nanofiber, graphene oxide-GO, metals and nonmetals) in several current research onto the surface of carbon fibers to improve the interfacial strength without degrading the mechanical properties of composites is a standard [6, 36]. Grafting of graphene oxide (GO) homogenously onto carbon fiber is an efficient technique for achieving hierarchical intralaminar and interlaminar reinforcement that introduce superior fiber-matrix bonding and enhanced overall mechanical properties to the composite material [37]. The novelty of using graphene oxide (GO) lies in its two-dimensional configuration coupled with its superior mechanical and thermal properties [38]. Furthermore, GO can be prepared in combination with several oxygen functional groups in substantial amounts at a reasonably reduced cost [39].

1.5 Radiation Treatment

Laser radiation treatment has also been successfully adapted for enhancing the fiber-matrix bond while additionally improving the mechanical properties of the matrix. In a study by Fischera et al. [40], CO_2 and UV lasers were tested for carbon fiber surface treatment which showed that both lasers equally improved the interfacial bond strength between the reinforcement and matrix.

1.6 Coating

Studies on the application of polydopamine (PDA) coating have shown it provides a flexible platform for functionalizing the surface of several substrates without degrading the mechanical properties. The impressive chemical properties such as the catechol functional group are mainly responsible for superior adhesiveness [41]. PDA coating treatment has mostly been conducted on glass fibers [42]; however, in a recent study by Kim and Song [43], the influence of PDA on surface modification of carbon fiber was assessed. They observed an increase in Young's modulus of PDA coated carbon fiber reinforced composite. Carboxyl functional group formation on fiber surfaces occur during oxygen plasma treatment, while hydrogen plasma treatment induces the formation of hydroxyl bonds facilitated by sp2 etching which provides the anchoring platform and also produces superior hydrophilic surfaces [44].

1.7 Layer-by-Layer (LbL) Self-assembly Technique

Most of the carbon fiber surface modification techniques for the inclusion of nanoscale materials are not environmentally friendly or green as toxic chemicals and catalysts are required [45]. In contrast, the layer-by-layer (LbL) self-assembly technique is generally considered a green approach because it is environmentally friendly, high flexibility for optimization and tuning, moderate processing conditions and excellent repeatability [45]. This approach basically operates on a cyclically based concept where alternating film deposition of materials that are oppositely charged takes place.

1.8 Covalent Functionalization

The two main limitations of untreated carbon fiber are the reduced interfacial adhesion of fiber and matrix and relatively high conductivity which is possibly detrimental to electric machines. Most of the available treatment methods for surface modification can resolve the interfacial adhesion issue; however, covalent functionalization is one of the few methods capable of decreasing carbon fiber conductivity as presented in the study by [46]. They achieved CF surface modification via covalent functionalization involving electrochemical reduction of phenyl-diazo species.

1.9 Microwave Irradiation

Microwave irradiation is another emerging tool for modifying the surface of carbon fiber made possible by carbon fiber's outstanding microwave absorbability. Wang et al. [47] developed a microwave irradiation method that was carried out in air atmosphere, however, some drawbacks reported included some level of burnout and degradation. A study by [48], avoided the limitations by conducting the irradiation via immersion of the carbon fiber. The microwave etching process resulted in large formation oxygen-containing groups on the surface of the carbon fibers.

2 Results and Discussion

A comparison of the different surface treatment is presented in Table 1. In comparing the various methods, electrochemical and wet oxidation treatments are the least preferred due to the environmental issues associated with chemical use and the time-consuming nature of the methods [49]. However, plasma treatment impacts both physical and chemical properties of carbon fibers without significantly altering the

Table 1 Carbon fiber surface morphological treatments

Surface morphology treatment	Treatment agent	Enhancement	Comment	Author
Plasma	Ammonia	(1) High surface energy (nanoporosity) (2) Enhanced wettability	Nitrogen functionalization was successfully conducted. Nitrogen functional groups did not contribute to improvement in fiber-matrix bonding	[30]
Plasma	Hydrogen	Improvement in mechanical properties such as modulus of elasticity, yield strength, tensile and compressive stresses	All mechanical properties significantly improved with 30% carbon fiber content increase	[49]
Electro-polymerization	Polymethacrylic acid	(1) 36% Improvement in interlaminar shear strength (ILSS) (2) A 3% polymer mass increase significantly enhanced the wettability	The process is cost-efficient and environmentally friendly	[35]
Chemical grafting	GO and CNTs	The CF-GO-CNTs composites showed significant improvement of 48% and 83% for the ILSS and IFSS, respectively	The excellent interfacial strength was attributed to good wettability and surface energy, improved fiber-matrix interlocking and interface chemical bonding	[9]

(continued)

Table 1 (continued)

Surface morphology treatment	Treatment agent	Enhancement	Comment	Author
Electrochemical grafting	Covalent attachment of small molecules	The functionalization of carbon fiber surface using small molecules improved the resistivity by 200% while the interfacial shear properties improved by over 19%	The reduction of carbon fiber conductivity was explored	[46]
Surface sizing	Sizing agents	Heat treatment at 130 °C and 150 °C resulted in significant improvement of 71.1% and 15.1%, respectively, on interfacial shear strength (IFSS)	Increased fibers-sizing interaction is detrimental to the sizing-matrix bonding	[15]
Layer-by-layer self-assembly method	GO/SiO_2 multilayers films	The interfacial shear strength (IFSS), interlaminar shear strength (ILSS) and flexural strength of composites improved by 86.1%, by 89.3% and 30.4%, respectively	The researchers claim that the approach can be classified as green and provides a viable alternative for the preparation of high interfacial properties	[45]
Coating	Polydopamine	The mechanical properties such as bending strength increased by 11% while significantly improved tensile strength and Young's modulus were recorded	The highest mechanical performance was observed at PDA concentrations of 1.5 and 1 g/L and TRIS concentration of 40 mM	[43]

(continued)

Table 1 (continued)

Surface morphology treatment	Treatment agent	Enhancement	Comment	Author
Chemical grafting	Nylon 6 (PA6)	The grafting of Nylon 6 (PA6) onto the carbon fiber surface, improved the tensile strength of the composites by 30.8%	Enhanced wettability, fiber-matrix interfacial adhesion, mechanical interlocking and chemical interaction was achieved when grafting was performed on oxidized carbon fibers	[56]
Microwave irradiation	Etching	Although the tensile strength of treated carbon fibers was slightly deteriorated, A significant improvement in the interfacial shear strength was observed while the tensile strength degraded slightly	The microwave irradiation was conducted with the carbon fibers immersed in water	[48]

bulk mechanical properties [8]. Although oxygen plasma treatment as reported by Montes-Moran et al. [50] and Fu et al. [51] shows high functional group density and improved interfacial shear strength, it can cause some fiber degradation. Argon plasma treatment also has detrimental effects on the fiber; however, argon/ammonia plasma improves fiber-matrix bonding without degrading the strength of the fiber [29]. The most preferred method, however, is sizing mainly because of its simplicity and flexibility [21]. Although a huge amount of research is going on, the surface modification techniques still require extensive research particularly in areas such as pre-treatments using infrared- and ultraviolet-lasers [52], solvent cleaning methods [53], acidic etching [54], peel-ply and abrasion [53] and plasma treatments [55].

3 Conclusion

An exhaustive review of the current methods for carbon fiber surface treatment has been conducted. A comparison of the techniques is carried out and tabulated. The advantages and disadvantages of the methods are also presented. All the methods still require further research and tuning to achieve optimal conditions. The chemical approaches are efficient but are mostly not environmentally friendly. It is envisaged that as technology advances more environmentally sensitive approaches will be developed.

References

1. Yuan X, Zhu B, Cai X, Liu J, Qiao K, Yu J (2017) Optimization of interfacial properties of carbon fiber/epoxy composites via a modified polyacrylate emulsion sizing. Appl Surf Sci 401:414–423. https://doi.org/10.1016/J.APSUSC.2016.12.234
2. Lin SP, Han JL, Yeh JT, Chang FC, Hsieh KH (2007) Composites of UHMWPE fiber reinforced PU/epoxy grafted interpenetrating polymer networks. Eur Polym J 43:996–1008. https://doi.org/10.1016/j.eurpolymj.2006.12.001
3. Zabihi O, Ahmadi M, Li Q, Shafei S, Huson MG, Naebe M (2017) Carbon fibre surface modification using functionalized nanoclay: a hierarchical interphase for fibre-reinforced polymer composites. Compos Sci Technol 148:49–58. https://doi.org/10.1016/j.compscitech.2017.05.013
4. Li W, Yao SY, Ma KM, Chen P (2013) Effect of plasma modification on the mechanical properties of carbon fiber/phenolphthalein polyaryletherketone composites. Polym Compos 34:368–375. https://doi.org/10.1002/pc.22385
5. Liu J, Tian Y, Chen Y, Liang J, Zhang L, Fong H (2010) A surface treatment technique of electrochemical oxidation to simultaneously improve the interfacial bonding strength and the tensile strength of PAN-based carbon fibers. Mater Chem Phys 122:548–555. https://doi.org/10.1016/J.MATCHEMPHYS.2010.03.045
6. Zhao Z, Teng K, Li N, Li X, Xu Z, Chen L, Niu J, Fu H, Zhao L, Liu Y (2017) Mechanical, thermal and interfacial performances of carbon fiber reinforced composites flavored by carbon nanotube in matrix/interface. Compos Struct 159:761–772. https://doi.org/10.1016/J.COMPSTRUCT.2016.10.022

7. Li F, Liu Y, Qu C-B, Xiao H-M, Hua Y, Sui GX, Fu S-Y (2015) Enhanced mechanical properties of short carbon fiber reinforced polyethersulfone composites by graphene oxide coating. Polym (Guildf) 59:155–165. https://doi.org/10.1016/J.POLYMER.2014.12.067
8. Tiwari S, Bijwe J (2014) Surface treatment of carbon fibers—a review. Procedia Technol 14:505–512. https://doi.org/10.1016/j.protcy.2014.08.064
9. Gao B, Zhang R, He M, Sun L, Wang C, Liu L, Zhao L, Cui H, Cao A (2016) Effect of a multiscale reinforcement by carbon fiber surface treatment with graphene oxide/carbon nanotubes on the mechanical properties of reinforced carbon/carbon composites. Compos Part A Appl Sci Manuf 90:433–440. https://doi.org/10.1016/j.compositesa.2016.08.012
10. Sui X, Shi J, Yao H, Xu Z, Chen L, Li X, Ma M, Kuang L, Fu H, Deng H (2017) Interfacial and fatigue-resistant synergetic enhancement of carbon fiber/epoxy hierarchical composites via an electrophoresis deposited carbon nanotube-toughened transition layer. Compos Part A Appl Sci Manuf 92:134–144. https://doi.org/10.1016/J.COMPOSITESA.2016.11.004
11. Morgan P (2005) Carbon fibers and their composites. Taylor & Francis. https://doi.org/10.1016/j.tube.2011.07.006
12. Paul R, Dai L (2018) Interfacial aspects of carbon composites. https://www.tandfonline.com/doi/full/10.1080/09276440.2018.1439632. https://doi.org/10.1080/09276440.2018.1439632
13. Sharma M, Gao S Mäder E, Sharma H, Wei LY, Bijwe J (2014) Carbon fiber surfaces and composite interphases. https://www.sciencedirect.com/science/article/pii/S0266353814002358. https://doi.org/10.1016/j.compscitech.2014.07.005
14. Drzal LT, Rich MJ, Koenig MF, Lloyd PF (1983) Adhesion of graphite fibers to epoxy matrices: II. The effect of fiber finish. J Adhes 16:3–152. https://doi.org/10.1080/00218468308074911
15. Wu Q, Zhao R, Ma Q, Zhu J (2018) Effects of degree of chemical interaction between carbon fibers and surface sizing on interfacial properties of epoxy composites. Compos Sci Technol 163:34–40. https://doi.org/10.1016/j.compscitech.2018.05.013
16. Kamae T, Drzal LT (2012) Carbon fiber/epoxy composite property enhancement through incorporation of carbon nanotubes at the fiber-matrix interphase—Part I: The development of carbon nanotube coated carbon fibers and the evaluation of their adhesion. Compos Part A Appl Sci Manuf 43:1569–1577. https://doi.org/10.1016/j.compositesa.2012.02.016
17. Qian J, Wu J, Liu X, Zhuang Q, Han Z (2013) Improvement of interfacial shear strengths of polybenzobisoxazole fiber/epoxy resin composite by n-TiO_2 coating. J Appl Polym Sci 127:2990–2995. https://doi.org/10.1002/app.37970
18. Qin W, Vautard F, Drzal LT, Yu J (2015) Mechanical and electrical properties of carbon fiber composites with incorporation of graphene nanoplatelets at the fiber-matrix interphase. Compos Part B Eng 69:335–341. https://doi.org/10.1016/j.compositesb.2014.10.014
19. Loos MR, Yang J, Feke DL, Manas-Zloczower I, Unal S, Younes U (2013) Enhancement of fatigue life of polyurethane composites containing carbon nanotubes. Compos Part B Eng 44:740–744. https://doi.org/10.1016/J.COMPOSITESB.2012.01.038
20. Pokharel P, Lee DS (2014) High performance polyurethane nanocomposite films prepared from a masterbatch of graphene oxide in polyether polyol. Chem Eng J 253:356–365. https://doi.org/10.1016/J.CEJ.2014.05.046
21. Wang Z, Huang X, Xian G, Li H (2016) Effects of surface treatment of carbon fiber: Tensile property, surface characteristics, and bonding to epoxy. Polym Compos 37:2921–2932. https://doi.org/10.1002/pc.23489
22. Baghery Borooj M, Mousavi Shoushtari A, Nosratian Sabet E, Haji A (2016) Influence of oxygen plasma treatment parameters on the properties of carbon fiber. J Adhes Sci Technol 30:2372–2382. https://doi.org/10.1080/01694243.2016.1182833
23. Moosburger-Will J, Jäger J, Strauch J, Bauer M, Strobl S, Linscheid FF, Horn S (2017) Interphase formation and fiber matrix adhesion in carbon fiber reinforced epoxy resin: influence of carbon fiber surface chemistry. Compos Interfaces 24:691–710. https://doi.org/10.1080/09276440.2017.1267513
24. Baghery Borooj M, Mousavi Shoushtari A, Haji A, Nosratian Sabet E (2016) Optimization of plasma treatment variables for the improvement of carbon fibres/epoxy composite performance by response surface methodology. Compos Sci Technol 128:215–221. https://doi.org/10.1016/j.compscitech.2016.03.020

25. Chang T (1999) Plasma surface treatment in composites manufacturing. J Ind Technol 15:1–7
26. Farrow GJ, Jones C (1994) The effect of low power nitrogen plasma treatment of carbon fibres on the interfacial shear strength of carbon fibre/epoxy composites. J Adhes 45:29–42. https://doi.org/10.1080/00218469408026627
27. Corujeira Gallo S, Charitidis C, Dong H (2017) Surface functionalization of carbon fibers with active screen plasma. J Vac Sci Technol A Vac Surf Film 35:021404. https://doi.org/10.1116/1.4974913
28. Jones C, Sammann E (1990) The effect of low power plasmas on carbon fibre surfaces. Carbon N Y 28:509–514. https://doi.org/10.1016/0008-6223(90)90046-2
29. Jang BZ (1992) Control of interfacial adhesion in continuous carbon and kevlar fiber reinforced polymer composites. Compos Sci Technol 44:333–349. https://doi.org/10.1016/0266-3538(92)90070-J
30. Moosburger-Will J, Lachner E, Löffler M, Kunzmann C, Greisel M, Ruhland K, Horn S (2018) Adhesion of carbon fibers to amine hardened epoxy resin: influence of ammonia plasma functionalization of carbon fibers. Appl Surf Sci 453:141–152. https://doi.org/10.1016/j.apsusc.2018.05.057
31. Bousquet A, Awada H, Hiorns RC, Dagron-Lartigau C, Billon L (2014) Conjugated-polymer grafting on inorganic and organic substrates: a new trend in organic electronic materials. https://www.sciencedirect.com/science/article/pii/S0079670014000380. https://doi.org/10.1016/j.progpolymsci.2014.03.003
32. Subramanian RV, Jakubowski JJ (1978) Electropolymerization on graphite fibers. Polym Eng Sci 18:590–600. https://doi.org/10.1002/pen.760180708
33. Harris B, Braddell OG, Almond DP, Lefebvre C, Verbist J (1993) Study of carbon fibre surface treatments by dynamic mechanical analysis. J Mater Sci 28:3353–3366. https://doi.org/10.1007/BF00354259
34. Kainourgios P, Kartsonakis IA, Dragatogiannis DA, Koumoulos EP, Goulis P, Charitidis CA (2017) Electrochemical surface functionalization of carbon fibers for chemical affinity improvement with epoxy resins. Appl Surf Sci 416:593–604. https://doi.org/10.1016/j.apsusc.2017.04.214
35. Semitekolos D, Kainourgios P, Jones C, Rana A, Koumoulos EP, Charitidis CA (2018) Advanced carbon fibre composites via poly methacrylic acid surface treatment; surface analysis and mechanical properties investigation. Compos Part B Eng 155:237–243. https://doi.org/10.1016/j.compositesb.2018.08.027
36. Luo W, Zhang B, Zou H, Liang M, Chen Y (2017) Enhanced interfacial adhesion between polypropylene and carbon fiber by graphene oxide/polyethyleneimine coating. J Ind Eng Chem 51:129–139. https://doi.org/10.1016/j.jiec.2017.02.024
37. Zhang RL, Gao B, Ma QH, Zhang J, Cui HZ, Liu L (2016) Directly grafting graphene oxide onto carbon fiber and the effect on the mechanical properties of carbon fiber composites. Mater Des 93:364–369. https://doi.org/10.1016/j.matdes.2016.01.003
38. Steurer P, Wissert R, Thomann R, Mülhaupt R (2009) Functionalized graphenes and thermoplastic nanocomposites based upon expanded graphite oxide. Macromol Rapid Commun 30:316–327. https://doi.org/10.1002/marc.200800754
39. Wu Y, Lin X, Shen X, Sun X, Liu X, Wang Z, Kim JK (2015) Exceptional dielectric properties of chlorine-doped graphene oxide/poly (vinylidene fluoride) nanocomposites. Carbon N Y 89:102–112. https://doi.org/10.1016/j.carbon.2015.02.074
40. Fischer F, Kreling S, Jäschke P, Frauenhofer M, Kracht D, Dilger K (2012) Laser surface pretreatment of CFRP for adhesive bonding in consideration of the absorption behaviour. J Adhes 88:350–363. https://doi.org/10.1080/00218464.2012.660042
41. Ran B, Jing C, Yang C, Li X, Li Y (2018) Synthesis of efficient bacterial adhesion-resistant coatings by one-step polydopamine-assisted deposition of branched polyethylenimine-g-poly(sulfobetaine methacrylate) copolymers. Appl Surf Sci 450:77–84. https://doi.org/10.1016/J.APSUSC.2018.04.184
42. Xiong S, Wang Y, Zhu J, Yu J, Hu Z (2016) Poly(ε-caprolactone)-grafted polydopamine particles for biocomposites with near-infrared light triggered self-healing ability. Polym (Guildf). 84:328–335. https://doi.org/10.1016/J.POLYMER.2016.01.005

43. Kim HJ, Song JH (2019) Improvement in the mechanical properties of carbon and aramid composites by fiber surface modification using polydopamine. Compos Part B Eng 160:31–36. https://doi.org/10.1016/j.compositesb.2018.10.027
44. Merenda A, Ligneris E des, Sears K, Chaffraix T, Magniez K, Cornu D, Schütz JA, Dumée LF (2016) Assessing the temporal stability of surface functional groups introduced by plasma treatments on the outer shells of carbon nanotubes. Sci Rep 6:31565. https://doi.org/10.1038/srep31565
45. Fu J, Zhang M, Jin L, Liu L, Li N, Shang L, Li M, Xiao L, Ao Y (2019) Enhancing interfacial properties of carbon fibers reinforced epoxy composites via Layer-by-Layer self assembly GO/SiO2 multilayers films on carbon fibers surface. Appl Surf Sci 470:543–554. https://doi.org/10.1016/j.apsusc.2018.11.168
46. Beggs KM, Randall JD, Servinis L, Krajewski A, Denning R, Henderson LC (2018) Increasing the resistivity and IFSS of unsized carbon fibre by covalent surface modification. React Funct Polym 129:123–128. https://doi.org/10.1016/j.reactfunctpolym.2017.06.016
47. Wang B, Duan Y, Zhang J, Zhao X (2016) Microwave radiation effects on carbon fibres interfacial performance. Compos Part B Eng 99:398–406. https://doi.org/10.1016/J.COMPOSITESB.2016.06.032
48. Yuan JM, Fan ZF, Yang QC, Li W, Wu ZJ (2018) Surface modification of carbon fibers by microwave etching for epoxy resin composite. Compos Sci Technol 164:222–228. https://doi.org/10.1016/j.compscitech.2018.05.043
49. Lee ES, Lee CH, Chun YS, Han CJ, Lim DS (2017) Effect of hydrogen plasma-mediated surface modification of carbon fibers on the mechanical properties of carbon-fiber-reinforced polyetherimide composites. Compos Part B Eng 116:451–458. https://doi.org/10.1016/j.compositesb.2016.10.088
50. Montes-Morán MA, Martínez-Alonso A, Tascón JMD, Paiva MC, Bernardo CA (2001) Effects of plasma oxidation on the surface and interfacial properties of carbon fibres/polycarbonate composites. Carbon N Y 39:1057–1068. https://doi.org/10.1016/S0008-6223(00)00220-7
51. Fu YF, Xu K, Li J, Sun ZY, Zhang FQ, Chen DM (2012) The influence of plasma surface treatment of carbon fibers on the interfacial adhesion properties of UHMWPE composite. Polym Plast Technol Eng 51:273–276. https://doi.org/10.1080/03602559.2011.617406
52. Reitz V, Meinhard D, Ruck S, Riegel H, Knoblauch V (2017) A comparison of IR- and UV-laser pretreatment to increase the bonding strength of adhesively joined aluminum/CFRP components. Compos Part A Appl Sci Manuf 96:18–27. https://doi.org/10.1016/J.COMPOSITESA.2017.02.014
53. Encinas N, Oakley BR, Belcher MA, Blohowiak KY, Dillingham RG, Abenojar J, Martínez MA (2014) Surface modification of aircraft used composites for adhesive bonding. Int J Adhes Adhes 50:157–163. https://doi.org/10.1016/J.IJADHADH.2014.01.004
54. Wetzel M, Holtmannspötter J, Gudladt H-J, Czarnecki JV (2013) Sensitivity of double cantilever beam test to surface contamination and surface pretreatment. Int J Adhes Adhes 46:114–121. https://doi.org/10.1016/J.IJADHADH.2013.06.002
55. Gude MR, Prolongo SG, Ureña A (2012) Adhesive bonding of carbon fibre/epoxy laminates: correlation between surface and mechanical properties. Surf Coatings Technol 207:602–607. https://doi.org/10.1016/J.SURFCOAT.2012.07.085
56. Hui C, Qingyu C, Jing W, Xiaohong X, Hongbo L, Zhanjun L (2018) Interfacial enhancement of carbon fiber/nylon 12 composites by grafting nylon 6 to the surface of carbon fiber. Appl Surf Sci 441:538–545. https://doi.org/10.1016/j.apsusc.2018.01.158

Preparation and Mechanical Property Analysis of Polymer Matrix Composite Containing Rice Husk and Saw Dust

Sarojrani Pattnaik, Arnab Sengupta, and Mihir Kumar Sutar

Abstract Nowadays, there is a great demand for natural/wood fiber composites in different areas. Rice husk (RH) and saw dust (SD) are profusely available in India. In the present study, RH and SD reinforced polymer matrix composite has been prepared where both materials shared equal volume fraction in four equal categories of polymer samples containing 30, 35, 40 and 45% of filler content. The epoxy resin (HSC 9221) and hardener (EH 411) are mixed in 2:1 ratio. Polymeric samples prepared for each category have been tested for tensile test, Vickers hardness test and surface roughness. The results indicated that the tensile strength increased with percent increase in reinforcing material from 30 to 40%, thereby it deceased slightly. The highest tensile strength obtained was 57.7 MPa at 40% filler content. The average hardness of composite was found to be more for 45% filler loading. However, the hardness of polymeric sample containing 40% filler loading at different locations was found to be more uniform than that obtained at 45% filler loading and the values ranged from 7 to 7.4 HV. The surface roughness (SR) was found to be maximum (0.79 μm) at 45% filler loading due to agglomeration of filler content at few places. It was found to be minimum (0.64 μm) at 35% filler loading.

Keywords Polymer matrix composite · Saw dust · Rice husk · Hardness · Tensile strength

1 Introduction

Nowadays, the application of polymer matrix composites (PMCs) in different sectors such as transportation, aerospace industries, packaging, etc., have been increased drastically owing to their good corrosion resistance, low density, translucency, high stiffness, etc [1]. Natural fibers are reinforced with PMCs to enhance their mechanical and electrical properties. These natural fibers are light in weight, cheaply available,

S. Pattnaik · A. Sengupta · M. K. Sutar (✉)
Mechanical Engineering Department, Veer Surendra Sai University of Technology, Burla, Odisha, India
e-mail: mihirsutar05@gmail.com

S. Vijayan et al. (eds.), *Trends in Manufacturing and Engineering Management*, Lecture Notes in Mechanical Engineering,
https://doi.org/10.1007/978-981-15-4745-4_23

biodegradable and contain no toxic products. Thus, they are frequently inserted in polymer matrix to form fiber-reinforced PMCs. Till date, commonly used natural fibers employed in PMCs are flax fiber, jute, date palm leaf fiber, hemp, sisal, coir, cotton, saw dust, rice husk, etc [2].

Yang et al. [3] used rice husk (RH) flour as filler material in preparing thermoplastic composite. The authors did four levels of filler loading and tested the morphology and mechanical properties of the developed composites. It was found that with increase of filler loading, tensile strengths of the composite decreased, whereas tensile modulus increased. Chen et al. [4] prepared high loading of RH containing recycled thermoplastic green composites. The authors found that the fire-retardancy and mechanical properties of the composites were greatly increased with the usage of rice husk at higher loadings. However, it adversely affected the dimensional stability of the composite. Monteiro et al. [5] studied the effects of RH ash and graphite as fillers separately in polypropylene (PP)-based composites. A good bonding of aforementioned fillers in the polymeric matrix was found.

Xu et al. [6] prepared perforated high-density polyethylene-based composites made from recycled rubber and fiber-board sawdust (SD) with different loadings (18, 40 and 42%) for sound performance testing. It was found that the acoustic properties of the prepared composite depended on hole-size, cavity depth and perforation ratio. Suarez et al. [7] studied the tensile properties plus the fracture phenomenon of sawdust containing polypropylene (PP) or PP together with maleated PP composites. The tensile strength of the composites containing increased amount of MAPP to the PP matrix was found to be enhanced and also, they exhibited improved bonding of sawdust to the PP matrix.

From the above literature survey, it can be seen that the natural fibers, namely rice husk or saw dust have been proved as a good source of filler materials in preparing polymer matrix composites. However, till date, the combination of both aforementioned natural products has not been utilized in preparing a composite material. Thus, the present work focuses on using both RH and SD as reinforcing materials for developing epoxy resin base composites.

2 Materials and Methods

The rice husk obtained from the winnowing process was sieved in order to remove the impurities from it. Saw dust was obtained from the carpentry shop. The epoxy resin used in the present study was HSC-9221 and the hardener used was EH-411. Silica powder was also used along with resin and hardener to act as support material at high temperatures. The composites were made using a compression mold and made up of mild steel.

First, the volume fraction of fibers (RH & SD) and resin was calculated. The volume fraction of fibers was equally divided for RH & SD. Then, the epoxy and hardener were mixed in 2:1 ratio and the silica powder was added in such a manner that it would be 33% of epoxy resin. Composites containing 30, 35, 40 and 45% fiber volume fractions were made, as shown in Fig. 1 and they were coded as Type 1, 2, 3 and 4, respectively.

Then, the specimens were prepared for conducting tensile tests from each of the composite types as shown in Fig. 2. The tensile test was performed on panel controlled digital universal testing machine (UTM) with hydraulic open type grip (Fig. 3). The sample hardness was measured at different positions using a digital microhardness tester. The surface roughness of the composites was measured using Taylor Hobson Talysurf surface roughness tester (shown in Fig. 4). Several readings were taken by moving the stylus tip on the surface of the workpiece, and finally, the average of the surface roughness readings of each of the composite types were computed for the sake of comparison.

(a) (b) (c) (d)

Fig. 1 Composites containing variable volume fractions of fibers **a** 30%, **b** 35%, **c** 40% and **d** 45%

Fig. 2 Tensile test specimens of composite types

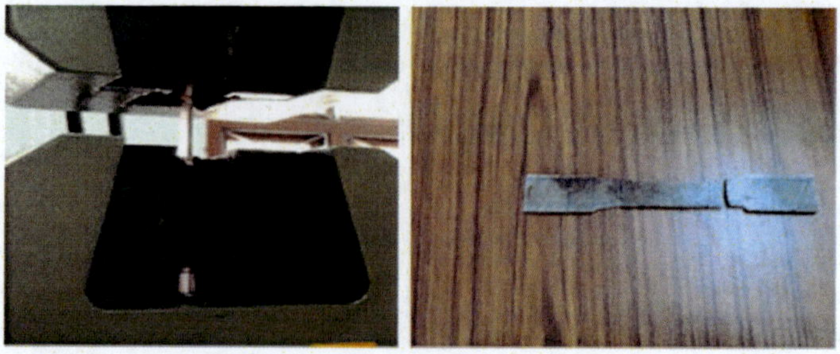

Fig. 3 Tensile test performed on prepared composite

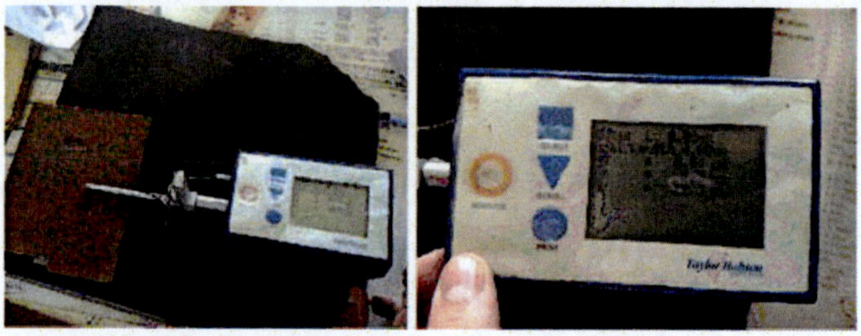

Fig. 4 Surface roughness testing of composite specimen using Talysurf

3 Results and Discussion

The tensile strengths of the composites of Types I, II, III and IV are shown in Fig. 5 and were found to be 51.7, 52.4, 57.7 and 53.7 MPa, respectively. Figure 5 depicts the tensile test results of each type of composites. It is seen that the tensile strengths of the composites increase with increase in filler content from 30 to 40 wt%, and thereafter, it decreases. The average microhardness test results of all composite types are depicted in Table 1. It is very obvious from the table that there is an increasing trend of hardness values with increase in filler percent from 30 to 45 wt%, and Type IV composite containing 45% filler material possessed maximum Vicker's hardness of 9.1 HV (Fig. 6), whereas average hardness value is 8.86 HV.

From the above results obtained for tensile strength and hardness tests, it is evident that up to 40 wt% of fillers loading, both the tensile strength and hardness of the composites increased. The Type III composite containing 40% fibers exhibited maximum tensile strength among all composite types because of highest interfacial adhesion between the fiber and the matrix which led to efficient load transfer between them. With rise in fiber content, the reinforcement in the matrix increases causing increase in hardness of the composites.

It is found that at 45 wt% filler content, the tensile strength of the composite decreased, whereas its microhardness is increased. The reason behind decline in tensile strength may be due to the weak chemical reaction at the interface to shift

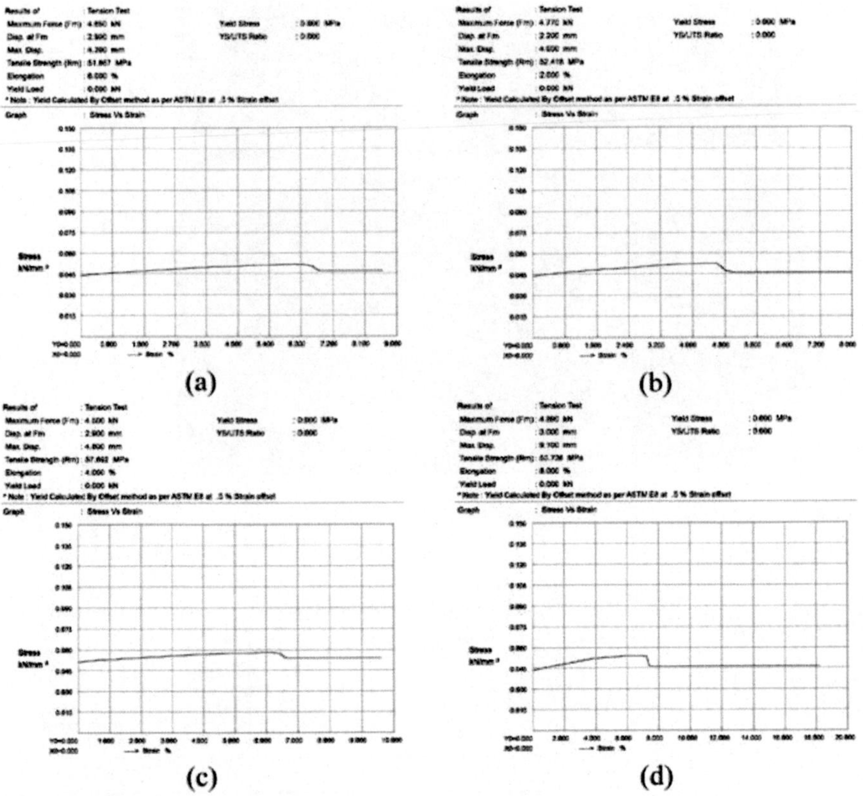

Fig. 5 Tensile test results of a Type I, b Type II, c Type III, d Type IV composite specimen

Table 1 Microhardness test results of all the composites

Sl.No	Volume fraction of fibers (%)	Vickers hardness (HV) avg. value
1.	30	5.56
2.	35	7.00
3.	40	7.26
4.	45	8.86

the tensile stress amid the matrix and the particulates. Also, there is a possibility of non-uniform distribution of reinforcing fillers in the matrix due to their increased wt. percent and these fillers would have somewhere agglomerated leading to delamination. Hardness test is done by pressing the filler particles and the epoxy matrix together so that both have a tight grip which facilitates easy transfer of pressure, causing increase in hardness of the composite.

The results of surface texture of all the prepared composites are depicted in Table 2.

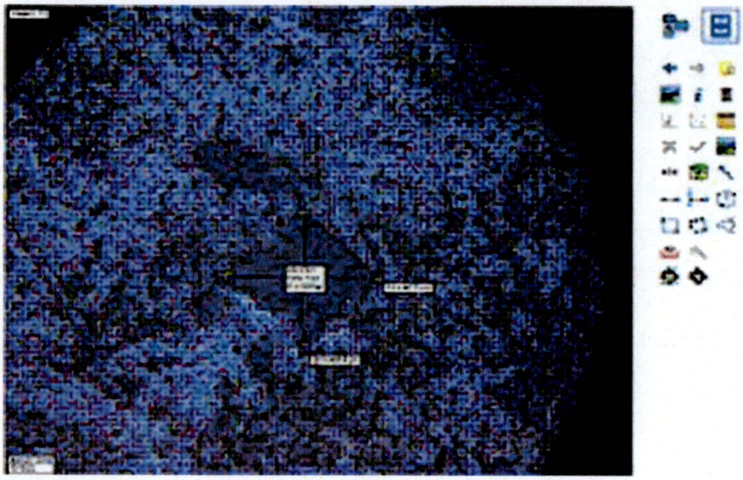

Fig. 6 Microhardness test conducted on composite of Type IV

Table 2 Surface roughness of all the composites

Type	Volume fraction of fibers (%)	Average roughness (Ra) (μm)
I	30	0.66
II	35	0.64
III	40	0.72
IV	45	0.79

It is seen that Type II composite containing 35% filler material showed maximum surface finish as regards to others, due to uniform distribution of fillers in the matrix. During the manufacture of the composites, few defects were detected. Few cavities were seen on the surface of the composite because of the sliding of the upper plate at the time of loading. Overflow of polymer also took place due to its high percent in the composition. A line was formed on one side of the composite due to slipping of the Mylar sheet.

4 Conclusions

From all the different types of tests performed on the fiber-reinforced epoxy composites containing saw dust and rice husk, final results are accomplished. The following observations are concluded from the present investigation:

- A good dispersibility of both sawdust and rice husk in the epoxy matrix had enhanced the tensile strength, hardness and surface texture of the composites.

- 40% of reinforcing fillers in the epoxy matrix gave the overall best results among all the prepared and tested composites.
- A very high surface finish of the composite can be obtained by careful handling of the specimen during the process.

References

1. Agarwal BD, Broutman LJ (1980) Analysis and performance of fiber composites. Wiley, New York, pp 3–12
2. Balakrishnan P, John MJ, Pothen L, Sreekala MS, Thomas S (2016) Natural fiber and polymer matrix composites and their applications in aerospace engineering. Adv Compos Mater Aerosp Eng. https://doi.org/10.1016/B978-0-08-100037-3.00012-2
3. Yang HS, Kim HJ, Son J, Park HJ, Lee BJ, Hwang TS (2004) Rice-husk flour filled polypropylene composites: mechanical and morphological study. Compos Struct 63:305–3012
4. Chen RS, Ahmad S, Gan S, Tarawneh MA (2019) High loading rice husk green composites: dimensional stability, tensile behavior and prediction, and combustion properties. J Thermoplast Compos Mater. https://doi.org/10.1177/0892705718815536
5. Monteiro AS, Barreira DAS, Bartolomei SS, Rene R. Oliveira and Esperidiana Augusta Barreto de Moura, Comparative Study of the Use of Rice
6. Xu X, Wang H, Sun Y, Han J, Huang R, Zhan X (2018) Sound absorbing properties of perforated composite panels of recycled rubber, fiberboard sawdust, and high density polyethylene. J Clean Prod. https://doi.org/10.1016/j.jclepro.2018.03.174
7. Suarez JCM, Coutinho FMB, Sydenstricker TH (2003) SEM studies of tensile fracture surfaces of polypropylene-sawdust composites. Polym Test 22:819–824

Investigation of the Mechanical Properties of Polyamide 6 Hybrid Nanocomposites with MWCNT and Copper Nanoparticles

T. Anand and T. Senthilvelan

Abstract The mechanical behaviour of polyamide6 (PA6)-based hybrid nanocomposites has been investigated. The polyamide6 (PA6) is blended with multi-walled carbon nanotube (MWCNT) and copper nanoparticles. The reinforcement materials in different proportions were blended with PA6 in a co-rotating twin-screw extruder and the specimens were made in an injection moulding machine. Mechanical tests were conducted on the prepared specimens to study the tensile, flexural and impact properties. The results suggested that Polyamide6 (PA6) hybrid nanocomposites infused with MWCNT and copper nanoparticles have significant increase in the tensile, flexural and impact properties. An atomic force microscope was used to examine the microstructure of the composites. Fractured surfaces of Polyamide6 nanocomposites after tensile and impact test were analysed through SEM images.

Keywords Polymer nanocomposites · Polyamide6 (PA6) · MWCNT · Copper nanoparticles · Atomic force microscope · SEM

1 Introduction

The developments in the field of engineering and technology majorly depend on the advancement of materials. Increasing demand for lightweight, strong and stiff structure has paved the way for the development of new materials. Composite materials have played a vital role in the process of optimization in materials from the beginning of the 1960s. Composite materials provide flexible designs and tailor-made materials for an optimum design.

T. Anand (✉) · T. Senthilvelan
Pondicherry Engineering College, Pondicherry University, Puducherry, India
e-mail: anand.tvel@gmail.com

S. Vijayan et al. (eds.), *Trends in Manufacturing and Engineering Management*, Lecture Notes in Mechanical Engineering,
https://doi.org/10.1007/978-981-15-4745-4_24

Nanomaterials as fillers or reinforcements [1, 2] in the composite can produce new physical properties and novel behaviours. Gheorghe Rusu et al. (2012) have studied the effect of TiO_2 reinforcement in nylon6 and found that the tensile and flexural characteristics were improved. Vijaya K. Rangari et al. (2012) have addressed the reinforcement of ZnO/CNTs with nylon6. This study revealed that the tensile stress and modulus were improved by 73% and 377.4%.

Polyamides (PA6) are semi-crystalline polymers synthesized by the ring opening polymerization of caprolactam, one of its monomers [3, 4]. The melting point of polyamide6 $(C_6H_{11}NO)_n$ is about 220 °C, crystalline density at 25 °C is 1.23 g/cm^3 and the molecular weight is 113.16 g/mol. Polyamides possess excellent mechanical properties, particularly toughness, and have good sliding and wear characteristics [5–8].

Carbon nanotubes are allotropes of carbon with a cylindrical nanostructure. These cylindrical molecules have extraordinary thermal conductivity, electrical and mechanical properties. Multi-walled carbon nanotubes (MWCNT) consist of multiple rolled layers of grapheme

Copper (Cu) is a ductile material with good thermal and mechanical properties [9]. Pure copper is soft and malleable. The atomic weight of copper is 63.546; melting point is about 1084.620 and it comes under the element category of a transition metal. The density of copper is 8.96 g/cm^3. Copper nanoparticles are used as a lubricant additive to reduce the friction and wear and to enhance the mechanical properties.

In this work, twin-screw extrusion and injection moulding processes were used to prepare polyamide6 nanocomposites with MWCNT and copper nanoparticles as reinforcements. The weight % of MWCNT was kept as constant and weight % of copper nanoparticles was varied. The mechanical behaviour of the composites was studied.

2 Experimental

2.1 *Materials*

The PA6 (commercial grade) material has been used in the present work. The multi-walled carbon nanotube is purchased from Applied Sciences Innovations private limited, Pune. The particle size of MWCNT is 25 nm and it is surface modified with COOH. Carbon coated copper nanoparticles of size 20 nm were purchased from Nanoshell Corporation, USA. A silane coupling agent γ-aminopropyltriethoxysilane has been used to create bonding between the matrix material and reinforcements.

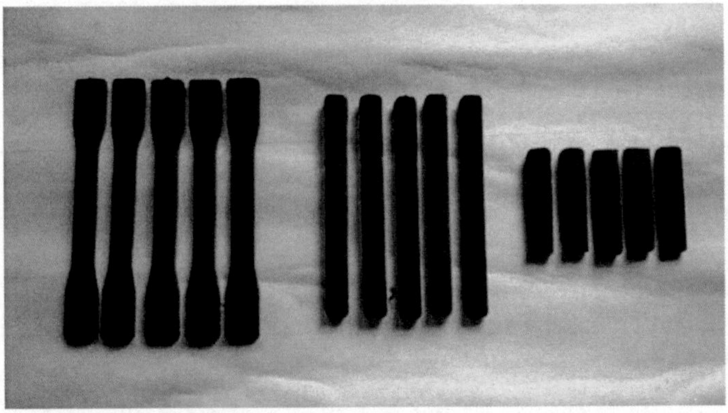

Fig. 1 Specimens for tensile, flexural and impact tests as per ASTM standards

2.2 *Preparation of Polyamide6/MWCNT/Nano-Copper Composites*

The materials are kept in an oven at a temperature of 80 °C for four hours to remove the moisture content present in them. The reinforcement materials are added to PA6 in weight percentage. The surface modified multi-walled carbon nanotube taken is 0.25 wt% and the carbon coated copper nanoparticles taken are 0.2, 0.4 and 0.6 wt% with PA6. The materials are blended in a co-rotating twin-screw extruder with a screw diameter of 25 mm and the rotating speed of the screw is kept at 180 rpm. The operating temperature is 250 °C. The blended material for each composition is pelletized in a pelletizer.

The blended composite mixture is then injection moulded using the MACI-FOLD injection moulding machine. The barrel temperature of the injection moulding machine is 290 °C and 10 MPa pressure. The specimens for the tensile, flexural and impact tests were prepared as per ASTM standards. Figure 1 shows the specimens prepared for the testing.

The operating temperature has been selected based on the melting point of the matrix material, whereas the screw diameter is kept constant. The rotating speed of the screw and operating pressure has been selected with reference to the machine catalogue depends on the matrix material. These parameters ensure proper blending of reinforcements with the matrix material [10].

2.3 *X-Ray Diffraction*

An XRD analysis [11] was employed to identify the dispersion of MWCNT and copper nanoparticles present in the matrix material PA6. X-ray diffraction analyses were performed on the prepared specimens using a Philips X-ray diffraction instrument with a radiation source of Cu-k-β using 30 kV and 100 mA. A step size of 0.02 degree, scan rate of 2 degree per minute and continuous scan mode are the parameters used for the analysis.

2.4 Atomic Force Microscopy

An atomic force microscope has been widely used for viewing the surface topology of the polymer composites. It is also applied to study the dispersion of the nano-reinforcements in the matrix material. AFM images were taken using the NTMDT atomic force microscope, Ireland. The frequency range is 5 MHz and the resolution of the atomic force microscope is 0.01 nm.

2.5 Mechanical Property Testing

The tensile test has been conducted in the specimen made as per ASTM D638(Type 1) with size 165 × 13 × 3 using the INSTRAN 3382, 100 KN universal testing machine with available load cells 100 N, 1 KN and 10 KN. The flexural test is conducted in the specimens made as per ASTM D 790 with size 127 × 12.7 × 6.4 mm using the AUTOGRAPH AG-IS 50 KN universal testing machine with the available range of loads 0.5,1,2.5,5,10,25 and 50 KN. The impact test is conducted in the specimen made as per ASTM D256 with size 63.5 × 12.7 × 6.4 mm using the izod impact testing machine [12]. A notch was provided on the impact test specimen with a depth of 2, 450 and 0.25 mm end radius. Five similar samples were used for each test and the average values are considered for the analysis. Figure 1 shows the prepared specimens for the tensile, flexural and impact tests as per ASTM standards.

3 Results and Discussion

The mechanical properties of PA6, multi-walled carbon nanotube and nano-copper hybrid composites are given in the following Table 1. The tensile strength, flexural strength and impact strengths of PA6 nanocomposites are illustrated and compared with that of pure PA6. The tensile strength, flexural strength and impact strength of PA6/0.25 wt%/0.6 wt% nano-copper was enhanced by 26.6, 39.47 and 139.46%, when compared to pure PA6.

In composite materials, the hardness increase with the percentage of the reinforcement materials. Figure 2 shows the shore–D hardness of PA6 and PA6 with various proportions of reinforcements. PA6 offered a shore–D hardness of 67 and PA6 with 0.25 wt% MWCNT and 0.6% nano-Cu offered a maximum hardness of 72.4 with 8.06% increase in hardness.

Table 1 Mechanical properties of PA6 and its composites

Sl.No	Description	Tensile strength (MPa)	% of increase in tensile strength	Flexural strength (MPa)	% of increase in flexural strength	Impact strength (J/m)	% of increase in impact strength
1	PA6	39.164	–	29.986	–	4837.2	–
2	PA6 + 0.25% MWCNT + 0.2% nano-Cu	43.029	9.869	35.609	18.75	5772.8	19.34
3	PA6 + 0.25% MWCNT + 0.4% nano-Cu	45.182	15.36	38.795	29.38	8589.5	77.69
4	PA6 + 0.25% MWCNT + 0.6% nano-Cu	49.595	26.6	41.823	39.47	11583.4	139.46

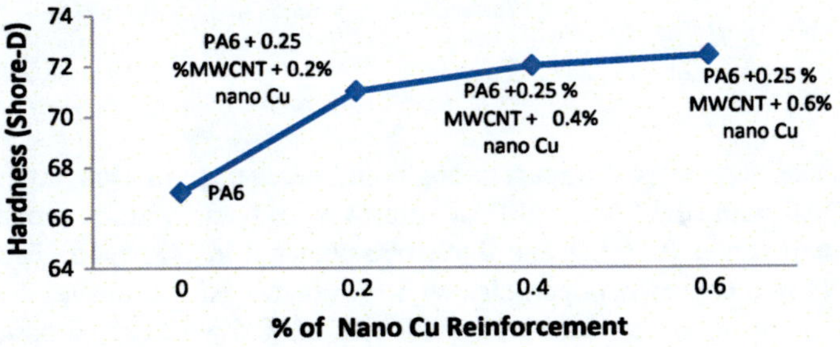

Fig. 2 Hardness (shore–D) of PA6 and its composites

The stress-strain behaviour [13] of PA6 and PA6 composites with 0.25 wt% MWCNT and 0.2, 0.4 and 0.6 wt% of copper nanoparticles are shown in Fig. 3. The stress value increases linearly within the elastic limit, beyond which the strain increases rapidly for a small stress value. The stress-strain behaviour of PA6 nanocomposites was more or less similar to that of virgin PA6.

Figure 4 shows the tensile modulus of PA6 and its composites. The tensile modulus of PA6 was increased for various proportions of reinforcements. The nanofillers dispersed evenly in the matrix material and created a strong interaction between the matrix material and reinforcements [14].

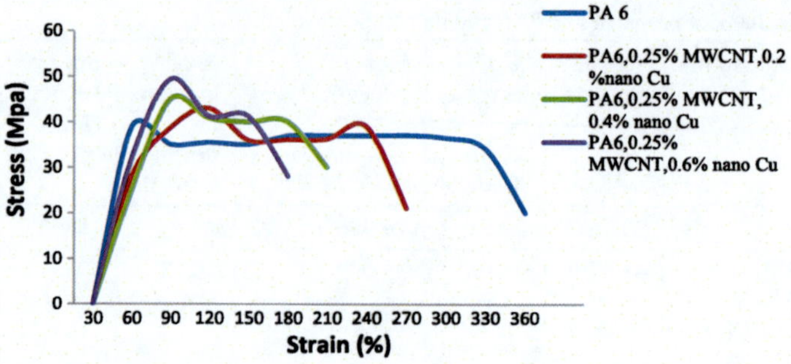

Fig. 3 Stress strain curves of PA6 and its composites

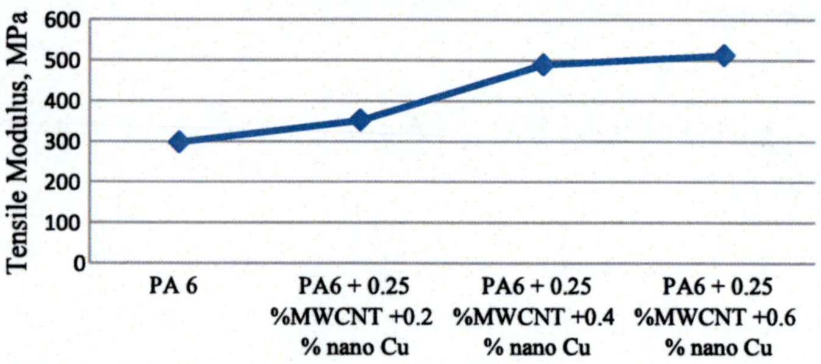

Fig. 4 Tensile modulus of PA6 and its composites

The tensile, flexural and impact strength and percentage of elongation at break for PA6, PA6 with MWCNT and PA6 with MWCNT and various proportions of copper nanoparticles (0.2, 0.4 and 0.6%) are depicted in Fig. 5a–d. The addition of MWCNT and copper nanoparticles with PA6 increased the tensile, flexural and impact strengths. PA6 with 0.25 wt% MWCNT and 0.6 wt% copper nanoparticles showed better mechanical properties than the other proportions. The deformability of PA6 nanocomposites decreases due to the reinforcement materials thus reducing the elongation at break [15].

Figure 6 shows the X-ray diffraction pattern of PA6 nanocomposites with various proportions of nano-reinforcements (PA6, 0.25 wt% MWCNT and 0.2 wt% Nano-Cu, PA6, 0.25 wt% MWCNT and 0.4 wt% Nano-Cu, PA6, 0.25 wt% MWCNT and 0.6 wt% Nano-Cu). The d-spacing was calculated by analysing the peaks of the XRD patterns using Bragg's law $n\lambda = 2d\sin\theta$, where θ is the scattering angle, n is a positive integer and λ is the wavelength of the incident wave and d is the interplanar distance [16]. An X-ray with a wave length of 1.54 A is used. The XRD pattern shows that PA6, MWCNT and nano-copper composites have diffraction peaks at $2\theta = 21.680$ and 43.40 corresponding to a d-spacing of 4.09 and 2.08 A.

The dispersion of nano-copper and multi-walled carbon nanotubes reinforced in different proportions with PA6 is studied by viewing the structure of PA6 nanohybrid

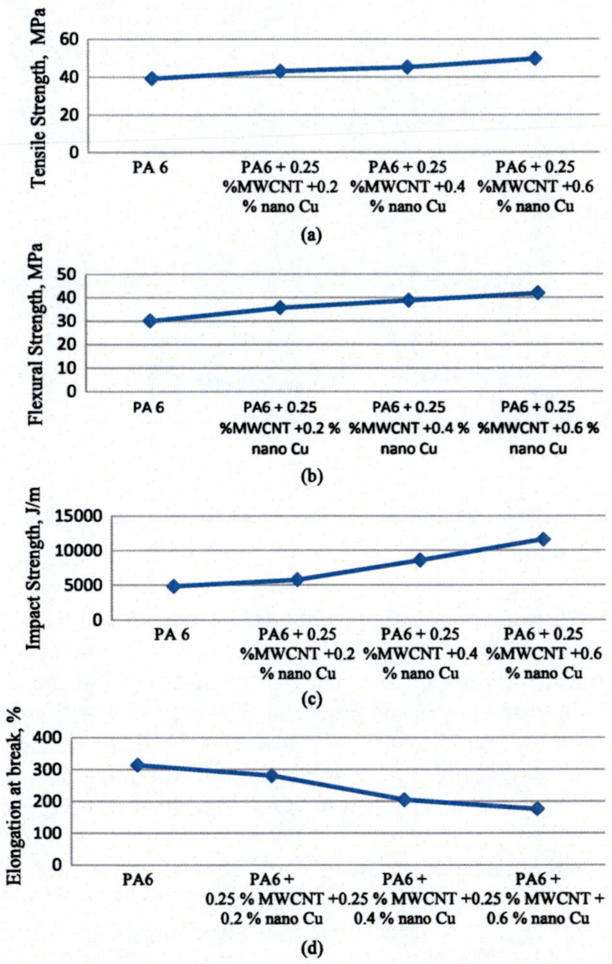

Fig. 5 **a** Tensile, **b** flexural, **c** impact strengths, **d** elongation at break of PA6 and its composites containing MWCNT and nano-copper

composites using AFM [17]. The height and phase images of the PA6 nanocomposites with different proportions of reinforcement are shown in Fig. 7.

The copper nanoparticles and multi-walled carbon nanotubes are evenly dispersed in the PA6 matrix with an average size of 61.49 nm. The microscopy images show, Fig. 7a PA6 with 0.25% MWCNT and 0.2% nano-Cu consists of an average particle size of 59.66 nm, Fig. 7b PA6 with 0.25% MWCNT and 0.4% nano-Cu consists of an average particle size of 61.293 nm, Fig. 7c PA6 with 0.25% MWCNT and 0.6% nano-Cu consists of an average particle size of 61.48 nm.

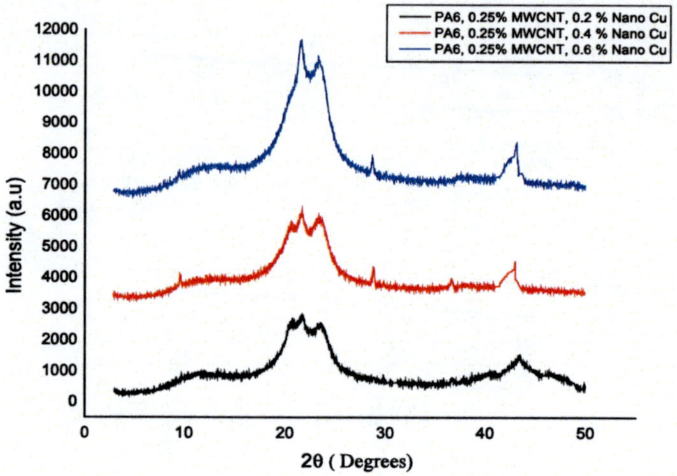

Fig. 6 X-Ray diffraction pattern of PA6 nano-hybrid composites with various proportions of nano-Cu and MWCNT

To study the fracture behaviour, fractographs of PA6 and PA6 with MWCNT and copper nanoparticles were taken using the scanning electron microscope. Figures 8 and 9 show the fracture surface of PA6 and PA6 composites after the tensile and impact tests. The microstructure of the fractured surface of PA6 appeared rough in Fig. 8a and PA6 with 0.25 wt% of MWCNT and 0.2 wt% of copper nanoparticles show small spherical dimples in Fig. 8b indicates a ductile fracture. Figure 8c, d shows smooth surface and some crazes were seen, indicating that the fracture was brittle in nature [18].

Figure 9a shows the fractured surface of PA6 after the impact test. The rough surface observed absorbs energy and that causes ductile fracture [19]. The nanoparticles, MWCNT and copper nanoparticles are found uniformly dispersed in the PA6 matrix material as shown in Fig. 9b, c. The addition of MWCNT and copper nanoparticles made the fractured surface of the composites smooth and brittle fracture was observed shown in Fig. 9d.

The incorporation of surface modified MWCNTs and carbon coated nano-copper particles into the PA6 matrix creates good interaction between the nano-reinforcements and polymeric chains [20–22]. The strong interaction between the nano reinforcements and PA6 matrix greatly enhanced the dispersion as well as interfacial adhesion. Hence, the mechanical properties of the composites could be improved.

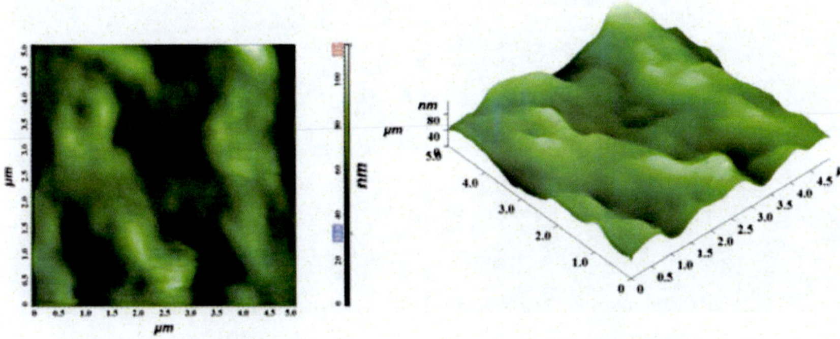

(a). Height images of PA6 with 0.25% MWCNT and 0.2% nano Cu

(b). Height images of PA6 with 0.25% MWCNT and 0.4 % nano Cu

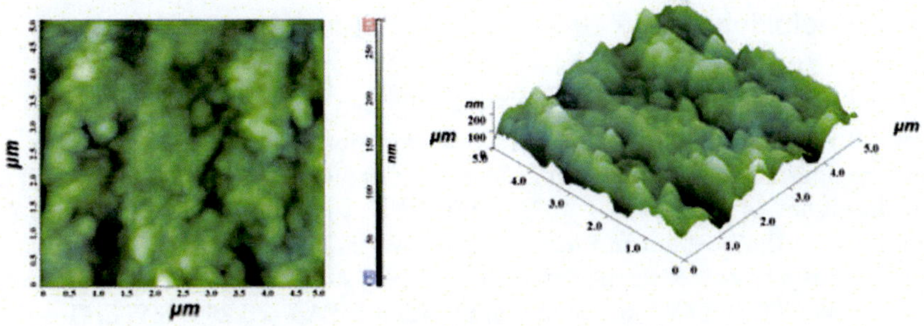

(c). Height images of PA6 with 0.25% MWCNT and 0.6 % nano Cu

Fig. 7 **a.** Height images of PA6 with 0.25% MWCNT and 0.2% nano-Cu. **b.** Height images of PA6 with 0.25% MWCNT and 0.4% nano-Cu. **c.** Height images of PA6 with 0.25% MWCNT and 0.6% nano-Cu

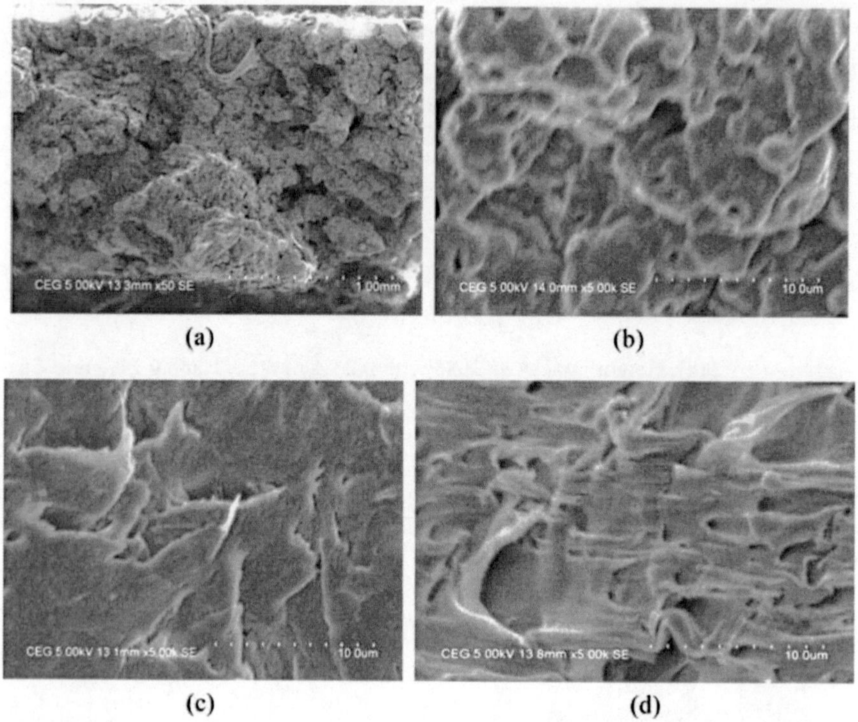

Fig. 8 Scanning electron microscopic images of the fractured surface after tensile test **a** PA6 **b** PA6, 0.25% MWCNT, 0.2% nano-Cu **c** PA6, 0.25% MWCNT, 0.4% nano-Cu **d** PA6, 0.25% MWCNT, 0.6% nano-Cu

4 Conclusion

PA6 nanocomposites reinforced with MWCNT and copper nanoparticles were successfully manufactured using twin-screw extrusion and injection moulding processes. Tensile, flexural and impact tests were conducted successfully on different compositions of PA6, PA6 + 0.25% MWCNT + 0.2% nano-Cu, PA6 + 0.25% MWCNT + 0.4% nano-Cu, PA6 + 0.25% MWCNT + 0.6% nano-Cu. The experiments showed that the tensile, flexural and impact strength values were improved by 26.6%, 39.47% and 139.46% in the composition of PA6 + 0.25% MWCNT + 0.6% nano-Cu compared to that of virgin PA6.

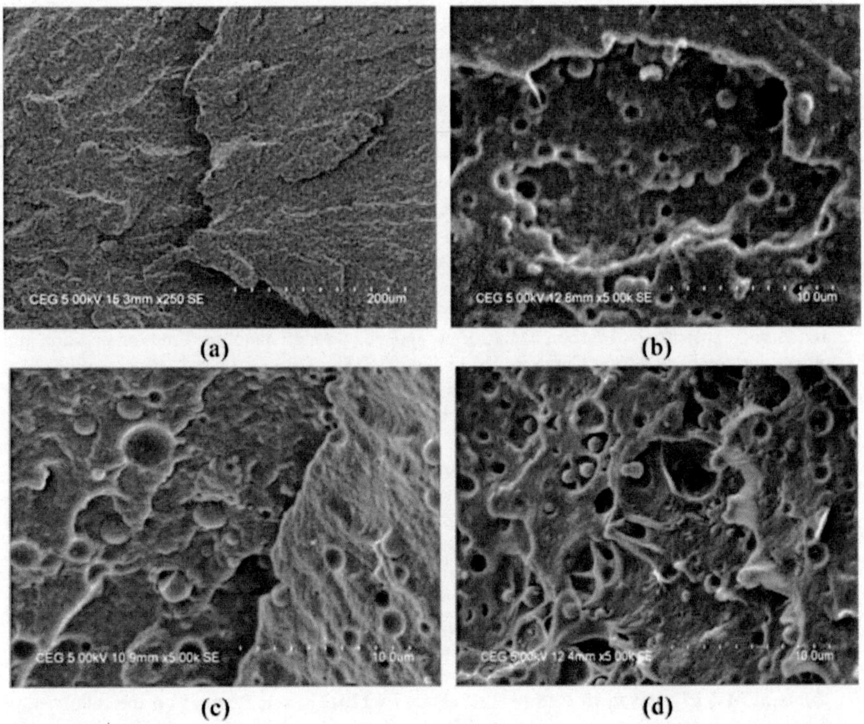

Fig. 9 Scanning electron microscopic images of the fractured surface after impact test **a** PA6 **b** PA6, 0.25% MWCNT, 0.2% nano-Cu **c** PA6, 0.25% MWCNT, 0.4% nano-Cu **d** PA6, 0.25% MWCNT, 0.6% nano-Cu

References

1. Rusu G, Rusu E (2012) Anionic Nylon 6/TiO_2 composite materials: effects of filler on the thermal and mechanical behaviour of the composites. Polym Compos 13:1557–1569
2. Rangari V, Mohammad GM (2012) Synthesis of hybrid ZnO/CNTs nano particles and their reinforcement in Nylon 6 polymer fibers. J Appl Polym Sci 129(1)
3. Liu Q, de. Wijjin JR, Van Blitterswijk CA (1977) Nano—Apatite/Polymer composites: mechanical and physiochemical characteristics. Biomaterials 18:1263–1270
4. Kransov AP, Aderikha VN, Afonicheva OV, Mit VA, Tikhonov NN, Yu A (2010) Vasil'kov: categorization system of nanofillers to polymer composites. J Friction Wear 31:68–80
5. Chiu FC, Kao GF (2012) Polyamide 46/multi walled carbon nanotube nano composites with enhanced thermal, electrical and mechanical properties. Compos Part A 43:208–218
6. Shen Z, Bateman S, Wu DY, McMohan P (2009) The effects of carbon nanotube on mechanical and thermal properties of woven glass fiber reinforced Polyamide 6 nano composites. Compos Sci Technol 69:239–244
7. Mohanty S, Nayak SK (2007) Mechanical, thermal and viscoelastic behaviour of Nylon 6/Clay nano composites with Cotreated Montmorillonites. Poly Plast Technol Eng 46:367–376
8. Liu TX, Liu ZH, Ma KX, Shen L, Zeng KY, He CB (2003) Morphology, thermal and mechanical behavior of Polyamide 6/layered—silicate nano composites. Compos Sci Technol 63:331–337

9. Luyt AS, Molefi JA, Krump H (2006) Thermal, mechanical and electrical properties of copper powder filled low-density and linear low-density polyethylene composites. Polym Degrad Stab 91(7):1629–1636
10. So HH, Cho JW, Sahoo NG (2007) Effect of carbon nanotubes on mechanical and electrical properties of polyimide/carbon nanotubes nano composites. Eur Polym J 43:3750–3756
11. Lee MW, Hu X (2003) Effect of fillers on the structure and mechanical properties of LCP/PP/SiO_2 in-situ hybrid nano composites. Compos Sci Technol 63(3):333–346
12. Chen GX, Kim HS, Park BH, Yoon JS (2006) Multi walled carbon nanotubes reinforced nylon 6 composites. Polymer 27:4760–4767
13. Amrishraj D, Senthilvelan T (2017) Acrylonitrile butadiene styrene composites reinforced with nano zirconia and PTFE. Polym Compos
14. Sahoo NG, Cheng HKF, Cai J, Li L, Chan SH, Zhao J, Yu S (2009) Improvement of mechanical and thermal properties of carbon nanotube composites through nanotube functionalization and processing methods. Mater Chem Phys 117 (2009) 313 – 320
15. Coleman JN, Khan U, Werner J, Blau Yurii K. Gun'ko (2006) Small but strong: a review of the mechanical properties of carbon nanotube—polymer composites. Carbon 44:1624–1652
16. Berti G, Angelo LD', Gatto A, Iuliano L. Mechanical characterization of PA—Al_2O_3 composites obtained by selective laser sintering. Rapid Prototyping. 16:124–129
17. Pollanen M, Pirinen S, Suvanto M, Pakkanen TT (2011) Influence of carbon nanotube—polymeric compatiblizer master batches on morphological, thermal, mechanical and tribologigal properties of polyethylene. Compos Sci Technol 71:1353–1360
18. Fu SY, Qiao Feng XI , Lauke B, Wing Mai Y (2008) Effects of particle size, particle/matrix interface adhesion and particle loading on mechanical properties of particulate—polymer composites. Composites: Part B 39:933–961
19. Saeed MB, Zhan MS. Adhesive strength of nano size particles filled thermoplastic polyimides. Int J Adhe Adhes 27(4):319–329
20. Kulkarni MV, Elangovan K, Reddy H, Prakash JN (2011) Investigations on the influence of filler on the impact properties of Polyamides and its composites. I-Managers J Mech Eng 1(4)
21. Yun Zheng L, Tak Lau K, Xin Zhao L (2009) mechanical and thermal properties of nano Al2O3/Nylon 6 composites. Chem Eng Commun 197:343–351
22. Chitte HK, Bhat NV, Karmakar NS, Kothari DC, Shinde GN (2012) Synthesis and characterization of polymeric composites embedded with silver nanoparticles. World J Nano Sci Eng 2:19–24

Preparation, Characterization, Image Segmentation and Particle Size Analysis of Cow Bone Powder for Composite Applications

O. M. Ikumapayi, Esther Titilayo Akinlabi, Paul A. Adedeji, and S. A. Akinlabi

Abstract Cow bone is a bio-waste material that has caused environmental unfriendliness to the people living around the place of production and deposition. In recent time, researchers have exploited various means of utilizing this bio-waste material and make it environmentally friendly and economically utilizable, especially in the area of applications, such as activated carbon, water purification, reinforcement in composites, filler and additives, and its efficacy has been traced to whether is in the form of macro-, micro- and nanoparticles. In this study, cow bone head (skull) was collected, washed and cleaned from meats, processed, sun-dried for six weeks, and it was then washed again with distilled water to remove impurities and contaminants, then placed inside the oven set at 50 °C to dry for 5 h to remove any trace of moisture content and to ensure absolute dryness. It was then milled into nanoparticle powder using vibratory disk milling machine and the milling times were 0 min which was taken to be 150 μm size, 20, 40 and 60 min. The morphological and physiological characterizations were carried out using scanning electron microscopy (SEM), energy dispersive X-ray (EDX) as well as X-ray fluorescence (XRF). MATLAB (R2015a) was employed for image processing into bimodal with foreground and background pixels using thresholding segmentation method. SEM images were taken at 1.00 ×, 2.00 × and 5.00 × which resulted to 50 μm, 20 μm and 10 μm, respectively. The image segmentation was employed to determine the foreground from a background of cow bone powder (CBP) and to enhance high resolutions, visual perception as well as to achieve better quality of the final output.

Keywords Cow bone powder · Image segmentation · Milling time

O. M. Ikumapayi (✉) · E. T. Akinlabi · P. A. Adedeji
Department of Mechanical Engineering Science, University of Johannesburg, Auckland Park Kingsway Campus, Johannesburg 2006, South Africa
e-mail: oikumapayi@uj.ac.za

S. A. Akinlabi
Department of Mechanical Engineering, Faculty of Engineering and Technology, Walter Sisulu University, Butterworth Campus, Butterworth 4960, South Africa

E. T. Akinlabi · S. A. Akinlabi
Department of Mechanical Engineering, Covenant University, Ota, Nigeria

S. Vijayan et al. (eds.), *Trends in Manufacturing and Engineering Management*, Lecture Notes in Mechanical Engineering,
https://doi.org/10.1007/978-981-15-4745-4_25

1 Introduction

Cow bone as organic bio-waste materials have been developed into different particles sizes from macro, micro as well as to nanoparticles. The development of microparticles from cow bones was carried out by first carbonized the cow bones at 750 °C using heat treatable furnace in the absence of oxygen and sieved using different mesh sizes of 150, 300 and 600 μm and these were then applied to polymer as reinforcement to form polymatrix composites [1] and it has also been used as reinforcement nanoparticle during friction stir welding of metal composites [2], in this study, cow bone was milling for 1 h and then carbonized at a temperature of 500 °C inside muffle furnace for 2 h and allowed for thermal stability inside the furnace to room temperature. It is a known fact that cow bones are waste products and making the environment unconducive to live in with the unbearable pollutants. Notwithstanding, researchers are making every effort to convert this agro-wastes to more useful engineering applications, especially for reinforcement purposes. Cow bone being natural animal fiber is expected to have good surface compatibility in addition to the structural compatibility requirements as biomaterials [3, 4].

It is essential to determine the particle size distribution as they affect the powders dissolution, dispersion and the flow properties. In the recent times, there have been tremendous works on the techniques to improve and address the dichotomy between the image backgrounds and foreground, especially when the object in a case is brighter with much variance [5, 6]. The image segmentation technique requires knowledge about difference intensity and object size for successful thresholding. To complement other studies, this study uses a sequence of an algorithm written in MATLAB to estimate the particle size at varying milling time. This is seen as a means of automating the process, such that particle size can be determined in real time during milling by using optimal imagery of samples. Thresholding becomes a potent tool for image segmentation when the variance of the object differs significantly from that of the background. This is often the case in SEM images and change monitoring applications [7]. The ideal threshold for segmentation is the intersection point of the image and its background. This is such that the intersection is far from the class with large variance [8].

2 Materials and Methods

2.1 *Material Collection and Preparation of Cow Bone Nanoparticle*

The skulls of the cow (cow head bone) were collected from the abattoir in the city of Johannesburg, South Africa as shown in Fig. 1a. It was then washed with de-ionized water, cleaned from meat and then sun-dried for about six weeks to allow for absolute dryness as shown in Fig. 1b. After 6 weeks in the sun, it was then washed again with

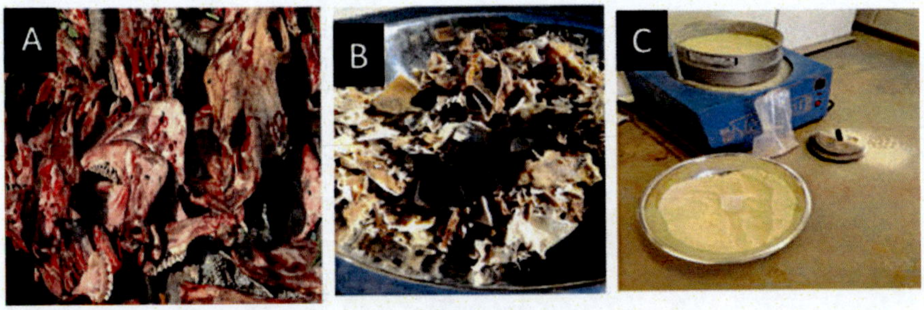

Fig. 1 **a** Fresh cow skull bone obtained, **b** dried cow bone and **c** cow bone powder (CBP) in a 150 μm ASTM standard sieve

de-ionized water to remove any impurities and contaminations that may be present. In order to ensure total dryness before crushing, it was dried again inside the electric oven set 50 °C to dry for 5 h for 7 days to ensure absolute dryness of moisture contents, this was further cleaned with acetone before crushing and pulverizing and sieved into 150 μm size using ASTM meshes standard range by employed KingTest Sieve, after which it was further milled with digital vibratory disk milling machine (VDMM) for 60 min before utilization as shown in Fig. 1c. The flowchart of the study is presented in Fig. 1.

2.2 *Characterizations of Cow Bone Powder*

The following characterizations were carried out on the cow bone powder scanning electron micoscopic (SEM), eergy dispersive X-ray (EDX) and X-ray fluorescence (XRF) to determine the elemental and chemical compositions as well as morphology of the powder (Fig. 2).

2.2.1 X-Ray Fluorescence (XRF) Spectroscopic Analysis of Cow Bone Powder

X-Ray fluorescence (XRF) spectroscopic analysis of cow bone powder (CBP) was carried out using PHILIP PW1404 XRF wavelength disperse spectrometer and it was revealed that CBP contains the following chemical composition 45.06% (CaO), 34.60% (P_2O_5), 0.065% (MgO), 0.25% (Fe_2O_3), 0.17% (Na_2O), 0.16% (Al_2O_3), 0.08% (Al_2O_3), 0.12% (K_2O), 0.06% (TiO_2), 0.02 (ZnO), 0.01% (MnO) and 16.28% (LOI).

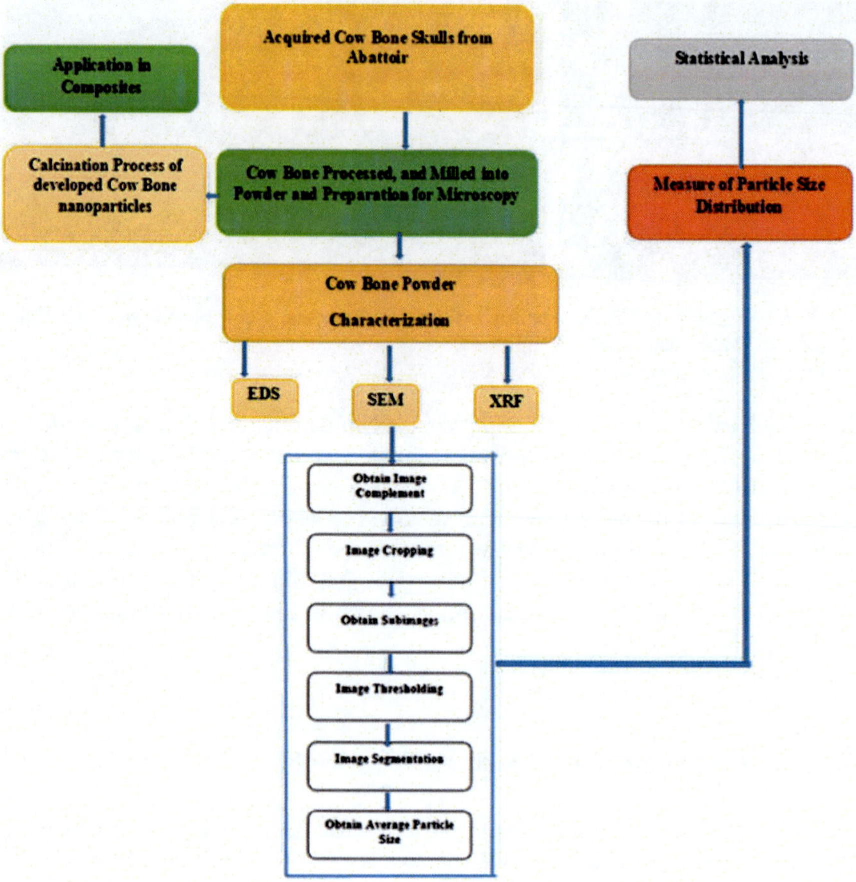

Fig. 2 Process flowchart of the study

2.2.2 Scanning Electron Microscopic (SEM) Analysis of Cow Bone Powder

Morphological and elemental characteristics of developed cow bone powder (CBP) were carried out using TESCAN model, type VEGA 3 LMH, scanning electron microscope (SEM) machine which was integrated with energy dispersive X-ray spectrometer (EDXS), this responsible for elemental composition analysis present in CBP. To ensure high conductivity and better resolution of the test samples (CBP), the sample was first sputter coated with a thin layer of carbon before the test and the test was conducted [9, 10] with the following parametric machine values, beam intensity of 12, accelerating voltage of 20 kV and a working distance of 15. SEM images were taken at 1.00 kx, 2.00 kx and 5.00 kx leading to 50 μm, 20 μm and 10 μm, respectively, for different milling times of 0 min, 20 min, 40 min and 60 min. The

SEM images at varying magnifications and milling time with respect to SEM image segmentation are shown in Fig. 3 while the elemental compositions are showed in Table 1.

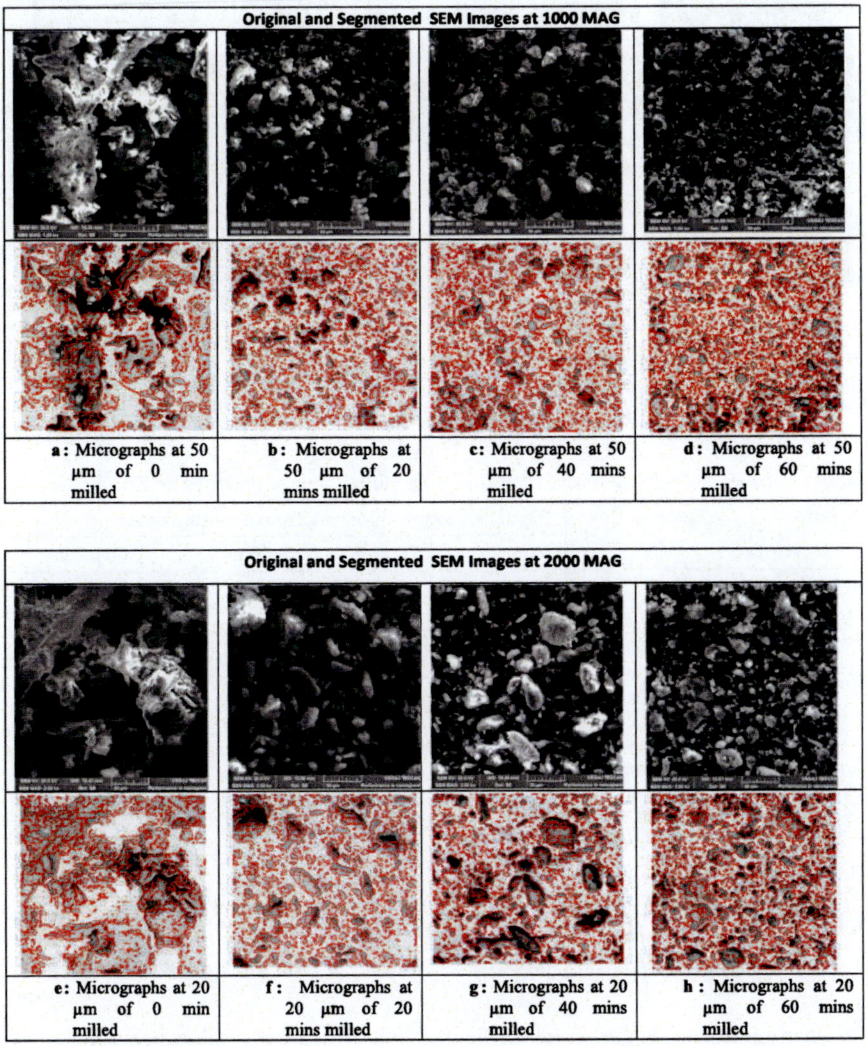

Fig. 3 SEM images and their segmentations for cow bone powder at varying magnification of 1.00, 2.00 and 5.00 kx at different milling time of 0, 20, 40 and 60 min

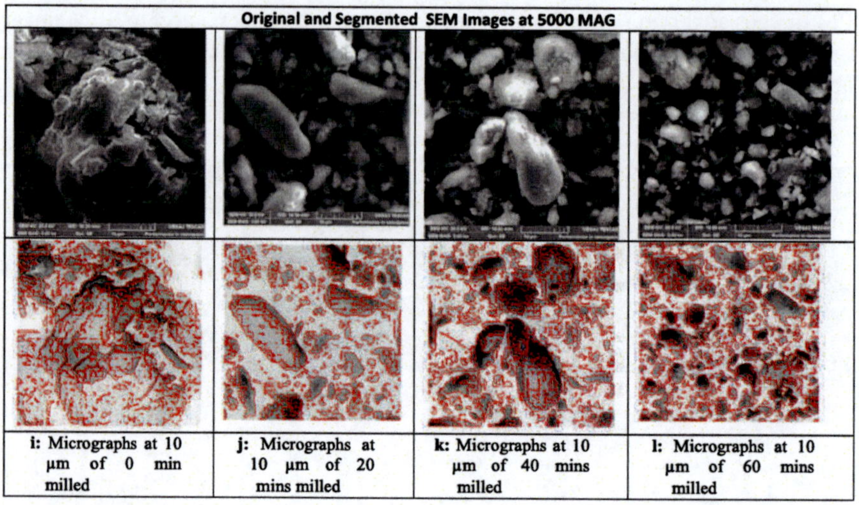

Fig. 3 (continued)

Table 1 Elemental compositions of CBP at varying milling time as analyzed by EDXS

Powder	Element	Elemental composition at different milling times			
		0 min	20 min	40 min	60 min
CBP	C	46.3	45.6	53.9	40.8
	O	27.2	30.7	30.0	32.1
	Ca	18.3	16.2	10.7	18.3
	P	7.5	6.8	4.8	8.0
	Mg	0.3	0.5	0.2	0.4

Note 0 min was taken at mesh size of 150 µm

3 Methodology

3.1 Method of Calcination of Cow Bone Powder for Composites Application

Calcination process is simply mean the method of heating a substance with a controlled temperature and controlled environment [11], i.e., thermal treatment process of material in the absence or limited supply of air in order to induce thermal stability, phase transition, thermal decomposition as well as removal of a volatile fraction from the material before using [12]. This is done to improve the strength, texture, thermal stability, etc. The milled cow powder cannot be applied directly as reinforcement to the workpiece that will be processed. If applied without any carbonization process, it will react with the metal and form slag within the processed metal as a result of a thermal generation that could lead to combustion of the particles, thereby leading to internal and surface defects. On this note, the processed powder was first undergone calcination process to ensure thermal stability. The calcination of the powder was carried out by putting each of CBP in pure graphite crucible and then put them carefully inside the muffle furnace set at 500 °C for two (2) hours, after the 2 h, the furnace was put off and the carbonaceous powder was allowed to cool to room temperature inside the furnace before it was then brought out for use, and then further grinded was performed on cow bone powder to ensure smoothness of the carbonaceous powder before use.

3.2 Image Segmentation and Thresholding

Distinguishing foreground pixels from the background pixels has been used to enhance visual perception of images [13–15]. One of the common techniques used to identify morphological maps [16] and image color intensity and is the threshold segmentation. The technique offers a simplicity in calculation and speed in operation [17]. Threshold segmentation recognizes image patterns aiming toward enhancing final image quality and resolutions [18]. This method does not perform well in cases where there is overlap in the grayscale value of the image [19, 20]. In this study, the SEM images, which are grayscale, were converted into binary images thus reducing data complexity.

Otsu's method [21], a nonparametric and an unsupervised thresholding technique, was used for image thresholding. The technique selects optimum thresholds by maximizing the intraclass variance of binary images. In this study, we assume that the SEM images are bimodal with foreground and background pixels.

Selecting from a set of local thresholds $t_1, t_2, t_3, \ldots t_n$, the thresholds subdivide the images into $n + 1$ classes of $C_1, C_2, C_3, \ldots C_n$. This is achieved by maximizing the objective function of a sum of all threshold variances [6];

$$J_1(t_1, t_2, t_3, \ldots t_n) = \sigma_0^2 + \sigma_1^2 + \sigma_2^2 + \sigma_3^2 + \cdots \sigma_n^2 \quad (1)$$

where

$$\sigma_0^2 = \omega_0(\mu_0 - \mu_T)^2, \omega_0 = \sum_{i=0}^{t_1-1} p_i, \mu_0 = \sum_{i=0}^{t_1-1} \frac{ip_i}{\omega_0}. \quad (2)$$

$$\sigma_1^2 = \omega_1(\mu_1 - \mu_T)^2, \omega_1 = \sum_{i=t_1}^{t_2-1} p_i, \mu_1 = \sum_{i=t_1}^{t_2-1} \frac{ip_i}{\omega_1}. \quad (3)$$

$$\sigma_2^2 = \omega_2(\mu_2 - \mu_T)^2, \omega_2 = \sum_{i=t_2}^{t_3-1} p_i, \mu_2 = \sum_{i=t_2}^{t_3-1} \frac{ip_i}{\omega_2} \quad (4)$$

$$\sigma_3^2 = \omega_3(\mu_3 - \mu_T)^2, \omega_3 = \sum_{i=t_3}^{t_4-1} p_i, \mu_3 = \sum_{i=t_3}^{t_4-1} \frac{ip_i}{\omega_3} \tag{5}$$

$$\sigma_n^2 = \omega_n(\mu_n - \mu_T)^2, \omega_n = \sum_{i=t_n}^{t_{n+1}-1} p_i, \mu_n = \sum_{i=t_n}^{t_{n+1}-1} \frac{ip_i}{\omega_n} \tag{6}$$

and $\sigma_0^2, \sigma_1^2, \sigma_2^2, \sigma_3^2, \ldots \sigma_n^2$. are the class variances, $\omega_0, \omega_1, \omega_2, \omega_3, \ldots \omega_n$ represents the class probabilities, $\mu_0, \mu_1, \mu_2, \mu_3, \ldots \mu_n$ represents the mean values of the segmented classes. The intensity of the whole image is

$$\mu_T \subseteq \mu_T = \omega_0\mu_0 + \omega_1\mu_1 + \omega_2\mu_2 + \omega_3\mu_3 + \ldots + \omega_m\mu_m\omega_0 + \omega_1 + \omega_2 + \omega_3 + \omega_n = 1.$$

The segmentation process was performed at different milling times and at different magnifications. A script for this was computed using MATLAB (R2015a) installed on a desktop computer workstation with configuration 64 bits, 32 GB RAM Intel (R) Core (TM) i7 5960X. The mean particle sizes were obtained at different milling time and different magnifications. A two-way analysis of variance was performed on the result to determine whether there is a statistically significant difference between the mean particle sizes across varying milling time and the across the magnifications.

3.3 *Results and Discussion*

Table 1 and Fig. 4 showed the variability in elemental compositions of cow bone powder as analyzed by EDXS. It was revealed that the following elements are present, calcium (Ca), carbon (c), phosphorous (P), oxygen (O) and little traces of magnesium (Mg). It can be observed that there is variation in composition of the elements as the milling progresses. The elemental composition varied along with the milling time from 0–60 min and this was also in agreement with the work carried out Omah et al. [1].

The mean particle sizes at each milling time and each magnification are as presented in Table 2. It is expedient to statistically establish whether the difference between the mean particle sizes at each milling time and magnification are statistically significant. To establish this, an analysis of variance was performed and presented in Table 3.

From Table 3, at a confidence interval of 0.05, the difference between the mean particle sizes obtained at different milling times is statistically significant. Similarly, the mean particle sizes obtained at different magnifications are statistically significant. However, the degree of significance in milling time is not as obvious as that from varying magnifications.

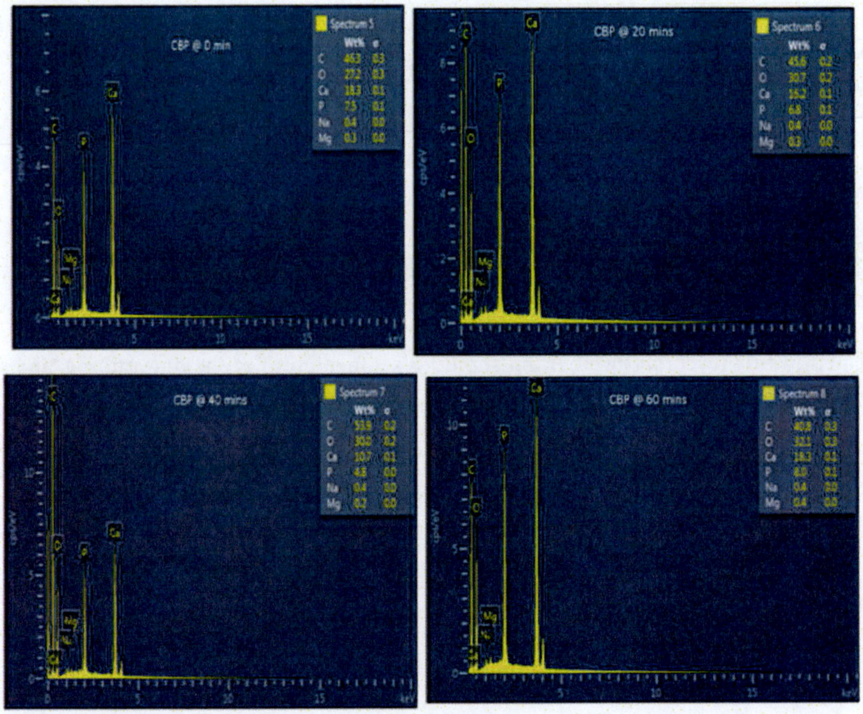

Fig. 4 EDS of the cow bone powder at varying milling time

Table 2 Mean particle size of segmented SEM images at different magnifications and milling times

	Magnification		
Milling time	×1000	×2000	×5000
0	2.70E+03	1.06E+03	507.996
20	2.98E+03	1.87E+03	526.284
40	3.54E+03	1.89E+03	727.674
60	4.03E+03	2.13E+03	781.975

Table 3 Statistical analysis of variance between mean particle sizes at varying milling time and different magnifications

Source of variation	SS	df	MS	*F*	*P*-value	*F* crit
Milling time	1,215,747	3	405,248.9	4.464	0.057	4.757
Magnification	14,487,754	2	7,243,877	79.799	4.76E−05	5.143
Error	544,659.7	6	90,776.62			
Total	16,248,160	11				

It can be inferred that the increasing the milling time of cow bone depending on the use, significantly affect its particle size. Particle sizes at different magnification are bound to vary due to change in image resolution.

4 Conclusion

The method used in this work permitted the manipulations of micrographs in order to have a true representation of the powder distribution and at the same time distinguishing foreground from background. Image processing analysis was investigated in this research through thresholding segmentation at varying milling times ($t = 0, 20, 40, 60$ min). It was revealed in the study that progressive milling leads to increase in particle size distributions and decrease in the area of particle size as noted in Table 2. It was revealed by EDX analysis that the main constituents of cow bone powder are calcium, oxygen, magnesium, carbon and phosphorous.

References

1. Omah AD, Okorie BA, Omah EC, Ezema IC, Aigbodion VS (2017) Measurement of dielectric properties of polymer matrix composites developed from cow bone powder. Int J Adv Manuf Technol 88:325–335. https://doi.org/10.1007/s00170-016-8759-1
2. Ikumapayi OM, Akinlabi ET (2019) Experimental data on surface roughness and force feedback analysis in friction stir processed AA7075-T651 aluminium metal composites. Data Br 23:103710. https://doi.org/10.1016/j.dib.2019.103710
3. Ahmad M, Lee SS, Lim JE, Lee SE, Cho JS, Moon DH et al (2014) Speciation and phytoavailability of lead and antimony in a small arms range soil amended with mussel shell, cow bone and biochar: EXAFS spectroscopy and chemical extractions. Chemosphere 95:433–441. https://doi.org/10.1016/j.chemosphere.2013.09.077
4. Aramide FO, Ibitoye SA, Oladele IO, Borode JO (2010) Pack carburization of mild steel, using pulverized bone as carburizer: Optimizing process parameters. Leonardo Electron J Pract Technol 9:1–12
5. Elbayoumi Harb SM, Isa NAM, Salamah SA (2015) Improved image magnification algorithm based on Otsu thresholding. Comput Electr Eng 46:338–355. https://doi.org/10.1016/j.compeleceng.2015.03.025
6. Khairuzzaman AKM, Chaudhury S (2017) Multilevel thresholding using grey wolf optimizer for image segmentation. Expert Syst Appl 86:64–76. https://doi.org/10.1016/j.eswa.2017.04.029
7. Hu QM, Hou ZJ, Wieslaw LN (2006) Supervised range-constrained thresholding. IEEE Trans Image Process 15:228–240
8. Xu X, Xu S, Jin L, Song E (2011) Characteristic analysis of Otsu threshold and its applications. Pattern Recognit Lett 32:956–961. https://doi.org/10.1016/j.patrec.2011.01.021
9. Abdulrahman A, Latiff AAA, Daud Z, Ridzuan MB, Jagaba AH (2016) Preparation and characterization of activated cow bone powder for the adsorption of cadmium from palm oil mill effluent. IOP Conf Ser Mater Sci Eng 136:1–11. https://doi.org/10.1088/1757-899X/136/1/012045
10. Adewole T (2015) Effect of cow bone ash particle size distribution on the mechanical properties of cow bone ash-reinforced polyester composites. Chem Mater Res 7:40–47

11. Mohammed A, Aboje AA, Auta M, Jibril M (2012) A comparative analysis and characterization of animal bones as adsorbent. Pelagia Res Libr Adv Appl Sci Res 3:3089–3096
12. Ihom PA (2013) Case hardening of mild steel using cowbone as Energiser. Afr J Eng Res 1:97–101
13. Zhu S, Xia X, Zhang Q, Belloulata K (2007) An image segmentation algorithm in image processing based on threshold segmentation. Third Int IEEE Conf Signal-Image Technol Internet-Based Syst 2007:673–678. https://doi.org/10.1109/SITIS.2007.116
14. Kuruvilla J, Sukumaran D, Sankar A, Joy SP (2016) A review on image processing and image segmentation. Int Conf Data Min Adv Comput 2016:198–203. https://doi.org/10.1109/SAPIENCE.2016.7684170
15. Ikumapayi OM, Akinlabi ET (2019) Image processing and particle size analysis of coconut shell. Int J Civ Eng Technol 10:2475–2482
16. Rohilla L, Garg V, Mallick SS, Setia G (2018) An experimental investigation on the effect of particle size into the flowability of fly ash. Powder Technol 330:164–173. https://doi.org/10.1016/j.powtec.2018.02.013
17. Zaitoun NM, Aqel MJ (2015) Survey on image segmentation techniques. Procedia Comput Sci 65:797–806. https://doi.org/10.1016/j.procs.2015.09.027
18. Aly A, Deris S, Zaki N (2011) Research review for digital image Segmentation techniques. Int J Comput Sci 3:99–106. https://doi.org/10.5121/ijcsit.2011.3509
19. Kaur D, Kaur Y (2014) Various image segmentation techniques: a review. Int J Comput Sci Mob Comput 3:809–814
20. Kaur A (2012) A review paper on image segmentation and its various techniques in image processing. Int J Sci Res 3:2319–7064 ISSN (Online Impact Factor)
21. Otsu N (1979) A threshold selection method from gray-level histograms. IEEE Trans Syst Man Cybern 9:62–66. https://doi.org/10.1109/TSMC.1979.4310076

Investigation of Mechanical and Chemical Properties of the Coir Fiber and Wood Powder Reinforced Hybrid Polymer Composite

T. Prabhuram, D. Elilraja, S. Prathap Singh, and Immanuel Durairaj

Abstract In many engineering applications, natural fibers are used because of high strength, low weight, and easy availability. In many composite studies reveals among the various natural fibers, the coir fiber is used as a reinforcing agent in polymer composites. In this paper, the mechanical and chemical properties of the hybrid polymer composites reinforced with coconut coir fiber and wood powder at different fractions were studied. The hand layup technique is used for sample preparation as per ASTM D638-03 standards. The mechanical and chemical properties such as tensile strength, flexural strength, and impact strength and water absorption capacity of the hybrid polymer composite were tested for different specimens prepared as per ASTM D638-03 standards. The results of the above tests prove that the coir fiber and wood powder have a significant influence on the mechanical properties of the hybrid polymer composite.

Keywords Hybrid polymer composite · Coir fiber · Wood powder · Hand layup technique

1 Introduction

Natural fiber-reinforced polymer composites have emerged as an environmentally friendly and alternative to synthetic fiber-reinforced composites. Natural fiber polymer composites play a major role in many engineering applications such as aviation, automotive (bumpers, battery cases, door panels, dashboards, door handles, and spoilers), construction, and packaging industries due to their unique properties including lightweight, ease of processing, availability, and cost-effectiveness. Rajesh et al. [1] checked the mechanical characterization and influence of chemical modifications and of natural fiber intra-ply fabric polymer composites. Omrani et al. [2] deal with state of the art on tribological behavior of polymer matrix composites reinforced with natural fibers. Prabhuram et al. [3] deal with mechanical

T. Prabhuram · D. Elilraja · S. Prathap Singh (✉) · I. Durairaj
St. Joseph's Institute of Technology, Chennai, India
e-mail: prathap.singh50@gmail.com

S. Vijayan et al. (eds.), *Trends in Manufacturing and Engineering Management*, Lecture Notes in Mechanical Engineering,
https://doi.org/10.1007/978-981-15-4745-4_26

properties of hybrid composite materials. Balaji et al. [4] used coir fiber and cotton fiber-reinforced unsaturated polyester composites for packaging applications. Kattimani et al. [5] conducted experimental investigation on coconut coir reinforced composite polymer. Prathapsingh et al. [6] investigate the tensile properties of alkali-treated pantandanus odoratissimus fiber-reinforced with polymer matrix composite. The main objective of this work is the preparation of coconut coir fiber and wood powder reinforced polymer composite with different weight fractions and compares the mechanical and chemical properties of the polymer composite.

2 Materials and Methods

2.1 Coconut Coir Fiber

Coir fiber is one of the naturally available fibers extracted from the coconut husk. The coir fiber is prepared from coconut husk available from rural areas of all over the southern states of India. The collected coir material is washed with normal water and dried in sunlight for several days. Hammering is done on dried coconut husk to extract coir fiber then it will be cut into the small pieces of 10 mm length approximately. Figure 1 shows the extracted coir fiber from coconut.

2.2 Wood Powder

The wood powder is collected from the local saw mill. The wood powder used in this work is extracted from the mango tree. The powder is sieved into the same size and dried in sunlight for several days. Figure 2 shows the processed wood powder.

Fig. 1 Extracted coconut coir fiber

Fig. 2 Wood powder of mango tree

2.3 Preparation of Composite

The hybrid composite specimen was prepared by hand layup technique which one of the common methodologies used for the preparation of composite materials. The coir fiber and wood powder act as the reinforcing material, whereas the epoxy resin acts as the base material. The hand layup technique may be divided into four stages such as mold preparation, gel coating, layup, and finishing. Molds were made of wood with a dimension of 30 × 30 × 0.5 cm. Gel coating is applied to the surface of the mold to separate the composite easily after curing. In layup stage, the required amount of coir and wood powder are measured along with resin and hardener with the ratio 10:1. Initially, the resin is mixed with coir fiber and wood powder and stirred for a few minutes to ensure uniform distribution of reinforcement material with the resin. After that the hardener is mixed with the resin and again stirred well. Then, the mixture is poured into the mold cavity and allowed to solidify for one day. After curing, the composite was taken away from the mold and dried is normal room temperature for a few hours. Table 1 shows the samples and the weight fractions of coir fiber and wood powder.

Table 1 Samples with weight fraction of coir fiber and wood powder

Samples	Percentage of epoxy resin (%)	Percentage of wood powder (%)	Percentage of coir (%)
A	70	30	0
B	70	0	30
C	70	15	15
D	70	20	10
E	70	10	20

3 Results and Discussion

3.1 Tensile Test

The tensile test was carried out to determine the tensile strength of the prepared hybrid composite material specimens. The specimen was prepared as per ASTM standard for tensile test. The prepared tensile test specimens were inspected under loading in the universal testing machine. The tensile strength of all the samples is tabulated in Table 2.

From Fig. 3, it was found that the composite with wood powder reinforcement shows lower tensile strength than the composite with coir reinforcement. The composites having the mixture of 15% of coir and 15% of wood powder reinforcement shows greater tensile strength than the individual reinforcement composites. This shows that the combination of coir and wood powder increased the tensile strength of the composites. Therefore, the increase in the coir percentage gives good tensile strength than the increase in wood powder percentage.

Table 2 Tensile strength of coir and wood powder reinforced polymer composites

Sample	Length (mm)	Width (mm)	Thickness mm)	Area (mm^2)	Load (kN)	Tensile strength (Mpa)
A	175	13.50	4.40	59.40	0.59	9.96
B	175	11.60	4.10	47.56	0.67	14.07
C	190	12.40	4.20	52.08	0.76	14.59
D	175	12.90	5.20	67.08	0.77	11.41
E	185	13.60	4.80	65.28	0.92	14.11

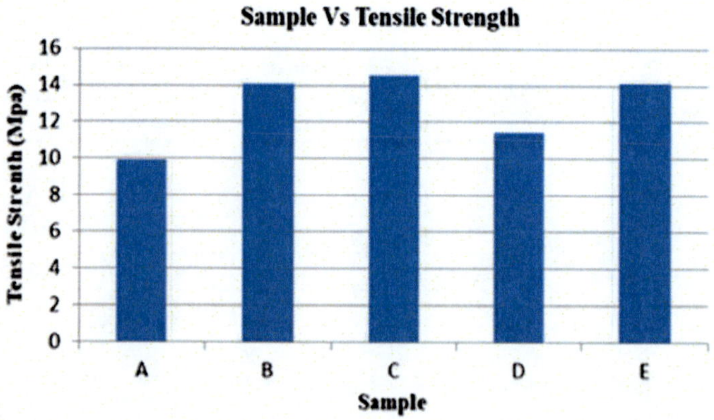

Fig. 3 Tensile strength comparison of samples

3.2 Flexural Test

The flexural test is carried out to determine the flexural properties of the composite material. The sample was prepared as per ASTM standard. The sample is kept on bending fixture and the compressive load is given and the curve generated till the failure of the sample takes place. Table 3 shows the flexural strength of all the samples.

From Fig. 4, it was found that the composite with wood powder reinforcement shows lower flexural strength, whereas the coir reinforced composite has greater flexural strength. Therefore, the increase in the coir percentage gives good flexural strength than the increase in wood powder percentage.

Table 3 Flexural strength of coir and wood powder reinforced polymer composite

Sample	Length (mm)	Width (mm)	Thickness (mm)	Area (mm^2)	Load (kN)	Flexural strength (N/mm^2)
A	100	14.50	4.10	59.45	0.06	36.92
B	100	14.50	4.60	66.70	0.08	39.11
C	100	14.10	4.60	64.86	0.07	35.56
D	100	15.50	4.30	66.65	0.09	47.10
E	100	15.50	4.30	66.65	0.10	52.34

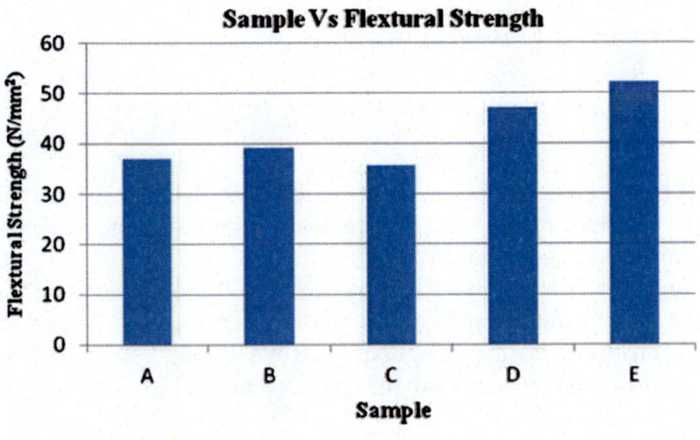

Fig. 4 Flexural strength comparison of samples

Table 4 Impact strength of coir and wood powder reinforced polymer composites

Sample	Length of span (mm)	Width (mm)	Thickness (mm)	Area (mm^2)	Impact strength (J)
A	80	15	5	75	2
B	80	15	5	75	4
C	80	15	5	75	2
D	80	15	5	75	2
E	80	15	5	75	2

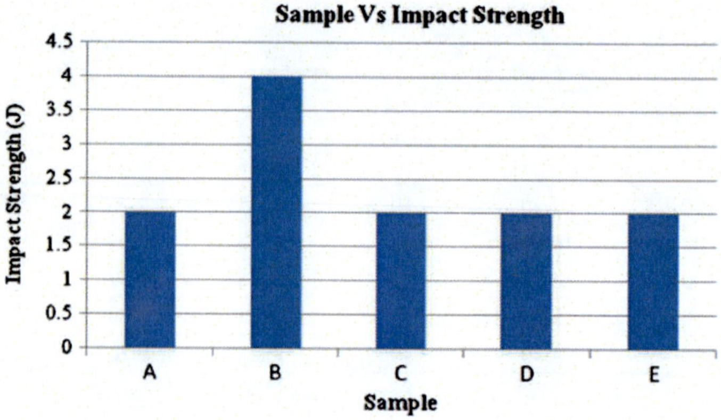

Fig. 5 Impact strength comparison of the samples

3.3 *Impact Test*

The impact test is the study of the ability of the material to withstand the sudden shock loads. This impact test was conducted in the impact testing machine. Table 4 shows the impact strength of all the samples.

From Fig. 5, it was found that the impact test result shows that there is a considerable increase in the impact strength for coir reinforced composite and there is no considerable change in impact strength of other specimens. It shows that the impact strength was improved by increasing the coir percentage.

3.4 *Water Absorption Test*

Water absorption test is used to determine the amount of water absorbed under room temperature conditions. Initially, the weights of the samples are measured using the weighing machine and then the samples were immersed in the normal water for 48 h. Then, the weights of the samples were measured again. The change in the weight of

Table 5 The water absorption capacity of coir and wood powder reinforced polymer composite

Sample	Initial weight (g)	Final weight (g)	% of water absorption
A	7.12	7.15	0.42
B	6.00	6.07	1.17
C	4.66	4.70	0.87
D	7.78	7.84	0.77
E	6.60	6.68	1.2

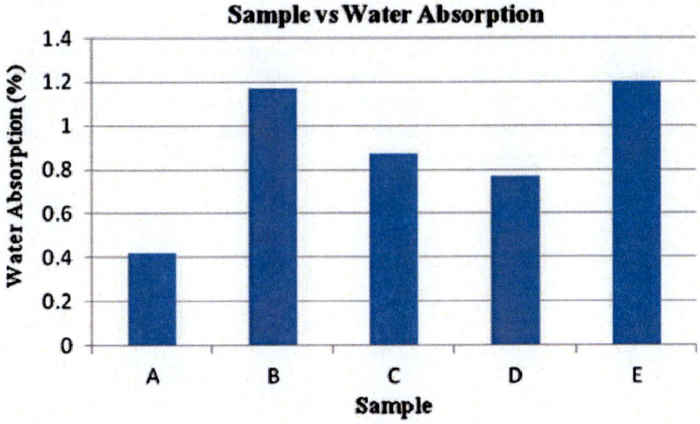

Fig. 6 Water absorption capacity comparison of the samples

the sample shows the water absorption capabilities of the composite. Table 5 shows the water absorption capacity of all the samples.

From Fig. 6, it was found that the composite absorbs a very minimum percentage of water. So, it is used in application such as storage tank and container fabrication.

4 Conclusions

The mechanical and chemical properties such as tensile strength, flexural strength, and impact strength, and water absorption capacity have been carried out on hybrid polymer composite samples. The results obtained from the above test conducted are concluded that the flexural strength of the composites increases with an increase in the percentage of coir than increases the percentage of wood powder. The coir reinforced composite has good impact strength than the wood powder reinforced composite. The water absorption capacity decreases when increasing amount of wood powder and a decreasing amount of coir fiber.

References

1. Rajesh M, Jeyaraj P (2017) Mechanical characterization of natural fiber intra-ply fabric polymer composites. Influence of chemical modifications. J Reinf Plast Compos 36(22):1651–1664
2. Omrani E, Menezes PL, Rohatgi PK (2016) State of the art on tribological behavior of polymer matrix composites reinforced with natural fibers in the green materials world. Eng Sci Technol Int J 19(2):717–736
3. Prabhuram T, Somurajan V, Prabhakaran S (2010) Hybrid composite materials. In: Frontiers in automobile and mechanical engineering (FAME), pp 27–31
4. Balaji V, Senthil Vadivu KP (2017) Mechanical characterization of coir fiber and cotton fiber reinforced unsaturated polyester composites for packaging applications. J Appl Packag Res 9(2):12–19
5. Kattimani P, Raikar N, Patil AN, Nandi P (2016) Experimental investigation on coconut coir reinforced composite polymer. Int J Engineering Trends Technol 41(3):128–133
6. Prathapsingh D, Elilraja D, Prabhuram T, Immanuel Durairaj J (2018) Investigation of tensile properties of Alkali treated pandanus odoratissimus fiber reinforced with polymer matrix composite. Int J Mech Prod Eng Res Dev 8(8):403–407

An Experimental Study on Hemp/Sisal Fiber Embedded Hybrid Polymer Composites

Akash, Shivakumar Rachoti, Vishwanath Patil, and K. G. Girisha

Abstract Environmental concerns encourage the use of artless fibers since bio-fiber embedded polymer composite crumble easily in nature and cheaper compare to petroleum-based fibers. Cellulosic fibers are gradually more gaining attention as their uses are diversified into structural and automobile parts for vehicle where low weight is necessary. The low density of cellulosic fibers is very beneficial in the structural and automobile industry. Cellulosic fibers offer many ecological and technical benefits for its use in incorporating composites. Current research paper investigates the mechanical strength (tensile, flexural, compression strength and shore-D hardness) and water absorption performance of different weight percentage of hemp/sisal fiber embedded polymer resin (epoxy resin) hybrid composites. Hemp and sisal fiber surfaces were subjected to 10 weight percentage of alkaline (NaOH) solution to improve the bonding characteristic between hydrophilic cellulosic fibers and hydrophobic polymer resin. Hybrid composite specimens were prepared by simple cold pressing technique. A 40 weight percentage of hemp/sisal fiber embedded hybrid composite sample discloses maximum tensile strength, flexural strength and compression strength. Shore-D hardness value and absorption of moisture of hybrid composite samples increases with increase in fiber loading. Morphological analyses are carried out to analyze the fractured surface of hybrid composite samples by using scanning electron microscope (SEM) and reported. The investigational result also discloses that the hemp and sisal fibers are promising reinforcements for use in cheaper bio-composites which have high strength to weight ratio.

Keywords Bio-fiber · Epoxy resin · Alkaline solution · Cold pressing · SEM

Akash (✉) · S. Rachoti · V. Patil
Sharnbasava University Kalaburagi, Kalaburagi, KA 585103, India
e-mail: akashbiradar@gmail.com

K. G. Girisha
Akshaya Institute of Technology, Tumkur, KA 585103, India

S. Vijayan et al. (eds.), *Trends in Manufacturing and Engineering Management*,
Lecture Notes in Mechanical Engineering,
https://doi.org/10.1007/978-981-15-4745-4_27

1 Introduction

Cellulosic fibers are increasingly gaining attention as their uses are diversified into structural and automobile parts for vehicles where low weight is required. The low density of cellulosic fibers is very helpful in the structural and automobile industry. Cellulosic fibers offer many ecological and technical benefits for its use in incorporating composites. Number of types of cellulosic fibers have been investigated for use in polymers, including bamboo, sisal, hemp, coir, flax, jute, banana leaf extracted, pineapple leaf extracted fiber etc., and the polymer matrix used for incorporating the fibers are classified as thermoplastics and thermosetting plastics. Cellulosic fiber incorporated composites are also claimed to offer ecological reward such as minimized dependence on non-renewable material/energy sources, lower green house gas emission, lower pollutant emissions, enhanced energy recovery and end of life biodegradability of components. Cellulosic fibers are renewable, discloses high specific properties, abundantly available with low density as well as they decompose in soil and less abrasive to costly molds and mixing tools [1]. Due to their poor resistance to moisture, incompatibility with hydrophobic polymer resin and their tendency to form aggregate during processing has reduced their potential use as reinforcement in polymer matrix composites.

The mechanical strength of cellulosic fiber reinforced polymer composites are lower compare to synthetic-based fibers [2]. To produce rough surface and reactive hydroxyl groups on cellulosic fiber surface for better bonding with polymeric resin, cellulosic fibers have to undergo chemical or/and physical treatment to modify the structure and surface. The hydrophobic epoxy resin—hydrophilic sisal and hemp fiber interfacial bond was improved by an alkaline treatment.

Hemp fiber and sisal fiber are most suitable cellulosic fibers for use in polymer composites, due to its greater specific strength and decompose in nature after use. Due to the more lignin content sisal and hemp fiber is long durable compared to other cellulosic fibers. With increasing emphasis on fuel efficiency, cellulosic fibers such as sisal and hemp fiber incorporated composite samples enjoying wider applications in structural, domestic and automobiles applications. Many aspects of the use of sisal and hemp fibers as incorporation in polymer resin composites are discussed in literature. Rout et al. [3] studied the mechanical properties of untreated and treated hemp fibers, discloses clear signs of the presence of poor bonding, long pulled out fibers without any polymer resin adhered to the fibers and poor mechanical strength were obtained compare to treated fiber embedded composites. Alkaline treatment to cellulosic fibers increases hydrophobic behavior with improving the resistance to fungi attack and also increases mechanical strength of polymer composites [4, 5]. According to author [5] 45 wt% of fiber loading is considerable reinforcing effect when the composite specimens are tested for tension. The utilization of cellulosic fiber alone, in a polymer resin is not enough in satisfactory tackling all the technical needs of a fiber embedded polymer composite. In an attempt to develop a better but inexpensive composite, a cellulosic fiber can be combining with another natural fiber in the same matrix so as to take the best favorable results of both natural fibers. The aim of this research work is to investigate the mechanical properties and moisture absorption behavior of sodium hydroxide treated hemp and sisal embedded epoxy resin polymer composites.

2 Experiments

2.1 *Materials*

Chopped hemp and sisal fibers were used as incorporation materials. These cellulosic fibers were procured from local dealers. Epoxy resin and corresponding hardener with brand name AW106 and HV953, respectively, were used as matrix phase to manufacture hybrid composites. Epoxy resin (AW 106) and hardener (HV953) at a ratio of 10:1 was mixed to prepare matrix material.

2.2 *Fiber Surface Treatment*

A 10 wt% of sodium hydroxide treatment removes cementing materials present in the cellulosic fiber, namely hemicelluloses, waxes, pectin and lignin, thus increase in the contact area of the cellulosic fiber. Increased fiber surface area leads to the better bonding of the matrix and cellulosic fiber, thus increasing the wettability and mechanical strength of the composites [6, 7]. Chopped hemp and sisal fiber were taken in separate trays, to these trays 10% NaOH solution was added and the both hemp and sisal fibers were soaked in the NaOH solution for 10 h. The hemp and sisal fiber were then washed thoroughly with distilled water to remove the excess of sodium hydroxide sticking to the hemp and sisal fibers. NaOH treated hemp and sisal fibers were dried in sun light for three days to remove completely moisture content from the fibers.

2.3 *Preparation of Hybrid Composites*

Simple cold pressing method was used to manufacture hybrid composite specimens in 250 * 250 * 6 mm metallic (cast iron) mold at high pressure. Hydraulic pressing machine was used to manufacture hemp/sisal fiber embedded epoxy resin hybrid composites. Thin layers of polyvinyl alcohol (PVA) were coated to both male and female molds to ensure that the sheet will not stick on to the metallic mold. Epoxy resin (Araldite AW 106) and hardener (HV 953) were mixed in a ratio of 10:1 in a bowl to prepare polymer matrix material. The quantity of fiber (hemp/sisal fiber) was varied from 10 to 50 wt% in steps of 10 wt%. A well-mixed mixture of reinforcement and matrix material was poured into PVA coated female mold die cavity. A PVA coated male die was positioned on the female die and pressure applied on male die up to 410.4 kg/cm^2 from hydraulic pressing machine for three hours. After drying, fabricated sample was cut to specimen to carry out tensile test, flexural test and compression test and water absorption behavior according to ASTM standard dimensions.

3 Results and Discussions

3.1 Mechanical Properties

Tensile test, compression test, flexural test and hardness test were conducted according to ASTM D-3039, ASTM D-695, ASTM 790-03 and ASTM D-2240, respectively. Three specimens were tested for individual test and average values are tabulated. Tensile, flexural and compression strength and shore-D hardness values of hybrid composite samples are shown in Figs. 1, 2, 3 and 4 respectively. The tensile strength of hemp/sisal fiber incorporated hybrid composites increases with increase in fiber loading up to 40 wt%, after that decreases is shown in Fig. 1. This indicates an effective and uniform stress transfer within composite after the incorporation of

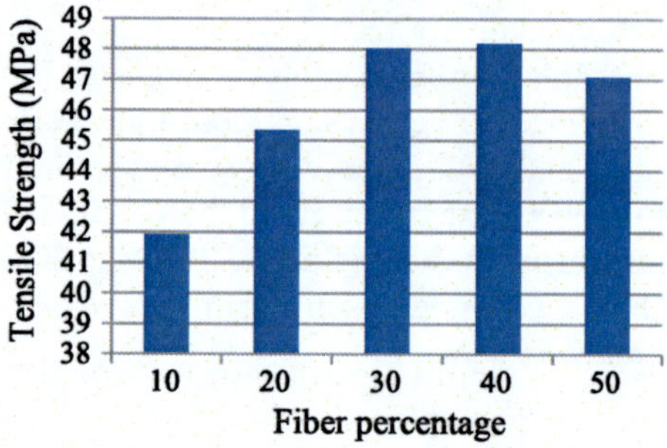

Fig. 1 Tensile strength

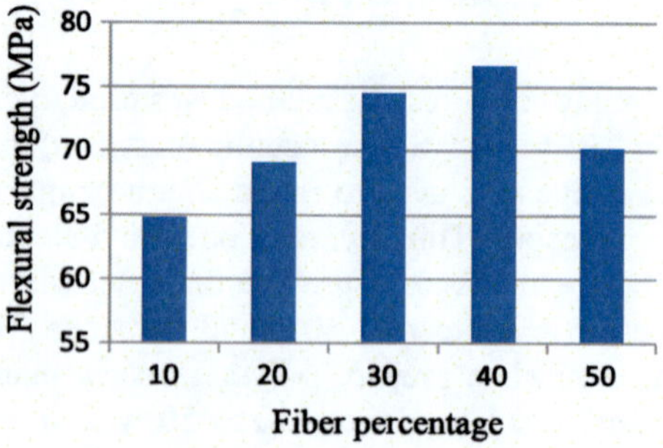

Fig. 2 Flexural Strength

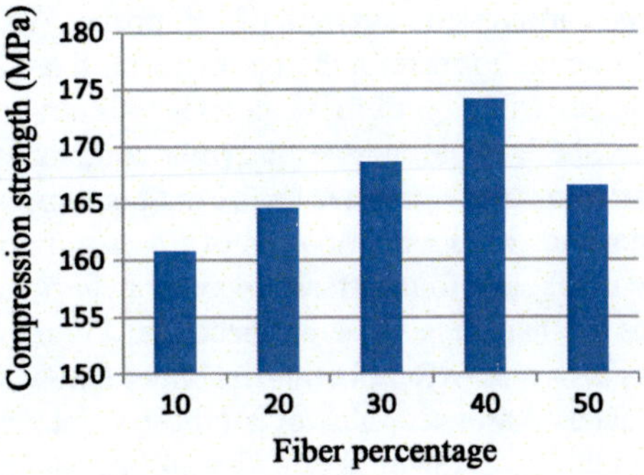

Fig. 3 Compression strength

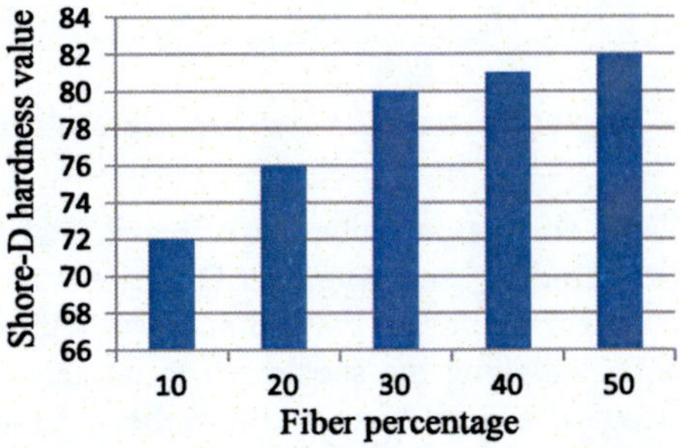

Fig. 4 Shore-D hardness number

cellulosic fibers in polymer matrix. At higher % of fiber loading, it is difficult for the polymer to penetrate the reducing spaces between fibers, leading to poor wetting, and hence, a reduction in the stress transfers efficiency through the fiber-matrix interface. Drop in tensile strength is attributed to the poor matrix/fiber interfacial bonding. The results highlight that the fiber weight fraction is the most dominant parameter that will affect the tensile strength [8].

Flexural strength of the fabricated hybrid composite specimens increases with increasing fiber loading up to 40 weight percentage, this enhancement in fiber/matrix strength is due to the ability of cellulose fiber to resist the bonding force. The lower flexural strength at lower cellulosic fiber content may be attributed to the lower loads transferred from the matrix to the fibers, thus resulting in lower load carried by fibers [9]. Significant increases in flexural strength at higher fiber content is because of the increase in stress transferred to the fibers as a result of increased adhesion at the fiber-matrix interface.

Compression strength of the fabricated hybrid composite specimens increases with increasing fiber loading up to 40 weight percentage this is due to the bonding at the crossing point between hydrophobic matrix and hydrophilic cellulosic fiber is too strong to transfer strength [10]. Higher cellulosic fiber concentration demands higher stress for the same deformation; therefore, the reinforcement of hemp fiber and sisal fiber into resin resulted in increases the compression strength. Further addition of hemp fiber and sisal fiber, i.e., 50 weight percentage lead to decrease the compression strength of composite. When the quantity of cellulosic fibers is not sufficient to restrain the epoxy resin, large stresses will be developed at low strains and distribution of these stresses will not be uniform [10, 11].

Shore-D hardness of polymer composite refers to its resistance to shape change after load/force applied on it. For fiber reinforced polymer composites, hardness value depends on the distribution of fibers in polymer resin. The shore-D hardness value was taken at four different spots for every sample and average result was considered as the hardness value of specimen. Figure 4 shows the variation of hardness value of different wt% of hemp/sisal fiber embedded hybrid composite. Figure 4 discloses that the hardness value of composites increases with increasing in fiber loading. The inclusion of hemp fiber and sisal fiber in epoxy resin has decreased the mobility of polymer chain in the rigid polymer composite [12].

3.2 *Moisture Absorption Behavior*

Water absorption properties of hemp/sisal fiber incorporated composite specimens were studied according to ISO 62:1999 standard. Manufactured hybrid composite samples were dried in oven at 60 °C and refrigerated to atmospheric temperature. After weighing the specimens to an accuracy of 0.1 mg, they were immersed in clean water in a bowl at atmospheric temperature. Daily once, the samples were taken out from the bowl, all the water particles from the surface were removed with cotton cloth and samples were weighed. Same process was done for 30 days. The % of moisture absorption of hemp fiber and sisal fiber incorporated epoxy resin composites increases with increasing fiber loading [13]. The % of moisture absorption was plotted against the time as shown in Fig. 5. Figure 5 also discloses the moisture absorption of hybrid composite increases with the increasing in the soaking time.

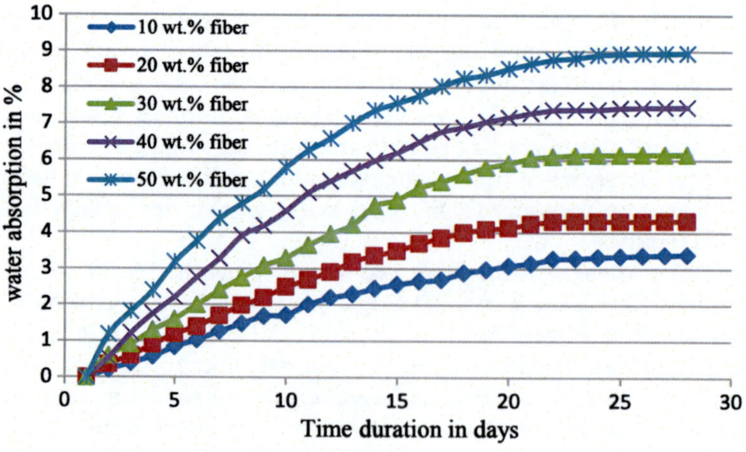

Fig. 5 Water absorption of hybrid composites

3.3 Morphological Analysis

The surface characteristics of the hemp/sisal fiber incorporated hybrid composite sample used for the investigation are studied through scanning electron microscope. Figure 6 discloses the SEM of fractured surface of fabricated hybrid composite sample. Figure 6 discloses the gap between fibers and matrix and fiber pullouts which indicates the weak bonding between the hydrophobic matrix and hydrophilic fibers. But some of the hemp and sisal fiber are fractured, and it is found to have strong interfacial bonding between matrix and fibers. The presence of voids in fractured samples, void space around the periphery of fibers confirms the de-wetting effect that is responsible for poor interface and de-bonding of matrix-fiber [14].

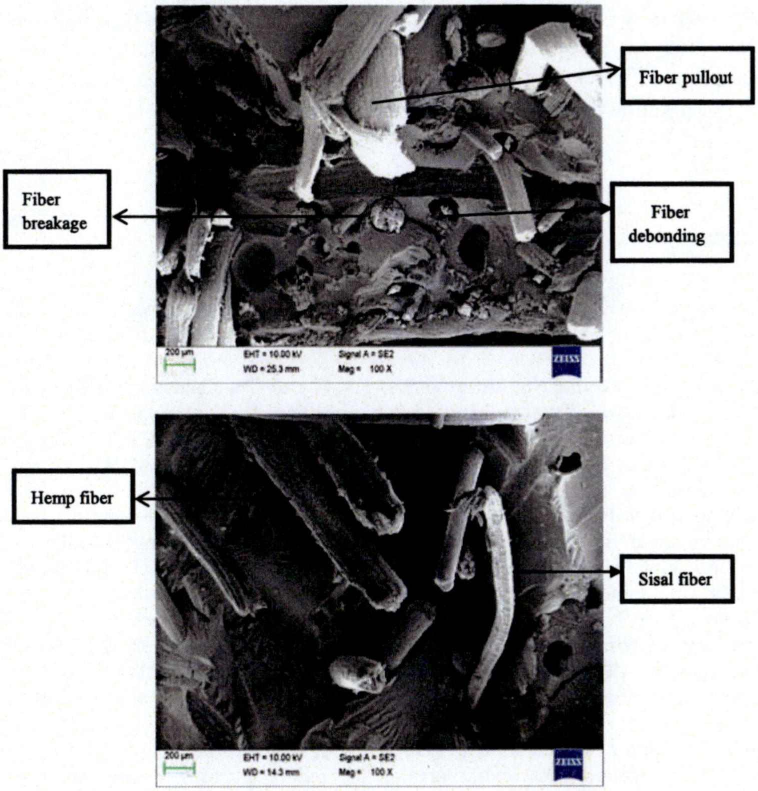

Fig. 6 SEM images of the hemp/sisal fiber reinforced hybrid composite samples

4 Conclusions

The study on mechanical, water absorption and fractured surface of hemp/sisal fiber incorporated epoxy resin hybrid composite led to the following conclusions.

- Tensile strength, flexural strength and compression strength of hemp/sisal fiber embedded hybrid composites increases with increasing fiber weight percentage in composite specimens up to 40 weight percentage, and then it decreases due to further loading the percentage of hemp/sisal fibers is incompatible with polymer resin and has tendency to form aggregates.
- Shore-D hardness value of hybrid composite increases with increasing in weight parentage of fiber due to addition of hemp/sisal fiber in polymer resin has decreased the mobility of polymer chain in the rigid composites.
- SEM images reveal that the hydrophilic cellulosic fibers (hemp fiber and sisal fiber) homogeneously mixed with hydrophobic polymer resin (epoxy resin).
- Absorption of water content increases with increase in the weight percentage of fibers of manufactured hybrid composite specimens.

References

1. Sanjeevamurthy GC, Rangasrinivas G, Manu S (2012) Mechanical performance of natural fiber-reinforced epoxy-hybrid composites. Int J Eng Res Appl (IJERA) 2(5):615–619. ISSN: 2248-9622
2. Akash, Venkatesh Gupta NS, Sreenivas Rao KV, Prasad CB, Khadka P (2015) Comparative valuation of mechanical and water absorption properties of pure epoxy resin, coir fiber/epoxy resin and hemp fiber/epoxy resin composite. Int J App Eng Res 10(55). ISSN 0973-4562
3. Rout J, Misra M, Mohanty AK, Nayak SK, Tripathy SS (2003) SEM observations of the fractured surfaces of coir composites. J Reinf Plast Compos 22:1083
4. Hill CAS, Khalil HPSA (2000) The effect of environmental exposure upon the mechanical properties of coir or oil palm fiber reinforced composites. J Appl Polym Sci 77:1322
5. Hill CAS, Khalil HPSA (2000) Effect of fiber treatments on mechanical properties of coir or oil palm fiber reinforced polyester composites. J Appl Polym Sci 78:1685
6. Gassan J, Bledzki AK (1999) Possibilities for improving the mechanical properties of jute/epoxy composites by alkali treatment of fibres. Compos Sci Technol 59:1303–1309
7. Sanjeevamurthy GC, Ranga Srinivas G (2012) Sisal/coconut coir natural fibers—epoxy composites: water absorption and mechanical properties. Int J Eng Innov Technol (IJJEIT) 2
8. Joseph K, Varghese S, Kalaprasad G (1996) Influence of interfacial adhesion on the mechanical properties and fracture behavior of short sisal fiber reinforced polymer composites. Eur Polym J 32(10):1243–1250
9. Ghosh R, Krishna AR, Reena G, Raju BL (2011) Effect of fibre volume fraction on the tensile strength of banana fibre reinforced vinyl ester resin composites. Int J Adv Eng Sci Technol 4(1):89–91
10. Akash, Sreenivasa Rao KV, Venkatesha Gupta NS, Arun kumar DS (2016) Mechanical properties of sisal/coir Fiber reinforced hybrid composites Fabricated by cold pressing method. IOP Conf Ser Mater Sci Eng 149:1–7

11. Venkatesha Gupta NS, Akash, Sreenivasa Rao KV, Arun Kumar DS (2016) Fabrication and evaluation of mechanical properties of alkaline treated sisal/hemp fiber reinforced hybrid composite. IOP Conf Ser Mater Sci Eng 149:1–7
12. Mir SS, Nafsin N, Hasan M, Hasan N, Hassan A (2013) Improvement of physico-mechanical properties of coir-polypropylene biocomposites by fiber chemical treatment. Mater Des 52:251–257
13. Akash, Girisha KG, Venkatesha Gupta NS, Sreenivas Rao KV (2017) A study on flammability and moisture absorption behavior of sisal/coir fiber reinforced hybrid composites. IOP Conf Ser Mater Sci Eng 2:1–6
14. Singh NP, Aggarwal L, Gupta VK (2015) Tensile behavior of sisal/hemp reinforced high density polyethylene hybrid composite. Mater Today Proc 2:3140–3148

Investigation on Chemical Isolation and Characterization of Cellulose from *Delonix regia* Fruit Fibers

Kalpit P. Kaurase and Dalbir Singh

Abstract Lignocellulosic fibers have gained popularity among the research fraternity in recent years due to its abundance in nature, biodegradability, high specific strength, etc. Cellulose can be extracted from biomasses by various chemical and mechanical methods. In this paper, *Delonix regia* fruit fibers are presented as a new source of cellulose, and chemical methods are used for extraction of cellulose from it. Chemically treated and untreated fibers were characterized by X-ray diffraction (XRD) and thermogravimetric analysis (TGA) to understand the effect of chemical treatment and properties of the cellulose yield. The results obtained demonstrated that a high yield of cellulose is obtained by chemical extraction methods followed and validated the suitability of *Delonix regia* fruit fibers for various applications including biocomposites and bionanocomposites.

Keywords Cellulose · Hemicellulose · Lignin · Bleaching · Alkali treatment · Lignocellulosic biomass · *Delonix regia* fruit fibers

1 Introduction

Lignocellulosic biomass is obtained from all types of plant materials, agricultural products, urban wastes, etc., and hence, it is abundantly available in nature. Lignocellulosic biomass in the raw and modified forms has been utilized from ancient times by human beings in various applications such as fodder for animals, food source, building houses, cloths, energy drinks and medicines. Plant biomass mainly consists of high energy materials like fat, protein, starch, etc., and plant cell walls which are made of cellulose, hemicellulose and lignin. Cellulose is the main structural component of cell wall which provides strength to the material, and it is very light in weight [1, 2]. Cellulose extracted from these lignocellulosic biomasses can be used in plenty

K. P. Kaurase (✉) · D. Singh
Hindustan Institute of Technology & Science, Hindustan University, Chennai, India
e-mail: kalpitkaurase@gmail.com

D. Singh
e-mail: dsinghcad@gmail.com

S. Vijayan et al. (eds.), *Trends in Manufacturing and Engineering Management*, Lecture Notes in Mechanical Engineering,
https://doi.org/10.1007/978-981-15-4745-4_28

of applications such as medical applications, food packing industry, biocomposites, bionanocomposites, etc.

Major component of lignocellulosic biomasses is water, but on dry basis, the main constituents are cellulose, hemicellulose and lignin which comprise almost 80% of the total weight, and remaining constituents are pectin, protein, lipid, ash, carbohydrate and wax. Percentage of these constituents varies in different plants, and also, in same plant, it varies in different parts and according to different climatic and growing conditions [3].

Cellulose and hemicellulose together are termed as holocellulose, and it consists of 65–70% of total dry mass. Cellulose is an organic compound with formula $(C_6H_{10}O_5)_n$ of *D*-glucopyranose units bonded together by β-glucosidic bonds. Glucose units altogether are frequently called as degree of polymerization (DP), and it varies from plant to plant. Cellulose is tasteless, odorless, hydrophilic white crystalline compound and insoluble in most of the organic solvents. It consists of almost 80% crystalline region and remaining amorphous region [4, 5].

Hemicellulose is a combination of polysaccharide polymers and consists of sugars of glucose, xylose, mannose, galactose, arabinose and rhamnose. It is a branched polymer and one of the most complex components of cell wall. It is amorphous compound having degree of polymerization less than that of cellulose [6].

Lignin is heterogeneous cross-linked phenolic polymers and serves the purpose of bonding cellulose and hemicellulose through strong hydrogen bonding and stable lignin–carbohydrate complex. There are basically three types, derived from phenyl propane, 4-hydroxy 3-methoxyphenylpropane, 3, 5-dimethoxy 4-hydroxy phenyl propane and 4-hydroxy phenyl propane [7].

Inorganic contents which are generally measured as ash content are also present in lignocelluloses. Proteins which are polymers of amino acids are available in high concentrations in some plants. Other contents of the lignocellulosic plants are termed as extractives which consist of waxes, terpenes, fats, fatty acids, phenols, etc.

Delonix regia is a flowering plant belonging to *Fabaceae* family of bean trees, and it is mostly found in tropical forests. In India, it is popularly known as Gulmohar plant and can be seen all across the country having very high population [8]. *Delonix regia* fruits are very hard having high strength and almost unutilized for any kind of applications in India and are potential biomass being wasted every year which can be used for cellulose extraction. The main objective of this research work is to extract cellulose for *D. regia* fruit fiber using chemical extraction methods and to characterize the treated and untreated fibers by using thermogravimetric analysis (TGA), X-ray diffraction (XRD) and differential scanning calorimetry (DSC) to verify the cellulose extraction process.

2 Materials and Methods

2.1 Materials

Delonix regia fruits were collected from the local fields of Raipur, Chhattisgarh, India. Chemicals used were: toluene, ethanol, sodium hydroxide, sodium hypochlorite, hydrogen peroxide and sulfuric acid. Various chemicals used were of analytical grades and obtained from Kasliwal Brothers, Raipur, Laxmi Scientific Works, Raipur and Lab Chemicals, Chennai.

2.2 Experimental Procedures

2.2.1 Sample Preparation

Green fruits of *D. regia* were cut from the *D. regia* tree, and seeds of fruit were removed. *Delonix regia* fruits were washed several times with deionized water and chopped into small pieces (approx. 2 cm × 2 cm) by chopper. Chopped fibers were dried in a convective oven at 80 °C for 24 h and grounded in a mill and sieved to attend uniform size in powder form and then washed and filtered to remove the dirt or waxing substances and improving swelling tendency as per NREL standard procedures [9]. Dirt-free and screened samples were stored in a polyethylene bags and kept in a desiccator for further use. Figure 1 shows the *D. regia* fruits biomass before and after size reduction.

2.2.2 Chemical Composition Determination

Size-reduced and screened samples of *D. regia* fruits were subjected to subsequential gravimetric composition analysis methods as per NREL Laboratory Analytical procedures (LAP) [9, 10].

Moisture Content
Screened sample (8 g) was subjected to convection oven drying at 105 °C for 24 h and weighed in balance and stored in a desiccator to avoid the moisture absorption. Total solid in the biomass is obtained by weight difference before and after oven drying and subtracting it from 100 gives moisture content of the biomass [9–11].

Extractive Content
Dry sample (5 g) was taken in a tared extraction thimble, and Soxhlet apparatus was assembled, and thimble was inserted in Soxhlet tube. Ethanol (190 ml) was added as solvent for extraction in the receiving flask and placed on Soxhlet apparatus. Temperature of the heating plate was kept at 80 °C, and distillation flask was kept in

Fig. 1 Raw biomass (*Delonix regia* fruit) before and after size reduction; **a** *Delonix regia* tree, **b** *Delonix regia* fruits (on tree), **c** *Delonix regia* fruits (Cut), **d** removed seeds, **e** dried and chopped fruits, **f** fruits after size reduction and screening

a water bath for the reflux time of 16 h. After complete extraction, sample was kept in Buchner funnel with cellulose filter paper and air dried at room temperature [9–11]. Weight difference before and after extraction process gives the extractive content.

Hemicellulose Content

Distilled water and NaOH (150 ml) were added to dry extracted sample (3 g) in Erlenmeyer flask and boiled for 4 h. Mixture was air cooled, vacuum filtered and washed to reach neutral pH after completion of boiling time. Residual content was dried in oven at 105 °C, and weight difference before and after treatment gives hemicellulose content [9–11].

Ash Content

Dried extracted sample (2 g) was placed in a tared crucible and ashed in a muffle furnace at 575 °C for 24 h. Crucible was removed and placed in a desiccator (to avoid moisture absorption) for initial cooling before weighing after complete ashing. Weight difference before and after complete ashing gives the ash content of the biomass [9–11].

Lignin Content

72% H_2SO_4 (3 ml) was added to dry extracted sample (1 g) and kept at room temperature for 2 h with continuous mixing by magnetic stirrer. After initial hydrolysis, distilled water (84 ml) was added and kept in autoclave for 1 h at 120 °C to complete hydrolysis. Then, mixture was air cooled at room temperature and vacuum filtered.

Acid insoluble lignin was calculated by drying residue at 105 °C and accounting for ash content. Acid soluble lignin was determined by acid digestion 320 nm. Lignin content of biomass was obtained by summing acid insoluble and acid soluble lignin content [11].

Cellulose Content

Cellulose content was estimated by considering that biomass contains moisture, ash, lignin, hemicellulose, extractives and cellulose and subtracting moisture, ash, lignin, hemicellulose, extractive percentage from 100%.

2.2.3 Cellulose Extraction

The following techniques were utilized for extracting cellulose from *D. regia* fruit fibers: Size-reduced and screened *D. regia* fruit fiber samples (20 g) were first subjected to extractive removal procedure by using Soxhlet apparatus. 2:1 (v/v) toluene/ethanol mixture was added to the round flask and placed on heating mantle, and Soxhlet apparatus was assembled for the total reaction time of 6 h at 90 °C, followed by washing with ethanol and drying [12]. Fibers were pulped chemically with 17.5% (w/w) NaOH and kept at room temperature overnight to improve swelling tendency of the fiber and partial removal of lignin and hemicellulose, successive cleaning with tap water, and filtration was carried out afterward. Fibers were delignified with 72% (v/v) sulfuric acid (H_2SO_4) at 70 °C for 1 h followed by washing with distilled water and centrifuged several times to reach neutral pH, and supernatant was collected [13]. Fibers were subjected to alkaline treatment with 2% (w/w) NaOH at 90 °C for 2 h in a water bath to remove hemicellulose. Fibers were bleached 5 times with 5% (v/v) sodium hypochlorite at 75 °C for 2 h followed by repetitive rinsing and filtration. Fibers were further bleached with 10% (v/v) hydrogen peroxide (H_2O_2) at 80 °C for 1 h, and filtrate was oven dried at 40 °C for 5 h to obtain highly purified yield of cellulose [14] (Fig. 2).

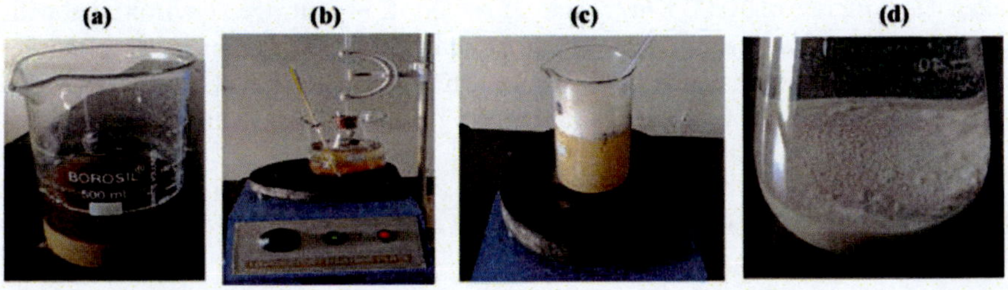

Fig. 2 Various stages of chemical extraction process **a** pulping, **b** delignification, **c** bleaching, **d** extracted cellulose

2.3 *Material Characterization*

Material characterization was conducted for raw *D. regia* fruit fiber samples and cellulose extracted by chemical treatments and termed as sample 1 and sample 2, respectively. Characterization was carried out using three techniques: X-ray diffraction (XRD), thermogravimetric analysis (TGA) and differential scanning calorimetry (DSC).

2.3.1 X-Ray Diffraction (XRD)

X-ray diffraction analysis was performed on Panalytical Xpert-Pro diffractometer using Cu-Kα radiation with wavelength of $\lambda = 1.5406$ nm at generator voltage of 40 kV and tube current of 30 mA. Samples were scanned at 2θ range of 20° to 100° with the scan step size of 0.026.

Crystallinity index (CrI) was calculated by the equation:

$$\text{CrI} = \left[\frac{(I_{002} - I_{\text{am}})}{2}\right] \times 100 \tag{1}$$

where I_{002} is the intensity of the peak at 2θ angle between 26° and 27° representing crystalline region, I_{am} is the intensity of the peak at 2θ angle between 21° and 22° representing amorphous region [15, 16].

2.3.2 Thermogravimetric Analysis (TGA)

Thermogravimetric analysis (TGA) was conducted by TGA-DTG Star^e Mettler-Toledo system equipped with high throughput autosampler and Star^e SW 13.00 software. For sample 1, initially, 7.9850 mg of weighed sample was taken in an aluminum crucible, and for sample 2, initially, 7.3000 mg of weighed sample was taken. Heating rate of 10.00 K/min from 25 to 700 °C in a nitrogen environment with flow rate of 70 ml/min was used for both the samples. Change in mass of the sample was recorded by microbalance system. First, differential curve (DTG) and TGA curve were obtained by Star^e SW 13.00 software by TGA 1 SF/1100/321 module [17].

2.3.3 Differential Scanning Calorimetry (DSC)

Differential scanning calorimetry (DSC) was conducted by DSC Star^e Mettler-Toledo system equipped with high throughput autosampler and Star^e SW 13.00 software. For sample 1, initially, 4.8000 mg of weighed sample was taken, and for sample 2, initially, 5.2000 mg of weighed sample was taken in an aluminum crucible. Heating rate of 10.00 K/min from 25 to 700 °C in a nitrogen environment with flow rate of

70 ml/min was used for both the samples. Difference in the amount of heat required to raise the temperature of the sample was recorded, and DSC curve was obtained by Star[e] SW 13.00 software by DSC 1/700/2861 module [18].

3 Result and Discussion

3.1 *Chemical Composition of the* Delonix regia *Fruit Fibers*

Chemical composition of the *D. regia* fruit fibers was determined by experimental procedures. All the experimental procedures were repeated thrice, and the average of results was determined. Table 1 shows the chemical composition of *D. regia* fruit fiber obtained from experimental procedures followed.

3.2 *X-Ray Diffraction (XRD)*

For analyzing the crystallinity index and crystallite size of the extracted cellulose with raw *D. regia* Fruit fiber sample, X-ray diffractometry was conducted. Figure 3 shows diffraction pattern of the extracted cellulose with raw *D. regia* Fruit fiber

Table 1 Chemical composition of *Delonix regia* fruit fiber (%w/w)

Cellulose (%)	Hemicellulose (%)	Lignin (%)	Extractives (%)	Moisture (%)	Ash (%)
66.9	11.6	6.5	4.5	8.3	2.2

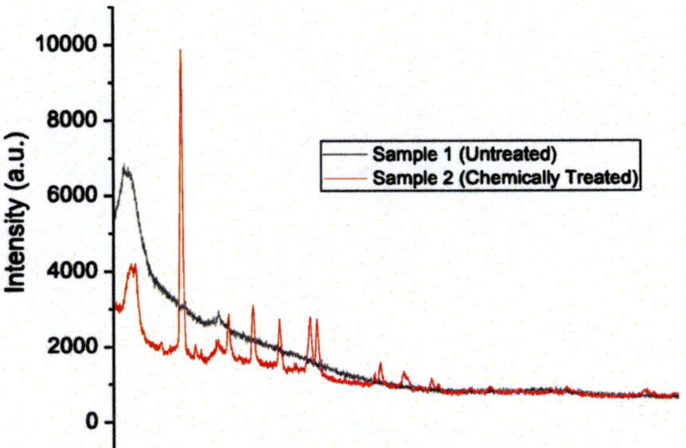

Fig. 3 XRD plot for raw (untreated) *Delonix regia* fruit fiber sample (sample 1) and chemically treated, i.e., extracted cellulose sample (sample 2)

sample. Two peaks were observed for sample 1 at 2θ values of 22.6855° and 186 34.8198° and most of the portion of curve is flat with the absence of sharp peaks which shows that the major the amorphous nature of the untreated material. For chemically treated sample, total 24 peaks were observed having narrow and sharp regions which represent crystalline nature of chemically treated material. Major peak for sample 2 was seen at 2θ value of 29.567°. Crystallinity index (CrI) was found to be 52.24% for raw *D. regia* fruit fiber sample (sample 1) and 73.41% for chemically treated sample (sample 2). Improvement in crystallinity index by chemical treatment is observed which clearly indicates that all the amorphous material such as hemicellulose, lignin and wax has been successfully removed leaving behind crystalline material, i.e., cellulose [19, 20].

3.3 *Thermogravimetric Analysis (TGA)*

Chemical structure of various constituents of lignocellulosic biomass such as waxes, lignin, hemicellulose and cellulose differs from each other, and hence, they degrade at different temperatures. Figure 4 shows the weight loss and its derivative during heating for raw (untreated) *D. regia* fruit fiber sample (sample 1), and Fig. 5 shows the weight loss and its derivative for chemically treated, i.e., extracted cellulose sample (sample 2). From these plots, we can observe that sample 1 and sample 2 has first degradation at 69.5 °C and 63.8 °C, respectively with reduction of almost 7.23% and 28.99% of the total mass which evaporation or loss of moisture content. The second degradation starts at 173.6 °C and 166.4 °C for sample 1 and 2, respectively, which represents degradation of hemicellulose, and lignin has started, and major weight loss is observed at 305.2 °C and 334.6 °C for sample 1 and 2, respectively, with 59.87% and

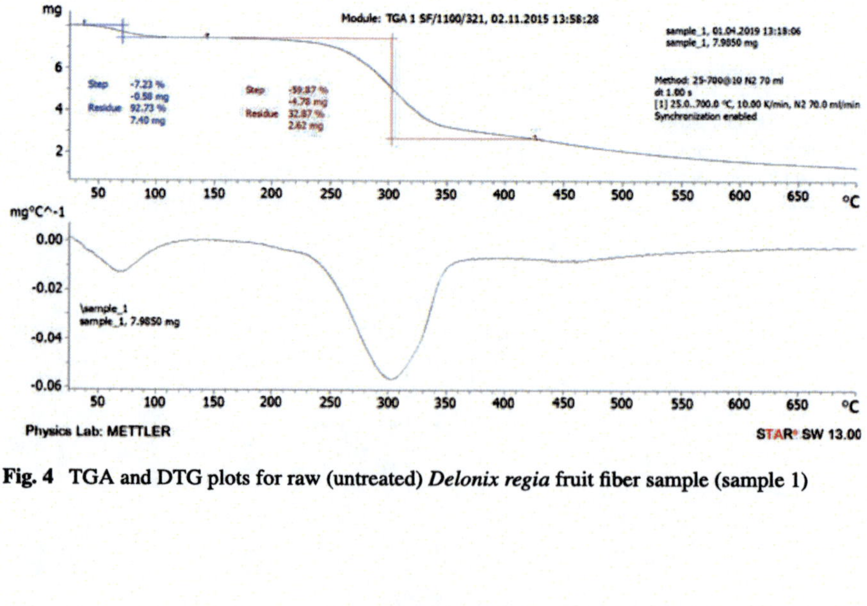

Fig. 4 TGA and DTG plots for raw (untreated) *Delonix regia* fruit fiber sample (sample 1)

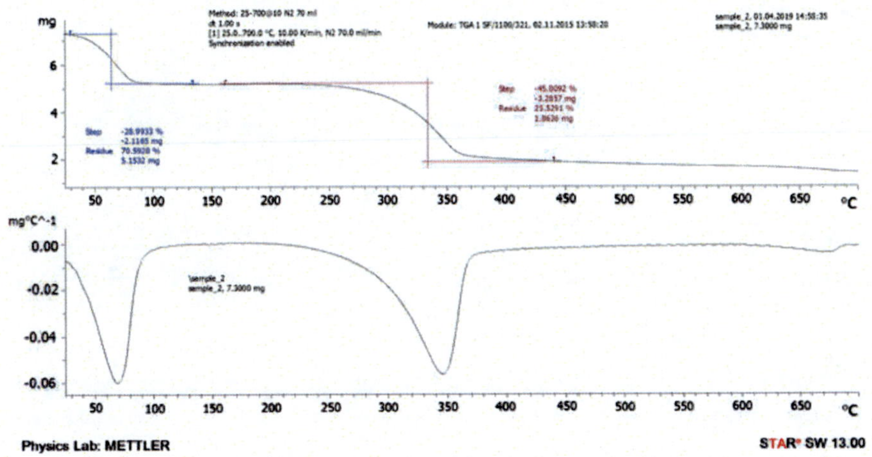

Fig. 5 TGA and DTG plots for chemically treated, i.e., extracted cellulose sample (sample 2)

45% reduction in total mass which represents the degradation of cellulose and lignin in sample 1 and degradation of cellulose in sample 2. This degradation continues further up to 600 °C till the ashing point after which degradation becomes negligible. The results indicate that weight loss at first degradation is less in untreated sample as compared to chemically treated sample due to fact that moisture absorption is less in untreated sample which is due to the presence of wax and other extractive content. Weight loss at the second degradation is more in untreated sample as compared to chemically treated sample because of the presence of lignin in untreated sample [21–23].

3.4 Differential Scanning Calorimetry (DSC)

Differential scanning calorimetry (DSC) was used to determine the thermal behavior of untreated sample (sample 1) and chemically treated sample, i.e., cellulose extracted (sample 2). From the results, we can observe that endothermic peak starts 20–130 °C which represents the energy consumed in evaporating the moisture content; low energy consumption is seen for untreated sample as compared to chemically treated sample because of the presence of wax and other extractive content. Further, endothermic peak was observed between 260 and 370 °C in untreated sample, due to energy consumption for degradation of lignin, hemicellulose and cellulose, whereas almost uniform rise in energy consumption 260–400 °C without any peak was observed in chemically treated sample which represents the presence of pure crystalline cellulose [24] (Figs. 6 and 7).

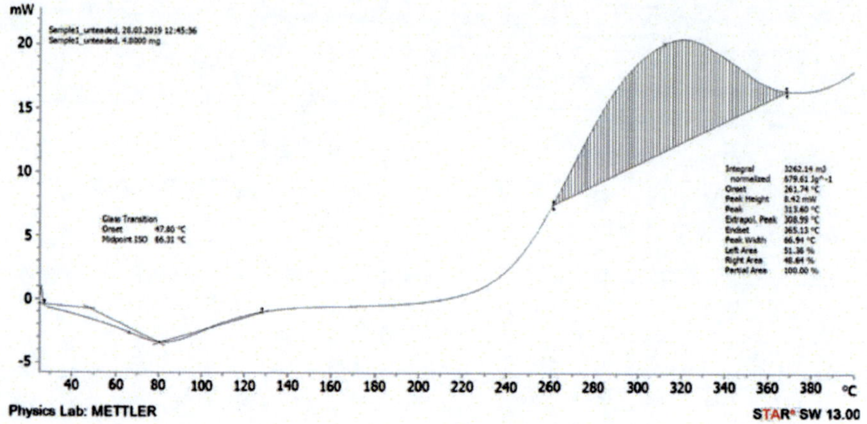

Fig. 6 DSC plot for raw (untreated) *Delonix regia* fruit fiber sample (sample 1)

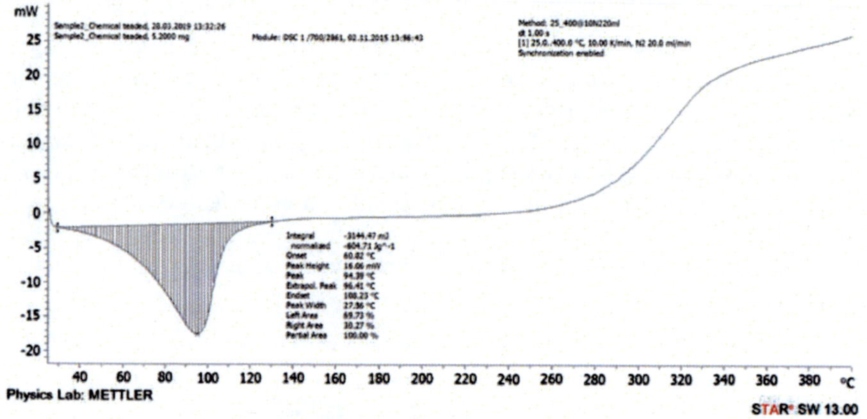

Fig. 7 DSC plot for chemically treated, i.e., extracted cellulose sample (sample 2)

4 Conclusion and Future Work

From the results obtained following conclusions can be drawn:

1. Cellulose content of the *D. regia* fruit fiber is 66.9% which is higher than most of popular cellulose sources, and due to high cellulose yield, it proves to be a potential candidate for cellulose and nanocellulose extraction and application in cellulose-based biocomposites and bionanocomposites,
2. XRD, TGA-DTG and DSC results proved that highly purified yield of cellulose is successfully extracted by the chemical route followed,

3. *Delonix regia* fruit fiber is suggested to be a suitable source for cellulose extraction, and in the future, further studies can be carried out for nanocellulose extraction and its application in cellulose-based biocomposites and bionanocomposites.

Acknowledgements I would like to thank Bombay Textile Research Association (BTRA), Mumbai (India), for providing necessary support and scientific help in compositional analysis. Also, I would like to thank National Institute of Technology, Raipur and Department of Physics, School of Sciences, Pt. Ravishankar Shukla University, Raipur for assisting in characterization.

References

1. Krasznai DJ, Champagne Hartley R, Roy HM, Champagne P, Cunningham MF (2018) Compositional analysis of lignocellulosic biomass: conventional methodologies and future outlook. Crit Rev Biotechnol 38(2):199–217
2. Han JS (1997) Chemical composition of fibers, Chapter 5
3. Watkins D, Nuruddin M, Hosur M, Tcherbi-Narteh A, Jeelani S (2015) Extraction and characterization of lignin from different biomass resources. J Mater Res Technol 4(1):26–32
4. Mohanty AK, Misra M, Hinrichsen G (2000) Biofibres, biodegradable polymers and biocomposites: an overview summary of contents 1. Introduction 2. Reinforcing biofibres 2.1 Chemical constituents and structural aspects 2.2 Properties of biofibres 2.3 Degradation properties of biofibres 2.4 Cost asp
5. Möller J (2014) Comparing methods for fibre determination in food and feed
6. Welker CM, Balasubramanian VK, Petti C, Rai KM, De Bolt S, Mendu V (2015) Engineering plant biomass lignin content and composition for biofuels and bioproducts. Energies 8(8):7654–7676
7. Ayeni A, Adeeyo O, Oresegun O, Oladimeji T (2015) Compositional analysis of lignocellulosic materials: Evaluation of an economically viable method suitable for woody and non-woody biomass. Am J Eng Res 4(4):14–19
8. Senthilkumar N, Murugesan S, Sumathi R, Babu DS (2015) Tree species in India, vol 6, no 2, pp 25–28
9. Templeton DW, Scarlata CJ, Sluiter JB, Wolfrum EJ (2010) Compositional analysis of lignocellulosic feedstocks. 2. Method uncertainties. J Agric Food Chem 58(16):9054–9062
10. Sluiter A et al (2011) Determination of structural carbohydrates and lignin in biomass determination of structural carbohydrates and lignin in biomass. Natl Renew Energy Lab 2011(July)
11. Giger-Reverdin S (1995) Review of the main methods of cell wall estimation: interest and limits for ruminants. Anim Feed Sci Technol 55(3–4):295–334
12. Sun RC, Fang JM, Mott L, Bolton J (1999) Extraction and characterization of hemicelluloses and cellulose from oil palm trunk and empty fruit bunch fibres. J Wood Chem Technol 19(1):167–185
13. Martins MA, Teixeira EM, Corrêa AC, Ferreira M, Mattoso LHC (2011) Extraction and characterization of cellulose whiskers from commercial cotton fibers. J Mater Sci 46(24):7858–7864
14. Sun JX, Sun XF, Zhao H, Sun RC (2004) Isolation and characterization of cellulose from sugarcane bagasse. Polym Degrad Stab 84(2):331–339
15. Ibrahim MM, Agblevor FA, El-Zawawy WK (2010) solation and characterization of cellulose and lignin from steam-exploded lignocellulosic biomass. Biol Syst Eng 5:397–418
16. Singh S et al (2015) Comparison of different biomass pretreatment techniques and their impact on chemistry and structure. Front Energy Res 2(February):1–12

17. Suryanto H, Marsyahyo E, Irawan YS (2013) Morphology, structure, and mechanical properties of natural cellulose fiber from Mendong grass (*Fimbristylis globulosa*). J Nat Fibers (November 2014):37–41
18. Morán JI, Alvarez VA, Cyras VP, Vázquez A (2008) Extraction of cellulose and preparation of nanocellulose from sisal fibers. Cellulose 15(1):149–159
19. Maheswari CU, Reddy KO, Muzenda E, Guduri BR, Rajulu AV (2012) Extraction and characterization of cellulose microfibrils from agricultural residue e *Cocos nucifera* L. Biomass Bioenerg 46:555–563
20. Reddy KO et al (2018) Extraction and characterization of cellulose single fibers from native african napier grass. Carbohydr Polym 188:85–91
21. Kalita RD, Nath Y, Ochubiojo ME, Buragohain AK (2013) Extraction and characterization of microcrystalline cellulose from fodder grass; *Setaria glauca* (L) P. Beauv, and its potential as a drug delivery vehicle for isoniazid, a first line antituberculosis drug. Colloids Surf B Biointerfaces 108:85–89
22. Jahan MS, Saeed A, He Z, Ni Y (2011) Jute as raw material for the preparation of microcrystalline cellulose. Cellulose 18(2):451–459
23. Nuruddin M et al (2011) Extraction and characterization of cellulose microfibrils from agricultural wastes in an integrated biorefinery initiative. Cellul Chem Technol 45(5–6):5–6
24. Ohwoavworhua F, Adelakun T, Okhamafe A (2009) Processing pharmaceutical grade microcrystalline cellulose from groundnut husk: extraction methods and characterization. Int J Green Pharm 3(2):97

Experimental Study and Analysis of Defragmented Carbon Nanotubes in Polyacrylonitrile Matrix

N. Arunkumar, Joven Job, D. Ananthapadmanaban, N. E. Arun Kumar, and N. Sathishkumar

Abstract The aim of this research is to elucidate the change in electrical properties of polyacrylonitrile (PAN) polymer when reinforced with defragmented multiwall carbon nanotubes. In conventional chopped fiber-reinforced polymer composites, uniform distributions of fibers throughout the matrix are critical for producing materials with superior physical and electrical properties. The previous methods have dispersed carbon nanotubes by aggressive chemical modification of the nanotubes or by the use of a surfactant prior to dispersion. Here, ultrasonic energy was used to uniformly disperse and defragment the multiwall nanotubes (MWNTs) in solutions and to incorporate them into composites without chemical pretreatment. A common solvent dimethylformamide (DMF) is used for the dissolution process. The film is formed by solvent casting method which involves evaporation of the homogeneous solution of CNT and polymer. Electrical characteristic studies were done on the film samples. The electrical properties such as conductivity and resistivities were found and compared to the original polymer matrix.

Keywords Carbon nanotubes (CNT) · Dimethylformamide (DMF) · Nanocomposite · Single-wall carbon nanotubes (SWNTs) · Multiwall carbon nanotubes (MWNTs) · Polyacrylonitrile (PAN) · Chemical vapor deposition (CVD)

1 Introduction

Recent theoretical and experimental studies suggest that carbon nanotubes have remarkable mechanical and electrical properties [1–4]. In this article, the electrical properties of carbon nanotubes [5] are applied to enhance the conducting property

N. Arunkumar (✉) · J. Job · N. E. Arun Kumar · N. Sathishkumar
Department of Mechanical Engineering, St. Joseph's College of Engineering, Old Mahabalipuram Road, Chennai, Tamil Nadu 600119, India
e-mail: narunkumar72@gmail.com

D. Ananthapadmanaban
Department of Mechanical Engineering, Sri Sivasubramaniya Nadar College of Engineering, Kalavakkam, Kanchipuram, Tamil Nadu 603110, India

S. Vijayan et al. (eds.), *Trends in Manufacturing and Engineering Management*, Lecture Notes in Mechanical Engineering,
https://doi.org/10.1007/978-981-15-4745-4_29

of a poor conductor [6]. The previous study indicated that most polymers are poor conductors [7]. Our broad objectives were to explore the development of advanced composites in which carbon nanotubes serve as the reinforcing element in a polymer matrix. Although several studies have focused on producing polymer nanotube composites, many practical challenges remain before their potential can be fully realized. Dispersing the nanotubes individually and uniformly into the matrix seems to be fundamental in producing composites with reproducible and optimal properties [8, 9]. The article focuses on dispersing defragmented MWNTs into the polymer matrix using solvent casting process [10, 11] and analyzing the change in the electrical properties of the nanocomposite material. Polyacrylonitrile (PAN) was chosen as the matrix material because its properties are well known; it is soluble in a broad range of solvents; and its clarity allows dispersion of MWNTs to be observed at the micron scale [12]. DMF (dimethylformamide) was chosen as a common solvent. MWNTs were prepared without any aggressive chemical pretreatment, purification, or modification using CVD method [13]. MWNTs and the polymer (PAN) solutions were dispersed and defragmented using a probe sonication process. The solution samples were evaporated to form a film using solvent casting process [14] which was later characterized by scanning electron microscopy (SEM) to determine dispersion and orientation of MWNTs within the PAN matrix. Composite samples were tested for its electrical properties such as conductivity and resistivity, and the results were plotted and tabulated.

2 Experimental

2.1 Synthesis of CNT (CVD Method)

See Fig. 1.

2.2 Process

The catalyst employed was a 1:1 mixture of misch metal-based alloy ($MmNi_3$) and graphene oxide (GO). Hydrogen decrepitation was employed for powdering the ingot of $MmNi_3$, and a modified Hummer's method [15] was used for preparing GO as described by the authors [16, 17]. The catalysts were grinded together in a mortar for a few minutes and then loaded into the center of a horizontal quartz tubular reactor, and hydrogen was allowed to enter at 240 °C for 30 min. Hydrogen reduction of GO to graphene has been observed under these conditions. The sample was heated further at 700 °C, before introducing a burst of acetylene (C_2H_2) as the carbon containing precursor, in addition to the argon (Ar) flow. The as grown samples were purified by refluxing in conc. HNO_3 at 60 °C for 6 h where the end hemispherical distorted

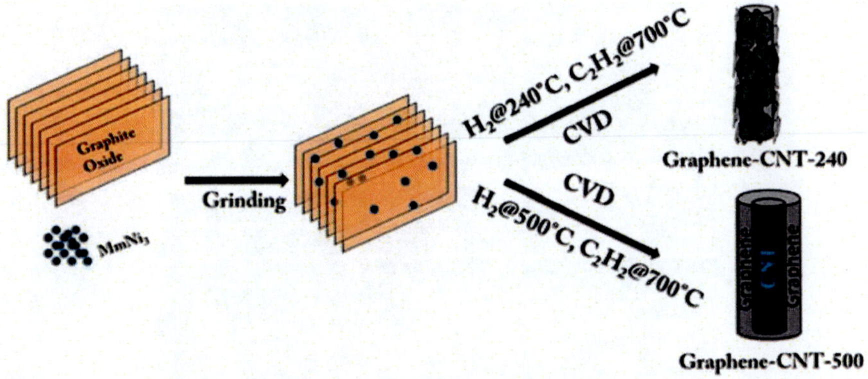

Fig. 1 Schematic illustration of the synthesis of graphene-coated CNT composites

carbon layers were removed. Acid-treated samples were washed with deionized water until pH 7, filtered using a 0.2 micron polytetrafluoroethylene (PTFE) membrane, and dried at 60 °C in a vacuum. The graphene-wrapped CNT samples grown at 240 °C and 500 °C of H_2 flow were labeled as graphene–CNT-240 and graphene–CNT-500, respectively. The growth mechanism of the graphene-wrapped CNT is now considered. At 500 °C, GO exfoliated into graphene sheets in the presence of hydrogen gas. The reaction was further carried out by allowing acetylene at 700 °C. When the catalyst particle ($MmNi_3$) is saturated with carbon, carbon leaves the particle in another form. Saturated carbon atoms nucleate into six membered rings and finally lead to CNT formation which is described in Fig. 1. The magnified images using FESEM and TEM are described in Fig. 2. This is the well-reported vapor liquid solid (VLS) mechanism for the growth of carbon nanotubes by chemical vapor deposition. The graphene nanosheets located at a distance of $\geq$0.35 nm from a nanotube get affixed to the surface of the nanotube by van der Waals interaction.

2.3 *Observation and Inferences*

The catalyst ($MmNi_3$) must be crushed into very fine powder. The growth of the carbon nanotubes takes place on the surface of the catalyst. Hence, finer the particles of the catalyst more surface area is formed for nanotube growth. The tubular structure of CNTs depends upon the insertion of gases at optimal temperature. Hydrogen is passed at 2400 °C to reduce the graphene oxide into graphene. The oxide group reacts with hydrogen and leaves as water vapor from the furnace. Graphene is the main source of carbon to form the carbon nanotubes. Acetylene is passed at 7000 °C to separate the carbon nanotubes from the misch metal catalyst.

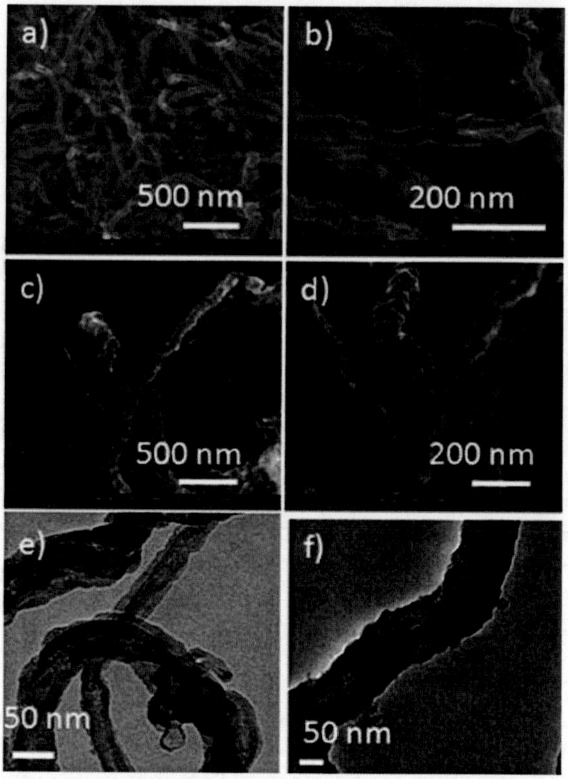

Fig. 2 **a** Low, **b** high-magnification FESEM images and **e** TEM images of graphene–CNT-500; and **c** low, **d** high-magnification FESEM images **f** TEM images of graphene–CNT-240

2.4 FESEM and TEM Images of Purified CNT

The CNTs formed by CVD method were checked for morphology using field electron scanning electron microscope (FESEM) and transmission electron microscope (TEM) at different magnifications. The tubular structures are the carbon nanotubes indicated in Fig. 2.

2.5 Carbon Nanotube Production Device by Chemical Vapor Deposition

Three independent temperature adjustors are applied to control the temperature at different positions of quartz tube furnace to gain high temperature balance so that the created carbon tube has high crystallizability. Various carbon nanotubes or nano-coil, micro-carbon coil, and hexagonal crystal lattice may be created and form film. Put silicon, quartz, porcelain, oxide, and other carriers in the external import device with catalyzer precursor which is prepared at advance, catalyst and carbon nanotube will be created at the same time, without the necessity to prepare catalyzer separately. The device can easily prepare highly dispersed CNT compound with CNT lamella structure/core structure of compound material. The catalyst is placed in the center of the furnace to ensure that there is uniform heating and uniform chemical vapor deposition. The gases and other chemicals are passed as mentioned earlier, and various formation processes of the carbon nanotubes take place at their respective temperatures.

2.6 *Preparation of Polyacrylonitrile Polymer (PAN)*

The polymer used in this research was purchased from Sigma-Aldrich. The polymer has 99% purity and a molecular weight ranging from 1,50,000 to 2,00,000 mol. Preparation method adopted to prepare this polymer is described by the authors of [18] in Fig. 3.

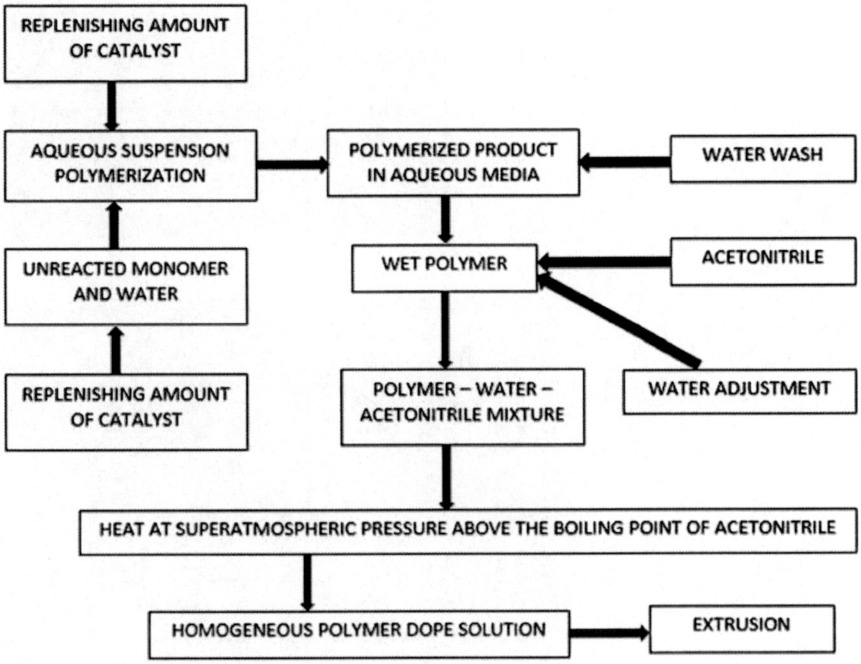

Fig. 3 Flowchart of the preparation process of polyacrylonitrile

2.7 Preparing the Polymer Solution

Polyacrylonitrile (PAN) powder was first weighed and placed in the beaker described in Fig. 4. It is then dissolved in DMF using a probe sonicator which produces ultrasonic waves to form a homogeneous polymer solution described in Fig. 5. The time used for obtaining a complete homogeneous solution of PAN is about 12 min having an OFF and ON cycle of 5 s and with a pulse of 50 Hz.

Fig. 4 Polyacrylonitrile polymer (PAN) weighed in a beaker before preparation of PAN solution

Fig. 5 Prepared PAN solution after probe sonication process

Fig. 6 Prepared CNT solution after probe sonication process

2.8 Preparing the Carbon Nanotube Solution

The multiwall carbon nanotubes were weighed using the electronic weighing balance. Solutions were created with wt% varying from 1 to 10. The solvent used in this application is the same dimethylformamide (DMF). 60 ml of the solvent was taken in 100 ml beaker, and the carbon nanotubes were added. The amount of carbon nanotubes added for every sample preparation varies from 10 to 100 mg, respectively. A total of 10 samples were created with weight percentage 1–10 %wt. The probe sonicator was used to disperse and defragment the carbon nanotubes in DMF. The complete dispersion takes place in 20 min and having an ON cycle of 15 s and OFF cycle of 5 s and with a pulse of 50 Hz. The final homogeneous solution of defragmented CNT and DMF is shown in Fig. 6.

2.9 Mixing the Polymer and Carbon Nanotubes Solutions

It was ensured that the solvent chosen (DMF) would be a common solvent for both PAN and CNTs, thereby assuring no chemical reactivity between solvents. The two solutions formed in 2.7 and 2.8 were mixed using a probe sonicator for about 30 min with a pulse of 50 Hz and an ON and OFF cycle of 5 s. A homogeneous solution of defragmented CNT and PAN was formed which is described in Fig. 7. Evaporation process is carried out by pouring the solution in a 4-inch petri dish and placing it on a hotplate as shown in Fig. 8. The hot plate is maintained at a constant temperature of 90 °C. A vacuum furnace can also be used for evaporation purpose.

The evaporation process takes place till all the solvent has evaporated and a thin film (50 μm thick) of composite material is formed in the petri dish as depicted in Fig. 9. The film was then peeled from the petri dish carefully without tearing as shown in Fig. 10. This method adopted for preparing a thin film by evaporating

Fig. 7 Prepared homogeneous solution of CNT and PAN by probe sonication process

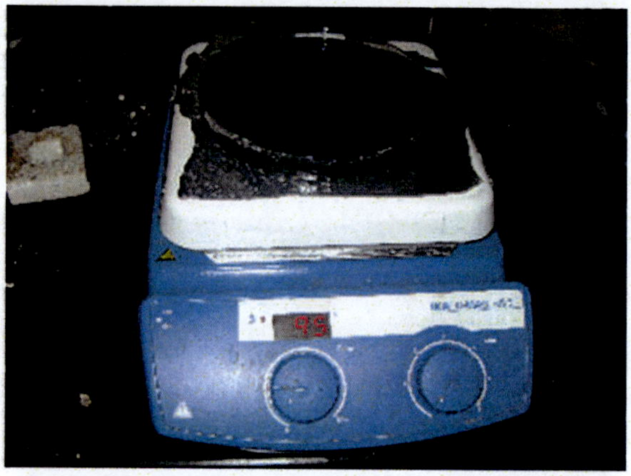

Fig. 8 Homogeneous composite solution placed on a hot plate for evaporation process

Fig. 9 Nanocomposite films which is formed after evaporation (solvent casting process)

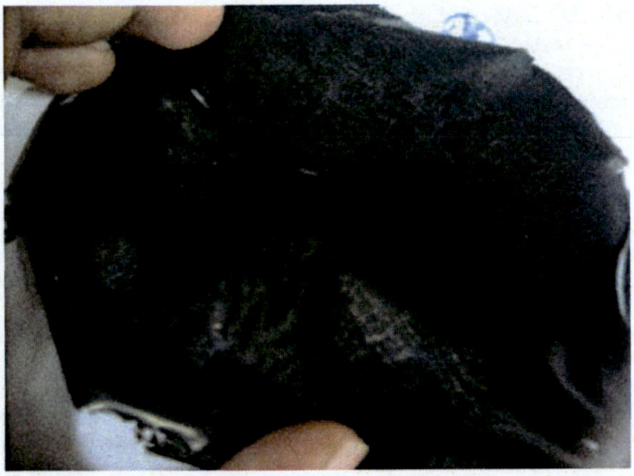

Fig. 10 Composite films when taken out from petri dish

the common solvent is known as solvent casting process explained in detail with an example by the authors of [19, 20]. The concentration of the nanotubes was varied with respect to the polymer concentration. The same process was repeated for all the samples and was checked for its morphology using a scanning electron microscope. SEM observations at low resolution were used to assess the degree of dispersion of the defragmented carbon nanotubes in the polymer matrix.

3 Result and Discussion

The tubular structure of carbon nanotubes has been distorted into flake-like structures during probe sonication. According to references [1–4], carbon nanotubes have excellent electrical conducting properties. Our discussion focused on using this property of CNTs to enhance the electrical conductivity of insulators, namely polymers by incorporating CNTs into them. Polymers have two major drawbacks which restrict its usage. One being its mechanical strength and other its electrical and thermal properties. By overcoming its drawbacks and with addition to its advantages (light in weight, high corrosion resistant, etc.), it would have a huge impact in replacing conventional materials in structural and electrical domains.

The samples were observed under a scanning electron microscope, and images of the morphology were obtained. The following are the SEM images for a few samples of the nanocomposite film. It is evident that the carbon nanotubes are completely defragmented into carbon layers or strands of carbon. This carbon is represented by the dark shaded area in the image. The lighter strands represent the polymer polyacrylonitrile strands which are spread throughout the matrix. Fig. 11 is the SEM image of 4 %wt CNT (which is 40 mg of CNT in 960 mg of polymer). The magnification was taken to 1 μm, and the analysis was conducted. While the defragmentation process using the probe sonicator, it is noticed that complete defragmentation into

Fig. 11 Low-resolution SEM images of composite films prepared using the probe sonicator

pure carbon will occur only during extensive use of the device. Figure 11, 12, and 13 indicate a lesser level of defragmentation of the CNTs which can be noticed by the flake-like structure indicating the graphene distorted rings merged with polymer. Hence, longer the sonication process, more the defragmentation of CNTs occurs. The flake-like structure indicated in Fig. 12 has graphene layers fused with the polymer which was the result of defragmenting the CNTs during solvent casting it with the polymer. The magnification was taken at 500 nm. Partial flakes and dispersion of PAN and defragmented CNTs are noticed in Fig. 13. The sample consisted of

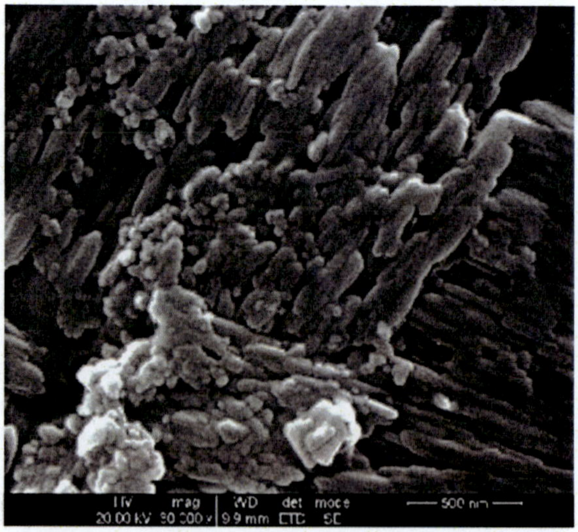

Fig. 12 SEM image of nanocomposite containing 5% wt CNT

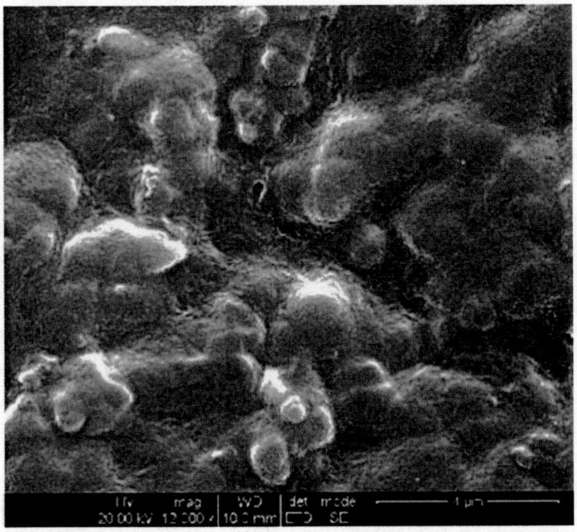

Fig. 13 SEM image of nanocomposite containing 8 %wt CNT

8 %wt CNT and was magnified at 1 μm. The broad flake-like structure represents the carbon/graphene structure merged with the polymer which is similar to Figs. 11 and 12 (Fig. 14 and Table 1).

Table 2 indicates the values which have been measured using a multimeter. This multimeter has a very high sensitivity of measuring resistivity of sample. The resistance and conductivity have been calculated using basic formula of electrical resistance and resistivity formula. The graph shows the variation of various electric properties of the nanocomposite with the variation of carbon nanotube weight percentage.

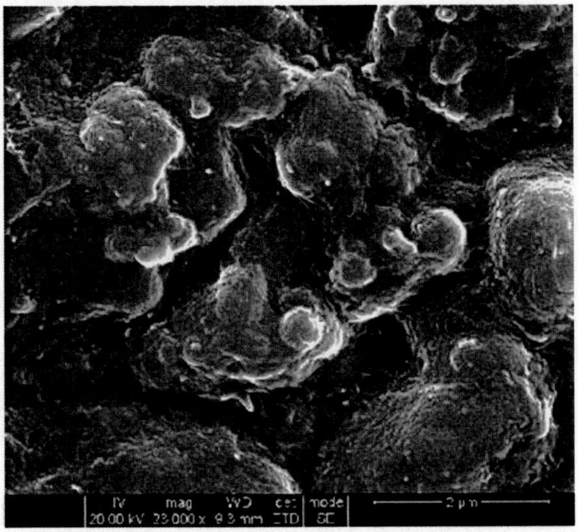

Fig. 14 SEM image of nanocomposite containing 9 %wt CNT

Table 1 Concentration of various samples

S. No.	Percentage of CNT (%)	Weight of CNT added (mg)	Weight of polyacrylonitile (mg)
1	2	20	980
2	3	30	970
3	4	40	960
4	5	50	950
5	6	60	940
6	7	70	930
7	8	80	920
8	9	90	910

Table 2 Electrical characteristics measured

S. No.	CNT (wt%)	Resistivity (Ωm)	Resistance (Ω)	Conductivity (S/m)
1	2	210.17	2526.45	3.96×10^{-4}
2	3	87.16	1045.43	9.57×10^{-4}
3	4	33.26	399.12	2.50×10^{-3}
4	5	8.66	103.92	9.62×10^{-3}
5	6	3.75	45	0.022
6	7	1.51	18.12	0.055
7	8	0.57	6.84	0.146
8	9	0.16	1.92	0.52

The electrical resistivity of the material decreases gradually with increase in concentration of CNT as shown in Fig. 15. Similarly, resistance decreases by a large extent with increase in CNT as shown in the graph. Conductance increases drastically with the increase in CNT as shown in the graph. Resistivity is measured in Ωm (ohm

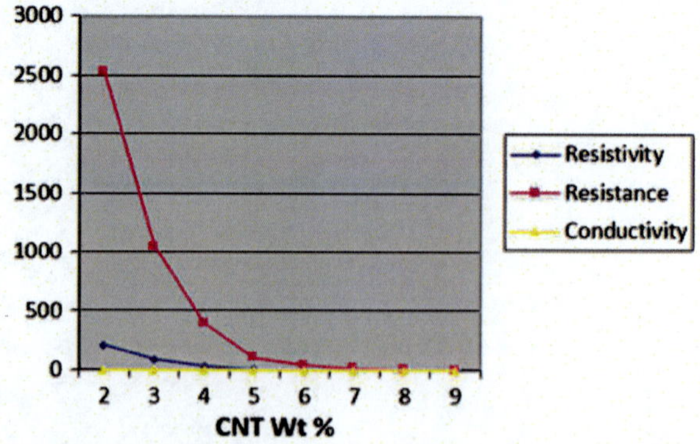

Fig. 15 Resistivity, resistance, and conductivity versus CNT concentration

Fig. 16 Resistivity versus CNT concentrations

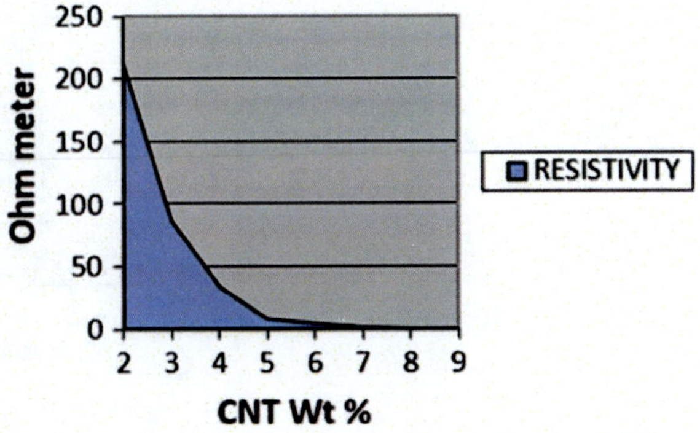

meter). Weight of the CNT in the composite is measured in mg. Figure 16 shows that with the increase in carbon nanotubes in the polymer matrix, the resistivity decreases. This happens because the free electrons travel through a defined path along the carbon nanotubes. Therefore, there is less opposition to the flow of electrons. Hence, there is decrease in resistivity with more number of CNTs added in the polymer matrix. Resistance is measured in ohms (Ω). Concentration of the CNT is measured in mg. Resistance is defined as the opposition to the flow of electric current. This opposition involves protons which oppose the flow of electric current by creating a barrier. When CNTs are added to the polymer matrix, the potential barrier space is reduced drastically and the concentration of current along the carbon nanotubes increases; hence, the opposition is less. Therefore, there is a decrease in resistance with more number of CNTs added in the polymer matrix as shown in Fig. 17. Conductivity is measured in siemens/meter. The concentration of the CNTs is measured in mg. Conductivity is defined as the flow of electrons through a material for a specified area of cross section and length. The conductivity of the material increases drastically with the increase in CNT concentration as depicted in Fig. 18. Hence, conductance also increases. The flow of electrons is aligned along the carbon nanotube walls and acts as a carrier of current which is present all through the polymer matrix.

Fig. 17 Resistance versus CNT concentrations

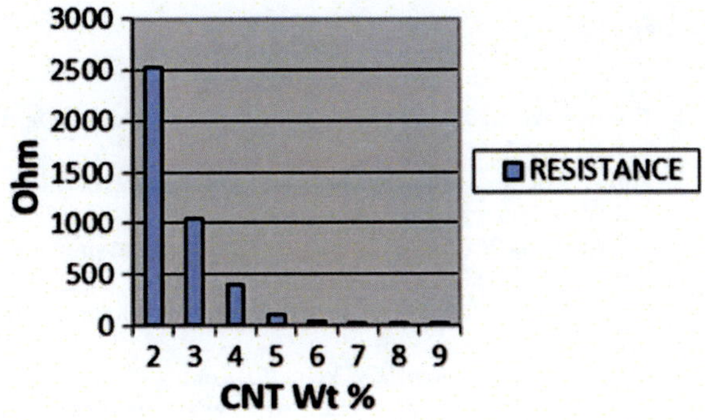

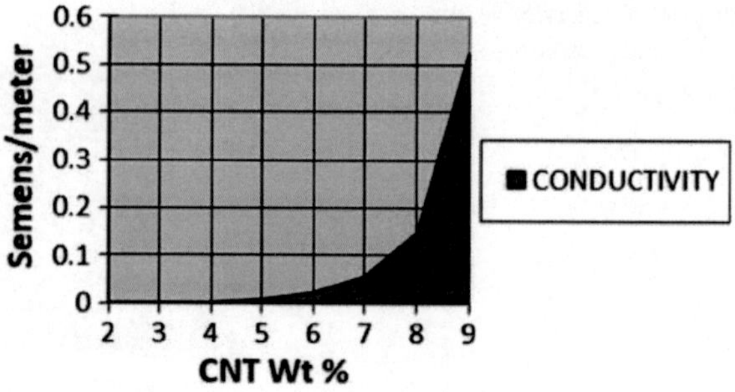

Fig. 18 Conductivity versus CNT concentrations

4 Conclusion

It has been experimentally proved that by inserting the carbon nanotubes into the polymer matrix, the electric properties of the polymer are improved drastically. The original resistance of the polymer was found to be in the range of 10^{16} Ω. After adding carbon nanotubes into the matrix, it was found to be in the range of 10 to 10^3 Ω. This proves that a significant increase in the electrical properties of a polymer can be improved by addition of carbon nanotubes. With further doping of carbon nanotubes, the conductance can be compared with the present materials which are used for electrical purposes such as copper and silver. The PAN nanocomposite is a unique material which consists of good electrical properties, high tensile strength which is inherited from the carbon nanotubes and also good corrosion resistant which is inherited from the polymer polyacrylonitrile.

Acknowledgements The authors would like to thank:

Dr. Ramaprabhu S, Dept of physics, Alternative Energy and Nanoscience Laboratory, IITM, for permitting to use the laboratory facilities and conducting the research.

Dr. Eswariah, Dept of physics, Alternative Energy and Nanoscience Laboratory, IITM, for his guidance through the course of the research work.

References

1. Bhushan B (2010) Introduction of carbon nano tubes. In: Springer handbook of nanotechnology, Part A, pp 47–118
2. Perie T, Brosse A-C, Tence-Girault S (2012) Mechanical and electrical properties of multi walled carbon nanotubes/ABC block terpolymer composites. Carbon 50(8):2918–2928
3. Bai J, Huang Y (2010) Fabrication and electrical properties of graphene nanoribbons. Mater Sci Eng: R: Rep 70(3–6):341–353
4. Shokrieh MM, Rafiee R (2010) Mechanics of composite materials. Chem Mater Sci 46(2):155–172
5. Marinho B, Ghislandi M, Tkalya E, Koning CE, de With G (2012) Electrical conductivity of compacts of graphene, multi-wall carbon nanotubes, carbon black, and graphite powder. Powder Technol 221:351–358

6. Inam F, Reece MJ, Peijs T (2012) Shortened carbon nanotubes and their influence on the electrical properties of polymer nanocomposites. J Compos Mater 46(11):1313–1322
7. Zhang Y, Rutledge GC (2012) Electrical conductivity of electrospun polyaniline and polyaniline-blend fibers and Mats. Macromolecules 45(10):4238–4246
8. Hou H, Reneker DH (2004) Carbon nanotubes on carbon nanofibers: a novel structure based on electrospun polymer nanofibers. Adv Mater 16:69–73
9. Rzepka M, Bauer E, Reichenauer G, Schliermann T, Bernhardt B, Bohmhammel K (2005) Hydrogen storage capacity of catalytically grown carbon nanofibers. J Phys Chem B 109:14979–14989
10. Greiner A, Wendorff JH (2007) Electrospinning: a fascinating method for the preparation of ultrathin fibers. Angew Chem Int Ed 46:5670–5673
11. Aravind SS, Eswaraiah V, Ramaprabhu S (2011) Facile synthesis of one dimensional graphene wrapped carbon nanotube composites by chemical vapor deposition. J Mater Chem 21:15179
12. Jeong JM, Choi MY, Ahn YT (2006) Morphology and properties of polyacrylonitrile nanocomposite prepared viain-situ polymeriszation with macroazoinitiator. Macromol Res 14(3):312–317
13. Lu C, Liu J (2006) controlling the diameter of carbon nanotubes in chemical vapor deposition Method by carbon feeding. J Phys Chem 110(41):20254–20257
14. Cantoro M, Hofmann S, Pisana S, Scardaci V, Parvez A, Ducati C, Ferrari AC, Blackburn AM, Wang K-Y, Robertson J (2006) Catalytic chemical vapor deposition of single-wall carbon Nanotubes at lowtemperatures. Nano Letters 6(6):1107–1112
15. Marcano DC, Kosynkin DV, Berlin JM, Sinitskii A, Sun Z, Slesarev A, Alemany LB, Lu W, Tour JM (2010) Improved synthesis of graphene oxide. ACS Nano 4(8):4806–4814
16. Arai T, Yamazaki H, Tsujimoto T (2006) Solvent casting process. US Patent No. 7,132,074
17. Eswaraiah V, Sankaranarayanan V, Ramaprabhu S (2011) Inorganic nanotubes reinforced polyvinylidene fluoride composites as low-cost electromagnetic interference shielding materials. Nanoscale Res Lett 6(1):137
18. Kim JH, Ganapathya HS, Hong S-S, Gal Y-S, Lim KT (2008) Preparation of polyacrylonitrile nanofibers as a precursor of carbon nanofibers by supercritical fluid process. J Supercrit Fluids 47:103–107
19. Safadi B, Andrews R, Grulke EA (2002) Multiwalled carbon nanotube polymer composites: synthesis and characterization of thin films. J Appl Polym Sci 84(14):2660–2669
20. Moniruzzaman M, Winey KI (2006) Polymer nanocomposites containing carbon nanotubes. Macromolecules 39(16):5194–5205

Design and Fabrication of Car Door Panel Using Natural Fiber-Reinforced Polymer Composites

Sodisetty V. N. B. Prasad, G. Akhil Kumar, P. Yaswanth Sai, and Shaik Varees Basha

Abstract For a lot of years, extensive research was carried on the use of natural fibers in automotive sector to meet the environmental regulations as well as to increase the profit margins of company. Natural fibers being of low cost, recyclable and lightweight provide industry with the best alternative for traditional materials such as steel, aluminum and synthetic fiber. Applications such as car door panels, sun visors, seat covers and rearview mirrors have provided excellent scope for usage of natural fiber-reinforced polymer composites. Various car manufacturers have successfully implemented natural fiber-reinforced polymers in some of their car models in recent years and have yielded desirable results in terms of lightness, cost saving and reduction of environmental pollution. Various factors such as combinations of fibers, layer sequence, fiber-to-matrix ratio and manufacturing method influence the properties of the composite specimens. The natural fibers used are flax, hemp and kenaf. Epoxy is selected as a matrix material. Hand layup method was used to fabricate the specimens as well the car door panel. The mechanical testing was carried out to determine the tensile strength, flexural strength, hardness and impact strength according to ASTM standards. The composite with most desirable mechanical properties is chosen to fabricate the final car door panel. This paper also discusses about the optimization of influencing parameters for drilling the natural fiber-reinforced epoxy plate using Taguchi method. Spindle speed, feed rate and drill bit diameter are selected as input parameters. These parameters which influence the thrust force are measured by using drill tool dynamometer.

Keywords Hand layup molding · Door panel · Mechanical testing · Taguchi method

S. V. N. B. Prasad (✉) · G. Akhil Kumar · P. Yaswanth Sai · S. V. Basha
SRM Institute of Science and Technology, Kattankulathur, Chennai, India
e-mail: prasadsodisetty.m@ktr.srmuniv.ac.in

S. Vijayan et al. (eds.), *Trends in Manufacturing and Engineering Management*, Lecture Notes in Mechanical Engineering,
https://doi.org/10.1007/978-981-15-4745-4_30

1 Introduction

Recently, the growing need of alternative materials in the field of aerospace, marines and construction, etc., had led to in-depth research in natural fiber-reinforced polymer composites [1, 2]. Natural fiber-reinforced polymer composites consist of polymer matrix reinforced with natural fibers such as flax, kenaf, sisal, hemp and jute. These fibers possess high specific strength and high modulus and are used in various applications [3]. Natural fibers offer advantages over synthetic fibers such as high specific strength, renewable, recyclable, less damage to processing equipment and easily available. Srinivas et al. [4] investigated the properties of various natural fiber composites such as jute, sisal, banana, bamboo, flax, kenaf, coir, hemp, curaua, cotton, abaca and Ramie and compared it with synthetic fibers such as glass, carbon and Kevlar. Hemp, flax, kenaf, abaca, etc., were found to have comparable properties of tensile strength and Young's modulus to those of synthetic fibers. Natural fibers were seen as potential replacements due to their lower cost and recyclable in nature. Various studies have been carried out on comparison of using hand layup method and vacuum infusion method for preparation of composites [5]. In the present study, hand layup method is selected based on various factors such as flexibility of design, low volume of production and good surface finish. It is also an easier method for production of heavy components.

Mechanical testing of natural fiber-reinforced polymer composites is quite essential to determine its tensile strength, flexural strength, impact strength, hardness and water absorption property. Numerous studies have been carried out on mechanical testing on natural fiber-reinforced polymer composites [6–12]. Mechanical properties of natural fiber-reinforced polymer composites depend on various factors such as type of fiber, fiber-to-matrix ratio, stacking sequence of fiber, fiber orientation, manufacturing method and number of layers. Elanchezhian et al. [13] evaluated the mechanical properties of kenaf, name and banana hybrid composites. The ratio of resin to hardener was found in the ratio of 10:1 to provide adequate interfacial bonding between fibers. It was found that kenaf fiber possessed the highest tensile strength and hardness among all the combinations and can be used for automotive components using car hood and mudguard. The ratio of hardener to resin should be 1:10 to provide suitable interfacial strength, and the critical ratio of fiber to matrix was found to be 40:60 for best mechanical properties. Girisha et al. [14] studied the properties of sisal, coconut fiber and ridge guard fiber. The mold was prepared by using hand layup method. The specimens of the three fibers with epoxy are made with varying the ratio of fiber to matrix. Increase in strength was observed with increase in fiber volume fraction. However, natural fibers are hydrophilic in nature because of cellulose content present in them. Hydrophilic fibers offer insufficient adhesion with hydrophobic polymer matrices. This may lead to improper adhesion and bond breakage between matrix and fibers. Various types of chemical treatment methods such as alkali treatment and saline treatments are available to remove the moisture from the natural fibers [15]. Alkali treatments such as treatment with 10% NaOH for 24 h improve the fiber–matrix adhesion due to removal of moisture. This along with increased surface roughness, increased aspect ratio and reduction in moisture absorption properties of fibers results in increased mechanical properties.

The applications of NFRCs in automobile industry include door panels, floor panels, sun visor, door cladding, mudguards and in aerospace applications such as aircrafts nose, doors, struts, outboard and indoor flaps and stabilizers [16]. Holbery et al. [17] studied the process of making thermoplastic polypropylene with natural fibers like compounding process, extrusion, kneading, injection molding, compression molding and sheet molding compound. The applications for these composites in automotive sector include door cladding, seatback lining, cushions, floor panel, floor body panels, head restraints and sound proofing. The improvement of these composites from plastic composites is increased stability, enhanced mechanical properties, excellent electrical properties, good impact strength. dos Santos et al. [18] postulated that the traditional materials used in automotive application can be replaced by natural reinforced composites. Experimentation was carried on modification of natural fiber to improve their properties like stretching, calendaring, electric discharge method, alkaline treatment and copolymerization. The fiber used for this study is curaua. The application of this natural fiber is sun visor, sun visor hinge pin and car interiors. The major problems which occur in drilling of natural fibers are fiber pullouts from matrix and delamination of composite. So, various parameters have to be optimized during drilling of natural fibers for higher surface finish and low thrust force [19]. Various experimental analyses were carried out to determine the torque and thrust force during drilling operation of NFRCs [20].

2 Methodology

Flax, hemp and kenaf fiber mats were selected as reinforcements for epoxy (LY556) matrix. Hy951 Hardener was used. Hand Layup was used for fabricating the composite specimens. Totally six different composite plates with layers each were fabricated. All the natural fibers were chemically treated with 10% NaOH solution to remove the moisture content present in the fibers. Flax fiber-reinforced epoxy composite plate with layers of flax fiber is fabricated with size 350 * 350 mm as shown in Fig. 1a. Similarly, with the same size, kenaf and hemp are fabricated as shown in Fig. 1b, c, respectively. Three hybrid plates are fabricated using three different combinations of fiber mats such as kenaf + hemp, flax + hemp and flax + kenaf as shown in Figs. 1d, e, f, respectively.

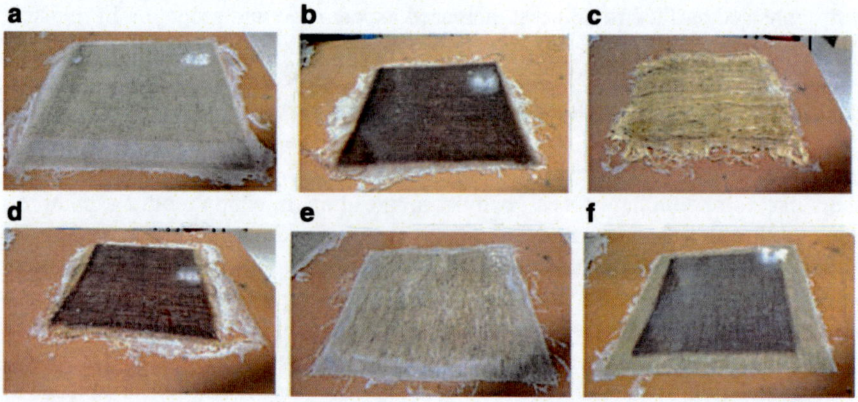

Fig. 1 **a** Flax [F_6], **b** Kenaf [K_6], **c** Hemp [H_6], **d** Kenaf + hemp [$(K/H)_3$], **e** Flax + hemp [$(F/H)_3$], **f** Flax + kenaf [$(F/K)_3$]

3 Experimental Testing

3.1 Tensile Test

Tensile test is carried out on all the six specimens as per ASTM D-3039 standard. In this test, the specimen is subjected to a constant tensile force until it fails. The stress–strain plot for various combinations of composite plates is shown in Fig. 2. Properties that can be determined by a tensile test are ultimate tensile strength, breaking strength, maximum elongation and area reduction.

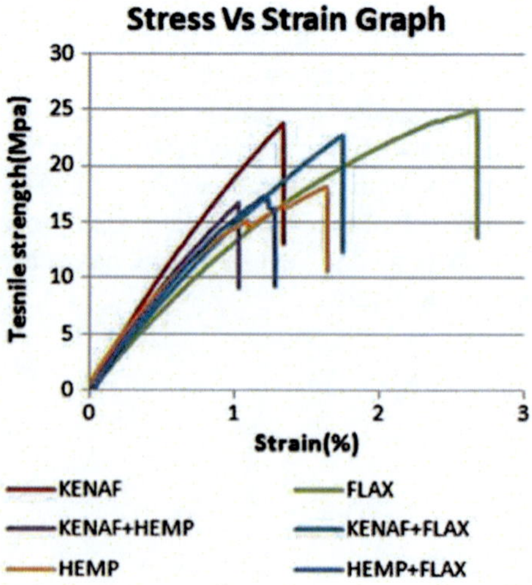

Fig. 2 Stress versus strain graph

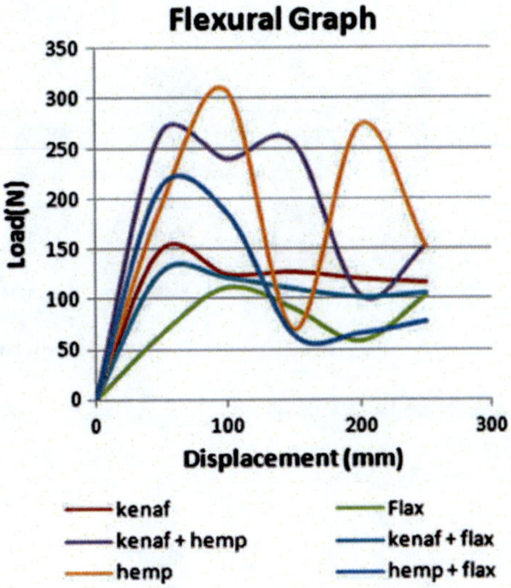

Fig. 3 Flexural testing graph

3.2 *Flexural Test*

Flexural test is carried out as per ASTM D-790 standard. The three-point flexural or bending test is used to determine the modulus of elasticity, flexural stress and flexural strain. The main advantage of a three-point bending test is the ease of the specimen preparation and its testing. The main disadvantage of this method is that the results are sensitive to specimen, the loading geometry and the strain rate of the specimen under the load. The load versus displacement graph is shown in Fig. 3.

3.3 *Izod Impact Test*

Impact test is carried out as per ASTM D-256 standard that was used for testing. The energy absorbed by the specimens is calculated from the height from which the arm swings to after hitting the specimen which is shown in Fig. 4.

3.4 *Hardness Test*

ASTM D-2583 standard was followed to carry out this testing. The Barcol hardness test studies the indentation hardness of materials by measuring the depth of penetration made by an indenter, loaded on a material sample and compared to the penetration in a reference material. The values of hardness are shown in Fig. 5.

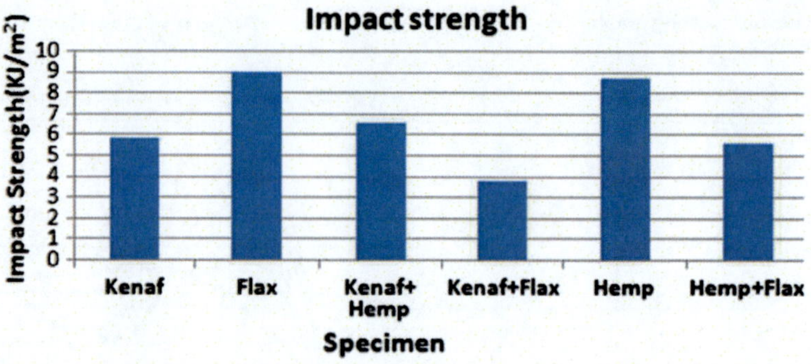

Fig. 4 Impact testing graph

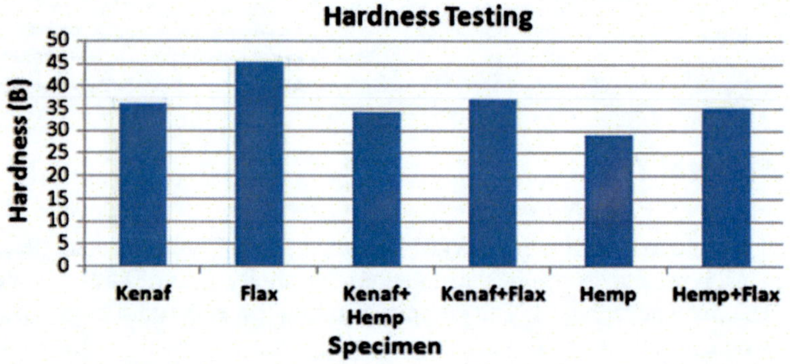

Fig. 5 Hardness testing graph

Mechanical testing was carried out on all the specimens; the results are shown in Table 1. The composite made up of all six flax fibers showed better mechanical properties over others. So, this combination of six flax fibers is chosen for fabrication of car door panel. It is observed that individual fibers possess higher mechanical properties than the hybrid combinations. This unusual trend is attributed to the presence of a large number of voids on the surface of hemp and kenaf fiber, which leads to poor interfacial bonding and hence poor mechanical properties among the hybrid fibers.

3.5 Comparison of Flax Fiber with ABS Plastic

Flax fiber-reinforced epoxy composite is selected among the five combinations. The properties of ABS plastic and the flax fiber-reinforced composite plate are compared as shown in Table 2.

Table 1 Mechanical testing results

Fiber combination	Tensile strength (MPa) (ASTM D-3039)	Flexural strength (MPa) (ASTM D-790)	Izod impact strength (kJ/m^2) (ASTM D-256)	Hardness (Barcol hardness) (ASTM D-2583)
Kenaf [K_6]	22.3	36.7	5.86	36.1
Flax [F_6]	28.0	62.7	9.0	45.1
Kenaf + hemp [$(K/H)_3$]	17.0	49.6	6.6	34.0
Kenaf + flax [$(F/H)_3$]	21.7	46.4	3.8	37.0
Hemp [H_6]	21.0	60.6	8.7	29.0
Hemp + Flax [$(H/F)_3$]	15.0	66.2	5.6	35.1

Table 2 Comparison between ABS plastic and flax

Properties	Flax fiber composite	ABS plastic
Flexural strength (MPa)	62.7	51.986470
Tensile Strength (MPa)	28.0	46
Izod impact strength (kJ/m^2)	9.0	9.01
Hardness (Barcol hardness)	45.1	97
Thickness of specimen (mm)	5	5

On comparing flexural strength of the flax fiber and ABS plastic, flax has more flexural strength. But impact strength is almost same for both flax and ABS plastic. So, flax can be used as alternative material for ABS plastic while fabricating the car door panel.

4 Machining

CNC machining center is used for drilling the natural fiber-reinforced epoxy composites using HSS twist drill tool. Drill tool dynamometer is connected to CNC machining center which is used to measure the thrust force as shown in Fig. 6. Thrust force values are measured for flax specimen. L9 orthogonal array is used in Taguchi analysis. Three influencing parameters are chosen in three levels (3 * 3). The three influential parameters are drill bit size, spindle speed and feed rate. The main objective function is to reduce the thrust force. So smaller is better is chosen while undergoing Taguchi analysis using Minitab software. Residual plots for thrust force

Fig. 6 Experimental Setup for drilling

are obtained after Taguchi as shown in Fig. 8 by selecting four in one. All the parameters are taken with three levels as shown in Table 3. Thrust force is the response variable recorded for all nine runs. General linear model was performed in analysis of variance using Minitab software to obtain the % of contribution of parameters on thrust force. Spindle speed plays a major role in varying the thrust force from the % of contribution in Table 5. Spindle speed, drill bit size and feed rate are contributing 13.33, 53.08 and 30.86, respectively. The signal-to-noise ratio shows the optimized parameters from the given parameters. The highest signal-to-noise ratio value will be the optimized value. At 5 mm drill bit size, 50 mm/min feed rate and 3000 rpm spindle speed are shown in Fig. 7. Response table for means is obtained as shown in Table 4.

4.1 Regression Equation

The following regression equation is obtained when analysis of variance is performed.

$$\begin{aligned}\text{Thrust force (N)} = {} & 63.31 - 18.52\,\text{Drill bit diameter (mm)\_ 5} \\ & + 11.73\,\text{Drill bit diameter (mm)\_ 6} + 6.79\,\text{Drill bit diameter (mm)\_ 7} \\ & - 12.64\,\text{Feed rate (mm/min)\_ 5(1)} + 0.55\,\text{Feed rate (mm/min)\_ 100} \\ & + 12.09\,\text{Feed rate (mm/min)\_ 150} + 9.21\,\text{Spindle speed (rpm)\_ 1000} \\ & - 3.00\,\text{Spindle speed (rpm)\_ 2000} - 6.21\,\text{Spindle speed (rpm)\_ 3000}\end{aligned} \tag{1}$$

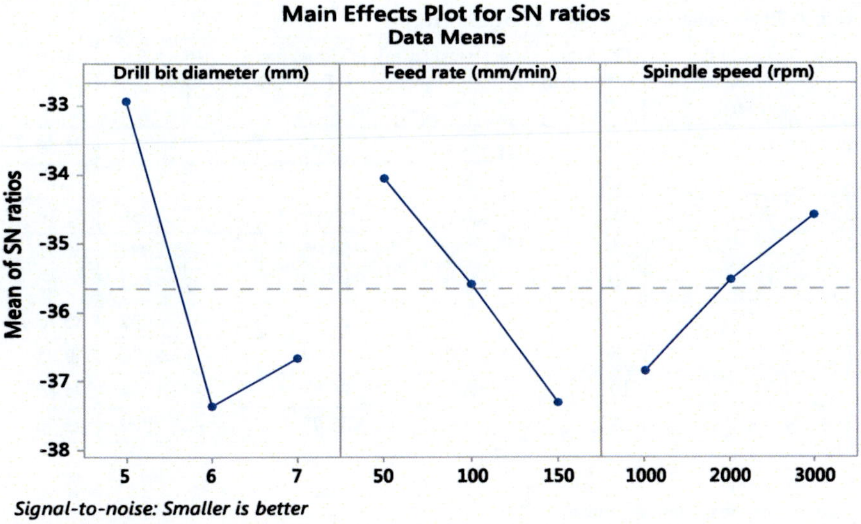

Fig. 7 Response graph

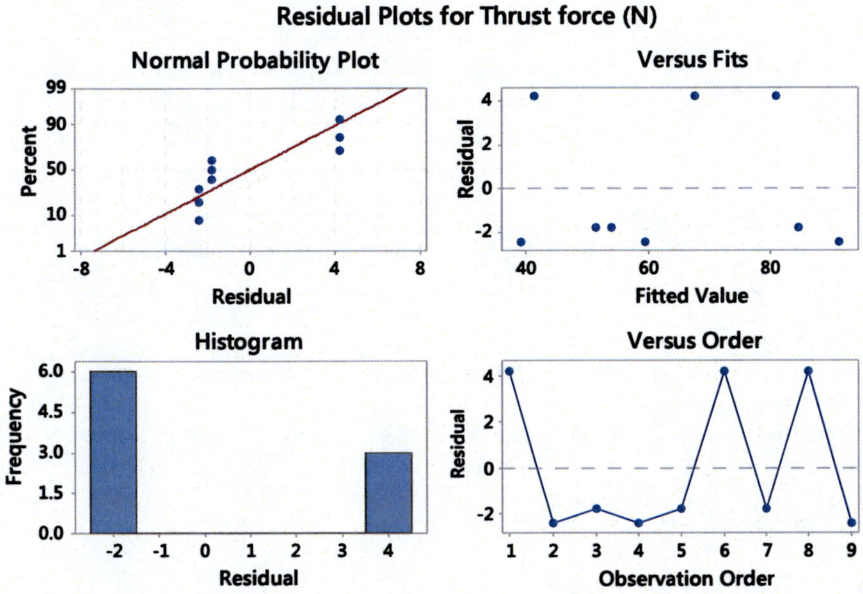

Fig. 8 Residual plots for thrust force

Table 3 Experimental design with orthogonal array

S. No.	Drill bit diameter (mm)	Feed rate (mm/min)	Speed (rpm)	Thrust force (N)	SNRA1	MEAN1
1	5	50	1000	45.59	−33.1774	45.59
2	5	100	3000	36.69	−31.2910	36.69
3	5	150	2000	52.10	−34.3368	52.10
4	6	50	2000	56.95	−35.1099	56.95
5	6	100	1000	83.02	−38.3837	83.02
6	6	150	3000	85.15	−38.6037	85.15
7	7	50	3000	49.48	−33.8886	49.48
8	7	100	2000	71.89	−37.1334	71.89
9	7	150	1000	88.95	−38.9829	88.95

Table 4 Response table for means

Level	Drill bit diameter (mm)	Feed rate (mm/min)	Speed (rpm)
1	44.79	50.67	72.52
2	75.04	63.87	60.31
3	70.11	75.40	57.11
Delta	30.25	24.73	15.41
Rank	1	2	3

Table 5 Analysis of variance

Source	DF	Seq SS	Adj SS	Adj MS	F-value	P-value	% Contribution
Drill bit diameter (mm)	2	1579.96	1579.96	789.98	19.46	0.049	53.08
Feed rate (mm/min)	2	918.49	918.49	459.24	11.31	0.081	30.86
Speed (rpm)	2	396.86	396.86	198.43	4.89	0.170	13.33
Error	2	81.19	81.19	40.60			2.73
Total	8	5726.01					100.00

4.2 *Confirmability Test*

Conformability test is performed for the optimized parameters of 5 mm drill bit size, 50 mm/min feed rate and 3000 rpm spindle speed. Least thrust force of 33.82 N is obtained.

5 Fabrication of Mold and Door Panel

Original door panel was purchased from the automobile spare parts vendor. To fabricate the mold using hand layup, firstly, applied wax on the specimen to clean the surface and then applied the PVA on the specimen for easy removal of mold after curing. Resin, catalyst and chalk powder are mixed and applied over the PVA. The glass fiber mat is placed, and resin mixture is transferred on to it. Repeated this procedure for three layers and cured it for 48 h. Mold is removed from the original door panel. The final mold obtained from hand layup is as shown in Fig. 9. Flax fiber is used to fabricate the final door panel. The same technique was employed to fabricate the door panel using hand layup method as shown in Fig. 10.

Fig. 9 Mold made of glass fiber

Fig. 10 Door panel fabricated by flax

6 Conclusion

Flax fiber-reinforced epoxy composite possesses higher values of tensile strength, impact strength and hardness among various combinations tested. The flexural strength of hemp and flax combination is higher than the individual flax fiber. Hence, flax fiber is chosen to fabricate the door panel. It is observed that individual fibers possess higher mechanical properties than the hybrid combinations. This unusual trend is attributed to the presence of a large number of voids on the surface of hemp and kenaf fiber, which leads to poor interfacial bonding and poor mechanical properties among the hybrid fibers. For low volume of production, hand layup method is most suitable. Natural fibers can be extensively used in automobile applications as alternatives for synthetic fibers and ABS plastic due to their comparable mechanical properties, low cost and recyclability. The optimized parameters to obtain less thrust force to drill flax fiber-reinforced epoxy specimen are drill bit of size 5 mm, feed rate of 50 mm/min and spindle speed of 3000 rpm. Spindle speed plays a major role in varying thrust force.

References

1. Haydar Zaman U, Mubarak Khan A, Ruhul Khan A, Arifur Rahman M, Lily Das R, Al-Mamun MD (2010) Role of potassium permanganate and urea on the Improvement of the mechanical properties of jute polypropylene composites. Fibers Polym 11(3):455–463
2. Ratna Prasad AV, Mohana Rao K, Gupta AVSSKS, Reddy BV (2011) A study on flexural Properties of Wild cane grass Fiber-reinforced polyester composites. J Mater Sci 46:2627–2634
3. Yang Yuqiu, Ota Tomoko, Morii T, Hamada H (2011) Mechanical property and hydrothermal aging of injection molded jute/polypropylene composites. J Mater Sci 46:2678–2684
4. Srinivas K, Lakshumu Naidu A, Raju Bahubalendruni MVA (2017) A review on chemical and Mechanical properties of natural fiber reinforced polymer composites. Int J Perform Eng 13(2):189–200
5. MohdYuhazri Y, Phongsakorn PT, Sihombing H (2010) A Comparison process between vacuum infusion and Hand lay-up method toward kenaf/polyester composites. Int J Basic Appl Sci IJBAS-IJENS 10(3)
6. Jamrichova Z, Akova E (2013) Mechnical testing of natural fiber composites for automotive industry. Univ Rev 7(3):20–25
7. Dan-mallam Y, Abdullah MZ, MegatYusoff PSM (2012) Predicting the tensile properties of woven kenaf/polyethylene terephthalate (PET) fiber reinforced polyoymethylene (POM) hybrid laminate composite. J Mech Civil Eng (IOSRJMCE)
8. Muthukumar V, Venkatasamy R, Sureshbabu A, Arunkumar D (2011) A study on mechanical properties of natural fiber reinforced Laminates of Epoxy (Ly 556) polymer matrix composites. Int J Prod Technol Manag Res 2(2)
9. Barile C, Casavola C, De Cillis F (2019) Mechanical comparision of new composite materials for aerospace applications. Elsevier J Compos Part B 162:122–128
10. Alagarraja K, Dhamodharan A, Gopinathan K Mathan Raj R, Ram Kumar K (2014) Fabrication and testing of fiber reinforced polymer composites material. IOSR J Mech Civil Eng (IOSR-JMCE)
11. Arun Kumar, Abishek T, Jegatheesh Raja S, Gokul Krishnan A (2016) A review on mechanical properties of natural fiber Reinforced hybrid polymer composites. Int J Adv Sci Tech Res 6

12. Rihayat T, Suryani S, Fauzi T, Agusnar H, Wirjosentono B, Alam PN, Sami M (2017) Mechanical properties evaluation of single and hybrid composites polyester reinforced bamboo, PALF and coir fiber. IOP Conf Ser Mater Sci Eng 334(2018):012081
13. Elanchezhian C, VijayaRamnath B, Kaosik R, Nellaiappan TK, Santhosh Kumar K, Kavirajan P, Sughan MU (2017) Evaluation of mechanical properties of kenaf based hybrid composite for automotive components replacement. ARPN J Eng Appl Sci 10(13)
14. Girisha C, Sanjeevamurthy, Rangasrinivas G (2012) Tensile properties of natural fiber reinforced epoxy hybrid composites. Int J Mod Eng Res (IJMER)
15. Suardana NPG, Piao Y, Lim JK (2011) Mechanical properties of hemp fibers and hemp/PP composites: effects of chemical surface treatment. Mater Phys Mech 11:1–8
16. Leao AL, Rowell R, Tavares N (1998) Application of natural fibers in automotive industry. Sci Technol Polym Adv Mater
17. Holbery J, Houston D (2006) Natural-fiber-reinforced polymer composites in automotive applications. J Mater
18. dos Santos PA, Giriolli JC, Amarasekera J, Moraes G, Natural fibers plastic composites for automotive applications. SABIC Innovative Plastics
19. Palani Kumar K (2011) Experimental investigation and optimization in drilling of GFRP composites. Measurement 44:2138–2148
20. Chandramohan D, Marimuthu K (2011) Drilling of natural fiber particle reinforced polymer composite material. Int J Adv Eng Res Stud 1(1):134–145

Investigation of Physical–Chemical Properties and Evaluation of Optimal Blend Ratio of Rice Bran Biodiesel: A Mathematical Regression Analysis Approach

M. Selvam, S. Palani, and P. Vaishnavi

Abstract Our research work mainly focusses on the examination of properties, both physical and chemical, of biodiesel blends among neat diesel at different volumetric proportions. Properties like viscosity; calorific value (CV); pour point (PP); cetane number; could point (CP); flash point and fire point are measured for each blending ratios, and it is examined against the relevant standards (USA ASTM D 6751; Europe EN:14214 as well as India IS:15607). Plot between the properties and different blending ratios is drawn, and regression equations (R2) are formulated from which biodiesel properties are predicted and compared by mineral diesel. From this investigation, the optimal blending ratio is projected considering the different properties of biodiesel.

Keywords Rice bran oil · Cetane number · Flashpoint · Calorific value · Mathematical regression analysis

1 Introduction

The fossil fuel is a great power source, but it has many drawbacks. It is the main contributor to the increase of global warming due to its emission which leads climatic changes and environmental effects like acid rain, depletion of the ozone layer, increase in greenhouse phenomena and formation of smog. This is a renewable source of energy, but the formation of fossil fuel takes over a million of years to renew when compared to other sources like solar and wind energy, so it is considered to be a non-renewable form of energy. Biodiesel be able to be extracted from a variety of raw materials similar to vegetable oil, animal fat, used cooking oil, etc. Biodiesel is said to be a good alternative because it is biodegradable, less toxic in nature, superior

M. Selvam (✉) · S. Palani
Anna University, Chennai, Tamil Nadu 600025, India
e-mail: mselvam@veltechmultitech.org

M. Selvam · P. Vaishnavi
Department of Mechanical Engineering, Vel Tech Multitech, Avadi, Chennai, Tamil Nadu 600062, India

S. Vijayan et al. (eds.), *Trends in Manufacturing and Engineering Management*, Lecture Notes in Mechanical Engineering,
https://doi.org/10.1007/978-981-15-4745-4_31

lubricity and the abundant availability of various raw materials around us. Biodiesel enhances the efficiency of diesel engines because it has high oxygen content, and it also has high carbon content which increases the combustion rate. Along with the efficiency, the emission characteristics of biodiesel are found to be better than diesel. It emits a relatively less amount of carbon monoxide, hydrocarbons, particulate matter than diesel. However, biodiesel tends to increase the emission of nitrogen oxide (NO_x), and also, utilization of biodiesel (B100%) within diesel engine will result in reduced power, torque and fuel economy. Many researchers have reported that various blends of diesel and biofuel could increase the power, torque and fuel economy of the vehicle.

In the recent environmental research scenario regarding biofuel, the investigation rice bran oil is considered for production of biodiesel due to the vast availability, and India produces of about 8 lakhs tonnes rice bran oil from rice husk (which contains nearly 10–23% of oil). Rice bran oil is removed from rice husk with help of huller and sheller mills. Such extracted rice bran oil consists of several compositions of fatty acids like 1.297% stearic, 14.883% palmitic, 42.013% oleic, 38.561% linoleic, 1.396% linolenic, 0.122% myristic. Production of biodiesel starts with transesterification process which involves to get fatty acid methyl ester through rice bran oil. Fatty acids from the oil are removed so as to minimize oil viscosity for meeting the standard value of mineral diesel.

Lin et al. (2009) produced biodiesel using transesterification process and optimum reaction conditions to analyze the fatty acid composition. The fuel properties obtained are well suited with the standards of ASTM D6751-02 and DIN V51606 [1]. Dennis et al. (2010) created biodiesel by means of catalyzed process of transesterification at atmospheric temperature and pressure which avoids the necessity of high temperature and pressure for animal fats, and waste cooking oils increase the molar ratio of triglycerides to alcohol of 6:1 [2]. Atabani et al. (2013) suggested *Sterculia foetida* oil biodiesel blend to increase the effectiveness of properties of maximum kinematic viscosity of 75.826 mm^2/s at 40 °C [3]. Evangelos (2013) found that the cold flow properties have higher impact on the biodiesel than that of conventional diesel [4]. Ashrafu et al. (2014) stated that the fatty acid composition plays a major role in the fuel properties such as viscosity, calorific value and cetane number. They also established that monosaturated fatty acid present in the biodiesel blend at minimum temperature increases the ignition quality and also increases the fuel properties [5].

Subhash et al. (2015) indicated that the biodiesel blend of B15 does not lead to the wall impingement, whereas the B20 has better boundary of wall impingement (within uncertainty ±1.3%). [6]. Bhaskor et al. (2016) carried out the three diverse compression ratios of 18, 17.5 and 17, at injection timing 23 °C at various loading conditions. In maximum load, the most extreme brake thermal efficiencies (BTEs) with respect to double fuel mode (DFM) are observed to be 20.27%, 19.97% and 18.39% for CRs of 18, 17.5 and 17 [7]. Mayank et al. (2016) found that the fatty acid content in crude rice bran oil is elevated with respect to oil quality since of over fatty content. The two-step transesterification process is followed which includes acid-catalyzed transesterification and base-catalyzed transesterification [8]. Baharak et al. (2016) found that the edible oil has various drawbacks like deforestation, increased vegetable oil price, to overcome which extraction can be done with non-edible oils(rice bran oil, rubber seed, mahua seed, silk cotton, babassu tree) [9]. Mohamed Shameer et al. [10] used the B20 blend because of its low emission and increased engine performance. Also, they used advanced injection timing as well as injection pressure for reducing emission [10].

Maria et al. (2017) found that rice bran is the by-product of rice which is under-utilized and could be used in biodiesel production and also rich in protein. They also stated that the rice bran oil is more captivating for its more nutraceutical characteristics and balanced fatty acid content [11]. Ertan et al. (2017) stated that the oxygenated fuel blends reduce NOx emission but increase CO_2, CO [12]. Evangelos et al. (2018) analyzed the properties of fuel with respect to the fatty acid composition of the biodiesel. They also analyzed the properties such as the cetane number, density, kinematic viscosity with respect to the fatty acid content present in the biodiesel. [13]. Ertan et al. (2019) stated that the fuel properties play a major role in the performance, combustion and injection characteristics of the diesel engine. They also stated that the combustion and performance of methyl and ethyl ester show better values when compared with diesel at various engine conditions such as engine speed and at different load conditions (BMEP 3.3, 5.0, 6.6, 8.3 bar) [14]. Jeyaprakash et al. (2019) investigated rice bran biodiesel in a four-stroke, single-cylinder, direct-injection engine which runs at 1500 rpm is made in which the biodiesel blend of 20%(B20) has greater values when compared with other blending ratio. Their results show that the rice husk contains about 16–20% of crude rice bran oil, and it has high in-cylinder pressure of about 72.1 bar and heat release rate of about 70.826 J/CA [15].

In our research work, consider rice bran oil for production of biodiesel due to the vast availability. Rice bran oil is removed from rice husk with help of huller and sheller mills. Such extracted rice bran oil consists of several compositions of fatty acids like stearic, palmitic, oleic, linoleic, linolenic, myristic. Production of biodiesel starts with transesterification process involves to get fatty acid methyl ester (FAME) from rice bran oil.

2 Extraction of Rice Bran Oil

Rice bran oil is edible oil which is used for cooking purpose. Rice husk is a main source of rice bran oil. India produces about 80 million tonnes of paddy from which 5 million tons of rice bran and to the extent of 8 lakh tonnes of rice bran oil are obtained. 2 lakh tonnes is currently produced and about 10–12% percentage is of edible grade. Around 5–10 kg of rice bran is produced from 100 kg of paddy. About 10–23% of rice bran oil is produced from rice bran. Rice bran is a mixture of substances which contains protein, fat, ash and crude fiber. 70% of the paddy is processed by huller mills, and about 30% is produced by sheller mills. Rice bran oil production comes under three production phases. The first and foremost is the primary phase which includes needing, feeding, separating of fogging material (killing insects that involves using fine pesticide spray). The separated material is fed into the cooker for steam processing, and then, the cooked material is fed into the palletizer to make material into pallets. Now, the pallets are allowed to cool down; then they are sent to the conveyor for the extraction. The second phase is started by pouring pallets into hexanoate in order to remove the hexane. Then, the final stage is the distillation process; the product from the distillation process is fed into the missile tank for storage, and now, the stored product is sent to the heater to evaporate the hexane in order to produce crude rice bran oil. The crude rice bran oil consists of dissolved gum which is removed by degumming process. Further, it is sent to bleaching section to refine the crude oil; following the bleaching section, it is sent to the de-waxing section where the wax is removed from the oil. At last, deodorizing process is carried out at high vacuum and temperature to remove the fatty acids. Thus, the required edible rice bran oil is obtained.

2.1 Production Methodology

The raw rice bran oil was extracted from conventional process. The oil was then filtered to remove minuscule residues. The raw rice bran oil was transformed to biodiesel by transesterification technique. Initially, the oil was heated to a temperature of about 55–60 °C. By the side, methanol (10%) and potassium hydroxide (KOH) were titrated and added to the oil. Then, the concoction is stirred thoroughly at 600 rpm for about an hour. The mixture is kept nearly for 24 h for the glycerol to settle down. At the end of the process, glycerol is settled down, and after its removal, fatty acid methyl ester is obtained. Warm water was added to the methyl ester and kept for 24 h in order to take away suspended impurities. The above said process is clearly shown in Fig. 1.

The properties like kinematic viscosity, density, flash point, fire point, cetane number, cloud point, pour point were tested using standard apparatus and examined for both raw rice bran oil and biodiesel.

3 Extraction of Rice Bran Oil

3.1 Acid Composition of Biodiesel and Crude Oil

Rice bran oil as well as rice bran biodiesel is composed of 17.892% and 18.849% of saturated fats and 82.102% and 81.152% of unsaturated fats, respectively. The level of free fatty acid of rice bran oil as well as rice bran biodiesel is evaluated by means of gas chromatography (GC) test. Gas chromatography (GC) is a typical kind of chromatography utilized for isolating and breaking down composites that can be vaporized without disintegration. The free fatty acid compositions are described in

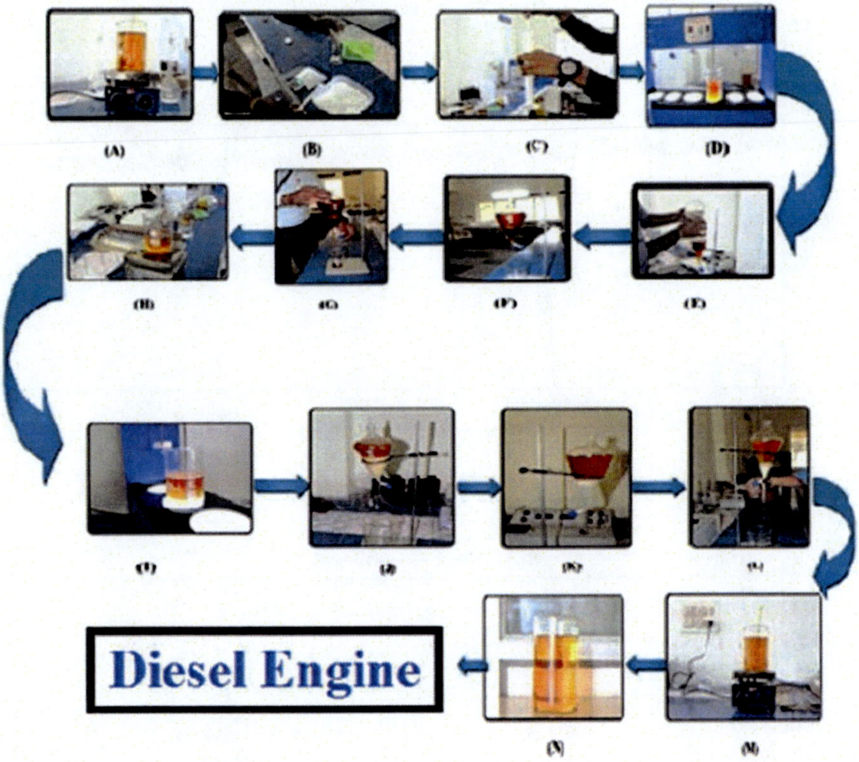

Fig. 1 Production methodology of raw rice bran oil

terms of C:D ratio, where C represents the number of carbon atoms and D represents the number of double carbon atom chains. The succeeding acids have the major part in rice bran oil and rice bran biodiesel. Rice bran oil and rice bran biodiesel consist of 42.013% and 41.414% of oleic acid C18:1 which is also used as an emulsifying agent as it withstands the structure of chemical compounds. The linoleic acid C18:2 has volumes of 38.236% and 38.032% in rice bran oil and biodiesel. The palmitic acid C16:0 serves as release agent and also used to increase the volatility. It is present about 14.883% and 15.897% in rice bran oil and biodiesel, respectively.

3.2 FTIR Analysis

The vibrational motion of the functional group is determined by the Fourier transformation infrared spectroscopy of the diesel sample and biodiesel blends of B30 as shown in Fig. 2 and B100 as shown in Fig. 3. In the infrared region, the light rays pass through the interferometer and the beam splitter to generate absorption frequencies.

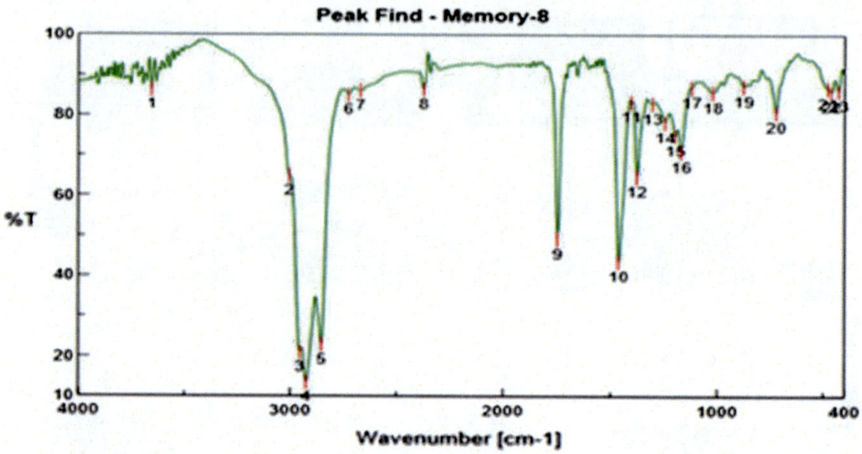

Fig. 2 FTIR analysis in the spectrum of B30

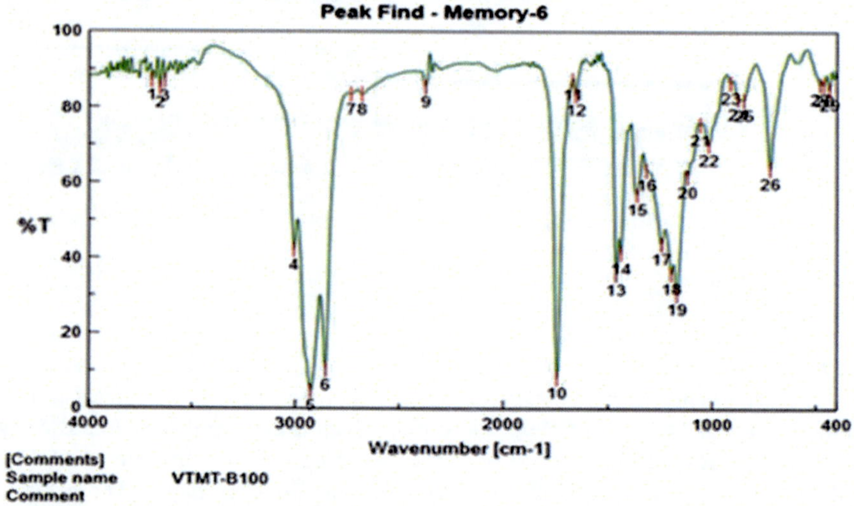

Fig. 3 FTIR analysis in the spectrum of B100

The peaks in the spectrum band are the absorbent of the vibration molecule and the concentration of the mixture. MIR (Mid-range) valDB; ues for this apparatus are in range of 400–4000 cm^{-1} and as a resolution of about 1 cm^{-1}. The highest peak is found for blend of B30. The double bond stretching vibration related to C–H alkyl methylene group which frame olefin and to extend fluctuation and reduce at methyl group and the second peak clearly at 1745 cm^{-1} C=O carbonyl group it frames esters the C–H group's tendency in $-CH_3$ and CH_2 assignment and as same as the B100

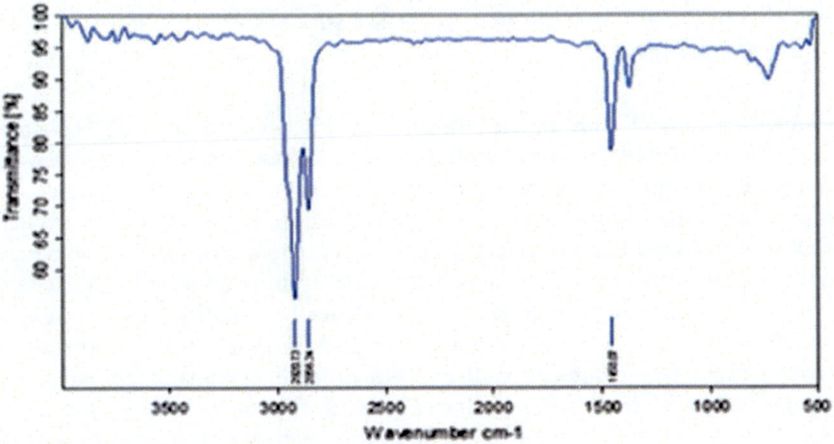

Fig. 4 FTIR analysis in the spectrum of D100

sample as shown in Fig. 3. In between 3007 and 2926 cm^{-1} stretch combined with 2855 cm^{-1} and yield stretch has shown $–CH_2$ and the C–H alkyl group is present at another peak got at 1743 cm^{-1} carbonyl C=O aldehyde are identified both sample bending has approximately at 723 cm^{-1}. The sample of diesel shown in Fig. 4 has similar functional groups at the approximate region of wave number in the spectrometric image, and the appropriate values were given as well as the functional group families and type of vibration motion.

3.3 CHN Elemental Analysis

The technique CHNS/O analysis is used to determine element weight percentage in fuels for diesel, rice bran biodiesel and rice bran oil that is mainly described in carbon, hydrogen, nitrogen, sulfur and oxygen, and the elemental analysis is synthesized by the Eagar smart data handling software automatic calculation. The high carbon diesel to RBD and RBO, than it shows the main element for high burn source oxygen RBO having more than other fuels and its sulfur content has very less in RBO and RB biodiesel than diesel which has 0.21% so that its concluded maximum Sulfur free and high burn source capacity elements are abundantly available in the raw material and proposed biodiesel.

3.4 Physical and Chemical Properties of Rice Bran Oil and Biodiesel

Blending the biofuel with mineral diesel is mostly preferred in many countries because the use of 100% of biodiesel in the engine may have some drawbacks like less availability, poor cold flow properties and comparatively reduced calorific value. For this purpose, a detailed investigation was done to analyze the deviation, and required information is gained in the physical and chemical properties of diesel and biodiesel blend. Rice bran oil is blended with mineral diesel at different blending ratios from 10 to 90% to determine the properties like kinematic viscosity, density, calorific value, fire point, flash point, cloud point, pour point. The above said properties are measured using the instruments compared with different bending ratios, and graphs are generated to determine the deviation of properties for each blending ratio for the same. The mathematical regression equations are determined from which values of properties for different blending ratios are established. This considerably reduces the time and cost.

3.5 Examination of Properties of Diesel and Biodiesel Blends

Graphs for different properties of rice bran biodiesel blends are established. Mathematical regression equation is developed for these properties using analysis software know as sigma plot software, which is mainly used to predict the values for the different properties of various blending ratios. Here, in the mathematical equation, '*x*' represents the blending ratio which is defined as the mixture of mineral diesel and biofuel.

3.5.1 Linear Regression

Equation:

$$Y = ax + b \tag{1}$$

'a' coefficient

$$A = \frac{\Sigma x_i \Sigma y_i - n \Sigma x_i y_i}{\Sigma x_i^2 - n \Sigma x_i^2} \tag{2}$$

'b' coefficient

$$b = \frac{\Sigma x_i \Sigma x_i y_i - \Sigma x_i^2 \Sigma y_i}{\Sigma x_i^2 - n \Sigma x_i^2} \tag{3}$$

Linear correlation coefficient

$$r_{xy} = \frac{n\Sigma x_i y_i - \Sigma x_i \Sigma y_i}{\sqrt{\left(n\Sigma x_i^2 - \left(\Sigma x_i^2\right)\left(n\Sigma y_i^2\right) - \left(\Sigma x_i^2\right)\right)}} \quad (4)$$

Coefficient of determination

$$R^2 = r_{xy}^2 \quad (5)$$

Standard error of the regression

$$\bar{A} = \frac{1}{n}\left|\frac{\Sigma y_i - \bar{y}_i}{y_i}\right| \times 100 \quad (6)$$

4 Results and Discussion

In our research work, rice bran oil properties have been evaluated and compared with various standards like American Standard (ASTM), European Standard (EN) and Bharat Standard (BIS).

4.1 Density

Density can be defined as ratio of mass per volume which is mostly communicated as kilogram per meter cubic. Density plays a major role in the engine output and performance since the change in density has a direct influence on the mass of the fuel, which has an effect on the fuel injection in the engine. Also, density has greater effect on emission like particulate matter (PM) and NO_x. When the density is increased by 10%, it leads to loss of combustion heat by 4%. Thus, in order to increase the heat of combustion, the density of rice bran is reduced. The main reason to reduce the density of biofuel is to reduce the viscosity which has influences on the fuel injection characteristics. The density of mineral diesel is in the range of about 820 kg/m^3. It is clearly detected that the density values increase in a linear manner in which the density increases with the increase in blending ratio because of the increase in the biofuel ratio in the biodiesel mixture. Thus, the density range for B30 is almost nearer to the density value and is also in range mentioned in the EN-590 (reference paper no). The density value reaches up to a range of 874 kg/m^3 for the B90 blends. These values are shown in Table 1.

Table 1 Properties of rice bran biodiesel for different blending ratios

Properties	Unit	B0	B10	B20	B30	B40	B50	B60	B70	B80	B90	B100
Density	kg/m^3	820	826	831	838	844	851	856	862	868	874	878
Kinematic viscosity	mm^2/s	2.4	2.6	2.8	3	3.3	3.5	3.7	3.9	4.2	4.4	4.75
Calorific value	kj/kg	43,000	42,850	42,250	42,050	41,750	41,250	40,950	40,650	40,050	39,700	39,500
Flash point	°C	55	68	79	89	100	112	122	135	145	156	165
Fire point	°C	62	73	85	98	111	125	136	149	161	175	182
Cetane number		49	49.9	50.6	51.3	52.1	53	53.9	54.7	55.5	56.1	57
Cloud	°C	1	2	4	5	6	8	9	10	12	13	14

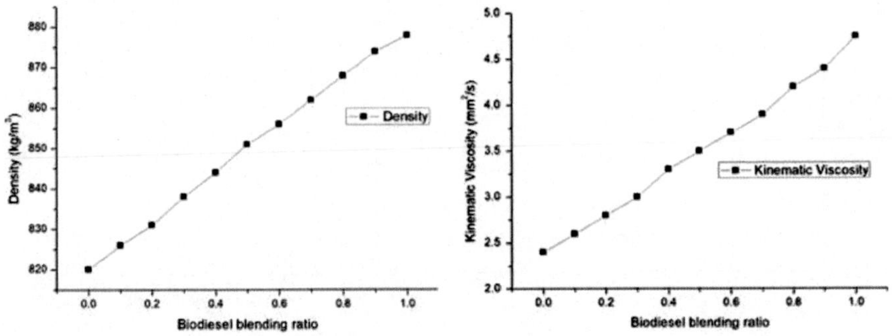

Fig. 5 Density vs Biodiesel blending ratio and kinematic viscosity vs Biodiesel blending ratio

4.2 Kinematic Viscosity

Viscosity is the oil's resistance to flow and shear. Thus, the kinematic viscosity was determined using Redwood Viscometer. This kinematic viscosity has an important influence in the fuel flow and fuel injection. Kinematic viscosity should be in the range of 2–5 mm^2/s which will affect the fluidity of the fuel at low temperature. When the kinematic viscosity increases, it will affect the combustion rate since the atomization of the fuel will be delayed. Spray cone angle and spray penetration are reduced with increase in viscosity and results in bigger droplets. High viscosity can reduce fuel flow rates which results in inadequate fueling of engine. On the other hand, low viscosity results in fuel leakage and loss of fuel system calibration. As in the density, the viscosity value also has a linear increase in value with the increase in the volumetric ratio of the rice bran biofuel with the mineral diesel as shown in Fig. 5. In this, the viscosity range for raw biodiesel (B100) is about 30.26 mm^2/s whereas the value for B30 blend is almost nearer at 3 mm^2/s which satisfies the engine specification and also helps to increase the combustion rate.

4.3 Calorific Value

The amount of heat liberated during the combustion at steady pressure and temperature is known as calorific value. It is one of the major factors that has an effect in the combustion rate and also affects the output power of the engine. The calorific value is measured using the bomb calorimeter, which is measured by the heat produced by combustion of a little quantity of oil. This is usually expressed in kilojoules' per kilogram. Figure 6 shows the variation of calorific value with the various blending ratios of rice bran biodiesel. In contrast to density and kinematic viscosity, the calorific value shows a decrease in value with increasing blending ratio. The calorific value for raw rice bran oil is almost around 36,283 kJ/kg which is less when compared to

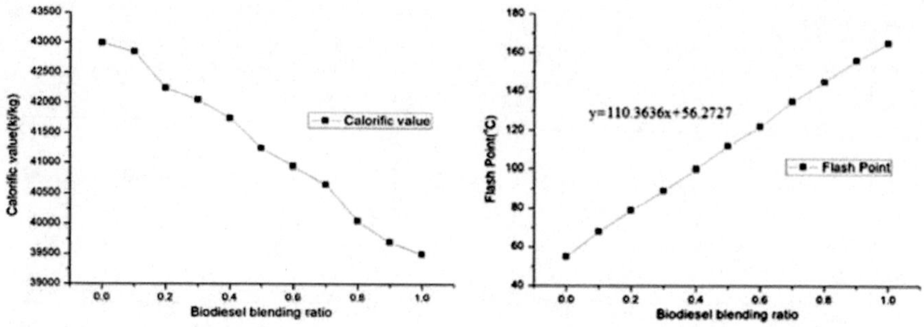

Fig. 6 Calorific value vs Biodiesel blending ratio and Flash point vs Biodiesel blending ratio

mineral diesel. But the biodiesel blends of B20 and B30 have a calorific value nearer to diesel.

4.4 Flash Point

The minimal temperature at which the fuel starts to vaporize and forms as an ignitable mixture in the presence of atmospheric air around the fuel surface is termed as flash point. Ignition source temperature does not have effect on flash and fire point, but its value is far higher than that of flash point and fire point. However, lower the flash point easier to ignite the fuel mixture. The flash point of diesel is 55, and flash point of biofuel is 165. In order to reduce the flash point value, the biofuel is blended with mineral diesel. The fuel is termed as flammable if the flash point value is less than 37.8 °C, and if value is greater than 37.8 °C, then the fuel is called combustible. Thus, crude rice bran oil and biodiesel come under combustible fuel. Figure 7 clearly shows that value of flash point gradually increases from B0 to B100.

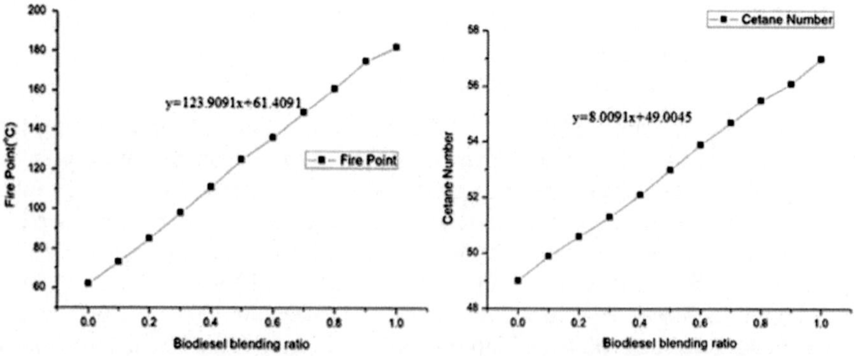

Fig. 7 Fire point vs Biodiesel blending ratio and Cetane number vs Biodiesel blending ratio

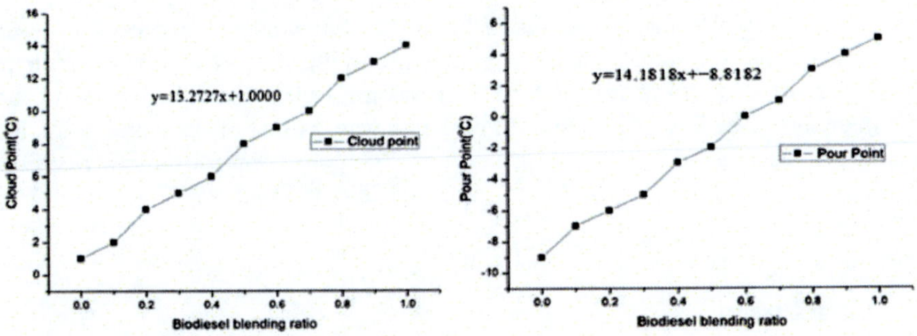

Fig. 8 Cloud point vs Biodiesel blending ratio and Pour point vs Biodiesel blending ratio

4.5 Fire Value

Fire point can be defined as the minimal temperature at which the fuel vapor starts to burn for a specified time after it is ignited. In general, the fire point will be 10 °C more than the flash point. In Fig. 8, same as that of the density and flash point, the fire point also has a linear increase in values for increasing values of blending ratio. It is observed that the raw rice bran oil has a fire point of about 242 °C. The blending ratio B30 and B20 has optimal fire point when compared to the mineral diesel.

4.6 Cetane Number

Cetane number is a marker or indicator which gives the info about the ignition delay and also indicates the ignition quality of the fuel. For the increasing value of cetane number, the ignition delay period reduces subsequently the rate of combustion increases and vice versa. As per the ASTM standards, the cetane number should have a minimal value of 40 to a maximum value of 60. If the value of cetane number falls below 40, it leads to smoke emission and fails to obey the emission laws. The cetane number value of biodiesel lies in the acceptable range, and it also enhances the combustion rate and reduces the ignition delay period alternatively. In the same way as before, it also shows a linear increase in value with increase in volumetric ratio of rice bran biofuel with the mineral diesel.

4.7 Cloud Point, Pour Point

Cloud point can be referred to as the temperature at which the wax crystal is created in the fuel which has a direct impact at the low condition application. Pour point is also an important property for the low temperature application. It is defined as the temperature at which the fuel losses its fluidity and flow characteristics. The cloud point also shows constant increase with increase in blending ratio where the cloud point of mineral diesel is around 1 °C. It is recommended that the cloud point and the pour should be low as possible. In the same way, as that of the cloud point, the pour point also shows an increase in readings. The pour point for the raw rice bran biodiesel is around +5 °C whereas for mineral diesel is about −9 °C.

5 Conclusion

In this investigation, the physical and chemical properties of rice bran oil are evaluated and compared with the mineral diesel as per ASTM-D-6751, EN-14214 and IS-15607 standards. From this investigation, it is clearly known that different properties of rice bran biodiesel are nearer to that of mineral diesel at lower blending ratios, and the deviation increases as the blending ratio increases. Mathematical regression calculation is used to determine the properties for different blending ratios. Sigma plot software is used to draw the graph using the values determined by regression values. From the graph, it is evaluated that B30 is optimum blending ratio. At this blending ratio, it is found that viscosity of oil is higher than that of mineral diesel, and other properties like density, cetane number, calorific value, flash point and fire point are closer to that of mineral diesel. Cold flow properties like cloud point and pour point do not show major deviation to the value of mineral diesel.

References

1. Lin L, Ying D, Chaitep S, Vittayapadung S (2009) Biodiesel production from crude rice bran oil and properties as fuel. Appl Energy 86:681–688
2. Dennis Leung YC, Xuan W, Leung MKH (2010) A review on biodiesel production using catalysed transesterification. Appl Energy 87:1083–1095
3. Atabani AE, Mahlia TMI, Masjuki HH, Badruddin IA, Yussof HW, Chong WZ, Lee KT (2013) A comparative evaluation of physical and chemical properties of biodiesel synthesized from edible and non-edible oils and study on the effect of biodiesel blending. Energy 58:296–304
4. Evangelos Giakoumis G (2013) A statistical investigation of biodiesel physical and chemical properties and their correlation with the degree of unsaturation. Renew Energy 50:858–878
5. Ashraful AM, Masjuki HH, Kalam MA, Rizwanul Fattah IM, Imtenan S, Shahir SA, Mobarak HM (2014) Production and comparison of fuel properties engine performance emission characteristics of biodiesel from various non-edible vegetable oils A review. Energy Convers Manag 80:202–228
6. Lahane S, Subramanian KA (2015) Effect of different percentages of biodiesel–diesel blends on injection spray combustion performance and emission characteristics of a diesel engine. Fuel 139:537–545
7. Bhaskor Bora J, UjjwalSaha K (2016) Experimental evaluation of a rice bran biodiesel e biogas run dual fuel diesel engine at varying compression ratios. Renew Energy 87:782–790
8. Chhabra M, Sharma A, Dwivedi G (2016) Performance evaluation of diesel engine using rice bran biodiesel. Egypt J Petrol 26(2):511–518

9. Sajjadi B, Aziz A, Raman A, Arandiyan H (2016) A comprehensive review on properties of edible and non-edible vegetable oil-based biodiesel Composition specifications and prediction models. Renew Sustain Energy Rev 63:62–92
10. Mohamed Shameer P, Ramesh K, Sakthivel R, Purnachandran R (2017) Effects of fuel injection parameters on emission characteristics of diesel engines operating on various biodiesel: a review. Renew Sustain Energy Rev 67:1267–1281
11. Maria Capellini C, Giacomini V, Maitê Cuevas S, Christianne Rodrigues EC (2017) Rice bran oil extraction using alcoholic solvents Physicochemical characterization of oil and protein fraction functionality. Ind Crops Prod 104:133–143
12. Alptekin E (2017) Emission injection and combustion characteristics of biodiesel and oxygenated fuel blends in a common rail diesel engine. Energy 119:44–52
13. EvangelosGiakoumis G, Christos Sarakatsanis K (2018) Estimation of biodiesel cetane number density kinematic viscosity and heating values from its fatty acid weight composition. Fuel 222:574–585
14. Alptekina LE, Sanlib H, Canakcia M (2019) Combustion and performance evaluation of a common rail DI diesel engine fueled with ethyl and methyl esters. Appl Thermal Eng 149:180–191
15. Dharmaraja J, Nguyen DD, Shobana S, Saratale GD, Arvindnarayan S, Atabani AE, Chang SW, Kumar G (2019) Engine performance, emission and bio characteristics of rice bran oil derived biodiesel blends. Fuel 239:153–161

Wear and Friction Behaviours of Stainless Steel (SS 316) Wire Mesh and Carbon Fibre Reinforced Polymer Composite

A. H. Ansari, V. Jayakumar, and S. Madhu

Abstract There is a growing demand for the development of new polymer matrix composite materials that exhibits high strength, high wear resistance, good rigidity and less weight. In this work, an attempt has been made to develop a novel carbon fibre reinforcement polymer (CFRP) composite with stainless steel (SS 316) wire mesh and to investigate its tribological characteristics with the objective of improving the wear resistance. The pin-on-disc experiment was performed on the fabricated CFRP-SS316 fibre metal laminate by using EN31 steel (high carbon alloy) pins. Wear rate and coefficient of friction have been examined by conducting four experiments on CFRP-SS316 laminate. The process parameters used for this investigation includes applied load, sliding velocity, sliding distance and speed. Experimental results showed that the wear rate increases on a CFRP-SS316 disc laminate when the load increases gradually from 10 N to 40 N. The minimum wear rate of 16.34 mm^3/N was obtained for 10 N applied load.

Keywords Carbon fibre reinforced polymer · Stainless steel 316 wire mesh · Wear rate · Coefficient of friction

A. H. Ansari
Department of Mechanical Engineering, Saveetha School of Engineering, Saveetha Institute of Medical and Technical Sciences, Chennai, Tamil Nadu, India

V. Jayakumar (✉)
Department of Mechanical Engineering, Amrita School of Engineering, Chennai, Amrita Vishwa Vidyapeetham, Tamil Nadu, India
e-mail: v_jayakumar@ch.amrita.edu

S. Madhu
Department of Automobile Engineering, Saveetha School of Engineering, Saveetha Institute of Medical and Technical Sciences, Chennai, Tamil Nadu, India

S. Vijayan et al. (eds.), *Trends in Manufacturing and Engineering Management*, Lecture Notes in Mechanical Engineering,
https://doi.org/10.1007/978-981-15-4745-4_32

1 Introduction

Polymer matrix composites potentially used in automobile, medical and engineering industries. They offer high stiffness, strength, abrasion resistance, creep resistance and non-flammability. These composites have been widely used for many applications because they are light in weight and ease of manufacturing. As compared with other engineering materials, polymer matrix composites have high brittleness. Nowadays, hybrid composites were fabricated which consist of different fibre with suitable matrix using various methods [1]. As compared with other engineering materials, polymer matrix composites have high brittleness. In order to increase the ductility, hybrid FRP composites (CFRPP and SRPP) were fabricated [2]. Carbon fibre is widely used as reinforcing materials in advanced polymer composites because of their unique properties. In order to control the surface characteristics, the carbon fibre has undergone for surface treatments [3]. Composite laminates have susceptible to damage induced by impact as compared with traditional metals [4]. Fibre reinforced metal laminate composites offer better damage tolerance to fatigue crack growth and impact damage [5].

Tensile behaviour of hybrid polymer composites fabricated with carbon fibre and three different sizes of stainless steel wire mesh were investigated and reported that addition of SS wire mesh increased the tensile strength [6]. Aluminium reinforced CFRP hybrid polymeric composite have been fabricated and investigated the friction and wear behaviour. This investigation concluded that the applied load is the significant parameter that affects the friction coefficient [7]. In the study, fibre reinforced polymer composite fabricated with aluminium thin sheets using embossing process. The effect of fibre orientation and polymeric matrix on the feasibility of hybrid joint was investigated [8]. Much previous research works investigated about thermosetting fibre reinforced polymer composite in terms of wear, tensile load, shear load, displacement and absorption energy were studied. Several studies on hybrid steel wire mesh reinforced fibre composite were also performed [9, 10].

Cryogenic-treated carbon fabric epoxy composite was prepared by hot compression moulding technique. Wear characteristics of treated and untreated carbon fabric/epoxy (CF/E) composite were investigated using pin-on-disc method at room temperature for varying load and sliding speed. In this work, morphology of worn surfaces of the wear test composite specimens was studied by scanning electron microscope (SEM) [11]. Carbon fibre reinforced epoxy composites containing 42 and 52 wt% carbon fibres were fabricated using moulding technique. Wear and friction characteristics were investigated using a pin-on-flat plate configuration. In this work, static and dynamic coefficients of friction and wear rates of the fabricated epoxy composites under different loading conditions were experimentally determined. Microstructures of the fabricated composites revealed that applied load and speed are the significant parameters that affect the wear properties [12]. Effect of the normal load and sliding speed on wear rate of two-body abrasive wear of aluminium reinforced polymer composite was investigated [13]. Recently, several studies on the wear behaviours of the hybrid composites were also performed [14–19].

The objective of this current research is to fabricate the stainless steel wire mesh (SS316) reinforced carbon fabric epoxy composite using hand layup process, and the tribological properties such as wear rate, frictional force and coefficient of friction were investigated on CFRP-SS316 composite sample under different applied load, sliding velocity, sliding distance and speed.

2 Materials and Method

2.1 Friction Material Processing

Stainless steel reinforced carbon fibre reinforced polymer composite is fabricated using hand layup process which is used for this investigation. Four layers of carbon fibre reinforced polymer composite arranged in unidirectional manner and stainless steel 316 wire mesh with the size of 50 microns material are combined with CFRP to fabricate a new composite material. Epoxy and hardener with 10:1 ratio is used as a matrix to strength both CFRP and SS316. The 30 N load was applied on the fabricated laminate for 24 h for uniform distribution of matrix and fibre. Figure 1a shows the fabricated CFRP-SS316 composite. Abrasive water jet machining is used for making the pin-on-disc specimen with inner and outer diameter 6 mm and 55 mm, respectively. Hence, a disc (CFRP-SS316) is formed with its initial weight 7.527 g. EN31 steel is a high carbon alloy manufactured by following forging, annealing, stress reviling, hardening and tempering to find the wear and frictional property of a composite. The height and thickness of the pin is 1 inch and 8 mm, respectively, as shown in Fig. 1b CFRP-SS316 sample and EN31 steel pins. Figure 1c demonstrated the microstructure fabricated CFRP-SS polymer composite.

2.2 Wear Testing Characterization

The pin-on-disc wear test is carried out to predict the wear performance of the fabricated carbon fibre reinforced SS polymer composite. A pin-on-disc wear tester is used to find the wear rate and coefficient of friction of the samples. A pin is located against a flat rotating disc specimen such that a circular wear path is described by the machine. The machine is used to evaluate wear and friction properties of materials under pure sliding conditions. The disc can serve as specimen, while the other as counter face. Pin with various geometry is used. Figure 2a shows the two-dimensional model of pin and disc. Figure 2b shows the CAD model of wear mechanism.

The disc placed above the flat rotating surface and a pin is just above the disc with sliding diameter as shown in Fig. 2a. The wear track formed depending upon the load applied on the material through pin as shown in Fig. 2b. This also shows the metal removal rate and wears depth of cut of material.

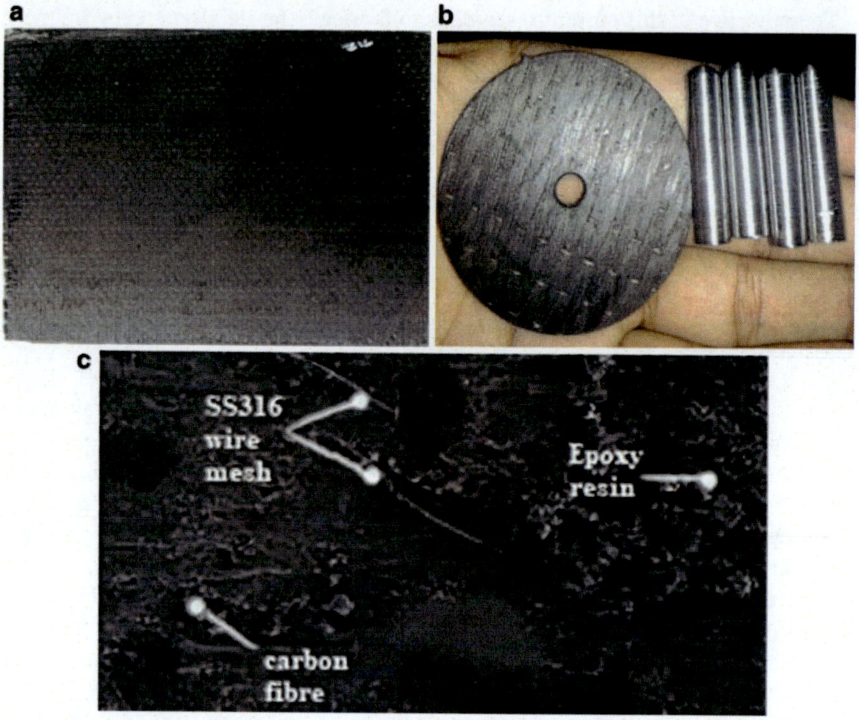

Fig. 1 **a** Fabricated CFRP samples. **b** Wear test specimen. **c** SEM image of the fabricated CFRP-SS polymer composite

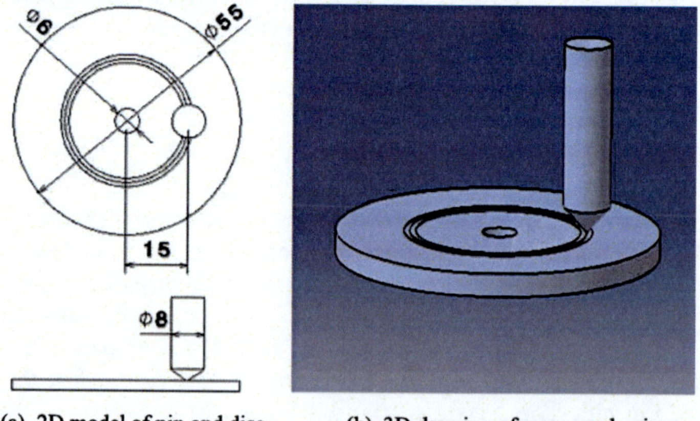

(a) 2D model of pin and disc (b) 3D drawing of wear mechanism

Fig. 2 Wear mechanism

2.3 *Pin-on-Disc Wear Parameters*

The wear test parameters used for this investigation is applying load (N), sliding velocity (m/s), sliding distance (m) and speed (rpm). The above three parameter values are considered depending upon the material properties. Table 1 shows the wear parameters used for this investigation, and Fig. 3a, b shows the pin-on-disc apparatus and sample.

The specific wear rate is calculated as loss in volume loss (ΔV) divided by applied load (P) and sliding distance (L) as shown in the following Eq. 1

$$k = \Delta m / P \cdot L \tag{1}$$

where 'Δm' is mass loss, 'P' is the applied load and 'L' is the sliding distance.

3 Results and Discussions

The wear tests have been done on CFRP-SS316 material specimen, and its average values are given in Table 2. With the help of software and arrangement made in the wear equipment, it was possible to record reading on every 1 s, and for the 16.664 min

Table 1 Wear test parameters

S. No.	Parameter	Range
1	Applied load (N)	10, 20, 30, 40
2	Sliding velocity (m/s)	1, 1.5, 2, 2.5
3	Sliding distance (m)	1000, 1500, 2000, 2500
4	Speed (RPM)	623, 637, 1038, 1274

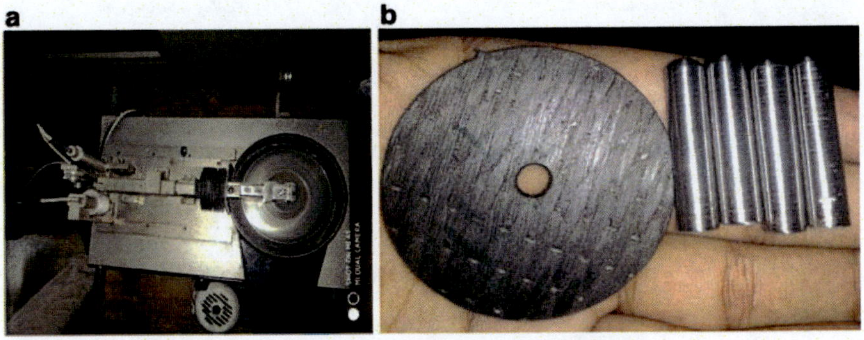

Fig. 3 **a** Pin-on-disc apparatus. **b** Wear test specimen

Table 2 Wear rate and COF of CFRP-SS316 polymer composite

Exp	Load (N)	Speed (rpm)	Sliding velocity (m/s)	Sliding distance (m)	Wear rate (mm^3/N-min)	Frictional force (N)	Coefficient of friction
1	10	623	1	1000	16.34	1.008	0.1008
2	20	637	1.5	1500	118.1	1.92	0.096
3	30	1038	2	2000	379.2	7.38	0.246
4	40	1274	2.5	2500	633.9	10.76	0.269

test duration, 998 readings were recorded for the wear rate and coefficient of friction. Table 2 shows the average values of wear rate and coefficient of friction.

3.1 *The Effect of Pin-on-Disc Parameters on Wear Rate*

Figure 4 shows the effect of pin-on-disc parameters on wear rate of CFRP-SS316 composite material. Figure 4a shows the effect of applied load on wear rate. From the figure, it was observed that the wear rate increases on CFRP-SS316 material when

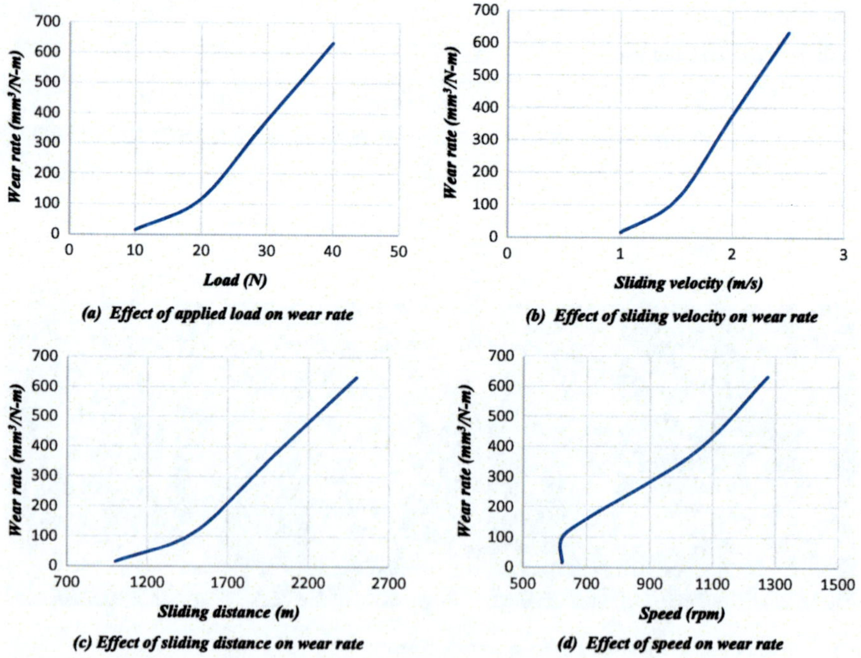

Fig. 4 Effect of pin-on-disc parameters on wear rate

the applied load increases. The minimum wear rate obtained from this experiment is 16.34 mm^3/N-m (10 N) and 633.9 mm^3/N-m was obtained for 40 N applied load as shown in Fig. 4a. Also, the minimum wear rate is obtained for sliding velocity of 1 m/s and sliding distance of 1000 m. The minimum speed is 623 rpm for less wear rate and for maximum speed 1274 rpm, the wear rate is high as shown in Fig. 4d. The applying load has a significant influence on wear rate. When the load increases, the friction between the pin and the work surfaces increases which in turn increase the wear rate of the fabricated composite [19].

The reason behind high wear loss of a material is that due to the applied load which is high on the material, there was no possibility of the fibre detachment due to deep embedding of fibres in the epoxy resin. The need of stainless steel 316 mesh inside the material is the wear rate that decreases due to less friction. The sliding velocity is a circumferential speed of the rotating face. When the sliding velocity increases on the material, the contact pressure between the pin on material increases and affects the unidirectional fibres for more material loss. The high speed also is one of the reasons, why the removal rate of material is high.

3.2 *The Effect of Pin-on-Disc Parameters on Coefficient of Friction (COF)*

The effect of wear parameters on COF for the fabricated CFRP-SS316 composite material is shown in Fig. 5. The effect of COF increases on CFRP-SS316 material when load increases. The minimum COF is 0.1008 for 10 N applied load and maximum 0.269 for 40 N applied load as shown in Fig. 5a. In the same way, the minimum COF is 0.1008 for sliding velocity and sliding distance (i.e., 1 m/s and 1000 m) and maximum 0.269 for 2.5 m/s and 2500 m as shown in Figs. 5b, c. For minimum speed, 623 rpm less COF was obtained, and for maximum speed 1274 rpm, COF is high as shown in Fig. 5d. The effect of coefficient of friction increases, because at low speed, the friction is high, whereas at high speed the friction is less.

4 Conclusions

The tribological properties of carbon fibre reinforced polymer—stainless steel 316—is investigated in this work. The effect of wear rate and coefficient of friction on load, sliding velocity, sliding distance and speed was investigated. The SS reinforced carbon fabric reinforced epoxy composite shows relatively good sliding wear resistance. The addition of stainless steel 316 material in carbon fabric improved the wear resistance and acts as solid lubricant during sliding. The developed composite material can be used in brake disc to improve the wear resistance.

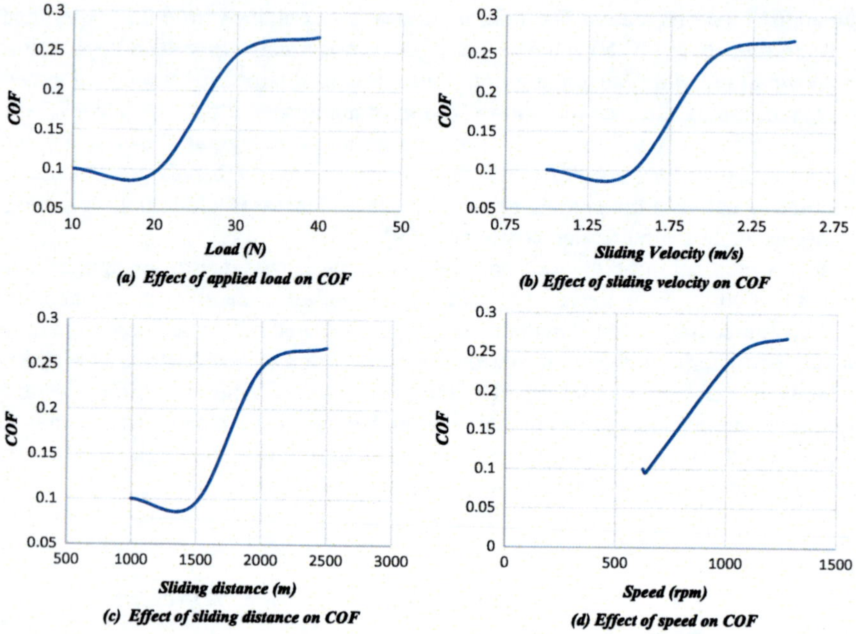

Fig. 5 Effect of pin-on-disc parameters on COF

References

1. Sathishkumar TP, Naveen J, Satheeshkumar S (2014) Hybrid fiber reinforced polymer composites—a review. J Reinf Plast Compos 33(5):454–471
2. Swolfs Y, Meerten Y, Hine P, Ward I, Verpoest I, Gorbatikh I (2015) Introducing ductility in hybrid carbon fibre/self-reinforced composites through control of damage mechanisms. Compos Struct 131:259–265
3. Jin F-L, Park S-J (2015) Preparation and characterization of carbon fiber-reinforced thermosetting composites: a review. Carbon Lett 16(2):67–77
4. Davim JP, Marques N, Baptista AM (2001) Wear 251:1100–1104
5. Sinmazçelik T, Avcu E, Bora MÖ, Çoban O (2011) A review: fibre metal laminates, background, bonding types and applied test methods. Mater Des 32:3671–3685
6. Truong GT, Choi K-K (2019) Tensile behavior of hybrid composites of carbon fibres—steel wire mesh reinforced polymer. Mech Adv Mater Struct
7. Caliman R (2016) Friction behaviour of aluminium composites mixed with carbon fibers with different orientations. Mater Sci Eng 145:022035
8. Huang Z, Sugiyama S, Yanagimoto J (2014) Adhesive–embossing hybrid joining process to fibre-reinforced thermosetting plastic and metallic thin sheets. Procedia Eng 81:2123–2128
9. Sakthivel Munisamy, Vijayakumar Subramani (2017) Influence of stainless-steel wire mesh on the mechanical behaviour in a glass-fibre-reinforced epoxy composite. Mater Technol/Mater Technol 51:455–461
10. Hasselbruch H, von Hehl A, Zoch H-W (2014) Properties and failure behavior of hybrid wire mesh/carbon fibre reinforced thermoplastic composites under quasi-static tensile load. Mater Des
11. Shivamurthy B, Murthy K, Anandhan S (2018) Tribology and mechanical properties of carbon fabric/MWCNT/epoxy composites. Adv Tribol, Article ID 1508145

12. Sahinand Y, De Baets P (2017) Tribological behaviour of unidirectional carbon fibre reinforced epoxy composites. Mater Sci Eng, 174
13. Odabas D (2018) Effects of load and speed on wear rate of abrasive wear for 2014 Al alloy. Mater Sci Eng, 295
14. Suresha B, Chandramohan G, Renukappa NM, Siddaramaiah (2007) Mechanical and tribological properties of glass-epoxy composites with and without graphite particulate filler. J Appl Polym Sci 103:2472–2480
15. Khalladi A, Elleuch K (2017) Tribological behavior of wheel rail contact under different contaminants using pin-on disk methodology. J Tribol 139:011102–011108
16. Rudas JS, Gomez LM, Toro A, Gutiérrez JM, Corz A (2016) Wear rate and entropy generation sources in a Ti6Al4V—WC/10Co sliding pair
17. Uvaraja VC, Natarajan N, Rajendran I, Sivakumar K (2013) Tribological behavior of novel hybrid composite materials using Taguchi technique. J Tribol 135:021101–021109
18. Shibo W, Niu C, Teng B (2017) Tribological behavior of polytetrafluoroethylene: effect of sliding motion. J Tribol 139:011301–011306
19. Xu S, Liu T, Tangpong XW, Zhong W-H (2012) Wear and friction of carbon nanofiber-reinforced HDPE composites. J Tribol 134:041602–041606

Tribological Properties of PEEK Reinforced with Synthetic Diamond Composite

K. Rajkumar, K. Vishal, and P. Sabarinathan

Abstract Tribological behaviour of synthetic diamond reinforced poly ether ether ketone was studied for polymer self-lubricate bearing components. Addition of reinforcement to the matrix improved its mechanical, thermal and tribo properties of composite. Synthetic diamond provides many advantageous properties which is responsible for higher hardness and lowered the thermal degradation. This paper deals with synthetic diamonds of various volume fractions reinforced with PEEK. This thermo polymer matrix composite was manufactured by powder metallurgy process. The characterization of synthetic diamond and polymer matrix composites was done by scanning electron microscopy. Wear and tribo coefficient of developed PEEK matrix composite was found by tribometer. Diamond particles were used to prevent severe galling during higher loading.

Keywords Wear · Synthetic diamond · Galling · Polymer matrix composite

1 Introduction

Tailor to the multifunctional properties of the material, one potential direction is to develop composite materials that possess combined properties of two or more materials. Metals are being replaced by polymeric materials, because of its inexpensiveness. They are also becoming materials of choice for many engineering applications. Polymeric materials have low density-to-strength ratio making them as more preferable. Semi-crystalline thermoplastics compose regions of both amorphous and ordered molecular structure which let them show a greater performance. As a result, these polymers have chemical resistance as well as wear resistance [1]. Polymer materials with low softening temperature have been replaced by high-temperature polymer like poly ether ether ketone (PEEK). It is known as a semi-crystalline high-performance thermoplastic with excellent mechanical properties. It is the most popular polymeric

K. Rajkumar (✉) · K. Vishal · P. Sabarinathan
Department of Mechanical Engineering, Sri Sivasubramaniya Nadar College of Engineering, Chennai, India
e-mail: rajkumark@ssn.edu.in

S. Vijayan et al. (eds.), *Trends in Manufacturing and Engineering Management*, Lecture Notes in Mechanical Engineering,
https://doi.org/10.1007/978-981-15-4745-4_33

material. It has greater elastic modulus, mechanical strength, wear resistance and inert to chemical reactivity. It is easy processable characteristics.

Recently, the usage of PEEK in the engineering and medical fields is increasing to multifold. It has good resistance against biological actions [3, 4]. It is a material for high temperature-bearing application since it has high creep property up to 250 °C [2]. Various researches have been done to test the tribo behaviour of PEEK and its composites. Various carbon allotropes are being reinforced with thermoplastics due to their high mechanical properties and low manufacturing cost. Some of such materials are graphite, diamond and graphitize carbon (long, short) fibre. Carbon fibre as a reinforcement in PEEK brings in good effect in its strength and tribo nature [5, 6]. PEEK composites are being used in many areas like electrical, automobile, biomedical and aerospace. Advantageous properties like lightness, self-lubrication and easy maintenance can be achieved by using this polymer during the application. Tribological performance of PEEK and PTFE has been researched. The results stated that PEEK is used for the production of précise-machined custom bearings which are suitable for market needs. PEEK's properties such as self-lubrication, high corrosion resistance and high melting temperature (340 °C) make it to be considered as one of the high-performance materials among polymers [7]. The stable operation of the polymer and polymer composites were influenced by the frictional wear [9]. Thermal conductivity of the composite was increased by the addition of the nano-particle [10]. Dispersion of zirconia fillers in the polymer matrix increases the mechanical property, thermal stability and certain combinations of their physical and mechanical properties. The amount of nano-filler affects the thermal stability of the polymer composite [11]. Diamonds exhibit properties like high thermal conductivity, high corrosion resistance, high wear resistance, low coefficient of thermal expansion and high hardness. Due to higher thermal conductivity and higher modulus of synthetic diamond, it is used as reinforcement in high temperature polymer [12]. By the addition of nano-sized TiO_2, the wear rate of the PEEK can be reduced [13]. Wear rate of the polymer matrix can be reduced in large sum by the addition of diamond as reinforcement. These literatures were reporting tribological properties of PEEK composite reinforced with ceramic and carbon nano-materials. Synthetic diamond is another form of high-strength crystalline carbon reinforcement which enhances wear resistance of polymer materials.

In this work, pure peek with different weight percentages of synthetic diamond were added in PEEK matrix, and its tribological properties were studied. Wear loss and friction coefficient of synthetic diamond reinforced PEEK were reported as function of sliding velocities.

Table 1 Properties of poly ether ether ketone (PEEK)

Properties	Values
Size	20 μm
Density	1.32 g/cm^3
Glass transition temperture	143 °C
Yield stress	107 MPa

2 Experimental Procedure

2.1 Materials

The received poly ether ether ketone of grade 2000FP powder form properties is shown in Table 1. In this, PEEK makes the matrix phase, and synthetic diamonds (SD) are the reinforcing material. The size of PEEK and synthetic diamond were 20 μm and 5 μm, respectively. Poly vinyl alcohol (PVA) was used as a binder agent.

2.2 Preparation of Composites

Powder metallurgy process was used to fabricate the polymer composite. The synthetic diamond was varied from 10 to 30 volume % to reinforce PMC. To get a homogeneous mixture, pure PEEK powder and different composition of synthetic diamond were mixed in ball milling technique. This mixture was mixed for three hours. The composites were prepared by pressing the powder hydraulically and sintering it. The powder was pressed at 100 MPa to get a cylindrical shape of diameter 10 mm and height 10 mm. During sintering process, temperature ramp of 100 °C was maintained. The polymer composites were sintered at 3250 °C and soaked for two hours. P/M process produces high potential homogeneous composite materials, it involves lesser time, and it is cost efficient. Cylindrical specimens were fabricated in the form of pellets, as shown in Fig. 1.

2.3 Micro-Structural Analysis

To confirm the structure of diamond, XRD was carried out for the synthetic diamond powder. From the fabricated samples, particle distribution in the composite was evaluated by optical microscope. SEM with 15 kV of accelerating voltage is applied for image observation. Before SEM analysis, the specimen was cleaned with alcohol and then gold sputtered over the specimen to make it surface conductive.

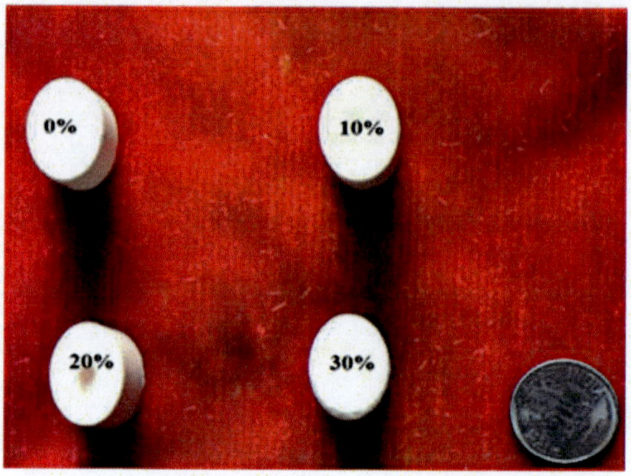

Fig. 1 Fabricated PEEK composites

2.4 Density

The density of the polymer matrix composite (PMC) was measured by Archimedes' principle. The fabricated composite was submerged in water, and its equivalent weight is calculated. Weight of the PMC was also calculated in the air. From these values, the density of the PMC was obtained.

2.5 Hardness

Hardness of PMC was measured by Vickers hardness testing machine NOVA 330/360. The diamond intender in the form of right pyramid with the angle of 136° was used. The hardness was measured with a loading time of 15 s. Load of 100 kgf is applied for about 10–15 s. The average value of 20 measurements was taken to measure the hardness of the polymer matrix composite.

2.6 Fourier Transform Infrared (FTIR) Spectroscopy

Fourier transform infrared (FTIR) spectroscopy was conducted in BRUCKER, Germany, 3000 Hypersion. It was observed that the sample absorbed the IR radiations that were passed through it. The usefulness of infrared spectroscopy is to evaluate the difference in chemical structures (molecules) of PMC.

2.7 Tribological Test

Pin-on-disc wear testing is a method of characterizing the wear rate and coefficient of friction, and frictional force between two sliding materials. The tests have been conducted under normal atmospheric condition in PIN ON DISC FRICTION & WEAR TEST RIG–TE-165-SPOD. The PMC pin specimen was normally loaded over the steel disc. Test was conducted on a flat surface of cylindrical pin of diameter 10 mm and height of 10 mm. Tribo test was conducted for a constant normal load of 100 N using different sliding velocities of 0.6, 1.2, 1.8, 2.4 m/s. Each test was conducted for a period of 30 min which ensures a steady-state coefficient of friction for all the tested samples. An average of three trials was done for each sample. All the tribological sliding tests were conducted at 20 °C in a dry-sliding condition.

3 Results and Discussion

3.1 Material Characterization

Figure 2 shows the XRD profile of synthetic diamond. From the figure, peeks at 430 and 750 show the presence of carbon which confirms that the compound is diamond. The crystal structure was found to be cubic in nature.

Figure 3 shows the typical optical image of the fabricated composite. The image clearly shows the synthetic diamond particles that were fairly dispersed. However, the presence of agglomerates is obvious. No micro-cracks are seen from the fabricated sample.

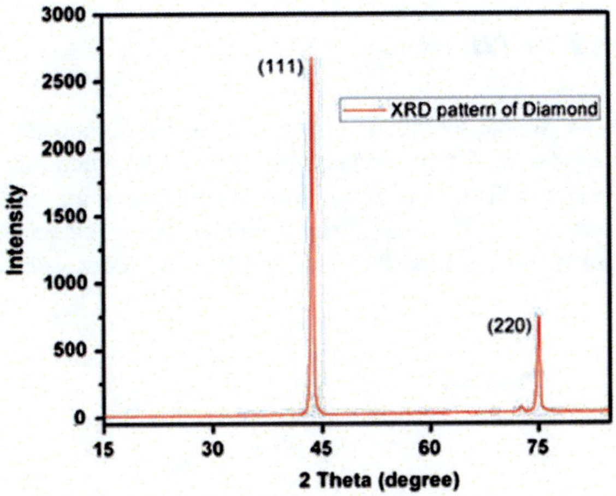

Fig. 2 XRD pattern of SD

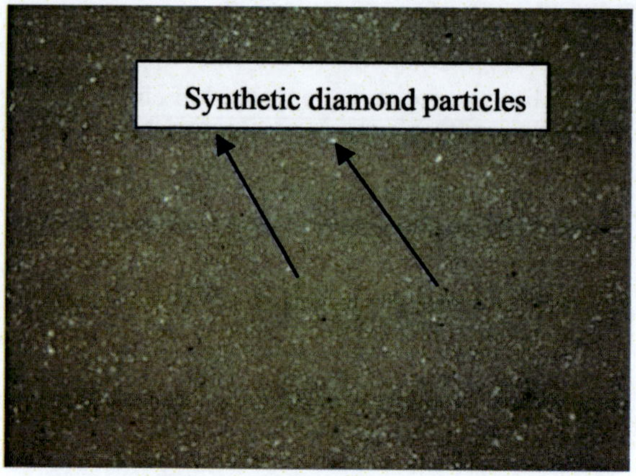

Fig. 3 Optical micrograph of PPEK-30% SD composite

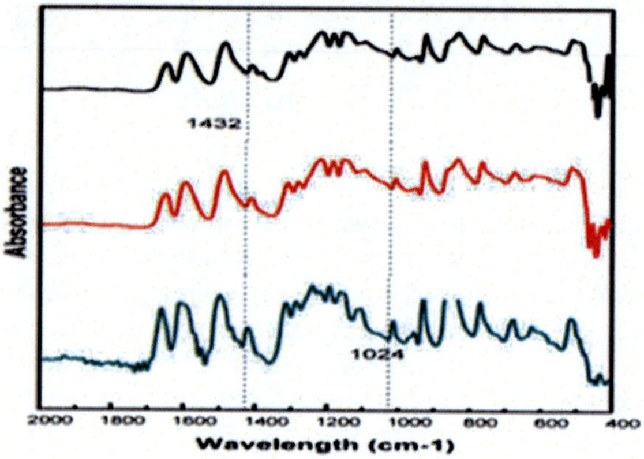

Fig. 4 FTIR spectrums (Black-10%SD, Red-20%SD and Green-30%SD)

3.2 FTIR

The sample of PEEK/synthetic diamond composite was analysed using the FTIR technique. A typical spectrum of PEEK/synthetic diamond composite can be seen in Fig. 4. The spectrum is trace distinctive of amorphous PEEK/synthetic diamond composite. The finger-print region of the spectrum (1432–1024 cm^{-1}) exhibits the distinctive spectral features typical to an IR-recorded spectrum of PEEK.

3.3 Density

The effect of synthetic diamond on the density of peek is shown in Fig. 5. Due to the high density of diamond, the density of the composite material increased as the reinforcement increased.

Fig. 5 Density of PEEK composites

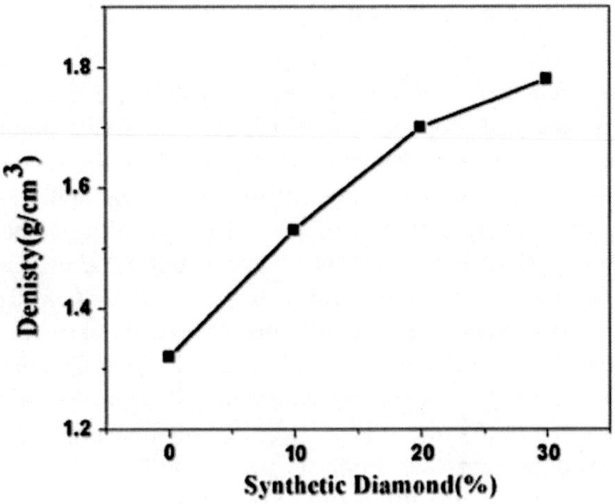

3.4 Mechanical Testing

As per the ASTM E92 standard, Vickers hardness test was conducted for the polymer composite. The hardness of peek and synthetic diamond is shown in Fig. 6. Hardness of the peek is improved when reinforcement is added, due to the high strength of synthetic diamond. Although there was an agglomeration of diamond particle beyond 10%, the hardness of peek composite was observed to increase slightly. This is because of the termination and pinning mechanism.

Fig. 6 Hardness of PEEK composites

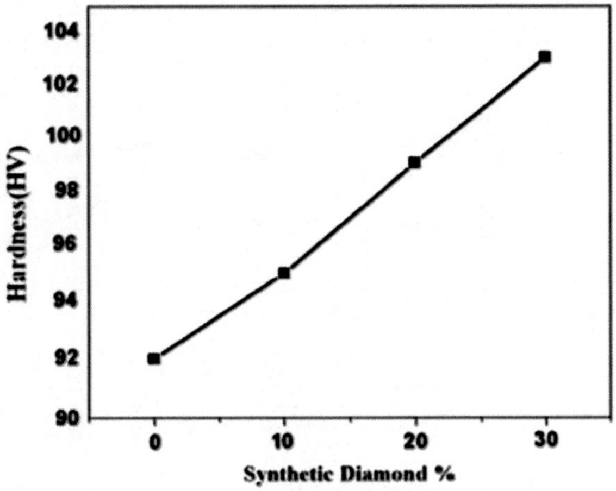

3.5 Tribo Analysis

As seen from the previous section, the 30% SD reinforced composite shows a high mechanical property. Therefore, the tribological studies are limited to this composition only when compared to pure PEEK. In this section, tribo behaviour of 30% SD reinforced composite was done with different sliding speed at constant normal load. The wear performance is influenced by sliding velocities according to the volume fractions and environment. Figure 7 shows the wear rate of virgin PEEK and PEEK composite. This figure shows that wear rate of fabricated composite and pure PEEK is decreasing with sliding velocity. This is because the contact region of the pin and the counterpart gets minimized [7]. PEEK composite shows a lower wear rate as compared to pure PEEK. SD in PEEK makes this composite stronger to sliding wear loss. As a result, the load-bearing constituent of SD provides a direct contact to counter surface. Greatest hardness of SD also shares the contact load and reduces tearing of PEEK [6]. It can be seen from SEM image Fig. 8a that PEEK surface was torn out along with some micro-cutting grooves. However, PEEK composite was observed to be the only counter surface with hard asperities scoring marks over the surface, as shown in Fig. 8b.

Under normal loading, there is a decrease in coefficient of friction at room temperature with sliding velocity increasing, as seen from Fig. 9. The graph clearly shows the reduction of coefficient of friction from 0.047 for the pure PEEK to 0.035 for the PEEK reinforced with synthetic diamond composite. The reduction of COF is around 50%. This is because of the addition of the SD which renders the rolling or mending effect of smaller-sized SD particles between the sliding surfaces [8].

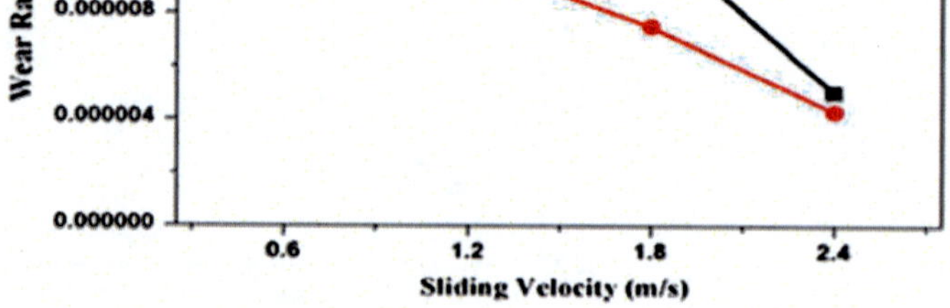

Fig. 7 Wear rate versus sliding velocity

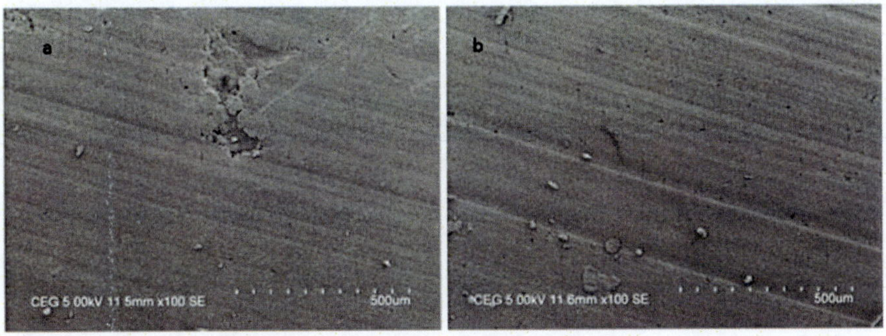

Fig. 8 Worn surface SEM images (**a**) pure PEEK (**b**) 30%SD-PEEK

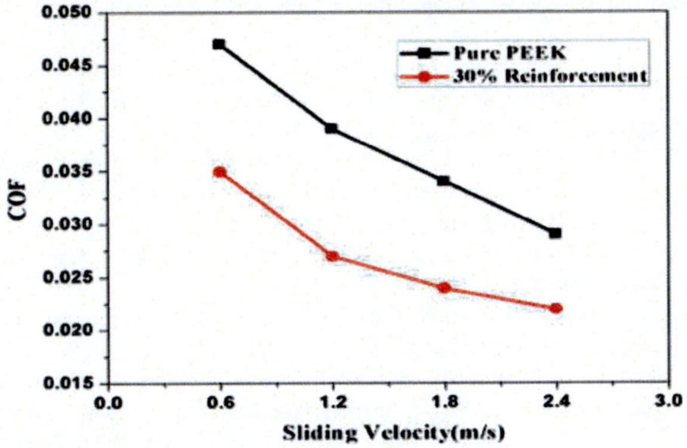

Fig. 9 Coefficient of friction versus sliding velocity

4 Conclusions

1. The effects of several volume fractions of fillers in a PEEK polymer matrix were studied. The mechanical and wear rate of PEEK and different composition of synthetic diamond powder have been studied in this paper.
2. It was observed that the dispersion of synthetic diamond particles in PEEK matrix is fair at low volume of fraction.
3. Density of PEEK composite increased due to the addition of synthetic diamond powder, due to the higher density of diamond.
4. The hardness of the PEEK polymer composite is higher with increase in the reinforcement when compared with pure PEEK.
5. Under constant loading condition of 100 N, tribological tests were performed at different sliding velocities of 0.6 m/s, 1.2 m/s and 1.8 m/s. Friction coefficients and wear rates were evaluated, and from scanning electron microscopy, worn surfaces were examined.

6. As the amount of the synthetic diamond powder increased in the polymer matrix, the wear rate of the composite decreased.
7. With the increase in reinforcement, coefficient of friction decreases. The coefficient of friction of PEEK-based composites was reduced by nano- and micro-particle which becomes more significant at higher contact pressure.

References

1. Simsiriwong J, Shrestha R, Shamsaei N, Lugo M (2015) Effect of microstructural inclusion on fatigue life of poly ether ether ketone (PEEK). J Mech Behav Biomed Mater 51:388–397
2. Werner P, Altstadt V, Jaskulka R, Jacobs O, Sandlerd JKW, Shaffer MSP, Windle AH (2004) Tribological behaviour of carbon-nanofibre-reinforced poly(ether ether ketone). Wear 257:1006–1014
3. Patel P, Richard Hull T, Richard E, Stanislav I, Richard N, Crowley S, Natallia S (2011) Investigation of the thermal decomposition and flammability of PEEK and its carbon and glass-fiber composites. J Polym Degrad Stab 96(1):12–22
4. Xie GY, Sui GX, Yang R (2011) Effects of potassium titanate whiskers and carbon fibers on the wear behavior of poly ether ether ketone composite under water lubricated condition. J Compos Sci Technol 71(2):828–835
5. Avci A, Arikan H, Akdemir A (2004) Fracture behavior of glass fiber reinforced polymer composite. J Cem Concr Res 34(1):429–434
6. Ovaert TC, Cheng HS (1991) Counter-face topographical effects on the wear of PEEK and a PEEK-carbon fiber composite. Wear 150(2):150–157
7. Koike H, Kida K, Santos EC, Rozwadowska J, Kashima Y, Kanemasu K (2012) Self-lubrication of PEEK polymer bearing in rolling contact fatigue under radial loads. Tribol Int 49:30–38
8. Avanzini A, Donzella G, Mazzu A, Petrogalli C (2013) Wear and rolling contact fatigue of PEEK and PEEK composites. Tribol Int 57:22–30
9. Koike H, Kida K, Mizobe K, Shi X, Oyama S, Kashima Y (2015) Wear of hybrid radial bearing (PEEK ring –PTFE retainer and alumina ball) under dry rolling contact. Tribol Int 90:77–83
10. Shang LG (2002) Characterisation of interphase nanoscale property variations in glass fibre reinforced polypropylene and epoxy resin composites. Composites 33(4):559–576
11. Mishra TK, Kumar A, Pandey VV, Kumar V (2013) PEEK composites reinforced with zirconia nano filler. Compos Sci Technol 72(1):1627–1632
12. Zhao Z, Liu Y, Feng W, Zhang Q, Cheng L, Zhang L (2017) Improvement on the thermal conductivity of Diamond/CVI-SiC Composites using large diamond particles. Diam Relat Mater 74:1–8
13. Kurdi A, Wang H, Chang L (2018) Effect of nano-sized TiO_2 addition on tribological behaviour of poly ether ether ketone composite. Tribol Int 117:225–235

Experimental Investigation on the Mechanical Properties of American Agave and Glass Fibre Reinforced Polypropylene Composites

M. Indra Reddy, M. Anil Kumar, and Vamsi Inturi

Abstract The utilization of natural fibres has become extensive instead of conventional synthetic fibres because of the issues with the greenhouse effect and consciousness towards the environment. Exploring natural fibres has several advantages such as economically viable, availability in fibrous form readily and ease of extraction from the plant leaves. In this study, an attempt has been made to investigate the mechanical properties such as tensile, impact and flexural strength of the composites made by reinforcing American agave fibre into a polypropylene resin. In addition to that, the mechanical properties of hybrid composites that are fabricated by reinforcing American agave and glass fibre as 1:1 proportion into a polypropylene resin are studied. The proportion of fibre content is varied from 10 to 30% by volume percentage, and the behaviour in the mechanical properties in each case is examined. It is observed that the American agave and glass fibre reinforced polypropylene hybrid composite has exhibited better mechanical properties than American agave fibre reinforced polypropylene composite.

Keywords American agave fibre · Polypropylene · Mechanical properties

1 Introduction

1.1 A Subsection Sample

Over the years, natural fibres are serving mankind and find their application in various fields. Eventually, natural fibres are contending with the man-made fibres, especially, as far as sustainability, quality and economy of production are concerned [1].

M. Indra Reddy (✉) · M. Anil Kumar
Department of Mechanical Engineering, S R K R Engineering College, Bhimavaram, India
e-mail: reddy.indra.m@gmail.com

V. Inturi
Department of Mechanical Engineering, BITS Pilani, Hyderabad Campus, Hyderabad, Telangana, India

S. Vijayan et al. (eds.), *Trends in Manufacturing and Engineering Management*, Lecture Notes in Mechanical Engineering,
https://doi.org/10.1007/978-981-15-4745-4_34

Because of their biodegradable nature, the utilization of natural fibres has become expensive. Alongside, they often have low densities, beneath processing costs and sustainable than the synthetic materials [2]. Conventionally, natural fibres have been used as reinforcements in polypropylene composites due to their recyclability nature [3]. The alkali and acrylic acid treatment of the banana fibres could enhance the water resistance property and improves the mechanical properties of the composites. However, the addition of compatibilizer will further intensify the above-mentioned properties. It was reported that the addition of compatibilizer improves the impact and flexural strength which further yields to alleviate the water absorption capacity [4]. Zaman et al. [5] considered irradiated banana fibres and treated them with monometer in MeOH solution blended with 2% benzyl peroxide. The authors have used thermal curing method and observed a significant enhancement in the mechanical properties because of the monometer treatment. Rana et al. [6] examined the influence of compatibilizer on jute fibre reinforced polypropylene composites. The authors observed that there is a reduction in the water absorption capacity and a sudden increase in the mechanical properties. This can be attributed to the fact that the addition of compatibilizer developed a linkage between the hydroxyl groups of jute and the carboxyl groups of the compatibilizer [6]. On the other hand, the properties such as modulus of composites and tensile strength are affected by fibre composition, liquid of immersion and number of immersion cycles. In the case of kenaf fibre reinforced polypropylene composites, there is a depression in the tensile strength values because of the number of immersion cycles and conditioning. However, continuous and cyclic immersion in bleach resulted to alter the tensile strength of the composites significantly. However, there is a minimal variation in the values of tensile modulus after a number of immersion cycles [7].

Oksman et al. [8] examined the effect of fibre morphology on the mechanical properties of natural fibres and on the fibre breakage of the composites due to the extrusion process. It was observed that the sisal composites had displayed the favourable impact properties and the longest fibres after the extrusion among the considered natural fibres. In general, the flexural stiffness of the composites increases with fibre content, being highest for flax-type composites. Perhaps, the addition of fibres has no effect on the flexural strength property because of the low compatibility [8]. Based on the exhaustive literature review, it has been found that the mechanical properties of American agave, polypropylene composites have not explored. The main objective of this study is to prepare the composite by incorporating the American agave and glass fibres at various volume percentages into a polypropylene resin matrix. Finally, the composites are examined and characterized to evaluate the tensile, impact and flexural properties.

2 Experimental Procedures

The following section elaborates the experimental procedure carried out, and the steps involved are:

1. Materials
2. Extraction of fibres
3. Fabrication of composites
4. Investigating the mechanical properties of composites.

2.1 Materials

The materials used in this experimental work are:

1. Reinforcement: American agave fibre and glass fibre
2. Matrix: Polypropylene resin
3. Hardener/Catalyst: Methyl ethyl ketone peroxide
4. Accelerator: Cobalt naphthenate.

2.2 Mater Extraction of Fibres

American agave fibre is extracted directly from the American agave plant. Initially, teen-aged American agave plants are taken, and then, the leaves are plucked. Later, the edges of the leaves are trimmed, and then, these edge-trimmed American agave plant leaves are allowed to soak in water for about 10–15 days. After soaking for sufficient time, the residual from leaves gets deposited in the water. Remaining leaf is taken into hand and rubbed, and final fibre is obtained (these fibres appear like threads) and then dried in sunlight. For removing additional organic waste from the fibre, we go for alkali treatment. In alkali treatment, we soak the fibre in 1% NaOH solution for sufficient time, and later, the fibres are again dried in sunlight to obtain the agave fibre. The American agave plant is shown in Fig. 1.

2.3 Fabrication of Composites

Polymer matrix composites are made using polypropylene resin matrix to evaluate the reinforcing ability of American agave fibres. The accelerator (1.5% by volume) and catalyst (1.5% by volume) are supplemented to the resin for curing at room temperature. The appropriate mixture of polypropylene resin and American agave fibres is filled up in the prepared mould and fed to injection moulding machine. In order to ensure good quality fibre reinforced composites, the deformation and movement of the fibre should be minimized. Hence, a compressive pressure of 0.05 MPa is enforced on to the mould, and the composite specimens are allowed to cure for 24 h. In addition to the curing at room temperature, post-curing at 70 °C for about two

Fig. 1 American agave fibre

hours is carried out onto the specimens. The fabricated samples are represented in Fig. 2. Five composite samples are prepared with five different volume percentages of American agave Fibres.

2.4 *Tensile Test*

In order to investigate the tensile properties of the American agave and glass fibre reinforced polypropylene composites, the specimens are prepared according to the ASTM D 638M standards (Standard, 2010). The composite specimens are prepared with a length of 165 mm, width of 12.7 mm and thickness of 3 mm. Three identical specimens are tested for each percentage volume fraction of fibre. These composite specimens are tested using a tensile testing machine operating with a cross-head speed of 2.5 mm/min.

2.5 *Flexural Test*

In order to examine the flexural properties of the specimens, three-point bend tests are carried out in accordance with ASTM D 790M (Standard, 2010) test standards. Three-point bend test is chosen, as it provides the flexibility in determining the centre point deflections with the equipment. In addition to that, it operates with the less amount of material for each test cycle. The specimens are made with 125 mm length, 12.7 mm width and 3 mm thickness. During the test, the outer rollers are kept 64 mm apart, and the specimens are examined under the strain rate of 0.2 mm/min. For each composition, three identical specimens are tested using tensile testing machine, and all the experiments are carried out at a cross-head speed of 2.5 mm/min. Flexural

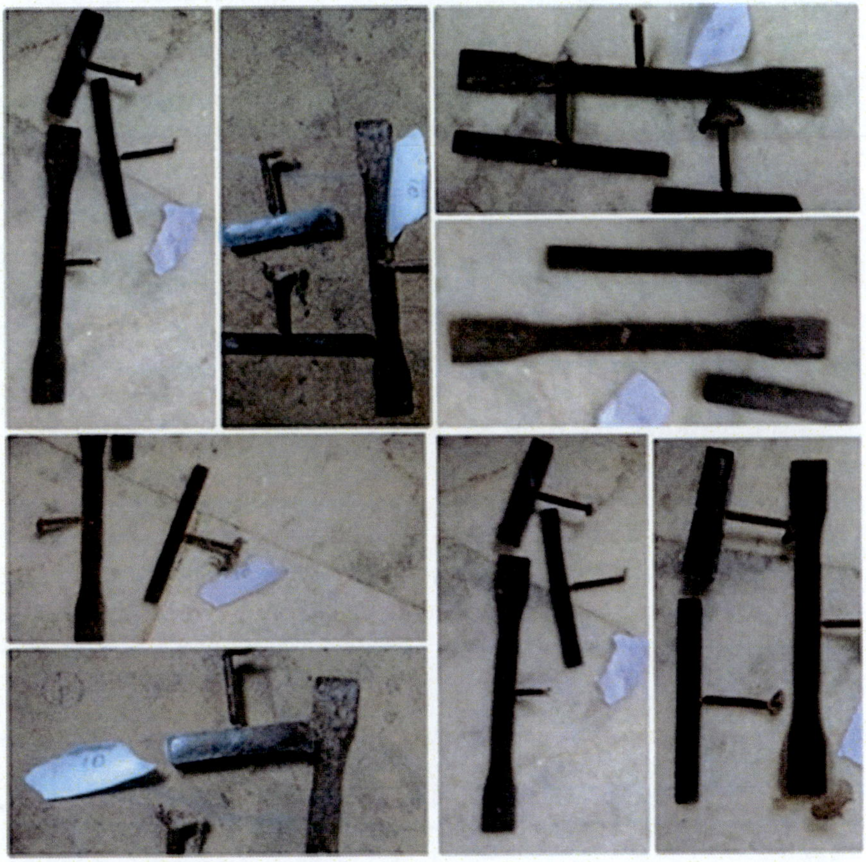

Fig. 2 Fabrication of composites

strength (σ) and modulus (E_f) are calculated using the following formulae:

$$\sigma = 3Pl/2bt^2$$
$$E_f = l3m/bt^3$$

where

σ Bending strength (N/mm^2)
P Maximum load (N)
L Distance between the supports (mm)
B Width of the specimen
T Thickness of the specimen
M Slope of the initial straight portion of the load-deflection curve.

2.6 *Impact Test*

The test specimens for carrying the impact test are prepared according to the standards of ASTM D256. These specimens are having a length of 55 mm, width of 10 mm and thickness of 10 mm. To achieve the notch onto the specimen, a sharp file having an included angle of 45° is drawn across the cut at an angle of 90° to the sample axis. These specimens are fractured in a plastic Izod impact testing machine (capacity—21.68 J). For each composition, three specimens are tested, and the value of the absorbed energy is noted. Finally, the impact strength for a specimen is calculated by dividing the measured absorbed energy value with the cross-sectional area.

3 Results and Discussions

3.1 *Tensile Strength*

The variation of mean tensile strength and tensile modulus of composite with varying fibre content is illustrated in Figs. 3 and 4. It can be observed that, the tensile strength is increasing with the increase in the fibre content of the polypropylene matrix. This can be attributed to the fact that as the polypropylene resin disseminates, it dispenses the applied stress to the American agave fibres which further resulted to higher strength value. The tensile strength of the composite with maximum fibre content (30% volume of the fibre) is found to be 27.03 MPa. The tensile strength and the tensile modulus of American glass fibre reinforced polypropylene composite considered in this study are far better than that of American agave fibre

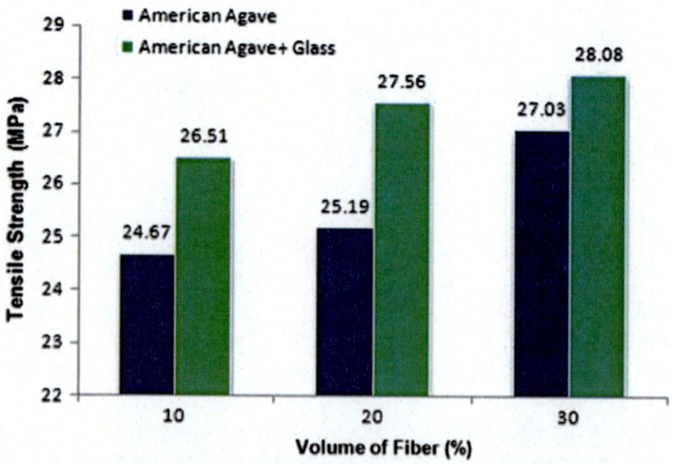

Fig. 3 Variation of tensile strength of composites with volume of fibre

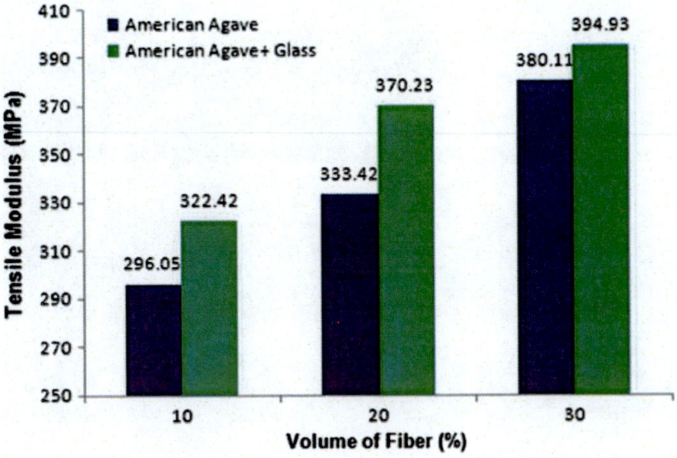

Fig. 4 Variation of tensile modulus of composites with volume of fibre

reinforced polypropylene composites. From the tensile test, we obtained load versus displacement graphs at 30% volume of fibre, which are shown in Fig. 5. The tensile tested samples are represented in Fig. 6.

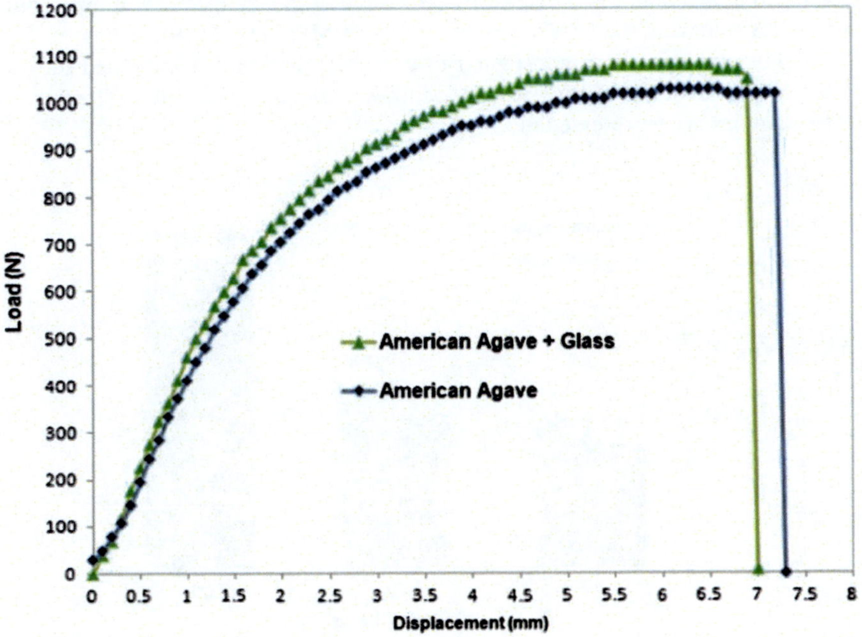

Fig. 5 Load versus displacement curve of composite at 30% volume of fibre

Fig. 6 Tensile tested specimens

3.2 *Flexural Strength*

Figures 7 and 8 describe the flexural behaviour of American agave fibre reinforced composites and American agave–glass fibre reinforced composites. The trend is similar to the trend observed for the case of tensile properties with the same above-mentioned conviction. The flexural strength value of the composite with maximum fibre (30% volume of the fibre) content is about 84 MPa. The American agave–glass fibre reinforced composite considered in this study has shown improved flexural strength value than that of American agave fibre reinforced composites. The flexural tested samples are represented in Fig. 9.

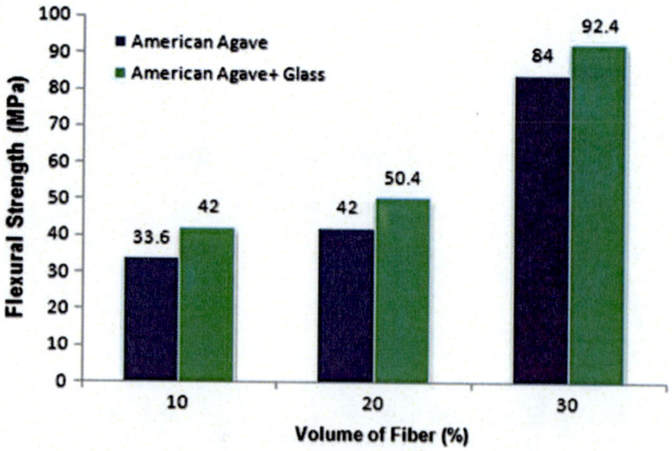

Fig. 7 Variation of flexural strength of composites with volume of fibre

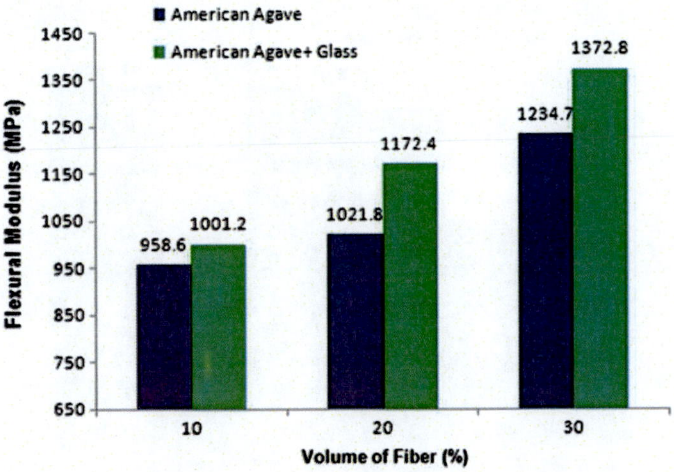

Fig. 8 Variation of flexural modulus of composites with volume of fibre

Fig. 9 Flexural tested specimens

3.3 *Impact Strength*

Figure 10 describes the results of Izod impact test. From below figure, it is evident that as the percentage of volume increases, the impact strength value increases. It can be noted that the impact strength value of the composite with maximum fibre (30% volume of the fibre) content is found to be 5.8 J. The impact strength of

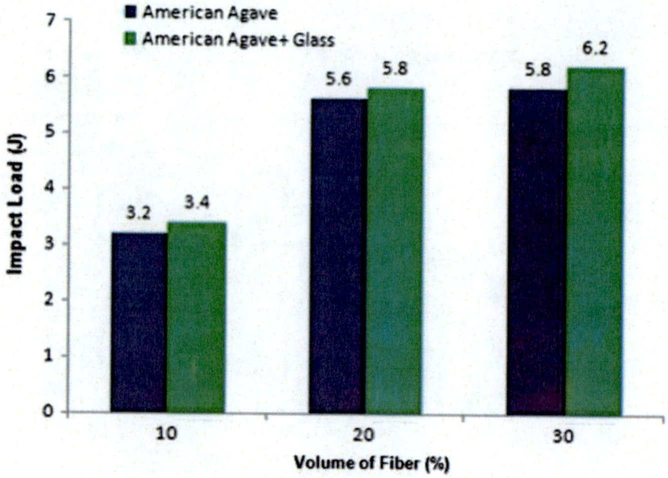

Fig. 10 Variation of impact strength of composites with volume of fibre

American agave–glass fibre composite is better than American agave fibre reinforced composites. The impact tested samples are represented in Fig. 11.

Fig. 11 Impact tested specimens

4 Conclusions

In this study, American agave fibre reinforced polypropylene composite and American agave–glass fibre reinforced polypropylene hybrid composites are fabricated with different volume of fibres, and the mechanical properties, namely tensile, impact and flexural properties, are examined. After evaluating the obtained experimental results, the following conclusions are made:

- The tensile strength and tensile modulus of the American agave and agave–glass fibre reinforced polypropylene composite increases with the increase in volume of fibre.
- The flexural strength and flexural modulus of the American agave and agave–glass fibre reinforced polypropylene composites increases with the increase in volume of fibre.
- The impact strength of the American agave and agave–glass fibre reinforced polypropylene composites increases with the increase in volume of fibre.
- American agave and glass reinforced polypropylene composites show better mechanical properties than American agave fibre reinforced polypropylene composites.

References

1. Kozłowski RM (ed) Handbook of natural fibres: types, properties and factors affecting breeding and cultivation. Elsevier
2. Heng JY, Pearse DF, Thielmann F, Lampke T, Bismarck A (2007) Methods to determine surface energies of natural fibres: a review. Compos Interfaces 14(7–9):581–604
3. Joshi SV, Drzal LT, Mohanty AK, Arora S (2004) Are natural fiber composites environmentally superior to glass fiber reinforced composites? Compos A Appl Sci Manuf 35(3):371–376
4. Prasad N, Agarwal VK, Sinha S (2016) Banana fiber reinforced low-density polyethylene composites: effect of chemical treatment and compatibilizer addition. Iran Polym J 25(3):229–241
5. Zaman HU, Khan MA, Khan RA (2013) Banana fiber-reinforced polypropylene composites: s study of the physico-mechanical properties. Fibers Polym 14(1):121–126
6. Rana AK, Mandal A, Mitra BC, Jacobson R, Rowell R, Banerjee AN (1998) Short jute fiber-reinforced polypropylene composites: effect of compatibilizer. J Appl Polym Sci 69(2):329–338
7. Haniffah WH, Sapuan SM, Abdan K, Khalid M, Hasan M, Hoque ME (2015) Kenaf fibre reinforced polypropylene composites: effect of cyclic immersion on tensile properties. Int J Polym Sci
8. Oksman K, Mathew AP, Långström R, Nyström B, Joseph K (2009) The influence of fibre microstructure on fibre breakage and mechanical properties of natural fibre reinforced polypropylene. Compos Sci Technol 69(11–12):1847–1853

Influence of Magnesium Hydroxide Fillers on Acoustic, Thermal, and Flame Retardant Properties of Pu Foam

L. Yuvaraj, S. Jeyanthi, Digvijay D. Kadam, and R. G. Ajai

Abstract Polyurethane foam is a versatile material for many applications like acoustic, thermal insulation, as well as for energy absorption. The main aim of this paper is to improve acoustic properties by adding weight percentage of 2, 4, and 6 of magnesium hydroxide ($Mg(OH)_2$) micro-particles in polyurethane (Pu) foam. To investigate the influence of micro-particles on acoustic properties, the samples are tested in an impedance tube to measure sound absorption coefficient. The experimental results are compared to finite element results predicted from Johnson–Champoux–Allard model. For thermal properties, the samples are experimented in thermal conductivity and fire retardant test. The results indicate that the significant improvement in the acoustic and thermal properties due to the addition of magnesium hydroxide.

Keywords Polyurethane foam · Magnesium hydroxide · Sound absorption coefficient · Thermal conductivity · Flame retardancy

1 Introduction

The noise produced by vehicles not only affects the quality of the automobile but also decreases the comfort of driving. Eliminating automotive interior noise has become an important and predominant topic in noise control engineering. This noise reduction depends on the coefficient of sound absorption. It is necessary to identify the coefficient of sound absorption of polyurethane at different frequency ranges. These porous materials such as polyurethane foam have been widely used in various fields like acoustic absorption thermal insulation, aeronautics, and building construction. Cao et al. [1] have discussed Delany–Bazley model for sound absorption of porous material. Ekici et al. [2] improved coefficient of sound absorption of polyurethane foam by the addition of tea leaf and luffa-cylindrical. Tao et al. [3] explained the

L. Yuvaraj · S. Jeyanthi (✉) · D. D. Kadam
School of Mechanical and Building Sciences, VIT, Chennai, Tamil Nadu 600127, India
e-mail: jeyanthi.subramanian@vit.ac.in

R. G. Ajai
Department of Mechanical Engineering, SVCE, Sriperumpudur, Tamil Nadu, India

S. Vijayan et al. (eds.), *Trends in Manufacturing and Engineering Management*, Lecture Notes in Mechanical Engineering,
https://doi.org/10.1007/978-981-15-4745-4_35

effect of rice straw and wheat straw fiber in polyurethane foam on sound absorption property. Chen et al. [4] added bamboo chips and bamboo stem in polyurethane foam in order to enhance the coefficient of sound absorption and also carried out airflow resistivity test. Tiuc et al. [5] have discussed the addition of textile waste in polyurethane foam with detailed microstructure studies they noticed improvement in the acoustic properties of developed hybrid foams. Benkreira et al. [6] investigated porosity, airflow resistivity, and sound absorption of porous material. Verdejo et al. [7] have discussed the effect of the addition of carbon nanotube in polyurethane foam on the acoustic property as well as on Young's modulus. Orfali et al. [8] added silicon oxide and carbon nanotubes in polyurethane foam to enhance the coefficient of sound absorption. Rasheed et al. [9] have explained compressive strength, thermal conductivity, and moisture absorption test of polyurethane foam with glass wool. Chen et al. [10] have discussed the effect of variation of reactants like percentage of blowing agent of polyurethane foam on sound-absorbing capacity.

Sripathy and Sharma [11] described the flammability and moisture absorption test of polyurethane foam. Mazraeh-shahi et al. [12] explained method to calibrate thermal diffusivity and thermal conductivity of porous material. Moretti et al. [13] explained the influence of granule size of aerogels on acoustic and thermal properties and also explained transmission loss. Xu et al. [14] discussed thermal stability and flame retardancy property of pure polyurethane and the addition of organoclay nano-composites. Nedkov et al. [15] and Ahmadinejad et al. [16] explained finite element modeling of the impedance tube and calibration of the coefficient of sound absorption. Wang et al. [17] have enhanced the coefficient of sound absorption by using rice hull in polyurethane foam. Matyka et al. [18] have described the concept of tortuosity, porosity, and the relationship between these factors of porous material. Lim et al. [19] discussed the sound-absorbing property of knife fibers and also discussed the effect of thickness of the sample and air gap behind sample on the coefficient of sound absorption. Othmani et al. [20] and Santoni et al. [21] presented different mathematical models for porous absorption. The explained concept of porosity, touristy, flow resistivity calibrated coefficient of sound absorption for recycled sugarcane wastes.

Putra et al. [22] explained the effect of air gap on sound-absorbing property of porous material by using pineapple-leaf fibers. Gle et al. [23] explained the acoustic properties of hemp concrete as a green building material. Piegay et al. [24] discussed sound absorption and transmission loss of flax wool and hemp wool samples, and carried out validation with Delany–Bazley and Tarnow perpendicular random model. Curtu et al. [25] used wood fiber, jute, and wood chips to prepare biodegradable composite samples and evaluated the acoustic properties of these samples. Hurrell et al. [26] explained mathematical models of airflow resistivity for non-woven materials. Berardi et al. [27] explained the comparative study of sound absorption of various fibers like coconut, straw, cane, sheep wool, cork, hemp, and kenaf. Kuranska et al. [28] explained microstructure details of pure and composite polyurethane foams and investigated their compressive strength performance. Or et al. [29] explained oil palm empty fruit bunch fibers as the sustainable acoustic absorber and carried out an experiment of sound absorption to explain the effect of air gap behind the sample and also explained the effect of density and thickness of sample on sound absorption property of porous material.

Fouladi et al. [30] explained the Biot–Allard model for sound absorption of coir fiber and projected comparative results of the impedance tube, Delany–Bazley model, and Biot–Allard model. Koruk et al. [31] experimented with pure luffa fiber, luffa fiber with perforated lines, and luffa composite with epoxy with varying volume fraction in order to evaluate the coefficient of sound absorption and transmission loss. Martellottaa et al. [32] explained the acoustic-absorbing property of samples made from olive pruning wastes and also projected the effect of air gap and the effect of thickness of samples.

In this study, the different weight percentage of magnesium hydroxide is added to Pu foam. The acoustic capabilities of the foam samples measured in impedance tube test, and other physical parameters are tested in the individual experimental setup. These physical parameters are given as input to the FEA analysis to predict the numerical sound absorption. For thermal properties, the sample is tested in thermal conductivity and flame retardant test.

2 Sample Preparation

The flexible foams are a simple mixture of isocyanides and polyol in the ratio of 100:65. First of all, pure foam with no additives is prepared by adding the two compounds in a circular mold with one face open and letting the reaction take place. The reaction is exothermic, and carbon dioxide gas is released during the blowing process. The mixture is allowed to expand and is then removed from the mold and is prepared according to the required dimensions. Now, three different samples were prepared with different weight percentages of magnesium hydroxide, i.e., 2%, 4%, 6% with average particle size, 6.8 (μm), reactants, and their weight percentage are mentioned in Table 1. Composition of magnesium hydroxide filled hybrid foam is mentioned in Table 2. Samples are shown in Fig. 1

Table 1 Reactants used for pure PUF

Chemicals	%wt
Isocyanate	43.55
Polyol	53.6
Blowing agent	1.6
Catalyst	0.85
Surfactants	0.4

Table 2 Composition for hybrid foam

Sample name	%wt of PuF	%wt of $Mg(OH)_2$
a	100	0
b	98	2
c	96	4
d	94	6

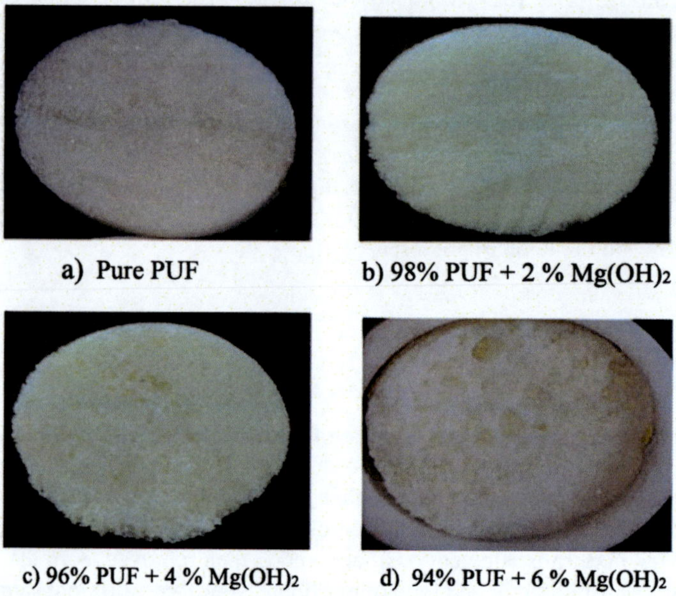

Fig. 1 Test samples of pure and hybrid polyurethane foam

3 Experimental Testing and FEA Analysis

3.1 Porosity Measurement

In order to characterize and better prediction of sound absorption coefficient, porosity is an essential physical parameter. Porosity is defined as the amount of fluid that occupied in the pores. Porosity can be tested by pressure/mass method introduced by Salissou et al. [33]. Foam sample placed in air-sealed container and the mass is calculated in high- and low-pressure condition. A simple formula is stated by Salissou which gives us porosity from measured values. The schematic working and the existing test setup are shown in Fig. 2.

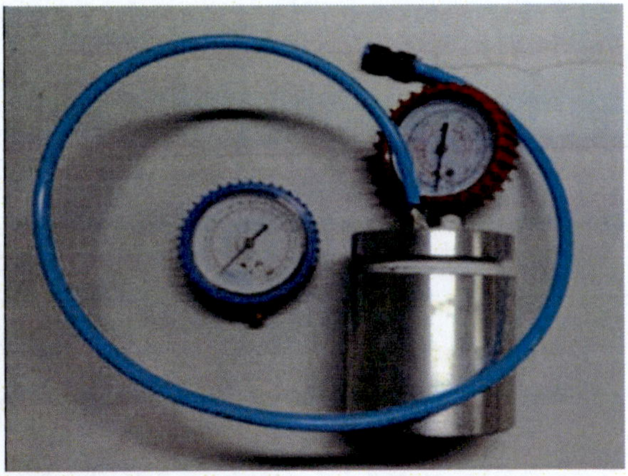

Fig. 2 Experimental setup for porosity measurement

3.2 *Flow Resistivity Measurement*

The airflow resistivity is major parameter for governing acoustic behavior of porous material. The airflow resistivity depends on the direction of airflow through the material. The flow resistance (R_f) is the ratio of the pressure difference between two sides of the sample to the air velocity, which is stably passed through a porous medium. The flow resistivity (σ) flows resistance per unit thickness of the sample. Experimental setup is shown in Fig. 3. By using this method, we can measure airflow resistivity of porous material ranging from 100 to 10,000 Ns/m^4, and air velocity can vary up to 50 mm/s. The pressure differential across samples range from 0.1 to 250 Pa.

$$R_f = \frac{\Delta P}{V} \tag{1}$$

Fig. 3 Experimental setup of flow resistivity measurement

Fig. 4 Experimental setup for tortuosity measurement

$$\sigma = \frac{R_f}{d}, \sigma = \frac{\Delta P}{V \times d} \tag{2}$$

The test samples with 100 mm diameter and 40 mm thickness are held between the holder and mounting plate. Compressed air with 50 mm/s is supplied to one side of the sample, and pressure difference (ΔP) is measured by a digital pressure gauge.

3.3 Tortuosity Measurement

Tortuosity is a dimensionless structural parameter, which indicates the effects of pore structure on the velocity of fluid flow. It is defined as the ratio of actual path length through the material to the linear path length. It shows how well the porous material prevents direct flow through the porous medium. The more complex the path, the more time a wave is in contact with the absorbent, and hence, the more energy dissipation and more absorbing capability. Using the below test setup, we can find the actual length of the twisted path. This experimental setup constitutes transmitter (T) and receiver (R) ultrasonic sensors, where the signal allow to pass from T to R and the travel time is measured. The experiment carried out in two conditions as presence and absence of sample and the existing test setup is shown in Fig. 4.

3.4 Sound Absorption Measurement

In this setup, confined acrylic tube with different diameter based on cut-off frequency is used to find the sound absorption coefficient as per ASTM E-1050. The test setup is

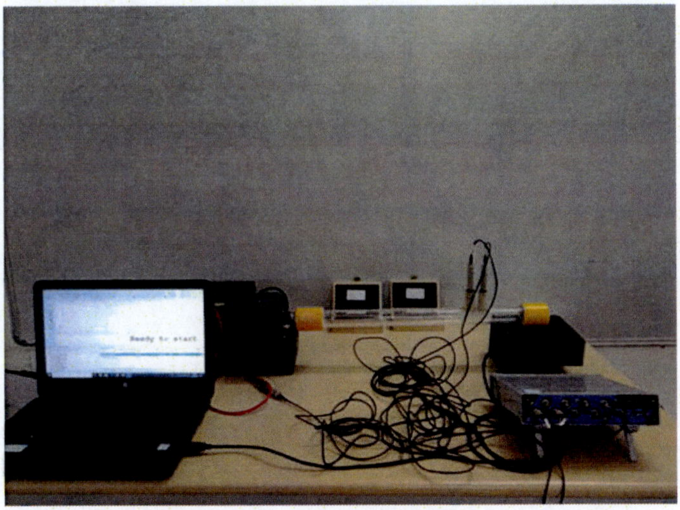

Fig. 5 Measurement of sound absorption by impedance tube

equipped with two microphones, loudspeaker, and DAQ system. A loudspeaker that excites the sound in quantified frequency and microphone senses the sound pressure at position one and two. The two-channel DAQ systems were used to record the time domain data for post-processing. The MATLAB® software is used to estimate the transfer function across two microphone data. Based on the transfer function, normal incidence sound reflection and absorption obtain following Eqs. (4) and (5)

$$H = \frac{P_1}{P_2} \tag{3}$$

$$R = \frac{H - e^{-jks}}{e^{jks} - H} e^{j2k(l+s)} \tag{4}$$

$$\alpha = 1 - |R|^2 \tag{5}$$

S is the distance between two microphones, and l is the distance between the speaker and first microphone (Fig. 5).

3.5 FEA Modeling of Impedance Tube

The acoustic body of impedance tube is presented with parameters like the speed of sound in air and volume density. For meshing 3D brick, acoustic fluid element is selected. The Johnson–Champoux–Allard model is selected for FEA analysis. Harmonic mass source (loudspeaker) is inserted at one end of the tube, and another

Fig. 6 FEA modeling of impedance tube

end sample is placed. Source of sound generates plane waves which incident normally on the sample to be tested. The frequency range is selected up to 5000 Hz. Mesh sensitivity was carried, and the maximum element length must satisfy the following equation (Fig. 6).

$$L = \frac{C}{8f} \tag{7}$$

where L is the maximum element length, C is the speed of sound in air, and f is the maximum frequency of the model. Parameters used for FEA modeling of impedance tube are listed below.

1. Impedance tube length = 600 mm
2. Impedance tube diameter = 33 mm
3. Speed of sound in air = 343 m/s
4. Density of air = 1.2 kg/m^3.

Other than above for FEA analysis, other parameters are also required like flow resistivity, touristy, and porosity which is listed in the result section.

3.6 *Measurement of Thermal Conductivity*

Thermal conductivity depends on density, porosity, shape, and size of pores. The addition of fillers should not increase the thermal conductivity, or at least, the material should maintain the existing value. So, it is essential to ensure the thermal conductivity of polyurethane filled with magnesium hydroxide. The principle used in heat flow meter method is represented by the schematic diagram in Fig. 7. This method is based on Fourier's law of heat flow. The sample is placed between two plates. One plate is heated, and the other is kept cool. The temperature gradient across sample and heat flux calibrated by the sensor and thermal conductivity is calculated by the equation of Fourier's law. Thickness of all sample is maintained as 30 mm.

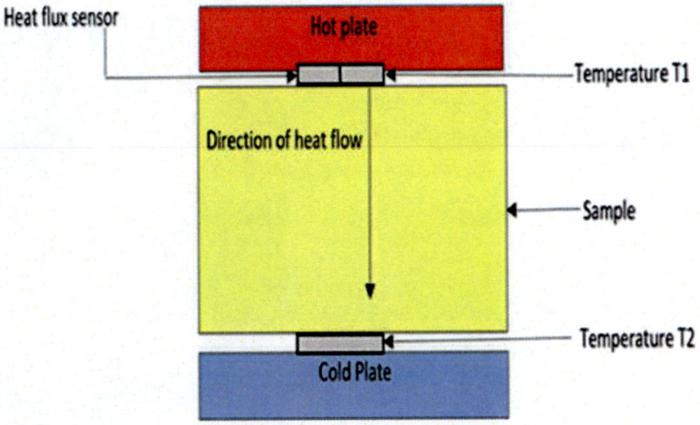

Fig. 7 Schematic representation of thermal conductivity measurement

$$Q = KA\frac{\Delta T}{d} \tag{7}$$

3.7 Flame Retardancy Test

Basically, polyurethane foam has excellent mechanical, acoustic, and thermal insulation properties, but its poor fire resistance leads to restriction for some application.

So, the samples underwent fire retardant test to study the influence of magnesium hydroxide as filler in Pu foam. The test is carried out by allowing to burn the sample for different periods of 10 s, 15 s, and 20 s (Fig. 8).

4 Results

4.1 Porosity, Flow Resistivity, and Tortuosity of Sample

The experimental test results of porosity, flow resistivity, and tortuosity are listed in Table 3. The results prove that the physical parameters are interconnected to each other. Addition of magnesium hydroxides increases the porous nature of Pu foam which means the number of pores is increased. The increasing number of pores leads to more number of cell walls, which makes the porous path as resistive and complex. One can notice that as the percentage of magnesium hydroxide increased, the values of porosity, flow resistivity, and tortuosity is also increased.

Fig. 8 Sample subjected to flame retardancy test

4.2 *Sound Absorption Results*

As mentioned previously, magnesium hydroxide enhances all parameters and the same resembles in the sound absorption result as shown in Fig. 9. The percentage of $Mg(OH)_2$ filler in polyurethane foam increases the coefficient of absorption which increases over all frequency ranges for all samples. However, in case of 6% sample, the sound absorption performance is poor between the range of 0–1000 Hz, and it is mainly due to the addition of more magnesium hydroxide causes the foam to behave rigid by compared to other samples. The other fact is that pores turn to void if excess addition of fillers and severely collapse the foam beyond 6% of magnesium hydroxide.

4.3 *Comparison of Experimental and FEA Result*

Figure 10 shows the comparison of experimental results to numerical results which is predicted from experimental results of porosity, flow resistivity, and tortuosity. There is a minimal gap notice at some region in the comparison plot which is mainly due

Table 3 Porosity, flow resistivity, and tortuosity factor of samples

Samples	Porosity	Flow resistivity (Ns/m^4)	Tortuosity factor
Pure PUF	0.857	3126.56	1.08
98% PUF + 2% $Mg(OH)_2$	0.869	3463.21	1.13
96% PUF + 4% $Mg(OH)_2$	0.914	3768.21	1.14
94% PUF + 6% $Mg(OH)_2$	0.943	3974.1	1.16

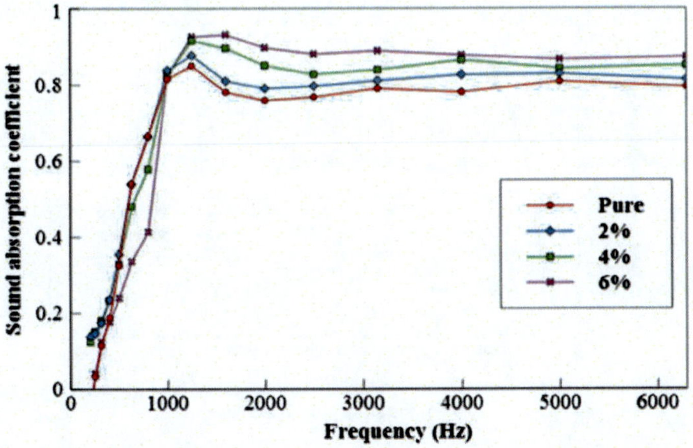

Fig. 9 Comparison of the coefficient of sound absorption of pure and hybrid polyurethane foam

to non ideal condition of experimental setup such as air leakage, loose fit of sample and so on.

4.4 Thermal Conductivity Measurement

There is no significant increase in thermal conductivity of pure and hybrid polyurethane foam. Thus, it maintains its insulation property with the addition of $Mg(OH)_2$ filler content in polyurethane foam (Table 4).

4.5 Flame Retardancy Test

The flame retardant property of samples b, c, and d increased significantly, based on the percentage of addition. The enhancement occurs due to the magnesium hydroxide is a non-flammable material. During burning, the magnesium hydroxide starts to liberate water vapors. The layer of water vapors starts to envelop the flame, and thus, flame retardancy property of these hybrid foam can be enhanced. Magnesium hydroxide $Mg(OH)_2$ acts as smoke suppressor as well as acid-free flame retardant (Fig. 11).

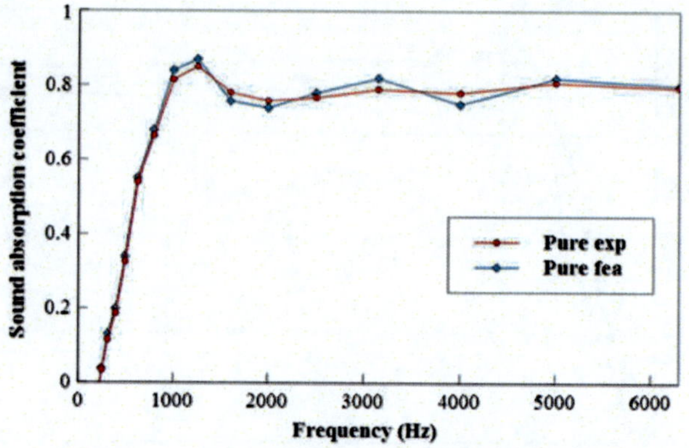

a) Pure Polyurethane foam

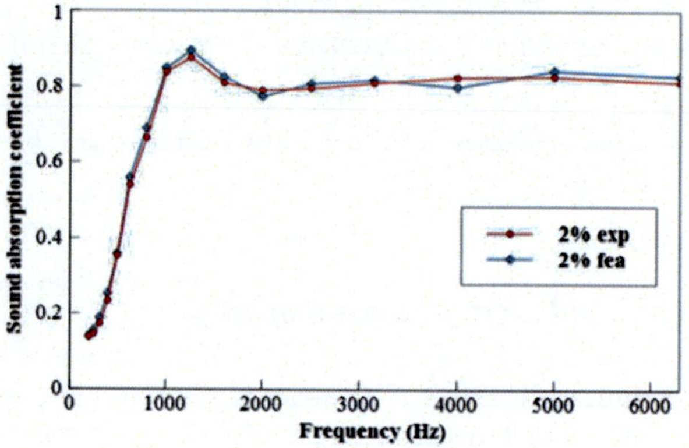

b) 98% PUF + 2 % $Mg(OH)_2$

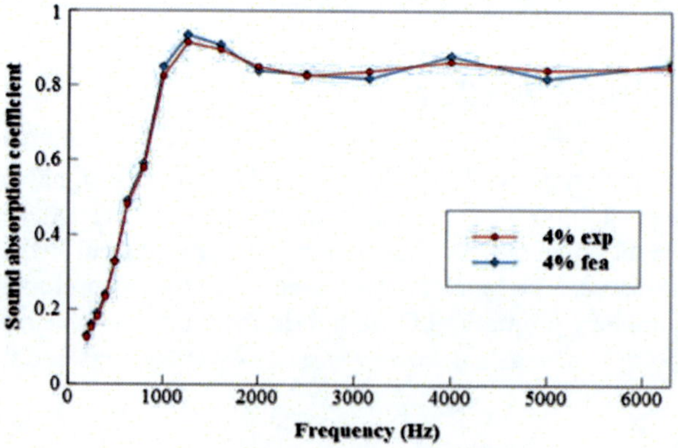

c) 96% PUF + 4 % $Mg(OH)_2$

Fig. 10 Comparison of experimental and FEA result of the coefficient of sound absorption of different foams

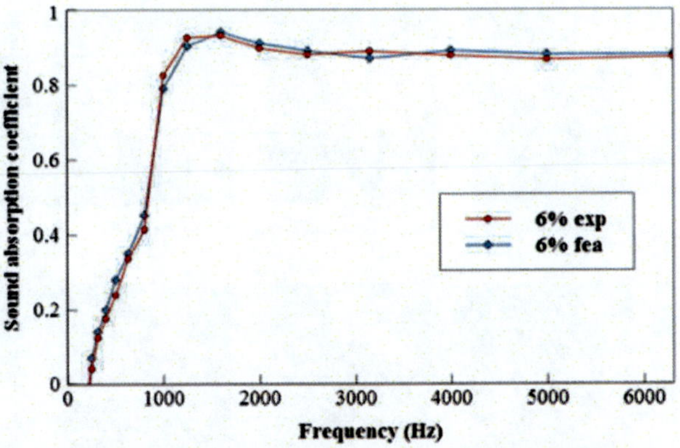

d) 94% PUF + 6 % $Mg(OH)_2$

Fig. 10 (continued)

Table 4 Parameters of measurement of thermal conductivity

Sample	(W/m)	T1 (°C)	T2 (°C)	K (W/m)
Pure PUF	58.01	80	30	0.035
98% PUF + 2% $Mg(OH)_2$	58.3	80.2	30.34	0.0358
96% PUF + 4% $Mg(OH)_2$	58.2	80.15	31.91	0.0362
94% PUF + 6% $Mg(OH)_2$	58.15	80.12	32.32	0.0365

5 Conclusion

It is concluded that this study investigated the acoustic and thermal properties different weight percentage addition of magnesium hydroxide as filler in Pu foam. The filler enhances the physical parameters of foam samples as well as thermal properties. The measured sound absorption results compared with numerical FEA showing fairly good agreement to each other. The insulating properties of foam samples remain constant, which is noticed in the result of a thermal conductivity test. Non-flammable nature of magnesium hydroxide improves fire retardant property of polyurethane foam.

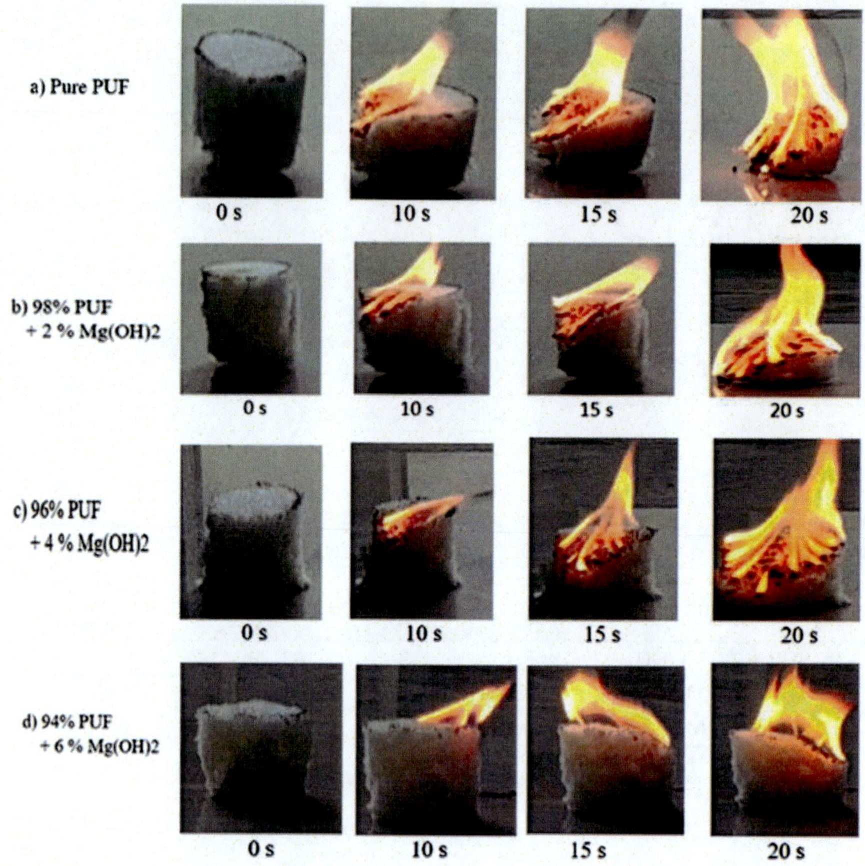

Fig. 11 Flammability Test of various samples

Acknowledgements We would like to thank the Department of Science and Technology (DST-SERB; File no: ECR/2015/000111) for providing us the necessary facilities and funds for conducting this study.

References

1. Cao L, Fu Q, Si Y, Ding B, Yu J (2018) Porous materials for sound absorption. Compos Commun 31(10):25–35
2. Ekici B, Kentli A, Küçük H (2012) Improving sound absorption property of polyurethane foams by adding tea-leaf fibers. Arch Acoust 37(4):515–520
3. Tao Y, Li P, Cai L (2016) Effect of fiber content on sound absorption, thermal conductivity, and compression strength of straw fiber-filled rigid polyurethane foams. BioResources 11(2):4159–4167

4. Chen S, Jiang Y (2018) The acoustic property study of polyurethane foam with addition of bamboo leaves particles. Polym Compos 39(4):1370–1381
5. Tiuc AE, Vermeşan H, Gabor T, Vasile O (2016) Improved sound absorption properties of polyurethane foam mixed with textile waste. Energy Procedia 85:559–565
6. Benkreira H, Khan A, Horoshenkov KV (2011) Sustainable acoustic and thermal insulation materials from elastomeric waste residues. Chem Eng Sci 66(18):4157–4171
7. Verdejo R, Stämpfli R, Alvarez-Lainez M, Mourad S, Rodriguez-Perez MA, Brühwiler PA, Shaffer M (2009) Enhanced acoustic damping in flexible polyurethane foams filled with carbon nanotubes. Compos Sci Technol 69(10):1564–1569
8. Orfali WA (2015) Acoustic properties of polyurethane composition reinforced with carbon nanotubes and silicon oxide nano-powder. Phys Procedia 70:699–702
9. Rasheed AK, Marhoon II (2015) Mechanical and physical properties of glass wool-rigid polyurethane foam composites. Alnahrain J Eng Sci 18(1):41–49
10. Chen S, Jiang Y, Chen J, Wang D (2015) The effects of various additive components on the sound absorption performances of polyurethane foams. Adv Mater Sci Eng
11. Sripathy M, Sharma KV (2013) Flammability and Moisture absorption test of rigid polyurethane foam. Int J Sci Eng Res 4(2)
12. Mazraeh-shahi ZT, Shoushtari AM, Bahramian AR (2015) A new method for measuring the thermal insulation properties of fibrous silica aerogel composite. Procedia Mater Sci 11:583–587
13. Moretti E, Merli F, Cuce E, Buratti C (2017) Thermal and acoustic properties of aerogels: preliminary investigation of the influence of granule size. Energy Procedia 111:472–480
14. Xu Z, Tang X, Zheng J (2008) Thermal stability and flame retardancy of rigid polyurethane foams/organoclay nanocomposites. Polym-Plast Technol Eng 47(11):1136–1141
15. Nedkov T, Pleshkova-Bekjarska S (2015) Modeling of 3D impedance tube with a complex termination impedance using finite element method. International Scientific Conference on Information, Communication and Energy Systems and Technologies, 2015 june, pp 170–173, Sofia, bulgaria
16. Ahmadinejad P, Monazzam MR (2018) 3D simulation of impedance tube: effect of mesh size on sound absorption characteristic. Int J Comput Sci Netw Secur 18(1):179–182
17. Wang Y, Zhang C, Ren L, Ichchou M, Galland MA, Bareille O (2013) Influences of rice hull in polyurethane foam on its sound absorption characteristics. Polym Compos 34(11):1847–1855
18. Matyka M, Khalili A, Koza Z (2008) Tortuosity-porosity relation in porous media flow. Phys Rev E 78(2):026306
19. Lim ZY, Putra A, Nor MJ, Yaakob MY (2018) Sound absorption performance of natural kenaf fibres. Appl Acoust 15(130):107–114
20. Othmani C, Taktak M, Zain A, Hantati T, Dauchez N, Elnady T, Fakhfakh T, Haddar M (2017) Acoustic characterization of a porous absorber based on recycled sugarcane wastes. Appl Acoust 120:90–97
21. Santoni A, Bonfiglio P, Fausti P, Marescotti C, Mazzanti V, Mollica F, Pompoli F (2019) Improving the sound absorption performance of sustainable thermal insulation materials: natural hemp fibres. Appl Acoust 150:279–289
22. Putra A, Or KH, Selamat MZ, Nor MJ, Hassan MH, Prasetiyo I (2018) Sound absorption of extracted pineapple-leaf fibres. Appl Acoust 136:9–15
23. Glé P, Gourdon E, Arnaud L (2011) Acoustical properties of materials made of vegetable particles with several scales of porosity. Appl Acoust 72(5):249–259
24. Piégay C, Glé P, Gourdon E, Gourlay E, Marceau S (2018) Acoustical model of vegetal wools including two types of fibers. Appl Acoust 129:36–46
25. Curtu I, Stanciu MD, Cosereanu C, Vasile O (2012) Assessment of acoustic properties of biodegradable composite materials with textile inserts. Materialeplastice 49:68–72
26. Hurrell AI, Horoshenkov KV, Pelegrinis MT (2018) The accuracy of some models for the airflow resistivity of nonwoven materials. Appl Acoust 15(130):230–237
27. Berardi U, Iannace G (2017) Predicting the sound absorption of natural materials: best-fit inverse laws for the acoustic impedance and the propagation constant. Appl Acoust 115:131–138

28. Kuranska M, Prociak A (2012) Porous polyurethane composites with natural fibres. Compos Sci Technol 72(2):299–304
29. Or KH, Putra A, Selamat MZ (2017) Oil palm empty fruit bunch fibres as sustainable acoustic absorber. Appl Acoust 119:9–16
30. Fouladi MH, Ayub M, Nor MJ (2011) Analysis of coir fiber acoustical characteristics. Appl Acoust 72(1):35–42
31. Koruk H, Genc G (2015) Investigation of the acoustic properties of bio luffa fiber and composite materials. Mater Lett 15(157):166–168
32. Martellotta F, Cannavale A, De Matteis V, Ayr U (2018) Sustainable sound absorbers obtained from olive pruning wastes and chitosan binder. Appl Acoust 141:71–78
33. Salissou Y, Panneton R (2007) Pressure/mass method to measure open porosity of porous solids. J Appl Phys 101:124913

Development and Analysis of GFRP Conical Springs

D. Vivek, R. Praveen, A. Sanjay Krishnan, Y. K. Sabapathy, D. Ebenezer, and M. Selvaraj

Abstract Increasing competition and innovation in automobile sector tends to modify the existing products or replace old products by new and advanced material products. A suspension system of vehicle is also an area where these innovations are carried out regularly. Nowadays, the automobile industry has shown much interest in using glass fibre reinforced plastics (GFRP) components replacing conventional steel components due to its 'high strength to low weight' ratio. Therefore, replacement of steel conical suspension springs (in heavy automobiles) with GFRP conical springs with an aim to reduce its weight was attempted. A manual pultrusion and braiding process is approached for fabricating the FRP open coil springs using e-glass and epoxy resin. Its total deformation, equivalent stress and equivalent elastic strain were analysed using GFRP conical spring at different static loads of 20, 40, 60 and 80 N.

Keywords Glass fibre reinforced plastics (GFRP) · Epoxy resins · Manual pultrusion · Creo 3.0 · ANSYS

1 Introduction

Generally, automobile parts are made up of steel and its alloys. Steel is an alloy consisting mainly of iron with carbon content of 0.2–2.1% by weight. Steel has many physical properties such as density, elasticity, strength and hardness. Among these properties, the strength property is very important, which can be achieved by the addition of more materials like manganese, niobium and vanadium. But, steel has some limitations like having more weight, and its corrosion resistance is low. By using FRP components, fuel efficiency and endurance of various components in machines can be achieved. FRP is slowly replacing steel, especially in the automobile sector. FRP is

D. Vivek · A. Sanjay Krishnan
Department of Mechanical Engineering, Sri Sivasubramaniya Nadar College of Engineering, Chennai, India

R. Praveen · Y. K. Sabapathy · D. Ebenezer (✉) · M. Selvaraj
Sri Sivasubramaniya Nadar College of Engineering, Chennai, India
e-mail: ebenezerd@ssn.edu.in

S. Vijayan et al. (eds.), *Trends in Manufacturing and Engineering Management*, Lecture Notes in Mechanical Engineering,
https://doi.org/10.1007/978-981-15-4745-4_36

a composite material made up of polymer matrix reinforced with fibres [6]. The fiber is usually glass (fiber glass), carbon (in carbon-fiber-reinforced polymer), aramid or basalt. The polymer is usually an epoxy, vinyl ester or polyester thermosetting plastic, though phenol formaldehyde resins are still in use. FRP composite material has excellent mechanical properties such as impact resistance, strength, stiffness and flexibility. Main advantages of this FRP are corrosion resistance, low weight and high formability [7, 8]. FRP materials are anisotropic and can be tailor made to suit any requirement to counter stresses in any desired direction and location. The components made of FRP are thermally and dimensionally stable and well suited to work at high temperatures. FRP is widely used in many industries like automobile suspension system, railway wagons, electrical applications, aerospace and marine, etc [8]. The conical springs are cone shaped springs to provide near constant spring rates with lower solid height than traditional compression springs. The improved solid height is achieved by nesting active springs within each other during compression. It can also achieve the objective of reducing the weight of the engine component, thus increasing shock absorption, reducing spring weight and improving strength of springs [9].

2 Literature Review

Conical compression springs designed by Krishna et al. [3] is having the characteristics of a cylindrical spring of much greater free length.

The spring is characterised by wire whose diameter diminishes from base to apex end and by coils whose pitch diminishes from base to apex end such that the ratio of mean coil diameter to wire diameter is constant from coil to coil. Further, each coil diameter is preferably dimensioned to fit within the next larger coil permitting a spring to be designed which exhibits a small solid compressed height [5].

Kudapane [4] invented a conical spring wherein the diameters of the coils vary progressively and the spaces separating adjacent coils vary in relation to the dimensions of the coils to provide pre-determined force displacement spring characteristics. A conical spring is provided wherein the diameters of the coils thereof vary progressively, and the spaces separating adjacent coils vary in relation to the dimensions of the coils to provide pre-determined force displacement spring characteristics has been referred.

Optimisation of steel helical spring by composite spring was presented by Mehdi Bakhshesh, and Majid Bakhshesh [2] proposed springs that can reserve high level of potential energy and have undeniable role in industries. Helical spring is the most common element that has been used in car suspension system. In their research, steel helical spring related to light vehicle suspension system under the effect of a uniform loading has been studied.

Arularasan and Sabapathy [1] fabricated GFRP helical spring with E-glass and epoxy resin using semi-mechanised pultrusion process, and braiding process has found that a compression load of 1205.327 N with the displacement corresponding as 101.986 mm was studied.

3 Experimentation

The equipments required for GFRP fabrication process consists of fibre rolls, tray, clamps, steel tube, PVC hose pipe and wooden mould. By using marker pen, the pitch is marked in the periphery of the wooden mould and hammered with number of nails in a marked area in order to mention the pitch gap. The E-glass fibres are made with different loops. Now, the resin and hardener are poured into a tray and thoroughly mixed until it attains heat reaction. This process is called exothermic reaction.

After, the fibres are dipped and completely submerged with resin and hardener solution. A steel tube is rigidly fixed by holding with clamps. A PVC hose tube is inserted over one side of steel tube and another side is kept open.

Steel wire is inserted into hose tube through steel tube. Now, steel wire is tied with fibres having more number of loops, and other side of steel wire ending is used for pulling. Slowly, the steel wire is pulled through steel tube.

The fibre with solution is doped into a hose tube. Finally, the hose tube containing solution and fibres are rolled on the mould with preferred pitch of 20 mm. After a period of 40–48 h, the hose tube is removed by hammering on mould. Cut the hose tube periphery, and FRP conical spring is manufactured. By using this data, the analytical solving is done by using software called ANSYS.

The procedure for analysing GFRP conical spring for different loads is explained in following, open the ANSYS software. Click and drag the static structural module. Select the engineering data and upload the composite material as epoxy E-glass with having unidirectional. Update the constant properties values like density (2000 kgm^{-3}), Young's modulus (x-direction = 4500 Mpa, y-direction = 10,000 Mpa and z-direction = 10,000 Mpa), shear modulus (xy = 5000 Mpa, yz = 3846.2 Mpa and xz = 5000 Mpa). Import the geometry from Creo 3.0 software in which it is converted into IGES file format. Generate the model in ANSYS Workbench software. Now, fix the one end of spring and other end as different loads (i.e., 20 N, 30 N, 60 N, 80 N) through setup option. Update and generate a solution which provides the total deformation, equal elastic strain and equivalent stress using von-Misses formulation. Finally, result is obtained and noted down the maximum values of total deformation, equal elastic strain and equivalent stress for plotting the graphs.

3.1 3D Modelling

Figure 1 describes how to create a tapered spring in Creo 3.0. The helical sweep option is used to create a tapered spring, define internal helix profile, select suitable axis and make profile for axis of revolution. Now, draw a cross line with preferred

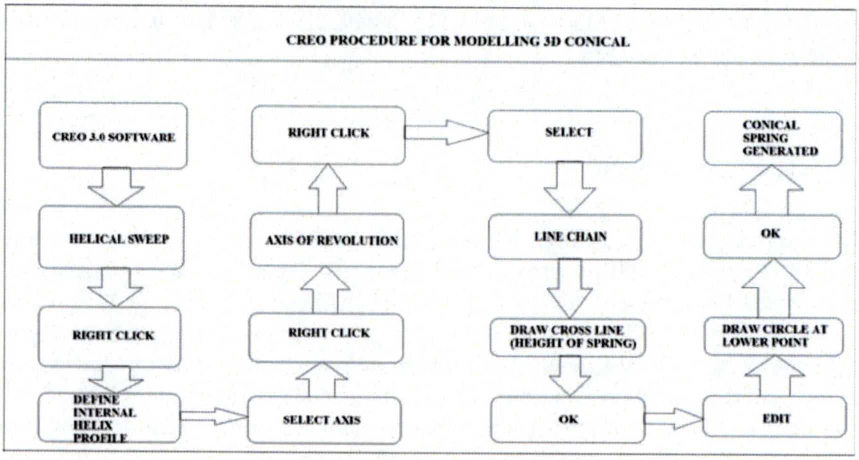

Fig. 1 Schematic line diagrams for creating tapered spring using Creo 3.0

dimensions, and after clicking ok, edit the profile to draw a wire diameter circle at lowest point and apply the pitch required in millimetre. Finally, the conical spring is generated and converted it into initial graphic exchange format (IGES) file. The generated 3D model is shown below in Fig. 2.

Fig. 2 3D model representations in Creo 3.0

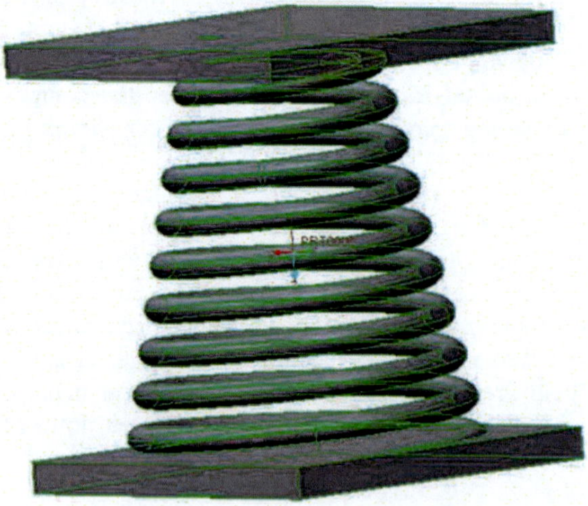

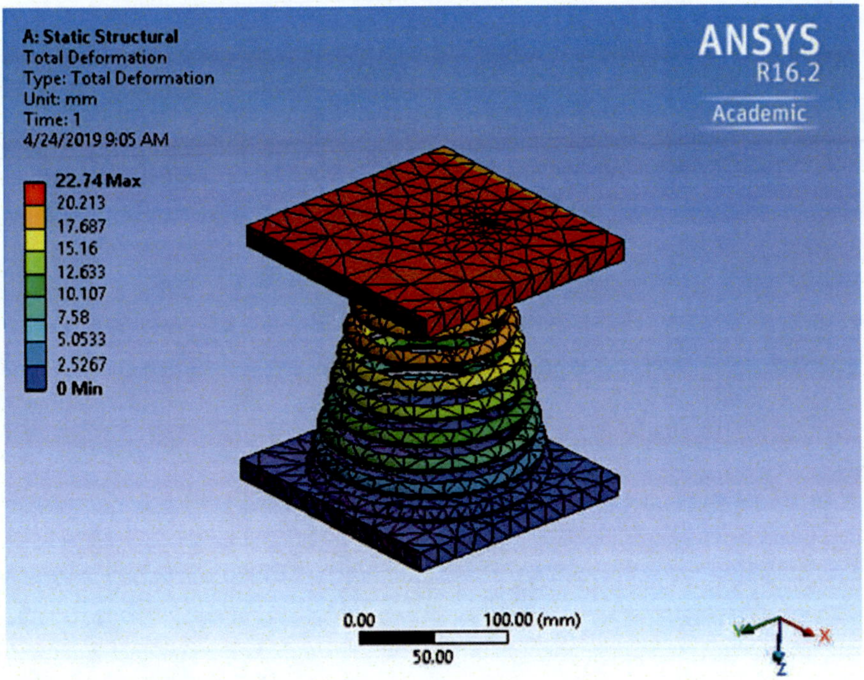

Fig. 3 Total deformation

4 Analysing of GFRP Conical Springs

4.1 Total Deformation

Total deformation is the elongation of conical spring. A compressive force is applied on the top and bottom of the plate. So, the acquired maximum total deformation for loads such as 20 N, 40 N, 60 N and 80 N are 22.74 mm, 45.48 mm, 68.22 mm and 90.96 mm. Figure 3 shows total deformation for 20 N load.

4.2 Equivalent Elastic Strain

The compressive loads such as 20 N, 40 N, 60 N and 80 N are applied on the spring plates. The maximum values acquired such as 1.172, 2.345, 3.518 and 4.691 are available. Figure 4 shows equivalent elastic strain for 20 N load.

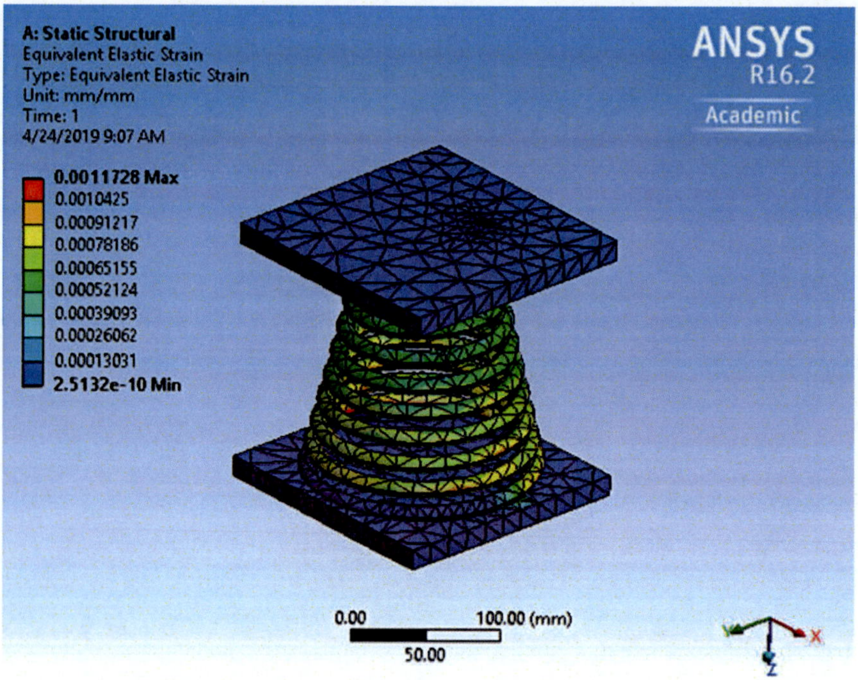

Fig. 4 Equivalent elastic strain

4.3 Equivalent Stress

Von-Misses equation is used to calculate for various loads such as 20 N, 40 N, 60 N and 80 N. The maximum values acquired are 10.708 MPa, 21.416 MPa, 32.124 MPa and 42.832 MPa. Figure 5 shows the equivalent stress for 20 N load.

4.4 Strain Energy

Strain energy for different loads such as 20 N, 40 N, 60 N and 80 N is maximum of 0.4419 MJ, 1.768 MJ, 3.9776 MJ and 7.0713 MJ. Figure 6 shows the strain energy for 20 N load (Table 1).

4.5 Graphical Analysis of Various Loads

Figure 7 shows the graphical representation between equivalent stress and equivalent

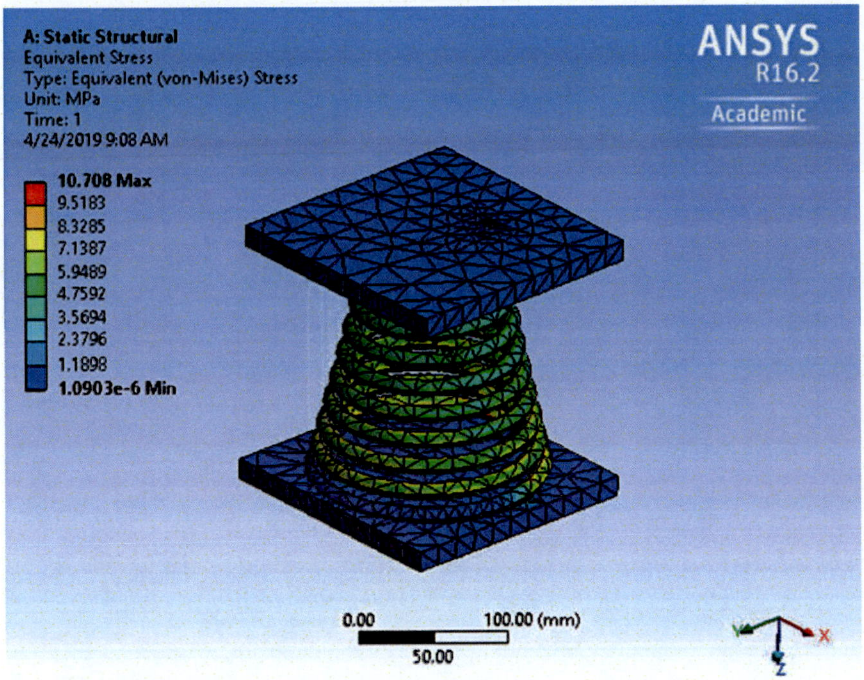

Fig. 5 Equivalent stress

elastic strain. In this graph, equivalent stress is noted in x-axis, and equivalent elastic strain is noted in y-axis. The maximum stress and strain values mentioned in the result are used in x-axis and y-axis. After jointing the points, the curve is gradually increased as shown in Fig. 7.

Figure 8 is a graphical representation between load and strain energy. The curve clearly shows that increasing in load causes gradual increase in strain energy.

4.6 *Fabrication of GFRP Conical Springs*

Figure 9a shows the wooden mould which is hammered with nails in a pre-determined dimensions before winding the hose tube. The pre-determined dimensions are also said to be pitch of spring. Figure 9b shows the final product after completely cut and removed hose tube.

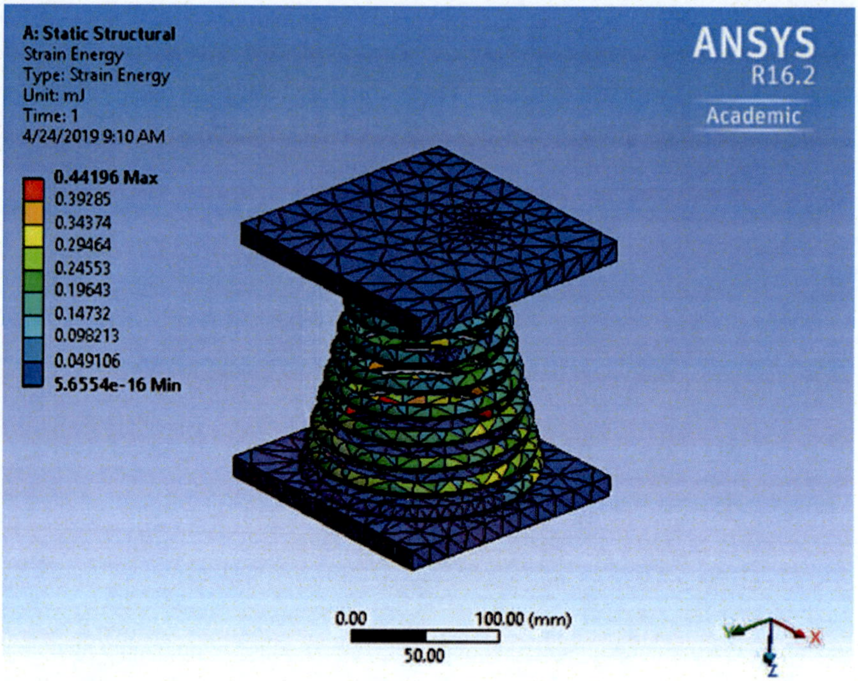

Fig. 6 Strain energy

Table 1 Software analytical results of FRP conical spring

Load (N)	Maximum total deformation (mm)	Strain energy (MJ)	Maximum equivalent stress (MPa)	Maximum equivalent elastic strain ($\times 10^{-3}$)
20	22.74	0.4419	10.708	1.172
40	45.48	1.7678	21.416	2.345
60	68.22	3.9776	32.124	3.518
80	90.96	7.0713	42.832	4.691

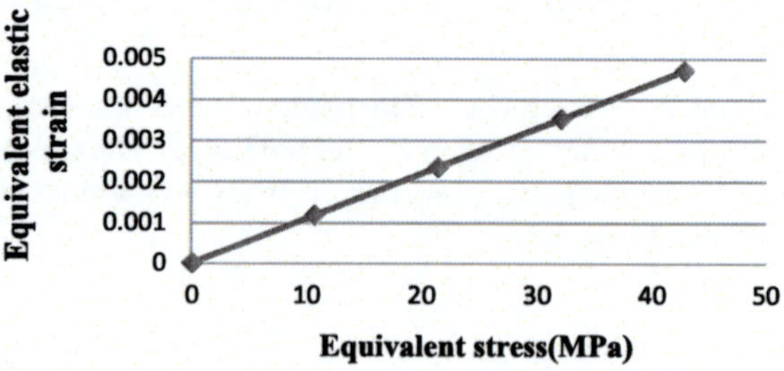

Fig. 7 Equivalent stresses versus equivalent elastic strain

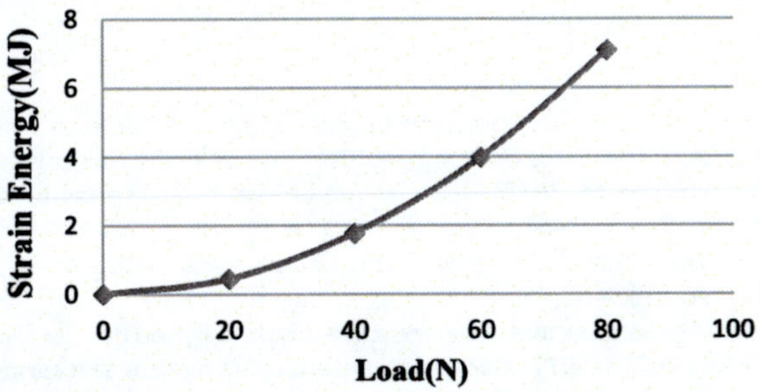

Fig. 8 Load versus strain energy

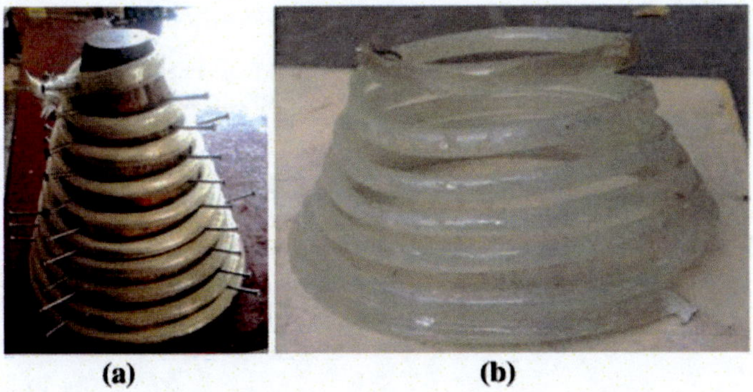

Fig. 9 **a** Wooden mould with conical spring, **b** fabrication of conical spring

5 Conclusion

In the present paper, conical springs or tapered springs using GFRP composites as epoxy resin and glass fibre were developed experimentally by the application of the manual pultrusion technique that can be used effectively to improve weight saving. In general, this study considered that the composite open coil springs can be used for light commercial vehicle, together with saving of weight of the component.

In case of automobiles, the weight reduction is one of the important factors and may not cause any technological impact. By using the geometric data, the conical springs are investigated by the aid of analysis software called ANSYS.

The result shows that for such loads (20 N, 40 N, 60 N and 80 N), total deformation decreases gradually, but elastic strain and stress are increased by applying compressive force. It is found that maximum values for each load are obtained.

References

1. Arularasan R, Sabapathy YK (2014) Fabrication and testing of FRP open coil springs. Appl Mech Mater 592–594:1065–1069. https://doi.org/10.4028/www.scientific.net/amm.592-594.1065
2. Bakhshesh M, Bakhshesh M (2012) Optimization of steel helical spring by composite spring. Int J Multidiscip Sci Eng 3:47–51
3. Krishna J, Manoj M, Reddy P, Kumar S (2014) Design and optimization of leaf spring using composites. IJMCA 2:163–168
4. Kudapane SA (2017) Modeling and finite element analysis of hybrid FRP leaf spring, pp 551–556
5. Mahanthi DL, Murali CVS (2017) Design and analysis of composite leaf spring for light weight vehicle. Int J Adv Eng Res Sci 4:147–152. https://doi.org/10.22161/ijaers.4.3.23
6. Pozhilarasu V, Parameshwaran Pillai T (2013) Comparison of performance of glass fibre reinforced plastic leaf spring with steel leaf spring. Int J Chem Tech Res 5:1339–1345
7. Sun W, Guan Z, Li Z (2017) Compressive failure analysis of unidirectional carbon/epoxy composite based on micro-mechanical models. Chinese J Aeronaut 30:1907–1918. https://doi.org/10.1016/j.cja.2017.10.002
8. Zarif Karimi N, Heidary H, Fotouhi M, Minak G (2017) Experimental analysis of GFRP laminates subjected to compression after drilling. Compos Struct 169:144–152. https://doi.org/10.1016/j.compstruct.2017.01.017
9. Zhou L, Zhao L, Liu F, Zhang J (2018) A micromechanical model for longitudinal compressive failure in unidirectional fiber reinforced composite. Results Phys 10:841–848. https://doi.org/10.1016/j.rinp.2018.08.003

Effect of Nano and Microfillers in Basalt/Epoxy Composites

M. M. Metro and M. Selvaraj

Abstract There is a growing demand emerging from industries towards usage of composite materials. Remarkable properties of basalt fibre reinforced polymer composites are being tailored for its application in various fields. Effect of epoxy/basalt fibre with nano and microparticles as filler has been studied in this work. Owing to easier manufacturing process, low cost and exceptional properties, basalt fibre finds its wider usage. Nano graphene platelets are added as filler to epoxy resin to enhance physical and mechanical properties of matrix. Synergistic effect of zinc borate with alumina trihydrate has also made them to incorporate with the resin as a flame retardant. Influence of nano titanium dioxide with graphene platelet nanopowder as reinforcement in basalt fibre-epoxy composite was also determined by appropriate tests. Mechanical properties such as tensile and flexure strength were investigated in accordance with ASTM standards. SEM image of failure pattern is also studied.

Keywords Basalt fibre · Epoxy · Nanographene platelets · Alumina trihydrate · Zinc borate · Titanium dioxide

1 Introduction

Recent research has its focus in energy conservation and environmental issues for green living. Booming polymeric composite materials attracts the inventors in research vicinity. Basalt fibre possesses high tensile strength when compared to glass fibre and better fracture strain than do carbon fibre. Abundant availability of basalt rock and its simple fabrication process makes the fibre a desired research component. Exceptional behaviour of basalt fibre in chemical, weather and heat resistance made itself a promising fibre [1]. Surface modification of basalt fibre improves interfacial interaction [2].

M. M. Metro · M. Selvaraj (✉)
Department of Mechanical Engineering, Sri Sivasubramaniya Nadar College of Engineering, Chennai, India
e-mail: selvarajm71@gmail.com

S. Vijayan et al. (eds.), *Trends in Manufacturing and Engineering Management*, Lecture Notes in Mechanical Engineering,
https://doi.org/10.1007/978-981-15-4745-4_37

Current research results indicate that composites with micro and nano-sized materials reinforced in polymer matrix perform better mechanical properties than do conventional fibre reinforced composites [3–6]. The fillers are added to polymer in order to reduce the cost and also to improve physical or chemical properties of polymer.

Addition of inorganic particulate fillers such as nano/microglass, CNT, Al_2O_3, $CaCO_3$, silicates into polymers improves stiffness and strength of polymer matrix, thus expanding their applications in diverse field. Particulate fillers also modify the physical and mechanical properties of polymer composites in many ways.

In the present work, basalt fibres were treated by alkaline agent, sodium hydroxide solution before fabrication of the composite. Effects of surface-treated basalt fabric with the addition of nano and microfillers on mechanical behaviour were studied by suitable tests. The commercially available nanoparticles are generally incorporated into polymers such as epoxy, polypropylene and polystyrene to form a matrix. Introduction of nano fillers like carbon nanotubes, graphene into micrometal matrix enhances mechanical properties and reduction in weight. High specific surface area and 2-D structure of graphene increase tensile strength of the composite [3, 4]. These nanoparticles enhance the mechanical, thermal and electrical properties of composite matrix [5, 6].

Nano graphene platelets (GnPs) are similar to carbon nanotubes but possess excellent performance characteristics when compared to other nano materials. Graphene nanoplatelets are two dimensional (2-D) thin sheets made up of tiny stacks of graphene, which replaces carbon nanotubes, carbon fibre, nanoclays in various composite applications. Carbon atoms of graphene nanoplatelets are arranged in a honeycomb structure with sp2 hybridization. Thickness of graphene nanoplatelet is of a single carbon [7]. Two-dimensional platelets have capability of sliding over one another, allowing higher loading and lower viscosities. GnPs are widely used as an ideal filler in industry owing to its excellent mechanical and physical properties, including exceptional conductivity and high specific strength. Incorporation of GnPs in very small quantity enhances the performance of composite [8]. Addition of graphene also has a considerable effect on flammability properties of the composites [9]. Outstanding performance of graphene nanoplatelets is due to stability of sp2 bonds that form a hexagonal lattice [10]. Surface modified CNT enhanced the performance of tensile strength and Young's modulus compared to plain CNT/basalt/epoxy composites [11, 12]. Silane modified CNT increases fracture toughness of basalt/CNT/epoxy by 40% than acid modified CNT composites [13]. Addition of graphene platelets nanopowder enhances thermo-mechanical and flame-retardant properties of polymer composite.

Flame retarders are substances that are added into polymer molecule or blended physically into the polymers. Non-halogen additives are used with polymers for its flame-retardant property. The most commonly used non-halogen additives are metal hydroxides, like aluminium trihydroxide (ATH) due to their low cost and toxicity, ease of handling and minimal corrosion effects. Alumina trihydrate (ATH), the most extensively used flame retardant, is in usage for decades [14]. Fire retardant property of alumina trihydrate flags an important role in epoxy resin. Alumina trihydrate itself has the potential to retard flame inevitably. Synergistic effect of alumina trihydrate could endure the flammable property of multicomponent epoxy nano composite [15].

Metal borates gained significant interest and extensive usage in industrial application due to their excellent structural chemistry [16]. Zinc borates are being used as a polymer additive for more than decades. Addition of zinc borate significantly improves the flame-retardant property. Synergistic effect of zinc borates in addition to polymer/metal hydroxide provides feasible properties [17].

Incorporation of aluminium trihydroxide and zinc borate in epoxy asphalt binders increases flame retardancy of neat cold-mix epoxy asphalt binders. Synergistic effect of aluminium trihydroxide and zinc borate in epoxy asphalt binders enhances glass transition temperature of the composite. Halogen-free flame retardants disperse uniformly and increase the tensile strength [18]. Metal fillers in polymers enhance mechanical performance of composite such as hardness. It also acts as a reinforcement and increases the interfacial bond between fillers and polymers [19]. Mechanical properties of metal matrix fabricated into epoxy depend mainly on size of particle. Tensile modulus of composite laminate fabricated with metal matrix reinforcement increases with increase in filler [20]. The size of particles added as fillers in epoxy concludes hardness value of fabricated composite. Aluminium particles when added to epoxy resin increases the hardness of composite with decrease in particle size [21]. Tensile modulus and density of metal matrix filled epoxy depend on rule of mixtures [22]. Interfacial strength between fibre and composite is improved by incorporation of titanium dioxide [23–25]. The interfacial shear strength is more for graphene–titanium than that of graphene–aluminium [25].

Several studies have been conducted to explore the application of basalt fibre as a reinforcing substance in various composites. However, performance analysis of graphene nano platelets with microfillers and graphene nano platelets with nano titanium dioxide on basalt/epoxy composite has not been studied elsewhere till date to the maximum knowledge of author.

2 Materials and Procedures

2.1 *Materials*

The commercially available woven-type basalt fabric with surface density of 300 g/m^2 was used as a reinforcing material in this study. Woven fabric has an average fibre diameter of 12 μm. A solution of sodium hydroxide (NaOH) was used for surface treatment of the fabric. Epoxy resin of bisphenol A diglycidylether (Araldite LY 556) and hardener (HY 951) were used for fabrication. Graphene platelet nanopowder of thickness ranging between 6 and 8 nm with surface area 150 m^2/g was used. The purity of carbon was 99.5%. Alumina trihydrate, Al(OH)3 or $Al_2O_3 \cdot 3H_2O$, a white amorphous powder with average particle size as 1.5–5 microns were used. Zinc borate ($2ZnO \cdot 3B_2O_3\_3.5H_2O$), a white amorphous powder with an average particle size less than 7 μm, was also used as microfiller. The weight ratio was considered by varied alumina trihydrate (ATH) and constant zinc borate. Titanium dioxide (TiO_2–rutile), an ultrapure white nanopowder with an average particle size 250 nm, was used with basalt/epoxy matrix and graphene platelet nanopowder.

2.2 Composite Preparation

Basalt fabrics were sprayed with 1 (N) NaOH solution and dried at room temperature. Surface-treated basalt fabrics were then cleaned with deionized water to maintain neutrality [25]. Finally, basalt fibres were made to dry in open air for 24 h at a room temperature of 30 °C. The nano filler, graphene platelet nanopowder of 0.2%, 0.4% and 0.6% by weight were added to resin and stirred using a magnetic stirrer of 600 rpm. ATH and zinc borate of 3% and 1%, respectively, by weight were then added to resin as microfillers followed by magnetic stirring. Epoxy resin with fillers was then mixed with hardener in the weight ratio of 10:1 and was mixed thoroughly using a mechanical stirrer continuously. Vacuum bagging procedure was adapted for fabrication of basalt fibre reinforced composite. Basalt fibre reinforced composite excluding nanofiller was considered as control specimen. The composite laminates were fabricated to the dimensions of 300 mm × 300 mm × 3 mm.

Similar procedure was repeated for graphene platelet nanopowder and titanium dioxide nanopowder reinforced basalt epoxy composite laminate. The schematic representation of fabrication process is illustrated in Fig. 1.

Various combinations of composite specimens are enlisted in Table 1. Basalt/epoxy composite laminates with nano and microfillers were fabricated with varying weight percentages of ATH as 0, 1%, 3% and 5% with a constant 1 wt% of zinc borate as microfillers. GnP nano fillers were also included by 0, 0.2%, 0.4% and 0.6%, respectively. Another set of basalt/epoxy composite laminates were fabricated with different nano fillers. Based on the previous work, 0.5 wt% of nano titanium dioxide was kept constant and varying graphene platelet nanopowder as 0, 0.2, 0.4 and 0.6 wt%, respectively.

2.3 Surface Morphology

The morphology and surface textures of basalt fibre reinforced with nano graphene platelet powder, titanium dioxide, alumina trihydrate and zinc borate were investigated by scanning electron microscope (SEM). However, long cylindrical smooth surface is the common morphological characteristics of fibres after treatment. Failure mechanisms related with dispersion of nano and microfillers in epoxy resin were also detailed using SEM images.

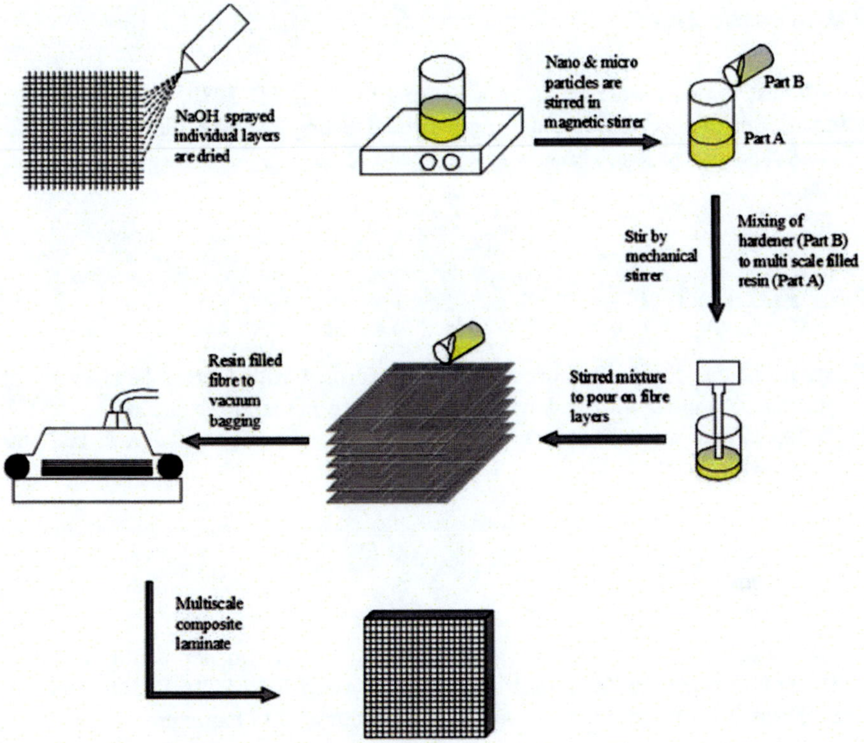

Fig. 1 Schematic diagram of fabrication process

Table 1 Combinations for composite specimen

Combination for composite specimen	% by wt of fillers as reinforcements
Epoxy + BF	0
Epoxy + BF + ATH + ZB	0, 1, 3 & 5 (ATH), 1(ZB)
Epoxy + BF + ATH + ZB + GnP	0, 1, 3 & 5 (ATH), 1(ZB), 0.2 (GnPs)
Epoxy + BF + ATH + ZB + GnP	0, 1, 3 & 5 (ATH), 1(ZB), 0.4(GnPs)
Epoxy + BF + ATH + ZB + GnP	0, 1, 3 & 5 (ATH), 1(ZB), 0.6(GnPs)
Epoxy + BF + TiO_2	0.5 (TiO_2)
Epoxy + BF + TiO_2 + GnP	0.5 (TiO_2), 0.2(GnPs)
Epoxy + BF + TiO_2 + GnP	0.5 (TiO_2), 0.4(GnPs)
Epoxy + BF + TiO_2 + GnP	0.5 (TiO_2), 0.6(GnPs)

2.4 *Tensile Test*

Tensile test samples were prepared according to ASTM D 3039, the international standard. Specimens were tested using universal testing machine, Shimadzu AG–X, of load capacity 50 kN. The cross head speed for tensile tests was maintained at 2 mm/min.

2.5 *Flexure Test*

Flexural test for composite specimen was performed with a cross head speed of 2 mm/min. Samples with and without GnPs were prepared according to ASTM D 790 specifications and subjected to three-point bending load. The load cell was equipped with 50kN.

2.6 *Impact Test*

Impact of composite specimens was tested using Charpy impact testing machine, AIT 300-N. Specimens for impact test were prepared for ASTM D 256 standard. The Charpy impact test machine was equipped with a 300 J hammer.

3 Results and Discussions

3.1 *Tensile Strength*

Basalt fibre reinforced composite specimens with multiscale fillers were subjected to standard tensile test. Variations in tensile properties of composites based on addition of nano and microfillers in matrix are discussed here. Incorporation of varied percentages of GnPs triggered the mechanical characterization of composites to certain extent.

Stress–strain values of varying weight % of GnPs in multiscale composite explain the deviation of strength after a sudden drop in load at 0.4 wt%. Tensile modulus of composite specimen fabricated with varying wt% of GnPs, ATH and constant zinc borate is shown in Fig. 2. It is clearly observed from Fig. 2 that tensile modulus of composite specimen with wt% of 0.4 GnPs with a combination of 3 wt% (ATH) and 1 wt% (Zinc borate) named as A3Z1 marks highest value, whilst A0Z1 without GnPs possesses the lowest as 18.97 MPa and 10.65 GPa, respectively. Incorporation of nano

Fig. 2 Tensile modulus versus wt% of nano filler

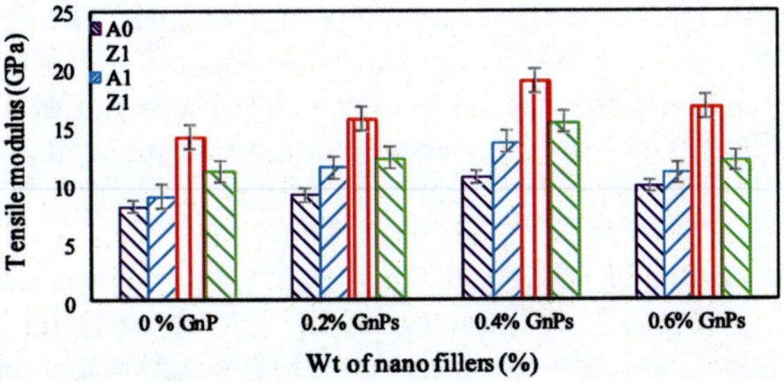

Table 2 Tensile test results of composite specimen with micro and nano fillers (A3Z1)

GnPs content (%wt)	Tensile strength (MPa)	Tensile modulus (GPa)	Tensile strain (mm)
Epoxy/Basalt	212	12.53	0.023
Epoxy/Basalt/ATH/ZB	216	14.12	0.027
0.2	234	15.76	0.029
0.4	244	18.97	0.030
0.6	219	16.81	0.028

filler along with microfillers enhances the tensile property of composite specimen. Nano fillers being able to slide over one another in its structural arrangement create a good interfacial bond with the matrix. Presence of microfillers in matrix grooves the surface and widens gap between fibre and matrix.

Table 2 shows the obtained result in terms of tensile properties for A3Z1 alone (i.e. 3% ATH with 1% zinc borate and varying GnPs as 0, 0.2%, 0.4%, 0.6% by weight). Tensile strength and modulus of control specimen were 216 MPa and 14.12 GPa, respectively. An enhancement in tensile strength and modulus was observed for 0.4 wt% GnPs with an increment of 14.03% and 40.89%, respectively, than that of control sample. It may be noted that tensile strength and modulus of hybrid multiscale composites with GnPs were observed to have higher values than composites without GnPs.

Weak Vander walls force lowers homogeneity of material due to the presence of multiscale fillers. Elongation at break decreased dramatically with almost linear tensile behaviour after 0.4% GnP. Tensile strength of multiscale composite specimen increases with increase of nano graphene platelets up to a certain percentage and then decreases with increase of it. Inclusion of GnPs up to 0.4 wt% exhibited a yielding behaviour which shows typical plastic, and the sudden fall for 0.6 wt% of GnPs exhibited a brittle behaviour. A decline in elastic modulus was observed, thereby indicating softening of material with an increase in nano filler of the composites. It was stated from previous study that the addition of GnPs at 0.1% by weight reaches critical value [26]. Addition of graphene enhances toughness of epoxy with an enhancement of 65% in fracture toughness [27, 28]. A decreasing trend for tensile strength was noticed after 0.4% of nano filler. This enhancement in tensile strength may be due to the inclusion of multiscale fillers into the polymers. Stiffness of composite specimen increases with 0.4 wt% of GnPs and decreases for 0.6 wt% of GnPs. Elasticity decreases from the addition of 0.6 wt% of GnPs to the polymer. However, there is a decrease in tensile strength and tensile strain values at break, by 10% and 6.67%, respectively, for same loading of GnPs. SEM images of tensile specimen of A3Z1 with 0.4 wt% of GnP are illustrated in Fig. 3.

From SEM image shown in Fig. 3, it is observed clearly that incorporation of nano fillers has made interfacial bond between fibre and matrix closer, thus reducing the gap. The surface roughness (Fig. 3a) is due to protrusion of microparticles on matrix. Presence of microfillers may result in stress concentrations around its surface creating weak bondage in matrix and hence makes the matrix to crack. Agglomeration of nanoparticles is also seen on the surface of composite laminates. These agglomerations might have caused interfaces to weaken and in turn to fracture. Fibres visualized without matrix coat in Fig. 3b is seen and may be due to error in fabrication. This space created by improper fill of matrix leads the composite specimen to fail.

Table 3 enlists tensile strength and modulus of composite specimen fabricated with 0.5 wt% of nano titanium dioxide and varied wt% of (0, 0.2, 0.4, 0.6%) graphene platelet nanopowder. It is observed that composite specimen with 0.4 wt% of GnPs marked maximum tensile strength and modulus, whilst e minimum is with exclusion of GnPs. Addition of 0.5 wt% of nano titanium dioxide has increased tensile strength and modulus of composite by 21.85% and 41.09%, respectively, than that of epoxy/basalt plain composite. Increase of 13.77% and 21.89% on tensile strength and modulus with 0.2 wt% of GnPs than 0.4 wt% of GnPs has an increase in strain rate with softening of composite. Strain at 0.4 wt% reaches ultimate and had sudden decline for 0.6 wt% making the composite brittle.

Ease of mobility of nano fillers into matrix makes the interaction better than that of multiscale fillers. Interfacial interaction of nano fillers with polymer forms a strong bondage and hence strong matrix. Mobility of polymer-coated nano fillers interlocks

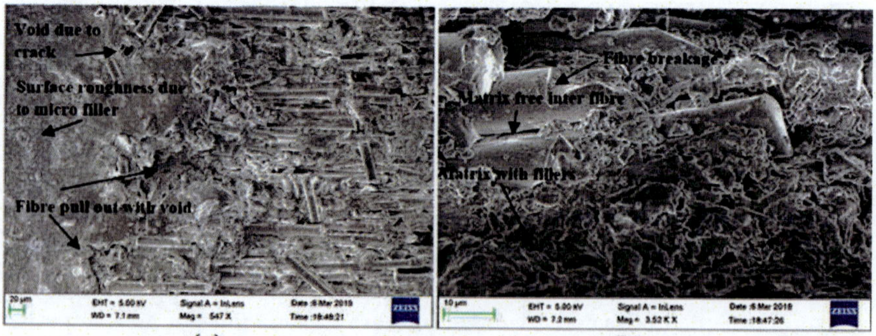

(a) (b)

Fig. 3 a and b SEM images of tensile specimen (A3Z1)

Table 3 Tensile test results of specimen with nano TiO_2 and GnPs

%wt of fillers	Tensile strength (MPa)	Tensile modulus (GPa)	Tensile strain (mm)
Epoxy/Basalt	212	12.53	0.023
Epoxy/Basalt/0.5 (TiO_2)	225	13.26	0.026
0.5 (TiO_2) & 0.2 (GnPs)	230	15.26	0.031
0.5 (TiO_2) & 0.4 (GnPs)	264	19.01	0.035
0.5 (TiO_2) & 0.6 (GnPs)	240	14.15	0.029

into fabric consequently resulting in a strong interaction of fibre/matrix. Synergistic effect of nanoparticles as filler into matrix has made composite stronger than unfilled specimen [29, 30]. Inclusion of titanium carbide enhances tensile property of composite [30].

3.2 *Flexural Strength*

Flexural test specimens were subjected to three-point bending tests. The flexural performance of multiscale composite specimen fabricated by varying wt% of GnPs, ATH with constant zinc borate was evaluated. Discrepancies in flexural modulus of composite specimen are depicted in Fig. 4. An enhancement in flexural performance of composite laminates was observed with an inclusion of 0.2 wt% nano filler with microfiller filled resin than that of epoxy/basalt composite. Multiscale composite with 3 wt% of ATH and 1 wt% of zinc borate in addition to 0.4 wt% of GnPs has

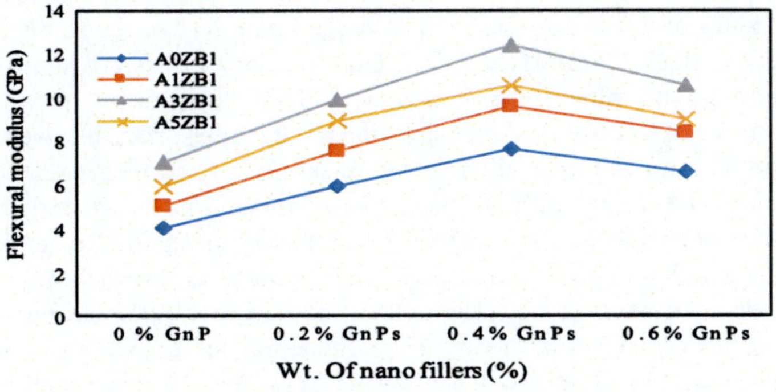

Fig. 4 Flexural values of composite specimen with micro and nano fillers

Table 4 Flexural values of composite specimen with micro and nano fillers (A3Z1)

% weight of GnPs	Flexural strength (MPa)	Flexural modulus (GPa)	Flexural strain (mm)
Epoxy/Basalt	208.92	4.03	0.022
Epoxy/Basalt/ATH/ZB	243.24	7.05	0.028
0.2	329.73	9.93	0.026
0.4	395.45	12.45	0.027
0.6	295.54	10.61	0.024

good effect on flexure modulus of composite specimen. A decremental effect in flexure strength and modulus after addition of 0.4% GnPs reveals that composite has better interfacial bond at 0.4% GnPs, thus making composite stronger. It is clearly observed that it makes the material more ductile beyond the addition of 0.4% GnPs by weight.

Delamination of composite formed on surface of specimen with 0.2 wt% GnPs is more than do 0.4 and 0.6 w% of GnPs composite laminates. The action of this type of behaviour features the brittle nature of the composite. This delamination might have formed due to weak interfacial adhesion of matrix and fibres.

Flexural values of tested composite specimen with micro and nano fillers for ATH and zinc borate with weight ratio of 3:1 and varied wt% of GnPs are enlisted in Table 4. Basalt/epoxy composite with microfillers attained more strength than basalt/epoxy composites by 15%. Presence of multiscale fillers increases surface roughness with the reaction of intraparticle fusion. Nano fillers increase the smoothness of composite surface.

Flexural modulus of multiscale composite has an incremental value with 0.4 wt% of GnPs to matrix. Addition of microfillers with nano matrix has improved flexural strength and modulus of composite by 18.13% and 22%, respectively, for 0.4 wt% of GnPs than that of composite without nano filler. Incorporation of nano material beyond 0.4 wt% makes the material more brittle. Inclusion of 0.6% of GnPs by weight has a decline in strength and modulus by 28.92% and 15.95%, respectively, from the values of composite specimen with maximum value. It was mentioned by earlier research that addition of 0.1 wt% of GnPs in epoxy with basalt fabric improved strength resulted in further decrease in interfacial bonding [26].

SEM images of flexural specimen describe the fracture surface as given in Fig. 5. Orientation of nano and microfillers along basalt fabric and resin indicates nature of fracture surface. Distribution of nano fillers formed a neat and smooth surface. Smooth delaminated surface is seen in Fig. 5a. Weak matrix interface between fibres causes matrix failure which in turn pulls out the fibres as shown in Fig. 5b. Non-uniformity in distribution of nano fillers clearly depicts poor interfacial bond between fibre and matrix which in turn ruptures fibre. Fibre pull outs are initiated due to weak interfacial bonding between epoxy and fibre. It is evident that more number of fibre pull outs result in the weak and brittle specimen.

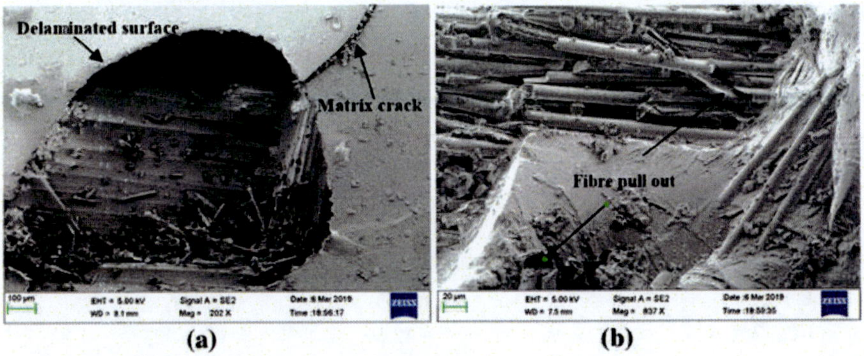

Fig. 5 **a** and **b** SEM images of flexural test specimen

SEM images of flexural specimen of nano graphene platelets and TiO_2 are illustrated in Fig. 6. The matrix that bounds fibre itself demonstrates clear and smooth matrix formed with inclusion of different nanoparticles. Presence of nanofillers grounds smooth matrix.

Flexural test results of composite specimen fabricated with 0.5 wt% of TiO_2 with varying wt% of GnPs are tabulated in Table 5. It is evident that specimen with 0.4 wt% has marked maximum, whilst epoxy/basalt scored the minimum. Increase in strain rate from plain specimen (epoxy/basalt) till composite specimen with 0.4 wt% GnPs measures higher tensile and flexure properties. Composite has proved its yield capacity till 0.4 wt% of GnPs with a limit. The specimen bounds its stiffness to 0.4 wt% of GnPs into 0.5 wt% of TiO_2 filled epoxy/basalt.

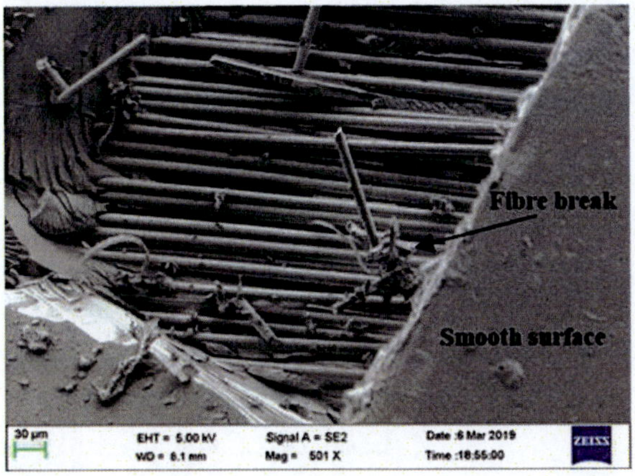

Fig. 6 SEM images of flexural test specimen fabricated with nano GnPs and TiO_2

Table 5 Flexural test results for composite specimen with TiO_2 and GnPs

% weight of GnPs	Flexural strength (MPa)	Flexural modulus (GPa)	Flexural strain (mm)
Epoxy/Basalt	208.92	4.03	0.022
Epoxy/Basalt/0.5 (TiO_2)	236.66	7.98	0.025
0.5 (TiO_2) & 0.2 (GnPs)	335.23	11.87	0.026
0. 5(TiO_2) & 0. 4(GnPs)	400.64	14.41	0.028
0.5 (TiO_2) & 0.6 (GnPs)	358.75	12.69	0.025

3.3 *Impact Strength*

Energy absorbed and lost by specimen during Charpy impact test gives impact resistance and impact strength of the specimen. The striking velocity of pendulum is 2.45 m/s. Impact strength of specimen subjected to impact energy of 300 J at room temperature is shown in Fig. 7. It is evident from Fig. 7 that incorporation of 0.4 wt% GnPs increased the impact strength of specimen with a reduction on further addition of GnPs. Addition of 0.6 wt% of GnPs declined in its impact strength. Impact strength of composite specimen fabricated with nano fillers is higher than composite with nano and microfillers. This might have been effected from interfacial interaction of fillers which enhances interfacial bonding between fibre and matrix. Impact strength of composite falls from 1517 to 1505 kJ/m^2 after the addition of 0.6 wt% GnPs into nano TiO_2 matrix. The composite fabricated with multiscale fillers recorded a value of 1507 kJ/m^2 for 0.4 wt% of GnPs from 1435 kJ/m^2 for plain epoxy/basalt composite. Though the impact strength values are similar, further addition of GnPs lowers impact strength and thus leads to delamination.

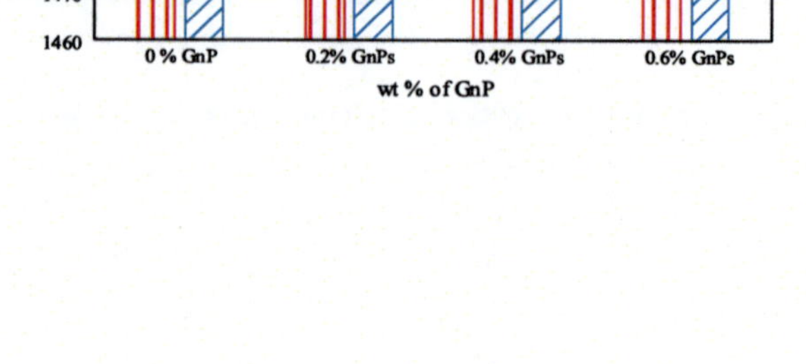

Fig. 7 Impact strength of multiscale and nano fillers

4 Conclusion

The experimental investigations on mechanical properties of basalt fibre reinforced composite were studied. The following are highlighted as remarks.

1. Incorporation of nano graphene platelets and microfillers such as aluminium trihydrate, zinc borate improved mechanical properties of multiscale composite with 0.4 wt% of GnPs.
2. Addition of 0.4 wt% graphene nano platelets to nano titanium dioxide enhances tensile strength of the composite. Increment in flexural strength signifies the strength of composite.
3. The impact strength of multiscale and nano composite remarks with 0.4 wt% of graphene nano platelets for higher loading.

References

1. Dhand V, Mittal G, Rhee KY (2015) A short review on basalt fiber reinforced polymer composites. Compos Part B Eng 73:166–180. https://doi.org/10.1016/j.compositesb.2014.12.011
2. Espano (2013) Investigation of the effect of different silane coupling agents on mechanical performance of basalt fiber composite laminates with biobased epoxy matrices. Polym Compos 34
3. Dasari BL, Morshed M, Nouri JM (2018) Mechanical properties of graphene oxide reinforced aluminium matrix composites, pp 136–144
4. Muhammad R, Fusheng P, Aitao T (2014) Effect of graphene nanoplatelets addition on mechanical properties of pure aluminum using a semi-powder method, pp 101–108
5. Park JH, Jana SC (2003) The relationship between nano- and micro-structures and mechanical properties in PMMA-epoxy-nanoclay composites. Polymer (Guildf) 44:2091–2100. https://doi.org/10.1016/S0032-3861(03)00075-2
6. Koricho EG, Khomenko A, Haq M (2015) Effect of hybrid (micro- and nano-) fillers on impact response of GFRP composite. Compos Struct 134:789–798. https://doi.org/10.1016/j.compstruct.2015.08.106
7. Sengupta R, Bhattacharya M, Bandyopadhyay S, Bhowmick AK (2011) A review on the mechanical and electrical properties of graphite and modified graphite reinforced polymer composites. Prog Polym Sci 36:638–670. https://doi.org/10.1016/j.progpolymsci.2010.11.003
8. Weihua K, Yuuki H, Lei Hua YI (2008) Thermal and mechanical properties of a poly(« -caprolactone)/graphite oxide composite. J Appl Polym Sci 107:1395–1400
9. Huang G, Gao J, Wang X (2012) How can graphene reduce the flammability of polymer nanocomposites? Mater Lett 66:187–189. https://doi.org/10.1016/j.matlet.2011.08.063
10. Papageorgiou DG, Kinloch IA, Young RJ (2017) Mechanical properties of graphene and graphene-based nanocomposites. Prog Mater Sci 90:75–127. https://doi.org/10.1016/j.pmatsci.2017.07.004
11. Lee JH, Rhee KY, Park SJ (2010) The tensile and thermal properties of modified CNT-reinforced basalt/epoxy composites. Mater Sci Eng, A 527:6838–6843. https://doi.org/10.1016/j.msea.2010.07.080
12. Liang M, Wong KL (2017) Study of mechanical and thermal performances of epoxy resin filled with micro particles and nanoparticles. Energy Procedia 110:156–161 https://doi.org/10.1016/j.egypro.2017.03.121

13. Kim MT, Rhee KY, Park SJ, Hui D (2012) Effects of silane-modified carbon nanotubes on flexural and fracture behaviors of carbon nanotube-modified epoxy/basalt composites. Compos Part B 43:2298–2302. https://doi.org/10.1016/j.compositesb.2011.12.007
14. Lynch TJ, Chen T, Riley D (2003) Advances in ATH benefit Aluminium hydroxide is a widely used flame retardant for reinforced plastics. Materials E, 44–46
15. Elbasuney S (2017) Novel multi-component flame retardant system based on nanoscopic. Powder Technol 305:538–545. https://doi.org/10.1016/j.powtec.2016.10.038
16. Schubert DM, Alam F, Visi MZ, Knobler CB (2003) Structural characterization and chemistry of the industrially important zinc borate, Zn [$B_3O_4(OH)_3$]. Chem Mater 4:866–871
17. Bourbigot S, le Bras M (2003) European meeting on fire retardancy and protection OF Materials, pp 327–335
18. Chen R, Gong J, Jiang Y (2018) Halogen-free flame retarded cold-mix epoxy asphalt binders: rheological, thermal and mechanical characterization. Constr Build Mater 186:863–870. https://doi.org/10.1016/j.conbuildmat.2018.08.018
19. Tanwyn M, Khushairi M, Sharif S (2015) Evaluation of mechanical properties of filled epoxy composite for improving mould performance—a review. Appl Mech Mater 735:13–18. https://doi.org/10.4028/www.scientific.net/AMM.735.13
20. Bhagyashekar MS (2014) J Reinf Plast Compos, 28–42. https://doi.org/10.1177/0731684408095034
21. Chung S, Im Y, Kim H, Jeong H (2005) Evaluation for micro scale structures fabricated using epoxy-aluminum particle composite and its application
22. Tilbrook MT, Moon RJ, Hoffman M (2005) Study on the mechanical properties of alumina—epoxy composites with an interpenetrating network structure. Mater Sci Eng A 393:170–178. https://doi.org/10.1016/j.msea.2004.10.004
23. Ma L, Zhu Y, Li X (2018) The architecture of carbon fiber-TiO_2 nanorods hybrid structure in supercritical water for reinforcing interfacial and impact properties of CF/epoxy composites. Polym Test 66:213–220. https://doi.org/10.1016/j.polymertesting.2018.01.021
24. Yi C, Bagchi S, Dmuchowski CM (2018) Direct nanomechanical characterization of carbon nanotube—Titanium interfaces. Carbon N Y. https://doi.org/10.1016/j.carbon.2018.02.069
25. Lipatov YV, Gutnikov SI, Manylov MS (2015) High alkali-resistant basalt fiber for reinforcing concrete. J Mater 73:60–66. https://doi.org/10.1016/j.matdes.2015.02.022
26. Bulut M (2017) Mechanical characterization of Basalt/epoxy composite laminates containing graphene nanopellets. Compos Part B 122:71–78. https://doi.org/10.1016/j.compositesb.2017.04.013
27. Silva F, Njuguna J, Sachse S (2013) The influence of multiscale fillers reinforcement into impact resistance and energy absorption properties of polyamide 6 and polypropylene nanocomposite structures. Mater Des 50:244–252. https://doi.org/10.1016/j.matdes.2013.02.041
28. Prasad V, Joseph MA, Sekar K (2018) Investigation of mechanical, thermal and water absorption properties of flax fibre reinforced epoxy composite with nano TiO_2 addition. Compos Part A 115:360–370. https://doi.org/10.1016/j.compositesa.2018.09.031
29. Mahesha CR, Mohan N, Rajesh M (2017) Role of nanofillers on mechanical and dry sliding wear behavior of basalt-epoxy nanocomposites. Mater Today Proc 4:8192–8199. https://doi.org/10.1016/j.matpr.2017.07.161
30. Thotsaphon T, Katsuyoshi K, Hisashi I (2019) Titanium matrix composite reinforced with in-situ formed TiC using carbon black nano particles via a wet process. Trans JWRI 38:13–17

A Feasibility Study on Microwave Joining of GFRP Composite Pipes with Interlayer Coupling Agents

K. Rajkumar, M. Dhananchezian, and S. Aravindan

Abstract This paper investigates the feasibility of energy, economic, and environment way to joining the GFRP pipes. These GFRP pipes are used to transport chemicals, fuel, oil, and natural gases. Microwave joining is economic and energy-efficient method for bonding thermoset composite with suitable intercoupling layers. Microwave joining was carried out in 2.45 GHz furnace with suitable susceptor and various coupling agents such as molten polypropylene and polyethylene, epoxy resin liquid, and sodium silicate. These interlayered GFRP pipes were tested to evaluate their lap shear strength and bond strength. Among the various coupling agents used, polyethylene interlayered coupling agent joints exhibited a good shear strength. The abstract should summarize the contents of the paper in short terms, i.e., 150–250 words.

Keywords Microwave joining · GFRP · Composite bonding · Inter layer

1 Introduction

In the electromagnetic spectrum, microwave energy is in the frequency range of 0.3–30 GHz. Microwaves are used to process the materials for many years. Its application is wide in material-processing industries in particular such as joining of thermoplastic composite (TPC) and joining of ceramic to ceramic [1–4]. The joining of thermoplastics using microwave heating has been exploited much in industry for a number of years. Thermoset materials having highly desirable properties are now readily available at economically attractive prices; use of such materials for the production of large-shaped articles has been highly restricted. This is particularly true if the articles have complex shapes or reinforced with glass fibers and too large to be molded

K. Rajkumar · M. Dhananchezian (✉)
Department of Mechanical Engineering, Sri Sivasubramaniya Nadar College of Engineering, Chennai, India
e-mail: dhananchezianm@ssn.edu.in

S. Aravindan
Department of Mechanical Engineering, Indian Institute of Technology, Delhi, India

S. Vijayan et al. (eds.), *Trends in Manufacturing and Engineering Management*, Lecture Notes in Mechanical Engineering,
https://doi.org/10.1007/978-981-15-4745-4_38

economically. An attempt has been made on solving this problem by microwave joining of glass fiber reinforced plastic pipe (GFRP). GFRP, an advanced composite material, is widely used in a variety of applications, including aircraft structural member, robots, and transportation of corrosive liquid [5, 6]. This study focuses on the effect of microwave exposure time, temperature, and different coupling agents on joining of GFRP pipes.

2 Materials and Microwave Joining

GFRP pipes were made using the resin composition of isophthalic (50%) and vinylester (50%). The volume fraction of the materials is 65:35 (resin/glass). The hand lay-up technique was used to manufacture the GFRP pipe with the dimensions of 30 mm and 55 mm of inner and outer diameters, respectively. The glass fiber orientation angle in matrix was 90°.

The term "joining" refers to a zone bounded by the GFRP substrates, in which the entanglement of polymer chains has progressed beyond simple, mechanical, and chain entanglement. The process comprises generating sufficient heat energy at interface of an assembly of thermosets substrates to bond them together using an "interface strip," referred to as a "coupling agent." Prior to joining, the surfaces of the pipes were cleaned, and surface contaminations were removed by fine SiC paper. The different coupling agents used were liquid sodium silicate, epoxy resin, molten polypropylene, and polyethylene. Pipe to be joined was placed in multimode cavity of industrial microwave furnace (2.45 GHz and 3.2 kW). Normally, GFRP thermoset substrates are transparent to the microwave radiation, and the only part that can be heated up is the coupling agent implanted at the joint. The microwave power and heating rate were 320 W and 10 °C/min, respectively. The temperature range fixed was based on the glass transition temperature T_g of the coupling agent. The lap shear strength of the processed joints was evaluated. Effect of microwave exposure was assessed through buckling strength testing of the GFRP pipes.

3 Results and Discussion

Initially, liquid sodium silicate was used as the coupling agent which was applied between the surfaces to be joined. The microwave power was adjusted to 320 W. The holding time, also referred to as "exposure time at maximum temperature," was varied between 5 and 10 min. Table 1 shows the results of microwave joining of GFRP pipe with different coupling agents. The samples prepared using sodium silicate as coupling agents gave a successful weld at 150 °C. When the temperature was increased to 200 °C, the weld exhibited bad strength, but joints have been made. But once the temperature was increased to 250 °C, both the pieces separated out on the application of slightest additional force and hence the joint failed at 250 °C. The

Table 1 Microwave heating history and different coupling agent

Coupling agent	Maximum temperature (°C)	Holding time (min)	Heating rate (°C/min)	Weld success	Shear strength (MPa)
Liquid sodium silicate	150	5	10	Yes	40.8
Liquid sodium silicate	150	10	10	Yes	40.8
Liquid sodium silicate	200	5	10	Yes	–
Liquid sodium silicate	200	10	10	Yes	–
Liquid sodium silicate	250	5	10	No	
Liquid sodium silicate	250	10	10	No	
Epoxy resin	150	10	10	No	
Epoxy resin	200	10	10	No	
Epoxy resin	250	10	10	No	
Polypropylene	150	10	10	No	
Polypropylene	200	10	10	No	
Polypropylene	250	10	10	No	
Polyethylene	250	10	10	No	
Polyethylene	350	2	12	No	
Polyethylene	300	2	12	Yes	122.4

variation in the holding time also had not affected the failure at 250 °C. Epoxy resin could not provide the joint since there was no bonding at all. Polypropylene paste was applied between the joints to be made. The samples failed to produce joint at any of the temperatures tested. Although there was some bonding between the two surfaces of joint but with a slight rotation, the joints separated away. Polyethylene thin film was then tested as a coupling agent. The polyethylene was cut in the shape of rings, and a double film was placed at the junction of two thermoset pipes.

The temperature was varied from 250 to 350 °C with 2–10 min holding time. The samples at 250 and 350 °C failed on application of slight force. Same samples processed at 300 °C and exhibited good strength. Figure 1 shows the successful joint made by using polyethylene as a coupling agent. Heat generated along the bond interface, i.e., coupling agent is transferred from inside to the outer surface. At the softening temperature, the coupling agent softens and joins together. When the temperature of coupling agent increases, it absorbs/couples with microwave energy. When amount of microwave energy is absorbed, heat generation also is increased. This cumulative effect leads to burnouts as noticed on samples. Samples prepared with sodium silicate at 250 °C suffered due to burnouts at the outer layer of the

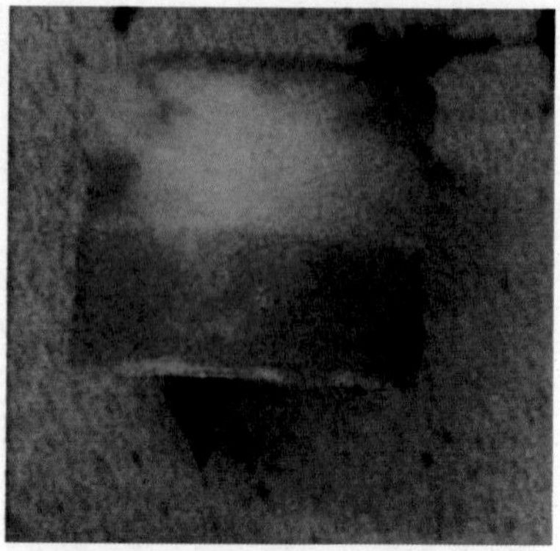

Fig. 1 Joint prepared with using polyethylene as coupling agent

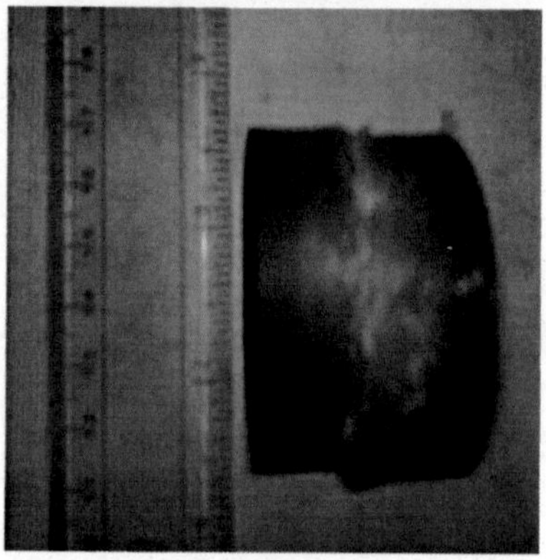

Fig. 2 Burnouts due to high temperature heating

substrates. Samples prepared with polyethylene at 350 °C suffered due to the burnouts on the whole of the outer surface of the substrate. Figure 2 shows the burnout sample of joint with polyethylene. From Table 1, it is understood that higher temperature and longer duration of microwave heating degrade the material properties. A very narrow weldability window is available, for joining these thermoset pipes.

3.1 Microstructure Analysis and Mechanism of Joint Formation

Figure 3a shows the SEM image of the joint made using liquid sodium silicate was a coupling agent (150 °C and 10 min). The SEM image shows well-defined boundary between the two pipes. The amorphous glass phase (sodium silicate) absorbs microwaves, and thus, a selective heating was observed [4]. Sodium silicate is converted into solid films or bonds between the two GFRP pipes by evaporation of water (dehydration). During microwave exposure, the viscosity of the glass drops down to

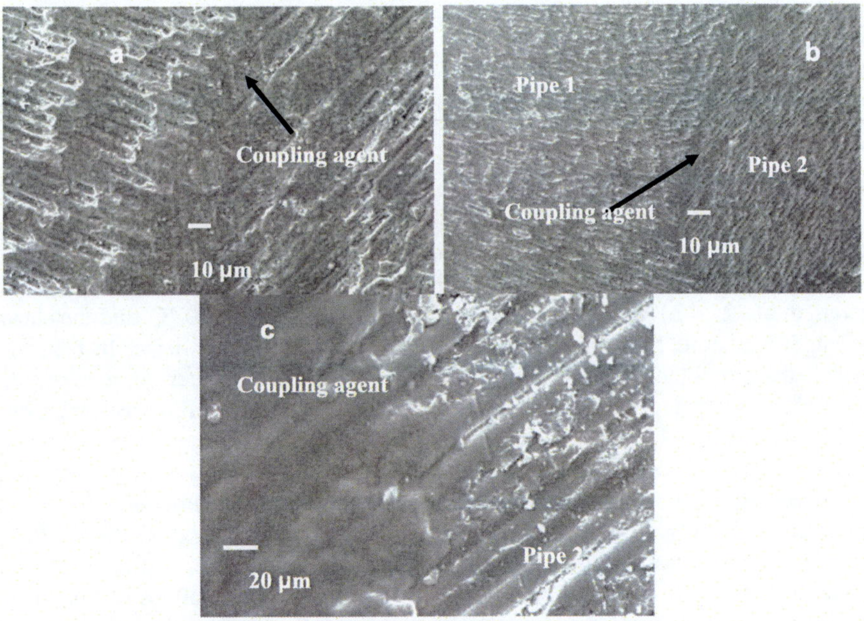

Fig. 3 SEM image of successful microwave joint with coupling agent **a** liquid sodium silicate, **b** polyethylene, **c** magnified image of polyethylene

the lower order and spreading of this material over the interface. Microwave exposure makes it easier to inter-diffusion of coupling agent over the GFRP pipes. Furthermore, attainment of crystalline state at the processing temperature can also facilitate effective bond. Sodium silicate films are having an affinity to moisture pickup, and then, degradation of joints is also possible. However, microwave heating can completely remove the water particles from the sodium silicate which is helpful to form good bond.

In the case of polyethylene coupling agent, the interface is observed very narrow as seen in Fig. 3b. To understand more clearly, the interface of the joint is magnified, and the corresponding image is shown in Fig. 3c. It is interesting to observe that there is no presence of voids or discontinuities at the interface. The role of the coupling agent is to absorb the microwave energy, generate local heat at the interface at the butt faces of thermoset material. The dielectric property of polymers to external electric field depends mainly on its polarity. Moreover, in unpolar polymers (polyethylene), only induced (electron or ion) polarizations play a role when interact with electromagnetic wave [7], whereas in polar polymer orientation, dipole polarization can contribute for heating. During microwave heating, due to localized interdiffusion occurred [8] between the coupling agent to adjacent substrate materials the bond formation occurred.

3.2 Joint Shear Strength and Compressive Strength

After the elimination of the failed samples, successful joints were tested for their joint strength. The mechanical strength of the joint was tested through shear strength. The successful joint was held in position with the help of the designed fixture. From Table 1, it can be understood that sodium silicate and polyethylene are typical coupling agents for bonding. Successful joints made from coupling agent such as sodium silicate with 150 °C and 5 min and 150 °C and 10 min exhibited same shear strength of 40.8 MPa. Polyethylene coupling agent with 300 °C and 2 min exhibited strength of 122.4 MPa. Polyethylene produced better joint at 300 °C, and the shear strength was quite high when compared with that of sodium silicate. In fact, the shear strength of the former was almost triple than that of the latter. At an exposure time of 10 min and higher temperature 350 °C, the substrate material was weakened, and when it was subjected to a shear strength test, failure occurred at the substrate material. The surface skin also been affected during the microwave exposure when exceeding the optimal parameters. The buckling test was carried out on microwave exposed at 150 °C and 10 min and unexposed samples. For carrying out the buckling test, the sample was placed in the compressive testing machine, and the load was steadily applied. Both the samples showed the same load of failure (2 kN). It is understood from this compressive strength data that the microwave exposure had not induced any detrimental effect on the GFRP pipe.

4 Conclusions

Microwave heating technology is successfully demonstrated for joining of thermoset substrate with coupling agent. Coupling agent such as liquid epoxy resin and polypropylene had not yielded positive results. Samples got successfully joined using polyethylene and liquid sodium silicate coupling agent. Joints made using polyethylene exhibited maximum shear strength compared to the liquid sodium silicate samples. Buckling strength of microwave exposed and unexposed thermoset samples had not exhibited any variation in compressive strength, which confirms that no thermal degradation occurred.

References

1. Ku HS, Siores E, Ball JAR, Horsfield B (1999) Microwave processing and permittivity measurement of thermoplastic composite at elevated temperature. J Mater Process Technol 89:419–424
2. Siores E (1994) Microwave technology for welding and joining. Mater World 2(10):526
3. Liu F, Turner I, Siores E, Groombridge P (1996) A numerical and experimental investigation of the microwave heating of polymer materials inside a ridge waveguide. J Microwave Power Electromagn Energy 31:71–82

4. Aravindan S, Krishnamurthy R (1999) Joining of ceramic composites by microwave heating. Mater Lett 38:245–249
5. Ramkumar J, Aravindan S, Malhotra SK, Krishnamurthy R (2004) An enhancement of the machining performance of GFRP by oscillatory assisted drilling. Int J Adv Manuf Technol 23:240–244
6. Aravindan S, Naveen Sait A, Noorul Haq A (2008) Machinability study of GFRP pipes using statistical techniques. Int J Adv Manuf Technol 37:1069–1081
7. Svorcik V, Ekrt O, Rybka V, Liptak J, Hnatowicz V (2000) Permittivity of polyethylene and polyethyleneterephtalate. J Mater Sci Lett 19:1843–1845
8. Rajkumar K, Aravindan S, Kulkarni MS (2012) Wear and life characteristics of microwave sintered copper-graphite composites. J Mater Eng Perform 21:2389–2397

Comparative Evaluation of Mechanical Properties of GFRP and Polymer Hybrid Composite

R. Raja, Sabitha Jannet, Allen Varughese, and Joby George

Abstract Synthetic fiber polymer laminates are becoming competitive material for engineering and structural applications. In this work, an attempt is made to synthesize a laminate with natural as well as synthetic fibers to get the advantage of cost-effective natural fibers. Coir is a natural fiber available in significant amount in south India. The present work aims at manufacturing a strong, cost effective, light and low impact on environment hybrid material. Both GFRP laminates and hybrid (GFRP + coconut Coir) laminates were fabricated by hand layup method and compression molding. Physical and mechanical properties of this laminates such as tensile strength, impact strength and density were evaluated. Microscopy images of this laminates were arrived at through scanning electron microscope. Even though the performance of hybrid (GFRP + coconut Coir) laminates is relatively lesser than GFRP laminates, these hybrid laminates may be adopted in the industries where cost is an important design factor.

Keywords GFRP · Coir · Impact strength · Density · SEM · Tensile strength

1 Introduction

Biocomposites are an attractive new age alternative for synthetic composites [1]. The increase in fiber reduces ductility but when treated with epoxy resin produces inhomogeneity however hardness increases [2]. Energy absorption and failure modes were significantly affected by strain rate [3]. Coir fibers obtained from disregarded coconut shells were used for reinforcement up to 80%. Up to 50% strength increased further addition resulted in agglomeration [4]. Highest tensile strength of 30.7 MPa was reported for cotton unsaturated polymer composite and tensile strength increased linearly [5]. It was found that 30 volume percentage of coir reinforcement displayed ballistic performance as good as Kevlar [6]. The coir reinforced epoxy matrix treated

R. Raja · S. Jannet (✉) · A. Varughese · J. George
Mechanical Engineering Department, Karunya Institute of Technology & Sciences, Coimbatore 641114, India
e-mail: sabithajannet@gmail.com

S. Vijayan et al. (eds.), *Trends in Manufacturing and Engineering Management*, Lecture Notes in Mechanical Engineering,
https://doi.org/10.1007/978-981-15-4745-4_39

with alkali showed 15% increase in impact strength when compared with untreated samples [7]. Coconut shell powder as filler with fiber reinforcement in epoxy exhibited reduction in wear rate due to the filler powder [8]. Epoxy resin was reinforced with untreated coconut fiber leading to increase in tensile strength and higher impact strength which was influenced by the high aspect ratio of the fibers [9]. The inclusion of cow dung and corn starch fibers increases the friction coefficient. Overall, 6% weight cow dung showed superior wear properties and thus proved that cow dung fiber is efficient in improving the friction properties of composites. [10] There was improved mechanical properties in treated fiber compared with raw fibers [11]. There was 25% and 28% increase in flexural and tensile strength, respectively, after the coir was treated in sodium carbonate for 96 h. Coir polyester achieved superior mechanical properties compared to coir epoxy derivatives composites. The solution also led to increase in the superfluous fiber roughness attributing to increased mechanical properties [12]. Coconut fibers pre-treated with alkali solution such as NaOH exhibited superior mechanical properties. And the length of the fibers subsequently affected the tensile properties of the composites [13]. The hygroscopic swelling and absorption effect on the mechanical properties of biocomposites revealed that the direction of the fiber played a major role in the improved mechanical properties [14]. Surface modification of fibers resulted in improved mechanical properties [15]. The flexural strength and tensile properties improved with longer coir [16]. The effect of immersion of coir in water leads to a drastic decrease was preceded with modulus stabilization [17]. The interaction effect between the curing time and coir reinforced volume decided the tensile strength properties of coir fiber reinforced epoxy composites [18].

2 Materials and Methods

Epoxy resins have to be cured for a minimum of 24 h at room temperature. 30 volume percentage of fiber reinforcement with polymer based materials are popular with hand layup method. Volume of fiber and matrix was first calculated. Epoxy content was of 30%, and glass fiber was 70%. Epoxy resin LY556 (density 1.15–1.20 g/cm^3 at 25 ^{0} C) and hardener HY951 (density 1 g/cm^3 at 20 °C) were used. A ratio of 10:1 by weight of resin and hardener were employed. OHP sheet is placed beneath the wooden mold. OHP sheet is used because it is a polymer plastic, so the resin will not stick to the sheet. The sides of the mold were covered using a tape. The glass fiber sheets were laid into the mold layer by layer, and the resin was applied gradually using a brush. This process was repeated until the required thickness was achieved. The mold was then covered using another OHP sheet, and weight was applied on it. Curing was done for first 24 h at room temperature and was then exposed to sunlight. Hybrid is made using the same method as hand layup GFRP. Alternate layers of coir and GFRP were laid, and resin was applied. Hybrid sample consisted of 30% epoxy, 35% glass fiber and 35% coconut fiber. Compression molding refers to a process where up to 2000 psi pressure and temperature from 120 to 200 °C is used in order to cure the part. This technique is used to rapidly cure large quantities fiberglass-reinforced polymer parts. Matched metal dies are placed in large molding press. The materials to be molded are placed between the hot dies and pressed. Fiber arrangement and volume percentage are similar to the hand layup GFRP. Only difference is that the fiber and resin are compressed in between hot dies at 200 °C. Fiber arrangement and volume percentage are similar to the hand layup hybrid. Only difference is that the fiber and resin are compressed in between hot dies at 200 °C. The size of the specimen for the impact test was as per ASTM A370. Brinell hardness test was carried out on a specimen with measurements 50 * 50 * 10 mm.

3 Results and Discussion

3.1 Density

From Fig. 1, it can be seen that the density of GFRP and hybrid produced using compression molding method is less dense when compared to hand layup method. It may be due to the removal of excess epoxy through vaporization or through side leaks when pressure is applied. It can therefore be inferred that GFRP and hybrid composite fabricated using compression molding method are suitable for production of low weight structural models and components (Fig. 2).

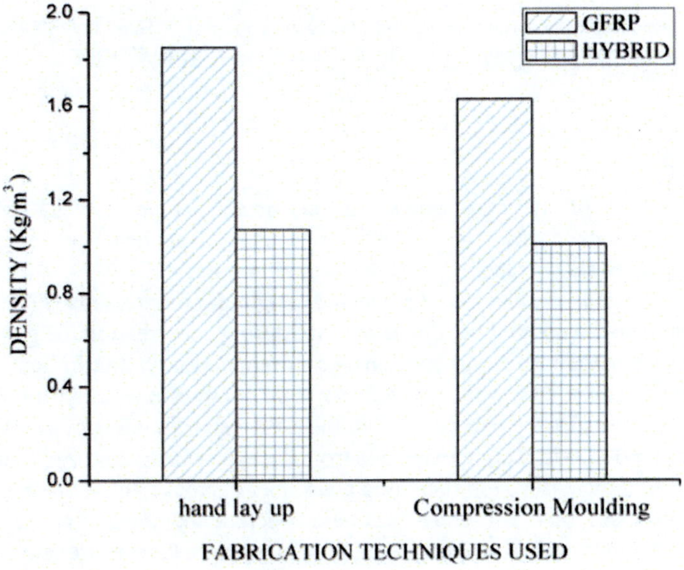

Fig. 1 Effect of GFRP and hybrid composite on the density

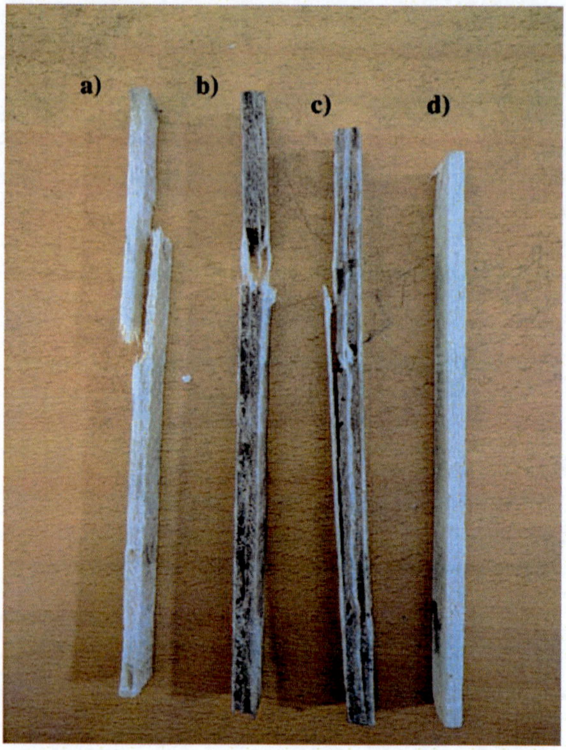

Fig. 2 Tensile samples according to ASTM D3039 after the test. **a** Hand layup GFRP. **b** Hand layup hybrid. **c** Compression molding hybrid. **d** Compression molding GFRP

3.2 Tensile Test

The tensile test of the composites was performed as per the ASTM D3039 (200 * 15 * 6) standards. The test was done using a universal testing machine. The above figure shows tensile stress samples after the test.

From Fig. 3, it is evident that the tensile strength of compressed GFRP is higher than that of hand layup method, this is because during compression the gap between different layers is reduced due to the high pressure applied as well as the heat treatment given, and this is also true when we look in hybrid composite. The hybrid composite possesses less tensile strength due to that the GFRP slab contains more glass fibers when compared to hybrid composite, and from the experiment, it is provident that the tensile strength increases with increased glass fiber content as glass fiber is stronger than natural fiber (coir). Therefore, we could conclude that the composite materials fabricated by compression molding method possess more tensile strength than those fabricated with usual hand layup method.

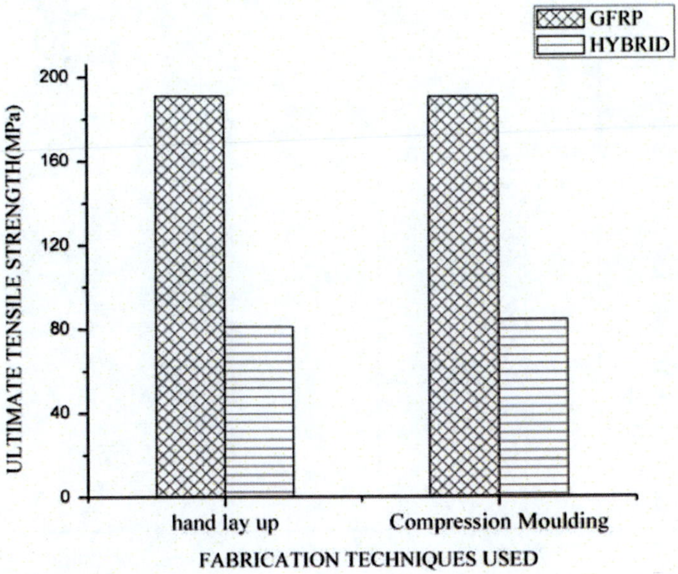

Fig. 3 Effect of GFRP and hybrid on the ultimate tensile strength (MPa)

3.3 Brinell Hardness

The typical test uses a 10 mm (0.39 in) diameter steel ball as an indenter with a 1000 kgf force. For softer materials, a smaller force is used; for harder materials, a tungsten carbide ball is substituted for the steel ball (Fig. 4).

From Fig. 5, it is evident that the indentation of compressed GFRP is lower than that of hand layup method, this is because during compression the gap between different layers is reduced due to the high pressure applied as well as the heat treatment given, and this is also true when we look in hybrid composite. Therefore, we could conclude that the composite materials fabricated by compression molding method possess more hardness than those fabricated with simple hand layup method.

3.4 Charpy Impact Test

The specimens were cut as per ASTM A370 standard. A V-notch is provided at 28 mm, i.e., at half the length of the specimen. The depth of the notch was around 2 mm, and 10 * 10 was its width and thickness, respectively.

From Fig. 6, it can be noticed that the impact strength of compression molding is found to be way higher than that of the hand layup method, and this is due to better adhesiveness of different layers in composite slabs. The compression molding uses pressure as well as temperature to give better compression of the material,

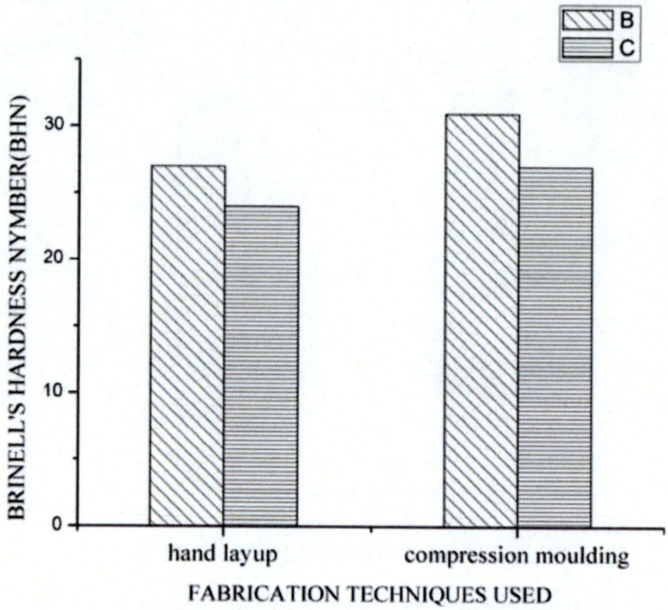

Fig. 4 Experimental values of Brinell hardness test obtained in this study are plotted in graph

Fig. 5 Charpy impact test specimens **a** Hand layup GFRP. **b** Hand layup hybrid. **c** Compression hybrid. **d** Compression GFRP

and as a result, the bonding between various atom particles increases on comparing with hand layup method. Among compression molded composite, the glass fiber composite provides better impact strength than hybrid composite because the fiber content in GFRP composite is found to be higher as well as the strength of glass fiber

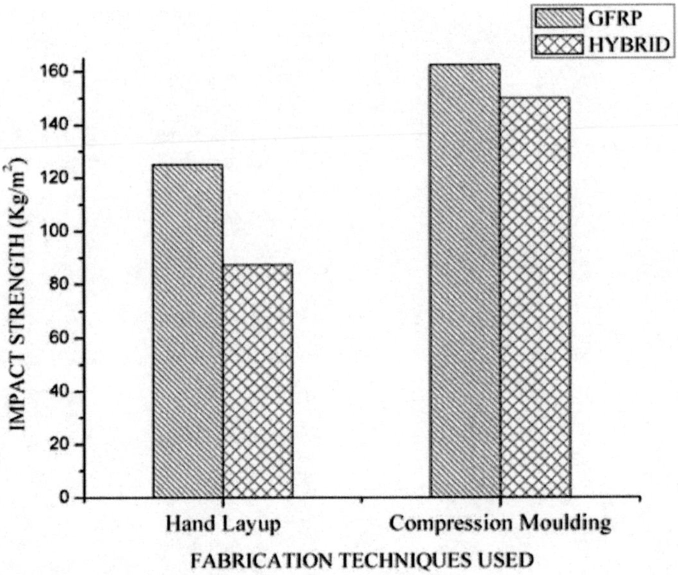

Fig. 6 Effect of fabrication technique and GFRP and hybrid polymer on the impact strength

strand is better when compared with a single coir fiber strand. Therefore, we can conclude that the compressed glass fiber composite possesses more impact strength when compared with other three samples.

Failures in composites are mainly due to voids, improper fiber bonding, poor adhesiveness between the fiber and matrix, etc. The SEM images of the four samples were taken at a magnification of 300× without any gold sputtering. The size of the image measured 50 microns. The surface topography of each sample was studied, and the conclusion obtained is as follows. In Fig. 7a and Fig. 7b, lots of voids are visible. Weak interface between the fiber and matrix due void leads to premature failure. These weak points lead to stress concentrations which cause de-bonding of fiber. Cracks propagate from voids to voids which eventually lead to matrix failure, fiber failure and ultimate failure of composites. Unlike the hand layup samples, the compression molding processed samples as shown in Fig. 7c, d have higher interfacial bonding between fire and matrix which will improve the ability of composite material to undergo loading. The matrix clings to individual strands of fibers which suggest that the fiber and matrix interfacial strength are exceptionally high. Thus, it can be concluded that compression molded composites have higher bond strength and can withstand higher loads without failing when compared to hand layup method.

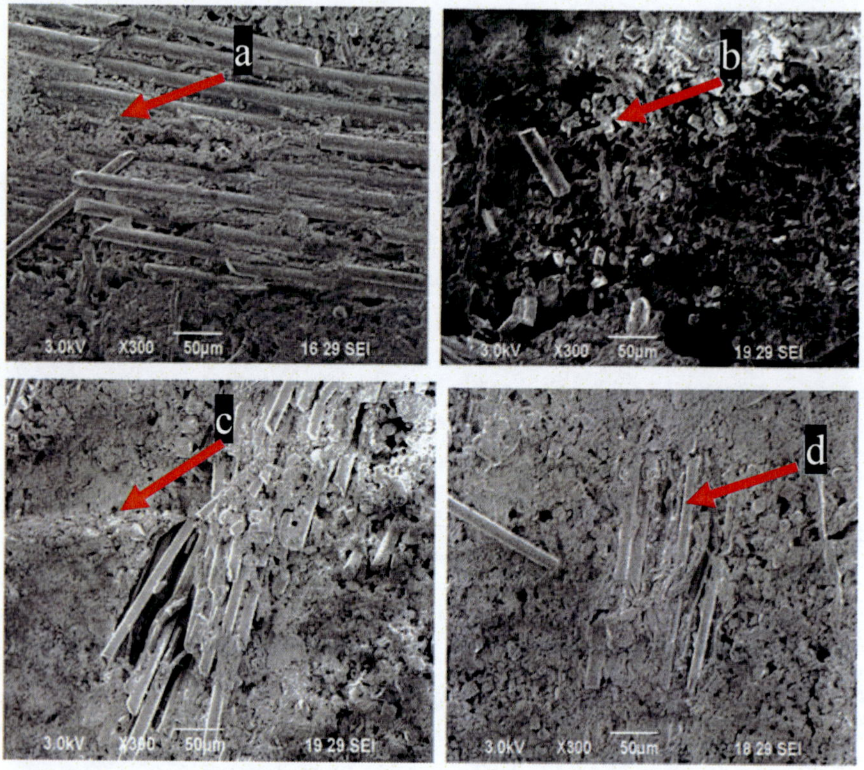

Fig. 7 SEM images of different samples. **a** SEM image of hand layup GFRP 1-Fiber breakage. **b** SEM image of hand layup hybrid. **c** SEM image of compression GFRP, 1-Fiber breakage. **d** SEM image of compressed hybrid 2-voids

4 Conclusion

In the present investigation, a glass fiber coir hybrid composite was fabricated using hand layup and compression molding method, and its mechanical properties were tested. It was observed that compression molding showed enhanced mechanical properties when compared to hand layup. The effect of the addition of coconut fibers has been studied, and the following conclusions have been drawn:

The tensile strength of hybrid is 44% of GFRP. Tensile strength of compression molded composite is slightly higher than hand layup. Hybrid is 87% as hard as GFRP, and when fabricated using compression molding, it showed slight increase in hardness. Compression molded hybrid was 92% as tough as the compression molded GFRP according to the impact test results.

Compression molding had also significantly raised the toughness of the composites. It was observed from the SEM images that the compression molded composites had better interfacial bonding between matrix and fibers.

The mechanical properties of hybrid are obviously lower than GFRP. But the addition of coconut fibers leads to cost and weight reduction of the regular glass fiber reinforced composites. Cost and strength of materials should always be maintained at an optimum level. In this case, hybrid materials are able to provide reasonable strength compared to purely synthetic materials at a lower cost, reduced weight and also efficient usage of natural eco-friendly resource.

References

1. Khan A, Ahmad MA, Joshi S (2015) A systematic study for electrical properties of chemically treated coir fiber reinforced epoxy composites with ANN model. Int J Sci Res 4:410–414
2. Deshpande S, Rangaswamy T, Engg M (2014) Effect of fillers on E-glass/jute fiber reinforced epoxy composites. Int J Eng Res Appl 4:118–123
3. Fazita MRN, Khalil HPSA, Izzati ANA, Rizal S (2019) 3 - Effects of strain rate on failure mechanisms and energy absorption in polymer composites. Elsevier Ltd. https://doi.org/10.1016/B978-0-08-102293-1.00003-6
4. Monteiro SN, Terrones LAH, Almeida JRMD (2008) Mechanical performance of coir fiber/polyester composites. Polym Test 27:591–595. https://doi.org/10.1016/j.polymertesting.2008.03.003
5. Balaji V, Senthil Vadivu K (2017) Mechanical characterization of coir fiber and cotton fiber reinforced unsaturated polyester composites for packaging applications. J Appl Packag Res 9(2), Article 2. https://scholarworks.rit.edu/japr/vol9/iss2/2
6. Santos F, Neves S, Sousa E, Pereira É, Júnior L (2017) Ballistic application of coir fiber reinforced epoxy composite in multilayered armor. Mater Res 20:23–28
7. Karthikeyan A, Balamurugan K (2012) Effect of alkali treatment and fiber length on impact behavior of coir fiber reinforced epoxy composites. JSIR 71(09):627–631. http://nopr.niscair.res.in/handle/123456789/14634
8. Kumar TV, Chandrasekaran M, Santhanam V (2017) Characteristics analysis of coconut shell husk. J Eng Appl Sci 12:2401–2406
9. Biswas S, Kindo S, Patnaik A (2011) Effect of fiber length on mechanical behavior of coir fiber reinforced epoxy composites. Fibers Polym 12:73–78. https://doi.org/10.1007/s12221-011-0073-9
10. Ma Y, Wu S, Zhuang J, Tong J, Qi H (2019) Tribological and physio-mechanical characterization of cow dung fibers reinforced friction composites: an effective utilization of cow dung waste. Tribiol Int 131:200–211. https://doi.org/10.1016/j.triboint.2018.10.026
11. Hasan M, Mir SS (2015) Manufacturing of coir fiber-reinforced polymer composites by hot, Chapter 15. https://doi.org/10.1007/978-3-319-07944-8
12. Cesar J, Luiz R, Machado L, Vieira G, Teixeira R, Freire S, Mano V, Hallak T (2018) Effects of sodium carbonate on the performance of epoxy and polyester coir-reinforced composites. Polym Test 67:533–544. https://doi.org/10.1016/j.polymertesting.2018.03.043
13. Nazeer A (2014) To Study the mechanical properties of coconut coir fiber reinforced with epoxy resin AW 106 & HV 953 IN. Int J Mod Eng Res 4:38–47
14. Péron M, Célino A, Castro M, Jacquemin F, Le Duigou A (2019) Study of hygroscopic stresses in asymmetric biocomposite laminates. Compos Sci Technol 169:7–15. https://doi.org/10.1016/j.compscitech.2018.10.027
15. Cunha G, Filho DO, Carlos R, Mota DS, Claudia A, Antunes M, Olimpio O, Filho DA (2019) Effects of hybridization on the mechanical properties of composites reinforced by piassava fibers tissue. Compos Part B 162:73–79. https://doi.org/10.1016/j.compositesb.2018.10.050
16. Obada DO, Kuburi LS, Dauda M, Umaru S, Dodoo-arhin D, Balogun MB, Iliyasu I, Iorpenda MJ (2018) Effect of variation in frequencies on the viscoelastic properties of coir and coconut

husk powder reinforced polymer composites. J King Saud Univ Eng Sci. https://doi.org/10.1016/j.jksues.2018.10.001
17. Habibi M, Laperrière L, Mahi H (2019) Effect of moisture absorption and temperature on quasi-static and fatigue behavior of nonwoven flax epoxy composite. Compos Part B 166:31–40. https://doi.org/10.1016/j.compositesb.2018.11.131
18. Romli FI, Nizam A, Shakrine A, Rafie M, Laila D, Abdul A (2012) Factorial study on the tensile strength of a coir fiber-reinforced epoxy composite. Procedia Soc Behav Sci 3:242–247. https://doi.org/10.1016/j.aasri.2012.11.040

Study of Hexagonal Boron Nitride Particulate as Vibration Behaviour Modifier of Alternate Stacked Glass–Natural Fibre Polymer Composite Laminate

K. Rajkumar and M. Selvaraj

Abstract The vibration behaviour of polymers can be modified by the particulates and natural fibres which are interesting study to the design engineers. A study has been carried out to investigate the tensile, flexural and vibrational behaviour of vinyl ester composites reinforcing with woven kenaf and glass fibres. Vibration behaviour of composite was further enhanced by the addition of hexagonal boron nitride (2&4% hBN) particles. The fibres were chemically treated to improve its adhesion to matrix. The tensile strength of fabricated composites was marginally reduced with increase in hBN particles. The vibration behaviours of composite were studied by electro-dynamic shaker. The experimental results were clearly showed that natural frequency of composite decreases with increasing particulate content. The hBN particles inclusion in the treated kenaf laminate has increased damping ratio by dissipating by more local vibration energy. This is due to combined effect of chemical treatment of fibre and local relaxing behaviour of secondary hBN particles. The two fold increase in damping ratio obtained by adding four times of hBN particle. Dynamic mechanical tests proved that hBN, a secondary particle, is improving vibration damping characteristics. Hence, it potentially benefits the composite as an engineered material.

Keywords Vibration energy · Electro-dynamic shaker · Natural frequency · Polymer composite

1 Introduction

In recent years, it is a known fact that the environment is completely messed up. As a result, the researchers are probing a solution to fix it using naturally acquired materials. There comes the natural fibre as an alternate for synthetic fibre. The studies, till

K. Rajkumar · M. Selvaraj (✉)
Department of Mechanical Engineering, Sri Sivasubramaniya Nadar College of Engineering, Kalavakkam, Chennai, Tamilnadu, India
e-mail: selvarajm71@gmail.com

S. Vijayan et al. (eds.), *Trends in Manufacturing and Engineering Management*, Lecture Notes in Mechanical Engineering,
https://doi.org/10.1007/978-981-15-4745-4_40

the date, have undoubtedly proved that the natural polymeric substances are substituted for synthetic polymers. In order to achieve the desired properties of the current applications, the evolution of composites with various natural fibres as reinforcement and fillers do meet the requirements. Various studies have reported a better result on the hybridization of natural fibres along with synthetic fibres reinforced with thermosetting polymers [1–3]. Woven fibres with particulates improve the properties of the composites. Glass fibres are hybridized with natural fibres like kenaf, sisal, jute, coir, banana, flax, date palm wood, basalt.

Researchers are attracted to the fibres acquired from nature to utilize in polymer composites due to their eco-friendly nature and sustainability [1]. The lignin, hemicellulose, cellulose-based composites are suitable for the electrical, mechanical, thermal and biodegradation applications [1, 3]. The elastic modulus of fibres and the interface properties of the matrix have a large effect on the stress and strain to failure initiation [2]. The thermal property of individual low conductivity kenaf fibres proves that a reduction in 16% of the effective thermal conductivity of the composite [3]. The addition of kenaf fibres reduces the degree of crystallinity of the recycled polypropylene/polyethylene sourced from rejected–unused diapers whose plastic and absorbent gel were separated [4]. The tensile strength and flexural strength of kenaf fibre were improved considerably when reinforced with vinyl ester, epoxy and polyurethane matrixes [5]. The double-layered kenaf fibres increase the thermal property and the tensile strength of the composite than single-layered kenaf fibre [6]. The thermal and dynamic mechanical properties of silane-treated pineapple leaf fibre and kenaf fibre have better than that of untreated fibres [7]. Kenaf fibres absorb sound and thus used for acoustic purpose. The sound absorption of kenaf fibres is similar to that of synthetic rock wool [8].

The dynamic flexural modulus has no significant effect on the radial configuration of the kenaf yarns, whilst the glass fibres have on damping factor. The bilateral configuration of glass yarns has considerable effects on the static flexural modulus, strength and the strain at break [9].

The addition of microsize calcium carbonate with kenaf fibres reinforced vinyl ester increases the tensile and flexural properties of the composite [10]. The flax/basalt hybrid reinforced with vinyl ester laminates enhances the durability and water repellence behaviour [11]. The inclusion of kenaf fibre with hexagonal boron nitride (hBN) and epoxy increases the thermal conductivity of the composite than with hBN and epoxy alone [12].

Hexagonal boron nitride also termed as white graphite is chemically stable and possesses good lubrication properties, resistant to oxides, high thermal conductivity and excellent electrical insulation [13]. Recently, the application of boron nitride nanomaterials (nanoparticles, nanotubes, nanosheets, etc.) in medical areas is increasing rapidly because of its unique properties and suitable to the other applications [14] also.

A study on the damping behaviour of composite is necessary since it reduces the vibration by dissipating energy. Various types of transducers have been used to measure the response of the vibrations of composite. An impact hammer attached with a force transducer is used as an exciter and a mini accelerometer to obtain the

Table 1 Physical and Mechanical properties of thermoset resin

Polymer	Vinyl ester
Density (gm/cm^3)	1.14
Tensile strength (MPa)	86
Elastic modulus (MPa)	3200
Elongation at break (%)	5–6

response of the simply supported specimen [15]. Flat coils were attached to the ends of a beam within the fields of permanent magnets. An alternating current through one coil was used as vibrators whilst the motion of the other coil as transducer. The composite specimen with high flexural modulus lowers specific damping capacity [16].

From the above review, it is concluded that the work presented in this paper has not been done anywhere in the past. This paper provides an overview of the characteristics of treated and untreated fabrics of kenaf and glass reinforced vinyl ester with boron nitride particle. The contribution of the boron nitride with the thermosetting polymer has a significant role on the mechanical and vibration properties of polymer laminates.

2 Materials and Methods

2.1 Materials

The E glass woven was chosen for its low cost and high performance in strength, stiffness, heat and chemical resistance. Kenaf being one of the lignocellulosic fibres was selected due to its low cost, low density, high specific strength and sustainability. Woven kenaf fibres were soaked in 6% (wt) of NaOH solution at 25 °C for 24 h maintaining a water solution ratio of 15:1. They were then washed well with distilled water and after neutralizing with dilute acetic acid. The kenaf fabrics were dried for 48 h. The boron nitride particle of size 5 μm is used due to its lamellar structure and high thermal conductivity. The vinyl ester resin of grade HPR 8711 with catalyst (1% MEKP), promoter (Benzyl peroxide) and accelerator (0.5% Cobalt Napthanate) to form a matrix was used. Table 1 refers the physical and mechanical properties of the thermoset polymer used.

2.2 Composite Fabrication

The hybrid composite laminates were fabricated at room temperature by hand layup technique. The treated and untreated fabrics of kenaf along with glass fibres of size 30×30 cm^2 were placed in the thermoset resin mixed with hexagonal boron nitride

Fig. 1 Specimen used for tensile test

of various proportions. It was then cured with the promoter, catalyst and accelerator. Five sets of hybrid composites were prepared by placing the fabrics of treated kenaf and glass alternately for 0, 2 and 4% of hBN. Similarly, the untreated kenaf fibres were arranged alternately with glass fibres for 2% and 4% hBN mixed with resin, respectively.

2.3 *Tensile Test*

The five samples prepared for each formulation were cut according to ASTM D-3039 (1995) with a size of 250*25*4 mm to carry out the tensile test in an universal testing machine (Instron 4206) at a feed rate of 0.1 mm/min. Figure 1 shows the test samples. During the tension test, the grips were moved apart at a constant rate to stretch the specimen. The results were averaged for the five specimen of each formulation for calculation purpose. Scanning electron microscope (SEM) was used to evaluate the fracture surface of the polymer matrix and to illustrate the fibrillose morphology of the kenaf fibres.

2.4 *Flexural Test*

The prepared flexural samples of each formulation with dimensions of 64*16*4 mm according to ASTM D790 were tested in universal testing machine. The bending strength of the composite was measured by a three point bending test.

2.5 Vibration Test

A multirecorder frequency processing unit was used to determine the vibration of the hybrid composite laminates. Vibration induced on the specimen by electro-dynamic shaker. The phase analyser (FFT analyser) has four input channels and was used as a data acquisition system. The accelerometer measures the acceleration of the vibrations. The impact hammer was used for impacts of varying amplitude. The real-time frequency analysis software, DEWESOFT, was used for the data analysis and verification. The one end of the specimen was hold by gripper, and another end was free, as cantilever mode. The specimens were excited at the free end by impact hammer, and the corresponding responses were recorded. This vibration test determines the natural frequency and damping ratio.

3 Results and Discussions

3.1 Tensile Strength

Figure 2 shows a comparative analysis of tensile strength for the composite laminates fabricated with treated and untreated kenaf/glass fabrics reinforced with 0, 2 and 4% of hBN. The fabricated specimen with treated kenaf fabric hybridized with glass fibres for the addition of 2% hexagonal boron nitride (hBN) mixed with vinyl ester resin has the highest tensile strength. The tensile strength gradually increases for the samples fabricated with the treated kenaf fibres hybridized with glass fibres matrix with 0–2% addition of hBN and decreases for 4% of hBN.

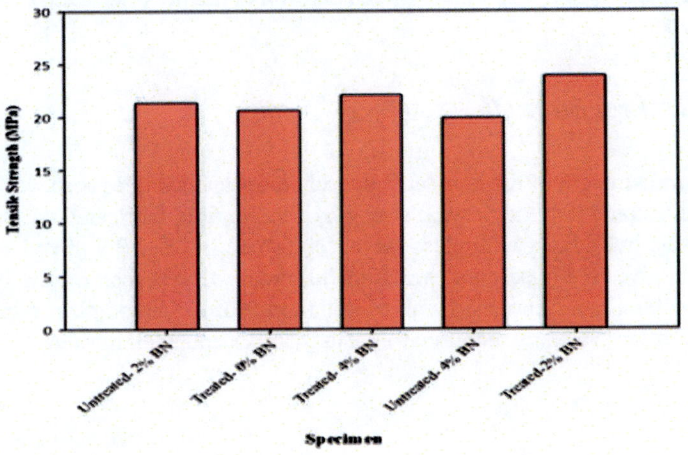

Fig. 2 Tensile strength of composites

Fig. 3 Flexural strength of the composites **a** Untreated **b** treated

The tensile strength obtained for the untreated kenaf fibre with the addition of 2% hBN gives the maximum value, whilst the addition of 4% hBN gives the minimum value. The notable improvement in tensile strength of the treated kenaf fibres hybridized with glass fibres can be due to strong interface adhesion. The proper addition of hBN, that is, 2% resulted a very good adhesion property for the fibres with the matrix than 4%. The fall of tensile strength value for 4% addition of hBN does not attribute high bond between fibre and matrix. Figure 3 shows fracture surface of untreated and treated kenaf fibre hybrid composites.

The untreated fibre composite shows an inefficient transfer of resin along the fibre direction leading to poor intact between fibre-matrix interfaces. The used NaOH treatment has removed impurities from the external surface of kenaf fibres, so their surface appears clean, and it gives good adhesion to the polymer matrix.

3.2 *Flexural Strength*

The flexural strength of the laminated hybrid composite samples is shown in Fig. 4. It is clearly noted that the flexural strength of the sample fabricated with untreated kenaf fabric hybridized with glass fabric reinforced in 2% hBN mixed with vinyl ester resin gives the maximum, whilst the minimal value is seen for 4% hBN. The composite laminates fabricated with treated fabrics of kenaf and glass reinforced in thermoset resin gradually increase from 0 to 2% hBN and a fall in value for 4% hBN.

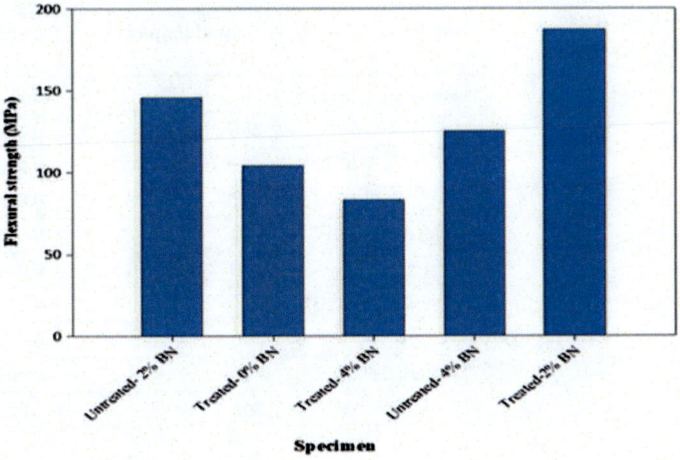

Fig. 4 Flexural strength of the composites

3.3 *Vibrational Behaviour*

The graphical output of vibrational behaviour polymer composite is shown in Fig. 5.

It is clearly seen from Fig. 5 that natural frequency of laminates decreases with increasing particulate content. Chemical treatment of kenaf fibre increases the natural frequency of laminates. The natural frequency of the laminate without hBN was higher due to continuous phase of matrix.

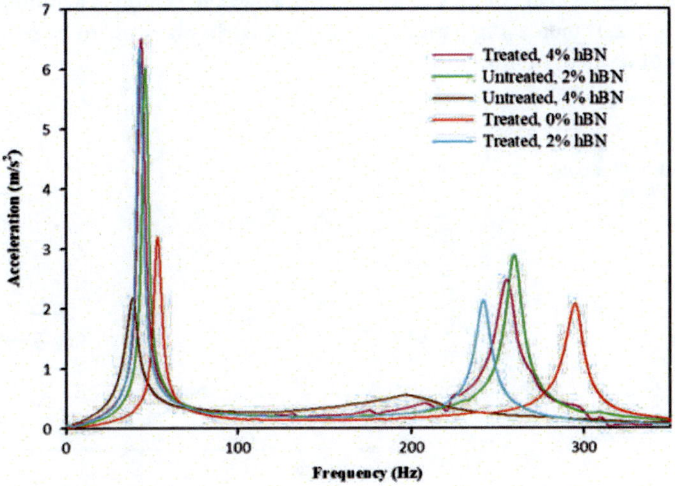

Fig. 5 Vibrational behaviour of hybrid composites

Table 2 Natural frequency and damping ratio

Samples	Natural frequency (Hz)	Damping ratio
Untreated, 2% hBN	50	0.033
	266	0.008
Treated, 0% hBN	46	0.043
	262	0.022
Treated, 4% hBN	34	0.076
	190	0.045
Untreated, 4% hBN	53	0.031
	296	0.013
Treated, 2% hBN	45	0.041
	245	0.025

The two modal frequencies of each sample are enlisted in Table 2. The damping ratio was obtained using the DEWESOFT software. In higher modal frequency, treated fibre laminate shows the higher damping than that of untreated kenaf fibre laminate. The strong adhesion of matrix and fibre mostly relaxes the vibration energy. The hBN particles inclusion in the treated kenaf laminate increases the damping ratio by dissipating by more vibration energy, as shown in Fig. 6. This is a synergizing effect by the strong adhesion between fibre-matrix and local relaxing behaviour of secondary hBN particles. The weak interface was observed from the untreated fibre reinforced composite, despite being the presence of hBN particle, leading to lower damping ratio. This is due that lower dissipation of vibration energy by the kenaf fibres.

The damping ratio of composite without hBN particle exhibited a minimum value. The four times of increasing particle content in matrix observed to be increased the two times of damping ratio.

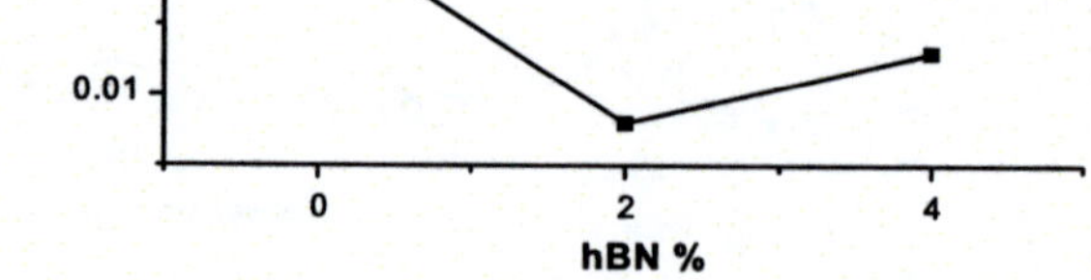

Fig. 6 Damping ratio of composite at high modal frequency

4 Conclusions

Vinyl ester hybrid composites were fabricated with kenaf fibre mat, glass fibre mat and different wt. fraction of hBN particles. The composite with treated fibre-2% hBN showed a higher tensile strength, but 4% hBN addition shows a lower value. Fracture surface of untreated fibre composite shows an incomplete resin transfer to fibre leading poor intact between fibre-matrix interfaces. The NaOH treatment has removed impurities from the surface of kenaf fibres paved the way for good adhesion property. It is clearly showed that the flexural strength of the treated kenaf fabric hybridized with glass-2% hBN exhibited a maximum, whilst the minimal value observed for 4% hBN content composite.

It is clearly seen that natural frequency of laminates decreases with increasing particulate content. Fibre treatment increases the natural frequency of laminates due to undisturbed continuous phase of matrix. The hBN particles inclusion in the treated kenaf laminate increases the damping ratio by dissipating by more vibration energy. This is a synergizing effect of strong adhesion between fibre-matrix and local relaxing behaviour of secondary hBN particles. The damping ratio of composite without hBN particle exhibited a minimum value. The four times increasing hBN particle content in the matrix increases the two times of damping ratio.

References

1. Mohammed L, Ansari MNM, Pua G, Jawaid M, Saiful Islam V (2015) A review on natural fiber reinforced polymer composite and its applications. Int J Polym Sci. Article ID 243947, 15 pages http://dx.doi.org/10.1155/2015/243947
2. Asp LE, Berglund LA, Talreja R (1996) Effects of fiber and interphase on matrix Initiated Transverse failure in polymer composites. Compos Sci Technol 56:657–665
3. Guillou J, Lavadiya DN, Munro T, Fronk T, Ban H (2018) From lignocelluloses to biocomposite: multi-level modeling and experimental investigation of the thermal properties of kenaf fiber reinforced composites based on constituent materials. Appl Thermal Eng 128:1372–1381
4. Taufiq MJ, Mansor MR, Mustafa Z (2018) Characterization of wood plastic composite manufactured from kenaf fibre reinforced recycled-unused plastic blend. Compos Struct 189:510–515
5. Sharba MJ, Salman SD, Leman Z, Sultan MTH, Ishakand Mohamad MR, AzmahHanim A (2016) Effects of processing method, moisture content and resin system on physical and mechanical properties of woven kenaf plant fiber composites. BioResources 11(1):1466–1476
6. Islam MR, Beg MDH (2016) Effects of reinforcing arrangement of kenaf fibers into unsaturated polyester for improved properties. J Mech Eng Sci 10(2):2020–2030
7. Asima M, Paridah MT, Saba N, Jawaid M, Alothman OY, Nasir M, Almutrai Z (2018) Thermal, physical properties and flammability of silane treated kenaf/pineapple leaf fibers phenolic hybrid composites. Compos Struct 202:1330–13338
8. Lim ZY, Putra A, Nor MJM, Yaakob MY (2018) Sound absorption performance of natural kenaf fibers. Appl Acoustics 130:107–114
9. Hashemi F, Brancheriau L, Thahir PM (2018) Hybridization and yarns configuration effects on flexural dynamic and static properties of pultruded hybrid kenaf/glass fibre composites. Compos Part A 112:415–422

10. Fairuz AM, Sapuan SM, Zainidin ES, Jaafar CNA (2016) Effect of filler loading on mechanical properties of pultruded kenaf fiber reinforced vinyl ester composites. J Mech Eng Sci 10(1):1931–1942
11. Ramraji K, Rajkumar K, Sabarinathan P (2019) Tailoring of tensile and dynamic thermo-mechanical properties of interleaved chemical-treated fine almond shell particulate flax fiber stacked vinyl ester polymeric composites. In: Proceedings of the Institution of Mechanical Engineers, Part L: Journal of Materials: Design and Applications, p 1464420719849616
12. Xia C, Garcia AC, Shi SQ, Warner N, Wu Y, Cai L, Rizvi HR, D'Souza NA, Nie X (2016) Hybrid boron nitride—natural fiber composites for enhanced thermal conductivity. Sci Rep 6:34726. https://doi.org/10.1038/srep34726
13. Kivnc M, Barutca B, Koparal AT, Goncu Y, Bostanci SH, Ay N (2018) Effects of hexagonal boron nitride nanoparticles on antimicrobial and antibiofilm activities, cell viability. Mater Sci Eng C. https://doi.org/10.1016/j.msec.2018.05.028
14. Gnanavelbabu A, Rajkumar K, Saravanan P (2018) Investigation on the cutting quality characteristics of abrasive water jet machining of AA6061-B4C-hBN hybrid metal matrix composites. Mater Manuf Process 33(12):1313–1323
15. Rao MD, Crocker MJ (2009) Vibrations of bonded beams with a single lap adhesive joint. J Sound Vibration 92(2):299–309
16. Kadioglu F (2009) Measurement of dynamic properties of composites in vibration by means of a non-contact mechanism. J Reinforced Plast Compos 28, No. 12/2009 https://doi.org/10.1177/0731684408089514

Study of Statistical Distribution and Morphology of Particles in a Polymer Matrix by Foldscope Imaging Technique

P. Kaythry, A. Madhan, and K. Rajkumar

Abstract Design and material engineers are focusing on strengthening and toughening of the polymer matrix by introducing secondary particles. However, particle distribution is a control parameter to ensure the properties of polymer materials. Further, the agglomeration of the fine particle affects polymer composite performance. The chosen process may affect the statistical distribution and aggregation of secondary particles in the polymer composite. There is a lack of simple and affordable technique to evaluate the particle distribution, morphology and degree of agglomeration of particles in the polymer matrix. In this paper, the origami crafted microscope called foldscope is used to find out the particle distribution in the polymer matrix. Initially, the fillers/particles are mixed in the epoxy matrix through an ultra-sonication technique. The experiments are conducted by varying the size of filler (30, 80 and 120 grit) with sonication parameters such as power rate (70, 80 and 100%) and processing time T_{on} (45, 90 and 135 s) with constant 3% filler in weight. With the above parameters, Taguchi L17 array was constructed, and the process parameters were optimized to the high degree of uniform distribution. From the particulate polymer composite, the images captured by foldscope were further analyzed by ImageJ technique for determination of the statistical distribution of particles. The processing parameters of 80% power, T_{on} 135 s for 120 grit size exhibited excellent particle distribution and higher tensile strength.

Keywords Polymer matrix · Foldscope · Ultra-sonication · Particle distribution

---

P. Kaythry (✉) · A. Madhan
Department of Electronics and Communication Engineering, Sri Sivasubramaniya Nadar College of Engineering, Chennai, India
e-mail: kaythryp@ssn.edu.in

K. Rajkumar
Department of Mechanical Engineering, Sri Sivasubramaniya Nadar College of Engineering, Chennai, India

S. Vijayan et al. (eds.), *Trends in Manufacturing and Engineering Management*, Lecture Notes in Mechanical Engineering,
https://doi.org/10.1007/978-981-15-4745-4_41

1 Introduction

The modern applications like aerospace, automobiles, renovation of building and bridges [1] are using polymer composite materials. The dispersion and interfacial adhesion between the matrix and filler are required to achieve the desired mechanical properties [2, 3]. The structural feature of polymer composite dictates by tensile and flexural strength.

Visual inspection is the first level evaluation to select any polymer composite in the applications [4]. It is used to identify the particle distribution, agglomeration, porosity and discontinuity by conventional microscopes like compound microscope, optical microscope and scanning electron microscope (SEM) [5–7]. These techniques require different sample preparation methods, which are more expensive and tedious. To overcome difficulties, foldscope is used for the visual study of the polymer matrix. It also provides flexibility over the other conventional microscopes. For using a foldscope, the images are captured by attaching a mobile phone camera over it [8]. It produces quality (3024 × 4032 pixels) microscopic images at any instance. Also, it does not require any special working conditions [9]. Hence, foldscope is a highly effective method for microscopic level image capturing. However, an image processing is further required to analyze the captured images [10, 11].

Specific to the composite, the dispersion of particles depends on the processing methods [3]. Generally, the particle distribution in a polymer matrix is done by centrifuging, vibratory table, magnetic particulate distribution and ultra-sonication method. Microscopic image defines the relative amount of mass, number of a particle according to size and mean distance between the particles in a spatial matrix [12]. In this work, an ultrasonic technique was used to disperse almond particles in the polymer matrix and correlated spatial distribution of particles to the tensile strength.

2 Materials and Methods

2.1 *Basic Components and Preparation*

In this work, dry almond shells were reinforced to obtain the polymer composite. Initially, the impurities from almond shells were removed. Then, it is crushed by using Willy's mechanical mill. The crushed particles were sieved to 30, 80 and 120 grit size by laboratory shaker and used as a filler material for the polymer matrix. Araldite HY550 epoxy resin and hardener HY951 were used as polymer matrix to fabricate the polymer composite. A constant filler weight of 3% is used to fabricate the composite laminate.

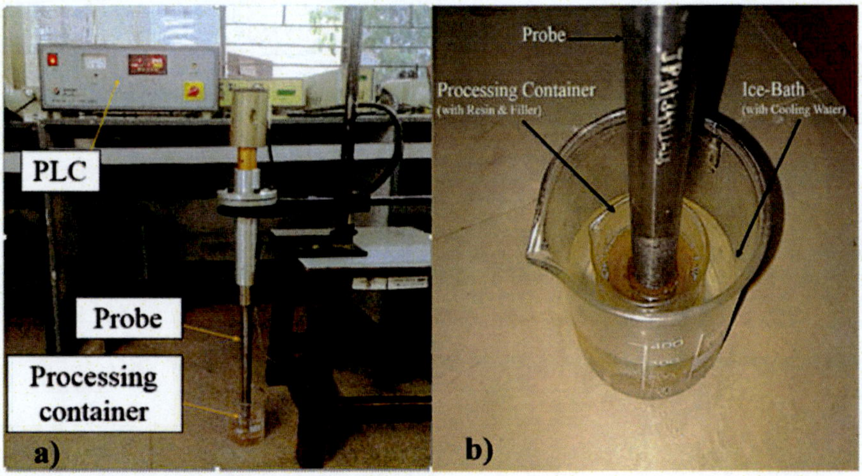

Fig. 1 **a** Sonication setup and **b** ice-bath setup

2.2 *Ultra-Sonication of Particle Dispersion*

Ultra-sonication is a technique used for dispersing particle in a solution. When the probe inserted into the solution, ultrasonic waves are formed [13]. It creates a large number of micro-bubbles, followed by high and low pressure in the solution. The difference in pressure leads to sudden and violent collapse of micro-bubbles. Under this circumstance, even dispersion of particles would be occurred [14].

Sonication and ice-bath setup are shown in Fig. 1a, b. It consists of a probe, processing container and programmable logic controller (PLC). The ultrasonic processor operates at 20 kHz frequency with the maximum power of 2000 W. The power is varied from 60, 70, 80 and 100% of maximum power. The variable operating time (T_{on}) and constant interval time ($T_{off} = 3$ s) were controlled by PLC to study the particle dispersion in the polymer composites. The sonication process increases the temperature of the resin, which changes viscosity. So, the processing container is surrounded by the ice-bath to minimize the temperature variation.

2.3 *Fabrication*

The hand layup method was used for the fabrication of polymer composite because of its low cost, maximum effectiveness and simplicity. The particulate laminate has dimensions of 300 mm × 300 mm × 4 mm, which is divided into three segments along its length. Each segment has a size of 300 mm × 100 mm × 4 mm. The almond shell powder of different grits was mixed with resin by ultra-sonication process. For the particle dispersion study, Taguchi experimental design was constructed, and the

Table 1 Selected parameters range and levels used for Taguchi design

Particle size in grit	Power % in 2000 W	T_{on} (s)
30	70	45
80	80	90
120	100	135

Table 2 Formation of L17 array–Box–Behnken method of experimentation

Experiment number	Particle size (Grit)	T_{on} (s)	Power (% in 2000 W)
1	80	135	70
2	120	135	80
3	80	90	80
4	120	90	70
5	80	45	70
6	80	45	100
7	30	90	100
8	80	90	80
9	80	90	80
10	30	135	80
11	80	90	80
12	120	90	100
13	30	45	80
14	80	90	80
15	80	135	100
16	30	90	70
17	120	45	80

range and level of the selected input parameters are given in Table 1. Table 2 shows the formation of L17 array by using a Box–Behnken method of experimentation. The fabricated almond particle dispersed in polymer composites is shown in Fig. 2.

2.4 *Tensile Test*

The tensile test was carried out to evaluate the tensile strength by universal testing machine (Instron 4206) at a feed rate of 0.2 mm/min. A standard dimension of 250 mm × 25 mm × 4 mm was prepared according to American Society for Testing and Materials (ASTM) D638 standard. Abrasive water jet machine (AWJM) was used to make the standard specimen. The prepared specimens and tested tensile samples are shown in Figs. 3 and 4, respectively.

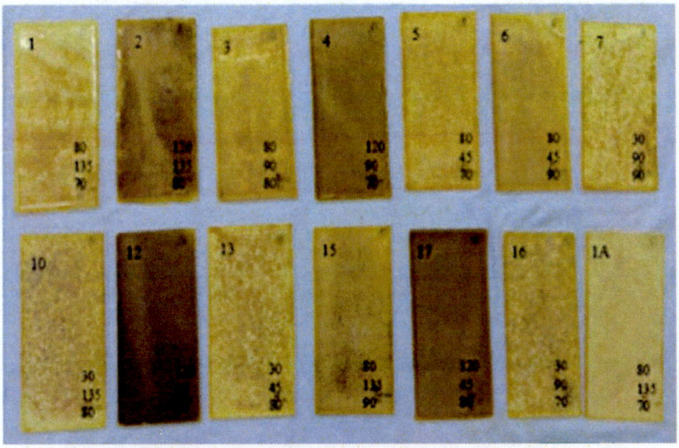

Fig. 2 Almond particle dispersed in polymer composite

Fig. 3 Prepared specimen for tensile test

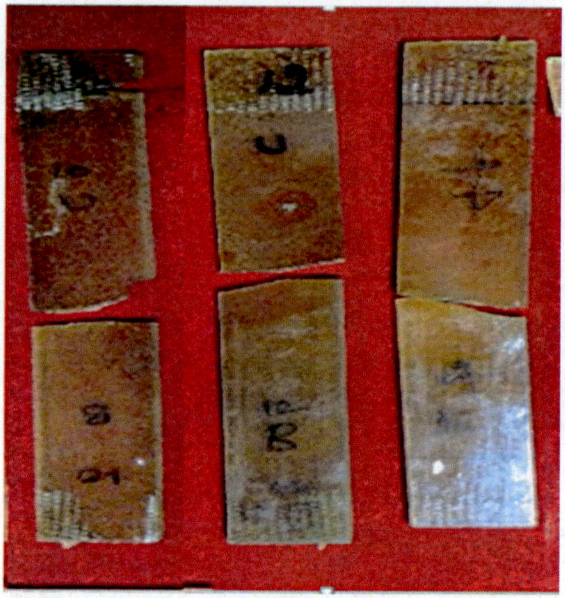

Fig. 4 Tensile tested samples of Foldscope

A foldscope is an optical microscope designed by Manu Prakash. It costs less than one US dollar. The foldscope weighs 8 grams and provides magnification from 140 $\times$ to 2000 $\times$. The foldscope can be attached to a smartphone with the help of magnets, permitting the user to take magnified pictures. It is compact and light, especially when compared with other microscopes. Due to its portability and maneuverability, the foldscope displayed greater ease of access.

The fundamental components like punched sheets, magnetic couplers and spherical glass lens, along with a battery that powers the LED as a light source are used to assemble a foldscope. The lens forces light waves to bend as they pass through an object. As the light waves bend, the object is magnified based on the principles of ray optics. The magnification usually depends on how curved the lens, the distance between the lens, object and mobile camera. The picture of an assembled foldscope is shown in Fig. 5. The linear movement along X- and Y-axis used to produce images over various point and fully constrained X–Y panning over a 20,620 mm^2 region. Z-axis movement is used to provide controlled space between optics stage relative to the sample-mounting stage by flexure-based focusing. This movement helps to capture clarity images and attached scale used to measure distance travelled across the sample slide with 0.5 mm resolution.

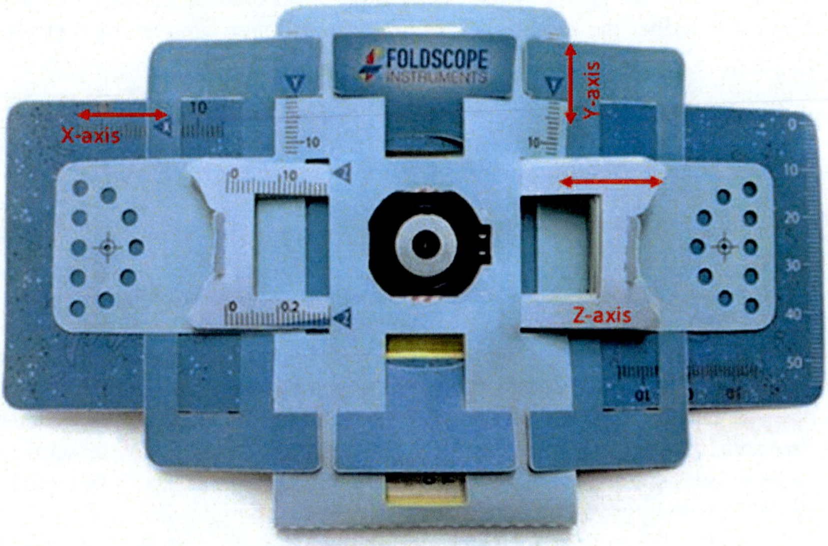

Fig. 5 An assembled foldscope [8]

2.5 Foldscope Image Analysis

Foldscope images of fabricated composites are captured and visually analyzed. ImageJ software is an image processing tool used to evaluate the degree of agglomeration, particulate counts, particle distribution, mean, the standard deviation of particles, histogram and an area of filler distribution [14]. Figure 6 shows the processed

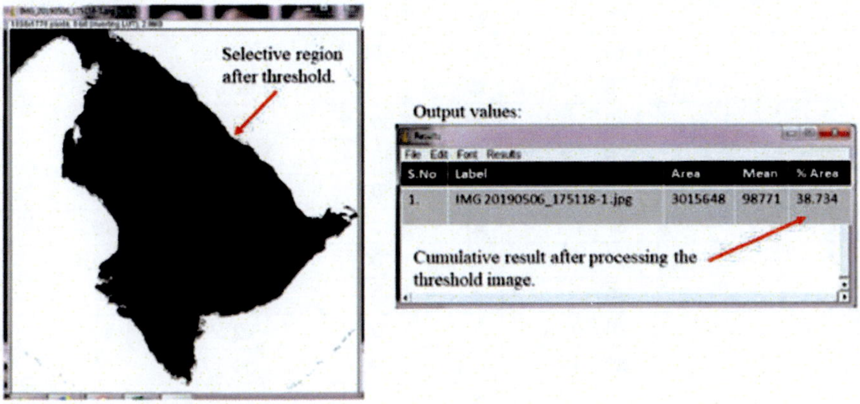

Fig. 6 Processed image

image of polymer composite which is an experiment number 10, as presented in Table 2. It describes the percentage of filler occupied area from the total available area.

3 Results and Discussion

3.1 Foldscope Visual Images

Table 3 represents the processing parameters and results for particle distribution and tensile strength. As the grit size increases, the mean particle size decreases. From the experiment number 2, 4 and 12, it is observed that smaller particle size shows a higher particle distribution with very less mean free path between the particles. Since the surface area to the volume of the particle is higher for smaller particles, the number of particles present in the fixed area is high. On the other hand, the experiment numbers 7, 10 and 13 with larger particle size reveal particle distribution with low surface area to the volume of the particle. As results, the number of particles in the fixed area was low. Thereby mean free path between the particles is also high. The foldscope images

Table 3 Image processing results of the tensile strength of composites

Experiment No.	Figure	Particle size (Grit)	T_{on} (s)	Power (% in 2000 W)	Particle distribution (% area)	Strength (MPa)
1	(1,2)	80	135	70	45.514	25.4
2	(1,1)	120	135	80	48.72	42.2
3	–	80	90	80	42.691	26.185
4	(2,1)	120	90	70	36.764	25.4
5	(2,2)	80	45	70	42.784	15.56
6	(3,2)	80	45	100	52.086	32.5
7	(2,3)	30	90	100	50.835	22.7
8	–	80	90	80	42.691	26
9	–	80	90	80	42.691	26
10	(1,3)	30	135	80	38.734	19.7
11	–	80	90	80	42.691	26
12	(3,1)	120	90	100	62.071	30.4
13	(3,3)	30	45	80	71.128	24
14	–	80	90	80	42.691	26
15	–	80	135	100	59.543	14.8
16	–	30	90	70	42.691	23.32
17	–	120	45	80	33.689	29

of polymer composites are given by a column–row matrix in Fig. 7. The grit size of 120, 80 and 30 is mentioned for each column. It also evidences that the strength of the composites partly depends on the degree of the particle distribution.

The statistical analysis of the experimental results is performed using Design Expert statistical software. A separate analysis of variance (ANOVA) is carried out for the particle distribution and tensile strength. The optimum level of variables for the uniform distribution is obtained using the Design Expert program. For a significant model, the P value should be < 0.05. From ANOVA for particle distribution in Table 4, the Model P value of 0.0098 implies the model is significant. There is an only chance of 0.98% error in Model F value occurs due to noise in experimentation. Power (*B*), Particle size × T_{on} (AC) were significant model terms.

Increasing the power of the sonication process increases the intensity of the sonic waves, which is shown in Fig. 8. There is no effect on the particle distribution when

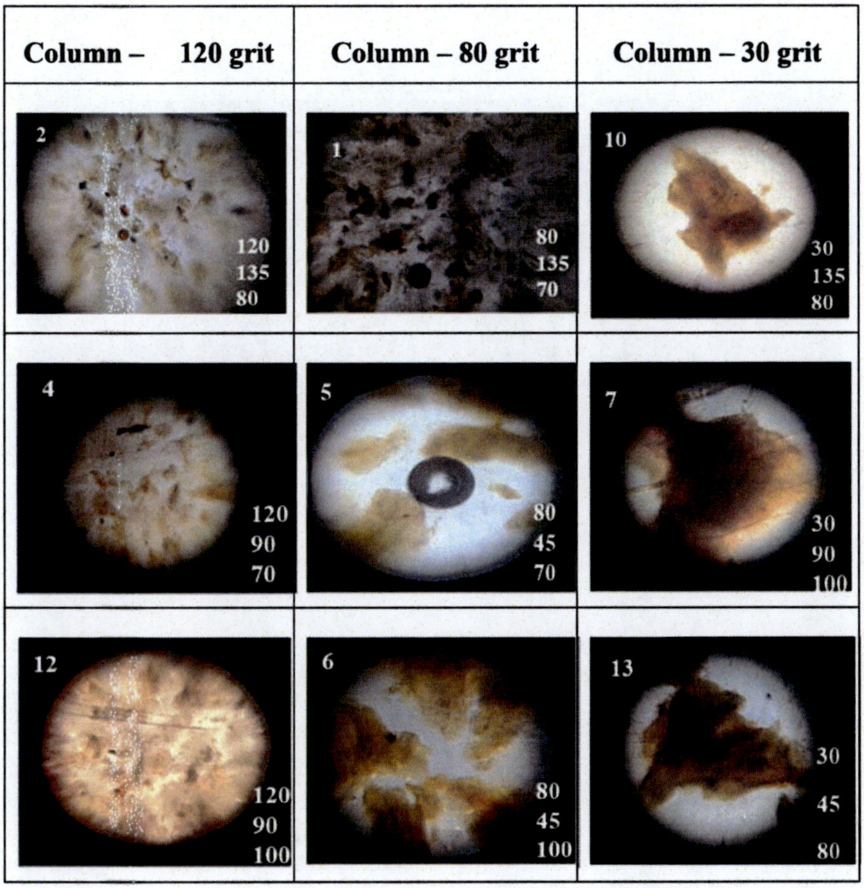

Fig. 7 Foldscope image of polymer composites with different process parameters

Table 4 Analysis of variance for particle distribution

Source	Sum of squares	df	Mean square	F value	P value Prob > F	
Model	1145.78	6	190.96	5.41	0.0098	Significant
A-Particle size	66.77	1	66.77	1.89	0.199	
B-Power	375.12	1	375.12	10.63	0.0086	
C-T_{on}	19.67	1	19.67	0.56	0.4725	
AB	68.47	1	68.47	1.94	0.1938	
AC	595.49	1	595.49	16.87	0.0021	
BC	5.59	1	5.59	0.16	0.6991	
Residual	352.91	10	35.29			
Pure error	1.45	4	0.36			
Cor total	1498.69	16				

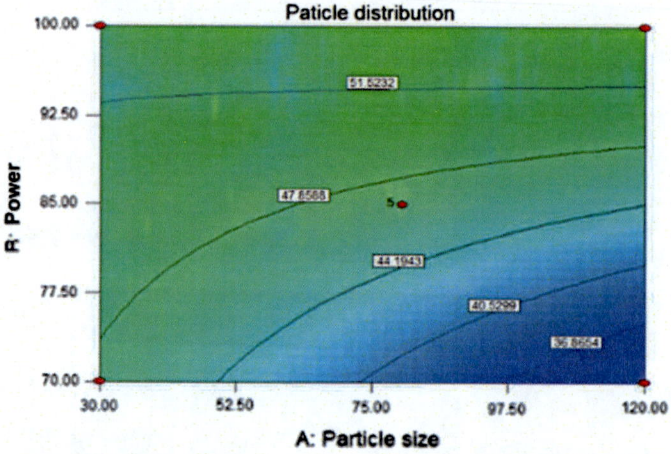

Fig. 8 Contour plot showing the effect of particle size and power on particle distribution

the threshold level of power is above 93%. Moreover, 120 grit particle size requires less energy to disperse the particle with high distribution percentage due to the mass–acceleration relationship. If the power is below 93%, then the maximum distribution of particles occurs for all range of particle size. For 30 grit particle size, it requires a higher power level of 85% for the dispersion of particulates to cover a maximum of 47.85% distribution area.

The 120 grit particle size with medium power (85%) gives the higher tensile strength. The higher particle size with all range of power reduces the strength of the composites due to mass–acceleration relationship as shown in Fig. 9.

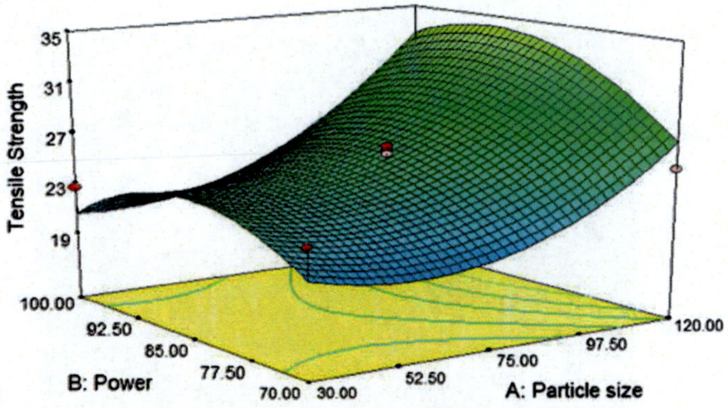

Fig. 9 Correlation of tensile strength of particle size and power

The ANOVA for tensile strength is given in Table 5. In this Model, P value of 0.0224 implies the model is significant. So, there is only a 2.24% chance that a model deviates, which could be due to the noise in the experimentation. Particle size (*A*), Particle size × T_{on} (AC), Power × T_{on} (BC) are significant model terms.

Figure 10 shows the relationship between tensile strength, particle size and power. The particle size affects the mechanical strength of the polymer matrix composites. In a smaller particle (120 grit), the lower mean free path of particles increases the cross-linking of polymer chain. Hence it results in higher tensile stress of particulates

Table 5 Analysis of variance for tensile strength

Source	Sum of squares	df	Mean square	*F* value	*P* value Prob > *F*	
Model	543.6581	9	60.40645	5.027942	0.0224	Significant
A-Particle size	173.7248	1	173.7248	14.46002	0.0067	
B-Power	12.6418	1	12.6418	1.052243	0.3391	
C-T_{on}	0.077477	1	0.077477	0.006449	0.9382	
AB	8.154704	1	8.154704	0.678758	0.4372	
AC	68.21096	1	68.21096	5.677552	0.0487	
BC	189.6129	1	189.6129	15.78246	0.0054	
A^2	55.42087	1	55.42087	4.612966	0.0688	
B^2	52.70503	1	52.70503	4.386913	0.0745	
C^2	0.302417	1	0.302417	0.025172	0.8784	
Residual	84.09905	7	12.01415			
Pure error	1.16392	4	0.29098			
Cor total	627.7571	16				

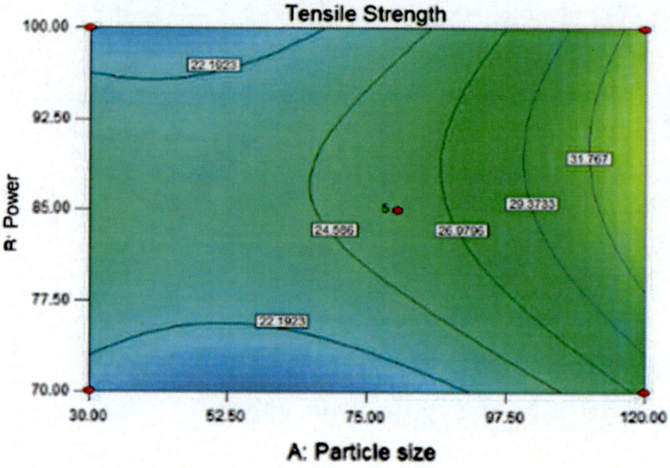

Fig. 10 Tensile Strength

in polymer composite. The degree of cross-linking is low for large particle size (30 grit), which leads to lower tensile strength.

Figure 11 shows the relationship between tensile strength, power and T_{on}. The tensile strength of composites increases with the T_{on} due to the higher degree of particulate disperse with all the range of particle size. It is directly derived that power of the sonication process increases the tensile strength of composite. This is because of higher degree particle distribution with all the range of particle size.

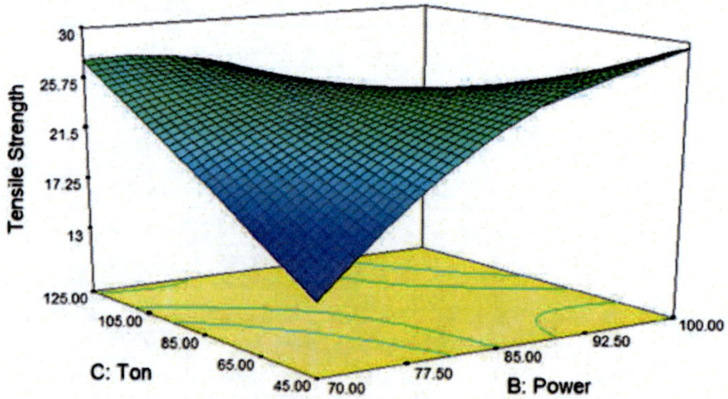

Fig. 11 Correlation of tensile strength of T_{on} and power

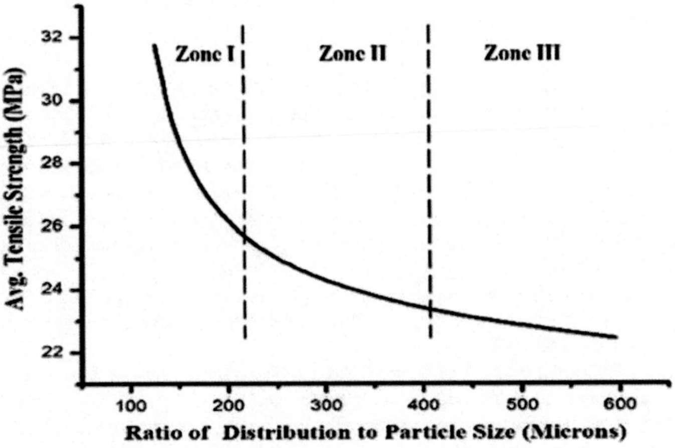

Fig. 12 Average tensile strength to ratio of distribution and particle size

3.2 *Tensile Strength-Particle Factional Image Area*

Figure 12 depicts the average tensile strength to the ratio of distribution and particle size. In zone-I, when the ratio of distributions to particle size is low, there is a steep rise in tensile strength due to the higher order of cross-linking. Zone-II is called as a transition zone, where the tensile strength begins to decrease. In zone-III, there is no improvement in the tensile strength due to the saturation in cross-linking.

4 Conclusions

Distribution of particles is analyzed by visual study through foldscope images for the various polymer composites with varying grit sizes and other process parameters. The foldscope captured images of polymer composites are evaluated using ImageJ software. Analysis of variance identified the desired parameters for the particle distribution and tensile strength of the polymer composites. From the results, the smaller particulates with a higher distribution increase tensile property even at lower T_{on} and power. Foldscope imaging is simple and least cost technique for characterization and visual inspection of polymer-based products.

Acknowledgements Authors are thankful to the Department of Biotechnology, Govt. of India—Foldscope Research Grant (BT/IN/Indo-US/Foldscope/39/2015) for supporting this work.

References

1. Friedrich K, Abdulhakim A, Almajid (2013) Manufacturing aspects of advanced polymer composites for automotive applications. Appl Compos Mater 20(2):107–128
2. Sabarinathan P, Rajkumar K, Gnanavelbabu A (2016) Mechanical properties of almond shell-sugarcane leaves hybrid epoxy polymer composite. Appl Mech Mater 852:43–48
3. Fu S-Y (2008) Effects of particle size, particle/matrix interface adhesion and particle loading on mechanical properties of particulate–polymer composites. Compos Part B: Eng 39(6):933–961
4. Ramakrishna S (2001) Biomedical applications of polymer-composite materials: a review. Compos Sci Technol 61(9):1189–1224
5. Saunders RA, Lekakou C, Bader MG (1999) Compression in the processing of polymer composites 1. A mechanical and microstructural study for different glass fabrics and resins. Compos Sci Technol 59(7):983–993
6. Ramraji K, Rajkumar K, Sabarinathan P (2019) Tailoring of tensile and dynamic thermo-mechanical properties of interleaved chemical-treated fine almond shell particulate flax fiber stacked vinyl ester polymeric composites. In: Proceedings of the Institution of Mechanical Engineers, Part L: Journal of Materials: Design and Applications, p 1464420719849616
7. Qaiss A, Bouhfid R, Essabir H (2015) Characterization and use of coir, almond, apricot, argan, shells and wood as reinforcement in the polymeric matrix in order to valorize these products. In: Agricultural biomass-based potential materials. Springer, Cham, pp 305–339
8. Cybulski JS, Clements J, Prakash M. Foldscope: origami-based paper microscope. PLoS ONE 9(6):98781. https://doi.org/10.1371/journal.pone.0098781
9. Kaythry P, Sangeetha P, Madhan A (2019) Study on shape and texture of different types of fine aggregate used in concrete using Foldscope In: Proceedings of national conference on 'MICROCOSMOS 2019', Port Blair, Andaman & Nicobar Islands, p 11
10. Igathinathane C (2008) Shape identification and particles size distribution from basic shape parameters using ImageJ. Comput Electron Agric 63(2):168–182
11. Papadopulos F (2007) Common tasks in microscopic and ultrastructural image analysis using ImageJ. Ultrastructural Pathol 31(6):401–407
12. Loganathan D, Gnanavelbabu A, Rajkumar K, Ramadoss R (2014) Effect of microwave heat treatment on mechanical properties of AA6061 sheet metal. In: Procedia engineering, vol 97, pp 1692–1697
13. Kaboorani A, Riedl B, Blanchet P (2013) Ultrasonication technique: a method for dispersing nanoclay in wood adhesives. J Nanomater 3
14. Singh AK, Sandun D, Fernando, Hernandez R (2007) Base-catalyzed fast transesterification of soybean oil using ultrasonication. Energy Fuels 21(2):1161–1164

Dynamic Mechanical Analysis of Flax Fiber Stacked Polyurethane Blend Epoxy Composites

K. Rajkumar, K. Ramraji, M. Rajesh, and M. Rajiv kumar

Abstract Dynamic mechanical properties of polymer blend composites reinforced with flax woven were investigated. Polyurethane prepolymer was used as a modifier with the varied composition of 5–20 wt% within the epoxy resin. The ultrasonic stirring process was employed to mix the polyurethane with epoxy resin. Polymer blend composite was fabricated by using hand layup technique followed compression pressing. Tensile and flexural test was carried out for the fabricated composite. Experimental results revealed that the increasing polyurethane composition gradually decreases the mechanical properties of the blended polymer composite. Additionally, polyurethane blending has reduced the brittleness of composites. Dynamic mechanical properties of fabricated blend composites like storage modulus, loss modulus, and loss factor (tan δ) were improved only with 10% polyurethane composition.

Keywords Polymer blend · Polyurethane · Mechanical properties · Flax fiber

1 Introduction

Natural fiber-reinforced polymer (NFRP) composites are recently playing an important role in designing of structural and semi-structural components due to its superior mechanical and dynamic mechanical properties. The use of natural fibers has highly accepted because of increase environmental concern and eco-friendly demands. Natural fibers are greatly reducing the application of glass fiber in the polymer composites [1–3]. Natural fiber is basically extracted from the plant, animal, and mineral, namely flax, kenaf, hemp, sisal, basalt, etc. [4]. Natural fibers are improving the damping characteristics of the composite [5]. Flax is a bast fiber with comparable mechanical properties to the glass fiber [6]. The natural fibers are normally hydrophilic nature which results in high moisture absorption and weak adhesion with the matrix. The

K. Rajkumar (✉) · K. Ramraji · M. Rajesh · M. Rajiv kumar
Department of Mechanical Engineering, Sri Sivasubramaniya Nadar College of Engineering, Chennai, Tamil Nadu, India
e-mail: rajkumark@ssn.edu

S. Vijayan et al. (eds.), *Trends in Manufacturing and Engineering Management*, Lecture Notes in Mechanical Engineering,
https://doi.org/10.1007/978-981-15-4745-4_42

chemical treatment method has reducing hydrophilic behavior of the natural fiber. Flax fiber woven was treated with 5% NaOH solution providing better mechanical properties [7]. However, its low degradation temperature (~200 °C) makes them incompatible to the thermoset polymer composite. This also restricts the selection of manufacturing method for natural fiber composite processing [8].

Epoxy (EP) resin is one of the most important matrixes for the polymer composites include surface coatings, insulating materials, and structural adhesives. But its high cross-linking network during curing produces undesirable brittle characteristic. This leads to poor resistance to crack growth, low toughness, and low impact strength [9, 10]. In order to change its brittle nature by blending with compatible polymers such as an elastomer or rubbery, thermoplastic is required. These are providing additional interpenetrating polymer networks (IPNs) with an epoxy cross-link network. Polyurethane and the epoxy resins are NCO (nitrogen, carbon, and oxygen) and OH group. Epoxy blending with polyurethane makes it strong and flexible. Also, it gives a stable epoxy structure by polyurethane IPNs [11, 12]. The properties of the blends are the same as those synthesized by batch polymerization. Moreover, it requires the least time and energy input for the polymerization process [13]. Polyurethanes (PUs) are a unique family of polymers that possess some of the properties of plastics and rubbers. PU/EP IPNs show excellent thermal and mechanical properties as derived from the advantage characteristic of both the PU and EP [14, 15]. During the past decades, the great deals of efforts have been made on the evaluation of mechanical properties, and morphology of PU/EP IPNs and optimal ratio of PU IPNs in EP [16].

Viscoelastic polymeric materials are used as a medium of attenuating sound and damping the vibration. Those capabilities of the polymer are determined by how much dissipating sound or the vibrational mechanical energy into heat. This is indicated by intensity and broadness of the loss modulus or tan δ peaks at a particular range of temperatures. As temperature or testing frequency increases, the loss modulus (E'') and loss tangent (tan δ) show an intensity peak when damping reaches a maximum value [17]. The storage modulus (E') or dynamic modulus is typically related to Young's modulus of a material. It depends on the stiffness of a material and determines the sample whether it is stiff or flimsy. E' is a material property that having an ability to store energy applied to it. The changes in strength and stiffness of epoxy thermoset are evident by blending with polyurethane resin. The damping characteristics of epoxy polymer composites can be tailored by optimization of polyurethane composition [18, 19].

This research work reports the optimum composition of polyurethane blend and its effect on the dynamic mechanical properties of flax fiber-reinforced epoxy polymer composite.

2 Materials and Methods

2.1 Materials

Flax fiber was purchased from the Go Green Fiber Ltd., Chennai. The properties of flax fiber are follows, fiber strength (88–1500 MPa), density (1.4 g/cm^3), specific modulus (60–80 GPa), and elongation at failure 1.2–1.6% [20]. The used epoxy resin grade was LY556 and of density 1.25 (g/cm^3), Young's modulus (3.5 GPa), tensile strength (50 MPa), and fracture toughness (0.5 MPa) [21]. Polyurethane properties are followed as density (1.2 g/cm^3), Young's modulus (0.01 GPa), tensile strength (20 MPa), and fracture toughness (0.1 MPa) [15]. Diethylenetriamine (DETA) was a curing agent of C4H13N3 purchased from the Vasavibala Chemical, Chennai. Methyl ethyl ketone (MEK), a grade of $C_2H_5{\cdot}CO{\cdot}CH_3$ as a compatizer was purchased from the Chen chem chemical, Chennai, India.

2.2 Fabrication of Blend Composites

The composite laminates were made by hand layup technique that followed a gravity compression loading. Flax woven was treated with 5% NaOH solution for three hours and then washed by running tap water. Treated woven fibers were dried under direct sunlight to remove moisture. The releasing agent (Mansion Wax) was properly spread inside of mold for easy separation. Polymer blend laminate has a dimension of 300*300*5 mm^3 and is fabricated. To prepare epoxy-polyurethane blend composite, epoxy resin and polyurethane were dissolved separately in methyl ethyl ketone (MEK). The polyurethane pre-polymer was acting as a slender and its concentrations (5, 10, 15, and 20%) in the epoxy resin as shown in Table 1. The polymer blend resin mixture was prepared by mixing epoxy-polyurethane resins at room temperature. The ultrasonic stirring process was employed to mix the polyurethane resin in the epoxy resin. The chosen ratio of 10% of diethylenetriamine (DETA) and 2% of methylethylketone (MEK) has been added. Then, a woven flax layer was an initial layer, above which flax layer by layer were stacked. Finally, blend mixture was applied between the flax layers for effective binding. The composite laminates were

Table 1 Polyurethane content of composite samples

Blend composites	% of polyurethane blend (by weight)
BC0	0
BC5	5
BC10	10
BC15	15
BC20	20

Fig. 1 Typical blend composite laminate

cured under dead weight loading for 24 h. The typical blend laminate is shown in Fig. 1.

2.3 *Characterization of Blend Composites*

The mechanical and dynamic mechanical properties were evaluated for the fabricated blend composites as given below.

Tensile and Flexural Strength

The specimens were cut by abrasive water jet machine for both tensile and flexural tests. The tensile property was recorded from the universal testing machine (UTM). The tensile specimen was confirming the ASTM-D638 standard and is shown in Fig. 2a. Loading on the sample was continued until the specimen breaks. The flexural property was recorded using the universal testing machine with a facility of three-point bending. The flexural specimen was prepared according to ASTM-D790 standard, as shown in Fig. 2b. Five specimens of tensile and flexural strength were tested for each blend compositions.

Dynamic mechanical analysis

DMA measurements were carried out in Du Pont 983 analyzer. Dynamic mechanical properties were measured in tensile mode at a fixed frequency of 1 Hz under nitrogen gas purging. The specimen was heated from 34 to 185 °C with a heating rate of 5 °C/min. The quantities of the storage modulus (E'), loss modulus (E''), loss factor (tan δ), and temperature were recorded in the data logger. These data were further used to construct the damping characteristics curve.

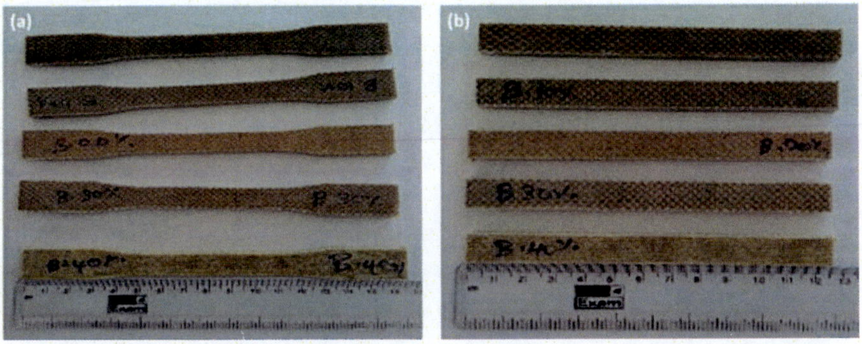

Fig. 2 Fabricated samples **a** Tensile and **b** flexural samples

3 Results and Discussion

3.1 Tensile and Flexural Strength

The tensile strength and modulus of blend composites are shown in Fig. 3a, b. This figure shows that tensile strength and modulus of blend composites increased with increase in polyurethane composition. An addition of polyurethane blend provides significant changes in the network structure. It reduces the tensile strength when polyurethane composition was exceeding 10%. Increases in polyurethane resin content above the 15 and 20% (BC15 and BC20) the failure of composite occurs under lower tensile stress. It can be seen that composite BC10 blend composition shows a higher tensile strength. The improvement of tensile strength was 25.5% which is more than that of base composite (BC0). This is due to saturation of polyurethane interpenetrating network with epoxy rings [16]. Similarly, BC10 blend composite shows a maximum tensile modulus. It was almost 44.4% greater than BC0 blend composite.

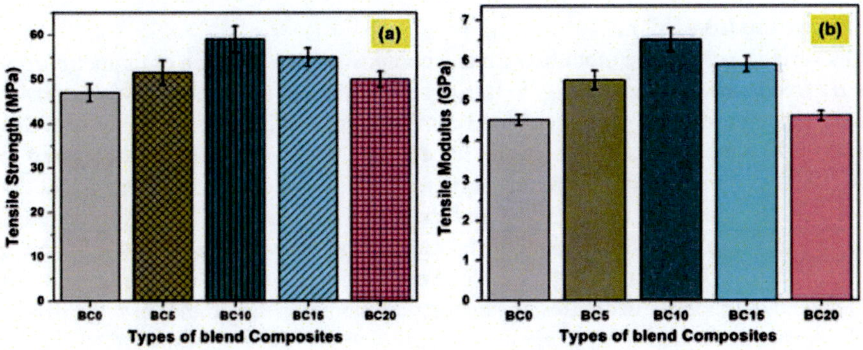

Fig. 3 **a** Tensile strength and **b** tensile modulus of blend composites

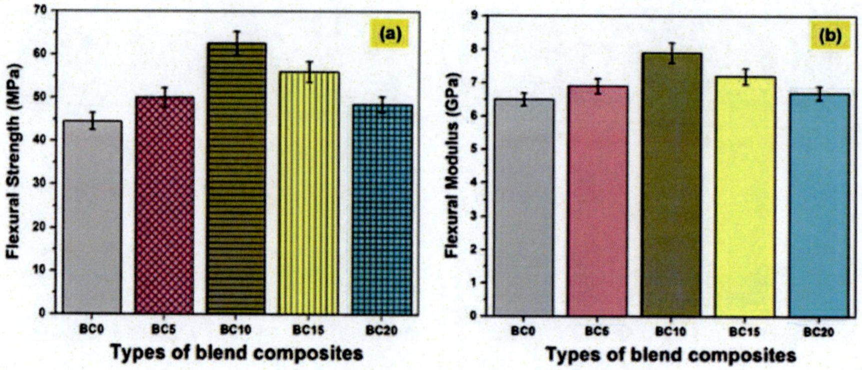

Fig. 4 **a** Flexural strength and **b** flexural modulus of blend composites

While, BC5 and BC15 blend composites observed higher tensile modulus by 22.2 and 31.1% than the BC0 blend composite.

The flexural strength and modulus of blend composites are shown in Fig. 4a, b. Figure 4a shows a flexural strength which increases with increasing polyurethane blend composition. This could be an addition of polyurethane blend which significantly increases interlinking networks of the epoxy matrix. IPNs are improving the strength and flexi of the composite. Additionally, flax fiber resisting bending stress induced upon it. Higher flexural strength was obtained from the BC10 blend composite. It is observed that BC10 was 40.4% higher than the BC0 blend composite. Moreover, BC5, BC15, and BC20 blend composites are showing 12.3, 25.8, and 9% higher than BC0 blend composite. Similarly, BC10 blend composite reaches a maximum flexural modulus and of 21.5% more than the BC0 blend composite.

3.2 *Effect of Composition on Dynamic Mechanical Properties*

Storage modulus (E')

The variation of storage modulus of blend composites as a function of temperature at 1 Hz frequency is shown in Fig. 5. It is observed that the storage modulus increases with an increase in polyurethane blend composition. The blend composites of BC15 and BC20 were gradually decreasing its storage modulus value when compared to BC10. By comparing the different polymer blend composites, it is found that the value of E' increases up to BC10 and then starts to decrease with polyurethane composition. In the glassy region, the maximum and lowest value of E' was found to be 1.30 E+010 Pa and 0.80 E Pa for blend composites BC0. The storage modulus of the blend composites decreases as temperature increases due to the loss in stiffness of fibers [17]. In the rubbery region, it is observed that the composite BC10 has the highest value of E' whereas composite BC20 has the lowest value. The lowest value of E' indicates an increase in molecular mobility in the blend composites matrix.

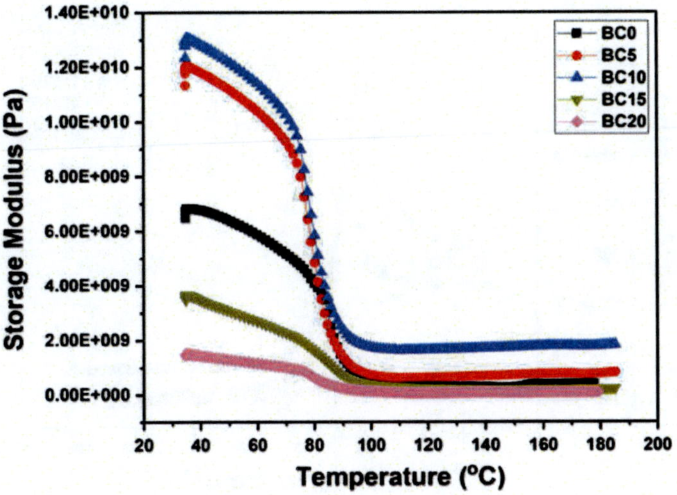

Fig. 5 Variation of storage modulus of blend composites at 1 Hz

The high value of E', specifically BC10, indicates a good adhesion between polymer blend composites. This fact shows that blend composite BC10 has a better interface bonding as compared to other blend composites. This figure also shows a shifting of glass transition temperature toward the left side as increasing in polyurethane composition in the matrix.

Loss modulus (E'')
The variation of loss modulus of blend composites as a function of temperature at 1 Hz frequency is shown in Fig. 6. Similar to storage modulus, the value of loss modulus of blend composite was also found to be increasing with increase in polyurethane composition up to 10%. It can be observed that the value of E'' increases up to the glass transition temperature (T_g) and then decreasing with temperature. In glassy region, the maximum values of loss modulus of blend composites BC0, BC5, BC10, BC15, and BC20 are found as 5.05E+7 Pa, 5.96E+7 Pa, 7.38E+7 Pa, 6.4E+8 Pa, and 1.87E+7 Pa, respectively. The peak height (maximum value) of loss modulus curve shows the T_g of corresponding composites. The blend composite BC10 shows the highest value of T_g (93 °C) which indicates better thermal stability than the other blend composites.

Loss factor (tan δ)
The variation of loss factor of blend composite as a function of temperature is shown in Fig. 7. The temperature corresponding to the tan δ peak is known as the glass transition temperature of the composite. The order of maximum value of tan δ for blend composites of BC10, BC5, BC0, BC15, and BC20 is found that 0.452, 0.448, 0.428, 0.417, and 0.378, respectively. The highest peak of tan δ value was found for BC10 composite which means that better damping property reached as compared

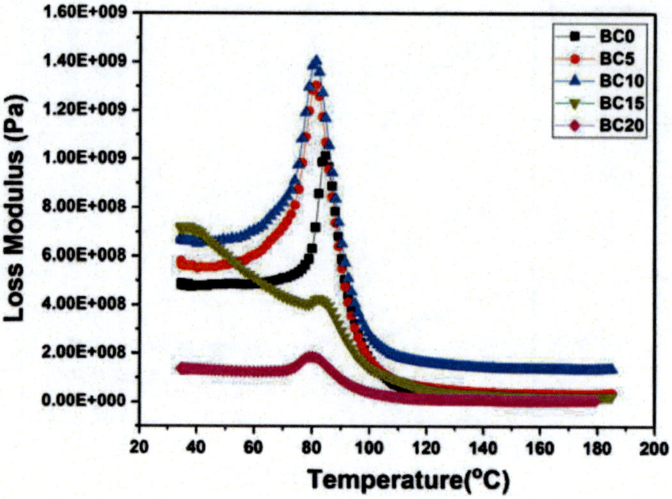

Fig. 6 Variation of loss modulus of blend composites at 1 Hz

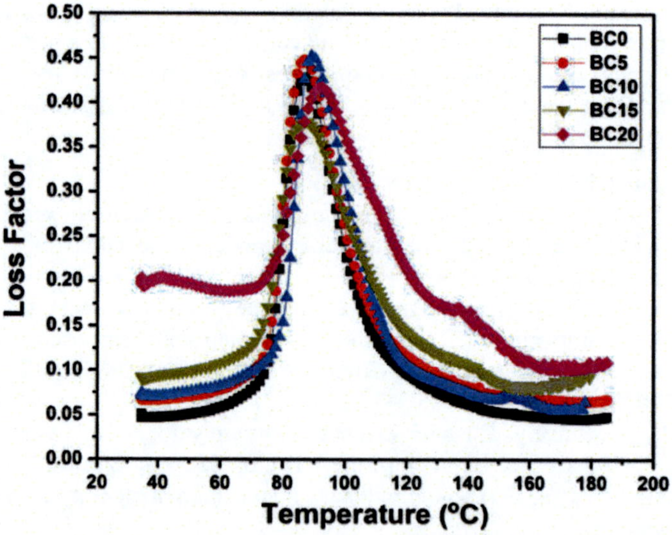

Fig. 7 Shows the variation of loss factor blend composites at 1 Hz

to other blend composites [19]. The lower value of tan δ is found for the blend composite BC20. Damping properties of different blend composite are shown in Table 2. The capability of damping was improving up to BC10 and begins to decrease as polyurethane composition increases. This clearly observed from Table 2. The similar trend is observed for the T_g of composites. It can be noted that the temperature

Table 2 Damping properties of different blend composite

Types of blend composite	$(\tan \delta)_{max}$	T_g (°C)	Temperature range ($\tan \delta > 0.3$)
BC0	0.428	86	78–94 (16 °C)
BC5	0.448	87	78–95 (17 °C)
BC10	0.452	93	83–102 (19 °C)
BC15	0.417	89	81–97 (16 °C)
BC20	0.378	90	82–98 (16 °C)

range of BC10 composite was 19 °C. This is indicative of damping characteristics which exist for a longer temperature range when compared to other composites.

4 Conclusions

Dynamic mechanical properties of polymer blend reinforced with flax fiber composite were studied and the following conclusions are drawn.

1. The tensile strength and modulus of BC10 composition reaches a maximum value and then decreases with further increase in blend composition. The tensile strength and its modulus improvement were 25.5 and 40.4% higher than that of base composite (BC0). This is attributed to the interpenetrating networks (IPNs) of polyurethane with epoxy cross-link networks.
2. The higher flexural strength and flexural modulus were obtained for the BC10 blend composite. The flexural strength and modulus were increased by 40.4 and 21.5% when compared to the BC0 composite.
3. It is found that the value of storage modulus increases with increase in polyurethane composition up to 10% in the matrix and then decreases with further increase in polyurethane composition. Due to the reduction in stiffness of matrix and flexi or rubbery nature, the loss modulus (E'') of blend composite BC10 shows the highest value of T_g (93 °C) which shows better thermal stability. The blend composite BC10 reveals a higher value of $\tan \delta$, which is indicative of higher dampness. Among the dynamic mechanical tested composites, the dampness of BC10 exists for a longer temperature range.

References

1. Ali M, Liu A, Sou H, Chouw N (2012) Mechanical and dynamic properties of coconut fiber reinforced concrete. Constr Build Mater 30:814–825
2. Yan LB, Chouw N, Yuan XW (2012) Improving the mechanical properties of natural fiber fabric reinforced epoxy composites by alkali treatment. J Reinforced Plastics Compos 31(6):425–437

3. Liang S, Gning PB, Guillaumat L (2012) A comparative study of fatigue behaviour of flax/epoxy and glass/epoxy composites. Compos Sci Technol 72(5):535–543
4. Ku H, Wang H, Pattarachaiyakoop N, Trada M (2011) A review on the tensile properties of natural fiber reinforced polymer composites. Compos Part B: Eng 42(4):856–873
5. Sgriccia N, Hawley MC, Misra M (2008) Characterization of natural fiber surfaces and natural fiber composites. Compos Part A: Appl Sci Manuf 39(10):1632–1637
6. Genc G, El Hafidi A, Gning PB (2012) Comparison of the mechanical properties of flax and glass fiber composite materials. J VibroEng 14(2):783
7. Ramraji K, Rajkumar K, Sabarinathan P (2019) Tailoring of tensile and dynamic thermo-mechanical properties of interleaved chemical-treated fine almond shell particulate flax fiber stacked vinyl ester polymeric composites. In: Proceedings of the Institution of Mechanical Engineers, Part L: Journal of Materials: Design and Applications, p 1464420719849616
8. Li X, Tabil LG, Panigrahi S (2007) Chemical treatments of natural fiber for use in natural fiber-reinforced composites: a review. J Polym Environ 15(1):25–33
9. May CA (1998) Epoxy resins. Marcel Dekker, New York
10. Lee L, Nevile K (1986) Handbooks of epoxy resins. McGraw-Hill, New York
11. Maxwell D, Young RJ, Kinloch AJ (1984) Hybrid particulate-filled epoxy-polymers. J Mater Sci Lett 3(1):9–12
12. Kinloch AJ, Shaw SJ, Hunston DL (1983) Deformation and fracture behaviour of a rubber-toughened epoxy: 2. Failure criteria. Polymer 24(10):1355–1363
13. Chen S, Tian Y, Chen L, Hu T (2006) Epoxy resin/polyurethane hybrid networks synthesized by frontal polymerization. Chem Mater 18(8):2159–2163
14. Kostrzewa M, Hausnerová B, Bakar M, Dalka M (2011) Property evaluation and structure analysis of polyurethane/epoxy graft interpenetrating polymer networks. J Appl Polym Sci 122(3):1722–1730
15. Chen CH, Chen MH (2006) Synthesis, thermal properties, and morphology of blocked polyurethane/epoxy full-interpenetrating polymer network. J Appl Polym Sci 100(1):323–328
16. Jin H, Zhang Y, Wang C, Sun Y, Yuan Z, Pan Y, Xie H, Cheng R (2014) Thermal, mechanical, and morphological properties of soybean oil-based polyurethane/epoxy resin interpenetrating polymer networks (IPNs). J Thermal Anal Calorimetry 117(2):773–781
17. Chern YC, Tseng SM, Hsieh KH (1999) Damping properties of interpenetrating polymer networks of polyurethane-modified epoxy and polyurethanes. J Appl Polym Sci 74(2):328–335
18. Jawaid M, Khalil HA, Hassan A, Dungani R, Hadiyane A (2013) Effect of jute fiber loading on tensile and dynamic mechanical properties of oil palm epoxy composites. Compos Part B: Eng 45(1):619–624
19. Wang C, Jia J (2014) Damping and mechanical properties of polyol cross-linked polyurethane/epoxy interpenetrating polymer networks. High Perform Polym 26(2):240–244
20. Mohammed L, Ansari MN, Pua G, Jawaid M, Islam MS (2015) A review on natural fiber reinforced polymer composite and its applications. Int J Polym Sci
21. Chawla KK (2012) Composite materials: science and engineering. Springer Science & Business Media

Material Processing Technology

Comparative Study of Ball Nose and Flat End Milling on A356 Alloy/SiC_p Metal Matrix Composite

K. Jayakumar

Abstract The end milling process is mainly used to quickly remove large amounts of material during heavier machining operations. It is most suitable for making slots, profiles and plunge cutting. The ball nose end milling is used to produce complex 3D sculptured surfaces for making moulds and dies. But, a variation of cutting force and surface roughness in both end milling processes is not well understood due to the constant change in cutting tool–workpiece engagement during these milling processes.

Aluminium alloy–SiC particle-reinforced metal matrix composites (MMCs) are used in many industries in different shapes through different machining processes. Therefore, the present study focuses on the comparative performance study of ball end and flat end milling on A356/SiC MMC.

Composite was prepared from A356 alloy powders (avg. size-50 μm) and SiC powders (avg. size-1 μm). MMC with 90 vol.% A356 alloy powder and 10 vol.% SiC powders were synthesized using vacuum hot pressing (VHP) process at 600 °C temperature with 25 MPa pressure. VHP is mainly used to achieve high densification with uniform distribution of reinforcement.

Both ball and flat end mill machining experiments were conducted as per Taguchi L_9 array. Experiments were carried out by varying cutting speed (100, 150 and 200 m/min), feed (0.1, 0.2 and 0.3 mm/rev) and depth of cut (0.4, 0.7 and 1 mm) for both machining processes. From the experimental results, it is observed that the flat end mill gave more material removal rate (MRR) in cm^3/min with low surface roughness (Ra) value than ball nose milling.

Keywords A356/SiC_p MMC · Vacuum hot pressing technique · Flat end milling · Ball nose end milling · Material removal rate · Surface roughness

---

K. Jayakumar (✉)
Department of Mechanical Engineering, Sri Sivasubramaniya Nadar College of Engineering, Kalavakkam, Chennai 603110, India
e-mail: kjayakumar@ssn.edu.in

S. Vijayan et al. (eds.), *Trends in Manufacturing and Engineering Management*, Lecture Notes in Mechanical Engineering,
https://doi.org/10.1007/978-981-15-4745-4_43

1 Introduction

Metal matrix composites (MMCs) are materials which combine metallic matrix and hard ceramic reinforcement to manufacture composite materials with superior properties to conventional metallic alloys. Among modern composite materials, particulate-reinforced metal matrix composites (PMMCs) are finding increased applications due to their favourable properties, including high mechanical properties and good wear and corrosion resistance. Silicon carbide particle (SiC)-reinforced aluminium-based MMCs are among the most common MMCs and commercially available due to their economic production. Increasing quantities of metal–matrix composites are being used to substitute conventional materials in many applications, particularly in the automobile, aerospace, electronics, armour and recreational industries, owing to increasing performance requirements [8]. In the modern manufacturing process, the objective is to produce low-cost, high-quality products in tiny time. Automated and flexible manufacturing systems are working for that purpose along with CNC machines that are capable of achieving high accuracy with low processing time. End milling is the second most common process (after turning) for metal cutting and especially for the finishing of machined flat parts, in many applications. Also, ball nose end milling is widely used for generating 3D sculptured surfaces for making moulds and dies.

Although SiC particle-reinforced aluminium matrix composites are regarded as one of the new engineering materials that have many applications, the abrasive particles reinforcement used in these materials makes them difficult to machine especially if the reinforcement is more than 10 vol. % of base matrix and also brittleness started increasing. If there is no solution to the machining problem, it is most likely that such composites would not find wide application. Many works have been carried about the machining of this material with turning, drilling, grinding and electric discharge grinding, etc. Kuram et al. [1] studied the effect of spindle speed, feed and depth of cut on tool wear, cutting force and surface roughness during microball nose end milling of Al 7075 material. They found the optimum process parameter for achieving higher machinability of the selected alloy. End milling study was carried out on Al 2124 alloy with 20% B_4C MMC to analyse the effect of machining parameter on surface roughness [2]. Conclusions were medium cutting speed, high feed and low depth of cut are suitable for machining the composite with low surface roughness value. Al 2014-T6 alloy was machined using high-speed ball nose end milling process by Dikshit et al. [3]. They identified surface roughness value which was influenced by cutting speed and axial depth of cut then the radial depth of cut and feed. Jayakumar [4] did an experimental investigation on ball nose end milling study on A356-based MMC and identified the best machining parameter for getting low cutting force and surface roughness value. Liu et al. [5] investigated microslot milling of 45% SiCp/Al composite using a single-edge PCD end mill of Ø 1 mm. They got optimum condition to get minimum Ra of 0.4 μm.

Optimization of parameters in the end milling of Al Al5059/SiC/MoS_2 aluminium hybrid metal matrix composites was carried out by Daniel et al. [6]. Major findings from the study increased in particle size of the composites tend to decrease in surface roughness, temperature, cutting forces and increased the MRR. Recently, ball nose end milling on Inconel 718 was conducted to study the effect of radial and axial depth of cut, feed and cutting speed on surface finish using minimum quantity lubrication [7]. It was identified that feed effects on the quality of surface finish whilst the radial depth of cut controls Ra variation.

From the available literature, it was found that ball nose end milling studies of A356 Al-based MMCs are in the initial stage and it needs further detailed analysis. Hence, the present study focuses on ball nose end milling of A356 alloy with 10 vol.% SiC MMC and its comparison with flat end milling machining process using the same process parameter. The output performance measures considered for both processes are surface roughness and MRR.

2 Experimental Details

2.1 *Synthesis of Composite*

A356 alloy powder (average size of 50 μm) and SiC particles (average size of 1 μm) of 10 vol. % were used as raw materials for processing of MMC. The major challenge in synthesizing the particulate MMCs is uniform mixing of finer reinforcement in the matrix. Vacuum hot pressing (VHP) shared with powder metallurgy (P/M) process was used in the present study to consolidate and process the composite with good densification (relative density—nearly 95%) and more isotropic in nature. Hence, the P/M process was used rather than stir casting of this alloy. The vacuum hot press used in this work is available at VSSC, Trivandrum, Kerala.

In sintering or VHP, at first, mixed powders were cold compacted at a pressure of 15 MPa. Then, the hot compaction was carried out at 600 °C temperature with 25 MPa pressure in a vacuum (10^{-5} mbar) to avoid oxidation of A356 alloy. The load and temperature were maintained to a dwell time of 30 min for effective densification of the powder compacts. The above-mentioned pressure and temperature were taken from the literature [4]. The hot pressed blank is shown in Fig. 1 which has nearly 125 mm in diameter and 20 mm thickness. Figure 2 shows the microstructure images of the composite taken from scanning electron microscope (SEM) at 1000 × . From the image, it was observed that the SiC particles are fairly well distributed on the matrix.

2.2 *Milling Study*

Figure 3a shows the experimental set-up. Rectangular piece of size 72 × 62 × 20 mm (Fig. 3b) was cut from the processed composite and end milling studies on

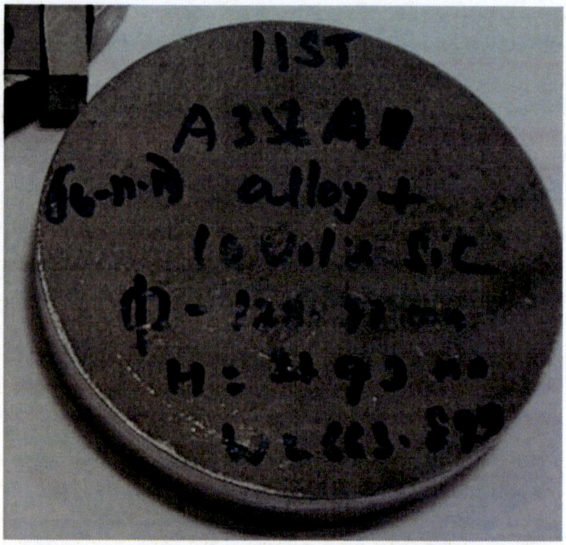

Fig. 1 Processed composite blank

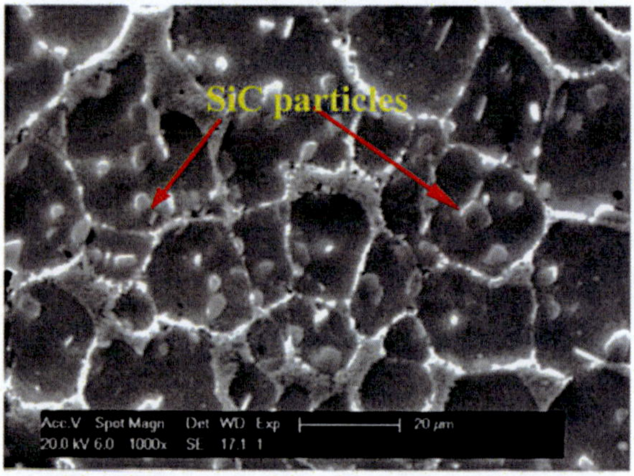

Fig. 2 Microstructure of MMC with 10% SiC

the composite material were carried in dry condition. HSS ball nose cutter of Ø 16 mm with two uncoated carbide inserts (R 390-11 T3 04E-NL H13A) and HSS end mill cutter of Ø 16 mm with two uncoated cemented carbide inserts (WIDIA XDHT 09 03 08) were used in this experimentation. The factors contributing to surface roughness and MRR during end milling of MMCs are work material (vol. % of reinforcement, particle size, type of ceramic particle and the base alloy chemical composition, etc.), machining parameters (up and down milling, cutting speed, feed and depth of cut), tool material and geometry, etc. Among these, cutting speed (100, 150, 200 m/min), feed (0.1, 0.2, 0.3 mm/rev) and axial depth of cut (0.4, 0.7, 1 mm) were considered as the process parameter for the present work.

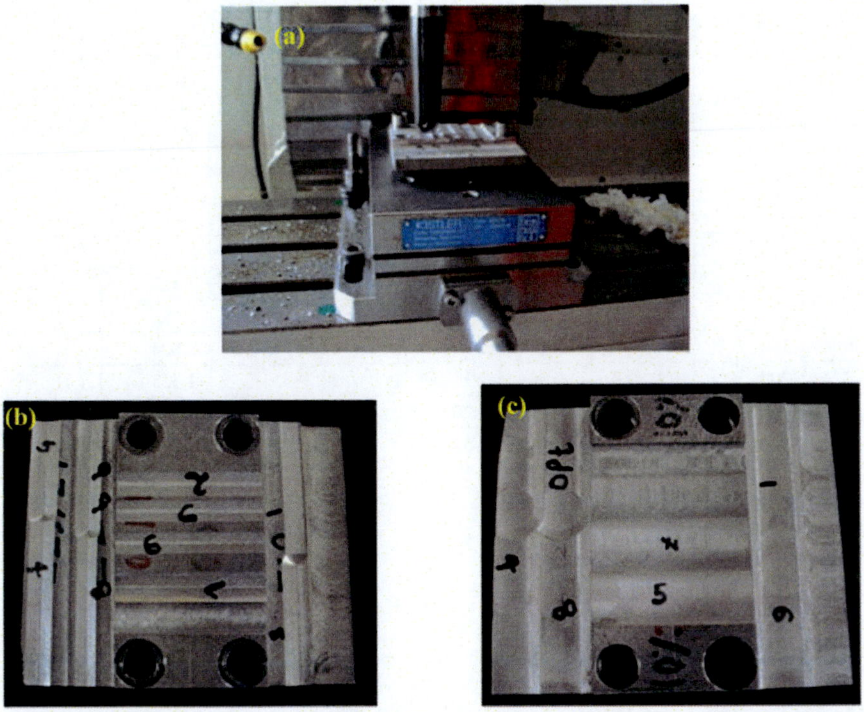

Fig. 3 **a** Ball nose end milling **b** machined workpiece after ball nose milling **c** machined workpiece after flat end milling

A good experimental design can decrease the number of experiments at the same time the error of the model does not change intensely. In the present work, Taguchi's L9 orthogonal array was used for three-level experiments with nine different conditions for flat end milling and ball nose end milling process for designing the experiments and is given in Table 1. During the experiments, inserts were indexed and changed at the end of five experiments to reduce the wear in the inserts. Slots were made for 16 mm width during both milling processes for each experiment.

The roughness (Ra) of the machined samples was measured using roughness tester (Model: Form Talysurf 50 Infra). Measurements were taken using the cut-off length (λ_c) of 0.8 mm and no. of samples of 5 using stylus with diamond tip of Ø 2 μm and an average of 3 readings taken at different locations in each slot was reported.

The weight of the workpieces before and after the experiments has been measured using a digital weighing balance (accuracy 1 mg). From the weight loss, actual density and machining time, MRR was calculated.

Table 1 Experimental results

Exp. No.	V_c-m/min	*f*-mm/rev	*d* of cut-mm	MRR (cm^3/min)-ball nose milling	MRR (cm^3/min)-flat end mill	Surface roughness (μm)-ball nose milling	Surface roughness (μm)-flat end mill
1	100	0.1	0.4	0.30	1.19	0.564	0.41
2	100	0.2	0.7	1.58	7.18	2.491	1.75
3	100	0.3	1	3.59	8.82	1.797	1.58
4	150	0.1	0.7	0.96	2.91	1.202	2.44
5	150	0.2	1	3.62	7.23	2.032	1.63
6	150	0.3	0.4	1.74	5.51	2.804	2.51
7	200	0.1	1	2.11	5.82	1.328	1.71
8	200	0.2	0.4	1.20	10.86	1.223	1.18
9	200	0.3	0.7	4.57	13.43	1.331	1.17

3 Results and Discussions

3.1 Comparison of MRR Between Ball Nose and Flat End Milling Processes

Material removal rate (MRR) is crucial in manufacturing industries for higher production and it can be attained by selecting proper machining parameters like cutting speed, feed and depth of cut. From Fig. 4, it was found that the MRR is increased with an increase in cutting speed, feed and depth of cut for both ball nose end milling and flat end milling processes. Increase in cutting speed caused the material to be soft due to the increase in temperature at the shear zone which reduced the strength of the composite workpiece that caused the effortless removal of chips. Increase in feed and depth of cut (DOC) increased the normal pressure, size of chip cut per tooth and chip load which improved the MRR. MRR is low for ball nose milling than flat end milling. The reason is that more compressive force is given to workpiece during ball nose end milling than flat end milling due to inadequate space available for the chip flow.

3.2 Comparison of Surface Roughness Between Ball Nose and Flat End Milling Processes

From Fig. 5a, it was found that surface roughness increased initially and then later it decreased for both milling processes. The reason is the development of built-up edge (BUE) whilst machining Al-based MMCs at lower cutting speed. BUE is break off when it becomes large at medium cutting speed. When this happens, broken

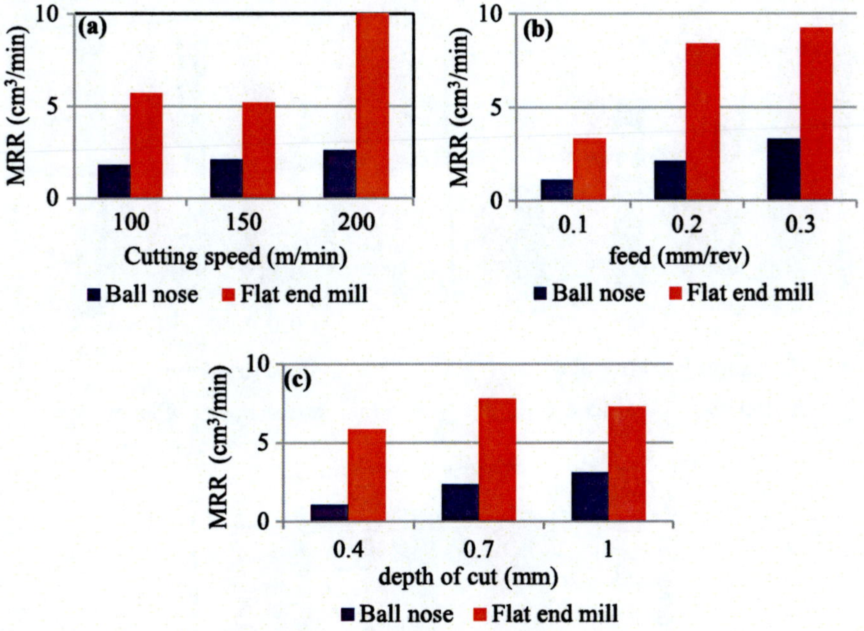

Fig. 4 **a** Cutting speed versus MRR (cm^3/min) **b** Feed versus MRR (cm^3/min) **c** Depth of cut versus MRR (cm^3/min)

BUE may pass off the underside of the chip and it randomly scratches the already machined surface and causes saw-toothed surface which finally makes the work surface as rough. As the cutting speed increases, the BUE vanishes, chip fracture decreases, thus the surface finish is improved. From Fig. 5b, the surface roughness of the composite is increased with a rise in feed for both milling cutters. The explanation is that, the feed rate has increased the distance between the two successive tool paths, which leads to increase in Ra value. As the depth of cut increased, the cutting process vibrations increases [8] and the development of high normal pressure on the rake faces of the tool resulted in the formation of BUE that worsened the surface finish (Fig. 5c). Surface roughness value was low for flat end mill cutting process than ball nose end milling process. Compared with the flat end mill profile, the surface produced by ball end mill is complex and a new cut is overlapped which generated a new surface roughness by removing the earlier cut thus the non-uniform feed marks being formed on the machined surface which increased surface roughness value.

Using design expert software's optimization option, cutting speed of 186 m/min, feed 0.29 mm/rev and 0.9 mm depth of cut were recommended as an optimum condition for ball nose end milling process to get high MRR and low surface value. Similarly, cutting speed of 200 m/min, feed 0.3 mm/rev and 0.83 mm depth of cut was recommended as an optimum condition for flat end milling process.

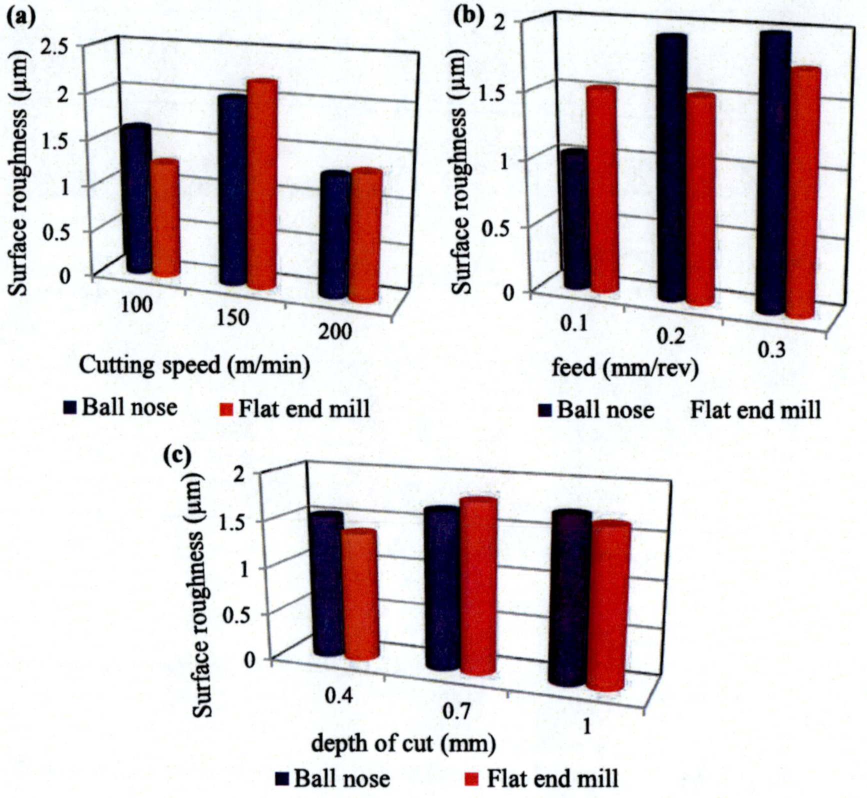

Fig. 5 **a** Cutting speed versus surface roughness (μm) **b** Feed versus surface roughness (μm) **c** Depth of cut versus surface roughness (μm)

4 Conclusions

In the present work, fairly homogeneous MMC of A356-10% SiC powders was synthesized using vacuum hot pressing technique.

The effects of cutting speed, feed and depth of cut on MRR and surface roughness during flat end milling and ball nose milling of Al-based MMC were investigated using Taguchi's L9 experimental design method.

Flat end milling process gave better responses than ball nose mill in terms of high MRR with low surface roughness value.

The optimum machining parameter was found to improve the machinability of the MMC for both milling processes.

References

1. Kuram E, Ozcelik B (2013) Multi-objective optimization using Taguchi based grey relational analysis for micro-milling of Al7075 material with ball nose end mill. Measurement 46(6):1849–1864
2. Boswell B, Islam MN, Davies IJ, Pramanik A (2017) Effect of machining parameters on the surface finish of a metal matrix composite under dry cutting conditions. Proc Inst Mech Eng Part B: J Eng Manuf 231(6):913–923
3. Dikshit MK, Puri AB, Maity A (2017) Optimization of surface roughness in ball-end milling using teaching-learning-based optimization and response surface methodology. Proc Inst Mech Eng Part B: J Eng Manuf 231(14):2596–2607
4. Jayakumar K (2018) Study of cutting force and surface roughness in Ball nose end milling of vacuum hot pressed A356 alloy/SiCp metal matrix composite. Mater Today: Proc 5(2):6526–6533
5. Liu J, Cheng K, Ding H, Chen S, Zhao L (2018) An investigation of the influence of phases' removal ways on surface quality in micro milling SiCp/Al composites. Proc CIRP 71:59–64
6. Ajith Arul Daniel S, Pugazhenthi R, Kumar R, Vijayananth S (2019) Multi objective prediction and optimization of control parameters in the milling of aluminium hybrid metal matrix composites using ANN and Taguchi-grey relational analysis. In: Defence technology
7. Kasima MS, Hafiza MSA, Ghanib JA, Haronb CHC, Izamshah R, Sundic SA, Mohamed SB, Othman IS (2019) Investigation of surface topology in ball nose end milling process of Inconel 718. Wear 426:1318–1326
8. Niu Z, Cheng K (2019) An experimental investigation on surface generation in ultraprecision machining of particle reinforced metal matrix composites. Int J Adv Manuf Technol 103:1–9

Numerical Modeling of Orthogonal Machining Process Using Smoothed Particle Hydrodynamics—A Parametric Study

Sarath Babu Thekkoot Surendran, C. S. Sumesh, and Ajith Ramesh

Abstract Some of the challenges faced during numerical modeling of machining processes using finite element method are: extreme deformations, complicated and discontinuous contact conditions between the workpiece and cutting tool, and the chances of self-contact due to chip curling. Another difficulty encountered in FEM is in the form of mass and energy losses. Lately, the smoothed particle hydrodynamics (SPH) method has developed as a potential alternative for modeling machining processes due to its ability to handle severe deformations while avoiding energy losses and mass losses encountered by traditional FE Model. This method has been implemented in commercial finite element package Abaqus, for solving problems involving localized severe deformations. Several control parameters are used in a typical SPH formulation. The purpose of this research work is to investigate the influence of the important factors such as the type of SPH formulation, mass scaling, particle density, artificial bulk viscosity, and the smoothing length in the numerical modeling of orthogonal machining of AISI 1045 steel. The challenges involved in accurately modeling this highly nonlinear problem is handled using the Abaqus/Explicit integration scheme along with the Johnson–Cook material model. Results from this parametric study are validated with the results from previously published literatures.

Keywords Finite element method (FEM) · Smoothed particle hydrodynamics (SPH) · Abaqus/explicit · Orthogonal machining · Johnson–cook material model · Smoothing length

1 Introduction

Numerical modeling normally gives close to accurate results for problems including nonlinear ones which analytical methods find difficult to handle. They are neither costly nor time-consuming as compared to experimental analysis. Machining is a

S. B. T. Surendran · C. S. Sumesh (✉) · A. Ramesh
Department of Mechanical Engineering, Amrita School of Engineering, Coimbatore, Amrita Vishwa Vidyapeetham, India
e-mail: cs_sumesh@cb.amrita.edu

S. Vijayan et al. (eds.), *Trends in Manufacturing and Engineering Management*, Lecture Notes in Mechanical Engineering,
https://doi.org/10.1007/978-981-15-4745-4_44

controlled material removal process in which a raw material is converted into desired shape and size. Orthogonal machining is a machining process in which the cutting edge of the tool is perpendicular to the direction of motion of the cutting tool. AISI 1045 steel used in this study is a medium carbon steel that has good machinability, good weldability, and high strength and impact properties both for hot rolled and normalized conditions. It is widely used in gears, shafts, connecting rods, crankshafts, hydraulic clamps, etc.

Orthogonal machining uses a wedge-shaped tool in which the cutting edge is perpendicular to the cutting direction. The chip is formed from the shear deformation along the shear plane. The important tool geometry angles are rake angle and clearance angle. The direction of chip flow is determined by the rake angle, whereas clearance between tool and workpiece is endowed by the clearance angle. This paper presents numerical modeling of orthogonal machining using SPH. Numerical modeling has many advantages over experimental and analytical methods. Numerical methods can give close to accurate results for problems even when analytical methods fail to solve the problem. It is cheaper compared to experimental analysis.

Smoothed particle hydrodynamics, abbreviated as SPH, is one among the most popular particle methods available. The SPH particles move in space following the conservation laws of continuum fluid dynamics. It has been proved stable to solve engineering problems. Hence, commercial FE packages like Abaqus/Explicit and LS-Dyna incorporated this method in their codes. In this research work, Abaqus/Explicit package is used for modeling.

In the literature, there are many researchers who have done research in the field of numerical modeling, orthogonal machining, smoothed particle hydrodynamics (SPH) and combination of these fields. Given below are details of some of the previous research work related to the machining area.

Niu et al. [1] introduced an improved SPH method and used it for modeling orthogonal cutting and predicted the cutting force and chip morphology under different cutting conditions. This modification is achieved by approximating density and kernel gradient. Olleak and El-Hofy [2] used SPH method in the modeling of high-speed machining. The paper also predicted the chip morphology and cutting forces during high-speed orthogonal machining of Ti6Al4V. Ramesh et al. [3] predicted the deformation and machining characteristics for high-speed orthogonal machining of Ti6Al4V using a detailed FE model. The influence of feed rate, spindle speed, and rake angle on output parameters such as cutting force and surface finish are studied. The paper also tackled the issue of self-contact using a unique model. Avachat and Cherukuri [4] studied the SPH control parameters in detail in modeling orthogonal machining. The paper evaluated the effects of various parameters like smoothing length, type of SPH formulation, and particle density. The influence of the above-mentioned parameters on stress distribution as well as chip morphology of orthogonal machining of AISI 1045 steel is also studied. Bagci et al. [5] presented the simulation and analysis of 3D machining using SPH. AISI H13 Steel was used in the model. The model accurately predicted the cutting forces. M.B. Liu and G.R. Liu [6] were the people who wrote the first book on SPH. In the year 2010, the authors published a paper discussing the SPH method and its latest development. This paper gives a very good background on SPH.

Kershah et al. [7] estimated cutting coefficients of AISI 1045 steel using finite element modeling. The paper also contains a parametric study which correlates cutting parameters to cutting coefficients. Madaja and Píškaa [8] presented a comparison of SPH results with experimental results and FE model result for orthogonal cutting simulation of Al 2024-T351. The study found that as particle spacing reduces, the particles tend to form segmented chips. Sumesh and Ramesh [9] studied the effect of process parameters like depth of cut, cutting speed, and feed rate on surface integrity of turned Al 6061–T6 alloys using RSM. Zetterberg [10] discussed the important parameters and different possibilities to form a chip in metal cutting simulation in Abaqus/Explicit. It emphasizes on how the flank wear and hardening implemented in the material model affects the formation of the chip. Villumsen et al. [11] predicted the cutting forces of orthogonal machining using the SPH method and compared with the existing literature to validate the model. The authors also carried out a sensitivity analysis of machining parameters. Haddag et al. [12] analyzed the effect of orthogonal machining parameters on chip segmentation of Al2024-T351 alloy. The paper also considered the effect of feed and rake angle on chip morphology and cutting force.

Barge et al. [13] developed a numerical model and studied the physical and process parameters dependence using finite element modeling. The paper discusses the numerical modeling of orthogonal machining. Zhang et al. [14] used digital image correlation technique to obtain the deformation field containing elastic as well as plastic parts. A new methodology has been proposed to obtain the cutting stress field. Limido et al. [15] evaluated SPH method for high-speed cutting modeling. The developed approach in the paper is used for finding the cutting forces in ball end milling. The paper used both a 2D model and a 3D model using the software LS-Dyna. Chen et al. [16] developed FE model for high-speed machining of Ti6Al4V alloy. Cleary et al. [17] used SPH for modeling thin-walled high-pressure die casting using SPH. The paper compared the results with the experimental data and found that the results were accurate. The study also performed flow analysis using the validated model. Hu et al. [18] used SPH for modeling high-pressure die casting. The paper compared SPH results with the finite element method and found that they are comparable and thus validated the model. The paper showed that the method is able to capture the fine details of the free surface motion, including plume shape, frequency, and phase of oscillation.

The above-mentioned research works have helped extensively in the presented research. In this paper, the influence of the important factors such as type of SPH formulation, particle density, smoothing length, mass scaling, and artificial bulk viscosity has been studied. The challenges involved in accurately modeling this highly nonlinear problem is handled using the Abaqus/Explicit integration scheme along with the Johnson–Cook material model.

2 Smoothed Particle Hydrodynamics

2.1 History

Smoothed particle hydrodynamics is a mesh-free Lagrangian particle method. This method mainly uses Lagrangian formulation along with particle approximation. A group of particles having material properties assigned individually represents the state of the system. The particles in this system behave depending upon the governing equations. A pre-defined mesh is not required to connect particles in SPH method, like in conventional FEM. In the year 1977, Lucy, Gingold, and Monaghan invented the method to solve astrophysical problems. The first book on SPH was published by Liu GR and Liu MB in the year 2003 [19]. The authors were working on high explosives and found that the existing grid-based methods were incapable of handling the phenomena due to excessive mesh distortion. The search for advanced methods ended up in SPH which later became popular. It was further extended to solid mechanics problems involving large deformations which lead to the incorporation of this method in the commercial FE packages like Abaqus/Explicit and LS-Dyna. SPH can naturally handle problems with extremely large deformations. Unlike other mesh-free methods which use nodes as only interpolation points, SPH particles carry material properties.

2.2 Mathematical Formulation

Function f (y) is approximated as

$$< f(y) >= \int_{\Omega} w(y - y\prime, h) f(y')\mathrm{d}\Omega \tag{1}$$

where
$< >$—the approximation
$w(y - y',h)$—the kernel function
h—the smoothing length
y'—the particle within the smoothing length

The conditions satisfied by the kernel function are:

$w(y - y',h) = 0$ for $| y - y'| \geq 2\,h$

The normalization condition:

$$\int_{\Omega} w(y - y',\ h)\mathrm{d}y' = 1 \tag{2}$$

As h tends to zero, the kernel function becomes the Dirac delta function,

i.e., $\lim_{h\to 0} w(y - y', h) = \delta(y - y', h)$

Thus, $\lim_{h\to 0} < f(y) > = f(y)$
So, for N discretized points, the function approximation becomes:

$$< f(y^i) > = \sum_{j=1}^{N} \frac{mj}{\ell j} f(y^j) w(y^i - y^j, h) \tag{3}$$

where
$\frac{mj}{\ell j}$—Volume of the particle j with mass mj

The value of a variable is computed by the contributions from the neighboring particles j for any particle i, where $j = 1, 2 \ldots N$. Here, the kernel function is nonzero. Cubic B-spline is the commonly used kernel function. It was proposed by Monaghan given by,

$$W(r, h) = \frac{c}{h^D} \times \begin{cases} \left(1 - \frac{3}{2}r^2 + \frac{3}{4}r^3\right), & 0 \le r \le 1; \\ \frac{1}{4}(2 - r)^3, & 1 \le r \le 2; \\ 0, & \text{Otherwise.} \end{cases} \tag{4}$$

where

$$r = \frac{|y - y'|}{h}$$

D = the number of spatial dimensions and C = scale factor
$C = 2/3$ for $D = 1$,
$C = 10/7\pi$ for $D = 2$,
$C = 1/\pi$ for $D = 3$.

3 Numerical Modeling of Orthogonal Machining Using SPH

The numerical model uses a tungsten carbide rigid tool. It has a rake angle of 8° and a clearance angle of 10°. The workpiece material used is AISI 1045 steel with dimensions 3.6 mm × 0.3 mm. The out-of-plane thickness is 0.04 mm. The material properties of workpiece are displayed in Table 1. Equation of state is used along with the material properties with US-UP formulation. The remaining parameters used in the model are as follows [4]:

(a) Speed of sound in the workpiece
(b) $c = 6.35 \times 10^6$ mm/s

Table 1 Material properties of AISI 1045 steel [4]

A (MPa)	*B* (MPa)	*n*	*C*	*m*	Room temperature (°C)	Melting temperature (°C)	Reference strain rate	Density (kg/m^3)	Shear modulus (GPa)
553	600.8	0.234	0.013	1	27	1460	1	7800	76.923

(c) slope $s = 0.0$
(d) Gamma = 0.0

The stress–strain relationship for the material is defined using Johnson–Cook (J-C) model. J-C model is found to provide a proper description of the material behavior which is unique to nonlinear large deformation processes by considering the high strain rates, temperature-dependent visco-plasticity, and large strains [20].

$$\sigma = (A + B\varepsilon^n)\left(1 + C\ln\left(\frac{\varepsilon}{\varepsilon 0}\right)\right)\left(1 - \left(\frac{T - Tr}{Tm - Tr}\right)\right) \tag{5}$$

where
ε^n—equivalent plastic strain

A, B, C, m and n—material constants

$\frac{\varepsilon}{\varepsilon 0}$—plastic strain rate
Tm—Reference melt temperature
Tr—Reference temperature

The model uses static-kinetic exponential decay as the friction formulation. The static coefficient of friction is taken as 0.5, dynamic coefficient value as 0.3, and decay coefficient value as 0.1. The default values of artificial bulk viscosity factor $Q1$ is changed from 0.06 to 0.5 and $Q2$ is changed from 1.2 to 1.5 [4].

Direct SPH continuum particle analyses technique has been used for validation instead of FE conversion to SPH. In the former technique, a node set was created and inertia mass was assigned to create mass elements in the .inp file. After completing the CAE model, the mass elements were renamed as PC3D to convert mass elements into SPH particles, in the .inp file. The solid elements of the workpiece were deleted. Then *mass definition was removed and the element set name of the continuum particle elements (PC3D) was copied to the *Solid Section element set name. Characteristic length was also defined in the *Solid Section. In the assembly module, a node-based surface for the workpiece was created to define the contact. *Section Controls was used to change the SPH to NSPH and also to define the smoothing length. Finally, the contact between the workpiece and the tool was defined in the .inp file. An additional damage model is not required while modeling using SPH method, which is another added advantage.

4 Results and Discussions

Following numerical controls were used for SPH modeling of the machining process. The influence of each parameter on the optimal modeling of the problem is investigated.

Fig. 1 Tool passing without interaction

4.1 *Influence of Mass Scaling*

Mass scaling is a numerical technique that is used to increase the time step by scaling up the density in specific elements. Initially, a mass scaling factor of 10,000 was used in the study. Initial results indicated the absence of any material removal, although the tool had contact with the workpiece (see Fig. 1). Hence, a reduced mass scaling factor of 100 was used and was found to initiate the material removal. However, the computational time increased to a significantly large value. Therefore, no mass scaling was used in this study for better results. It has been understood that an increased mass scaling can lead to high inertial effects that can corrupt the results.

4.2 *Need of Direct SPH*

In this work, initially, only the bottom and left faces of the workpiece were encastered. However, the results indicated that boundary conditions were not transferred to the SPH particles. Figure 2 shows the end results of the above-mentioned condition. This problem was resolved by using smoothed particle hydrodynamics instead of finite element conversion to SPH particles [21]. In the SPH analyses, a node set had to be

Fig. 2 Encastered workpiece moving along with the tool

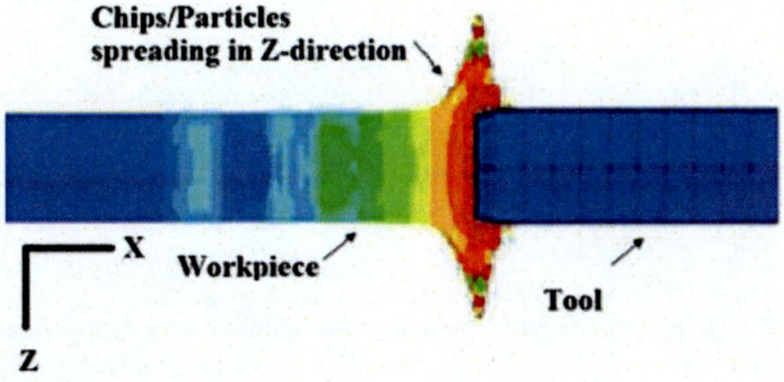

Fig. 3 Particles spreading

created and inertia mass had to be assigned to the nodes to create mass elements in the .inp file. Later, these mass elements had to be replaced with continuum particle elements (PC3D).

4.3 Contact Issues

Another major problem that was encountered was the inability to use general contact algorithm. The reason being due to the extensive nodal penetration of tool nodes to the workpiece surface. This was tackled by using the traditional master–slave, surface-to-surface contact algorithm.

4.4 Necessity of Additional Non-default Constraints

The initial simulation results indicated the movement of SPH particles in the *Z*-direction instead of flowing over the tool rake (see Fig. 3). This issue was resolved by assigning encaster boundary conditions to the bottom most and left most nodes, and a displacement/rotation boundary condition to the rest of the nodes of the workpiece where the displacement along the *Z*-direction and rotations about *Y* and *X* axes were constrained.

4.5 Effect of Smoothing Length

Smoothing length is the radius of the sphere of influence within which any particle may affect the neighboring particle. Initially a value of 0.005 was assumed for smoothing length based on the calculation of half of the characteristic length [4]. This, however, leads to non-physical results (see Fig. 4). Then, a higher value of 0.02 was assumed which resulted in the termination of the simulation with an error.

Fig. 4 Non-physical result due to low value of smoothing value

Fig. 5 Non-physical result due to constant smoothing length

The error indicated that the smoothing length was too large as the number of particles exceeded the maximum specified. Between these two extreme values, a suitable trade-off value was reached as 0.01, which was used as a constant smoothing length.

Figure 5 shows the results of constant smoothing length that lead to breaking up of workpiece or flow of chips over the tool. This issue was resolved by using a variable smoothing length option in the .inp file. Maximum number of particles within the smoothing length was maintained at a default value of 140, since the computational memory required to run the simulation was too high even for a slight increase from the default value.

Influence of Artificial Bulk Viscosity and SPH formulation

The von Mises stress contour plot was found to be uniformly distributed for $Q1 = 0.5$ than the default value of $Q1 = 0.06$. Another important parameter was the type of SPH formulation. The default SPH formulation was found to be inadequate and inconsistent for modeling the process, and it leads to non-physical results (See Fig. 6). Thus, NSPH and XSPH were used simultaneously instead of default SPH formulation.

5 Validation

Although the numerical model was developed, the initial results were not found to match with the literature since the simulation had many issues which are mentioned in the previous section. After making necessary modifications based on the trial-and-error simulations, the model was finally developed successfully. This work was able

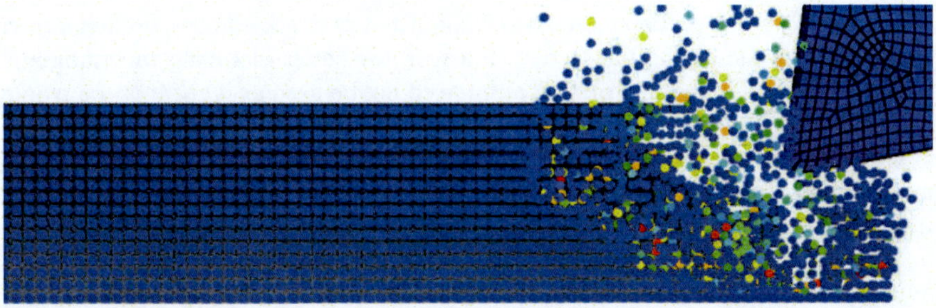

Fig. 6 Non-physical result due to default SPH formulation

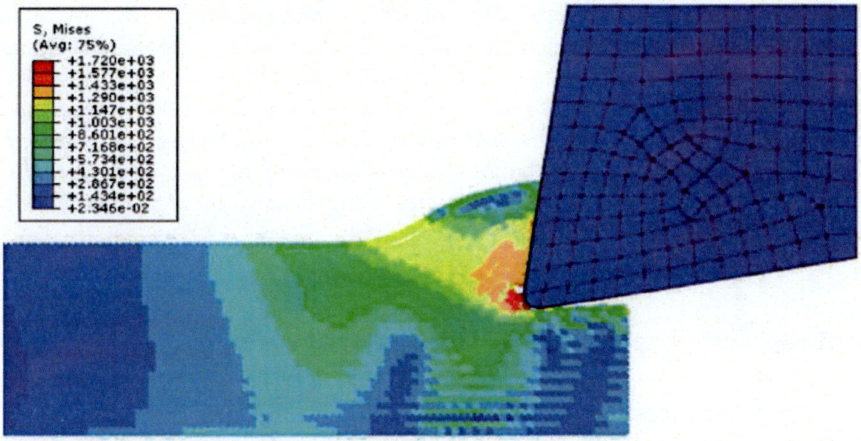

Fig. 7 Validated model

to reproduce the results obtained from the literature, Chinmay et al. [4] and Kershah et al. [7]. The average stress in the shear zone was found to be matching with the literature. The von Mises stress contours of the validated model are shown in Fig. 7.

6 Conclusions

This paper discusses the development of a numerical model for simulating orthogonal turning of AISI 1045 steel using SPH method. The FE package Abaqus/Explicit was used. The results obtained are found to be in good agreement with the results found in the literature.

By varying SPH control parameters and boundary conditions, parametric studies were performed. All the simulations were conducted by modifying the .inp file, except for FE conversion to SPH results.

Influence of mass scaling factor was studied and as expected computational time decreased with increasing the factor. However, for better modeling of orthogonal machining, it was found that mass scaling need not be used as it changes the whole dynamics.

Initially, the encaster boundary conditions given to the workpiece were not transferred during the conversion of elements to SPH particles. This issue was solved using the direct SPH method instead of FE conversion to SPH method.

The significance of using variable smoothing length was also investigated. When low smoothing length was used, the particles were found to fly out of the domain, since there were a less number of particles in the sphere of influence. On the other hand, the maximum number of particles in the sphere of influence was maintained at a default value of 140. Even a slight increase from the default value was found to increase the computational memory required to run the simulation.

It was found that when the correct SPH parameters were used, the computational results obtained were in good agreement with the results from conventional FEM. It was also found that SPH can naturally handle self-contact, mesh distortion, and material separation. However, the SPH method has some limitations. The values of some SPH parameters used in the model are obtained by trial-and-error methods. Also, computational time was found to increase significantly due to particle sorting and neighbor search.

References

1. Niu W, Mo R, Liu GR, Sun H, Dong X, Wang G (2017) Modeling of orthogonal cutting process of A2024-T351 with an improved Sph method. Int J Adv Manuf Technol 95:905–919
2. Olleak AA, El-Hofy HA (2015) Prediction of cutting forces in high speed machining of Ti6Al4 V using Sph method. In: Proceedings of the ASME 2015 International Manufacturing Science And Engineering Conference MSEC 2015 June 812, 2015, Charlotte, North Carolina, USA
3. Ramesh A, Sumesh CS, Abhilash PM, Rakesh S (2015) Finite element modelling of orthogonal machining of hard to machine materials. Int J Mach Machinability Mater 17(6):543–568
4. Avachat CS, Cherukuri HP (2015) A parametric study of the modeling of orthogonal machining using the smoothed particle hydrodynamics method. In: Proceedings of the ASME 2015 International Mechanical Engineering Congress and Exposition IMECE 2015, 13–19 Nov 2015, Houston, Texas
5. Bagci E (2011) 3-D Numerical analysis of orthogonal cutting process via Meshfree method. Int J Phys Sci 6(6):1267–1282
6. Liu MB, Liu GR (2010) Smoothed particle hydrodynamics (Sph): an overview and recent developments. Arch Comput Methods Eng 17:25–76
7. Kershah T (2000) Prediction of cutting coefficients during orthogonal metal cutting process using FEA approach
8. Madaja M, Píškaa M (2013) On the Sph orthogonal cutting simulation of A2024-T351 alloy. In: 14th CIRP Conference on Modeling of Machining Operations, Procedia Cirp, vol 8, pp 152–157
9. Sumesh CS, Ramesh A (2018) Numerical modelling and optimization of dry orthogonal turning of Al 6061 T6 alloy. Periodica Polytechnica Mech Eng 62(3):196–202

10. Zetterberg M (2014) A critical overview of machining simulations in Abaqus. Degree Project. In: Solid mechanics, Second Level Stockholm, Sweden
11. Villumsen MF, Fauerholdt TG (2008) Simulation of metal cutting using smooth particle hydrodynamics. Ls-Dyna Anwenderforum, Bamberg
12. Haddag B, Atlati S, Nouari M, Barlier C, Zenasni M (2012) Analysis of the cutting parameters influence during machining Aluminium alloy A2024-T351 with uncoated carbide inserts. Eng Trans 60(1):31–39
13. Barge M, Rech J, Hamdi H, Bergheau J-M (2005) Numerical modelling of orthogonal cutting numerical, physical and process parameters dependence. In: Proceedings of the 2nd International Conference on Manufacturing Engineering (ICMEN), 5–7 Oct 2005, Kallithea-Chalkidiki, Greece
14. Zhang D, Zhang X-M, Xu W-J, Ding H (2016) Stress field analysis in orthogonal cutting process using digital image correlation technique. J Manuf Sci Eng 139:031001-1 (March 2017)
15. Limido J, Espinosa C, Salaun M, Mabru C, Chieragatti R, Lacome JL (2011) Metal cutting modelling Sph approach. Int J Mach Machinability Mater 9(¾)
16. Chen G, Ren C, Yang X, Jin X, Guo T (2011) Finite element simulation of high-speed machining of Titanium alloy (Ti-6Al-4V) based on ductile failure model. Int J Adv Manuf Technol 56:1027–1038
17. Cleary PW, Savage G, Ha J, Prakash M (2014) Flow analysis and validation of numerical modelling for a thin walled high pressure die casting using Sph. Comp Part Mech. https://doi.org/10.1007/s40962-017-0144-9
18. Hu MY, Cai JJ, Li N, Yu HL, Zhang Y, Sun B, Sun WL (2017) Flow modeling in high-pressure die casting processes using Sph model. Am Foundry Soc. https://doi.org/10.1007/s40962-017-0144-9
19. Liu GR, Liu MB (2003) Smoothed particle hydrodynamics: a Meshfree particle method. World Scientic, Singapore
20. Dassault Systmes Simulia Corporation. ABAQUS Analysis user's guide (2013)
21. Dassault Systmes Simulia Corporation. ABAQUS Analysis user's guide (2016), Section 15.2.2

Acoustic Emission-Based Grinding Wheel Sharpness Monitoring Using Machine Learning Classifier

J. Revant, Rahul Sree Kumar, K. Rameshkumar, and D. S. B. Mouli

Abstract Grinding is one of the important secondary manufacturing processes used to improve the dimensional accuracy, surface finish and geometric form of the component. Grinding wheel consists of abrasive particles, which perform the metal removal function. The sharpness of the grinding wheel is one of the important factors for achieving the required surface geometry in the component. In this study, a simple device used to measure the sharpness of the abrasive particles of the grinding wheel is designed and fabricated. Aluminium oxide grinding wheel conditions are established using the sharpness of the abrasive grinding wheel. Grinding process is monitored using acoustic emission (AE) Sensor. AE features are extracted in time domain and dominated features which contain useful information about the grinding wheel that are identified. A correlation between grinding wheel condition and AE feature is established using ANN-based machine learning classifier.

Keywords Grinding · Acoustic emission (AE) · Machine learning classifiers · ANN

1 Introduction

Grinding is an abrasive machining process carried out using grinding wheel. The grinding wheel is embedded with randomly distributed abrasive particles like aluminium oxide, silicon carbide, diamond or cubic boron nitride. Abrasive particles of different grit sizes are used in the machining process based on metal removal rate and surface quality requirements. The grinding process starts with sharp abrasive particles, and the abrasive particles will become dull and lose its cutting ability due to wear. This phenomenon is called as the glazed condition of the grinding wheel. The wear is attributed due to attritious wear or progressive wear [1]. Also, during grinding of soft materials, workpiece chip will clog into the grinding wheel and

J. Revant · Rahul Sree Kumar · K. Rameshkumar (✉) · D. S. B. Mouli
Department of Mechanical Engineering, Amrita School of Engineering, Amrita Vishwa Vidyapeetham, Coimbatore 641112, India
e-mail: k_rameshkumar@cb.amrita.edu

S. Vijayan et al. (eds.), *Trends in Manufacturing and Engineering Management*, Lecture Notes in Mechanical Engineering,
https://doi.org/10.1007/978-981-15-4745-4_45

deteriorates the cutting ability of the abrasive wheel and called as clogging of the grinding wheel. Once the grinding wheel becomes glazed or clogged, it is necessary to subject the grinding wheel for dressing operation. The dressing is a grinding wheel conditioning operation, which will expose the abrasive grits and make the grinding wheel free from glazing and clogging. Truing operation is performed for new grinding wheels especially for a precision grinding process for ensuring concentricity of grinding wheel with the work surface. Dressing and truing operations are integrated as a single process for the conventional grinding wheels.

1.1 Sharpness Evaluation of the Grinding Wheel

The sharpness of the abrasive particles in a grinding wheel plays a prominent role in determining the capacity of the abrasive wheel to remove the material and achieving the required finish in the work surface. The sharpness of the wheel face is also directly related to heat generation and grinding forces developed during grinding. Nakayama et al. [2] proposed a procedure for evaluating the abrasive wheel sharpness. Alexandre et al. [3] proposed a test bench for measuring the sharpness of the abrasive wheel to study the dresser condition. In this study, the methodology proposed in [1, 3] is used to design and fabricate the sharpness measurement device which is used to establish the conditions of the grinding wheel.

1.2 Grinding Process Monitoring

Sensor-based machine condition monitoring is increasingly used in industries to predict the fault condition in advance to prevent the catastrophic failure of the machine tool. Vibration, motor current, optical, force, temperature, and acoustic emission sensors are commonly used for monitoring the conditions of the machines or processes. Among these sensors, the AE sensor is found suitable for precision machining processes such as grinding. AE is happening at higher frequencies in the range of 100 kHz to 1 MHz. It is possible to eliminate the external disturbances with much ease and only the signals relevant to the faulty conditions are utilized using the AE-based condition monitoring techniques.

AE-based condition monitoring of the grinding process has been investigated by many researchers. Neuro-fuzzy classification algorithms were proposed by Lezanski [4] using AE, force and vibration sensors. Liao et al. [5, 6] monitored grinding wheel conditions using hidden Markov models, genetic clustering algorithm, boosted classifiers and neural networks. Support vector machines (SVM) are used by Subrahmanya and Shin [7] for grinding condition monitoring. Yang and Yu [8] proposed AE-based grinding wheel wear monitoring system using SVM algorithms with AE signature. Advancements in sensor technology, hardware development, signal processing methods and feature extraction techniques were discussed in detail by Roth et al. [9]. ANN models were developed by Moia et al. [10] for monitoring dressing operation using AE signals. Dias et al. [11] identified a range of frequency for the occurrence of roundness, cylindricity and surface roughness. In a cylindrical grinding process, Arun et al. [12] predicted grinding wheel conditions in a cylindrical grinding process using classifiers such as SVM, decision trees and artificial neural network (ANN). Off late, decision tree, Naive Bayes, SVM and artificial neural network models are used to predict the tool conditions in high-speed precision machining process [13–15]. In a surface grinding process, dresser conditions of aluminium oxide (Al_2O_3) grinding wheel was studied by Alexandre et al. [3] with AE and fuzzy methods.

In this paper, the grinding wheel sharpness measurement device has been designed and fabricated. This device is used to identify the conditions of the grinding wheel. A grinding wear plot was established for the complete grinding cycle starting from sharp wheel condition to the wornout condition. Grinding experiments were carried out using an Al_2O_3 grinding wheel and a correlation is established between the features extracted from AE and grinding wheel sharpness condition. A statistical prediction model is established using feed-forward back-propagation neural network.

2 Grinding Wheel Sharpness Measurement Test Bench

The principle involved in measuring the sharpness is shown in Fig. 1. Cylindrical steel disc is subjected to ground with the surface of the abrasive wheel with the normal force, F_N. During the grinding test, the cylindrical disc is pushed by the normal force (F_N) generated due to the cutting action. The normal force intensity reduces when the contact area is increased. The displacement of the cylindrical disc is measured with the help of transducer. The amount of displacement depends on

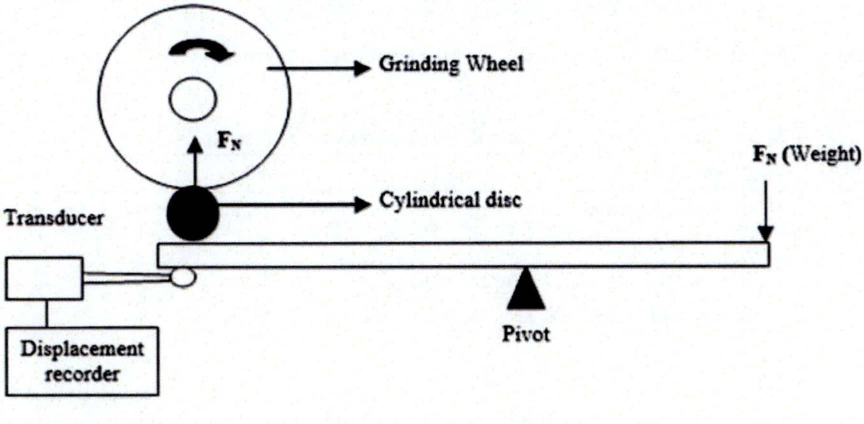

Fig. 1 Principle of grinding wheel sharpness measurement [2]

the sharpness of the grinding wheel face. A relationship between 'displacement' and 'time' is established to arrive at the sharpness of the grinding wheel. The sharpness of the grinding wheel is estimated using Eq. (1).

Abrasive grinding wheel sharpness,

$$K = \frac{2b\sqrt{8r}}{3F_N} a^{2/3} \tag{1}$$

where '**b**' is the width of the disc, '**r**' is the radius of the disc, '$\mathbf{F_N}$' is the normal force applied to the disc over the grinding wheel, '**a**' is the gradient of the characteristic curve of grinding wheel, i.e. regression line obtained from the plot of 'displacement' versus '$t^{(2/3)}$'. A device used for measuring the grinding wheel sharpness has been designed and fabricated based on the principle of sharpness measurement as shown in Fig. 1. The photograph of the device is shown in Fig. 2. The specification of the device is as shown in Table 1. The grinding wheel grade used in this study is 'AA60-K5-V8'. The details of the grinding wheel are shown in Table 2. The sharpness measurement device is fixed over the surface grinding machine table using a fixture.

Fig. 2 Photograph of the grinding wheel sharpness measurement device

Table 1 Specification of the grinding wheel sharpness measurement device

S. No.	Parameters	Value
1	Overall dimension	100 mm × 60 mm × 80 mm
2	Disc diameter	25 mm
3	Distance between disc centre and pivot point	55 mm
4	Counterweight added to keep the disc in contact	105 g

Table 2 Specification of the grinding wheel (AA60-K5-V8)

S. No	Parameters	Value
1	Grinding wheel size	OD—200 mm; ID—31.75 mm; Thickness—20 mm
2	Grinding wheel material	Aluminium oxide (White)
3	Grain size/Grit	Medium
4	Grade	Medium
5	Type of bond	Vitrified

2.1 Grinding Wheel Sharpness Evaluation

Grinding tests were conducted in a surface grinding machine by fixing the sharpness measurement device over the grinding table. The grinding wheel sharpness is evaluated for all the states of the abrasive wheel using the sharpness measurement device fabricated in this study. The grinding wheel sharpness is quantified for sharp and dull-wornout conditions separately. Time–displacement plots are drawn, and the gradient of the regression lines are computed. Time–displacement plots are shown in Figs. 3 and 4. Figure 3 shows the regression plot of the sharp condition, and Fig. 4 shows the dull-wornout condition. Sharpness parameters of grinding wheel of three different conditions are shown in Table 3.

Grinding experiments were repeated to validate the sharpness of the grinding wheel. From the results, it is observed that the sharpness is lower in the dull or wornout condition due to clogging or natural wear of the abrasive particles in the grinding wheel. Lower sharpness value indicates that the grinding wheel has to be dressed for improving its cutting ability. Sharpness values of the grinding wheel are indicated in Table 3 for sharp and dull conditions of the grinding wheel.

Grinding wheel wear plot is drawn as a function of metal removal rates of the grinding wheel and work material. Experiments were conducted starting with a dressed grinding wheel with sharp abrasive edges. Experiments were continued until the

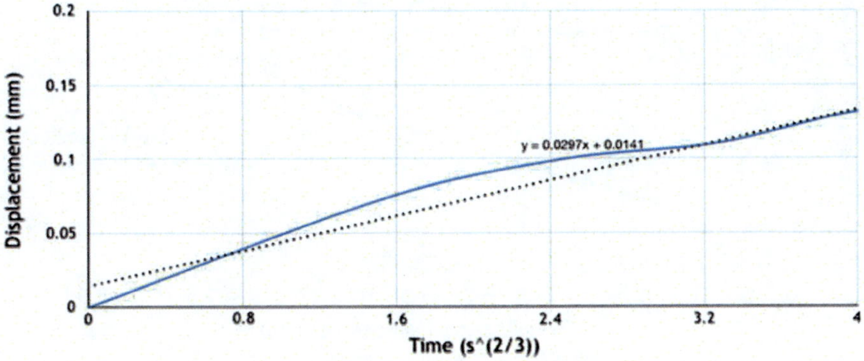

Fig. 3 Time–displacement plot of sharp grinding wheel condition

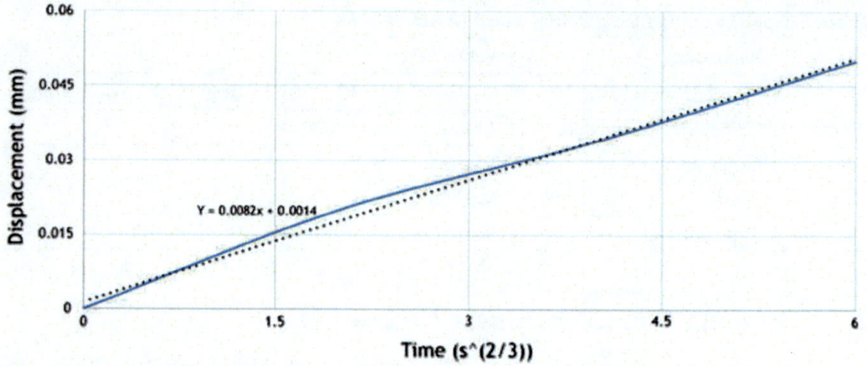

Fig. 4 Time–displacement plot of wornout grinding wheel condition

Table 3 Sharpness parameters

Parameters	Sharp condition	Dull condition
Disc width 'b'	4 mm	4 mm
Disc radius 'r'	12.5 mm	12.5 mm
Applied force 'F_N'	1.02 N	1.02 N
Regression gradient 'a'	0.4596	0.0145
Sharpness 'K'	4.505 mm^3/N s	1.063 mm^3/N s

wheel became a dull and glazed condition. During the grinding process at different intervals of time, the metal removal rates in workpiece and grinding wheels are recorded and grinding wheel wear plot is established and shown in Fig. 5. Three

Fig. 5 Grinding wheel wear plot indicating grinding wheel conditions

grinding wheel conditions, viz. Sharp-initial, sharp-intermediate and dull-wornout are also indicated in Fig. 5. Once the grinding wheel becomes dull, it is subjected to dressing operation.

3 AE Monitoring of Grinding Process

The main objective of AE monitoring of the grinding process is to predict the grinding wheel condition based on the sharpness parameter using the features extracted from the AE signature. The experimental set-up used in this study is shown in Fig. 6. The experimental set-up for AE measurement consists of (a) surface grinding machine, (b) AE sensor, (c) AE sensor attachment, (d) AE sensor data acquisition system, (e) AE win software, (d) machine learning classifier and (e) signal processing hardware.

AE sensor used in this study is supplied by M/s Physical Acoustic Corporation, USA. Micro 30 D sensor used in this study is capable of capturing AE signature in the frequency range 100–350 kHz. An AE attachment is used to hold the AE sensor to capture the acoustic emissions from the grinding wheel. AE sensor output is connected to a pre-amplifier and to the data acquisition system. The signal is converted into digital form and stored in the computer memory for further processing. AE features are extracted using AE-Win software supplied by M/s Mitras Corporation, USA. After extracting the features, those having wealthy information about the grinding wheel conditions are selected. The selected features are used for developing a statistical model using artificial neural network (ANN) classifier for predicting the grinding wheel condition.

Fig. 6 Experimental set-up for AE measurement

3.1 *Feature Selection*

Grinding experiments were carried out from the sharp-initial condition to the dull-wornout condition. AE signatures were acquired for the complete grinding cycle. AE signals were captured at a sampling rate of 1 MSPS with a threshold of 40 dB. AE features, viz. Count, RMS, amplitude, rise time, energy and duration are extracted from the raw signal. The features having useful information about the grinding wheel conditions are selected by plotting the feature value versus grinding time. It is observed that the AE features RMS and Count have a strong correlation with the grinding wheel conditions as shown in Fig. 7. RMS and Count feature data are used to train the feed-forward back-propagation neural network classifier for predicting the grinding wheel condition using a sharpness parameter.

4 Neural Network Modelling

Neural networks or multi-layer perceptrons have the ability to extract useful information from the AE sensor feature data and have the capability to extract patterns and detect the conditions of the grinding wheel. RMS and Count data extracted from the AE signature are provided as an input to neural network with the known condition of the surface grinding wheel. Training of the neural network is carried out using back-propagation algorithms. After training, the neural network model is tested with AE signal feature data for which the condition of the grinding wheel is not known. Training and testing procedure of neural network classifiers is shown in Fig. 8.

Multi-layer perceptron used in this study consists of three layers namely (a) input layer, (b) hidden layer and (c) output layer. A supervised learning technique, back-propagation algorithm, is used for training the network. Input layer consisting of neurons is based on the input data. In this case, 1363 AE data (RMS and Count) are used for training and testing the network. Among 1363 AE data, 320 are belonging to the sharp-initial state of the abrasive wheel. 636 samples are belonging to the sharp-intermediate state of the abrasive wheel. The balance 407 samples belong

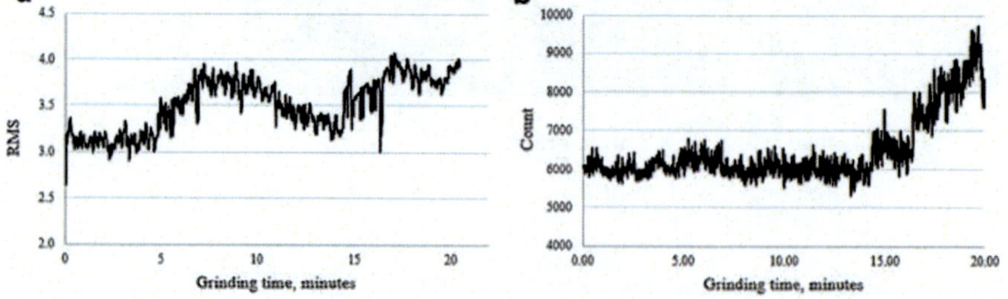

Fig. 7 **a** Grinding time versus RMS. **b** Grinding time versus Count

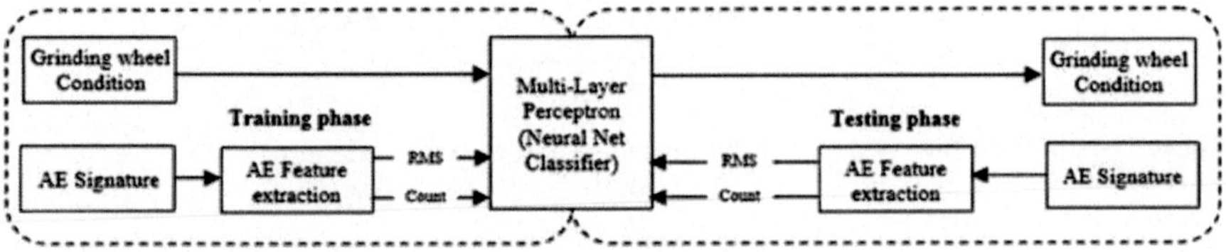

Fig. 8 Training and testing procedure in neural net classifier

to the dull-wornout state of the abrasive wheel. The samples were extracted by conducting the experiment in appropriate grinding wheel conditions established in this study. Experiments were conducted to optimize the number of hidden layers. It is found that the no. of hidden layers as '3' that yields good results. The sigmoidal transfer function is used. Learning rate and momentum are taken to be 0.3 and 0.2, respectively. A '10' fold cross-validation is used for training and testing of the neural network classifiers. In a '10'-fold cross-validation process, the data sets are divided into '10' groups and perform the training on nine groups and validate the results using the remaining group. To reduce the variability, 10 times of validation experiments are performed using different groups of training data leaving out the validation (test) data. The results are averaged over 10 times of validation. Neural network classifier is implemented using the Weka platform [16]. Performance of the classifier is studied using classification accuracy, misclassification rate, Kappa and ROC statistics.

4.1 Performance Measures

The performance of the classifier is studied using a confusion matrix. Performance measures, viz. (a) classification accuracy, (b) misclassification rate, (c) Kappa statistics and (d) receiver operating characteristic (ROC) are extracted from the confusion matrix. These matrices are used for evaluating the ANN classifier considered in this study. Classification accuracy of a classifier is computed using Eq. (2).

$$\text{Classification accuracy, \%} = \frac{\text{Total No. of correctly classified instances}}{\text{Total No. of instances}} \times 100 \quad (2)$$

Landis and Koch [17] proposed statistics called 'Kappa' for evaluating the performance of the classifiers. A 'Kappa' value of 0.81–1 is referred to as a 'perfect classifier'. A 'Kappa' value of 0.61–0.80 is a 'substantial classifier', 0.41–0.60 is a 'moderate' one, 0.21–0.40 is termed as 'fair' and 0–0.20 is called as a 'slight' classifier. The 'Kappa' statistic is computed using the expression shown in Eq. (3). Receiver operating characteristic (ROC) of the classifier is found using true positive and false positive rates obtained from the confusion matrix.

$$\text{Kappa} = \frac{p_o - p_e}{1 - p_e} = \frac{(\text{Observed accuracy} - \text{Expected accuracy})}{(1 - \text{Expected accuracy})} \tag{3}$$

where ‘p_o’ is the observed accuracy and ‘p_e’ is the expected accuracy.

5 Results and Discussions

AE features RMS and Count were given as an input to ANN classifiers for training. Training and testing data are acquired from the grinding wheel conditions arrived using the grinding wheel sharpness data. The confusion matrix obtained after training and testing the AE features and the ANN classifier performance is shown in Table 4. The ANN classifier is able to predict the grinding wheel conditions with an overall accuracy of 90.201%. Among 1363 instances, 1230 instances were classified correctly according to its conditions. Misclassifications of AE features are more in the features which were acquired in the boundary regions of ‘sharp-initial’ and ‘sharp-intermediate’. ANN is unable to differentiate the very minute variations in the sharpness of the abrasive particles in the boundary regions. There are 21 instances of ‘sharp-initial’ misclassified as ‘sharp-intermediate’ and 50 instances of ‘dull-wornout’ are misclassified as ‘sharp-intermediate’ condition.

Kappa statistics value of 0.8468 justifies the good performance of the ANN classifiers. Receiver operating characteristic (ROC) curve statistics provided by the ANN classifiers also demonstrate the diagnostic ability of the classifier and show that the ANN is able to distinguish the grinding wheel conditions with good accuracy. ROC statistics were arrived by using the true positive rate and false positive rates of the classifier. It is observed that ANN is able to classify the grinding wheel condition with good accuracy. The time taken for building the ANN model for grinding wheel

Table 4 Confusion matrix: Performance of ANN classifier

Grinding wheel conditions				
		Predicted condition		
		Sharp-Initial	Sharp-Intermediate	Dull-Wornout
Actual condition	Sharp-Initial	298	21	1
	Sharp-Intermediate	35	578	23
	Dull-Wornout	3	50	354
Correctly classified instances		1230		
Overall accuracy, Avg., %		90.2421		
Incorrectly classified instances		133		
Misclassification rate, Avg., %		9.7579		
Kappa statistics		0.8468		
ROC area		0.973		

condition prediction is 0.53 s using 'Intel i5 processor' with 4 GB internal memory. ANN model proposed in this study can be extended for grinding wheel condition prediction in real time.

6 Conclusions

A grinding wheel sharpness measurement device has been designed and fabricated in this study. This device will be useful in automated manufacturing environments for detecting the grinding wheel conditions. AE sensor is used for capturing the signals of the grinding process and grinding wheel conditions are established using the grinding wheel wear plot. Time domain AE features such as RMS and Count were found to have a good association with the abrasive wheel conditions. A statistical model using ANN is developed for estimating the grinding wheel condition. AE sensor with ANN-based machine learning classifiers is able to predict the grinding wheel conditions with an accuracy of 90.2%. Kappa and ROC statistics are also justifying the effectiveness of ANN classifier in predicting the conditions of the grinding wheel. Further studies can be taken up to develop an AE-based real-time grinding wheel condition monitoring system using the data and knowledge generated in this study.

Acknowledgements This research is supported by Directorate of Extramural Research and Intellectual Property Rights (ER & IPR), Defense Research and Development Organization (DRDO), ERIP/ER/0803740/M/01/1194, 13 January 2010.

References

1. Jackson MJ, Davim JP (2011) Machining with abrasives. Springer, New York, pp 4–5
2. Nakayama K, Takagi JI, Irie E, Okuno K (1980) Sharpness evaluation of grinding wheel face by the grinding of steel ball. CIRP Ann 29(1):227–231
3. Alexandre FA, Lopes WN, Dotto FRL, Ferreira FI, Aguiar PR, Bianchi EC, Lopes JC (2018) Tool condition monitoring of Aluminum oxide grinding wheel using AE and fuzzy model. Int J Adv Manuf Technol 96(1–4):67–79
4. Lezanski P (2001) An intelligent system for grinding wheel condition monitoring. J Mater Process Technol 109(3):258–263
5. Liao TW, Ting CF, Qu J, Blau PJ (2007) A wavelet-based methodology for grinding wheel condition monitoring. Int J Mach Tools Manuf 47(3):580–592
6. Liao TW (2010) Feature extraction and selection from acoustic emission signals with an application in grinding wheel condition monitoring. Eng Appl Artif Intell 23(1):74–84
7. Subrahmanya N, Shin YC (2008) Automated sensor selection and fusion for monitoring and diagnostics of plunge grinding. J Manuf Sci Eng 130(3):031014
8. Yang Z, Yu Z (2012) Grinding wheel wear monitoring based on wavelet analysis and support vector machine. Int J Adv Manuf Technol 62(1–4):107–121
9. Roth JT, Djurdjanovic D, Yang X, Mears L, Kurfess T (2010) Quality and inspection of machining operations: tool condition monitoring. J Manuf Sci Eng 132(4):041015

10. Moia DFG, Thomazella IH, Aguiar PR, Bianchi EC, Martins CHR, Marchi M (2015) Tool condition monitoring of aluminum oxide grinding wheel in dressing operation using acoustic emission and neural networks. J Brazilian Soc Mech Sci Eng 37(2):627–640
11. Dias EA, Pereira FB, Ribeiro Filho SLM, Brandão LC (2016) Monitoring of through-feed centerless grinding processes with acoustic emission signals. Measurement 94:71–79
12. Arun A, Rameshkumar K, Unnikrishnan D, Sumesh A (2018) Tool condition monitoring of cylindrical grinding process using acoustic emission sensor. Mater Today: Proc 5(5):11888–11899
13. Krishnakumar P, Rameshkumar K, Ramachandran KI (2018) Acoustic emission-based tool condition classification in a precision high-speed machining of titanium alloy: a machine learning approach. Int J Comput Intell Appl 17(03):1850017
14. Krishnakumar P, Rameshkumar K, Ramachandran KI (2018) Feature level fusion of vibration and acoustic emission signals in tool condition monitoring using machine learning classifiers. Int J Progn Health Manag 9:1–15
15. Krishnakumar P, Rameshkumar K, Ramachandran KI (2018) Machine learning based tool condition monitoring using acoustic and vibration data in high speed milling. Int J Intell Decis Technol 1:1–18
16. https://www.cs.waikato.ac.nz/~ml/weka/
17. Landis JR, Koch GG (1977) The measurement of observer agreement for categorical data. Biometrics 159–174

Turning Process Characteristics of Aluminium Matrix Hybrid Composite Using Grey Relational Surface Methodology

A. Gnanavelbabu, V. Arunachalam, K. T. Sunu Surendran, R. Saranraj, and K. Rajkumar

Abstract Aluminium metal matrix composites have generated huge attention in modern industries because of its high strength-to-weight ratio, low density and corrosion resistance. Turning is a widely used material removal process to produce cylindrical products. Presence of solid lubricant reinforcements such as graphite (Gr), carbon nanotube (CNT) and hexagonal boron nitride (hBN) could increase the feasibility of turning on aluminium metal matrix composites. Aluminium alloy 6061 reinforced with B_4C (5–15 vol.%) and hBN (10 vol.%) particulates were fabricated using stir casting process. The effect of process parameters such as feed, speed, depth of cut and volume percentage of boron carbide has been investigated to explore the impact on tangential force, cutting power and tool wear. Experiments were conducted using Taguchi's L_{27} orthogonal array (OA). The factors affecting the responses were optimized by applying grey-response surface methodology (G-RSM) using grey relational grade (GRG) calculated by normalized S/N ratio and grey relational coefficient (GRC).

Keywords Aluminium hybrid composites · Turning · Grey relational grade · Response surface methodology

1 Introduction

Metal matrix composite (MMC) has been considered as an important class of material, as it progressively replaces conventional materials utilized in structural engineering, aeronautics, mechanical engineering and in many other sectors [1]. MMCs are considered for engineering applications where high specific strength, stiffness and lightweight are important. Common matrix metals include aluminium, magnesium,

A. Gnanavelbabu (✉) · V. Arunachalam · K. T. Sunu Surendran · R. Saranraj
Department of Industrial Engineering, CEG, Anna University, Chennai 600025, India
e-mail: agbabu@annauniv.edu

K. Rajkumar
Department of Mechanical Engineering, Sri Sivasubramaniya Nadar College of Engineering, Chennai 603110, India

S. Vijayan et al. (eds.), *Trends in Manufacturing and Engineering Management*, Lecture Notes in Mechanical Engineering,
https://doi.org/10.1007/978-981-15-4745-4_46

copper and titanium in its pure or alloy forms. Aluminium metal matrix composites (AMMCs) are suitable material for aerospace, defence and automotive applications. Generally, aluminium (6 series) alloys have been considered broadly because of its merits such as strength, castability, formability, corrosion resistance and low cost comparing to other aluminium alloys.

Ceramic reinforcements such as silicon carbide (SiC), boron carbide (B_4C) and titanium carbide (TiC) are common reinforcements in AMMC [2, 3]. Incorporation of such reinforcements improved the specific properties of the metal matrix phase [4, 5]. In addition, graphite, graphene, carbon nanotube (CNT) and hexagonal boron nitride (hBN) are commonly used solid lubricants in developing AMMCs. A. Gnanavelbabu et al. investigated the cutting quality characteristics of AA6061-B_4C-hBN hybrid composites by abrasive water jet machining and reported favourable results [6]. C. Muthazhagan et al. found a positive influence of solid lubricant on the mechanical properties of aluminium-B_4C composites. Also, they found that the addition of B_4C conversely increased the hardness of the composites [7].

Design of experiment (DOE) plays a vital role in reducing the number, time and cost of the experiments. Optimizations of process parameters are important for improving the quality characteristics of the responses. W.H. Yang et al. optimized the cutting parameters for turning operations for steel using Taguchi's orthogonal array, S/N ratio ANOVA [8]. A. J. Makadia et al. investigated the influence of tool geometry (nose radius) and cutting parameters (cutting speed, feed and depth of cut) on the surface finish obtained in turning of mild steel by applying Taguchi-response surface methodology (T-RSM) [9]. R. Adalarasan et al. investigated the quality characteristics of laser cutting parameters using grey-response surface methodology (G-RSM) and found the optimum level of cutting parameters by applying grey relational analysis and G-RSM [10]. A. Dey et al. predicted the optimal combination of processing parameters by hybrid approach of grey-based response surface methodology (G-RSM) and found the significant model using grey-based approach [11].

It is observed from the literature that optimization of multiple responses based on orthogonal array with GRA tends to take advantage of both the Taguchi method and the GRA. Hence, an approach named GRSM which combines with GRA is presented to find out the optimal parameter setting. This paper presents a detailed experimental analysis on the turning process of AA6061-B_4C-hBN hybrid composites by applying G-RSM for optimizing the machining quality characteristics such as tangential force, cutting power and tool wear.

2 Materials and Methods

Aluminium alloy 6061 is the most commonly used alloy which has a better combination of mechanical properties and it was chosen as the matrix material.B_4C and hBN were the reinforcement particulates of size 10–25 μm and 5 μm, respectively. The composition of the fabricated hybrid composites is shown in Table 1. Aluminium hybrid composites were fabricated using two-stage stir casting process.

Table 1 Composition of fabricated composites

Composite	hBN (vol.%)	B_4C (vol.%)	AA6061 (vol.%)
Comp A	10	5	85
Comp B	10	10	80
Comp C	10	15	75

Table 2 Factors and levels of the input parameters

Factors	Speed (RPM) [A]	Feed (mm/rev) [B]	Depth of cut (mm) [C]	Reinforcement (vol.%) [D]
Low	60	0.05	0.2	5
Medium	120	0.10	0.4	10
High	180	0.15	0.6	15

The fabricated composites comp A, comp B and comp C were cut into required dimensions for performing turning operation. Taguchi's L_{27} orthogonal array was used to collect the experimental data and to analyse the characteristics of turning parameters of AMMC. Turning parameters namely speed, feed, depth of cut and solid lubricant are chosen for controlling the responses and each parameter is designed with three levels, denoted by, low, medium and high as shown in Table 2.

CNC turning centre with 360–480 V and 40/60 Hz was used to perform the experiments. Polycrystalline diamond (PCD) was used as the cutting tool. PCD tools are preferred instead of polycrystalline cubic boron nitride (PCBN) tools as they exhibit high fracture resistance and low adhesion property. Also, it improves the machining productivity and tool life characteristics. Each experiment was carried out using a new cutting edge every time to obtain accurate response values which are tangential force, cutting power and tool wear. The values of tangential force (F_z) and tool wear (T_w) are obtained directly from CNC turning machine, and the cutting power (P_c) is calculated using Eq. 1.

$$P_c = \frac{D * F * S * K_c}{60 * 10^3 * \eta} \tag{1}$$

where P_c–cutting power, F—feed, S—speed, D—depth of cut, K_c—Specific cutting force and η—machine coefficient. The results of CNC turning process are shown in Table 3.

The grey relational analysis was performed for the multi-performance optimization, in which the optimization processes are gone through, until the grey relational grade (GRG) obtained. Higher the value of GRG, closer the parameter to its optimum value. The resultant GRG value reflects the single optimization value of the multiple experimental performances. Within the range of sequences to acquire exact solutions, one has to pre-process the data which are related to group of sequences called as grey relational generation by Deng, J.L. (1989). Converting an original sequence

Table 3 Experimental data

Trail	A (RPM)	B (mm/rev)	C (mm)	D (vol.%)	F_z (N)	P_c (kW)	T_w (mm)
1	60	0.05	0.2	5	5	1.571	0.071
2	60	0.05	0.2	10	10	3.143	0.045
3	60	0.05	0.2	15	8	2.514	0.1
4	60	0.1	0.4	5	9	2.829	0.152
5	60	0.1	0.4	10	11	3.457	0.123
6	60	0.1	0.4	15	12	3.771	0.121
7	60	0.15	0.6	5	11	3.457	0.179
8	60	0.15	0.6	10	15	4.714	0.145
9	60	0.15	0.6	15	13	4.086	0.162
10	120	0.05	0.4	5	13	8.171	0.209
11	120	0.05	0.4	10	13	8.171	0.175
12	120	0.05	0.4	15	7	4.4	0.185
13	120	0.1	0.6	5	12	7.543	0.222
14	120	0.1	0.6	10	14	8.8	0.2
15	120	0.1	0.6	15	12	7.543	0.201
16	120	0.15	0.2	5	16	10.057	0.256
17	120	0.15	0.2	10	15	9.429	0.223
18	120	0.15	0.2	15	13	8.171	0.225
19	180	0.05	0.6	5	11	10.371	0.289
20	180	0.05	0.6	10	18	16.971	0.245
21	180	0.05	0.6	15	11	10.371	0.262
22	180	0.1	0.2	5	14	13.2	0.302
23	180	0.1	0.2	10	19	17.914	0.269
24	180	0.1	0.2	15	12	11.314	0.292
25	180	0.15	0.4	5	15	14.143	0.322
26	180	0.15	0.4	10	20	18.857	0.28
27	180	0.15	0.4	15	14	13.2	0.315

into a decimal sequence between 0.00 and 1.00 is known as pre-processing raw data. The pre-processing starts with calculating signal-to-noise ratio (S/N). S/N ratio was calculated to reduce the variability in the process. The objective of this study is to minimize the tangential force, cutting power and tool wear for better characteristic performance. Hence, the S/N ratio was calculated by considering "smaller the better" condition for all three responses. The obtained sequence normalized with the "smaller is better" condition.

Further, GRC was calculated to express the relationship between the best and actualized S/N ratio. Finally, the average of grey relational coefficients was taken to get grey relational grade (GRG). Further, response surface methodology (RSM) was applied to optimize and generate ANOVA and regression equations. Surface plots and contour plots were generated using G-RSM. Statistical analysis of the experimental data was performed using design expert software.

3 Results and Discussions

The current optimization is focused on the improvement of minimum tangential force, cutting power and tool wear in turning process. From GRC values of F_z, P_c and T_w, grey relational grade (GRG) was calculated. The experiments of grey relational analysis were ranked based on the higher order of their GRG values. The GRG values for all experiments are shown in Table 4. Experiment no. 26 has the highest GRG (0.9611). Therefore, experiment no. 26 is the optimal process parameters setting for minimum tangential force, cutting power and tool wear simultaneously (i.e. the best multi-performance characteristics) among the twenty-seven experiments. Since there is a possibility for a certain degree of uncertainties on GRG, RSM model is employed on GRG for further optimizing the responses.

The GRG representing the multiple responses was taken as a single performance and used in the generation of response surfaces. The RSM method involves developing equation of quadratic model using the decision variables. The generated regression equation creates the way to clear about the behaviour of system by quantifying the association and thereby exploring the individual and interaction effects of turning parameters on the responses. The GRG values are entered in the design expert software and the mathematical model are obtained as shown in Eq. 2.

$$\begin{aligned}\mathbf{GRG} = {} & +0.66 + 0.15 * \mathrm{A} + 0.048 * \mathrm{B} + 0.011 * \mathrm{C} - 0.014 * \mathrm{D} \\ & + 0.015 * \mathrm{A} * \mathrm{B} - 0.032 * \mathrm{A} * \mathrm{C} - 0.20 * \mathrm{A} * \mathrm{D} \\ & - 0.27 * \mathrm{B} * \mathrm{C} - 1.025\mathrm{E} - 003 * \mathrm{B} * \mathrm{D} + 6.007\mathrm{E} - 003 * \mathrm{C} * \mathrm{D} \\ & + 0.023 * \mathrm{A}^2 + 0.013 * \mathrm{B}^2 - 0.069 * \mathrm{D}\end{aligned} \tag{2}$$

The coefficient of the regression equation is determined and contour and surface plots are plotted for the p values lower than the significance level (α) equals to 0.05. The results of response surface regression and analysis of variance for the grey relational grades are shown in Table 5.

Based on the results from ANOVA, the response surfaced models generated are significant as shown in Table 6.

Referring to the ANOVA table, the regression model p value showed 0.0001, which indicated that the model was more significant. Thus, the model is more suitable to predict the values. The significance of the regression fitted model is determined by incorporating the coefficient correlation. The obtained R-sq. and R-sq. (adj) for the model are 94.24 and 88.48, respectively. Thus, the statistical measure of value shows that the model is fitted closely. Response surface optimizer in design is used to find the parameters which provide the best results. The goal is to maximize the GRG

Table 4 GRG calculation

Trail	S/N ratio (normalized)			GRC			GRG
	F_z	P_c	T_w	F_z	P_c	T_w	
1	0	0	0.231	0.333	0.333	0.394	0.3523
2	0.5	0.279	0	0.5	0.409	0.333	0.4158
3	0.339	0.189	0.405	0.430	0.381	0.456	0.4214
4	0.423	0.236	0.618	0.464	0.395	0.567	0.4723
5	0.568	0.317	0.510	0.536	0.422	0.505	0.4866
6	0.631	0.352	0.502	0.755	0.435	0.501	0.5028
7	0.568	0.317	0.701	0.536	0.422	0.626	0.5243
8	0.792	0.442	0.594	0.706	0.472	0.552	0.5754
9	0.689	0.384	0.653	0.616	0.484	0.585	0.5483
10	0.689	0.663	0.780	0.616	0.597	0.694	0.6343
11	0.689	0.663	0.690	0.616	0.597	0.617	0.6101
12	0.242	0.414	0.718	0.397	0.460	0.639	0.4956
13	0.631	0.631	0.811	0.575	0.575	0.725	0.6225
14	0.742	0.693	0.757	0.660	0.619	0.673	0.6501
15	0.631	0.631	0.760	0.575	0.575	0.676	0.6070
16	0.839	0.747	0.883	0.756	0.664	0.810	0.7407
17	0.792	0.721	0.813	0.706	0.641	0.728	0.6904
18	0.689	0.663	0.817	0.616	0.597	0.732	0.6463
19	0.568	0.759	0.955	0.536	0.675	0.909	0.6996
20	0.923	0.957	0.861	0.868	0.921	0.782	0.8604
21	0.568	0.759	0.895	0.536	0.675	0.826	0.6764
22	0.742	0.856	0.967	0.660	0.776	0.938	0.7886
23	0.963	0.979	0.908	0.931	0.960	0.845	0.9147
24	0.631	0.794	0.950	0.575	0.708	0.909	0.7271
25	0.792	0.884	1	0.706	0.812	1	0.8356
26	1	1	0.928	1	1	0.875	0.9611
27	0.742	0.856	0.988	0.660	0.776	0.978	0.8009

values so that the response values can be optimized. The optimal parameter settings are shown in Table 7 at which our objective function can be satisfied.

Figure 1 illustrates the plot of actual versus predicted values of responses and it is observed that all the points of the experimental runs are spotted randomly inside the reliable range of residuals.

Surface plots and contour plots for GRGs are given in the Figs. 2 and 3, respectively. The response surface plots are drawn based on the generated regression equation associating the dependent variable GRGs with the independent variables such as tangential force, cutting power and tool wear. Increased level of speed and feed

Table 5 Response surface regression

Term	Coefficient	SE Coefficient	T value	P value
Constant	0.6883	0.0340	20.26	0.000
Speed	0.1647	0.0125	13.13	0.000
Feed	0.0619	0.0177	3.49	0.004
DOC	−0.002	0.0177	−0.13	0.898
% OF hBN	−0.013	0.0125	−1.08	0.298
Speed*Speed	0.0094	0.0217	0.43	0.672
Feed*Feed	−0.006	0.0251	−0.03	0.980
DOC*DOC	−0.013	0.0251	−0.54	0.595
% of hBN* % of hBN	0.068	0.0217	−3.15	0.008
Speed*Feed	−0.012	0.0251	−0.48	0.638
Speed*DOC	−0.004	0.0251	−0.19	0.852
Speed*% of hBN	−0.020	0.0154	−1.32	0.210
Feed*% of hBN	−0.001	0.0154	−0.07	0.948
DOC*% of hBN	0.0060	0.0154	0.39	0.702

improved the GRG values and responses as shown in Fig. 2. It is observed that speed and feed are directly proportional to the responses, while the volume percentage of hBN is inversely proportional to the responses. From the surface plot, the maximum value of GRG is obtained for Comp B with high cutting speed, high feed and medium depth of cut.

The contour plots (Fig. 3) shows the graph of GRG and process parameters, where the dark green colour indicates higher the GRG value and light green colour indicates lower the GRG value. They hold values in the contour plots which are 120 rpm, 0.1 mm/rev, 0.4 mm and 10% for speed, feed and depth of cut and volume percentage of hBN, respectively. Lower the cutting speed lead to striations and cause rough surface finish while higher the cutting speed caused thermal damage to the material. Hence, a moderate level (120 rpm) of cutting speed is desired for turning operation.

4 Conclusion

The G-RSM optimization tool was applied for the optimization of CNC turning process parameters on AA6061-B4C-hBN hybrid composites and the following conclusions were drawn.

1. The derived second-order polynomial equation has shown good correlation between the predicted and experimental values.

Table 6 ANOVA for GRG

Source	Sum of squares	DOF	Mean square	F value	P value	Remarks
Model	0.60	13	0.046	16.35	<0.0001	Significant
A Speed	0.082	1	0.082	29.00	0.0001	
B Feed	8.732E − 003	1	8.732E − 003	2.96	0.1093	
C DOC	5.787E − 004	1	5.787E − 004	0.20	0.6587	
D % hBN	3.324E − 003	1	3.324E − 003	1.17	0.2984	
AB	3.491E − 004	1	3.491E − 004	0.12	0.7312	
AC	1.160E − 003	1	1.160E − 003	0.41	0.5333	
AD	4.918E − 003	1	4.918E − 003	1.74	0.2104	
BC	8.403E − 004	1	8.403E − 004	0.30	0.5952	
BD	1.260E − 005	1	1.260E − 005	4.450E − 003	0.9478	
CD	4.332E − 004	1	4.332E − 004	0.15	0.7021	
A^2	1.371E − 003	1	1.371E − 003	0.48	0.4989	
B^2	4.368E − 004	1	4.368E − 004	0.15	0.7009	
C^2	0.00	0				
D^2	0.028	1	0.028		0.0076	
Residual	0.037	13	2.833E − 003	9.94		
Cor total	0.64	26				
Std.dev.—0.053	R^2—0.9424		Adj. R^2—0.8848		Adq. precision—14.078	

Table 7 GRSM results at 10% hBN

Input	Optimum level
Speed	180
Feed	0.15
Depth of cut	0.3374
% hBN	8.6364

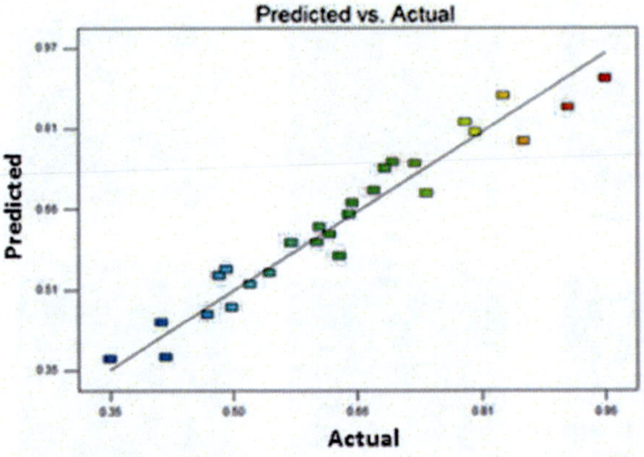

Fig. 1 Predicted versus actual response plot

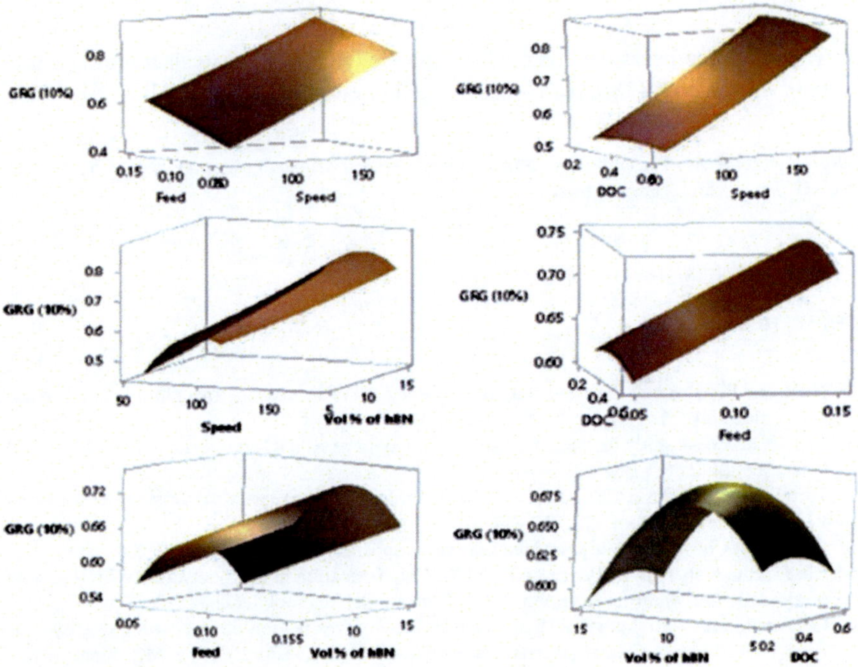

Fig. 2 Surface plot for GRG

2. Experiment no. 26 has the highest GRG value (0.9611) and it is the optimal process parameters setting to obtain the best multiperformance characteristics.
3. Effect of speed is much more prominent than the effects of feed rate and depth of cut, on the tangential force, cutting force and tool wear.

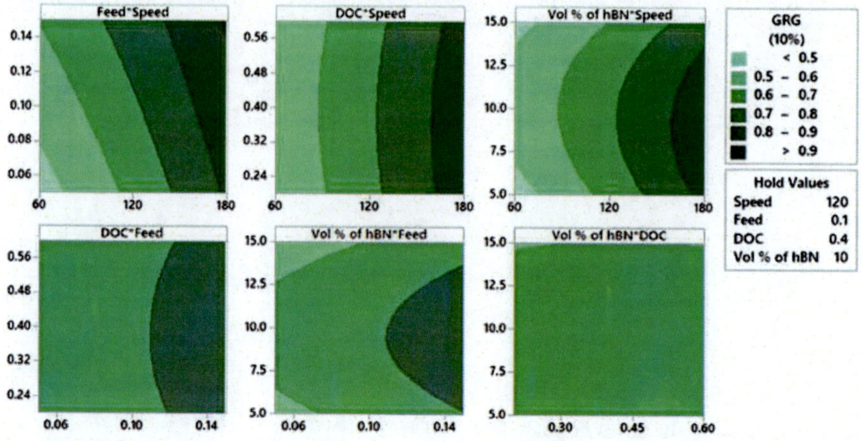

Fig. 3 Contour plot for GR

4. The optimal process parameters from the grey relational analysis for turning process were speed—180, feed—0.15, depth of cut—0.3374 and %B_4C—8.6364.

Acknowledgements The authors gratefully acknowledge the financial support from "DST-PURSE Phase II" for research consumables.

References

1. Nicholas JC, Boswell B, Davies J, Islam MN (2017) Review of machining metal matrix composites. Int J Adv Manuf Technol 90:2429–2440
2. Rai RN (2006) A study on the machinability behavior of Al-TiC composite prepared by in situ technique. Mater Sci Eng 428:34–40
3. Torralba JM, Da Costa CE, Velasco F (2003) P/M aluminum matrix composites: an overview. J Mater Process Technol 133:203–206
4. Chawla N, Chawla KK (2013) Metal matrix composites, 2nd edn. Springer, New York
5. Lin JT, Bhattacharyya D, Ferguson B (1998) Chip formation in the machining of SiC-particle-reinforced aluminium-matrix composites. Compos Sci Technol 58(2):285–291
6. Gnanavelbabu A, Rajkumar K, Saravanan P (2018) Investigation on the cutting quality characteristics of abrasive water jet machining of AA6061-B4C-Hbn hybrid MMC. Mater Manuf Process (45):1313–1323
7. Muthazhagan C, Gnanavelbabu A, Bhaskar GB, Rajkumar K (2014) Influence of graphite reinforcement on mechanical properties of Al-boron carbide composites. Adv Mater Res 495:398–402
8. Makadia AJ, Nanavati JI (2013) Optimization of tool geometry and cutting parameters for turning operations based on response surface methodology. In: Proceedings of AMPC, Chennai, pp 6–8
9. Yang WH, Tarng YS (1998) Design optimization of cutting parameters for turning operations based on the Taguchi method. J Mater Process Technol 84:122–129

10. Adalarasan R, Santhanakumar R, Rajmohan M (2015) Optimization of laser cutting parameters for Al6061/SiCp/Al_2O_3 composite using grey based response surface methodology (GRSM). Measurement 73:596–606
11. Dey A, Debnath S, Pandey KM (2017) Optimization of electrical discharge machining process parameters forAl6061/cenosphere composite using grey based hybrid approach. Transform Nonferrous Metals Soc China 27:998–1010

Corrosion Behaviour of Hot Rolled AA8015 in Natural Seawater at 1.37 μm Surface Roughness

O. Olaogun and Esther Titilayo Akinlabi

Abstract Aluminium alloy 8015 (AA8015) is a general-purpose alloy having a wide range of applications that include household appliances and marine ship structures. Electrochemical testing of hot rolled AA8015 with a particular roughness immersed in natural seawater solution was investigated. The AA8015 under investigation in this research is manufactured by a continuous casting method and subjected to mechanical hot rolling process at a temperature above its recrystallization temperature into 7 mm thickness plate. A roughness average of 1.37 μm with three cold-mounted samples was accomplished on an automatic dual multi-purpose grinding-polishing machine with the aid of 320 grit silicon carbide waterproof paper sheet. Corrosion experiments via electrochemical reactions were conducted on the corrosion samples immersed in natural seawater with the aid of a computerized controlled Ivium potentiostat in an open electrochemical polarization cell arrangement at ambient temperature condition. Observation of the surface morphologies of the corroded samples was achieved by a high megapixel camera and scanning electron microscope (SEM). Findings show asymmetric polarization curves for all the investigated samples, showing an actively corroding region. Consequently, the SEM examination confirms corrosion occurring at localized areas in the form of pitting.

Keywords AA8015 · Electrochemical corrosion · Mechanical hot rolling · Surface roughness

---

O. Olaogun (✉) · E. T. Akinlabi
Department of Mechanical Engineering Science, University of Johannesburg, Johannesburg, South Africa
e-mail: yinka.olaoluwa@live.com

E. T. Akinlabi
Department of Mechanical Engineering, Covenant University, Ota, Nigeria

O. Olaogun
Department of Mechanical Engineering, Kwara State University, Malete, Nigeria

S. Vijayan et al. (eds.), *Trends in Manufacturing and Engineering Management*, Lecture Notes in Mechanical Engineering,
https://doi.org/10.1007/978-981-15-4745-4_47

1 Introduction

Aluminium alloy 8015, among other aluminium alloys, is equally known for its excellent corrosion resistant property owing to the very thin oxide film layer formed on the metallic surface. Hot rolled aluminium alloy 8015 is distinctly based on the fact that it has undergone mechanical hot rolling process achieved with heat. The hot rolled aluminium alloy 8015 product has the best applications where dimensional tolerances are not as important as the overall material strength and likewise, surface finish is not a major concern [1].

Corrosion of aluminium alloys has significant records by various scholars [2–12]. The presence of intermetallic particles in aluminium alloys, usually on the surface, causes a weak thin oxide film layer [13]. Research shows that these intermetallic particles act as a cathodic role to the aluminium matrix in corrosive environments containing chloride ions [14]. These intermetallic alloying element particles have a varying degree of solid solubility in the aluminium matrix. Polmear et al. [15] reveal the solid solubility of individual alloying elements in aluminium. Findings show that zinc, silver, magnesium and copper have high solid solubility compared to iron, nickel, manganese and silicon, thereby suggesting that intermetallic alloying elements of the latter are the main causes of localized corrosion in aluminium alloys.

Electrochemical corrosion studies of aluminium alloys in natural seawater have been reported in the literature [16–18], and all reveal the mode of corrosion by pitting. Information on electrochemical corrosion of hot rolled aluminium alloy 8015 is yet to be reported in the literature.

This present work was motivated to investigate the corrosion behaviour of hot rolled aluminium alloy 8015 in natural seawater at 1.37 μm surface roughness average. Electrochemical corrosion test using Ivium potentiostat was used to analyse the open-circuit potential (OCP), polarization measurement and corrosion rate of the aluminium alloy 8015. The aluminium alloy 8015 specimen samples, before and after electrochemical corrosion, were analysed with TESCAN VEGA scanning electron microscope having energy dispersive spectrometry to observe the microstructure and morphologies of the corroded surface. The details of the corrosion behaviour are discussed in the present study.

2 Methods

2.1 Material and Corrosion Sample Preparation

The electrochemical experiments were performed with three specimen samples of aluminium alloy 8015 prepared by a continuous casting method and mechanical hot rolling process. This alloy under study was obtained industrially from Tower

Aluminium Rolling Mill Nigeria Plc, Sango-Ota, Nigeria. The elemental composition as determined by Bruker Elemental Optical Emission Spectrometry is shown in Table 1.

Prior to the hot rolling process, the aluminium alloy 8015 slab formed by solidification of metallic solution of pure aluminium and the alloying elements were fed into the furnace and heated to above recrystallization temperature up to 55 °C. The molten alloy was continuously casted and solidified using roll casters into a continuous strip of approximately 20 mm thickness. Further mechanical deformation process in a hot rolling mill in tandem reduced the thickness to 7 mm plate.

The hot rolled aluminium alloy 8015 specimen samples having thickness of 7 mm were prepared for electrochemical corrosion testing according to the ASTM standard G1-03 [19]. The specimen samples were square-cut into dimensions 10 mm by 10 mm. Insulated copper wire with SANS 60227-5 specification, cut into short lengths, was adhered to the cut specimen samples using epoxy glue for connectivity. Subsequent cold mounting was achieved and further grinding with 320-grit SiC paper was done. Surface roughness measurement using HOMMEL-ETAMIC TURBO WAVE V7.53 roughness and contour metrology before and after sample preparation reads 2.50 μm and 1.37 μm, respectively.

2.2 Electrochemical Corrosion Test Procedure

Potentiodynamic electrochemical experiments were conducted in a conventional three-electrode open glass-cell configuration containing 200 ml natural seawater solution using a CompactStat Potentiostat, an Ivium Technologies product, with corrosion analysis software. This instrument controls the driving mechanism required for electrochemical reactions that takes place on the working electrode. The hot rolled aluminium 8015-alloy with 1.37 μm surface roughness serves as working electrode, a cylindrical coiled platinum wire as counter-electrode, and Ag/AgCl as a reference electrode. The three-electrode cell was utilized to allow the potential taking place between working and reference electrodes to be controlled. The electrochemical experiments were performed at room temperature of 20 ± 2 °C within the succeeding order: Open-circuit potential (OCP) versus time for an hour duration and polarization measurements over a potential range starting from 250 mV lower than the OCP value and ending at 250 mV directly above the OCP value at a scan rate of 0.67 mV/s. These experiments procedures were followed according to ASTM standard G69 and G5 for corrosion potential and polarization measurements, respectively [20, 21]. The hot rolled metal alloy surface before and after the electrochemical test was subjected to macro- and micro-observation by a high pixel camera and scanning electron microscope (SEM), respectively.

Table 1 Measured element composition of aluminium alloy 8015

	Si	Fe	Cu	Mn	Mg	Cr	Zn	Ti	Al
Weight %	0.4780	1.3380	0.1460	0.0740	0.0350	0.00670	0.0720	0.0110	Bal.

3 Results and Discussion

3.1 Open-Circuit Potential and Potentiodynamic Polarization

The progression of the open-circuit potential (OCP) as a function of time for the hot rolled aluminium alloy 8015 samples (at 1.37 μm surface roughness) exposed to natural seawater electrolyte solution is presented in Fig. 1.

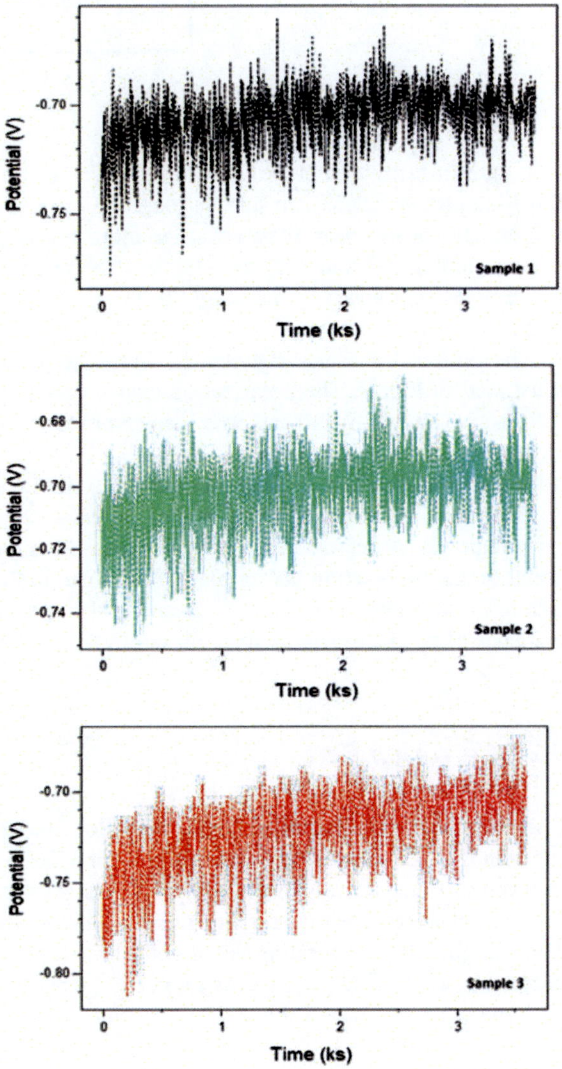

Fig. 1 Display of open-circuit potential (OCP) versus time for hot rolled aluminium alloy 8015 ($R_a \approx$ 1.37 μm) in natural seawater

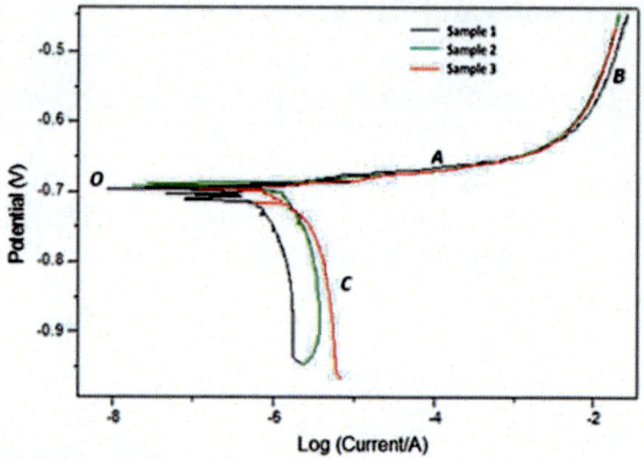

Fig. 2 Polarization curves of hot rolled aluminium alloy 8015 samples immersed in natural seawater at $R_a \approx$ 1.37 μm

Figure 1 illustrates that the OCP variation for the three hot rolled aluminium alloy 8015 samples with 1.37 μm surface roughness slightly shifted to the lower negative values. This observation shows that the inert thin-film oxide layers of the samples exhibit similar resistance to seawater electrolyte attack. The recorded OCP values for samples 1, 2 and 3 against Ag/AgCl are −0.693 V, −0.696 V and −0.714 V, respectively.

The polarization curves generated from the potentiodynamic experiments are illustrated in Fig. 2. The polarization curves show the anodic scan (progresses in the positive direction) and cathodic scan (progresses in the negative direction). Point O denotes the open circuit or rest potential. At this point, the measured current is relatively zero, confirming that the sum of the anodic and cathodic reaction rates on the hot rolled AA8015 electrode is zero. In the anodic scan, the potential increases up to region B. At region A, the steeply rise of the anodic current confirms dominant metal oxidation reaction taking place. Likewise, in the cathodic scan, a fast reduction reaction is observed. Rapid decrease in the applied potential in region C shows evidence of no change in the reaction rate.

3.2 *Corrosion Rate*

Corrosion rate of the hot rolled aluminium alloy 8015 at 1.37 μm was analysed from the polarization curve using Tafel analysis. The Tafel analysis for the three samples is given in Fig. 3. The straight lines that best fit the data at high potentials were achieved at selected potential range markers on both the anodic and cathodic curves. The straight lines were extrapolated until it intersects. Table 1 shows the computed corrosion potential (E_{corr}), corrosion current density (I_{corr}) polarization resistance (R_p) and corrosion rate (Table 2).

Fig. 3 Tafel analysis of hot rolled aluminium alloy 8015 at $R_a \approx 1.37$ µm. **a** Sample A1; **b** Sample A2; **c** Sample A3

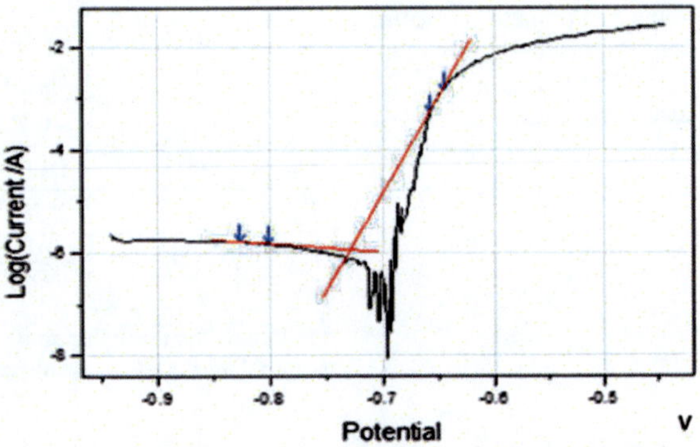

(a)

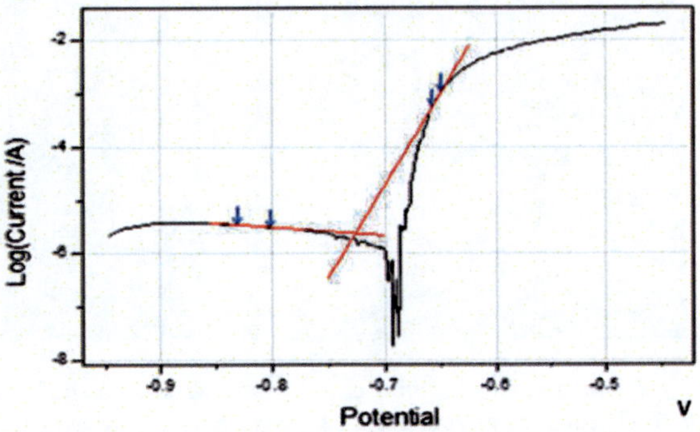

(b)

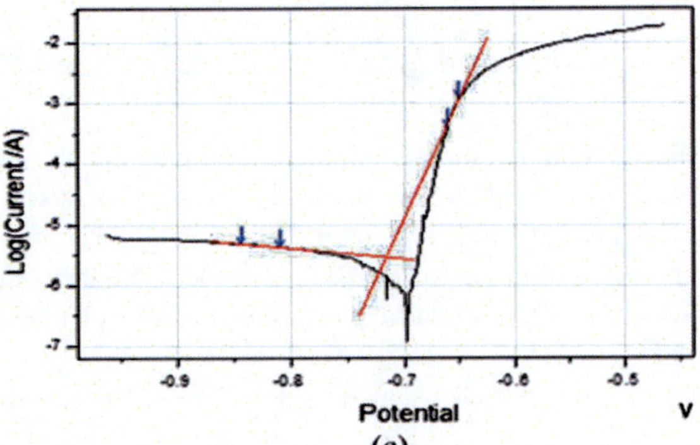

(c)

Samples 2 and 3 computed corrosion rate is selected to determine the average corrosion rate of 0.029385 mm/y. This confirms that the hot rolled aluminium alloy-8015 of 7 mm plate will have a service life of 238 years in natural seawater before it is dissolute completely.

Table 2 Computed corrosion rate of hot rolled aluminium alloy 8015 at 1.37 μm surface-roughness

	Sample A1	Sample A2	Sample A3
	Tafel data	Tafel data	Tafel data
E_{corr} (V)	−0.7281	−0.7264	−0.715
I_{corr} (A/cm^2)	1.12E − 06	2.43E − 06	2.97E − 07
R_p (Ω)	9879	4968	3520
Corrosion rate (mm/y)	0.01217*	0.02646	0.03231

* indicates the corrosion rate value asterik in the table was not use to calculate the mean corrosion rate

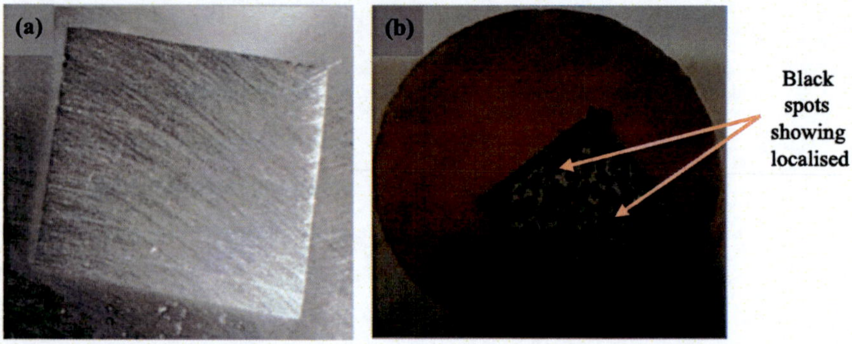

Fig. 4 Macrographs showing photographic observation for hot rolled aluminium alloy 8015 sample at 1.37 μm surface roughness. **a** Image before corrosion of hot rolled sample; **b** images after corrosion for hot rolled sample

3.3 *Visual and Microstructural Observation*

Figure 4 shows the visual camera images of the prepared hot rolled alloy surface before and after electrochemical corrosion. A larger surface area of the aluminium alloy surface is attacked in the form of localized corrosion with the existence of black spots observed on the surface of the alloy. Further microstructural observation using TESCAN VEGA scanning electron microscope revealed pitting corrosion, as shown in Fig. 5.

4 Conclusions

This research has provided an electrochemical study on corrosion behaviour of aluminium alloy 8015 at 1.37 μm in natural seawater solution. From the experimental investigation, it is shown that:

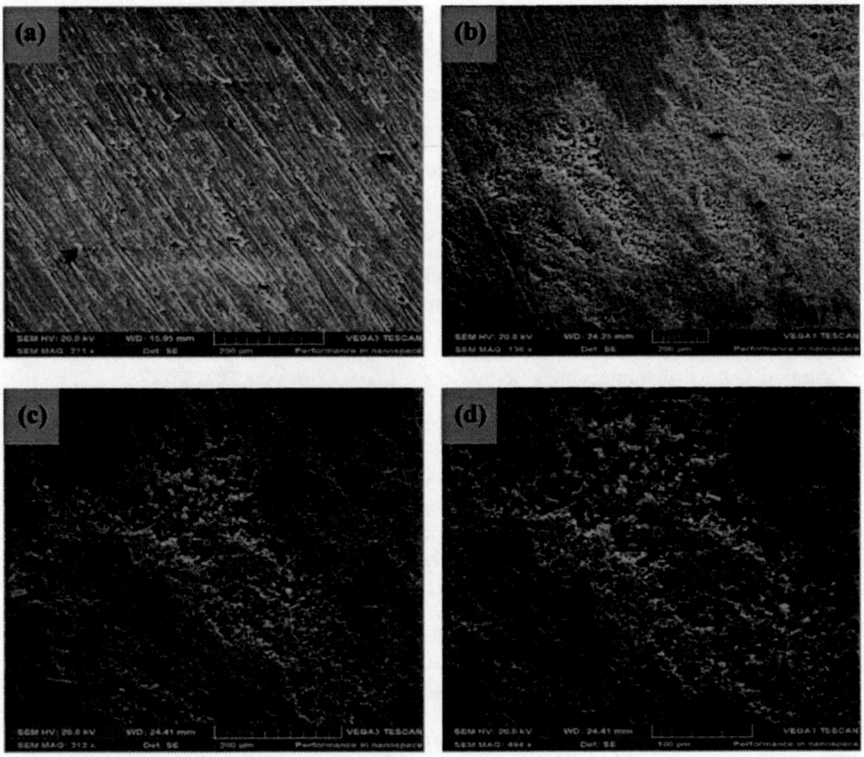

Fig. 5 SEM images of corroded hot rolled aluminium alloy 8015 at 1.37 μm surface roughness. **a** Image before corrosion at 271- × magnification; **b–d** images after corrosion at amplified magnifications

- The potentiodynamic polarization curves for the three samples are similar, showing an actively corroding region. This confirms precision of the experimental investigation.
- Mean corrosion rate of the hot rolled aluminium alloy 8015 is at 1.37 μm, and surface roughness under natural seawater condition at room temperature of 20 ± 2 is given as 0.029385 mm per year.
- Microstructural analysis of the hot rolled aluminium alloy 8015 surface after electrochemical corrosion reveals localized corrosion in the form of pitting.

Acknowledgements The Council for Scientific and Industrial Research (CSIR), South Africa, is acknowledged for provision of financial support.

References

1. Dieter GE, Kuhn HA, Semiatin SL (2003) Handbook of workability and process design. ASM International, USA
2. Gundersen JTB, Aytaç A, Ono S, Nordlien JH, Nişancıoğlu K (2004) Effect of trace elements on electrochemical properties and corrosion of aluminium alloy AA3102. Corros Sci 46(2):265–283
3. Laferrere A, Parson N, Zhou X, Thompson G (2013) Effect of microstructure on the corrosion behaviour of extruded heat exchanger aluminium alloys. Surf Interface Anal 45(10):1597–1603
4. Shahidi M, Gholamhosseinzadeh MR (2015) Electrochemical evaluation of AA6061 aluminum alloy corrosion in citric acid solution without and with chloride ions. J Electroanal Chem 757:8–17
5. Krüger L, Schwarz F, Mandel M, Hockauf M (2015) Electrochemical corrosion studies of ultrafine-grained aluminium alloy EN AW-6063. Mater Corros 66(3):226–232
6. Dai X, Wang H, Ju L, Cheng G, Cong H, Newby BZ (2016) Corrosion of aluminium alloy 2024 caused by Aspergillus Niger. Int Biodeterior Biodegrad 115:1–10
7. Aribo S, Fakorede A, Ige O, Olubambi P (2017) Erosion-corrosion behaviour of aluminum alloy 6063 hybrid composite. Wear 376–377:608–614
8. Tian G, Jin G, Zhang W, Huang Z, Yang Z (2017) Investigation on electrochemical corrosion characteristic of 2a14 aluminum alloy in nitric acid. Surf Rev Lett 24(Supp 01):1850016
9. Deepa P, Padmalatha R (2017) Corrosion behaviour of 6063 aluminium alloy in acidic and in alkaline media. Arab J Chem 10:S2244
10. Donatus U, Thompson GE, Omotoyinbo JA, Alaneme KK, Aribo S, Agbabiaka OG (2017) Corrosion pathways in aluminium alloys. Trans Nonferrous Metals Soc China 27(1):55–62
11. Zhang X, Zhou X, Hashimoto T, Liu B (2017) Localized corrosion in AA2024-T351 aluminium alloy: transition from intergranular corrosion to crystallographic pitting. Mater Charact 130:230–236
12. Liu B, Zhang X, Zhou X, Hashimoto T, Wang J (2017) The corrosion behaviour of machined AA7150-T651 aluminium alloy. Corros Sci 126:265–271
13. Seri O (1994) The effect of NaCl concentration on the corrosion behavior of aluminum containing iron. Corros Sci 36(10):1789–1803
14. Verma AS, Sumankant S, Yashpal NM (2015) Corrosion behavior of aluminum base particulate metal matrix composites: a review. Mater Today: Proc 2(4–5):2840–2851
15. Polmear I, John D, Nie J-F, Qian M (2017) 2-physical metallurgy of aluminium alloys. Light alloys. Elsevier Ltd, pp 31–107
16. Al-Moubaraki AH, Al-Rushud HH (2018) The red sea as a corrosive environment: corrosion rates and corrosion mechanism of aluminum alloys 7075, 2024, and 6061. Int J Corrosion 2018:1–15
17. Lameche-Djeghaba S, Benchettara A, Kellou F, Ji V (2014) Electrochemical behaviour of pure aluminium and Al–5%Zn alloy in 3% NaCl solution. Arab J Sci Eng 39(1):113–122
18. Kim S, Jang S, Han M, Park J, Jeong J, Chong S (2013) Mechanical and electrochemical characteristics in sea water of 5052-O aluminum alloy for ship. Trans Nonferrous Metals Soc China 23(3):636–641
19. ASTM International. (a). ASTM G1-03 standard practice for preparing, cleaning, and evaluating corrosion test specimens. ASM International
20. ASTM G5-14 Standard Reference test method for making potentiodynamic anodic polarization measurements. ASTM G5-14 Standard Reference test method for making potentiodynamic anodic polarization measurements
21. ASTM International. (c). ASTM G69-97 (2009) Standard test method for measurement of corrosion potentials of aluminum alloys. ASM International

Influence of Tool Profiles on Heat Transfer Analysis in Al-6061 Alloy Using Friction Stir Welding

M. Bala Chennaiah, A. Sreenivasulu, and K. Ravi Kumar

Abstract The following paper describes the thermal analysis on friction stir welded aluminum alloy AA 6061. The tool for this investigation is made up of high-speed steel material. We considered tool profile as tapered and the pin is with threaded and without threaded and also discusses the process parameters and their effects of thermal analysis of similar welded aluminum alloys. The joint is fabricated at a tool rotational speed of 825 rpm, feed rate of 32 mm/min and angle of 3 degree for the joining process. There is no surface-level cracks in a welded region by using liquid penetrant testing (NDT) method. The heat transfer model and evolution of temperature during the process of FSW are calculated theoretically, and the values of heat rate and peak temps are evaluated. These are compared with practical analysis values based on ANSYS values.

Keywords Aluminum alloys · Tapered tool · Heat transfer · Temperature distribution · Tensile · Hardness

1 Introduction

Friction stir welding (FSW) is a process of joining of metals in solid state. This process is efficient in energy, eco-friendly and comfortable to do. It is considered that the most significant development in the metal joining process in the twenty-first decade. 6061 Aluminum alloys are widely used for structural and marine applications. They have the advantage of being lightweight at the same time exhibiting good tensile strength. Non-consumable tool is used to join two faces of workpieces without melting the workpiece material. In Friction stir welding (FSW) major role play the tool material, tool profile, rotational speed, axial force and also temperature distribution perpendicular to the tool moment. Heat is generated by friction between the rotating tool and the workpiece and plastic deformation of workpiece, which leads to a softened region near the FSW tool. While the tool is traversed along the

M. Bala Chennaiah (✉) · A. Sreenivasulu · K. Ravi Kumar
ME Department, V.R. Siddhartha Engineering College, Vijayawada, India
e-mail: chennai303.mech@vrsiddhartha.ac.in

S. Vijayan et al. (eds.), *Trends in Manufacturing and Engineering Management*, Lecture Notes in Mechanical Engineering,
https://doi.org/10.1007/978-981-15-4745-4_48

joint line, it mechanically intermixes the two pieces of metal, and forges the hot and softened metal by the mechanical pressure, which is applied by the tool, much like joining clay. Thus, a joint is produced in solid state. It is primarily used on wrought or extrude and particularly for structures which need very high weld strength.

2 Literature Review

A Daneji et al. [1] studied about " Influence of tool profile and processing parameters on welding defects and mechanical properties of friction welding of AA 6061." In this study, he took rotational speed, welding speed, pin height and pin thread pitch as input parameters. However, the weld quality-related defects and mechanical properties were treated as output parameters. A cylindrical tool of external thread is taken as tool geometry, rotational speed of 1750 rpm and weld speeds of 6, 9, 12 mm/min. From the experimental results, he concludes that welding speed plays a key role toward the heat generation and lead to the generation of wormhole in the welding nugget in case of wrong selection. The thread pitch has a controlling influence on the stirring action and it can contribute toward the material accumulation toward the retrieving side of the welding.

M.S. Srinivasa Rao et al. [2] studied about "Study of weld characteristics during friction stir welding of aluminum alloy 6061." They selected AA6061 of dimensions 200*60*6 mm and two types of tools such as taper threaded tool and half grooved tool. They used 900 rpm, feed 16 mm/min, and 2 degrees of tilt angle as process parameters for the welding process. They observed that there is no influence on tool profile on hardness value. Hardness is mainly influenced by rotational speeds. The maximum Brinell's hardness numbered 69 BHN occurred at 900 rpm welded using the half grooved tool.

Ch Mohana Rao et al. [3] studied on "Friction Stir Welding of Aluminium Alloys 6061-To-6061 Similar Metals." Friction stir welding is performed to join 4 mm thick plates of 6061 aluminum to aluminum with varied parameters (like tool rotation speed (rpm), welding speed (mm/min), and the joining conditions are characterized. All welds were defect-free. Microstructure of weld and microhardness were shown at center of the weld. Tensile strength was good. As tensile strength increases from 710 to 1120 rpm, hardness decreases from 710 to 1120 rpm.

Abdul Arif et al. [4] studied on "Temperature Validation for Friction Stir Welding of Dissimilar Aluminum Alloys." The results of thermal analysis which are simulated using ANSYS® are used to evaluate the temperature distribution in the workpiece geometry. From ANSYS software, evaluated temperature distribution in dissimilar plates and compared actual models. Values of temperatures received by the finite element method are close to the real temperature distribution in the welded construction. % of error between experimental and simulation results do not exceed 5%. The analysis results are closely matched with experimental results.

CM Chen et al. [5] studied the FEM of friction stir welding—thermal and thermo-chemical analysis. They took the AA6061-T_6 as base metal of dimensions

240*50*6 mm, AIRKOOL STEEL of shoulder and diameter of 24 mm, 6 mm is used as a tool material. They took the rotational speed of 500 rpm and weld speeds of 140 mm/min, 350 mm/min for welding. In the experimental results, they found that if there is an increase in speed from 140 mm/min to 350 mm/min, there will be a decrease in temperature of the weld plate, especially in the weld zone. They calculated the heat generation during welding and used it as an input for ANSYS analysis of temperature, and they got the highest temperature 425 degrees at weld speed of 140 mm/min and rotation of 500 rpm (Figs. 1 and 2).

W. Tang et al. [6] "Studied about heat input and temperature distribution in friction stir welding". They took AA6061 as a base material of dimensions 300*100*6.5 mm, rotational speeds of 300 rpm,1200 rpm and travel speed was 2 mm/s. In the studied they observed that grain size of the weld zone is notably smaller and more equiaxed than that of parent metal as a result of recrystallization. They used thermocouples for measuring the temperature along the weld zone. The maximum temperature obtained is 450°. Perumalla Janki Ramulu et al. [7], " Experimental Study On Temperature

Fig. 1 Before the FSW process

Fig. 2 After the FSW process

Evolution During Frication Welding Of 6061-T6 Aluminum Alloy." they were considered AA 6061 as a base metal of thickness 2.1 mm, three types tools of shoulder diameters 12 mm, 15 mm, 18 mm and pin with equal probe lengths are used for entire process. They used thermocouples to measure the temperature distribution. They conducted the experimental work taking rotational speed of 1300, 1400, 1500 rpm and weld speeds of 90,100,120 mm/min and tool tilt angle of 2.5°. They concluded that increase in shoulder diameter and rotations speeds increase the temperature.

Summary of Literature Survey: Work done has been done on FSW of Al6061 plates of different thickness, different researches concluded that tool rotational speed, welding speed, axial force, pin profile, pin diameter, shoulder profile, shoulder diameter, tool tilt angle are the important process parameters which determine the quality of weld. Before the start of the experiment this values has to be chosen carefully. Improperly selection this process parameters leads to weaker joints containing several defects like tunnel defect voids, cracks etc.

3 Experimental Procedure

Aluminum plates of 6 mm thick, Al 6061 alloy base metal were cut to the required size (150 m × 50 mm) by power hacksaw cutting and milling. Square butt joint configuration (300 mm × 300 mm) was prepared, and single-pass welding procedure was followed to fabricate the joints. The chemical composition and mechanical properties of base metals are presented in Tables 1 and 2. The non-consumable tool is made of high-speed steel (HSS) material. The shoulder and pin of the tool are designed and machined by using the CNC machine. The shoulder diameter profile of the tool (20 mm) is flat and pin length (5.6 mm) is tapered with thread, and the other one is tapered without thread. Then, both tools are made by using heat treatment process. Machining process is done by the universal milling machine and the plates are tightened by clamping force of 2 kN by varying process parameters that are tool rotational speed, feed rate, and tilt angle. The process parameters are shown in Table 3. The plates were successfully welded using friction stir welding technique. The welded samples were first visually inspected to element any defective workpieces. All the samples passed this test. Next, the samples were subjected to non-destructive tests. Only one sample (Al6061 and Al6061 without threaded tool)

Table 1 Mechanical properties of Al 6061

Property	Yield strength	Tensile strength	Elasticity modulus	Brinell hardness
Value	276 MPa	310 MPa	68.9 GPa	95 HB

Table 2 Heat value at tool rotational speed

S. No.	Tool rotational speed	Amount of heat (kW)
1.	700	5.18

Table 3 Heat and temperature distribution on plates

Plate	Heat value (kW)	Temperature (°C)
6061-6061	3.626e3	399.714
6061-6061	2.59e3	362.285

Table 4 Theoretical and experimental temperature values

Plate	Theoretical temperature (°C)	Practical temperature (°C)	% error
6061-6061	385.65	399.714	3.51
6061-6061	385.65	362.285	6.05

was found, and all others were non-defective. In the present work, 6 mm thick commercial grade aluminum plates have been welded by using different types of tool (tapered and tapered with thread). From the numerical analysis, it is observed that the tool with tapered threaded pin profile generates less temperature than the tool with tapered pin profile because for the same base diameter the cylindrical pin has more contact area with the workpiece than the contact area of pin with conical pin profile (Table 4).

3.1 Heat Transfer Analysis

A 3D finite element transient thermal model was developed in the present work to analyze the heat transfer and temperature distribution in the FSW process. The actual conditions were accommodated as far as possible in the present thermal model. The following assumptions were made in developing the thermal model of a FSW process (Figs. 3 and 4).

The temperature calculation is based on Fourlier's equation:

$$\rho c \frac{\mathrm{d}T}{\mathrm{d}t} = \mathrm{div}(k \cdot \mathrm{grad}T) + q \mathrm{in}\ \Omega \tag{1}$$

where q is the power generated by friction between the tool and the top of the workpiece and by the plastic deformation work of the central weld zone, T is the temperature, n is the conductivity, r is the material density, and c is the heat capacity. The main heat source in FSW is generally considered to be the friction between the rotating tool and the welded plates, and the "cold work" in the plastic deformation of material in the vicinity of the tool. Considering an element at the contact surface between the tool shoulder and the top surface of workpiece, the rate of heat generation derived from the friction in the element at radius r is:

$$d\dot{q} = 2\pi\omega.r^2\mu(T)p(T)\mathrm{d}r \tag{2}$$

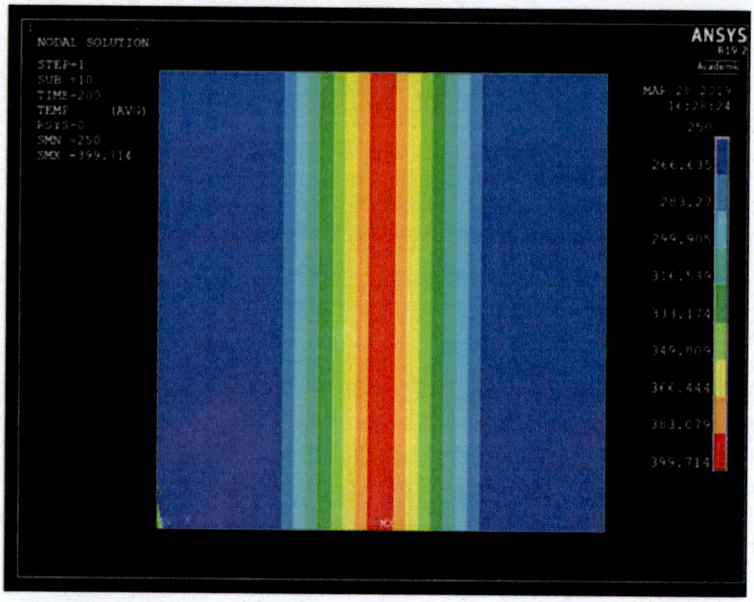

Fig. 3 6061 with thread tool—heat distribution

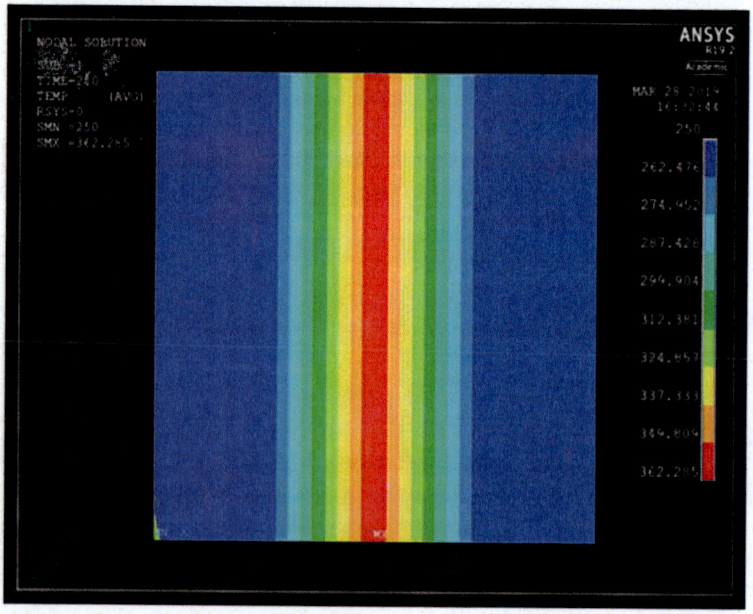

Fig. 4 6061 without threaded tool—heat distribution

The rate of heat generation (caused by the friction) over the entire interface of the contact will be:

$$\dot{q} = \int_{r_0}^{R_0} 2\pi\omega.r^2\mu(T)p(T)\mathrm{d}r = \frac{2}{3}\pi\omega\mu(T)p(T)\left(R_0^3 - r_0^3\right) \quad (3)$$

The rate of heat generation at the interface between the shoulder and the top of the workpiece surface is a function of the coefficient of friction m, angular velocity w, and radius r. As the *m* (*T*) and *p* (*T*) are dependent on the local temperature and the radius *r*, Eq. (3) is difficult to evaluate. As the temperature increases, the friction coefficient is expected to decrease, and the work of the plastic formation increases. In this model, the p is from the experimental measurement. A constant value of the friction coefficient is used to approximate the comprehensive effect of both factors of thermal and plastic effects during FSW, and the predicted temperature history is verified by the measured ones. The heat generation in the vicinity of the probe is considered the same as that at the shoulder periphery with a radius equal to that of the probe.

3.2 *Heat Analysis*

From the considering, the Fourier heat equation

$$\rho c\frac{\mathrm{d}T}{\mathrm{d}t} = \mathrm{div}(k\cdot \mathrm{grad}T) + q\mathrm{in}\Omega$$

$$q = \frac{2}{3}\pi\omega\mu(T)p(T)\left(R_0^3 - r_0^3\right)$$

We assume that
Shoulder diameter $R_0 = 20$ mm
Probe diameter $r_0 = 7$ mm
We taken $\mu = 0.5$

$$\omega = \frac{2\pi N}{60}$$

At 700 rpm,

$$q = \frac{2}{3}\pi\frac{2\pi 700}{60}0.5\frac{5000}{\pi\left(3.5*10^{-3}\right)^2}(T)p(T)\left(0.01^3 - 0.0035^3\right)$$

$= 5185.7\,\mathrm{W}$
$= 5.18\,\mathrm{kW}$

For 6061-6061(P = 5.18 kW),

$$5.18 \times 10^3\,\mathrm{W} = 5.67 \times 10^{-8}\tfrac{\mathrm{W}}{\mathrm{m^2K}} \times \frac{\pi}{4}\left(2.0 \times 10^{-3}\right)^2 \times T^4$$

$$\frac{5.18 \times 10^3 \times 4}{5.67 \times 10^{-8} \times \pi \times 20^2 \times 10^{-6}} = T^4$$

$$T_{\mathrm{the}} = 385.65\,^\circ\mathrm{C}$$

4 Conclusions

In this experiment, the two process parameters of FSW, i.e., feed rate and tool rotational speed are varied to find the effect on the thermal properties of friction stir welded Al 6061 alloy plates of 6.0 mm thickness of those parameters. A combination of tool rotational speeds and welding speeds were employed in this experiment. As a result, two joints were fabricated. The following are the results that are derived from these experimentations.

- We chosen two types of tools profiles: one of them is tapered with thread and another type is tapered without thread tool.
- Then, the highest temperature is generated at taper with thread tool, the amount of increase temperature in %, i.e., 6.05%

Acknowledgements We sincerely thanks to VR Siddhartha Engineering College management providing necessary resources and Dr. N. Vijaya Sai, Head of the Department, Mechanical Engineering, VRSEC for his continued encouragement to completion of this work.

References

1. Daneji A, Ali M, Pervaiz A (2018) Influence of tool geometry and processing parameters on welding defects and mechanical properties for friction stir welding of 6061 alloy. Mater Sci Eng 346
2. Srinivasa Rao MS, Ravi Kumar BVR, Manzoor Hussain M. Experimental study of weld characteristics during friction stir welding (FSW) of Aluminum Alloy (Aa6061-T6). Int J Res Eng Technol. ISSN: 2319-1163
3. Mohana Rao C, Mallikarjuna Rao K (2017) Friction stir welding of Aluminium Alloys 6061-To- 6061 similar metals. Int J Mech Eng Technol (IJMET) 8(1):264–269
4. Arif A et al (2013) Temperature validation for friction stir welding of dissimilar Aluminum Alloys. In: International Conference on Advances in Mechanical Engineering, ICAME-2013
5. Chen CM, Kovacevic R (2003) Finite element modeling of friction stir welding thermal and thermo mechanical analysis, by Received 12 May 2003; Accepted 10 June 2003

6. Tang W, Guo X, Mcclure J. Heat input and temperature distribution in friction stri welding. University of Texas at El pasa
7. Ramulu PJ, Ganesh Narayana R (2014) Experimental study on temperature evolution during frication welding of 6061-T6 Aluminum Alloy. (AIMTDR 2014) 12th–14th Dec 2014

Tensile, Hardness, and Impact Properties of Amorphous Al–Si–Mg Cast Alloys

Olayide R. Adetunji, Adeniran S. Afolalu, Mufutau A. Mustapha, Oluwasegun J. Adelakun, Samson O. Ongbali, and Abiodun A. Abioye

Abstract This research work focused on tensile, hardness, and influence properties of amorphous Al–Si-Mg cast alloys. Enhancement of aluminum alloy with Mg and Si was performed after which heat treatments were carried out on the produced samples. The result showed the percentage of Mg as 0.14, 4.14, 0.14, 0.32, and 0.20%, for as-cast aluminum, annealed, age-hardened, hardened, and glassy-phase samples, respectively. The percentages of Si were 2.01%, range of 1.94–4.10% and 3.37% for as-cast, heat-treated, and glassy-phase aluminum alloys, respectively. The impact tests showed an increase in impact value from 5.423 J of as-cast sample to 8.135 J of age-hardened sample. Hardened sample has the lowest impact value of 2.712 J while glassy-phase sample has the highest impact value of 10.847 J. The ultimate tensile strength of annealed sample is higher compared to as-cast while glassy-phase has the highest value. The SEM of the annealed alloy sample contains bright and dark regions indicating α and β phases, respectively.

Keywords Mechanical property · Amorphous al alloy · SEM characterization

1 Introduction

Aluminum is a substance having a unique significant as a result of the proportion of its capacity to weight [1], that is, simplicity in recovering the substance through a process of recycling, opposing deterioration of materials, high tendency of transmitting heat, and electricity. Utilization of alloys of aluminum, composite, and titanium was adopted in reducing aircraft weight. Wide variety of solutions were assumed in the design, operation, and manufacturing of automobiles in which the alloy of aluminum constitutes a part of it [2]. In most instances, alloys of wrought and cast

A. S. Afolalu (✉) · O. J. Adelakun · S. O. Ongbali · A. A. Abioye
Mechanical Engineering Department, College of Engineering, Covenant University, Ota, Nigeria
e-mail: Sunday.afolalu@covenantuniversity.edu.ng

O. R. Adetunji · M. A. Mustapha
Mechanical Engineering Department, Federal University of Agriculture, Abeokuta, Nigeria

S. Vijayan et al. (eds.), *Trends in Manufacturing and Engineering Management*, Lecture Notes in Mechanical Engineering,
https://doi.org/10.1007/978-981-15-4745-4_49

aluminum are applicable in joining the roof of automobile, engine, and side panels to the underbody, [3]. Steel or iron was previously used for automobile; but discovery has tested alloys of aluminum proven to result into lighter part which might occur to be complicated most times [4]. As a result of the increase in part design or a material property, it is most likely to generate a specific part that will perform the activities of other parts, thereby making it easy to minimize the parts to be produced [5].

The particular combination of metals and another element type leading to Al–7Si–0.4 Mg is normally utilized in automobile part, aircraft; as a result of the Mg in the combination, deterioration of materials is opposed [6]. Cast Al–Si–Mg alloys are applicable in automobile and aviation industry as a result of their excellent characteristics and the ratio of its capacity to weight [7]. Mechanical properties required for the aforementioned application are excellent opposition to weakening of materials, high proof strength, and high resistant to fracture [8]. Heat treatment of cast Al–Si–Mg alloys is normally examined in the aspect of homogeneous solid solution, undesirable low-temperature changes been hindered, and accelerated changes occurring in the properties [9], and this method for strengthening materials is utilized in other to achieve a preferred combination of its mechanical properties [10, 11]. This research work therefore aimed to produce amorphous Al–Si–Mg cast alloys with an improved mechanical property.

2 Materials and Methodology

The material, tools, and equipment required for the production of the aluminum alloys and heat treatments included aluminum ingot, silicon ingot, magnesium ingot, alloy additives, heating furnace, vernier calipers, and hacksaw.

2.1 *Procedure for Producing Aluminum Alloy Casting*

The aluminum (Al), magnesium (Mg), and silicon (Si) ingots were properly dried with acetone. In other to easily detach the mold from the casting, graphite coatings are made available within the mold. The casting of aluminum alloy was carried out at research center in the country. The molds were prepared toward a particular high temperature. Electric induction furnace was used for the melting of metal, and the alloy was inverted twice and was allowed to cool in sand. The alloys produced were categorized into as-cast sample and heat-treated samples. The cast samples were analyzed to determine the percentage of elemental composition in the alloys.

2.2 Procedure for Heat Treatment

In other to enhance the mechanical properties of cast alloy, heat treatment process was done on mechanical test samples. Heat treatment process comprises mainly of age-hardening, annealing, hardening, and glassy phase which were carried out at the research center.

2.3 Chemical Composition Analysis of the Test Samples

Percentage weight elemental composition of the alloys produced was carried out in a local aluminum rolling mill. The equipment used for the analysis was optical emission spectrometer. The test was carried out following the procedure below:

The samples were prepared by forming into the shape accommodated by the machine. This involved cutting and filling to shape, and prepared samples as described above were then sparked using argon gas after which the sample was inserted into metal analyzer for proper analysis.

2.4 Materials and Equipment for Mechanical Testing of the Samples

Various mechanical tests were carried out on the samples; hardness and impact tests were carried out at the university laboratory. Tensile test was carried out in another university. The equipment used included meter rule, testing equipment apparatus, hardness testing equipment, tensile strength testing equipment, and inspection equipment apparatus.

Hardness Impact and Tensile Tests. Rockwell hardness testing machine was used for Brinell hardness test using heat-treated samples and control samples. The surface was prepared by grinding and polishing. 20 mm length of no specific diameter was sectioned out from the sample and fixed on the machine using a dwell time of 15 s.

2.5 Procedure for Microstructure Examination

The SEM is majorly applicable in the evaluation of samples that are thick by putting together electrons which are emitted from the sample under experiment so as to obtain secondary electrons. In addition, other signals including electron beam were also collected. This was done using Phenom ProX scanning electron microscope, model number-800-07334.

3 Results and Discussion

The findings of this research are contained in Tables and Figures.

3.1 *Results of Aluminum Casting*

The casting produced was grouped into two for the purpose of comparison on the test to be carried out: The first group is as-cast (i.e., non-heat-treated samples), and the second is heat-treated sample. Both groups of samples were subjected to chemical analysis to determine the percentage weight composition of elements in the aluminum alloy produced (Fig. 1).

3.2 *Results of Chemical Analysis*

The result of chemical analysis is shown in Table 1.

3.3 *Results of Scanning Electron Microscopy Test*

The scanning process as earlier explained was followed by the EDX analysis with the acquired microstructure using the Phenom identification software. The scanning electron micrograph in Fig. 2 shows the microstructure of annealed alloy sample. The bright region indicates the primary solid solution, and the dark region shows the non-equilibrium eutectic solid solution between the grains. The energy dispersive X-ray as shown in Fig. 3 reveals the chemical composition. Figure 4 shows the SEM of age-hardening test carried out on the alloy samples. The emergence of a scattered solid form from a solution in combining two or more metals is the aim of the solid-hardening process. The fine solid in the combination of the two or more metals as seen in the micrograph prevents abnormality separation in mobility by forcing the abnormality to pass through the solid particles or encompasses round it. Preventing abnormality in mobility makes combination of two or more metals to be strengthened (Figs. 5 and 6).

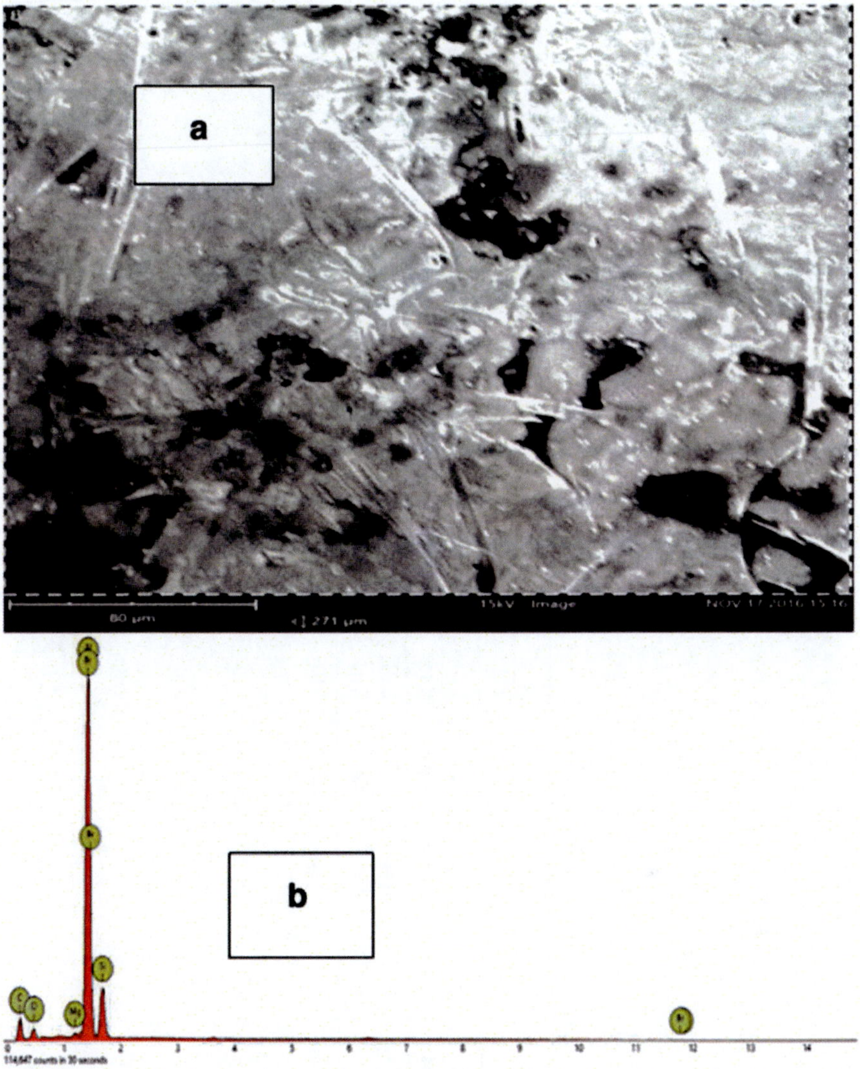

Fig. 1 **a** SEM image of annealed test sample **b** Energy dispersive spectrometer

3.4 Results of Mechanical Tests

Mechanical tests carried out on the samples were impact test, hardness test, and tensile test, and the results are as given below. These results are given in Tables 2 and 3.

Table 1 Percentage chemical composition of element in the cast sample

Heat-treat	% Elemental Composition											
	Mg	Si	Mn	Cu	Zn	Ti	Fe	Na	B	Al	Sn	Pb
As-cast	0.14	2.01	0.36	0.42	0.30	0.01	0.59	0.00	0.00	96.31	0.01	0.02
Annealing	4.14	3.94	0.90	0.24	2.48	0.12	3.00	0.12	0.059	81.51	2.81	0.68
Age-Hardening	0.14	1.94	0.21	0.40	0.31	0.01	0.72	0.01	0.00	96.24	0.01	0.02
Hardening	0.32	4.10	0.12	0.51	0.41	0.01	2.81	0.00	0.00	91.68	0.02	0.03
Glassy-Phase	0.20	3.37 0.12		0.42	0.37	0.01	2.36	0.00	0.00	93.12	0.01	0.03

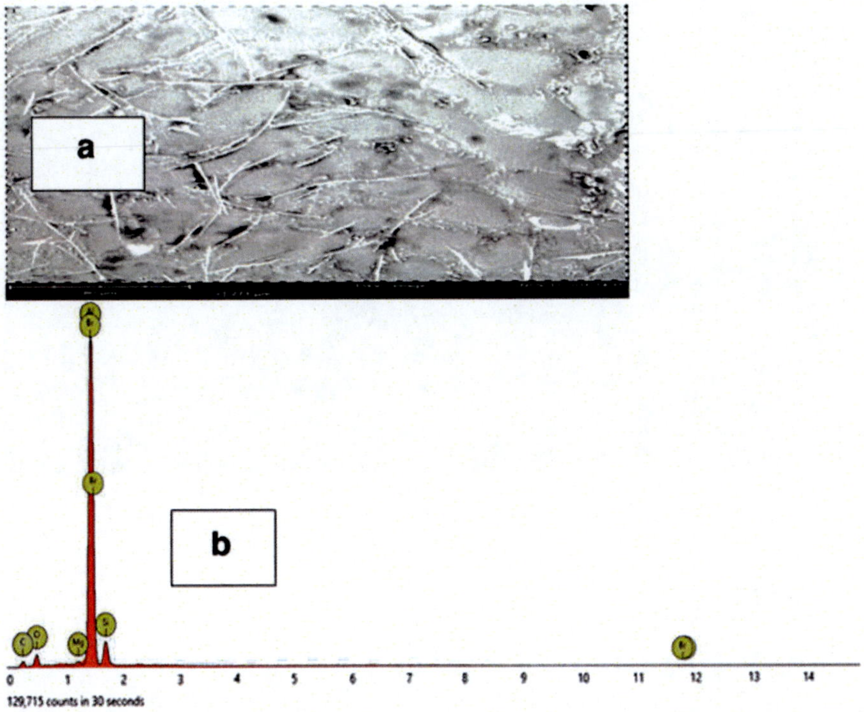

Fig. 2 **a** SEM image of the As-cast sample **b** Energy dispersive spectrometer (EDS)

3.5 Discussions of Results

The results show the percentage weight of elements in the as-cast, heat-treated, and glassy-phase aluminum alloys. The as-cast aluminum alloy shows a percentage of Mg 0.14%, heat-treated samples show annealed sample to be 4.14%, age-hardened sample 0.14%, hardened sample 0.32%, and glassy phase shows 0.20%. The percentage of Si in the as-cast Al alloy is 2.01 by wt% as the range is between 1.94% and 4.10% in the heat-treated aluminum alloy while glassy phase is 3.37%.

The impact value of as-cast sample is 5.423 J just as it is in annealed sample; age-hardened sample experiences an increase in the impact value to 8.135 J, and hardened sample has aÀ reduced impact value of 2.712 J. Glassy phase has the highest impact value of 10.847 J.

Comparatively, the results of hardness (HRB) value of as-cast sample is 68, and this value remained in hardened sample and increased to 70 in the annealed sample while the value reduced to 66 in age-hardened sample. Glassy phase only shows the lowest rate of reduction to 62. This result conformed to those of other researchers like Adeyemi et al. [10].

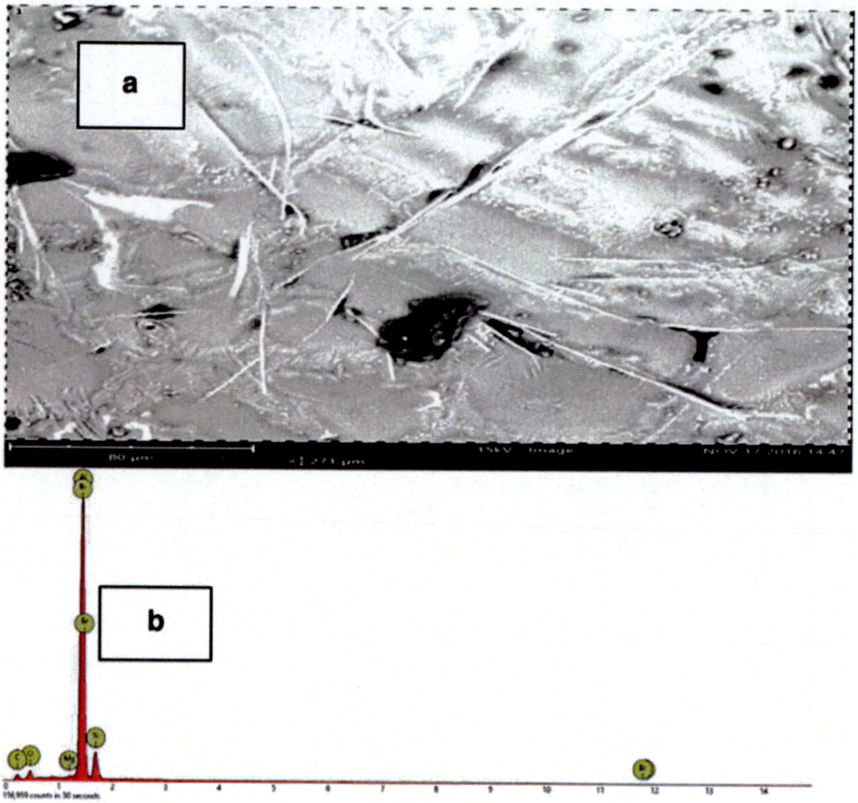

Fig. 3 a SEM image of age-hardened sample b Energy dispersive spectrometer

The tensile yield strength of as-cast is 82 MPa, and the ultimate tensile strength is 94.22 MPa while the percentage elongation is 3.62. The test also shows tensile yield strength in annealed sample to be 98 MPa, ultimate tensile strength increased to 108.24 MPa, and the percentage elongation is 2.81. In age-hardened sample shows tensile yield strength increased to 114 MPa, ultimate tensile strength is 91.81 MPa, and percentage elongation reduced to 1.60. The results show the tensile strength to be 116 MPa, ultimate tensile strength as 87.98 MPa, and percentage elongation is 1.62. Glassy-phase sample shows tensile yield strength to be 116 MPa, ultimate tensile strength to be 115.87 MPa, and percentage elongation to be 2.68.

The bright region in the SEM of the annealed alloy sample indicates the primary solid solution which accounted for solid solution hardness got in the result, and the dark region shows the non-equilibrium eutectic solid solution between the grains which accounted for the higher tensile and impact toughness values. This is in line with the findings of Chen et al. [11, 12]. The energy dispersive X-ray reveals the

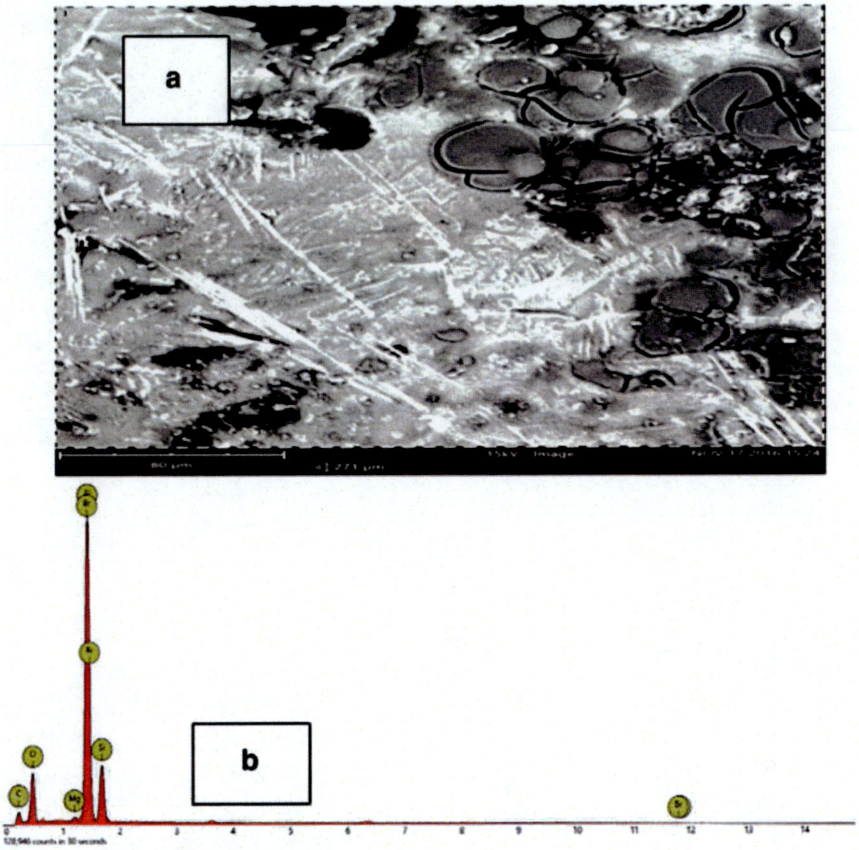

Fig. 4 **a** Micrograph of hardened test **b** Energy dispersive spectrometer (EDS)

traces of the chemical composition while Al (65.8%), O (11.6%), and Br (14.9%) were predominant; traces of Mg, C, and Si were seen.

4 Conclusions

The results of analysis show that the produced Al–Si–Mg alloys exhibit excellent casting characteristics evident from the results of microstructural test. Mechanical properties including impact toughness, hardness, and tensile were of good values. Glassy-phase alloy had the best impact toughness value followed by age-hardened one. The best hardness value was obtained for hardened alloy and followed by as-cast, age-hardened, and lastly glassy phase. The best yield and ultimate tensile strength

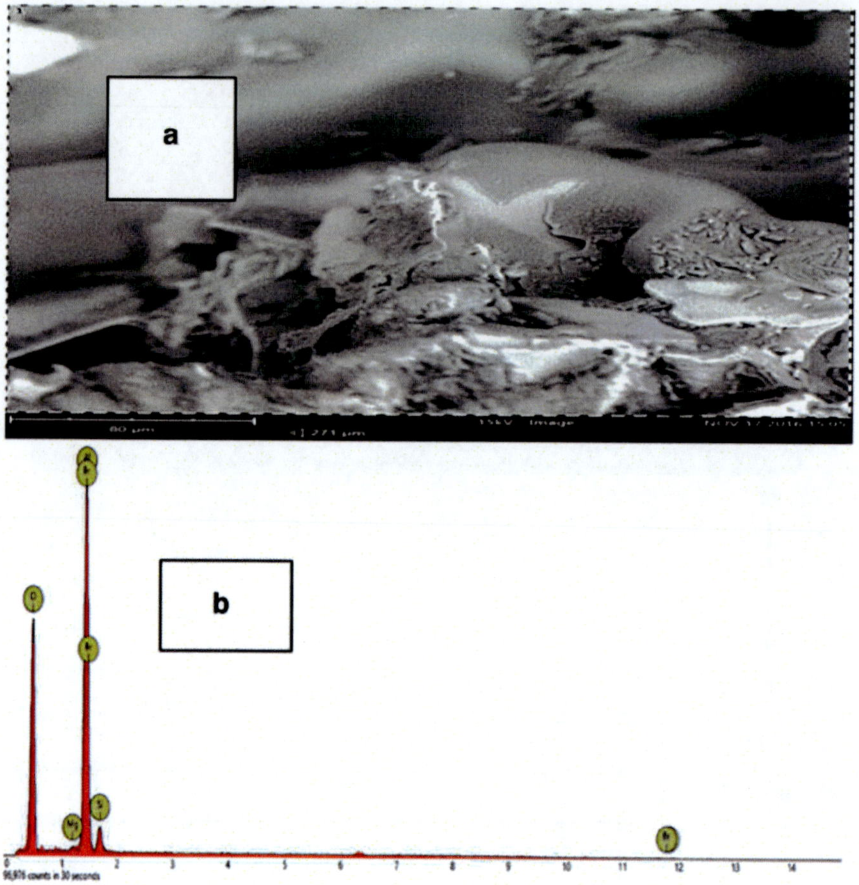

Fig. 5 **a** SEM image of glassy-phase test sample **b** Energy dispersive spectrometer

were obtained for glassy-phase alloy, followed by age-hardened, annealed, hardened, and as-cast. As-cast alloy had the best ductility followed by annealed one.

Acknowledgements We acknowledge the financial support offered by Covenant University in the actualization of this research work for publication.

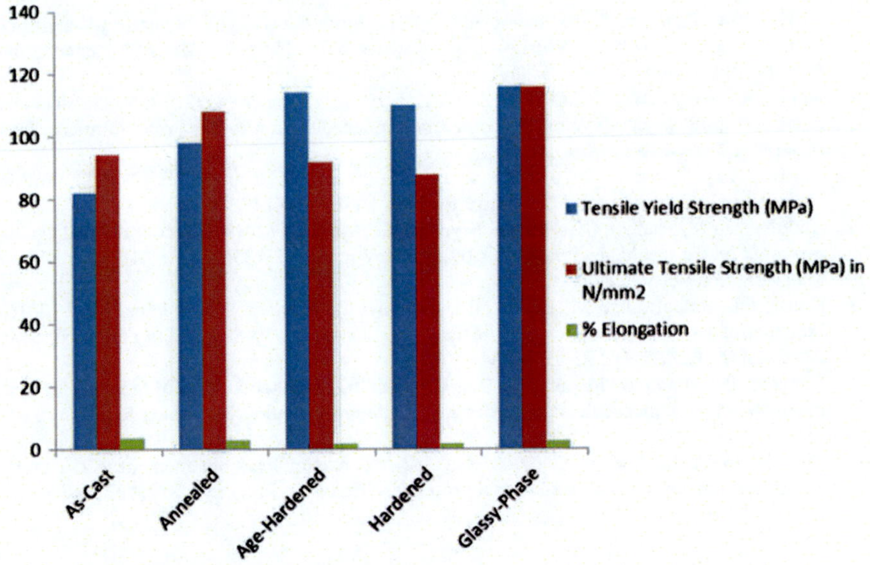

Fig. 6 Comparison of the results of tensile test samples

Table 2 Summary of results for impact tests

S/N	Sample	Impact value (J)	Hardness test (HRB)
1	As-Cast	5.423	68
2	Annealed	5.423	70
3	Age-Hardened	8.135	66
4	Hardened	2.712	68
5	Glassy-Phase	10.847	62

References

1. Adetunji OR, Olukuade ML, Simka W, Sowa M, Adesusi OM, Okediran IK (2017) Production and characterization of amorphous Aluminium-Copper alloy for aerospace applications. Eur J Eng Res Sci 2(2):1–5
2. Afolalu SA, Abioye AA, Dirisu JO, Okokpujie IP, Ajayi OO, Adetunji OR (2018) Investigation of wear land and rate of locally made HSS cutting tool. In: AIP conference proceedings, vol 1957, no 1. AIP Publishing, p 050002
3. Afolalu SA, Abioye OP, Salawu EY, Okokpujie IP, Abioye AA, Omotosho OA, Ajayi OO (2018) Impact of heat treatment on HSS cutting tool (ASTM A600) and its behavior during machining of mild steel (ASTM A36). In: AIP conference proceedings, vol 1957, no 1. AIP Publishing, p 050003
4. Afolalu SA, Okokpujie IP, Salawu EY, Abioye AA, Abioye OP, Ikumapayi OM (2018) Study of the performances of nano-case treatment cutting tools on carbon steel work material during turning operation. In: AIP conference proceedings, vol 1957, no 1. AIP Publishing, p 050001

5. Al-Khazraji KK, Hanna WA, Muhammed OS (2011) Study the effect of polymeric quenching of wear resistance of eutetic modified Aluminum-Silicon alloy. J Minerals Mater Characterization Eng 10(10):941–958
6. Abioye A, Abioye OP, Ajayi OO, Afolalu SA, Fajobi MA, Atanda PO (2018) Mechanical and microstructural characterization of ductile iron produced from fuel- fired rotary furnace. Int J Mech Eng Technol 9(1):694–704
7. Raj KS, Amit T, Das S (2016) Microstructure and mechanical properties of Al-Si alloy in As-cast and heat-treated condition. Am J Eng Res (AJER) 5(8):133–137
8. Abdulwahab M, Madugu IA, Asuke F, Fayomi OSI, Ayeni FA (2013) Effect of thermal ageing treatment on the mechanical properties of antimony-modified A356.0-type Al-Si-Mg alloy. J Mater Environ Sci 4(1):87–92
9. Umaru OB, Abdulwahab M, Hafsat NM, Maidawa H, Murtada AO, Suleiman TM (2016) Influence of a low temperature ageing on the properties of Al-6.5%Si-0.35%Mg alloy. Nigerian J Technol (NIJOTECH) 35(3):550–554
10. Adeyemi DI, Bolaji A, Mosobalaje OA, Oluyemi JO, Moshood DS (2013) Effect of heat treatment on some mechanical properties of 7075 Aluminum alloy. Int J Mater Res 16(1):190–194
11. Chen R, Wang L, Apelian D (2000) fludized bed reactor heat treatment of A356 alloy: microstructure analysis and mechanical properties. Trans Am Foundry Soc 2(4):289–304

Experimental Investigation on the Influence of Tool Geometry in Minimum Chip Thickness of Microendmilling Using Cutting Force Analysis

M. Prakash, M. Kanthababu, A. Arul Jeya Kumar, and V. Prasanna Venkadesan

Abstract Minimum uncut (t_m) chip thickness is the promising factor of the microendmilling process, which is affected by many factors. The tool geometry is one among them. In this work, the effect of different tool geometries, two flutes and four flutes of microendmill (500 μm diameter), are investigated, during slot milling of copper alloy (BSS 249) by considering the surface roughness (R_a) and cutting forces. The results show that the feed rate has more impact on the R_a. A higher order of R_a value is observed with the two flutes microendmill than that of with four flutes microendmill, due to the closer tool path generation of the subsequent passes of the four flutes. It is also observed that the transition of chip formation from interrupted chips to continuous chips has occurred with two flutes and four flutes. From the cutting forces analysis, it is found that the feed force (F_y) is greater than that of the transverse forces (F_x) irrespective of the tool geometry. The behaviour of cutting forces also helps to identify the minimum uncut chip thickness (t_m). The angular shifting of the cutting forces is also observed concerning the tool geometry. The material removal mechanism reveals that the ploughing effect is more dominant than that of the shearing effect with two flutes microendmill. The present investigations have shown that four flutes microendmill is more suitable for tool-based microendmilling in order to achieve better R_a with minimum cutting forces.

Keywords Microendmill · Tool geometry · Surface roughness · Cutting forces · Uncut chip thickness

M. Prakash (✉) · A. Arul Jeya Kumar
SRM Institute of Science and Technology, Kattankulathur, India
e-mail: prakashmuniyandi@gmail.com

M. Kanthababu
CEG, Anna University, Chennai, India

V. Prasanna Venkadesan
Indian Institute of Information Technology Design & Manufacturing, Kancheepuram, India

S. Vijayan et al. (eds.), *Trends in Manufacturing and Engineering Management*, Lecture Notes in Mechanical Engineering,
https://doi.org/10.1007/978-981-15-4745-4_50

1 Introduction

Recently, with more demand for accurate and precise micro-components, the importance of tool-based micromachining processes has been increased in the number of fields, such as medical, aerospace and automobile [1]. Among the tool-based micromachining processes, microendmilling is used for manufacturing the micro-components like microsurgical equipment, micro-impellor and micro-nozzle. Microendmilling uses microendmills with diameter in the sub-millimetre range. The edge radius of a typical microendmill is in the order of a few microns, and also, the uncut chip thickness (t) is usually lesser than the edge radius. Due to the size of the microendmill, it is very difficult to achieve the quality in the finished components. Previous investigations have indicated that the tool life in the tool-based micromachining is unpredictable, and also it fails prematurely [2–4].

Microendmilling distinguished from macro-regime machining in terms of size effect and t [5]. The t is greatly influenced by the chip formation process in the microendmilling. If the t is less than the minimum uncut chip thickness (t_{m}) (i.e. 0.33 of nose radius (r_{e})), ploughing is found to be dominating than that of the shearing [4], which resulted in the increase of cutting forces. New by et al. [5] made an attempt to calculate the uncut chip thickness (t) based on the feed per flute and cutting edge radius, which is shown in Eqs. 1 and 2.

$$\text{Uncut chip thickness } (t) = t_x \sin\alpha - \frac{n}{2\pi r_e} t_x^2 \sin\alpha\cos\alpha + \frac{1}{2r_e} t_x^2 \cos^2\alpha \quad (1)$$

where

$$t_x = \frac{F \times 1000}{S \times n} \quad (2)$$

where t—uncut chip thickness (μm), t_x—feed per flute (μm), F—feed rate (mm/min), S—rotational speed (rpm), n—number of flute, α—tool rotational angle, r_{e}—tool nose radius.

Generally, in tool-based micromachining processes the quality of the machined surface is mostly affected by the characteristics of cutting forces, which are mostly affected by the tool geometry [6]. Therefore, there is a need for more investigation on the effect of cutting forces generated during microendmilling. However, few attempts were made by researchers to study the effect of cutting forces in conventional/macro-regime endmilling [6–11].

Literature review related to the studies on the effect of tool geometry on the cutting forces is briefly presented here. Fang et al. [6] carried out the experimental investigations to study the effect of different tool geometries such as two flutes, triangle type and semicircular type and its failure modes using cutting force signals in the microendmilling of brass material. They have observed that the tool rigidity is much higher with the triangular-based microendmill, than that of the two flutes microendmill. However, the machining quality is found to be poor with triangular-based microendmill than that of two flutes types of microendmill. Filiz et al. [7] investigated the effect of two flutes of the microendmill with cutting forces during microendmilling of copper. They have found that the t_m and associated ploughing effects induce variations in the cutting forces. Li et al. [8] carried out experimental analysis on the tool wear and surface roughness using two flutes 0.1 mm diameter microendmill in microendmilling of OFHC Copper. They have observed that, the effect of the t_m on the surface roughness is found to be lower, when the feed per tooth is lower than that of the t_m. Lai et al. [9] developed an analytical model based on the finite element simulations for microendmilling. It is resulted that the edge radius is the major factor for the t_m. Bissacco et al. [10] found that during microendmilling the effect of machining process parameters is not similar to that of macro-regime milling due to the size effect and the reduced stiffness of the microendmill. These differences are mainly due to the changes in physical phenomena, t_m and microstructure of the work material.

From the literature, it is observed that during microendmilling, the tool geometry plays a significant role in terms of generation of the t_m subsequently cutting forces and surface roughness. It is also necessary to understand the microendmilling mechanism with respect to the tool geometry [11–14]. From the literature review, it is also observed that the effect of cutting forces in microendmilling is not yet fully explored. In addition to the cutting forces, the other factors that are affected by the tool geometry in microendmilling are entry and exit of the flutes, which occurs repeatedly during the cutting process, run out of the tool tip, t_m, ploughing, indentation, elastic recovery, etc. [15–30].

Therefore, in this work investigations are carried out to identify the effect of different tool geometries such as two flutes and four flutes of the microendmill by analysing the behaviour of uncut chip thickness (t) along with the cutting forces (transverse force (F_x) and feed force (F_y)).

2 Experimental Set-up

Microendmilling was performed using MIKROTOOLS made multi-process machine tool driven by a 100 W AC servo drive motor that can provide a rotational speed up to 5000 rpm. The *X*-*Y* table has optical linear scale with resolution of 0.1 μm. Two flutes carbide microendmill (Model: E9661; Make: AXIS, India) and four flutes carbide microendmill (Model: E9671; Make: AXIS, India) of 500 μm diameter each coated with aluminium titanium nitride (AlTiN) of 400 nm thickness are used for the experimentation (Fig. 2). The workpiece material used in the work is copper alloy (BSS 249), and its chemical composition is copper 55.70%, lead 3.30% and zinc 38.58%. Vickers's hardness of the workpiece is 146.10 HV. The workpiece is mounted on the worktable of the machine, and the microendmill is clamped to the spindle unit of the machine (Fig. 1).

Fig. 1 Experimental set-up closer view

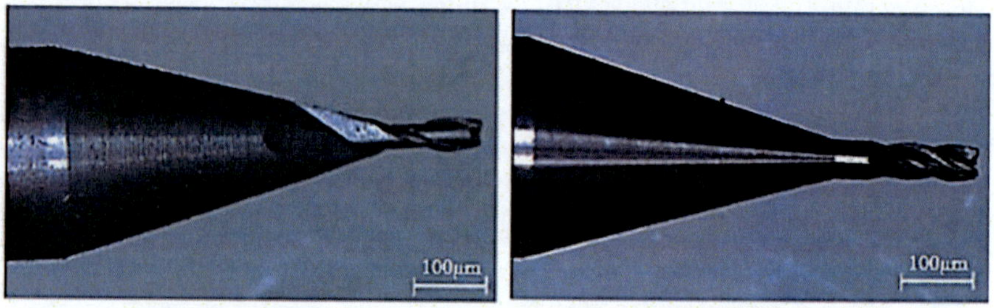

(a) Two flute **(b) Four flute**

Fig. 2 Photograph of the microendmills

Experiments are conducted as per the design of experiments approach (DoE), using the L_{16} orthogonal array, and optimal parameters based on the surface roughness such as rotational speed of 2800 rpm, feed rate of 10 mm/min and depth of cut (DOC) of 50 μm are obtained [27]. The above-optimized cutting parameters are used to analyse the effect of different tool geometry, namely two flutes and four flutes.

The surface roughness (R_a) is measured on the machined surfaces at three different locations (entry, middle and exit along the length of cut) in the feed direction using non-contact instrument Talaysurf CCI Lite, with 0.983 μm cut of length and 333 μm traverse length. The average value is recorded and given in Table 1.

Table 1 Allocation of input parameters in L_{16} OA and experimental results

Ex. No.	Speed [*S*] (rpm)	Feed [*F*] (mm/min)	Surface roughness [R_a] (μm)		Cutting Forces (*N*)			
					Two flutes		Four flutes	
			Two flutes	Four flutes	[F_x]	[F_y]	[F_x]	[F_y]
1	700	5	0.194	0.145	0.037	0.771	0.400	0.470
2	700	10	0.163	0.126	0.583	0.771	0.819	0.444
3	700	15	0.169	0.159	1.050	0.875	1.500	0.695
4	700	20	0.219	0.204	2.376	2.498	2.910	1.320
5	1400	5	0.192	0.106	0.057	0.955	0.545	0.389
6	1400	10	0.122	0.101	0.117	0.300	0.491	0.529
7	1400	15	0.108	0.093	0.844	0.910	1.140	0.623
8	1400	20	0.174	0.108	0.759	0.774	1.107	0.292
9	2100	5	0.163	0.079	0.121	0.903	0.538	0.011
10	2100	10	0.088	0.071	0.390	0.827	0.809	0.381
11	2100	15	0.104	0.063	0.084	0.449	0.611	0.351
12	2100	20	0.114	0.086	0.229	0.690	0.617	0.354
13	2800	5	0.125	0.062	0.137	0.938	0.336	0.293
14	2800	10	0.091	0.075	0.178	0.675	0.468	0.388
15	2800	15	0.076	0.066	0.134	0.487	0.425	0.375
16	2800	20	0.101	0.067	0.001	0.419	0.382	0.257

The cutting forces in transverse (F_x), feed (F_y) and axial (F_z) direction were measured using piezoelectric cutting force dynamometer (Fig. 1). The dynamometer (Make: Kistler, Model: 9256C2) is mounted on the fixture unit of the machine and then connected to a charge amplifier, which in turn connected to a PC for acquiring the data using data acquisition system with DynoWare software (Fig. 1). The directions of cutting forces (F_x, F_y and F_z) are measured using the dynamometer which is shown in Fig. 3. F_x, F_y and F_z are measured along the *X*-, *Y*- and *Z*-axes, respectively. The cutting force signals are measured at the sampling rate of 1000 Hz with the regular intervals of 120 s. During microendmilling (slot milling), it is observed that both downmilling and upmilling occurs, which is also similarly observed by Kang et al. [20]. The schematic indicates the direction of cutting forces measured during downmilling and upmilling is shown in Fig. 4a, b, respectively. Figure 4 indicates that ploughing occurs during the rotation of the microendmill from A to B and E to D regions during upmilling and downmilling, respectively, whereas shearing occurs during the rotation of the microendmill, from B to C and C to D regions during upmilling and downmilling, respectively.

From the preliminary studies, it is observed that the axial force is found to be non-significant in all the machining conditions due to the lower order of depth of cut (i.e. 50 μm), which is also similarly observed by Bissacco et al. [10]. Therefore, the

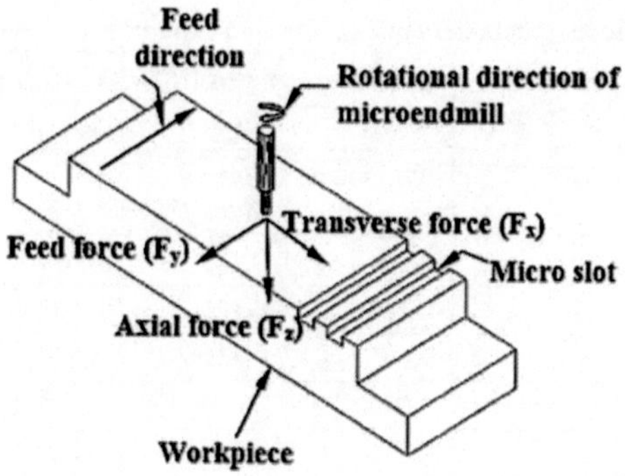

Fig. 3 Schematic diagram of the cutting force components in microendmilling

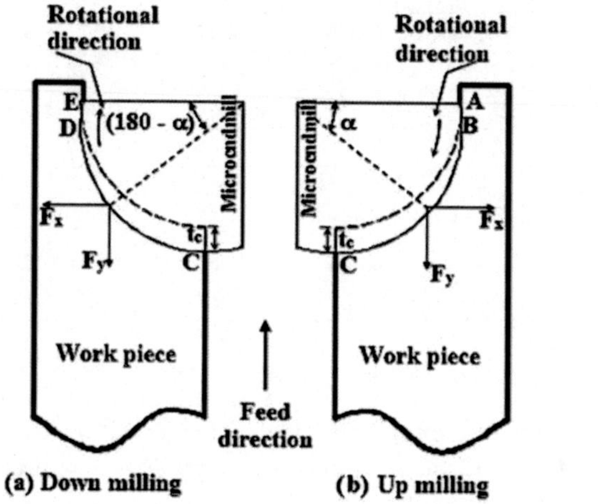

A to B: Ploughing ($t < t_m$), B to C: Upmilling ($t \geq t_m$),
C to D: Downmilling ($t \geq t_m$), E to D: Ploughing ($t < t_m$)

t – Uncut chip thickness, t_c – Maximum uncut chip thickness,
t_m – Minimum uncut chip thickness, α – Tool rotational angle

Fig. 4 Schematic of the direction of cutting forces in the workpiece in microendmilling

axial force is not considered for further analysis. The resultant force is also calculated by considering transverse and feed force using Eq. 3, which is given as

$$\text{Resultant force } (F_r) = \sqrt{F_x^2 + F_y^2} \tag{3}$$

3 Results and Discussion

The effect of two different microendmill geometries such as two flutes and four flutes during microendmilling on the surface roughness (R_a), the uncut chip thickness (t), and cutting forces (F_x and F_y) is analysed. They are briefly presented below.

3.1 *Surface Roughness* (R_a)

Experiments are carried out by varying the rotational speed and feed rate with constant depth of cut of 50 μm. The effect of two different tool geometries such as two flutes and four flutes of the microendmill at various cutting parameters on the R_a is shown in Fig. 5. Typical surface topography of the machined surface with two flutes and four flutes of microendmill is shown in Fig. 6. From Figs. 5 and 6, it is observed that R_a values are found to be lower in the four flutes microendmill than that of two flutes microendmill. This may be due to the formation of continuous chips with four flutes than that of the two flutes (Fig. 7). This may also be due to the formation of closer tool path (i.e. due to the increase in number of flutes in the four flutes microendmill) for subsequent passing of microendmill in the feed direction as illustrated by Bao et al. [16]. In the case of microendmilling with two flutes, higher R_a is observed due to interrupted chip formation (Fig. 7). It is also observed that R_a is found to be decreased with the increase in the rotational speeds up to the feed rate of 12 mm/min both with the two flutes and four flutes microendmill. However, R_a is found to be increased as the feed rate increases beyond 12 mm/min. There is no significant variation in R_a at the higher rotational speed above 1400 rpm with increase in feed rate beyond 12 mm/min. Therefore, in order to achieve a better R_a, four flutes microendmill

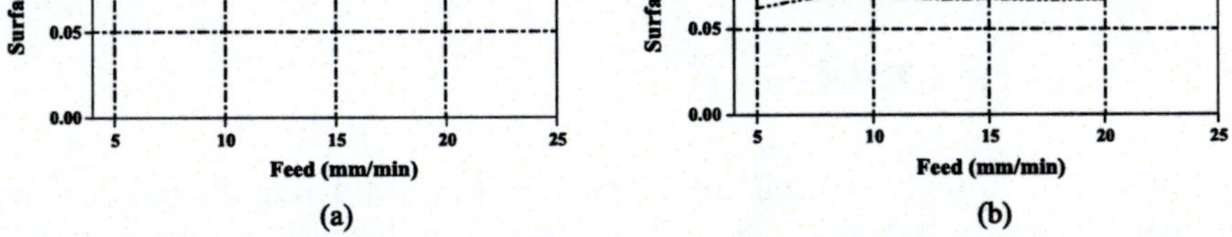

Fig. 5 Tool geometry effect on surface roughness in microendmilling. **a** Two flutes microendmill. **b** Four flutes microendmill

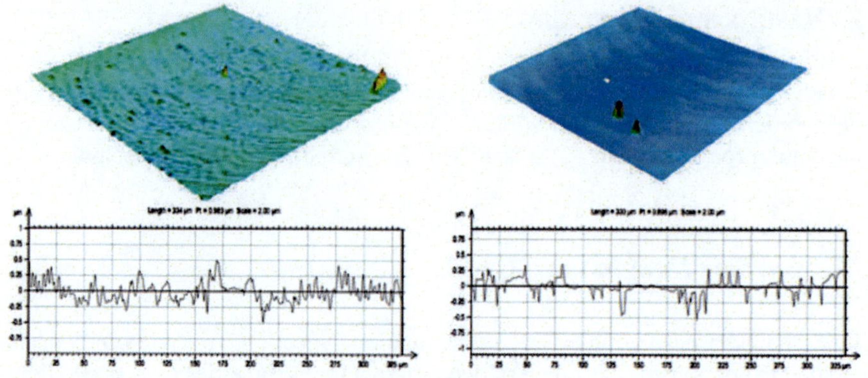

(a) (b)

Fig. 6 Typical surface topography of the machined surface in microendmilling. **a** Two flutes microendmill. **b** Four flutes microendmill

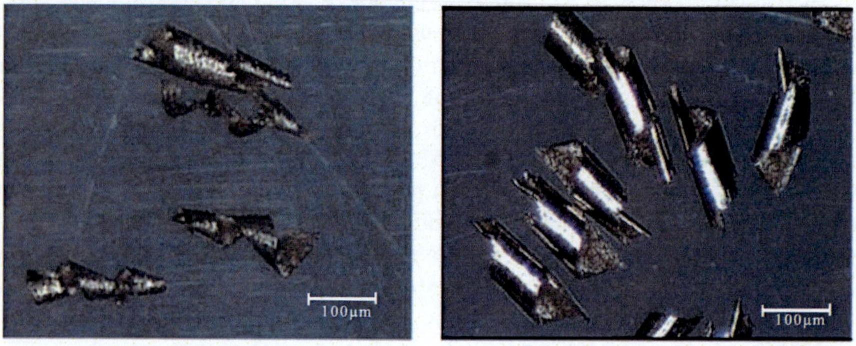

(a) (b)

Fig. 7 Typical chips obtained in microendmilling. **a** Two flutes microendmill. **b** Four flutes microendmill

with higher rotational speed along with moderate feed rate are found to be highly desirable.

3.2 *Uncut Chip Thickness* (t)

The behaviour of the uncut chip thickness (t) is analysed during the half rotation for both two flutes and four flutes microendmill by varying the rotational speed. Figure 8 indicates the two regions of material removal mechanism in microendmill based on the uncut chip thickness (t) such as ploughing (regions A to B and D to E)

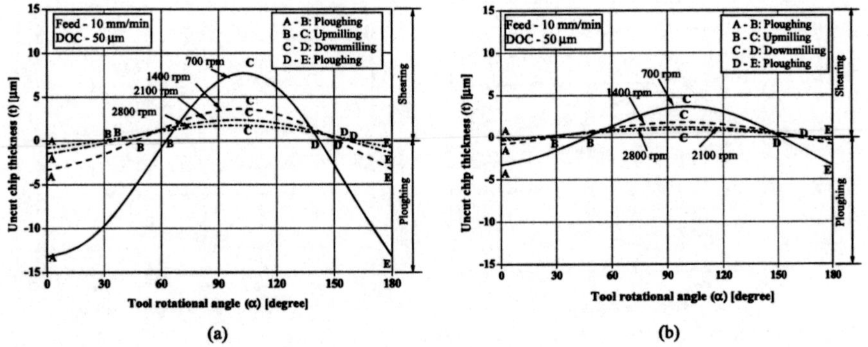

Fig. 8 Behaviour of uncut chip thickness of different tool geometry with varying speed in microendmilling. **a** Two flutes microendmill. **b** Four flutes microendmill

and shearing (regions B to C and C to D). During shearing (regions B to C and C to D), both upmilling and downmilling occurred.

Figure 8a indicates that in the case of two flutes microendmill, with the increase in the rotational speed and there is a decrease in the thickness value of uncut chip thickness (t). However, ploughing (regions A to B and D to E) is found to be more dominant than that of the shearing (regions B to C and C to D). This is also reflected in the formation of interrupted chip (Fig. 6a), in which the chips are not formed until the uncut chip thickness (t) reaches the minimum uncut chip thickness (t_m) during the subsequent passes of microendmill along the feed direction. This indicates that the chip formation did not occur during the single pass of the microendmill, and therefore, subsequent passes are required for the chip formation process.

Figure 8b indicates the behaviour of uncut chip thickness (t) with four flutes microendmill with the increase in the rotational speed. From the Fig. 8b, both ploughing (regions A to B and D to E) and shearing (regions B to C and C to D) are observed during the half tool rotation of the microendmill, irrespective of the rotational speed (Fig. 8b). However, the size of uncut chip thickness (t) is found to be lesser than that of the two flutes microendmilling. It is also observed that in the four flutes microendmill, the shearing (regions B to C and C to D) is found to be more dominating than that of the ploughing (regions A to B and D to E), which is also reflected in the formation of continuous chip (Fig. 7b).

Figure 9a, b shows the effect of feed rate with respect to two flutes and four flutes microendmill on the uncut chip thickness (t) during the half rotation of the microendmill. Both ploughing (regions A to B and D to E) and shearing (regions B to C and C to D) are observed during the half tool rotation of the microendmill, irrespective of the rotational feed rate (Fig. 9). From Fig. 9a, it is observed that the uncut chip thickness (t) is found to increase with the increase in the feed rate. It is also observed that with the increase in the feed rate, ploughing effect dominates the shearing effect. However, in the case of microendmilling with four flutes microendmill (Fig. 9b), it is found that the size of uncut chip thickness (t) and the effect of ploughing are

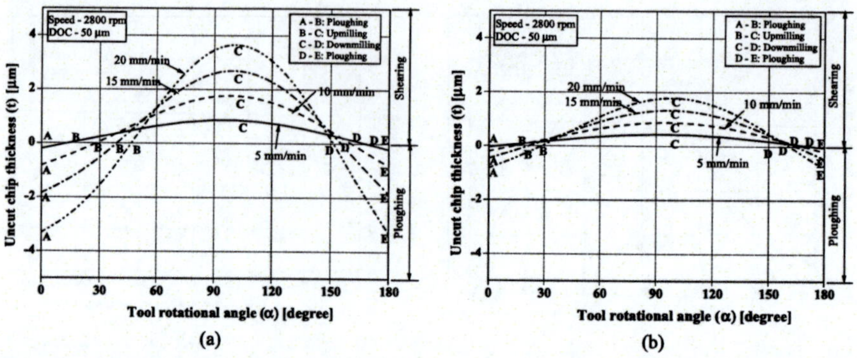

Fig. 9 Behaviour of uncut chip thickness of different tool geometry with varying feed in microendmilling. **a** Two flutes microendmill. **b** Four flutes microendmill

lesser than that of with the two flutes microendmill. This may be due to the closer trochoidal tool path generation of the four flutes microendmill than that of two flutes microendmill, which is also similarly observed by Newby et al. [5]. These results indicate that the tool path generation in microendmilling significantly depends on the number of flutes and the feed rate, which is also similarly observed by Tansel et al. and Bao et al. [14, 16]. Higher feed rate with more number of flutes results in continuous chip formation, which is also similarly observed in the conventional endmilling [11–13].

3.3 Cutting Forces

Figures 10 and 11 show the typical cutting forces in the transverse and feed directions,

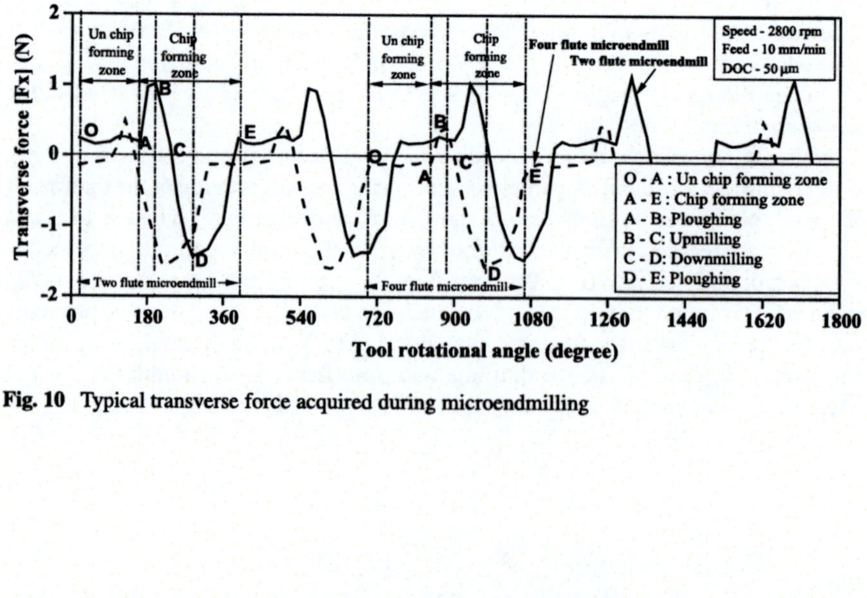

Fig. 10 Typical transverse force acquired during microendmilling

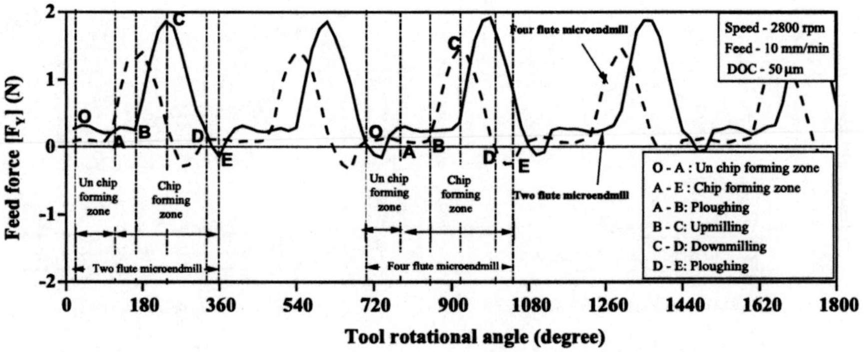

Fig. 11 Typical feed force acquired during microendmilling

respectively, for two flutes and four flutes microendmill. From Figs. 10 and 11, it is observed that the signature of the transverse force and feed force signals for both two flutes and four flutes microendmill is found to be in increasing and decreasing trend both in positive and negative directions with respect to the tool rotational angle. However, from Figs. 10 to 11, it is observed that, at the earlier stages with four flutes microendmill there is the significant angular shift in cutting forces at the region of unchip forming zone (O to A), where no chips are formed, than that of with the two flutes microendmill. This may be due to the closer trochoidal tool path of the four flutes microendmill than that of two flutes microendmill. This is well agreed with the result of Bao et al. [16]. It is also observed from Figs. 10 and 11, that the signature of the cutting force signal during unchip forming zone (O to A) and chip forming zone (A to E) is found to be different from the four flutes microendmill than that of two flutes microendmill, due to the change in frequency of contact of two flutes microendmill than that of four flutes microendmill.

From Fig. 10, it is observed that the direction changes in the transverse force from upmilling (B to C) to downmilling (C to D) with both two flutes and four flutes microendmill as shown in Fig. 4. However, the magnitude of the transverse force during the upmilling is lower in the case of four flutes microendmill. But in the case of downmilling, there is no significant variation is observed. Transverse force due to ploughing (regions A to B and D to E) is more dominant in the case of two flutes microendmill. From the above, it is observed that, the microendmilling with two flutes is mostly affected by ploughing than that of shearing.

From Fig. 11, it is observed that increasing and decreasing trends of feed force from upmilling (B to C) and downmilling (C to D) in both microendmills as shown in Fig. 4. However, the magnitude of feed force during upmilling is lesser in the case of four flutes microendmill than that of two flutes microendmill.

Figure 12a, b shows the cutting forces behaviour with respect to the uncut chip thickness (t) and tool geometry. Feed force is found to be the maximum when the uncut chip thickness (t) reaches the maximum limit. From Fig. 12, it is also observed that transverse force approaches nearly zero, when the uncut chip thickness (t) is

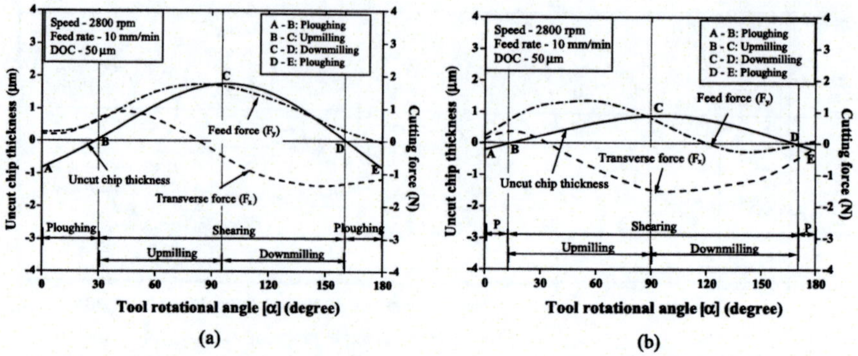

Fig. 12 Typical behaviour of cutting forces with the uncut chip thickness of different tool geometry in microendmilling. **a** Two flutes microendmill. **b** Four flutes microendmill

found to be the maximum level. The transverse force is found to change the direction from positive direction to negative direction during downmilling (regions C to D) while using two flutes and four flutes microendmill (Fig. 12a, b). By comparing Fig. 12a, b, it is observed that, shearing mode of material removal mechanism is dominant with four flutes than the two flutes microendmill. This may be due to the increase in contact area between the tool and workpiece. Figure 12 also indicates that the transverse force and feed force are found to be lesser in the case of four flutes microendmill due to the reduced uncut chip thickness (t).

Figure 13 shows the tool geometry effect on the resultant forces (F_r) with various rotational speed and feed rate with the constant depth of cut (50 μm). It is observed that, the rotational speed and feed rate are compounded irrespective of the tool geometry at the lowest speed of 700 rpm, as similar to the conventional endmilling. The same trend is observed by Lai et al. [9]. In conventional endmilling, increased rotational speed and feed rate resulted in the increasing trend of the cutting force

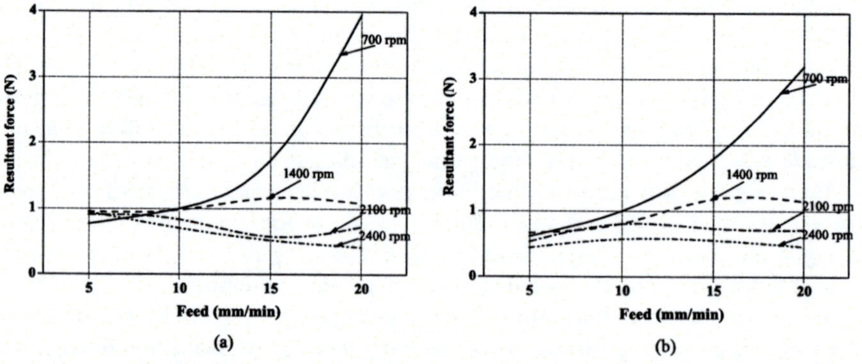

Fig. 13 Tool geometry effect on cutting forces in microendmilling. **a** Two flutes microendmill. **b** Four flutes microendmill

values [11]. At lower speed of 700 rpm along with increased feed rate, higher cutting forces are observed due to the dominance of rubbing and ploughing. However, when the rotational speed increases above 2100 rpm the resultant forces tends to decrease. In the case of two flutes microendmill, the cutting forces are found to be higher than that of four flutes microendmill due to the effect of ploughing than that of the shearing [28].

4 Conclusions

This work presented the investigations on the effect of tool geometries of microendmill with two flutes and four flutes during microendmilling of copper alloy (BSS 249) on the surface roughness (R_a), the uncut chip thickness (t) and cutting forces (feed and transverse forces). The following conclusions are drawn from the present investigations.

- Four flutes microendmill resulted in lesser R_a than that of the two flutes microendmill, due to the closer tool path generation of the flutes in the microendmill in the feed direction and also due to the formation of continuous chip. Therefore, to achieve the better R_a during microendmilling, four flutes microendmill with higher speed and with moderate feed rate is recommended.
- Based on the behaviour of uncut chip thickness (t), it is found that both ploughing and shearing are observed during the half tool rotation of the microendmill, irrespective of the rotational speed and feed rate, in both two flutes and four flutes microendmills. However, the size of uncut chip thickness (t) and the ploughing effect is found to be lesser while using four flutes. Shearing is found to be dominant during microendmilling with four flutes microendmill due to the increase in contact area of tool–workpiece interface.
- From the cutting forces analyses, it is found that the signature of the cutting force signals for both two flutes and four flutes microendmill are similar with respect to the tool rotational angle. However, significant angular shift is observed in the case of four flutes microendmill due to the closer trochoidal tool path generation of successive passing of microendmill. Transverse and feed forces are found to be low in the case of four flutes microendmill due to the reduced uncut chip thickness (t) in the same parametric level of microendmilling. Cutting forces are found to be higher with two flutes microendmill due to the dominance of ploughing than that of the shearing.

References

1. Masuzawa T (2000) State of the art of micromachining. Ann CIRP 49:473–488
2. Alting L, Kimura F, Hansen HN, Bissacco G (2003) Micro engineering. CIRP Ann Manuf Technol 52:635–657
3. Dornfeld D, Min S, Takeuchi Y (2006) Recent advances in mechanical micro machining. CIRP Ann Manuf Technol 55:745–768
4. Chae J, Park S, Freiheit T (2006) Investigation of micro cutting. Int J Mach Tools Manuf 46:313–332
5. Newby G, Venkatachalam S, Liang SY (2007) Empirical analysis of cutting force constants in micro-end-milling operations. J Mater Process Technol 41–47:192–193
6. Fang FZ, Wu H, Liu XD, Liu YC, Ng ST (2003) Tool geometry study in micromachining. J Micromech Microeng 13:726–731
7. Filiz S, Conley CM, Wasserman MB, Ozdoganlar OB (2007) An experimental investigation of micro-machinability of copper 101 using tungsten carbide micro-endmills. Int J Mach Tools Manuf 47:1088–1100
8. Li H, Lai X, Li C, Feng J, Ni J (2008) Modelling and experimental analysis of the effects of tool wear, minimum chip thickness and micro tool geometry on the surface roughness in micro-end-milling. J Micromech Microeng 18. https://doi.org/10.1088/0960-1317/18/2/025006
9. Lai X, Li H, Li C, Lin Z, Ni J (2008) Modelling and analysis of micro scale milling considering size effect, micro cutter edge radius and minimum chip thickness. Int J Mach Tools Manuf 48:1–14
10. Bissacco G, Hansen HN, De Chiffre L (2005) Micromilling of hardened tool steel for mould making applications. J Mater Process Technol 167:201–207
11. Koshy P, Dewes RC, Aspinwall DK (2002) High speed end milling of hardened AISI D2 tool steel (~ 58 HRC). J Mater Process Technol 127:266–273
12. Matsumura T, Hiramatsu T, Shirakashi T (2005) A study on cutting force in the milling process of glass. J Manuf Process 7(2):102–108
13. Li HZ, Zhang WB, Li XP (2001) Modelling of cutting forces in helical endmilling using predictive machining theory. Int J Mech Sci 43:1711–1730
14. Tansel I, Rodriguez O, Trujillo M, Paz E, Li W (1998) Micro-end-milling: I. Wear and breakage. Int J Mach Tools Manuf 38:1419
15. Friedrich CR, Coane PJ, Vasile MJ (1997) Micromilling development and applications for micro-fabrication. J Microelectron Eng 35:367
16. Bao WY, Tansel IN (2000) Modeling micro-end-milling operations. Part II: tool run-out. Int J Mach Tools Manuf 40:2175–2192
17. Fang FZ (2002) Cutting edge effect in micromachining. Presentation to the Cutting Committee Meeting (Part II) in 52nd CIRP General Assembly (Spain)
18. Mian AJ, Driver N, Mativenga PT (2011) Identification of factor that dominate the size effect in micro-machining, Int J Mach Tools Manuf 51:383–394
19. Yun HT, Heo S, Lee MK, Min BK, Lee SJ (2011) Ploughing detection in micromilling process using the cutting force signal. Int J Mach Tools Manuf 51:377–382
20. Srinivasa YV, Shunmugam MS (2013) Mechanistic model for prediction of cutting forces in microendmilling and experimental comparison. Int J Mach Tools Manufa 67:18–27
21. Sooraj VS, Mathew J (2011) An experimental investigation on the machining characteristics of microscale endmilling. Int J Adv Manufa Technol 56:951–958
22. Kang IS, Kim JS, Seo YW (2010) Investigation of cutting force behaviour considering the effect of cutting edge radius in the micro-scale milling of AISI 1045 steel. Proc IMechE 225:163–171
23. Malekian M, Park SS, Jun MBG (2009) Modeling of dynamic micro-milling cutting forces. Int J Mach Tools Manuf 49:586–598
24. Park SS, Malekian M (2009) Mechanistic modeling and accurate measurement of microend-milling forces. CIRP Ann Manuf Technol 58:49–52

25. Cuba Ramos A, Autenrieth H, Straub T, Deuchert M, Hoffmeister J, Schulze V (2012) Characterization of transition from ploughing to cutting in micromachining and evaluation of minimum chip thickness of cut. J Mater Process Technol 212:594–600
26. Afazov SM, Zdebski D, Ratchev SM, Segal J, Liu S (2013) Effect of micro-milling conditions on the cutting forces and process stability. J Mater Process Technol 213:671–684
27. Prakash M, Kanthababu M (2013) In-process tool condition monitoring using acoustic emission sensor in microendmilling. Int J Mach Sci Technol 17(2):209–227
28. Chen W, Teng X, Huo D, Wang Q (2017) An improved cutting force model for micro milling considering machining dynamics. Int J Adv Manuf Technol 93(9–12):3005–3016
29. Sahoo P, Patra K (2019) Mechanistic modeling of cutting forces in micro-end-milling considering tool run out, minimum chip thickness and tooth overlapping effects. Mach Sci Technol 4;23(3):407–430
30. Chen W, Huo D, Teng X, Sun Y (2017) Surface generation modelling for micro end milling considering the minimum chip thickness and tool runout. Proc CIRP 1(58):364–369

Temperature and RF Power Effect on the Morphology and Structural Properties of TiC Thin Film Grown by RF Magnetron Sputtering

Olayinka Oluwatosin Abegunde, Esther Titilayo Akinlabi, and Oluseyi Philip Oladijo

Abstract In this research study, the effect of temperature and radio frequency (RF) power on the surface morphology and structural properties of titanium carbide (TiC) thin film was studied. TiC thin film was deposited on the surface of commercially pure titanium (CpTi) alloys to enhance its surface properties. During the deposition process, the sputtering temperature and RF power were varied while other sputtering parameters were kept constant. Field emission scanning electron microscope was used to analyze the surface profiling of the film, and atomic force microscope was used to study the surface roughness. Raman spectroscopy was used to determine the film composition, and grazing incidence X-ray diffractometer was used to analyze the phase composition. Nanoindentation was carried out to understand the hardness and Young's modulus properties of the coating. The results reveal the dependency of the thin film properties on the independent process parameters.

Keywords RF power · Temperature · Thin film · TiC

1 Introduction

TiC coatings are extensively applied for modifying and enhancing the surface properties of materials. This is due to the superior quality of their properties such as high hardness, low coefficient of friction, low wear rate, good electrical conductivity, excellent thermal resistance and stability, high modulus, excellent biocompatibility adhesion and chemical inertness [1–6].

O. O. Abegunde (✉) · E. T. Akinlabi · O. P. Oladijo
University of Johannesburg, Johannesburg 2006, South Africa
e-mail: oabegunde@uj.ac.za

E. T. Akinlabi
Covenant University, Ota, Nigeria

O. P. Oladijo
Botswana International University of Science and Technology, Palapye, Botswana

S. Vijayan et al. (eds.), *Trends in Manufacturing and Engineering Management*, Lecture Notes in Mechanical Engineering,
https://doi.org/10.1007/978-981-15-4745-4_51

Numerous techniques like laser deposition, electrochemical deposition, chemical vapor deposition, and physical vapor deposition are available for making surface coating depending on the area of application and coating thickness [7–9]. Among these several techniques, RF magnetron sputtering, a variant of physical vapor deposition, has proven to be one the most promising coating techniques due to advantageous factors like deposition at low temperature, dense and uniform coating, easy to control the process parameters and ability to be used on both conductive and insulating materials [10, 11]. During RF magnetron sputtering process, several process parameters like sample holder configuration, sputtering power, vacuum temperature and pressure, sputtering bias potential and current, deposition time and substrate–target distance determine the behavior of the resulting thin film properties [12–17]. The distribution of the temperature and RF power is two crucial parameters focused on in this research work. Distribution of the temperature within the vacuum chamber is important for the growth of thin film and affects the coalescence and deposition rate. Thinner film tends to form a region with high-temperature profile compared to region with low-temperature distribution [18, 19]. RF magnetron power is another vital process parameter in growing thin films. RF power determines the intensity of the plasma formation, which influences the bombardment of the target and coalescence of the adatoms on the surface of the substrate [13, 20]. Therefore, optimizing and controlling the interaction between these process parameters can help to achieve film of desired and specific properties.

In this research study, the effect of the RF power and temperature was studied on the evolving properties (microstructure, roughness, hardness, Young's modulus, structure, and Raman) of TiC thin film deposited on commercially pure titanium substrate.

2 Experimental Setup

2.1 RF Magnetron Sputtering

TiC thin films were grown on commercially pure titanium (CpTi) using HHV TF500 versatile RF magnetron sputtering coater. Before deposition of the film, the substrate (CpTi) was ground using silica carbide paper and then polished to remove any surface impurities. Further, cleansing of the surface was done by using acetone to remove any form of lubricant or contaminant on the surface. The purity of the TiC target used is 99.99%. During the sputtering process, the RF power and temperature were varied while other sputtering parameters were kept constant. Three runs of experiments were conducted, and the experimental matrix is shown in Table 1. The sputtering apparatus was first evacuated to a base pressure of 1.13×10^{-5} mbar. The reason for the low pressure is to provide a long mean free path for a collision between the target and substrate and allows for the control and minimization of contaminants. After evacuation, the system is refilled with argon to a partial pressure of 2.5×10^{-3} mbar.

Table 1 Experimental matrix of the process parameters for thin film deposition

S/N	RF Power (W)	Temperature (°C)
A	150	80
B	200	90
C	250	100

The flow rate of argon was kept constant at 12 sccm while the sputtering time was maintained at 2.5 h. The working pressure was constant throughout the experiments. Maintaining constant pressure helps to place many argon atoms in the path of the ions and ejected coatings.

2.2 *Characterization* Techniques

A veeco Di2100 atomic force microscope was used to evaluate the surface topographies of the samples. Image scan size at $5 \times 5\ \mu m^2$ was obtained in tapping mode. All analysis was performed in ambient temperature. Nanoscope software was used for capturing and analyzing the images from the surface of the samples. Field emission scanning electron microscope (ZEISS Gemini*2, Germany) (FESEM) capable of capturing nanoscale images effortlessly at very high magnification was used to observe the surface morphology evolution. The FESEM can take images at very high magnification and images were taken at 50,000× magnification. Raman analyses were performed on TiC thin films using an alpha300R (WITec) confocal laser Raman microscopy coupled with frequency-doubled Nd-YAG laser of 532 nm wavelength. A magnification of ×50 on Nikon objective lens was used in capturing the Raman spectra intensity. A laser power of 2 mW at room temperature was used to prevent the burning of the film surface. Before the acquisition of the Raman spectra, beam centering and Raman spectra calibration were done using standard Si(111) samples. The Raman spectrum of the substrate was obtained and used to compare with the TiC thin films deposited. Grazing incidence X-ray diffractometer (PANalyticals's Xpert Pro with Cu K-alpha and wavelength 1.540598 A) at a very low angle of incidence that allows penetration into the thin film was used to study the structural properties of the thin film. Young's modulus and hardness thin film of the nanomechanical properties were obtained using Hysitron TI 950 Triboindenter. The control mode used was load controlled, and the load pattern was modeled to follow a trapezoidal geometry of loading dwelling and unloading profile with dwelling time of 15 s at peak load.

Berkovich tip was used as the diamond indenter with a tip radius of curvature of 100 nm. From the analyzed load-displacement curves, Oliver and Pharr analysis method was used to calculate Young's modulus of the thin film [21–23]. The depth of penetration for all the samples was limited to 10% of the coating thickness to eliminate the effect of the substrate on the thin film properties [24, 25]. The hardness and

Young's modulus reported corresponding to average of 10 measurements performed on the thin film coating.

3 Result and Discussion

Figure 1 depicts the microstructural evolution of the surface morphology of the coating at three different process parameters. Three different distinct features and growth modes are noticeable at each process parameter. For sample A, the microstructural growth pattern is similar to island growth which shows an early stage of film growth and nucleation on the surface and the adherence bond between the adatoms is stronger than the bonding between the adatoms and substrate. The coatings are sparingly distributed on the surface with granular grain shape. The coating becomes dense and homogenous distributed as the power and temperature increased. This might be associated with the fact that at sufficient temperature and power, the rate of nucleation and growth of the film is rapid. In sample B, the images reveal a columnar structure as expected. It is of no doubt that these columnar structures usually associated with ceramic coating Amina et al. [2]. For sample C, two layers of coating are noticeable. The topmost layer is coarse with high porosity while the latter layer coating is evenly distributed. This implies that further increased in RF power and temperature leads to thicker coating surface and formation of a new layer with the nucleation mechanism pattern of the Stranski–Krastanov mechanism.

The 3D morphology of the surface roughness is shown in Fig. 2, and the statistical

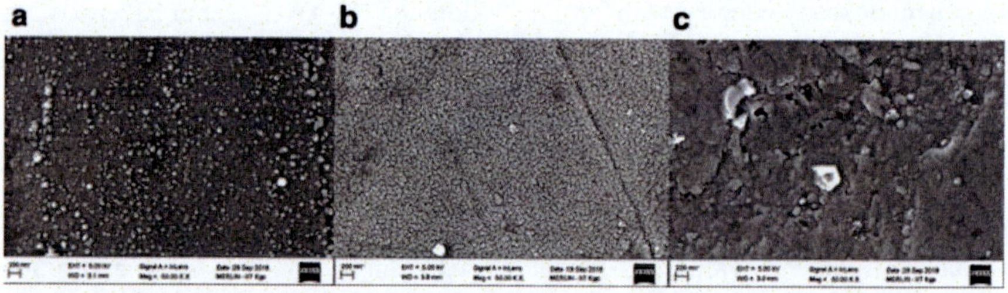

Fig. 1 FESEM images taken at a 50,000× magnification of the surface morphology

Fig. 2 3D images from the AFM surface profiling at scan size of 5 × 5 μm^2

Table 2 Statistical details of the AFM results

Experimental run	Z-range (nm)	RMS (nm)	Mean roughness (nm)	Skewness	Kurtosis
A	397.25	51.93	39.68	0.263	3.898
B	268.35	30.58	20.48	1.646	8.316
C	440.66	53.467	40.330	0.451	4.223

details are summarized in Table 2. The 3D AFM structure for sample 1 exhibits extended columnar grain growth which agrees with the result of the FESEM images. The AFM morphology of samples 2 shows evenly distributed coating, and for sample 3, the surface becomes uneven.

The surface roughness reduced from 39.68 nm for sample A to 20.48 nm in sample B. This phenomenon might due to the homogenous morphology of sample B. The roughness increases for sample C which has the highest roughness and most uneven surface due to incomplete formed multilayer surface. The hardness and Young's modulus are presented in Table 3. Similarly, the graph is shown in Fig. 3. The highest hardness value is associated with sample B that exhibits uniform coating of the film. Sample C has the least hardness value. Although for thin film, the hardness is not

Table 3 Hardness and Young's modulus results from nanoindentation

Experimental run	Hardness (GPa)	Young's Modulus (GPa)	Wear resistance H^3/E^2 (GPa)	Plasticity Index H/E
A	5.00 ± 2.04	134.19 ± 21.39	0.007	0.037
B	6.08 ± 2.90	139.03 ± 25.67	0.012	0.044
C	4.09 ± 1.99	113.21 ± 34.44	0.005	0.036

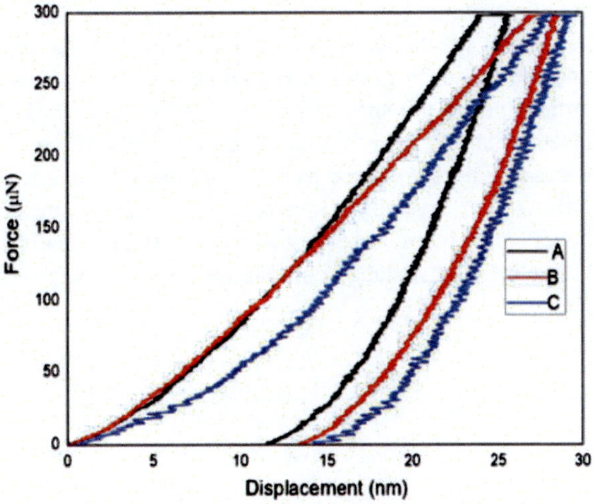

Fig. 3 Force versus displacement curve of the nanoindentation plot

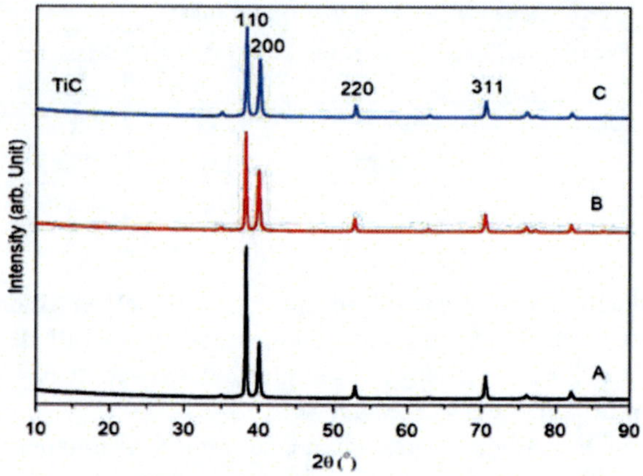

Fig. 4 XRD spectra of TiC thin film coating on CpTi

enough to conclude the resistance of the surface to deformation or failure, the wear resistance which is the ratio of H^3/E^2 and plasticity index (H/E) is used to study the film resistance. The wear resistance and plasticity index were highest at B. The wear resistance and plasticity increase with an increase in RF power and temperature and then decrease as the RF power and temperature were raised from 200 W and 90 °C to 250 W and 100 °C, respectively.

The crystallographic structures from the XRD are shown in Fig. 4. The patterns of the TiC peaks obtained exhibit a preferential orientation of 110, and other planes found are 200, 220, and 311. TiC films have a rock salt-type (NaCl type) face-centered cubic (FCC) with diffraction peaks intensity at 110, 200, 220, and 311 plane reflection [26]. This agrees with the result, and different peak intensities were noticed due to the RF power and temperature of the coating. The highest peak intensity was found at 110 for all the samples. 110 spectra are known to result in high hardness value [2]. Sample A produces the highest peak value at 110 dues to the geometrical factor on the slip system [14].

Raman spectroscopy is responsible for determining the crystal structure quality, and the response graph is shown in Fig. 5. Both the D and G bands for TiC are noticeable on the spectra graph. The D band for TiC was found around 500–700 (1/cm). Sample B tends to have smaller Raman intensity peaks compared to A and C. Raman peaks are caused by lattice mismatch, chemical composition, defects, short-range order, crystalline structure, and internal stresses. Since sample B possesses better homogenous and uniformly distributed coating, it can be concluded that the lattice is better matched compared to A and C with less defect.

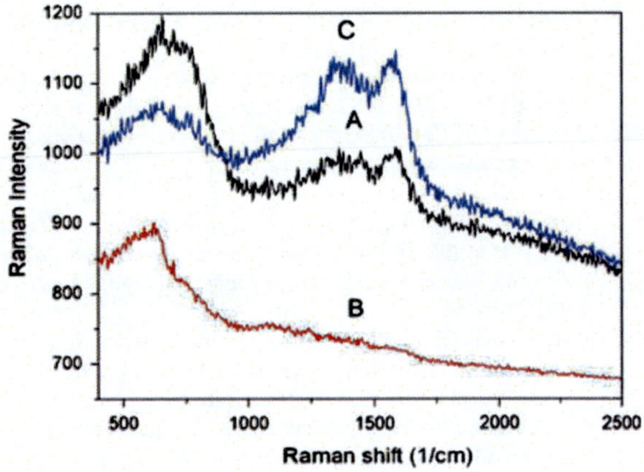

Fig. 5 Raman spectra of TiC thin film coating on CpTi substrate

4 Conclusion

The observation from the characterization shows how RF power and temperature contribute to the properties of TiC thin film coating. Three distinct growth types were noticed depending on the process parameters, and the lowest surface roughness was noticed at the film with the dense and uniformly distributed coating. The XRD revealed the preferential plane of 111 has the highest peak, and the hardness properties show substantial dependence on the process parameters.

Acknowledgements The authors will like to acknowledge CSIR South African and the University of Johannesburg for funding this project.

References

1. Kulkarni M et al (2014) Biomaterial surface modification of titanium and titanium alloys for medical applications. Nanomedicine 111:111
2. Djafer ZA et al (2014) Deposition and characterization of titanium carbide thin films by magnetron sputtering using Ti and TiC targets. Appl Surf Sci 312:57–62
3. Longo G et al (2010) Effect of titanium carbide coating by ion plating plasma-assisted deposition on osteoblast response: a chemical, morphological and gene expression investigation. Surf Coat Technol 204(16–17):2605–2612
4. Brama M et al (2007) Effect of titanium carbide coating on the osseointegration response in vitro and in vivo. Biomaterials 28(4):595–608
5. Galevsky G et al (2015) Titanium carbide: nanotechnology, properties, application. In: IOP conference series: materials science and engineering, p 012017
6. Bugaev SP et al (2003) Amorphous hydrogenated carbon films deposited by a closed-drift ion source. Laser Part Beams 21(2):285–289
7. Abegunde OO et al (2019) Overview of thin film deposition techniques. AIMS Mater Sci (2):174

8. Zhang S et al (2003) Recent advances of superhard nanocomposite coatings: a review. Surf Coat Technol 167(2–3):113–119
9. Guo HC et al (2017) Recent progress of atomic layer deposition on polymeric materials. Mater Sci Eng C 70:1182–1191. https://0-doi-org.ujlink.uj.ac.za/10.1016/j.msec.2016.01.093
10. Alexeeva OK (2016) Application of the magnetron sputtering for nanostructured electrocatalysts synthesis. Int J Hydrogen Energy 41(5):3373–3386
11. Wu F et al (2016) Characteristics of the structure and properties of ZnSnO3 films by varying the magnetron sputtering parameters. Acta Metallurgica Sinica (English Letters) 29(9):827–833
12. Toku H et al (2010) Influence of process parameters on the growth of pure-phase anatase and rutile TiO_2 thin films deposited by low temperature reactive magnetron sputtering. Braz J Phys 40(3):340–343
13. Ghoranneviss ZG (2016) Effects of various deposition times and RF powers on CdTe thin film growth using magnetron sputtering. J Theor Appl Phys 10(3):225–231
14. Zhang X et al (2013) Effects of substrate temperature on the growth orientation and optical properties of ZnO: Fe films synthesized via magnetron sputtering. J Alloys Compd 574:149–154
15. Fancey K (1995) A coating thickness uniformity model for physical vapour deposition systems: overview. Surf Coat Technol 71(1):16–29
16. Kim K, Kim HW, Lee CM (2003) Effect of growth temperature on ZnO thin film deposited on SiO_2 substrate. Mater Sci Eng B 98(2):135–139
17. Gennardo D (2010) Design, construction, and optimization of a magnetron sputtering system for urania deposition
18. Dong H et al (2006) Study and analysis on the film uniformity. Piezoelectr Acoustoopt 28:578–580
19. Tian M, Liu D (1991) Handbook of thin film science and technology
20. Li FWJ, Qi Y, Guo WDY (2016) Characteristics of the structure and properties of $ZnSnO_3$ films by varying the magnetron sputtering parameters. 29(9):827–833
21. Oliver WC, Pharr GM (1992) An improved technique for determining hardness and elastic modulus using load and displacement sensing indentation experiments. J Mater Res 7(6):1564–1583
22. Oliver WC, Pharr GM (2004) Measurement of hardness and elastic modulus by instrumented indentation: advances in understanding and refinements to methodology. J Mater Res 19(1):3–20
23. Fang T, Jian S, Chuu D (2004) Nanomechanical properties of TiC, TiN and TiCN thin films using scanning probe microscopy and nanoindentation. Appl Surf Sci 228(1–4):365–372
24. Bec S et al (1996) Improvements in the indentation method with a surface force apparatus. Philos Mag A 74(5):1061–1072
25. Mani et al (2005) Effects of residual stress on the mechanical and structural properties of TiC thin films grown by RF sputtering. Surf Coatings Technol 194(2–3):190–195
26. Chun S (2010) Bias voltage effect on the properties of TiN films by reactive magnetron sputtering. J Korean Phys. Soc 56(4):1134–1139

Effect of Temperature on the Surface Characteristics of Anodized Aluminium Tubes

A. L. Sriram Sudhan and A. Brusly Solomon

Abstract Thin porous coating was found to be vital for boiling heat transfer enhancement in various heat transfer devices that use a phase change mechanism. However, the formation of a stable thin porous surface is a challenging process. It is known that the aluminium tubes were used to fabricate different kinds of heat pipes which are used in space applications. Therefore, developing nanoporous coating in an aluminium tube is essential to improve the performance of heat pipes. Anodization is an electrochemical process that applied mostly for aluminium materials to protect the surface from corrosion. This anodized aluminium surface can also be utilized for heat transfer applications by controlling the morphology of the surface through adjusting the process parameters of the anodization. One of the most critical parameter that controls the morphology of coating surface is the temperature of the anodizing cell. Therefore, in this study, the effect of cell temperature on the coating thickness, pore size and contact angle is analysed while maintaining the electrolyte concentration, cell voltage, electrolyte flow rate and anodizing time as constant. An aluminium tube of the outer and inner diameter of 19 mm and 16.8 mm, respectively, with a length of 350 mm was used for anodization. The anodized aluminium tubes were characterized by analysing the surface morphology, coating thickness, pore size and contact angle. It was found that the formation of micro/nanoporous structure in the inner layer of aluminium tube with the pore size ranging from 2.05 μm to 3.6 μm and contact angle less than 10° between the temperature range of 20–35 °C when 10% H_2SO_4 solution is used as the electrolyte.

Keywords Anodization · Temperature · Nanoporous surface · Coating thickness · Capillary rise · Heat transfer

---

A. L. Sriram Sudhan · A. Brusly Solomon (✉)
Micro and Nano Heat Transfer Lab, Centre for Research in Material Science and Thermal Management, Department of Mechanical Engineering, Karunya Institute of Technology and Sciences, Coimbatore, India
e-mail: abruslysolomon@gmail.com

S. Vijayan et al. (eds.), *Trends in Manufacturing and Engineering Management*, Lecture Notes in Mechanical Engineering,
https://doi.org/10.1007/978-981-15-4745-4_52

1 Introduction

The capillary rise is a critical parameter for an efficient operation of heat transfer device such as heat pipe. To enhance the capillary rise, anodization (electrochemical process) was performed on the inner surface of the aluminium tube. The anodization process prevents corrosion and also increases the number of nucleation sites by forming micro/nanopores [1, 2]. By controlling the anodizing parameters, surface morphology can be modified for efficient heat transfer applications [3, 4]. Singh et al. [5] studied the thermal performance of anodized flat thermosyphon with a heat input range of 50–300 W, various fill ratio (40%, 60% and 100%) and different inclination angles (0°, 45°, 90°). It was found that the thermal resistance was reduced by more than 20% in the anodized surface when compared to the non-anodized surface. Also, it was noticed that the pore density was increased up to 90% compared to the non-anodized surface due to the anodization. Solomon et al. [6] investigated the thermal performance of anodized two-phase thermosyphon by varying the heat input in the range of 50–300 W with the combination of R600a and R290 as working fluid. It was found that the reduction in thermal resistance of 17% to 23% while varying the inclination angle from 0° to 90°. Further, the heat transfer coefficient was increased up to 33% for anodized thermosyphon. Solomon et al. [7] studied the thermal performance of anodized grooved heat pipe with various parameters such as heat input in the range of 25–250 W, fill ratio of 10–70% and inclination angle of 0°–90° using R600a as working fluid. It was found that the formation of micro/nanostructures enhances the heat transfer performance by 39% while reducing the thermal resistance of 30% compared to the non-anodized surface. Due to the anodization process, the generation of nucleation site is more; thus, it makes the surface more wettable and leads to capillary rise and heat transfer. Weng et al. [8] studied the transfer performance of the anodized heat pipe by developing anodic aluminium oxide nanotubes in the inner surface of the heat pipe. It was found that heat transfer was enhanced by 4.12% and the reduction in thermal resistance was 58.68% by anodization. From the above literature studies, it is evident that the anodized coating formed in the inner surface of the heat pipe enhances the thermal performance of the heat pipe. This paper aimed at modifying the surface morphology such as pore size, coating thickness and surface wettability, which are key for heat transfer enhancement. Also, the effect of temperature on the pore size, coating thickness and contact angle of anodized surface is studied and discussed.

2 Experimental Analysis

2.1 Anodization of Aluminium Tube

The anodizing facility consists of three systems, namely the electrolyte flow system, chilling unit and DC power supply. The aluminium tube of OD 19 mm, ID 16.8 mm and length of 350 mm is used as an anode, and stainless-steel rod of 10 mm OD is used as a cathode. The cathode is placed inside the anode, and it is supported by Teflon sleeves at both ends, as shown in Fig. 1. Initially, etching is done by circulating NaOH (Merck) solution of 5% for about 1 min and then washed with de-ionized water. Secondly, desmutting is done by circulating 70% of nitric acid solution (Merck) for 2 min, and the excess material is removed from the surface of aluminium and then cleaning is done with DI water. Finally, the aluminium tube is cleaned by passing de-ionized water for 5 min. After the cleaning process, the electrolyte solution is prepared with 10vol.% H_2SO_4 solution (Merck) and another solution with 7% H_2SO_4 + 3% oxalic acid. For anodization, initially 1 V was set, and then the voltage was raised 1 V per minute up to 15 V, and then the voltage was kept constant. The anodization is performed at different temperatures ranging from 20 to 35 °C with H_2SO_4 solution and with a temperature range from 10 to 30 °C for a hybrid solution having 7% H_2SO_4 and 3% oxalic acid solution. More detailed procedure for anodizing the inner tube of aluminium tube is presented in our previous study [4]. After the anodization process, the anodized aluminium tubes were characterized using SEM micrograph, and from the SEM image, the pore size was calculated by using a reference scale provided in the same image.

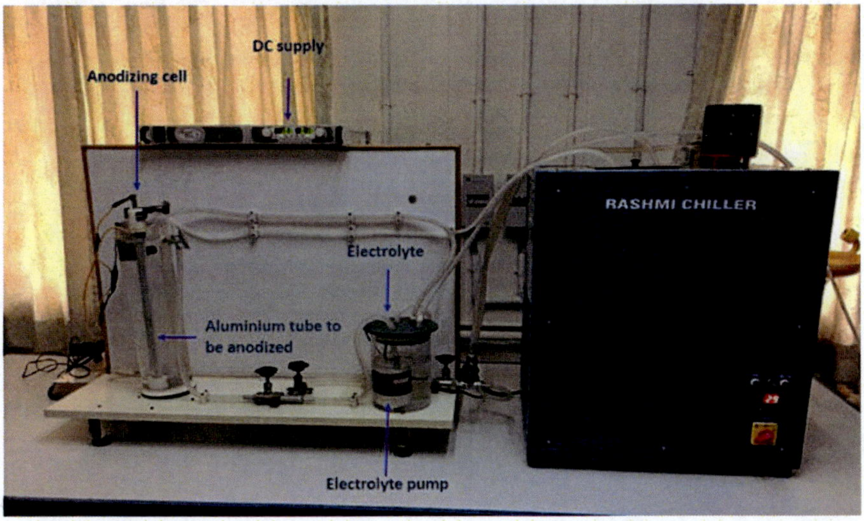

Fig. 1 Anodization cell set-up

3 Results and Discussions

3.1 *Anodization of Aluminium Tube*

Variations of anodizing current with respect to the anodizing time at different temperatures while using H_2SO_4 as electrolyte solution during the anodization process is shown in Fig. 2. During the voltage increment, a thick anodized layer is formed, and when the voltage is maintained at 15 V, the pore formation starts. After the pore formation started, the current consumption increases slightly, as shown in Fig. 2. Therefore, as the temperature is maintained, the increase in current leads to more prominent pore formation. Also it is noticed that the pore size is increasing with the increase in temperature of the anodizing cell. Therefore, it was decided to decrease the temperature further to minimize the pore size. Moreover, the literature suggesting that adding oxalic acid reduces the pore size formation. Therefore, oxalic acid of 3% is combined with the 7% H_2SO_4 solution is considered while decreasing the temperature. The variation of current at different temperature with respect to the anodizing time using 7% H_2SO_4 + 3% oxalic acid as electrolyte solution is shown in Fig. 3. It was observed that the current increases with the increase in time for all anodizing temperatures as similar to H_2SO_4 as an electrolyte solution. The thickness of the coating for this combination of electrolyte is skinny, and it is around 3 to 5 μm. Though the thickness of the anodized layer is thin, there were no pores found on the anodized surface. From these experiments, we can conclude that using H_2SO_4 solution as an electrolyte at an anodizing temperature of 20 °C, the pore size (2.053 μm) and coating thickness (6.446 μm) are found to be less which is suitable for enhanced heat transfer applications.

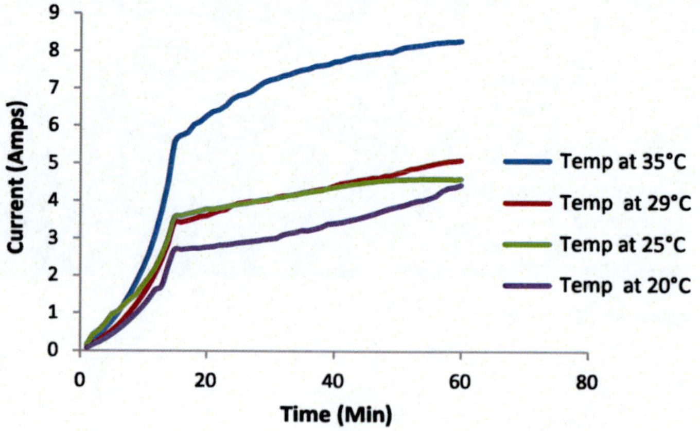

Fig. 2 Variation of current with respect to the anodizing time at different temperatures using H_2SO_4 as an electrolyte

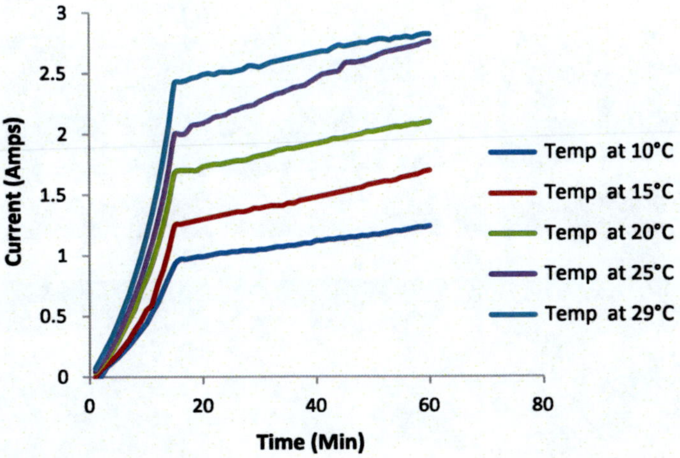

Fig. 3 Variation of current with respect to anodizing time at different temperature using 3% oxalic acid + 7% H_2SO_4 as electrolyte

3.2 Characterization of Anodized Aluminium Tube

In our previous study [4, 7], anodization results for the electrolyte flow rate of 380 ml/min, the anodizing voltage of 1 to 15 V and the concentration of 10 vol% of H_2SO_4 were discussed. In the present study, the surface morphology of anodized aluminium tube using H_2SO_4 as electrolyte solution with an increase in temperature is shown in Fig. 4a–d and Table 1. It was found that the pore size is increasing with the increase in anodizing temperature. The pore size is found to increase suddenly from 2 to 3.6 μm when the temperature is increased from 29 to 35 °C. The maximum experimental temperature was limited at 35 °C due to the formation of larger pore size, which is against our requirement of smaller pore size. To find the thickness of coating formed at the surface of the inner layer, SEM photograph for 10% of H_2SO_4 as electrolyte solution is shown in Fig. 4 (e, f, g and h). A sponge-like substance was formed in the inner layer of the aluminium tube, which represents the coating thickness. This sudden decrease in coating thickness may happen due to the dissolution of the aluminium oxide layer in the inner layer of the tube. A similar dissolution pattern is common in anodization practices [9]. This dissolution occurs with consuming more current of 9 A while forming coating at the inner side of the aluminium tube and if it is continuing the dissolution may take place. With these above discussions, anodization concerning increasing in temperature may result in more dissolution. The anodizing temperature of 20 °C with a minimum pore size of 2.053 μm and minimum coating thickness of 6.44 μm is found to be optimum, which will be suitable for enhanced heat transfer applications.

Surface morphology of anodized aluminium tube while using 7% H_2SO_4 + 3% oxalic acid as electrolyte solution is given in Fig. 5a–e. It was observed that the

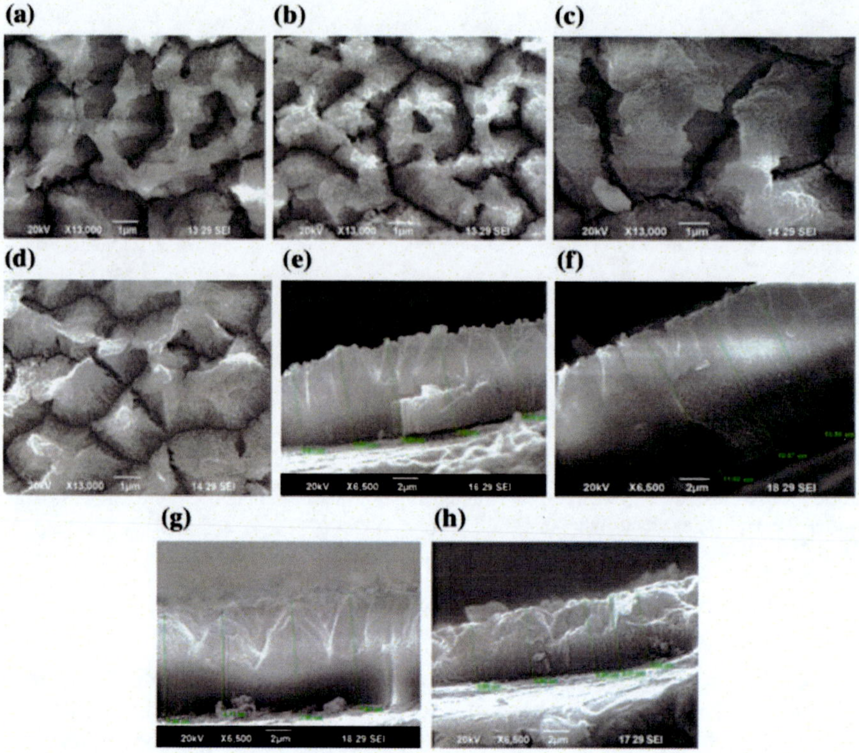

Fig. 4 SEM image of anodized aluminium tube **a** at 20 ° C **b** at 25 °C **c** at 30 °C **d** at 35 °C, coating thickness of anodized aluminium tube **e** at 20 °C **f** at 25 °C **g** at 30 °C **h** at 35 °C

pore size is 4.54 nm for the anodizing temperature of 29 °C. When the anodization temperature decreases from 29 to 10 °C, there was no pore size noticed. This combination of 7% H_2SO_4 + 3% oxalic acid electrolyte consumes less current (about two amps) for anodizing compared to 10% H_2SO_4 electrolyte. Due to less consumption of current, the coating layer formed on the inner surface of the aluminium tube is found to be very thin with cracks and breaks. Though the coating layer is very thin, it is necessary to estimate the coating thickness from the SEM photographs, and it is shown in Fig. 5 f–j. With the decrease in anodizing temperature, the coating thickness also decreases up to a temperature of 15 °C and then starts increasing for 10 °C and it is around 4.7 μm. As a result, this combination of 7% H_2SO_4 + 3% oxalic acid of electrolyte will not be suitable for anodization since there was no pore formation.

To find the wettability of the anodized surface, the contact angle of the surface was measured using Drop Shape Analyzer (DSA 25, Kruss GmBH) with a needle diameter of 0.652 mm. The contact angle for the tube anodized using 7% H_2SO_4 + 3% oxalic acid as electrolyte solution is given in Fig. 6a–e. With the increase in temperature, the contact angle was found to be decreasing, and it is in the range of 98.4–135.6°. Further, it was found that the minimum contact angle obtained at a

Table 1 Anodization conditions

Experiment number	Electrolyte flow rate (ml min^{-1})	Electrolyte Solution	Anodizing Temperature (°C)	Final cell voltage (V_{final})	Acid concentration (V/V %)	Anodizing time (minutes)	Average coating thickness (μm)	Average pore size (μm)	Average contact angle (θ)
Uncoated	–	–	–	–	–	–	–	–	110
1	380	H_2SO_4	20	15	10	60	6.446	2.053	<10
2	380	H_2SO_4	25	15	10	60	7.465	2.167	<10
3	380	H_2SO_4	29	15	10	60	10.97	2.333	<10
4	380	H_2SO_4	35	15	10	60	4.84	3.61	<10
5	380	3% oxalic acid + 7% H_2SO_4	29 °C	15	10	60	5.425	0.00453	98.4
6	380	3% oxalic acid + 7% H_2SO_4	25	15	10	60	4.038	No pore size sited	104.8
7	80	3% oxalic acid + 7% H_2SO_4	20	15	10	60	3.851	No pore size sited	110.5
8	380	3% oxalic acid + 7% H_2SO_4	15	15	10	60	3.604	No pore size sited	113.8
9	380	3% oxalic acid + 7% H_2SO_4	10	15	10	60	47.032	No pore size sited	135.5

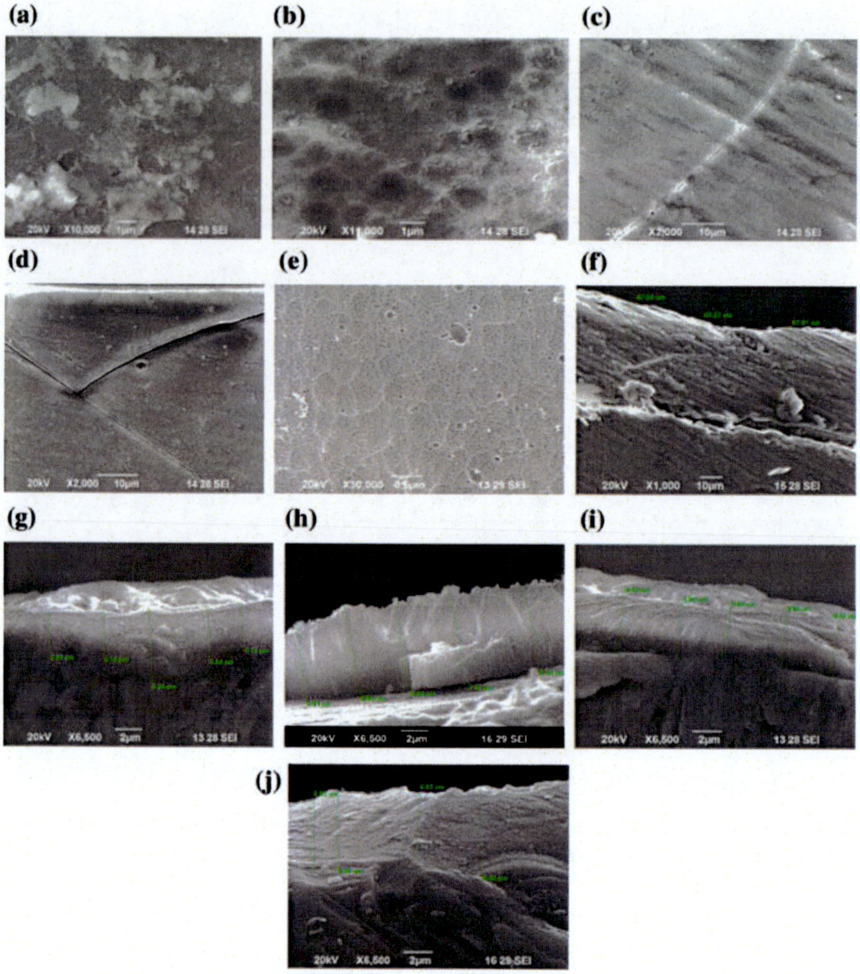

Fig. 5 SEM image of anodized aluminium tube **a** at 10 °C **b** at 15 °C **c** at 20 °C **d** at 25 °C **e** at 29 °C, Coating thickness of anodized aluminium tube **f** at 10 °C **g** at 15 °C **h** at 20 °C **i** at 25 °C **j** at 29 °C

higher temperature is 29 °C. The minimum contact angle implies that the anodized surface is hydrophilic, where the solid–liquid interaction is more, which leads the surface to produce capillary rise. The contact angle for the inner surface of aluminium tube coated using 10% of H_2SO_4 as an electrolyte solution is much less than 10°, which cannot be measured as the liquid drop is spreading quickly.

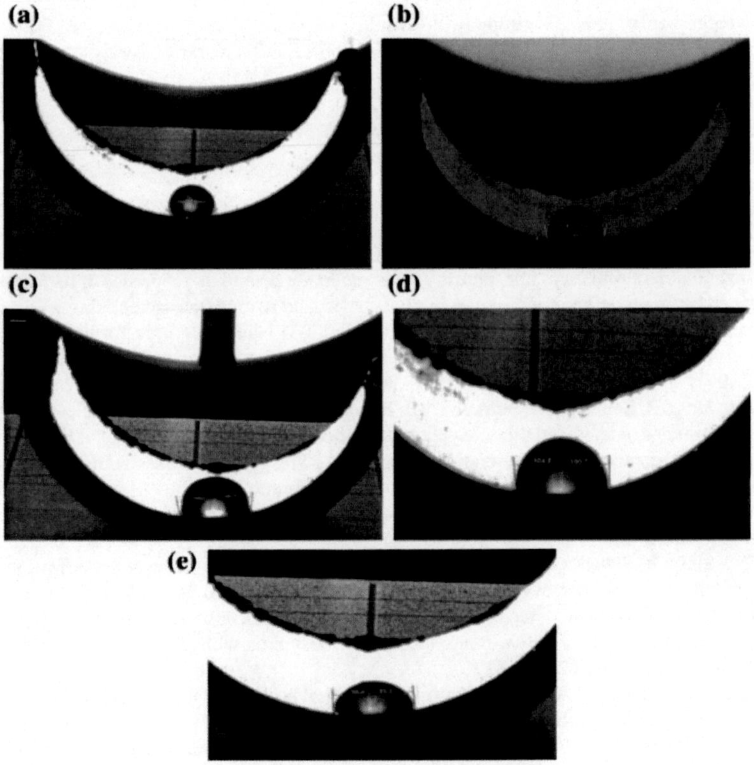

Fig. 6 Contact angle measurement of anodized aluminium tube **a** at 10 °C **b** at 15 °C **c** at 20 °C **d** at 25 °C **e** at 29 °C

4 Conclusion

The effects of anodization on the aluminium tube with varying temperature for two different electrolytes were analysed and discussed. From this study, the following conclusions arrived.

- The pore size and coating thickness were found to be increasing with an increase in temperature for anodization using 10% of H_2SO_4 as an electrolyte solution
- When the cell temperature was varied from 10 to 30 °C, there were no pores found in the anodization process with 7% H_2SO_4 + 3% oxalic acid as an electrolyte solution
- H_2SO_4 of 10% electrolyte solution with an anodizing temperature of 20 °C is found to be the optimum conditions due to the minimum the pore size and minimum coating thickness.

It was noticed that the anodized surface with the above conditions seems to be more hydrophilic, which will be suitable for enhanced heat transfer applications.

Acknowledgements The authors would like to thank DST–SERB (DST/SERB/YSS/2015/001084) for the financial support for this research work.

References

1. Lee W, Kim J, Splitting PW, Template AO (2009) Properties of nanostructures obtained by anodization of aluminum in phosphoric acid at moderate potentials properties of nanostructures obtained by anodization of aluminum in phosphoric acid at moderate potentials
2. Thamida SK, Chang H, Thamida SK, Chang H (2002) Nanoscale pore formation dynamics during aluminum anodization nanoscale pore formation dynamics during aluminum anodization, vol 240
3. Renjith Singh R, Selladurai V, Solomon AB, Balakrishnan T (2015) Effect of nucleation's site on the performance of anodized thermosyphon. In: Proceedings of the 23rd National Heat and Mass Transfer Conference and 1st International ISHMT-ASTFE Heat and Mass Transfer Conference, IHMTC2015, 17–20 December, 2015, Thiruvananthapuram, India
4. Solomon AB, Noel M (2015) IHMTC2015-30 anodization and evaluation of an aluminium thermosyphon with, pp 1–9
5. Renjith Singh R, Selladurai V, Ponkarthik PK, Solomon AB (2015) Effect of anodization on the heat transfer performance of flat thermosyphon. Exp Therm Fluid Sci 68:574–581
6. Solomon AB, Roshan R, Vincent W, Karthikeyan VK, Asirvatham LG (2015) Heat transfer performance of an anodized two-phase closed thermosyphon with refrigerant as working fluid. Int J Heat Mass Transf 82:521–529
7. Solomon AB et al (2017) Characterisation of a grooved heat pipe with an anodised surface. Heat Mass Transf und Stoffuebertragung 53(3):753–763
8. Weng H, Yang M-H (2018) Heat transfer performance enhancement of gravity heat pipes by growing AAO nanotubes on inner wall surface. Inventions 3(3):42
9. Abd-elnaiem AM, Mebed AM, Gaber A (2014) Effect of the anodization parameters on the volume expansion of anodized aluminum films, vol 8, no 2013, pp 10515–10525

Experimental Investigation on Heat Pipe-Assisted Cooling During Milling Process of AISI 1040

I. Kantharaj, D. J. Hiran Gabriel, Julius Benedict Prakash, and S. Mohanasundaram

Abstract The cutting temperatures during milling operation are high owing to the hardness of work specimens and due to friction at the tool–work and tool–chip interfaces. This elevated temperature results in the deformation of the cutting edges during dry milling. The most common method employed to reduce the heat that is generated is the application of cutting fluids in the cutting zone. The tool life during milling is prominently enhanced by providing a higher quantity of cutting fluid at the cutting zone. Though the cutting fluid effectively removes heat from the tool, its sudden application causes thermal shocks in the tool. This thermal shock makes the tool brittle specifically at the cutting edges resulting in tool wear. Also, coolant application has a negative impact on the environment as well as the machine operator's health. Moreover, tool cost is higher than the cost of safekeeping and clearing of cutting fluids. Machining with heat pipe is a newly established technique to lower the problem involved with cutting fluid. Thus, in this experiment, a customized milling tool is fabricated such that it can house a ring-shaped heat pipe. The experiments were conducted in conditions of dry milling with no coolant and heat pipe-assisted cooling. The results obtained in the experiments confirm that milling with heat pipe-assisted cooling exhibited reduced cutting force and cutting temperature than dry machining. Thus, heat pipe-assisted cooling can be suggested as an alternative for conventional methods.

Keywords Dry milling · Cutting fluid · Heat pipe

I. Kantharaj (✉) · J. B. Prakash · S. Mohanasundaram
Department of Mechanical Engineering, Karunya Institute of Technology and Sciences, Coimbatore, Tamil Nadu 641114, India
e-mail: ikantharaj@gmail.com

D. J. Hiran Gabriel
Department of Mechanical Engineering, PSG College of Technology, Coimbatore, Tamil Nadu 641004, India

S. Vijayan et al. (eds.), *Trends in Manufacturing and Engineering Management*, Lecture Notes in Mechanical Engineering,
https://doi.org/10.1007/978-981-15-4745-4_53

1 Introduction

Metal machining plays a significant part in all manufacturing industries. Machining industry has incorporated continuous changes in technology and processes to increase the effectiveness of manufacturing [1]. Metal machining is a cutting process that removes the unwanted material from the work material to produce the desired shape and size [2]. Machining includes turning, drilling, boring, milling processes with single or multi-point cutting tools to remove excess material from the work to attain a required geometry during the metal removal operation [3]. Temperature rise and friction at the tool–work interface generate difficulties concerning the quality of finished surface and tool life [4]. According to several authors, the contact between workpiece and tool is the source friction produced during the metal removal process. The tool wear is due to thermal stresses produced during metal cutting and affects the surface finish [5]. During metal removal, heat energy produced at the tool–chip interface and it is carried away by coolants. To overcome these issues, cutting fluids can be used. Cutting fluid is a lubricant as well as a coolant for metal machining processes. A typical cutting fluid contains oil, oil with water emulsion, gel, aerosols or mist with air. It is widely used to increase the level of productivity and surface quality of the specimen during machining [6]. In machining, the cutting fluid, along with cutting parameters like depth of cut, cutting speed and feed rate, plays a vital role. Though the principal role of the cutting fluids is to cool the primary cutting zone and lubricate the interfaces, it also reduces the consumption of power, prevents corrosion on the surface of the machined component and also prolongs the tool life [7]. The global demand for lubricant was high at 39.4 million tons a year in 2015. It is predicted to be in the range of 43.9 million tons in 2022. Most of the lubricants used are engine oils, hydraulic lubricants and gear oils. A metal cutting fluid represents 5% of the global market of lubrication. Though the cutting fluid has several advantages, it also suffers from certain shortcomings such as cost, environmental issues, operator safety and health [8]. Around 17% of the total cost of machining is ascribed to use of coolants in the machining industry [9]. In recent years, alternatives for cutting fluids have been established to overcome the downsides of cutting fluids. The alternatives include cryogenic cooling, dry machining, minimum quantity lubrication (MQL), solid lubrication, gaseous cooling and nanofluids [10]. Dry machining and MQL are being widely used among these alternatives. The tool wear was found to be higher at elevated temperatures of dry machining [11]. During dry machining, achieving the required geometrical accuracy was difficult due to the above-mentioned shortcomings [12]. MQL applied in the form of mist during machining, can be inhaled by the metalworking operator, which can be carcinogenic [13]. Embedding the heat pipe with the tool during machining is the better solution to avoid above-said problems and to remove the heat produced in the primary cutting zone without applying coolant [14–16]. In order to evaluate the efficiency of heat pipe-assisted milling over dry milling, experiments were conducted and the results are discussed below.

2 Materials and Methods

2.1 Investigation Parameters

Milling is the most commonly employed metal removal process in the industry. The main advantage of the process is that it can create a wide variety of features on the component by removing the material from the component. In milling, a large amount of heat is produced at cutting zone due to the shearing of the workpiece by the milling cutter. In dry machining, the elevated temperature results in high tool wear and surface finish of the workpiece being poor. Thus, to remove the heat generated, a large amount of cutting fluid is employed in the cutting zone. Though cutting fluid aids in better surface finish and increased tool life, it adds up to the indirect cost of the product. Also, cutting fluid makes the work floor unclean and hazardous to the operator. Thus, it is suggested to implement heat pipe in the process for the effective transfer of heat from the tool.

The main objective of the experimental work is to study dry machining and heat pipe-assisted cooling in the milling process. The parameters considered for the comparison are temperature, cutting force and surface roughness. For a proper investigation, a nine-level experimental run was designed with variation in metal cutting parameters such as spindle speed, feed and depth of cut. The preset values of the nine-level experimental are given in Table 1.

2.2 Tool and Heat Pipe Fabrication

For the experimental purpose, a new tool was designed and fabricated. The material used for the fabrication of the tool is HSS. To embed the ring-shaped heat pipe, a housing position is made using EDM. The metal removal process using EDM was

Table 1 Parameters used in the investigation for each run of the experiment

Run	Speed (m/min)	Feed (mm/rev)	Depth of cut (mm)
S1	250	16	0.25
S2	250	20	0.5
S3	250	25	0.75
S4	500	20	0.25
S5	500	25	0.5
S6	500	16	0.75
S7	710	25	0.25
S8	710	16	0.5
S9	710	20	0.75

Fig. 1 Fabricated milling cutter with cavity

Table 2 Dimensions of the heat pipe

Parameter	Dimension (mm)
Outer diameter	70
Inner diameter	54
Wall thickness	0.5
Height	35

carried out in a way that does not weaken the tool as well as enabling the heat pipe to reach the maximum possible distance near to the cutting zone. A pre-shaped electrode in the shape of the required cavity is used in the process. The fabricated tool is shown in Fig. 1.

The heat pipe was made using thin copper plates of 0.5 mm thickness. Special care was taken in order to maintain the ring shape since any deformation in the shape would lead to mismatch with the housing slot on the tool. Distilled water is the working fluid used in the heat pipe. The heat pipe filling is done using a vacuum pump and is sealed properly. The dimensions of the heat pipe are given in Table 2.

The fabricated heat pipe shown in Fig. 2 is fitted in the housing cavity made on the tool. The fit was made in such a way that the heat pipe does not come off the tool. The assembly of milling tool with the heat pipe is shown in Fig. 3.

2.3 Experimental Set-up

AISI 1040 steel workpiece was used to carry out experiments with heat pipe-assisted cooling method. The parameters measured during the experiments were temperature, surface roughness and cutting force. The cutting force during milling was measured using a Kistler 9257B dynamometer. Mahr TR100 surface roughness tester of type Mahr Surf GD 25 was used in measuring the surface roughness of the workpiece.

Fig. 2 Fabricated heat pipe

Fig. 3 Milling cutter embedded with heat pipe

Among the various parameters measured using the surface roughness tester, roughness average (Ra) is considered in the study. The temperature during the milling operation is measured using Amprobe IR-750 an infrared thermometer.

The fabricated heat pipe is housed on the custom-made milling tool. A heat pipe is a tool that transfers heat using the ideas of phase transition and thermal conductivity to transfer heat efficiently. A heat pipe is always fabricated using a high thermally conductive material like copper. The part of the heat pipe which is exposed to the higher temperature is the hot interface and part that is exposed to the atmosphere is the cold interface. At hot interfaces, the liquid which is in contact with the highly conductive material is converted to vapour. This vapour reaches the cold interface of the heat pipe where the temperature is low. The vapour at the cold interface due to the low temperature condenses to become liquid again giving away the latent heat to the atmosphere. This condensed liquid is then returned to the hot interface by capillary action, centrifugal force, or gravity and the cycle restarts.

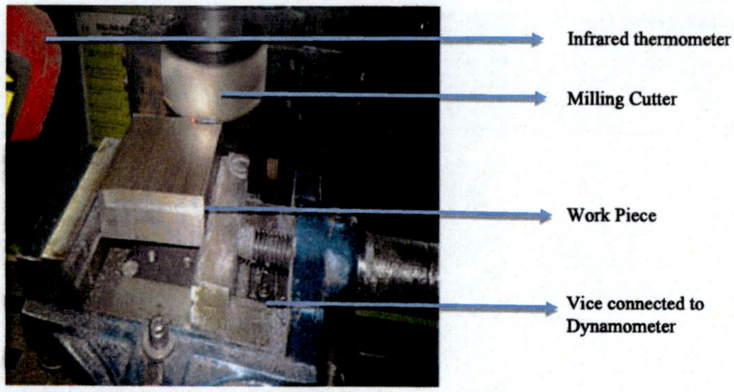

Fig. 4 Experimental set-up

The set-up of the experiment consists of a milling machine to which an arbour is fixed. The tool is mounted in the machine using the arbour. The dynamometer is fixed with the work table along with the vice of the machine for the measurement of the cutting forces. The workpiece is mounted on the vice and checked for levelling. The infrared thermometer is used for the measurement of the cutting temperature. Surface roughness tester measures the value of surface roughness. The experiments were conducted for both the conditions of dry and heat pipe-assisted cooling with cutting parameters remaining the same. The experimental set-up is shown in Fig. 4.

3 Results and Discussion

The effect of the milling process was studied under the conditions of dry machining and heat pipe-assisted machining. The cooling performance of the heat pipe-assisted cooling in milling is analysed by comparing the values of surface roughness, cutting temperature and cutting forces.

It is clear from Table 3 that irrespective of speed of spindle, depth of cut and feed values of surface roughness, cutting force and cutting temperature obtained during milling with heat pipe cooling has an improvised result in comparison with that of dry machining. The cutting temperature of dry machining is so high since there is no external agent such as coolant for heat removal. Due to an increase in temperature and cutting forces, dry machining results in the improper surface finish of the component. This elevated temperature of the tool leaves burnt marks on the machined surface of the component.

Figure 5 shows a comparison of cutting temperatures obtained during each run, from the graph it is clear that in heat pipe-assisted milling machining happens at a lower temperature. The heat pipe housed on the milling cutter effectively removes heat from the cutting zone due to its thermal conductivity and phase transition properties. This reduced temperature during the milling process aids in the production of

Table 3 Observation during the experiments

Run	Cutting temperature (°C)		Cutting force (N)		Surface roughness (μm)	
	Dry milling	Heat pipe assisted milling	Dry milling	Heat pipe assisted milling	Dry milling	Heat pipe assisted milling
S1	37.4	31.5	90.6	77.6	1.143	1.234
S2	42	35.6	109.4	103.3	1.101	1.201
S3	58.9	42.3	215.7	135.6	1.144	1.164
S4	38.2	34.7	223.8	130.3	0.954	1.098
S5	43.2	38.6	196.95	181.9	0.871	0.997
S6	66.3	49.1	170.27	146.1	0.852	0.894
S7	39.8	35.2	219.5	142.6	0.857	0.942
S8	52.2	39.1	274.9	189.8	0.582	0.64

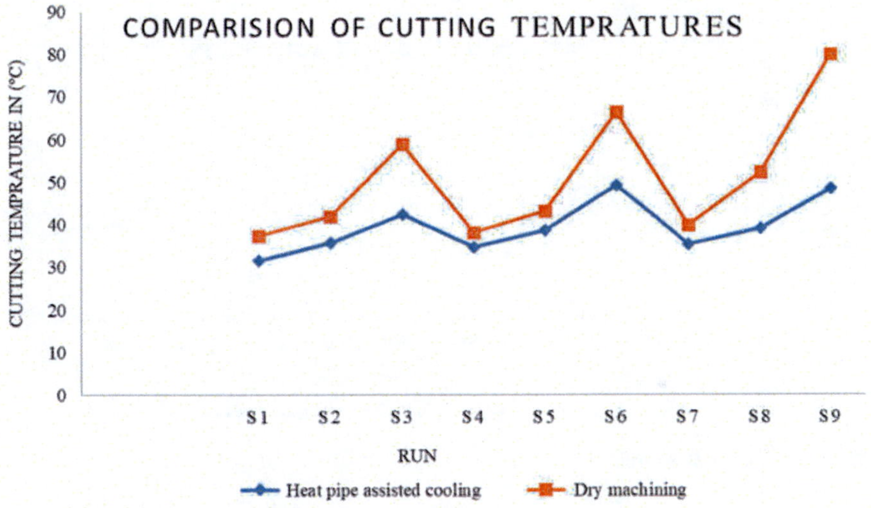

Fig. 5 Comparison of cutting temperature

good quality of surface finish on the component along with a considerable reduction in cutting force which is shown in Fig. 6. The surface finish of the obtained component is measured for each run of the experiment, and a comparison is made between dry and heat pipe-assisted milling plotted in the form of a graph as shown in Fig. 7. Since there is no usage of cutting fluid in the heat pipe-assisted milling process, it causes no hazard to the environment as well as the operator.

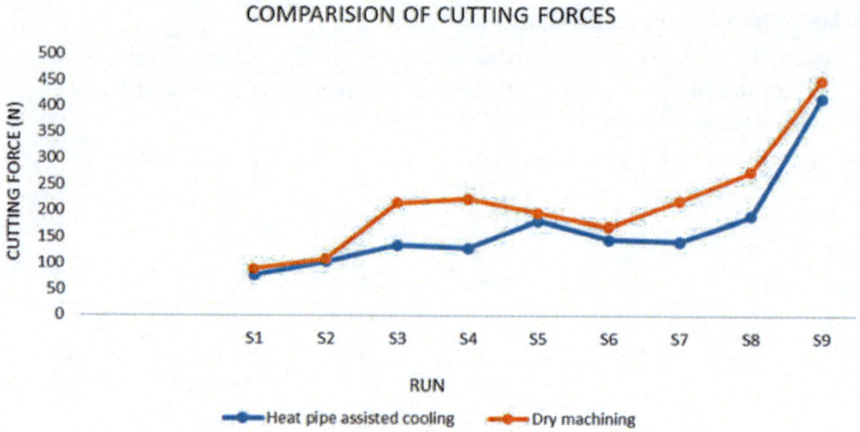

Fig. 6 Comparison of cutting forces

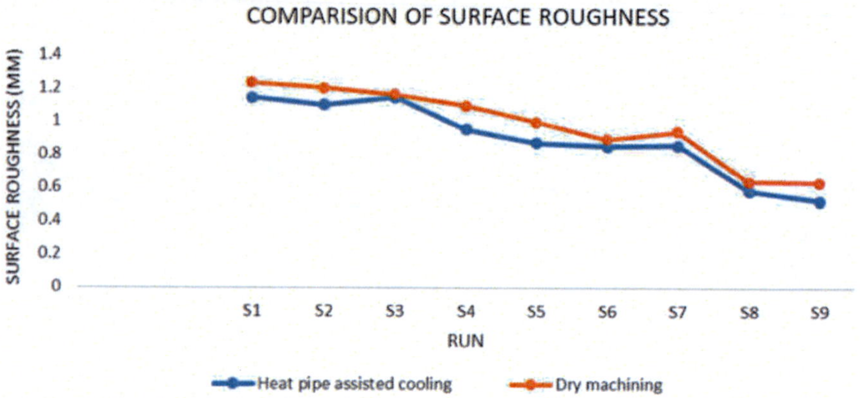

Fig. 7 Comparison of surface roughness

4 Conclusion

For the milling process under study, for each experimental run, parameters such as surface roughness, cutting force and temperature are considered and a comparison is made between both dry machining and heat pipe-assisted machining. From the results obtained, it is observed that the heat pipe-assisted milling process exhibited improved surface finish in comparison with dry machining, due to the lower cutting temperature observed during the machining process. Also, there is a considerable reduction in cutting force, thus conserving energy. Since heat pipe-assisted cooling does not use cutting fluid, it eliminates the process of safekeeping and disposal of cutting fluid, thereby reducing the cost of the final product. Thus, heat pipe-assisted milling will serve as a better alternative for conventional dry milling process. This process also aids in the promotion of greener manufacturing environment since it does not consume any coolant for the process.

References

1. Shokoohi Y, Paiva JM, Fox-Rabinovich G, Bork CAS, Veldhuis SC (2019) Evaluation of the superabsorbent coolant as a new approach to semi-dry machining. Int J Adv Manuf Technol 101:2037–2050
2. Kalpakjian S (2001) Manufacturing engineering and technology. Pearson Education India
3. Liew PJ, Shaaroni A, Sidik NAC, Yan J (2017) An overview of current status of cutting fluids and cooling techniques of turning hard steel. Int J Heat Mass Transf 114:380–394
4. Sharma VS, Dogra M, Suri N (2009) Cooling techniques for improved productivity in turning. Int J Mach Tools Manuf 49:435–453
5. Chan C, Lee W, Wang H (2013) Enhancement of surface finish using water-miscible nano-cutting fluid in ultra-precision turning. Int J Mach Tools Manuf 73:62–70
6. Arulraj GA, Wins KLD, Raj A (2014) Artificial neural network assisted sensor fusion model for predicting surface roughness during hard turning of H13 steel with minimal cutting fluid application. Proc Mater Sci 5:2338–2346
7. Liang L, Liu X, Li X-Q, Li Y-Y (2015) Wear mechanisms of WC–10Ni3Al carbide tool in dry turning of Ti6Al4V. Int J Refract Metal Hard Mater 48:272–285
8. Adler D, Hii W-S, Michalek D, Sutherland J (2006) Examining the role of cutting fluids in machining and efforts to address associated environmental/health concerns. Mach Sci Technol 10:23–58
9. Brinksmeier E, Meyer D, Huesmann-Cordes AG, Herrmann C (2015) Metalworking fluids—mechanisms and performance. CIRP Ann 64(2):605–628
10. Anton S, Andreas S, Friedrich B (2015) Heat dissipation in turning operations by means of internal cooling. Proc Eng 100:1116–1123
11. Das SR, Dhupal D, Kumar A (2015) Experimental investigation into machinability of hardened AISI 4140 steel using TiN coated ceramic tool. Measurement 62:108–126
12. Debnath S, Reddy MM, Yi QS (2016) Influence of cutting fluid conditions and cutting parameters on surface roughness and tool wear in turning process using Taguchi method. Measurement 78:111–119
13. Shokrani A, Dhokia V, Newman ST (2012) Environmentally conscious machining of difficult-to-machine materials with regard to cutting fluids. Int J Mach Tools Manuf 57:83–101
14. Ghosh S, Rao PV (2015) Application of sustainable techniques in metal cutting for enhanced machinability: a review. J Clean Prod 100:17–34
15. Robinson Gnanadurai R, Varadarajan AS (2016) Investigation on the effect of cooling of the tool using heat pipe during hard turning with minimal fluid application. Eng Sci Technol Int J 19(3):1190–1198
16. Stephenson DA, Agapio JS (2016) Metal cutting theory and practice, 3rd edn. CRC Press, Boca Raton

Development of a Modified Magnetic Moulding Set up for Improved Heat Transfer

B. Anand Ronald, M. Harshal, K. Barath Varadaraj, and G. Gopinath

Abstract Magnetic moulding is a relatively lesser-known process used for casting of metallic materials. The conventional magnetic moulding set up has certain shortcomings, due to heat transfer from the molten metal to the container wall through the steel balls; simultaneously heat is generated from the container wall due to Joule Thomson heating from the current-carrying copper coils, leading to the possibility of stagnation in heat transfer and thereby retarding the heat carried away to the steel balls. In order to overcome this, the current work deals with, the conventional magnetic moulding set up modified with water circulating jacket, so as to enhance the heat transfer, which is expected to improve the solidification rate and hence have control over the grain size of the cast products.

Keywords Magnetic moulding · Heat transfer · Water jacket

1 Introduction

Metal casting is one of the widely used primary manufacturing processes. Currently, there is a greater need for manufacturing processes that can reduce lead time, cost of manufacturing without compromising on the quality of products. Also, many efforts are made towards production strategies/processes that reduce damage to the environment. Magnetic mould casting is an innovative process having a great potential to replace conventional casting methods due to various advantages associated with it. In magnetic moulding steel balls are used instead of moulding sand and an electromagnetic field, surrounding the moulding flask acts as a binding agent [1]. This process is similar to the lost foam process because the pattern used is an expendable polystyrene pattern. Once the cast metal is solidified the electromagnetic field is interrupted so that the steel balls would collapse by itself; the part is recovered and the steel balls could be reused in the subsequent castings [2]. Compared to the moulding

B. Anand Ronald (✉) · M. Harshal · K. Barath Varadaraj · G. Gopinath
Department of Mechanical Engineering, Sri Sivasubramaniya Nadar College of Engineering, Chennai, Tamil Nadu 603110, India
e-mail: anandronaldb@ssn.edu.in

S. Vijayan et al. (eds.), *Trends in Manufacturing and Engineering Management*, Lecture Notes in Mechanical Engineering,
https://doi.org/10.1007/978-981-15-4745-4_54

sand (mixture of sand, clay and water) used in sand casting, this mould has no moisture content to generate steam and vapour, which generally results in casting defects like blow holes and pores when they are trapped in the mould itself [3]. Therefore, replacing the sand with alternate materials with better thermal conductivity, like steel balls improves the heat transfer rate and reduces the solidification time of casting. This paper deals with the development of a modified magnetic moulding set up to overcome the heat transfer limitations in the conventional magnetic moulding set up.

2 Modified Magnetic Moulding Set up

The type of process employed for casting has an effect over properties like surface finish, microstructure, hardness, toughness, etc. of the resultant product. Magnetic mould casting is an application of electromagnetism to the process of casting. Steel shots constitute the mould that is formed by application of magnetic field on them. The application of magnetic field induces temporary magnetic force between steel shots giving strength to mould [4]. This reduces the time elapsed in the ramming process. Also, breaking of mould becomes easier by using magnetic field, as it is required to switch off the supply to turn down the magnetic field and mould breaks.

The conventional magnetic moulding set up (Fig. 1) consists of a mild steel container inside which the steel balls are filled with the pattern at the center. The pattern is made using Expendable Polystyrene pattern (EPS) and it is cut according to the cast product size and shape required. Polystyrene is in solid-state (glassy) at room

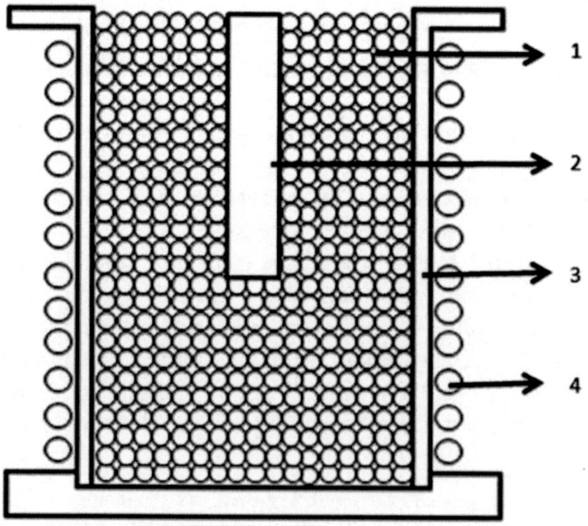

Fig. 1 Schematic diagram of the conventional magnetic moulding setup. 1. Steel shots. 2. Expendable polystyrene pattern (EPS). 3. Container. 4. Copper coil

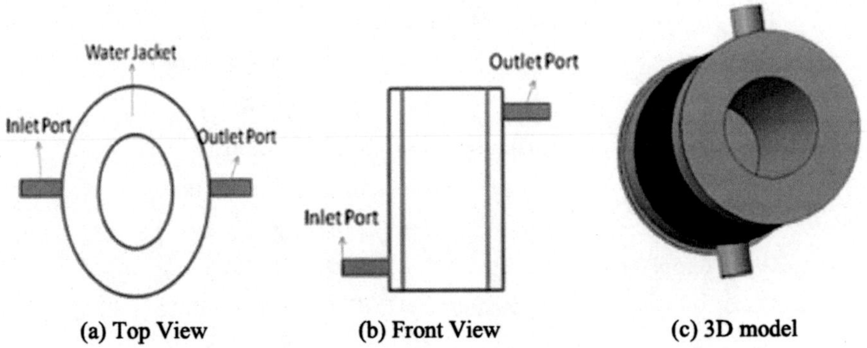

Fig. 2 Schematic diagram of the modified magnetic moulding set up

temperature, but when heated above 100 °C it evaporates. Copper coils of 22 gauge were wound surrounding the mild steel container. The number of turns of the copper wire was 500 with 8 A DC power supply given.

When the molten metal is poured over the polystyrene pattern, the molten metal evaporates the polystyrene, and the molten metal takes the shape of the pattern, transfers the heat to the surrounding steel balls, which act as mould wall. This contributes to nucleation and thereby encourages directional solidification of the mould material. In the conventional magnetic moulding set up, the heat that is dissipated from the molten metal moves towards the steel balls by conduction and then to the mild steel container wall. Simultaneously, certain amount of heat is generated due to Joule heating/Ohmic heating, from the current-carrying copper coils that are wound around the container. This can lead to stagnation of heat in the steel balls and hence lead to reduction in solidification time for the cast material. To overcome this stagnation of heat, a modified magnetic moulding set up was developed with a water jacket in the container wall.

The water jacket was developed such that room temperature water enters inlet port available at the bottom, the heat is transferred to the water during circulation and the hot water exits through the top port. The schematic diagram of the modified magnetic moulding set up is shown in Fig. 2.

3 Calculations

Calculation of amount of heat from aluminium

The amount of heat from the molten aluminium is calculated as follows:

Mould Dimension:

$$\text{Diameter} = 30\,\text{mm} = 0.03\,\text{m}$$
$$\text{Radius} = 15\,\text{mm} = 0.015\,\text{m}$$

$$\begin{aligned}
\text{Volume of Mould} &= \pi r^2 h \\
&= \pi \times (0.015)^2 \times (0.06) \\
&= 4.24 \times 10^{-5}\,\text{m}^3 \\
\text{Mass of Aluminium} &= \text{Volume} \times \text{Density} \\
&= 4.24 \times 10^{-5} \times 2700 \\
&= 0.11448\,\text{kg} \\
\text{Temperature Difference } \Delta T &= 700\,^\circ\text{C} - 30\,^\circ\text{C} \\
\text{Required} &= 670\,^\circ\text{C}
\end{aligned}$$

Specific Heat of Aluminium $C_p = 0.91\,\text{kJ/kg K}$

Amount of heat from the mould

$$\begin{aligned}
Q_1 &= mC_p\Delta T \\
&= 0.1148 \times 0.91 \times 670 \\
&= 69.99\,\text{kJ}
\end{aligned}$$

Therefore, $\mathbf{Q_1 = 69.99\,kJ}$

Calculation of amount of heat produced in the coil

According to Joule's law,

$$\text{Heat produced in the coil } H = I^2RT$$

To find Current (I), Selecting the Output Current (I) from the transformer as

$$I = 3A$$

To find Time (T), Time of current supply to the coil $T = 5\,\text{min}$

$$5 \times 60 = 300\,\text{s}$$

To find Resistance (R), Resistance of the coil $R = \frac{\rho l}{A}\,\Omega$

$$\begin{aligned}
\text{Resistivity of copper coil} &= 1.72 \times 10^{-8}\,\Omega\text{m} \\
\text{Length of the wire} &= \text{No of turns} \times \text{Circumference of the outer cylinder} \\
&= 500 \times \pi \times 144 = 226.080\,\text{m} \\
\text{Area of the coil} &= \pi r^2 = 6.28 \times 10^{-6}\,\text{m}^2 \\
\text{Resistance of the copper coil } R &= 0.6192\,\Omega \\
\text{Heat produced in the coil } H &= I^2RT \\
H &= 3^2 \times 0.6192 \times 300 \\
&= 1671.83\,\text{J} \\
&= 1.67183\,\text{kJ}
\end{aligned}$$

$$\mathbf{Q_2 = H = 1.67183\,kJ}$$

Amount of magnetic induction produced in current-carrying coil

$$\text{Magnetic Induction } B = \frac{\mu I}{2\pi r}\,\text{T}$$

To find Magnetic Permeability (μ),
Magnetic Permeability for materials

$$\text{Magnetic Permeability of MS cylinder} = 1.262344 \times 10^{-6}\,\text{H/m}$$
$$\text{Magnetic Permeability of Water} = 1.256627 \times 10^{-6}\,\text{H/m}$$
$$\text{Magnetic Permeability of Steel shots} = 5.044567 \times 10^{-6}\,\text{H/m}$$
$$\text{Average Permeability of materials} = 2.521179 \times 10^{-6}\,\text{H/m}$$

To find Current (I),
Selecting the Output Current (I) from the transformer as

$$\text{Current } I = 3A$$

To find Radius of the coil (r),
Selecting copper coil of grade (SWG 19)

$$\text{Radius of the copper coil} \quad r = 1\,\text{mm} = 0.001\,\text{m}$$

$$\begin{aligned}\text{Magnetic Induction } B &= \frac{\mu I}{2\pi r} = \frac{2.52117 \times 10^{-6} \times 3}{2 \times \pi \times 0.001} \\ &= 1.20377 \times 10^{-3}\,\text{T} \\ \text{Actual Magnetic Induction } B &= N \times B \\ &= 500 \times 1.20377 \times 10^{-3} \\ &= 0.60188\,\text{T}\end{aligned}$$

Mass flow rate of water at time 't' second

$$\begin{aligned}\text{Total Amount of Heat from the setup } Q_{\text{total}} &= Q_1 + Q_2 \\ &= 69.99 + 1.67183 \\ &= 71.66183\,\text{kJ} \\ \mathbf{Q_{total}} &= \mathbf{71.66183\,kJ}\end{aligned}$$

Mass flow rate of water to decrease 5 °C at 30 s

Fig. 3 Actual photograph of the inner and outer concentric cylinders during fabrication

$$\text{Total Amount of Heat from the setup } Q_{\text{total}} = \dot{m} C_{\text{p}} \Delta T$$

$$\dot{m} = \frac{Q_{\text{total}}}{\Delta T \times C_{\text{p}} \times 30}$$

$$= 0.119\,\text{kg/s}$$

$$= \mathbf{0.119\,l/s}$$

Thus the Mass Flow rate determined is equal to 0.119 l/s.

4 Fabrication of the Mild Steel Container

Two concentric cylinders are used so that there can be a passage of fluid in between the walls of the closed cylinder. A cylindrical mild steel pipe is chosen for this purpose. As per the design calculation, we require a 2 mm thick-walled cylinder so that magnetic field can penetrate. From the determined flow rate for proper removal of heat the diameters of the cylinders are selected as 80 and 140 mm (outer diameter). So in order to achieve the required dimensions 88 and 148 mm mild steel pipe (Outer diameter) of thickness 4 mm is purchased and is turned using four jaw lathe to 80 and 140 mm diameter with 2 mm thickness (Fig. 3).

5 Conclusions

A Modified Magnetic Mould casting setup was successfully fabricated. Since the modified set up has a water jacket, it is expected that the modified set up will give better heat transfer, which can be ascertained by evaluating (at the outlet port) the amount of heat carried away. Also after solidification, the mechanical (tensile strength) and metallurgical properties (fine grain size) of the cast material will give insight into the improvement due to modification of the existing magnetic moulding set up.

References

1. Wittmoser A (1975) The new third generation of molding processes. AFS Trans 63–72
2. Suganth Kumar KJ, Senthilkumaran S, Rajendra Boopathy S, Ramesh A (2008) Theoretical and experimental investigation of mold strength in magnetic molding compacts. J Mater Process Technol 205:235–242
3. Rajkolhe R, Khan JG, Gajanan SS (2014) Defects, causes and their remedies in casting process: a review. Int J Res Advent Technol 2(3). E-ISSN: 2321-9637
4. Anand Ronald B, Arun Prakash C, Suba Karthik M (2016) Influence of steel shots size on tensile properties of magnetic moulded MMC. Appl Mech Mater 852:118–122

Quality Assessment Studies on AA7075 Plate in Hot Rolling Process

K. Jayakumar

Abstract Rolling is an important material-forming process and more than 90% of ferrous and nonferrous metals and alloys are processed to their usable shapes by rolling. In the rolling process, permanent deformation is achieved by subjecting the material to high compressive stress by allowing the material to pass through the gap between two rotating cylindrical rolls. In 7075 aluminum alloy, zinc as the primary alloying element. It is strong, with a strength comparable to many steels, and has good fatigue strength and average machinability. In order to improve further the strength of the AA7075 alloy, the hot rolling study was carried out on a 6.62 mm thickness plate. Hot rolling was carried out as per L9 orthogonal array by varying roller speed (750 and 850 rpm), the number of passes (10 and 20) and preheating the workpiece at 350 and 500 °C. From the experimental results, it was observed that the hardness value was increased from 42.6 to 48.3 HRB with an increase of all the selected process parameters. The thickness of the rolled plate was reduced from 6.62 to 5.53 mm at a higher number of passes. Dye penetrant inspection was carried out to see the surface-level defects or microstructural changes. The rolled plate with low roller speed, less number of passes, and higher preheating temperature showed more surface-level defects or microstructural deformation.

Keywords AA7075 alloy · Hot rolling study · Hardness · Thickness variation · Dye penetrant inspection

1 Introduction

AA7075 alloy is mainly used in Aircraft fittings, Gears and shafts, Fuse parts, Missile parts, Worm gears, keys, aircraft, aerospace and defence applications, bike frames, all-terrain vehicle (ATV) sprockets, etc. It has lower resistance to corrosion than

K. Jayakumar (✉)
Department of Mechanical Engineering, Sri Sivasubramaniya Nadar College of Engineering, Kalavakkam,
Chennai 603110, India
e-mail: kjayakumar@ssn.edu.in

S. Vijayan et al. (eds.), *Trends in Manufacturing and Engineering Management*, Lecture Notes in Mechanical Engineering,
https://doi.org/10.1007/978-981-15-4745-4_55

many other aluminum alloys but has significantly better corrosion resistance than the 2000 series aluminum alloys. Its relatively high cost limits its use. Rolling is the most important material-forming process by which properties of the materials can be improved further by grain refinement. In hot rolling, the workpiece is heated nearer to its recrystallization temperature and hence the compressive force required for deformation is comparatively less and porosity of the rolled metal can be minimized. Also, the smaller capacity rolling stand can be used for rolling operation. Much research work has been conducted on the rolling of different aluminum alloys. Panigrahi et al. [1] did rolling studies on AA7075 alloy at cryogenic and room temperature and they compared its mechanical properties from tensile tests and hardness testes results. The conclusion from the study was rolling with cryogenic temperature is good. The effects of rolling parameters on the mechanical behavior of AA6061 alloy was carried out by Rajabi et al. [2]. They concluded that higher rolling temperature and strain rate were improved the tensile strength of the material. Guo et al. [3] suggested that the true stress-strain curves at 300–500 °C temperature and strain rates of 0.01–10 s^{-1} were obtained through isothermal compression tests, providing a data basis for the hot processing map of AA7075 alloy. From 2015 onwards, the accumulative roll bonding process (ARB) came to replace the conventional rolling to further improve the properties of the AA alloys. Alvandi et al. [4] studied the mechanical properties of fine structured AA7075 alloy by ARB process at room temperature. Results showed that the tensile and hardness of the sample were improved by ARB compared with alloy with a coarse grain sample. Guo et al. [5] evaluated the microstructure of hot-rolled AA7075 alloy through Friction Stir Processing (FSP). Major conclusions were the maximum superplastic elongation of 776.4% was achieved at 500 °C and strain rate 10^{-3} s^{-1} for the hot-rolled AA7075 subjected to one pass FSP. Microstructure and deformation behavior of Al–Fe–Mn alloys with different Fe contents during Cold Rolling was tried by Pan et al. [6].

From the available literature, it was found that different researchers have tried to improve the mechanical properties of the AA7075 alloy by different rolling processes. But, they evaluated the improvement in properties by destructive testing. However, the present work focused on the improvement of mechanical properties of AA7075 by hot rolling process and assessment of the mechanical properties by destructive and nondestructive testing through dye penetrant testing.

2 Experimental Details

The AA7075-T651 alloy with the chemical composition in maximum weight % of 6.1 Zn, 2.9 Mg, 2 Cu, 0.5 Fe, 0.4 Si, 0.28 Cr, 0.2 Ti, and Al balance (91.4%) was taken as raw material. The initial size of the plate is 300 mm × 300 mm × 6.62 mm. To improve further the mechanical properties of the alloy, eight sample pieces of 20 × 70 × 6.62 mm were cut from the plate and used for the further rolling study. Before the rolling process, the thickness and hardness of a parent sample were measured

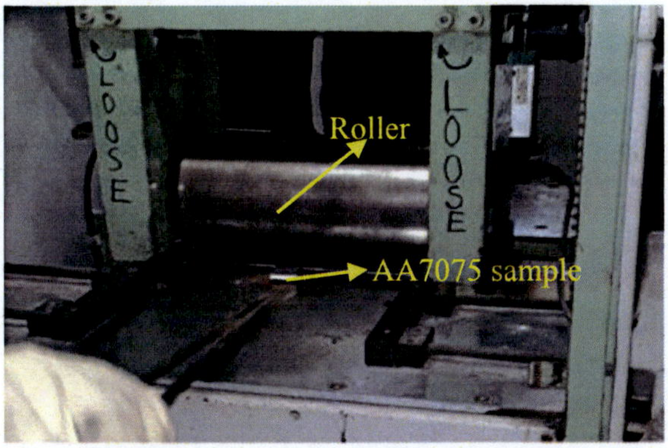

Fig. 1 Rolling of AA7075 sample

using digital vernier caliper and Rockwell hardness tester with B scale respectively. Initial hardness value obtained as 42.6 HRB and thickness as 6.62 mm.

The hot rolling mill (Maximum Temperature = 1200 °C) facility which is available in the Department of Metallurgical and Materials Engineering at IIT Madras was used in the present work. Figure 1 shows the experimental setup with the rolling of aluminum plate. The hot rolling study was conducted by varying roller rotational speed, the number of passes and preheating temperature of the workpiece. The rolling parameter was selected based on trial and error as well as machine constraints. Rolling pressure or load was kept as constant with 0.5 ton. Eight experiments were conducted as per three factors with two level-full factor experimental designs due to facility constrain.

After rolling, the hardness test was conducted for all eight pieces to check the improvement of hardness. Rockwell B scale hardness test conditions are Load-100 kg, Indenter-1/16 in. or 1.6 mm diameter ball indenter. Hardness was measured in each plate at three different locations and the average is considered. Dwell time of 10–15 s was maintained during the hardness test to allow for elastic recovery. Thickness variation of the rolled samples was also measured and Table 1 lists the experimental design with results. During thickness measurement, thickness variation was observed in all samples from middle to edges. Average of thickness variation from all samples is 0.03 mm.

Figure 2 shows the samples after the rolling process in which the first sample which is present in the right end is a not rolled one and the remaining eight pieces are rolled one as per eight experimental conditions.

To check the internal defect of the rolled samples, Dye penetrant (DP) based nondestructive testing (NDT) was carried. DP test is sensitive to small surface discontinuities/defects, gives real-world results and affordable one than other NDTs. There are three types of liquids that were used in the DP test and they are cleaner, penetrant, and developer.

Table 1 Rolling experimental condition and results

S. No.	Roller speed (rpm)	No. of passes	Temperature (°C)	Hardness (HRB)	Thickness (mm)
1	750	10	350	45.6	6.11
2	850	10	350	44.5	6.34
3	750	20	350	46.6	5.98
4	850	20	350	45.3	5.53
5	750	10	500	44.3	6.31
6	850	10	500	46.3	6.22
7	750	20	500	47	5.8
8	850	20	500	48.3	6.23

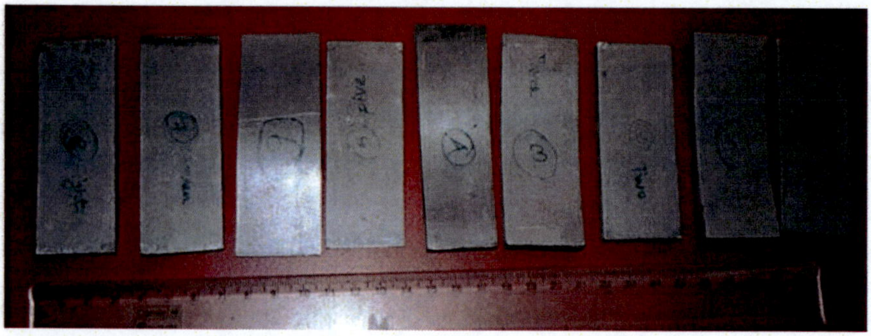

Fig. 2 Samples after rolling process

The cleaner is used to remove the dust particles on the workpiece surface and the penetrant is used to fill the surface cracks. Dwell time of 15 min was maintained after applying penetrant and Fig. 3 shows the rolled plate after applying the developer.

A developer is a nonaqueous wet type one and they are acetone, isopropyl alcohol, or a propellant that is a combination of these two.

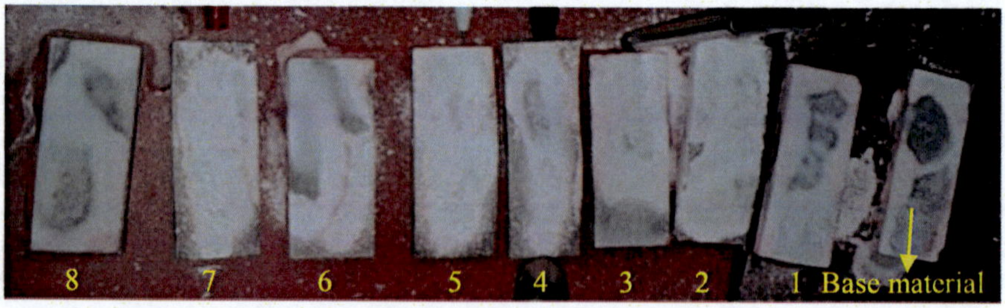

Fig. 3 Rolled specimens after applying developer

3 Results and Discussions

3.1 *Effect of the Rolling Parameter on Hardness and Thickness Variation*

The effect of a rolling parameter on hardness variation is shown in Fig. 4 in which the number of passes in hot rolling is the most sensitive factor than the other two factors.

Main effect graphs were obtained from Minitab software based on the experimental results. The hardness of the sample was increased for an increase in all the selected process parameters. As the number of passes and temperature during rolling is increased, the work hardening and dislocation density are increased which caused the grain refinement and fragmented precipitates which improved the hardness.

As the rolling speed increased, the hardness of the strip was also increased which is due to the decrease of time taken for temperature drop which caused the grain refinement during the application of compressive load in the rolling process.

Similar to hardness, the thickness of the rolled plates was increased owing to the increase of roller speed and preheating temperature of a workpiece (Fig. 5). The thickness of the rolled strip is increased for an increase in rolling speed since the time for temperature drop was less which caused more deformation. The thickness of the AA7075 strip was decreased because of the increase in the number of passes. The reason is that as the no. of passes increased, the edges got tore instead of deformation which finally decreased the thickness.

An increase in the thickness of the sample as the increase in temperature was observed due to more deformation during rolling instead of compression.

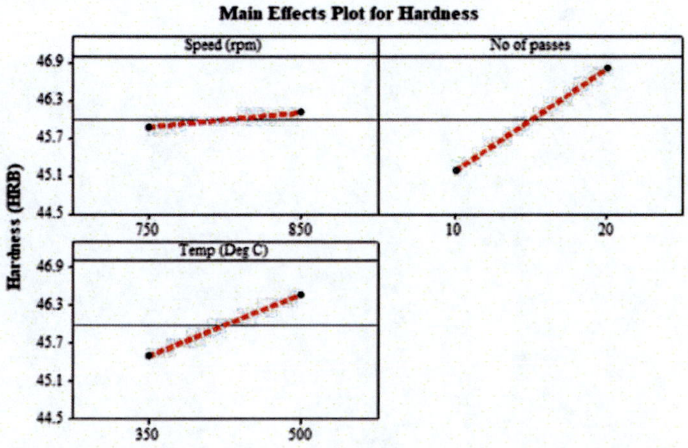

Fig. 4 Main effect plot for hardness variation

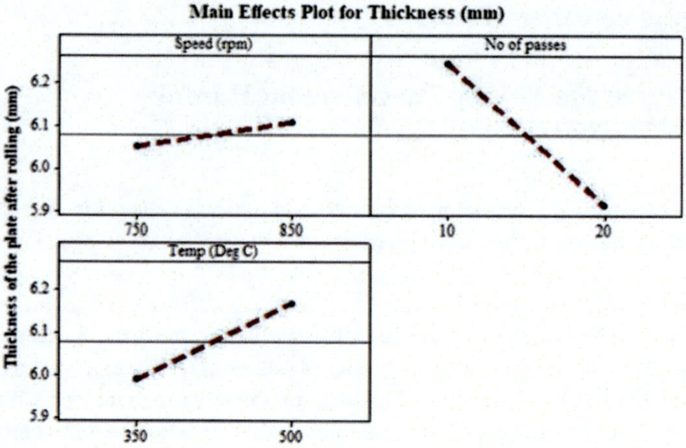

Fig. 5 Main effect plot for thickness variation

3.2 *Effect of the Rolling Parameter on Surface Defects*

From the DP test after applying developer, sample number 3 and 4 are shown in Fig. 6 has more surface-level microcracks. The reason is that the rolling process was carried out with more number of passes (20 numbers), low temperature (350 °C) with high speed (850 rpm) in those experiments. Due to these effects, the material is not able to deform and got slightly torn on the edges which caused microcracks.

Overall, from these rolling studies, it was observed that the experiment number 7 and 8 can be recommended for getting good mechanical properties of the AA7075 strip.

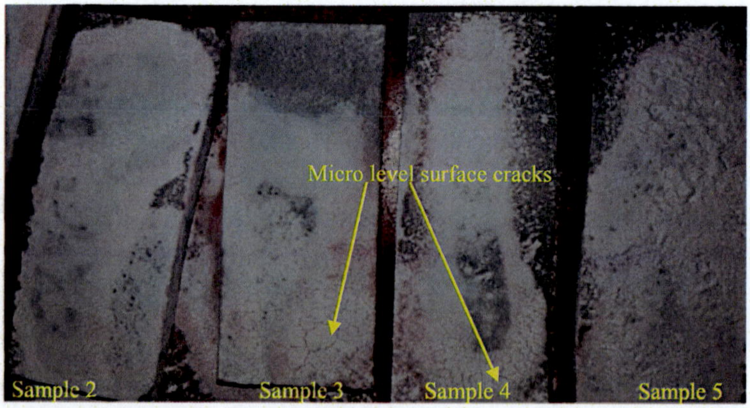

Fig. 6 Rolled samples with surface-level defects

4 Conclusions

- In this study, the hardness, thickness variation, and surface-level defects of AA7075 aluminum alloy plate through the hot rolling process were investigated.
- The hardness of the AA7075 sample is increased from 42.6 to 48.3 HRB by hot rolling process.
- The thickness of the sample is reduced from 6.62 to 5.53 mm through the hot rolling process. The number is passes of the sample during the rolling process was identified as the dominant factor for both responses.
- Dye penetrant inspection was carried out to see the surface-level defects or microstructural changes from that microcracks or defects were observed for the third and fourth experiments.

References

1. Panigrahi SK, Jayaganthan R (2008) A comparative study on mechanical properties of Al 7075 alloy processed by rolling at cryogenic temperature and room temperature. Mater Sci Forum 584:734–740
2. Rajabi F, Zarei-Hanzaki A, Eskandari M, Khoddam S (2013) The effects of rolling parameters on the mechanical behavior of 6061 aluminum alloy. Mater Sci Eng A 578:90–95
3. Guo L, Yang S, Yang H, Zhang J (2015) Processing map of as-cast 7075 aluminum alloy for hot working. Chin J Aeronaut 28(6):1774–1783
4. Alvandi H, Farmanesh K (2015) Microstructural and mechanical properties of nano/ultra-fine structured 7075 aluminum alloy by accumulative roll-bonding process. Procedia Mater Sci 11:17–23
5. Guo ML, Tan MJ, Liu FC, Song X, Chua BW (2016) Microstructure evolution in hot rolled 7075 Al via friction stir processing. In: AIP conference proceeding, vol 1769, no 1, pp 1–6
6. Pan Y, Shen Y, Zhao P (2018) The microstructure and deformation behavior of Al-Fe-Mn alloys with different Fe contents during cold rolling. Metals 753(8):1–6

Artificial Neural Network and Genetic Algorithm-Based Models for Predicting Cutting Force in Turning of Hardened H13 Steel

K. Leo Dev Wins, B. Anuja Beatrice, D. S. Ebenezer Jacob Dhas, and V. S. Anita Sofia

Abstract The manufacturing sector in the modern era is striving hard to reduce the cost of production by employing innovative techniques. One such technique is hard turning where the workpiece is heat treated to the requisite hardness, and the final size and shape of the component are obtained directly through hard turning process. Hard turning is generally carried out with a huge amount of cutting fluid to enhance the output performance. Since petroleum-based emulsions are easily available in the market at reasonable price, they are widely used in industries. Petroleum-based cutting fluids create a number of environmental and health issues. In this perspective, pure dry turning is a logical substitute as it does not possess the harmful effects connected with the cutting fluids. The feasible tool life and surface quality are often disturbed while carrying out the machining operation under pure dry condition. Under such circumstances, the concept called minimal cutting fluid application (MCFA) performs itself as a possible solution. This paper investigates the effect of applying cutting fluid using MCFA technique at the critical contact zones while hard turning of H13 steel. An artificial neural network (ANN) model was developed for the prediction of the main cutting force, and its ability to predict cutting force (F_z) was analyzed. An effort is made to optimize the cutting parameters to accomplish minimum cutting force using genetic algorithm.

Keywords Hard turning · Artificial neural network · Minimal cutting fluid application · Genetic algorithm

---

K. Leo Dev Wins (✉) · D. S. Ebenezer Jacob Dhas
Karunya Institute of Technology and Sciences, Coimbatore, Tamil Nadu, India
e-mail: leo@karunya.edu

B. Anuja Beatrice · V. S. Anita Sofia
Sri Krishna Arts and Science College, Coimbatore, Tamil Nadu, India

S. Vijayan et al. (eds.), *Trends in Manufacturing and Engineering Management*, Lecture Notes in Mechanical Engineering,
https://doi.org/10.1007/978-981-15-4745-4_56

1 Introduction

The manufacturing sector is increasingly adopting newer methods and techniques to reduce the cost of production and manufacturing lead time to ensure productivity growth and delivery of top quality products in the most effective and economical manner [1]. In this endeavor, lot of researchers took it as a promising area of research to device new prospects in manufacturing processes. By virtue of rigidity, good compressive strength, and high wear resistance, hardened steels meet industry demands in the fabrication of high-performance components. They are highly used in the field of automotive and die tool industry [2]. In the recent years, the concept of hard machining has gained great importance in metal cutting. The increasing need for more efficient machining techniques and the timely emergence of high-quality cutting tool materials such as CBN have led to the wide usage of hard turning process. Reduced equipment costs, few process steps, and flexibility are the influencing factors for the selection of hard turning compared to conventional practices [3].

Hard turning process necessitates large quantities of cutting fluids. Cooling and lubrication of workpiece and cutting tool during hard turning operation are normally executed by free flow of cutting fluids to reduce friction and heat developed during machining. The high temperature at the cutting zone during hard turning produces harmful fumes which not only create adverse effects to the workers but also causes environmental pollution [4]. With the advent of new standards regarding the environment, health and safety, some of the ingredients in many of the cutting fluids were identified as problematic. Because of such adverse effects created by cutting fluids, a number of government agencies have established regulations and standards for the need of reducing particulate exposure [5].

Cutting fluid usage, storage, and disposal affect the cost-effectiveness of the machining processes. Industries spend 7–17% of the manufacturing cost on cutting fluids [2]. Under such situation, dry machining is a viable alternative but it requires machine tools and cutting tools that can take enormous cutting forces. Hence, such a scheme cannot be executed as such in the existing shop floor. Enormous amount of vibration is produced which may also lead to premature tool failure [6].

For reducing harmful effects linked with the use of cutting fluids, minimal quantity lubrication (MQL) technique was developed. MQL method also improves surface finish and dimensional accuracy [7] but adds the risk of mist in the shop floor [8].

Alternative to this problem, Varadarajan et al. developed a new fluid application system called minimal cutting fluid application system (MCFA) in 2002 [9], by excluding the role of compressed air from the system. In this method, cutting fluid is delivered at high velocity and is pumped directly to the cutting zone with the help of an injector in a pulsating manner, and this process created very less oil mist [10]. The results proved that the MCFA reduces cutting force and vibration and improves surface quality and tool life when compared to wet cooling and dry machining [11].

Analytical models often use unfavorable assumptions and simplifications to get superior and proper results. In real-time applications, artificial intelligence-based modeling approaches are preferred, in that ANN is found to be attractive, reliable,

Table 1 Chemical composition of H13 steel (weight %)

C	Mn	Si	P	S	Cr	Mo	Fe
0.430	0.214	1.08	0.0330	0.0290	5.02	1.13	Balance

and viable [11]. Özel and Karpat used ANN model for predicting flank wear and surface finish during hard turning [12]. An ANN model was developed by Leo et al. for predicting surface finish based on ANN during surface milling of AISI 4340 steel of hardness 43 HRC with minimal fluid application [13].

Genetic algorithm with the basic principle of genetics and evolution is one of the optimization methods. It mimics biological evolution for solving search and optimization problems [14]. Computer programs based on genetic algorithm approach helps to produce optimal solutions [15].

The literatures review revealed that, lot of research works have been reported on the optimization of surface finish and cutting force during hard turning of different steels but no work is reported on the prediction of cutting force using ANN during hard turning of H13 steel with minimal cutting fluid application. In this research, hard turning process was assisted with minimal cutting fluid application technique, which reduced the usage of cutting fluid to a minimum amount of 8 ml/min. Based on the earlier researches in the optimization while machining of hardened steels [16], an attempt was also made to achieve minimum cutting force by optimizing the cutting parameters using genetic algorithm. The prediction model derived out of GA was found to be highly accurate with more than 99% accuracy.

2 Selection of Workpiece and Cutting Tool

H13 tool steel of 45 HRC was used as the workpiece material in this research. It is one of the toughest tool steels. It has high toughness and red hardness. Its good hardenability makes it useful in making extrusion tools and dies.

Accordingly, a cylindrical rod of 65 mm diameter with 355 mm length was used for the experimentation. Table 1 shows the percentage composition of elements in the workpiece material. The tool used for hard turning was SNMG120408 (MT TT 5100) hard metal inserts and tool holder used was PSBNR2525M12.

3 Experimental Setup

Experimental setup used an all-geared Kirloskar Turnmaster-35 lathe with variable feed and speed drive. Figure 1 shows the experimental setup. Variable DC controllers were used to control the feed rate and speed independently.

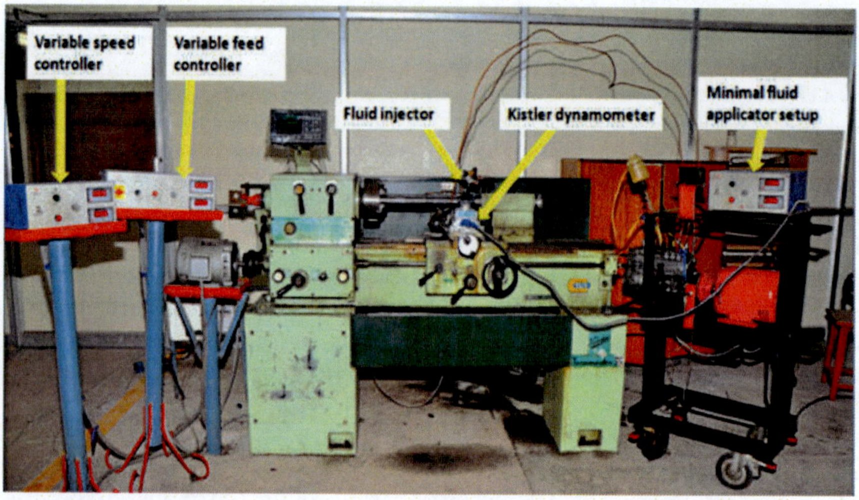

Fig. 1 Experimental setup

Kistler multi-component piezoelectric dynamometer of type 9257B was used for the measurement of cutting force. Screening experiments were conducted to identify the main factors that have major effect on the output parameter, and thorough analysis was done to fix the upper and lower limits of the factors. Accordingly, trial runs were conducted by changing one of the input parameters while keeping other parameters constant. Accordingly, spindle speed of the lathe was fixed between 77 and 115 m/min. In line with this factor, feed rate of the lathe was fixed between 0.05 and 0.1 mm/rev and the depth of cut was fixed between 0.5 and 1 mm. The fluid application parameters were kept constant. Accordingly, pressure at the injector was kept at 100 bar, frequency of pulsing at 500 pulses/min, quantity of application at 8 ml/min, and the cutting fluid concentration as 20% of oil in water [17].

Five-level factorial central composite rotatable design with full replications was used for the design of experiment because these designs can include information from a correctly planned factorial experiment. Table 2 presents five levels of parameters.

Table 3 shows the design matrix with the experimental values and results.

Table 2 Five levels of cutting parameters

Cutting parameters	Levels				
Cutting speed (mm/min)	77	85	96	107	115
Feed rate (mm/rev)	0.05	0.06	0.07	0.09	0.1
Depth of cut (mm)	0.5	0.6	0.75	0.88	1

Table 3 Experimental data

Run order	Type of data	Cutting speed (v) (m/min)	Feed rate (f) (mm/rev)	Depth of cut (d) (mm)	Cutting force (F_z) (N)
1	Training	85	0.06	0.6	211.8
2	Training	107	0.06	0.6	248.2
3	Training	85	0.09	0.6	250
4	Training	107	0.09	0.6	260.7
5	Training	85	0.06	0.88	228
6	Training	107	0.06	0.88	240
7	Testing	85	0.09	0.88	318
8	Training	107	0.09	0.88	321
9	Training	77	0.07	0.75	245.3
10	Training	115	0.07	0.75	254
11	Training	96	0.05	0.75	239.3
12	Training	96	0.1	0.75	345
13	Training	96	0.07	0.5	202.7
14	Testing	96	0.07	1	281.5
15	Training	96	0.07	0.75	276.5
16	Training	96	0.07	0.75	277.3
17	Training	96	0.07	0.75	276.5
18	Training	96	0.07	0.75	277.1
19	Training	96	0.07	0.75	277.5
20	Testing	96	0.07	0.75	278.1

4 Network Algorithm for ANN

It has been proved by many researchers that the artificial neural network (ANN) model has better accuracy and yields more admissible results when compared with the simple tools such as Taguchi method and regression analysis. Artificial neural network seems to be the most preferable algorithm for developing a predictive model for complex nature of the hard turning of H13 steel. Artificial neural network was developed and trained using quickprop (QP) algorithm, and cutting force values were predicted. The neural network contains one input layer and an output layer. The hidden layer was changed into single and multi-layer types with the changes in number of neurons, and cutting force was predicted with the help of best configuration which gave better coefficient of determination with least MSE error through trial-and-error method [18]. "Tansig" was considered as the transfer function. Equation (1) shows the sigmoid function used in this experimentation.

$$F(x) = \frac{1}{1 + e^{-x}} \tag{1}$$

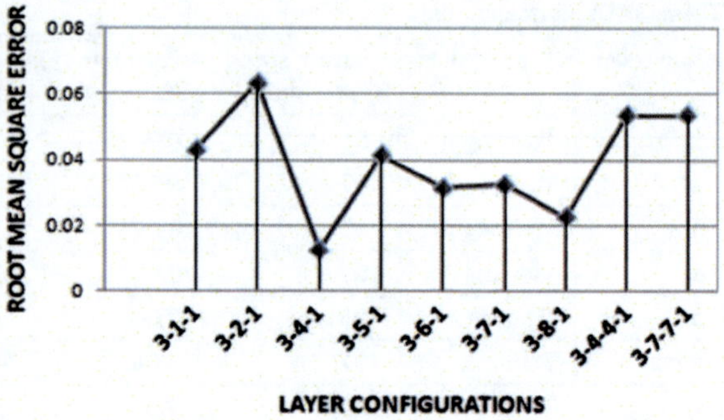

Fig. 2 ANN layer configurations with limiting RMSE values

In order to get the best possible network structure, ANN model was trained with 80 different network configurations. Some of the network configurations which gave lower values of RMSEs are presented in Fig. 2. Based on the results, 3-4-1 network structure was selected with multilayer normal feedforward algorithm.

This ANN model consists of one input layer with three neurons corresponding to cutting speed, feed rate, and depth of cut, one hidden layer with four neurons, and an output layer corresponding to cutting force as shown in Fig. 3.

The regression coefficient (R) value of 0.99996 was obtained for the 3-4-1 network configuration, which is very close to unity, and it shows a close relationship between observed and predicted data. Table 4 shows the comparison between test patterns with experimental results for validating the ANN model. The comparison results revealed

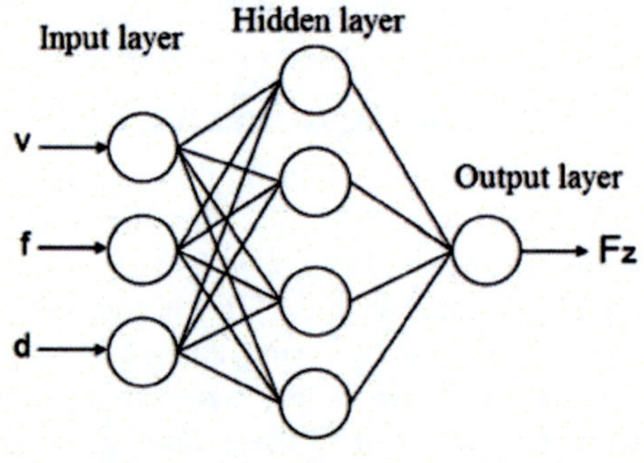

Fig. 3 The 3-4-1 network of the ANN model

Table 4 Comparison between testing data and experimental results

Run order	Testing data			Cutting force (F_Z) (N)		
	v (mm/min)	f (mm/rev)	d (mm)	Experimental results	Prediction by ANN	% Error
1	85	0.09	0.88	318	312.5	1.73
2	96	0.07	1	281.5	276.3	1.85
3	96	0.07	0.75	278.1	273.6	1.62

Table 5 Optimum machining parameters using hybrid ANN-GA approach

Machining parameter	Optimum value
Cutting speed (m/min)	77.00109
Feed rate (mm/rev)	0.0600000325
Depth of cut (mm)	0.610000123

Table 6 Confirmatory test results

Optimum cutting force predicted through GA approach (F_z) (N)	Confirmatory experimental result (F_z)	% Error
195.4946	198.076	1.303

the good concurrence between the predicted values and experimental results. No significant interaction effect between the variables was found.

5 Optimization Using Hybrid ANN-GA

In this present investigation, an effort is made to find the optimum values of cutting parameters that will lead to minimum cutting force. To solve the optimization problem for this study, a hybrid ANN-GA approach was used. Predictive model was trained and developed [19]. The developed artificial neural network (ANN) was used as the fitness function for genetic algorithm (GA).

The machining parameters leading to minimum cutting force are the optimization problem in this case. The constraints in this case are cutting speed (77–115 m/min), feed (0.05–0.1 mm/rev), and depth of cut (0.5–1 mm). The GA parameters used were population size of 30, mutation rate of 0.1, crossover rate of 0.8, and of uniform crossover type. The optimum machining parameters which led to minimum cutting force using hybrid ANN-GA approach are given in Table 5. The optimum cutting force value obtained was 195.4946 N.

Confirmatory tests were conducted to check the adequacy of the optimum parameter combinations, and the experimental results showed good agreement with the predictions. The confirmatory test results are shown in Table 6.

6 Conclusions

- The developed ANN model can be very much helpful in setting up the machining parameters to reduce the force needed to achieve the best results.
- The new scheme is well suitable for the industry as it can be directly practiced in the shop floor with no modifications in the existing setup.

- Hybrid ANN-GA approach can be effectively used for optimizing cutting parameters to achieve minimum cutting force.
- MCFA technique can be used in promoting a green environment in the metal cutting industries, by minimizing risks involved in the use of huge quantity of cutting fluid.

References

1. Enabling manufacturing competitiveness and economic sustainability. In: Proceedings of the 4th international conference on changeable, agile, reconfigurable and virtual production (CARV2011), Montreal, Canada (2011)
2. Attanasio A, Gelfi M, Giardini C, Remino C (2006) Minimal quantity lubrication in turning: effect on tool wear. Wear 260(3):333–338
3. Aronson RB (1995) Why dry machining. Manuf Eng 33–36
4. NIOSH (2001) Metal working fluids—recommendation for chronic inhalation studies. National Institute for Occupational Safety and Health, Cincinnati, OH USA
5. Marano RS, Smolinnski JM, Esingulari CWM (1997) Polymer additives as mist suppressant in metal cutting fluids. J Soc Tribol Lubr Eng 25–32
6. Sam Paul P, Varadarajan AS (2013) Performance evaluation of hard turning of AISI 4340 steel with minimal fluid application in the presence of semi-solid lubricants. Proc Inst Mech Eng Part J J Eng Tribol 227(7):738–748
7. Itoigawa F, Childs THC, Nakamura T, Belluco W (2006) Effects and mechanisms in minimal quantity lubrication machining of aluminium alloy. Wear 260:339–344
8. Dhar NR, Kamruzzaman M, Ahmed M (2006) Effect of minimum quantity lubrication (MQL) on tool wear and surface roughness in turning AISI-4340 steel. J Mater Process Technol 172:299–304
9. Varadarajan AS, Philip PK, Ramamoorthy B (2002) Investigations on hard turning with minimal cutting fluid application (HTMF) and its comparison with dry and wet turning. Int J Mach Tools Manuf 42:193–200
10. Robinson GR, Varadarajan AS (2012) Investigation on the effect of an auxiliary pulsing jet of cutting fluid on the top side of the chip during hard turning with minimal fluid application. Int J Mach Mach Mater 12:321–336
11. Chryssolouris G, Guillot M (1990) A comparison of statistical and AI approaches to the selection of process parameters in intelligent machining. ASME J Eng Ind 112:122–131
12. Özel T, Karpat Y (2005) Predictive modeling of surface roughness and tool wear in hard turning using regression and neural networks. Int J Mach Tools Manuf 45:467–479
13. Leo Dev Wins K, Varadarajan AS, Ramamoorthy B (2007) Artificial neural network assisted sensor fusion model for predicting tool wear online during hard turning. In: 9th international symposium on measurement and quality control, IIT Madras, pp 267–270
14. Bhoskar T, Kulkarni O, Kulkarni N (2015) Genetic algorithm and its applications to mechanical engineering: a review. Mater Today 2(4–5):2624–2630
15. Goldberg DE (2003) Genetic algorithm in search. Optimization and machine learning. Pearson Education Asia, New Delhi
16. Kaur M, Bhattacharya A, Singh M, Batish A (2014) Hard turning: parametric optimization using genetic algorithm for rough/finish machining and study of surface morphology. J Mech Sci Technol 28(5):1629–1640
17. Raj A, Leo Dev Wins K, Varadarajan AS (2016) Optimization of fluid application parameters during hard turning of AISI H13 tool steel using minimal cutting fluid application. Int J Res Mech Eng 4(3):190–196

18. Zhang G, Patuwo BE, Hu MY (1998) Forecasting with artificial neural network; The state of the art. Int J Forecast 14:35–62
19. Mia M, Dhar NR (2017) Prediction and optimization by using SVR, RSM and GA in hard turning of tempered AISI 1060 steel under effective cooling condition. Neural Comput Appl 1–22

Experimental Study of the Effect of TiN–Zn Coated High-Speed Steel Cutting Tool on Surface Morphology of AL1060 Alloy During Machining Operation

I. P. Okokpujie, C. A. Bolu, O. S. Ohunakin, and Esther Titilayo Akinlabi

Abstract In machining operations, surface morphology is a significant parameter in developing mechanical components, as the level of the surface roughness gives the consumer the notion of whether the product is well developed or not. Also, the surface roughness of the mechanical component will affect the component during machining operations. The application of coatings in cutting tools is very significant, as coatings will help in improving the cutting tool life as well as protect it from rapid wear and corrosion rate. This research work focuses on the investigation of the significance of TiN–Zn coatings on high-speed steel (HSS) when compared with non-coated HSS cutting tool on the surface morphology, machining parameters and machining time on 1060 aluminum alloy during the machining operation. Computer numerical control (CNC) program and excel software was employed to investigate the effects the Coated cutting tool, various depth of cut and such as 1, 1.5, 2, 2.5, and 3 mm and at a constant cutting speed of 1250 rpm, constant feed rate of 10 mm/rev, with machining time varies from 15 to 500 s on the machined surface morphology of the 1060 aluminum alloy. After each machining time, the average surface roughness was determined, from three-point of the work piece, that is the starting point, middle point, and the end point. The result shows that the TiN–Zn coated HSS cutting tool on the surface morphology of 1060 aluminum alloy was found to reduce surface roughness with about 15.5% and both minimum surface roughness of 0.96 and 1.06 μm was achieved for the TiN–Zn coated and uncoated HSS cutting tool, respectively, with cutting time of 15 s at 1 mm depth of cut, which is viable for machining operations.

I. P. Okokpujie (✉) · C. A. Bolu · O. S. Ohunakin · E. T. Akinlabi
Department of Mechanical Engineering, Covenant University, Ota, Ogun State, Nigeria
e-mail: imhade.okokpujie@covenantuniversity.edu.ng

O. S. Ohunakin
Faculty of Engineering and the Built Environment, University of Johannesburg, Johannesburg, South Africa

E. T. Akinlabi
Department of Mechanical Engineering Science, University of Johannesburg, Auckland Park, Kingsway Campus, Johannesburg 2006, South Africa

S. Vijayan et al. (eds.), *Trends in Manufacturing and Engineering Management*, Lecture Notes in Mechanical Engineering,
https://doi.org/10.1007/978-981-15-4745-4_57

Keywords Coatings · Machining · Surface morphology · Aluminum 1060 alloy · HSS cutting tool

1 Introduction

Coating and its application in cutting tool used for metal to metal machining are very significant, as the coatings help to improve the wear rate of the cutting tool life, prevent the cutting tool from corrosion attack, increases the electrical, thermal, and mechanical characteristic of the cutting tool material, which enables the cutting tool to withstand high temperature at the cutting zone and vibration during machining operations [1–3]. Machining operations are greatly affected by the cutting tools. The continuous failures of cutting tools during machining operations can give rise to high cost of production of mechanical parts [4, 5]. Coating is the application of additives on the surface of working materials for improved mechanical properties. The state-of-the-art technology during the machining process is the use of coated cutting tool. Improvements in covering hardware and procedures currently empower us to deliver a wide scope of various hard nitride and oxidic films and to store them on different apparatus substrates as monolayer, multilayer, or composite coatings [4, 6–13]. Ghani et al. [14] examined the wear rate of carbide cutting tool coated with titanium nitride (TiN) and the uncoated carbide cutting tool, using different values of machining speed, feed rate, and depth of cut. From the result, the authors confirm that the uncoated cutting tool, crack with little or no time and the wear rate with the machining parameters were very high, when compared with the carbide cutting tool coated with TiN. Huang [15] also carried out tool evaluation in associations of tool life centered on flank wear at standard machining conditions, such as machining speed, depth of cut, and feed rate. They discovered that machining speed plays a great role in analyzing the tool life evaluation, due to the fact that the increase of the cutting speed led to gradual reduction of the tool wear, followed by depth of cut and feed rate.

According to Schulz et al. [16] who worked on the chemical vapor deposition (CVD) and physical vapor deposition (PVD) of TiN, TiC, and TiAl on cutting tool, the result shows that the cutting tool life was improved significantly when compared with the uncoated cutting tool. Fahad [17] confirmed that the application TiN coating on the cutting tool leads to a great reduction of heat from getting to the cutting tool, the heat rate was analyzed and the coated cutting decrease the heat by 17% for traditional cutting speed and 60% for high-speed machining region.

The comparison of different coatings on cutting tools and uncoated cutting tools were also studied by Astrand et al. [18] on heat distribution in the cutting tool. This proves that coating has great influence in tool flank face and racks on the tool wear. Grzesik [19] and Corduan et al. [20] also confirm that the multilayer coatings on cutting tool done by physical vapor deposition performed better when compared with the chemical vapor deposition method. Bayraktar et al. [21] produced a permanent Al-35Zn alloy using mold casting method and carried out cutting evaluations on titanium aluminum nitride (TiAlN) coated carbide tool with ordinary carbide tool without coatings on A-35Zn alloy during end milling operation. The study employs CNC vertical milling machining center, using a constant feed rate, depth of cut and machining speed. From their result it is clearly seen that the TiAlN increases the cutting force and the surface roughness when compared with the uncoated cutting tool during the machining operation, the authors clamped that the coatings applied reduce the contact area between the cutting tool and the working material, which also reduces the sharpness of the cutting teeth. Quadrini [22], Qin [23] and Heaney [24] contradict this study, saying surface roughness and cutting force can be optimized with the application of coated cutting tool with smaller grain size.

According to Tamerabet et al. [25] the elimination of the lubrication process from machining operation, which is dry machining process, gives a clear picture to study and know all the factors that can lead to cutting tool damages. To optimize the cutting condition, there is need to know the wear pattern, contact angles, and the factors in the operating environment, this idea is to select the important or the most influential factor values that can help to eliminate the wear behavior. The work of the authors focusses on the application of multilayer coating on the cutting tool and the optimization of the cutting parameters on tool wear and temperature at the tool-chip cutting region for two different cutting tools of KC810 and KC910 during dry turning of SAE 1030 carbon steel. The study is to achieve the optimized cutting tool conditions using Taguchi techniques. The result shows that machining speed is the most significant parameter on the wear rate study and the temperature distribution. From the observations of the authors, the KC810 applied to perform better during the turning of carbon steel with a longer lifetime of 228 min during operation. Cutting tool also has a great impact on surface roughness of the mechanical component if not properly optimized and controlled. From the study carried out from [26–32] shows that vibration and surface roughness can also be affected by coated cutting tool and non-coated cutting tool. Therefore, the need to study cutting tool performance on surface roughness is very significant.

However, this experimental study is aimed at analyzing the performance of coated TiN–Zn HSS cutting tool on surface morphology during the machining operation of Aluminum 1060 alloy.

2 Materials and Methods

The material used for this research work are coated HSS cutting tool, non-coated HSS and AL1060 alloy rectangular bar size of length of 2000 mm, width of 50 mm and thickness of 5 mm. The machining experiment was carried out using WARCO X6323A vertical milling machine, Vernier caliper, Steel rule, Mitutoyo surface roughness tester. After each experiment, the surface morphology is studied by measuring the surface with the Mitutoyo surface roughness tester, in order to determine the effects of the coated cutting tool on the surface morphology of AL1060 alloy. Tables 1 and 2 shows the chemical composition, mechanical, thermal and physical properties of AL1060 alloy.

Table 1 Chemical composition of 1060 aluminum alloy [20]

Element	Fe	Si	Cu	V	Zn	Mg	Mn	Ti	AL
Weight %	0.4	0.25	0.05	0.05	0.1	0.03	0.03	0.03	Reminder

Table 2 Physical, mechanical, and thermal properties of AL1060 alloy

Properties	Units
Density	2700 kg/m^3
Melting start (solidus)	649 °C
Tensile strength	130 MPa
Yield strength	94 MPa
Elastic modulus	80 GPa
Poisson's ratio	0.33
Thermal expansion	23.6 (10^{-6}/°C)
Thermal conductivity	234 (mK)

Table 3 The parameter employed for this investigation

Cutting speed (m/min)	Feed rate (mm/rev)	Axial depth of cut (mm)
1250	10	1, 1.5, 2, 2.5 and 3
1250	10	1, 1.5, 2, 2.5 and 3
1250	10	1, 1.5, 2, 2.5 and 3
1250	10	1, 1.5, 2, 2.5 and 3

2.1 Experimental Set-up for the Turning Operation of AL1060 Alloy

Table 3, present the parameters employed for the machining analysis, of different axial depth of cut, while the cutting speed and feed rate were kept constant. Figure 1, shows the experimental set up of the SIEG 3/10/0016 milling machine employed for this research work. After each machining process, the surface morphology is measured with the surface roughness tester and the machining process was carried out between 15 and 500 s, under the dry machining conditions.

3 Result and Discussion

The result obtained in this experimental study is presented in Figs. 1, 2, 3, 4 and 5. Figure 1 presents the effects of the coated HSS cutting tool and the uncoated HSS cutting tool on the surface morphology of Aluminum 1060 alloy, with different cutting time ranging from 15 to 480 s with 1 mm depth of cut during the machining operations. From Fig. 1, it is clearly seen that there was continues increase of the surface roughness with an increase in cutting time, for both the TiN–Zn coated cutting tool and the uncoated cutting tool. This trend is being seen as the same in the experimental result of varying the depth of cut from 1 to 3 mm in this study. However, due to the significant mechanical properties of the coatings applied, there

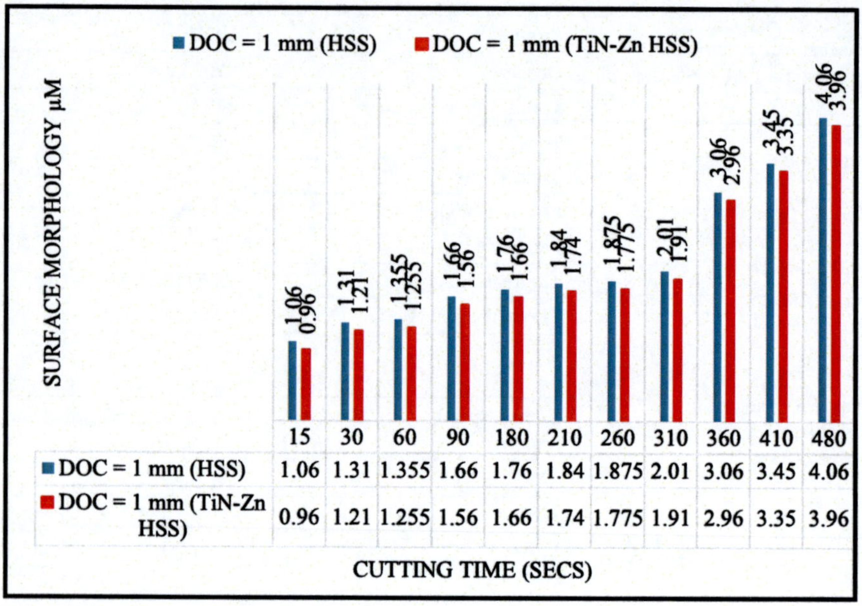

Fig. 1 Surface morphology variation versus cutting time with depth of cut of 1 mm, machining speed 1250 rpm and feed rate 10 mm/rev for TiN–Zn coated HSS and non-coated HSS

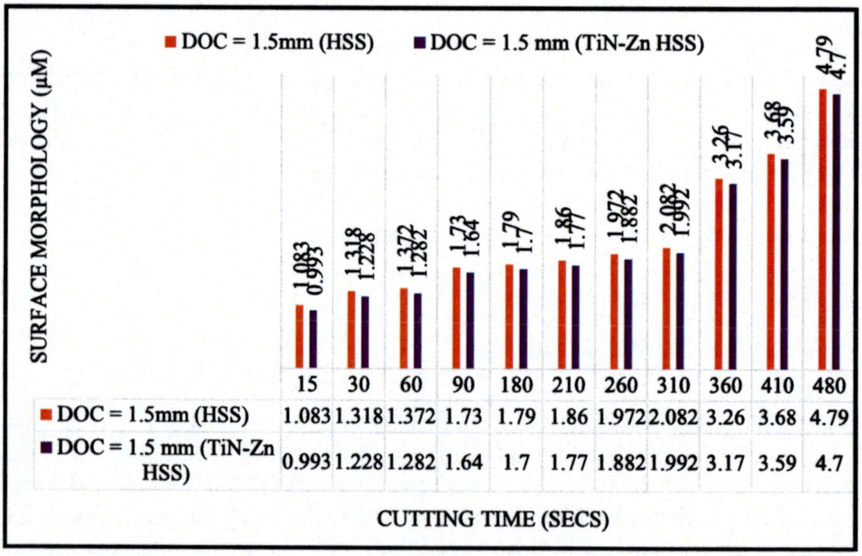

Fig. 2 Surface morphology variation versus cutting time with depth of cut of 1.5 mm, machining speed 1250 rpm and feed rate 10 mm/rev for TiN–Zn coated HSS and non-coated HSS

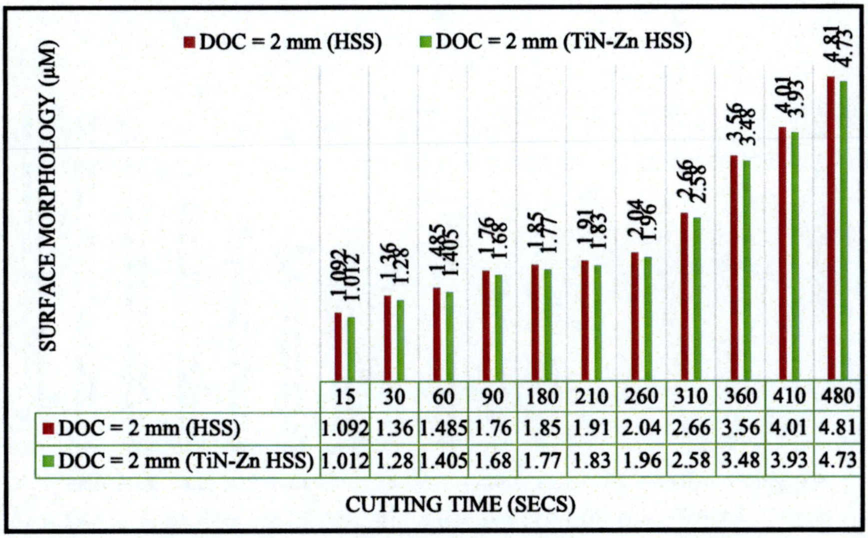

Fig. 3 Surface morphology variation versus cutting time with depth of cut of 2 mm, machining speed 1250 rpm and feed rate 10 mm/rev for TiN–Zn coated HSS and uncoated HSS

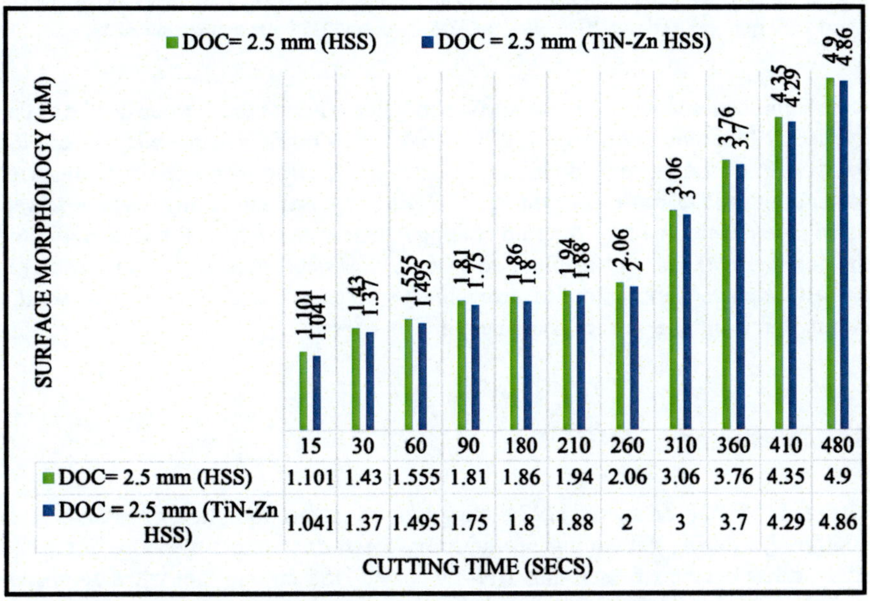

Fig. 4 Surface morphology variation versus cutting time with depth of cut of 2.5 mm, machining speed 1250 rpm and feed rate 10 mm/rev for TiN–Zn coated HSS and non-coated HSS

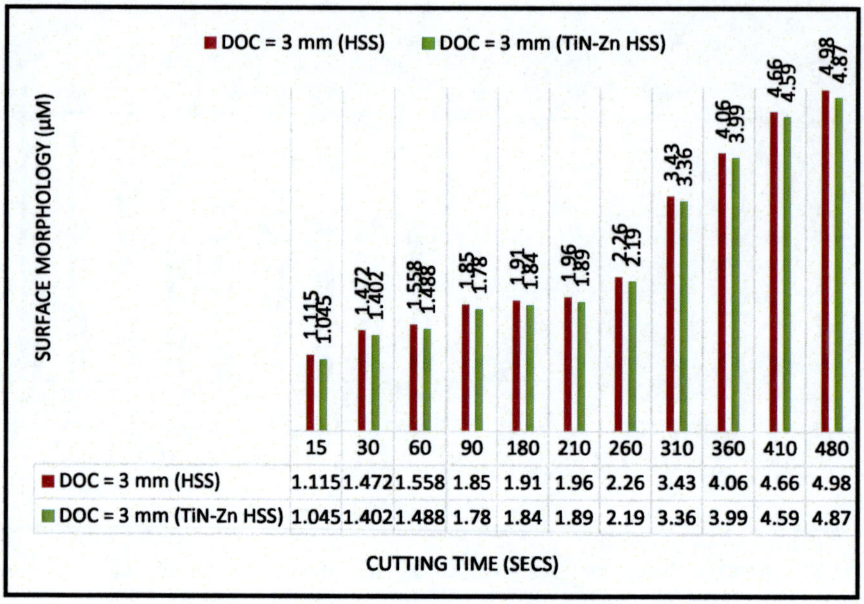

Fig. 5 Surface morphology variation versus cutting time with depth of cut of 3 mm, machining speed 1250 rpm and feed rate 10 mm/rev for TiN–Zn coated HSS and non-coated HSS

were significant reductions in the surface roughness across the various depths of cut applied in this study, from Figs. 1, 2, 3, 4, and 5. This result is in line with Fahad [16] but it contradicted the study done by Bayraktar et al. [20] applying TiAlN coatings on carbide cutting tool during machining Al-35Zn alloy. From the study it was observed during the development of the said coatings, there was no critical consideration of the coating thickness put in place so when it was deposited on the cutting tool it caused the coated cutting tool to have a blunt surface, which definitely increases the cutting force and the surface roughness during operations.

4 Conclusion

The study of TiN–Zn coated HSS cutting tool on surface morphology of aluminum 1060 alloys during machining operation has been done experimentally. From the study, it has been confirmed that TiN–Zn coated HSS cutting tool when compared with the uncoated HSS cutting tool during machining operations has significant effects on the surface morphology of the aluminum 1060 alloy during the machining operation. The following finding was observed

- The experimental result proves that cutting time has a great effect on surface morphology, as the cutting time and the depth of cut increases there was continuous increase in the surface morphology (roughness).
- The minimum surface morphology of 0.96 and 1.06 μm was achieved for the TiN–Zn coated and uncoated HSS cutting tool at a cutting time of 15 s at 1 mm depth of cut.
- From the observation, during the machining operation, the TiN–Zn coated HSS cutting tool was able to perform at high temperature at the cutting zone without loosening its hardness.

Acknowledgements The author wishes to appreciate the covenant university management for their sponsorship.

References

1. Ucun İ, Aslantas K, Bedir F (2013) An experimental investigation of the effect of coating material on tool wear in micro milling of Inconel 718 super alloy. Wear 300(1–2):8–19
2. Okokpujie IP, Ohunakin OS, Bolu CA, Okokpujie KO (2018) Experimental data-set for prediction of tool wear during turning of Al-1061 alloy by high speed steel cutting tools. Data Brief 1(18):1196–1203
3. Ezugwu CA, Okonkwo UC, Sinebe JE, Okokpujie IP (2016) Stability analysis of model regenerative chatter of milling process using first order least square full discretization method. Int J Mech Appl 6(3):49–62
4. Nwoke ON, Okonkwo UC, Okafor CE, Okokpujie IP (2017) Evaluation of chatter vibration frequency in CNC turning of 4340 alloy steel material. Int J Sci Eng Res 8(2):487–495
5. Afolalu SA, Salawu EY, Okokpujie IP, Abioye AA, Abioye OP, Udo M, Adetunji OR, Ikumapayi OM (2017) Experimental analysis of the wear properties of carburized HSS (ASTM A600) cutting tool. Int J Appl Eng Res 12(19):8995–9003
6. Ogundimu O, Lawal SA, Okokpujie IP (2018) Experimental study and analysis of variance of material removal rate in high speed turning of AISI 304L alloy steel. In: IOP conference series: materials science and engineering, vol 413, no 1, p 012030. IOP Publishing, UK
7. Afolalu SA, Okokpujie IP, Salawu EY, Abioye AA, Abioye OP, Ikumapayi OM (2018) Study of the performances of nano-case treatment cutting tools on carbon steel work material during turning operation. In: AIP conference proceedings, vol 1957, no 1, p 050001. AIP Publishing, New York
8. Afolalu SA, Abioye AA, Idirisu J, Okokpujie IP, Ikumapayi OM (2018) Abrasion wear of cutting tool developed from recycled steel using palm kernel shell as carbon additive. Prog Ind Ecol Int J 12(1/2):206–218
9. Okokpujie IP, Ikumapayi OM, Okonkwo UC, Salawu EY, Afolalu SA, Dirisu JO, Nwoke ON, Ajayi OO (2017) Experimental and mathematical modeling for prediction of tool wear on the machining of aluminium 6061 alloy by high speed steel tools. Open Eng 7(1):461–469
10. Okokpujie IP, Salawu EY, Nwoke ON, Okonkwo UC, Ohijeagbon IO, Okokpujie KO (2018) Effects of process parameters on vibration frequency in turning operations of perspex material, pp 700–707
11. Afolalu SA, Abioye OP, Salawu EY, Okokpujie IP, Abioye AA, Omotosho OA, Ajayi OO (2018) Impact of heat treatment on HSS cutting tool (ASTM A600) and its behaviour during machining of mild steel (ASTM A36). In: AIP conference proceedings, vol 1957, no 1, p 050003. AIP Publishing, New York

12. Afolalu SA, Abioye AA, Dirisu JO, Okokpujie IP, Ajayi OO, Adetunji OR (2018) Investigation of wear land and rate of locally made HSS cutting tool. In: AIP conference proceedings, vol 1957, no 1, p 050002. AIP Publishing, New York
13. Afolalu SA, Oluwaseyifunmi O, Ongbali SO, Abioye AA, Okokpujie IP, Oloyede OR (2018) The study of the impact of nano carbon additives on ASTM A53 mild steel during machining. In: IOP conference series: materials science and engineering, vol 413, no 1, p 012028. IOP Publishing, UK
14. Ghani JA, Choudhury IA, Masjuki HH (2004) Wear mechanism of TiN coated carbide and uncoated cermets tools at high cutting speed applications. J Mater Process Technol 10(153):1067–1073
15. Huang Y, Liang SY (2005) Effect of cutting conditions on tool performance in CBN hard turning. J Manuf Process 7(1):10–16
16. Schulz U, Peters M, Bach FW, Tegeder G (2003) Graded coatings for thermal, wear and corrosion barriers. Mater Sci Eng A 362(1–2):61–80
17. Fahad M, Mativenga PT, Sheikh MA (2012) A comparative study of multilayer and functionally graded coated tools in high-speed machining. Int J Adv Manuf Technol 62(1–4):43–57
18. Åstrand M, Selinder TI, Fietzke F, Klostermann H (2004) PVD-Al_2O_3-coated cemented carbide cutting tools. Surf Coat Technol 188:186–192
19. Grzesik W, Van Luttervelt CA (2005) Analytical models based on composite layer for computation of tool-chip interface temperatures in machining steels with multilayer coated cutting tools. CIRP Ann 54(1):91–94
20. Corduan N, Himbart T, Poulachon G, Dessoly M, Lambertin M, Vigneau J, Payoux B (2003) Wear mechanisms of new tool materials for Ti-6AI-4V high performance machining. CIRP Ann 52(1):73–76
21. Bayraktar S, Hekimoglu AP, Turgut Y, Haciosmanoglu M (2018) A performance comparison study of uncoated and TiAlN coated carbide end mill on machining of the Al-35Zn alloy. In: IOP conference series: materials science and engineering, vol 295, no 1, p 012013. IOP Publishing, UK
22. Quadrini F, Tagliaferri V, Polini R (2008) Effect of coating roughness on performance of small CVD diamond coated tools. Mach Sci Technol 12(3):390–404
23. Qin F, Chou YK, Nolen D, Thompson RG (2009) Coating thickness effects on diamond coated cutting tools. Surf Coat Technol 204(6–7):1056–1060
24. Heaney PJ, Sumant AV, Torres CD, Carpick RW, Pfefferkorn FE (2008) Diamond coatings for micro end mills: enabling the dry machining of aluminum at the micro-scale. Diam Relat Mater 17(3):223–233
25. Tamerabet Y, Brioua M, Tamerabet M, Khoualdi S (2018) Experimental investigation on tool wear behavior and cutting temperature during dry machining of carbon steel SAE 1030 using KC810 and KC910 coated inserts. Tribol Ind 40(1)
26. Okonkwo UC, Okokpujie IP, Sinebe JE, Ezugwu CA (2015) Comparative analysis of aluminium surface roughness in end-milling under dry and minimum quantity lubrication (MQL) conditions. Manuf Rev 2(30):1–11
27. Okokpujie IP, Okonkwo UC (2015) Effects of cutting parameters on surface roughness during end milling of aluminium under minimum quantity lubrication (MQL). Int J Sci Res 4(5):2937–2942
28. Nwoke ON, Okokpujie IP, Ekenyem SC (2017) Investigation of creep responses of selected engineering materials. J Sci Eng Dev Environ Technol (JOSEDET) 7(1):1–5
29. Okokpujie IP, Ajayi OO, Afolalu SA, Abioye AA, Salawu EY, Udo M (2018) Modeling and optimization of surface roughness in end milling of aluminium using least square approximation method and response surface methodology. Int J Mech Eng Technol (IJMET) 9(1):587–600
30. Dunmade I, Udo M, Akintayo T, Oyedepo S, Okokpujie IP (2018) Lifecycle impact assessment of an engineering project management process—a SLCA approach. In: IOP conference series: materials science and engineering, vol 413, no 1, p 012061. IOP Publishing, UK
31. Okokpujie I, Okonkwo U, Okwudibe C (2015) Cutting parameters effects on surface roughness during end milling of aluminium 6061 alloy under dry machining operation. Int J Sci Res 4(7):2030–2036

32. Chuan SP, Ghani JA, Tomadi SH, Hassan CH (2012) Analysis of Ti-base hard coating performance in machining process: a review. J Appl Sci (Faisalabad) 12(18):1882–1890

Experimental Investigation of Surface Roughness in End Milling of AA6061 Alloy with Flooded Cooling and Minimum Quantity Lubrication (MQL) Technique

D. Nathan, D. Elilraja, T. Prabhuram, and S. Prathap Singh

Abstract This article presents an experimental investigation on the effect of surface roughness during end milling of aluminium alloy 6061 grade in minimum quantity lubrication (MQL) technique. The MQL technique is becoming increasingly more popular due to the safety of the environment. The cutting speed, depth of cut, feed rate and MQL flow rate are selected input parameters in this study. This experiment was conducted based on the Orthogonal Array (L_{27}). The comparative effectiveness of surface roughness was investigated between MQL technique end milled components and flooded lubrication end milled components. The findings of this study show that surface roughness values have a significant difference upon application of the MQL technique.

Keywords Minimum quantity lubrication · High-speed machining · Surface roughness · Orthogonal array

1 Introduction

Nowadays, modern industries are looking for ways to reduce the use of lubricants in metal cutting operations due to ecological and economical limitations. Lubrication in the machining process becomes one of the most important challenges. Almost conventional flooded lubrication with water-based lubricants in high-speed machining has a negative effect on tool life [1]. Although various lubricants are used in the industries, unfortunately, these cutting fluids have many negative characteristics like costly; need more quantity; they need a periodical replacement; and, lead to health and environmental issues [2, 3]. Minimum Quantity Lubrication otherwise called "micro lubrication" is the latest technique used for delivering coolant at the tool–workpiece interface. In order to improve the effectiveness of high-speed machining,

D. Nathan (✉)
Dr. Mahalingam College of Engineering and Technology, Pollachi, India
e-mail: nathan14576@gmail.com

D. Elilraja · T. Prabhuram · S. Prathap Singh
St. Joseph's Institute of Technology, Chennai, India

S. Vijayan et al. (eds.), *Trends in Manufacturing and Engineering Management*, Lecture Notes in Mechanical Engineering,
https://doi.org/10.1007/978-981-15-4745-4_58

MQL came into existence because it assists in lowering temperature in chip tool interface as well as tool–workpiece interface and maintains the sharpness of the cutting edges. Vegetable oil or synthetic easter oil are used instead of mineral oil which can improve lubrication properties in MQL [4]. The exposure to flooded metal cutting fluid may lead to adverse health effects and safety issues, including toxicity, cancer, dermatitis, and respiratory disorders [5]. Comparative experiments were performed to investigate machining of titanium alloy (Ti–6Al–4V) and difficult-to-cut alloys under different lubrication conditions such as dry, MQL, and high-pressure air [6]. Although dry machining without using any cutting fluids may result in tremendous environmental and economic benefits, it suffers a rise in temperature at workpiece tool interface and hence reduced tool life. The goal of MQL is to use the minimum amount of cutting fluids during machining, typically of a flow rate of 50–500 ml/h. Out of total machining costs, 7–17% of the cost is spent for cutting fluids while 15–20% of total machining cost is used for coolants compared to the tool cost which ranges from 2 to 4% [7–9]. Thus minimum use of lubricants in machining can be a prime indicator of sustainable manufacturing. End milling is one of the most widely used machining operations in the industry because of its high MRR and good surface quality [10]. Usually, in the end milling process, abundant amounts of liquid coolant are applied. Since liquid coolant provides intermittent cooling in end milling causes an increase in temperature variations at the tool–workpiece interface and hence buildup edge [11]. MQL has been effectively applied in increasing tool life while machining components [12]. Also, the MQL technique possesses many advantages over dry, air and flooded lubrication besides requiring some accessories and attachment like chip removal system in machine tool [13]. MQL parameters such as lubricant flow rate, nozzle angle, and distance between the nozzle tip and the tool cutting edge have remarkable effects on the machining performance also in their work optimal condition of nozzle distance, MQL flow rate and nozzle angle using Taguchi Method [14]. The influence of dry, flood and minimum quantity lubrication on diameter error and circularity was studied in aluminium, mild steel 1030 and alloy steel 4340 and found that circularity has a significant effect [15]. Significant parameters for end milling of AA6061T6 alloy end milling using an HSS tool under flood cooling were studied in the statistical method and Genetic Algorithm [16]. Due to high adhesive properties and smearing effects on the cutting tool edges, effective lubrication is required for aluminium alloys [17]. However, very few investigations found employing MQL into the end milling of aluminium alloys [18]. Although most of the researches have been concerned mainly with the turning, drilling and grinding processes involved with MQL. The role of MQL is still to be explored in end milling [19]. This leads to further exploration of the effect of MQL in end milling of aluminium alloys. Hence in this study, end milling of aluminium alloy is investigated by comparing the surface roughness of the components machined under flooded cooling and MQL cooling.

2 Materials and Methods

2.1 Materials

Aluminium alloy (AA6061) workpiece of dimensions 50 mm × 50 mm × 25 mm is machined using "STM Manford VL 610" CNC vertical milling machine. Table 1 shows the chemical composition of AA6061 aluminium alloy. Table 2 describes the properties of AA6061 aluminium alloy. A twist fluted PVD carbide coated HSS end milling cutter of diameter 12.5 mm has been used as a cutting tool. End milling of the AA6061 aluminium workpiece has been carried out at various spindle speed, table feed, and depth of cut values. Same machining parameter combinations were used in end milling under the flooded cooling condition and MQL condition. "ST KOOL 74AL" Water-soluble oil was used as the flood coolant with a concentration of 10%, emulsion (pH-9). Flood coolant was used at 2 bar pressure and 0.5 LPM flow rate during cutting. A Portable surface tester "Mitutoyo-Surf test SJ-210" is used for surface roughness measurement. According to ISO 2488:1996 procedure, surface roughness measurement was carried out along the feed direction for both flooded cooling condition and MQL condition end milling operation.

2.2 Selection of Design Matrix

In order to analyze the machining process, a large number of random experiments have to be done. The Taguchi Method can be used to study the whole parameter space with a minimum number of experiments. Depending on the number of control parameters and their levels the suitable Orthogonal Array can be selected. L_{27} orthogonal array was prepared for machining parameters combinations of speed, feed and depth of cut to study the effect of surface roughness on end milling operation. Table 3 depicts the set of cutting conditions based on L_{27} Orthogonal Array.

2.3 MQL Setup

A custom-designed MQL set up was fabricated for this study. Compressed air at 2 bar pressure is passed through the FRL unit. Then at the mixing chamber, compressed air is mixed with coolant at 6.588 l/h. A nozzle of diameter 2 mm sprayed out the coolant air mixture at the tool–workpiece zone as shown in Fig. 1. Figure 2 describes the arrangement of MQL setup.

Table 1 Chemical composition of AA6061 by weight %

Al	Si	Fe	Cu	Mn	Cr	Zn	Ti	Each others	Other total
Available	0.4–0.8	0.7 max	0.15	0.8–1.2	0.04–0.35	0.25 max	0.15 max	0.05 max	0.15 max

Table 2 Properties of AA6061 aluminium alloy

Properties	Value
Density (ρ)	2.70 g/cm^3
Young's modulus (E)	68.9 GPa
Tensile strength (σ_t)	124–290 MPa
Elongation (ε) at break	12–25%
Poisson's ratio (ν)	0.33
Melting temperature (T_m)	585 °C
Thermal conductivity (K)	151–202 W/m-K
Specific heat capacity (c)	897 J/kg-K

Table 3 Machining parameters

Exp. No.	Speed (rpm)	Feed (mm/min)	Depth of cut (mm)
1	700	200	0.4
2	700	200	0.8
3	700	200	1.0
4	700	500	0.4
5	700	500	0.8
6	700	500	1.0
7	700	1000	0.4
8	700	1000	0.8
9	700	1000	1.0
10	1500	200	0.4
11	1500	200	0.8
12	1500	200	1.0
13	1500	500	0.4
14	1500	500	0.8
15	1500	500	1.0
16	1500	1000	0.4
17	1500	1000	0.8
18	1500	1000	1.0
19	2000	200	0.4
20	2000	200	0.8
21	2000	200	1.0
22	2000	500	0.4
23	2000	500	0.8
24	2000	500	1.0
25	2000	1000	0.4
26	2000	1000	0.8
27	2000	1000	1.0

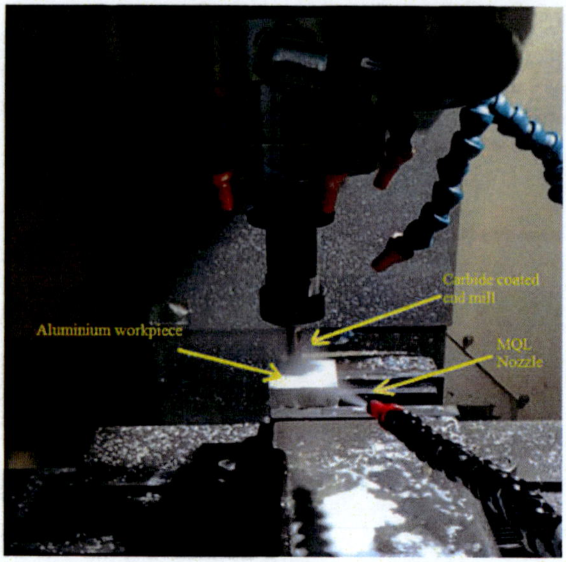

Fig. 1 Photograph of minimum quantity lubrication arrangement

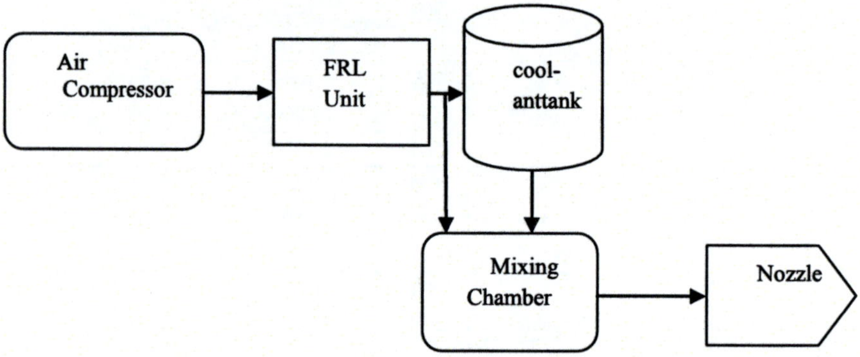

Fig. 2 Block diagram of MQL setup

3 Results and Discussions

3.1 Effect of Depth of Cut and Feed at 700 RPM

Figure 3 describes that, at 700 rpm, the highest surface roughness values were observed at 1000 mm/min feed with flood cooling. Lowest surface roughness values were given by 200 mm/min with flood cooling. Comparing the various depth

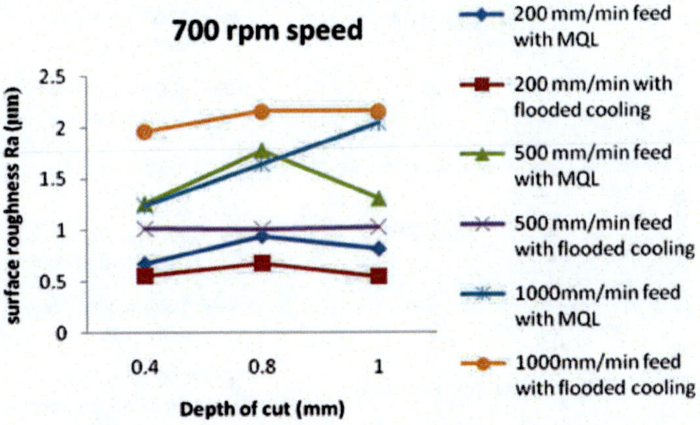

Fig. 3 Influence of depth of cut and feed with surface roughness at 700 rpm

of cut, the lowest depth of cut of 0.4 mm produces low surface roughness. Irrespective of depth of cut, almost minimum variation in surface roughness observed in 500 mm/min feed with flood cooling. Even under flood cooling higher feed rate results in lesser heat dissipation at the workpiece and cutting edge interface subsequently increases the surface roughness values. At higher feed rate with MQL, it is evident that heat dissipation is still lower as the depth cut increases which results in a higher gradient of surface roughness. Hence, the strong influence of flood cooling at this speed, feed, and depth of cut combinations were observed.

3.2 *Effect of Depth of Cut and Feed at 1500 RPM*

Figure 4 reveals that, at 1500 rpm, lower feed of 200 mm/min MQL end milling operation produces lowest surface roughness because of sufficient heat dissipation with MQL was achieved at the tool chip interface. On the other hand, the higher feed rate of 1000 mm/min with MQL gives the highest surface roughness by sticking phenomena of aluminium chips due to insufficient exposure of lubricant at the tool–workpiece interface. At lower feed rate, irrespective of the depth of cut it is evident that flood cooling produces almost constant surface roughness values as the heat dissipation and consequently low friction at tool chip interface.

3.3 *Effect of Depth of Cut and Feed at 2000 RPM*

From Fig. 5, at highest cutter speed and feed rate, as the temperature at the tool–workpiece interface became high, it is evident that MQL produces the highest surface

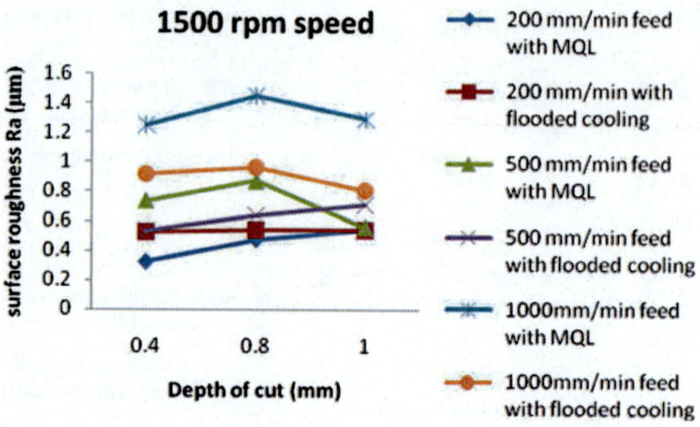

Fig. 4 Influence of depth of cut and feed with surface roughness at 1500 rpm

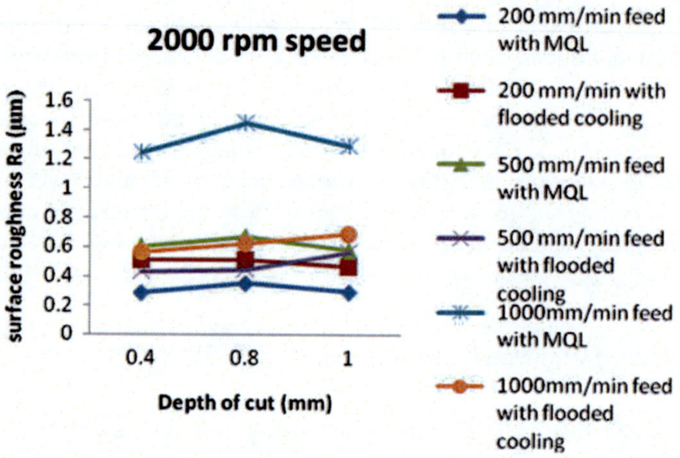

Fig. 5 Graph showing influence of depth of cut and feed with surface roughness at 2000 rpm

roughness values due to insufficient flow of lubricant at the tool–workpiece interface. Alternately, no distinct change in the surface roughness values with the influence of MQL and flooded cooling has been observed in respect of feed rate and depth of cut.

3.4 Analysis of Variance (ANOVA)

Analysis of Variance is performed to investigate the influence and contribution of each process parameter under MQL conditions and flooded cooling for surface roughness in end milling of AA6061 alloy. Table 4 indicates the significant parameters and

Table 4 ANOVA for surface roughness under MQL

Source	DF	Adj SS	Adj MS	*F*-value	*P*-value	% Con
Speed	2	2.1023	1.05117	33.85	0	40.70
Feed rate	2	2.8406	1.42032	45.74	0	55.09
DOC	2	0.2129	0.10645	3.43	0.052	4.13
Error	20	0.621	0.03105			
Total	26	5.7769				

Table 5 ANOVA for surface roughness under flood cooling

Source	DF	Adj SS	Adj MS	*F*-value	*P*-value	% Con
Speed	2	2.49834	1.24917	14.35	0	53.41
Feed rate	2	2.16466	1.08233	12.44	0	46.30
DOC	2	0.01307	0.00654	0.08	0.928	
Error	20	1.74067	0.08703			
Total	26	6.41675				

contribution of each parameter under the MQL condition. From Table 4 it is evident that feed rate is the most significant parameter with 55.09% contribution and followed by the speed with 40.7% contribution of the total effect. The linear model expresses that 89.25% variability in surface roughness.

Table 5 describes the result of ANOVA under flood cooling for surface roughness with linear model prediction. From Table 5 it is observed that speed gives a major contribution of 53.41% followed by 46.3% of contribution by feed rate in minimizing surface roughness. The model has 72.87% variability in surface roughness.

Figure 6 describes the comparison of surface roughness values of the end milled specimens both under minimum quantity lubrication and flood cooling for various experiments. From the comparison graph, it is found that a considerable amount of deviation in the surface roughness values has been noted. The trend line deviates much in MQL cutting conditions compared with flood cooling conditions.

From literature, MQL increases tool life and reduces surface roughness in machining, but this work gives contradictory results. However, other researchers claimed that MQL application has given poor tool life compared to flood cooling conditions in some work tool material combinations. Many researchers argued that cooling method or machining performance depends on workpiece tool material. This study supports the argument and hence AA6061 aluminium alloy and carbide coated tool worsen surface roughness in end milling operation.

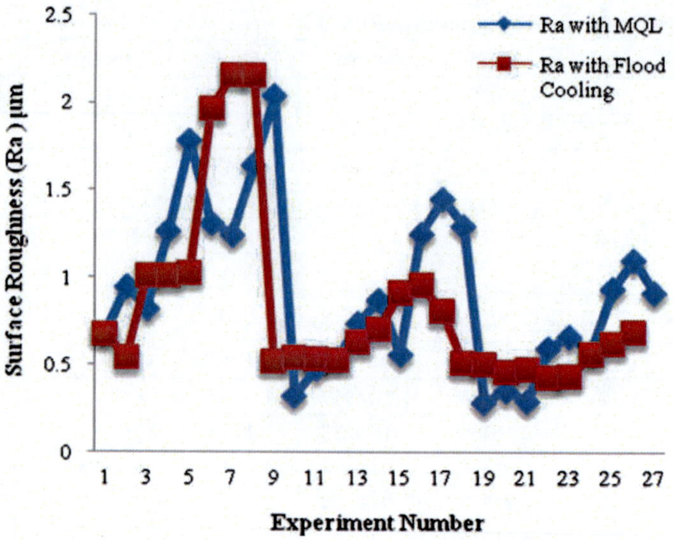

Fig. 6 Comparison of surface roughness values between MQL cooling and flood cooling

4 Conclusions

End milling of AA6061 aluminium alloy with a carbide-coated end mill tool under MQL and flooded cooling conditions were performed successfully by varying cutter speed, feed, and depth of cut. Within the limitations of current research, it was concluded that the application of MQL does not give favorable results in reducing surface roughness compared with the flooded cooling condition. The surface roughness of AA6061 machined with MQL was higher than that obtained with flood cooling method irrespective of the other cutting conditions. ANOVA was employed to investigate the significant parameters for surface roughness under both MQL and flood cooling conditions. As future work, other MQL settings such as distance from the nozzle tip to the tool–workpiece interface, angle of impinging spray, coolant flow rate and compressed air pressure are to be optimized.

References

1. Dhar NR, Islam S, Kamruzzaman M (2007) Effect of minimum quantity lubrication (MQL) on tool wear, surface roughness and dimensional deviation in turning AISI-4340 steel. G.U. J Sci 20(2):23–32
2. Rozzi JZ, Sanders JK, Chen W (2010) The experimental and theoretical evaluation of an indirect cooling system for machining. J Heat Transf 133(3):1–10

3. Wichmann H, Stache H, Schmidt C, Winter M, Bock R, Herrmann C, Bahadir M (2013) Ecological and economic evaluation of a novel glycerol based biocide-free metalworking fluid. J Cleaner Prod 43:12–19
4. Woods S (2005) Going green. Cutting Tool Eng 57(2)
5. Abdalla HS, Baines W, McIntyre G, Slade C (2007) Development of novel sustainable neat-oil metalworking fluids for stainless steel and titanium alloy machining. Part 1—formulation development. J Adv Manuf Technol 34:21–33
6. Liu ZQ, Chen M, An QL (2015) Investigation of friction in end-milling of Ti-6Al-4V under different green cutting conditions. Int J Adv Manuf Technol 78:1181–1192
7. Rahman MM, Yusoff AR, Kadirgama K (2012) Investigation of flow behavior in minimum quantity lubrication nozzle for end milling processes. Int J Autom Mech Eng 6:768–776
8. Puvanesan M, Rahman MM, Najiha MS, Kadirgama K (2014) Experimental investigation of minimum quantity lubrication on tool wear in aluminum alloy 6061-T6 using different cutting tools. Int J Autom Mech Eng 9:1538–1549
9. Najiha MS, Rahman MM (2015) Experimental study on minimum quantity lubrication in end milling of AA6061-T6 using TiAlN coated carbide tools. Int J Autom Mech Eng 11:2771–2785
10. Lakshmi VVK, Subbaiah KV (2012) Modelling and optimization of process parameters during end milling of hardened steel. Int J Eng Res Appl 2:674–679
11. Boswell B, Voges E (2012) The effect of combined cold air and minimum liquid cooling on end milling. Adv Mech Eng 2(1):1–7
12. Boubekri N, Shaikh V, Foster PR (2010) A technology enabler for green machining: minimum quantity lubrication (MQL). J Manuf Technol Manage 21:556–566
13. Kadu MB, Hambire U (2017) Review on MQL in green environments used for machining process. Int J Eng Res Appl 7(10:3):23–26
14. Safiei W, Rahman MM, Rusdan SA (2018) Experimental investigation of MQL optimum parameters in end milling of AA6061-T6 using Taguchi method. Int J Eng Technol 7(4.35):186–189
15. Islam MN (2013) Effect of additional factors on dimensional accuracy and surface finish of turned parts. Int J Mach Sci Technol 17(1). https://doi.org/10.1080/10910344.2012.747936
16. Najiha MS, Rahman MM, Yusoff AR (2013) Modeling of the end milling process for aluminum alloy AA6061T6 using HSS tool. Int J Autom Mech Eng 8:1140–1150
17. Tosun N, Huseyinoglu M (2010) Effect of MQL on surface roughness in milling of AA7075-T6. Mater Manuf Process 25:793–798
18. Boswell B, Islam MN (2013) The challenge of adopting minimal quantities of lubrication for end milling aluminium. In: Yea G-C (ed) Lecture notes in electrical engineering, IAENG transactions on engineering technologies. Springer Science + Business Media Dordrecht, Australia
19. Boswell B, Voges E (2011) The effect of combined cold air and minimum liquid cooling on end milling. In: Ma M (ed) International conference on mechanical, industrial, and manufacturing engineering. Information Engineering Research Institute, Melbourne, Australia, pp 54–57

Material Joining Technology

Performance Study of Dissimilar Alloy Joints of SS321 and SS347 Under MIG Welding Process

S. Nishanth, S. Prem, T. Prakash, J. Siva, M. Sathish Kumar, and N. Siva Shanmugam

Abstract An attempt is made in this research work to study the influence of filler wire materials on the dissimilar alloys between SS321 and SS347 are made using Robotic MIG welding process Preliminary experimental welding trials are conducted by varying the MIG welding process parameters like current, welding speed and stick out the distance. The prepared metal coupons of two different stabilized stainless steels are welded using ER321 as well as ER347 filler wires. The performance of the joint is studied by mechanical testing and microstructural characterization. The results of mechanical testing showed that the welded joint with ER347 filler wire exhibited improved tensile strength and hardness value than the joint made by ER321 filler wire.

Keywords SS321 · SS347 · Dissimilar metal welding · MIG

1 Introduction

The highest demand for enhanced material performance in both military and civilian sectors is the main driving force for advances in new alloys and technologies for newer designs. In this work type 321 stainless steel, an austenitic alloy is stabilized with titanium, making it resistant to intergranular attack caused by chromium carbide precipitation is joined using gas metal arc welding with dissimilar 347 which resembles the same chemical composition using two different filler wires ER321 and ER347 the specimens are prepared and welded in specific conditions and tested for mechanical properties along with microstructure.

S. Nishanth · S. Prem · T. Prakash · J. Siva · M. Sathish Kumar (✉)
Department of Mechanical Engineering, SNS College of Engineering, Coimbatore, Tamil Nadu, India
e-mail: Sathishmfg14@gmail.com

N. Siva Shanmugam
Department of Mechanical Engineering, National Institute of Technology Tiruchirappalli, Tiruchirappalli, Tamil Nadu, India

S. Vijayan et al. (eds.), *Trends in Manufacturing and Engineering Management*, Lecture Notes in Mechanical Engineering,
https://doi.org/10.1007/978-981-15-4745-4_59

It is identified that stabilized austenitic stainless steel such as 347 and 321 are widely used in components designed for high-temperature applications like nuclear reactors, boilers, superheaters and chemical reactors [1]. This material can be selected for high temperatures services due to its high creep and intergranular corrosion resistance. However, for services into the sensitization range of temperatures (450–850 °C) [2]. Austenitic stainless steel type AISI 321, has an incredible resistance to corrosion and oxidation [3].

2 Experimental Procedure

2.1 Material Selection

The grade SS347 is chosen as one of the dissimilar parent alloys which are stabilized austenitic stainless steel with the addition of niobium. The intermittent temperature is about 800–1650 °F (427–897 °C). It is nonmagnetic. The grade SS321 is another alloy to be welded with 347 which is a stabilized austenitic stainless steel which is similar to SS304 but with a titanium addition of at least five times the carbon content. The intermittent temperature is about 800–1500 °F (427–816 °C). It is normally magnetic. This grade is equivalent to SS304 in the annealed condition and stronger if weldments in these grades have not subjected to annealing or if this grade is put into general service in the 420–900 °C range. Table 1 gives the material composition of the alloys.

2.2 Applications of These Dissimilar Alloys

This alloy has been extensively used in Expansion joints Spiral Welded tube for burner pipes and flues, Bellows, Woven or welded screens for high-temperature mineral processing Furnace parts, Heating element tubing, Heat Exchangers. Due to the above-said reasons, their weld characteristics have been studied. Stabilized austenitic stainless steel (SS) such as type 347 and 321 are widely used in components designed for high-temperature applications like nuclear reactors, boilers, superheaters and chemical reactors [1].

2.2.1 Experiment

The experiment is carried out in robotic gas metal arc welding machine shown in Figs. 1 and 2. The specifications of the GMA welding setup is DM 350 MIG power source of OTC Daihen. GMA welding torch was attached with a 6 axis OTC Daihen robot. The experimental study denotes that MIG weld process provides higher quality

Table 1 Composition of 321/347 grades

Grade		C	Mn	Si	P	S	Cr	Mo	Ni	N	Other
321	Min	–	2.00	0.75	0.045	0.030	17.0	–	9.0	0.10	T=5(C+N)
	Max	0.08	–	–	–	–	19.0	–	12.0	–	0.70
347	Min	0.08	2.00	0.75	0.045	0.030	17.0	–	9.0	0.10	Nb=10(C+N)
	Max	–	–	–	–	–	19.0	–	13.0	–	1.0

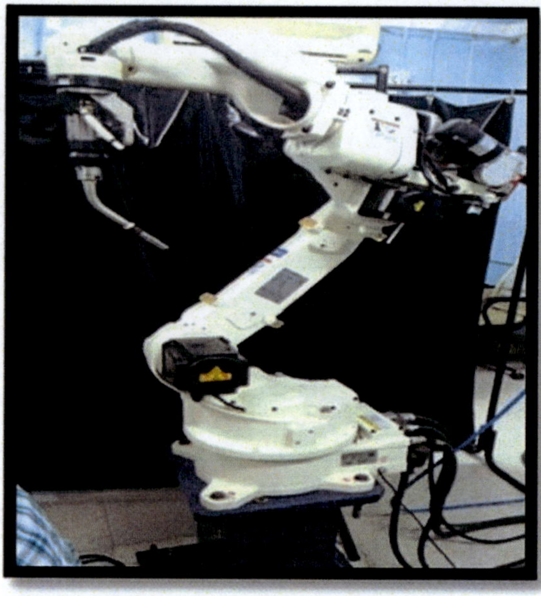

Fig. 1 Robot setup

Fig. 2 Welding process

welds in a wide variety of metal and alloys. Therefore, it is most commonly used to join stainless steel and other metals. The helium and argon are shielding gases that are used to protect the welding zone from the atmosphere [4].

The specimen preparation is carried out to get sound welding joints.

The specimen size is 100 × 50 × 6 mm (two pieces). The butt weld joint is preferred. The specimen is set to resist the degrees of freedom in a welding fixture available. The welding is carried out with parameters arrived from the previous literature. The welding current is chosen as 140 A for welding. The filler wire of two types one is ER321 and ER347 is taken. The specimens were welded using ER321 and ER347 separately and subject to mechanical test.

3 Results and Discussion

3.1 Tensile and Hardness Properties

The tensile tests were carried out at room temperature, specifically, the methods of determination of yield strength, ultimate tensile strength, elongation, and reduction of area. The specimens were prepared according to ASTM E8. The tests were carried out in x-direction and with Tinius Olsen H50KL tensile testing machine. The Fig. 3 shows the specimen size as per the standards.

Some of the literature findings reveal that the tensile test sample fractures occurred at the base metal of SS321 austenitic stainless steel, which shows that the strength of the welded joint was better than that of base metal [5].

The tensile test provides data on the strength and ductility of materials under uniaxial tensile stresses. The results of the tensile test, which was conducted for dissimilar welds of low alloy steel to austenitic stainless steel, showed that fracture happened in the HAZ of low alloy steel [6].

In this work, it is studied that the tensile strength of the base metals SS321 and 347 is about 515 MPa. After welding the dissimilar combination, the specimens were subjected to tensile test. The results of the tensile tests were shown in Table 2. The comparison is to find the best suitable filler wire for the dissimilar alloy combinations

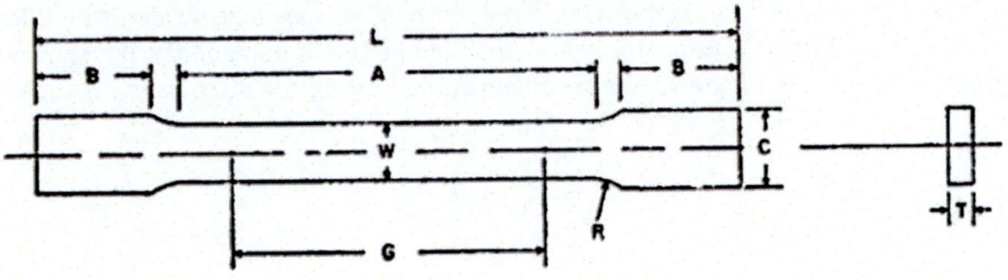

Fig. 3 Test specimen for tensile test

Table 2 Tensile test results

S. No.	Filler/welding current	Ultimate stress (MPa)	Yield stress (MPa)	Total elongation (%)
1	ER321/140 A	330	255	17.6
2	ER347/160 A	363	343	20.9

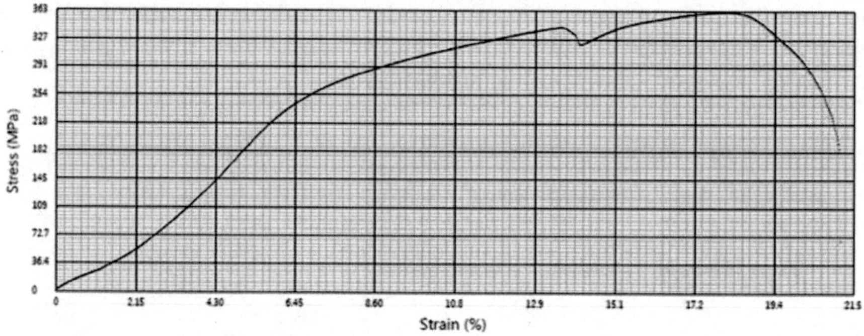

Fig. 4 Stress–strain curve for specimen welded with ER347

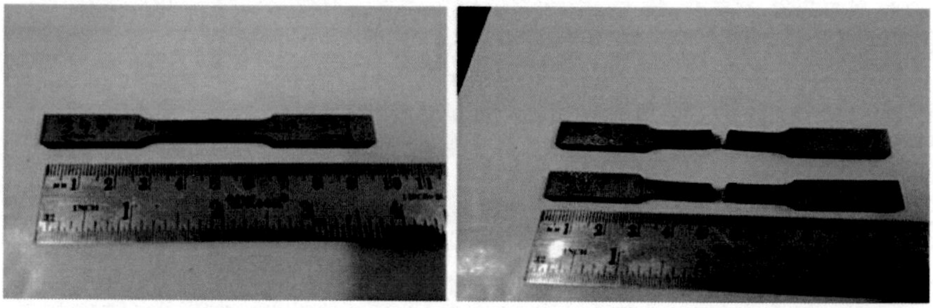

Fig. 5 Tensile test

having high tensile strength. It is found that the weld specimen of the combination of filler wire 347 and welding current 160 A has high tensile strength.

The stress–strain curve is shown in Fig. 4 for the weld specimen welded with filler ER347. The ultimate tensile strength is obtained at 18% strain and the upper yield point occurs at 13% of strain. The specimen before testing and after testing is shown in Fig. 5.

3.1.1 Hardness Test

The hardness measurements were performed to determine the strength along with hardness. Vickers hardness provides medium indentations. The specimen is subjected

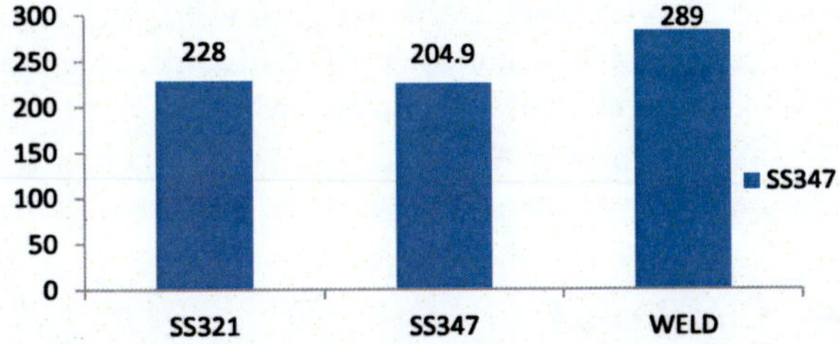

Fig. 6 Vickers hardness test

to Vickers hardness test. Hardness is lower in the HAZ region compared to the weld metal and base metal regions from welding technique [7].

Figure 6 shows the values obtained in the hardness test. Vickers scale is considered. The specimen welded with ER347 has hardness value of 289 HV and the specimen welded with ER321 has 257 HV in the welded region. The base metal always gave the highest hardness value for all the welding parameters [8].

3.2 *Microstructure Properties*

The microstructure properties were studied using SEM analysis. The Fig. 7 shows a

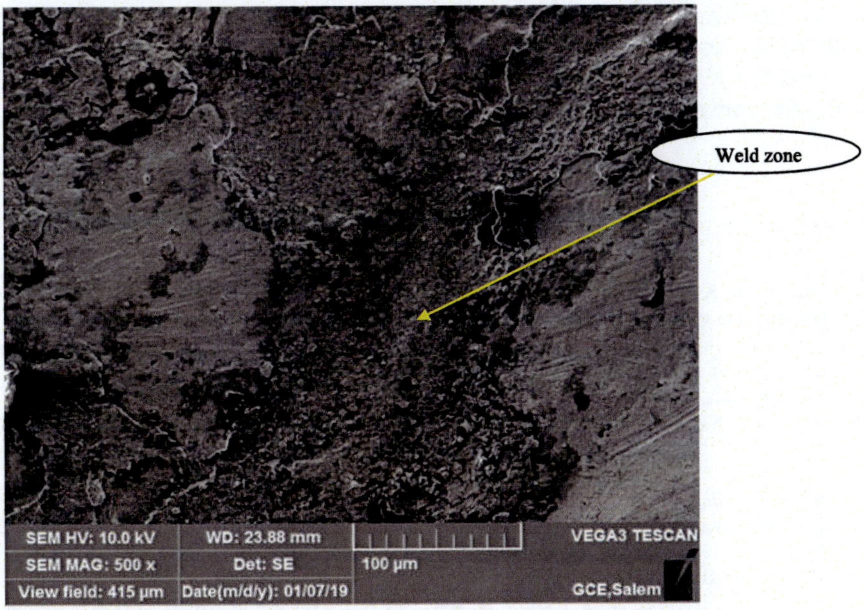

Fig. 7 SEM images

uniform flow of filler material and good weld joint. The scanning electron microscopy was used for the examination of weld microstructure. In the interdendritic region where the small white precipitates which were sometimes segregated clearly from the field were observed [9].

4 Conclusion

The present investigation on dissimilar alloy deduces the following conclusions.

The butt welding joint is best suitable for testing tensile strength. The filler wire ER321 and ER347 are used to weld the joint which improves strength and hardness to a possible extent. The tensile strength of a specimen with ER347 is higher than the specimen welded with ER321. The hardness tested with specimen (welded with ER347) is more when compared to (ER321 filler wire) weld specimen. The neck formation and failure of specimen occur at the ultimate strength of 363 MPa in the specimen welded with ER347. The neck formation and failure of the specimen occur at ultimate strength of 330 MPa in the specimen welded with ER321. The hardness of the weld joint is more than the base metal signifies the lesser tendency to abrasion. The results of mechanical testing showed that the welded joint with ER347 filler wire exhibited improved tensile strength and hardness value than the joint made by ER321 filler wire. This is due to the fact that the presence of intermetallic compounds, i.e. NbC (revealed during metallographic study) in the weld metal is observed to be more in joints produced with ER347 filler wire. In the SEM analysis, the specimen welded with ER347 resembles a more uniform grain structure than the specimen welded with ER321. In the future, the welding current and speed may be optimized along with filler wire to improve joint properties.

References

1. Chandra K, Kain V, Tewari R (2012) Microstructural and electrochemical characterisation of heat-treated 347 stainless steel with different phases. Corros Sci 67:118–129
2. Moura V, Kina AY, Tavares SS, Lima LD, Mainier FB (2007) Influence of stabilization heat treatments on microstructure, hardness and intergranular corrosion resistance of the AISI 321 stainless steel. J Mater Sci 43:536–540
3. Nakhodchi S, Shokuhfar A, Iraj SA, Thomas BG (2015) Evolution of temperature distribution and microstructure in multipass welded AISI 321 stainless steel plates with different thicknesses. J Pressure Vessel Technol 137
4. Kang BY, Prasad YK, Kang MJ, Kim HJ, Kim IS (2009) The effect of alternate supply of shielding gases in austenite stainless steel GTA welding. J Mater Process Technol 209:4722–4727
5. Mohan Kumar S, Siva Shanmugam N (2018) Studies on the weldability, mechanical properties and microstructural characterization of activated flux TIG welding of AISI 321 austenitic stainless steel, vol 5, no 10. IOP Publishing Ltd Materials Research Express

6. Arivazhagan N, Surendra Singh S, Prakash S, Reddy GM (2011) Investigation on AISI 304 austenitic stainless steel to AISI 4140 low alloy steel dissimilar joints by gas tungsten arc, electron beam and friction welding. Mater Des 32:3036–3050
7. Lakshminarayanan K, Shanmugam K, Balasubramanian V (2009) Effect of welding processes on tensile and impact properties, hardness and microstructure of AISI 409M ferritic stainless joints fabricated by duplex stainless steel filler metal. J Iron Steel Res Int 16:66–72
8. Gulenc B, Develi K, Kahraman N, Durgutlu A (2005) Experimental study of the effect of hydrogen in argon as a shielding gas in MIG welding of austenitic stainless steel. Int J Hydrogen Energy 30(13–14):1475–1481
9. Hajiannia I, Shamanian M, Kasiri M (2013) Microstructure and mechanical properties of AISI 347 stainless steel/A335 low alloy steel dissimilar joint produced by gas tungsten arc welding. Mater Des 50:566–573
10. Green G, Higginson R, Hogg S, Spindler S, Hamm C, Najorka J (2015) Analysis of ferrite formed in 321 grade austenitic stainless steel. Mater Sci Technol 31:418–425

Effect of Friction Stir Welding Tool on Al–SiC Composites by Varying Tool Pin Profile and Tool Material

P. Jayaseelan, T. V. Christy, and S. J. Vijay

Abstract Efficient joints can be accomplished through the friction stir welding (FSW) process with the selection of appropriate tool profile, tool material and *D*/*d* ratio. This experiment to weld Aluminium 6061 with 10% silicon carbide reinforcement was carried out in a CNC FSW machining set-up, the tool materials being H13 and OHNS, the tool profiles plain pentagon, plain taper and threaded cylinder and the *D*/*d* ratio, a common 2. The different welding parameters chosen were 1000 rpm tool rotation speed and 25 mm/min travel speed with an axial load varying between 3.8 and 4.3 kN. The joint produced with the threaded cylinder tool, compared with the rest of the tools, exhibited the maximum tensile strength of 150 MPa and hardness of 41 HRC and was found to be defect-free with a uniform grain size distribution.

Keywords FSW · Profile · Material · *D*/*d* ratio · Al–SiC · Tensile strength · Hardness

1 Introduction

Friction stir welding was first developed in December 1991 by "the Welding Institute" (TWI). Friction stir welding procedure is a process of solid-state connection without any external filler. This entire process is carried out in the CNC machine setup. Large alloys, like highly alloyed 2xxx and 7xxx series, are not weldable because their microstructure and porosity in the area of melting are poorly solidified. This entire process is carried out in the CNC machining setup [1]. When there are high

P. Jayaseelan · S. J. Vijay
Department of Mechanical Engineering, School of Engineering and Technology, Karunya Institute of Technology and Sciences, Coimbatore, India
e-mail: jeevaglory@gmail.com

S. J. Vijay
e-mail: vijayjoseph.2001@gmail.com

T. V. Christy (✉)
Professor of Mechanical Engineering, PRIST, Vallam, India
e-mail: drchristytv@gmail.com

S. Vijayan et al. (eds.), *Trends in Manufacturing and Engineering Management*, Lecture Notes in Mechanical Engineering,
https://doi.org/10.1007/978-981-15-4745-4_60

levels of thermal stress and solidification of shrinkage during the welding process, Aluminium is susceptible to Solidification cracking or hot cracking can be found out. FSW has several benefits in the welding process which can be listed: FSW does not need a joint preparation for plates, the result is high-quality welding with increased tensile strength, expectation fatigue and corrosion resistance [2]. In order to improve material mechanical properties, combining with various alloying elements shows high performance, and heat treatable alloys. In this experimental process, we used Aluminium Silicon Carbide alloy as a weldment. The four major parameters determine the quality and strength of joint are mainly classified into (i) Speed of the Rotation, (ii) Traverse speed, (iii) Plunging depth, (iv) Tilt angle. Then, the selection of appropriate tool profile, tool material, tool geometry is risky and challenging one [3]. By determining the proper ratio for the shoulder and pin, the tool was selected [4]. Palin pentagon, Plain Tapering and Threaded cylinder with various instruments are the tool profiles used in this experimental work. After the welding process once completed, the scans of the weldment are important to check the different zones and its welded quality. The areas discussed here are (i) Base Material (BM) or Parent Zone (PZ) (ii) Heat affected zone (HAZ) (iii) Thermo-Mechanically affected zone (TMAZ) (iv) Weld zone (WZ) or Nugget Zone (NZ). In this experimental work, we finally discussed the various Tensile strength and hardness for the welded specimen in different tool profiles and the values are tabulated, and the micrographs for the weldment zones are shown.

2 Experimental Procedure

The experiment was initially started by selecting the proper tool profiles which namely Plain Pentagon, Plain tapered, threaded cylinder tool pin and the material used for the tools are H13, OHNS are selected for use. Hence the *D*/*d* ratios for the tool profiles are listed in Table 1. Hence, the listed tool profiles are allowed to weld in the Al–SiC weldment are measured the tensile strength and hardness values are compared. The tools used in this experiment are shown in Fig. 1. The experiment was therefore performed with a composite aluminium metal matrix. Al-10% SiC with the dimensions 100 mm × 50 mm × 6 mm is produced under the in situ casting process with the composition of the metal plates. As we discussed before, the composition of the Aluminium alloy with other alloys possess greater mechanical strength [5]. If the material composition is balanced, then the defect-free weld can be obtained [6]. The Al-10% SiC MMC microstructure is displayed in Fig. 2.

Table 1 FSW tools descriptions

Tool No.	Tool profile	Tool material	*D*/*d*
1	Plain pentagon	H13	2
2	Plain tapered	OHNS	2
3	Threaded cylinder	H13	2

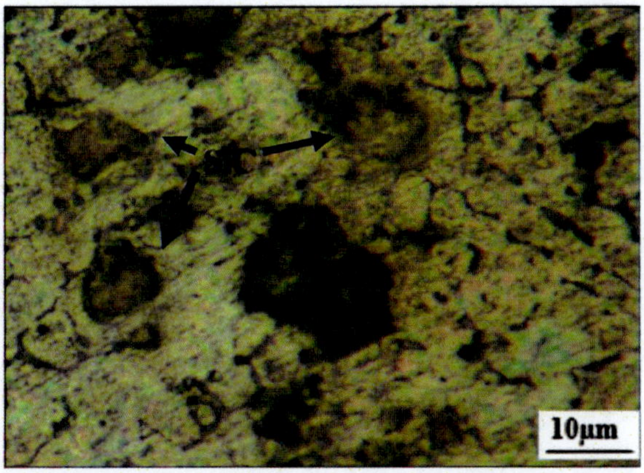

Fig. 1 Al-10% SiC MMC optical micrograph

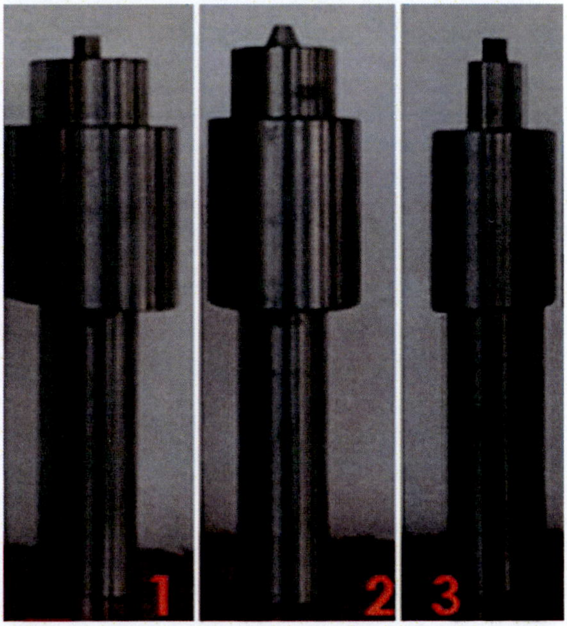

Fig. 2 Fabricated FSW tools

Then, the proper selection of the tool material is important that the tool should withstand such mechanical properties [7]. The selection of the material is important so that the good quality of the weld can be achieved.

H13 and OHNS with proper tool geometry are the tools used in this experiment which have a strong temperature resistance to thermal fatigue and resistance to wear.

The parameters used in this experiment are (1) 1000 rpm tool rotation speed; (2) 25 mm/min traversing speed; (3) 3.8–4.3 kN axial load. The parameters are fixed after we did some trial experiments which yielded defect-free weld [8]. The parameter used

Table 2 Welding parameters

Tool rotational speed (RPM)	Traverse speed (mm/min)	Axial load (kN)
1000	25	3.8–4.3

in this experiment is tabulated as shown in Table 2. Then the experiment was followed with this fixed parameter for all the tool profiles, then the results are discussed. The connection between the number of revolutions and the speed of soldering directly affects the toughness of the fracture and the energy needed to initiate and propagate the crack [9].

3 Results and Discussion

Specimens are welded with above-mentioned parameters and it was prepared for testing the hardness and tensile strength. The test was conducted as per ASTM E08 standard. Table 3 shows the tensile strength of the welded Aluminium Silicon carbide (Al–SiC) is welded with different tool profiles (Plain Pentagon, Plain Tapered, Threaded Cylindrical). And it was shown that Tool number 3 (Threaded Cylinder—H13) yields high tensile strength than the other tool profiles. And it is clearly stated that it yields 150 MPa, 118 MPa, 127 MPa with respect to Threaded cylinder—H13, Plain Tapered—OHNS, Plain Pentagon—H13 respectively. As we know that it yields high tensile strength only if the grains are evenly distributed (uniformly). That the condition is satisfied in the case of Threaded Cylinder—H13 and its microstructure are given in Fig. 3. The welding temperature and welding speed are mainly determined for the UTS of the friction stir welds in EN AW-2219-T87 [10]. Other than the H13, Threaded cylinder, tools profile, and materials made like H13—plain pentagon and OHNS—plain tapered couldn't even uniformly distribute than the H13 Threaded cylinder profile. So, the other two tools yield slightly lower than the H13—Threaded Cylinder profile. The optimal combination of process parameters for the high-speed steel (HSS) tool [11].

After the test, the maximum hardness for the welding specimens (Al–SiC) was determined by a hardness test. The hardness value for the welded specimen is shown in Table 4. It is clearly stated that again H13—Threaded Cylinder shows the maximum hardness value of 41 HRC compared to OHNS—Plain tapered and H13—Plain

Table 3 Tensile strength of weldments

Tool No.	Tool profile	Tool material	Tensile strength (in MPa)
1	Plain pentagon	H13	127
2	Plain tapered	OHNS	118
3	Threaded cylinder	H13	150

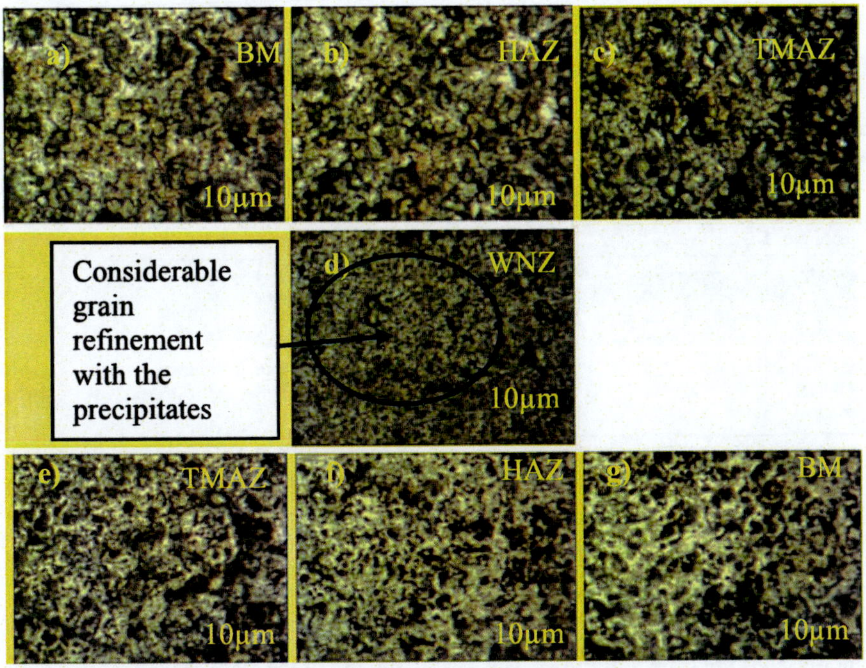

Fig. 3 Composite FSW micrographs using the threaded cylinders tool H13

Table 4 Hardness of weldments

Tool No.	Tool profile	Tool material	Hardness (in HRC)
1	Plain pentagon	H13	40
2	Plain tapered	OHNS	26
3	Threaded cylinder	H13	41

Pentagon of 26 HRC and 40 HRC, respectively. Hence, the Mechanical characterization results that H13—Threaded cylinder profile yielded high strength. Hardness increases during welding at a high tool rotation and reduces wear [12]. The tool material does contribute significantly to joint characteristics [13].

In current work, Al 10% SiC using the H13—Threaded cylinders show, according to the metallurgical characterization, that in the Weld zone microstructure in Fig. 3(d) the uniform distribution of particles is clearly indicated. Through examining of the weld zone, the uniform distribution of particles with a defect-free weld is obtained for H13—Threaded cylinder than the other two tool profiles.

As we know, metallurgical characterization is divided into four major categories. This includes the Weld Zone (WZ), THM Zone (TMAZ), the Thermal Area (HAZ), the THM Zone (NZ), or BA (BM) Zones. Therefore Al–SiC composite welded

Table 5 Macrostructure of weldments

Tool profile	Macrograph of FSW joint
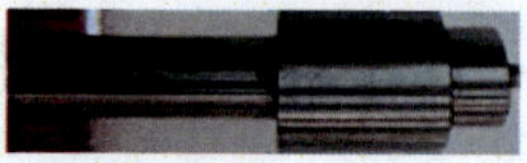 Plain pentagonal H13 Tool No. 1	 4.2 kN, 25 mm/min, 1000 rpm
 Cylindrical tapered OHNS Tool No. 2	 3.5 kN, 25 mm/min, 1000 rpm
 Threaded cylinder H13 Tool No. 3	 4.3 kN, 25 mm/min, 1000 rpm

micrographs are shown in Fig. 3. The grain structure of the HAZ is well known to be almost the same as base metal (BM). Since, when compared to Thermo-Mechanically Affected zone (TMAZ) grain size are smaller than the other grains due to its plastic deformation near the weld zone (WZ). The heat is allowed to pass through the TMAZ. The welding zone with a maximum heat generation is thus smaller than TMAZ and BM in grain size. In different areas of the welded joint, the asymmetry of the sold joints and the changes in metallurgical transformations that occur around the pine and below the tool during their joint movement affect the effect strength value [14].

Also, the macrostructure of the Welded zone of Al–SiC is shown along with the Tensile Strength (in MPa) as shown in Table 5. As we see that, weldment made using the Tool No 3 has exhibits defect-free joint and yield high tensile strength with desired hardness.

4 Conclusion

Significant difficulties in achieving high-quality welds are the right choice for the FSW tool. As already stated, the material, pin profile and ratio of *D*/*d* affect soldering quality significantly and therefore require a thorough investigation. Detailed investigations were carried out using tools of various profiles, such as friction stir welding aluminium metal matrix composites (Al–SiC). Simple flap, cylinders, materials such as simple tapering and threading. FSW, OHNS and H13 and *D*/*d* ratio 2 are allowed on commercial CNCs. The results show that Al-MMC joints can produce superior tensile strength and durability in a threaded cylinders H13 tool with a ratio of *D*/*d* 2. The ongoing work in the production of quality welds using the FSW process is considered an important contribution.

References

1. Barath VR, Vignesh RV, Padmanaban R (2018) Analysing the strength of friction stir welded dissimilar aluminium alloys using Sugeno fuzzy model. IOP Publishing, Coimbatore
2. Bosneag A, Constantin MA, Nitu E, Iordache M (2018) Friction stir welding of three dissimilar aluminium alloy: AA2024, AA6061 and AA7075. IOP Publishing, Romania
3. Palani K, Elanchezhian C, Avinash K, Karthik C, Chaitanya K, Sivanur K, Reddy KY (2018) Influence of friction stir processing parameters on tensile properties and microstructure of dissimilar AA8011-H24 and AA6061-T6 aluminium alloy joints n nugget zone. IOP Publishing, Kanchipuram
4. Jayaseelan P, Christy TV (2019) Behaviour of threaded cylinder tool in the friction stir welding of Al-ZrB2 and Al-SiC composite. Int J Mech Eng Technol 10(1):1751–1761
5. Daneji A, Ali M, Pervaiz S (2018) Influence of tool geometry and processing parameters on welding defects and mechanical properties for friction stir welding of 6061 aluminium alloy. IOP Publishing, Dubai
6. Chand S, Vineetha S, Madhusudhan D, Krishna CS, Devi GK, Hemarao K, Naidu GG (2018) Effect of FSW welding speed on microstructure and microhardness of Al-0.84Mg-0.69Si-0.76Fe alloy at moderate rotational tool velocity. IOP Publishing, Andhra Pradesh
7. Jayaseelan P, Christy TV, Rubesh, Prabhu G, Srinivasan R (2019) Behaviour of threaded cylinder tool in the friction stir welding of Al–ZrB_2 composite. J Comput Theor Nanosci 16(2):719–724
8. Vijayan D, Abhishek P (2018) Multi objective process parameter optimization of friction stir welding using NSGA-II. IOP Publishing, Kanchipuram
9. Jayaseelan P, Christy TV (2019) Behavior of threaded cylinder tool in the friction stir welding of Al-SiC composite. In Materials today: Proceedings (Vol. 16)
10. Zhang H, Zhao C, Cui B, Yan X, Dou C, Shao T, Pan R (2018) The corrosion behaviour of cold spray coating on 2219 aluminium alloy joints prepared by friction stir welding. IOP Publishing, Beijing
11. Milcic M, Vuherer T, Radisavljevic I, Milcic D, Kramberger J, Andjelkovic B (2018) Mechanical behavior of Al 2024 alloy welded by friction stir welding. IOP Publishing, Beograd
12. Chernykh IK, Vasil'ev EV, Matuzko EN, Krivonos EV (2018) Upgrading weld quality of a friction stir welded aluminium alloys AMG6. IOP Publishing, Omsk
13. Constantin MA, Bosneag A, Nitu E, Iordache M (2018) Orientation of process parameter values of TIG assisted FSW of copper to obtain improved mechanical properties. IOP Publishing, Romania
14. Jayaseelan P, Christy TV, Vijay SJ, Nelson R (2019) Effect of tool material, profile and D/d ratio in friction stir welding of aluminium metal matrix composites. Materials Research Express, IOP Science Publishing

Fabrication of AISI 304 Austenitic Stainless Steels with TiN Addition Using Spark Plasma Sintering Method

Babatunde Abiodun Obadele, Ramokone Marcia Mafafo, Walter Roberto Tuckart, and Peter Apata Olubambi

Abstract The spark plasma sintering (SPS) is a type of electric discharge sintering technique used to consolidate metallic/ceramic powders. In this study, spark plasma sintering was used to fabricate 304 stainless steel with TiN addition. The best parameters that influence the sintering process such as sintering temperature, heating time and pressure were investigated. Results show that by varying the amount of TiN, the sintered properties of the composites could be significantly affected. The relative density decreased with the increment of TiN. Microhardness values vary from 270 to 350 HV0.1.

Keywords Spark plasma sintering · 304 stainless steel · Density · Hardness

1 Introduction

In modern industrial applications, engineering materials are being subjected to harsh demanding environments that require materials to possess good mechanical properties, good oxidation and corrosion resistance. Engineering materials such as titanium, aluminium and stainless steels are finding wider application in such environment [1]. Austenitic stainless Steels (ASS) type 304 is widely used in these engineering environments. 304 ASS has excellent corrosion resistance properties in various environments; as a result, austenitic stainless steels find application in various industrial application which include thermal power plants, power generation plants, marine,

B. A. Obadele (✉) · R. M. Mafafo · P. A. Olubambi
Department of Chemical Engineering Technology, University of Johannesburg, Doornfontein Campus, Johannesburg, South Africa
e-mail: obadele4@gmail.com

B. A. Obadele
Department of Chemical, Materials and Metallurgical Engineering, Botswana International University of Science and Technology, Palapye, Botswana

W. R. Tuckart
Tribology Group, Universidad Nacional del Sur/CONICET, Av. Alem 1253, 8000 Bahía Blanca, Argentina

S. Vijayan et al. (eds.), *Trends in Manufacturing and Engineering Management*, Lecture Notes in Mechanical Engineering,
https://doi.org/10.1007/978-981-15-4745-4_61

urban water, chemical, aircraft gas turbine engines and food industries [2]. Apart from corrosive environment application, high impact velocity of particles is frequently encountered in components used in these industries. Examples are coal-fired boilers used in the power generation industries [3]. Although austenitic stainless steels are well known for their good corrosion properties, they are famous for poor mechanical properties such as friction and wear. According to Sun and Bell [4], when ASSs are in contact with each other or other materials, they experience severe metallic wear which results from the adhesion junction and severe plastic deformation occurring between the surfaces of the materials in contact with each other. Furthermore, ASS fails mainly because of wear rather than corrosion-related material loss problems.

According to More et al. [5], 50% or more investigators focuses on steel as their target material because steel is used in many industrial applications, however, it has been noticed that steel material can be effectively replaced or alloyed with some other materials such as titanium. Mainly because titanium is the second-largest target material and they have good wear properties. Titanium nitride (TiN), which has relatively good wear properties, is intentionally alloyed to 304 ASS to improve the hardness and wear properties of the resulting material.

There are several methods that can be used to consolidate 304L ASS with TiN. Such methods include hot isostatic pressing (HIP), hot pressing (HP), microwave sintering and spark plasma sintering (SPS) [6, 7]. This study focuses on ASS grade 304 reinforced with TiN under high-temperature conditions.

2 Experimental Procedures

Commercial as-received 304-austenitic stainless steel and TiN were chosen as the starting powders. The two feedstock powders were characterized for particle size distribution (PSD) using Microtrac S3500 analyzer. It uses the three red laser diodes (tri-laser) that provides accurate and repeatable particles sizes [8]. Powders were weighed according to the proportions as shown in Table 1. The weighed powders were then mixed (under dry environments) using Turbula Shaker Mixer T2F for 8 h at a speed of 72 rpm. The function of mixing is to achieve uniformly blended particles without altering the properties required for the next steps [9]. Admixed powders were then characterized using SEM-EDX and the phases formed were analyzed using XRD. The admixed powders were also characterized by PSD.

Table 1 Chemical compositions of admixed powders

Number	Powder	Weight ratio
1	304-ASS	100
2	304-ASS/2TiN	98:02
3	304-ASS/4TiN	96:04
4	304-ASS/6TiN	94:06

Table 2 Spark plasma sintering conditions

Sintering conditions	Weight ratio
Sintering temperature (°C)	1100
Holding time (min)	10
Heating rate (°C/min)	100
Vacuum pressure (MPa)	50

The admixed powders were then consolidated using spark plasma sintering technique in which specimens with dimensions of 35 mm × 1.5 mm were produced. The substrates were machined using the robofil 240-SL electrical discharge machine (EDM) wire cut. Scanning electron microscopy (SEM, Tescan Vega 3) which has the energy-dispersive X-ray spectrometer (EDS) attached to it was used to analyze the morphology of the feedstock and admixed powders to confirm the element present in the powders. Phase identification of the powders was also analyzed with X-ray diffraction (XRD Rigaku Ultima IV). Admixed powders were sintered using spark plasma sintering technology (H-HPD25-FCT Systeme GmbH Germany) in a 40 mm graphite die and the thickness between the graphite foil (used to separate the graphite die and the punches) and the powder was 6.5 mm. The densification process took place at varying operating parameters under a heating rate of 100 °C/min from room temperature up to the desired temperature (1100 °C) and holding time of 10 min. The temperature was monitored using the optical pyrometer throughout the SPS process. A constant vacuum pressure of 50 MPa was applied in each test. The sintering parameters used are presented in Table 2. After sintering, the samples were sand-blasted to remove any excess graphite material on them. Furthermore, the densities of the sintered samples were accurately weighed using the Archimedes for five times using Mittler electronic balance and the average densities were recorded.

The sintered samples were ground and polished to 3 μm diamond finish. The samples were further polished with fumed silica using a polishing suspension of 0.2 μm to achieve a mirror-like surface. The morphology of the sintered specimen was investigated using scanning electron microscope model TESCAN Vega 3 which is equipped with Energy-dispersive spectroscope. The Rigaku Ultima IV X-ray diffractometer XRD equipped with a graphite-monochromatic Cu Kα radiation source at 40 kV and 30 mA was used to perform the XRD analysis. Collection of diffractograms took place in the 2θ range between 3° and 90° with a step size of 0.01 [8]. The purpose of the XRD analysis is to identify the phases present/formed in the composites. The effect of adding TiN on the hardness of the resulting composite was carried using the INNOVATEST Falcon 500 Vickers microhardness tester. The specimens were indented under a load of 100 g and dwell time of 10 s on 10 different positions. The average values were calculated and recorded.

3 Results and Discussion

The SEM morphologies and EDS analysis of the starting 304 ASS and TiN powders used in this study are given in Fig. 1. From Fig. 1a, it could be seen that 304 ASS particles are spherical in shape, smooth and non-porous with a few satellites. This is typical for powders that were produced by gas atomization [10]. From Fig. 1b, it could be seen that the TiN powder show an irregular shape and it is porous. The TiN powder could have been produced by the sol-gel method. This method is used to produce metal-oxides or metal chlorides (usually known as precursors), especially that of silicon and titanium. The morphology of particles produced by this technique is porous. EDS reveals Fe, Cr and Ni in the 304 ASS powder as the major elements and Ti and N as the main elements in the TiN starting powder.

The XRD spectra of as-received 304 ASS and TiN powders are given in Fig. 2. In Fig. 2a, the XRD pattern shows the phases present in the starting 304 ASS powder. The phases that have emerged contain elements such as Cr, Ni, and Fe. Similarly, in Fig. 2b, the diffractometer reveals that there is only TiN present in the powder. The XRD results of the starting powders obtained in this study are in line with the EDS analysis.

The particle size distribution curves are presented in Fig. 3. The PSD graph represents the amount of materials in percentage passing through a specific screen size in microns. From the graph, it could be observed that for pure 304 SS, 304 SS/2TiN, 304

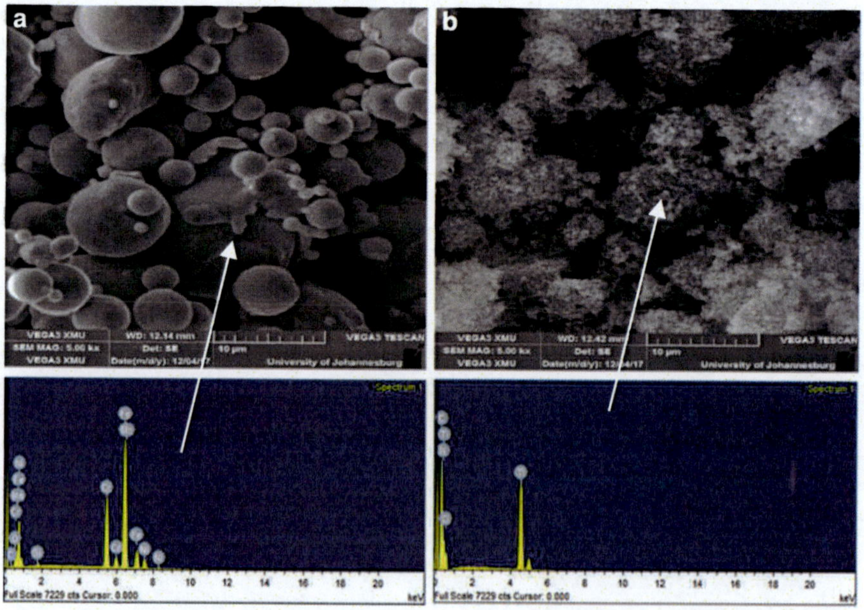

Fig. 1 SEM morphologies of **a** as-received 304 ASS and **b** TiN particles with EDS analyses

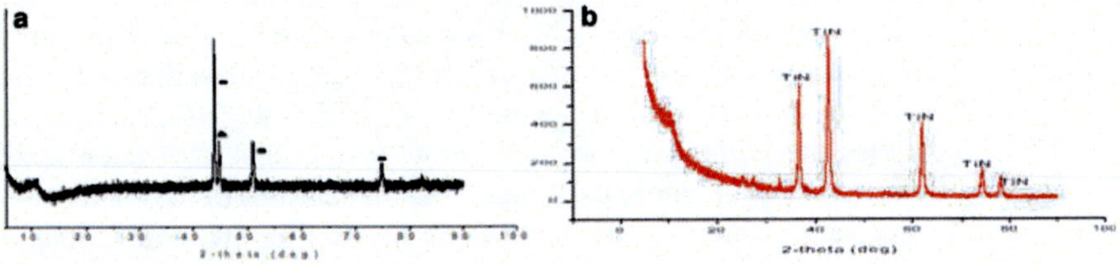

Fig. 2 XRD pattern of the feedstock **a** 304 ASS and **b** TiN powders

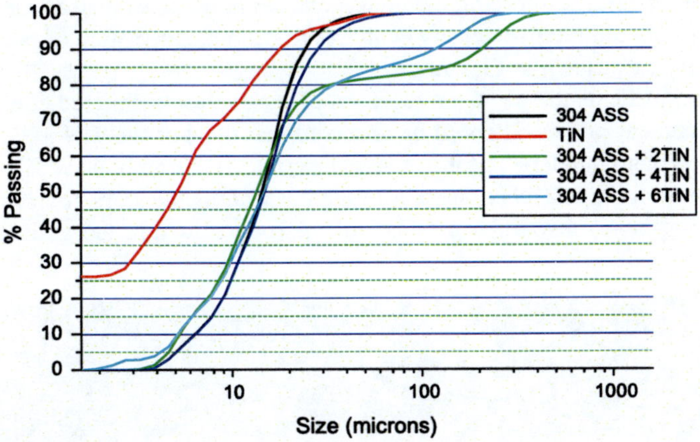

Fig. 3 PSD of 304 stainless steel matrix composite

SS/4TiN and 304 SS/6TiN, 50% (D50) of the particles pass through approximately 15 μm sieves. In the case of TiN, 50% (D50) of the particles pass through 5.5 μm. From SEM analysis of the powders (see Fig. 1), TiN particles were identified by smaller particle size whereas 304 ASS particles were classified by larger particles. The PSD curves confirm that the particles sizes of TiN powder to be smaller than that of 304 ASS powder.

Relative densities for the SPS composites used in the study are shown in Fig. 4. From the graph, it could be observed that the densities of the composites decrease

Fig. 4 Relative density of composites after SPS process

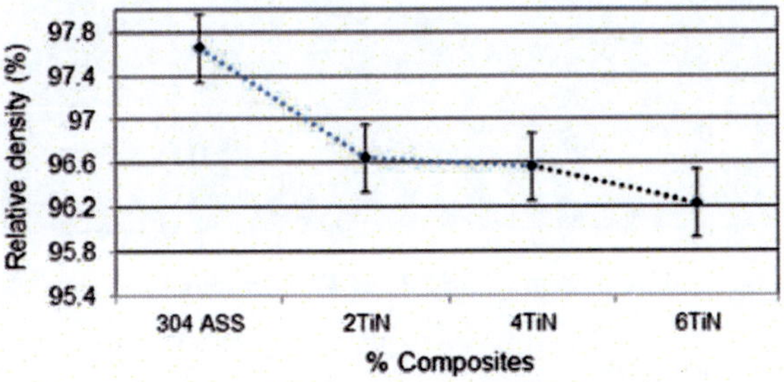

with an increasing amount of TiN reinforcing material in the steel matrix. The reason for this trend may be due to the lower density of TiN (5.43 g/cm^3) than that of 304 SS (8.03 g/cm^3). Relative densities of the composites were 97.65%, 96.64%, 96.56% and 96.23% for compositions without and with TiN (0 wt%, 2 wt%, 4 wt% and 6 wt%) respectively. Sulima et al. [11] reported that the apparent densities of the composites decrease with increasing content of the reinforcing material in the matrix. This is so because of the lower densities the reinforcing material usually possess. On their studies on the effect of SPS parameters on densification and properties of steel matrix composites, they indicated that the apparent densities of the steel matrix decrease as the content of TiB2 reinforcing material increases in the steel matrix. On the other hand, a decrease in relative density could result from the formation of pores around the TiN phase as shown in Fig. 5.

Figure 5 presents the SEM micrographs of spark plasma sintered composites of different content of TiN. It could be observed that most of the TiN are distributed within the grain boundaries of the matrix material. As the amount of TiN increases from 5 to 6%, more of TiN (white phase) segregate along the grain boundaries. SEM images (Fig. 5a–d) reveal the presence of micropores within the grain and along

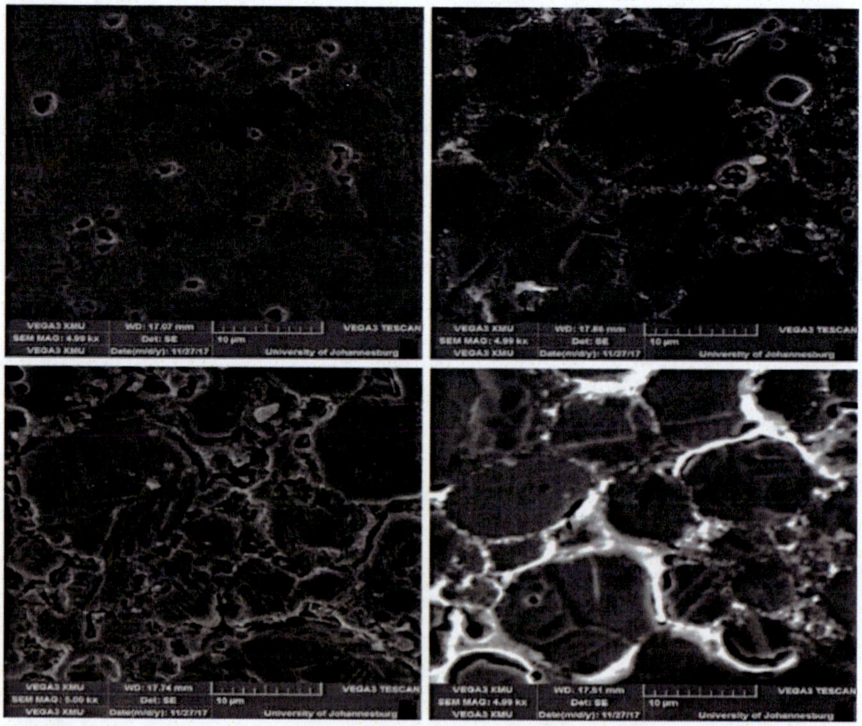

Fig. 5 SEM micrograph of SPS 304 ASS without and with 2%, 4% and 6% TiN

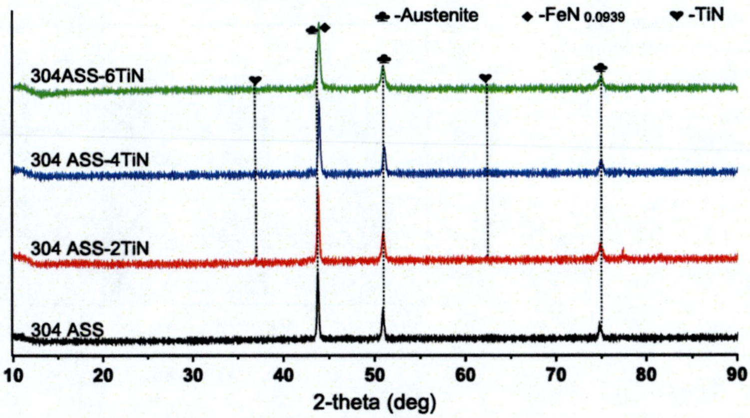

Fig. 6 X-ray diffractograms of 304 SS composite after SPS process

grain boundaries of 304 ASS. The decrease in sintered relative density (see Fig. 4) correlates with SEM images presented in Fig. 5.

The XRD analysis of the SPS stainless steel composites with various amount of reinforcement (TiN) in wt% sintered at a sintering temperature (1100 °C), heating rate (100 °C/min) and holding time (10 min) is given in Fig. 6. The peaks were scanned at room temperature under argon atmosphere. The XRD reveals the presence of the following phases; TiN with a chemical formula TiN0.90, phases containing iron, chromium and nickel such as FeCr0.29Ni0.16C0.06, austenite. The main elements of 304 ASS which are Fe, Ni and Cr, are revealed, these elements lie close to each other on the periodic table. Therefore, they possess similar properties and lattice parameters. In situations like this, it is practically impossible to accurately determine the phase composition of the spark plasma sintered 304 ASS composites because the XRD analysis of these composites can only prove the solid solution of iron [11, 12]. There has been no peak shift during sintering. However, there has been a slight change in peak intensities. The change in peak intensities could result from the sintering temperature, time and heating rate. The XRD patterns before and after spark plasma sintering process are very similar. There has been no significant transformation or phase change during spark plasma sintering. However, cubic nitride (chemical formula, FeN0.0939) has been detected in the major peak of ASS 304 with 2, 4 and 6 wt% TiN at peak position $2\theta = 43.884°$. This finding could be due to the intense phenomena that take place during the SPS process, which include surface activation, diffusion, surface melting as well as necking between sintered powder particles and plastic flow. An increase in mass transportation is obtained during sintering and this improves the bonding between powder particles. The iron nitride that has emerged in the studied material after sintering is formed as a result of iron and nitrogen diffusing in the grain boundary of the steel matrix. It is very crucial to mention that the SPS process involves a stage where the material undergoes non-equilibrium conditions

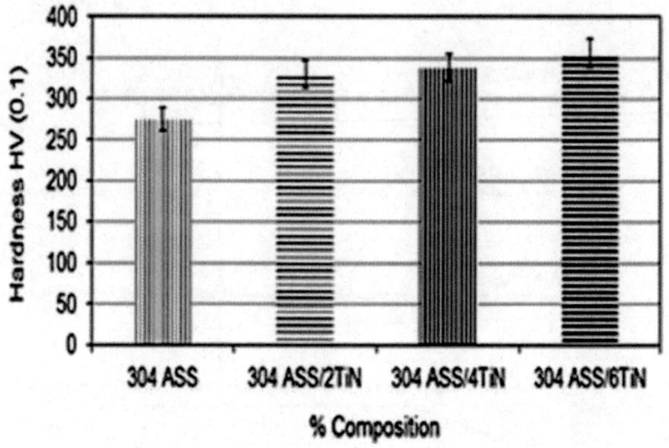

Fig. 7 Microhardness of 304 SS composites before erosion studies

and this has a great impact on the reactions as well as the diffusions which take place during SPS [12].

Microhardness profiles of ASS 304 without and with TiN additions are given in Fig. 7. As expected, the microhardness values increase as the amount of TiN in the specimen increases. This could be due to the hard phase TiN presence in the matrix material. ASS 304 with 6% TiN shows the highest microhardness value of about 350 $HV_{0.1}$ while ASS 304 without and with 2% and 4% TiN microhardness values of 275, 330, 338 $HV_{0.1}$ respectively.

4 Conclusion

In this study, the effect of TiN on the microstructure and hardness behaviour of 304 austenitic stainless steel has been evaluated. Based on the results obtained, the following conclusion could be drawn; SPS results given in terms of relative densities show a decrease in relative density with increasing content of TiN. Also, an increase in TiN value in the matrix increased the hardness values.

Acknowledgements The authors are grateful to Tribology Group, Universidad Nacional del Sur, Bahía Blanca, Argentina for the use of equipment and facilities and The Institute of Nanoengineering Research (INER), Tshwane University of Technology, Pretoria for the use of sintering and characterization facilities.

References

1. Obadele BA, Lepule ML, Andrews A, Olubambi PA (2014) Tribocorrosion characteristics of laser deposited Ti–Ni–ZrO_2 composite coatings on AISI 316 stainless steel. Tribol Int 78:160–167

2. Bregliozzi G, Di Schino A, Ahmed S-U, Kenny J, Haefke H (2005) Cavitation wear behaviour of austenitic stainless steels with different grain sizes. Wear 258:503–510
3. Samandi M, Shedden B, Smith D, Collins G, Hutchings R, Tendys J (1993) Microstructure, corrosion and tribological behaviour of plasma immersion ion-implanted austenitic stainless steel. Surf Coat Technol 59:261–266
4. Sun Y, Bell T (2002) Dry sliding wear resistance of low temperature plasma carburised austenitic stainless steel. Wear 253:689–693
5. More SR, Bhatt DV, Menghani JV (2017) Recent research status on erosion wear—an overview. Mater Today Proc 4:257–266
6. Smuk O (2004) Microstructure and properties of modern P/M super duplex stainless steels (Doctoral dissertation, Materialvetenskap).
7. Brytan Z, Dobrzański L, Grande MA, Rosso M (2009) Characteristics of vacuum sintered stainless steels. J Achievements Mater Manuf Eng 33:126–134
8. Dada OR (2015) Characterization and beneficiation of foundry greensand waste from ferrous foundries. University of Johannesburg
9. Obadele BA, Masuku ZH, Olubambi PA (2012) Turbula mixing characteristics of carbide powders and its influence on laser processing of stainless steel composite coatings. Powder Technol 230:169–182
10. Obadele BA, Ige OO, Olubambi PA (2017) Fabrication and characterization of titanium-nickel-zirconia matrix composites prepared by spark plasma sintering. J Alloy Compd 710:825–830
11. Sulima I, Putyra P, Hyjek P, Tokarski T (2015) Effect of SPS parameters on densification and properties of steel matrix composites. Adv Powder Technol 26:1152–1161
12. Oke SR, Ige OO, Falodun OE, Obadele BA., Shongwe MB, Olubambi, PA (2018) Optimization of process parameters for spark plasma sintering of nano structured SAF 2205 composite. *Journal of materials research and technology, 7*(2), 126–134.

Effect of Process Parameters on Bead Width of 202 Grade Stainless Steel Gas Tungsten Arc Welded Plates Using Response Surface Methodology

R. Sudhakaran, P. S. Sivasakthivel, M. Subramanian, S. Mahendran, M. Sathish Kumar, and R. Vijayakumar

Abstract Welding quality can be accomplished by meeting quality prerequisites, for example, bead geometry by detecting either directly or indirectly the different procedure parameters engaged with the procedure. Lacking weld bead measurements may add to the disappointment of a welded structure. This paper presents an improvement of a scientific model for foreseeing bead width for gas tungsten arc welding of 202-grade stainless steel plates. This model depends on the connection between procedure parameters and bead width and can be utilized to foresee the impact of procedure parameters on bead width. To assemble the required information for demonstrating, real tests were done dependent on a central composite rotatable plan with 31 test runs. The procedure factors considered here incorporate welding current (I), welding speed (V), gas flow rate (Q) and welding gun angle (θ). The sufficiency of the model was tried utilizing ANOVA procedure. The outcomes demonstrate that the proposed model can foresee bead width with sensible precision. Primary and cooperation impacts of the control factors are exhibited in a graphical structure that helps in choosing rapidly the procedure parameters to accomplish the ideal outcomes.

Keywords Gas tungsten arc welding · Central composite rotatable design · ANOVA · Bead width

R. Sudhakaran (✉) · S. Mahendran · M. Sathish Kumar
Department of Mechanical Engineering, SNS College of Engineering, Coimbatore, Tamil Nadu, India
e-mail: absudha@yahoo.com

P. S. Sivasakthivel
School of Mechanical Sciences, SASTRA University, Thanjavur, India

M. Subramanian · R. Vijayakumar
Department of Mechanical Engineering, SNS College of Technology, Coimbatore, Tamil Nadu, India

S. Vijayan et al. (eds.), *Trends in Manufacturing and Engineering Management*, Lecture Notes in Mechanical Engineering,
https://doi.org/10.1007/978-981-15-4745-4_62

1 Introduction

With the expanded utilization of mechanized welding, the determination of welding process parameters and welding methodology must be increasingly explicit to guarantee that the weld bead parameters of good quality are gotten at least expense with high repeatability. GTAW quality is emphatically portrayed by the weld bead geometry. This is on the grounds that weld bead geometry assumes an imperative job in deciding the mechanical properties of the weld [1]. Subsequently, the input welding factors which impact the bead width should consequently be appropriately chosen to acquire a satisfactory bead width and henceforth a great joint. EL Rayes et al. [2] considered the impact of different crossbreed welding parameters on bead geometry. They conducted experiments on 316 L austenitic stainless steel work piece. They changed arc and laser capacity to contemplate their effect on different bead measurements. They found that arc control impacts bead width. Esme et al. [3] improved weld bead geometry in the TIG welding process utilizing Gray relational investigation and Taguchi technique. They researched the multi-response optimization of TIG welding process for an ideal parametric blend to yield positive bead geometry of welded joints utilizing the Taguchi technique. They directed sixteen exploratory runs dependent on an orthogonal array of Taguchi strategy to advance the target work for example bead width. Thao and Kim [4] built up a connection demonstrate for foreseeing weld bead geometry for lab joint in gas metal arc welding process. They led tests dependent on full factorial plan with two dimensions of five procedure parameters to get bead geometry utilizing a GMA procedure. They created linear, curvilinear and interaction models for the procedure and found that forecast capacities of interaction models are solid than the linear and curvilinear models. From the investigation, it was induced that welding voltage, arc current, welding velocity, and welding gun angle have vast noteworthy impacts on bead width. Siva et al. [5] advanced the weld bead parameters of nickel-based overlay saved by plasma exchanged arc surfacing. They led tests dependent on a five-factor, five-level central composite rotatable design and created numerical models utilizing the regression method. They found that bead width expanded when the welding current is expanded and diminished when the welding speed is expanded. Mostafa and Khajavi [6] optimized welding process parameters for friction controlled arc welding process. They built up a scientific model for foreseeing weld penetration as an element of welding process parameters. The outcomes demonstrated that weld penetration achieves its most extreme esteem when welding current, arc voltage, nozzle to plate distance and electrode to work angle are greatest and welding speed is least. Raja Dhas and Kumanan [7] proposed a clever strategy, versatile neuro fluffy derivation framework (ANFIS) to anticipate the weld bead width in the submerged arc welding process for a given arrangement of welding parameters. They structured the analyses utilizing Taguchi's standards and utilized the outcomes to build up various multiple regression models. Information got from the model was utilized to prepare the smart network. They found that the proposed ANFIS is adaptable, precise in anticipating bead width. Menaka et al. [8] assessed bead width amid welding by infrared thermal imaging. A ton of work has been completed to foresee weld bead width for gas metal arc welding, GTAW, Friction stir welding process for various evaluations of stainless steels yet no work has been accounted for anticipating bead width for 202-grade stainless steel which can be utilized as a substitute for 304-grade stainless steel in many applications like food processing equipment, kitchen hardware, sink and apparatus and medicinal instruments and so on. Increasingly over the collaboration impacts of procedure parameters on bead width are not intricately talked about. Thus an endeavor has been made to connect GTAW process parameters, for example, welding current (I), welding speed (V), gas flow rate (Q) and welding gun angle (θ) with bead width. A factually planned examination dependent on central composite rotatable design was utilized for the improvement of numerical model [9]. The direct and intuitive impacts of procedure parameters on bead width were examined. The created model was helpful to decide quantitatively bead width of 202 stainless steel GTAW plates.

2 Experimental Procedure

The trials were structured and dependent on four-factor five-level central composite rotatable plans with full replication method [10]. These trials were directed according to the design matrix utilizing Lincoln V 350 PRO electric computerized welding machine. A servo motor driven controller was utilized to keep up uniform welding speed. The fundamental trial set up utilized comprised of a travelling carriage with a table for supporting the specimens. A power source was kept prepared. A welding gun was held stationary in an edge over the table and was given a connection for setting the expected nozzle to plate distance and welding gun angle individually. The nozzle to plate distance was kept consistent at 2.5 mm all through the experimentation procedure. A high recurrence connection was utilized to produce the arc at this separation. Test plates of size (100 mm × 30 mm × 5 mm) were cut from 202 stainless steel plates and one surface was cleaned to evacuate oxide scale and earth before welding. The concoction synthesis of AISI stainless steel plate is given in Table 1.

Argon gas with flow rates between 5 and 25 l/min was utilized for protecting the weld bead. The reason for utilizing the protecting gas was to shield the weld region from climatic gases, for example, oxygen, nitrogen, carbon dioxide and water vapor. The welding machine and controller utilized for leading the trials is shown in Figs. 1 and 2.

Table 1 Chemical composition of stainless steel 202 grade

SAE designation	%C	%Mn	%Si	%Cr	%Ni	%P	%S	%N	%Fe
202	0.15	9.25	0.49	17.1	4.1	0.06	0.03	0.25	70.01

Fig. 1 Gas tungsten arc welding machine

Fig. 2 Manipulator used for welding

3 Plan of Investigation

3.1 Identification of Procedure Parameters

The autonomously controllable procedure parameters impacting profundity of bead width were recognized to empower the completing of exploratory work and building up the scientific model. These are welding current (*I*), welding speed (*V*), shielding gas flow rate (*Q*), and welding gun angle (θ).

Table 2 Welding parameters and their levels

Parameter	Unit and notation	Levels				
		−2	−1	0	1	2
Welding current	A (*I*)	70	80	90	100	110
Welding speed	mm/min (*V*)	170	180	190	200	210
Shielding gas flow rate	l/min (*Q*)	5	10	15	20	25
Welding gun angle	Degrees (θ)	50	60	70	80	90

3.2 Finding the Limits of the Procedure Parameters

The working extents of each picked factor were settled by coordinating fundamental runs. This was finished by contrasting one of the components while keeping whatever is left of them as relentless characteristics. The working extent of every method parameter was constrained by exploring the dot for a smooth appearance with no conspicuous distortions, for instance, surface porosity, undercut, etc. The maximum furthest reaches of a given factor were coded as (+2) and lower limit was coded as (−2). The coded qualities for centre qualities were resolved using the Eq. (1) [11]

$$X_i = \frac{2(2X - (X_{\max} - X_{\min}))}{X_{\max} - X_{\min}} \tag{1}$$

where X_i is the required coded estimation of a variable X and is any estimation of the variable from $X_{\min}$ to $X_{\max}$. The chosen procedure parameters with their points of confinement and documentation are given in Table 2.

3.3 Development of the Design Matrix

The plan framework picked to lead the investigations was a five-level, four-factor central composite rotatable design comprising of 31 sets of coded conditions and involving a half replication 24 = 16 factorial structure in addition to eight star focuses and seven centre focuses. All welding factors at the middle level (0) comprise the inside focuses while the mix of each welding factors at either its lower level (−2) or its more elevated amount (+2) with the other two factors at the moderate level establish the star focuses. Subsequently, the 31 test runs permit the estimation of straight, quadratic and two way intuitive impacts of the procedure factors on bead width [12]. Examinations were led at random to maintain a strategic distance from schematic errors crawling into the exploratory methodology.

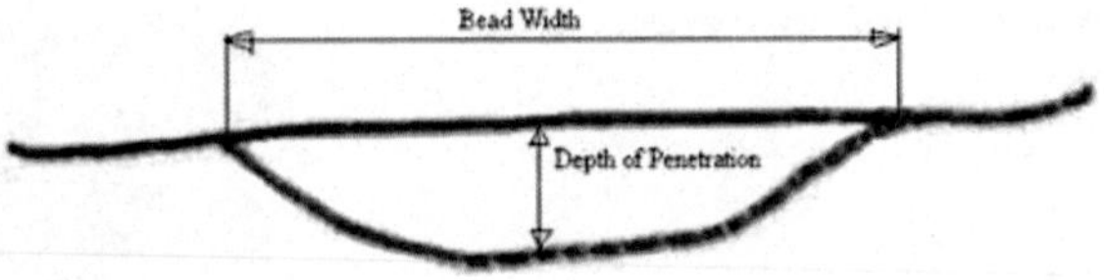

Fig. 3 Bead profile of the welded specimen

3.4 Recording the Responses

To measure the bead width the following processes were carried out on the specimens (1) Segmenting, (2) Granulating, (3) Polishing, (4) Etching and (5) Profile tracing.

Segmenting: The transverse areas of each weld were cut utilizing band saw and care was taken to keep away from twisting of the delicate austenitic stainless steel.
Granulating: It was performed so as to expel the cold work from cutting. The granulating was done at rates of roughly 300 RPM.
Polishing: The specimens were then harsh cleaned by hand. So as to acquire better edge evenness, the specimens were polished utilizing silicon carbide grating papers of evaluations 100, 220, 400, 600 and 800 individually. The specimens were then cleaned utilizing a rough—slurry of alumina (Al_2O_3) and water (H_2O) on a polishing machine.
Etching: After cleaning, etching was done on the specimens. Etching was important for analyzing the microstructure of the weld bead. The etchant utilized was Marble's reagent which is a blend of HCl (50 ml), $CuSO_4$ and H_2O (50 ml). The polished countenances of every specimen were swabbed utilizing the etchant for around 50–60 s to uncover the weld bead.
Profile Tracing: The bead profiles of the specimens were followed utilizing a reflective type optical profile projector.

The followed bead profiles were examined so as to decide the bead width. The bead width was estimated with the assistance of AUTOCAD. The followed bead profile is shown in Fig. 3. The welded specimens are shown in Fig. 4.

The design matrix and measured value of bead width are shown in Table 3.

The welding conditions continue as before for trial keeps running from 25 to 31 yet the reaction shifts marginally. This is because of the impact of obscure and eccentric factors called noise factors that creep into the tests. To represent the effect of these obscure factors on the reaction rehashed runs were incorporated into the design.

3.5 Recording the Responses

The development of a numerical model was done dependent on the regression technique to foresee the bead width. The response surface capacity speaking to bead width can be communicated as $\alpha = f\ (I,\ V,\ Q,\ \theta)$ and the relationship chose is a

Fig. 4 Welded specimens

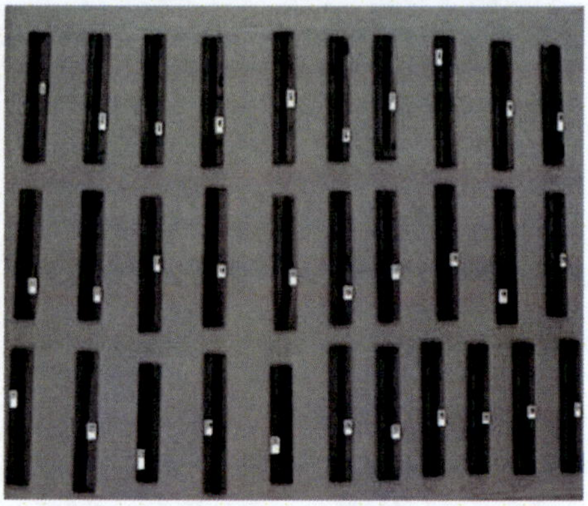

second request reaction surface for k factors is given by Eq. (2) [9]

$$Y = b_0 + \sum_{i=1}^{k} b_i X_i + \sum_{\substack{i,j=1 \\ i \neq j}}^{k} b_{ij} X_i X_j + \sum_{i=1}^{k} b_{ii} X_i^2 \tag{2}$$

Factual programming bundle (Minitab Release 14) [11] was utilized to figure the estimations of these coefficients. An underlying numerical model was created utilizing the coefficients got from the above conditions.

3.6 Testing the Coefficients for Significance

The estimation of the regression coefficients gives a thought about what degree the control parameters influence the response quantitatively. The less critical coefficients are dispensed with alongside the responses with which they are related without giving up a significant part of the exactness. This is finished by utilizing a student's t-test [13]. As per this test, when the determined value of 't' comparing to the coefficient surpasses the standard classified an incentive for the likelihood rule kept at 0.75, the coefficient ends up noteworthy. The final numerical model was created utilizing just the noteworthy coefficients. The final scientific model as controlled by the above examination is given by Eq. (3).

$$\begin{aligned} W = {} & 3.612 + 0.384I - 0.150V - 0.211Q + 0.416\theta - 0.191Q^2 \\ & + 0.098IV - 0.157IQ - 0.056VQ + 0.182V\theta \end{aligned} \tag{3}$$

The square numerous estimations of 'R' of the full model and the diminished model are exhibited in Table 4. It is apparent from the table that the full model has the higher estimation of balanced square multiple 'R' than the diminished model and

Table 3 Design matrix and responses

S. No.	Welding process parameters in coded values				Bead width (mm)
	I	*V*	*Q*	$\theta°$	
01	−1	−1	−1	−1	3.0268
02	1	−1	−1	−1	3.7861
03	−1	1	−1	−1	2.3194
04	1	1	−1	−1	3.5200
05	−1	−1	1	−1	3.0144
06	1	−1	1	−1	3.3449
07	−1	1	1	−1	2.0593
08	1	1	1	−1	2.9645
09	−1	−1	−1	1	3.5187
10	1	−1	−1	1	4.3151
11	−1	1	−1	1	3.3384
12	1	1	−1	1	4.7879
13	−1	−1	1	1	3.4434
14	1	−1	1	1	3.7218
15	−1	1	1	1	3.3701
16	1	1	1	1	3.8312
17	−2	0	0	0	2.7917
18	2	0	0	0	4.4517
19	0	−2	0	0	4.0100
20	0	2	0	0	3.3366
21	0	0	−2	0	3.4124
22	0	0	2	0	2.4511
23	0	0	0	−2	2.9074
24	0	0	0	2	4.6083
25	0	0	0	0	3.6754
26	0	0	0	0	3.7935
27	0	0	0	0	3.4325
28	0	0	0	0	3.4425
29	0	0	0	0	3.4788
30	0	0	0	0	3.7886
31	0	0	0	0	3.7345

Table 4 Comparison of square multiple '*R*' values and standard error of estimate for full and reduced models

Response	Adjusted square multiple *R*		Standard error estimate	
	Full model	Reduced model	Full model	Reduced model
Bead width	0.950	0.959	0.140	0.126

the diminished model has lesser estimations of the standard error of estimate than that of the individual full model. Thus the diminished model is superior to the full model.

3.7 *Checking the Adequacy of the Model*

The adequacy of the model was tried utilizing the Analysis of Variance Technique (ANOVA). According to this procedure [14], the determined estimation of the *F*—the proportion of the model created ought not to surpass the standard estimation of *F*-proportion for an ideal dimension of certainty, i.e., 95% and the determined estimation of *R*-proportion of the model created ought to surpass the standard organized estimation of the *R*-proportion for a similar certainty level. In the event that these conditions are satisfied, the model is viewed as sufficient. The aftereffects of ANOVA are exhibited in Table 5. It is obvious from the table that the model is sufficient.

4 Results and Discussion

The scientific model created can be utilized to foresee bead width by substituting the estimations of the individual procedure parameters. The impact of the procedure parameters on the bead width was concentrated utilizing the created model. The direct impact of procedure parameters was contemplated utilizing the created model. The direct impact was concentrated by keeping all the procedure parameters at the middle level with the exception of the parameter whose direct impact was examined. The interaction of the parameters is considered by keeping every one of the parameters at the middle level aside from the parameters whose interaction impacts are contemplated. The direct impact of the considerable number of parameters and the interaction impacts of welding process parameters which have solid interaction on bead width are examined underneath.

4.1 *Direct Effect of Welding Current and Welding Speed on Bead Width*

Figure 5 demonstrates the direct impact of welding current on bead width. It is obvious from the figure that bead width increments essentially as the welding current increments. This is because of the way that heat input increments as the welding current increments. The expansion in heat input outcomes in preheating of the work piece amid forwarding welding. Subsequently, additionally liquefying of the base metal happens. This builds the spreading of warmth which results in an increment of

Table 5 Results of ANOVA analysis.doc

Parameter	First-order term		Second-order term		Lack of fit		Error term		*F* ratio	*R* ratio	Remarks
	SS	DOF	SS	DOF	SS	DOF	SS	DOF			
Bead width	9.298	4	2.201	10	0.175	15	0.161	16	0.433	47.525	Model is adequate

SS Sum of Squares, *DOF* Degrees of Freedom
Mean Sum of Squares = Sum of Square Terms/DOF
F ratio = Ms of Lack of Fit/Ms of Error Terms
R ratio = Ms of First-Order Term and Second-Order Term/Ms of Error Term
F ratio (15, 6, 0.05) = 0.433
R ratio (9, 6, 0.05) = 4.10

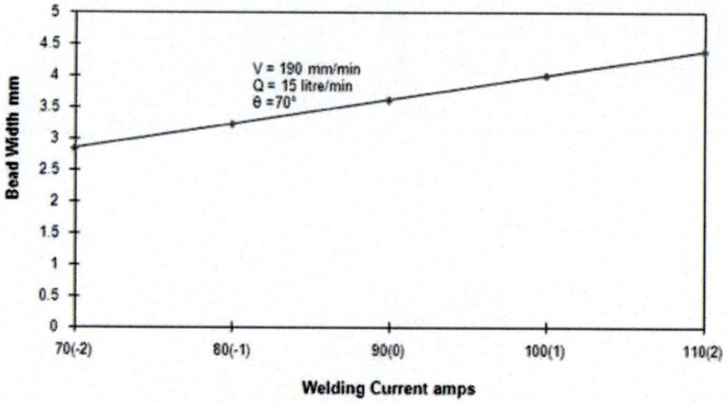

Fig. 5 Direct effect of welding current on bead width

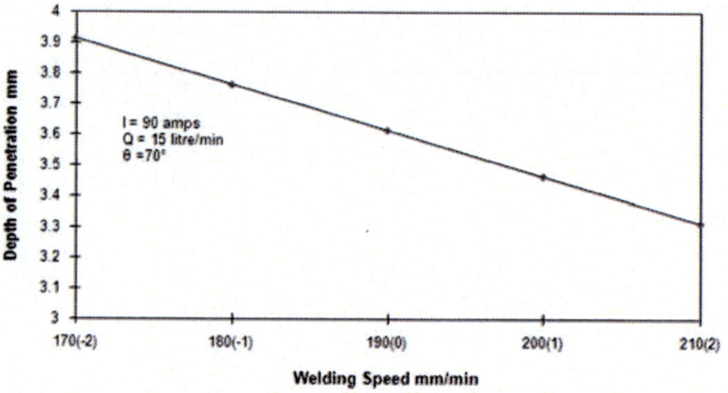

Fig. 6 Direct effect of welding speed on bead width

bead width. Figure 6 demonstrates the direct impact of welding speed on bead width. Welding speed is one of the primary factors that control heat info and bead width. Bead width and measurements of the warmth influenced zone diminish with the expansion in welding speed. This is because of the way that heat input is contrarily corresponding to welding speed. Henceforth, there is an abatement in bead width as the welding speed increments.

4.2 *Direct Effect of Shielding Gas Flow Rate and Welding Gun Angle on Bead Width*

Figure 7 demonstrates the direct impact of shielding gas flow rate on bead width. At the point when the gas flow rate is changed from lower level to more elevated amount there is a peripheral increment in bead width up to (−1) level and afterward

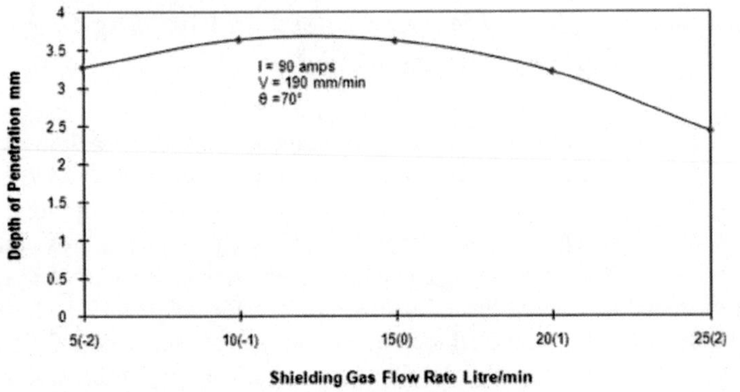

Fig. 7 Direct effect of shielding gas flow rate on bead width

there is a diminishing pattern up to (+2) level. This is because of the way that at higher gas flow rates more warmth is diverted by the gas. These outcomes in less bead width and diminishing in measurements of the warmth influenced zone. Thus there is an abatement in bead width for comparing increment in shielding gas flow rate. Figure 8 demonstrates the direct impact of welding gun angle on bead width. From the figure, bead width increments as the welding gun angle increments. This is because of the way that at higher gun angles the preheating of the base metal is high and the base metal gets more introduction to the arc. Thus there is an expansion in bead width for comparing increment in gun angle.

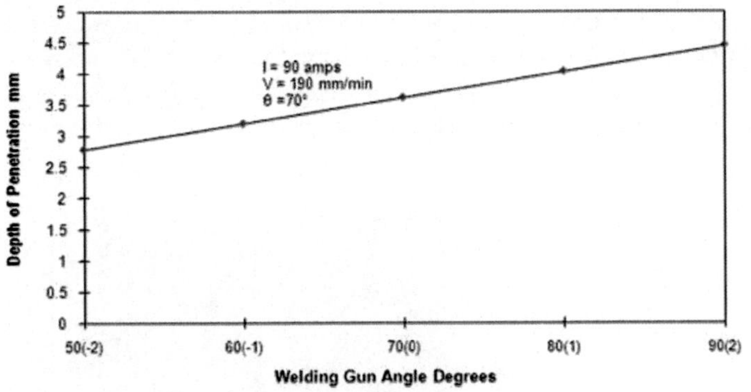

Fig. 8 Direct effect of welding gun angle on bead width

4.3 Interactive Effect of Welding Speed and Welding Current, Welding Gun Angle on Bead Width

Figure 9 speaks to the contour diagram for interactive effect of welding speed and welding current on bead width. From the contour plot, when the welding speed is at its lower limit for example (−2) level, bead width is in the scope of 3–3.5 mm for increment in welding current from (−2) to (0) level. It increments and is in the scope of 3.5–4 mm for the welding current from (0) to (+2) level. At the point when the welding speed is at (−1) level, bead width is in the scope of 2.5–3 mm for the welding current up to (−1) level. It increments and is in the scope of 3–3.5 mm for the welding current up to (+1) level. It is in the scope of 3.5–4 mm for the welding current at (+2) level. For the following dimension of welding speed for example (0) level, bead width is in the scope of 2–2.5 mm for the welding current up to (0) level. It is in the scope of 3–3.5 mm for the welding current at (+1) level and increments to 4 mm for the welding current at (+2) level. At the point when the welding speed is at (+1) level, bead width is under 2 mm for the welding current at (−2) level. The bead width builds like (0) level for different dimensions of welding current. A similar pattern is watched for welding speed at (+2) level. The expansion in bead width for increment in welding current for all dimensions of welding speed is because of the beneficial outcome of welding current which is more noteworthy than the negative impact of welding speed. Figure 10 demonstrates the contour plot for interactive of welding speed and welding gun angle on bead width. From the contour plot, when the welding gun angle is at (−2) level, bead width is in the scope of 4–4.5 mm for all dimensions of welding speed. At the point when the welding gun angle is at (−1) level, bead width is in the scope of 3.5–4 mm for the welding speed up to (−1) level. It at that point diminishes and is in the scope of 3–3.5 mm for the welding speed at (0) level. It further reduces and is in the scope of 2.5–3 mm for the welding speed at (+1) level. It tumbles to under 2.5 mm for the welding speed at (+2) level. At the point when the welding gun angle is at (0) level, a similar pattern is watched for bead width for

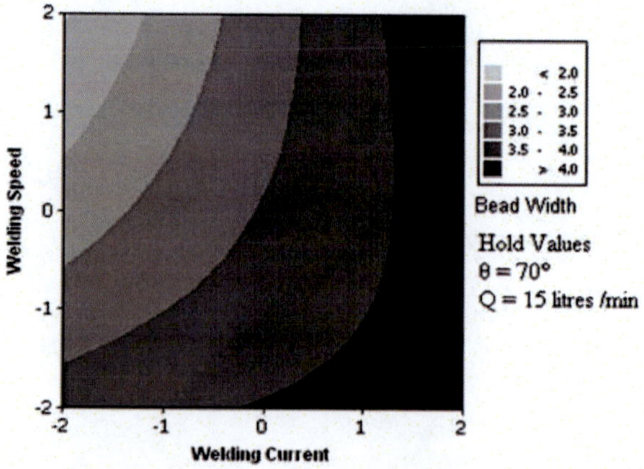

Fig. 9 Interactive effect of welding speed and welding current for bead width

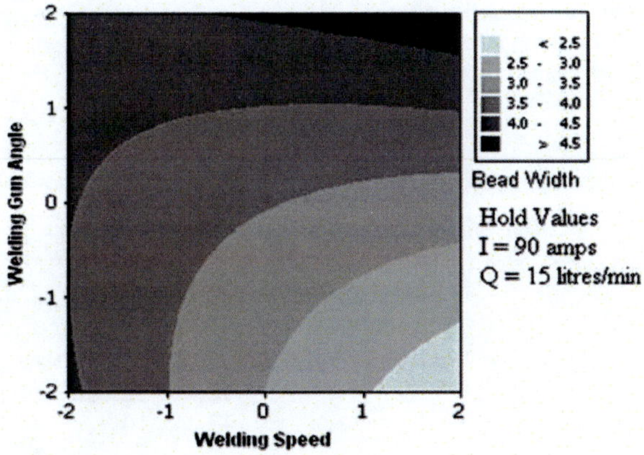

Fig. 10 Interactive effect of welding speed and welding gun angle for bead width

all dimensions of welding speed. For the following dimension of welding gun angle for example (+1) level, bead width is in the scope of 4–4.5 mm for all dimensions of welding speed. At the point when the welding gun angle is at its maximum farthest point for example (+2) level, bead width is in the scope of 4–4.5 mm for the welding speed up to (0) level. It at that point insignificantly increments to 4.5 mm for the other two dimensions of welding speed. From the above investigation, it tends to be derived that bead width increments for increment in welding gun angle for all dimensions of welding speed. This is because of the beneficial outcome of welding gun angle which is more critical than the negative effect of welding speed.

4.4 *Interactive Effect of Welding Current, Welding Speed and Shielding Gas Flow Rate on Bead Width*

Figure 11 speaks to the shape outline for the interactive effect of welding current and shielding gas flow rate on bead width. From the contour plot, when the shielding gas flow rate is at (−2) level and welding current is differed from (−2) to (+2) level bead width increments from 1.5 to 5 mm. At the point when the shielding gas flow rate is at (−1) level, bead width increments from 2 to 5 mm for increment in welding current from (−2) to (+2) level. A similar pattern is watched for gas flow rate at (0) and (+1) level. At the point when the shielding gas flow rate is expanded to its furthest restrict for example (+2) level, bead width is in the scope of 2–3 mm for increment in welding current from (−2) to (+2) level. From the above investigations, it may be surmised that bead width increments when the welding current increments for all dimensions of gas flow rate. This is because of the way that the beneficial outcome of welding current is more critical than the negative effect of shielding gas flow rate on bead width. Figure 12 demonstrates the contour plot for interactive effect of shielding gas flow rate and welding speed on bead width. As appeared in the contour plot, when the shielding gas flow rate is at (−2) level bead width is in the scope of 2–2.5 mm for

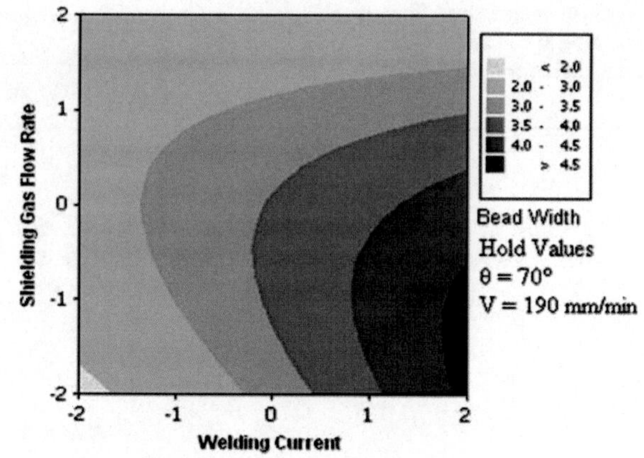

Fig. 11 Interactive effect of welding current and shielding gas flow rate for bead width

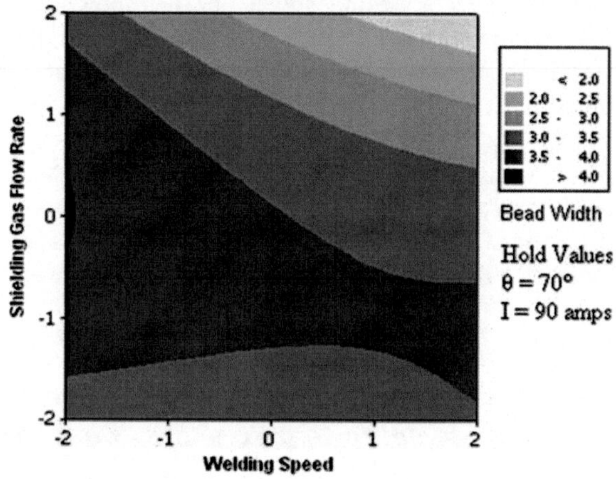

Fig. 12 Interactive effect of welding speed and shielding gas flow rate for bead width

increment in welding speed from (−2) to (+2) level. At the point when the shielding gas flow rate is at (−1) level, bead width is in the scope of 3.5–4 mm for increment in welding speed from (−2) to (+2) level. At the point when the shielding gas flow rate is at (0) level bead width is more noteworthy than 4 mm for welding speed at (−2) level. It at that point diminishes and is in the scope of 3.5–4 mm for different dimensions of welding speed. For the following dimension of shielding gas flow rate for example (+1) level, bead width is in the scope of 3.5–4 mm for the welding speed up to (+1) level. It at that point diminishes and is in the scope of 3–3.5 mm for the welding speed up to (0) level. It further declines and is in the scope of 2.5–3 mm for the welding speed up to (+1) level. It is in the scope of 2–2.5 mm for the welding speed at (+2) level. At the point when the shielding gas flow rate is at its maximum breaking point for example (+2) level, bead width is in the scope of 3–3.5 mm for the welding speed at (−2) level. It is in the scope of 2.5–3 mm for the welding speed at (−1) level. It further declines and is in the scope of 2–2.5 mm for the welding speed at (0) and (+1) levels. It goes underneath 2 mm for the welding speed at (+2) level. From the above investigation, it very well may be induced that bead width diminishes when the welding speed increments for all dimensions of gas flow rate. This is because of the way that both welding speed and shielding gas flow rate have a negative effect on bead width.

5 Conclusions

The accompanying ends were touched base at from the present examination.

1. The regression display created in this work from the experimental information can be utilized to control the procedure so as to accomplish the ideal weld quality in butt-welded plates.
2. The percentage of error got for profundity of entrance among predicted and experimental esteem falls inside the breaking point of 95% certainty level. Subsequently, the created regression display is fit for making the forecast of bead width with sensible exactness.
3. Central composite design can be advantageously used to examine the immediate and intuitive impacts of various blends of procedure parameters inside the scope of the examination.
4. Out of four procedure parameters chose for examination, welding current and welding gun angle has constructive outcome on bead width where as welding speed and shielding gas flow rate has negative impact on bead width.
5. Out of the distinctive blends of procedure parameters the following combinations, i.e., welding current and welding speed, welding current and shielding gas flow rate, welding speed and shielding gas flow rate and welding speed and welding gun angle have solid interactions on bead width.
6. The greatest bead width acquired from experimental investigations was 4.78 mm when the procedure parameters, for example, welding current, welding speed, and welding gun angle were at 100 A, 200 mm/min and 80° individually and shielding gas flow rate was at 10 l/min.
7. The least bead width acquired from experimental investigations was 2.06 mm when welding current and welding gun angle were at 80 A and 60° separately whereas welding speed and shielding gas flow rate were at 200 mm/min and 20 l/min individually.

References

1. Giridharan PK, Murugan N (2009) Optimization of pulsed GTA welding process parameters for the welding of AISI 304L stainless steel sheets. Int J Adv Manuf Technol 40(5–6):478–489
2. EL Rayes M, Walz C, Sepold G (2004) The influence of various hybrid welding parameters on bead geometry. Weld J 147–153

3. Greyjevo OGTZ, Esme U, Bayramoglu M, Kazancoglu Y, Ozgun S (2009) Optimization of weld bead geometry in TIG welding process using grey relation analysis and Taguchi method. Mater Technol 43(3):143–149
4. Thao DT, Kim IS (2009) Interaction model for predicting bead geometry for lab joint in GMA welding process. Comput Mater Sci Surf Eng 1(4):237–244
5. Siva K, Murugan N, Raghupathy V (2009) Modelling, analysis and optimisation of weld bead parameters of nickel based overlay deposited by plasma transferred arc surfacing. Surf Eng 1:174–182
6. Mostafa NB, Khajavi MN (2006) Optimisation of welding parameters for weld penetration in FCAW. J Achievements Mater Manuf Eng 16(1):13–17
7. Raja Dhas JE, Kumanan S (2007) ANFIS for prediction of weld bead width in a submerged arc welding process. J Sci Ind Res (India) 66(4):335–338
8. Menaka M, Vasudevan M, Venkatraman B, Raj B (2005) Estimating bead width and depth of penetration during welding by infrared thermal imaging. Insight Non-Destr Test Cond Monit 47(9):564–568
9. Cochran WG, Cox GM (1963) Experimental designs. Asia Publishing House, India
10. Vel Murugan V, Gunaraj V (2005) Effects of process parameters on angular distortion of gas metal arc welded structural steel plates. Weld J 84(11):165–171
11. Mini tab release 14, Minitab Inc (2004)
12. Montgomery DC (2001) Design and analysis of experiments. Wiley, New York
13. Yang LJ, Bibby MJ, Chandel RS (1993) Linear regression equations for modeling the submerged-arc welding process. J Mater Process Tech 39(1–2):33–42
14. Murugan N, Gunaraj V (2000) Prediction and optimization of weld bead volume for the submerged arc process—part 1. Weld Res 331–338

Feasibility Study of TIG Welding of AA6063-AA7075 Alloys

D. Nathan, S. Ashwin Kannan, and P. Krishna Kumar

Abstract The aim of this paper is to study the feasibility of welding dissimilar aluminum alloys AA6063 and AA7075 using destructive and nondestructive testing. Dissimilar Aluminum alloys AA6063 and AA7075 plates were butt-jointed using the tungsten inert gas (TIG) welding method. The feasibility of welding of dissimilar aluminum alloy can be confirmed by both strength calculations as well as identification of welding defects. The feasibility of welding of dissimilar aluminum alloy joint was studied through visual appearance, microstructures, tensile strength, hardness, liquid penetrant test and ultrasound test. Microstructures at different zones of dissimilar TIG joints were identified. The average tensile strength and hardness values were calculated. Using the liquid penetrant test and ultrasound test, welding defects were identified. It can be concluded that TIG welding of a dissimilar aluminum alloy is feasible.

Keywords Aluminum alloys · TIG welding · Nondestructive testing · Liquid penetrant test

1 Introduction

Automotive manufacturing environments have been focusing on light metal such as alloy without compromising its functionality. Aluminum alloys are widely used in the industry due to their promising properties like high strength to weight ratio, high thermal conductivity, low density and high corrosion resistance. Alloys have different varieties. Out of these varieties, the 6xxx series aluminum alloys are heat treatable, have high strength and excellent in corrosion resistance. Structural members and architectural elements where strength and stiffness are significant employ 6xxx series alloys due to its high extrudability. Other applications of 6xxx series include

D. Nathan (✉)
Dr. Mahalingam College of Engineering and Technology, Pollachi, India
e-mail: nathan14576@gmail.com

S. Ashwin Kannan · P. Krishna Kumar
St. Joseph's Institute of Technology, Chennai 600119, India

S. Vijayan et al. (eds.), *Trends in Manufacturing and Engineering Management*, Lecture Notes in Mechanical Engineering,
https://doi.org/10.1007/978-981-15-4745-4_63

weld structural members such as truck and marine frames, pipelines and railroad cars. Besides, the 7xxx series alloys are also heat treatable with the highest strength of all aluminum alloys. AA7075 is also wide applications in aircraft as well as in automotive industries [1]. The joining of dissimilar alloys becomes a prime demand in the automotive and aircraft industries for products with improved quality in terms of strength and ductility but light in weight. In order to produce vehicles with economical fuel consumption and manage pollution, many parts are manufactured by combining alloys that have lightweight and good corrosion resistance such as aluminum alloys [2].

Tungsten Inert Gas (TIG), also known as Gas Tungsten Arc Welding (GTAW) is the preferred welding process of alloys because of the comparatively easier applicability and superior economy [3]. The TIG welding process is also suitable to weld thin plates, resulting in high-quality and defect-free welded joints [4]. The arc is established between the tip of a non-consumable tungsten electrode and the work piece with a shielding gas applied to protect oxidation of the weld pool area [5].

In the TIG welding, problems may occur during the welding process where the welds may fail in the soft and over age heat-affected zone [6]. The limitations exist to join dissimilar aluminum alloys that are mainly related to the presence of an oxide layer. This oxide layer weakens the grain boundary cohesion and consequently the weld becomes prone to inter-granular cracking experienced during welding-pool solidification. Also, thermal conductivity and composition with alloying elements are one of the major causes these materials are difficult to join [7, 8]. Hence, feasibility of welding dissimilar alloys needs to be studied deeply before applied to the fabrication of engineering products.

In this study, the weldability of dissimilar aluminum alloys 6061 and 7075 was studied using both destructive and nondestructive testing. Visual inspections, microstructure identification, tension test and hardness test are conducted to investigate the effect of the welding process.

2 Materials and Methods

2.1 Materials

AA6063 and AA7075 plates of dimensions 200 mm × 50 mm × 10 mm were used in this study. The chemical compositions and mechanical properties are tabulated in Tables 1 and 2, respectively. Alloy plates are prepared by polishing with 180 grit emery paper at the faying surface and cleaned with ethanol. Lincoln ER5356 aluminum welding wire is used as filler material. After surface preparation, manual TIG welding of the dissimilar alloy was done. Welding parameters which are tungsten electrode diameter of 3.2 mm, 60° electrode tip angle and argon shielding gas were

Table 1 Chemical compositions of AA6063 and AA7075 by weight%

Grade	Al	Cr	Cu	Fe	Mg	Mn	Si	Ti	Zn	Other, each
AA6063	Max 97.5%	Max 0.1%	Max 0.1%	Max 0.35%	0.45–0.9%	Max 0.1%	0.2–0.6%	Max 0.1%	Max 0.1%	Max 0.05%
AA7075	87.1–91.4%	0.18–0.28%	1.2–2%	Max 0.5%	2.1–2.9%	Max 0.3%	Max 0.4%	Max 0.2%	5.1–6.1%	Max 0.05%

Table 2 Properties of AA6063 and AA7075

Grade	AA6063	AA7075
Physical properties		
Density ($\times 1000$ kg/m^3)	2.69	2.8
Mechanical properties		
Young's modulus (GPa)	68.9	70–80
Tensile strength (MPa)	241	230
Yield strength (MPa)	214	105
Elongation (%)	12%	17%
Poisson's ratio	0.3	0.33
Shear strength (MPa)	152	150
Fatigue strength (MPa)	68.9	160
Thermal properties		
Coefficient of thermal expansion ($\times 10^{-6}$/°C)	25.6	23.2
Thermal conductivity (W/mK)	201–218	130

Table 3 Welding parameters and its level

Parameters	Level 1	Level 2	Level 3
Welding current (A)	150	175	200
Welding voltage (V)	18	20	22
Gas flow rate (L/min)	5	7	9

fixed. The other parameters such as welding current welding voltage and gas flow rate are varied. The various welding parameters and its level are tabulated in Table 1 (Table 3).

2.2 *Experimentation*

Tungsten Inert Gas Welding is a multi-factor metal joining technique. The various process parameters include welding current, welding voltage, gas flow rate, electrode diameter, nozzle gap, etc. influence on weld bead geometry, weldment quality and also mechanical-metallurgical properties of the weldment. A limited number of experimental runs have been done in this study to use three levels of welding process parameters namely welding current, welding voltage, gas flow rate. Single side 60° V groove has been made before welding. Experiments have been carried out to create a single pass butt joint of dissimilar aluminum plates by varying the process parameters in the TIG welding. The tensile test specimens were prepared as per ASTM standards.

Table 4 Experimentation matrix

Exp. No.	Welding current (A)	Welding voltage (V)	Gas flow rate (L/min)
1	150	18	5
2	150	20	7
3	150	22	9
4	175	18	7
5	175	20	9
6	175	22	5
7	200	18	9
8	200	20	5
9	200	22	7

2.3 Orthogonal Array

Taguchi's orthogonal arrays are standard experimental design tables in order to perform only a small number of experimental trials to find the influence of the main factors on the response. The minimum number of experiments to be conducted is calculated from the formula given below

$$N = 1 + V(L - 1) \tag{1}$$

N Number of experiments needed
V Number of input parameters
L Number of levels of each parameter.

In this study,
$V = 3$ and $L = 3$, Hence
$N = 1 + 4(3 - 1) = 9.$

Taguchi experimental design of experiments suggests L_9 orthogonal array, where nine experiments are sufficient to optimize the parameters. Hence at least nine experiments are to be conducted. Based on this L_9 orthogonal array (OA) is to be selected which has at least nine rows, i.e., nine experimental runs (Table 4).

3 Results and Discussion

3.1 Tension Test Effect of Gas Flow Rate

Destructive testing was done on three specimens (specimen 2, specimen 4 and specimen 9). The results are as shown in Fig. 1. The result shows the specimen 2, welded

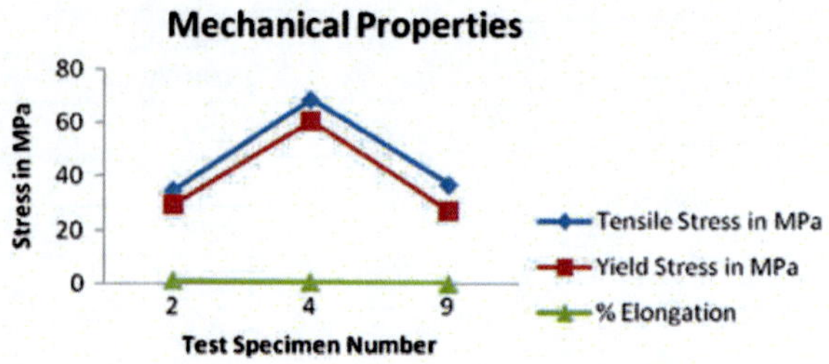

Fig. 1 Tensile test results

Table 5 Results of tensile test

Sample Id	2	4	9
Tensile strength in MPa	34.86	68.83	36.99
Yield stress in MPa	29.95	61.03	27.49
Elongation in %	1.50	0.90	0.24
Fracture location	At weld	At weld	At weld

at low welding current, low welding voltage and medium gas flow rate has a low tensile strength.

The result of the testing done on specimen 4 which is welded under medium welding current, medium welding voltage and low gas flow rate gives out high tensile strength. The result depicts the specimen nine welded low welding current, low welding voltage and medium gas flow rate brings very low tensile strength (Table 5).

3.2 Hardness Test

The Vickers Hardness test was done on all specimens under 10 kgf load with a dwell time of 10 s. Before doing hardness test the specimens were prepared by polishing with help of different grades of emery paper in order to get clear indentation marks. Hardness measurement was carried from either side of base metal towards the center of the weld. The results of specimens 2, 4, and 9 were plotted in a graph as shown in Fig. 2. The result shows the specimen 2, welded at low pulsed current, low pulsed voltage and medium gas flow rate has a high hardness structure in HV.

The result of the testing done on specimen 4, which is welded under medium pulsed current, medium pulsed voltage and low gas flow rate gives out medium Hardness structure in HV. The result depicts the specimen 9, welded high pulsed current, high pulsed voltage and high gas flow rate bring very low Hardness structure in HV.

Distribution of microhardness values was analyzed and depicted in Fig. 3. The graph shows that hardness value decreases at HAZ on both sides of welding due to the coarsening of grain structure. Hardness value has an increasing trend near the

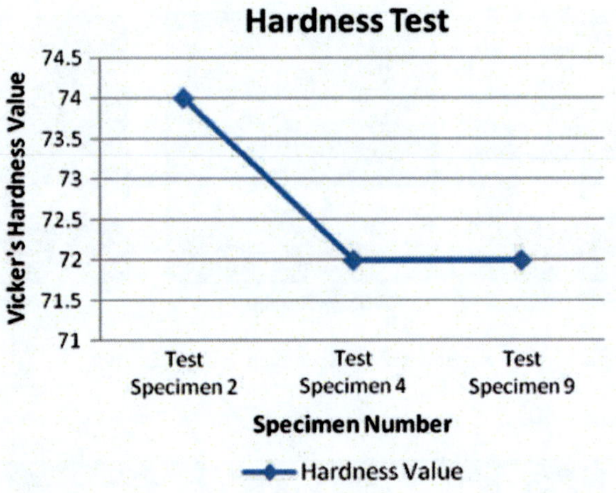

Fig. 2 Hardness test results

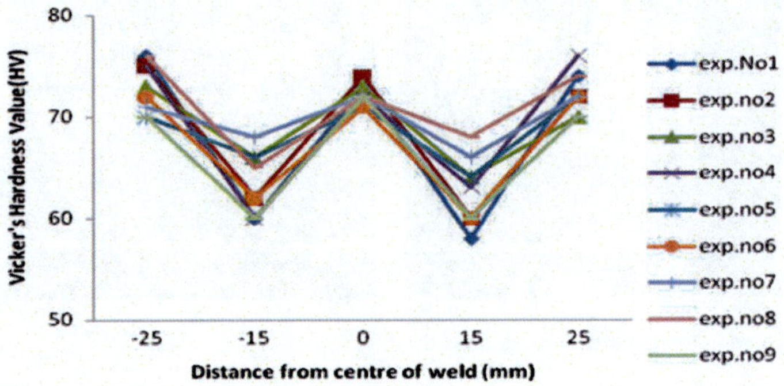

Fig. 3 Nugget profile of different experiments

center of the weld bead on both sides of the welded region. It is found to be AA 7075 side has higher hardness distribution compared to the AA 6063 side.

3.3 Microscopic Examinations

Microscopic examinations of TIG-welded plates with different welding parameters were studied. It is found the microstructure of the test specimen 9 shows a uniform grain structure with no distinct heat-affected zone. In specimen 4, the structure of the HAZ is coarse-grained. The presence of large grains due to the overheating zone causes weakening the strength properties of the joint. The microstructure of the test specimen 4 reveals distinct grain coarseness and heat-affected zone. In the

a) Specimen 2 b) Specimen 4 c) Specimen 9

Fig. 4 Microscopic view of **a** specimen 2, **b** specimen 4, **c** specimen 9

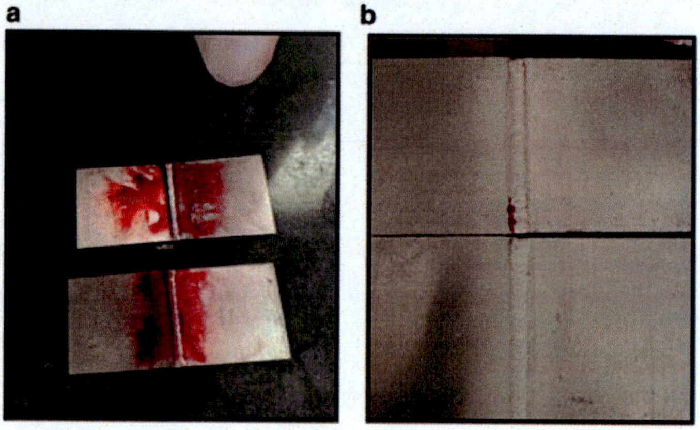

Fig. 5 **a** Liquid penetrant test, **b** crack detection

test specimen, 2 grain distortion was found clearly. A small amount of intermetallic phases in the form of fine precipitates is present in the grains and grain structure found to be coarse (Fig. 4).

3.4 Liquid Penetrant Test

Liquid penetrant test results are depicted in Fig. 6. A maximum flaw length of 10 mm was found in the test number 1 is an unacceptable value. Test number 3, 4, 5, 6, 7 and 8 have an average flaw length of 2.91 mm. Test number 2 and 9 have no defects (Fig. 5).

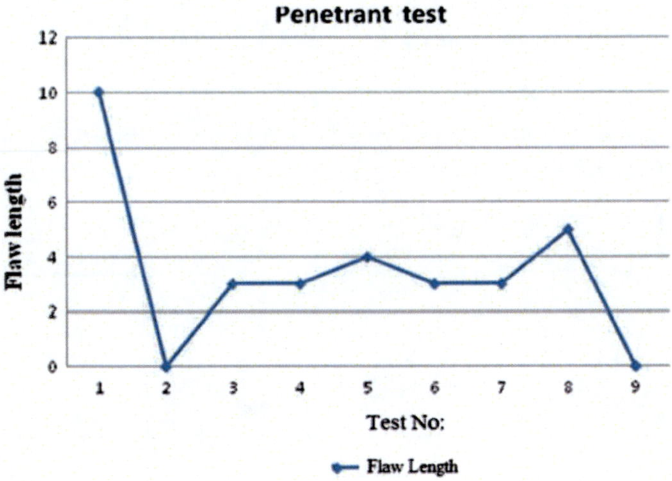

Fig. 6 Liquid penetrant test result

3.5 *Ultrasonic Test*

The calibrated ultrasound flaw detector can calculate the depth of the discontinuity and the thickness of the component accurately. Figure 7 shows the ultrasonic testing on TIG-welded specimen.

From Fig. 8, it is evident that the maximum flaw length of 3.8 mm occurs in test number 5 at the gas flow rate of 9 L/min and welding of 175 A. Test numbers 2, 8,

Fig. 7 Ultrsonic testing of welded specimen

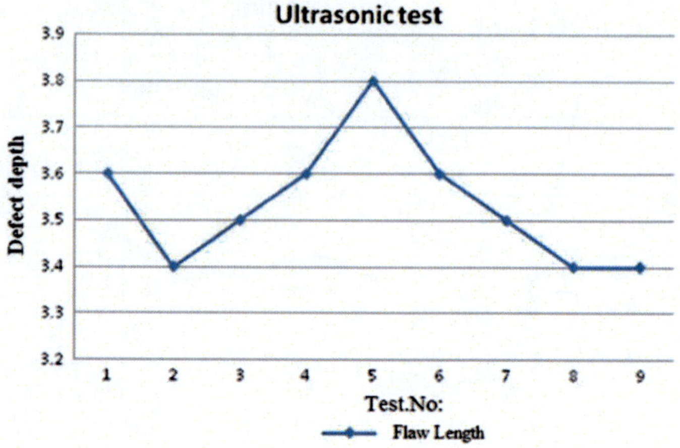

Fig. 8 Ultrsonic test result

and 9 show minimum flaw length of 3.4 mm. Average flaw length of 3.5 mm found in test 3 and 7. Hence, it is evident that all the test specimens have flaws in the weld.

3.6 *Effect of Gas Flow Rate*

From literature, gas flow rate in the TIG welding process of Aluminum alloy has a significant effect on weld quality. Shielding gas has a greater influence on weld strength, porosity, ductility, impact strength and resistance to corrosion. Argon is used as shielding gas in this work. Welding defects such as porosity, lack of penetration and fusion are taken care of by suitable shielding gas at a certain flow rate. Hence the gas flow rate is considered as one of the welding parameters in this study. Observations show a significant variation in tensile strength with respect to gas flow rate.

4 Conclusion

This investigation found the feasibility of various TIG welding by estimating mechanical properties and microscopic study of the weld. Taguchi's L_9 Orthogonal Array has been implemented in the experimentation. Tensile testing, hardness testing and microscopic examinations have been done. Liquid penetrant test and ultrasonic inspection were successfully conducted on welds produced at various parameter combinations. It can be concluded that based on the various parameters of welding the liquid penetrant testing and ultrasonic technique successfully detected the defects present in the welds and can be said to be appropriate in this regard. This study shows the idea to choose suitable welding parameters for the better welding of dissimilar alloy.

References

1. Fukuda T (2012) Weldability of 7000 series alloy materials. Weld Int 26:256–269
2. Hayat F (2012) Effect of aging treatment on the microstructure and mechanical properties of the similar and dissimilar 6061-T6/7075-T651 RSW joints. Mater Sci Eng A 556:834–843
3. Kumar TS, Balasubramanian V, Sanavullah MY (2007) Influences of pulsed current tungsten inert gas welding parameters on the tensile properties of AA 6061 alloy. Mater Des 28:2080–2092
4. Nascimento MP, Voorwald HJC, PayãoFilho JDC (2011) Fatigue strength of tungsten inert gas-repaired weld joints in airplane critical structures. J Mater Process Technol 211:1126–1135
5. American welding. Society 261–269 (2012)
6. Mutombo K, Du Toit M (2011) Corrosion fatigue behaviour of alloy 6061-T651 welded using fully automatic gas metal arc welding and ER5183 filler alloy. Int J Fatigue 33:1539–1547
7. Lakshminarayanan AK, Balasubramanian V, Elangovan K (2007) Effect of welding processes on tensile properties of AA6061 alloy joints. Int J Adv Manuf Technol 40:286–296
8. Song JL, Lin SB, Yang CL, Ma GC, Liu H (2009) Spreading behavior and microstructure characteristics of dissimilar metals TIG welding–brazing of aluminum alloy to stainless steel. Mater Sci Eng A 509:31–40

Effects of Processing Parameters on Temperature Distributions, Tensile Behaviour and Microstructure of Friction Stir Welding of Dissimilar Aluminium Alloys

Olatunji P. Abolusoro and Esther Titilayo Akinlabi

Abstract Research has shown that there is a correlation between process parameters, temperature variations, tensile strength, microstructures and the durability of the welds obtained in friction stir welding (FSW). The need to institute this correlation is crucial in order to achieve a weld free of defects and having sound mechanical behaviours for industrial applications. This work examines process parameters effects on variations of weld temperature, tensile strength and microstructures in dissimilar FSW of 6101-T6 and 7075-T651 aluminium alloys with a plate thickness of 6 mm. The welding was done with rotational speeds of 1250, 1550 and 1850 rpm and traverse speeds of 50 and 110 mm/min. The results obtained indicate that processing parameters significantly affect the temperature distributions in the weld. Increase in rotational speed increases the temperature but cause a decrease in tensile strength. While the increase in travel speed cause a reduction in temperature which results to increase in the tensile strength. Highest tensile strength of 143 MPa was obtained at 1250 rpm and 50 mm/min but better mixing of both materials was achieved at 1550 rpm and 50 mm/min.

Keywords Aluminium · Friction stir welding · Temperature · Processing parameters

1 Introduction

One of the metals that are broadly used in many engineering applications is aluminium. The extensive use of this alloy emanates from its properties. One of which is its high strength to weight ratio. FSW is a technology that has been in use for joining both similar and dissimilar aluminium alloys. The unique advantages of this

O. P. Abolusoro (✉) · E. T. Akinlabi
Department of Mechanical Engineering Science, University of Johannesburg, Johannesburg 2006, South Africa
e-mail: abolusoroolatunji@yahoo.com

E. T. Akinlabi
Department of Mechanical Engineering, Covenant University, Ota, Ogun State, Nigeria

S. Vijayan et al. (eds.), *Trends in Manufacturing and Engineering Management*, Lecture Notes in Mechanical Engineering,
https://doi.org/10.1007/978-981-15-4745-4_64

welding technology which include non-requirements of filler metals as the case in non-solid state welding technology and environmentally friendly have endeared it to its wide usage in industries for joining aluminium alloys [1–3]. There are parameters that govern the friction stir welding process. Two most prominent of these factors are the rotation and travel speed of the tool pin on the workpiece. These two parameters affect largely the heat generated during the welding and consequently affect temperature distributions, tensile and microstructures of the weld [4–7]. Aissani et al. [8] reported the influence of process parameters on tensile behaviour. Lombard et al. [9] optimized the rate of feed and rotation of tool effects on tensile strength and concluded that tool rotational speed impact more on tensile behaviour. Many other researchers have reported the relationship between these factors, tensile and microstructures [10–14] Also, the influence of these processing parameters on temperatures distributions have been examined by some researchers. Umasanka and Vijay studied the relationship between tool rotational speed, mechanical impact strength and temperature on AA6101-T6 and AA6351-T6 alloys welded using FSW. Their result indicates that temperature goes higher with an increase in rotational speed and that the impact strength depends on the pressure and the duration of impact [15]. Yoshima et al. welded dissimilar ADC12 and AA6061 alloy and examined the temperature distributions during the welding. The temperature measurements were done at various offset values. Their results indicate that temperature goes higher with an increase in tool offset and the temperature at any point depends on its distance from the welding tool which generates the heat [16]. Modelling of temperature distributions in FSW of one of the chosen alloys for this work (AA7075) has been attempted. Kandasamy et al. analyzed temperature changes in FSW of AA7075 aluminium alloy using finite element approach. The temperature measurement was done using k-type thermocouple on the workpiece and hand infrared thermometer on the tool. Their result demonstrated close agreement with the experimental values obtained [17]. Many researchers have utilized various approaches to investigate temperature changes during FSW. Some of these methods include thermal cameras, thermocouple inserted in the welding tool or in the workpiece [18–22]. Other methods like ultrasound and neutron-based techniques have also been used [23, 24]. However, the most commonly used method observed in literature is that of a thermocouple embedded in workpiece [15, 16, 25, 26]. This particular method was employed in this study. Thermocouple positions have also been reported to affect the accuracy of temperatures measured. While close proximity of the thermocouple to the rotating tool pin or the stir zone gives a better estimate of the weld temperature, the likely destruction of the thermocouple wires as the pin progresses limits the use of this method [15]. Hence in this study, the thermocouple was embedded in the workpiece in an area close to the heat-affected zone. Although numerous researches on FSW of dissimilar aluminium alloys have been carried out, however, welding parameter's relationship with microstructures, tensile and temperature variations especially on FSW of 7075-T651 and 6101-T6 have not been fully reported. This paper, therefore, presents this relationship for more understanding.

2 Experimental Procedures and Processing Parameters

2.1 Elemental Composition and Properties of the Sample

Dissimilar aluminium alloys 6101-T6 and 7075-T651 were used for the experiment. The elemental constituents of the two materials are indicated in Table 1 while Table 2 shows the mechanical properties.

2.2 Process Parameters

Three tool rotational speeds and two feed rates were engaged in this study as presented in Table 3.

2.3 Experimental Procedures

The tool for the experiment was fabricated using high carbon steel H13. It has shoulder diameter of 22 mm with a tapered pin profile of big diameter 7.3 mm, small diameter 5.5 mm and pin length of 5.6 mm as illustrated in Fig. 1. The two alloy plates of 7075-T651 and AA6101-T6 were cleaned and arranged in butt configuration with AA6101-T6 on the advancing side and 7075-T651 on the retreating side as illustrated in Fig. 2. The FSW was performed along the rolling direction of the two alloy samples with the tilt angle at 2^0. Two thermocouple wires were embedded in the workpiece 2 mm away from the welding zone on both sides of the plates and are placed 40 mm apart. The thermocouples were connected to a temperature data capturing machine for in-process temperature measurements. The thermocouple arrangement is shown in Fig. 3. Four different experiments were carried out using the parameters shown in Table 3. Three tensile samples cut according to ASTM E8 standard were tested for each weld and the average for each weld taken for evaluation. The microstuctural analysis was done on the nugget zone of each weld using an optical microscope.

Table 1 Elemental constituents of the two materials

Alloy	Si	Fe	Cu	Mn	Mg	Cr	Zn	Ti	Al
6101-T6	0.53	0.14	0.01	0.002	0.600	0.001	0.003	0.008	Others
7075-T651	0.40	0.50	1.70	0.300	2.40	0.22	5.50	0.20	Others

Table 2 Mechanical properties of the alloy

Alloy	Tensile strength (Mpa)	Ultimate tensile strength (Mpa)	Elongation (%)	Thermal conductivity (W/m °C)
6101-T6	172	180	15	167
7075-T651	462	575	11	130

Table 3 Processing parameters

Rotational speed (rpm)	Travel speed (mm/min)
1250	50
1550	50
1850	50
1850	110

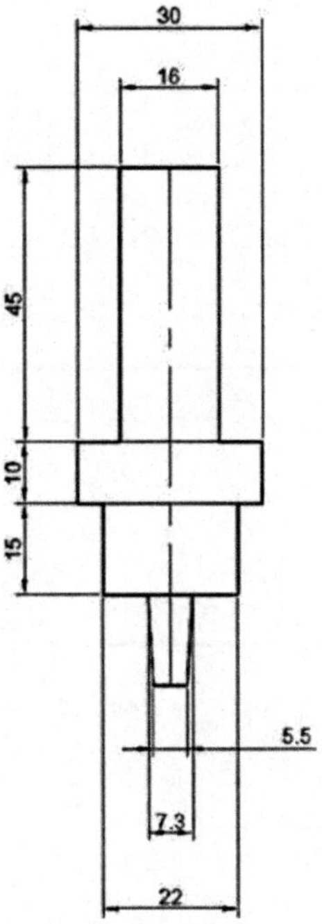

Fig. 1 Schematic diagram of the welding tool

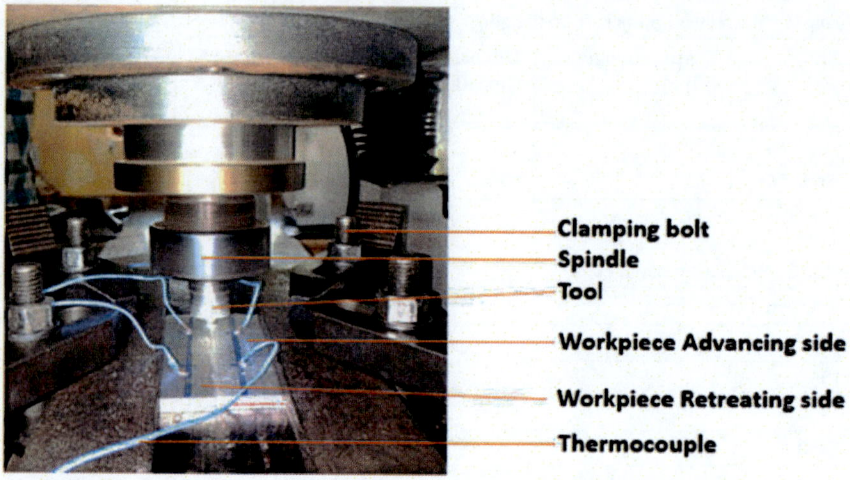

Fig. 2 Workpiece retreating side

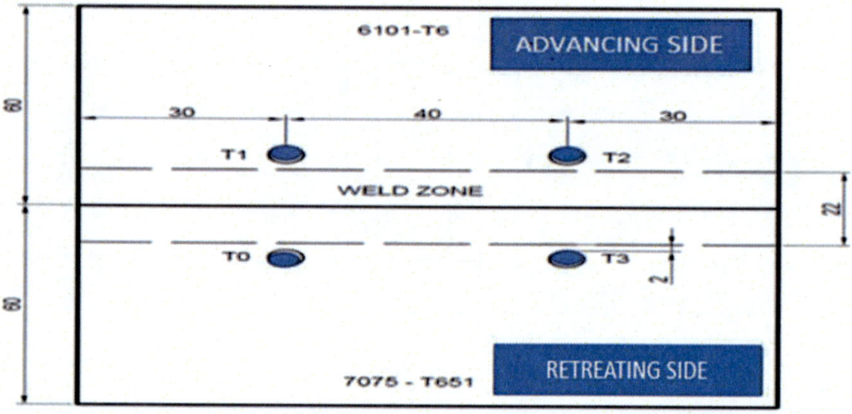

Fig. 3 Schematic diagram of the thermocouple arrangements T0, T1, T2, T3 on the alloy plates

3 Results and Discussion

3.1 Weld Appearance

The weld joints were carefully observed after the welding and were found to be externally free form defects although little flashes are noticed in some of the welds as shown in the photograph in Fig. 4.

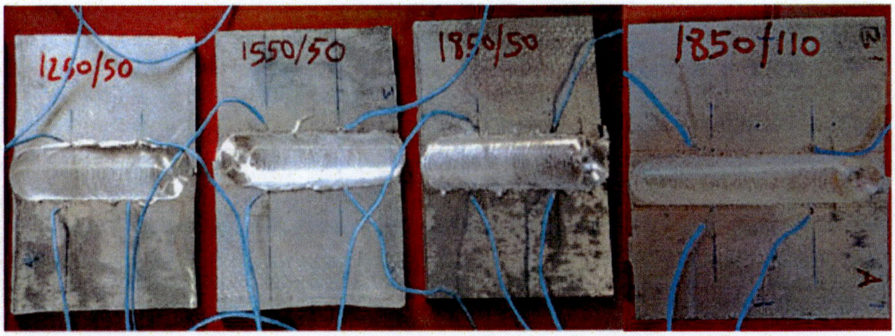

Fig. 4 Welded plates with embedded thermocouple

Fig. 5 The highest temperature profile of the processing parameters used

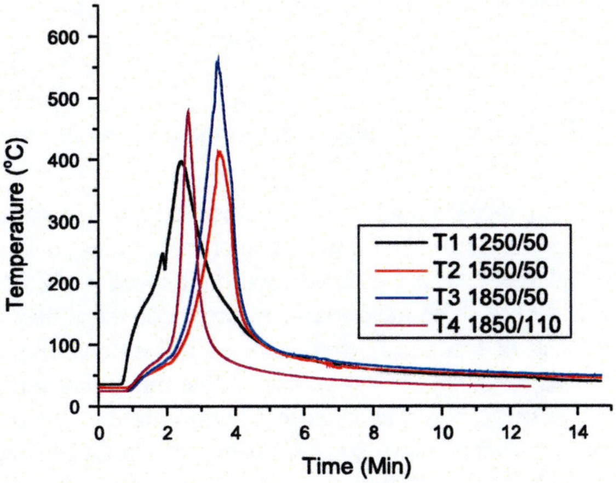

3.2 Temperature Profile of the Weld

Heat is produced as a consequence of the friction between the workpiece and the rotating tool. This generated heat influenced the thermo-mechanical welding process and impact on the mechanical behaviours of the joints. The highest heat takes place at the workpiece and rotating tool interface. The temperature profile of the welding process with time is presented in Fig. 5. The temperatures recorded at the 6101-T6 side of the weld which is the advancing side were lower than those at the 7075-T651 side which is the retreating side. This is as a result of different thermal conductivity influence of the welded plates. The 6101-T6 alloy on the advancing side of the welds has a higher thermal conductivity than 7075-T651 placed on the retrieving side. This implies that the rate of heat transfer and heat losses is higher in 6101-T6 than in 7075-T651, hence the aluminium alloy 7075-T651 on the retreating side retains more heat than 6101-T6 and therefore gave higher temperature values than the 6101-T6 alloys during the welding. The temperature profile also indicates that the temperature generally increases as the welding progresses and get to the peak in the middle of the weld.

3.3 Effects of Tool Rotational Speed on the Temperature Distribution

The highest temperature of about 560 °C was recorded at a rotational speed of 1850 rpm while the least was about 365 °C at a rotational speed of 1250 rpm. The temperature generally increases across the studied samples as the tool rotational speed increases. This is due to higher heat generation as the rotating tool speed goes higher own to increase in frictional movement between the rotating tool and the workpiece.

3.4 Effects of Welding Speed on the Temperature Distribution

Varying the travel speed also influenced the temperature distributions. At 1850 rpm and 50 mm/min, the highest temperature recorded was 560 °C while at 110 mm/min travel speed with the same rotational speed of 1850 rpm, the highest temperature was 480 °C. The lower temperature recorded at the higher travel speed could be as a result of less heat generated own to less dwelling time of the rotating tool as it moves faster on the workpiece. While the higher temperature obtained at the lower travel speed was due to more dwelling time of the rotating tool as it moves slower on the workpiece. Also, the ratio of the rotational speed to the welding speed of the tool decreases from 37 to 17 rpm min/mm as the speed rises from 50 mm/min to 110 mm/min. This decrease in the ratio leads to a reduction in heat input and consequently lowers the temperature of the weld.

3.5 Tensile Behaviour of the Welds at Constant Travel Speed

At a steady travel speed of 50 mm/min, the ultimate tensile strength reduces as the rotational speed rises. This is shown in Fig. 6. This emanates from more heat generated as the rotational speed increases as shown in the temperature profile in Fig. 5. The excessive heat cause deterioration of the precipitates at the welding zone which leads to poor grain bonding and consequently decreases the mechanical strength.

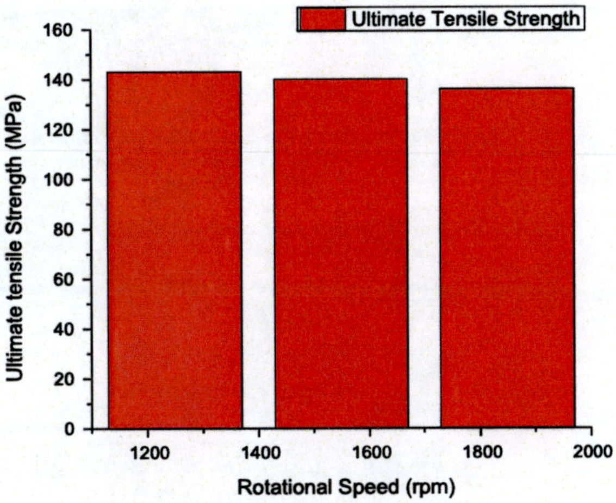

Fig. 6 Effect of rotational speed on ultimate tensile strength

3.6 Tensile Behaviour of the Welds at Constant Rotational Speed

At a steady rotational speed of 1850 rpm, the rising of feed rate from 50 to 110 mm/min causes the ultimate tensile strength to increase by 4 MPa. The increase follows a drop in the temperature as shown in Fig. 5. This implies that there was an improvement in grain bonding as the temperature drops and consequently an improvement in the ultimate tensile strength of the joint. This is shown in Fig. 7.

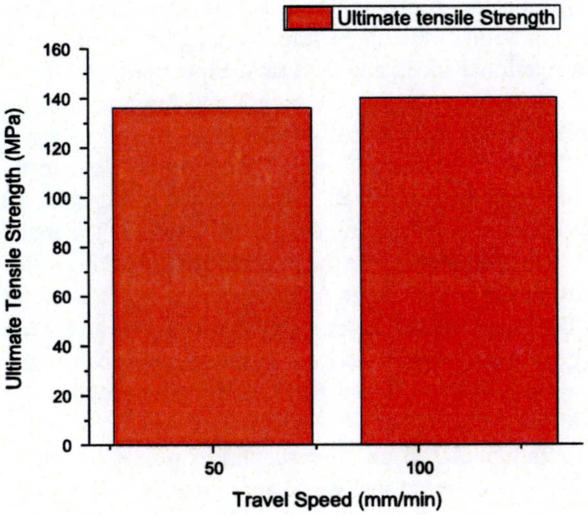

Fig. 7 Effect of travel speed on the ultimate tensile strength

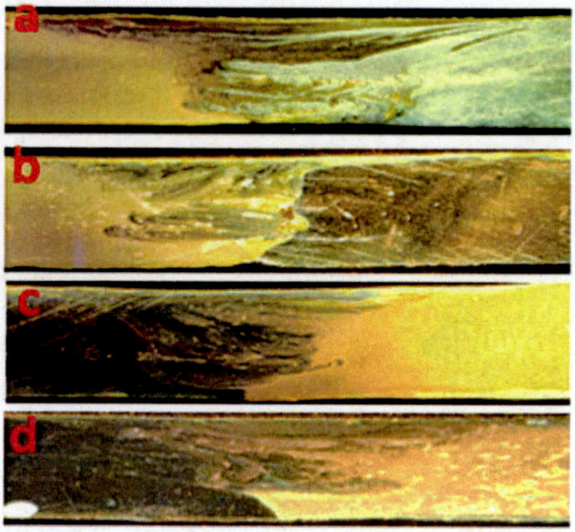

Fig. 8 Macrographs of the welded zone. **a** 1250 rpm/50 mm/min, **b** 1550 rpm/50 mm/min, **c** 1850 rpm/50 mm/min, **d** 1850 rpm/110 mm/min

3.7 Macrostructure and Microstructure of the Welds

The macrographs of all the welds (Fig. 8a–d) show that there was sufficient heat for plasticization for all the welds. The welds exhibited onion ring structures at all the rotational speed. Lamellae structure pattern of material flow was observed in all the parameters. Larger material movement is observed at the 1850 rpm and 110 mm/min. However, full penetration of the alloys into one another occurred more at 1550 rpm and 50 mm/min rotational speed and feed rate respectively.

The microstructure varies across the length and the breath of the weld and can be divided into unmixed, mechanically mixed and mixed regions as reported by some authors [27–29]. The unmixed region is observed close to the tool shoulder areas. Fine-grain structure is observed here as shown in Figs. 9a, 10a, 11a and 12a as a result of dynamic recrystallization during the welding. The mechanically mixed region (Figs. 9b, 10b, 11b and 12b) is the interface region of both alloys at the fusion zone. It contains the microstructures of both alloys. Weck's reagent etchant used revealed 7075-T651 as a lighter colour and the 6101-T6 as brown colour. At the mixed region, the tool pin stirring mechanism moved plastically deformed materials in alternate layers of 7075-T651 and 6101-T6 to form lamellae structural pattern (Figs. 9c, 10c, 11c and 12c.

Material penetration of the aluminium alloy into each other was highest at 1550 rpm and 50 mm/min rotational speed and travel speed, respectively.

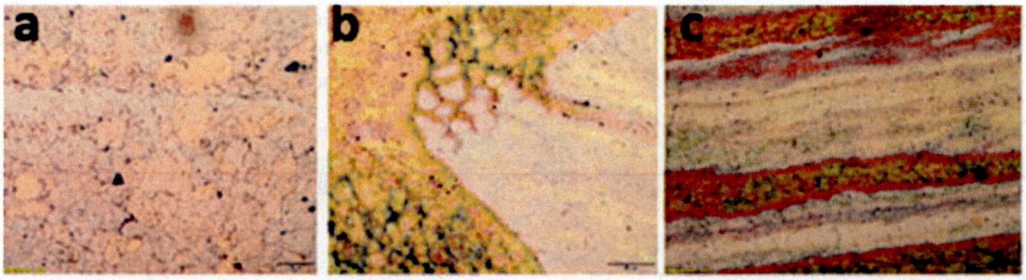

Fig. 9 Microstructures of weld nugget zone at 100× for 1250 rpm/50 mm/min, **a** unmixed, **b** mechanically mixed, **c** mixed regions

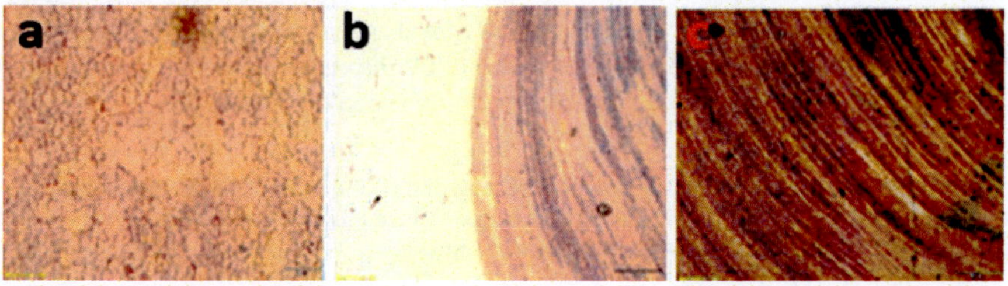

Fig. 10 Microstructures of weld nugget zone at 100× for 1550 rpm/50 mm/min, **a** unmixed, **b** mechanically mixed, **c** mixed regions

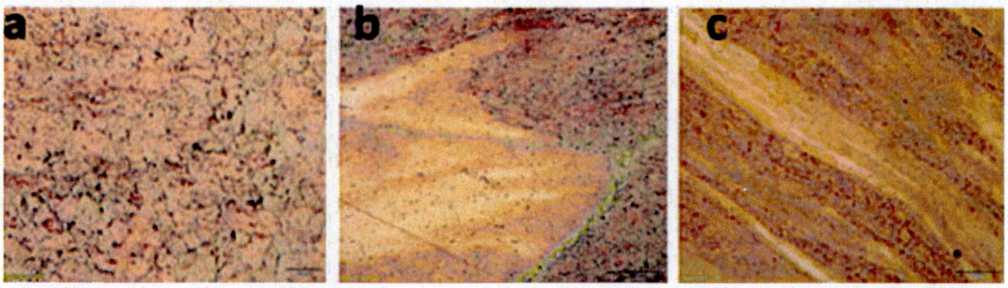

Fig. 11 Microstructures of weld nugget zone at 100× for 1850 rpm/50 mm/min, **a** unmixed, **b** mechanically mixed, **c** mixed regions

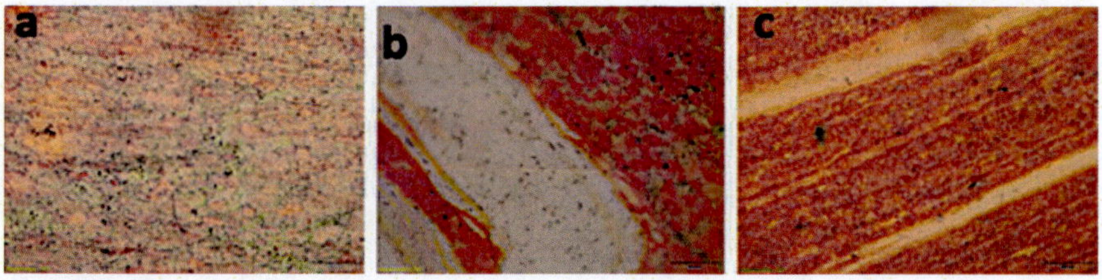

Fig. 12 Microstructures of weld nugget zone at 100× for 1850 rpm/110 mm/min, **a** unmixed, **b** mechanically mixed, **c** mixed regions

4 Conclusions

Conclusions from the results obtained are as follows:

1. Increase in the tool rotational speed result in an increase in the temperature generated during the welding process.
2. Increase in travel speed decreases the temperature.
3. Temperature profile varies relative to welding time
4. Tensile strength decreases as the tool rotational speed increases.
5. There is an increase in tensile strength as the welding speed increases.
6. Material distributions in the welding zone are not even. The volume of material swept into one another was more in 1850 rpm at a welding speed of 110 mm/min. However, the highest penetration of the alloy into one another occurred at the 1550 rpm rotational speed and at the welding speed of 50 mm/min.

Acknowledgements The authors appreciate the head and all the scholars of the Friction stir laboratory of the India Institute of Technology Kharagpur where this experiment was carried out and to the University of Johannesburg South Africa for sponsorship.

References

1. Padhy GK, Wu CS, Gao S (2017) Friction stir based welding and processing technologies - processes, parameters, microstructures, and applications: a review. J Mater Sci Technol 34:1–38
2. Sidhu MS, Chatha SS (2012) Friction stir welding—process and its variables: a review. Int J Emerg Technol Adv Eng 2(12):275–279
3. Abolusoro OP, Akinlabi ET (2019) Wear and corrosion behaviour of friction stir welded aluminium alloys—an overview. Int J Mech Prod Eng Res Dev 9(3):967–982
4. Mishra RS, Mahoney MW, Sato Y, Hovanski Y (2016) Friction stir welding and processing VIII
5. Silva ACF, De Backer J, Bolmsjö G (2017) Temperature measurements during friction stir welding. Int J Adv Manuf Technol 88(9–12):2899–2908
6. Khandkar MZH, Khan JA, Reynolds AP (2003) Prediction of temperature distribution and thermal history during friction stir welding: input torque based model. Sci Technol Weld Join 8(3):165–174
7. Yan F, Zhang Y, Fu X, Li Q, Gao J (2019) A new calculating method of frictional heat and its application during friction stir welding. Appl Therm Eng 153(January):250–263
8. Givi MKB, Asadi P (2014) Advances in friction-stir welding and processing
9. Lombard H, Hattingh DG, Steuwer A, James MN (2008) Optimising FSW process parameters to minimize defects and maximize fatigue life in 5083-H321 aluminium alloy. Eng Fract Mech 75(3–4):341–354
10. Prabha KA, Putha PK, Prasad BS (2018) Effect of tool rotational speed on mechanical properties of aluminium alloy 5083 weldments in friction stir welding. Mater Today Proc 5(9):18535–18543
11. Aydin H, Bayram A, Esme U, Kazancoglu Y, Guven O (2010) Application of grey relation analysis (Gra) and Taguchi method for the parametric optimization of friction stir welding (FSW) process. Appl Grey Relat Anal 44(4):205–211

12. Abd Elnabi MM, Elshalakany AB, Abdel-Mottaleb MM, Osman TA, El Mokadem AA (2019) Influence of friction stir welding parameters on metallurgical and mechanical properties of dissimilar AA5454-AA7075 aluminum alloys. J Mater Res Technol 8(2):1684–1693
13. Rajakumar S, Balasubramanian V (2012) Establishing relationships between mechanical properties of aluminium alloys and optimized friction stir welding process parameters. Mater Des 40:17–35
14. Kumar SR, Rao VS, Pranesh RV (2014) Effect of welding parameters on macro and microstructure of friction stir welded dissimilar butt joints between AA7075-T651 and AA6061-T651 alloys. Procedia Mater Sci 5:1726–1735
15. Das U, Toppo V (2018) Effect of tool rotational speed on temperature and impact strength of friction stir welded joint of two dissimilar aluminum alloys. Mater Today: Proc 5:6170–6175
16. Takayama Y, Akutsu Y, Choshiro N, Kato H, Watanabe H (2010) Temperature measurement during friction stir welding of dissimilar aluminum alloys, pp 0–5
17. Kandasamy J, Sairam J, Hussain MM (2018) ScienceDirect ICMPC_2018 comparative analysis of temperature variation in friction stir welding. Mater Today Proc 5:18798–18804
18. Sun YF, Fujii H, Sato Y, Morisada Y (2019) Friction stir spot welding of SPCC low carbon steel plates at extremely low welding temperature. J Mater Sci Technol 35:733–741
19. Silva-Magalhães A, De Backer J, Martin J, Bolmsjö G (2019) In-situ temperature measurement in friction stir welding of thick section aluminium alloys. J Manuf Process 39:12–17
20. Hamilton C, Dymek S, Sommers A (2010) Characteristic temperature curves for aluminum alloys during friction stir. Weld J
21. Covington JL, Robison W, Webb BW (2006) Experimental characterization of tool heating during friction stir welding. Mech Eng 179–184
22. Buffa G, Donati L, Fratini L, Tomesani L (2006) Solid state bonding in extrusion and FSW: Process mechanics and analogies. J Mater Process Technol 177(1–3):344–347
23. Feng Z et al (2007) In situ neutron diffraction measurements of temperature and stresses during friction stir welding of 6061-T6 aluminium alloy. Sci Technol Weld Join 12(4):298–303
24. Fehrenbacher A, Duffie NA, Ferrier NJ, Pfefferkorn FE, Zinn MR (2014) Effects of tool-workpiece interface temperature on weld quality and quality improvements through temperature control in friction stir welding. Int J Adv Manuf Technol 71(1–4):165–179
25. Upadhyay P, Reynolds AP (2010) Effects of thermal boundary conditions in friction stir welded AA7050-T7 sheets. Mater Sci Eng A 527(6):1537–1543
26. Miles M, Karki U, Hovanski Y (2014) Temperature and material flow prediction in friction-stir spot welding of advanced high-strength steel. JOM 66(10):2130–2136
27. Palanivel R, Koshy Mathews P, Murugan N, Dinaharan I (2012) Effect of tool rotational speed and pin profile on microstructure and tensile strength of dissimilar friction stir welded AA5083-H111 and AA6351-T6 aluminum alloys. Mater Des 40:7–16
28. Ouyang JH, Kovacevic R (2002) Material flow and microstructure in the friction stir butt welds of the same and dissimilar aluminum alloys. J Mater Eng Perform
29. Shigematsu I, Kwon YJ, Suzuki K, Imai T, Saito N (2003) Joining of 5083 and 6061 aluminum alloys by friction stir welding. J Mater Sci Lett

Optimizing the Parameters for Friction Stir Welding of an Aluminium Alloy

M. Selvaraj

Abstract A novel slip factor accounted for thermal model is being used to predict the temperature with time and the effect of welding and rotations on heat input for weld length peak-temperature of the friction stir welding process. Material used in this study is aluminium alloy of AA6061 grade which is used in aerospace applications. For various rotations and weld velocities, friction stir welding experiments were conducted on aluminium alloy plates. The plates were fixed with thermocouples at different locations from weld centerline to measure the temperatures during the welding process. The temperature with time and locations predicted by the model along the transverse direction are closely matching with the experimental results. The heat input for unit length of weld and peak-temperature increases as rotation increases and decreases as weld velocity increases. The energy required for welding for length is minimum at the combination of lower rotation and higher weld velocity in the defect-free zone.

Keywords Friction stir welding · Aluminium alloy · Thermal model · Energy required · Temperature

1 Introduction

Heat flow modeling during the FSW process was done by Rajamanickam et al. [1] and Riahi and Nazari [2] using a heat flow model to obtain the thermal history. Song and Kovacevic [3] presented the isotherms and temperature with time of FSW process by simulating the welding process. Woo et al. [4] and Chao and Tang [5] developed a heat flow model for the FSW process, numerically predicted temperature with time. Prasanna et al. [6] modeled the FSW process of steel, predicted the temperature with time. Nandan et al. [7] presented a 3D thermal and material flow model and predicted the thermal history, velocity and strain rate during the FSW process. Hamilton et al. [8, 9] done simulation of FS welding of aluminium alloy

M. Selvaraj (✉)
Sri Sivasubramaniya Nadar College of Engineering, Chennai, Tamil Nadu, India
e-mail: selvarajm71@gmail.com

S. Vijayan et al. (eds.), *Trends in Manufacturing and Engineering Management*, Lecture Notes in Mechanical Engineering,
https://doi.org/10.1007/978-981-15-4745-4_65

to predict the thermal cycle. 3-dimensional model for FSW considering energy due to friction and plastic deformation is developed by Emam and Domiaty [10] and predicted change of peak normalized temperature with the effective energy. Selvaraj et al. [11, 12] proposed a thermal model for FS welding of steels assuming slip factor as a function of temperature to predict the temperature with time and energy required per unit weld length during the welding process.

The literature survey indicates that most of the thermal studies concentrate on predicting temperature distribution and temperature with time during FSW process. Most of the researchers have not predicted the energy required for different combinations of weld velocity and rotation. In some combinations of weld, parameters require to weld with minimum energy. It will save a lot of energy. In this paper, attention has been given to predicting the energy-efficient rotational and weld velocity in the FS welding process using a mechanics-based heat flow model.

2 Experiment Work

This study uses AA6061 aluminium alloy which is used in applications like aerospace and marine. The material was cut into a plate of 150 × 75 × 6 mm size. Two plates are aligned and fixed by clamps. Aligned plates were welded using an M2 steel tool as shown in Fig. 1. The tool and welding parameters are given in Table 1. The welding experiments were done by an FSW machine with 12 kW motor as shown in Fig. 2.

Fig. 1 FSW tool dimensions

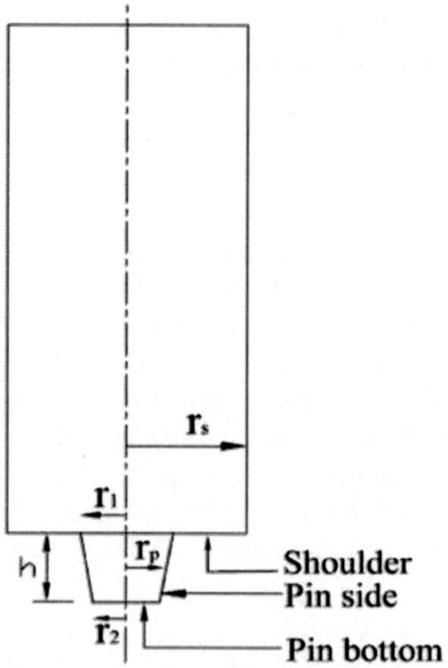

Table 1 FSW parameters

Parameters	Value
Radius of shoulder, r_s	9 mm
Max. radius of pin, r_1	4 mm
Min. radius of pin, r_2	2 mm
Length of pin, h	5.5 mm
Axial load, F_n	15.5 kN
Rotations, N	300–2000 rpm (7 speeds)
Translational velocity, V	300, 600, 900, 1400 and 1800 mm/min

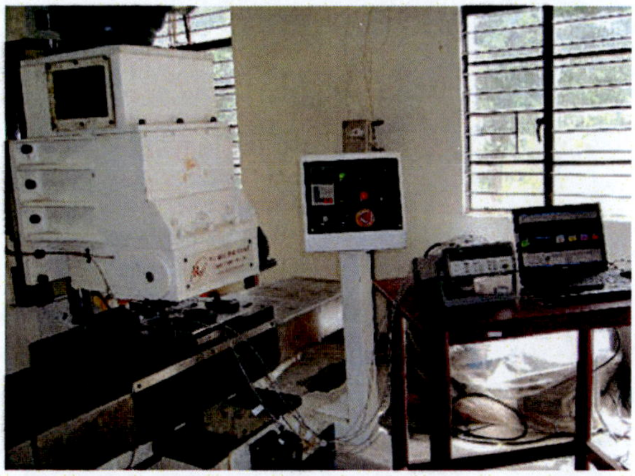

Fig. 2 Experimental setup

One mm diameter four K type thermocouples were fixed at different locations. They are at 4, 10, 15 and 25 mm from the center of weld on the advancing side. The temperature during the process was measured using data logger.

3 Thermal Model

3.1 Mathematical Model

Shoulder heat
Friction heat at shoulder

$$Q_{\text{sh-fr}} = \frac{2\pi}{3}\mu\, p_n\, \omega\, (r_s^3 - r_1^3) \tag{1}$$

Pressure in axial direction,

$$p_n = \frac{F_n}{\pi r_s^2} \tag{2}$$

Tool angular velocity,

$$\omega = \frac{2\pi N}{60} \tag{3}$$

The friction coefficient (μ) is linearly varying from 0.5 to 0.4 between the room and solidus temperature [13].

Shear deformation heat at shoulder

$$P_{\text{sh-pl}} = \frac{2\pi}{3}\tau\omega(r_s^3 - r_1^3) \tag{4}$$

Shear stress,

$$\tau = \frac{\sigma_y}{\sqrt{3}} \tag{5}$$

where σ_y is the material yield stress which is shown in Table 2.

The expression for heat generated at the shoulder is obtained by merging the heat generation due to friction and shear deformation with stick factor.

Shoulder heat

$$Q_{\text{sh}} = \delta.Q_{\text{sh-pl}} + (1-\delta)Q_{\text{sh-fr}} \tag{6}$$

Total heat

Calculated pin heat at side Q_{ps} and bottom Q_{pb} are

Pin heat,

$$Q_{\text{pin}} = Q_{\text{ps}} + Q_{\text{pb}} \tag{7}$$

Total tool heat,

$$Q_{\text{T}} = Q_{\text{sh}} + Q_{\text{ps}} + Q_{\text{pb}} \tag{8}$$

Shoulder and pin heat flux

The heat over weld length

$$Q_l = \frac{Q_{\text{T}}}{V} \times 60 \tag{9}$$

Table 2 Properties of 6061-T6 aluminium alloy [2]

Temperature (°C)	26	38	93	147	205	260	316	372	425
Yield strength (MPa)	276	275	265	249	219	159	66	34	17
Density (kg/m^3)	2700	2686	2686	2666	2656	2656	2631	2621	2601
Thermal con. (W/m °C)	167	171	178	185	191	202	205	215	225
Heat capacity (J/kg °C)	896	921	979	1005	1029	1051	1079	1105	1134

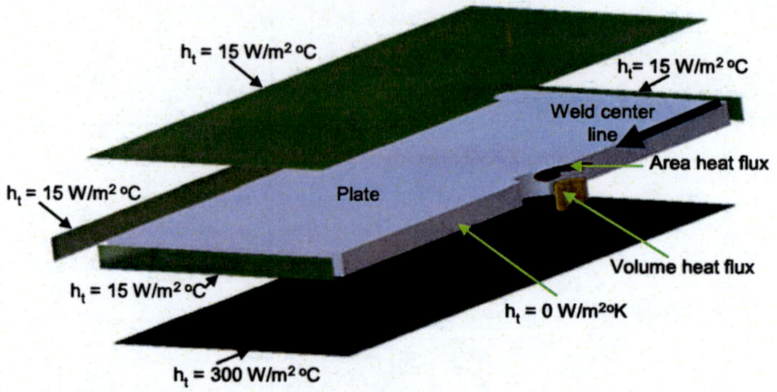

Fig. 3 FE 3D model with boundary conditions and heat flux

By assuming 90% of heat transferred to workpiece [14], shoulder heat flux expression arrives linear with radius of the tool [13].

$$q_{\mathrm{sh}} = 0.9 \times \frac{3Q_{\mathrm{sh}}r}{2\pi(r_{\mathrm{s}}^3 - r_{\mathrm{p}}^3)} \quad \text{for } r_{\mathrm{p}} \le r \le r_{\mathrm{s}} \tag{10}$$

Heat around the pin is taken as volume heat flux in the shape of hollow cylinder based on the shear layer thickness of 2 mm as stated by Chen et al. [15].

$$q_{\mathrm{pin}} = 0.9 \times \frac{Q_{\mathrm{pin}}}{\pi[(r_{\mathrm{p}} + 2)^2 - r_{\mathrm{p}}^2]h} \tag{11}$$

3.2 *Finite Element Model*

See Fig. 3.

4 Results and Discussion

4.1 *Thermal History*

The variation of temperature with respect to time during FS welding of aluminium alloy for different rotations and weld velocities at four different locations from the weld centerline is shown in Fig. 4. For all input parameters, the thermal history predicted by the model matches closely with that of experimental history. It shows that this model is reliable to predict the temperature distribution during FS welding.

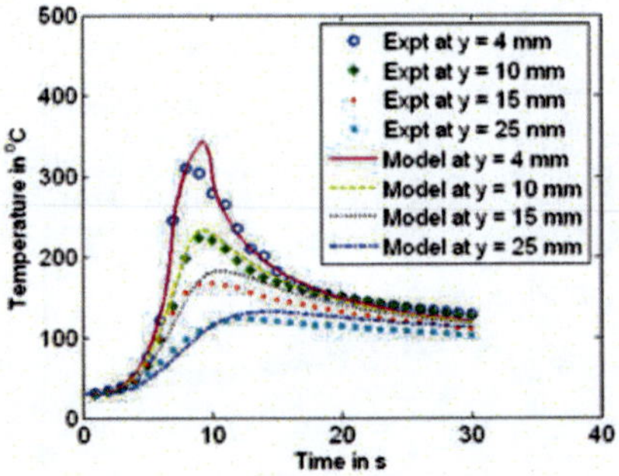

(a) Weld velocity 300 mm/min & rotation 300 rpm.

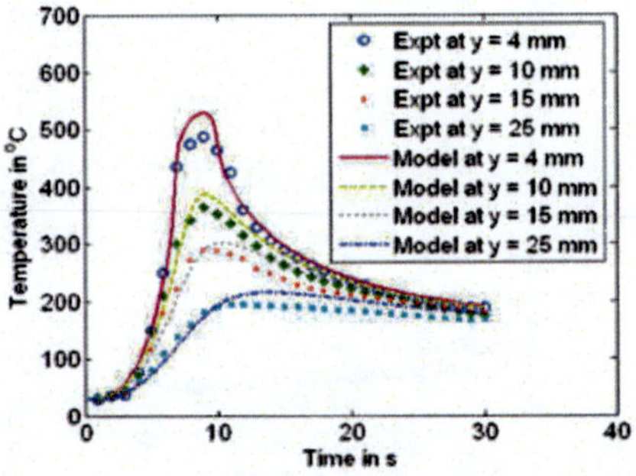

(b) Weld velocity 300 mm/min & rotation 1200 rpm.

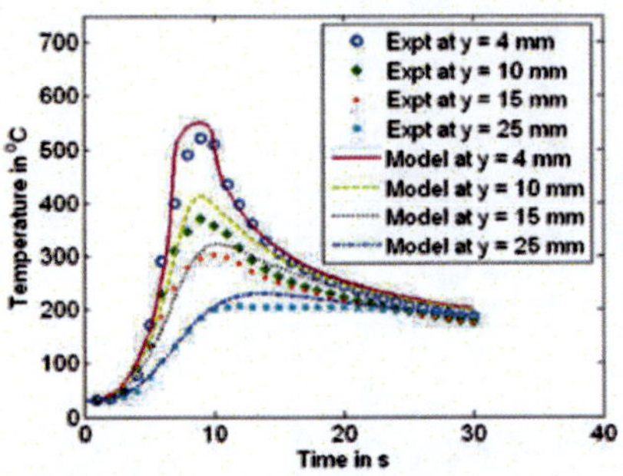

(c) Weld velocity 300 mm/min & rotation 2000 rpm.

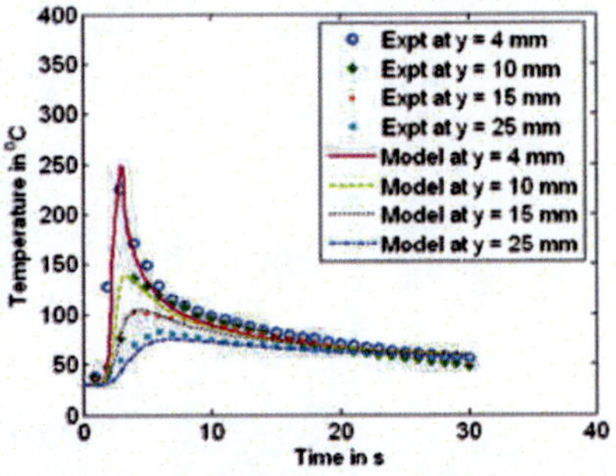

(d) Weld velocity 900 mm/min & rotation 300 rpm.

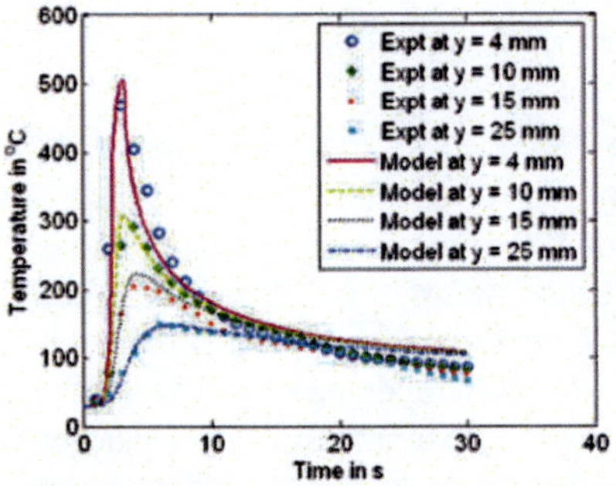

(e) Weld velocity 900 mm/min & rotation 1200 rpm.

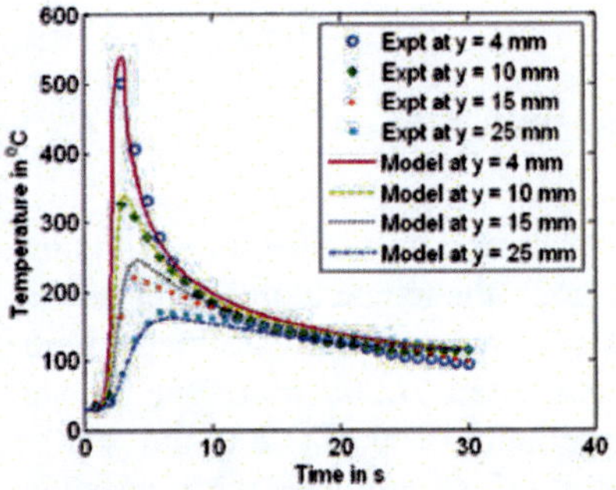

(f) Weld velocity 900 mm/min & rotation 2000 rpm.

Fig. 4 Temperature with time at various locations from the weld centerline for different Weld velocity and rotations

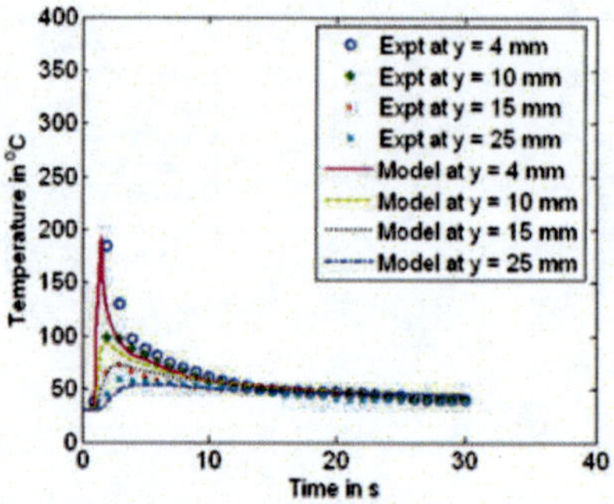

(g) Weld velocity 1800 mm/min & rotation 300 rpm.

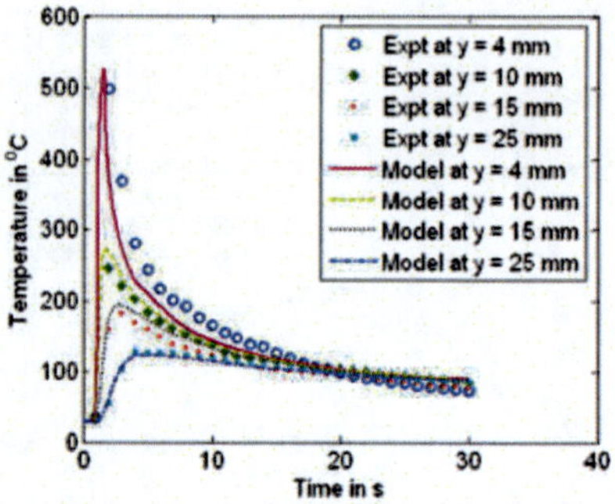

(h) Weld velocity 1800 mm/min & rotation 1200 rpm

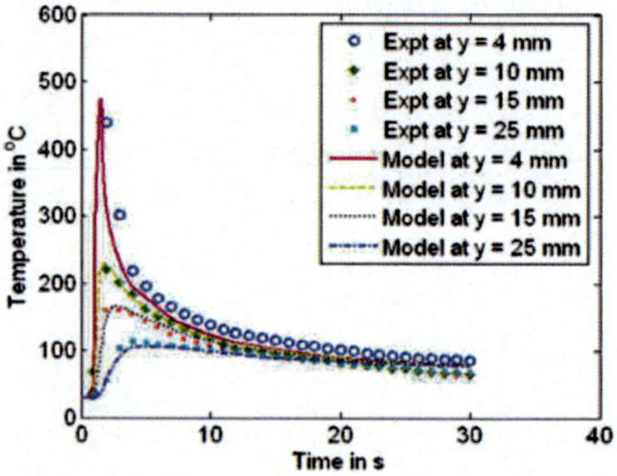

(i) Weld velocity 1800 mm/min & rotation 2000 rpm.

Fig. 4 (continued)

Analyzing the temperature with time of 300 mm/min weld velocity and different rotations, the temperature increases and drops sharply for 300 rpm than 1200 rpm. It also shows that the peak-temperature is high and wider cycle is obtained for 1200 as compared to 300 rpm. But the variation of temperature with time and peak-temperature between 1200 and 2000 rpm is not significant. For a constant weld velocity (300 mm/min), when speed is increased from 300 to 1200 rpm, heat generation increases and leads to variation in temperature with time. The variation in temperature with time between 1200 and 2000 rpm is not significant because variation in heat generation is not significant. The same kind of pattern is visible in the other weld velocities also.

Analyzing the effect of various weld velocity for a constant rotation (300 rpm), the rise and dip in temperature are very sharp and peak-temperature decreases as weld velocity increases. As weld velocity increases, the heat source quickly approaches and passes away the measuring location and it leads to a sharp rise and drops in temperature.

4.2 *Effect of Weld Parameters on Energy Required and Peak-Temperature*

The energy required for length of weld and peak-temperature for various weld velocity and rotations are illustrated in Figs. 5 and 6. It is noted that the heat input for length of weld and peak-temperature increases as rotation increases but decreases as weld velocity increases. It is also noted that at lower rotations slope of the curve is more and it reduces as rotation increases. This is due to reduction in heat generation rate as rotation increases. The heat input for length of weld is inversely proportional to the weld velocity. The heat input for length of weld and peak-temperature

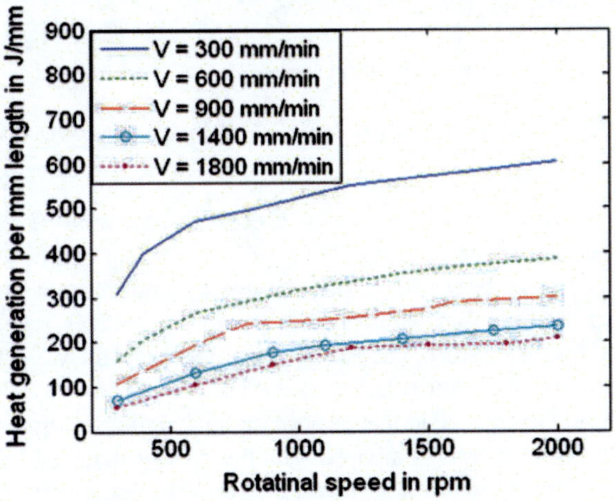

Fig. 5 Energy required per mm length of weld for various weld velocity and rotations

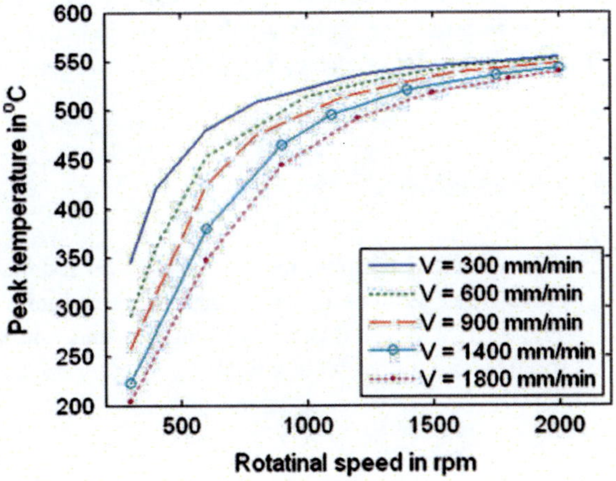

Fig. 6 Simulated peak-temperature for various weld velocity and rotations

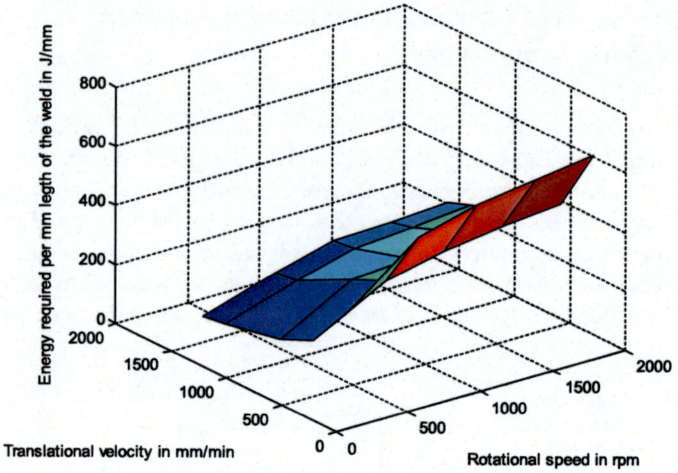

Fig. 7 Energy required for length of weld for various weld velocity and rotations

follow the same trend. It is clear that heat input for length of weld influences the peak-temperature (Fig. 7).

In the defect-free zone, the energy required per unit weld length is 188.2 J/mm at the combination of lower rotation (1200 rpm) and higher weld velocity (1800 mm/min). The energy required per unit weld length is 549.6 at the combination of higher rotation (1200 rpm) and lower weld velocity (300 mm/min). The energy required at the combination of higher rotation and lower weld velocity is 2.9 times higher as compared with the energy required per unit weld length the combination of lower rotation and higher weld velocity. It shows that the energy required for welding for length is minimum at the combination of lower rotation and higher weld velocity in the defect-free weld zone. Hence it is vital to see the combination of lower rotation and higher weld velocity in order to weld with minimum energy. This will save a lot of energy required for welding.

5 Conclusions

1. The predicted temperature with time and location in transverse direction has closely merged with that of experimental values.
2. The energy required per weld length is minimum at the combination of lower rotation and higher weld velocity.

References

1. Rajamanickam N, Balusamy V, Reddy GM, Natarajan K (2009) Effect of process parameters on thermal history and mechanical properties of friction stir welds. Mater Des 30:2726–2731
2. Riahi M, Nazari H (2010) Analysis of transient temperature and residual thermal stresses in FS welding of aluminum alloy 6061-T6 via numerical simulation. Int J Adv Manuf Technol 55:143–152
3. Song M, Kovacevic R (2003) Thermal modeling of FS welding in a moving coordinate system and its validation. Int J Mach Tools Manuf 43:605–615
4. Woo W, Choo H, Withers PJ, Feng Z (2009) Prediction of hardness minimum locations during natural aging in an aluminum alloy 6061-T6 friction stir weld. J Mater Sci 44:6302–6309
5. Chao YJ, Tang W (2003) Heat transfer in FS welding—experimental and numerical studies. Trans ASME 125:138–145
6. Prasanna P, Rao BS, Mohana Rao GK (2010) Finite element modeling for maximum temperature in FS welding and its validation. Int J Adv Manuf Technol 51:925–933
7. Nandan R, Roy GG, Lienert TJ, Debroy T (2007) Three-dimensional heat and material flow during FS welding of mild steel. Acta Mater 55:883–895
8. Hamilton C, Dymek S, Sommers A (2008) A thermal model of FS welding in aluminum alloys. Int J Mach Tools Manuf 48:1120–1130
9. Hamilton C, Sommers A, Dymek S (2009) A thermal model of FS welding applied to Sc-modified Al–Zn–Mg–Cu alloy extrusions. Int J Mach Tools Manuf 49:230–238
10. Emam SA, Domiaty AE (2009) A refined energy-based model for friction-stir welding. World Acad Sci Eng Technol 53:1016–1022
11. Selvaraj M, Murali V, Rao SK (2013) Thermal model for FS welding of mild steel. Multidiscip Model Mater Struct 9:49–61
12. Selvaraj M (2014) A temperature dependent slip factor based thermal model for FS welding of stainless steel. Sadhana-Acad Proc Eng Sci 38:1393–1405
13. Soundararajan V, Zekovic S, Kovacevic R (2005) Thermomechanical model with adaptive boundary conditions for FS welding of Al 6061. Int J Mach Tools Manuf 45:1577–1587
14. Dickerson T, Shi Q, Shercliff, HR (2003) Heat flow into FS welding tools. In: 4th international symposium on FS welding, Park City Utah, USA
15. Chen Z, Pasang T, Qi Y (2008) Shear flow and formation of Nugget zone during FS welding of aluminium alloy 5083-O. Mater Sci Eng, A 474:312–316

Machining of ZE41 Magnesium Alloy in WEDM Using Taguchi Approach

G. Selvakumar, S. Ram Prakash, and E. Caleb Kovilpillai

Abstract This study explains the effects of process parameters of Wire Electrical Discharge Machining (WEDM) on the machinability and surface characteristics of Magnesium ZE41 alloy. ZE41 alloy has a plethora of applications in the structural industry, aerospace industry, military equipment, video cameras, and vibration testing equipment. In this study, Pulse on time (Ton), Pulse off time (Toff), Current (I) and Wire Tension (WT) were considered as the input parameters to conduct the experiments for the performance measures Material removal rate (MRR) and Surface roughness (Ra). Experimentation works were conducted by using Taguchi's L9 orthogonal array. Analysis of variance (ANOVA) was performed to identify the significant parameters. XRD test was carried out to identify the phase constituents of the material. Further, Surface morphology of the machined surface is analyzed using SEM micrographs.

Keywords ZE41 magnesium alloy · WEDM · Material removal rate · Surface roughness · Process parameters

1 Introduction

The demand for alloy materials is rising abound in recent years due to its appealing properties like high strength to weight ratio, ductility, formability, and high resistance to chemical attack. Alloying is an effective technique to enhance the mechanical properties of the materials by combining the beneficial properties of two or more metals [1]. It is the most predominantly used material in all engineering applications since pure metals have limitations in its properties. Magnesium is the 6th widely available lightweight material in the earthly resources [2]. Magnesium alloys have renowned properties like good castability, formability, high sound dampening capacity, and low density [3]. Among aluminum, steels, and copper-base alloys, magnesium alloy

G. Selvakumar · S. Ram Prakash (✉) · E. Caleb Kovilpillai
Department of Mechanical Engineering, Sri Sivasubramaniya Nadar College of Engineering, Kalavakkam, Chennai 603110, India
e-mail: ramprakashnec@gmail.com

S. Vijayan et al. (eds.), *Trends in Manufacturing and Engineering Management*, Lecture Notes in Mechanical Engineering,
https://doi.org/10.1007/978-981-15-4745-4_66

is the supreme material for structural applications due to its high strength to weight ratio. However, it has poor hardness, wear, and corrosion resistance. Pure magnesium is highly susceptible to corrosion and weakens the material behavior which can be enhanced by adding alloying elements [3]. Magnesium and its alloys have a plethora of applications in biomedical engineering like screws and plates for bone fixation, tissue engineering scaffolds, as well as on cardiovascular stents [4], automobile, aerospace, electronics, and structural industries [5]. Machining of magnesium alloy at the required shape with the superior surface finish is necessary to perform its intended functions. So, it is indeed crucial to machine the material with minimal surface roughness and higher material removal rate. It is identified that nontraditional machining is the best way out to cut intricate shapes with a better surface finish. Amongst all other nontraditional machining processes, WEDM is the apt machining process to process magnesium alloys. The material removal mechanism for WEDM is the melting and vaporization of the material due to the electrical discharges. It has the potential to cut any intricate shapes with great accuracy of electrically conductive material [6]. A material with high electrical and thermal conductivity would be preferred as wire electrodes. Brass wire and zinc-coated brass wire are commonly used wire electrodes in the recent era due to its utmost performance [7].

Janardhan and Samuel [8] developed a pulse discrimination algorithm to study the effect of machining parameters at spark gap and revealed that the surface roughness (Ra) and roundness error are majorly affected by the occurrence of arc regions, width of arc and average ignition delay time. Weingartner et al. [9] investigated the shape and size of craters produced during single discharge experiments and developed an electro-thermal model. Further, the experimental results are then compared with the simulation results. Mohammadi et al. [10] discussed the effects of machining parameters for roundness error and surface quality of a cylindrical component and further presented the mathematical model equation for predicting the roughness using regression analysis.

Few researchers have attempted on processing ZE41 magnesium alloy in WEDM but there is a gap in investigating the surface integrity and morphology of the machined surface. The objective of the present work is to study the effect of process parameters of WEDM on machining of ZE41 alloy for the responses MRR and surface roughness. Surface morphology of the machined surface was examined through SEM micrographs. Besides, the phase constituents of the alloy were investigated using XRD pattern.

2 Materials and Methods

The experimental work is performed in Ratnaparki CNC WEDM machine, model: smart cut 2530. The machine setup contains a machine tool, control system, and circulation type dielectric unit with a capacity of 250 L. The chemical composition of the ZE41 magnesium alloy used in this experimentation work is shown in Table 1. The phase constituents of the ZE41 magnesium alloy have been identified by XRD

Table 1 Chemical compositions of ZE41 Magnesium alloy

Material	Zinc	Rare earth	Zirconium	Magnesium
Magnesium ZE41 alloy	3.5–5	0.8–1.7	0.4–1	Balance

Table 2 Parameters and their levels

S. No.	Process parameters	Level 1	Level 2	Level 3
1	Pulse on time—Ton (μs)	0.1	0.2	0.3
2	Pulse off time—Toff (μs)	30	40	50
3	Peak current—I (A)	1	3	5
4	Wire tension—WT (N)	18	20	22

spectrum. Copper is used as target material, generator voltage and tube current were set as 45 kV and 40 mA, respectively. In Fig. 1, from the obtained XRD peaks, it is evident that magnesium occupies the dominant phases and few traces of zinc are also detected when compared with JCPDS data.

Zinc-coated brass wire is used as wire material with the diameter 250 μm and deionized water is used as dielectric medium. Taguchi's L9 orthogonal array was employed to conduct the experiments. The controllable factors considered for this experiment were based on previous literature and trial experiments. Table 2 discusses the controllable parameters and their levels considered for conducting the experiments. The constant parameters used for these experiments were wire feed: 3 m/min, Gap voltage:30 V, Water Pressure: 9 bar.

MRR is calculated by the following expression:

$$\text{MRR}\ (\text{mm}^3/\text{min}) = \text{cutting speed (mm/min)} \times \text{kerf width (mm)} \times \text{thickness (mm)} \tag{1}$$

Surface roughness is measured by MitutoyoSurftest, Model: SJ 210 with the cut-off length as 0.25 mm. While Taguchi's parametric analysis approach, the better performance of WEDM for MRR is indicated by choosing the problem type as "larger is better" and "smaller is better" is chosen for surface roughness. Table 3 represents the experimental results for the L9 orthogonal array. In Taguchi's approach, the WEDM performance analysis for MRR is marked as "Larger is better". Besides, "smaller is better" is chosen for surface roughness. The analysis of variance (ANOVA) method has been deployed to figure out the contribution of individual factors for the responses MRR and Ra (Fig. 1).

Table 3 Experimental results

S. No.	Ton (μs)	Toff (μs)	I (A)	WT (N)	MRR (mm^3/min)	Ra (μm)
1	0.1	30	1	18	2.822	2.83
2	0.1	40	3	20	2.405	2.64
3	0.1	50	5	22	2.202	2.63
4	0.5	30	3	22	10.797	3.56
5	0.5	40	5	18	9.867	2.95
6	0.5	50	1	20	9.602	3.51
7	1	30	5	20	11.020	3.70
8	1	40	1	22	10.606	3.32
9	1	50	3	18	11.557	3.12

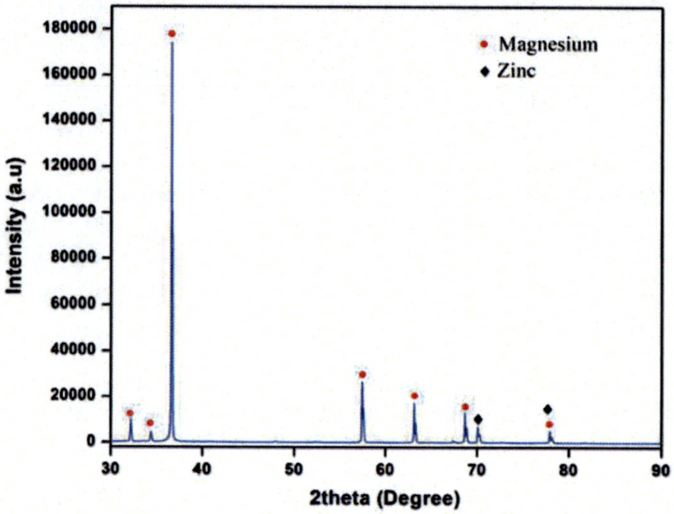

Fig. 1 XRD pattern of ZE41 magnesium alloy

3 Results and Discussion

From Fig. 2 and Table 4, it is revealed that the pulse on time is the most predominant factor which majorly influences the MRR. Next to pulse on time, current plays a major role in deciding MRR. The rest of the factors have minor influences over MRR. The increase in pulse on time and current increases the material removal rate, this is due to the fact that the amount of discharge energy rises up drastically which in turn melt and vaporize large lump of material and produces huge crater. Therefore, an increasing trend is observed in Fig. 3 for both pulse on time and current.

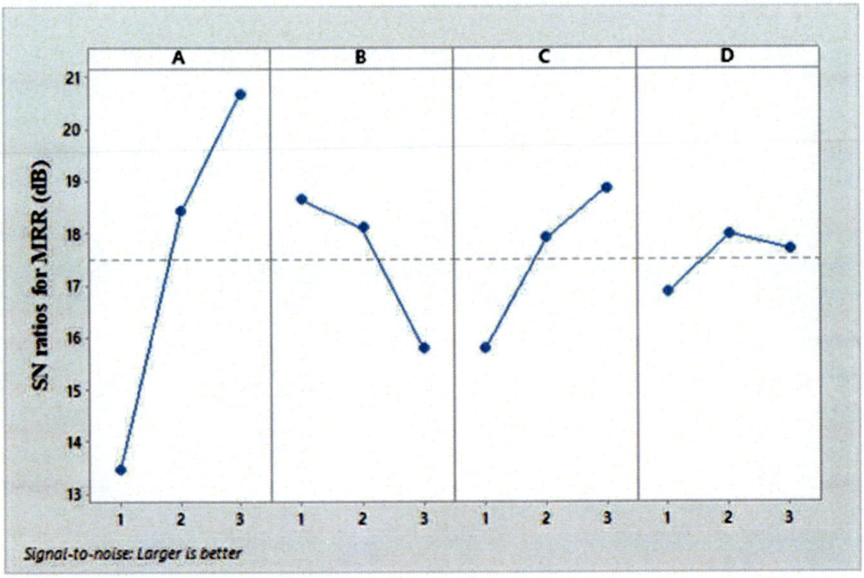

Fig. 2 Effect of parameters on MRR

Table 4 ANOVA result

Factor	DF	MRR		Ra	
		AdjSS	% of contribution	AdjSS	% of contribution
Ton	2	77.67	69.36	6.3157	53.80
Toff	2	12.256	10.94	0.777	6.62
I	2	14.13	12.62	2.6744	22.78
WT	2	1.042	0.93	0.1478	1.26
Error	0	6.883		1.8333	
Total	8	111.981		11.7481	

From Fig. 3 and Table 4, it is revealed that the pulse on time plays the major role followed by the current on determining the surface roughness of the machined surface. The amount of discharge energy applied is directly proportional to the longevity of the pulse and the magnitude of discharge energy per pulse decides the crater size. An increase in pulse on time increases the discharge energy which tends to produce large craters and creates an uneven surface. Hence, increase in pulse on time and current diminish the surface quality of the material [11].

Figure 4 explains the surface morphology of the machined surface. In WEDM, material removal takes place in the form of craters. The size of the crater depends on the intensity of the electrical discharge. As the discharge energy increases, the more will be the crater size. It is observed from the SEM images that the machined surface

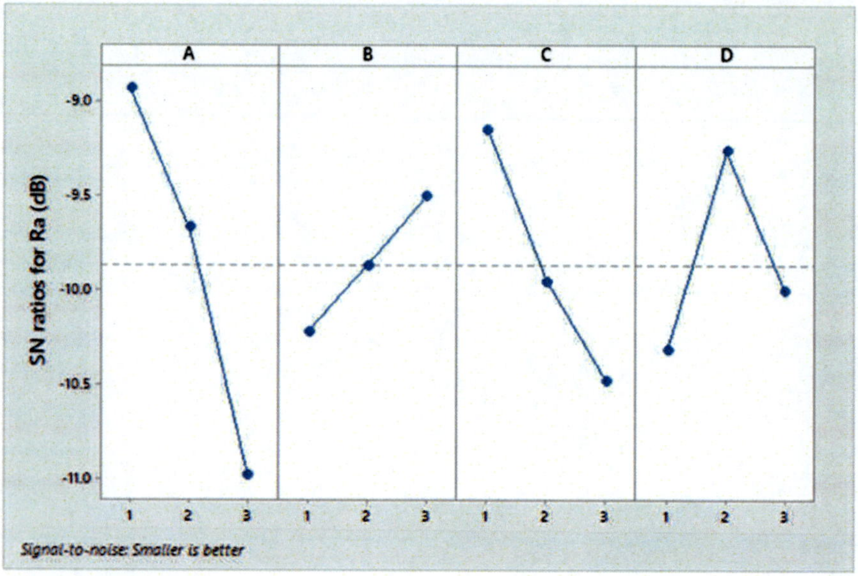

Fig. 3 Effect of parameters on Ra

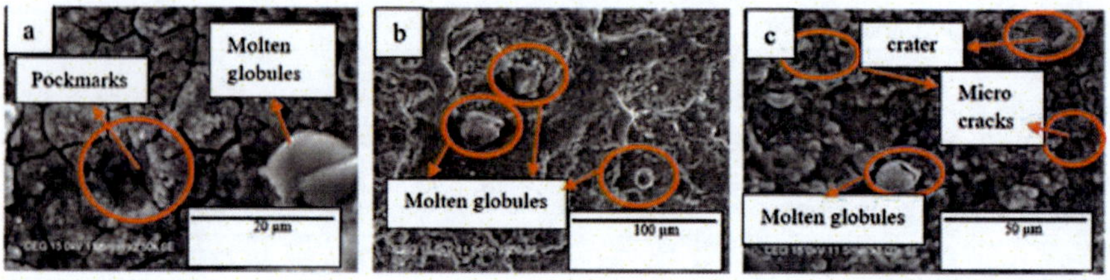

Fig. 4 **a** Microcracks on the machined surface, **b** presence of molten globules, **c** crater and microcrack formation of the machined surface

contains microcracks, microvoids, craters, and molten globules. In some regions, it shows the accumulation of fragmented debris on the machined surface. Further, it witnesses the random presence of molten globules due to the rapid cooling effect of the deionized water. During spark emission, gas bubbles are formed which later explodes and produces microcracks on the surface. The presence of microcracks and voids deteriorates the mechanical strength of the material. Proper flushing along with appropriate pulse off time should be provided to reduce the recast layer on the machined surface.

4 Conclusion

In this work, WEDM machining of ZE41 alloy was performed and the effect of process parameters was studied for the responses MRR and Ra. The XRD analysis has been conducted to identify the phase constituents of the material. Pulse on time and current were identified as the prime factors which influence the material removal rate and surface roughness. The surface morphology of the machined surface has been studied with the SEM micrographs. Molten globules, microcracks, pockmarks, and microvoids were observed in the machined region.

References

1. Chen J, Tan L, Yu X, Etim IP, Ibrahim M, Yang K (2018) Mechanical properties of magnesium alloys for medical application: a review. J Mech Behav Biomed Mater 87:68–79
2. Singh K, Singh G, Singh H (2018) Review on friction stir welding of magnesium alloys. J Magn Alloys 6:399–416
3. Dinodi N, Nityananda Shetty A (2013) Electrochemical investigations on the corrosion behaviour of magnesium alloy ZE41 in a combined medium of chloride and sulphate. J Magn Alloys 1:201–209
4. Hohna S, Virtanena S, Boccaccini AR (2019) Protein adsorption on magnesium and its alloys: a review. Appl Surf Sci 464:212–219
5. Luo AA (2013) Magnesium casting technology for structural applications. J Magn Alloys 1:2–22
6. Selvakumar G, Thiruppathi Kuttalingam KG, Ram Prakash S (2018) Investigation on machining and surface characteristics of AA5083 for cryogenic applications by adopting trim cut in WEDM. J Braz Soc Mech Sci Eng 40:267
7. Gnanavelbabu A, Saravanan P, Rajkumar K, Karthikeyan S, Baskaran R (2018) Optimization of WEDM process parameters on multiple responses in cutting of Ti-6Al-4V. Mater Today: Proc 5:27072–27080
8. Janardhan V, Samuel GL (2010) Pulse train data analysis to investigate the effect of machining parameters on the performance of wire electro discharge turning (WEDT) process. Int J Mach Tools Manuf 50:775–788
9. Weingartner E, Wegener K, Kuster F (2012) Wire electrical discharge machining applied to high-speed rotating workpieces. J Mater Proc Technol 212:1298–1304
10. Mohammadi A, Tehrani AF, Emanian E, Karimi D (2008) A new approach to surface roughness and roundness improvement in wire electrical discharge turning based on statistical analyses. Int J Adv Manuf Technol 39:64–73
11. Selvakumar G, Sarkar S, Mitra S (2013) An experimental analysis of single pass cutting of aluminium 5083 alloy in different corner angles through WEDM. Int J Mach Machinabil Mater 13(2/3)

Comparative Study of Friction Stir Welding and Underwater Friction Stir Welding on Magnesium ZE41 Alloy

S. Cyril Joseph Daniel and A. K. Lakshminarayanan

Abstract 5-mm-thick ZE41 magnesium alloy plates were subjected to normal friction stir welding (in air) and under water friction stir welding (UWFSW). A comparative study was made to investigate the process parameter and resultant properties of microstructure and tensile behaviour. The friction stir welding (FSW) was carried out at two different speeds 660 and 1220 rpm in order to obtain the influence of rotation speed on the performance of underwater joints. And the tool transverse speed and the tool tilt angle were fixed at 40 mm/min and 2.5°, respectively. From this investigation, it can be found that the UWFSW joint made using the higher tool rotational speed of 1220 rpm exhibited good tensile properties when compared to FSW joints without any defects like tunnel-type defects or hot cracking. The higher cooling rate in the UWFSW caused reduced thermomechanically affected zone (TMAZ) in the weld. The joint welded at UWFSW made the grain size fine when compared to the grain size of joints at the air weld condition. It is also noted that the hardness was increased in the UWFSW specimen due to the fine grains present in the stir zone. For air weld, the fracture occurred in the weld region and for UWFSW the fracture moved away from the stir zone.

Keywords Friction stir welding · Magnesium ZE41 · Underwater · Microstructure · Mechanical properties

1 Introduction

Magnesium alloys have brought more and more attention lately in response to the rising eco-awareness. Due to the fact that magnesium possesses the lowest density among all structural metals (such as Al, Mg and Ti), the purpose of magnesium alloys will reduce the energy consumption and has very high vibration damping capacity.

S. Cyril Joseph Daniel (✉) · A. K. Lakshminarayanan
Department of Mechanical Engineering, Sri Sivasubramaniya Nadar College of Engineering, Chennai,
Tamil Nadu 603110, India
e-mail: cjdaniel1994@gmail.com

S. Vijayan et al. (eds.), *Trends in Manufacturing and Engineering Management*, Lecture Notes in Mechanical Engineering,
https://doi.org/10.1007/978-981-15-4745-4_67

In addition, magnesium also has the advantage of recyclability, further decreasing the squandering of natural resources. Based on the lower density and recyclability, magnesium alloys have gradually become the highly potential metallic material. The light alloys with lower density, higher strength and good corrosion resistant properties magnesium are widely used. Magnesium alloys are the lightweight material among the structural metals, which have two-thirds density (1.74 g/cm^3) of the aluminium alloys (2.7 g/cm^3) and also four times lighter the steels which have the density of 7.86 g/cm^3. It has a good ductility and excellent casting ability.

Friction stir welding is the solid-state welding process which is recently adopted in all possible processes. In FSW process, the non-consumable tool moves between two plates which are to be welded together by the effect of friction between tool and the workpiece. Moreover, the stirring action and pressure applied play the major role in FSW process. The lower heat generation in FSW stands unique and avoids many defects due to higher temperature on fusion welding. Expanded details about the process of FSW were discussed in many previous publications [1–5].

Nowadays, there are many methods like GTAW, AGTAW, GMAW, SMAW and laser beam welding followed for welding of magnesium ZE41 alloy [1, 6]. Some researchers had tried different materials with FSW in underwater condition and got better mechanical properties [7]. But mechanical properties and microstructural distribution for joining of similar magnesium ZE41 underwater condition have not been studied fully. In this paper, similar magnesium ZE41 alloy was welded together using both FSW at underwater condition and FSW at air condition. The microstructure distribution, tensile strength and fracture region were investigated to understand a comparative study of welding of underwater FSW and FSW of magnesium ZE41 alloy.

2 Experimental Procedure

As per the requirement, the magnesium ZE41 alloy (200 mm × 40 mm × 5 mm) was prepared and checked for defect-free condition. Their chemical composition and mechanical properties are shown in Table 1. The material was completely cleaned to make sure the absence of the foreign particles which can cause defects during the welding. The as-received alloys have coarse grains, so the faces of the plates were machined and finished to get the flat surface. The plates were placed together on the welding direction, and tightly clamped to the table to avoid distortion during the welding. The process was done at a XL6026AV machine, which is a modified form

Table 1 Chemical properties of magnesium ZE41

Grade		Zn	Ce	Zr	Mg
%	Min	3.5	0.8	0.4	Remaining
	Max	5	1	1	

of vertical friction stir welding machine with the 5HP capability shown as in Fig. 5. The machine has 360 cm of table width and 1370 cm of table length (Fig. 5), and the chamber for underwater welding was designed and fabricated with dimensions of 400 mm × 280 mm × 75 mm length, width and height, respectively. This set-up is designed to hold the material to be welded and capable of carrying the water to make the material immersed condition (Figs. 1, 2, 3 and Table 2).

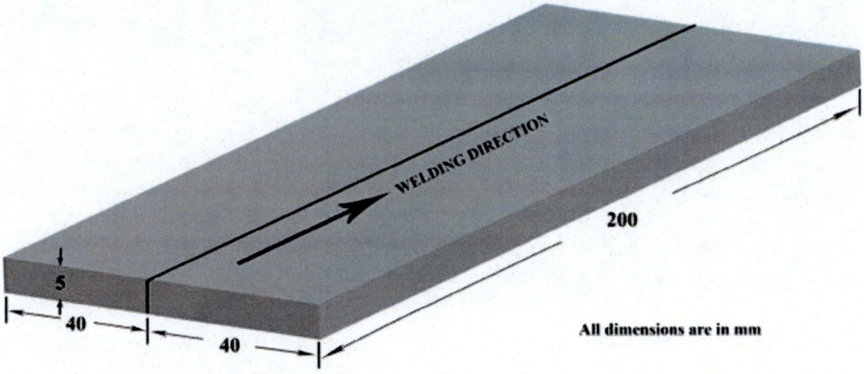

Fig. 1 Schematic of weld specimen

Fig. 2 Joint to be fabricated (Mg ZE41)

Fig. 3 Chamber for underwater FSW

Table 2 Mechanical properties of magnesium ZE41

Tensile strength	Modulus of elasticity	Poisson's ratio	Elongation	Brinell's hardness	Melting point
218 MPa	44.12 GPa	0.35	4.50%	55–70	532–638 °C

Figure 4 shows that the non-consumable tool was made up of high carbon steel with dimension of shoulder diameter (D) 15 mm, insert pin diameter (d) 5 mm and pin length (L) of 4 mm, respectively. Two plates were welded at rotational speed of 1220 rpm and travel speed of 40 mm/min. The tool's tilt angle used is 2.5° to get effective material flow during tool transverse. Various parameters used in this study were indicated in Table 3.

After the FSW process at both conditions, the samples were cut perpendicular to the direction of welding using electrical discharge machine. The cross section was observed by the optical microscopy (OM). Afterwards, the prepared samples were mounted and polished using SiC papers with different grit sizes. Polished surfaces were swabbed with the etching solution (picric acid, ethanol, acetic acid and water of 4.2 g, 70 ml, 40 ml and 10 ml, respectively) until the macrostructure was revealed. The etched samples were cleaned with ethanol for few seconds. Microstructure images were obtained by the optical microscope (Figs. 6, 8, 9 and 10).

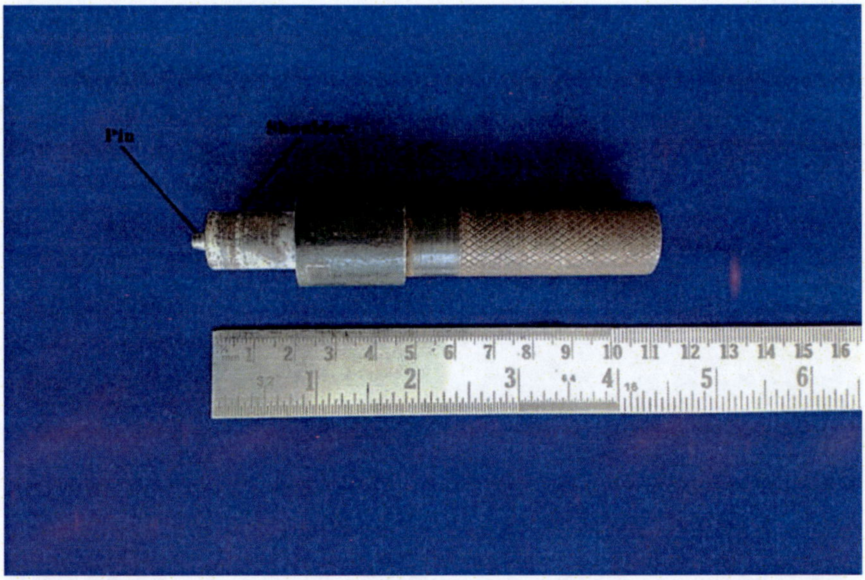

Fig. 4 FSW tool

Table 3 FSW process parameter and tool nomenclature

Parameter	Value	
Tool rotation speed (rpm)	660	1220
Tool transverse speed (mm/min)	40	
Axial force (KN)	5	
Tool tilt angle (°)	2.5	
Shoulder diameter (mm)	15	
Shoulder length (mm)	20	
Pin diameter (mm)	5	
Pin length (mm)	4	
D/d ratio of tool	3	
Tool profile	Taper cylindrical	

3 Results and Discussions

3.1 Appearances and Microstructures

Figure 6 shows the optical micrograph of ZE41 base alloy. The α-Mg matrix has T-Phase: Mg_7Zn_3-RE precipitates which are distributed over the matrix eutectically. There are some more finely distributed particles in the matrix; they are Zr_4Zn_3 or Zn_2Zr_3 zirconium-rich particles. These zirconium-rich particles have grain size of

Fig. 5 XL6026AV semi-automatic milling machine

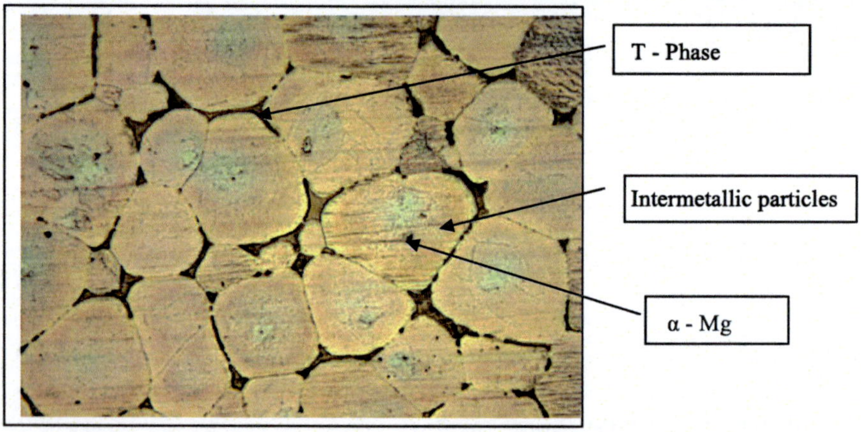

Fig. 6 Optical micrograph of ZE41 alloy

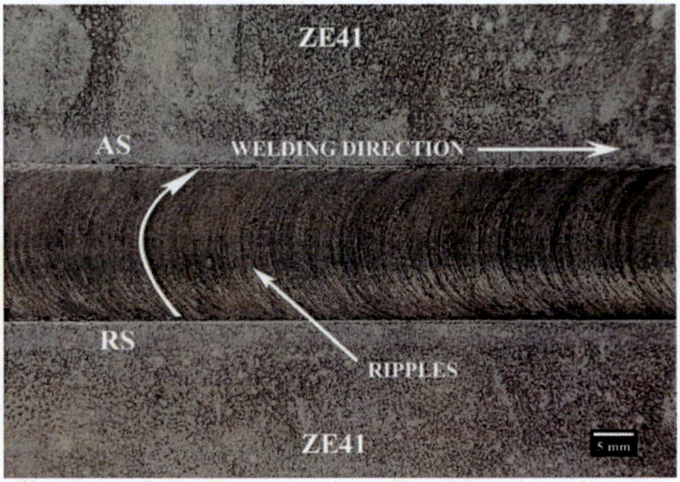

Fig. 7 Welded joint of ZE41 alloy

around 4 μm [8, 9]. From the Mg–Zr phase diagram, below 653 °C, the α-Zr particles and α–Mg particles are present in the magnesium matrix and they are not dissolved in the base material [10, 11]. The brighter contrast of the grain boundaries shows that the presence of higher atomic number of the secondary-phase particles in the grain boundary. The T-phases in the Mg matrix show a white- and light-coloured zone in the SEM micrograph of ZE41 alloy, and similar observations are reported by Neil et al. [12].

Figure 7 shows the FSW joint of ZE41. It also showed no defects on surface when the rotational speed of tool was 1220 rpm and transverse speed was 40 mm/min. And the spiral formation on upper surface of the joints can be seen. It shows that material flow during the welding process and the spiral obtained due to the relation between tool rotation and the substrate movement. Figures 8 and 9 represent the macrograph and its microfeatures of underwater FSW and FSW, respectively. The boundary between stir zone (SZ) and the thermomechanically affected zone (TMAZ) is called delineation. From Fig. 8, it can be seen that the delineation in the advancing side is sharper and the delineation in the retreating side is gradually changed. It evidences the state of plastic flow during welding process. In the stir zone, the friction effect is produced by the welding tool and the material on the advancing side is squeezed to the back of the pin [6].

Heat dissipation rate was high on the underwater friction stir welding since water at room was in contact with magnesium plate; thus, the thermal gradient developed during welding was dissipated from plate. So the frictional heat between tool and the specimen was absorbed by the surrounding water. So that there was no enough time given for growth of recrystallized grains than the air weld. On the other side, the heat dissipation rate in the normal air atmosphere was low where sufficient time was given for grain refinement than the underwater condition. So it had enough time

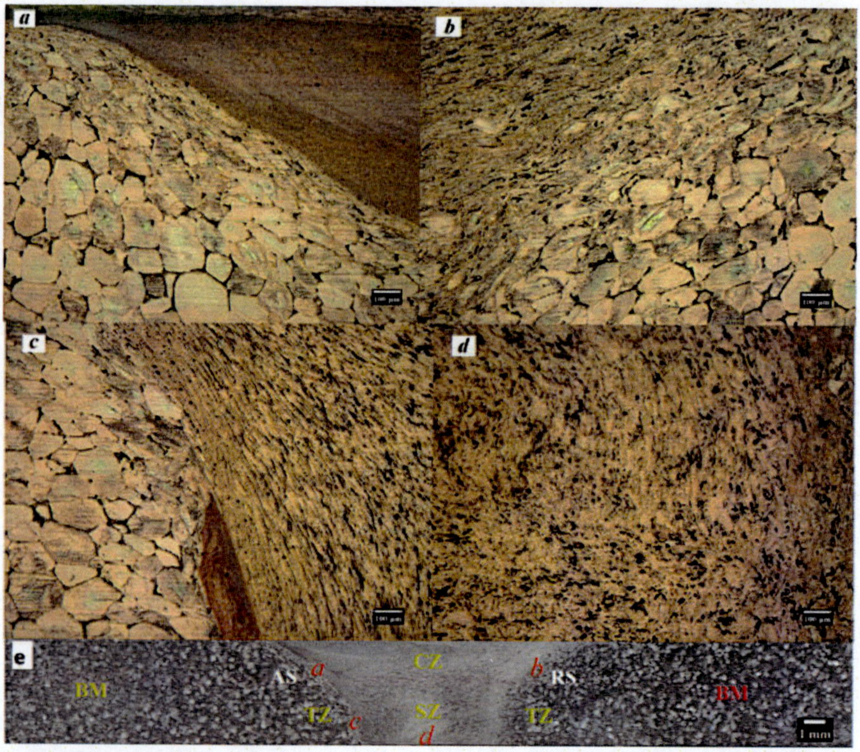

Fig. 8 Micrograph and macrograph of underwater welded sample: **a** advancing side, **b** retreating side, **c** SZ and base interface, **d** stir zone and **e** macroimage

for grain growth near the interface. So the geometry of stir zone in the air weld condition was quite higher than the underwater condition.

The macrostructure of friction stir welded ZE41 alloy is shown in the figure in which grain refinement occurred heavily in stir zone. The FSW weld joint reveals the bathtub-like curve which denoted the stir zone, a narrow region of interface like structure that divides the stir zone from HAZ in TMAZ (Figs. 8 and 9). This TMAZ appeared as a resultant of mechanical stirring effect occurring between tool and workpiece. The retreating and advancing side of the weld metal can be identified by grain refinement pattern. The sudden grain refinement was observed in advancing side due to the initial area of contact was at advancing side during the welding. Then the plasticized material was wiped towards the rotating direction of the tool. It is called as retreating side where the grain growth took place eventually.

Figure 8d shows that the stir zone of the underwater welding has smaller cross section when compared to the stir zone of the air weld FSW, because the higher heat dissipation in underwater limited the grain growth at the interface region.

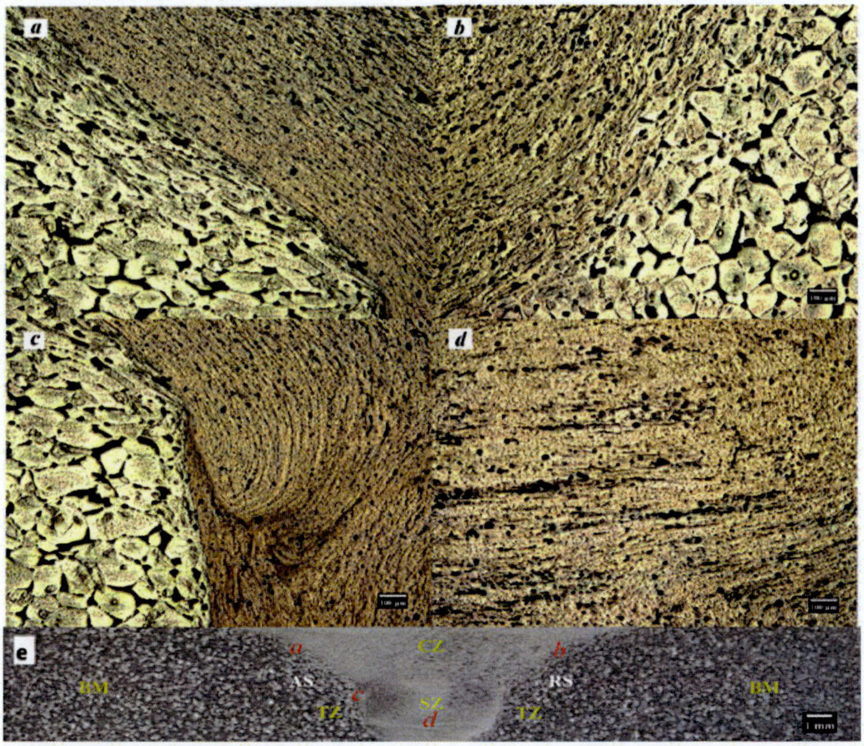

Fig. 9 Micrograph and macrograph of air welded sample: **a** advancing side, **b** retreating side, **c** SZ and base interface, **d** stir zone and **e** macroimage

Figures 10a and b show that the average grain size measured for underwater welded specimen is 25 μm and for the air welded specimen is 40 μm. This is due to the cooling rate was higher in the underwater FSW when compared to air FSW.

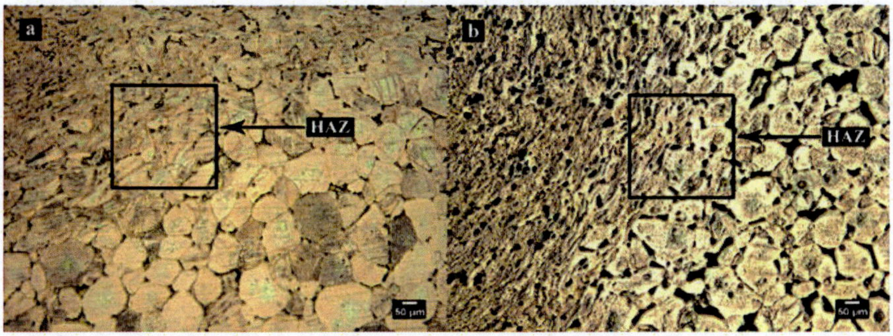

Fig. 10 Heat-affected zone: **a** underwater condition and **b** air welded condition

Table 4 Micro- and macrograph of tensile samples: (a) air welded conditions and (b) underwater welded conditions

S. No.	Macrograph	Micrograph	Observation
1.		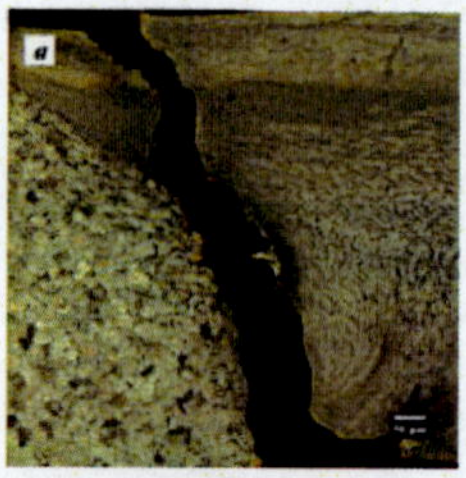	Failure occurs in the TMAZ
2.		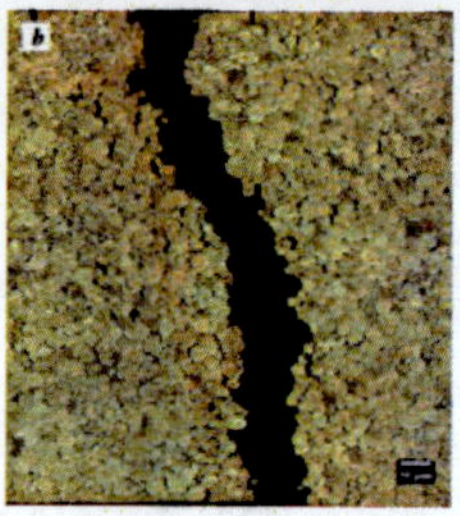	Failure occurs in the base

3.2 *Tensile Properties*

The tensile strength of underwater friction stir welded ZE41 alloys is up to 189 MPa and for air weld condition is 168 MPa. Maximum elongation of air welded sample is 20% and for underwater weld is 8%. The normal elongation for magnesium ZE41 is 4%. In air weld condition, the heat input at the stir zone is higher and it causes excessive grain growth. Therefore, the sample exhibits lower tensile strength when compared to the weld done at underwater condition.

From Table 4, we can see that the failure occurs in the stir zone for air weld specimen and the fracture is located at the base metal for underwater welded specimen. It denotes that the weld strength of the underwater specimen is higher than the strength of air weld specimen. The fracture location of friction stir weld underwater has less elongation than the air weld FSW. This is because of the formation of precipitation on the welding region decelerates the elongation of grains in interface.

4 Conclusions

In this study, the investigation of UWFSW and FSW on MgZE41 alloy was experimentally conducted in which the process parameter, resultant properties, microstructure and tensile behaviour were studied.

- Heat dissipation rate was high on the underwater friction stir welding since water at room was in contact with magnesium plate; thus, thermal gradient developed during welding was dissipated from plate. So the frictional heat between tool and the specimen was absorbed by the surrounding water. So that there was no enough time given for the growth of recrystallized grains than the air weld.
- On the other side, the heat dissipation rate in the normal air atmosphere was low. And so sufficient time was given for grain refinement than the underwater condition. So it had enough time for grain growth near the interface. So the stir zone in the air weld condition was quite higher than the underwater condition.
- The tensile strength of underwater friction stir welded ZE41 alloys was up to 189 MPa and for air weld condition was 168 MPa. Maximum elongation of air welded sample was 20%, and for under water weld was 8%. The normal elongation for magnesium ZE41 is 4%.
- The failure occured in the stir zone for air weld specimen, and it was located at the base metal in underwater welded specimen.
- The heat input at the stir zone was higher, and it caused excessive grain growth. Therefore, the sample exhibited lower tensile strength when compared to the weld done at underwater condition.

References

1. Rajasekaran R, Lakshminarayanan AK, Vasudevan M, Vasantharaja P, Role of welding processes on microstructure and mechanical properties of nuclear grade stainless steel joints. Proc IMechE Part L: J Mater: Design Appl. https://doi.org/10.1177/1464420719849448
2. Esparza JA, Davis WC, Murr LE (2003) Microstructure-property studies in friction-stir welded, thixomolded magnesium alloy AM60. J Mater Sci: Mater Electron 38(5):941–952
3. Zhang HJ, Liu HJ, Yu L (2011) Microstructure and mechanical properties as a function of rotation speed in underwater friction stir welded aluminum alloy joints. Mater Des 32:4402–4407
4. Esparza JA, Davis WC, Trillo EA, Murr LE (2002) Friction-stir welding of magnesium alloy AZ31B. J Mater Sci Lett 21(12):917–920
5. Zhao Y, Wang Q, Chen H, Yan K (2014) Microstructure and mechanical properties of spray formed 7055 aluminum alloy by underwater friction stir welding. Mater Des 56:725–30
6. Zhao Yong, Zhengping Lu, Yan Keng, Huang Linzhao (2015) Microstructural characterizations and mechanical properties in underwater friction stir welding of aluminum and magnesium dissimilar alloys. Mater Des 65:675–681
7. Huda Z, Edi P (2013) Materials selection in design of structures and engines of supersonic aircrafts: a review. Mater Des 46:552–60
8. Mofid MA, Abdollah-Zadeh A, Malek Ghaini F (2012) The effect of water cooling during dissimilar friction stir welding of Al alloy to Mg alloy. Mater Des 36:161–167
9. Koganti R, Lakshminarayanan AK, Ramprabhu T (2019) Understanding the effect of tool rotational speed on microstructure and mechanical properties of friction stir processed ZE41 grade magnesium alloy. In: Advances in materials and metallurgy, Lecture notes in Mechanical engineering. Springer Nature Singapore Pte Ltd. https://doi.org/10.1007/978-981-13-1780-4_41
10. Sanchez C, Nussbaum G, Azavant P, Octor H (1998) Structural material properties, microstructure and processing. Mater Sci Eng A 252:269

11. Neil WC, Forsyth M, Howlett PC, Hucthinson CR, Hinton BRW (2009) Corrosion of magnesium alloy ZE41—the role of microstructural features. Corros Sci 51:387–394
12. Genghua C, Datong Z, Wen Z, Cheng Q (2015) Microstructure evolution and mechanical properties of Mg–Nd–Y alloy in different friction stir processing conditions. J Alloy Compd 636:12–19

Characterizing the Tensile Deformation Behavior of Friction Stir Welded Dissimilar Joints Using Acoustic Emission Technique

A. Venkatakrishna, A. K. Lakshminarayanan, K. Radhika, and R. Rajasekaran

Abstract In the present investigation, P91 and 316LN dissimilar weld joints were fabricated at the tool speeds of 600 and 900 rpm with 0.5 and 1 mm offset conditions. For all mentioned welding conditions, the tensile testing was carried out at the constant strain rate of $3.3 \times 10^{-4}\ S^{-1}$. A non-destructive acoustic emission (AE) technique was used to analyze the tensile deformation behavior at elastic and plastic deformation stages. The AE signals were recorded until the crack initiation and the deformation behavior is compared with base metals. The obtained results were useful to identify the deformation characteristics and joint integrity in different tool speed and offset conditions.

Keywords Dissimilar weld joints · Tensile deformation · Acoustic emission · Strain rate · AE signals

1 Introduction

The dissimilar weld joints of P91 steel and low carbon and nitrogen enhanced stainless steel 316LN were widely used in pressure vessels, piping materials, and steam generator applications in fast-breeder nuclear reactors [1, 2]. During sudden start-ups and shutdowns of the nuclear reactor, high cyclic thermal stresses were generated due to the mismatches in thermal expansion coefficients of the dissimilar weld metals. This cyclic stresses are superimposed on the residual welding stresses and leads to service failure/premature failure during long time service of the welded structure [3]. Therefore, by online monitoring, the material behavior through a wide range of non-destructive testing techniques, the real-time information on material integrity can be predicted and the failures of the welded structures can be controlled in advance.

Acoustic emission technique (AE) is a non-destructive technique that used to monitor the structural integrity, dislocation kinetics, crack initiation and inclusions

A. Venkatakrishna (✉) · A. K. Lakshminarayanan · K. Radhika · R. Rajasekaran
Department of Mechanical Engineering, Sri Sivasubramaniya Nadar College of Engineering, Chennai, Tamil Nadu 603110, India
e-mail: venkatakrishnaa@ssn.edu.in

S. Vijayan et al. (eds.), *Trends in Manufacturing and Engineering Management*, Lecture Notes in Mechanical Engineering,
https://doi.org/10.1007/978-981-15-4745-4_68

in various types of materials [4]. Due to its simplicity in testing, the deformation characteristics have been analyzed in different kinds of materials such as brass [5], copper [6], aluminum [7], and magnesium [8]. But only a few researchers [9–15] have shown interest in characterizing the tensile behavior in a wide variety of steels and stainless steels. During tensile deformation, the variation in AE signals will depend on the physical, chemical, and the structural properties of the base materials. The acoustic signals will generate high activity due to high precipitate content, low dislocation densities, different grain sizes, and variation in carbon content in different weld thermal cycles [9]. Khamedi et al. [10] investigated the acoustic behavior of DP steel under tensile loading. The rate of deformation and modes of failure were comparatively correlated with the use of AE frequency signals and the counts. Barat et al. [11] studied the tensile deformation behavior of AISI-type 304LN SS in low temperature, it is depicted that the AE counts were gradually increased due to the evolution of martensite during low-temperature deformation. Han et al. [12] studied tensile deformation behavior of Q345 steel weld and base metals. It is reported that the crack initiation in weld was comparably higher than the base metal. It is due to the presence of oxide inclusions and coarse grain structures in the weld samples. Raj et al. [13] studied the systematic variation and a high source of AE signals that generated during the prior stage of deformation in austenitic stainless steels; it is due to the material dislocations and multiplication of existing dislocation, i.e., Frank-Read grain boundary during the tensile straining. Venkataraman et al. [14] monitored the AE activity of 316 austenitic stainless steel, and the AE count rate is varied with respect to different strain rates during tensile testing. In maximum straining condition, the generated AE counts and signal frequency were high due to high deformation rate and the source activation of acoustic signals. Haneef et al. [15] investigated the tensile behavior of grade 316 austenitic stainless steel. The AE parameters such as RMS voltage and dominant frequency were monitored in different strain rates from $1.4 \times 10^{-3}\ S^{-1}$ to $1.4 \times 10^{-2}\ S^{-1}$. The RMS voltage was increased with the increase in the strain rate. The dominant frequency was decreased during necking, but it is increased due to the increased dislocation velocity at high strain rates.

From the above literature, it is identified that by using AE technique, the tensile deformation behavior was characterized in various kinds of metallic and non-metallic materials. But only a few studies were available on characterizing the weld deformation behavior using the AE technique. Moreover, no work has been carried out on monitoring the tensile deformation behavior of ferritic–martensitic–austenitic dissimilar weld joints using a friction stir welding process. In this present study, an attempt was made to study the acoustic emission characteristics during the tensile deformation of P91-316LN dissimilar joints. The weld integrity in different tool speed and offset conditions was characterized, and the differences in AE parameters were compared with base metal studies.

Table 1 Physical properties of the base metals

Base metals	Density (kg/m^3)	Specific heat (J/kg K)	Melting point (°C)	Thermal conductivity (W/mK)
P91	7.77×10^{-9}	620	1420	26–28
316LN	7.99×10^{-9}	470	1450	16.2–16.5

2 Experimental Details

2.1 Materials and Weld Specimens

The materials used in the present study were P91, a nuclear grade ferritic/martensitic steel, and 316LN, a corrosion-resistant austenitic stainless steel. The physical properties of the base metals and its chemical compositions were listed in Tables 1 and 2.

The normalized and tempered condition of P91 steel and solution annealed 316LN was received in the form of 10-mm-thick plates. With the use of an electrical discharge machine (EDM), the plates were sliced to the thickness of 3 mm. The sliced plates were fabricated to the length of 125 mm and a breadth of 60 mm. Before conducting the FSW trials, the machined plates were cleaned with methyl alcohol (CH_3OH) solution, and the centerline of the weld plates was tightly clamped to each other. The P91 steel plate was arranged in the retreating side, and the 316LN SS plate was arranged in the advancing side. By using tungsten lanthanum oxide tool, FSW process was carried out. The weld trials were carried out at the rotational speeds (ω) of 600 and 900 rpm with 0.5 and 1 mm offset towards the P91 side. For all FSW trials, constant welding speed (v) of 50 mm/min was maintained. The process parameters used for the FSW process were shown in Table 3. The representation of the FSW tool and its schematic view of the FSW process were shown in Fig. 1.

2.2 Acoustic Emission Monitoring and Testing Procedure

The weld joints and the base metals were machined according to ASTM E8M-09 standards for testing the tensile properties of the weld joint. The graphical representation of the tensile sample is shown in Fig. 2a. The tensile deformation behavior was characterized by using 100 kN hydraulic tensile testing machine (Make: INSTRON, Model: 8801), and the testing was accomplished at the constant strain rate of 3.3×10^{-4} S^{-1}.

Two piezoelectric AE sensors (Make: Physical Acoustics, Model: micro 30) with a frequency range of 150–400 kHz was fixed at the shoulder regions of the tensile samples. A very small quantity of silicon grease gel was used as a sensor couplant in gauge transition regions. To obtain proper signals without any disturbances,

Table 2 Chemical compositions of the base metals measured in (wt%)

	C	Cr	Mo	Ni	Si	Mn	N	Cu	P	V	Nb	Fe (bal)
P91	0.09	9.04	0.95	0.42	0.28	0.37	0.04	0.15	0.01	0.23	0.09	88.3
316LN	0.02	18.0	2.74	12.3	0.54	1.97	0.08	0.19	0.04	–	0.005	63.9

Table 3 FSW process parameters

Parameters	Notations	Units	Process variables
Tool and pin profile			Taper cylindrical
Tool pin offset	x	mm	0.5, 1.0
Average axial load	F	kN	32 kN
Rotational speed	N	rpm	600, 900
Welding speed	v	mm/min	50

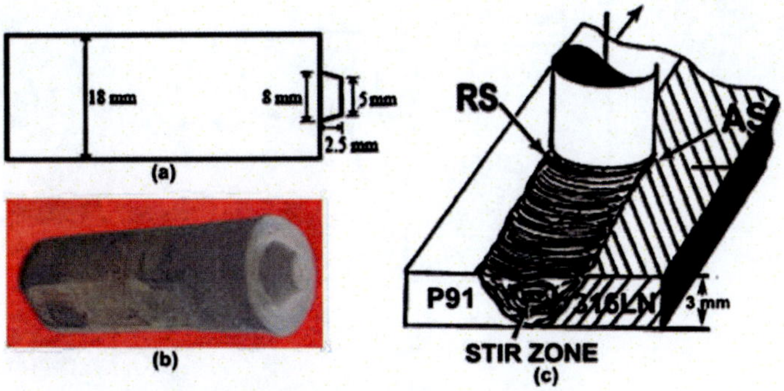

Fig. 1 FSW tool, dimensions, and the welding process

petroleum-based lubricant gel was applied between the sensor interface and around the specimen surface. During tensile loading, the AE sensor starts sensing the deformation levels and generates the acoustic signals in the shoulder region. These signals were captured and processed by pre-amplifier and 100–300 kHz range bandpass filter. An average threshold frequency was set to eliminate the external noises that occurred during tensile testing. The AE signals produced during tensile deformation were simultaneously recorded and processed by using an AE workstation (Make: Physical electronics), and the tensile deformation results were evaluated by using AEwin disp software v2.19. Finally, the fractured surfaces of the deformed tensile samples were studied by using scanning electron microscopy (Make: Hitachi, Model: 3400 N). The graphical and pictorial views of the AE experimental setup were shown in Fig. 2b.

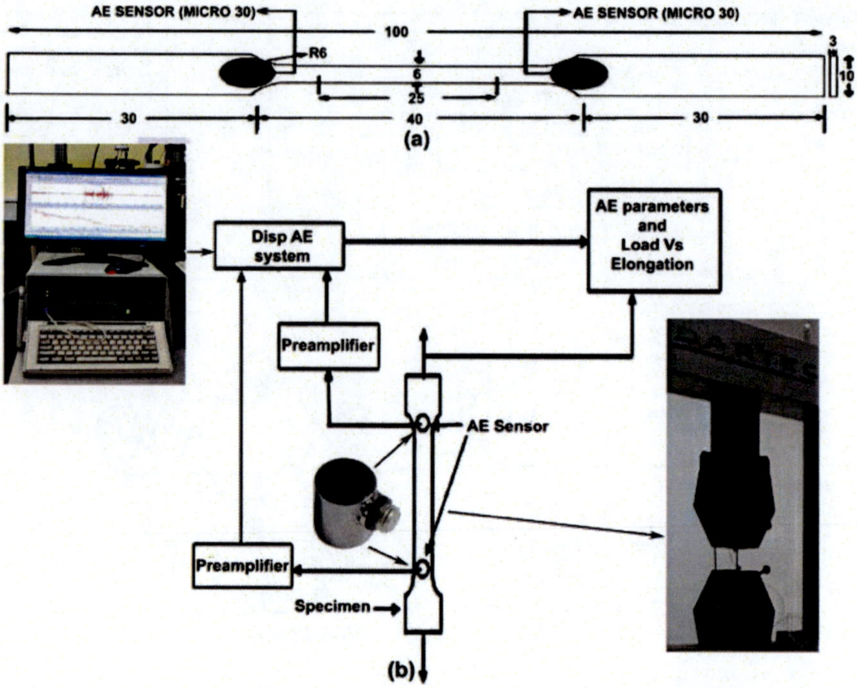

Fig. 2 Schematic of tensile sample and AE experimental setup with block diagram

3 Results and Discussion

3.1 Tensile Properties and Joints Characteristics

The tensile properties of the base metals were illustrated along with the acoustic emission characteristics as shown in Fig. 3. The tensile and yield strength of P91 steel were high around 36 and 150 Mpa. But the rate of elongation was low around 32% as compared to 316LN SS. In P91 steel (Fig. 3a), high tensile strength was observed due to co-precipitation of MX and $M_{23}C_6$ precipitates in ferrite and coarse martensite grain boundaries [16]. In 316LN (Fig. 3d), the elongation percentage was increased due to the effect of work hardening. Dissimilar joints between these two materials exhibited different tensile properties in different tool speed and offset conditions as shown in Figs. 4 and 5. In both 600 and 900 rpm weld conditions, the tool was offset toward the P91 side. The tensile strength of 600 rpm 0.5 condition (Fig. 4a) was reduced due to inadequate heat input and insufficient mixing of 316LN material toward the P91 area. Therefore, the tensile strength of 600 rpm and 0.5 mm weld condition was reduced around 110 Mpa than the base metal. Similarly, by increasing the offset up to 1 mm toward P91 side (Fig. 4d), the weld region was not softened

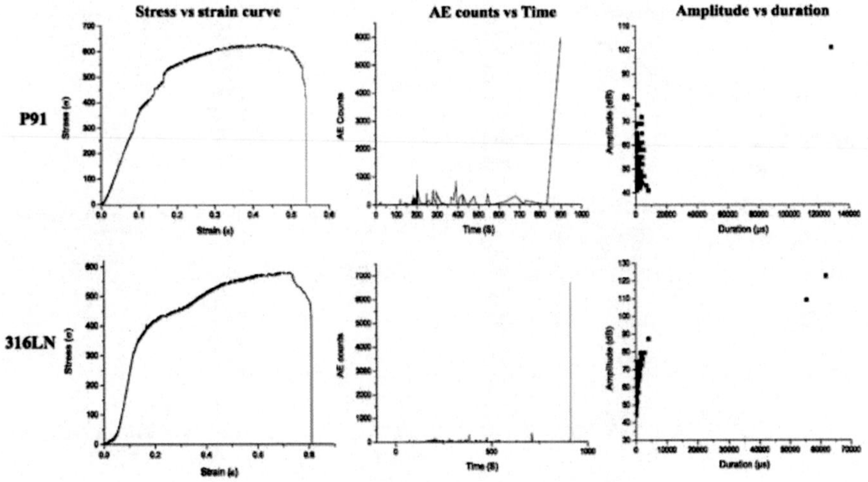

Fig. 3 Tensile properties of the base metals and its AE characteristics

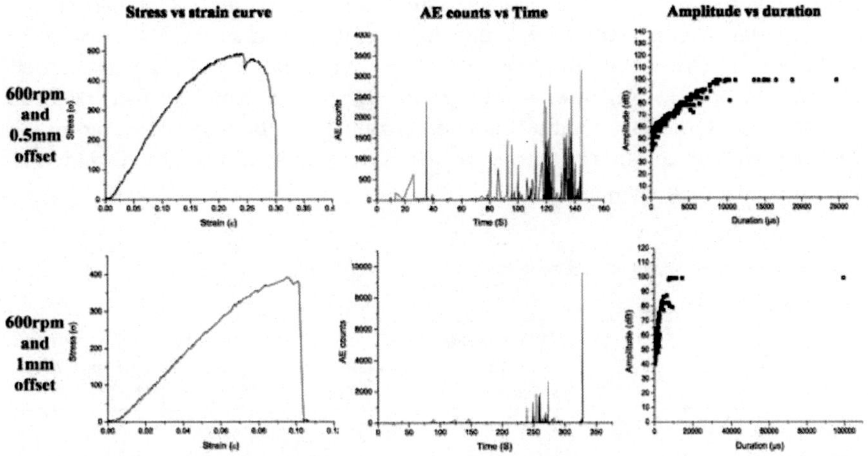

Fig. 4 Tensile properties of 600 rpm joints and its AE characteristics

properly. So, the defects such as micropores and tunnel defect were developed around the 316LN HAZ side. Therefore, the tensile strength was drastically decreased around 315 Mpa. Moreover, in both offset conditions, the fracture was observed in the weld region.

In 900 rpm weld with 0.5 offset condition (Fig. 5a), material mixing proportions were good due to increased heat input. The obtained tensile strength was almost equal to 316LN base metal properties. When compared to both base materials, a slight drop in strength around 25 Mpa was noticed. It is due to the coarsening effect

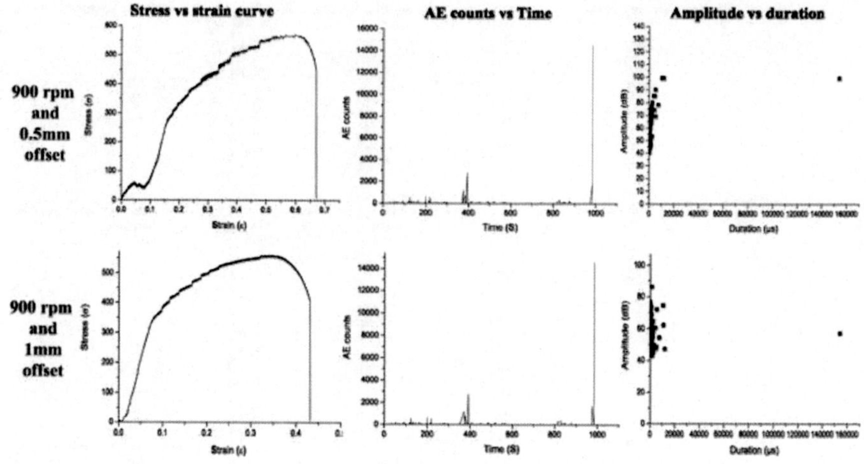

Fig. 5 Tensile properties of the 900 rpm joints and its AE characteristics

of grains during welding. The yield strength was increased up to 66% than 316LN, and the elongation percentage was increased up to 21% than the P91 base metal. But in 1 mm offset condition (Fig. 5d), the mixing ratio of 316LN was reduced; therefore, the tensile properties are comparatively reduced than 0.5 offset condition. Moreover, in both offset conditions, ductile mode of fracture was observed in P91 HAZ region. The variation in tensile properties for both base metals and weld joints was illustrated in Table 4.

Table 4 Tensile test results of the base metals and the weld joints

Strain rate 3.3×10^{-4} S^{-1}						
P91		316LN	600 rpm and 0.5 mm offset	600 rpm and 1 mm offset	900 rpm and 0.5 mm offset	900 rpm and 1 mm offset
Tensile strength (Mpa)	618	582	479	303	569	554
Elongation (%)	54	82	31	12	75	60
Yield strength (Mpa)	431	290	262	212	356	304

3.2 *Acoustic Emission Characteristics During Tensile Deformation*

AE behavior during tensile testing was characterized in four regions; they are: (1) 0–3% elastic deformation region, (2) 3–10% yield region, (3) UTS and necking region, and (4) final fracture after UTS [15]. During tensile testing, the transient elastic AE waves were generated from the shoulder region of the tensile sample and the rate of deformation is identified in the form of counts. The emission characteristics during AE analysis differ in three stages—they are (1) crack initiation, (2) crack propagation, and (3) the final stage of failure. In the crack initiation stage, the emission characteristics will be low, but in crack propagation stage, the emission characteristics will be increased, and during the fracture stage, the AE signal and emission characteristics will be in maximum peak level [17]. Similar to these emission characteristics, the variation of AE counts with respect to time was identified in P91 and 316LN base metals and its weldments.

In P91 base metal (Fig. 3b), the AE counts have been increased at the beginning stage of plastic deformation, but sudden rise and fall of signals were observed till the yield region; it is due to the movement of dislocations during straining. In crack initiation region, the dislocation densities were increased and necking formation was observed, and the amplitude signals were slightly increased with respect to the acoustic signals. But at the stage of crack propagation, the dislocation motion and its velocity were increased and the rise in AE signals was observed. With the increase of AE counts at maximum straining condition, the duration of acoustic signals (Fig. 3c) has increased to 70,000 μS. Generally, the AE signals generated during deformation were mainly depended upon the microstructure of the metal, strain rate, and presence of any inclusions [18]. Due to single-phase austenite microstructure in 316LN base metal, AE signals are observed at the constant strain levels corresponding to the phenomenon of macroyielding (Fig. 3e). But compared to P91 steel, the AE counts were less during crack initiation. In the crack propagation region, AE signals were generated due to the effect of dislocation motion and twinning. At the final stage of fracture, the burst AE signals were observed due to the increase in source activation signals at the constant strain rate. Therefore, the duration of acoustic with respect to amplitude (Fig. 3f) was increased up to 140,000 μS.

In 600 rpm weld joint with 0.5 offset condition (Fig. 4b), the AE counts were slightly increased until the homogenous plastic deformation stage. But at the beginning stage of crack propagation, the AE counts were gradually increased until the end of the tensile test. Due to high dislocation movements in the weld region, the stress concentration was increased and the material was failed in an earlier stage as compared to base metals. Therefore, the amplitude of AE signals was increased up to 100 dB and the AE burst signals were increased until the minimum duration of 25,000 μS (Fig. 4c).

In 1 mm offset condition (Fig. 4d), the AE counts were increased at the beginning stage of plastic deformation. Sudden rise and fall of AE peaks were continuously monitored in the crack initiation region. But at the crack propagation region, the AE counts have increased again till the occurrence of a failure in the tensile specimen. The sample was failed without necking, so it is concluded that the sample is failed due to the presence of increased dislocations in the weld region. Therefore, the material was failed in an earlier stage as compared to 600 rpm 0.5 offset weld condition. Similarly, the AE signals were constantly increased up to 100 dB, and the AE burst signals were increased until the minimum duration of 25,000 μS (Fig. 4e).

In 900 rpm weld joint with 0.5 offset condition (Fig. 5b), the AE counts were very low until the final stage of elastic deformation, but a sudden increase of AE signals was noticed due to the decohesion of inclusions during crack initiation. The recorded AE counts were too weak for a long time up to 800 s of testing time (i.e., up to necking), because the rate of deformation of this weld joint was high as compared to other welds. Due to improved tensile properties, AE counts was increased upto 160,000 μs during crack propagation stage of the tensile sample. Similarly, in 1 mm offset condition (Fig. 5f), a peak of AE emission was noticed due to the presence of coarse grains in the plastic deformation region. But other than the obtained AE signals, the counts were almost the same as the 0.5 mm offset condition. But the amplitude frequency signals were reduced up to 40 dB than the 0.5 mm offset condition; this change is attributed with the effect of moving dislocations associated with the tensile deformation.

4 Conclusion

Using the AE technique, tensile deformation behavior has been successfully monitored for P91-316LN FSW joints that welded at the tool speed of 600 and 900 rpm with 0.5 and 1 mm offset conditions. From the experimental investigations, some important results were observed and it is listed below as follows:

1. In 600 rpm welds, AE signals were generated in yielding stage of deformation—it is mainly associated due to the increased fraction of dislocation density in the weld region.
2. Due to the improper coalescence of dissimilar materials in 600 rpm weld condition, the ductility of the joint has been comparably reduced in 1 mm offset condition. Hence, AE counts and signal duration were found to be increased at the beginning stage of plastic deformation.
3. In 600 rpm weld joints, crack initiated in weld region, and for 900 rpm joints, the fracture occurred in P91-HAZ region for both offset conditions. The 900 rpm joints exhibited good tensile properties as compared to 600 rpm joints. But compared to base metals, the strength of 900 rpm joints was reduced due to the presence of inclusions and coarse grains in the weld joints.
4. During necking, frequencies and amplitude signals were found to be increased in 900 rpm weldments; it is attributed to the increase of the material dislocations during deformation at the constant strain level.

5. The amplitude signals and AE waves generated during tensile deformation were varied in both 600 and 900 rpm weld joints. In 600 rpm condition, the peaks were totally varied due to the presence of defects in 1 mm offset condition. But in 900 rpm, slight changes in AE counts and amplitude were observed during deformation in both offset conditions.

Therefore, it is concluded that by increasing the tool speed with a low offset value, the defect formation in dissimilar metals has been reduced and the joint integrity has been improved. Moreover, by using the AE technique, overall material life can be predicted in earlier stages of deformation; therefore, the joint failures can be controlled in advance.

Acknowledgements Authors are thankful to Mr.T Haneef, Scientific Officer D, and Dr. C.K Mukhopadhyay Head, NDTD/MMG, IGCAR, Kalpakkam, India, for providing the acoustic emission test facility in their laboratory. The authors are thankful to Mr. Thirukumaran, Research fellow, NDTD/MMG, IGCAR, for providing constant encouragement and support during the experimental analysis.

References

1. Zinkle SJ, Ghoniem NM (2011) Prospects for accelerated development of high-performance structural materials. J Nucl Mater 417:2–8
2. Chai M, Duan Q, Hou X, Zhang Z, Li L (2016) Fracture toughness evaluation of 316LN stainless steel and weld using acoustic emission technique. ISIJ Int 56(5):875–882
3. Bhaduri AK, Venkadesan S, Rodriguez P (1994) Transition metal joints for a steam generators—an overview. Int J Press Vessels Pip 58:251–265
4. Ríos-Soberanis CR (2011) Acoustic emission technique, an overview as a characterization tool in materials science. J Appl Res Technol 9(3):367–379
5. Pattnaik A.B. Jha B.B and Sahoo R.: Effect of strain rate on acoustic emission during tensile deformation of α-brass. Materials Science and Technology 29(3),294–299 (2013)
6. Hunaidi O, Chu WT (1999) Acoustical characteristics of leak signals in plastic water distribution pipes. Appl Acoust 58:235–254
7. Skalski VR, Lyasota IM (2014) Features of acoustic-emission signals during the initiation of a fatigue failure in a welded joint of an aluminum alloy of the Al Cu-Mn system. Russ J Nondestr Test 50(2):120–126
8. Máthis K, Chmelik F, Trojanová Z, Lukác P, Lendvai J (2004) Investigation of some magnesium alloys by use of the acoustic emission technique. Mater Sci Eng A 387–389:331–335
9. Wadley HNG, Scruby CB (1991) Cooling rate effects on acoustic emission-microstructure relationships in ferritic steels. J Mater Sci 26(21):5777–5792
10. Khamedi R, Fallahi A, Refahi Oskouei A (2010) Effect of martensite phase volume fraction on acoustic emission signals using wavelet packet analysis during tensile loading of dual phase steels. Mater Design 31(6):2752–2759
11. Barat K, Bar HN, Mandal D, Roy H, Sivaprasad S, Tarafder S (2014) Low-temperature tensile deformation and acoustic emission signal characteristics of AISI 304LN stainless steel. Mater Sci Eng A 597:37–45
12. Han Z, Luo H, Cao J, Wang H (2011) Acoustic emission during fatigue crack propagation in a micro-alloyed steel and welds. Mater Sci Eng A 528(25–26):7751–7756
13. Raj B, Jayakumar T (1991) Acoustic emission during tensile deformation and fracture in austenitic alloys. ASTM-STP-1077, p 218

14. Venkataraman B, Mukhopadhyay CK, Raj B (2004) Effect of variation of strain rate on thermal and acoustic emission during tensile deformation of nuclear grade AISI type 316 stainless steel. Mater Sci Technol 20:1310–1316
15. Haneef T, Lahiri BB, Bagavathiappan S, Mukhopadhyay CK, Philip J, Rao BPC, Jayakumar T (2015) Study of the tensile behavior of AISI type 316 stainless steel using acoustic emission and infrared thermography techniques. J Mater Res Technol 4(3):241–253
16. Zhang S, Melfi T, Narayanan BK (2016) Effects of precipitates on mechanical properties of P91 submerged arc welds. Sci Technol Weld Joining 21(2):147–156
17. Yu J, Ziehl P, Zárate B, Caicedo J (2011) Prediction of fatigue crack growth in steel bridge components using acoustic emission. J Constr Steel Res 67(8):1254–1260
18. Ono K (2005) Current understanding of mechanisms of acoustic emission. J Strain Anal Eng Design 40(1):1–15

Study of Infrared Thermography on Tensile Behavior of Laser Beam Welded 316LN Austenitic Stainless Steel

R. Rajasekaran, A. K. Lakshminarayanan, A. Venkatakrishna, and K. Radhika

Abstract In this investigation, the tensile behavior of laser beam welded 316LN austenitic stainless steel was studied by infrared thermography (IRT) and compared with the base metal tensile behavior. Initially, microstructural characterization of base metal and weld bead was carried out by optical microscopy (OM). Base metal comprises an average grain size of 60 μm and LBW exhibited refined grains at the fusion zone. Very narrow HAZ around 15 μm was recorded at LBW interface. Tensile samples were prepared as per ASTM standard E8, and the test was conducted at room temperature of 26.5 °C. A strain rate of 4.4×10^{-4} s^{-1} was used during the tensile test of the base metal and LBW samples. Temperature variations of the base metal and weld bead were recorded by an infrared camera at different stages of deformation. Compared to LBW sample base metal sample deformed more and displayed the percentage of elongation as 64.14%, whereas LBW sample displayed less percentage of elongation around 53.2% due to lesser deformation as compared to the base metal. This is attributed to grain refinement during rapid solidification of LBW process at fusion zone. High temperatures were recorded on the base metal (39.2 °C) and center of the LBW fusion zone (30.8 °C) before just the time of fracture. This high temperature of the base metal over LBW sample indicates that the base metal deformed more compared to laser beam welded sample.

Keywords Infrared thermography technique (IRT) · Laser beam welding (LBW) · 316LN austenitic stainless steel · Stain rate · Deformation

1 Introduction

Austenitic stainless steels are used widely in nuclear power plants for their excellent mechanical and corrosion resistance properties. Many structural components, except the nuclear core structures, are built by 316LN grade austenitic stainless steel for

R. Rajasekaran (✉) · A. K. Lakshminarayanan · A. Venkatakrishna · K. Radhika
Department of Mechanical Engineering, Sri Sivasubramaniya Nadar College of Engineering, Chennai, Tamil Nadu 603110, India
e-mail: rajasekaranr@ssn.edu.in

S. Vijayan et al. (eds.), *Trends in Manufacturing and Engineering Management*, Lecture Notes in Mechanical Engineering,
https://doi.org/10.1007/978-981-15-4745-4_69

its superior mechanical and corrosion resistance property at high temperatures [1]. During the structural component fabrication process the joining of 316LN together is done using conventional GTAW process. Many research works are going on joining of 316LN using high energy density and solid-state welding techniques [2, 3]. Joining of 316LN using the LBW process could improve mechanical properties in the fusion zone. Improvement in toughness property by LBW process on 409 M ferritic stainless steel was reported in earlier research work [4]. During the service, 316LN joints may experience different types of load/stress conditions. This phenomenon leads to deformation in that particular material as well as joints. Hence, the study of the deformation behavior of 316LN base and its joints is essentially one.

Infrared thermography is a contactless nondestructive technique used in many engineering applications due to the accurate monitoring of temperature variations in fast inspection rates [5]. Assessment of material deformation is successfully carried out on metals, polymers, and composite materials using infrared thermography technique [6–8]. This is followed by the concept of infrared radiations from a deforming object detected by IR camera, and the radiations are converted in terms of temperature patterns. These temperature patterns are stored and analyzed using computer workstations. Deformation distribution in the material and its severity is evaluated by the intensity of the temperature patterns. In this experimental investigation, IRT has been used for studying tensile deformation of laser beam welded 316LN austenitic stainless steel.

2 Experiment

2.1 Material

Austenitic stainless steel 316LN was taken to study about IR thermography during the tensile test. The details of the chemical composition of 316LN austenitic stainless steel are given in Table 1. Plate form of the base material with a length of 300 mm, the width of 75 mm, and a thickness of 3 mm was prepared to make weld joint. Edge preparation was made by square butt configuration. Argon with a purity level of 99.995% was used as a purging gas during the welding process. The following process parameters were used for fabricating LBW joint: (i) laser power: 2.5 KW and (ii) welding speed: 1500 mm/min.

2.2 Microstructural Evolution

Metallographic samples of base metal and weld joints (transverse side) were polished up to 2500 girt emery sheet. Further samples were polished by using 0.5-μm diamond paste to eliminate micro-sized impurities. After polishing, etching process was done

Table 1 Chemical composition of 316LN austenitic stainless steel

	Weight % of elements									
	C	Cr	Ni	Mo	Mn	Si	S	P	N	Fe
316LN	0.026	17.8	11.7	2.4	1.6	0.45	0.009	0.026	0.095	Balance

by a solution which contains 1:3 ratios of HNO_3 to HCl as per ASTM standard designation E407. Microstructures of base metal and the LBW fusion zone were captured by olymphus BX 51 M optical microscope.

2.3 *Infrared Thermography*

Figure 1 depicts experimental setup of IRT during the tensile test. Figure 2 depicts samples of base metal and LBW after the tensile test. Tensile test of the base metal and LBW samples were performed by Dartec, UK, 100KN hydraulic servo motor computer-controlled universal testing machine (UTM). 2 mm/min crosshead speed

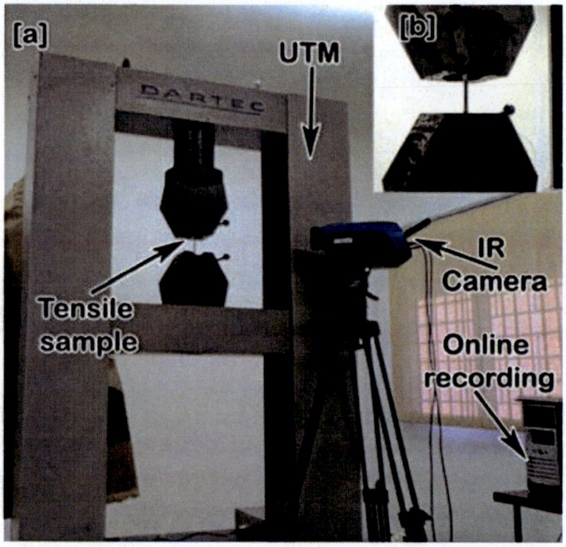

Fig. 1 Experimental setup of IRT during tensile test, **a** universal testing machine and IR camera and **b** tensile sample of LBW

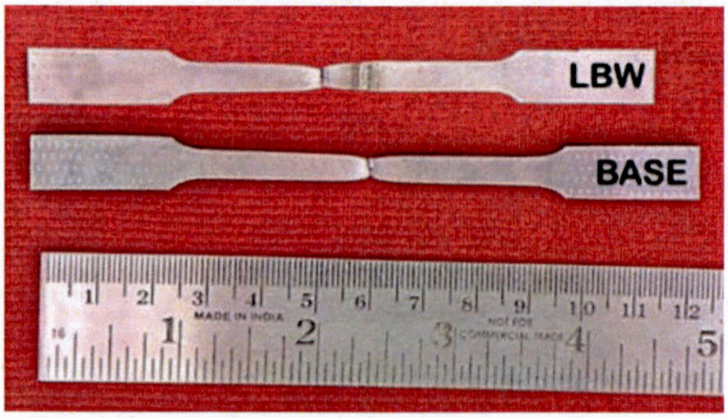

Fig. 2 Samples after the tensile test

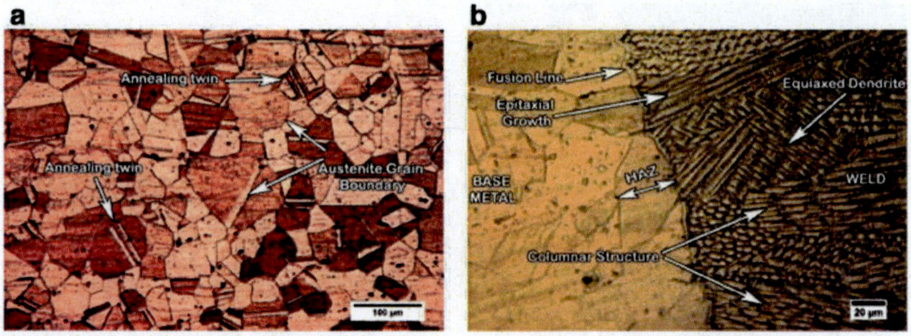

Fig. 3 Microstructures, **a** 316LN base, **b** LBW: fusion zone–base metal interface

was maintained during the tensile test. Thermographic images were recorded by the IR camera model: CEDIP silver 420 with 25 m/C temp sensitivity.

3 Result and Discussion

3.1 Microstructural Characterization

Figure 3 depicts the microstructural morphologies of 316LN base metal and LBW joint. The base metal consists complete austenite phase with grain size around 60 μm. Annealing twins were clearly observed in the base metal microstructure. LBW fusion zone exhibited the primary austenite mode of solidification. Following morphologies were recorded at the fusion zone of the LBW process viz: (i) epitaxial grain growth, (ii) equiaxed dendrite, and (iii) columnar structure. Higher welding speed of LBW process resulted in the formation of very narrow weld bead and HAZ around 15 μm [9].

3.2 Deformation Study by IR Thermography

Thermographic images of the base metal at different stages during tensile deformation are shown in Fig. 4a. The highest temperature around 39 °C was recorded at stage III (before fracture) in the base metal. Temperature curve and its different stages are explained in Sect. 3.3. Temperature evolution curves along the axial length of the base metal sample at different stages are shown in Fig. 4b. Necking was observed almost center of the gauge portion in the base metal. This is clearly correlated with the peak point of the temperature evolution curve.

Thermographic images of the LBW sample at different stages during tensile deformation are shown in Fig. 5a. The highest temperature around 32 °C was recorded

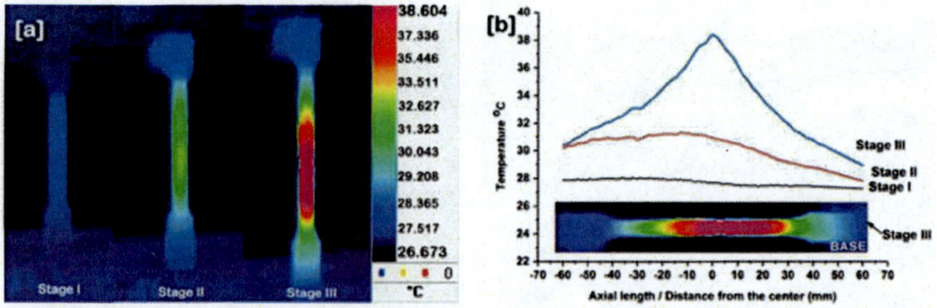

Fig. 4 **a** Thermographic images during tensile deformation of base metal and **b** evolution of temperatures in axial length of the base metal

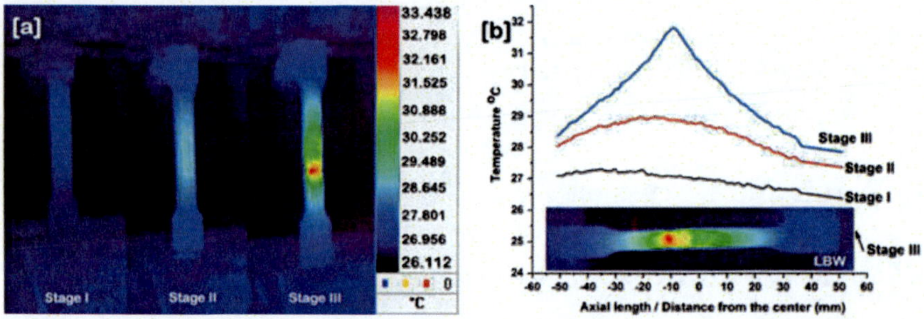

Fig. 5 **a** Thermographic images during tensile deformation of LBW joint and **b** evolution of temperatures in axial length of LBW sample

at stage III (before fracture) in LBW sample. Temperature evolution curves along the axial length of the LBW sample (Fig. 5b) revealed that the position of the peak temperature is away from the center of the gauge portion (LBW fusion zone).

3.3 *Tensile Behavior Versus Peak Temperature*

Transverse tensile properties of the base metal and the LBW joint are given in Table 2. Base metal showed more percentage of elongation as compared to the LBW tensile sample. This phenomenon is mainly attributed to the grain refinement at the fusion zone during the solidification process. Failure of LBW sample occurred at the base metal side around 8 mm from the fusion zone. This failure location indicating that the LBW joint contains more tensile strength compared to the base metal. Further micro-tensile samples [10] were extracted from LBW fusion zone along weld direction to asses UTS of LBW joint. From UTS evaluation, it was seen that the LBW joint exhibited 11.7% more strength (688 MPa) compared to the base metal UTS value.

Table 2 Tensile properties of base and LBW joint

S. No.	Sample specification	Crosshead speed (mm/min)	Strain rate (s^{-1})	Yield (MPa)	UTS (MPa)	% of Elongation
1.	316LN Base	2	4.4×10^{-4}	319	614	64.14
2.	LBW	2	4.4×10^{-4}	320	619	53.2

Evolution of peak temperature with the stress plot of the base metal and the LBW samples is shown in Figs. 6 and 7, respectively. This temperature curve is divided into four main stages based on the deformation behavior of the material. They are (i) linear temperature drop stage, (ii) temperature rise stage, (iii) rapid temperature rise stage, and (iv) final temperature decline stage [11, 12]. The considerable temperature drop

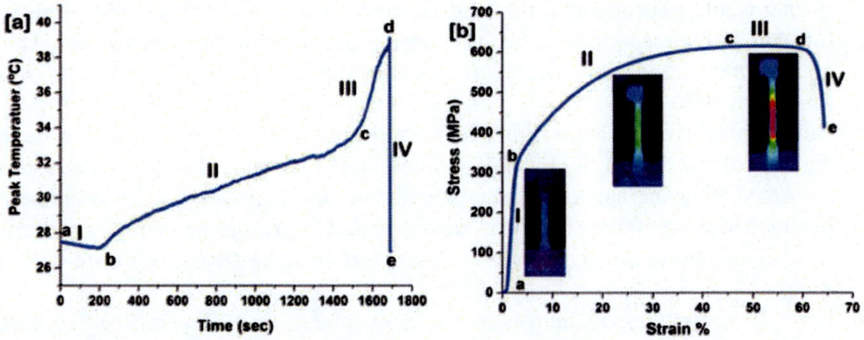

Fig. 6 **a** Evolution of peak temperature of the base metal during deformation with respect to time and **b** stress versus strain curve of base metal

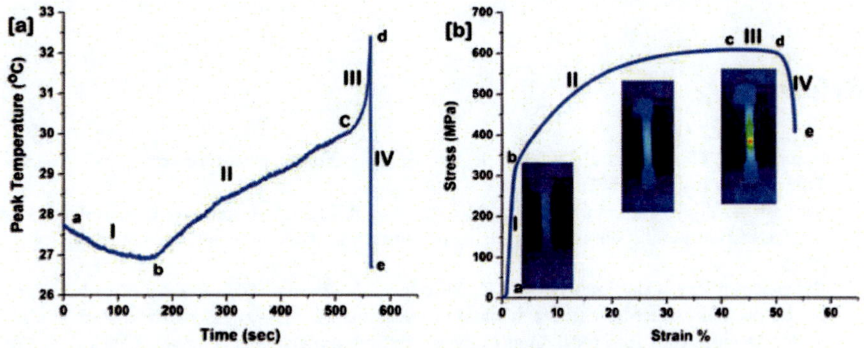

Fig. 7 **a** Evolution of peak temperature in LBW sample during deformation with respect to time and **b** stress versus strain curve for LBW sample

was observed in stage I due to the thermoelastic effect of the material. This effect occurred up to yield point of the base metal and LBW sample. When the plastic deformation started, the continuous rise in temperature was noticed on both base metal and LBW sample. This is attributed to the generation of heat due to plastic deformation effect [13]. Sudden rise in temperature occurred during the necking and cracking stage of the samples. This is attributed to the local plastic deformation of the initiated crack tip during necking. On stage III, the base metal sample exhibited 39.2 °C, whereas the LBW sample exhibited a peak temperature of 32.4 °C. But the fusion zone of the LBW sample experienced peak temperature of 30.8 °C before fracture (stage III). Lesser deformation in the gauge portion of the LBW sample compared to the base metal sample is the reason for this temperature difference.

4 Conclusions

- Higher welding speed and solidification behavior of LBW process influenced the formation of refined grains at fusion zone compared to the base metal grains. This grain refinement increased strength in the fusion zone of the LBW process over the base metal.
- During tensile deformation study, peak temperatures were recorded just before to the time of fracture for both LBW and base metal samples. This phenomenon is attributed to the increase of temperature with respect to increasing deformation in metals. Moreover, failure locations were revealed in terms of temperature patterns in advance before the occurrence of fracture during the tensile test on different samples.
- From IR thermography images, peak temperatures of 39.2 and 30.8 °C were observed at base metal and LBW fusion zone, respectively. This increased temperature at the base metal indicates that the base metal has undergone more deformation compared to LBW sample.

References

1. Mannan SL, Chetal S et al (2003) Selection of materials for prototype fast breeder reactor. Trans Indian Inst Met 56(2):155–178
2. Vignesh A, Vijay Prakaash VG, Lakshminarayanan AK (2015) Improving wear resistance of AISI 316LN austenitic stainless steel using friction stir processing. Appl Mech Mater 787:421–425
3. Rajasekaran R, Lakshminarayanan AK, Vasudevan M, Vasantharaja P, Role of welding processes on microstructure and mechanical properties of nuclear-grade stainless steel joints. J Mater: Design Appl. In: Proceedings of the institution of mechanical engineers, part L. SAGE. https://doi.org/10.1177/1464420719849448
4. Lakshminarayanan AK, Balasubramanian V (2012) Characteristics of laser beam and friction stir welded AISI 409M ferritic stainless steel joints. J Mater Eng Perform 21:530–539

5. Bagavathiappan S, Lahiri BB, Saravanan T, Philip J, Jayakumar T (2013) Infrared thermography for condition monitoring—a review. Infrared Phys Technol 60:35–55
6. Punnose S, Mukhopadhyay A, Sarkar R, Alam Z, Das D, Kumar V (2013) Determination of critical strain for rapid crack growth during tensile deformation in aluminide coated near-α titanium alloy using infrared thermography. Mater Sci Eng, A 576:217–221
7. Górka Jacek, Janicki Damian, Fidali Marek (2017) Thermographic assessment of the HAZ properties and structure of thermo mechanically treated steel. Int J Thermophys 38(183):1–21
8. Kumar Jalaj, Baby Sony, Kumar Vikas (2008) Thermographic studies on IMI-834 titanium alloy during tensile loading. Mater Sci Eng, A 496:303–307
9. Lakshminarayanan AK, Balasubramanian V (2012) Influences of welding processes on microstructure and mechanical properties of modified 12 wt% Cr ferritic stainless steel. Int J Manuf Res 7(4):331–352
10. Lakshminarayanan AK, Balasubramanian V (2012) Evaluation of microstructure and mechanical properties of laser beam welded AISI 409M grade ferritic stainless steel. J Iron Steel Res Int 19(1):72–78
11. Haneef Thodamrakandy et al (2015) Study of the tensile behavior of AISI type 316 stainless steel using acoustic emission and infrared thermography techniques. J Mater Res Technol 4(3):241–253
12. Yan Z, Zhang H, Chen P, Wang W (2017) Anisotropy of fatigue behavior and tensile behavior of 5A06 aluminum alloy based on infrared thermography. J Wuhan Univ Technol Mater Sci Ed 32(1):155–161
13. Kutin Marina, Ristic Slavica, Burzic Zijah, Puharic Mirjana (2010) Testing the tensile features of steel specimens by thermography and conventional methods. Sci Tech Rev 60(1):66–70

Production and Operation Management

Evaluation of Ergonomics Issues in Repetitive Scrap Handling Work in Automobile Industries

Aravind Babu Dasari and Dhinesh Balasubramanian

Abstract Ever since the origin of the evolution of human being either named as ape or man, striving to survive with safety, by becoming wise on live hazards, he has been innovating as many ways and means for the safe living/working. To this day, the struggle for safety is existing but in a sophisticated manner, as he grew from being wise to being an intellectual. This paper discusses the ergonomics issues in an automobile industry handling the hard substance, among the engineering industries, as the manual scrap handling system is one of the most critical safety concern aspects which includes more manual effort, due to its impact on the manpower expenses, medical expenses, personal protective equipment expenses, and time consumption. Ergonomic assessment, as a tool and method for analyzing human activities and their interactions with the surrounding environment, is thus crucial for designing operations and workplaces that achieve high safety. In the engineering industry, however, the constant repetitive work environment and laborious tasks cause traditional approaches to ergonomic analysis, such as manual observations and measurements to require substantial time and effort to yield reliable results. This study mainly concentrates on scrap handling because of its repetitive actions with heavy loads and also explores the adaptation and integration of various existing methods for data collection, analysis, and output representation potentially available for comprehensive ergonomic analysis. The proposed framework integrates the 3DSSP's (Static Strength Prediction) localized fatigue report for the inputs such as measurement of body angles using goniometer and NIOSH (National Institute of Occupational Safety and Health) weight lifting equation. The proposed framework is demonstrated through a case study using data from on-site scrap handling system through 3DSSP software.

A. B. Dasari (✉) · D. Balasubramanian
Mepco Schlenk Engineering College, Sivakasi, India
e-mail: daravindb6@gmail.com

D. Balasubramanian
e-mail: dhineshbala91@gmail.com

S. Vijayan et al. (eds.), *Trends in Manufacturing and Engineering Management*, Lecture Notes in Mechanical Engineering,
https://doi.org/10.1007/978-981-15-4745-4_70

Keywords 3DSSP software · NIOSH weight lift equation · AS-IS model · Goniometer · Localized fatigue report · Manual observation-based ergonomics analysis

1 Introduction

The engineering industry is labor- and machine-intensive, and worker activities can significantly affect the success of the engineering industry and scrap handling operations. Labor is one of the most crucial resources and has the highest direct impact on the outcomes in the industry, including time and cost. Furthermore, labor operations in scrap handling involve physically demanding motions and tasks that frequently expose workers to risk in their working environments, leading to a rate of injuries and chronic injuries that are among the highest of any engineering industry's scrap handling system.

The risk involved in the repetitive manual scrap handling system gets its peak with respect to the duration of the work, work atmosphere, distance between the loading and unloading region, and loading to unloading pathway of the scrap movement. The repetitive manual scrap handling system leads to the considerable effects on the expenses such as personal protective equipment (PPE) and medical expenses.

Golabchi et al. [1] state that the constantly changing work environments and laborious tasks subject traditional approaches to ergonomic analysis such as manual observations and measurements and his proposed framework used sensing for data collection, action recognition, and simulation modeling, by using different techniques that could facilitate the assessment of manual operations and could enhance the implementation of ergonomic practices during the work by reducing time, complexity and effort required, improvement achieved by proposing automated, simple and reliable analysis of manual operation.

Munck-Ulfsfält et al. [2] inferred that the ergonomics work is not a separate entity, but it is based on a strategy and much easier to get the management and other employees to realize, accept, and understand the risk involved in ergonomics work. Training on the load ergonomics concerned will be the key to success.

Dul et al. [3] argued that by contributing the shared goals of business performance, ergonomists would also be better able to reach their traditional health and safety objectives and also discussed the organizational strategies and business goals to which ergonomics contribute.

Leung et al. [4] studied the impact of many organizational stressors and stress on construction work safety behaviors and injury incidents. To achieve this, they studied a number of workers, also by using factor analysis, identified the major organizational stressors (inappropriate safety equipment, unfair reward and treatment, lack of goal setting and poor physical treatment, lack of provision of training), two types of stressors (physical and emotional), and safety behaviors and concluded that setting personal, long-term, and specific work goals with the workers would reduce the injury incident rate directly.

Golabchi et al. [5] discussed the approach that enabled modeling of manual activities at human motion level in order to provide standard task durations in addition to initial insight into the level of ergonomic risks associated with a planned production scenario.

Golabchi et al. [6] proposed the concept by making the approach with motion data from the 3D model of the jobsite to evaluate the risk factors that probably could produce more physical loads on the human body through biomechanical analysis. The framework was proposed after analyzing the jobsite ergonomic safety.

Tran et al. [7] introduced the metric-based learning approach for human activity identification, with the abilities to reject invisible actions, and discussed the vital role of invisible actions when applying human activity recognition to real-world applications.

Wei et al. [8] stated that their proposed model could generate physically realistic animations that react to external forces or changes in physical quantities of human bodies and interaction environments and evaluated their system by comparing against ground-truth motion data and alternative methods.

Case et al. [9] discussed the digital human modeling practices for the last four decades which is the first usable tool and still existing for the designing of the ergonomics-related issues.

This current study deals with the procedure of analyzing and evaluating the issues in the ergonomics in the automobile industry, analyzing the number of machines, plant layout, and the frequency of the scrap produced from different operations, gathering the anthropometric details through personal interview, video recording the manual scrap handling process for analyzing AS-IS model through MOST method, measuring the body angles using goniometer as input to the 3DSSP software for analyzing the load on the body parts and their ability of bearing the load, evaluating the RWL (Recommended Weight Limit) and LI (Lifting Index) for safe lifting weight, and giving the remedial action according to the NIOSH evaluation technique.

2 Case Study and Analysis

The main activities carried out to complete the task include picking up s45c grade steel scrap from the trolley and throw it into the scrap dump yard. As the first step, the work was video graphed to extract time lapses and types of activities and then angles of the worker's body joints were measured during his bending, picking, twisting, and throwing work postures. By using goniometer, the body angles were measured.

Details of the Machinery in the Automobile Industry

Figure 1 the graph gives information about the number of machines present in the industry for a particular operation which produces the scrap and also gives the information about the frequency of the scrap unloading per shift for different operations.

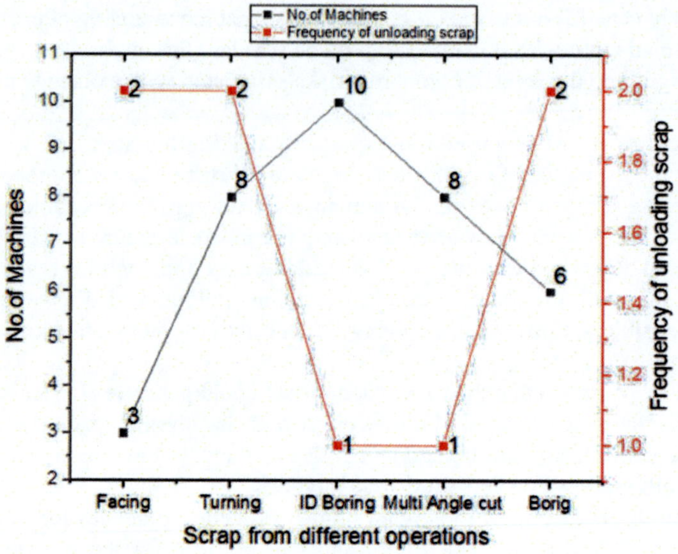

Fig. 1 Number of machinery and frequency of scrap with respect to the operation

2.1 *Safety Assessment Module*

The biomechanical analysis component of the framework enables the evaluation of an operation by examining the loads and their exertion on the human body and comparing them to safe limits. The results can be used along with productivity analysis output to improve the operation and select the optimal remedial action.

Table 1 gives the details about the size of the different types of scrap produced from a different type of operations in the automobile industry after its particular operation. S45C grade low-carbon steel is the scrap produced from different machines.

Table 2 gives the details about the number of exertions taken to unload the scrap from the trolley with respect to the scrap produced from their particular operation.

Figure 2 shows the exact replica of the manual scrap unloading procedure followed in the automobile industry. The practice involves bending, holding, picking, and throwing operations. The whole scrap is unloaded from the trolley of the above shown size, which is made of MS material weighing about 100 kg (Fig. 3).

2.2 *Activities Sequence and Duration*

Table 3 gives the details about the time lapse used for the MOST method of analysis, which represents the duration in seconds taken for the single exertion to get completed

Table 1 Different types of scrap from different operations for manual scrap handling

S. No.	Operation	Scrap sample
1	Facing	
2	Turning	
3	Turning and ID boring	
4	VMC-multi-angle cut	
5	Boring	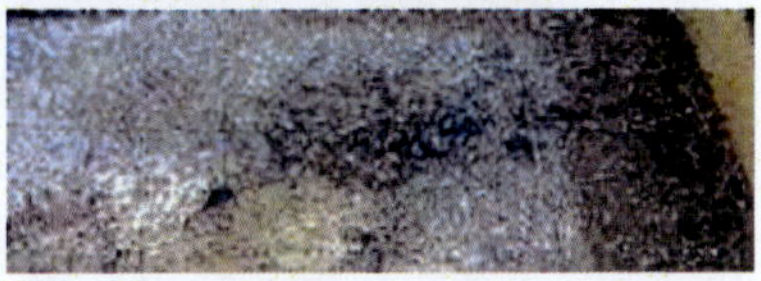

Table 2 Number of exertions for unloading different types of scrap

S. No.	Operation name	No. of exertions for unloading
1	Facing	550
2	Turning	460
3	Turning and ID boring	225
4	VMC-multi-angle cut	220
5	Boring	420

in the sequence, the actions for the scrap handling being identical for the all scraps produced from different operations.

AS-IS Model

The AS-IS model of the plant layout gives the detailed description of the unloading work pathway from its source to destination, besides the details like distance. As per the above plant layout, the sources of the unloading trolley are not the same, but the

Fig. 2 Manual operation done by the worker

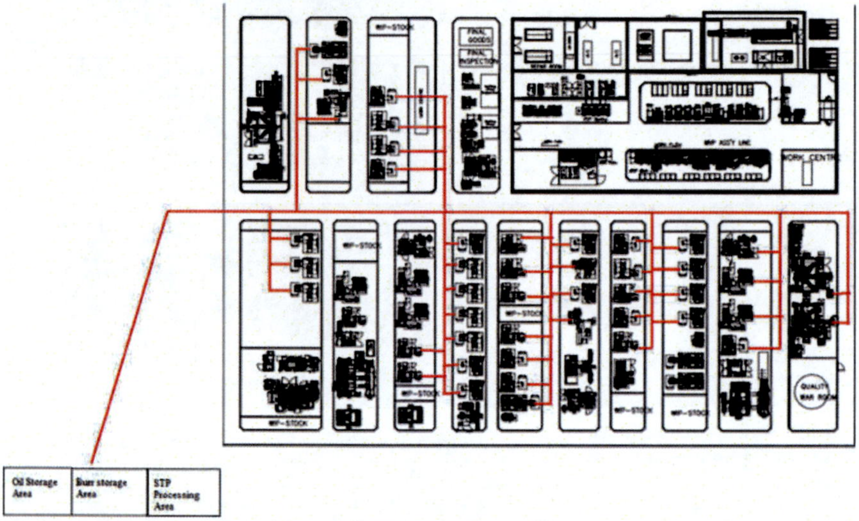

Fig. 3 AS-IS model representation of the factory layout, for scrap collection

destination of unloading the trolley is the same. The pathway is drawn connecting the individual source to the destination.

Analysis of the Direct and Indirect Expenses Due to the Inappropriate Ergonomics

Figure 4 gives the details about the average monthly usage of the expensive gloves for 3 scrap unloading workers used for handling the scrap unloading process; because of the variation in the method of unloading (crowbar and hand unloading), irrespective of unloading time, the unloading system depends on the shape of the scrap. Around

Table 3 Details of the action sequence and duration of the manual scrap lifting operation

S. No.	Activity sequence	Duration
1	Bend 	@00 h:00 m:01 s
2	Hold 	@00 h:00 m:04 s
3	Pickup 	@00 h:00 m:05 s
4	Throw 	@00 h:00 m:06 s

45–50 expensive gloves are used on an average for the work of scrap unloading. The average life span of the gloves used is hardly 2 days.

Figure 5 gives the monthly medical expenses in a year for 3 scrap unloading workers; the inference from the graph is that on an average the monthly medical

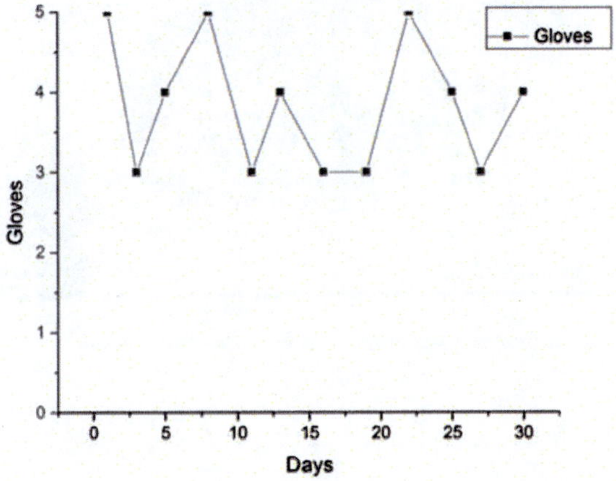

Fig. 4 Number of gloves used by the scrap handling worker during a month

expenses which is been referred from the monthly medical expenses bill of the particular workers are approximately 1500 Indian rupees.

Figure 6 gives the information about the average of different types of medical expenses incurred in a month; the inference from the pie chart is that the medical expenses are major for the hip and torso pains.

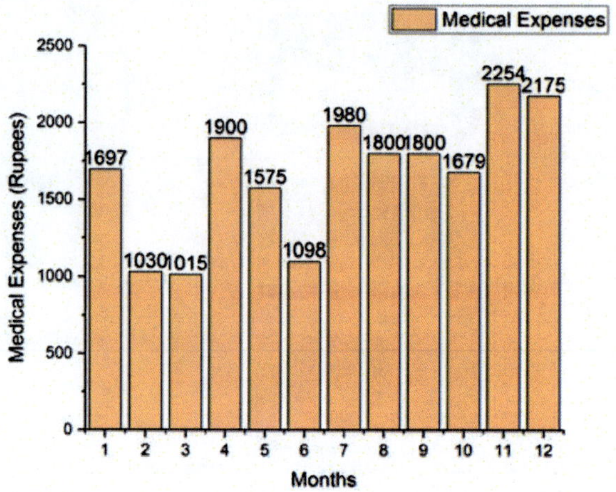

Fig. 5 Medical expenses of scrap handling workers during a year

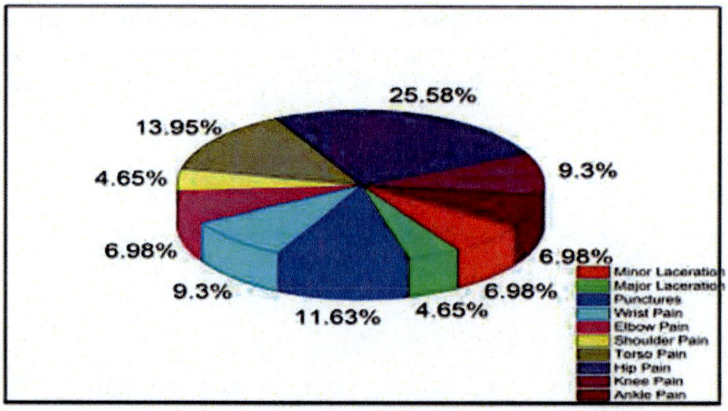

Fig. 6 Pie chart for the different types of medical expenses for a month

3 Methodology

This study proposes and tests an integrated framework that couples data acquisition and 3D static simulation program with an analysis of manual operations to enable effective evaluation of those manual operations for a comprehensive ergonomic analysis. The three objectives are (1) exploring the data associated with the different angles of the body, (2) examining possibility and applicability of sharing data among those different methods, and (3) testing, feasibility, and effectiveness of integrating the various methods. The framework is composed of the following modules: simulation, as-is modeling, and safety assessment. This analysis was held by gathering information about the conditions in the work environment through manual measurement.

The workers' normal activities at the work site were video graphed, and then extracts were obtained of the type of work, the sequence of operation, and duration of tasks and were used to build a simulation and get analysis report of the particular operation. The simulation made through 3DSSP software served to evaluate the static strength of the operation and gave the information about the strength or pressure applied on the particular parts of the body because of the particular operation and helped to evaluate the duty cycle based on the number of and duration of exertions. This 3DSSP 7.0 simulation program has the advantage of built-in range manipulator for duty cycle limit, exertion limits, etc., according to American Conference of Governmental Industrial Hygienists (ACGIH) 2016 (Fig. 7).

3.1 *Manual Observation-Based Ergonomic Analysis*

To carry out a thorough ergonomic analysis, information about the effects of physical activities on a worker's body needs to be available. Main contributors to those effects are the type, duration, and sequence of manual tasks. A complete ergonomic analysis involves evaluating the motions and postures of working people and the physical attributes of job site to assess present work conditions and propose the new designs for manual operations (safe conditions) and working spot (work spot dimensions). To carry out assessments, the ergonomists need to go through the three stages: (i) data collection, (ii) data analysis, and (iii) interpretation and presentation of results.

Ahead of data collection, the ergonomist has to plan the analysis process and define the strategy based on certain conditions of work being analyzed. The required inputs for each relevant data are collected, traditionally by observing the subjects (posture, anthropometry), their motions during work (leaning, bending), and the work environment particulars (workbench, tools, equipment).

After the completion of data collection, the ergonomist uses the gathered data to conduct an ergonomic evaluation using tools such as ergonomics assessment checklist (RULA, ROSA) and time and motion studies (MTM, MOST), giving the details

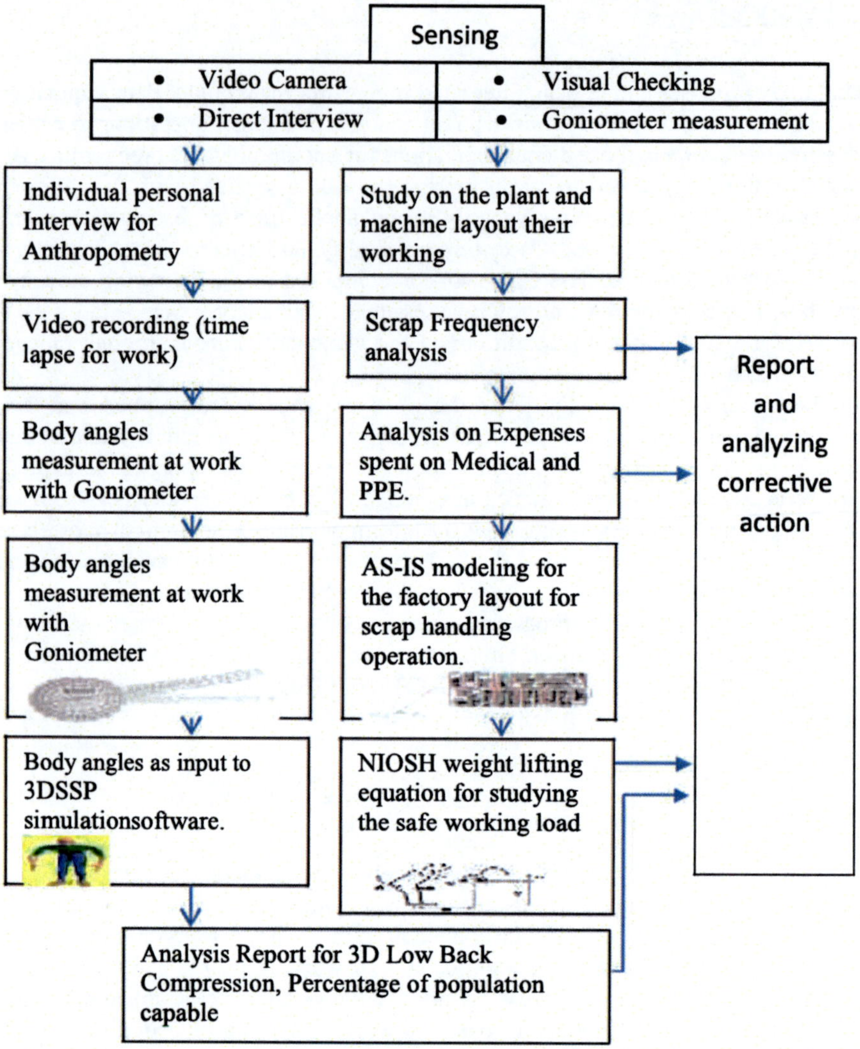

Fig. 7 Methodology for the evaluation and analysis of ergonomics in the automobile industry

like worker's posture (body angles/joint angles) along with the frequency and duration of exposure allowing posture-based tools to report on the level of ergonomic risk associated with the task.

This work involves three stages of the ergonomics analysis. They are (i) data acquisition, (ii) analysis (simulation modeling), and (iii) representation (result/corrective action).

3.2 MTM Method Study

Methods Time Measurement (MTM) is the motion time system that is primarily used in industrial settings to analyze the methods to perform any manual task or operation and as a product of analysis, set the standard time that a worker should complete that particular task.

3.3 MOST Method Study

MOST systems are designed to provide the optimal combination of speed, detail, and accuracy of the work/operation. It was designed to be much faster than the conventional work measurement techniques such as time study [10].

Integrated Ergonomic Analysis

To carryout thorough ergonomic analysis, the availability of information regarding the effects of physical activity is imperative. Physical activity type, duration, and sequence of the manual task are the main contributors to the physical effect.

Ergonomists rely on human judgment and estimates in acquiring data when designing the new operations for new or prospective workplaces, which can lead to uncertain information. This issue can be addressed by linking simulation model.

Predetermined motion time systems (PMTSs), which enable calculating the standard duration of the physical/manual task based on the involved movements, as this simulation environment allows not-yet-existing scenarios to be conveniently modeled and explored.

Simulation Model

To create a simulation model of manual operation and analyze its operational activities, either human observation or sensing methods have to be used to gather the required inputs. Human observation typically requires time, effort, and expertise, but adding the human observation along with simulation software enables more facts regarding the ergonomics issues of the particular work.

The developed simulation model using 3DSSP software represents the current status of an ongoing operation which is also used to serve as a base model to evaluate other scenarios for operation in terms of safety to find the most reliable.

NIOSH Weight Lifting Equation

The NIOSH weight lifting equation is an important tool mostly used by occupational health and safety experts to assess the manual material handling risks associated with the workplace while performing lowering tasks and lifting.

It offers the recommendation regarding the weight limit (RWL); the recommended weight lifting equation is as follows [11]:

$$\text{RWL} = \text{LC}(51) \times \text{HM} \times \text{VM} \times \text{DM} \times \text{AM} \times \text{FM} \times \text{CM}$$

- H = Horizontal position of an object relative to the body.
- V = Vertical position of an object relative to the floor.
- D = Distance of the object moved vertically.
- A = Asymmetry angle/twisting requirement.
- F = Duration of lifting and frequency.
- C = Coupling/quality of worker grip on an object (M stands for multiplier).

As-Is Modeling Module

The as-is modeling module that includes main components such as the current conditions of the existing workplace (building and objects) has to be modeled, and the path that each worker's 3D animation will use in the virtual representation is identified through path planning.

The above table gives detailed information about the different types of operations and the number of exertions took to unload the scrap totally from the trolley. This exertion count for unloading each trolley made a path for analyzing the ergonomics in the automobile industry, which includes the anthropometry data of the individual and manual body angle measurement using the goniometer.

The manual measurements taken from goniometer are to be taken as the input to the most useful ergonomic analysis software (3DSSP), which eventually gives the resultant details such as analysis summary report and localized fatigue report.

Details of Weight and Number of Exertions for Unloading Different Operations

Figure 8: the graph gives details about the weight of the scrap lifted manually and the number of exertions taken for the employee to unload the scrap present in the trolley. The scrap produced from facing operation is less in weight compared to the weights of other produced scraps, due to the low weight and the irregular size, the holding capacity of this #05 operation scrap is considerably low.

4 Analysis Using 3DSSP Software for All the Scrap Handling Operations

Table 4 gives the details regarding the anthropometry details of the worker performing manual scrap handling taken prior to the starting of the 3DSSP analysis (Fig. 9).

The simulation of the images is done using the 3DSSP software from the input parameters taken through goniometer.

Table 5 gives the details regarding the input parameters for the 3DSSP software being taken manually using the goniometer, though it is a time-consuming process but the most effective and accurate process of taking the body angles of the worker during the working posture (Fig. 10).

The simulation of the images is done using the 3DSSP software from the input parameters taken through goniometer (Table 6).

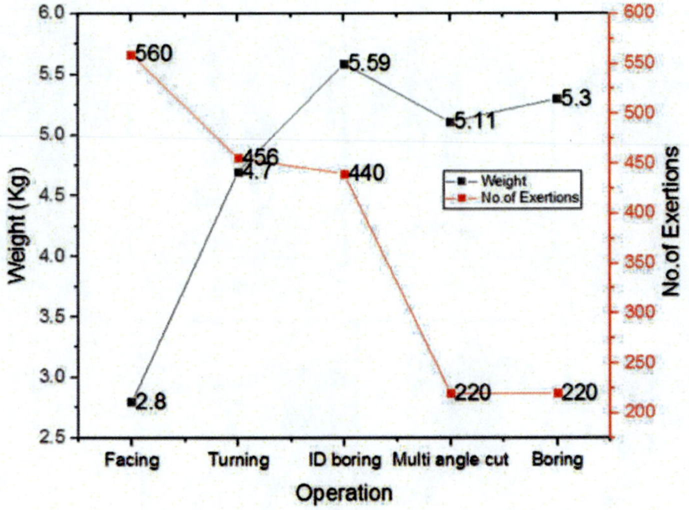

Fig. 8 Details of weight & number of Exertions for unloading different operations

Table 4 Anthropometry data of the employees working in manual scrap handling

S. No.	Details	Person 1	Person 2	Person 3
1	Gender	Male	Male	Male
2	Height (cm)	165	163	167
3	Weight (kg)	54	52	54
4	Shoe height (cm)	2.5	2.5	2.5
5	Age	48	46	48

By using the proposed body angles measuring algorithm on the human body with a goniometer, recognized scrap handling procedure which is induced from different operations is obtained.

Inference from the Analysis Summary Report for the 3D Low Back Compression
Figure 11 gives details obtained from the analysis summary report which discusses the 3D low back compression (Newton) with respect to the scrap produced from different operations.

Figure 12 gives the details of the low back compression for the different steps followed in handling (unloading) the scrap produced from #05 operation.

Inference of the Analysis Summary Report for the Percent of the Population Capable
Figure 13 gives details in percentage about the capability of the population to execute the particular operation; the analysis summary gives the report for the following body parts such as wrist, elbow, shoulder, torso, hip, knee, and ankle. The inference

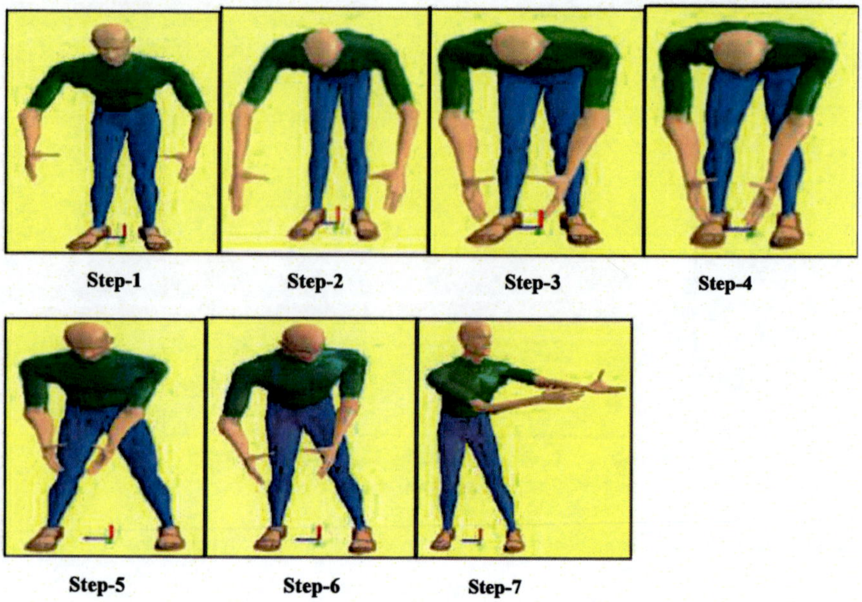

Fig. 9 The above figures are the steps an unloading person follows from pick to drop

that can be derived from the analysis summary report obtained from the 3DSSP software is that while manually handling this type of scrap nearly 96.5% of people can effectively and capably handle the wrist load, almost 100% of people can capably handle the elbow load, on an average 98.5% population can capably handle the shoulder load, around 93.75% of the population can manage the torso load, only 90.25% of population can capably manage the hip load, on an average 95.5% of population can handle the knee loads, and 96.25% of population can capably handle the ankle load while handling the scrap. As a result, the load on the hips while handling (unloading) the scrap is comparatively more than other body parts loads.

Table 7 gives information about the percent of the population capable of bearing the loads on the individual body part. The inference from the table is that only 90.25 and 93.75% of the population are capable of handling the load on the hip region and the torso region, respectively.

The results of the localized fatigue analysis report from 3DSSP software are used to give awareness about the work duration limits and also threshold duty cycle limits of the particular unloading work.

Table 5 Input parameters for the steps involved in the manual scrap handling of the facing operation scrap

Parameters	Step-1				Step-2			
	Left		Right		Left		Right	
	Horizontal	Vertical	Horizontal	Vertical	Horizontal	Vertical	Horizontal	Vertical
Hand (°)	106	−55	106	−55	106	−53	106	−56
Fore arm (°)	106	−55	106	−55	106	−53	106	−56
Lifting weight (N)	0				0			
Trunk angles	Flexion	Axial rotation	Lateral bending		Flexion	Axial rotation	Lateral bending	
Head	44	0	0		12	0	−1	
Angles	45	0	0		12	0	0	
Parameters	Step-3			Step-4				
	Left		Right	Left			Right	
	Horizontal	Vertical	Horizontal	Vertical	Horizontal	Vertical	Horizontal	Vertical
Hand (°)	110	−53	110	−54	106	−52	110	−50
Fore arm (°)	110	−53	110	−54	106	−52	110	−50
Lifting weight (N)	0				27.5			
Trunk angles	Flexion	Axial rotation	Lateral bending		Flexion	Axial rotation	Lateral bending	
Head	12	1	−1		6	−1	−1	
Angles	12	0	0		7	0	0	
Parameters	Step-5			Step-6				
	Left		Right		Left		Right	
	Horizontal	Vertical	Horizontal	Vertical	Horizontal	Vertical	Horizontal	Vertical
Hand (°)	106	−52	110	−50	118	−66	120	−41
Fore arm (°)	106	−52	110	−50	118	−66	120	−41
Lifting weight (N)	27.5				27.5			
Trunk angles	Flexion	Axial rotation	Lateral bending		Flexion	Axial rotation	Lateral bending	
Head	50	0	0		43	25	−4	
Angles	33	0	0		26	0	0	

Parameters	Step-7			
	Left		Right	
	Horizontal	Vertical	Horizontal	Vertical
Hand (°)	43	−12	162	13

(continued)

Table 5 (continued)

Parameters	Step-7			
	Left		Right	
	Horizontal	Vertical	Horizontal	Vertical
Fore arm (°)	43	−12	162	13
Lifting weight (N)	0			
Trunk angles	Flexion	Axial rotation	Lateral bending	
Head	60	40	−6	
Angles	75	10	−4	

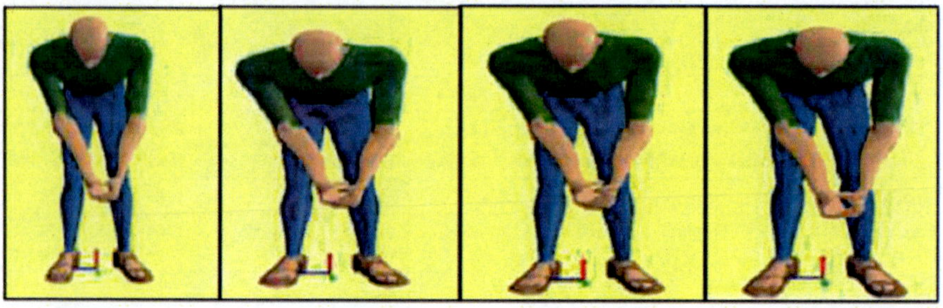

#10 scrap Handling **#20 Scrap Handling** **#50 Scrap Handling** **#60 Scrap Handling**

Fig. 10 The figures are the unloading postures of the scrap produced from #10, #20, #50, #60 operations

4.1 NIOSH Weight Lifting Equation Analysis

Table 8 gives the total information regarding the maximum and minimum weights that the scrap unloading worker handles for different scrap operations; minimum average load and maximum average load of the handling material are used for the calculation of the NIOSH weight lifting equation (Fig. 14).

From Table 9, it is clear that the Lifting Index (LI) of this scrap unloading work is in the highest risk factor, it badly needs the corrective remedial action to reduce the Lifting Index value for the safe lifting work, and it should be less than 1.0.

5 Results

5.1 3DSSP Analysis Output

Table 10 gives the details about the center of gravity of the particular position of the scrap unloading work for each step; this is obtained based on the weight of the load that the scrap handling person picks and the position of the torso; in the above figures,

Table 6 Input parameters for the steps involved in the manual scrap handling of the other operation scrap

Parameters	Turning				Turning and ID Boring			
	Left		Right		Left		Right	
	Horizontal	Vertical	Horizontal	Vertical	Horizontal	Vertical	Horizontal	Vertical
Hand (°)	117	−54	115	−51	115	−51	109	−48
Fore arm (°)	117	−54	115	−51	115	−51	109	−48
Lifting weight (N)	46.1				46.1			
Trunk angles	Flexion	Axial rotation	Lateral bending		Flexion	Axial rotation	Lateral bending	
Head	19	0	0		18	−1	1	
Angles	18	0	0		16	0	0	
Parameters	Multi-angle cut				Boring			
	Left		Right		Left		Right	
	Horizontal	Vertical	Horizontal	Vertical	Horizontal	Vertical	Horizontal	Vertical
Hand (°)	115	−51	109	−48	115	−51	109	−48
Fore arm (°)	115	−51	109	−48	115	−51	109	−48
Lifting weight (N)	52				54.9			
Trunk angles	Flexion	Axial rotation	Lateral bending		Flexion	Axial rotation	Lateral bending	
Head	18	−1	0		18	−1	3	
Angles	16	0	0		16	0	0	

the red dot represents the center of gravity; if the red dot is out of the green region, the particular position is not acceptable to work and that at any time the worker may fall down or cause to himself heavy strain to the hip and the torso region.

Table 11 gives the details about the center of gravity for the other scrap handling operations; this kind of analysis is made according to the ACGIH 2016 standards of the ergonomic analysis.

The 3D Analysis Summary Report

Figures 15 and 16 gives the result about the body parts which are subjected to high strain for the population of 25%, The entire analysis is done in the software with 25% strength population; among the 25% of the population, the hip and the torso are the major parts that are affected badly and among that population 90.25% and 93.75%, respectively, are capable of handling the loads on the particular operation.

NIOSH Weight Lifting Equation

The NIOSH weight lifting equation confirms that (ref Table 9), the Lifting Index (LI) value is out of the limit, i.e., exceeding 1 by 0.11, thereby indicating that it critically needs engineering and administrative control for remedial action.

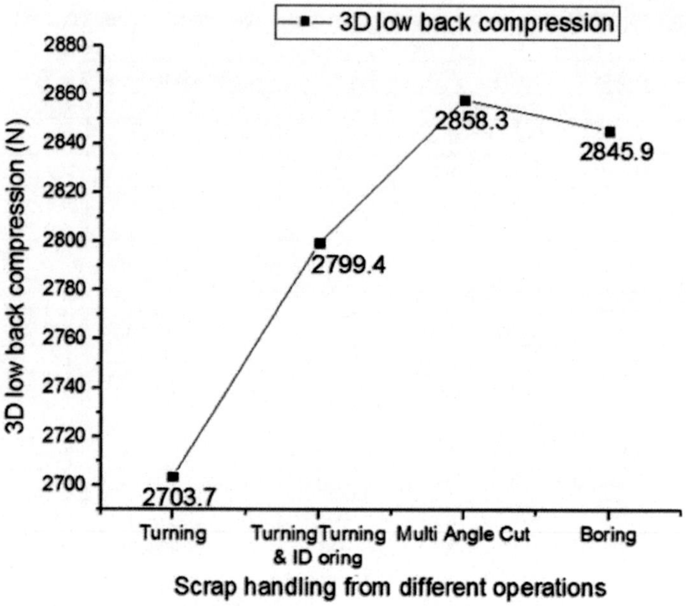

Fig. 11 3D low back compressions for different scrap handling operations

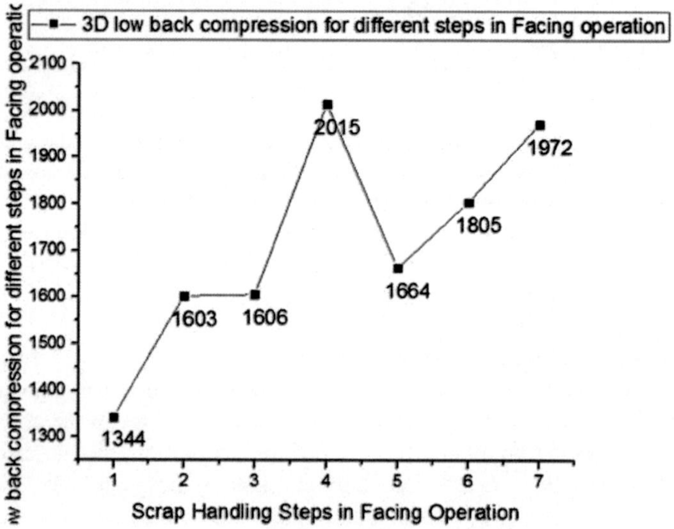

Fig. 12 3D low back compressions for handling facing operation scrap

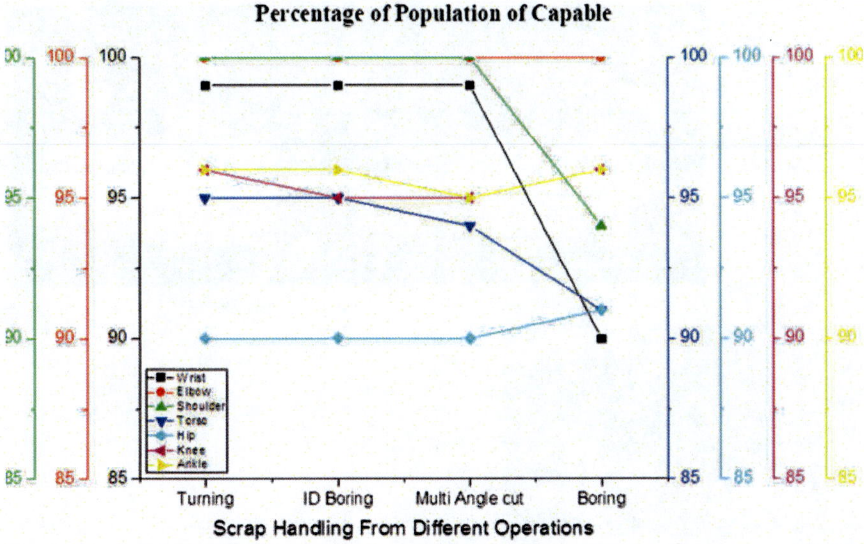

Fig. 13 Percentage of population capable to handle the loads

Table 7 Percent of population capable of handling the scrap with respect to body parts

S. No.	Load on body parts for different operation	Percent of population capable (%)
1	Wrist	96.5
2	Elbow	100
3	Shoulder	98.5
4	Torso	93.75
5	Hip	90.25
6	Knee	95.5
7	Ankle	96.25

Table 8 Calculation of the NIOSH lifting equation

S. No.	Operation name	Min weight lift/exertion (kg)	Max weight lift/exertion (kg)
1	Facing	2.0	2.8
2	Turning	3.0	4.7
3	Turning and ID boring	4.0	5.11
4	Multi-angle cutting	4.0	5.3
5	Boring	5.0	5.59

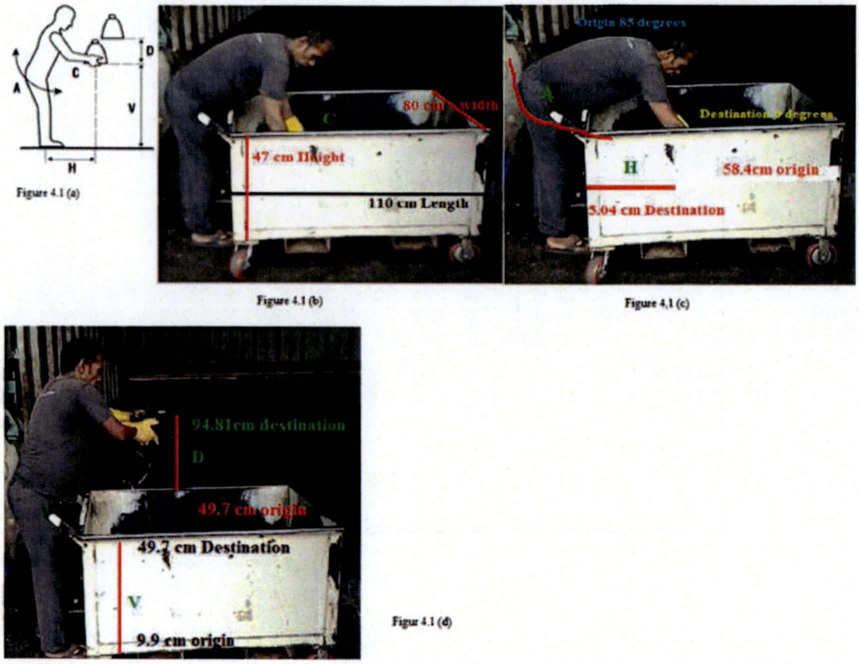

Fig. 14 **a** Represents NIOSH weight lifting equation details, **b** represents the dimensions of the trolley, **c**, **d** represent the real-time action with NIOSH weight lifting standards

6 Conclusions

This evaluation and analysis study on the ergonomics in manual scrap handling system of a particular automobile industry has enabled to draw the following conclusions

3DSSP Software Analysis

- Confirms that the load is heavy on the hip and torso of the scrap handling person.
- The unacceptable level of the center of gravity of the scrap handling person eventually leads to fall of person and strongly supports the heavy load on the hip and torso region of the worker.

NIOSH Lifting Equation

- The NIOSH weight lifting equation indicates that the Lifting Index (LI) is above the acceptable value of 1, i.e., (1.11). So the scrap handling operation critically needs the engineering and administrative controls as corrective action.

Table 9 Calculation of NIOSH lifting equation

Lifting task	Unloading stock origin	Multiplier origin	Unloading stock destination	Multiplier destination
Horizontal location (25.34–63.5) cm **H**	58.42	HM = 0.43	50.8	HM = 0.50
Vertical location (0–177.8) cm **V**	9.9	VM = 0.80	49.76	VM = 0.92
Travel distance (25.4–177.8) cm **D**	45.11	DM = 0.92	45.11	DM = 0.92
Angle assymetry (0–342.3) **A**	0	AM = 1.00	0	AM = 1.00
Coupling (1 = good, 2 = fair, 3 = poor) **C**	3	CM = 0.90	3	CM = 0.90
Frequency (0.2–15 lifts/min) **F**	7.5	FM = 0.22	7.5	FM = 0.22
Average weight lifted in kg	3.6		3.6	
Maximum weight lifted in kg	4.6		4.6	
Duration (1, 2, 8 h)	8	8 h	8	8 h
Recommended Weight limit RWL = HM * VM * DM * AM * CM * FM	3.3			4.3
Lifting index (LI) = load/RWL	1.11			0.84

MOST Method

- The MOST method of analysis for the scrap lifting operation produced from all the operations has the quick and shorter time duration of handling the scrap from its origin to the destination.
- It also serves the same to the plant layout; the traveling distance of unloading trolley to bring from origin to destination is the shortest and quickest as per the time duration and the plant layout.

Table 10 Details on the center of gravity for the steps involved in facing operation scrap unloading

S. No.	Step posture	Center of gravity	%MVC (maximum voluntary contraction)	%DC (maximum exertion duty cycle)
1		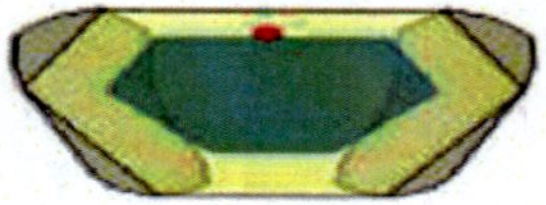 Critical	Are within the limits	Neck Flexion and torso flexion are out of limits
2	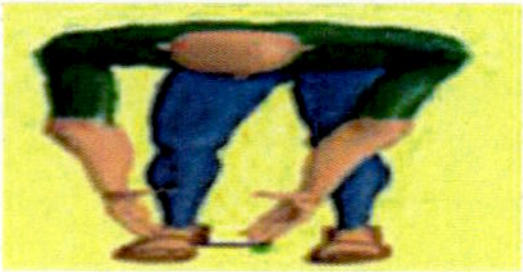	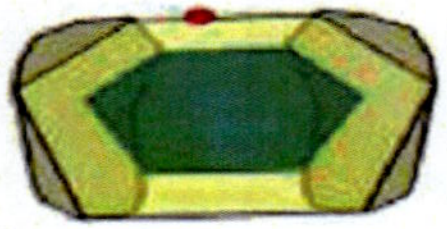 Unacceptable	Are within the limits	Wrist flexion, elbow flexion, shoulder abduction, and torso flexion are out of limits
3		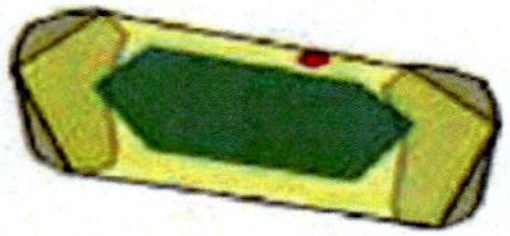 Critical	Are within the limits	Wrist ulnar/radius, elbow flexion, shoulder abduction, neck flexion, lateral bending, and torso are out of limits
4	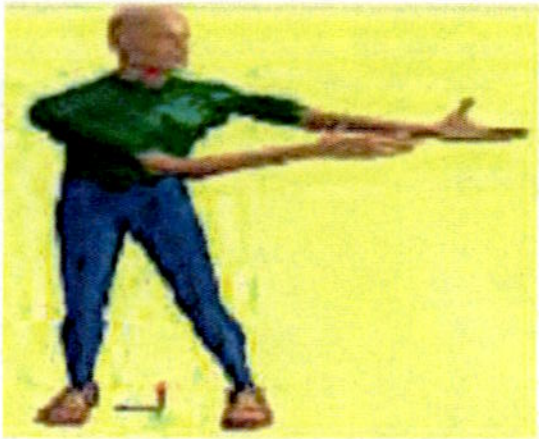	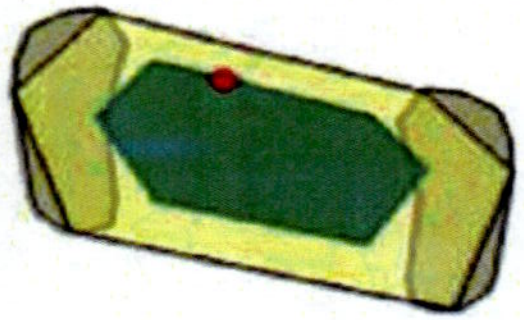 Acceptable	Are within the limits	Wrist flexion, Ulnar/radius, elbow flexion, shoulder rotation backward/forward, abduction, neck lateral bending, Torso flexion and rotation are out of limits

Table 11 Details on the center of gravity for the steps involved in boring operation scrap unloading

S. No.	Step posture	Center of gravity	%MVC	%DC
1	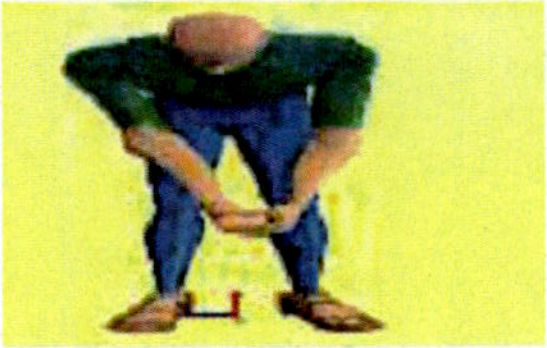	 Acceptable	Are within the limits	Wrist flexion and ulnar/radius, elbow flexion, shoulder rotation backward, frontward, neck flexion, torso flexion are out of limits
2		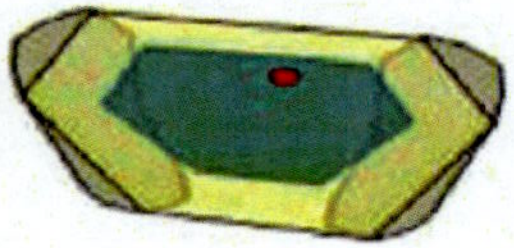 Acceptable	Are within the limits	Wrist flexion and Ulnar/radius, elbow flexion, shoulder rotation backward, frontward, neck flexion, torso flexion are out of limits
3		 Acceptable	Are within the limits	Wrist flexion and ulnar/radius, elbow flexion, shoulder rotation backward, frontward, neck flexion, torso flexion are out of limits

(continued)

Table 11 (continued)

S. No.	Step posture	Center of gravity	%MVC	%DC
4		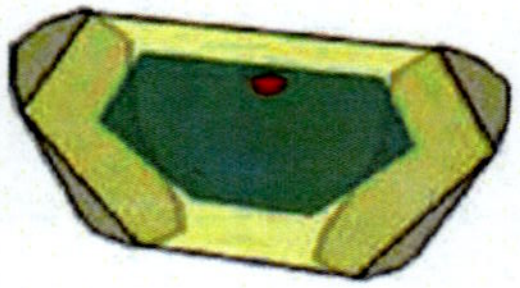 Acceptable	Are within the limits	Wrist flexion and ulnar/radius, elbow flexion, shoulder rotation backward, frontward, neck flexion, torso flexion are out of limits

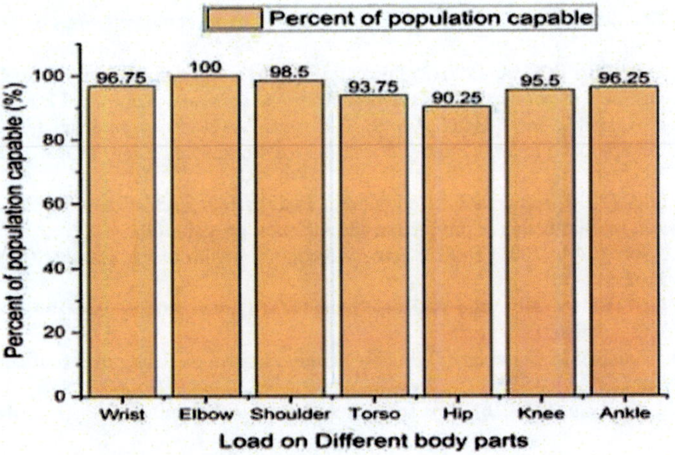

Fig. 15 Average percent of the population capable

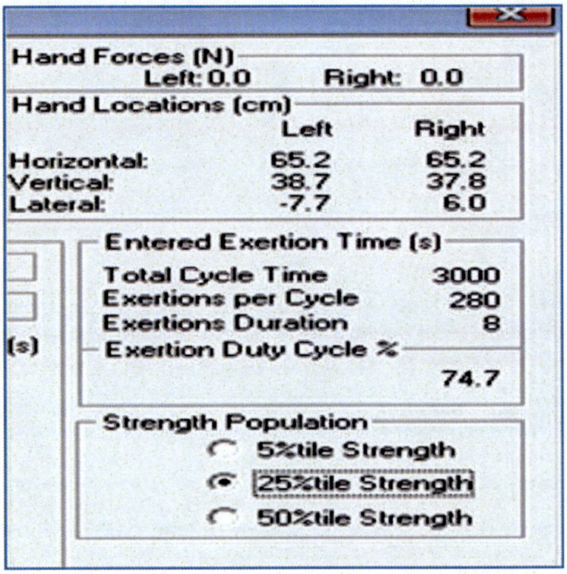

Fig. 16 Indication for the percentage of population considered for the analysis

References

1. Golabchi A et al (2018) An integrated ergonomics framework for evaluation and design of construction operations. Autom Constr 95:72–85
2. Munck-Ulfsfält U (2003) Corporate ergonomics programme at Volvo car corporation. Appl Ergon 34:17–22

3. Dul J, Neumann WP (2009) Ergonomics contributions to company strategies. Appl Ergon 40(4):745–752
4. Leung MY, Chan IYS, Yu J (2012) Preventing construction worker injury incidents through the management of personal stress and organizational stressors. Accid Anal Prev 48:156–166
5. Golabchi A, Han S, Seo J, Han S, Lee S, Al-Hussein M (2015) An automated biomechanical simulation approach to ergonomic job analysis for workplace design. J Constr Eng Manage 141(8)
6. Golabchi A (2015) Integration of ergonomic analysis into simulation modelling of manual operations. In: Simulation in production and logistics, pp 491–500
7. Tran D, Sorokin A (2008) Human activity recognition with metric learning. ECCV, Part I, LNCS 5302:548–561
8. Wei X (2011) Physically valid statistical models for human motion generation. ACM Trans Graph 30(3): Article 19
9. Case K, Marshall R, Summerskill S (2016) Digital human modeling over four decades. Int J Digit Hum 1(2):112–131
10. Senthil J, Haripriya G (2016) Time analysis with most technique. Int J Chem Sci 14(S2):519–526
11. https://ergo-plus.com/niosh-lifting-equation-single-task/

Reduction of Terminal Rejections and Failure Cost through Kaizen at Shop Floor of a Wire Harness Manufacturing Company

Pramod Kumar and Jaiprakash Bhamu

Abstract In the cutthroat market, demand for higher-quality products is currently the most important task in front of organization and management looks for better effectiveness. In manufacturing units, high cost, declining profit boundaries, discrepancy of quality of products, and delivery of such products in the competitive market are the factors those have encouraged many organizations to focus on ways to increase the involvement of employees to improve quality and productivity. Kaizen is one of the employee contribution methods intended to bring together all level of workforce in an organization for setting principles of excellence and achieving better consequences. In this paper, it is tried to analyze the performance of Kaizen in a wire harness manufacturing unit which is considered for the study, and the paper presents a case study of the selected company to reduce rejections due to high contact gap in terminal that further confides decrement in failure cost.

Keywords Kaizen, root cause analysis · Cost-effectiveness · Process improvement · Productivity · Rejection level

Abbreviations

OEE	Original equipment efficiency
PDCA	Plan-do-check-act
AWP	Advance wire process
NG	Not good
BOP	Begin of process
DOL	Direct on line
WIP	Wire in process

P. Kumar (✉) · J. Bhamu
Government Engineering College, Bikaner, India
e-mail: pramodfaujdar1984@gmail.com

S. Vijayan et al. (eds.), *Trends in Manufacturing and Engineering Management*, Lecture Notes in Mechanical Engineering,
https://doi.org/10.1007/978-981-15-4745-4_71

1 Introduction

In this study, effective lean tool such as Kaizen has been adopted for process improvement in a wire harness manufacturing company and Kaizen is used for betterment of cost-effectiveness at the shop floor in the company through teamwork. The study deals with various aspects of Kaizen as a lean manufacturing tool and how improvements can be made by adopting practices of Kaizen in wire harness manufacturing companies. OEE method was applied to reduce the cost by using Kaizen technique to a semiconductor industry which results in cost reduction of 7% annually for the use of gases and chemicals [1]. Kaizen was adopted to reduce operational cost and improve efficiency [2]. Kaizen-based case study was performed in Romania for improvement through PDCA cycle which results in reduction in service interventions, labor costs, and cost of replaced parts [3]. Kaizen signifies small improvements that have been made in the quality as a result of continuing efforts. Kaizen method concerned to work out the part mismatch problems in automobile assembly production line [4].

2 Methodology of the Study

Step-by-step Kaizen procedure has been attempted to solve the problem by data collection, root cause analysis, selection of the best solution method, corrective actions, and documentation. Identifying the problems, analyzing them, signifying solutions, implementing them, and getting feedback are all sequential procedures that help to resolve work-related problems properly and efficiently. The sequential method of Kaizen is followed to solve work-related problems. The study is carried out over a period of three months, during which team of Kaizen worked together to identify the major problem and found a suitable solution at the shop floor of the company. This study was performed on the basis of different Kaizen tools and problem-solving techniques for collecting and analyzing the data. The sequential procedure that we followed is shown by a flowchart in Fig. 1.

3 Case Study

3.1 Company Background

The ABC Company is a professionally managed company known for its quality and attitude toward perfection in the field of auto-electrical and electronic components. Appreciating the continually evolving needs of the automotive industry, the company has constantly developed new products for its customers and has endeavored to remain at par with the leading technology available worldwide. The selected company is among the largest manufactures of wiring harnesses in India. A well-equipped

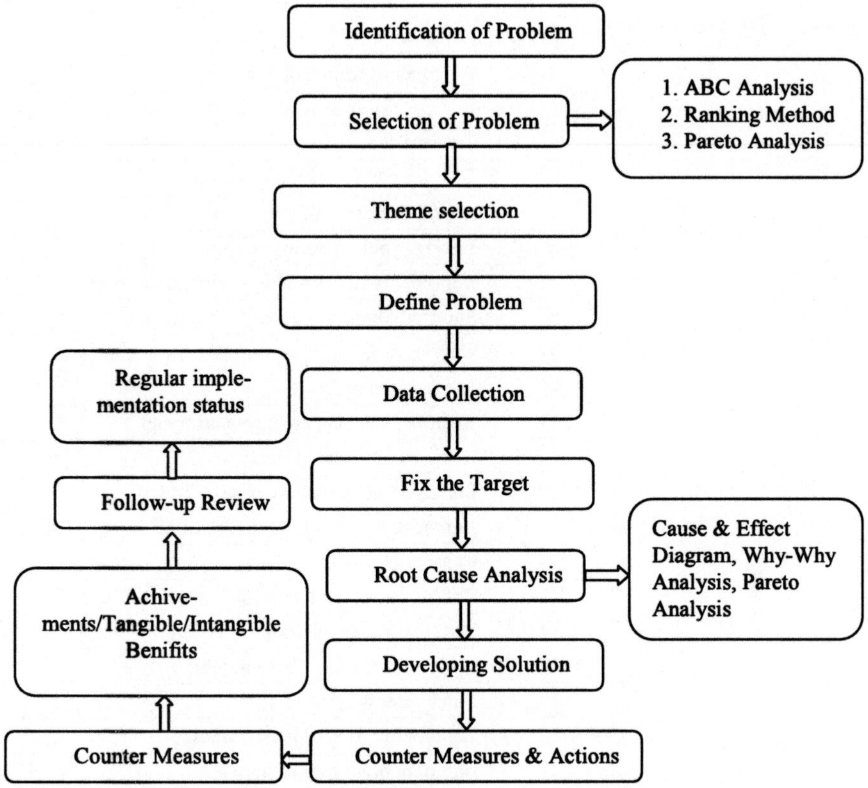

Fig. 1 Flowchart for problem solving case study

design and manufacturing facility combined with strong engineering capability ensures production of world-class products.

3.2 *Problem Recognition*

Team members identified problems by visiting the shop floor of production department of the company, and after discussion with shop floor workers, 52 problems (shown in Table 1) have been recognized in crimping section of WIP area of the company.

Table 1 Problem recognition

S. No.	Problem
1	Improper overlapping in taping
2	Improper storage system on shop floor
3	Tape wastage
4	Coupler interchange at coupler assembly stage
5	Improper storage of tape
6	Plating problem during storage
7	Improper issuing system of processed wires
8	Short molding in coupler
9	Cutting variation in terminal during crimping
10	Mixing in coupler
11	Improper fitment gap during crimping
12	Mixing of processed wires
13	Improper identification of job card
14	Improper storage of processed wires
15	Improper fitment of fuse case
16	Cutting variation in terminals during crimping
17	Terminal damaged during handling
18	Lock broken in 2p male coupler
19	Band harness missing
20	Improper pulling of band harness
21	Improper branching of taping
22	Time loss in opening of tape
23	Missing of protector cord
24	Wire interchange at coupler assembly stage
25	Terminal damage at wire assembly stage
26	Excess tool setup time at coupler insertion fixture
27	Excess tool setup time in continuity testing stage
28	Testing pins wear out
29	Terminal damaged at coupler insertion stage
30	Excess changeover time of taping fixture
31	Roller bearing of conveyor broken
32	Improper height of hangers
33	Improper identification of wire spool at AWP
34	Improper record storage of used job cards
35	Improper storage of raw material at coupler insertion stage
36	Improper storage of fixtures

(continued)

Table 1 (continued)

S. No.	Problem
37	Improper display of work instruction at shop floor
38	Improper storage of master sample
39	Time loss in changeover of terminal
40	Terminal damage during taping
41	Break down of 2.8 m terminal
42	Time loss in opening the tape
43	Tape wastage during taping
44	Tongue revert back
45	Contact gap in terminal
46	Improper knurling in terminal
47	Terminal face damaged
48	Leg height variation in terminal
49	Improper identification system of sleeves
50	Terminal blackout
51	Excess machine setup time of continuity tester
52	Conveyor hanger broken

3.3 *Problem Selection*

It is required to select the problem which affects the most and due to which more rejections are occurring. So to identify this problem in this study, three basic methods are followed, i.e., ABC analysis, Pareto analysis and rating method.

ABC Analysis
After ABC analysis, it is observed that out of 52 problems, 43(shown serial no. 1–43 in Table 1) problems are in category of A, which have minimum involvement of other department and only 5 problems (serial no. 44–48 in Table 1) are under B category in which involvement of other department is necessary and 4 problems (serial no. 49–52 in Table 1) are signified in C category which is needed management support for implementing solutions. The problems which have minimum involvement of other department are considered in 'A' category and which have involvement of other department is necessary are in 'B' category. In 'C' category problems management support may be needed in implementing solutions.

Pareto Analysis
Pareto analysis is basically a statistical technique used for selection of the problem which has significant overall affect on operations. Pareto analysis not only shows you the most important problem to solve, also gives you a score showing how severe the problem is. Out of 52 problems, it is decided to select a major problem that is responsible for increasing rejection level of terminal. After discussion and deep

inspection, the team members decided to take one problem out of 5 problems from B category problems. With the help all members' opinion, all these five problems are assigned score of severity of rejections. After plotting the score on chart, it is found that highest rejections terminal are occurring due to improper gap between leg of terminal so this problem is first selected to solve. Pareto chart is shown in Fig. 2.

Rating Method

Rating method is also an important technique which is also utilized for selection of the major problem for responsible for most of rejections of terminal. In this method, rating from 1 to 5 is decided based on three factors, i.e., occurrence, detection and extent of impact. 1 rating is for LOW, and 5 is for HIGH. Problem selected on the

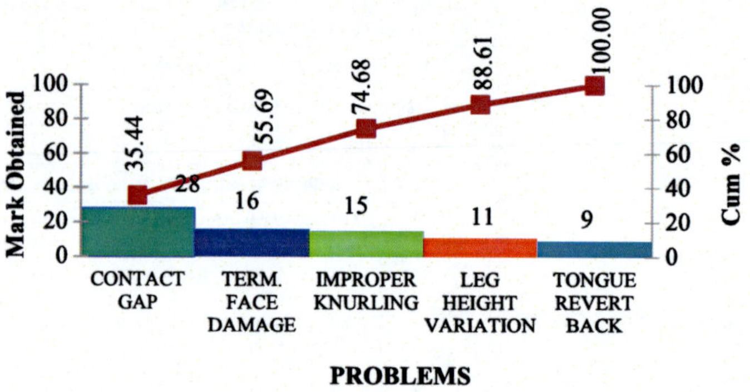

Fig. 2 Pareto chart analysis

Photo 1 Contact gap NG

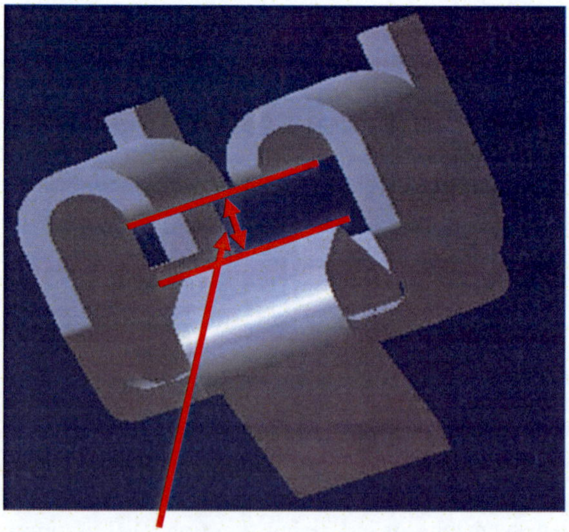

basis of highest priority numbers is calculated by multiplying the rating of given three factors.

For the occurrence factor, 5 rating is given for most frequently occurring problem and 1 for the problem which have lowest frequency of occurring. In detection, 1 rating is allotted for the problem that can be easily detected at the stage of generation time and 5 for the problem which cannot be detected. In extent of impact, 5 rating is allotted for the problem having impact on safety and 1 rating given to the problem which is having no significant impact (Table 2).

Table 2 Severity score of impact

S. No.	Problem	Severity
1	Contact gap in terminal	28
2	Terminal face damage	16
3	Improper knurling in terminal	15
4	Excessive tool setup time in crimping section	11
5	Tongue revert back	9

Table 3 Rating for three factors

Rating	Occurrence
5	Every hour
4	Every day
3	Every week
2	Every month
1	6 or more month
Rating	**Detection**
1	Can be detected at the stage of generation (fool proofing)
2	Can be detected at the next stage using gauge/jig
3	Can be detected by sample inspection
4	Can be detected by 100% inspection
5	Can't be detected
Rating	**Extent of impact**
5	Impact on safety
4	Functionality is affected
3	Sec. functionality is affected
2	Inconvenience to customer/user
1	Insignificant impact

4 Defining the Problem

The problem that creates major rejections is found the improper contact gap or height in 1.6F female type of terminal designed to slip inside a socket contact. The gap should be as per standards decided by the company, but at most of working station, this was not found as per the standards. As per the quality standards, this gap should not be greater than 2 mm (shown in Photo 2), but after measuring this gap, it was found 5–6 mm or up to 7 mm (shown in Photo 1). Due to this improper gap, the rejections levels were very high. As per data collected for 6 days in month of April 2018, the rejection was 66% because of this improper contact gap.

4.1 Impact of the Problem

The problem that was occurring before the improvement affected so many things in the company, i.e., increase in-house rejections, decrease in vendor rating, decrease in customer satisfaction, loss in productivity, increase in rework cost, and operator fatigue.

Photo 2 Contact gap ok

Table 4 Priority score calculation

S. No.	Problem	Occurrence	Detection	Extent of impact	Priority No.
		1 Low, 5 high	5 Low, 1 high	1 Low, 5 high	O * D * E
		O	D	E	
1	Tongue revert back	2	4	4	32
2	Contact gap in terminal	3	4	4	48
3	Improper knurling in terminal	1	2	3	6
4	Terminal face damage	2	4	2	16
5	Leg height variation in terminal	2	5	1	10

4.2 *Objectives*

The objectives of our team are to ensure 100% customer satisfaction, increase vendor rating, reduce in-house rejections, increase productivity, increase confidence level of operator, and reduce the cost of failure. As per (Table 4) rating method, the priority number of 'contact gap in terminal' is high so it is decided to solve this problem on priority basis.

5 Process Flow

The flow process of wire harness manufacturing section of the company is shown in Fig. 3. The raw material is received at raw material store and then goes through BOP; after that, the material is transferred to main store of the company. After issuing from main store, it reaches to wire assembly section and then transfers to coupler assembly. In coupler assembly, coupler is fixed to wires and this semi-finished product is transferred for board tapping process, and after that, this is electrically tested and cover connector is inserted. After complete assembly, the harness is sent to final inspection stage, and after inspection, it is ready for dispatch (Table 3).

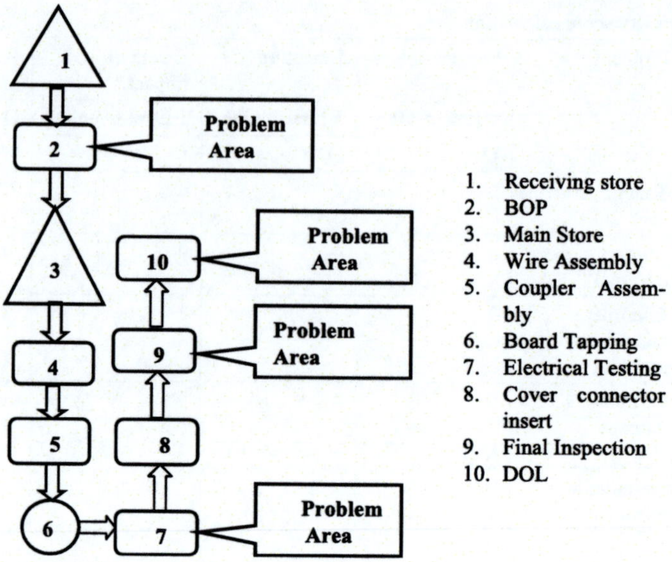

Fig. 3 Process flow diagram for indicating the problem

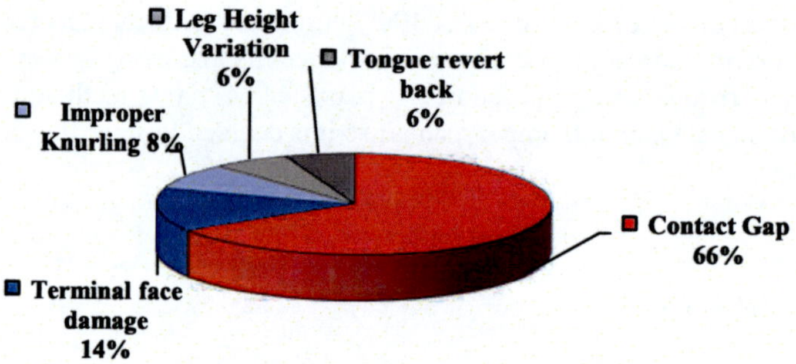

Fig. 4 Data representation on pie chart (refer Table 5)

6 Data Collection

For analysis of the impact of the five B category problems, data are collected for non-conforming pieces of terminal. After the analysis of six-day observations, it was found that 10 pieces are non-conforming due to leg height variation, 13 are due to improper knurling, 22 are due to face damaged, 9 are due to tongue revert back, and highest non-conforming pieces are due to improper contact gap of the terminal. This analysis is also shown in pie chart; according to that, 66% pieces of terminal are non-conforming due to contact gap problem. Total rejection level at different stages of production process is shown in Fig. 5. Out of total received quantity 1,120,000,

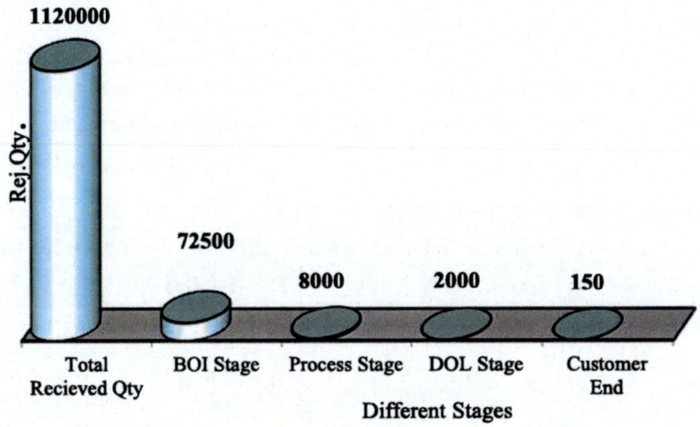

Fig. 5 Rejections faced at different stages during April 2018

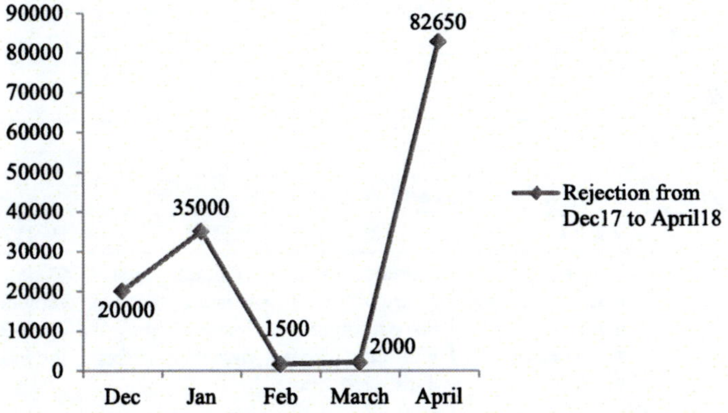

Fig. 6 Month-wise rejection from December 2017 to April 2018

72,500 pieces are rejected at BOI, 8000 pieces are rejected during process stage, 2000 at DOL, and 150 are rejected at customer end. Five month rejections status is shown in Fig. 6. It is observed that the rejection is continuously increasing due to the present problem as in the month of April 2018 this is found 82,650 pieces (Fig. 4).

7 Root Cause Finding

Why-Why analysis method is used to move past symptoms and understand the true root cause of a problem. It is said that only by asking 'Why?' Successively, it can investigate into a problem deeply enough to understand the ultimate root cause. With the help of this technique, it is observed that first tool setup rolled is due to insufficient training given to operator, latch/lever stopper wear problem is due to not analyzing tool life properly, the problem of improper setting of lever stroke length occurs due to inadequate tool design, and gap induced during rework of wire interchange is the problem which is due to inadequate rework method (Tables 4 and 5).

Table 5 Data collection for problem analysis

Category of problem	Data collection of 2nd week of March 2018						Total NC pcs	Priority
	Day-1	Day-2	Day-3	Day-4	Day-5	Day-6		
Leg height variation	1	2	2	1	3	1	10	4
Contact gap of term	15	16	12	27	19	24	**103**	**1**
Improper knurling	2	1	3	2	2	3	13	3
Terminal face damaged	3	2	3	4	5	5	22	2
Tongue revert back	1	2	1	2	1	2	9	5

Table 6 Why-Why Analysis for root cause finding

Problems	WHY1	WHY2	WHY3	WHY4
First tool setup piece rolled	First setup non-conforming pieces not separated by operator	Operator is not aware of the disposition of first setup pieces during job setup pieces	Operator is not aware of the impact of the first setup pieces	Training is not provided to operator about the impact of first setup pieces
Latch/lever stopper wear out	Stopper wear out before its life	Improper preventive maintenance	Preventive maintenance schedule is not adequate	Actual tool life is not analyzed properly
Improper setting of lever stroke length	Die setup length is more than required	Stopper screw is loose during production	Screw locking is not adequate for high-speed press	Tool design is not adequate
Gap induced during rework of wire interchange	Contact gap increased during terminal releasing	Pin use for terminal releasing from coupler	There is no other option to release the terminal	Rework method is not adequate

Table 7 Data analysis

S. No.	Causes	Observations	No. of failures	
			Nos.	In (%)
1	First tool setup piece rolled	15 times	9	53
2	Latch/lever stopper wear out	15 times	5	29
3	Improper setting of lever stroke length	15 times	2	12
4	Gap induced during rework of wire interchange	15 times	1	6

8 Data Analysis

After the root analysis of causes of failure, it is found that major causes are there due to which rejections level is high that are shown in Table 7. Observations are taken 15 times, and out of that, 9 times failure occurred due to tool setup problem and 5 times due to wear out lever stopper, only 2 failure are because of improper setting of lever stroke length and 1 time due to gap induced during rework of wire interchange. So it is observed from data that 53% failures are due to tool setup problem (Table 6).

9 Developing Solution

Theme that got maximum percentage in ABC analysis, rating method, and Pareto analysis is directly related to company's esteemed customer and fit/function at customer end, internal customer harassment, overall quality rating, time loss in productivity and rejection increase/rejection cost increase, so it is must to eliminate the root causes of this problem that are identified during root cause finding stage. Training is provided to the operator about the impact of first set pieces, and customer complaints were displayed at the notice board. Fresh pieces were used while reworking the wire interchange, and training is provided to operator for change of rework method that concludes to adequate method of reworking. Tool life is analyzed again and displayed, and fixed punch is used at the place of lever/latch. To implement the improvements, following are the resistances that are responsible for further step to overcome the causes of failures.

Before the implementation of the counter measures, the rejection level found 82,650 pieces at the end of April 2018, it is observed that after implementation of countermeasure 1, rejection level reduced to 55,600 pieces, after countermeasure 2, it reduced to 45,600 pieces, after countermeasure 3, it is found 40,000 pieces and after implementation of countermeasure 4, the rejections level found at zero level. This is great achievement of our objectives (Table 8).

Table 8 Countermeasures adopted for elimination of root causes

S. No.	Root causes	Countermeasures
1	Training is not provided to operator about the impact of first setup pieces	Training is provided to operator about the impact of first setup pieces Customer complaint is displayed at shop floor
2	Rework method is not adequate	Fresh pieces are used while reworking the wire interchange Training is provided to operator for change of rework method
3	Actual tool life is not analyzed properly	Tool life is analyzed again and displayed
4	Tool design is not adequate	Fixed punch is used in place of lever/latch

10 Calculation for Cost-Effectiveness

Monthly rejection qty. of terminal = 10,150 pcs.
Cost of reject w/p term = 10,150 * 0.90 paisa = Rs. 9135.

Internal failure cost
Operator cost for processing = 2 min.* 8000 = 267 h.*20 = Rs. 5340.
Operator cost (DOL) = 2000 * 2 min. = 67 h * 20 = Rs. 1340.
Total = Rs. 6680.

External failure cost
Customer visit cost (5-times) = 200 km * 5 = 1000 = Rs. 1000.
Operator cost for rework at customer end = 6 days * 160 = Rs. 960
Staff cost = 16 h * 72 = Rs. 1152
Total cost save = 9135 + 5340 + 1340 + 1000 + 960 + 1152 **= Rs. 18,927/month**
Total cost saving per year = Rs. 227,124 (Table 9).

Table 9 Probable resistance

S. No.	Problem resistance	Step to overcome
1	Supplier may not accept this as they might think that this will be an extra burden to them	Counseling the supplier and to convince them that this will help them for a smooth and enjoyable work
2	Operator will not agree to implement the change method of rework	Counseling the operator about the impact of the problem and convince them that this will help to avoid the occurrence of the problem at customer end
3	Shutdown of the machine is required for tool modification, likely to lead production loss	A system is developed to overcome the problem by planning the tool modification on turn basis

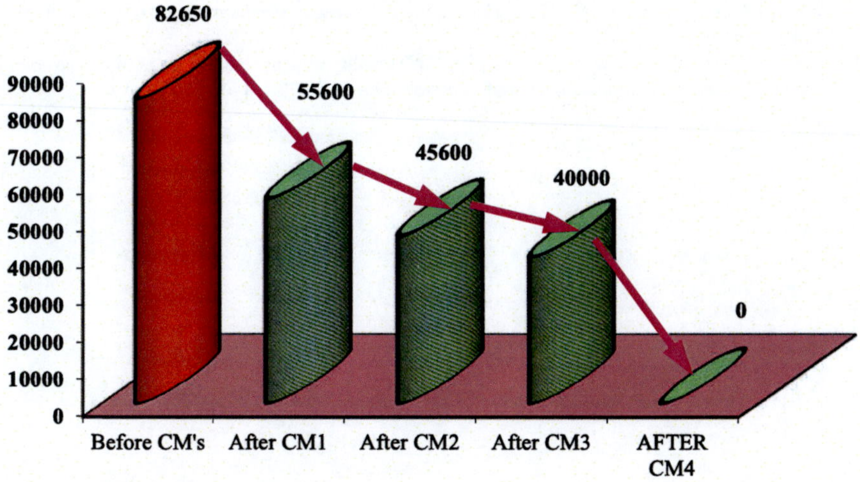

Fig. 7 Effectiveness of countermeasures

11 Conclusion

The theme of this was to reduce material cost and failure cost by solving the contact gap problem of the terminal in WIP area by adopting Kaizen lean tool. In this regard, total 52 problems are identified, out of those 43 in A, 5 in B, and 4 in C category.

This problem affects that total 82,650 pieces were rejected maximum in month of April 2018. In this case, to remove the contact gap, a modified mechanism with fixed punch installed in place of existing mechanism. After improvement in mechanism, the rejection level reduced to zero and the total cost saved Rs. 227,124/year. At the closing stages of this work, it concludes that Kaizen technique is very effective to solve the work-related problems in the organizations. This way of improving the production processes and getting cost-effectiveness is very precious for building far above the ground productivity of the company (Fig. 7).

References

1. Kikuchi K, Kikuchi T, Takai T (2007) Method of overall consumable effectiveness. In: IEEE international symposium on semiconductor manufacturing, Santa Clara, USA, pp 1–4
2. Modarress B et al (2005) Kaizen costing for lean manufacturing: a case study. Int J Prod Res 43(9):1751–60

3. Veres C, Marian L, Moica S (2017) Case study concerning effects of Japanese Management Model. Procedia Eng 181:1013–1020
4. Chandrasekaran M, Kannan S, Pandiaraj P (2008) Quality improvement in automobile assembly production line by using kaizen. Manuf Technol Today 7(3):33–38

Solving the Flexible Job Shop Scheduling Problem Using a Hybrid Artificial Bee Colony Algorithm

Rylan H Caldeira, A. Gnanavelbabu, and J. Joseph Solomon

Abstract In this work, a hybrid artificial bee colony algorithm is proposed for solving the flexible job shop scheduling problem (FJSP) which is a classification of the classical job shop scheduling problem (JSP) considered to NP-hard in nature. In FJSP, an operation can be processed on a set of capable machines with different processing times, thereby dealing with a routing and sequencing problem. The objective considered is to minimize the makespan. The basic artificial bee colony (ABC) algorithm stresses on the balance between global exploration and local exploitation. However, the drawback of the basic ABC algorithm is that it converges prematurely and may get trapped in the local optima. Hence to improve its exploration capability in local space, it is hybridized using a Tabu search (TS) algorithm. At first, initial solutions are generated with certain quality and diversity as food sources using multiple strategies in combination. Crossover and mutation operations are carried out for machine assignment and operation sequencing separately generating new neighboring solutions. Lastly, a local search strategy based on TS is proposed to enhance the local search capability. Kacem's and Brandimarte's benchmark instances are used to compare the performance of the proposed approach to five other well-known algorithms in the literature. Experimental results revealed the superiority of the proposed approach in solving FJSP.

Keywords Flexible job shop scheduling · Artificial bee colony algorithm · Tabu search · Makespan

---

R. H. Caldeira · A. Gnanavelbabu (✉) · J. Joseph Solomon
Department of Industrial Engineering, CEG Campus, Anna University, Chennai, Tamil Nadu 600025, India
e-mail: agbabu@annauniv.edu

S. Vijayan et al. (eds.), *Trends in Manufacturing and Engineering Management*, Lecture Notes in Mechanical Engineering,
https://doi.org/10.1007/978-981-15-4745-4_72

1 Introduction

The process of converting the resources into a finished or semi-finished product by adding value to it is called manufacturing. There are various types of manufacturing systems based on their size and activity. They are job shop, batch production, mass production, flow production, and assembly lines. This work deals with job shop since it is an extensively adopted type of production system. A job shop is a type of manufacturing process in which small batches of a variety of customized products are produced. JSP is the process of scheduling these jobs to machines. FJSP which is an extension of the JSP is an NP-hard combinatorial optimization problem. These problems are complex as each operation can be processed on more than one machine with different processing times. It consists of two sub-problems, routing, and scheduling. FJSP is usually solved by adopting a hierarchical or integrated approach. Brandimarte [1] proposed a hierarchical strategy for solving the scheduling problem by using TS. In recent times, hybrid meta-heuristics have become a popular tool for solving the FJSP as the hybridization of two or more meta-heuristics have proven to improve the performance of each meta-heuristic while eliminating the drawbacks of the other. ABC is one of the popular meta-heuristics. ABC provides a structured search framework that is characterized by individual improvement, cooperation, and competition of the population, and thus, it can perform global exploration efficiently (large-scale search). But the original ABC algorithm falls back in its ability to perform exploitation effectively by converging to a solution in earlier iterations, thus being incapable of exploring the search space effectively. In order to improve the local search capability of the ABC algorithm, a local search technique based on TS is introduced which greatly enhances the exploitation capability of the algorithm.

A beer froth ABC was proposed by Sharma et al. [2] in which to maintain proper harmony amid exploration and exploitation capabilities a beer froth phenomenon inspired position update was incorporated. A novel ABC based on the cosine similarity is proposed by Xiang et al. [3], a method based on cosine similarity for choosing a better neighbor individual to overcome the insufficiency of slow convergence in ABC. Levy flight ABC proposed by Sharma et al. [4] illustrated that there is a higher probability of skipping the optimal solution owing to large step sizes and a search strategy based on Levy flight was integrated with ABC which balances the exploitation and exploration capability. A hybrid ABC for FJSP proposed by Thammano and Phu-ang [5] where dispatching rules and the harmony search algorithm were used to create the initial solutions after which one of the two search techniques was randomly selected with a probability that is proportional to their fitness values. Simulated annealing algorithm or filter and fan algorithm was employed to escape from the local optimum. A hybrid ABC for the JSP proposed by Zhang et al. [6] aimed at minimizing the total weighted tardiness in JSP. Considering the high complexity, a novel ABC algorithm was proposed for solving the problem. A neighborhood property of the problem was discovered, and then, a tree search algorithm was devised to enhance the exploitation capability of ABC. An effective ABC for the FJSP proposed by Wang et al. [7] stresses the balance between global exploration and local exploitation. A local search strategy based on critical path theory was proposed to enhance the local intensification capability of the onlooker bees.

A hybrid ABC for solving the FJSP proposed by Li et al. [8] considered the total flow time criterion, where TS heuristic is introduced to perform a local search for employee, onlooker, and scout bee phases. A best-so-far ABC proposed by Banharnsakun et al. [9] biases the solution direction toward the best-so-far solution rather than a neighboring solution as proposed in original ABC method and also uses the set theory to describe a mapping of their proposed method to the problem in the combinatorial optimization domain. Among different approaches, ABC algorithm proposed by Karaboga and Gorkemli [10] is a widely employed swarm intelligence algorithm for scheduling FJSP problems. ABC algorithm was first proposed to solve the multi-variable and multi-modal continuous functions. Localization and multi-objective evolutionary optimization for FJSP proposed by Kacem et al. [11] present two new approaches to jointly solve routing and scheduling by localization and then by evolutionary approach. Mehrabad and Fattahi [12] proposed a TS algorithm to solve the FJSP to minimize makespan. Hurink et al. [13] performed TS for solving the FJSP.

In this work, we propose a hybrid ABC with TS to improve the exploration capability in local space (exploitation). The proposed algorithm consists of the basic ABC phases, initialization, employee bee phase, and the scout bee phase. Here, solutions are initialized as two separate chromosomes to facilitate individual crossover. The two chromosomes are machine assignment and operation sequencing. Machine assignment follows the random rule, maximum time remaining rule, and most number of remaining operations rule, whereas operation sequencing follows the random rule, global minimum processing time rule, and work-time considered rule. These rules are adopted so that the population generated is diverse and feasible.

Employee bee phase initiates with the modified precedence operation crossover operator (MPOX) for crossover in operation sequencing. A uniform crossover is used for machine assignment. For half of the population chosen at random maximum workload reduction, mutation operator is used for modifying machine assignment. The mutation operator is used to enhance the exploration capability of the algorithm. The solution obtained is now introduced into a TS local search. This search performs an iterative search on the entire population to get new and better solutions in each step. TS is chosen as it prevents the local search from converging on a solution prematurely, thereby improving the local exploration capability of the algorithm. When the solution does not improve, a random solution is generated on behalf of the scout bee phase which is compared with the TS solution. A new better population is obtained from both of these solutions and taken for the next ABC iteration until the termination criteria are satisfied.

2 Problem Definition

The formulation of FJSSP is as follows: There is a set of '*m*' machines $c = 1, 2, \ldots, m$ and a set of '*n*' jobs $a = 1, 2, \ldots, n$. Each job '*a*' consists of a sequence of '*k*' operations. Processing of each operation O_{ab} ($b = 1, 2, \ldots, k$) of job '*a*' has to be done on single machine 'M_c' out of a set of given available machines M_{ab} (*for* M_c € M_{ab}, M_{ab} € M). Each operation needs a single machine picked from a set of capable machines. Further, FJSP requires to set starting and ending times of each operation. The FJSP needs to find both allotment and an operations sequence on the machines so that given criteria are satisfied. If there is at least one operation M_{ab} € M, it is partial flexibility FJSP (P-FJSP), whereas if $M_{ab} = M$ for every operation, it is total flexibility FJSP (T-FJSP). The objective is to minimize the makespan.

3 Framework of the Proposed Approach

3.1 *Artificial Bee Colony Algorithm*

The ABC algorithm is a meta-heuristic designed by observing the foraging behavior of honey bee, introduced by Karaboga [1] in 2005. Here, all possible solutions to the optimization problem are considered as food sources, and the honey bees search to find the best solution. There are three types of bees in the population which are involved in finding the food source. They are

- **Employer bees**. For all employer bees, generate new candidate solutions and retain the candidate solutions possessing better objective function value.
- **Onlooker bee phase**. For all onlooker bees, select an employed bee randomly. Produce a new candidate solution. Compute fitness of individual, and if the fitness of the new solution is better than the existing solution, replace the older solution.
- **Scout bees**. If any source is exhausted, replace it by randomly generated solution by scout and memorize the best solution until stopping criteria are met. Old is substituted by an arbitrarily picked source within a specified region.

3.2 *Solution Representation*

The problem is solved hierarchically, and thus, solutions are represented as separate vectors each for operation sequencing and machine assignment which is termed as the two-vector solution representation (see Fig. 1). To guarantee an initial population with a certain quality, different heuristics are adopted to generate a population.

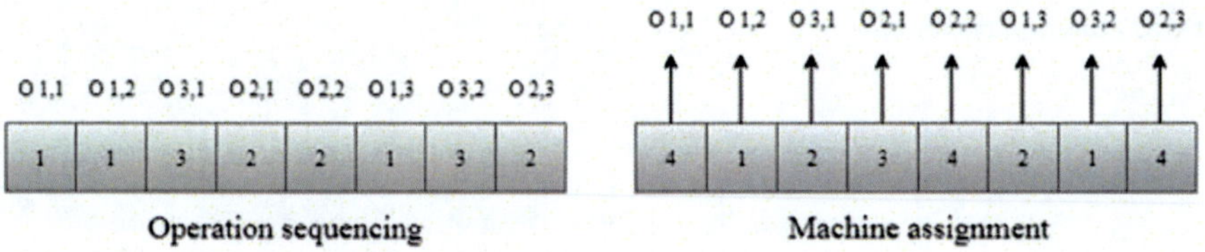

Fig. 1 Two-vector solution representation

For machine assignment, the following heuristics are employed:

- Global minimum processing time rule (20%);
- Workload considered rule (20%);
- Random rule (60%).

For operation sequencing, the following heuristics are employed:

- Random rule (80%);
- Most time remaining rule (10%);
- Most no. of operations remaining rule (10%).

3.3 Crossover and Mutation

In this phase, the two vectors are modified separately by crossover and mutation operators. Machine assignment vector undergoes a uniform crossover (see Fig. 3), where the new vector u_{new} is derived from the initial vector u_i and the vector with the best solution u' by generating a binary vector and then a maximum workload reduction mutation operator is used to reduce the assignment of the machine with maximum workload.

Operation sequencing vector undergoes a modified precedence operation crossover (MPOX) operation (see Fig. 2), where a random subset of half the jobs is generated, and they are given priority from one vector, whereas the other jobs are

Fig. 2 Modified precedence operation crossover

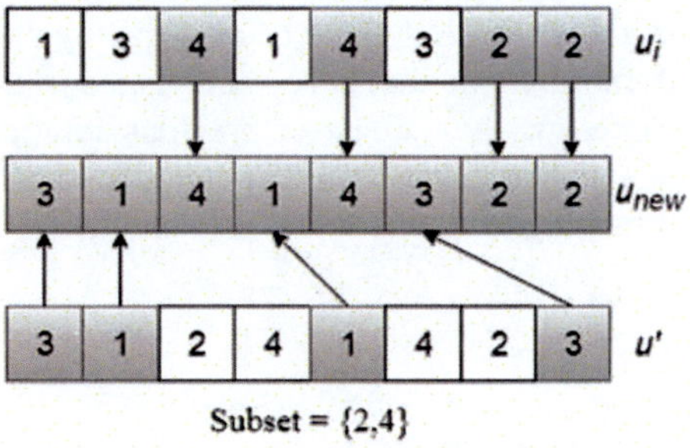

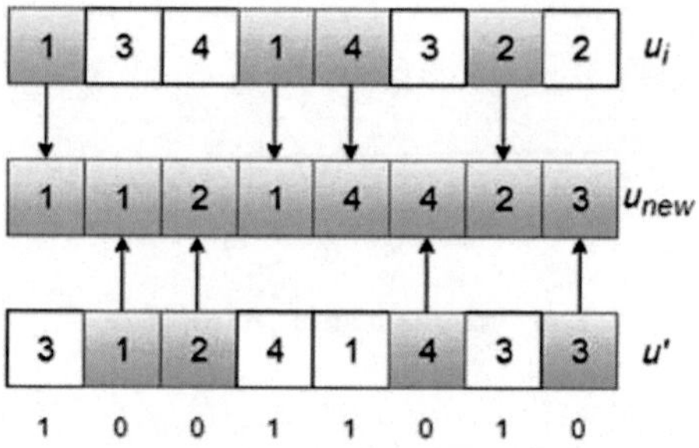

Fig. 3 Uniform crossover

assigned from another vector from left to right. Then, a maximum workload reduction operator is applied which identifies the machine with the maximum workload and re-assigns possible operations to other machines with lesser workloads. This acts as the mutation operator which is performed on only half of the population (Fig. 3).

3.4 Tabu Search

The TS is a local search heuristic that uses a neighborhood search procedure to iteratively move from one possible solution to x to another improved solution x' in the neighborhood of x until a terminating criterion is fulfilled. Local search techniques usually get stuck in less efficient areas and do not explore the search space completely. TS carefully explores the neighborhood of each solution as the search progresses. So to prevent the heuristic from returning to a recently visited solution, a short-term memory of recent moves within the search space is recorded, and this prevents future moves from repeating these recent moves. That is if a new potential solution has been previously visited within a short time period, it is marked as Tabu (forbidden) and the algorithm does not consider that solution. Thus, the solutions admitted to the new neighborhood, $N * (x)$, are determined through the use of memory structures. Using these memory structures, the search progresses iteratively moves from the current solution x to an improved solution x' in $N * (x)$.

These memory structures form a list known as the Tabu list which will contain a set of rules and banned solutions used to filter solutions that will be admitted to the neighborhood $N * (x)$ to be explored by the search. So a Tabu list is a short-term set of the solutions that have been visited in the recent past (less than n iterations ago, where n is the number of previous solutions to be stored which is also called the Tabu tenure). So a Tabu list consists of solutions that have changed by the process of moving from one solution to another.

3.5 Proposed Hybrid ABC

The parameters for the algorithm like population size, number of iterations, and bee colony size are set. The initial population solution representation is done by using two-vector notations employing population initialization rules.

In the employee bee phase, the machine assignment vector undergoes uniform crossover and a maximum workload reduction mutation operation where the mutation is performed on only half of the population, whereas the operation sequencing vector undergoes a modified operation crossover operation. Then, for the whole modified population, a TS local search is performed which provides the best solution in every iteration without converging predominantly, thus improving the exploration capability of the ABC algorithm in local search space. The flowchart of the proposed hybrid ABC is illustrated (see Fig. 4).

4 Experimental Results

The proposed approach was coded in MATLAB software on an i3 processor with a speed of 1.8 GHz on a Windows 10 operating system. Computational experiments are conducted on all ten instances of Brandimarte [1] and five instances of Kacem et al. [11] since these instances are widely solved in literature and hence extensive comparison of solution methodologies can be carried out. Kacem's instances range from sizes of 4 jobs and 5 machines to 15 jobs and 10 machines with a single instance of 8 jobs and 8 machines with partial flexibility, while the remaining have total flexibility. Brandimarte's instances consist of problems ranging from sizes of 10 jobs and 6 machines to 20 jobs and 15 machines with medium flexibility.

From Table 1, it can be seen that the proposed hybrid approach obtains best solutions in ten out of fifteen benchmark instances. In comparison with other algorithms, it can be seen that the proposed algorithm outperformed the algorithm proposed by Xing et al. [14] in seven problems. The performance of the hybrid ABC was superior to TLBO proposed by Buddala et al. [15] in six problem instances. The proposed approach gave better results to two problems each compared to AIA proposed by Bagheri et al. [16] and HTSA proposed by Li et al. [17]. The proposed ABC outperformed BBO proposed by Rahmati et al. [18] in eight problem instances. The convergence of Brandimarte's MK09 problem is shown (see Fig. 5). Gantt chart obtained using the proposed approach is shown for Kacem's 10 * 15 problem instance (see Fig. 6).

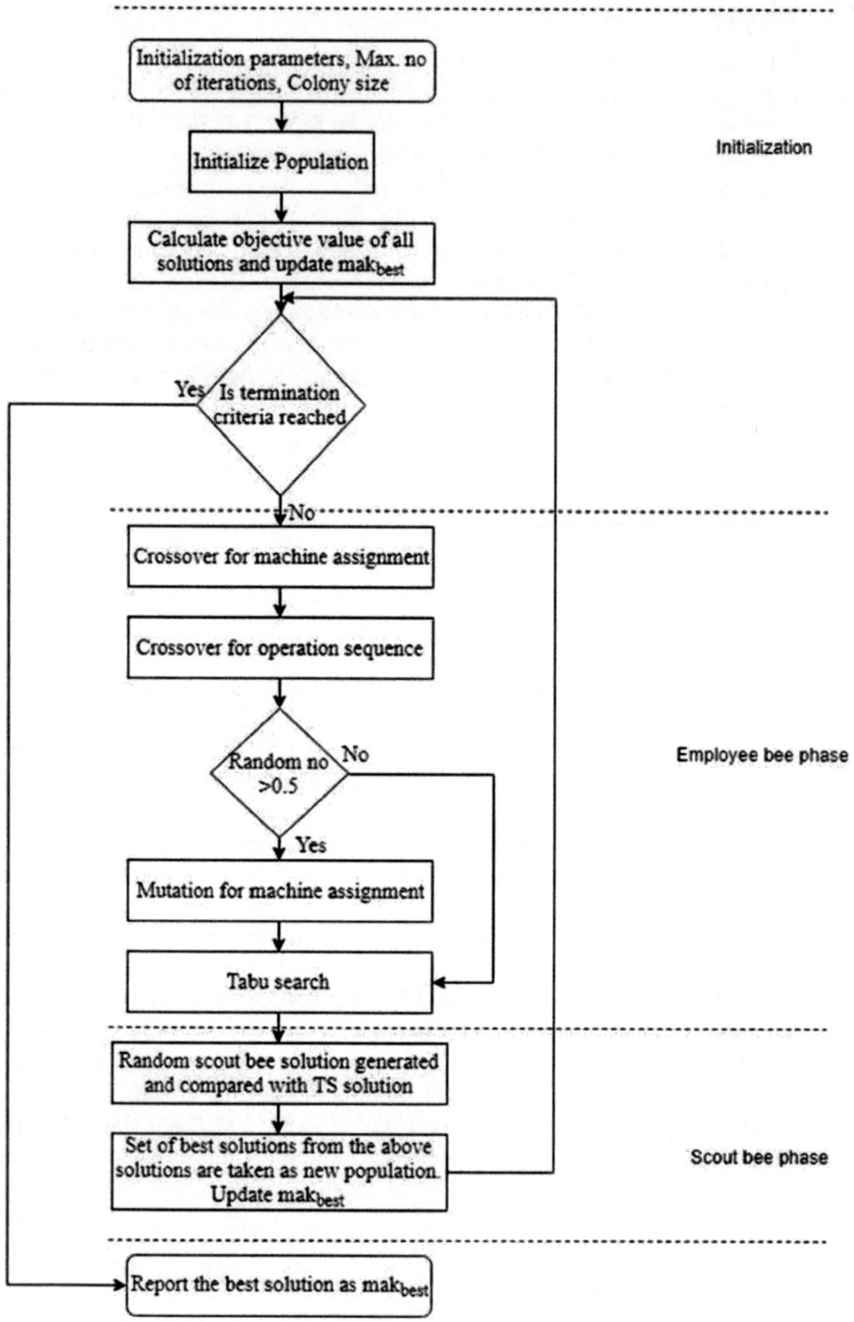

Fig. 4 Flowchart of the proposed hybrid ABC

Table 1 Results of benchmark problems

S. No.	Problem size	Xing's algorithm [14]	HTSA [17]	AIA [16]	BBO [18]	TLBO [15]	Proposed ABC
1	4 × 5	11	11	NA	11	11	**11**
2	8 × 8	14	14	14	14	14	**14**
3	10 × 7	11	11	NA	NA	11	**11**
4	10 × 10	7	7	7	7	7	**7**
5	15 × 10	11	11	11	12	12	**11**
6 (MK01)	10 × 6	42	40	40	40	40	**40**
7 (MK02)	10 × 6	28	26	26	28	28	27
8 (MK03)	15 × 8	204	204	204	204	204	**204**
9 (MK04)	15 × 8	63	61	60	64	63	**60**
10 (MK05)	15 × 4	181	172	173	173	172	**172**
11 (MK06)	10 × 15	60	65	63	66	65	**63**
12 (MK07)	20 × 5	148	140	140	144	144	**140**
13 (MK08)	20 × 10	523	523	523	523	523	523
14 (MK09)	20 × 10	308	310	312	310	311	**307**
15 (MK10)	20 × 15	212	214	214	230	214	214

Note The significance of the "Bold" is to indicate that the proposed problem solution is better than benchmark problem solutions

5 Conclusion and Future Scope

This paper considers the NP-hard FJSP with the makespan criteria, and the same is solved by a proposed hybrid algorithm where a TS is incorporated in the ABC algorithm. The effectiveness of using TS as a local search is highlighted by the improved exploration and convergence capability of the ABC algorithm in the local search space. The proposed hybrid ABC algorithm showed superior performance in comparison with other well-known meta-heuristics on Kacem's and Brandimarte's instances.

This work can be extended to solve other scheduling problems with constraints.

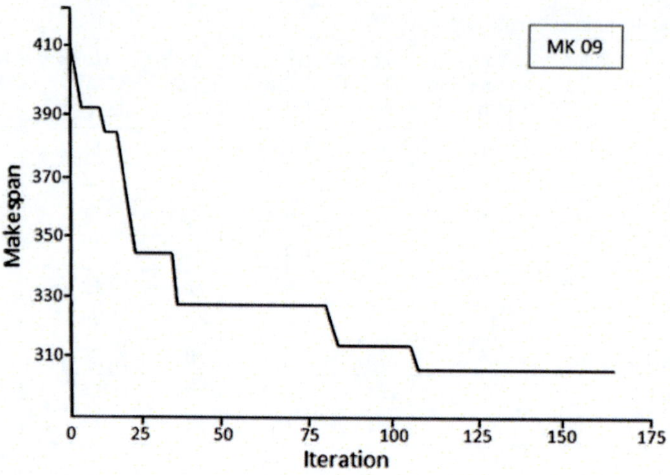

Fig. 5 Convergence of MK09 instance

Fig. 6 Gantt chart of Kacem's 10 × 15 instance

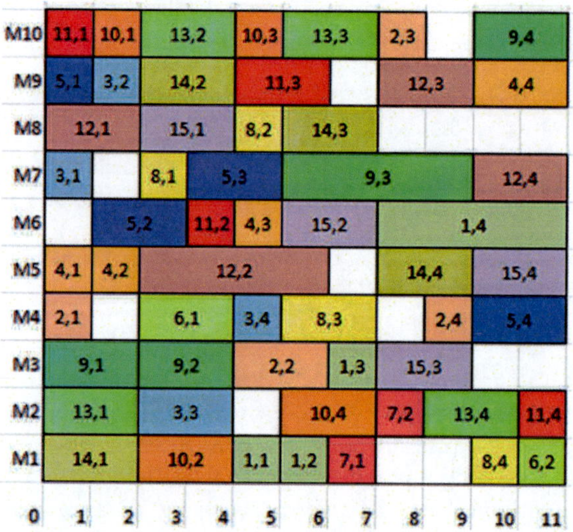

References

1. Brandimarte P (1993) Routing and scheduling in a flexible job shop by tabu search. Ann Oper Res 41(3):157–183
2. Sharma N, Sharma H, Sharma A (2018) Beer froth artificial bee colony algorithm for job-shop scheduling problem. Appl Soft Comput 68:507–524
3. Xiang W et al (2018) A novel artificial bee colony algorithm based on the cosine similarity. Comput Ind Eng 115:54–68

4. Sharma H, Bansal JC, Arya KV, Yang XS (2016) Lévy flight artificial bee colony algorithm. Int J Syst Sci 47(11):2652–2670
5. Thammano A, Phu-ang A (2013) A hybrid artificial bee colony algorithm with local search for flexible job-shop scheduling problem. Procedia Comput Sci 20:96–101
6. Zhang R, Shiji S, Cheng W (2013) A hybrid artificial bee colony algorithm for the job shop scheduling problem. Int J Prod Econ 141(1):167–178
7. Wang L, Zhou G, Xu Y, Wang S, Liu M (2012) An effective artificial bee colony algorithm for the flexible job-shop scheduling problem. Int J Adv Manuf Technol 60(1–4):303–315
8. Li JQ, Xie SX, Pan QK, Wang S (2011) A hybrid artificial bee colony algorithm for flexible job shop scheduling problems. Int J Comput Commun Control 6(2):286–296
9. Banharnsakun A, Sirinaovakul B, Achalakul T (2012) Job shop scheduling with the best-so-far ABC. Eng Appl Artif Intell 25(3):583–593
10. Karaboga D, Gorkemli B (2012) A quick artificial bee colony-qABC-algorithm for optimization problems. In: International symposium on innovations in intelligent systems and applications. IEEE, pp 1–5
11. Kacem I, Hammadi S, Borne P (2002) Approach by localization and multiobjective evolutionary optimization for flexible job-shop scheduling problems. IEEE Trans Syst Man Cybern Part C (Appl Rev) 32(1):1–13
12. Saidi M, Fattahi P (2007) Flexible job shop scheduling with tabu search algorithms. Int J Adv Manuf Technol 32(5–6):563–570
13. Hurink J, Jurisch B, Thole M (1994) Tabu search for the job-shop scheduling problem with multi-purpose machines. Oper Res Spektrum 15(4):205–215
14. Xing LN, Chen YW, Yang KW (2009) An efficient search method for multi-objective flexible job shop scheduling problems. J Intell Manuf 20(3):283–293
15. Buddala R, Mahapatra SS (2018) An integrated approach for scheduling flexible job-shop using teaching–learning-based optimization method. J Ind Eng Int 1–12
16. Bagheri A, Zandieh M, Mahdavi I, Yazdani M (2010) An artificial immune algorithm for the flexible job-shop scheduling problem. Futur Gener Comput Syst 26(4):533–541
17. Li JQ, Pan QK, Liang YC (2010) An effective hybrid tabu search algorithm for multi-objective flexible job-shop scheduling problems. Comput Ind Eng 59(4):647–662
18. Rahmati SHA, Zandieh M (2012) A new biogeography-based optimization (BBO) algorithm for the flexible job shop scheduling problem. Int J Adv Manuf Technol 58(9–12):1115–1129

Linear Programming in Market Management Using Artificial Intelligence

Shetty Vignesh Uday, Hamritha, and Gaurav Chaudhary

Abstract The intelligence exhibited by machines in decision making using mathematical algorithms in contrast to the human intelligence is known as artificial intelligence (AI). It is an area of engineering science that focuses on the modeling of intelligent systems that work and act in response like humans. The tasks that AI includes are face recognition, voice recognition, planning, learning, and predicting. Linear programming problem (LPP) is a tool to solve the real-life problems using the standard mathematical approach that gives us the optimal solution. This paper aims at providing the investors an aid in decision making while advertising their business with combination of the tools that are available. Neural network model will be formulated and trained on the basis of dataset that includes past data of variables such as amount of investment, type of business, demographical area, and consumer attributes. The output nodes include the variables based on the type of advertisement. After training and testing are finished, user provides the model with the inputs and the output predicts the optimal percentage of marketing of product in different fields. This way the model will help us to make sound decisions in allocating the total investment among the variables.

Keywords Linear programming problem · Marketing · Artificial intelligence · Productivity

1 Introduction

Artificial intelligence is widely used in decision making and prediction using mathematical algorithms. Artificial neural network (ANN)-based system works similar

S. V. Uday (✉) · Hamritha
Department of Industrial Engineering and Management, Ramaiah Institute of Technology, Bangalore 560054, India
e-mail: shettyvignesh01@gmail.com

G. Chaudhary
Department of Electronics and Communication Engineering, Ramaiah Institute of Technology, Bangalore 560054, India

S. Vijayan et al. (eds.), *Trends in Manufacturing and Engineering Management*, Lecture Notes in Mechanical Engineering,
https://doi.org/10.1007/978-981-15-4745-4_73

to a human brain in an individual's body. This is an easier method and guarantees consistency across the marketing chain [1]. An artificial neural network is a network of many interconnected nodes connected by unidirectional communication channels, called weights that carry numeric data which adjusts itself as per the dataset that contains past data of variables such as amount of investment, type of business, demographical area, and consumer attributes.

Linear programming enables us to make sound decisions in several fields based on the given constraints. There are several ways to solve a linear programming itself depending upon the type of the problem [2]. This is one of the oldest and most dependable methods which have already proved giving the most optimized results in several fields like scheduling, food and agriculture, energy and transportation problems. It is fundamentally used for allocating the limited resources since they are not abundantly available in the nature.

One of the most upcoming technologies of this era is the concept of artificial intelligence, where the system is trained and tested for predicting, categorizing, and collecting the data using facial recognition, theoretical data, speech recognition, virtual reality, image processing, visual audio manipulation, etc [3].

The computer-based system simulates the problems on par with human processing system. There are several ways where the system could be classified depending on the processing technique and applications.

A product to be sold needs to have a solid marketing strategy, existing marketing scenario has been completely evolved, and compared to the past years, there are several types of marketing methods like digital marketing, direct marketing, inbound marketing, cause marketing, guerrilla marketing, relationship marketing, multi-level marketing, personalization marketing, database marketing, etc. But the confidence interval of the product promising the returns plays a vital role in the type of marketing strategy that is adopted.

So, this paper in the whole discusses the algorithm and crucial part of the choosing appropriate marketing techniques using artificial-based system. The basic objective function of linear programming is to maximize the reach, frequency, and quality of exposure within the constraints of cost and availability of variables. Each of these depends on the product, demography, investment, customer behavior, and economy [4].

Henceforth, as the constraints increase, the factors on which it depends will contain huge set of data. As the data size increases, it is difficult to study and analyze the data manually. Therefore, linear programming problem can be framed based on type of the problem.

The real-world linear programming problem has a large set of variables and is quite difficult to solve manually which can be resolved with the help of programs and solve the problem based on the input set of constraints and obtain feasible solutions. This can be illustrated using limited set of constraints but when making use of software which are available in the market for solving the problem constraints considered are many. Algorithms are written for solving problem in the area of marketing, and also it would help in the prediction of the market returns of the product.

2 Methodology

The formulation of a linear programming problem considers only few variables and can be solved manually, but on real time applications, the variables considered are to be many. The main focus is not to solve a linear programming problem, but utilizing the linear programming system by a computer-embedded program that can solve the figures of large variables and generate the best suitable answer based on the past data and also configures the future prediction of the product chosen by the people.

In this system, input is taken from the end user like minimum investment, maximum investment, area, etc., and these inputs are fed to the system with '*n*' variable, and from the past records, it gives the output how the amount should be invested in each format of marketing. So, data in the system enables the customer to decide on the investment and thus helps to make a smart business.

2.1 Common Terminologies Used in Linear Programming

Decision Variables The decision variables are those which will decide the output of a problem statement. Decision variables need to be identified to solve any problem. For the defined problem, decision variables are "add input constraint names the ones that are visible" investment, type of business, city, time (in months), and assets.

Objective Function It is the prime function which is to be worked upon depending upon the problem. In this problem, the company wishes to know the better option in investing money in advertisement ("out of 4 options, add name of output nodes here") digital marketing, newspapers, pamphlets, and banners.

Following is the algorithm can be considered for obtaining the desired output:

The model comes under multiple class neural network classification that involves building neural networks that map the input to the network and output containing more than two classes. In this case, dependency on the input and output is not known. Therefore, three hidden layers are used with reLu as activation function for hidden layer and Softmax for output layer (Fig. 1).

Supervised Learning During the training of ANN under supervised learning, the input vector is presented to the network, which will give an output vector. The obtained result vector is compared with the desired result vector. An objection signal is generated if there is a difference between the actual result and the desired result vector.

The weights are adjusted (added or subtracted with value) until the actual output is matched with the desired output. It is a non-recurrent network that has processing units or nodes in layers, and all the nodes in a layer are connected to the nodes of the previous layer and also to node of next layer. The connection has different weights upon them (chosen randomly before training). These weights change as per

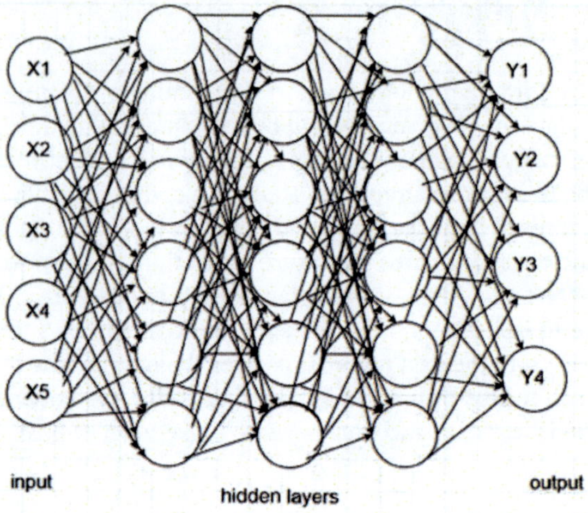

Fig. 1 Basic neural network model

the dataset using back-propagation algorithm to give us an output with acceptable error.

Activation functions are used to map the output of a layer between 0 to 1 and −1 to 1 depending on the type of function used. Rectified linear activation function—reLu—is the mostly used activation function (Fig. 2).

$f(z)$ is zero, if z is less than zero. $f(z)$ is z, if z is greater than zero. The function and its derivative both are monotonic.

Softmax activation function is used for final layer, where the Softmax calculates a probability for every possible class. By applying the standard exponential function to each element 'z' of the input vector, normalize these values by dividing the sum of all these exponentials, as shown in the formula below.

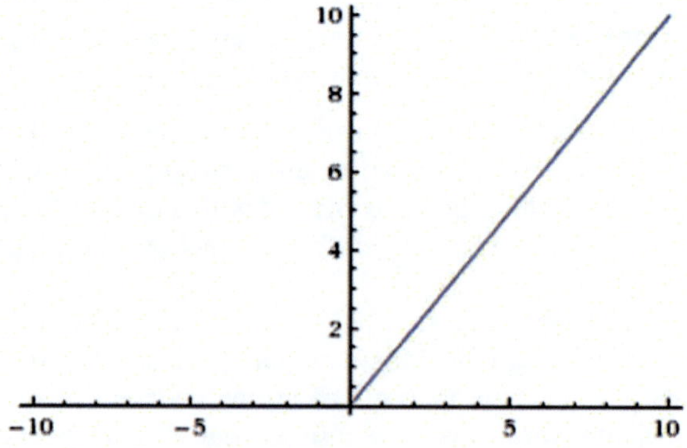

Fig. 2 Graphical representation of input output of reLu

$$\sigma(z) = \frac{e^{\beta z_j}}{\sum_{k=1}^{K} e^{\beta z_k}} \tag{1}$$

Training Algorithm For training, use reLu activation function for hidden layers and Softmax for the output layer. The training of neural network will have the following four phases.

Step 1: Feed-forward phase:

- Initialize to start the training with following inputs
  - Weights (choose random weights)
  - Learning rate α (should be small)
- Calculate activation function (a) of unit i in layer j using below relation

$$a_i^j = g\left(\sum_{i=1}^{n} x_i H_i^j\right) \tag{2}$$

where i is layer of which the activation function is being calculated.

Step 2: Calculate the error and adjust weights as follows:

- Cost function in neural network for multiple output will be:

$$J(\theta) = -\frac{1}{m}\left[\sum_{i=1}^{m} t^{(i)} \log(y_\theta(x_i))_k + (1 - t_k^i) \log(1 - y_\theta(x_i))_k\right] \tag{3}$$

where m is total number of nodes in a layer, t is desired output, and $(y_\theta)_i$ is ith output.

Step 3: Back propagation:

Intuition: δ_j^i = error of node j in layers i.

Intuition for output layer is:

$$\delta_j^{(m-1)} = y_j^{(m)} - t_j^{(m)}$$

where $\delta_j^{(2)}$ is intuition of the output layer

In most cases, back propagation is calculated by the backend library itself

Step 4: Update the weights in the network until dataset classified correctly or stopping criterion satisfied.

$$H^{d+1} = H^d - \propto \frac{\partial J}{\partial h} \tag{4}$$

where $\propto$ = learning rate (should be low), d is number of iterations, and J is loss function

Table 1 Results obtained for the input variables

Output class	Probability (%)
Digital marketing	0.84
Newspaper	0.10
Pamphlets	0.05
Banners	0.01

Testing Stage Simulate the network with new dataset. By using various test cases, neural network will be trained and that network will be used to solve real cases by linear programming problem.
Outcome Data is captured to get the best possible way of advertising, calculated by the neural network

3 Results and Discussion

Considering the marketing issues and availability, it is difficult to obtain the solutions considering all the variables. Hence, by using a formulated mathematical model, it is easier to obtain the probability of variables. Hence, the objectives can be met and profits can be maximized in real-life applications.

As marketing strategy is being one of the most important criteria of a product, it is appreciated if it reaches the end user with necessary requirements in short duration.

The above model will help the marketing team to meet these demands by training the algorithm. The probability of each variable will be obtained, and with the help of these values, suitable constraints and variables can be identified. This model helps the user decide on expenditures and types of marketing.

To demonstrate the working of the trained neural network model, the input is taken as [200,000, 'non-profit', 'Mumbai', 15, 3,500,000].

Passing the above as input, the probability obtained is as follows (Table 1).

The above output obtained will help marketing team on understanding the types of marketing and the priority.

4 Conclusion

This paper gives a complete set of algorithms for the system to train and learn the marketing environment and predict the most accurate outputs using the system of linear programming. As the time passes, large datasets will be collected and the system would have more variables to match and work upon; this will keep system improving.

In the future, the scope of obtaining the feedback in the system can be incorporated which will help the system constantly to update and develop. The developed system will help both the marketing firms to develop individually and also reach to the customers based on their preferences. Each of the products marketed varies from each other, based on the type of advertisement which varies from the demographical area. Based on place, people, and economy, a customized marketing plan can be formulated considering all possible variables.

Manually, only few variables can be formulated, as the complexity increases, and when the real-life solutions are needed to meet the customer satisfaction, ANN would give the best solutions to the problem within least time which gives a better impact on the marketing firm.

References

1. Kennedy MP, Chua LO (1988) Neural networks for nonlinear programming. IEEE Trans Circuits Syst CAS 35:554–562
2. Ghasabi-Oskoei H, Mahdavi-Amiri N (2006) An efficient simplified neural network for solving linear and quadratic programming problems. Appl Math Comput J 452–464
3. Wen U-P, Lan K-M, Shi H-S (2009) A review of Hopfield neural networks for solving mathematical programming problems. Eur J Oper Res 675–687
4. Gao X, Liao L-Z (2010) A new one-layer, neural network for linear quadratic programming. IEEE Trans Neural Netw 21(6)

Queueing Theory an Index Factor for Production Inventory Control in Automotive Industry—A Review

Sunday A. Afolalu, Segun Oladipupo, Samson O. Ongbali, Abiodun A. Abioye, Ademola Abdulkareem, Mfon O. Udo, and Oluseyi O. Ajayi

Abstract In this paper, various approaches to inventory control within the automotive industry were reviewed using queueing theory. Different models used in past literature were stated and the model considered to be most effective in this review is dock management modeling. This model was used to analyze inventory control and how effectively automotive industries can minimize inventory by getting the component needed in the assembly line just in time, this helps to reduce additional costs for warehouse maintenance and capital tied down in form of excess inventory. Analytical and simulation models are the mathematical models that are considered in this review as they are used in several papers by different authors.

Keywords Queueing theory · Index factor · Inventory · Automotive · Model

1 Introduction

The resultant effect of a queue could be a result of server factors of which the inability of servers not to meet up with the demand of the customers is one of them, this increases waiting time and length of the queue [1]. Other factors that affect a queueing system are service time, arrival pattern, number of servers, etc. Queues can be found in various settings which could either be in the manufacturing or service organization; in this case, the major concern in the manufacturing sector taking the automotive industry as a case study. In recent times, there has been an increase in variety of products and mix models in order for the manufacturers to effectively compete in the global market. For manufacturers to stay relevant, flexibility in manufacturing is an important concept that must be embraced [2]. However, flexibility has its own cons which are high initial investment which requires large inventory and this is a non-negligible cost [3].

S. A. Afolalu (✉) · S. Oladipupo · S. O. Ongbali · A. A. Abioye · M. O. Udo · O. O. Ajayi
Mechanical Engineering Department, College of Engineering, Covenant University, Ota, Nigeria
e-mail: sunday.afolalu@covenantuniversity.edu.ng

A. Abdulkareem
Department of Electrical and Information Engineering, Covenant University, Ota, Nigeria

S. Vijayan et al. (eds.), *Trends in Manufacturing and Engineering Management*, Lecture Notes in Mechanical Engineering,
https://doi.org/10.1007/978-981-15-4745-4_74

Manufacturing car, parts are stamped from steel rolls and the entire body is then welded to produce a shell, and then each composition is finally welded in the body on the moving assembly line [4]. It was observed in the automotive industry that inventory management is a key factor in maximizing productivity and minimizing waste [5]. Inventory can be in different forms like raw material, work in progress, finished products, and goods in the warehouse between customers and suppliers. It is important for all industries to manage inventory properly as it is determinant on how productive and profitable the business can be one of the most successful models used to tackle inventory problems in this industry is just-in-time manufacturing that makes use of the supermarket rule of taking exactly what is needed in the assembly line from the warehouse [6].

The efficiency of the production system can be determined using the batch size and throughput. The batch size and the throughput must be aligned to optimize the system in order to make sure that the system is efficient [7]. Queueing network was used in this paper to study how throughput and batch size affects the optimal utilization of resources, especially machine resources in manufacturing systems. This study is focused on plants in the manufacturing industry that carry out assembly operations. The result of this study shows that utilization is a function of batch size and throughput, which means an increase or decrease in batch size or throughput directly affects the increase and decrease in utilization. If the demand requirement is higher than the capacity bottlenecks will definitely occur. For the manufacturing industry to remain relevant, sustainable, and compete strongly they must always find a way of improving efficiency in the production process. The production line where production is carried out is divided into three parts, namely single, batch, and mixed-model line [8]. The machines in the production lines have a cycle time in which they operate. In a manufacturing system, the efficiency of the production operation is based on how the machines utilize the resource in particular cycle time. Some operations did not make full use of the resources, this resulted in increased cycle time which also affects the throughput rate at which a product enters and leaves the production line. However, the ideal batch size and throughput for a given production can be defined with the help of the queueing theory in order to optimally use resources. In the manufacturing system, modeling can be done by using either an analytical model or a simulation model. Queueing network is one of the analytical models used in evaluating manufacturing system [9].

2 Reviewed Literature

In the automotive assembly, the techniques and emerging technologies used in vehicle assembly plants are discussed in this paper. Assembly processes such as handling, joining, etc., that involves the use of technology, as well as supporting systems were considered in this paper. A summary of existing technologies is provided. Mass customization requirements have been a challenge for the assembly systems as they are unable to keep up, this means that upgrading technology is necessary to achieve high flexibility. For operation and system, design flexibility is a key factor that should be integrated into future assembly plants [10].

The automotive industry faces great challenges such as advancement in technology, new competitors, and industry restructuring, and sourcing globally. The quick change from mass production to mass customization is dependent on the fact that customers wants certain features incorporated into their products and manufacturers have to provide many variants using fewer resources in a relatively short period [11]. The ability of the manufacturer to provide several variants per model and bring onboard new models at a fast rate is restricted by the current equipment used in mass production and also current technologies that are not capable of supporting product variability. In automotive assembly increased complexity requires an encompassing view of the main attributes of manufacturing that should be considered when manufacturing decisions relating to cost, quality, flexibility, and time are taken [12].

In the automotive industry, the typical framework of a plant comprises four phases - stamping, body shop, painting, and final assembly. Assembly is carried out mainly in the body shop or final assembly. The assembly of the body in white (BIW) uses high level of automation, while the final assembly stage introduces a combination of hybrid human/machine systems. Four approaches can thus be identified when designing the assembly system; (1) flexible assembly (2) manual assembly, (3) semi-automated assembly, (4) fixed assembly.

Generally, different vehicles in an automotive assembly make use of the same assembly line known as mixed-model assembly line (MMAL). The MMAL are distinguished because of their ability not to hold large inventory when assembling different models of a given product [13]. Demand for more secured and faster ramp-up processes which are cost-intensive is one of the major problems modern assembly system has to tackle. Steady increase in number of products demanded underpins the goal, which is a result of improved innovation and increased market launches of product variant and new product. Groover [9] further expounded that the present trend that Original Equipment Manufacturers (OEMs) in automotive industry follow is incorporating equipment, product and process standardization.

One of the probabilistic models used for predicting and making a decision under uncertainty are Bayesian networks. The inventory, production quantity and deliver quantity of any manufacturing organization such as the automotive industry varies according to the several unexpected events [14]. Thus, predicting inventory in production is sacrosanct in order to cope with the variations in the market. This paper considers the way to adjust production for the car parts process through the use of a dynamic Bayesian network. Material in a production system (known as factory automation) is delivered to clients and stored in inventory for subsequent shipment through each manufacturing, sub-assembly and final product process [15].

To utilize a system like factory automation, attributes of production such as production rate of the company varies because of the irregular changes found generally in the operating capabilities and also the strength of the labor force. There are random changes to product order which means managers of each process have to decide on the output of products by using their know-how and experience to tackle this issue [16]. Taking a look at the inventory in a Kanban way, the inventory is considered to be a cost. Kanban is used to schedule for just-in-time manufacturing, it operates on the supermarket principle of picking the things necessary for production at the moment from the store. From a production point of view, Kanban controls the logistical and supply chain [17]. In a manufacturing system, putting uncertainties into consideration is of upmost importance because it helps proper planning against unforeseen events. Production planning models that put uncertainties into consideration are better off as they generate excellent planning than the models that don't account for uncertainties [18]. This paper reviews existing production planned literature under uncertainty. According to Pearl [22], uncertainty can be defined as the distinction between the information already gained and the amount of information actually needed to carry out the task. There are many ways in which uncertainty have an impact on production processes. It is classified into two groups, uncertainty about the environment and system uncertainty. Environmental uncertainty is those beyond uncertainties in demand and supply; this type of uncertainty is not influenced by the production process itself. System uncertainty refers to uncertainty that is influenced directly by the production process, such as failure of the production system, quality uncertainty, lead time uncertainty, operation yield uncertainty and changes to the structure of the product. The typology of uncertainty is used in this paper [19].

The supply chain of automotive industry is characterized for a properly organized structure, where the partial exchange of forecasting amongst manufacturers and their first-tier suppliers are requirement demand; this means that suppliers in most cases must be ready to handle a variety of potential requirements. In order to handle such varieties, it important to consider technological tools for efficiently managing the decision-making in this area [20].

The task of logistics planning in the automotive industry was studied in this paper using the macroscopic simulation model [21]. In specific terms, macroscopic simulation model lies between simulation models that are object-based discrete-event or flow-based continuous simulation models. On a broader view, processes of logistic flow are represented using macroscopic model via a piecewise flow rate that is constant rather than modeling flow objects individually. The manufacture of assembled products with various components such as cars, engines, etc. involves the collaborative effort of several factories. The individual components are to be designed and organized properly in order to meet customers' expectations [22]. The main concern here is the decision problems (medium and long-term), which contains supply chain macroscopic models. As stated by Lee et al., continuous worldview is used at this level, as representing the individual flows of numerous parts becomes quite tedious. The macroscopic models are based on the paradigm of Forrester system dynamics that are currently used by large French automotive companies [23]. In this article, the supply chain presented is made of two production areas with a total of five existing plants. The results show the real advantages that can be achieved. Through ferrous metal separation, 75% of the material weights are successfully recovered by the automotive recycling infrastructure once the lifecycle of a vehicle is complete [24]. Nevertheless, the rate at which automotive manufacturers make use of nonmetallic and nonferrous material has increased and this a major challenge for the automotive recycling industry [25]. Recycling planning models were proposed in this paper for automotive shredders in order to make a short-term strategic decision with regards to the level at which materials should be processed and reprocessed through multiple passes. A model known as mixed-integer linear programming is used to determine whether to separate or combine materials for shipment [26].

3 The Problem

3.1 Material Management

Generally, in supply chain materials move through the distribution center from the supplier and finally arrive at the moor of the assembly plant. The arriving components or materials are then taken immediately to the assembly line or the inventory. Prompt arrival of the components determines the correct feeding of the assembly line because of any mistake in leaving a part out result to heavy cost for production re-scheduling [27].

3.2 Feasible Alternatives

There are two ways of addressing any of the delays from the supplier, through means of transport or dock, they discussed as follows: Transport management: this is taking into consideration when the supplier doesn't deliver on time or problems arise while transporting the goods. In this type of scenario is an alternative which is to take advantage of faster means of transportation in order to reduce the lead time of goods arriving at the plant. It should be noted that change to a faster means of transportation will lead to additional cost for faster delivery [28].

Dock management takes care of any form of disruptive event that can occur during delivery of goods.

4 Solution

4.1 Dock Management Model

Askin et al. [28] stated that incoming disruptive events managed by this model led to a reduction in the total cost of updating the dock work, which helps to identify the most appropriate dock management solution. The main objective here is to remember the actual production time that displays the time when the assembly line requires an incoming order. Based on this time, calculating when the order at the dock should be ready is possible at maximum. The time difference between production time and time required to transfer orders for the assembly line from the warehouse is known as due time [29]. This transfer can be managed in two ways, which are the Standard method and strategies that are faster which include advantageous transport and an increase in price which is used to make up for the delay experienced that arrives after due time. Two moments name acceptable due time_S1 and _S2 are defined by the two strategies mentioned above, these are calculated as the difference between respective transfer time and actual production time. Any delayed order arriving after these moments are transferred for production re-scheduling [30]. Also, the time between the specific window time when the production re-scheduling is required and due time is known as buffer time. Arrivals after the buffer time are placed within the re-scheduling time. Orders that takes place after the estimated arrival time is due to disruption during order delivery [31].

A specific amount of time known as the unloading duration is used to unload and transfer goods to the warehouse after the order arrives at the dock. After unloading, the order is then placed on the assembly line. If the estimated readiness time is before the due time, then there is no additional cost. If the time of readiness after the due date is estimated, then the buffer time or cost of re-scheduling will be additional as a matter of time [32].

5 Conclusions

A number of manufacturing industries like the automotive industry took the use of queueing theory in inventory management, leading to reductions in lead times to maximize profit, mathematical models that include analytical models and simulation models were evaluated in various literature, meanwhile in this paper, the model considered to be effective is dock management, the use of the model result in proper inventory keeping of materials necessary for the assembly line in order to avoid stoppage during production [33–35].

Acknowledgements We acknowledge the financial support offered by Covenant University in actualization of this research work for publication.

References

1. Balaji K, Kumar VSS (2014) Multicriteria inventory ABC classification in an automobile rubber components manufacturing industry. Procedia CIRP 17:463–468
2. Başak Ö, Albayrak YE (2015) Petri net based decision system modeling in real-time scheduling and control of flexible automotive manufacturing systems. Comput Ind Eng 86:116–126
3. Bhatnagar R, Lin B (2018) The joint transshipment and production control policies for multi-location production/inventory systems. Eur J Oper Res

4. Bruniaux R, Pierreval H, Caux C (1999) Simulating supply chains with system dynamics. In: Proceedings of the ESS'99 European simulation symposium, Nuremberg, Germany, 26–28 Oct, pp 209–213
5. Bruniaux R (2000) Simulation continue de réseaux de sites industriels: application aux chaines logistiques dans le secteur automobile. Ph.D. Dissertation, in French, Université Blaise Pascal, Clermont-Ferrand, France
6. Bruniaux R, Caux C (2007) A continuous simulation approach for supply chains in the automotive industry. Simul Model Pract Theory 15(2):185–198
7. Boysen N, Emde S, Hoeck M, Kauderer M (2015) Part logistics in the automotive industry: decision problems, literature review and research agenda. Eur J Oper Res 242(1):107–120
8. Dalmijn WL (1990) Practical applications of eddy current separators in the scrap recycling industry. In: van Linden JHI, Stewart DL Jr, Sahai Y (eds) Proceedings of the second international symposium on recycling of metals and engineered materials, pp 303–314
9. Groover MP (1987) Automation, production systems, and computer integrated manufacturing. Prentice-Hall, Englewood Cliffs, pp 36–40, 404–407
10. Ho C (1989) Evaluating the impact of operating environments on MRP system nervousness. Int J Prod Res 27:1115–1135
11. Michalos G, Makris S, Papakostas N, Mourtzis D, Chryssolouris G (2010) Automotive assembly technologies review: challenges and outlook for a flexible and adaptive approach. CIRP J Manufact Sci Technol 2(2):81–91
12. Hu SJ, Zhu X, Wang H, Koren Y (2008) Product variety and manufacturing complexity in assembly systems and supply chains. CIRP Ann Manuf Technol 57(1):45–48
13. Lang S, Reggelin T, Wunder T (2017) Mesoscopic simulation models for logistics planning tasks in the automotive industry. Procedia Eng 178:298–307
14. Marsudi M, Shafeek H (2013) Production line performance by using queuing model. IFAC Proc Vol 46(9):1152–1157
15. Marsudi M, Dzuraidah AW, Che Hassan CH (2009) Application of spreadsheet and queuing network model to capacity optimization in product development. World Acad Sci Eng Technol 34:1068–1071
16. Salawu EY, Okokpujie IP, Afolalu SA, Ajayi OO, Azeta J (2018) Investigation of production output for improvement. Int J Mech Prod Eng Res Dev 8(1):915–922
17. Afolalu AS, Enesi YS, Kehinde O, Samuel UA, Ikechi VI, Remilekun RE (2018) Failure mode and effect analysis a tool for reliability evaluation. Eur J Eng Res Sci 3(4):65–68
18. Salawu EY, Ajayi OO, Inegbenebor AO, Afolalu SA, Ongbali SO (2018) Pareto analysis of product quality failures and cost effects in bottling machines—a lean thinking solution for alcohol industry. Int J Mech Eng Technol 9(11):2380–2388
19. Afolalu SA, Ayuba SU, Ihebom IV, Elewa RR, Oluyemi K (2018) Role of reliability management tools and documentations—a review. Glob J Eng Sci Res Manage 5(3):39. ISSN 2349-4506
20. Meyer A, Sejdovic S, Glock K, Bender M, Kleiner N, Riemer D (2017) A disruption management system for automotive inbound networks: concepts and challenges. EURO J Transp Logist 1–32
21. Mula J, Poler R, García-Sabater JP, Lario FC (2006) Models for production planning under uncertainty: a review. Int J Prod Econ 103(1):271–285
22. Pearl J (1986) Fusion, propagation and structuring in belief networks. Artif Intell 29(3):241–288
23. Lee YH, Cho MK, Kim SJ, Kim YB (2002) Supply chain simulation with discrete continuous combined modeling. Comput Ind Eng 43(1–2):375–392
24. Rabelo L, Magdy H, Lertpattarapong C (2004) Analysis of supply chains using system dynamics, neural networks and eigenvalues. In: Proceedings of the 2004 winter simulation conference, Dec, Washington, DC, USA
25. Abioye AA, Atanda PO, Abioye OP, Akinlabi SA, Akinlabi ET, Bolu CA, Afolalu SA, Ajayi OO, Ohijeagbon IO (2018). A review on automotive industries and foundries in Nigeria. In: IOP conference series: materials science and engineering, vol 413, no 1. IOP Publishing, p 012003

26. Ramírez-Granados M, Hernàndez JE, Lyons AC (2014) A discrete-event simulation model for supporting the first-tier supplier decision-making in a UK's automotive industry. J Appl Res Technol 12(5):860–870
27. Remiel FM, Cauvin A, Ferrarini A (2014) Conceptual design and simulation of an automotive body shop assembly line. IFAC Proc Vol 47(3):760–765
28. Askin RG, Mitwasi MG, Goldberg JB (1993) Determining the number of kanbans in multi-item just-in- time systems. IIE Trans 25(1):89–98
29. Sethi SP, Yan H, Zang Q (2002) Optimal and hierarchical controls in dynamic stochastic manufacturing systems: A survey. Manuf Serv Oper Manage 4:133–170
30. Shin J, Kim S, Lee JM (2015) Production and inventory control of auto parts based on predicted probabilistic distribution of inventory. Dig Commun Netw 1(4):292–301
31. Shiina T, Birge JR (2003) Multistage stochastic programming model for electric power capacity expansion problem. Jpn J Ind Appl Math 20:379–397
32. Spengler T, Ploog M, Schroter M (2003) Integrated planning of acquisition, disassembly and bulk recycling: a case study on electronic scrap recovery. OR Spectrum 25:413–442
33. Ajayi OO, Agarana MC, Akinwumi II, Okokpujie PI, Salawu EY, Abioye AA, Afolalu AS, Leramo RO (2019) Modelling rate of traffic-induced building vibrations in Sango-Ota, Nigeria: an assumption based analysis. WIT Trans Built Environ 182:303–313
34. Weiss EN, Tucker C (2018) Queue management: elimination, expectation, and enhancement. Bus Horiz 61(5):671–678
35. Williams JAS, Wongweragiat S, Qu X, McGlinch JB, Bonawi-tan W, Choi JK, Schiff J (2007) An automotive bulk recycling planning model. Eur J Oper Res 177(2):969–981

Design and Analysis of ASRS Using AGV for Rapid Inventory Storage System

Pranav Santhosh Nair, Sourabh Nair, Soukhin Sarkar, Arockia Selvakumar Arockia Doss, and D. Dinakaran

Abstract Non-competitive market pricing is easy to use and maintain but it is a major issue in automated guided vehicles (AGVs) that are available in the current market. Space is a very valuable commodity in urban areas. The current AGV available in the market needs optimization and is specifically designed for one particular form of storage system. In order to reduce the complexity of the system, hard locating method approach is used. Hard locating is a process that assists the operator in avoiding mistakes. In this work, an AGV for rapid inventory storage system is proposed. The AGV consists of a 5-degrees of freedom (DOF) robotic arm with a two-fingered gripper is used for picking and placing the boxes. The gripper is equipped with a piezoelectric sensor to sense the size of the object. By using this approach, the existing rack can be used without much alteration. This system has the capability to increase the utilization of the warehouse. This paper presents an optimized automated storage and retrieval system (ASRS) with the help of AGV with reduced cost and complexity.

Keywords Automated guided vehicle · Automated storage and retrieval system · Warehouse optimization · Piezoelectric sensor

1 Introduction

The design and implementation of a 7-channel line sensor for line following robot were having advanced functionalities such as software-based automatic sensor calibration and individual threshold per sensor channel. The system was implemented using light-emitting diodes, phototransistor and an on-board microcontroller Arduino which communicates with any navigation control system using serial communication [1]. The optimization problem of AGV in smart warehouse dispatching mechanism

P. S. Nair · S. Nair · S. Sarkar · A. S. Arockia Doss (✉)
School of Mechanical Engineering, Vellore Institute of Technology, Chennai, India
e-mail: arockia.selvakumar@vit.ac.in

D. Dinakaran
Centre for Robotics and Automation, HITS, Chennai, India

S. Vijayan et al. (eds.), *Trends in Manufacturing and Engineering Management*, Lecture Notes in Mechanical Engineering,
https://doi.org/10.1007/978-981-15-4745-4_75

is studied with genetic algorithm [2]. The AGV logistics and distribution components which are used for AGV in discrete manufacturing process are explained [3]. AGV is a kind of simple mobile robot, which integrates light, sound, electricity and computer, and it is applied in the fields of manufacturing industry and logistics, such as automobile industries [4]. An AGV with low-cost camera was used to capture images to figure out the surrounded conditions and decide the next move. Region-based convolutional neural network-based deep learning algorithm was used to analyse the images [5]. In smart factory, as one of key features for our industry, it requires automation within a manufacturing plant such as a shop floor. Automated guided vehicle (AGV) systems may be the best idea, whose accuracy depends on some factors. An AGV with radio frequency identification (RFID)-enabled positioning system is used in for smart factory [6]. The robotic gripper was designed and fabricated taking into account the economics, simplicity and reliability. Suitable-end effectors can be installed for particular application [7]. The portable robotic arm with 4-degrees of freedom working model was able to imitate the movement of a human arm with imperceptible delay and had a payload capacity of 150 gm that is achieved [8]. The ASVS/RS rack design is focused on avoiding collision and deadlock in the highroad traffic network, by restricting the layout and traffic regulations of trunk rails, storage rails, lifters, etc. [9]. The ASRS is basically used to store or retrieve the loads in the warehouse or out of it. The computer will calculate wherein the storage area the item can be retrieved from and does the retrieval operation. The components are stored into or retrieved from the racks, and the computer updates the inventory system [10].

Automated storage and retrieval systems have many potential applications. The most common ASRS application is the storage and retrieval of inventory component, but other ASRS applications can greatly increase the efficiency of an operation as well [11]. An automated guided vehicle (AGV) is a robot that follows marker (wires) on the floor, magnets or lasers for navigation. With the rise of online shopping, the requirement of large warehouses has increased exponentially. To manage these large warehouses, an army of workers is needed, which is not very economical. A better option would be to implement an automated storage and retrieval system by using automated guided vehicle which will greatly help the institution in economic terms and increase the storage density of the warehouse. The objective of the proposed system is to find the perfect balance between the two systems such that the overall efficiency of the warehouse storage can be optimized. Implementing human workforce, wherein large ever-growing demand for storage places, human efficiency may not update with the technology. Hence to increase the efficiency and storage density of the packages, implementation of ASRS becomes vital. The ASRS reduces the probability of human-caused errors, and the AGV increases the efficiency and the work duration. This is cheaper to handle, and very less maintenance is required. The integration and optimization of these two complex systems are presented and detailed in this proposed work.

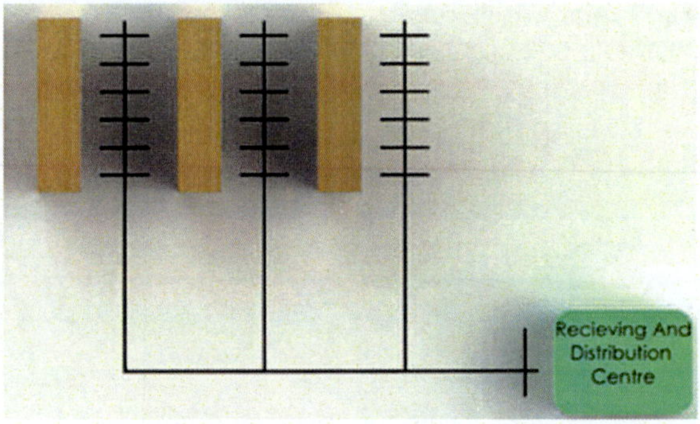

Fig. 1 Warehouse layout

2 Proposed System

Figure 1 shows the warehouse layout. In front of each box levels, a line or a junction is created to position the bot in front of the box. The boxes will be kept in the rack using locating strips so that the position of the box will be known and fixed; hence, the arm could be hardcoded to reach the designated position of the boxes. Based on the proposed fixing position of the boxes, the complexity, both mechanically and electrically, is reduced. Base of the AGV is kept as three-fourth of the track width (distance between two columns) and not too short of the wheelbase that will misbalance the centre of gravity of the bot while handling the load. The standby dimensions of the bot are 800 * 630 * 900. These two package sizes (12.5″ × 9.75″ × 3.25″ and 18″ 21.5″ × 5.25″) are the most common size used in the shipping industry. The rack is a simple 3-level rack and used to store objects of various sizes in inventory systems. The rack system is designed to help with the hardcoded gripper mechanism and ensure zero error and arrangement of boxes. The arena is designed such that the bot will have to go through minimum number of iterations. It will receive a set of location addresses via Bluetooth, using the location instructions; it will take the necessary turns, will grab the package with the help of gripper and get back to its original position.

3 Design and Analysis

The 5-DOF manipulator is mounted on the chassis which is having four wheels driven by DC motors. The end effector is driven by a DC motor for open and close operation of the jaw, and it is designed to have a piezoelectric sensor to adapt to the size of the object. The line following guidance system is designed for free movement of the bot without any interference. The precise measurement of the centre of the bot to the corner of the columns is kept in mind while marking the path. The gripper is flexible with its gripping length; i.e. it can lift packages of various dimensions.

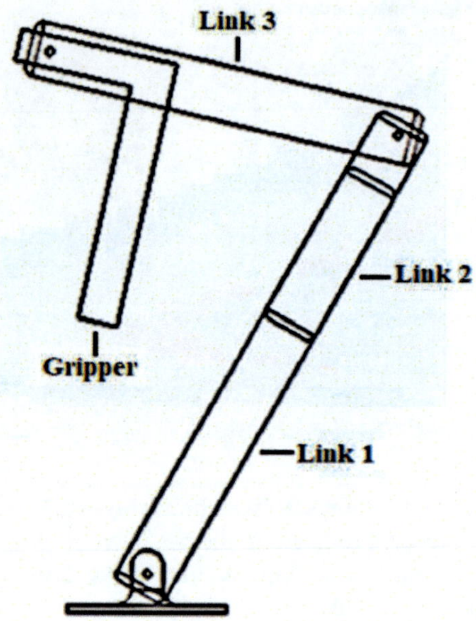

Fig. 2 Arm assembly ideal condition

Linear ball screw is used for moving the movable arm of the gripper. Figures 2 and 3 show the standby and fully extended position of the arm, respectively. The robotic arm consists of 5 DOF, which is controlled using electric linear actuators and high torque DC motors. The arm along with gripper has a lifting capacity of 100 kgs; i.e. it can lift package weighing up to 100 kg. The AGV components details are tabulated with specifications in Table 1.

The electric linear actuator proposed in this work is a simple gear and screw mechanism. The electric motor rotates the main gear connected to the motor shaft which is connected to the gear which is in turn connected to the screw. The screw moves up or down depending on the direction of motion of the motor. The robotic arm consists of three links as shown in Figs. 4 and 5. Link 2 slides over link 1 in order to extend the overall length of primary link from 500 to 750 mm. Because of this extendable link, the arm is able to lift objects placed at a height of 1.7 m. The sliding motion is controlled by a square gear DC motor of torque 30 kgcm and 500 RPM. The inclination of all the links is controlled by linear actuators of sizes 4 and 8 in.

4 Simulation

Structural analysis of the chassis of the bot is done to ensure that it can handle the stress while handling maximum load and has enough lifetime before service is needed. Kinematics is the field of focus while designing the arm. The load acting on the joints, torque acting on the juncture and the material required to suit the conditions

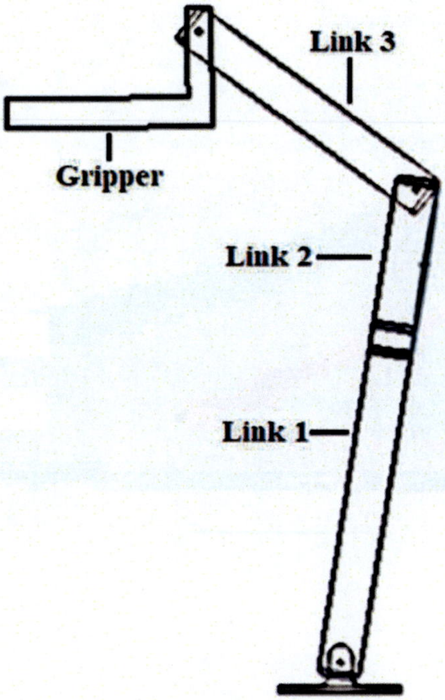

Fig. 3 Arm assembly extended condition

Table 1 Details of AGV components

S. No.	Part name	Dimensions (mm)
1	Base chassis	800 * 550 * 210
2	Link 1	500 * 70 * 96
3	Link 2	340 * 125 * 70
4	Link 3	550 * 70 * 2.5
5	Linear actuator 4 cm	205 (min) 305 (max)
6	Linear actuator 8 cm	320 (min) 520 (max)
7	Wheel diameter	100

are calculated. The gripper force directly depends upon the power of the motor used to move the jaws; keeping this in considerations, the gripper is designed. In order to check the safety of the components, the components are simulated in the ANSYS 16.0 and the displacements of base and the links are calculated and tabulated in Table 2.

Figure 6 shows the total deformation of base chassis for the payload of 2500 N; the deformation is found to be 0.000511 m; and it is found to have FOS of 4.8927. Figure 7 shows the total deformation of Link 1 for the payload of 2500 N; the deformation is found to be 0.000144 m; and it is found to have FOS of 1.9249. Figure 8 shows

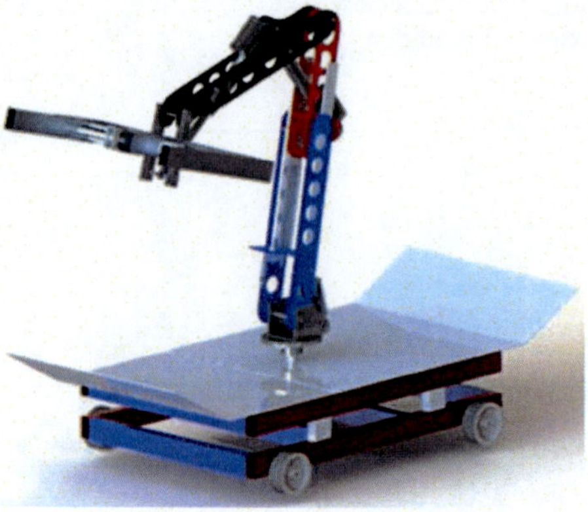

Fig. 4 Prototype made for static analysis

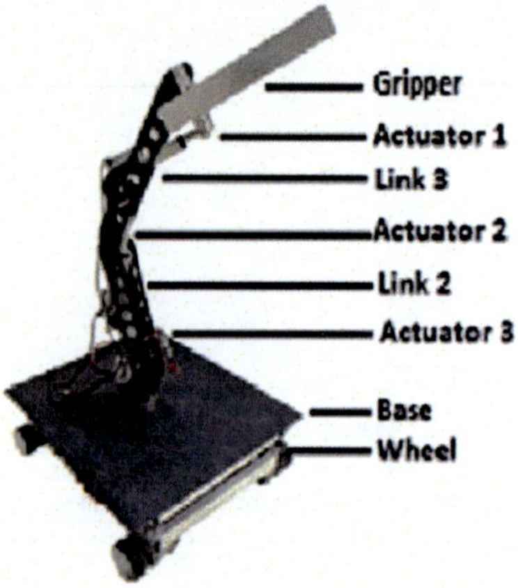

Fig. 5 Line following robot using IR sensors

Table 2 Details of AGV components

S. No.	Part name	Force applied (N)	Displacement (m)	Factor of safety
1	Base	2500	0.000511	4.8927
2	Arm Link 1	2500	0.000144	1.9249
3	Arm Link 2	2500	0.000144	4.3403
4	Arm Link 3	1000	0.00028158	1.9169

the total deformation of Link 2 for the payload of 2500 N; the deformation is found to be 0.000144 m; and it is found to have FOS of 4.3403. Figure 9 shows the total deformation of Link 3 for the payload of 1000 N; the deformation is found to be 0.00028158 m; and it is found to have FOS of 1.9169.

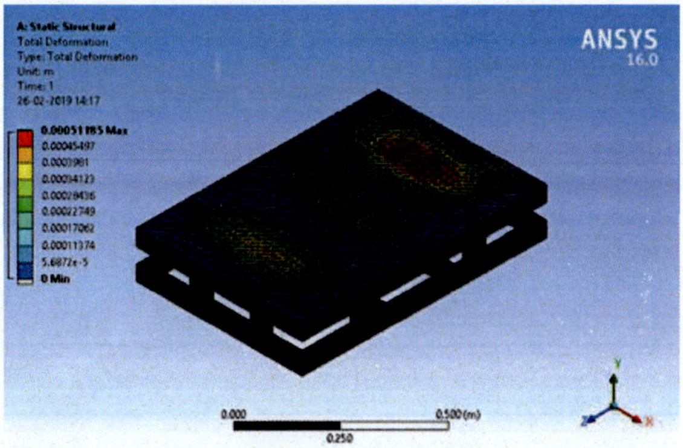

Fig. 6 Deformation results of base chassis

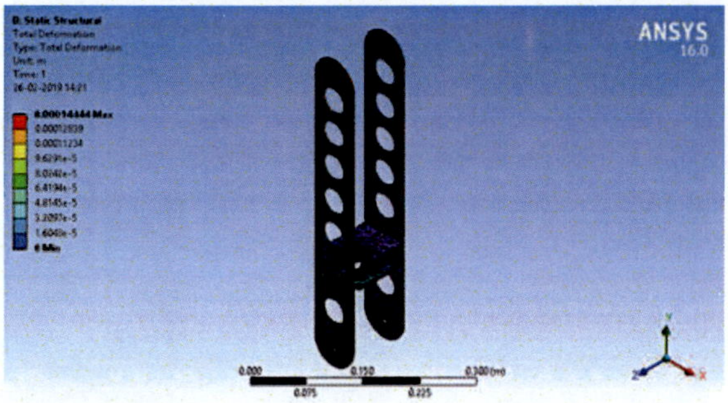

Fig. 7 Deformation results of Link 1

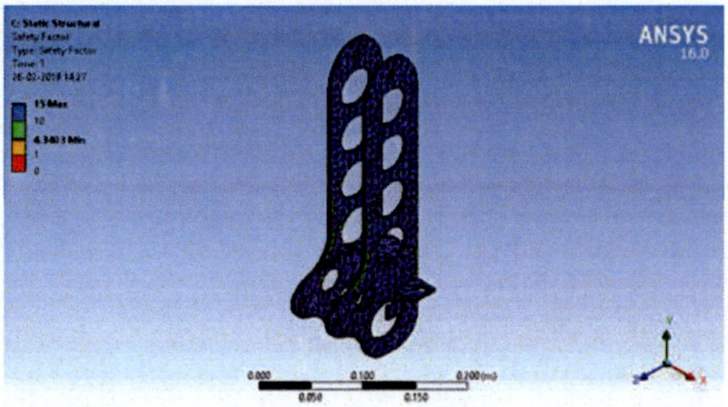

Fig. 8 Deformation results of Link 2

5 Control System Design

Figure 10 shows the line following sensors implemented in the AGV. The basic manoeuvrability is found to be perfect, and further, it is preceded with the line following mechanism. Two IR sensor arrays are placed one behind the other to follow a straight line. Black tapes are used to mark the line. The operator will be using a Bluetooth transmitter to send the coordinates to the Bluetooth receiving

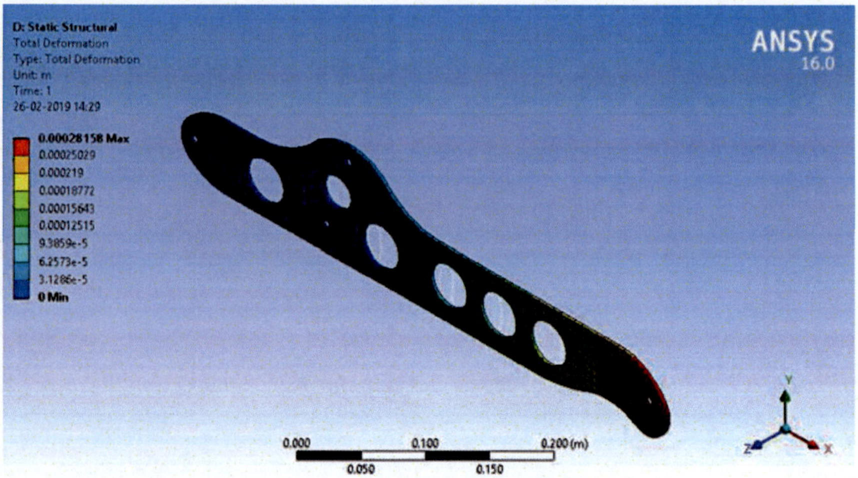

Fig. 9 Deformation results of Link 3

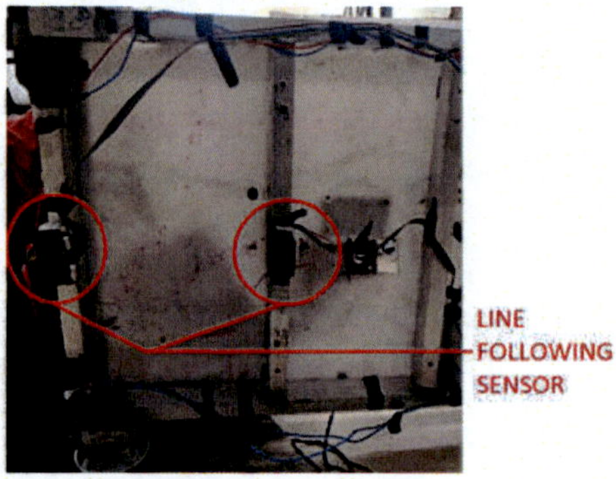

Fig. 10 Line detection sensors

module situated on the AGV. The code will be received by the master Arduino. The code consists of the coordinates of the rack where the package is placed or is to be placed. The master Arduino will in turn control 2 other slave Arduinos which controls both arm and base, respectively. The base controller Arduino will manage the motor RPM, and the line following mechanism is to guide the bot to the designated position according to the coordinates received from the operator. The arm controller Arduino will be hardcoded to reach the hardcoded position of the package. In Fig. 11, the control system of the AGV is shown. The Bluetooth module will receive the command from the user which will in turn send the received command to the master Arduino. According to the system layout, the master Arduino will give commands to the slave Arduino 1 and slave Arduino 2 which controls the arm and base of the AGV, respectively. The command received by the slave Arduino 1 will

Fig. 11 Control system unit

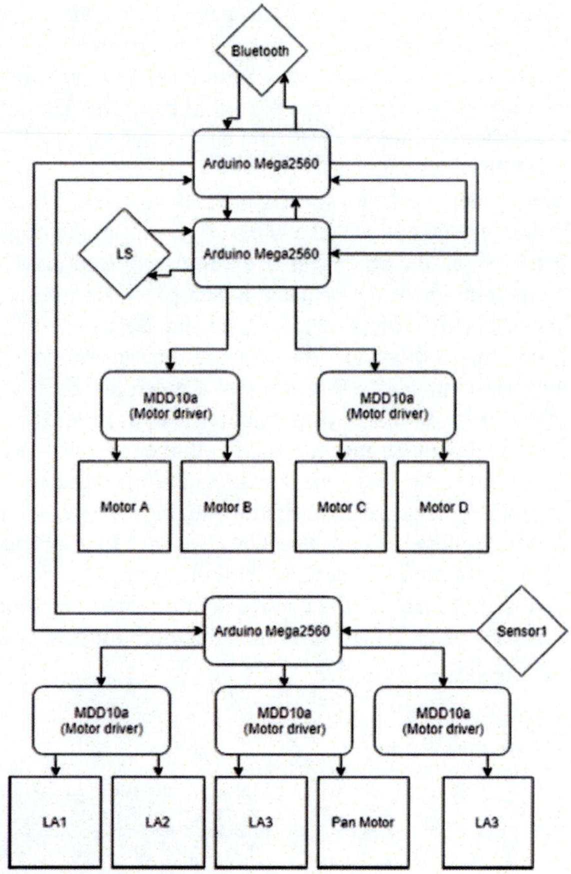

make the arm position itself in the desired position. The slave Arduino 2 will make the base of the bot manoeuvre and position itself using the line following algorithm.

The code sent by operator consists of three basic parts. This helps in keeping the code simple and prevents data skipping by Arduino due to excessive data values. An example of the code that will be sent to the master Arduino situated on the bot by the means of Bluetooth is given in Table 3.

The Bluetooth receiver is embedded on the Arduino which is connected to the transmitter which is in turn controlled by the operator. The code is kept small so as to prevent data skipping which might occur due to large data values. The master Arduino is connected to two other slave Arduino which controls the arm and the

Table 3 Code to Arduino

1^A	3^B	2^C

base, respectively. The first two parts of the code are sent to the base slave Arduino. The part A of the code is used to identify the rack location. The bot understands the location of the rack, and the motors start to spin following the line. There are two-line following sensors placed in line. The first sensor is used to implement the line following mechanism. The second sensor acts as the counter which counts the junctions. The value of A is being 1, the bot turns in alley 1 and so on. The part B is used to identify the location of the rack column. This is used in similar method of line following and counter like in part A to reach the designated location. After reaching the location, the arm slave Arduino comes into play. The part C is used to identify the height where the package is placed in the column. The arm will position itself according to the location of the package. Because the position of the package is placed according to the strips, the arm could be hardcoded. This enables swift movement and less complexity. The arms of the gripper are fitted with a piezoelectric sensor which helps in identifying whether the package is firmly gripped by the gripper or not. The bot undergoes minimum number of iterations, simplifying the process. The bot receives the code, goes to the junction, takes a left turn facing the rack, positions its gripper in accordance with the height and picks up the package. After the package is firmly gripped, the arm will retrieve back to its original position and the bot returns back to the retrieval and distribution centre.

Figure 12 depicts the control layout of the Arduinos (4, 5, 6 from left) and motor drivers (1, 2, 3, 7 from left) and Bluetooth module connected to the master Arduino (4 from left).

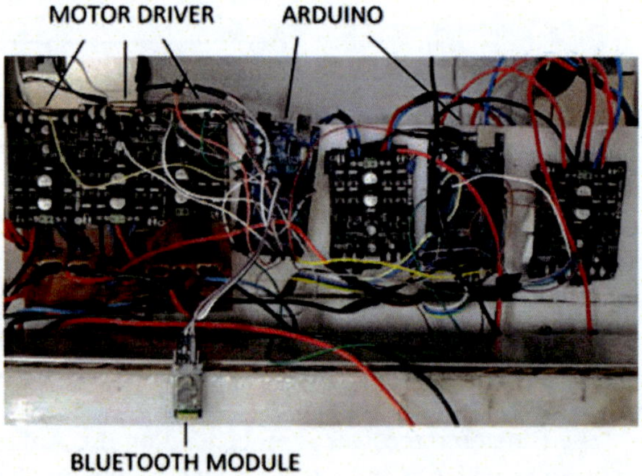

Fig. 12 Control set-up

6 Summary

An AGV for rapid inventory storage system is fabricated and tested for industrial application. Based on the design, structural analysis and experimental results the following points are summarized,

- The line following sensors (line following) that make the AGV follow a single path which makes the movement of the AGV much faster than the existing AGVs.
- This AGV can perform pick/place operations up to a height of 6 ft from the ground using the robotic arm without taking out the whole rack for a single object.
- The AGV is very cost-effective. In general, the market value of an AGV is approximately 5–10 lakhs. But the total cost of this AGV is around 0.30–0.35 lakh.
- Piezoelectric sensor allows AGV to avoid using visual sensor in assisting bot to pick up the package, which in turn helps us in reducing the cost.
- As the system is hardcoded, the system follows the written code command and any decision-making cannot be made by the AGV.

References

1. Patil S, Wagh A, Sawant M, Panda S, Bhopale A (2016) Design and implementation of advanced auto calibrating line following sensor for coloured surfaces with a white line. In: 2016 IEEE 1st international conference on power electronics, intelligent control and energy systems (ICPEICES), Delhi, pp 1–6
2. Liu S (2008) Optimization problem for AGV in automated warehouse system. In: 2008 IEEE international conference on service operations and logistics, and informatics, Beijing, pp 1640–1642
3. Qi BY, Yang QL, Zhou YY (2015) Application of AGV in intelligent logistics system. In: Fifth Asia international symposium on mechatronics (AISM 2015), Guilin, pp 1–5
4. Shengfang L, Xingzhe H (2006) Research on the AGV based robot system used in substation inspection. In: 2006 international conference on power system technology, Chongqing, pp 1–4
5. Cheong H-W, Lee H (2018) Concept design of AGV (automated guided vehicle) based on image detection and positioning. Procedia Comput Sci 139:104–107
6. Lua S, Xub C, Zhongc RY, Wangd L (2017) A RFID-enabled positioning system in automated guided vehicle forsmart factories. J Manuf Syst 44:179–190
7. Chouhan R, Kanwal F, Ali S, Ali N (2014) Design and development of a prototype robotic gripper. In: 2014 international conference on robotics and emerging allied technologies in engineering (iCREATE), Islamabad, pp 317–320
8. Bora P, Nandi V (2015) Low cost shadow function based articulated robotic arm. In: 2015 international conference on energy, power and environment: towards sustainable growth (ICEPE), Shillong, pp 1–4
9. Fu X, Zhang B, Yu H (2015) Rack design for an autonomous-steerable-vehicle storage and retrieval system of a warehouse. In: 2015 7th international conference on intelligent human-machine systems and cybernetics, Hangzhou, pp 112–116

10. Farah Hanani MK, Zulkhairi MY, Mohamad Zaihirain MR, MohdAswadi A, Ismail A (2016) Development of automated storage and retrieval system for flexible manufacturing system. J Eng Technol 4(1):43–50
11. Cheonga H-W, Leeb H (2018) Requirements of AGV (automated guided vehicle) for SMEs (small and medium-sized enterprises). Procedia Comput Sci 139:91–94

Optimization Techniques

Multi-response Optimization of Inconel 825 Process Parameters Using LN_2 Cooled Zinc-Coated Brass Wire in CNC Wire-Cut EDM

Midthur A. Salman Khan, C. Nandakumar, B. Mohan, and R. Senthil Kumar

Abstract Inconel 825 is a nickel-based superalloy which is mostly used in most of the industries where high corrosion resistance and high temperature withstanding are needed. Some of these industries are Petrochemical, Chemical, Missiles and Nuclear power plants. Inconel 825 has possessed high anti-oxidation nature even at high temperatures, i.e. about 650–760 °C. Inconel 825 cannot be machined in the traditional machines because of its high strength at elevated temperatures, high strain hardened and low thermal conductivity. WEDM is an alternative machine which is used for machining of Inconel 825, because WEDM can machine any material without considering its hardness if the material is electrically conductive. Liquid Nitrogen cooled (LN_2) Zinc-coated brass is taken as a cathode and the Inconel 825 is taken as an anode during the machining process. Taguchi's L_{27} orthogonally array is used in the experimentation method. Five parameters of WEDM i.e., Wire Tension, Wire Feed, Pulse on Time, Pulse off Time, Servo Voltage are taken as inputs and MRR and Surface finish are taken as the output responses. In this paper, an attempt is made to study the effect of various parameters and optimize the result. Mathematical modelling of the output responses is developed.

Keywords Nickel-based superalloy (Inconel 825) · Taguchi method · Material removal rate · Surface roughness · Liquid cooled nitrogen

---

M. A. Salman Khan (✉)
Department of Mechanical Engineering, KG Reddy College of Engineering and Technology, Hyderabad 501504, India
e-mail: m.a.salman087@gmail.com

C. Nandakumar
Department of Production Technology, MIT Campus, Anna University, Chennai 600044, India

B. Mohan
Department of Mechanical Engineering, CEG Campus, Anna University, Chennai 600025, India

R. Senthil Kumar
Department of Mechanical Engineering, Mohamed Sathak A.J. of Engineering, Chennai 603103, India

S. Vijayan et al. (eds.), *Trends in Manufacturing and Engineering Management*, Lecture Notes in Mechanical Engineering,
https://doi.org/10.1007/978-981-15-4745-4_76

1 Introduction

Wire Electric Discharge Machining (WEDM) is one of the non-conventional machines, which is used in machining of the materials without considering the strength of the material. It can be used to machine components with complex shapes, profile and sharp edges which cannot be machined in non-traditional machining process. Hence it has wide application areas like aerospace industries, medical implants, electronic industry, automobile industry, etc. [1]. Wire electrode is made of Copper, Brass, Zinc-coated Brass and Tungsten of about 0.05–0.3 mm diameter. Thermoelectric energy is the working principle of WEDM. Series of spark are generated between the wire and workpiece whose temperature is around 10,000–20,000 °C [2]. Dielectric fluid is continuously passed between the wire electrode and workpiece. The wire electrode moves close to the workpiece until the applied voltage between the wire and the workpiece increases and causes the dielectric to get ionized and breakdown of dielectric happens. The gap between both the electrodes should be in the range of 10–100 μm. Short-circuiting will happen if the gap reduces less this range. Dielectric fluid is also used to flush these tiny particles evaporated. There will no direct contact between the tool and the workpiece which reduces the mechanical stress and vibrations problems during the machining [3]. When cryogenic treatment is made on the material there will be a change in its crystal structure. Surface finish will be improved when deep cryogenic treatment is carried out on the material, which further improves the fatigue strength, corrosion and wear resistance of OHNS Die Steel [4]. It was observed that the cutting temperature of Ti–6Al–4V has reduced to 36 and 47% when CO_2 and LN_2 are used as cryogenic in comparison with wet machining. Cryogenic machining produces better chip control and minimizes the tool wear more than dry, wet machining [5]. Navjot Singh et al. [6], investigated effect on AISI D3 Die Steel using two different Cryogenic Treated Wires in Wire Electrical Discharge Machining by designing the experiments in Taguchi's L_9 orthogonal array. Better Material removal rate was obtained in cryogenically treated zinc-coated diffused brass wire than cryogenic treated plain brass wire. Parveen et al. [7], tried to optimize the WEDM process parameters when H-13 is machined using Taguchi L_9 orthogonal array method. MRR, Wire wear rate and surface flatness lead to improvements in the machining efficiency. Patil et al. [8], have studied surface finish (Ra) of Ceramic materials reinforced Al matrix composite and developed the semi-empirical model. The surface roughness of Al/SiC_p composite reduces with an increase in volume fraction of SiC and in Al/Al_2O_3 surface roughness decreases with the increase in volume fraction of aluminum. A. Thakur et al. have investigated the influence of cutting speed (V_c) (51, 84, and 124 m/min) on various machining characteristics like chip morphology, chip thickness ratio, tool wear, surface, and sub-surface integrity during dry turning of Inconel 825 and have concluded that the surface and sub-surface integrity obtained with coated tool were superior to that while machining Inconel 825 with uncoated tool. Rajyalakshmi and Venkata Ramaiah [9] have applied an effective approach, Taguchi grey relational analysis, to experimental results of wire-cut electrical discharge machining (WEDM) on Inconel 825 with consideration of multiple response measures along with the orthogonal array design of experiment, to find the best combination of parameters. Kumari et al. [10], have investigated the extent of machinability in purview of process performance surface integrity on of Inconel 825 Superalloy by using different tool material and flushing conditions. Moreover, effects of peak discharge current on EDM responses were also discussed. Aspinwall et al. [11], have employed Roughing and finishing (multiple trim cut) strategies on two high specification machines with pulse generators designed to provide minimum workpiece integrity damage. Results include productivity; 3D topographic maps of workpiece surfaces, microstructural and microhardness depth profile data. Average recast thickness was less than 11 μm were achieved. Yildiz and Nalbant [12] have reviewed the application of coolant in a cutting process which can increase tool life, dimensional accuracy and decrease cutting temperatures, surface roughness, amount of power consumed in a metal cutting process using nitrogen, as a cryogenic coolant. Molinari et al. [13], have studied the effect of deep cryogenic treatment (−196 °C) on the properties of some tool steels by means of both field tests on real tools and laboratory tests and concluded that the deep cryogenic treatment on quenched and tempered high-speed steel tools increases hardness and reduces tool consumption. Neeraj Sharma et al. [14], have used cryogenic treated high carbon high chromium tool steel (D-2 tool steel) and investigated the effect of parameters on surface roughness for WEDM. They employed Mathematical modelling of the process is carried with the help of Response surface methodology (RSM) and the central composite rotatable design (CCRD) for planning the experiments.

2 Research Methodology

A Computer Numerically Controlled (CNC) WEDM is used for the experimentation whose worktable can withstand a maximum load of 300 Kgs and can accommodate a workpiece of 200 mm height. The worktable can move in *X*, *Y* and *Z* directions whereas the wire upper guide can move in *U* and *V* directions. A standard wire of 0.25 mm can be passed through upper and lower guides. Deionized water is passed between the wire and workpiece, which behaves as a dielectric fluid between the wire electrode and workpiece electrode. In this research to make the Zinc-Coated Brass wire is used as wire electrode which is cryogenically cooled using Liquid Nitrogen (LN_2). Stainless steel box in which a hollow copper tube is at the center through with the Zinc-coated brass wire is passed. This box is insulated from inside using thermocol to avoid the dissipation of heat from the box. The cryogenic liquid which is stored in a separated cylinder is pumped into the stainless steel box continuously during the machining. 3D model of the cryogenic setup is shown in Fig. 1.

Inconel 825 of 250 × 250 × 5 mm is used as workpiece. Experiments were designed using Taguchi's L_{27} orthogonally array. The process parameters for machining are Wire Feed (WF), Wire tension (WT), Pulse on Time (T_{on}), Pulse off Time (T_{off}) and Servo Voltage which are taken at three different levels. The input parameters and its levels are shown in Table 1.

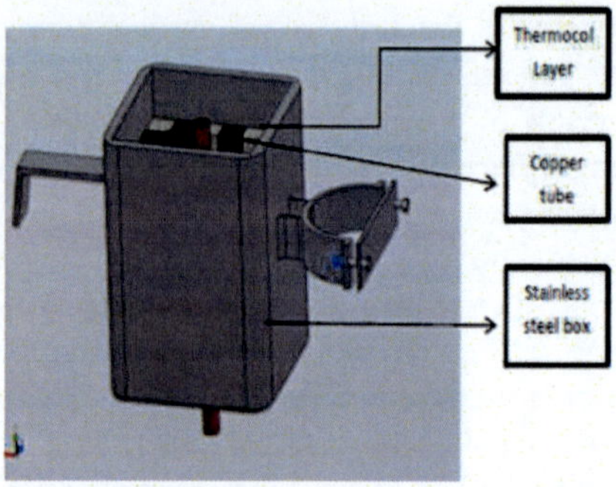

Fig. 1 3D modelling of cryogenic setup

Table 1 Machine process parameters and its levels

Symbol	Parameter	Unit	Coding levels		
			−1	0	1
A	Wire tension	g	850	1150	1300
B	Wire feed	mm/min	7	8	9
C	Pulse on	μs	7	8	9
D	Pulse off	μs	6	7	8
E	Servo voltage	V	4	5	6

Gear-shaped machining profile is selected for machining. Using Taguchi 27 experiments were conducted with five parameters and its three different levels. Taguchi method has reduced the number of test using orthogonal array and effects of the actors that cannot be controlled can be minimized. It even provides an efficient and systematic approach to obtain the optimum parameters. Material Removal Rate (MRR) and Surface Roughness are the responses recorded for 27 experiments. The material Removal rate is calculated using the formula given below.

$$\text{MRR} = \ldots \tag{1}$$

Volume of Material removed (V_m) = Perimeter × Thickness of workpiece (T_m = 5 mm) × Thickness of Wire (T_w = 0.3 mm).

2.1 Calculation of Perimeter

For calculating the perimeter of the Spur gear few assumptions are made to make the calculation simpler and the Fig. 2 shows the gear profile and Fig. 3 shows the Dedendum triangle reference, Figs. 4 and 5 gives details about the Dedendum

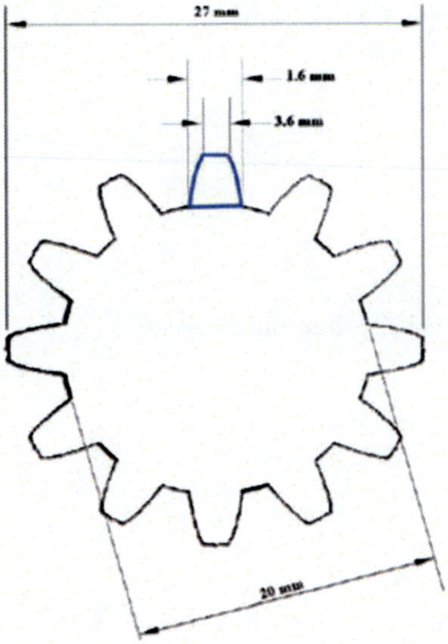

Fig. 2 Spur gear profile

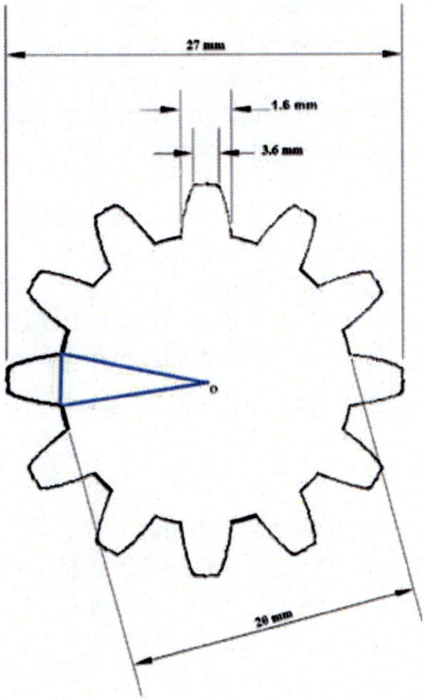

Fig. 3 Dedendum triangle reference

Fig. 4 Gear tooth dimensions

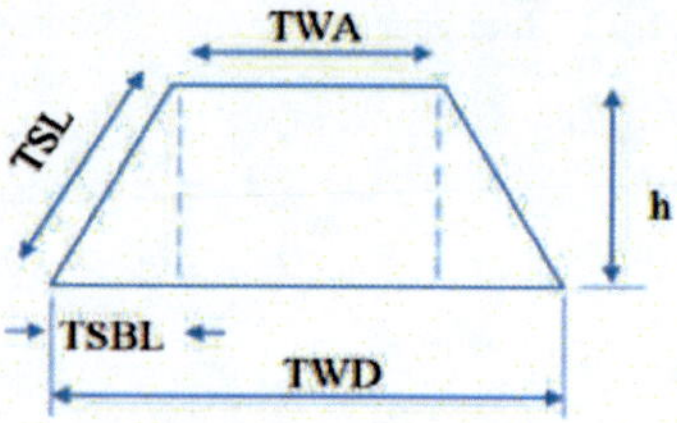

Fig. 5 Dedendum triangle

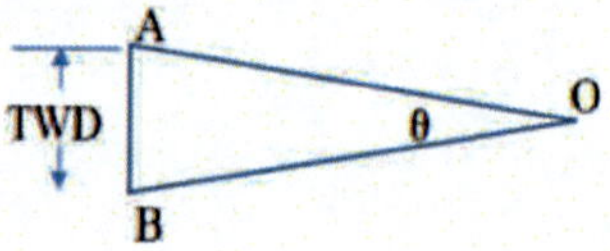

triangle.

$$P_p = P_d + \text{TTSL} + \text{TTWA} = 132.89\,\text{mm}.$$

where P_d = Perimeter Covered at Dedendum
TTSL = Total Tooth Side Length.
TTWA = Tooth Width at Addendum.
where θ = Angle of dedendum width (∠AOB).
O = Centre of Gear.

$$P_d = C_d - (L_{twd} \times N)$$

P_d =22.21 mm.
where C_d = Dedendum Circumference ($\pi \times 20.1$).
L_{twd} = Arc length from TWD length.
N = No. of tooth.

$$\text{TTSL} = \text{STSL} \times N \times S$$

TTSL = 91.4832 mm.
where STSL = Single Tooth Side Length (mm).
S = No. of Sides in a Tooth (2).

$$\text{STSL} = \sqrt{}(h^2 + \text{TSBL}^2)$$

STSL = 3.8118 mm
where h = Gear Tooth Height (3.685 mm)
TSBL = Tooth Side Bottom Length (0.97 mm).

$$L_{twd} = \pi d$$

L_{twd} = 3.411636 mm

where d = Dedendum diameter (20.1 mm)
θ = Angle of arc by TWD
$\theta = \tan - 1 = 19.45°$.
Surface roughness of the specimen is calculated using TaylSurf Surface roughness machine. The surface roughness is calculated at three sides of the specimen and the average of these values is taken as consideration. Responses for the 27 experiments are tabulated in Table 2. Machined specimens are shown in Fig. 6.

Table 2 Responses obtained for the experiments

S. No.	Wire feed WF (mm/min)	Wire tension WT (g)	Pulse on time T_{on} (μs)	Pulse on time T_{off} (μs)	Servo voltage S_v (V)	Material removal rate MRR (mm^3/min)	Surface roughness (μm)
1	7	850	7	6	4	3.24	2.65
2	7	850	7	6	5	3.35	2.80
3	7	850	7	6	6	3.45	2.98
4	7	1150	8	7	4	3.35	2.60
5	7	1150	8	7	5	3.56	2.70
6	7	1150	8	7	6	3.70	2.86
7	7	1300	9	8	4	3.95	2.50
8	7	1300	9	8	5	4.05	2.65
9	7	1300	9	8	6	**4.35**	2.80
10	8	850	8	8	4	3.25	2.81
11	8	850	8	8	5	3.45	2.95
12	8	850	8	8	6	3.67	3.05
13	8	1150	9	6	4	3.56	2.68
14	8	1150	9	6	5	3.79	2.84
15	8	1150	9	6	6	4.05	3.01
16	8	1300	7	7	4	3.85	**2.15**
17	8	1300	7	7	5	4.05	2.29
18	8	1300	7	7	6	4.21	2.41
19	9	850	9	7	4	3.65	2.85
20	9	850	9	7	5	3.79	2.98
21	9	850	9	7	6	3.97	3.15
22	9	1150	7	8	4	3.59	2.31
23	9	1150	7	8	5	3.90	2.43
24	9	1150	7	8	6	4.10	2.65
25	9	1300	8	6	4	3.89	2.26
26	9	1300	8	6	5	3.95	2.39
27	9	1300	8	6	6	4.25	2.50

Fig. 6 Samples specimen after machining

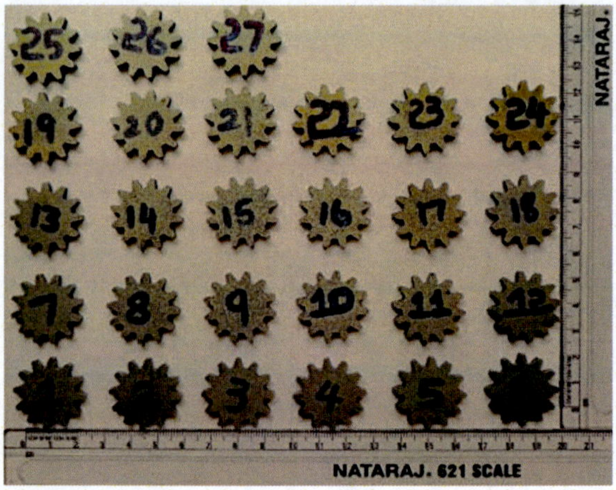

3 Evaluation of Experiments and Its Analysis

The effect of input parameters on the responses is studied by analyzing Taguchi design in Minitab software. The optimum solutions of the controlled factors were provided by Signal to noise (S/N) ratios. The highest value of MRR and lowest value of Surface Roughness are required. The S/N ratios graphs for MRR and Surface Roughness are shown in Figs. 7 and 8 which show the optimal machining parameters for obtaining smaller surface roughness and large MRR.

The parameters best level is one which is having the highest value in S/N response table. The optimal value of MRR is obtained at Wire feed 9 mm/min, Wire tension 1300 g, Pulse on time 9 μs, Pulse off time 7 μs and Servo voltage 6 V. Similarly for best surface roughness is obtained Wire feed 9 mm/min, Wire tension 1300 g, Pulse on time 7 μs, Pulse off time 7 μs and Servo voltage 4 V.

Modelling of variables is the relation between input parameters and the responses are obtained using regression equations. The linear regression equation of the Material Removal Rate and Surface Roughness are obtained in Eqs. 2 and 3 which gives the linear relation between input parameters with its responses. The response obtained by Eqs. 2 and 3 will give the liner variation of response with inputs. The R^2 value for the linear regression equation of MRR is found to be 98.9% and that of Surface roughness is found to be 99.2%

$$\text{MRR} = 0.1854\,\text{WF} + 0.001009\,\text{WT} + 0.0198\,T_{\text{on}} + 0.0029\,T_{\text{off}} + 0.1963\,S_{\text{v}} \quad (2)$$

$$\text{Ra} = 0.0238\,\text{WF} - 0.000806\,\text{WT} + 0.2355\,T_{\text{on}} + 0.0726\,T_{\text{off}} + 0.1947\,S_{\text{v}} \quad (3)$$

The Quadratic regression equation of MRR and Ra is obtained by using Eqs. 4 and 5 which will help to obtain the response for any value of the input instead of predefined values. Its predicted R^2 values are 96.9 and 98.9%.

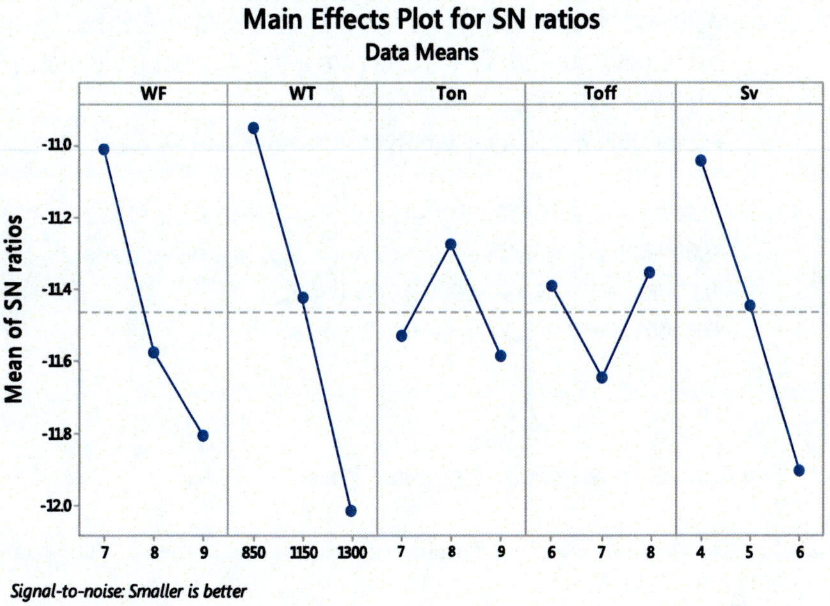

Fig. 7 Signal to noise ratio for MRR

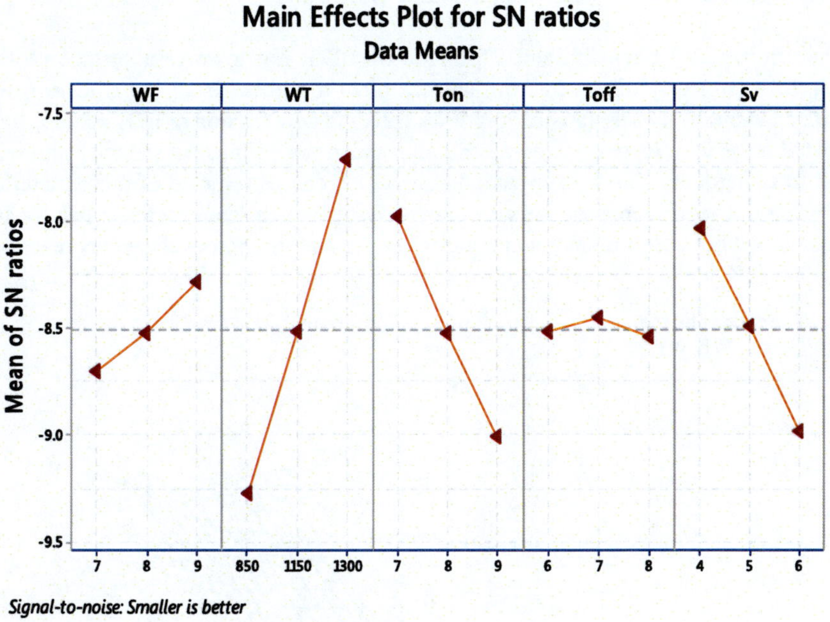

Fig. 8 Signal to noise ratio for surface roughness

$$\begin{aligned}\text{MRR} = {} & 0.805\,\text{WF} + 0.00728\,\text{WT} - 1.157\,\text{Ton} + 0.199\,T_{\text{off}} - 0.237\,S_{\text{v}} \\ & - 0.001050(\text{WF} \times \text{WT}) + 0.0909(\text{WF} \times T_{\text{on}}) - 0.0433(\text{WF} \times T_{\text{off}}) \\ & + 0.0250(\text{WF} \times S_{\text{v}}) + 0.000236(\text{WT} \times T_{\text{on}}) \\ & + 0.000067(\text{WT} \times S_{\text{v}}) + 0.0042(\text{Ton} \times S_{\text{v}}) + 0.0167(T_{\text{off}} \times S_{\text{v}}) \end{aligned} \quad (4)$$

$$\begin{aligned}\text{Ra} = {} & -0.276\,\text{WF} - 0.001193\,\text{WT} + 1.385\,T_{\text{on}} - 0.515\,T_{\text{off}} + 0.182\,S_{\text{v}} \\ & + 0.000408(\text{WF} \times \text{WT}) - 0.0902(\text{WF} \times T_{\text{on}}) + 0.0639(\text{WF} \times T_{\text{off}}) \\ & - 0.00083(\text{WF} \times S_{\text{v}}) - 0.000387(\text{WT} \times T_{\text{on}}) \\ & - 0.000017(\text{WT} \times S_{\text{v}}) - 0.00167(T_{\text{off}} \times S_{\text{v}}) \end{aligned} \quad (5)$$

3.1 Variations in Material Removal Rate

The Changes which are obtained because of the input responses in the Material removal rate are studied.

3.1.1 Surface Plot of Wire Tension and Wire Feed

The effect of Wire tension and Wire feed on MRR is shown as a surface plot in Fig. 9. It is clearly observed from surface plots that there is an increase in MRR with an increase in Wire tension and Wire feed. As wire tension of LN_2 cooled Zinc-coated brass wire increases, there is an increase in the stiffness of the wire and wire becomes hardened due to cryogenic liquid which causes removal of more material. When Wire feed increases the wire moves faster which leads to less time for the breakdown between the wire and material hence material removal rate decreases.

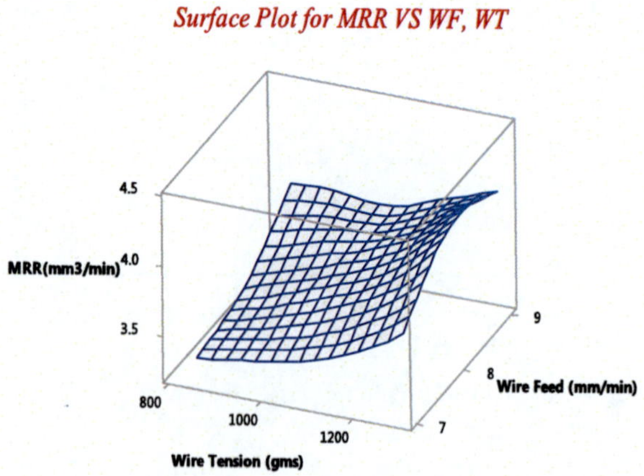

Fig. 9 Surface plot for MRR versus WF, WT

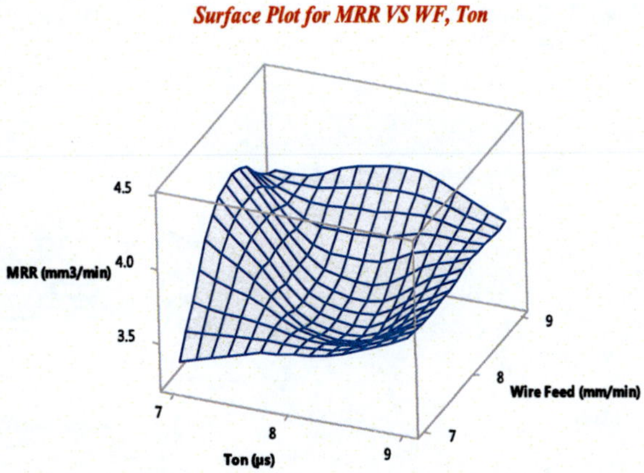

Fig. 10 Surface plot for MRR versus WF, T_{on}

3.1.2 Surface Plot of Wire Feed and Pulse On Time Versus MRR

The effect of Wire feed and Pulse on time are shown on surface plot in Fig. 10, in which it is observed that when pulse on time through the cryogenic cooled zinc-coated brass wire was increased, the duration of time increases that leads to evenly distribution of energy over the wire and removes more amount of material from the surface. This leads to increase in the amount of energy through the wire. Hence, the MRR increases.

3.1.3 Surface Plot of Wire Feed and Pulse Off Time Versus MRR

Surface Plot of MRR against Wire feed and Pulse off time is shown in Fig. 11. The figure clearly depicts that the wire feed is prominent than pulse off time. With

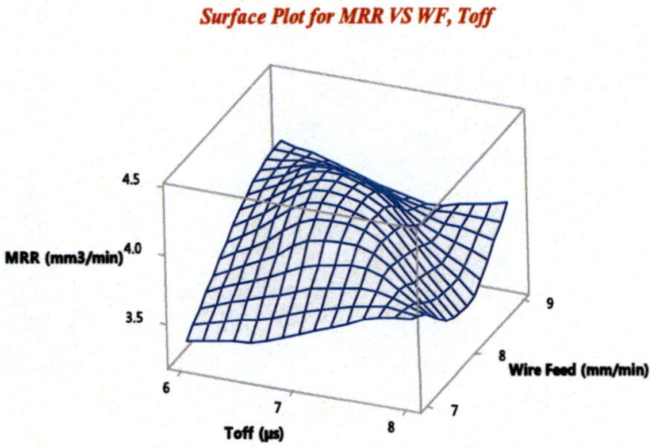

Fig. 11 Surface plot for MRR versus WF, T_{off}

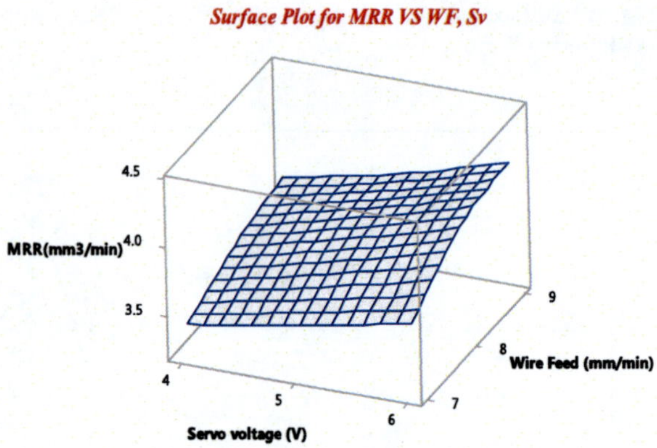

Fig. 12 Surface plot for MRR versus WF, S_v

increase in Pulse off time, there is an increase in duration for cooling the wire and the energy passed through the wire will be reduced. This leads to a decrease in MRR.

3.1.4 Surface Plot of Wire Feed and Servo Voltage Versus MRR

The variation in MRR with Wire Feed and Serve voltage is shown in Fig. 12. If Servo voltage is increased the dielectric will breakdown easily which in turn causes large energy between the electrodes, i.e. wire and the material, thus MRR increases.

3.1.5 Surface Plot of Wire Tension and Pulse On Time Versus MRR

The variations in MRR with the Wire tension and Pulse on time are shown on surface plot in Fig. 13. It is observed that MRR increases with the increase of pulse on time and Wire tension. Highest MRR is observed when Pulse on time is low and Wire

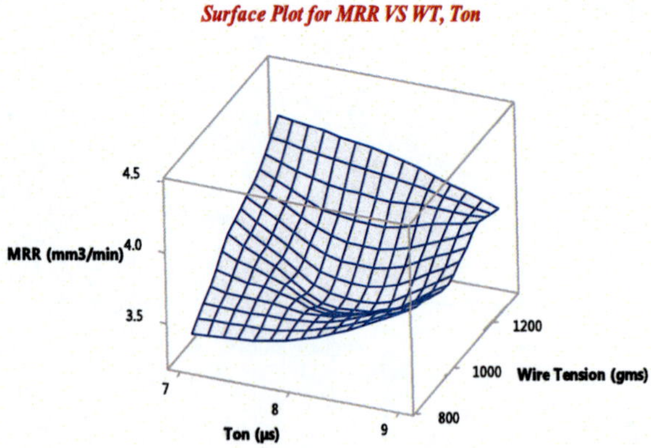

Fig. 13 Surface plot for MRR versus WT, T_{on}

Fig. 14 Surface plot for MRR versus WT, T_{off}

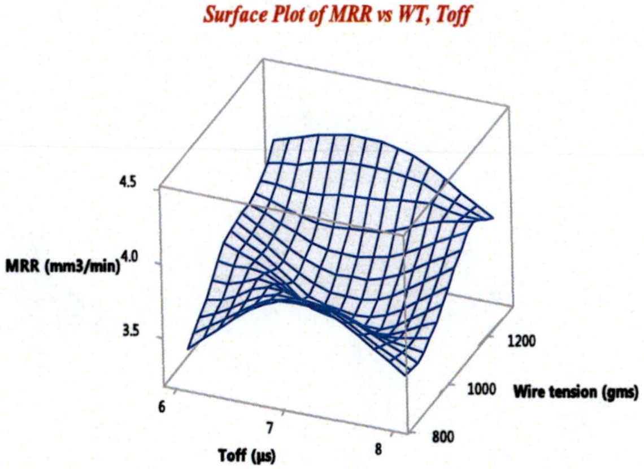

tension is high which indicates that wire tension is more significant compare to Pulse on time.

3.1.6 Surface Plot of Wire Tension and Pulse Off Time Versus MRR

The variation in MRR with Wire tension and Pulse off time is shown in Fig. 14, which clearly depicts that the MRR increases with increase in Wire tension and decreases with an increase in Pulse off time. As wire tension increases the vibrations of the wire between the upper and lower guide decreases which causes the energy to distribute evenly, thus MRR increases. When pulse off time increases more delay will be there in the production of breakdown hence MRR decreases.

3.1.7 Surface Plot of Wire Tension and Servo Voltage Versus MRR

The variation in MRR with wire tension and Servo voltage is shown on surface plot Fig. 15. It is observed that MRR increases with increase in both wire tension and Servo voltage. The MRR is highest in Fig. when both servo voltage and wire tension are highest.

3.1.8 Surface Plot of Pulse On Time and Pulse Off Time Versus MRR

The variation in MRR with Pulse on time and pulse off time is shown in Fig. 16 which clearly depicts that the MRR is high when Pulse on time is Medium and Pulse off time is less. When Pulse on time increases beyond a certain limit the wire will break because of short circuit. When Pulse off time is less, the sparks are generated within short duration of time, and hence MRR is more when Pulse off time is less.

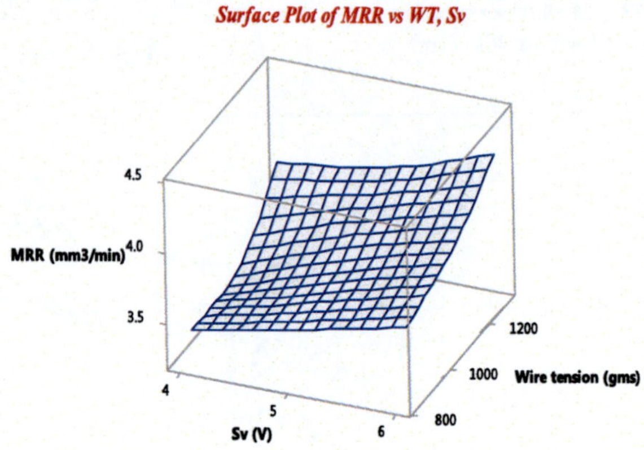

Fig. 15 Surface plot for MRR versus WT, S_v

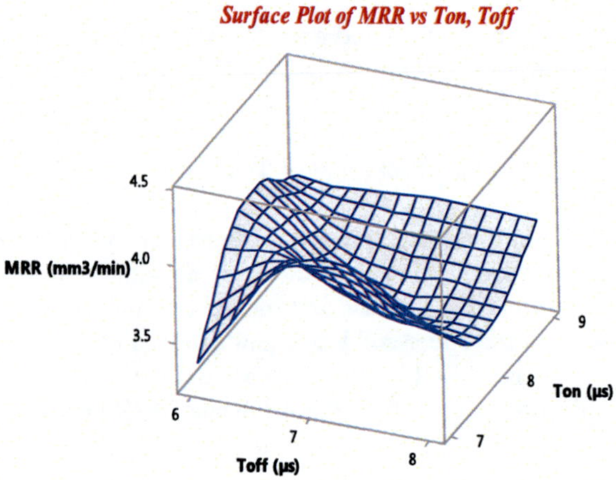

Fig. 16 Surface plot for MRR versus T_{on}, T_{off}

3.1.9 Surface Plot of Pulse On Time and Servo Voltage Versus MRR

MRR variations with Pulse on time and Pulse off time are shown in surface plot Fig. 17. The figure clears depicts that the MRR increases with an increase in both Pulse on time and servo voltage. The variation in MRR with Pulse on time and Servo voltage is linear. Highest MRR is observed when both Pulse on time and Servo Voltage are high.

3.1.10 Surface Plot of Pulse Off Time and Servo Voltage Versus MRR

The Surface plot of Pulse off time and Servo voltage on MRR is shown in Fig. 18. The variation in figure clearly depicts that the highest MRR is obtained when Pulse

Fig. 17 Surface plot for MRR versus WT, S_v

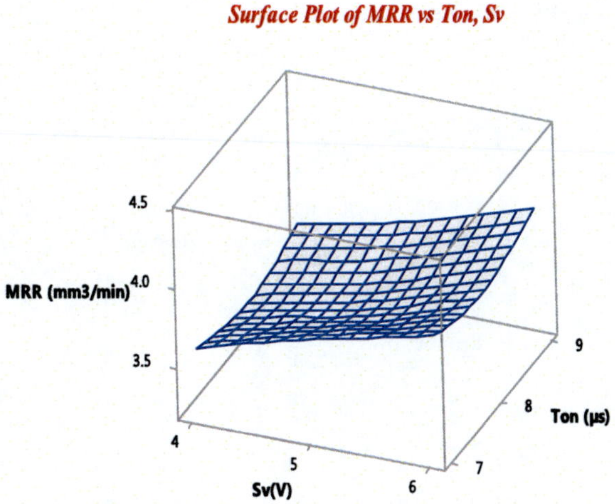

Fig. 18 Surface plot for MRR versus T_{on}, T_{off}

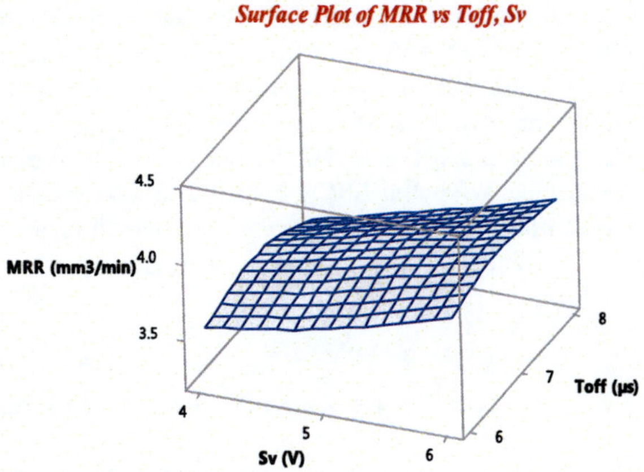

off time is less and Servo voltage is more because the delay in time for the generation of sparks will be less and breakdown occurs faster.

3.2 Variations in Surface Roughness

The effect of various input parameters on surface roughness is observed by using surface plots.

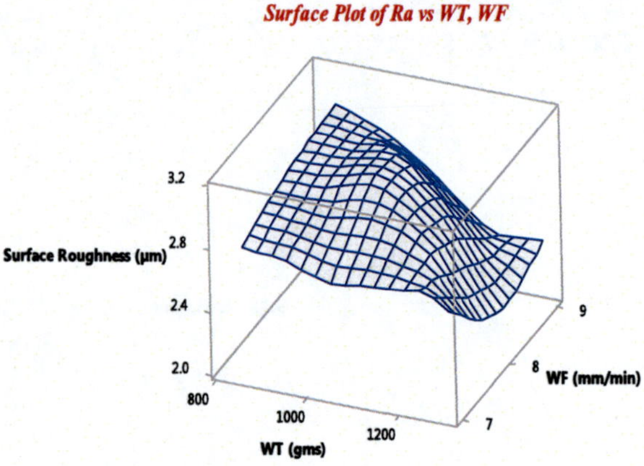

Fig. 19 Surface plot for Ra versus WF, WT

3.2.1 Surface Plot for Ra Against Wire Tension and Wire Feed

The effect of wire tension and Wire feed on surface roughness is shown on surface plot in Fig. 19. It is observed from the figure that Surface roughness is least when both wire tension and wire feed are high. When wire tension of cryogenic cooled brass wire is increased the stiffness of wire increases and vibrations are reduced which leads to even spark generation from the wire and material is removed evenly from the surface. When this Stiff wire is fed continuous, i.e. wire feed increases the newer wire comes in the spark generation are which further helps to remove the material evenly. Hence, Surface roughness decreases when both Wire tension and Wire feed are high.

3.2.2 Surface Plot for Ra with Wire Feed and Pulse On Time

The variation in surface roughness against Wire feed and Pulse on time is shown in Fig. 20. The figure depicts that Surface roughness increases with increase of both Wire feed and Pulse on time. As Pulse on time increases, the duration for which energy distortion between wire and workpiece increases which leads to more amount of material removal from the same surface hence surface roughness increases. When Wire Feed is increased the break down between newer wire and work piece takes place continuously, which leads to more amount of energy gets distorted at once. Thus there is an increase in surface roughness.

3.2.3 Surface Plot for Ra with Wire Tension and Pulse On Time

A surface variation for Ra with Wire tension and Pulse on time is shown in Fig. 21. It is observed that Surface roughness increases with increase in pulse on time and decreases with increase in wire tension. When Pulse on time increases, there is

Fig. 20 Surface plot for Ra versus WF, T_{on}

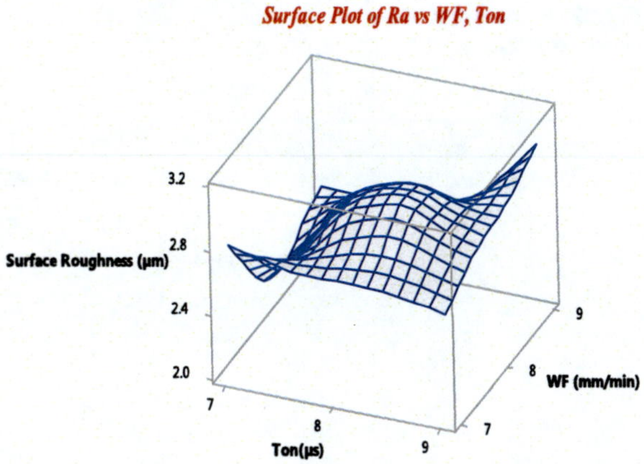

Fig. 21 Surface plot of Ra versus WT, T_{on}

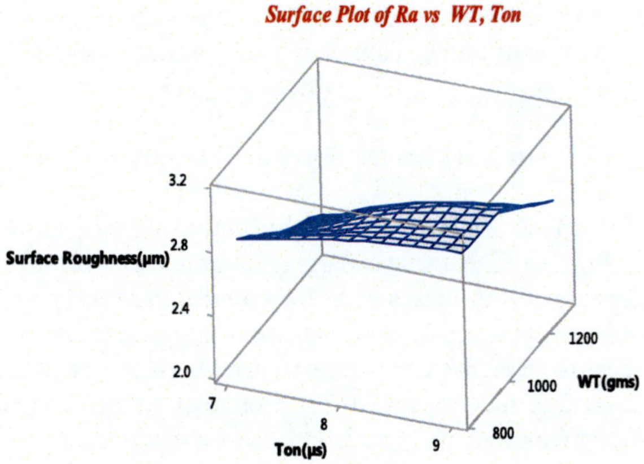

increase in time of spark generation which causes more amount of material getting eroded from the surface of the work piece. Thus Surface roughness increases with increase in Pulse on time. When Wire tension is increased there is reduction in vibrations of the wire which further leads to even distribution of energy between work piece and the wire, hence surface roughness decreases. It is clearly observed from the figure the surface roughness is least when Pulse on time is less and wires tension is high.

3.2.4 Surface Plot for Ra with Wire Tension and Pulse Off Time

The changes in Surface plot for surface roughness with Wire tension and Pulse off time are shown in Fig. 22. It is observed from the figure that Surface roughness decreases with increase both Pulse off time and Wire tension, the least surface roughness is obtained when both wire tension and pulse off time are high. When Pulse off

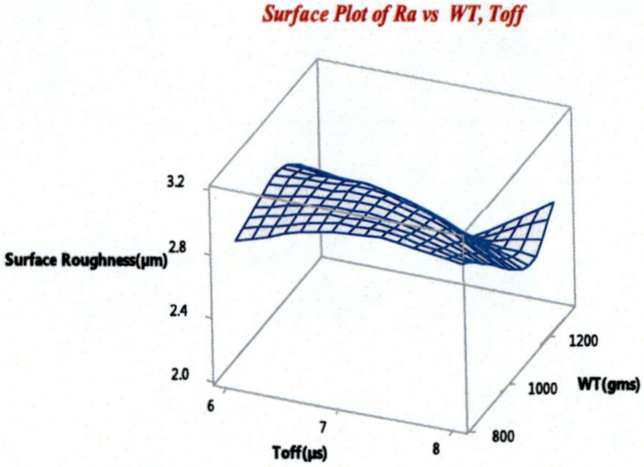

Fig. 22 Surface plot of Ra versus WT, T_{off}

time is high the duration for wire to get cool increases and increase in the delay in formation of spark. Thus Surface roughness will decrease.

3.2.5 Surface Plot for Ra with Wire Tension and Servo Voltage

The effects of Wire tension and Servo voltage on Surface Roughness are shown in Fig. 23. The surface roughness increases with an increase in Servo voltage and decreases with increase in Wire tension. As Servo voltage increases the amount of voltage between the wire and work piece increases the breakdown increases which leads to an increase in energy transfer between the electrodes, which leads to increase in surface roughness. The least value of surface roughness is observed in the figure when wire tension is high and servo voltage is less.

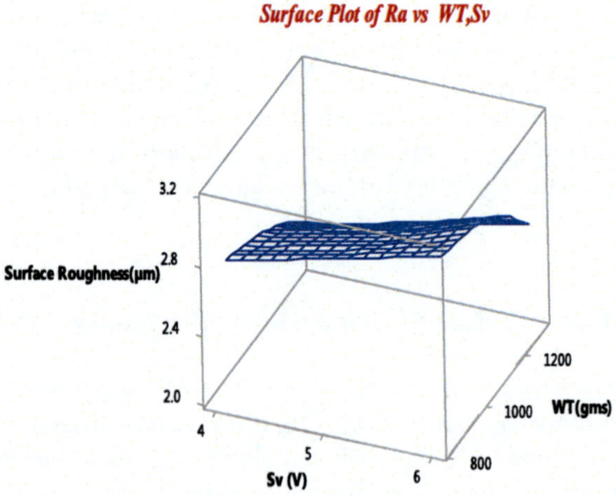

Fig. 23 Surface plot of Ra versus WT, S_v

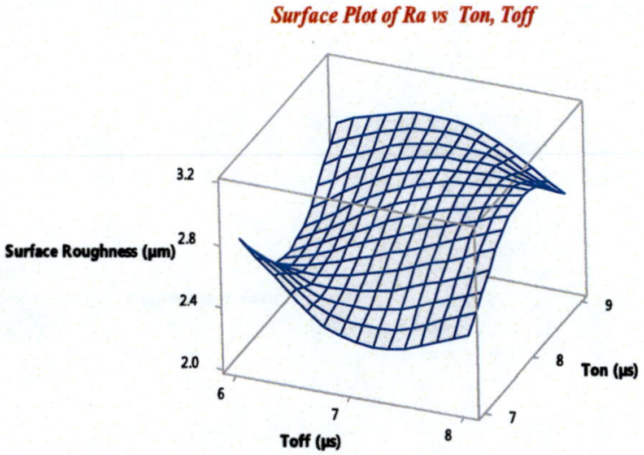

Fig. 24 Surface plot of Ra versus T_{on}, T_{off}

3.2.6 Surface Plot of Ra with Pulse On Time and Pulse Off Time

The variations in surface plot of Ra with Pulse on time and Pulse off time is shown in Fig. 24. The figure depicts that surface roughness decreases with increase in Pulse off time and increases with increase in Pulse on time. The least surface roughness in the figure is obtained when Pulse off time is high and Pulse on time is low.

3.2.7 Surface Plot of Ra with Pulse On Time and Servo Voltage

The Surface plot of surface roughness with pulse on time and servo voltage is shown in Fig. 25. The figures clearly indicate the surface roughness increases with increase

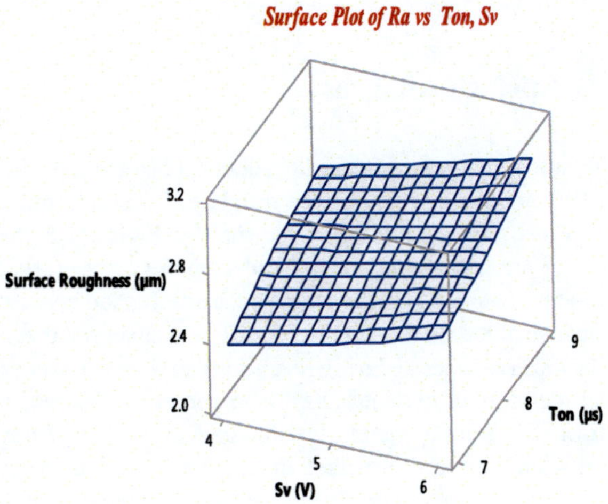

Fig. 25 Surface plot of Ra versus T_{on}, S_v

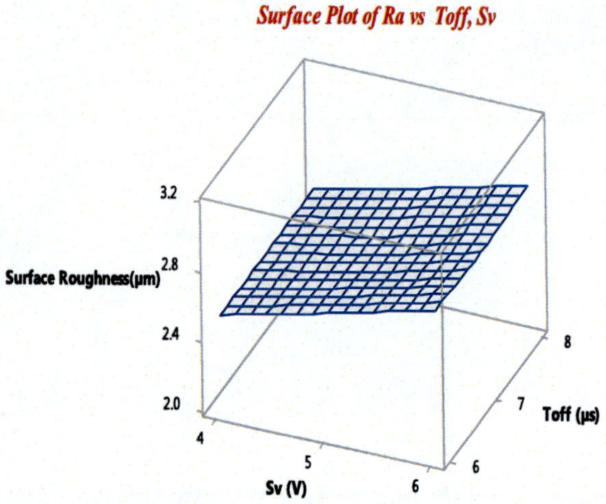

Fig. 26 Surface plot of Ra versus S_V, T_{off}

in both servo voltage and pulse on time. The highest value of the surface plot is obtained when both servo voltage and pulse on time are high.

3.2.8 Surface Plot of Ra with Pulse Off Time and Servo Voltage

The variation in the surface roughness with Pulse off time and Servo voltage is shown in Fig. 26. When Servo voltage increases there will be an increase in Surface roughness and when Pulse off time increases the surface roughness will be decreased. The least value in surface plot is obtained when servo voltage is less and Pulse off time is high.

4 Surface Analysis

To study the surface Scanning Electron Microscopic images are taken for the highest MRR and the lowest Surface roughness experimental specimens.

From Fig. 27 it is observed that for high MRR the surface has damaged more. There are few cracks, debris and globules are observed. The debris is formed when molten material of the work piece is not flushed properly and gets re-solidifies on the surface of the workpiece. In Fig. 27, it is observed that more debris is formed because the molten material during machining gets re-solidifies quickly due to decreases in the temperature of dielectric as it is in contact with cryogenic cooled wire. Globules are formed when the material which gets evaporated from the surface of the workpiece falls back on to the surface during flushing. Due to the difference in the impact of pulse discharge energy the craters are formed on the work piece.

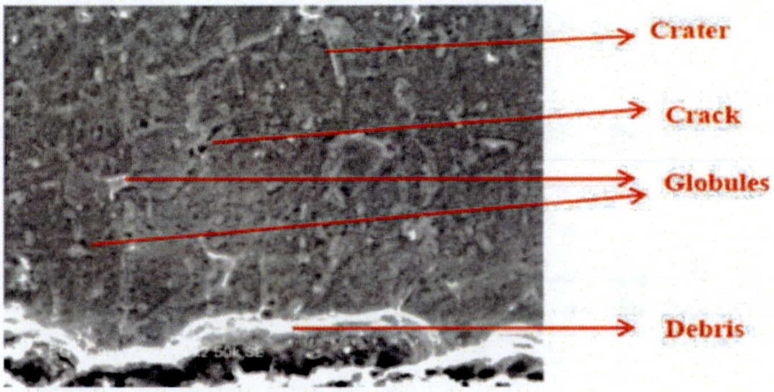

Fig. 27 SEM image for high MRR specimen

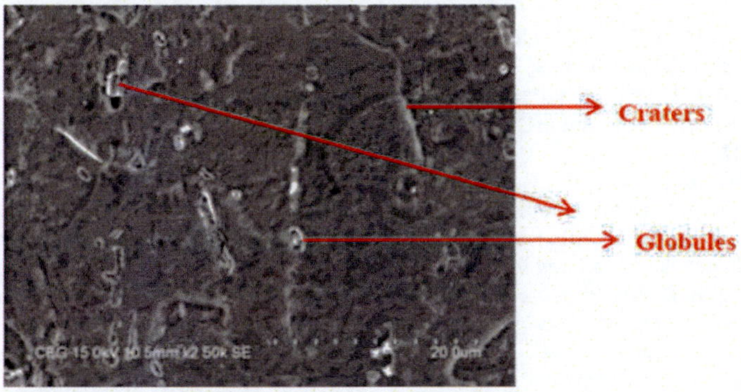

Fig. 28 SEM image for high MRR specimen

From Fig. 28, it is observed that better surface is observed. There are less amount globules and craters are observed. Surface roughness is improved as the as less energy is discharged during this experiment which causes less material to discharge and less molten material gets accumulated and fewer globules are formed.

5 Conclusions

- MRR of Inconel 825 varies from 3.24 to 4.35 mm^3/min.
- Surface Roughness varies from 2.15 to 3.01 μm.
- Highest MRR of 4.35 mm^3/min is obtained when the input parameters are Wire Feed (WF)–7 mm/min, Wire Tension (WT)–1300 g, Pulse on time (T_{on})–9 μs, Pulse off time (T_{off})–8 μs and Servo Voltage (S_v)–6 v.

- Least Surface Roughness of 2.15 μm is observed when input parameters are Wire Feed (WF)–8 mm/min, Wire Tension (WT)–1300 g, Pulse on time (T_{on})–7 μs, Pulse off time (T_{off})–7 μs and Servo Voltage (S_v)–4 v.
- Wire Tension, Pulse on time and Servo voltage are the most significant factors for MRR.
- Wire tension, Pulse off time and Wire Feed are the most significant factors for surface roughness.
- The Mathematical modelling for both MRR and Surface Roughness is found to be in good agreement with the experimental values.

References

1. Singh T, Misra JP, Singh B (2017) Empirical modelling of wear ratio during WEDM of nimonic 263. Mater Today 5(11):23612–23618
2. Majumder H, Maity K (2018) Prediction and optimization of surface roughness and micro-hardness using grnn and MOORA-fuzzy-a MCDM approach for nitinol in WEDM. Measurement 118(1):1–13
3. Ajay BT, Mayur Vitthal P, Rajyalakshmi G (2017) WEDM machining on aerospace Materials for improving material properties. Mater Today: Proc 4(8):9107–9116
4. Singh A, Kumar S (2015) Surface roughness and microhardness evaluation for EDM with Cu–Mn powder metallurgy tool. Mater Manuf Process 31(4):514–521
5. Dilip Jerold B, Pradeep Kumar M (2013) The Influence of cryogenic coolants in machining of Ti–6Al–4V. J Manuf Sci Eng 135(3):1–8
6. Singh N, Kumar P, Goyal K (2013) Effect of two different cryogenic treated wires in wire electrical discharge machining of AISI D3 DIE steel. J Mech Eng 43(2):54–60
7. Kumar P, Bhushan B, Gupta RK (2013) Response of wire electrical discharge machining for H13 using Taguchi L_9 array. Int J Eng Res Appl 3(5):907–912
8. Patil NG, Brahmankar PK (2009) Some studies into wire electro-discharge machining of alumina particulate-reinforced aluminum matrix composites. Int J Adv Manuf Technol 48(5–8):537–555
9. Rajyalakshmi G, Venkata Ramaiah P (2013) Multiple process parameter optimization of wire electrical discharge machining on Inconel 825 using Taguchi grey relational analysis. Int J Adv Manuf Technol 69(5–8):1249–1262
10. Kumari S, Datta S, Masanta M, Nandi G, Pal PK (2018) Electro-discharge machining of Inconel 825 super alloy: effects of tool material and dielectric flushing. Silicon 10(5):2079–2099
11. Aspinwall DK, Soo SL, Berrisford AE, Walder G (2008) Workpiece surface roughness and integrity after WEDM of Ti–6Al–4V and Inconel 718 using minimum damage generator technology. CIRP Ann—Manuf Technol 57(1):187–190
12. Yildiz Y, Nalbant M (2008) A review of cryogenic cooling in machining processes. Int J Mach Tools Manuf 48(9):947–964
13. Molinari A, Pellizzari M, Gialanella S, Straffelini G, Stiasny KH (2001) Effect of deep cryogenic treatment on the mechanical properties of tool steels. J Mater Process Technol 118(1–3):350–355
14. Sharma N, Singh A, Sharma R, Deepak (2014) Modelling the WEDM process parameters for cryogenic treated D-2 tool steel by integrated RSM and GA. Procedia Eng 97(1):1609–1617

Optimization of Process Parameters During EDM on Inconel Alloy 625

K. Jayakumar

Abstract Inconel 625 is a Nickel–Chromium–Molybdenum based superalloy which is extensively used in numerous engineering applications like, aerospace, nuclear and marine fields, aeronautical, chemical, petrochemical industries, oil and gas extraction industries. It has high strength and outstanding corrosion resistance properties. The main problem with using this material is their low machinability while machining with conventional machining processes due to its high mechanical properties. From different unconventional machining processes (UCMP), the Electrical discharge machining (EDM) process is the most important and versatile UCMP and is suitable for machining any difficult to machine materials and it is most appropriate for the manufacturing of dies, molds, and press tools. Normally machining industries targets the achievement of high-quality products, relating to the surface finish, workpiece dimensional accuracy, high material removal rate (MRR) at low cost. Hence, the present research work deals with the optimization of process parameters in EDM of Inconel 625 alloy for minimizing the surface roughness (Ra) and maximizing MRR. EDM studies were conducted as per Taguchi's L_9 orthogonal array by varying process parameters such as current, voltage, pulse on time (T_{on}) and electrode bottom profile. Surface roughness and MRR values were measured. Effects of process parameters on the selected machinability responses were analyzed. Optimum process parameters were found for the optimization of EDM on Inconel alloy 625.

Keywords Inconel alloy 625 · EDM studies · Material removal rate · Surface roughness

K. Jayakumar (✉)
Department of Mechanical Engineering, Sri Sivasubramaniya Nadar College of Engineering, Kalavakkam, Chennai 603110, India
e-mail: kjayakumar@ssn.edu.in

S. Vijayan et al. (eds.), *Trends in Manufacturing and Engineering Management*, Lecture Notes in Mechanical Engineering,
https://doi.org/10.1007/978-981-15-4745-4_77

1 Introduction

Inconel 625 is a nickel-based superalloy that has high strength and resistance to elevated temperatures, corrosion and oxidation. Its capability to survive high stress and a wide range of temperatures, both in and out of the water, makes it a fitting best for nuclear and marine uses. Originally Inconel 625 was developed in the 1960s for the use of making material for steam-line piping. Later, some modifications were made to be for more creep-resistant and weldable. Because of this, the uses of Inconel 625 has expanded into the chemical processing industry, marine, and nuclear applications. However, the same properties pose a great challenge during their machining by conventional machining [1]. Non-conventional machining methods with no direct contact between tool and workpiece would be a better option to machine such materials. Electric discharge machining (EDM) is one of the most frequently used non-conventional machining techniques to machine almost any electrically conductive material into complex and intrinsic shapes. EDM is an electrothermal process where material removal takes place through the controlled spark erosion. EDM is the most widely used machining route for Inconel superalloys because of its less setup cost and excellent accuracy of the machining performance in terms of quality of the end product. EDM drilling is most widely performed to machine superalloys in the aerospace industry, gas turbine, and rocket engine [2].

Due to the technological development, researchers have also worked on EDM of Inconel alloys (Inconel 625). Dhanabalan et al. [3] conducted EDM studies on Inconel 625 to analyze the effect of input parameters namely peak current, pulse on time (T_{on}), and pulse off time (T_{off}) on the form tolerances and MRR with circular, square and hexagonal shaped electrodes.

Torres et al. [4] analyzed the effect of EDM parameters on surface roughness (Ra), MRR and electrode wear (EW) of an Inconel 600 alloy by varying current, duty cycle, pulse duration and polarity. They found that +ve positive polarity leads to higher MRR whereas −ve polarity leads to lower Ra values. The same authors [5] continued the machining of Inconel 600 alloy using copper infiltrated graphite (C–Cu). They suggested that the use of negative polarity leads to a higher MRR, higher EW and higher Ra value. Moreover, they found the optimal condition to obtain a maximum MRR (Material Removal Rate) of 30.49 mm^3/min was: 8 A, 100 μs and 0.6, respectively, for the current intensity, pulse time and duty cycle. Dikshit et al. [6] recently used EDM for machining rectangular shape in Inconel 625. Experiments were conducted by varying peak current, pulse on time and pulse off time as per central composite design. They found optimum process parameter conditions for maximizing MRR and minimizing Ra value.

From literature, it was found that the effect of EDM process parameter on different machinability of Inconel alloys was studied but the effect of electrode bottom profile on the machinability of Inconel 625 is missing and it needs a detailed study. The present work focused on the effect of EDM process parameter and electrode bottom profile on Inconel 625 in terms of MRR and surface roughness.

2 Experimental Details

Inconel 625 of size 60 × 60 × 6 mm was chosen as workpiece material. Major constituents in weight % of Inconel 625 are Ni (55%), Cr (23%), Mo (10%) and Fe (5%). A cylindrical Copper (Cu) electrode of diameter 6 mm was used as electrode material. Cu can machine all hard materials with low tool wear and it is affordable.

EDM studies were conducted in the Die-Sinking EDM setup (Make: Ratnaparkhi Electronics EDM 500X300 ZNC; Country: India). Commercial EDM oil was circulated as a dielectric medium. The workpiece is connected to positive polarity and the electrode is connected to negative polarity during machining. Experiments were conducted using four major controllable process parameters and individual parameters were varied at three levels. The selected process parameters were current (I), voltage (V), pulse on time (μs) and electrode bottom profile (flat, chamfer and fillet). Pulse off time was maintained as a constant of 20 μs for all experiments due to the machine constraint. Fillet profile was created in a lathe machine with 1 mm front radius and the chamfer profile was machined with 1 mm × 45° dimension.

Taguchi L_9 orthogonal array (OA) design was used to conduct experiments and Table 1 shows the experimental conditions and results. The EDM hole depth was kept as constant (5.9 mm) and machining time was noted for each experiment to calculate MRR. Figure 1 shows the close view of the EDM setup with workpiece and electrode. Figure 2 shows the machined workpiece.

The machining responses of MRR and average surface roughness (Ra) at the machined surface were measured for each experimental condition. MRR was calculated based on weight difference between the workpiece before and after machining and time taken for each machining. Surface roughness (Ra) of the drilled holes surfaces was measured using Mitutoyo SJ210 series surface roughness tester. Cut off length and number of sample was taken as 0.8 mm and 4, respectively. Roughness measurements were made at three different areas of each hole surface by keeping the

Table 1 EDM experimental design and experimental results

Sl. No.	Current (A)	Voltage (V)	Ton (μs)	Electrode	MRR (g/min)	Surface roughness (μm)
1	15	50	20	Flat	**0.092**	**5.471**
2	15	65	30	Chamfer	0.146	7.617
3	15	80	40	Fillet	0.164	9.132
4	20	50	30	Fillet	0.167	9.787
5	20	65	40	Flat	0.178	9.194
6	20	80	20	Chamfer	0.129	6.958
7	25	50	40	Chamfer	0.159	**11.910**
8	25	65	20	Fillet	0.154	6.779
9	25	80	30	Flat	**0.210**	11.318

Fig. 1 EDM setup for machining Inconel 625

Fig. 2 Workpiece after EDM

workpiece in a different orientation and the average of three readings was considered. Figure 3 shows the arrangement for surface roughness measurement and results of Ra is listed in Table 1.

Fig. 3 Arrangement for measurement of surface roughness

3 Results and Discussions

3.1 Effect of EDM Parameter on MRR

Figure 4 shows the main effect plots of MRR. With an increase of current, voltage and Ton time, the MRR is increased. As current and voltage increased, the power and

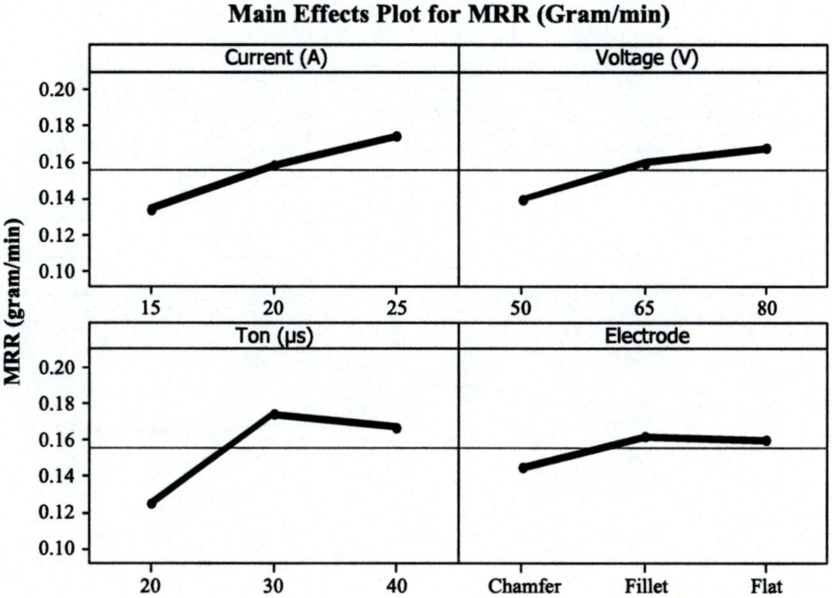

Fig. 4 Main effect plots for MRR

number of sparks produced were increased this improved the MRR. When the Ton increased, MRR increased at the start but later decreased with further raise in Ton. It is the result of an additional reduction on the gap between the work and electrode developing an electrical short circuit instead of spark which reduced the MRR.

MRR obtained from the study for an electrode with chamfer and the flat electrode is low compared with the fillet profile of the electrode. The reason is that the fillet profile has more front curvature surface area than the other two profiles which increased the spark discharges per unit area of the electrode and hence improved the MRR. Also, for the same EDM process parameters, the density of the electric discharges is high; faster separation of the metal from the workpiece and effective flushing of the molten metal happened with fillet profile than chamfer and flat profile [7].

3.2 *Effect of EDM Parameter on Surface Roughness*

The effect of process parameters on surface roughness is shown in Fig. 5.

The effect of process parameters on surface roughness is shown above plots. From that, it was observed that the electrode bottom profile has a little effect than the other three parameters. As power (current and voltage) and Ton of the EDM electric pulse increased, sparks become more energetic, causes more melting and affect the surface

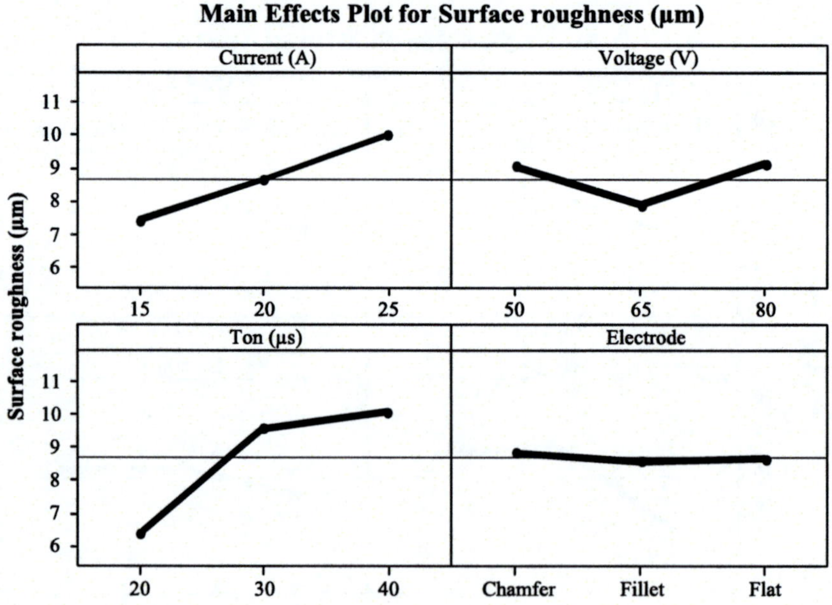

Fig. 5 Main effect plots for surface roughness

over a longer period, leading to larger and deeper craters which increased the surface roughness value [8, 9].

Amount of spark discharged from the electrode bottom is decreased from chamfer to fillet and flat profile, i.e. amount of spark discharged from fillet bottom profile is more than chamfer and flat profile, which created the melting boundary as deeper and wider cavities and increased the roughness from 5.471 to maximum of 11.910 μm.

From the experimental runs, experiment number 8 with fillet bottom profile can be considered as the optimum condition for better MRR and lower surface roughness.

Design Expert software's optimization module was used for finding optimum parameter combination. From that, it was found that the process parameter combination of current-15 A, voltage-55 V, Ton 40 μs with flat bottom profile electrode was found as the overall optimum parameter for getting maximum MRR of 0.21 g/min and minimum surface roughness value of 6.34 μm with the highest desirability of 0.93.

4 Conclusions

In this present study, the influence of EDM parameters on Inconel alloy 625 has been carried out with a copper electrode using three different bottom profiles. The results showed that, for MRR and surface roughness, Ton and current were identified as the most influencing factor.

For getting low surface roughness value, an electrode with a flat profile is recommended than the other two profiles and for high MRR; electrode with fillet profile is suggested.

Application of Design Expert software has determined the optimal setting of process parameters for concurrently attaining multi-response features (high MRR and low Ra value) in relation to the overall EDM performance.

References

1. Gangadharudu Talla S, Gangopadhyay S, Biswas CK (2017) Influence of graphite powder mixed EDM on the surface integrity characteristics of Inconel 625. Particulate Sci Technol 35(2):219–226
2. Mishra DK, Rahul, Datta S, Mahapatra SS, Masanta M, Mahapatra SS (2019) Through hole making by electro-discharge machining on Inconel 625 super alloy using hollow copper tool electrode. Proc Inst Mech Eng, Part B: J Eng Manuf 233(2):348–370
3. Dhanabalan S, Sivakumar K, Sathiya Narayanan C (2014) Analysis of form tolerances in electrical discharge machining process for Inconel 718 and 625. Mater Manuf Process 29(3):253–259
4. Torres C, Luis J, Puertas I (2015) Analysis of the influence of EDM parameters on surface finish, material removal rate, and electrode wear of an INCONEL 600 alloy. Int J Adv Manuf Technol 80(1–4):123–140

5. Salcedo AT, Arbizu IP, Pérez CJL (2017) Analytical modelling of energy density and optimization of the EDM machining parameters of Inconel 600. Metals 166(7):1–21
6. Dikshit MK, Anand J, Narayan D, Jindal S (2019) Machining characteristics and optimization of process parameters in die-sinking EDM of Inconel 625. J Braz Soc Mech Sci Eng 302:1–14
7. Manohar M, Selvaraj T, Sivakumar D, Gopinath S, George KM (2014) Experimental study to assess the effect of Electrode bottom profiles while machining Inconel 718 through EDM process. Procedia Mater Sci 6:92–104
8. S Ahmad, M A Lajis (2013) Electrical discharge machining (EDM) of Inconel 718 by using copper electrode at higher peak current and pulse duration. IOP Conf Ser: Mater Sci Eng 50:1–7
9. Goyal A, Pandey A, Sharma P (2017) Machinability of Inconel 625 aerospace material using cryogenically treated WEDM. Solid State Phenom 266:38–42

Optimization of Laser Trepanning Parameters for Mild Steel by Taguchi Response Surface Methodology (T-RSM)

A. Gnanavelbabu, V. Arunachalam, K. T. Sunu Surendran, V. Dharaniya, and K. Rajkumar

Abstract Laser trepanning plays a vital role in the machining of holes as it needs to eliminate material only from a thin annular region, not the entire circle. The key benefits of trepanning are material saving and less energy consumption. In this study, responses from laser trepanning on mild steel were taken for optimization. Responses like hole-top circularity (HTC), hole-bottom circularity (HBC) and hole taper (HT) were recorded according to the variations made in the parameters such as laser power, gas pressure, speed of spot movement and pulsing frequency as per Taguchi's L_{27} orthogonal array. To optimize the operating parameters, Taguchi–response surface methodology (T-RSM) was applied. Three-dimensional response surface graphs and ramp function graph for desirability were plotted for the obtained results. Optimal level of parameters obtained was laser power-2716.83 W, gas pressure-7 bar, speed of spot movement-1200 mm/s, pulsing frequency-2000 Hz and trepanning hole-20 mm.

Keywords Laser trepanning · Mild steel · Optimization · Response surface methodology (RSM)

1 Introduction

Laser-induced material removal is getting wide attention for precision hole drilling on difficult to cut superior alloys [1]. The laser drilling is categorized into two types: trepan drilling and percussion drilling. Laser trepan drilling is more popular than percussion drilling, and it involves trimming around the boundary of the hole to be generated [2]. Laser trepanning provides better dimensional accuracy [3]. The major

A. Gnanavelbabu (✉) · V. Arunachalam · K. T. Sunu Surendran · V. Dharaniya
Department of Industrial Engineering, CEG, Anna University, Chennai 600025, India
e-mail: agbabu@annauniv.edu

K. Rajkumar
Department of Mechanical Engineering, Sri Sivasubramaniya Nadar College of Engineering, Chennai, Tamil Nadu 603110, India

S. Vijayan et al. (eds.), *Trends in Manufacturing and Engineering Management*, Lecture Notes in Mechanical Engineering,
https://doi.org/10.1007/978-981-15-4745-4_78

advantage of laser trepan drilling is its less energy consumption for producing hole with least material wastage [4].

The statistical relationship between the factors and responses in an experiment can be determined using various optimization tools. The Taguchi DOE is a reliable and effective method for evaluating several design parameters by reducing the number of experiments [5]. Avanish Kumar Dubey et al. made robust design using Taguchi Orthogonal array and performed multi-objective optimization of laser beam cutting parameters on aluminium alloy sheet using ANOVA [6]. Response surface methodology (RSM) is an emerging optimization tool to characterize the process performance. Murthy B. R. N. et al. integrated Taguchi and response surface methodology to set optimum combination of process parameters for drilling of GFRP [7]. Sivaraos et al. by applying Taguchi and RSM methods compared the CO_2 laser machining on PVC foam by considering the parameters such as cutting speed, laser power, frequency, duty cycle, gas pressure on kerf width. They concluded that both the techniques predicted near values of average errors and RSM produced promising results over Taguchi method [8]. R. Adalarasan et al. studied the effect of CO_2 laser cutting parameters on surface roughness and kerf width of AA 6061/Al_2O_3 composite using L_{27} orthogonal array. Desirability analysis was generated by RSM and to obtain optimum setting of laser cutting parameters [9]. Rupesh Goyal et al. applied RSM to find the optimum parameters setting of laser trepan drilling (LTD) to improve the hole circularity and hole taper on superalloy sheets. They considered pulse width, pulse frequency, gas pressure and trepanning speed as the input parameters, more circular hole due to increased peak power for better penetration and higher pulse width and lower gas pressure reduced the hole taper [10].

From the literature review, not much work has been reported on the optimization of laser trepanning parameters on mild steel using Taguchi–response surface methodology (T-RSM) method. Therefore, in the present work, optimization of laser trepanning parameters for optimizing the responses such as hole-top circularity, hole-bottom circularity and hole taper on mild steel is carried out by T-RSM.

2 Materials and Methods

Mild steel is soft, and hence, it can be machined easily. In this work, a mild steel plate thickens 10 mm was used for laser trepanning. The non-uniformity in melting of work material, incomplete ejection melt, improper penetration, etc., affect the laser trepan hole quality characteristics. Hole quality characteristics of laser trepanning depend on hole circularity on top/bottom and hole taper. The uneven surface in hole diameter at top and bottom is known as hole taper. The responses such as hole top circularity (HTC) and hole bottom circularity (HBC) and hole taper (HT) can be calculated by Eqs. (1–3).

$$\text{HTC} = \frac{\text{Minimum hole diameter at top}}{\text{Maximum hole diameter at top}} \tag{1}$$

Table 1 Experimental parameters and levels

Symbol	Parameters	Level 1	Level 2	Level 3
A	Laser power (W)	2500	2750	3000
B	Gas pressure (bar)	5	6	7
C	Speed of spot movement (mm/s)	1000	1100	1200
D	Pulsing frequency (Hz)	1000	1500	2000
E	Trepanning hole diameter (mm)	10	15	20

$$\text{HBC} = \frac{\text{Minimum hole diameter at bottom}}{\text{Maximum hole diameter at bottom}} \tag{2}$$

$$\text{HT} = \left[\tan^{-1} \times \left(\frac{\text{hole diameter at top-hole diameter at bottom}}{2 \times \text{plate thickness}}\right)\right] \times \left(\frac{180}{\pi}\right) \tag{3}$$

Taguchi DOE as a robust statistical design is used for the optimization of process parameters. Laser trepanning on mild steel was performed as per the structured Taguchi's L_{27} orthogonal array with 5 factors and 3 levels as shown in Table 1.

The S/N ratio and ANOVA are applied to study the contribution of input parameters in various processes. Taguchi transforms the objective function rate into S/N ratio of mean standard deviation. It is essential to minimize the responses such as HBC, HTC, HT always. Therefore, "smaller is better" (Eq. 4) for S/N ratio was considered for this experiment. *P*-values of individual input process parameters were obtained by ANOVA using design expert software.

$$\text{S/N} = -10 * \log\left(\sum (Y^2)/n\right) \tag{4}$$

where Y^2 = average observed data, n = number of observations, y = observed data.

Response surface methodology (RSM) is a popular statistical analysis tool for the optimization of independent variable and dependent variables. It analyses the relationships between several explanatory variables and one or more response variables. The 3D surface graph generated by RSM helps in visualizing the parameter effects on the responses. The desirability value range between 0 and 1 indicates the nearness of the outputs towards the target.

3 Result and Discussion

Laser trepanning on mild steel was performed successfully at the defined level of parameters, and the corresponding responses obtained are shown in Table 2.

Table 2 Laser trepanning parameters and responses (L_{27} OA)

S. No.	A (W)	B (bar)	C (mm/s)	D (Hz)	E (mm)	HTC (Min/Max)	HBC (Min/Max)	HT (°)
1	2500	5	1000	1000	10	0.9977	0.9896	1.3
2	2500	5	1000	1000	15	0.9951	0.9845	1.7
3	2500	5	1000	1000	20	0.989	0.9818	1.2
4	2500	6	1100	1500	10	0.9996	0.9916	1.3
5	2500	6	1100	1500	15	0.9925	0.9914	0.2
6	2500	6	1100	1500	20	0.9915	0.9897	0.3
7	2500	7	1200	2000	10	0.9927	0.9906	0.3
8	2500	7	1200	2000	15	0.9926	0.9848	1.3
9	2500	7	1200	2000	20	0.9918	0.9845	1.2
10	2750	5	1100	2000	10	0.995	0.9921	0.5
11	2750	5	1100	2000	15	0.9926	0.9918	0.1
12	2750	5	1100	2000	20	0.9844	0.9783	1
13	2750	6	1200	1000	10	0.9985	0.9931	0.9
14	2750	6	1200	1000	15	0.9959	0.9881	1.3
15	2750	6	1200	1000	20	0.9948	0.9825	2
16	2750	7	1000	1500	10	0.9934	0.9903	0.5
17	2750	7	1000	1500	15	0.9931	0.9863	1.1
18	2750	7	1000	1500	20	0.9903	0.9768	2.2
19	3000	5	1200	1500	10	0.998	0.9931	0.8
20	3000	5	1200	1500	15	0.9942	0.9901	0.7
21	3000	5	1200	1500	20	0.9904	0.9898	0.1
22	3000	6	1000	2000	10	0.9981	0.9921	1
23	3000	6	1000	2000	15	0.9908	0.9898	0.2
24	3000	6	1000	2000	20	0.9894	0.9866	0.5
25	3000	7	1100	1000	10	0.9957	0.9941	0.3
26	3000	7	1100	1000	15	0.9941	0.9876	1.1
27	3000	7	1100	1000	20	0.9885	0.9847	0.6

3.1 *Analysis of Variance (ANOVA)*

The ANOVA for hole top circularity is given in Table 3. The model *F*-value 6.93 implies the model which is significant (S). *P*-values less than 0.050 indicate significant models at 95% of confidence level. In this case, B, D, E, AC, BE are significant model terms. The obtained values for the HTC from ANOVA were S.D = 0.001715, mean = 0.993322, $R^2 = 0.890$, Adj. $R^2 = 0.762$, pred. $R^2 = 0.459$.

The ANOVA for hole bottom circularity is given in Table 4. The model *F*-value 3.354 implies that the model is significant at 95% of confidence level. In this case, E,

Table 3 ANOVA for HTC

Source	SS	DOF	MS	*F*-value	*P*-value	Significant
Model	2.85×10^{-04}	14	2.04×10^{-05}	6.931993	0.0009	S
A	4.24×10^{-06}	1	4.24×10^{-06}	1.441922	0.2530	
B	1.98×10^{-05}	1	1.98×10^{-05}	6.748194	0.0233	S
C	5×10^{-07}	1	5×10^{-07}	0.170023	0.6874	
D	4.29×10^{-05}	1	4.29×10^{-05}	14.57943	0.0024	S
E	1.91×10^{-04}	1	1.91×10^{-04}	64.87228	0.0001	S
AB	6×10^{-06}	1	6×10^{-06}	2.04027	0.1787	
AC	2.09×10^{-05}	1	2.09×10^{-05}	7.109208	0.0206	S
AD	0	0				
AE	2.8×10^{-06}	1	2.8×10^{-06}	0.95326	0.3482	
BC	3.83×10^{-06}	1	3.83×10^{-06}	1.301428	0.2762	
BD	0	0				
BE	2.05×10^{-05}	1	2.05×10^{-05}	6.984808	0.0215	S
CD	0	0				
CE	5.74×10^{-06}	1	5.74×10^{-06}	1.952142	0.1877	
DE	3×10^{-08}	1	3×10^{-08}	0.010201	0.9212	
A^2	3.49×10^{-06}	1	3.49×10^{-06}	1.185381	0.2976	
B^2	0	0				
C^2	0	0				
D^2	0	0				
E^2	1.67×10^{-07}	1	1.67×10^{-07}	0.056674	0.8158	
Residual	3.53×10^{-05}	12	2.94E-06			
Corr. total	0.000321	26				

where *SS* sum of squares, *DOF* degree of freedom, *MS* mean square, *S* significant

BC, A^2 are significant model terms. The obtained values for the HBC from ANOVA were S.D = 0.002996, Mean = 0.987989, $R^2 = 0.796$, Adj. $R^2 = 0.559$, pred. $R^2 = -0.062$.

The ANOVA for hole taper is given in Table 5. The model *F*-value of 55.57 implies that the model is significant and significant model terms are A, B, C, D, E, AB, AE, BE. The corresponding values were S.D = 0.001248, Mean = 0.257444, $R^2 = 0.984809$, Adj. $R^2 = 0.967087$, pred. $R^2 = 0.905739$.

3.2 3D Surface Graphs

To find the effects of interactions of input parameters, 3D surface graphs of HTC, HBC, and HT were shown in Figs. 1a, b, 2a, b and 3a, b, respectively.

Table 4 ANOVA for HBC

Source	SS	DOF	MS	*F*-value	*P*-value	Significant
Model	0.00042139	14	30.09×10^{-05}	3.354396731	0.0211	S
A	1.068×10^{-05}	1	10.68×10^{-05}	1.190646047	0.2966	
B	3.528×10^{-05}	1	35.28×10^{-05}	3.9317609	0.0707	
C	3.0680×10^{-05}	1	30.68×10^{-05}	3.419178252	0.0892	
D	3.0508×10^{-07}	1	30.51×10^{-07}	0.033999408	0.8568	
E	2.8720×10^{-04}	1	2.872×10^{-04}	32.00691368	0.0001	S
AB	1.4727×10^{-05}	1	14.73×10^{-05}	1.641205561	0.2244	
AC	2.8167×10^{-05}	1	28.17×10^{-05}	3.139019237	0.1018	
AD	0	0				
AE	4.8×10^{-07}	1	4.8×10^{-07}	0.053493346	0.8210	
BC	4.8676×10^{-05}	1	4.868×10^{-05}	5.424621489	0.0381	S
BD	0	0				
BE	1.4008×10^{-06}	1	1.401×10^{-06}	0.156115128	0.6997	
CD	0	0				
CE	3.8533×10^{-06}	1	38.53×10^{-07}	0.429432691	0.5246	
DE	4.8×10^{-07}	1	4.8×10^{-07}	0.053493346	0.8210	
A^2	7.1501×10^{-05}	1	71.50×10^{-05}	7.968385739	0.0154	S
B^2	0	0				
C^2	0	0				
D^2	0	0				
E^2	1.0417×10^{-06}	1	1.042×10^{-06}	0.11608799	0.7392	
Residual	1.0768×10^{-04}	12	89.73×10^{-07}			
Corr. total	5.2907×10^{-04}	26				

Table 5 ANOVA for HT

Source	SS	DOF	MS	*F*-value	*p*-value	Significant
Model	1.212×10^{-03}	14	8.6569×10^{-05}	55.5690936	0.0001	S
A	2.692×10^{-04}	1	2.6924×10^{-04}	172.822494	0.0001	S
B	5.69×10^{-05}	1	5.6889×10^{-05}	36.5170877	0.0001	S
C	2.69×10^{-05}	1	2.6889×10^{-05}	17.2600297	0.0013	S
D	8.643×10^{-06}	1	8.6429×10^{-06}	5.54786669	0.0364	S
E	2.222×10^{-05}	1	2.2222×10^{-05}	14.2644874	0.0026	S
AB	2.017×10^{-05}	1	2.0167×10^{-05}	12.9450223	0.0037	S
AC	6.66×10^{-07}	1	6.6667×10^{-07}	0.42793462	0.5253	
AD	0	0				
AE	8.333×10^{-06}	1	8.3333×10^{-06}	5.34918276	0.0393	S
BC	5×10^{-07}	1	5×10^{-07}	0.32095097	0.5815	
BD	0	0				
BE	8.333×10^{-06}	1	8.3333×10^{-06}	5.34918276	0.0393	S
CD	0	0				
CE	3×10^{-06}	1	3×10^{-06}	1.92570579	0.1905	
DE	2.083×10^{-06}	1	2.0833×10^{-06}	1.33729569	0.2700	
A^2	1.93×10^{-06}	1	1.9286×10^{-06}	1.23795373	0.2877	
B^2	0	0				
C^2	0	0				
D^2	0	0				
E^2	6.667×10^{-07}	1	6.6667×10^{-07}	0.42793462	0.5253	
Residual	1.869×10^{-05}	12	1.5579×10^{-06}			
Corr. total	1.231×10^{-03}	26				

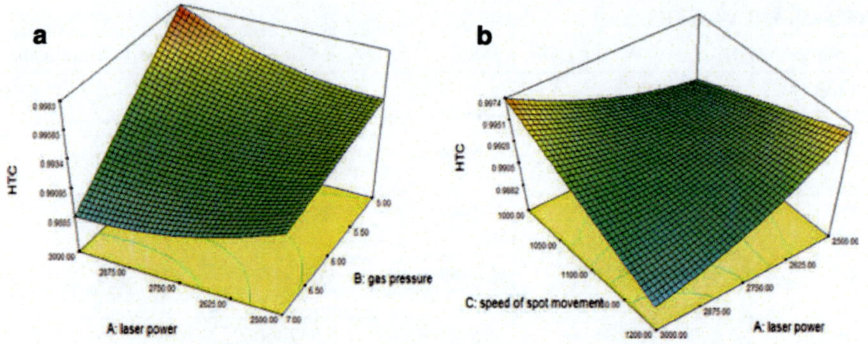

Fig. 1 **a** Interaction between A and B, **b** interaction between A and C

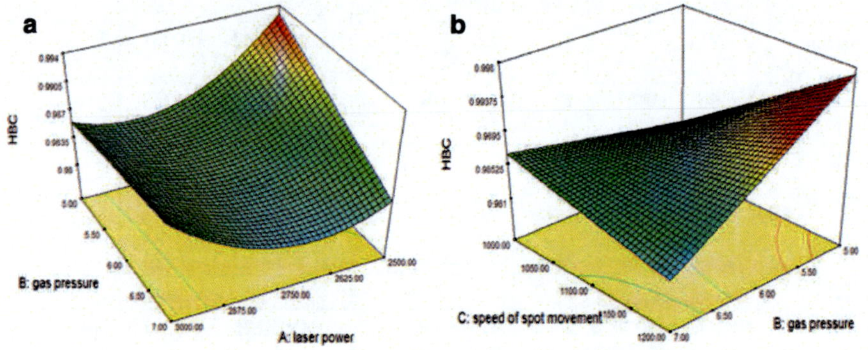

Fig. 2 **a** Interaction between A and B, **b** interaction between B and C

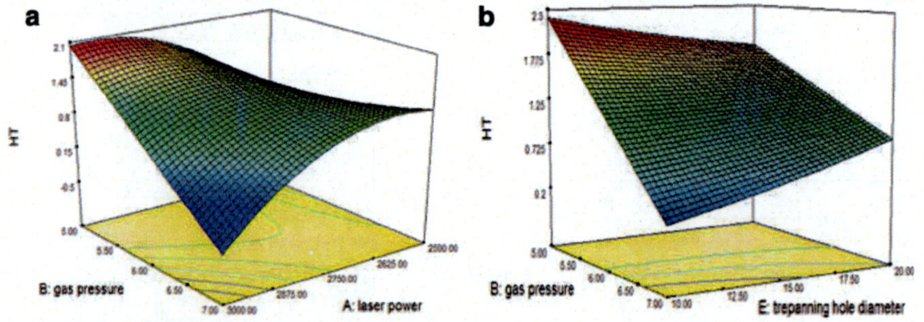

Fig. 3 **a** Interaction between B and A, **b** interaction between B and E

Figure 1a shows laser power (*X*-axis), gas pressure (*Y*-axis), and HTC (*Z*-axis). It inferred from the graph that higher the laser power and higher the gas pressure, minimizes the HTC value (0.9781). When laser power is maximum and gas pressure is minimum, and HTC is maximum. Figure 1b shows laser power, speed of spot movement and HTC are on *X*-, *Y*-, *Z*-axes, respectively. At higher speed on spot movement and higher laser power, a minimal optimal value 0.9781 is achieved. Hence, higher the laser power and higher speed on spot movement and minimized the HTC.

Figure 2a represents laser power, gas pressure and HBC on *X*-, *Y*-, *Z*-axes, respectively. The minimum HBC value 0.9772 was obtained at higher gas pressure and nominal laser power. The maximum HBC values are obtained at the lower level of gas pressure and laser power. Figure 2b represents gas pressure at *X*-axis, speed of spot movement at *Y*-axis and HBC at *Z*-axis. An increase in speed of spot movement and increase in gas pressure resulted in a minimum HBC.

Figure 3a represents *X*-axis laser power, *Y*-axis gas pressure and *Z*-axis hole taper. At higher laser power and higher gas pressure, the hole taper was minimum. Maximum HT obtained at higher laser power and lower gas pressure. Figure 3b represents gas pressure, trepanning hole and taper hole at *X*-, *Y*-, *Z*-axes, respectively. Hole taper was reduced significantly at high gas pressure and low trepanning hole. Lower trepanning value minimizes the hole taper.

3.3 Ramp Function Graph of Desirability

Ramp function graph for desirability is shown in Fig. 4. The desirability value obtained is 0.992. It proves a higher closeness interval between the target data and the responses.

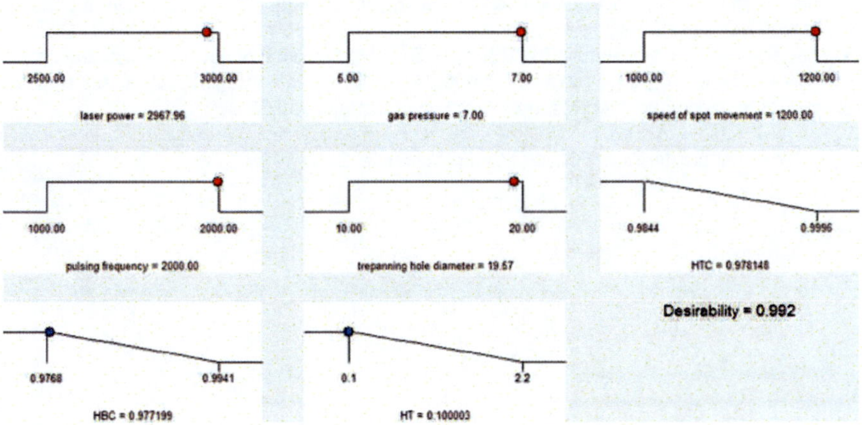

Fig. 4 Ramp function graph of desirability

4 Conclusion

- Laser trepanning on mild steel plate was successfully carried out as per Taguchi's L_{27} orthogonal array, and the parameters were optimized using response surface methodology.
- Optimal combination of parameters obtained through RSM was laser power-2716.83 W, gas pressure-7 bar, speed of spot movement-1200 mm/s, pulsing frequency-2000 Hz and trepanning hole 20 mm.
- The ramp graph indicates a desirability value of 0.992 proving a higher degree of closeness between the target data and the responses.

References

1. Ashkenasi D, Kaszemeikat T, Mueller N, Dietrich R, Eichlera HJ, Illing G (2011) Laser trepanning for industrial applications. Phys Procedia 12:323–331
2. Choudhury IA, Chong WC, Vahid G (2012) Hole qualities in laser trepanning of polymer materials. Opt Lasers Eng 50:1297–1305
3. Shi C, Ren N, Wang H, Xia K, Wang L (2018) Effects of ultrasonic assistance on micro hole drilling based on Nd: YAG laser trepanning. Opt Laser Technol 106:451–460
4. Kumar S, Dubey VK, Pandey AK (2013) Computer-aided genetic algorithm based multi-objective optimization of laser trepan drilling. Int J Precision Eng Manuf 14:1119–1125
5. Chien W-T, Hou S-C (2006) Investigating the recast layer formed during the laser trepan drilling of Inconel 718 using the Taguchi method. Int J Adv Manuf Technol 33:308–316
6. Dubey AK, Yadava V (2008) Robustparameter design and multi-objective optimization of laser cutting for aluminum alloy sheet. Int J Adv Manuf Technol 38:268–277
7. Murthy BRN, Rodrigues LLR, Anjaiah D (2012) Process parameters optimization in GFRP drilling through integration of Taguchi and response surface methodology. Res J Recent Sci 6:7–15
8. Sivaraos, Milkey KR, Samsudin AR, Dubey AK, Kidd P (2014) Comparison between Taguchi method and response surface methodology (RSM) in modelling CO_2 laser machining. Jordan J Mech Ind Eng 8:35–42
9. Adalarasan R, Santhanakumar M, Thileepan S (2016) Selection of optimal machining parameters in pulsed CO_2 laser cutting of Al6061/Al_2O_3 composite using Taguchi-based response surface methodology (T-RSM). Int J Adv Manuf Technol 93:305–317
10. Goyal R, Dubey AK (2014) Quality improvement by parameter optimization in laser trepan drilling of superalloy sheet. Mater Manuf Process 29:1410–1416

Parametric Optimization of Cracked Cantilever Beam Using Genetic Algorithm

Mihir Kumar Sutar, Sarojrani Pattnaik, and Pawan Kumar Modi

Abstract This paper presents a smart crack detection system using genetic algorithm (GA) for a cantilever beam. The accuracy of prediction is optimized by performing parametric optimization of the first three natural frequencies. The three modes of natural frequencies are obtained from the experimental procedures and are optimized to reduce the percentage of error in predicting the relative crack position and crack depth from a fixed position. Comparison of analytical method is in agreement with the experimental values as the error percent between the experimental and predicted values is below 10%.

Keywords Cracked cantilever beam · Genetic algorithm · Natural frequency · Parametric optimization

1 Introduction

Cracks generated in structural and engineering parts lead to their failure, and these cracks influence the vibration characteristics of the machines. The occurrence of crack in a machine accounts for decline in natural frequencies, and crack length rises as the amplitude of vibration increases. This concerned issue has been addressed by many researchers around the globe, and many methodologies have been suggested by them.

Mahfouz and Banerjee [1] introduced cracks of diverse lengths at three different sites along a cantilever steel beam, which was exposed to various amplitudes and frequencies by the usage of an electromagnetic shaker. The crack length and crack propagation data have been linked with statistical moments, frequency spectra, and

M. K. Sutar (✉) · S. Pattnaik
Mechanical Engineering Department, Veer Surendra Sai University of Technology, Burla, Odisha, India
e-mail: mihirsutar05@gmail.com

P. K. Modi
Electrical Engineering Department, Veer Surendra Sai University of Technology, Burla, Odisha, India

S. Vijayan et al. (eds.), *Trends in Manufacturing and Engineering Management*, Lecture Notes in Mechanical Engineering,
https://doi.org/10.1007/978-981-15-4745-4_79

wavelet coefficient data. Khatir et al. [2] used experimentally determined natural frequencies along with the particle swarm optimization (PSO) technique to detect an open crack on a cantilever beam made up of steel. Genetic algorithm (GA) is an optimization method which is bio-inspired, and it supports the ideas of natural genetics and selection theories as proposed by Charles Darwin. It was developed at the University of Michigan by John Holland and his co-workers. It provides robust search in complex spaces [3]. GA is widely used in the field of engineering and manufacturing. Pathan et al. [4] successfully built up a real-coded constrained GA along with finite element (FE) method for optimizing the damping action of a unidirectional composite laminate. Pajand and Tavakoli [5] pioneered a proficient process for detecting cracks in concrete gravity dams using GA and finite element modeling (FEM). It was found that this hybrid method could predict accurately and quickly the displacement values in concrete gravity dam analysis.

Pawar and Ganguli [6] used genetic fuzzy approach to determine the location of damage in cantilever beam and helicopter rotor pads. They used FEM to find out the variation in beam frequencies due to structural damage. A fuzzy model was developed and its rule-base, as well as membership functions, was optimized by GA so that there was an automatic rule generation. The adopted method proved effective in identifying the structural damage in both the beam and pads. Wang and Chen [7] applied GA to choose optimal support location of beams with three different boundary conditions. They considered both elastic and rigid supports, and it was found that GA's performance was better as compared to other optimization tools. Cho [8] used GA to optimize the design of a composite cylindrical shell which is most prone to buckling and failure on application of axial load, torsional load, or both. It was found that the GA optimized shell design possessed better mechanical properties as regards to conventional ones. Lin and Lee [9] applied GA along with regression analysis to optimize the stacking sequence of laminated composite structures. Khatir et al. [10] proposed a methodology using GA and firefly algorithm (FA) to identify the location of damage in simple and continuous beams and two-dimensional frame structure.

The current analysis includes an experimental investigation of obtaining the three modes of natural frequencies, e.g. first natural frequency (*fnf*), second natural frequency (*snf*), and third natural frequency (*tnf*), at different relative crack length (RCL) and relative crack depth (RCD) values. These experimental data are used further to develop a GA controller that can be used to predict the RCL and RCD. Comparison between the experimental and predicted values indicates a minimal percentage of error, and thus, the controller is acceptable. This will be helpful to predict the defects present in beam in the form of crack, without any experimentation, thus minimizing the cost of experiment. The current analysis focuses on the health monitoring of beam structural element for their safe functioning.

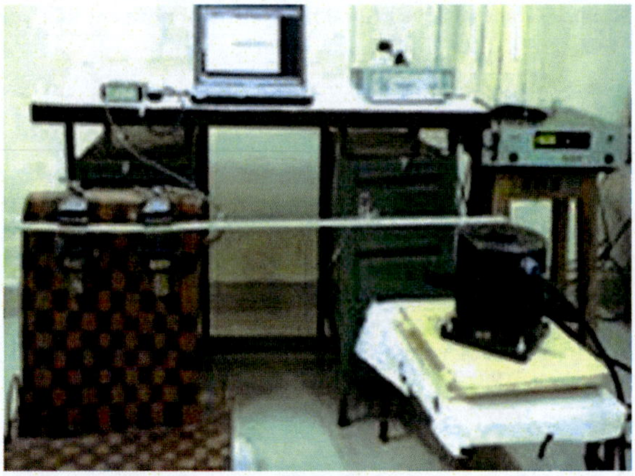

Fig. 1 Experimental setup for modal frequency determination

2 Experimental Procedure

A cantilever beam with single crack was used for the experimentation. The relative crack positions from the fixed end were varied at an interval of every 100 mm, and the three modal frequencies were recorded. The results obtained in the experimentation were further used as the input training data for the GA controller. The GA controller was able to predict the RCL and RCD from the input of the three modal frequencies. Number of experimental tests has been carried out on an 800 mm long aluminium cantilever beam having a cross section of 50 mm width and 6 mm thickness. Complete experimental setup is shown in Fig. 1. The three modes of natural frequencies recorded during the experimentation for different RCL and RCD are enlisted in Table 1.

3 Genetic Algorithm

GA is an optimization technique based on iteration process, and every iteration is composed of stages, namely reproduction, crossover, and mutation. Reproduction is nothing but selection of good parents from initial population who is having the best fitness value so to produce offspring for the subsequent generation. Crossover is done by choosing random mating points, and after that the new generation figured out possess characteristics comparable to that of parents. Mutation is the intermittent arbitrary alteration of the worth of a string position.

In this study, the fitness of all selected output variables is calculated on the basis of fitness function based on multi-variable linear regression model. This model consists of "n" independent input variables (x) whose relationship with the dependent variable (response y) is given by Eq. 1:

Table 1 Three relative natural frequencies at different RCL and RCD from experimentation

Sl. No.	Inputs			Outputs	
	Relative 1st natural frequency "fnf" (*A*)	Relative 2nd natural Frequency "snf" (*B*)	Relative 3rd natural Frequency "tnf" (*C*)	Relative crack length (RCL)	Relative crack depth (RCD)
1	0.9848	0.9958	0.9974	0.205	0.073
2	0.9673	0.9874	0.9942	0.43	0.084
3	0.9624	0.9948	0.9982	0.568	0.158
4	0.9689	0.9921	0.9948	0.4317	0.2774
5	0.9755	0.9976	0.9972	0.391	0.188
6	0.9655	0.9837	0.9961	0.3952	0.279
7	0.9785	0.9993	0.9987	0.387	0.189
8	0.9732	0.9916	0.9963	0.3772	0.1486
9	0.9852	0.9984	0.9967	0.23	0.24
10	0.9859	0.9989	0.9974	0.236	0.248
11	0.9724	0.9961	0.9817	0.545	0.287
12	0.9755	0.9869	0.9936	0.4827	0.2585
13	0.9824	0.9872	0.9917	0.447	0.406
14	0.981	0.9809	0.9932	0.495	0.424
15	0.9842	0.9926	0.9948	0.2345	0.2017
16	0.9834	0.9685	0.9974	0.535	0.534
17	0.9689	0.9921	0.9948	0.4317	0.2774

$$y = a + bx_1 + cx_2 + dx_3 \tag{1}$$

The regression equations for the selected responses, as given by Eqs. (2–3), are further used as fitness function for process optimization using GA.

$$y_1 = 24.158 - 7.8717x_1 - 6.9487x_2 - 9.2338x_3 \tag{2}$$

$$y_2 = 13.615 + 4.596x_1 - 10.371x_2 - 7.6128x_3 \tag{3}$$

where y_1 and y_2 are responses RCL and RCD; x_1, x_2, and x_3 are input parameters *A*, *B*, and *C*, respectively. The flowchart of the GA process is shown in Fig. 2.

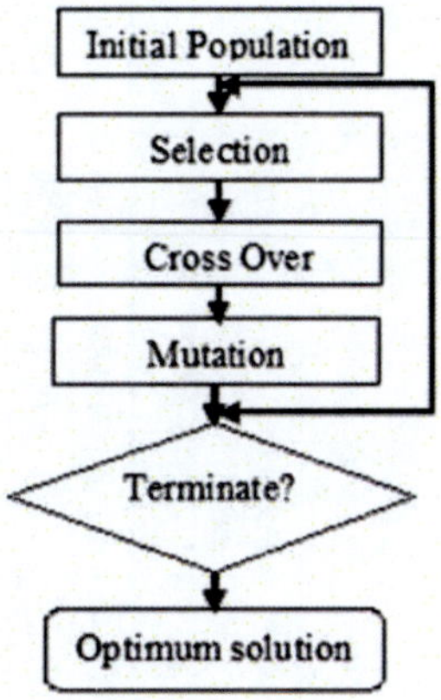

Fig. 2 Genetic algorithm flowchart

4 Confirmatory Experiments

Further confirmatory experiments were conducted at the obtained optimal settings as predicted by GA. Then, the error percentage between the experimental and predicted results was computed on the basis of mean absolute percentage deviation (MAPD), as given by Eq. 4.

$$\text{MAPD} = \frac{100}{n}\sum_{t=1}^{n}\left|\frac{(E_t - P)_t}{E_t}\right| \quad (4)$$

where E_t and P_t are the experimental and predicted values. In Table 2, a comparison has been made between the experimental and the predicted responses of the cracked cantilever beam. It is seen that the percentage of error is found to be less than 10% in all the cases, which shows that GA can be successfully used to forecast the responses of the cracked cantilever beam.

From the confirmatory experimentation, it was observed that the GA predictions are well within a limit of 10%, and the model is able to predict the relative crack length and relative crack depth properly.

5 Conclusions

A genetic algorithm approach for predicting the relative crack position and its depth is presented in the current analysis. The following conclusions are drawn from the analysis:

- An aluminium cantilever beam of 800 mm length, 50 mm width, and 6 mm thickness is taken for the current analysis.

Table 2 Assessment of the experimental and predicted responses of the cracked cantilever beam by GA

S. No.	Relative 1st natural frequency "fnf"	Relative 2nd natural frequency "snf"	Relative 3rd natural frequency "tnf"	Experimental values		Predicted values		% error	
	A	*B*	*C*	RCL	RCD	RCL	RCD	RCL	RCD
1	0.9850	0.9993	0.9987	0.2532	0.184	0.238	0.180	5.64	2.43
2	0.9757	0.9992	0.9986	0.2971	0.142	0.313	0.147	−5.38	−4.01
3	0.9834	0.9991	0.9986	0.2752	0.075	0.254	0.072	7.66	3.06
4	0.9828	0.9992	0.9986	0.247	0.169	0.257	0.164	−4.29	3.00
5	0.9640	0.9991	0.9986	0.3894	0.112	0.405	0.107	−4.18	4.44
6	0.9852	0.9993	0.9987	0.2447	0.115	0.237	0.119	2.86	−2.93
7	0.9697	0.9993	0.9986	0.3716	0.075	0.360	0.072	3.06	3.44
8	0.9735	0.9992	0.9986	0.3512	0.106	0.330	0.112	5.80	−5.54
9	0.9821	0.9993	0.9987	0.2791	0.163	0.262	0.170	6.12	−4.28
10	0.9786	0.9992	0.9987	0.2773	0.135	0.289	0.130	−4.47	3.91
11	0.9678	0.9992	0.9986	0.3867	0.117	0.375	0.115	2.81	1.69
12	0.9724	0.9992	0.9986	0.3276	0.118	0.339	0.125	−3.63	−5.47
13	0.9774	0.9992	0.9986	0.2885	0.181	0.300	0.176	−4.05	2.97
14	0.9707	0.9992	0.9986	0.3411	0.142	0.353	0.137	−3.48	3.57
15	0.9662	0.9992	0.9986	0.3672	0.151	0.388	0.158	−5.74	−4.61
16	0.9750	0.9993	0.9985	0.3408	0.167	0.319	0.173	6.30	−3.82
17	0.9640	0.9991	0.9986	0.3899	0.104	0.405	0.100	−4.05	3.45
Average % of error								5.86	7.82

- Relative crack locations are intentionally provided in the beam with a variation of crack depth to obtain the first three relative natural frequencies.
- The data obtained from the experimental procedure is then used as training data to prepare the GA model to predict the structural health of the cantilever beam by giving input of the three relative natural frequencies.
- Prediction of the GA model is found to be satisfactory within a limit of 10%.
- The average prediction error for RCL is found to be 5.86%.
- The average prediction error for RCD is found to be 7.82%.

References

1. Mahfouz IA, Banerjee A (2017) Crack detection and identification using vibration signals and fuzzy clustering. Procedia Comput Sci 114:266–274
2. Khatir S, Dekemele K, Loccufier M, Khatir T, Wahab MA (2018) Crack identification method in beam-like structures using changes in experimentally measured frequencies and particle swarm optimization. CR Mec 346(2):110–120
3. Griesbach JD, Etter DM Fitness-based exponential probabilities for genetic algorithms applied to adaptive IIR filtering, IEEE transactions. In: Thirty-second Asilomar conference on signals, systems and computers, 1–4 Nov, 1998
4. Pathan MV, Patsias S, Tagariellia VL (2018) A real-coded genetic algorithm for optimizing the damping response of composite laminates. Comput Struct 198:51–60
5. Pajand MR, Tavakoli FH (2015) Crack detection in concrete gravity dams using a genetic algorithm. Proc Inst Civ Eng—Struct Build 168(3):192–209
6. Pawar PM, Ganguli R (2003) Genetic fuzzy system for damage detection in beams and helicopter rotor blades. Comput Methods Appl Mech Eng 192:2031–2057
7. Wang BP, Chen JL (1996) Application of genetic algorithm for the support location optimization of beams. Comput Struct 58(4):797–800
8. Cho HK (2018) Optimization of laminated composite cylindrical shells to maximize resistance to buckling and failure when subjected to axial and torsional loads. Int J Precis Eng Manuf 19(1):85–95
9. Lin CC, Lee YJ (2004) Stacking sequence optimization of laminated composite structures using genetic algorithm with local improvement. Compos Struct 63(1):339–345
10. Khatir A, Tehami M, Khatir S, Abdel WM (2018) Multiple damage detection and localization in beam-like and complex structures using co-ordinate modal assurance criterion combined with firefly and genetic algorithms. J Vibro Eng 20(1):832–842

Study of the Influence of Reinforcement Parameters on Thermal Conductivity of Magnesium-Based MMCs Through Taguchi's Orthogonal Array Approach

M. R. Shivakumar, N. V. R. Naidu, M. Jai Surya, and D. Indhuja

Abstract Light metal alloy-based composites (MMCs) find applications in automotive and aerospace sectors, mainly due to their high strength-to-weight ratio. Such composites exhibit enhanced mechanical properties too. However, these MMCs inherently have inadequate thermal properties. In the present investigation, the thermal conductivity of pure magnesium/soda–lime glass composite was analyzed through Taguchi's orthogonal array approach, and reinforcement parameters were optimized to realize maximum possible thermal conductivity. Analysis of variance was applied to find out the contribution of individual reinforcement parameters to the thermal conductivity of the composite.

Keywords Magnesium composites · Thermal conductivity · Taguchi's orthogonal array

1 Introduction

Magnesium and its alloys have found certain specific uses in many industries. The density of pure magnesium (1.70 g/cm^3) is just 70% of the density of pure aluminum, and this aspect enhances its strength-to-weight ratio. The immediate need for fuel economy and cost-savings in aerospace and automobile sectors has mandated increasing uses of magnesium components [1]. Although the mechanical properties (such as UTS and yield strength) of magnesium alloys are somewhat near to those of aluminum alloys, their thermal properties (specific heat and thermal conductivity) are much lower than most other non-ferrous and ferrous alloys. The mechanical properties of magnesium alloys can be enhanced by incorporating ceramic reinforcements (to result in MMCs) [2, 3]. But, such reinforcements are likely to lead to adverse

M. R. Shivakumar (✉) · N. V. R. Naidu · M. J. Surya · D. Indhuja
Industrial Engineering and Management, M S Ramaiah Institute of Technology, Bangalore, Karnataka 560054, India
e-mail: mrshivakumar@msrit.edu

S. Vijayan et al. (eds.), *Trends in Manufacturing and Engineering Management*, Lecture Notes in Mechanical Engineering,
https://doi.org/10.1007/978-981-15-4745-4_80

effects on the thermal properties. There is an immediate need to analyze such negative effects in order to make the Mg alloy MMCs more appropriate for specific end-users.

The following are the different processing routes that are suitable for processing Mg alloy-based composites [3]:

1. Stir casting,
2. Squeeze casting, and
3. Powder metallurgy methods.

Of the above, the stir casting method appears to be very popular due to its simplicity, economy of operation, and applicability even to small casting units [4, 5].

The effect of reinforcement on the properties of composites can be analyzed by varying the reinforcement parameter levels. Optimum parameter levels can be determined by adopting statistical methods. Design of experiments (DOE) is a method for scientifically applying statistics in conducting experimental runs. The analysis involving experimental trials coupled with a methodically planned design procedure is helpful to lessen the research effort, financial resources, and facilities. Taguchi simplified the DOE approach with orthogonal array (OA) that significantly lowers the number of experimental trials needed. Taguchi's approach to DOE is easy to adopt, apply, and analyze. Taguchi's signal-to-noise ratio (S/N ratio) is invariably used to analyze and optimize the responses [6]. Analysis of variance (ANOVA) is a tool used to identify the significance of input parameters and to determine the effect of each input parameter on the properties of the composites [7].

2 Material Selection and Processing

Magnesium and its alloys exhibit good castability, machinability, formability, and weldability. In view of this and in the light of the ever-increasing applications of magnesium components in various industries, an experimental investigation on Mg-based MMCs was planned, with magnesium as the base material (matrix material) and soda–lime glass powder as the reinforcement (soda–lime glass powder is an easily available, inexpensive ceramic material) [8].

Magnesium ingots (of German origin) of purity more than 99.50% were procured from a standard metal supplier. The certified composition (Table 1) is provided by the manufacturer along with the supply.

The chemical composition of soda–lime glass is given in Table 2. It is noted that the composition is cited in the literature (8).

Table 1 Chemical composition of magnesium

Elements	Mg	Fe	Si	Mn	Cu	Ni
Percentage by weight	99.92	0.025	0.017	0.026	0.0027	0.0006

Table 2 Composition of soda–lime glass

Elements	SiO_2	Na_2O	CaO
Percentage by weight	70–73	12–16	10–15

Table 3 Levels of the reinforcement parameters

Parameter	Level		
	1	2	3
A: weight of glass particles added (%)	1.5	3.0	4.5
B: glass particle size (μm)	75	125	210
C: temperature of glass particles (°C)	260	380	500

Magnesium ingots were charged into a graphite crucible covered with Magrex-60 flux and melted in an electric resistance furnace. When the melt reached 700 °C, which is well above the melting temperature of magnesium, dross was skimmed off from the surface of the melt after treating with degassing tablet hexachloroethene. The melt was stirred (using the custom made mechanical stirrer) at a fixed stirring speed of 400 RPM to form a sufficiently large vortex. Preheated soda–lime glass powder (varying weight percentages, varying particle sizes and varying preheat temperatures) was added into the vortex. After adding the particles, the stirring impeller was driven for some more time till the temperature of the melt dropped to 640 °C. Soon after this, the melt mixture was poured into the preheated die.

The three levels considered for each reinforcement parameter in the experiments are shown in Table 3.

For the purpose of experimentation, based on the degrees of freedom for the factor levels, the minimum number of trials needed (as per total degrees of freedom) is 9. Taguchi's L_9 OA was selected suitably. As per the design layout of experimentation, nine castings were produced.

3 Test Results and Signal-to-Noise Ratio Analysis

The principle of comparative cut bar method (ASTM E1225 test method) was adopted in the thermal conductivity measurement; this is perhaps the most widely used method for axial thermal conductivity assessment. Cylindrical test specimen of size 8 mm diameter and 50 mm length was sandwiched between two reference specimens as shown in Fig. 1. As the test set-up is heated to predetermined temperatures, the temperatures at different marked spots on the test specimen were measured accurately using the temperature sensors located between the reference specimens and the test specimen. These were then used to compute the thermal conductivity K of the specimen under test.

If K_R is the thermal conductivity of the reference specimens, the thermal conductivity of the unknown sample (K_S) can be derived from the following equation:

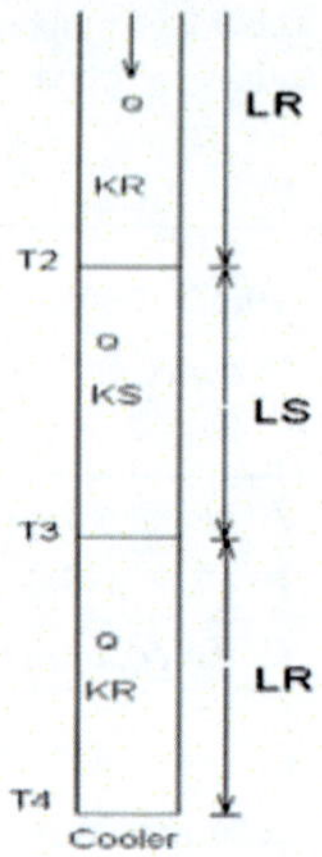

Fig. 1 Test specimen location (in between the reference specimens)

$$K_s = K_R \times \frac{L_s}{L_R} \times \left[\frac{(T_1 - T_2) + (T_3 + T_4)}{2(T_2 + T_3)}\right] \quad (1)$$

L_s	Length of test specimen
L_R	Length of reference specimens
T_1, T_2, T_3, and T_4	Temperatures as shown in Fig. 1.

The experimental observations were further examined through S/N ratio. These S/N ratios, which are log functions of desired output, serve as objective functions for optimization and help in data analysis and prediction of optimum results. The S/N ratio provides a measure of the impact of noise factors on performance. There are different types of S/N ratios, viz. smaller-the-better, nominal-the-best and larger-the-better. Larger the S/N ratio, the more robust the product is against noise. As the objective of this investigation was to maximize the thermal conductivity, "larger-the-better" concept was adopted in this work.

For "larger-the-better" characteristics,

$$\text{S/N ratio} = -10 \log 10(1/n) \sum (1/yi2) \quad (2)$$

where "*n*" is the number of observations and "*y*" is the observed data.

Thermal conductivity test results and S/N ratios with design layout of experimentation are shown in Table 4. The S/N ratio response values and the range for each parameter are shown in Table 5.

The plot of the main effect for S/N ratios is shown in Fig. 2.

This plot indicates that among all the factors, weight % (Delta = 1.63) is the most significant factor affecting the thermal conductivity. The optimal parameter levels were identified from the main effect plot. The optimum parameter levels are:

Table 4 Experimental results and S/N ratios

Test run	*A*	*B*	*C*	Observed values of thermal conductivity (W/m K)			Mean thermal conductivity (W/m K)	S/N ratio (dB)
1	1.5	75	260	143	150	151	148	43.3975
2	1.5	125	380	145	148	145	146	43.2858
3	1.5	210	500	146	148	144	146	43.2854
4	3.0	75	380	144	135	134	138	42.7631
5	3.0	125	500	137	135	139	137	42.7326
6	3.0	210	260	134	134	137	135	42.6053
7	4.5	75	500	128	120	118	122	41.7113
8	4.5	125	260	121	120	125	122	41.7232
9	4.5	210	380	122	118	123	121	41.6515

Table 5 Main effects (thermal conductivity) for S/N ratio

Level	A	B	C
1	43.32	42.62	42.58
2	42.70	42.58	42.57
3	41.70	42.51	42.58
Delta	1.63	0.11	0.01
Rank	1	2	3

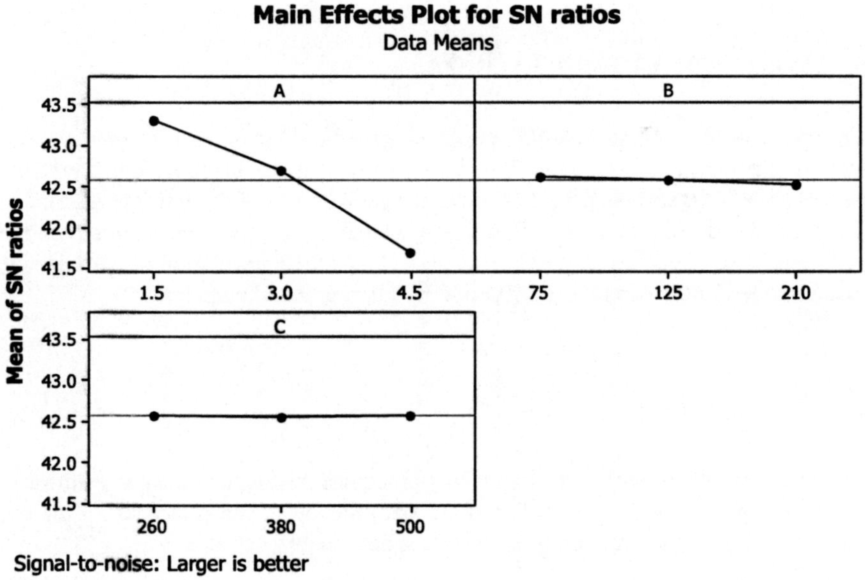

Fig. 2 Plot of main effects (data mean) for S/N ratios

Table 6 Results of confirmation experiment

Factor	*A*	*B*	*C*	Thermal conductivity (W/m K)			Mean thermal conductivity (W/m K)
Value	1.5	75	260	148	150	151	150

Table 7 ANOVA for test results

Source	DF	Seq SS	Adj SS	Adj MS	*F*	% of contribution
A	2	950.00	950.00	475.00	475.00	99.2
B	2	6.00	6.00	3.00	3.00	0.60
C	2	0.00	0.00	0.00	0.00	0.00
Error	2	2.00	2.00	1.00		0.20
Total	8	958.00				

Weight %—1.5%
Particle size—75 μm
Preheat temperature—260 °C.

Confirmation tests were conducted corresponding to the optimum parameter levels, and the results are shown in Table 6. From the results, it is seen that mean thermal conductivity was maximum at the optimum parameter levels.

4 Analysis of Variance (ANOVA)

The objective of applying ANOVA is to investigate the effect of reinforcement parameters on the thermal conductivity. Test results were analyzed using ANOVA, with the help of MINITAB software. Table 7 shows the contribution of the control parameters. The results showed that the contribution of weight % of glass reinforcement on the properties of composites is a high, 99.2%, whereas the contribution of particle size is negligible (0.60%) and no contribution by glass preheat temperature.

5 Conclusions

Consistent production of magnesium/soda–lime glass powder composites (with varying levels of weight per cent and particle size) has been established using the stir casting route, and the following observations are made:

- Thermal conductivity of these Mg alloy composites is affected by the weight percentage as well as the particle size of the reinforcement.

- Taguchi's orthogonal array approach has substantially reduced the number of experiments to be performed to determine the thermal conductivity of the composites with the entire range of process variables.
- The optimum combination that results in the maximum thermal conductivity is 1.5% weight percentage of the reinforcement (soda–lime glass powder) of particle size 75 μm.
- Preheating the soda–lime glass powder (prior to adding to the liquid melt) does not have any effect on the thermal conductivity of the composites.
- ANOVA tests (confirmatory tests) reveal that the weight percentage of soda–lime glass powder alone contributes 99.2% to the variation of thermal conductivity of the composites.

Acknowledgements The authors wish to thank Prof. S. Seshan, Indian Institute of Science Bangalore, for his involvement in this work. Financial support received from Visvesvaraya Technological University, Belagavi, is gratefully acknowledged.

References

1. Watarai H (2006) Trend of research and development for magnesium alloy. Sci Technol Trends Q Rev 8
2. Luo A (1995) Processing microstructure, and mechanical behavior of cast magnesium metal matrix composites. Metall Mater Trans A 26A:2445–2455
3. Koti V, George R, Shakiba A, Shivananda Murthy KV (2018) Mechanical properties of copper nanotubes reinforced with uncoated and nickel coated carbon nanotubes. FME Trans 46:623–630
4. Shivakumar MR, Naidu NVR (2019) Characterization of mechanical and thermal properties in soda–lime glass particulate reinforced LM6 alloy composites ICEMMM 2018. In: Advances in materials and metallurgy. Springer, pp 149–157
5. Madhoo G, Shilpa M (2017) Optimization of process parameters of stir casting technique using orthogonal arrays. Int J Adv Res Methodol Eng Technol 1(2)
6. Jayalakshmi, S. Kailas V, Seshan S, Fleury E (2006) Tensile strength and fracture toughness of two magnesium metal matrix composites. J Ceram Process Res 7(3)
7. Taguchi G, Chowdhury S, Wu Y (2003) Taguchi's quality engineering handbook. Wiley, Hoboken, NJ
8. Hasanuzzaman M, Rafferty A, Sajjia M, Olabi AG (2016) Reference module in materials science and materials. Published by Elsevier

Finite Element Modelling and Optimisation of Sheet Hydroforming for Cryo-rolled AA5083 Sheets

Akhil B. Raj, A. Arun, and Ajith Ramesh

Abstract Hydroforming is a manufacturing process that is used to form complex geometries by applying fluid pressure. The punchless hydroforming process is the most popular in the race, considering the absence of any punch that helps in reducing the tooling costs. Based on the part geometries formed, the punchless process can be classified into three categories, namely sheet hydroforming, shell hydroforming and tube hydroforming. Sheet metal hydroforming is a hydroforming process that uses hydrostatic fluid pressure for deforming the blank into a die cavity of the desired shape. Owing to the inert advantages, like lower tooling costs, reduced processing steps, remarkable precision, waste reduction and weight reduction, this process finds extensive use in applications requiring a high strength-to-weight ratio, as in complex automobile parts. The presented work involves the development of a nonlinear 2D finite element model for the sheet hydroforming process of AA5182 (aluminium alloy) using the FE package Abaqus/Explicit and validation of the numerical results using the available literature (Daryl in Logan: a first course in the finite element method. Cengage Learning Products, Canada 2007 [2]). The model is first validated by reproducing the research work carried out by Bharatkumar Modi et al., considering the input parameters, like blank holding force (BHF), sheet thickness and internal pressure, and the corresponding output parameters, like material thickness reduction and die corner radius. The research is further extended by replacing the current blank material AA5182 using cryo-rolled AA5083. The effect of varying BHF, at varying annealing temperature of blank considering varying frictional coefficients, is also studied. This work also investigates the optimisation of the process using a well-established design of experiments (DOE) technique—response surface methodology (RSM).

Keywords Sheet hydroforming · Blank holding force (BHF) · Die corner radius · Finite element modelling · RSM

A. B. Raj · A. Arun (✉) · A. Ramesh
Department of Mechanical Engineering, Amrita School of Engineering, Coimbatore, Amrita Vishwa Vidyapeetham, India
e-mail: a_arun@cb.amrita.edu

S. Vijayan et al. (eds.), *Trends in Manufacturing and Engineering Management*, Lecture Notes in Mechanical Engineering,
https://doi.org/10.1007/978-981-15-4745-4_81

1 Introduction

Forming is one of the most critical manufacturing processes where the material is deformed and takes the desired shape with improved material strength. Hence, a lot of research is carried out in this field to improve the process. Deep drawing is one such forming processes, which uses a die and a punch for forming. The problem with this kind of conventional forming process is that it deteriorates the surface quality of the material. Also, the material thickness is non-uniform throughout the formed surface. To improve this, the punch used in conventional processes is replaced with hydrostatic fluid pressure. This kind of process is generally known as hydroforming processes.

Hydroforming can be broadly classified into three: i. shell hydroforming processes, ii. tube hydroforming process and iii. sheet hydroforming process [6]. The current research work focuses on the effect of multiple parameters on a punchless sheet hydroforming process.

Figure 1 shows a punchless sheet hydroforming set-up. As shown in zone A (portion under the blank holder) due to tangential compression and radial tensions, the sheet deforms in deep drawing mode [8]. In zone B, which indicates the die corner radius, the state of stress is three-dimensional. This is because of the superimposing of deep drawing and stretch bending components. The central portion of the blank, i.e. the pole region (zone D), is subjected to uniform hydrostatic pressure due to which it deforms in stretching mode, which leads to thinning of the blank. Initially, the region between the die corner radius and the central portion (zone C) is slightly deformed. During the corner formation, it undergoes additional stretching once the draw-in is completed.

Blank holding force (BHF) is one of the most critical parameters in the sheet hydroforming process. A varying BHF can yield better formability in terms of minimum corner radius and lower thinning compared to a constant BHF [8]. Many other parameters are also studied to improve formability. Varying blank annealing temperatures of around 300 °C have given better results for modern lightweight alloys [7, 9, 13]. The annealing temperature is highly dependent on the blank material used for

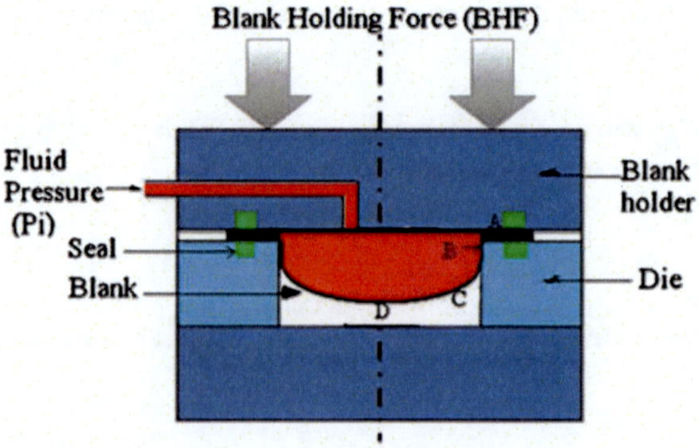

Fig. 1 Diagram showing hydroforming of a cup-shaped part [8]

the process. Other vital parameters that have significant effects on formability are friction and lubrication between the die and the blank surface. Teflon-coated die has found to be producing less material thinning during the sheet hydroforming process [3].

In recent years, there has been a tremendous demand for the need for lightweight and high-strength alloys in the automotive and aerospace industries. Aluminium has been found as one of the best solutions for this. However, aluminium's low strength and formability when compared to carbon steels have restricted its usage. Hence, cryo-rolled Al alloy sheets that display relatively higher strength due to its ultra-fine grain structure have occupied the space [4]. Annealing is found to enhance the formability of cryo-rolled sheets. Cryo-rolling followed by hydroforming has a very high potential to produce parts having complex shapes with lighter weight and high strength [9, 11].

In the present work, the experimental results obtained in the literature by Modi and Ravi Kumar [8] are first replicated using finite element modelling (FEM). The research is further extended by replacing the current blank material AA5182 using cryo-rolled AA5083. The effect of varying BHF, at varying annealing temperatures along with a variable frictional coefficient, is also studied. This work also investigates the optimisation of the process using a well-established design of experiments (DOE) technique—response surface methodology (RSM) [14, 15].

2 Methodology

A nonlinear 2D FE model for sheet hydroforming process is developed using Abaqus/Explicit package [1]. The effects of constant and varying BHF are studied. The initial model is a replica of experimental work undergone by Modi and Ravi Kumar [8].

FE model is a powerful tool for analysing the hydroforming process and for saving cycle time in designing and manufacturing. FEM is one of the powerful tools [2, 5]. The optimisation technology will secure an increased interest in practical application. Table 1 shows the material property of the blank AA5182.

FE model is a powerful tool for analysing manufacturing processes and for saving product design cycle time. The optimisation technology will secure an increased interest in practical applications due to its effectiveness. Table 1 shows the material property of the blank AA5182.

Table 1 Material properties of blank material AA5182

Young's modulus (MPa)	Ultimate tensile strength (MPa)	Strain hardening index, *n*	Strength coefficient, *K* (MPa)
152.5	281.2	0.35	594.8

The arbitrary Lagrangian–Eulerian (ALE) scheme is used to model material flow. The mesh follows a Lagrangian approach at the region of relatively lower deformation and solves for displacements. However, in the area of the domain that corresponds to extensive material flow, the mesh adapts into a Eulerian approach and addresses for velocities [10, 12].

A surface-to-surface interaction is defined between the die and the bottom portion of the blank. The die is set as the master and the blank as the slave. The friction for die–blank interface is taken as semifluid film lubrication, and coefficient of friction used is 0.3 [8]. In the model, the die is fixed in all degree of freedom and the blank is allowed to flow in all the three directions.

In problems involving nonlinear geometry, the quality and the level of refinement of mesh, especially at complicated contours, have a significant role to play in getting accurate results. A four-node bilinear plane stress quadrilateral with reduced integration and hourglass control CPS4R is used for the mesh around 18,000 elements used for the discretisation. It is a planar element with only one integration point. However, discrete rigid elements: R2D2, are used to discretise the die geometry. It is a two-dimensional linear rigid element.

Figure 2 shows the assembly model of blank and die that displays the applications of fluid pressure and BHF. Figure 3 shows the variation of applied fluid pressure concerning time. The fluid pressure is initially applied at a gradual rate, and once the time reaches 0.125, there is a steady increase in the applied pressure. The BHF, on the other hand, is used in two different ways, i.e. constant BHF, as shown in Fig. 4, and varying BHF, as shown in Fig. 5.

The obtained results are validated with the numerical results available in the literature [8]. The research work is further extended by replacing AA5182 with cryo-rolled AA5083 in the validated model. Table 2 displays the material properties of cryo-rolled AA5083 material.

The variation of input parameters like varying BHF and pressure is plotted in Figs. 6 and 7. The effect of other parameters, like friction coefficient and blank annealing temperature, is studied. Friction coefficient ranging from 0.04 to 0.25 is considered for the simulation [3]. Also, the annealing temperature ranging from

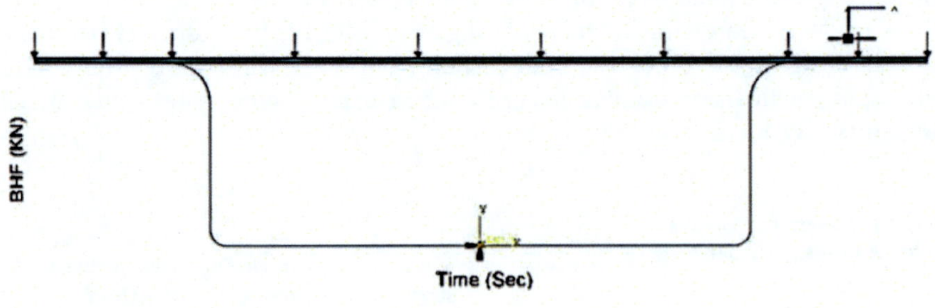

Fig. 2 Meshed 2D assembly model

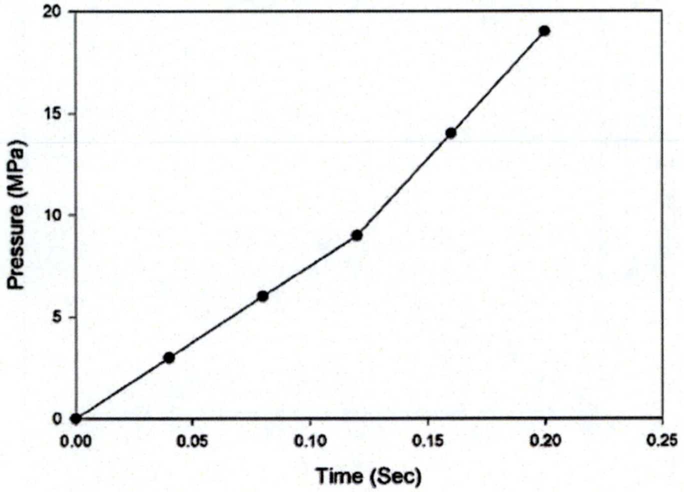

Fig. 3 Plot for fluid pressure versus time [8]

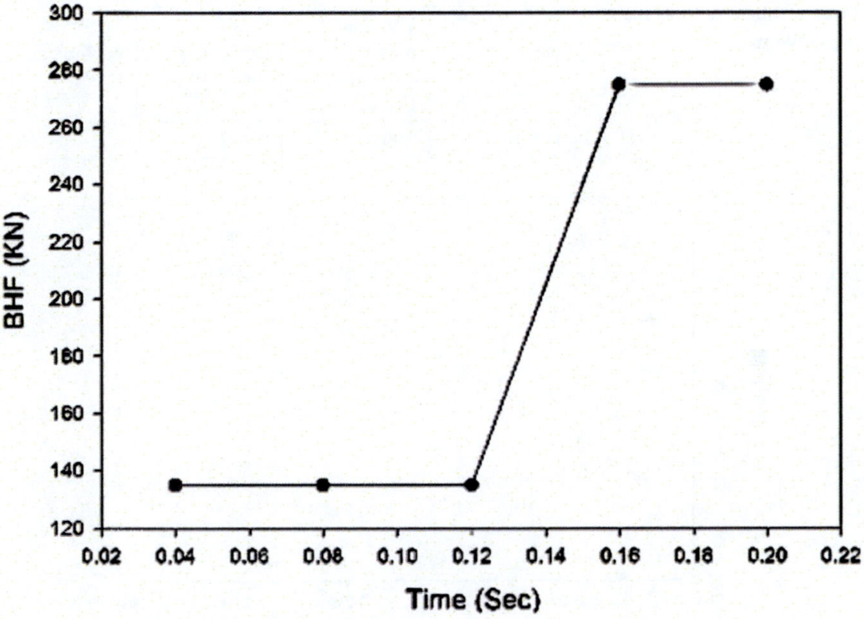

Fig. 4 Constant BHF versus time—AA5182 [8]

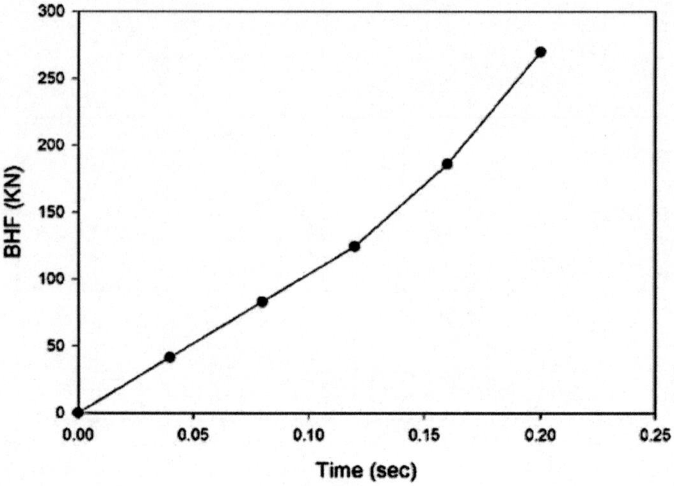

Fig. 5 Variable BHF versus time—AA5182 [8]

Table 2 Material properties of AA5083

Young's modulus	Ultimate tensile strength (MPa)	Strain hardening index (n)	Strength coefficient, K (MPa)
186.1	309.8	0.243	533

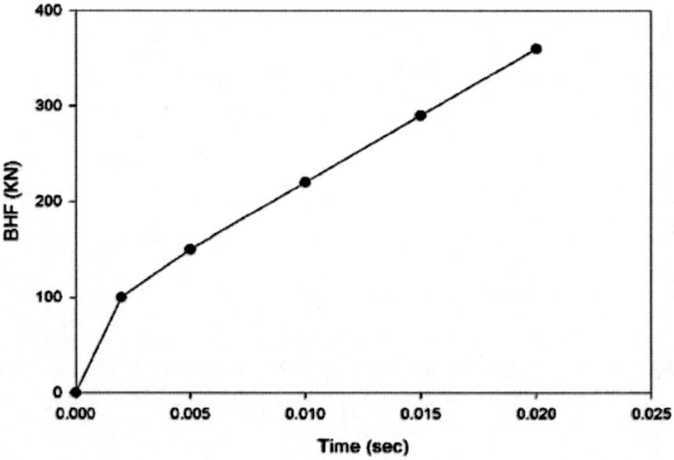

Fig. 6 Varying BHF versus time—cryo-rolled AA5083 [3]

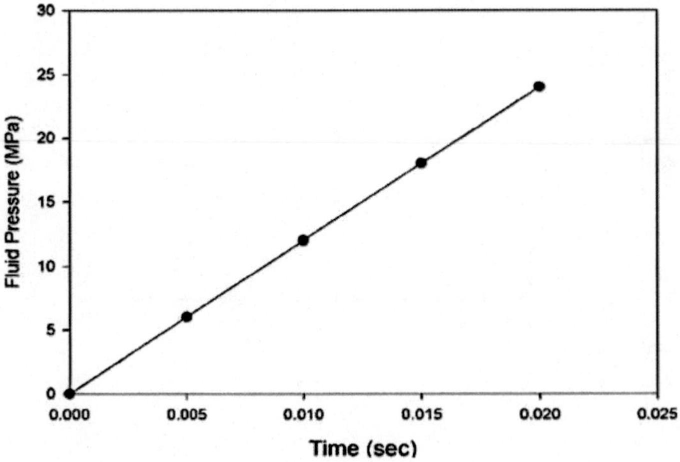

Fig. 7 Fluid pressure versus time—cryo-rolled AA5083 [3]

150 to 300 °C is considered, and their effects on output parameters like thickness reduction and corner radius are investigated [13].

3 Results and Discussion

The first part of this work is focused on validating the developed model. The influence of input process parameters like fluid pressure, BHF, friction at the bottom portion of blank and die, etc., on output parameters like corner radius and material thickness is investigated. Table 3 shows the comparison of output parameters obtained under two conditions, i.e. constant BHF and varying BHF. Comparing experimental (from journal paper) and simulation results, the percentage of difference in the case of constant BHF for maximum thickness reduction is 3.7 and for variable BHF is 5.95. For a minimum corner radius, the difference is 4.24 and 0.6 for constant BHF and varying BHF, respectively.

Table 3 Comparison of experimental (from the literature) and FEM results for sheet hydroforming for AA5182

Pressure BHF load path in hydroforming	Max. thickness reduction (mm)		Min. corner radius (mm)	
	Experimental [8]	FE simulation	Experimental [8]	FE simulation
Constant	0.8	0.77	37.7	36.1
Variable	0.84	0.79	33.4	33.2

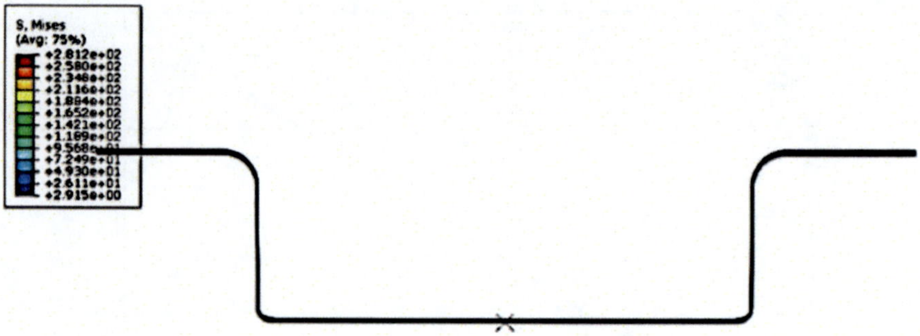

Fig. 8 Formed diagram

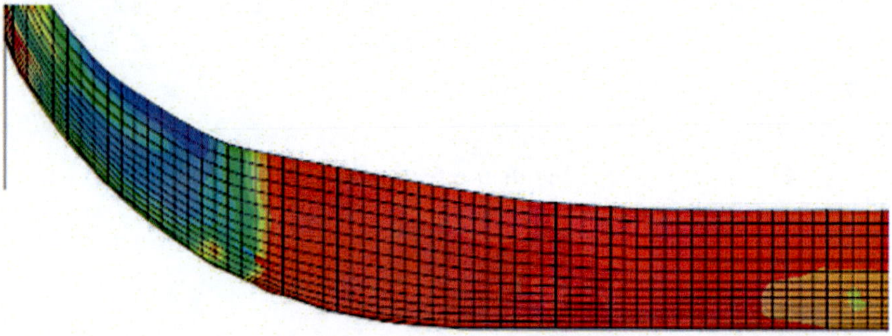

Fig. 9 Von Mises stress measured immediately after deformation in case of varying BHF

Figure 8 shows the forming diagram. Von Mises stress distribution in the material after the hydroforming process is analysed, and it is found that higher stress concentration is at the bottom corners and along with the diagonally opposite faces of the blank. The instantaneous stress distribution for the blank when the fluid pressure and BHF (variable and constant BHF) are applied is shown in Fig. 9. Spring-back behaviour of the deformed blank is analysed, and percentage relaxation in case of hydroforming using constant BHF technique is 38.3% and that in case of variable BHF technique is 32.52%. Thus, the spring-back effect is less in case of hydroforming using variable BHF technique. Hence, better results are achieved with variable BHF technique when compared to the constant blank holding force (BHF) technique. The deformed shape obtained from the model is almost in agreement with the journal paper.

Further, the work is extended by replacing the blank material with cryo-rolled AA5083. Simulations are carried out with varying coefficient friction and varying blank annealing temperature. Optimisation of output parameters is carried out using response surface methodology (RSM), and ANOVA table is generated. Figure 10 shows the main effect plot for maximum thickness reduction, and Fig. 11 shows the main effect plot for a minimum corner radius. In both cases, the coefficient

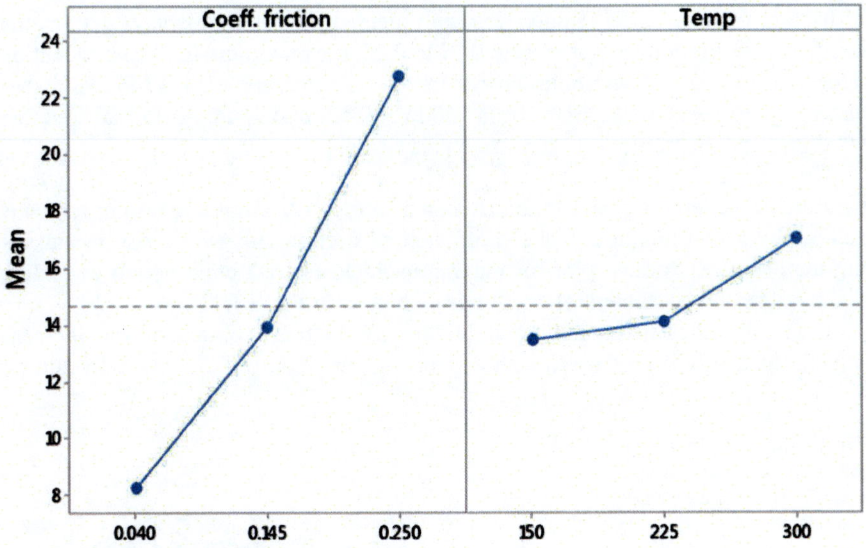

Fig. 10 Main effect plot for maximum thickness reduction

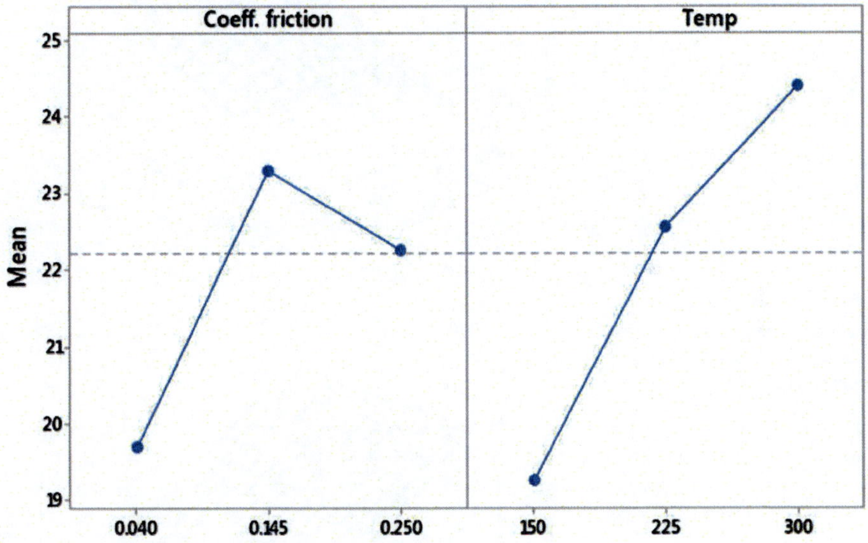

Fig. 11 Main effect plot for minimum corner radius

of friction and annealing temperature are reflected. The trend displays that, as the coefficient of friction increases from 0.04 to 0.25, the maximum thickness reduction is decreasing and the minimum corner radius is increasing until 0.145. Similarly, annealing temperature increases from 150 to 300 °C, maximum thickness reduction is falling, and the minimum corner radius is rising.

Three-dimensional surface plot for minimum corner radius versus annealing temperature and coefficient of friction is shown in Fig. 12. Minimum corner radius is achieved with an increase in the coefficient of friction and with a rise in annealing temperature. Surface plot for maximum thickness reduction versus annealing temperature and coefficient of friction is displayed in Fig. 13.

The trend indicates that maximum thickness reduction is obtained by an increase in both the coefficient of friction and annealing temperature. The optimal value for

Fig. 12 Surface plot for minimum corner radius vs annealing temperature and coefficient of friction

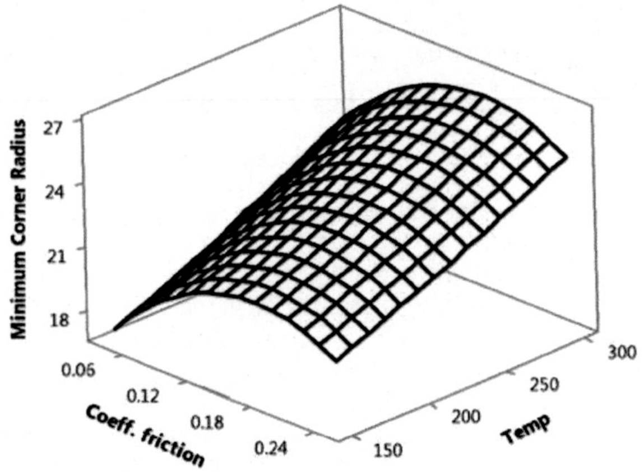

Fig. 13 Surface plot for maximum radius vs annealing temperature and coefficient of friction

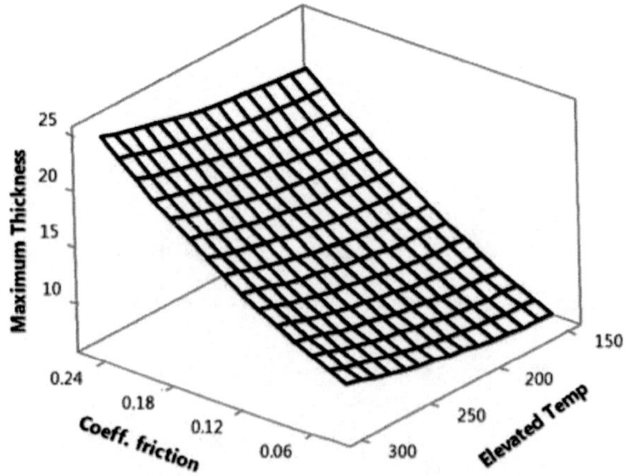

Table 4 Optimised results vs simulation results

Maximum thickness reduction (mm)			Minimum corner radius (mm)		
Optimised result	Simulation result	Percentage of difference	Optimised result	Simulation result	Percentage of difference
21.15	20.4	3.54	20.59	21.3	3.33

the coefficient of friction is 0.25 and for blank annealing temperature is 155 °C. Simulations are carried out using the optimum value, which is illustrated in Table 4. The percentage of difference is around 3.5% only.

4 Conclusion

A detailed finite element model for the punchless hydroforming process is developed and is validated with the available literature. The work is further extended by changing the blank material from AA5182 to cryo-rolled AA5083 considering the extent of applications. Cryo-rolling followed by hydroforming has better potential to provide complex sheet metal parts using lightweight, high-strength aluminium alloy sheets. The effect of input parameters like the coefficient of friction and annealing temperatures for blank is studied, with varying BHF, on output parameters like minimum corner radius and maximum thickness reduction. It has been observed that minimum corner radius and lower thinning are achieved in the case of variable BHF technique than in the case of constant BHF. It indicates that the variable BHF technique is superior in forming complex sheet metal parts in sheet hydroforming. The optimisation studies show an optimum value of 0.25 for the coefficient of friction and annealing temperature; it is 155 °C.

References

1. ABAQUS (2016) ABAQUS documentation. Dassault Systèmes, Providence, RI, USA
2. Daryl L (2007) Logan: a first course in the finite element method. Cengage Learning Products, Canada
3. Feyissa FT, Ravi Kumar D (2018) Enhancement of drawability of cryorolled AA5083 alloys sheets by hydroforming, J Mater Sci Technol 2238–7854 2018. Brazilian Metallurgical, Materials and Mining Association, pp 7–13, Elsevier Editora Ltd, pp 7–13 (2018)
4. Feyissa F, Ravi Kumar D (2018) Characterization of microstructure, mechanical properties and formability of cryorolled AA5083 alloy sheets. ASM Int, JMEPEG 27:1614–1627
5. Huges TJR (2000) The finite element method. Dover Publications Inc., Mineola, New York
6. Lang LH, Wang ZR, Kang DC, Yuan SJ, Zhang SH, Danckert J, Nielsen KB (2004) Hydroforming highlights: sheet hydroforming and tube hydroforming, J Mater Process Technol 151(1–3 spec. iss.):165–177
7. Mahabunphachai S, Koç M (2010) Investigations on forming of aluminium 5052 and 6061 sheet alloys at warm temperatures. Mater Des, Elsevier 31:2422–2434

8. Modi B, Ravi Kumar D (2013) Development of a hydroforming setup for deep drawing of square cups with variable blank holding force technique. Int J Adv Manuf Technol 66:1159–1169
9. Raja Satish D, Feyissa FT et al (2018) Formability of cryorolled aluminum alloy sheets in warm forming. Int J Mater, Mech Manuf, 6(2):123–126
10. Ramesh A, Sumesh CS, Abhilash PM, Rakesh S (2015) Finite element modeling of orthogonal machining of hard to machine materials. Int J Mach Machinability Mater 17:543–568
11. Rao PN, Singh D, Jayaganthan R (2013) Effect of annealing on microstructure and mechanical properties of Al 6061 alloy processed by cryorolling. Mater Sci Technol 29(1):76–82
12. Sumesh CS, Ramesh A (2018) Numerical modelling and optimization of dry orthogonal turning of Al6061 T6 alloy. Period Polytech Mech Eng 62(3):196–202
13. Taye F, Das P, Ravi Kumar D, Ravi Sankar B (2014) Characterization of mechanical properties and formability of cryorolled aluminium alloy sheets. In: 5th International and 26th All India manufacturing technology, design and research conference (AIMTDR), pp 511.6–511.11
14. Wang L, Chan LC, Lee TC (2007) Process modelling of controlled forming with time variant blank holder force using RSM method. Int J Mach Tools Manuf 47:1929–1940
15. Zhu X, Liu D, Yang Y, Hu Y, Zheng Y (2016) Optimization on cooperative feed strategy for radial-axial ring rolling process of Inco718 alloy by RSM and FEM. Chin Soc Aeronaut Astronaut Beihang Univ 29(3):831–842

Selection of Parameterization Method for Fitting of Freeform Curves Using Uniformly Spaced Data

G. Rajamohan

Abstract The term "parameterization" refers to the mapping of measured data points in Cartesian space into parametric space for the fitting of freeform curves and surfaces, etc. Among the conventional parameterization methods, the chord length and centripetal parameterization methods are generally considered adequate for most of the applications. However, the choice between these two parameterization methods is not obvious. The present work attempts to analyze the measured data points in order to select between the chord length and centripetal parameterization methods for achieving better results while fitting the freeform curves. The chord length between two adjacent measured data points and the discrete curvature at each point have been used to develop an approach to select between the two parameterization methods considered. For validation purposes, the measured data points with uniform spacing have been taken from some assumed example freeform curves of varying complexities. The proposed approach has been found to correctly select the appropriate parameterization method in all examples considered.

Keywords B-splines · Centripetal parameterization · Chord length parameterization · Freeform curves and surfaces

1 Introduction

Reverse engineering is carried out to transform the engineering objects, for which the original part drawings are not available, into computer-aided design (CAD) models in order to improve their design, manufacture and analysis. In reverse engineering, the data points from the desired engineering objects are acquired using appropriate measuring devices, such as 3D scanners. The data points are acquired in Cartesian space, whereas the CAD models are created in parametric space. The mapping of data points from Cartesian space to parametric space is known as parameterization. The accuracy of fitted CAD model depends on adequate parameterization.

G. Rajamohan (✉)
National Institute of Foundry and Forge Technology, Ranchi 834003, India
e-mail: grajamohan.nifft@gov.in

S. Vijayan et al. (eds.), *Trends in Manufacturing and Engineering Management*, Lecture Notes in Mechanical Engineering,
https://doi.org/10.1007/978-981-15-4745-4_82

Parameterization also has applications in computer-aided geometric design (CAGD), such as texture mapping, re-meshing, curve and surface fitting. Among the available conventional parameterization methods, the chord length and the centripetal methods are generally used. The present work considers the fitting of parametric freeform curves to measured data points using the B-splines representation.

2 Literature Review

The problem of fitting of freeform curves (and surfaces) to measured point clouds has been the subject of research for many years. The existing parameterization methods can be broadly categorized into conventional (single-step) methods, and evolutionary and nature-inspired algorithms (optimization-based).

The literature reveals various conventional parameterization methods, such as uniform, chord length, centripetal, etc. A summary of pros and cons of different parameterization methods can be found in [1]. These parameterization methods work fairly well with uniformly spaced data points. The literature also reveals that no single parameterization method is capable of handling all types of data points (e.g., collinear data points). While most of the reported research is focused on open curves and surfaces, few works based on closed curves and surfaces have been reported [2].

The conventional parameterization methods may not lead to optimal allocation of parameters for data points with nonuniform spacing. This has led the researchers to focus on developing algorithms based on some soft-computing techniques, such as the evolutionary and nature-inspired algorithms. The algorithms reported in this direction are based on artificial neural networks [3–7], genetic algorithms [8–14], swarm intelligence [15–18], etc.

The chord length and centripetal parameterization methods are generally adequate for most practical purposes [19]. The objective of this work is to develop and validate an approach to choose between chord length and centripetal parameterization methods in order to fit more accurate freeform curves for the given set of measured data points. The B-splines have been used to represent the freeform curves.

3 Description of the Parameterization Methods

A brief description of the chord length and centripetal parameterization methods is presented in this section. The parameter u is taken to be in the range [0, 1].

3.1 Chord Length Parameterization Method [19]

The chord length parameterization method uses the chordal approximation for curves. Let d be the total chord length, computed using Eq. (1).

$$d = \sum_{i=1}^{N} |Q_i - Q_{i-1}| \tag{1}$$

Equation (2) can be used to compute the u parameters, i.e., $\bar{u}_i$.

$$\bar{u}_0 = 0;\ \bar{u}_N = 1;\ \bar{u}_i = \bar{u}_{i-1} + \frac{|Q_i - Q_{i-1}|}{d},\ i = 1, \ldots, (N-1) \tag{2}$$

where, Q_i ($i = 0,\ldots,N$) are the data points. This parameterization method is widely used and is generally considered adequate.

3.2 Centripetal Parameterization Method [19]

This parameterization method uses a slight variation to the chordal approximation for curves. Let d be the total chord length, computed using Eq. (3).

$$d = \sum_{i=1}^{N} \sqrt{|Q_i - Q_{i-1}|} \tag{3}$$

Equation (4) can be used now to compute the u parameters, i.e., $\bar{u}_i$.

$$\bar{u}_0 = 0;\ \bar{u}_N = 1;\ \bar{u}_i = \bar{u}_{i-1} + \frac{\sqrt{|Q_i - Q_{i-1}|}}{d},\ i = 1, \ldots, (N-1) \tag{4}$$

where, Q_i ($i = 0,\ldots,N$) are the data points. This method gives better results than chord length parameterization method when the data points take very sharp turns, i.e., when the curvature changes rapidly.

4 Proposed Approach

The chord lengths and discrete curvatures are used in the proposed approach to select between the chord length and centripetal parameterization methods. The chord length is the distance between two adjacent measured points and is computed using the distance formulae. The discrete curvatures are calculated at all measured points using the algorithm described in the subsection below. The calculated chord lengths and discrete curvatures are normalized in the range [0, 1], and standard deviations are calculated. The attribute, i.e., the chord length or discrete curvature, having higher standard deviation is used to select the most suitable parameterization method. In case of chord length, the chord length parameterization is selected, and in case of discrete curvature, the centripetal parameterization method is selected.

4.1 Computation of Discrete Curvature [20]

Let Q_i ($i = 0, 1,\ldots, N$) represent an ordered set of ($N + 1$) measured data points. Assuming the sequential connection of these points to form an open polygon, the curvature at each measured point Q_i (denoted as K_i) is the relative curvature of the curve passing through three adjacent points, viz. Q_{i-1}, Q_i and Q_{i+1} (Fig. 1a). The curvature at the interior points (Q_1, Q_2, …, Q_{N-1}) can be expressed using circular curvature as $K_i = (2 * \Delta Q_{i-1}\, Q_i\, Q_{i+1})/(L_i\, L_{i+1}\, S_i)$, where $L_i = \|Q_i - Q_{i-1}\|$, $L_{i+1} = \|Q_{i+1} - Q_i\|$, $S_i = \|Q_{i+1} - Q_{i-1}\|$ and $\Delta Q_{i-1}\, Q_i\, Q_{i+1}$ is the signed area of triangle. The

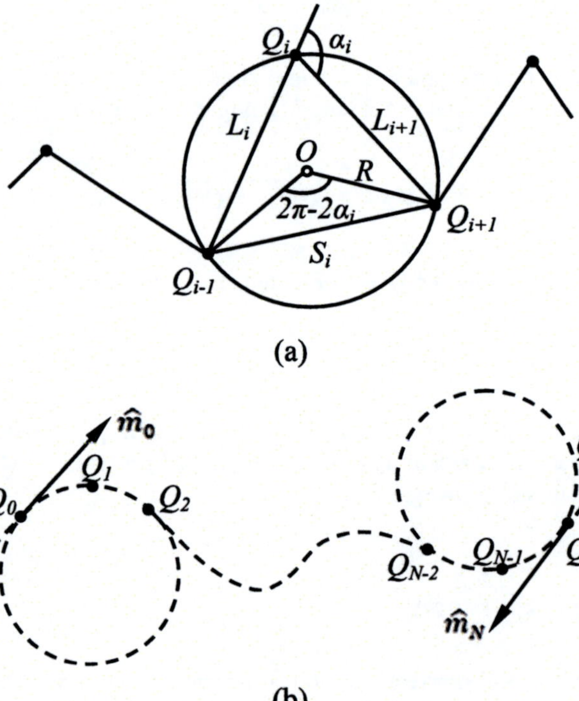

Fig. 1 Estimation of discrete curvature; **a** curvature of a polygon; **b** additional point supplement for two end points

curvature at endpoints Q_0 and Q_N can be computed using the procedure described below:

Step 1: Compute the tangent vectors at endpoints Q_0 and Q_N by applying the circular end conditions. The points Q_0, Q_1 and Q_2 are selected for computing the tangent vector $\hat{m}_0$ at Q_0, and points Q_{N-2}, Q_{N-1} and Q_N are selected for computing the tangent vector $\hat{m}_N$ at Q_N. Let $a_0 = Q_1 - Q_0$, $b_0 = Q_2 - Q_0$ and $c_0 = a_0 \times b_0$. The tangent vector $\hat{m}_0$ is then computed as $\hat{m}_0 = (r_0 \times c_0)/\|r_0 \times c_0\|$, where r_0 is computed as $r_0 = \{\|a_0\|^2 (b_0 \times c_0) + \|b_0\|^2 (c_0 \times a_0)\}/(2 \|c_0\|^2)$.

Step 2: As shown in Fig. 1b, two additional points Q_0^- and Q_N^+ are added, respectively, along the direction of $-\hat{m}_0$ and $-\hat{m}_N$, satisfying $\|Q_1 - Q_0\| \geq \| Q_0^- - Q_0\|$ and $\|Q_{N-1} - Q_N\| \geq \| Q_N^+ - Q_N\|$.

Step 3: The curvatures at Q_0 and Q_N are now calculated in the same way as interior points with Q_0^-, Q_0, Q_1 and Q_{N-1}, Q_N, Q_N^+, respectively.

5 Results and Discussion

The accuracy of freeform curves (and surfaces) fitted using the measured data points depends mainly on parameterization. Among the available conventional parameterization methods, the chord length and centripetal methods are mostly used owing to their simplicity and ease of implementation. The present work is an attempt to select between these two methods to fit better freeform curves for given measured data points and to validate the selection approach. The selection is based on the standard deviation of discrete curvatures and chord lengths, computed using the given measured data points. The B-splines representation has been used to represent the freeform curves. The programs were written in C++ language.

In order to make a comparison between the centripetal and chord length parameterization methods, the measured data points have been selected from known freeform curves. The curves are first discretized in parametric space with fine spacing of 0.005, giving 201 points in the range assumed, i.e., [0, 1]. Measured data points have been taken with uniform spacing between them, as the measuring machines typically generate such data easily. Three runs have been made with the numbers of measured data points as 11, 21 and 51. The discrete curvature and chord lengths have been computed using these data points, and standard deviations of the respective measures have been computed. The chord length parameterization method is chosen when the standard deviation of chord length is higher and vice versa.

Both the centripetal and the chord length parameterizations have been implemented in order to validate the selection by using the proposed approach. The measured data points and the respective parameters were used to fit the freeform curves of degree 3 using interpolation technique [19]. The normal deviations between the fitted freeform curves and the corresponding point on the known freeform curves are then calculated. The maximum normal deviation has been used to compare the chord length and centripetal parameterization methods. The selected parameterization method is supposed to yield lower value of normal deviation. The normal deviations (mm) obtained using different example freeform curves are shown in Tables 1, 2 and 3 for the numbers of measurements points of 11, 21 and 51, respectively. The boldfaced entries in column 2 of these tables indicate the mismatch in selected method using the proposed approach.

As evident from the tables, the prediction accuracy of the proposed approach is less with fewer data points, and it gets better with increase in the number of data points. It may be because the discrete curvature calculation algorithm gives better

Table 1 Normal deviations (mm) for measurement data points size of 11

Example No.	Selected method	Centripetal parameterization	Chord length parameterization
#1	Chord length	0.003163	0.003038
#2	**Chord length**	0.032463	0.049099
#3	Centripetal	0.008274	0.043202
#4	Centripetal	0.998697	2.588100
#5	**Chord length**	0.016959	0.026198
#6	**Chord length**	0.430049	0.635121

Table 2 Normal deviations (mm) for sparse measurement data points size of 21

Example No.	Selected method	Centripetal parameterization	Chord length parameterization
#1	Chord length	0.000122	0.000087
#2	Centripetal	0.003633	0.008180
#3	Centripetal	0.000687	0.002746
#4	Centripetal	0.039825	0.154892
#5	Centripetal	0.002091	0.003382
#6	Centripetal	0.054421	0.084704

Table 3 Normal deviations (mm) for dense measurement data points size of 51

Example No.	Selected method	Centripetal parameterization	Chord length parameterization
#1	Chord length	0.000024	0.000022
#2	Centripetal	0.000146	0.000210
#3	Centripetal	0.000033	0.000069
#4	Centripetal	0.000561	0.002775
#5	Centripetal	0.000112	0.000112
#6	Centripetal	0.000587	0.000774

results with dense points. The examples curves #2, #5 and #6 particularly have higher curvature variations, while the remaining curves have moderate curvature variations. Considering this aspect, it appears that the curvature calculation algorithm yields better values even with smaller numbers of measured points if the variation in curvature is not severe. With dense measured data points (Tables 2 and 3), the proposed approach has been able to accurately estimate the suitable parameterization method in all cases considered. The fitting accuracy also improves significantly. In most cases, the centripetal method is being suggested as the most suitable method. This may be because the freeform curves typically have more curvature variations along their length.

6 Conclusions

Many industrial and technological applications require the fitting of parametric freeform curves and surfaces to measured points. The parameterization method used plays crucial role on the accuracy of fitted freeform curves and surfaces. An approach has been proposed in this paper to select between the chord length and centripetal methods for fitting of freeform curves using the given sets of measured data points. The approach has been validated using multiple example freeform curves of varying complexities. The following conclusions are drawn:

- The accuracy of determination of most suitable parameterization method has been found to be lower with small number of measured data points, and it improves with increase in the number of data points.
- The proposed approach works well with uniformly spaced, dense measured data points. Such a scenario being typically the case in applications such as reverse engineering, the proposed approach can be of practical use.

The following may constitute the scope for future work:

- In curvature calculation, higher order approximation between the adjacent points may be tried, instead of straight line approximation, for better curvature estimates.
- The present work uses the interpolation technique to fit freeform curves for uniformly spaced data points. Use of approximation technique may be attempted.

References

1. Haron H, Rehman A, Adi DIS, Lim SP, Saba T (2012) Parameterization method on B-spline curve. Math Prob Eng. https://doi.org/10.1155/2012/640472
2. Park H (2001) Choosing nodes and knots in closed B-spline curve interpolation to point data. Comput Aided Des 33(13):967–974
3. Bishop CM, Roach CM (1992) Fast curve fitting using neural networks. Rev Sci Instrum 63:4450–4456

4. Gu P, Yan X (1995) Neural network approach to the reconstruction of freeform surfaces for reverse engineering. Comput Aided Des 27(1):59–64
5. Hoffmann M, Varady L (1998) Freeform surfaces for scattered data by neural networks. J Geom Graph 2:1–6
6. Echevarria G, Iglesias A, Galvez A (2002) Extending neural networks for B-spline surface reconstruction. Lect Notes Comput Sci 2330:305–314
7. Krause FL, Fischer A, Gross N, Barhak J (2003) Reconstruction of freeform objects with arbitrary topology using neural networks and subdivision techniques. Ann CIRP 52(1):125–128
8. Sarfraz M, Raza SA (2001) Capturing outline of fonts using genetic algorithm and splines. In: Proceedings of the international conference on information visualization. ISBN: 0-7695-1195-3
9. Renner G, Ekart A (2003) Genetic algorithms in computer aided design. Comput Aided Des 35:709–726
10. Yoshimoto F, Harada T, Yoshimoto Y (2003) Data fitting with a spline using a real coded genetic algorithm. Comput Aided Des 35:59–64
11. Kumar GS, Kalra PK, Dhande SG (2003) Parameter optimization for B-Spline curve fitting using genetic algorithms. Proc Congr Evoluti Comput (CEC'03) 3:1871–1878
12. Sun YH, Xiang WJ, Xia DS (2010) Parameter optimization for B-Spline curve fitting based on adaptive genetic algorithm. J Comput Appl 30:1878–1882
13. Zhao L, Jiang J, Song C, Bao L, Gao J (2013) Parameter optimization for Bezier curve fitting based on genetic algorithm. Adv Swarm Intell 7928:451–458
14. Hasegawa AY, Tormena C, Parpinelli RS (2014) Bezier curve parameterization using a multi-objective evolutionary algorithm. Int J Comput Sci Appl 11(2):1–18
15. Galvez A, Cobo A, Pey JP, Iglesias A (2008) Particle swarm optimization for Bezier surface reconstruction. In: Proceedings of the 8th international conference on computational science (part II)
16. Galvez A, Iglesias A (2011) Efficient particle swarm optimization approach for data fitting with free knot B-Splines. Comput Aided Des 43:1683–1692
17. Galvez A, Iglesias A (2012) Particle swarm optimization for non-uniform rational B-Splines surface reconstruction from clouds of 3D points. Inf Sci 192:174–192
18. Galvez A, Iglesias A (2013) Firefly algorithm for Bezier curve approximation. In: Proceedings of the 12th international conference on computational science and its applications, pp 81–88
19. Piegl L, Tiller W (1997) The NURBS book. Springer, Berlin
20. Liu GH, Wong YS, Zhang YF, Loh HT (2002) Adaptive fairing of digitized point data with discrete curvature. Comput Aided Des 34:309–320

Energy Engineering

Experimental Study of an Axial Turbine for Wave Energy Conversion

Kumud Kumar, Tapas K. Das, R. Srikanth, and Abdus Samad

Abstract With the growing menace of increase in population and pollution, renewable sources of energy are being preferred over the fast depleting fossil fuels. Ocean wave energy is one such form of renewable source with a huge potential all around the globe. Wave energy conversion (WEC) devices are used to harness the energy from ocean waves. An oscillating water column (OWC) is a WEC device, which uses an axial turbine to convert wave energy into useful electrical energy. The experimental setup designed and assembled at Wave Energy and Fluids Engineering Laboratory, IIT Madras tested the turbine subjected to bidirectional airflow. The objective of the preliminary experiment is to measure the rotational speed and pressure drop across the turbine. The variable parameters are stroke length and time period of the oscillating piston, which simulates different ocean wave conditions. A comparative study of the parameters is carried out and reported in this article.

Keywords Wells turbine · Wave energy · Experimental study · OWC

Nomenclature

α	Angle of attack (degree)
F_{ax}	Axial force component (N)
F_{tan}	Tangential force component (N)

K. Kumar · T. K. Das (✉) · R. Srikanth · A. Samad
Wave Energy and Fluids Engineering Lab, Ocean Engineering Department,
IIT Madras, Chennai, India
e-mail: mech.tapas@gmail.com

K. Kumar
e-mail: kumud142k@gmail.com

R. Srikanth
e-mail: srikanthrangisetty@gmail.com

A. Samad
e-mail: samad@iitm.ac.in

S. Vijayan et al. (eds.), *Trends in Manufacturing and Engineering Management*,
Lecture Notes in Mechanical Engineering,
https://doi.org/10.1007/978-981-15-4745-4_83

b	Blade width (mm)
c	Blade chord length (mm)
D	Drag force (N)
L	Lift force (N)
N_b	Number of blades
V	Relative velocity (m/s)
σ	Turbine Solidity
h	Hub-to-tip ratio
SL	Stroke length (m)
CT	Cycle time
R_t	Rotor tip radius (mm)
R_h	Rotor hub radius (mm)
t_c	Tip clearance (mm)
AR	Aspect ratio
N	Revolutions per minute
ΔP	Differential Pressure drop (kPa)
T	Time (s)

Abbreviation

OWC	Oscillating water column
WEC	Wave energy conversion
CAD	Computer aided design
VFD	Variable frequency drive
DAQ	Data acquisition system

1 Introduction

Wave energy is a form of renewable energy source which is environment-friendly and has the potential to be economically accessible in the near future. Renewable sources of energy are unlimited and eco-friendly; they don't produce harmful pollutant gases such as carbon dioxide, nitrogen oxides, or sulfur oxides. For countries that are surrounded by ocean such as India, Japan, Australia, England, to mention a few, and remotely situated islands, wave energy conversion principle is very promising for energy needs. Wave energy is the energy harnessed from water waves. When the wind blows over the water surface, it transfers the energy to waves. Wave power is directly proportional to wind speed; more the speed of wind more will be the wave energy. The global wave energy potential is estimated to be 8000–80,000 TW/yr [1]. This energy potential can be harnessed by using wave energy conversion technology to do useful work such as the generation of electricity, desalination of seawater, and pumping water to reservoirs.

Wave energy is harnessed by using wave energy conversion devices such as point absorbers, overtopping device, oscillating water columns, and many more. Oscillating water column (OWC) is a WEC device that is easy to install in both onshore and offshore conditions [2]. An OWC plant consists of an air chamber with an opening above the water surface, as shown in Fig. 1. The incident water waves force the water column to oscillate up and down like a piston-cylinder arrangement inside the chamber. The air trapped above the water surface is subjected to compression and decompression simultaneously, and this results in the generation of bidirectional air flow. The energy from this bidirectional air flow can be harnessed by using an air turbine designed by professor Alan Arthur Wells at Queens University, Belfast in the 1970s. This air turbine is known as the Wells turbine [3].

Plenty of research has been carried out on both numerical and experimental analysis of the Wells turbine [4]. While many experimental works on Wells turbine studied steady-state characteristics of the turbine for different blade geometries [5–7], researchers also carried out laboratory experiments to study the turbine characteristics for bidirectional airflow similar to the conditions in OWC [8, 9]. The Wave Energy and Fluids Engineering Lab (WEFEL) at IIT Madras developed a test rig with a piston-cylinder arrangement to emulate the bidirectional airflow similar to an OWC. Previously a scaled impulse turbine model was tested in this test rig [10]. In the present work, a preliminary experiment carried out on monoplane Wells turbine subjected to bidirectional air flow through the piston-cylinder arrangement. The

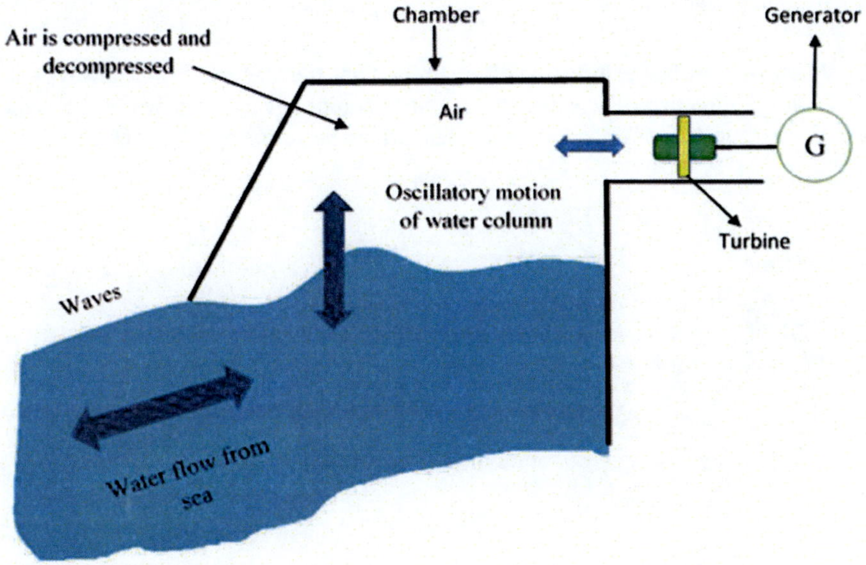

Fig. 1 Schematic of an OWC

measured performance parameters are rotational speed and the pressure difference across the turbine. The stroke length and the time period of the oscillating piston are varied to create different bidirectional airflow profiles similar to different wave conditions.

2 Working Principle of the Wells Turbine

Oscillating water columns utilize Wells turbine for converting wave energy to electrical energy. Wells turbine is an axial flow bidirectional turbine especially used in OWC type wave energy converters. The turbine is made up of symmetrical aerofoil type blades fixed around the hub (Fig. 2). The aerofoil blades are positioned around the hub with the chord planes perpendicular to the axis of rotation. The turbine transmits wave energy to the coupled electric generator through rotation.

The Wells turbine working principle is explained in [11]. When the flow is incident on the turbine blade at an angle of attack (α), a lift force is generated normal to the direction of relative velocity. Simultaneously a drag force is experienced in the direction of the relative velocity. Resolving these lift and drag forces (Fig. 2) gives components of forces in the plane of rotation (tangential force) and normal to the plane of rotation (axial force), F_{tan} and F_{ax}, respectively.

$$F_{\text{tan}} = L \sin\alpha - D \cos\alpha \tag{1}$$

$$F_{\text{ax}} = L \cos\alpha + D \sin\alpha \tag{2}$$

Whatever may be the direction of incoming flow, the tangential force always acts in the same direction. Therefore, the turbine rotation always remains in the same direction regardless of flow direction. This turbine performs optimally up to a value

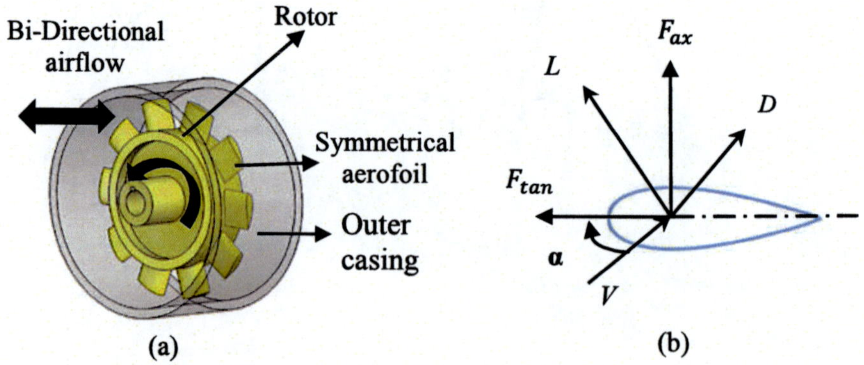

Fig. 2 **a** CAD model of Wells turbine with shroud, **b** forces acting on the blade

of angle of attack (α), further increase in its value results in flow separation around the blades. This angle at which flow separation occurs is called as stall angle. Beyond this angle, the magnitude of lift and drag forces reduces; as a result, efficiency decreases. One main drawback of Wells turbine is that it is difficult to achieve its operational speed due to its poor starting characteristics [12].

3 Turbine Design Parameters

The aerodynamic performance of the Wells turbine depends upon various parameters. These parameters with optimum values are discussed in this section. All the parameters are inter-related. Based on the analytical studies, turbine geometric parameters are given in [11]. All such important parameters are enlisted in Table 1. One of the most important parameters is solidity, which measures the mutual interaction between the blades. With the increase in solidity, turbine blades tend to come closer, thereby hindering flow passage and as a result efficiency of turbine reduces. Previous experiments show that with an increase in the hub-to-tip ratio beyond certain value results in loss of efficiency [11]. The aspect ratio and tip clearance play an important role in turbine performance. The Wells turbine shows high sensitiveness towards tip clearance.

A Monoplane Wells turbine was modeled and then fabricated for testing in the laboratory. The 3D CAD model and the actual turbine is shown in Fig. 3. The turbine is modeled on the recommended values of various design parameters. It is fabricated by rapid prototyping, and the material used is ABS Plastic (Acrylonitrile Butadiene Styrene). The turbine geometrical dimensions are listed in Table 2.

Table 1 Recommended parameters

Parameter	Analytical description	Remarks
Solidity (σ)	$N_b c/\pi R_t$	Efficiency reduces for $\sigma > 0.5$
Aspect ratio (AR)	b/c	AR ~0.5 is recommended
Hub-to-tip ratio (h)	R_h/R_t	Recommended value h ~0.6
Tip clearance (t_c)	% of c	≤2% of c is recommended

Fig. 3 Actual Wells turbine model

Table 2 Actual turbine parameters

Parameter	Description	Parameter	Description
Aerofoil profile	NACA0021	Chord length	36 mm
Number of blades	10	Rotor solidity	0.69
Hub radius	68 mm	Aspect ratio	0.8
Hub-to tip ratio	0.69	Tip clearance	2 mm

4 Experimental Setup and Procedure

The experiment is carried out at a laboratory facility in IIT Madras. This experimental setup consists of a piston-cylinder arrangement which is driven by a motor and gearbox. Motor rotations are controlled by a variable frequency drive (VFD). The measured parameters include differential pressure drop across the turbine rotor and the rotational speed. Table 3 lists the instruments used for measuring the parameters. The signals from the instrument are recorded by a data acquisition system (DAQ). The input signals are processed and stored as data in a readable format using LAB VIEW software.

The VFD runs the motor to generate the required cycle time (CT) of the piston, which relates to ocean wave time period. The power from the motor shaft is transferred to the piston rod through worm wheel gearbox (gear ratio 1:40). Ocean wave height is replicated by piston stroke length (SL). The frequencies corresponding to different CT of the piston are: 22.5 Hz relates to CT = 4 s and 18.5 Hz to CT =

Table 3 Instrument specifications

Instrument	Parameter	Range	Accuracy	Output signal
ABB 266DSH	Differential pressure	±10 kPa	0.06%	4–20 mA
Rotational speed frequency converter (Pepperl + Fuchs)	Rotational speed	0–5000 rpm	±5% of reading	4–20 mA

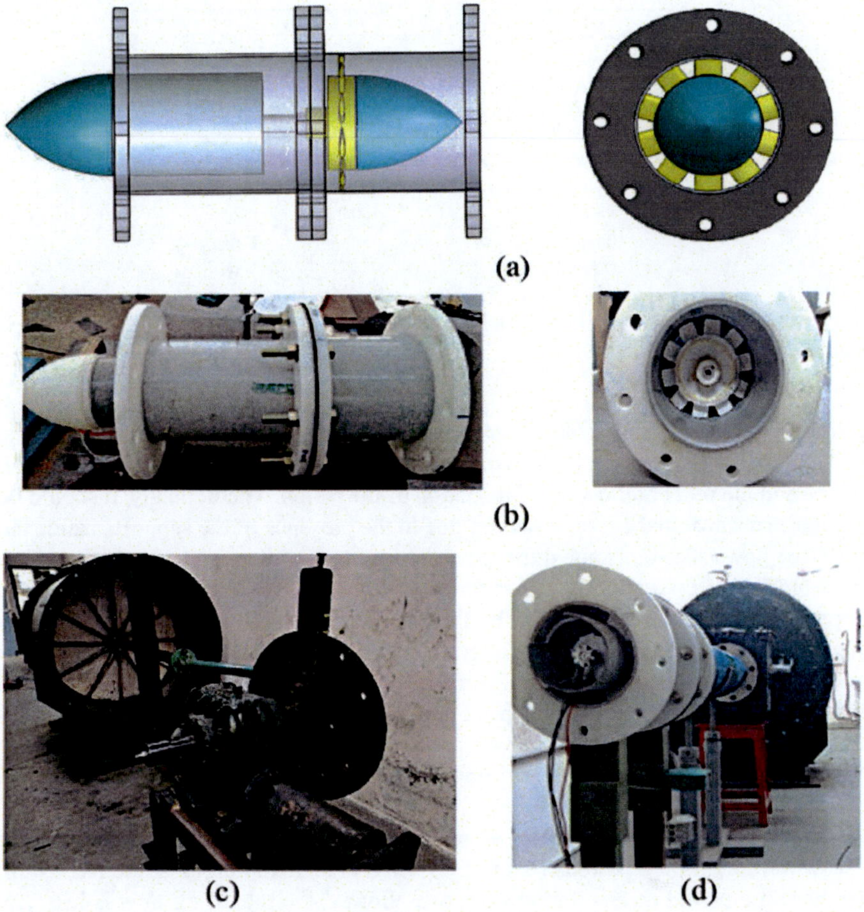

Fig. 4 **a** CAD model of the turbine assembly, **b** actual turbine assembly, **c** the piston-cylinder system, **d** complete setup

5 s. The oscillating motion of the piston creates the bidirectional air flow through the duct. Once the piston starts reciprocating, data is recorded after some time until the flow is fully developed. The sampling rate of data taken is 100 ms, and it is recorded for three minutes. The complete experimental setup is shown in Fig. 4.

5 Results and Discussions

All the experiment tests data is recorded by DAQ system, and the comparative results are plotted as the function of time. Although each data is collected for three minutes duration, the present result is shown for only one full cycle for the ease of understanding. Figure 5 shows the variation of pressure drop and rotational speed of the

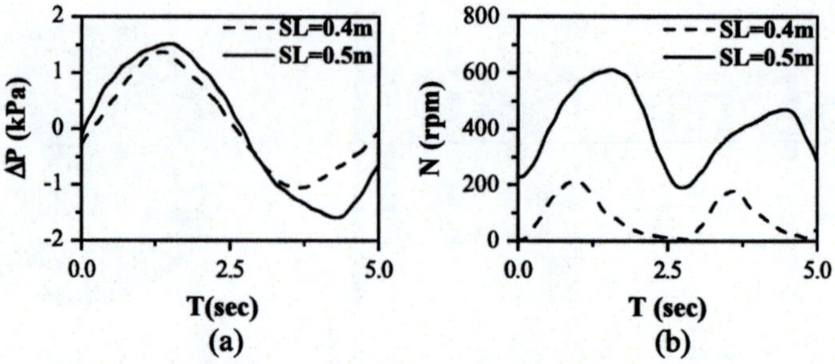

Fig. 5 Variation of **a** pressure drop and **b** rotational speed for different stroke length at same time period $T = 5$ s

turbine for two different stroke lengths (0.4 and 0.5 m) for a fixed time period of 5 s. In Fig. 5a, the first half cycle shows differential pressure during compression of the piston and the other half cycle is for suction of the piston. Theoretically, it should be similar to a sinusoidal curve; however, due to the presence of the supports inside the duct, the flow faces blockage during the suction stage. As a result, the compression peak pressure drop is slightly higher than the suction side pressure drop. The turbine experiences more air flow rate when the stroke length is longer. This is visible in Fig. 5(b) for the rotational speed. The rotational speed for SL = 0.5 m is much higher compared to smaller stroke length. For SL = 0.4 m, the turbine rotational speed is almost zero during the transition from compression to the suction stage. This situation interrupts the continuous power production from the generator. However, for SL = 0.5 m, the turbine rotates continuously during both suction and compression stage of the piston motion.

Figure 6 shows the variation of pressure drop and rotational speed of the turbine for two different time periods (4 and 5 s) when the stroke length is fixed at 0.4 m. Higher time period means the piston moves slowly; the rate of air flow is less. So, the pressure drop is also less when the time period is higher. The same is visible in Fig. 6a. The variation of rotational speed is shown in Fig. 6b. When the time period is

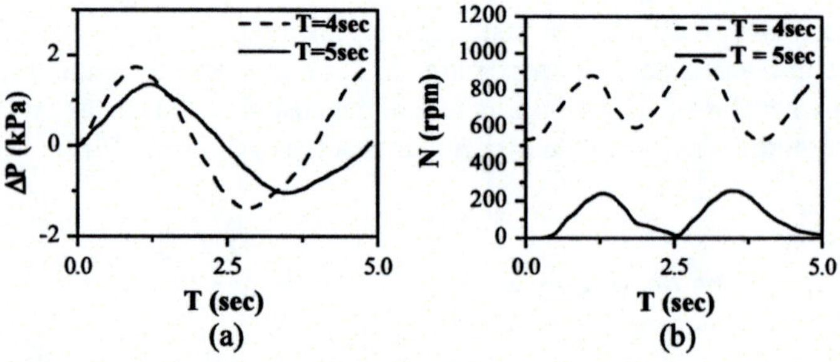

Fig. 6 Variation of **a** pressure drop and **b** rotational speed for different time period at same stroke length SL = 0.4 m

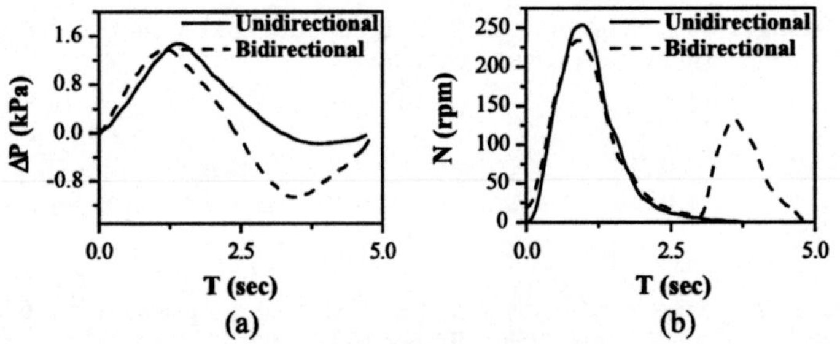

Fig. 7 Variation of **a** Pressure drop and **b** rotational speed for unidirectional and bidirectional condition at SL = 0.4 m, $T = 5$ s

less, the piston moves faster, and the transition from compression to suction happens very fast. So, the turbine speed does not fall to zero and oscillates between 600 and 1000 rpm approximately. However, this rotational speed is less compared to the design speed of the Wells turbine of similar dimension. This is due to the generator support, which does not allow uninterrupted airflow to the turbine. For $T = 5$ s, the turbine almost stops at the transition from compression to the suction stage.

Figure 7 shows the comparison between unidirectional and bidirectional flow at SL = 0.4 m and $T = 5$ s. The piston-cylinder arrangement can be made unidirectional by opening a valve. In this configuration, the air flows through the turbine only during the compression stage; whereas, during suction, the air flows through the valve. During the compression stage, the values for pressure drop and rotational speed are almost the same for both unidirectional and bidirectional flow. However, for unidirectional flow, the turbine does not rotate during the suction stage, and the values of pressure drop and rotational speed remain zero throughout this stage.

6 Conclusion

A preliminary experimental investigation on the performance of monoplane Wells turbine is carried out for bidirectional air flow, and the results are compared for two different time periods and stroke lengths of the piston. Due to limited instrument availability, only the pressure drop and the rotational speed was measured. The turbine experienced higher pressure drop and higher rotational speed when the stroke length was high and the time period is low. The generator supports created blockage of airflow to the turbine, which was the main reason for low rotational speed. Also, the turbines assembly had a large structural vibration at higher rotational speeds. So, the results had large deviations from the actual turbine performance. From this preliminary experience, a modified system is designed, and further tests will be carried out in future.

Acknowledgements The authors thank the Department of Ocean Engineering, IIT Madras for providing financial support to carry out the experiment.

References

1. Sannasiraj SA, Sundar V (2016) Assessment of wave energy potential and its harvesting approach along the Indian coast. Renew Energy 99:398–409
2. Falcao AFDO, Henriques JCC (2016) Oscillating-water-column wave energy converters and air turbines: a review. Renew Energy 85:1391–1424
3. Raghunathan S, Setoguchi T, Kaneko K (1994) Aerodynamics of monoplane Wells turbine—a review. Int J Offshore Polar Eng 4(1):68–75
4. Shehata AS, Xiao Q, Saqr KM, Alexander D (2017) Wells turbine for wave energy conversion: a review. Int J Energy Res 41(1):6–38
5. Gato LMC, Curran R (1996) Performance of the biplane Wells turbine. Trans ASME 118:210–215
6. Curran R, Gato LMC (1997) The energy conversion performance of several types of Wells turbine designs. Proc Inst Mech Eng Part A J Power Energy 211(2):133–145
7. Takasaki K, Takao M, Setoguchi T (2014) Effect of blade shape on the performance of Wells turbine for wave energy conversion. Int J Mech Aerospace, Ind Mechatronics Eng 8(12):2080–2083
8. Paderi M, Puddu P (2013) Experimental investigation in a Wells turbine under bi-directional flow. Renew Energy 57:570–576
9. Gato LMC, Webster M (2001) An experimental investigation into the effect of rotor blade sweep on the performance of the variable-pitch Wells turbine. Proc Inst Mech Eng Part A J Power Energy, 215(5):611–622
10. George A et al (2018) Experimental analysis of turbine-chamber coupling for wave energy conversion. Int J Energy Res 42(15):4770–4782
11. Raghunathan S (1995) The Wells air turbine for wave energy conversion. Prog Aerosp Sci 31(4):335–386
12. Inoue M, Kaneko K, Setoguchi T, Raghunathan S (1986) Simulation of starting characteristics of the Wells turbine. In: 4th fluid Mechanics, plasma dynamics and laser conference, 1986

Investigations into Nonlinear Energy Sinks for a Stochastic Dynamical Oscillator

Anuroop Parvathaneni, Dhritimaan Sharma, Dhruv Vashishtha, Pradeep V. Malaji, and J. Venkatramani

Abstract The paper deals with Nonlinear Energy Sinks (NES), utilizing piezoelectric transduction mechanism, focusing on the degree of effect the auxiliary nonlinear stiffness has on the performance of the NES and the performance of NES with the primary system subjected to random excitation. Hence, a parametric sweep of the auxiliary nonlinear stiffness over a broad range of values has been done and the variations in primary vibration suppression and voltage generation by the NES have been observed for its corresponding values. It has been conducted with the NES attached to a linear primary system and then an essentially nonlinear one. Comparison of results and validation of the performance of NES for both the cases have been performed. Following that, performance of the NES has been investigated when a linear primary system is subjected to random excitation. Two separate cases have been utilized to randomize the excitation. Results regarding vibration control and voltage generated have been derived for both and compared to those obtained for deterministic excitation. By and large, it is found that NES is successful in protecting a primary system and broadening the operation bandwidth, while satisfyingly generating voltage, irrespective of the type of excitation on the primary system.

Keywords Nonlinear energy sink (NES) · Energy harvesting · Vibration suppression

A. Parvathaneni · D. Sharma · D. Vashishtha · J. Venkatramani (✉)
Shiv Nadar University, Tehsil Dadri, Greater Noida, U.P., India
e-mail: j.venkatramani@snu.edu.in

P. V. Malaji
BLDEA's V P Dr. P G Halakatti College of Engineering and Technolog, Vijayapur, Karnataka 586103, India

S. Vijayan et al. (eds.), *Trends in Manufacturing and Engineering Management*, Lecture Notes in Mechanical Engineering,
https://doi.org/10.1007/978-981-15-4745-4_84

1 Introduction

Protection of mechanical equipment from excessive vibration is vital as vibrational damage can severely limit an equipment's service life. On the other hand, vibration being such a ubiquitous resource, utilization of vibrational energy for energy harvesting can be very useful for small portable equipment or for equipment present in isolated regions [1, 2]. Conventional energy harvesters have a limitation of harvesting energy at resonance [3]. Other options consisted of multimodal techniques [3, 4] or a switch to nonlinear techniques [3, 5]. While tuned Mass Dampers (TMD) and linear vibration absorbers have been utilized for control of primary structure [6], limitations such as negligible frequency robustness and poor performance for random vibration render their utilization an infeasible task [6]. Attachment of auxiliary systems as energy harvesters for vibration suppression has an advantage of vibration control and useful energy harvesting [7]. Further investigations into nonlinear alternatives have led to the development of nonlinear energy sink (NES). While it is successful in suppressing primary systems' responses, it can also be coupled with a piezoelectric vibrational energy harvester (PVEH) [6] to perform simultaneous energy harvesting. NES has also been found to provide certain unique features such as Strongly Modulated Response (SMR) [8] and Targeted Energy Transfer (TET) [8]. Such features allow vibration control in a broadband manner with simultaneous energy harvesting, making NES a lucrative and more advantageous system when compared to its counterparts. Also, while it has yet not been fully investigated, NES's ability to increase operational bandwidth also allows it to effectively suppress primary vibrations, unlike the linear systems which falter when operated upon by frequency varying excitation [6].

NES allows passive control of vibration of a primary system and consists of a mass and viscous damper with essential nonlinearity. Essential nonlinearity refers to an absence of linear stiffness. The presence of essential nonlinearity or quasi-essentially nonlinearity [9] is vital as only then can the NES resonate at any frequency and attain its objective. Quasi-essential nonlinearity refers to a situation in which the linear stiffness of the NES is considerably decreased in comparison to its nonlinear stiffness, allowing very accurate emulation of essential nonlinearity. It is employed in physical models as attaining essential nonlinearity is extremely challenging [8]. NES provides the Targeted Energy Transfer (TET) [8] feature which refers to its ability to irreversibly transfer energy from the primary system and dissipating it within itself. However, rather than dissipation, the vibrational energy can be utilized for minor energy needs. Efforts have been made to convert it into electrical energy and various transduction models have been introduced to achieve desirable results. By coupling the NES with a transduction model, the voltage can be generated instead of limiting to only vibration control. Electrostatic, electromagnetic, magnetostrictive and piezoelectric mechanisms [8] can be utilized with each providing certain advantages over the other. Most of the study to date has been focused on electromagnetic transducers [5]. But it has been proposed that piezoelectric vibrational energy harvesters (PVEH) [5] based on Nonlinear Energy Sinks can also reduce force transmissibility while harvesting energy on a broad frequency scale [9]. It can also be said that it occupies lesser space while providing better power density [5]. Another area lacking in investigation has been the performance of an NES for random excitations. Most of the studies done to date have concentrated on harmonic excitation upon the primary system which severely limits an NES's potential to operate satisfyingly on frequency varying environments while in the meantime, also fails to emulate real-life scenarios to maximum accuracy. Random vibration has indeed started replacing its harmonic counterpart when realistic scenarios are concerned and as it is much more capable of accurately emulating them. Hence, a study is required to understand if an NES can indeed perform satisfyingly when subjected to random vibrations as only then, practical application of an NES can be considered on a larger scale.

This work proposes the effect of cubic coefficient of the nonlinear stiffness of the NES spring on voltage generated by a NES and the structural response of the NES and the primary spring-mass system. It also investigates NES behaviour for a nonlinear primary system. The work further investigates into the behaviour of the NES when the primary system is acted upon by random excitations and analyses the difference between the voltage generated and structural response of the NES and primary system when acted upon by random excitations and when acted upon by deterministic harmonic excitations. The present study utilizes numerical simulations with the help of MATLABODE45 function to address these issues. The values of the parameters used in the study are taken from previous literature [9]. A parametric sweep of the value of the cubic nonlinear spring stiffness is performed for four different initial conditions [9]. The interplay and relation between the cubic hardening nonlinear spring stiffness of the NES and the newly introduced nonlinear spring stiffness of the primary system are evaluated. The modelling of random vibrations is done by introducing a deterministic harmonic (sinusoidal) excitation, which depending on the random variable, was split into two cases to introduce the random vibration, forcing magnitude or the frequency of forcing. The RMS voltages generated in both the cases are evaluated and compared with the deterministic harmonic excitation as well as with the results validated from literature [9]. The time response plots and time voltage plots hence obtained are presented in this paper.

The paper has been divided into three sections. Section 2 presents the mathematical model used in this study, followed by results and discussions in Sect. 3. The salient findings from this study are summarized in Sect. 4.

2 Mathematical Modelling

See Fig. 1.

To study the performance of the piezoelectric vibrational energy harvester (PVEH) based on the NES, a model is developed consisting of two parts: a primary system and a PVEH attached to a NES. The governing equations of the linear primary system with free vibration are given as

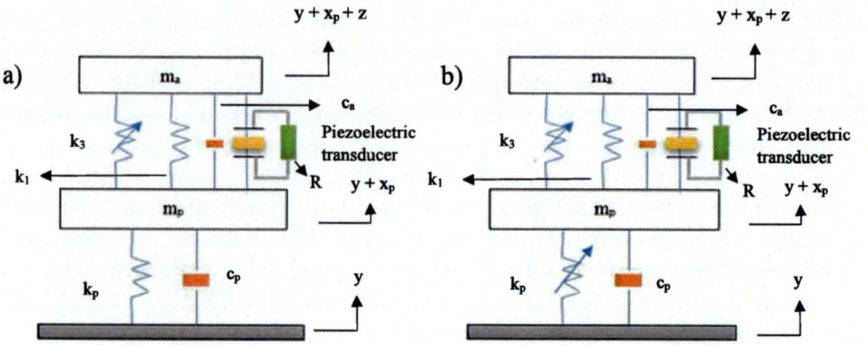

Fig. 1 **a** NES attached to a linear primary system; **b** NES attached to a nonlinear primary system

$$m_p\ddot{x}_p + c_p\dot{x}_p + k_p x_p - \left[c_a\dot{z} + \left(k_1 z + k_3 z^3\right) + \theta V\right] = 0 \tag{1}$$

$$m_a\ddot{z} + (1+\mu)\left[c_a\dot{z} + (k_1 z + k_3 z^3\right) + \theta V] - \mu[c_p\dot{x}_p + k_p x_p] = 0 \tag{2}$$

$$\frac{V}{R} + C^S\dot{V} - \theta\dot{z} = 0 \tag{3}$$

The governing equations of the nonlinear primary system with free vibration are given as

$$m_p\ddot{x}_p + c_p\dot{x}_p + k_p x_p^3 - \left[c_a\dot{z} + \left(k_1 z + k_3 z^3\right) + \theta V\right] = 0 \tag{4}$$

$$m_a\ddot{z} + (1+\mu)\left[c_a\dot{z} + (k_1 z + k_3 z^3\right) + \theta V] - \mu[c_p\dot{x}_p + k_p x_p] = 0 \tag{5}$$

$$\frac{V}{R} + C^S\dot{V} - \theta\dot{z} = 0 \tag{6}$$

For forced vibration, Eqs. (1) changes to Eq. (7) while Eqs. (2) and (3) are also utilized

$$m_p\ddot{x}_p + c_p\dot{x}_p + k_p x_p - \left[c_a\dot{z} + \left(k_1 z + k_3 z^3\right) + \theta V\right] = F\sin(\omega t) \tag{7}$$

In the governing equations, $x_a(t)$ and $x_p(t)$ represents the displacement of NES and the primary mass, respectively. z refers to the relative displacement between the NES and primary system and is represented as $z = x_a - x_p$. m_a and m_p represent the mass of the NES and the primary system respectively. c_a and c_p represent the damping coefficient of the NES and primary system. k_1 and k_3 represent the linear and nonlinear stiffness of the NES spring while k_p represents the stiffness of the primary mass. Mass ratio is denoted by $\mu = m_a/m_p$. The capacitance and the electromechanical coupling coefficient of the PVEH are represented as C^S and θ. R and V represent the resistive load and voltage.

These above governing equations can also be represented in the state space form, and numerical simulation is done with the help of MATLAB software using the ODE45 function to solve the above equations. The initial conditions used are: $x_p(0) = X$, $x_a(0) = X$, $\dot{x}_p(0) = 0$, $\dot{x}_a(0) = 0$. X represents the initial displacements given to the system. Four different values of X are used in the numerical simulation, $X = 0.52$ mm, $X = 1.42$ mm, $X = 2.65$ mm, $X = 4.45$ mm. Only the first value has been used when working with the Eqs. (2), (3) and (7) in Sect. 3.3.

Here, the base excitation has been considered to be zero as the base is fixed. The values of the parameters used in the study are taken from previous literature [9].

3 Results

The focus of this study was to investigate the steady-state performance in addition to the transient performance of the 2DOF mass-spring system. First, the results of previous literature were validated [9] which showed that NES can indeed increase frequency robustness while also decreasing the primary vibrations with the primary system being subjected to harmonic excitation. It also displayed satisfying results regarding voltage generation, irrespective of the initial displacement utilized. Hence the NES was capable of solving the drawbacks present in linear techniques.

Following that, parametric sweep of the auxiliary nonlinear stiffness has been done for linear and nonlinear primary systems. Logical inferences have been drawn from the two sets of results and have been compared to understand the difference in behaviour of NES as per the nature of the primary system. Following that, performance of the NES, when the primary system is subjected to random excitation, has been investigated. Two cases have been used to randomize the excitation. Results for both voltage generation and primary vibration suppression have been considered. To understand if NES remains effective for such a case, the results hence obtained have been compared to the behaviour of the NES when attached to primary system with deterministic excitation.

3.1 Parametric Sweep of k_3 *with a Linear Primary System*

In this section, Eqs. (1)–(3) have been used. A parametric sweep of k_3 has been performed as the focus is on understanding if auxiliary nonlinear stiffness plays any role in voltage generation and primary vibration suppression for a linear primary system. It can be observed that the value of voltage generated is directly proportional to the value of displacement of auxiliary mass. Initial displacement (X) or k_3 doesn't affect the relation between the two in any way as can be seen in Fig. 2.

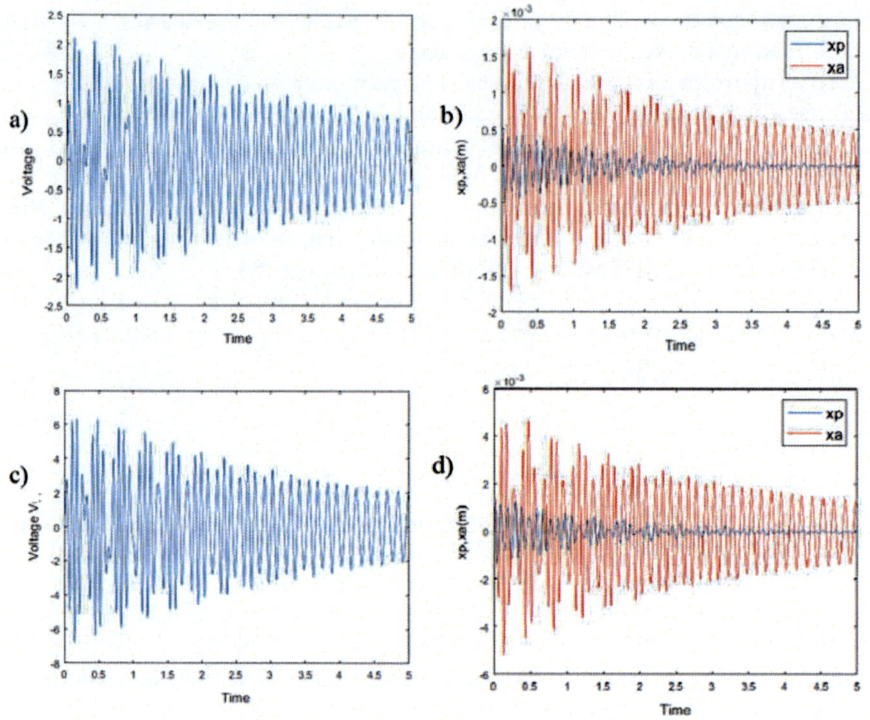

Fig. 2 **a** V versus t plot for $X = 0.52$ mm; **b** x versus t plot for $X = 0.52$ mm; **c** V versus t plot for $X = 1.42$ mm; **d** x versus t plot for $X = 1.42$ mm

For $X = 0.52$ mm, the maximum value of voltage can be obtained for a k_3 value of 2×10^8 and the maximum displacement of auxiliary mass is also obtained for the same k_3 value.

For $X = 1.42$ mm, however, the maximum value of voltage and x_a are obtained at k_3 value of 2×10^7 with a similar observation relation being observed for the two other initial displacement values 2.65 and 4.45 mm as well. Hence, k_3 doesn't affect the interrelation between auxiliary mass displacement and voltage generated in any way.

For constant k_3 value, greater initial displacements showed higher voltage values as is observable in Fig. 3. Similar increment in magnitude can be observed for the other two higher initial displacement values as well.

Also, irrespective of the value of initial displacement, negligible change is observed in the pattern of voltage generated (with respect to time) when k_3 lies between 2×10 and 2×10^4. There is a negligible change in magnitude as well. Variations occur only when k_3 values higher than 2×10^5 are considered. It can be observed in Fig. 4 where $X = 0.52$ mm has been considered with similar observations being made for the other initial displacement values as well.

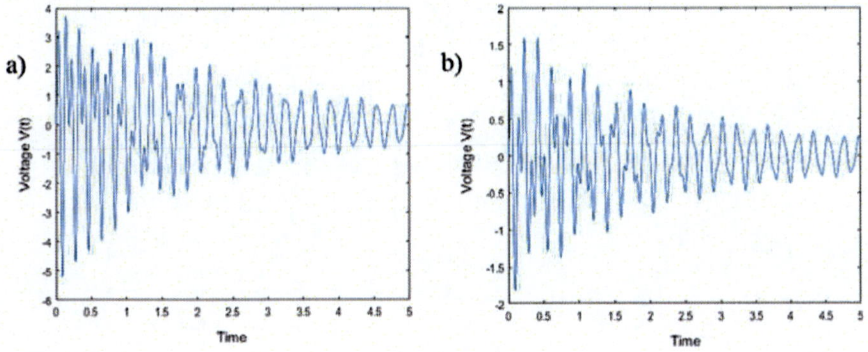

Fig. 3 **a** V versus t plot for $X = 1.42$ mm with $k_3 = 2 \times 10^6$; **b** V versus t plot for $X = 0.52$ mm with $k_3 = 2 \times 10^6$

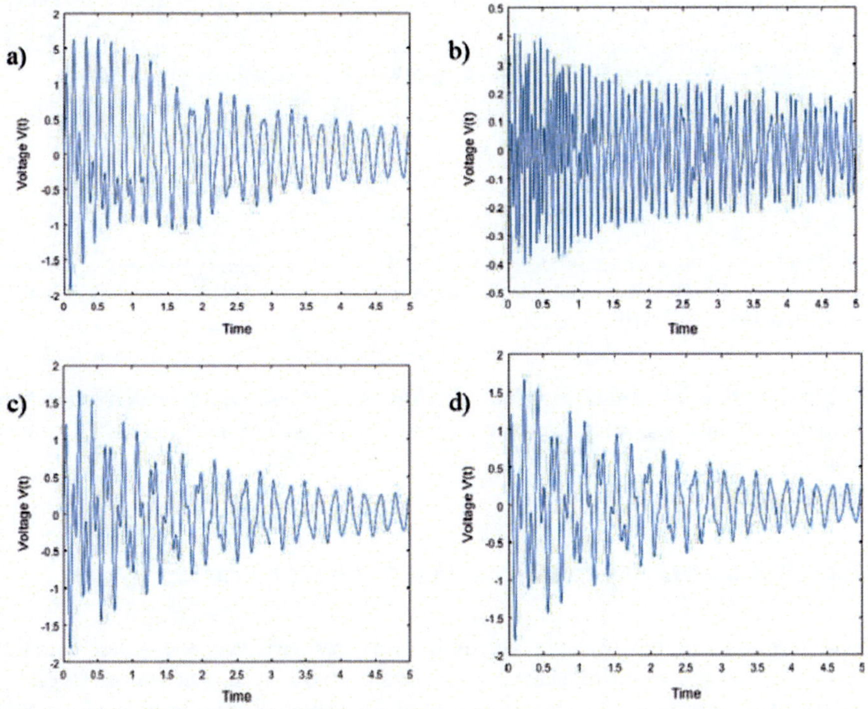

Fig. 4 **a** V versus t plot with $k_3 = 2 \times 10^7$; **b** V versus t plot with $k_3 = 2 \times 10^{10}$; **c** V versus t plot with $k_3 = 2 \times 10$; **d** V versus t plot with $k_3 = 2 \times 10^4$

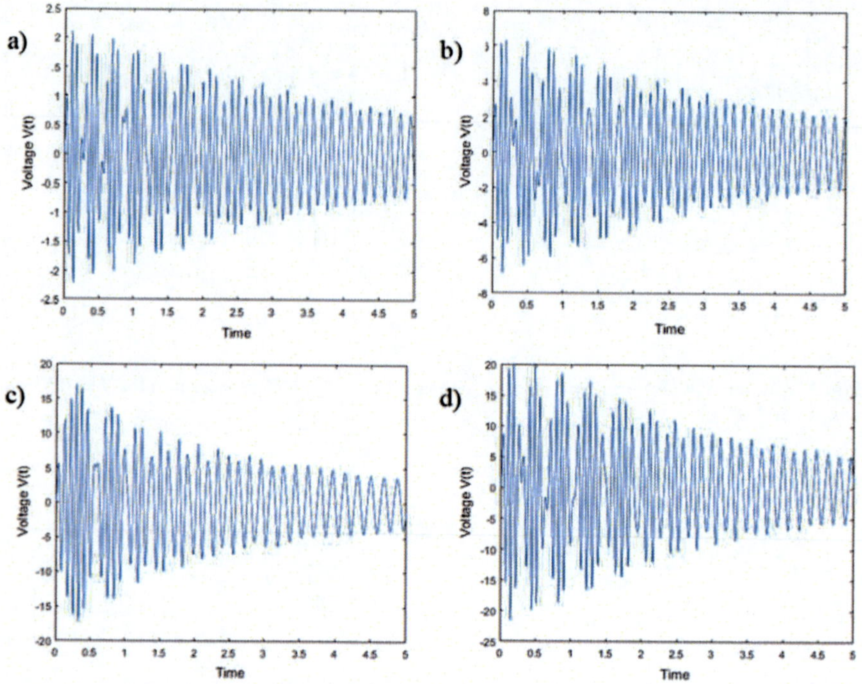

Fig. 5 **a** V versus t plot with $X = 0.52$ mm and $k_3 = 2 \times 10^8$; **b** V versus t plot with $X = 1.42$ mm and $k_3 = 2 \times 10^7$; **c** V versus t plot with $X = 2.65$ mm and $k_3 = 2 \times 10^6$; **d** V versus t plot with $X = 4.45$ mm and $k_3 = 2 \times 10^6$

Noteworthily, no optimum value of nonlinear stiffness could be derived as the system exhibited maximum voltage for different X values at different k_3 values (Fig. 5).

3.2 *Parametric Sweep of* k_3 *with a Nonlinear Primary System*

In this section, Eqs. (4)–(6) have been used. Two cases have been considered with the value of k_3 being varied while keeping a constant value of k_p and vice versa. Its effect on voltage generation and primary vibration suppression have been investigated. Here, k_p refers to the nonlinear stiffness of the primary system. The primary system considered is essentially nonlinear in nature as no primary linear stiffness has been considered.

As can be observed in Fig. 6, for a nonlinear primary system too, primary vibration has been reduced for both the aforementioned cases. Irrespective of the value of k_p or k_3 considered, with respect to the constant value chosen, primary vibration can be seen to be dying down.

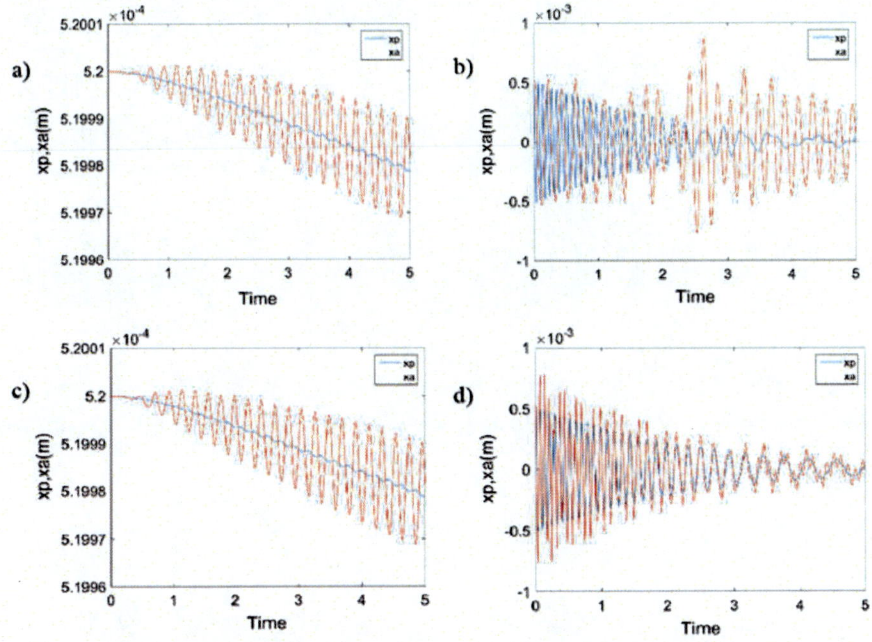

Fig. 6 **a** x versus t plot with $k_3 = 2 \times 10$ and $k_p = 4 \times 10$; **b** x versus t plot with $k_3 = 2 \times 10$ and $k_p = 4 \times 10^{10}$; **c** x versus t plot with $k_3 = 2 \times 10^{10}$ and $k_p = 4 \times 10$; **d** x versus t plot with $k_3 = 2 \times 10^{10}$ and $k_p = 4 \times 10^{10}$

For $X = 0.52$ mm–

Similar observations have been made for the other three initial displacement values as well.

Also, a pattern can be observed. Irrespective of k_3 value, the pattern and magnitude of voltage generated is determined by k_p as can be seen in Figs. 7 and 8.

For $X = 0.52$ mm–

As can be seen, k_3 is irrelevant to the magnitude and pattern of voltage generated. However, what can be observed throughout is that even for nonlinear primary systems, voltage is generated by NES and hence, NES can be concluded to be useful for such systems as well.

3.3 Random Vibrations

The modelling of random vibrations was done by introducing a deterministic harmonic (sinusoidal) excitation, which depending on the random variable, was split into two cases to introduce the random vibration, forcing magnitude F or the frequency of forcing ω. The modelling of random vibration was done with the help

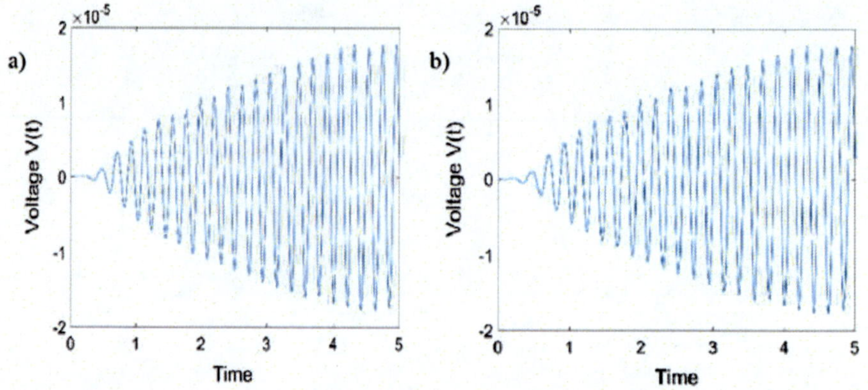

Fig. 7 **a** V versus t plot for $k_3 = 2 \times 10$ for $k_p = 4 \times 10$; **b** V versus t plot for $k_3 = 2 \times 10^{10}$ for $k_p = 4 \times 10$

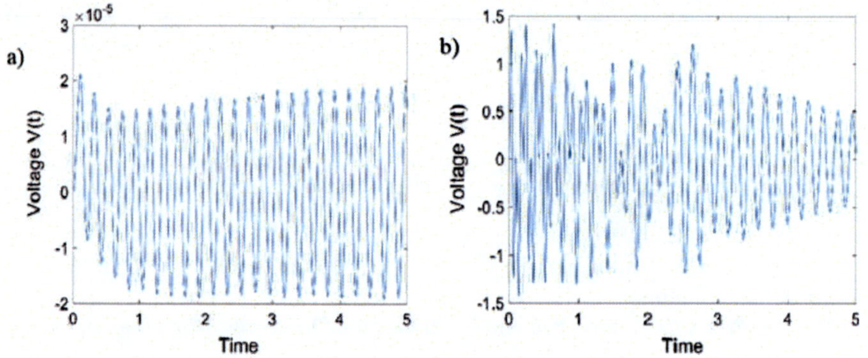

Fig. 8 **a** V versus t plot for $k_3 = 2 \times 10$ for $k_p = 4 \times 10^4$; **b** V versus t plot for $k_3 = 2 \times 10$ for $k_p = 4 \times 10^{10}$

of rand() function in MATLAB which generates a random number between 0 and 1. The two cases are (1) Random forcing, where F is the random variable in the harmonic excitation and $F = F_{\text{mean}} + \sigma$*random and (2) Random frequency, where w is the random variable in the harmonic excitation and $\omega = \omega_{\text{mean}} + \sigma$*random. In this formula, σ, represents the noise intensity and random represents the random number generated by the MATLAB function rand().

Using the governing Eqs. (2), (3) and (7) and values considered from previous literature [9], one case of initial displacement $X = 0.52$ mm was considered. Arbitrary values of F, F_{mean}, ω, ω_{mean} are considered in both the cases and σ is varied from 0.1 to 0.9 in order to study the effect of noise intensity on the voltage generated and displacements of the NES and primary system. Values of parameters used in case 1 are $F_{\text{mean}} = 1$, $\omega = 5$,σ varying from 0.1 to 0.9 and in case 2 is $F = 1$, $\omega_{\text{mean}} = 5$, σ varying from 0.1 to 0.9. Using the same F and ω values, the comparison between

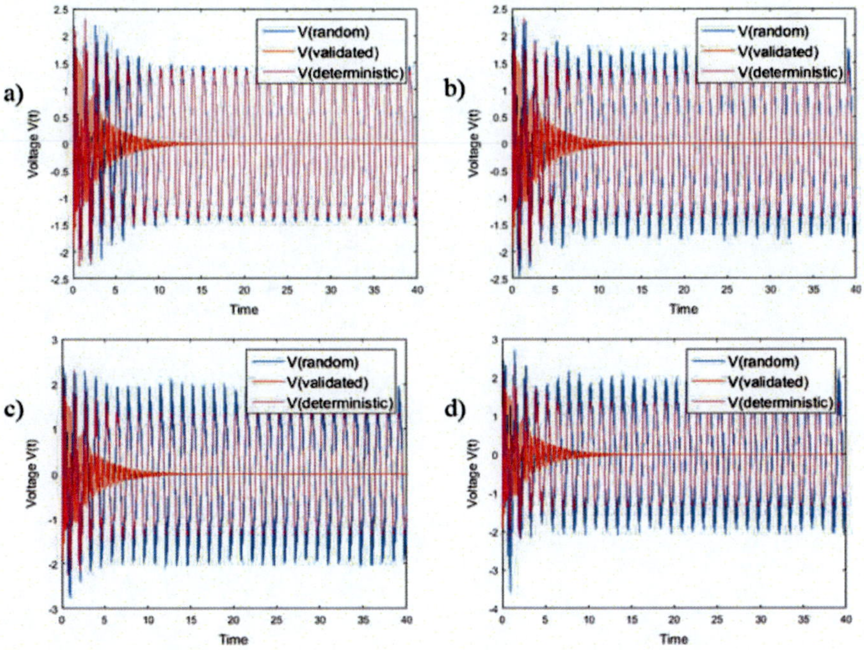

Fig. 9 Simulation results of voltage generated in case 1 from PVEH with random and deterministic excitation for $F_{\text{mean}} = 1$, $\omega = 5$ and $X = 0.52$ mm: **a** for $\sigma = 0.1$; **b** for $\sigma = 0.5$; **c** for $\sigma = 0.7$ and **d** for $\sigma = 0.9$

the voltage generated by both the deterministic harmonic excitation and the random excitation in both the cases for different σ values is presented in Figs. 9 and 10. In case 1, random F, the difference in magnitude between the random and deterministic voltage generated increases with an increase in σ as shown in Fig. 9. In case 2, random ω, the voltage generated dies down faster with increase in σ as shown in Fig. 10. The RMS voltages generated in both the cases are evaluated and compared with the deterministic harmonic excitation as well as with the results validated from literature [9]. This is represented in Tables 1 and 2. The RMS voltage generated increases with increase in σ in case 1 while it decreases with increase in σ in case 2 as shown in Tables 1 and 2.

The structural response of the system is shown in Figs. 11 and 12. It can be observed in Figs. 11 and 12, that NES displacement is more than the displacement of the primary mass, and even at higher values of σ the NES protects the primary system from excessive vibrations in both the random cases as shown in Figs. 11 and 12.

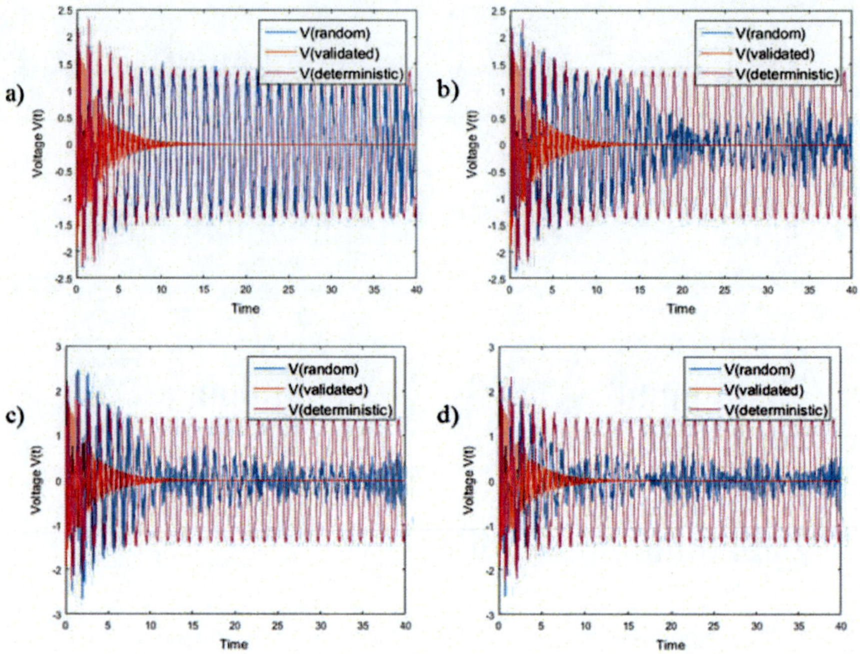

Fig. 10 Simulation results of voltage generated in case 2 from PVEH with random and deterministic excitation for $F_{\mathrm{mean}} = 1$, $\omega = 5$ and $X = 0.52$ mm: **a** for $\sigma = 0.1$; **b** for $\sigma = 0.5$; **c** for $\sigma = 0.7$ and **d** for $\sigma = 0.9$

Table 1 Comparison of RMS voltage generated in case 1: random F

σ value	RMS voltage (validated)	RMS voltage (deterministic)	RMS voltage (random)
$\sigma = 0.1$	0.2077	1.0093	1.0414
$\sigma = 0.3$	0.2077	1.0093	1.0896
$\sigma = 0.5$	0.2077	1.0093	1.1470
$\sigma = 0.7$	0.2077	1.0093	1.1945
$\sigma = 0.9$	0.2077	1.0093	1.2582

Table 2 Comparison of RMS voltage generated in case 2: random ω

σ value	RMS voltage (validated)	RMS voltage (deterministic)	RMS voltage (random)
$\sigma = 0.1$	0.2077	1.0093	0.8883
$\sigma = 0.3$	0.2077	1.0093	0.5858
$\sigma = 0.5$	0.2077	1.0093	0.4919
$\sigma = 0.7$	0.2077	1.0093	0.4221
$\sigma = 0.9$	0.2077	1.0093	0.3953

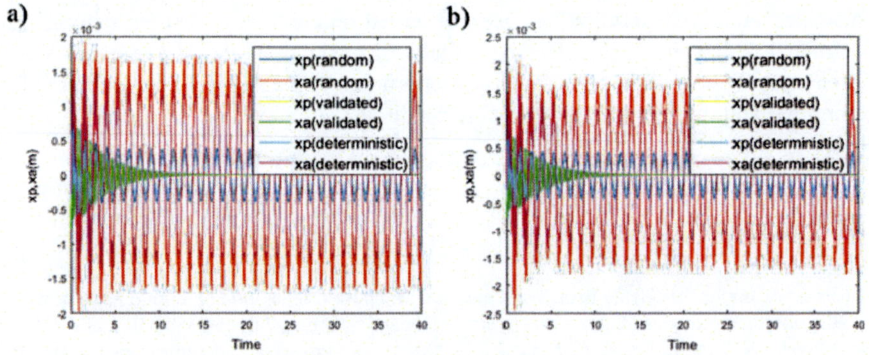

Fig. 11 Simulation results of the structural response of the system for case 1 with random and deterministic excitation for $F_{mean} = 1$, $\omega = 5$ and $X = 0.52$ mm: **a** for $\sigma = 0.7$ and **b** for $\sigma = 0.9$

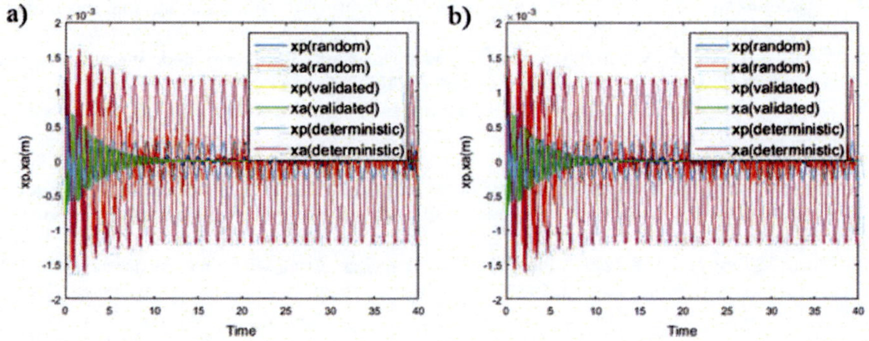

Fig. 12 Simulation results of the structural response of the system for case 2 with random and deterministic excitation for $F = 1$, $\omega_{mean} = 5$ and $X = 0.52$ mm: **a** for $\sigma = 0.7$ and **b** for $\sigma = 0.9$

4 Conclusion

Based on the NES principle, a piezoelectric harvesting device has been modelled. Validation of previous literature [9] was done. The transient response and the steady-state response of the system, regarding energy harvesting and vibration suppression, were analyzed using numerical simulations. Investigations into effect of auxiliary nonlinearity when connected to a primary system with deterministic harmonic excitation were made. All in all, the NES successfully suppressed primary vibration and continued to generate voltage over a wider operational bandwidth, regardless of the primary system being linear or nonlinear. The degree of effect auxiliary nonlinearity has on the variation in voltage generation, voltage pattern and its magnitude have been understood. Random excitation to the system is also modelled and the structural response as well as the voltage generated have been presented. It was observed that NES continues to suppress primary vibration just as in deterministic forcing but sustained voltage generation has been observed, in contrast to that for deterministic model which showed decreasing voltage. Investigations into the effect of noise intensity (σ) showed that it is directly proportional to the rms voltage generated for random F and indirectly proportional to random ω.

References

1. Ertuk A, Inman DJ (2011) Broadband piezoelectric power generation on high-energy orbits of the bistable duffing oscillator with electromechanical coupling. J Sound Vib 330:2339–2353
2. Liuyang X, Tang L, Liu K, R. Mace B., On the use of piezoelectric nonlinear energy sink for vibration isolation and energy harvesting. In: Conference on smart materials, adaptive structures and intelligent systems, pp 1–6 (2018)
3. Malaji PV, Ali SF (2018) Analysis and experiment of magneto-mechanically coupled harvesters. Mech Syst Signal Process 108:304–316
4. Rajarathinam M, Ali S (2018) Energy generation in a hybrid harvester under harmonic excitation. Energy Convers Manage 155:10–19
5. Kumar KA, Ali SF, Arockiarajan A (2017) Magneto-elastic oscillator: modeling and analysis with nonlinear magnetic interaction. J Sound Vib 393:265–284
6. Xiong L, Tang L, Liu K, Mace B (2018) Broadband piezoelectric vibration energy harvesting using a nonlinear energy sink. J Phys D Appl Phys 51:1–2
7. Malaji PV, Rajarathinam M, Jaiswal V, Ali SF, Howard IM (2019) Energy harvesting from dynamic vibration pendulum absorber. Recent Adv Struct Eng 2:467–478. Springer Singapore
8. Kremer D, Liu K (2014) A nonlinear energy sink with an energy harvester: transient responses. J Sound Vib 333:4859–4861
9. Zhang Y, Tang L, Liu K (2017) Piezoelectric energy harvesting with a nonlinear energy sink. J Intell Mater Syst Struct 28(3):307–322

A Novel Banana Leaf Waste-Based Activated Carbon for Automobile Emission Control

A. John Presin Kumar, S. Sivakumar, R. Balaji, and Mukesh Nadarajan

Abstract Recycling of plant and vegetable wastes is very much needed for improving the economy and minimizing wastages of materials. This can be a very beneficial act toward the progress of society overall. Environmental pollution issues especially from automobiles are also much decreased by this. Activated carbon is better known for its characteristics and applications in the areas of gas absorption, water purification and as composite reinforcement fillers. In this current research work, an attempt is made to economically synthesize activated carbon from dried banana leaf waste taken from a local vegetable market. This is characterized by measuring moisture content, ash content, bulk density, yield of charcoal, fixed carbon and hardness. Gas adsorption test is conducted using Atlon gas analyzer for various engine speeds. Its effect on being porous is analyzed by means of comparing its performance with the solid ones. Recycled banana leaves are used as the natural resource, and heat activation is used for converting the so-produced carbon into activated carbon. Potassium hydroxide is used for synthesizing porous activated carbon. Comparison is made between solid and porous activated carbons for better adsorbing capabilities for automobiles. Scanning electron microscopy was done to analyze the morphology of particles.

Keywords Activated carbon · Potassium hydroxide · Porous structure · Fixed carbon · Bulk density

---

A. John Presin Kumar (✉) · S. Sivakumar · R. Balaji · M. Nadarajan
Department of Mechanical Engineering, Hindustan Institute of Technology & Science, Padur, Kelambakkam, Chennai, Tamil Nadu 603103, India
e-mail: johnpk@hindustanuniv.ac.in

S. Sivakumar
e-mail: ssivak@hindustanuniv.ac.in

R. Balaji
e-mail: sridharabalaa@gmail.com

M. Nadarajan
e-mail: revamukesh@gmail.com

S. Vijayan et al. (eds.), *Trends in Manufacturing and Engineering Management*, Lecture Notes in Mechanical Engineering,
https://doi.org/10.1007/978-981-15-4745-4_85

1 Introduction

Activated carbon is a type of carbon processed to have minute, low-volume pores. These pores can increase the surface area of activated carbon particles for better adsorption capability. Activated carbon particle is shown in Fig. 1. Thus due to its high degree of microporosity, one gram of activated carbon has a greater surface area. The amount of activation level may be varied. A desired activation level sufficient for useful application may be attained from heating.

Many varieties of applications are possible for activated carbons. Such type of industrial applications involves the use of activated carbon as gas absorbents, in metal finishing for purification of electroplating solutions and as fillers in composites. Activated carbon technique is a main purification technique for removing impurities from nickel plating solutions. Activated carbon can also be effectively used as reinforcement fillers in fabricating various composites such as wood or polymer or metal composites [1, 2]. Though there are many advantages of activated carbons, they also have few drawbacks. For instance, the price of activated carbons is considerably high due to immense energy cost during their synthesis. Therefore, it is always welcome to create an economical method of producing activated carbon based on natural resources.

Every day, an average of over 5000 useful products are developed from plant wastes in our country. Banana trees always play a big role in our personal life as well as domestic life. The Indian government in earlier established a target to raise the forested area to one-third of India's land area. But increased pressure from humans stopped this initiative. Examples of these are harvesting wood for fuel and other uses, and clearing the land for agriculture, and others have prevented the government from attaining its objective. Wood recycling is the process of converting waste timber into new and usable products. This is adopted in many countries along the globe. Through recycling of wood only, we have developed re-useable products, artificial wood, etc., which are environment friendly. Banana leaves after their usage in banana mandis (market) go simply as scrap or wastes as shown in Fig. 1. Dried banana leaves are a good source of carbonaceous material. Figure 2 shows a useful product such as a bin, made out of banana leaves (Fig. 3).

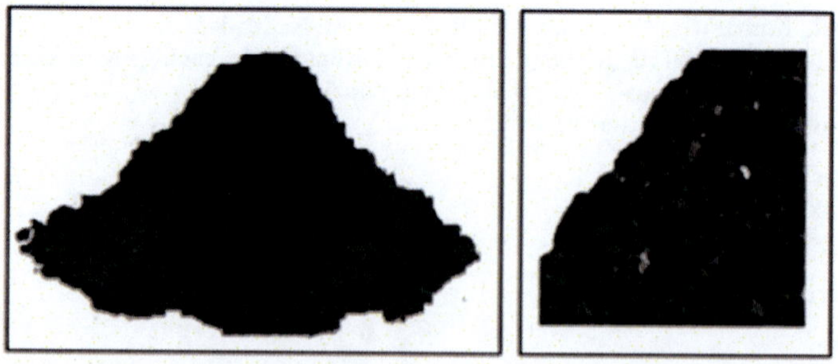

Fig. 1 Activated carbon powder

Fig. 2 Dried banana leaf

Fig. 3 A bin made of banana leaf

The literature survey indicates that activated carbons can be prepared from a variety of raw materials such as coconut shell, rice husk, walnut shell, banana stem and animal bones [3–5]. The aim of this paper is to synthesize activated carbons from dried banana leaves by heat and chemical activation and to investigate their important characteristics, especially as an efficient gas adsorbing agent. Two forms of activated carbon were compared for characterization like solid and porous structures. Porous structure was achieved by using potassium hydroxide KOH during chemical activation. Figure 4 shows the overall main steps involved in this current work.

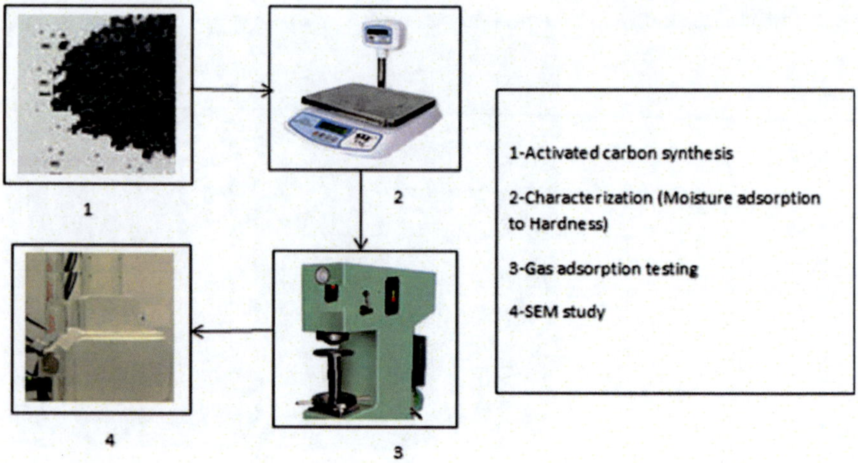

Fig. 4 Steps involved in the methodology

2 Preparation of Activated Carbon

Figure 1 shows the steps involved in the methodology clearly. The dried banana leaf samples were collected from a local banana and vegetable market in Chrompet, Chennai, Tamil Nadu, India. These were crushed into small pieces of about 2- to 3-mm diameter for providing large surface area. Then, these particles are sieved to get rid of dust particles and other impurities. These crushed pieces were put in an oven, and the temperature was maintained at 105 °C and soaked for 24 h. The dried samples were taken in airtight plastic bags to avoid re-absorption of moisture from the atmospheric air. About 103 g of samples was taken in a crucible with tightly closed tightened lid for ensuring airtight environment. The crucible is placed in the furnace carefully. The furnace was switched on, heated to a temperature of 460 °C and held constant for 30 min. The charcoal obtained from pyrolysis process was kept in an airtight bag in the activation unit, and the furnace was heated to a temperature of 750 °C. Steam was allowed to flow to it for 30 min by keeping the temperature constant. Activated carbon thus obtained has very less porosity which is ignorable and thus considered solid activated carbon particles.

In order to produce porous structured activated carbon [2], chemical activation was performed using KOH at various concentrations like 0.0, 0.3, 0.5 and 0.7 M preferably at 0.5 M. All samples were carbonized at a temperature of 600 °C using a furnace followed by physical activation using CO_2 gas at a temperature of 850 °C which produces porous activated carbon. Thus, both solid and porous activated carbon particles are obtained. These different forms of activated carbon were compared later for gas adsorption performances.

3 Characterization of the Activated Carbon

The activated carbon produced was characterized for its moisture content, ash content, bulk density, yield of charcoal, fixed carbon, hardness and gas adsorption capability.

3.1 Moisture Content (MC)

Moisture content of activated carbon was calculated by finding the difference in weight of sample before and after drying. The weight of the empty crucible was noted down first. Then, total weight of sample and crucible was noted as W_s. The sample was dried in an oven at 105 °C for a time period of 24 h, and then it was allowed to cool down. The sample along with crucible was weighed again W_f, and the value of $((W_s - W_f * 100))/(W_s - W_e)$ gives the percent of moisture content [1].

3.2 Ash Content (AC)

Dried sample of activated carbon was weighed and recorded as W_d. These were placed in a crucible, and the total weight of crucible plus dry sample was noted and recorded as W_s. Finally, it was placed in a furnace at 500 °C for a time period of 6 h [1]. It was removed and its content reweighed and recorded W_f. These steps were repeated twice for all the activated carbon samples, and the estimated value of the ash content was calculated as: $((W_s - W_f * 100))/(W_d)$.

3.3 Bulk Density

For calculating this, the activated carbon sample was transferred into a cylinder of 10 ml capacity and the weight of the sample along with cylinder was recorded as W_s [2]. The estimated value of the bulk density was calculated as: weight of sample (W_s)/weight of equal volume of water (W_w).

3.4 Yield of Charcoal (YC)

The yield of charcoal [1] that is available in the activated carbon was estimated as:

$$\mathrm{WS} - \mathrm{AC} + \mathrm{MC} * 100/\mathrm{WS}$$

where WS is the original weight of dry sample, AC is the ash content and MC is the moisture content.

3.5 *Fixed Carbon (Fc)*

The fixed carbon content in each activated carbon sample was estimated [1] using:

$$\mathrm{YC} - \mathrm{AC} - \mathrm{MC} * 100/\mathrm{YC}$$

where YC is the yield of charcoal, AC is the ash content and MC is the moisture content.

3.6 *Hardness (H)*

This involves the following steps. First, 2 g of activated carbon sample was measured in a standard sieve of size 40. The sieve was made to shake along with particles for a time period of 1 h using mechanical shaker at 160 rpm with four steel balls of 6.00-mm diameter inside the sieve [1]. Finally, the quantity of activated carbon retained inside the sieve after 1 h was expressed in percentage as hardness.

3.7 *Gas Adsorption*

To obtain the gas adsorption characteristics of activated carbon, batch adsorption experiments were conducted [4]. This involves taking into account the effects of the following parameters like initial concentration, contact time, volume and weight of adsorbent. The adsorbed amounts of the solution are given by the equation [4] as

$$Q = (\mathrm{CO} - \mathrm{CT}) * V/W$$

where CO is the initial concentration of solution, CT is the final concentration of solution at time T, V is the volume of solution and W is the weight of adsorbent used.

3.8 *Scanning Electron Microscopy*

The surface morphology was obtained using scanning electron microscopy (SEM).

3.9 Engine and Gas Analyzer

This current research work was conducted in our institute Hindustan Institute of Technology and Science, Chennai. Figure 5 shows single cylinder water cooled naturally aspirated compression ignition engine having maximum horse power (hp) of 10 hp. The output shaft is connected to an electrical dynamometer which has maximum capacity of 24 amps. The engine used a mechanical injector, and it is a constant engine which is designed to run at 1500 rpm [6, 7].

The engine specification is given in Table 1.

Atlon 5 gas analyzer shown in Fig. 6 was used to measure the emission parameter

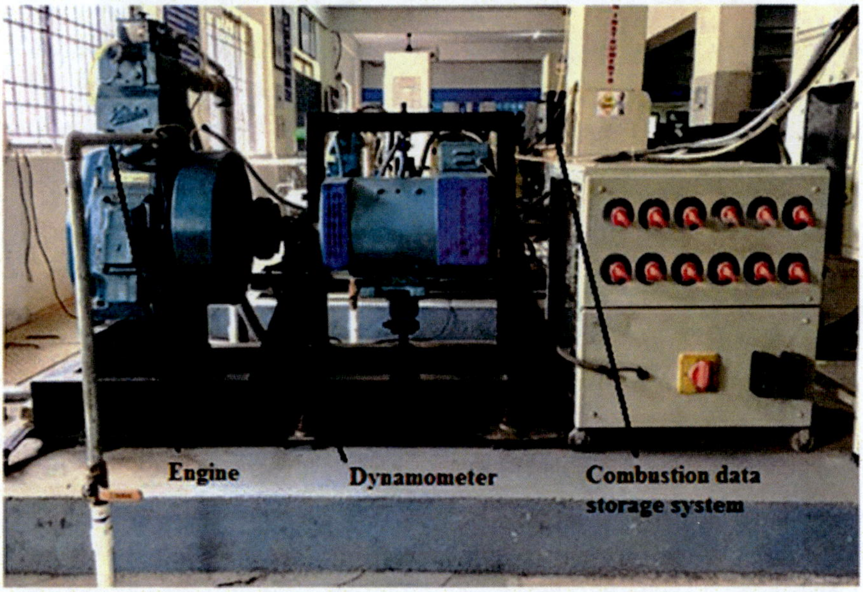

Fig. 5 Kirloskar DM 10 engine

Table 1 Engine specification

Make and model	Kirloskar DM 10
Bore/stroke	102 * 116 mm
Speed	1500
B.H.P	10 hp
Swept volume	661 cc
Compression ratio range	17:5:1
Fuel	Diesel
Torque	0.048 KN m

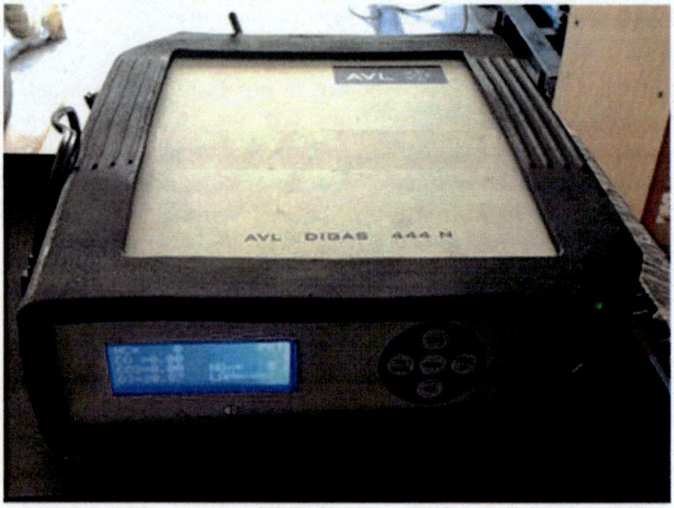

Fig. 6 Atlon gas analyzer

Table 2 Gas analyzer specification

S. No.	Make	AVI
1	Model	DIGAS 444
2	Display	LCD
3	Interface	RS 232
4	Power consumption	25 W
5	Operating voltage	230 V 55 Hz

of the exhaust gas. This gas analyzer is capable of measuring 5 parameters: carbon monoxide (CO), carbon dioxide (CO_2), hydrocarbon (HC), oxides of nitrogen (NO_x) and oxygen (O_2) whose specification is given in Table 2.

4 Results and Discussion

The technical analyses of 5 samples are shown in Fig. 3. It should be noted that all the samples were representative of activated carbon particles produced under similar conditions [5, 8–11], namely AC1–AC5. These samples were categorized based on the different combinational values of temperature and soaking duration. Temperatures were taken like 105, 110, 115, 120 and 125 °C, and soaking times were taken like 30, 24, 20, 15 and 10 h. The properties were found to be showing similar values in all the samples indicating isotropic nature [12, 13]. These are shown in Tables 1 and 2 for comparing the solid and porous samples, respectively.

Table 3 clearly shows various characteristics of activated carbon without porosity inside them. Being solid particles, these activated carbon particles may have lesser surface area compared to the porous ones.

From Table 4 values for solid activated carbon, it was clear that the sample AC3 showed promising results indicating the better preparation temperature of 115 °C and an optimum soaking time of 20 h for all properties other than moisture and ash removal.

By comparing the two tables, the values of hardness, bulk density and amount of carbon were slightly better than that of solid activated carbon. Similarly, from Table 4 values for porous activated carbon, it was clear that the sample AC3 showed promising results indicating the better preparation temperature of 115 °C and an optimum soaking time of 20 h for all properties other than moisture and ash removal.

The adsorption of total unburnt gases was first analyzed using the Atlon gas analyzer, and comparison is made among various performance curves drawn for varying engine speeds against activated carbon samples prepared at different activation temperatures. This performance curve is illustrated in Fig. 7. It was found that the performance was better for full speed of the engine (1500 rpm) and also at a higher activation temperature of 810 °C. The temperature for 15 different samples chosen was varying at 30 °C from its predecessor.

The adsorption capacity toward NO_x, CO_2 and HC was analyzed for various activated carbon samples. These samples indicate the different temperature and soaking

Table 3 Characteristics of solid activated carbon

Sample No.	Moisture content %	Ash content %	Bulk density g/cm^3	Fixed carbon %	Hardness
AC1	11.2	12.2	0.775	88.2	7.34
AC2	11.8	12.6	0.728	88.7	7.11
AC3	10.8	12.1	0.639	89.1	7.56
AC4	10.3	11.9	0.788	88.9	7.23
AC5	11.9	12.3	0.774	89.0	7.44

Table 4 Characteristics of porous activated carbon

Sample No.	Moisture content %	Ash content %	Bulk density g/cm^3	Fixed carbon %	Hardness
AC1	10.9	11.3	0.776	89.3	7.35
AC2	10.8	11.6	0.738	89.7	7.31
AC3	10.5	11.1	0.798	89.9	7.66
AC4	10.2	10.9	0.791	88.2	7.25
AC5	10.9	11.4	0.771	89.1	7.64

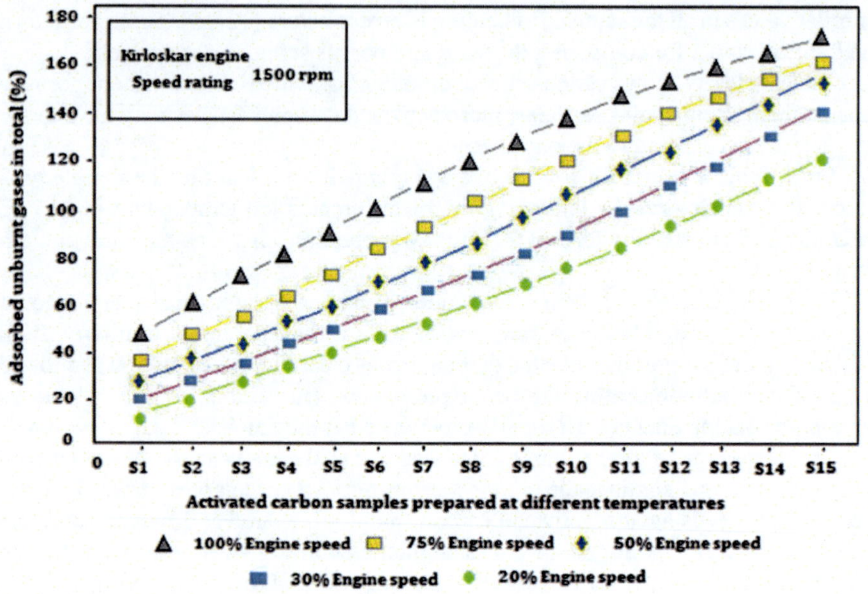

Fig. 7 Gas adsorption comparison for different activation temperatures

time durations in oven. The sample showing better gas adsorption indicated clearly the appropriate temperature and soaking time duration which were optimum for preparing activated carbon.

The following graph in Figs. 8 and 9 for solid and porous activated carbon, respectively, showed the performance characteristic of activated carbon samples from AC1 to AC5 in observing the harmful gases. For each sample, the average values of 5 trials were plotted. The sample AC3 showed promising characteristic in adsorbing all three types of gases (Table 5).

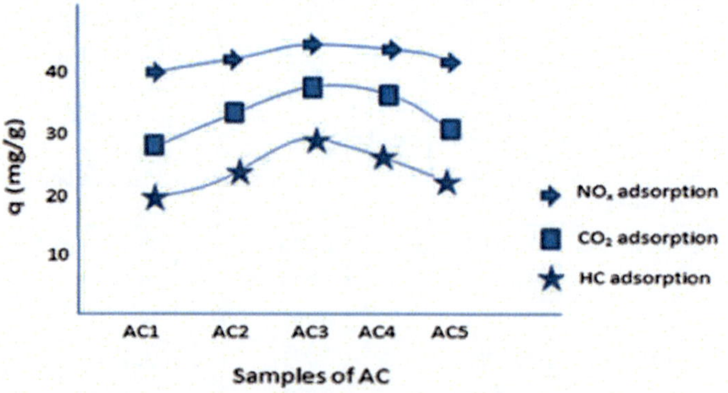

Fig. 8 Gas adsorption curves for solid activated carbon

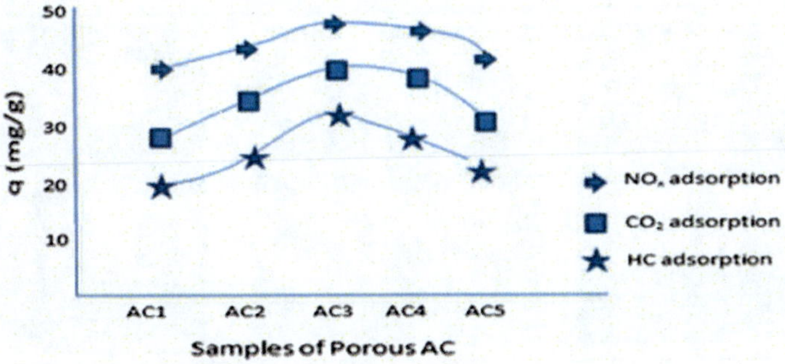

Fig. 9 Gas adsorption curves for porous activated carbon

Table 5 Comparison of gas adsorption %

	Solid activated carbon	Porous activated carbon
% of gas adsorption	66	70

The effect of higher to lower loading on the gas adsorption performance is shown as various trials in Fig. 10. Porous activated carbon gave better results in all trials compared to the solid ones having the values of 68%, 55% and 48%, respectively, whereas the solid ones were having 65%, 53% and 42%, respectively, for trial 1, trial 2 and trial 3, respectively. The effect of chemical and heat activation on the surface morphology of activated carbon particles from banana true stems is clearly shown in Fig. 11. SEM micrographs show that the samples prepared have more pores. The

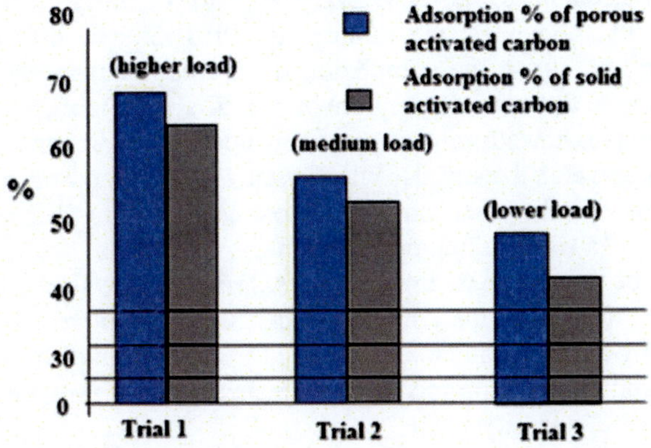

Fig. 10 Overall gas adsorption performance

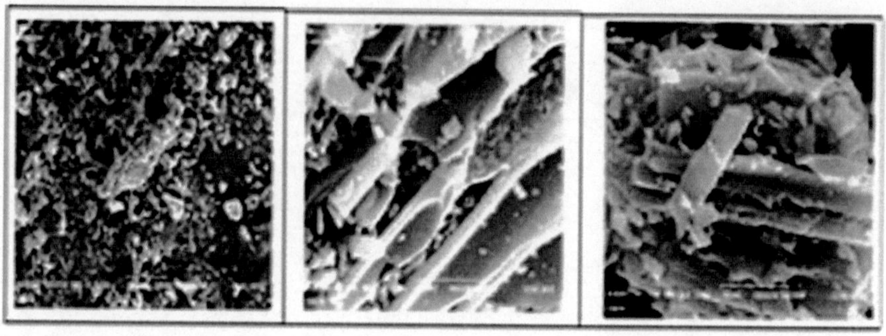

Fig. 11 SEM micrographs of activated carbon particles

sample contains fiber structures such as an agglomeration of fibers. Inclusion of more pores and agglomeration are important and needed properties for successfully activated carbon particles.

5 Conclusion

- The effectiveness of dried banana leaves in providing activated carbon was analyzed in this work by means of comparing the performance among two of its forms such as solid and porous ones. It was found initially in an overall performance observation made for different activated carbon samples prepared at different activation temperatures that the gas adsorption was better for full speed of the engine (1500 rpm) and also at a higher activation temperature of 810 °C. The temperature for 15 different samples chosen was varying at 30 °C from its predecessor.
- In the case of solid ones, the sample AC3 had removal efficiency of gases greater than 66% upon the completion of five consecutive adsorption cycles corresponding to the oven temperature of 115 °C and 20 h of soaking time.
- The same sample showed promising performance in other properties like hardness, fixed carbon and bulk density also whereas the moisture removal and ash removal had been better for the other sample AC4 corresponding to the oven temperature of 120 °C and 15 h of soaking time duration.
- In the case of porous ones, the sample AC3 had removal efficiency of gases greater than 70% upon the completion of five consecutive adsorption cycles corresponding to the oven temperature of 115 °C and 20 h of soaking time.
- The same sample showed promising performance in other properties like hardness, fixed carbon, bulk density.
- Here also, the moisture removal and ash removal have been better for the other sample AC4 corresponding to the oven temperature of 120 °C and 15 h of soaking time duration.

- Dried banana leaves had shown good ability to be a better resource for activated carbon synthesis. The porosity providing effect of potassium hydroxide solution was also very clear from the results.

Acknowledgements We would like to express our sincere gratitude and thanks to Dr. A. Velayudham, Scientist 'G', Ministry of Defence, CVRDE, Avadi, Chennai, India, and Dr. Ashish Selokar, Accendere, CL Educate Ltd., New Delhi, India, for their valuable suggestions. We express our sincere gratitude and thanks for their helps in the whole duration of executing this research work.

References

1. Sunday NJ, Okechukwu NS, Elom N, Anthony O, Michael PT (2018) Quantitative characterization of activated carbon from cow, donkey, chicken and horse bones from Ezzangbo in Ebonyi State, Nigeria. Amer J Appl Chem Sci Publ Group 6(5):169–174
2. Taer E, Taslim R, Mustika WS, Kurniasih B, Agustino, Afrianda, Apriwandi A (2018) Int J Electrochem Sci 13:8428–8439
3. Lebedeva MV, Ayupov AB, Yeletsky PM, Parmon VN (2018) Rice husk derived activated carbon/polyaniline composites as active materials for supercapacitors. Int J Electrochem Sci 13:3674–3690
4. Haji Azaman SA, Afandi A, Hameed BH, Mohd Din AT (2018) J Appl Sci Eng 21(3):317–330
5. Subramani T, Revathi PK (2015) Production of activated carbon from agricultural raw waste. IOSR JEng (IOSRJEN) 05(05):54–63
6. Udhaya Kumar V, Bharath P, John Presin Kumar A (2017) Experimental Investigation of performance and emission characteristics of ethanol blended diesel with zinc oxide nanoparticles. Int J Eng Tech 9(3):1524–1531
7. John Presin Kumar A, Bharath V, Balakumar S, Godwin John J (2014) A study on the effect of the dual fuel mixture of diesel and LPG, CNG gases on diesel engines. Int J Ind Prod Eng Tech 4(1):11–13
8. Aziz NAA, Mohamed NAA, Mohamad M, Amini MHM, Abdul Aziz MS, Yusoff H, Ismael Rizman Z (2015) ARPN J. Eng Appl Sci 10(1):376–386
9. Chandramohan D, John Presin Kumar A (2017) Experimental data on the properties of natural fiber particle reinforced polymer composite material. Elsevier, Data Brief 13:460–468
10. Chandramohan D, John Presin Kumar A (2017) Data-enabled discover. Appl, Cross Mark, Springer Inter Publ 1–9
11. Tsyntsarski B, Marinov S, Budinova T, FerhatYardim M, Petrov N (2013) Synthesis and characterization of activated carbon from natural asphaltites. Fuel Processing Tech. Elsevier, 116:346–349
12. Abdul Khalil HPS, Noriman NZ, Ahmad MN, Ratnam MM, NikFuaad NA (2007) J Reinf Plast Comp 26(3):305–320
13. Flores A, Cagiao ME, Ezquerra TA, BaltaCalleja FJ (2001) J. Appl. Pol. Sci. 79:90–95
14. SubhashreePradhan, Project report, B. Tech-Chem Engineering, Department of Chemical Engineering, NIT, Rourkela, pp 1–9 (2010)
15. Abubakar M, Alechenu AA, Manase A, Mohammed J (2012) Adv Appl Sci Res 3(5):3089–3096
16. Cobb A, Warms M, Maurer EP, Chies S (2012) IntJ Serv Learn Eng 7(1):93–104
17. Aziz NAA, Mohamed M, Mohamad M, Amini MHM, Aziz MA, Yusoff H, Rizman ZI (2015) ARPN J Eng Appl Sci 10(1):376–386

Analyzing Different Methods to Increase the Natural Period of a Compact Wave Energy Converter

Vishnu Vijayasankar and Abdus Samad

Abstract In this article, we present the numerical analysis performed on a compact mechanical-based direct drive wave energy converter (WEC), in order to improve its performance on longer wave periods. For this work, the numerical modeling of the point absorber is done with the help of an open-source code written in MATLAB, WEC-Sim and the hydrodynamic parameters of the WEC are determined using the BEM code AQWA. From the past experience, it is a known fact that the range of frequencies for which a point absorber can perform satisfactorily is less if they are small-sized. In this work, we make an effort to increase the operational bandwidth of a point absorber and to match the natural period of the system with the period of sea waves, without increasing its size (diameter). During the process, the variation in different hydrodynamic coefficients and performance parameters, triggered by the proposed design alteration is studied. In the end, it has been concluded that the proposed modification was able to improve the operational bandwidth of the device and the resonant frequency of the system was reduced, as expected.

Keywords Point absorber · Direct drive · Wave energy converter · BEM

1 Introduction

Traditional wave energy converter (WEC) systems like oscillating water columns (OWC) use a turbine rotating at a high speed to generate electricity [1–3]. However, since the frequency of oscillations imparted by ocean waves are particularly low, the complex processes involved in compressing the fluid to attain such a high velocity proves to be extremely inefficient and less feasible especially if the power requirement is less or moderate, as in the case of a self-powered navigational or data buoy. A

V. Vijayasankar · A. Samad (✉)
Wave Energy and Fluids Engineering Lab, Department of Ocean Engineering, Indian Institute of Technology Madras, Chennai, India
e-mail: samad@iitm.ac.in

V. Vijayasankar
e-mail: vishnuvijayasankar@gmail.com

S. Vijayan et al. (eds.), *Trends in Manufacturing and Engineering Management*, Lecture Notes in Mechanical Engineering,
https://doi.org/10.1007/978-981-15-4745-4_86

substitute for these systems would be direct drive systems [4]. The mode of power conversion in a direct drive system can be mechanical based (using gears) or electrical based (using linear generators). Even though in the case of a linear generator, there are only fewer moving parts, the complexity involved in handling a voltage which varies both in amplitude and frequency makes it difficult to implement.

In the paper, we consider a mechanical-based direct drive wave energy converter that uses a power take-off (PTO) similar to the design proposed by Ai et al. [5]. The aforementioned PTO is coupled to a heaving buoy so that together they will perform like a point absorber WEC. These types of WECs are capable of extracting energy from a wave of crest length greater than its width [6]. A typical point absorber comprises a buoy whose width is less compared to the incident wavelength. In the present work, the main focus is to improve the performance of such a point absorber WEC in long period waves without increasing its size or the cost of manufacturing. This study uses the Boundary Element Method (BEM) code Ansys AQWA to determine the hydrodynamic coefficient for different test cases which subsequently is fed to an open-source MATLAB based code that is tailor-made for simulating wave energy converters (WEC-Sim) [7] to analyze their response in simulated wave conditions.

2 Hydrodynamics of Buoys

In the case of a point absorber, it is convenient to assume the motion of the buoy is in the heave direction only [8]. Let us consider the buoy makes a displacement z in the heave direction then, its equation of motion can be written as [9]:

$$m\ddot{z} = F_e + F_r + F_{\mathrm{rs}} + F_{\mathrm{damp}} \tag{1}$$

where m is the structural mass of the system, F_e is the excitation force imparted to the buoy by the incoming wave, F_r is the wave radiation force, F_{rs} is the hydrostatic restoring force and F_{damp} is the hydrodynamic damping force.

The excitation force is composed of two different force components, the diffraction force (F_d) and the Froude–Krylov force (F_{fk}) [10]. The radiation force consists of radiation damping and added mass terms.

$$F_r = -\big(-\omega^2 A(\omega) + j\omega b_d(\omega)\big)z(\omega) \tag{2}$$

where, ω is the angular frequency, $A(\omega)$ is the frequency-dependent added mass and $b_d(\omega)$ is the frequency-dependent radiation damping coefficient. The normalized value of added mass and radiation damping can be obtained as:

$$A'(\omega) = \frac{A}{\rho} \tag{3}$$

$$B'(\omega) = \frac{b_d}{\rho\omega} \tag{4}$$

where ρ is the density of the fluid.

2.1 Hydrodynamic Resonance

Resonance occurs when a system is forced to oscillate at a frequency near to its natural frequency. It is desirable for a wave energy converter to operate near resonance in order to extract more power [11]. The resonance of a floating buoy occurs for an angular frequency;

$$\omega_n = \sqrt{\frac{\rho g \pi r_b^2 + k}{a + m}} \tag{5}$$

$$T_n = \frac{2\pi}{T} \tag{6}$$

where r_b the radius of the buoy, and m is the mass of the buoy.

In many coastal areas like Chennai, the periods of the most percentage of the waves lies in the range of 4–9 s. So, for any point absorber to work efficiently in those conditions, their natural period should lie somewhere near the same range. However, small-sized bodies usually possess a natural period lower than that. From Eqs. (5) and (6), it is evident that the natural period varies with the structural mass, added mass, and the buoy radius. Since the added mass varies as the cube of diameter [12], the natural period can be increased by increasing the diameter of the buoy. However, this idea contradicts the main objective of this work, i.e. to propose a compact low-cost design to increase the natural period of the buoy. This leaves us with two options, either to increase the structural mass of the buoy or to increase the added mass of the buoy by some other means.

2.2 Theoretical Added Mass

Newman [12] quoted that an approximate value of the added mass formed by a semi-infinite cylinder with radius r can be found by the formula $a = 2.064\ \rho r^3$. In 1966, *Sarpkaya* [13] observed that the added mass formed by a circular plate is approximately equal to the mass of water in an imaginary sphere enclosing the plate. Based on these two observations, the following equations were derived to find the added mass of the system of a floating cylindrical buoy with and without a heave plate.

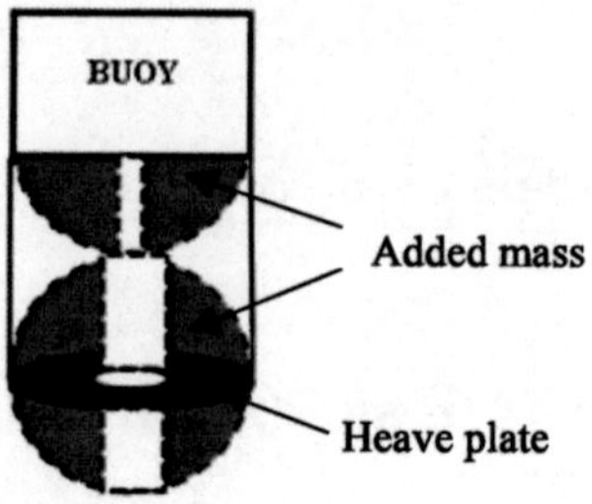

Fig. 1 Theoretical added mass of the system

The theoretical added mass of the buoy without heave plate:

$$A'(\omega) = \rho\pi\left(\frac{2}{3}R_b^3 - r_0^2 R_b\right) \tag{7}$$

The theoretical added mass of the buoy without heave plate:

$$A_b(\omega) = \rho\left[\left(\frac{8}{3}r_p^3 - 2\pi R_0^2 r_p\right) + \pi\left(\frac{2}{3}R_b^3 - r_0^2 R_b\right)\right] \tag{8}$$

whereR_b is the radius of the buoy, r_p is the radius of the heave plate and r_0 and R_0 are the radii of holes in the buoy and heave plate respectively (Fig. 1).

2.3 Wave Power

Wave power is the power content of the incoming ocean wave per crest lest length of the wave. Wave power depends on the wave height and wave period and can be mathematically represented as:

$$P_{\text{wave}} = \frac{\rho g^2 T H^2}{32\pi} \tag{9}$$

where g is the acceleration due to gravity (9.81 m/s^2), T is the wave period and H is the wave height (Fig. 2).

Fig. 2 CAD model **a** Simple buoy **b** buoy with heave plate

Table 1 Buoy specifications

Buoy specifications	Value
Height (cm)	40
Outer diameter (cm)	60
Inner diameter (cm)	15
Material	SS

3 Methodology

3.1 Geometry and Meshing

Hydrodynamic software packages such as the ANSYS AQWA provide a modeling platform sufficient for creating basic shapes and devices. The geometry is discretized using the panel meshes generated by AQWA. Figure 2a, b shows the geometry of two different buoy configurations and their corresponding meshes which are used for the present study.

The buoy dimensions are fixed during the entire study and its details are given in Table 1.

The mass properties of the geometry like the moment of inertias and the center of mass can is obtained from SOLIDWORKS. These mass properties have to be fed to the AQWA solver before starting the analysis.

3.2 Finding Hydrodynamic Coefficients

The surface model generated by the design modeler is analyzed by the AQWA solver to determine the hydrodynamic coefficients. During the initial phase of analysis, the mass properties of the geometry, the range of required wave frequencies and wave directions are to be specified. For the present work, 99 different wave frequencies with directions ranging from $-180°$ to $+ 180°$ were chosen.

3.3 WEC-Sim Code

The response of the buoy for various wave conditions be studied using the WEC-Sim code. WEC-Sim requires four types of inputs from the user

1. Hydrodynamic data generated by the AQWA.
2. Geometry files of the body.
3. Wave conditions like regular/irregular, wave height, wave period, etc.
4. A Simulink model of the hydrodynamic system.

3.4 Response Amplitude Operator

The Response Amplitude Operator of a body in any direction is the ratio of its motion amplitude in that direction to the incoming wave amplitude. A floating body can have RAOs for heave, pitch, and roll motions. In the process of designing a point absorber, the RAO in heave direction plays a significant role to optimize the power absorption. RAO for heave motion can be expressed as:

$$\text{RAO} = \frac{\zeta_a}{\eta_a} \tag{10}$$

where, ζ_a is the heave amplitude of the buoy and η_a is the incoming wave amplitude.

3.5 Test Cases

In the present work, the buoy is subjected to two main test cases; the first one in which the experiments are done for different masses of the buoy and the second one in which the experiments are done for the buoy attached with heave plates of different diameters. Variation in mass refers to increasing or reducing the mass of buoy without changing its dimension. For testing the effect of mass on the resonance, three masses (30, 40 and 50 kg) were chosen and for testing the effect of heave plate diameter, three different diameters (30, 60, and 90 cm) were used. One important point to be noted is that the AQWA solves the potential flow theory and therefore does not take into account the drag forces. So, in order to get the exact results, one has to find the damping coefficients due to drag experimentally and manually include it in the AQWA code. However, experimentation is out of the scope of the present work.

4 Results and Discussions

The results of all the test cases developed for the buoy are discussed in this section. As mentioned in the previous sections, the numerical experiments are first carried out for different buoy masses without the heave plate and then by connecting heave plates of different diameters to the buoy after fixing the buoy mass.

4.1 Mesh Dependency

Mesh dependency study is done to find an optimum mesh resolution required to get an accurate numerical solution at a moderate computational cost. Three different mesh resolutions; fine, medium, and coarse were considered for this purpose. The specifications of the different meshes are shown in Table 2.

In the present work, since the added mass plays a vital role in approaching the conclusions; the values of added mass obtained for different mesh sizes were compared to find an appropriate mesh. As seen in Fig. 3, the added mass values obtained for the medium quality mesh were the nearest to the theoretically calculated value of added mass (Sect. 2.3) and it required less computational time compared to the fine mesh. Hence, the medium-sized mesh was used for all further analyses.

Table 2 Types of meshes used to study mesh convergence

	Max. element size (m)	No. of elements
Coarse	0.024	11,103
Medium	0.019	16,911
Fine	0.016	24,615

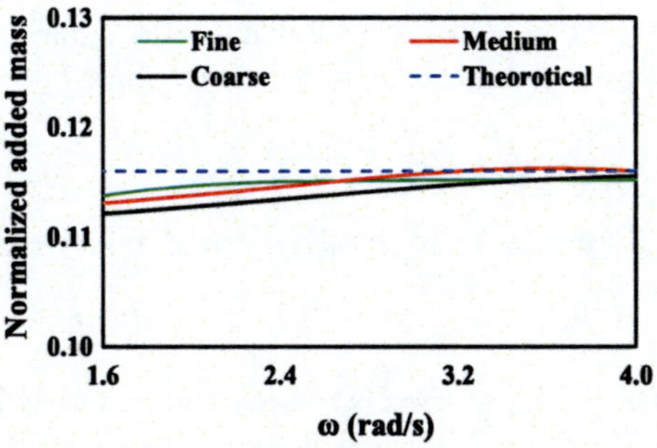

Fig. 3 Normalized added mass obtained for different mesh resolutions

Table 3 Natural periods and RAOs obtained for different buoy masses

Buoy mass (kg)	Nat. period (s)	Max. RAO	% Diff in natural period
30	1.02	1.64	0
40	1.07	2.45	4.90
50	1.16	2.26	13.72

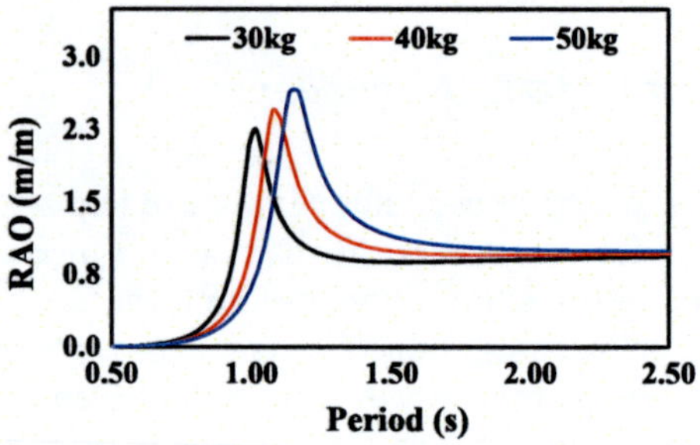

Fig. 4 RAO for different buoy masses

4.2 Numerical Experiments for Different Buoy Masses

The same buoy design is tested for three different masses: 30, 40, and 50 kg. The value of natural frequencies obtained for different buoy masses is listed in Table 3. It can be noted that the natural period of the buoy keeps on increasing with an increase in buoy mass. However, since the dimensions of the buoy are kept constant, the maximum buoyancy that this particular buoy can provide is also constant. This limits the maximum mass possible to around 55 kg. Figure 4 shows the variation of RAO with wave periods for different buoy masses. The peaks are obtained near the resonant period of corresponding buoy configurations.

As discussed in Sect. 2, added mass is independent on the structural mass of the buoy. It simply represents the added inertia that develops during motion. Therefore, added mass remains more or less constant for all the three buoy masses.

4.3 Numerical Experiments for Different Heave Plate Diameters

In the second test case, the buoy mass is conveniently fixed as 30 kg (excluding the weight of heave plate) and the analysis is done on the buoy fitted with heave plates of three different diameters, 30, 60, and 90 cm.

It can be seen that the maximum percentage increase in the natural period for the second test case (89.10%) is much higher than the same in the first test case (13.72%).

Table 4 shows the natural periods for the buoy connected with three different heave plates. Table 4 also shows an enhancement in the peak power absorbed as the size of

Table 4 Naturalperiods and RAOs obtained for different buoy masses

Plate diameter (cm)	Nat. period (s)	Max. RAO	% Diff in natural period	Max. power
30	1.10	2.61	0	19.37
60	1.46	5.26	32.73	36.24
90	2.08	7.50	89.10	79.85

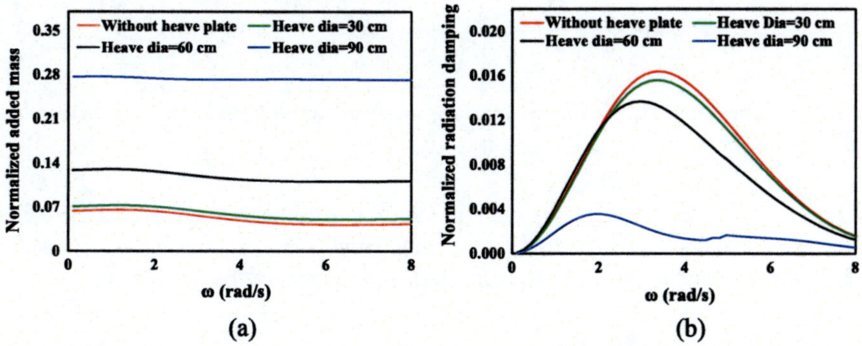

Fig. 5 **a** Normalized added mass versus angular frequency. **b** Normalized radiation damping versus angular frequency

the heave plate increases. Even though this estimate does not include the power lost due to drag, it shows the dependence of the power absorbed on the diameter of the heave plate.

Added mass

As shown in Fig. 1, the added mass is heavily dependent on the size of the heave plate connected at the bottom of the buoy. As the diameter of the heave plate increases, the added mass increases significantly (See Fig. 5a) since the added mass varies as the cube of the diameter of the plate.

Radiation damping

Figure 5b indicates that the radiation damping tends to decrease as the heave plate diameter increases. Since the radiation damping accounts for only 20–30% of the total damping [14], this doesn't necessarily mean that the total damping of the system decreases with an increase in heave plate diameter.

Excitation force

It is evident from Fig. 6a that the excitation force acting on the buoy drops as the size of the heave plate is increased. This is occurring due to the dependence of excitation on added mass.

As explained earlier, the heave excitation force is composed of two counteracting components, the diffraction force (F_{diff}) and the Froude–Krylov force (F_{fk}). The

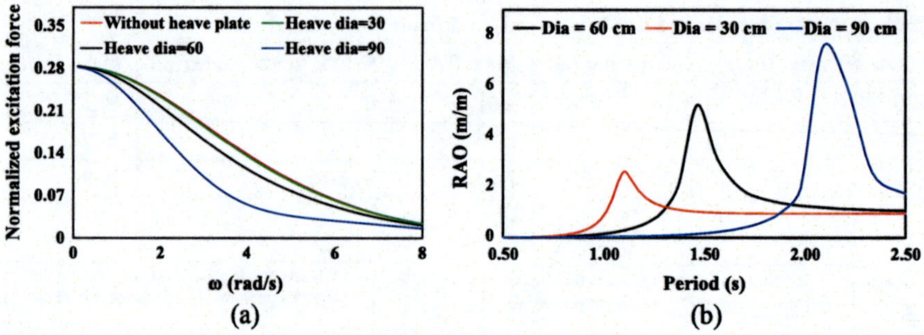

Fig. 6 **a** Normalized excitation force versus angular frequency. **b** RAO for different heave plate diameters

diffraction component increases with an increase in added mass, while at the same time the more decisive Froude–Krylov component remains more or less constant. Thus, the net excitation force drops for higher diameters of the heave plate.

Response amplitude operator
The peak value of RAO tends to increase as the heave plate diameter is increased (See Fig. 6b). This can be explained as a combined effect of two possibilities; (i) as seen in Eq. (9), the energy content of the wave increases and the wave period increases and (ii) the resonance occurs at higher wave periods for larger heave plate diameters.

5 Conclusion

The following conclusions were drawn from the numerical analyses carried out for the two aforementioned test cases:

1. The resonant period of the buoy increased for both the test cases, but the percentage of increase is different in the two cases.
2. Increasing the mass of the buoy has little effect on its resonant period when compared to the addition of a heave plate.
3. In the first test case, an increase in mass has virtually no effect on the added mass and radiation damping of the buoy.
4. In the second test case, an increase in the heave plate significantly increased the natural period of the buoy as much as by 89.10%. Also, the excitation force decreased as the heave plate diameter is increased.

Summing up, the present work was able to successfully fulfil the objective of improving the performance of the point absorber in longer wavelengths, without increasing its size.

References

1. Falcão AFdO (2010) Wave energy utilization: a review of the technologies. Renew Sustain Energy Rev 14(3):899–918
2. Clément A et al. (2002) Wave energy in Europe: current status and perspectives. Renew Sustain Energy Rev 6(5):405–431
3. Cruz J (2008) Ocean wave energy current status and future prepectives. Springer, Berlin
4. Baker NJ, Mueller MA (2001) Direct drive wave energy converters. Rev des Energies Renouvelables 1–7
5. Ai J, Zuo L, Lee H, Liang C (2015) Design, fabrication, simulation and testing of a novel ocean wave energy converter. In: ASME international manufacturing science and engineering conference, pp 1–7
6. Budar K, Falnes J (1975) A resonant point absorber of ocean-wave power. Nature 256(5517):478–479
7. WEC-Sim | Open Energy Information. (Online). Available: https://openei.org/wiki/WEC-Sim. Last accessed 06 Dec 2018
8. Chakrabarti SK (1987) Hydrodynamics of offshore structures. Computational Mechanics Publications, Southampton
9. Falnes J (2002) Ocean waves and oscillating systems: linear interaction including wave energy extraction. Cambridge University Press, Cambridge, UK
10. Haslum HA (2000) Simplified methods applied to nonlinear motion of spar platforms. Nowergian University of Science and Technology, Ph.D
11. Arthur Pecher JPK (ed) (2017) Handbook of ocean wave energy, vol 7. Springer, Aalborg, Denmark
12. Newman JN (2017) Marine hydrodynamics, 40th Anniversary. The MIT Press, Cambridge
13. Sarpkaya T (2010) Wave forces on offstructure structures, First Edit. Cambridge University Press, USA
14. Vugts JH (1968) The hydrodynamic coefficients for swaying, heaving and rolling cylinders. Int Shipbuild Prog 15(167):251–275

Comparison of Hydrogen Yield from Ball-Milled and Unmilled Magnesium Hydride in a Batch System Hydrogen Reactor

J. A. Adeniran, R. S. Fono-Tamo, Esther Titilayo Akinlabi, and T. C. Jen

Abstract Energy plays a crucial role in the economic development of a nation and its sustenance. In view of the important role of energy to man; energy demand has been on the increase as the human population increases. Interests in solid-state hydrogen generation especially from lightweight metals have increased lately due to high gravimetric and volumetric hydrogen storage/reserve. In this study, effect of ball milling on hydrogen yield was examined by comparing hydrogen generation from unmilled MgH_2 and ball-milled MgH_2 in a hydrolysis reaction carried out in a batch system hydrogen reactor. Furthermore, the effects of acetic acid concentration and MgH_2 weight on hydrogen generation was investigated. The reaction was carried out at 30 °C with three substrates weights 0.2 g, 0.4 g and 0.6 g respectively. Ball-milled MgH_2 performed better than unmilled MgH_2 by recording higher hydrogen yield when compared to unmilled MgH_2 components with a hydrogen yield of about 0.0194 L relative to 0.0131 L using 0.6 g MgH_2 obtained in unmilled MgH_2. The results from the XRD spectra validates the reduction of crystallite size of ball-milled MgH_2 compared to the unmilled MgH_2. The crystallite size reduces from an average of 52.4 μm to about 92.25 nm after one-hour ball milling. This development enhances reaction kinetics by increasing the reaction surface area. Similarly, the fracturing of the substrate crystals during ball milling increases the nucleation reaction in the particles thereby increasing the hydrogen release phenomenon.

Keywords Magnesium hydride · Ball milling · Hydrolysis · Hydrogen generation

1 Introduction

Energy remains a pivot for the economic development and sustenance of nations of the world. With the importance of energy also comes challenges such as high cost and environmental pollution from some of the conventional energy generation

J. A. Adeniran (✉) · R. S. Fono-Tamo · E. T. Akinlabi · T. C. Jen
Department of Mechanical Engineering Science, University of Johannesburg, Auckland Park Kingsway Campus, Johannesburg 2006, South Africa
e-mail: nikjocrown2000@yahoo.com

S. Vijayan et al. (eds.), *Trends in Manufacturing and Engineering Management*, Lecture Notes in Mechanical Engineering,
https://doi.org/10.1007/978-981-15-4745-4_87

technologies. It is widely believed that energy technologies of the future are those with sustainability attributes, that is minimal environmental harm, readily accessible materials, relatively cheap and safe handling.

Hydrogen is one of the veritable renewable energy techniques with increasing interests from researchers and policymakers in recent years. Among the different hydrogen generation/storage methods, solid-state hydrogen storage and generation especially from metal hydrides have attracted renewed interests. The development is due to the attributes such as high gravimetric and volumetric hydrogen potentials of most of these categories of materials [1]. Other alluring qualities of metal hydrides include less harmful environmental reaction by-products, relatively cheaper cost of experiments, comparative safe handling of the substrates. Magnesium hydride is among the metal hydrides with increased interests from the experts for hydrogen storage and generation especially for onboard vehicular application. Magnesium hydride is readily available due to the abundant nature of its main component (magnesium) regarded as the fourth most abundant element on earth [2]. Similarly, magnesium hydride has high volumetric hydrogen storage estimated to be about 7.6 wt% [3, 4], the reaction products are considered to be of less harmful effect on the environment and the cost of the material is comparatively affordable [5]. However, the two major challenges facing the use of MgH_2 like most metal hydrides are sluggish reaction kinetics and thermodynamic limitations [6]. Researchers have explored different interventions towards ameliorating the limitations of slow reaction kinetics and thermodynamics. Among such interventions the applications of catalysts such noble metals, organic and inorganic acids have been reported to enhance hydrogen yield [7–10]. Ball milling of metal hydrides for hydrogen generation results in reduction of crystallite size of the powder, ensures increase in reaction surface area of the materials [11]. Similarly, the fracturing of the crystals leads to creation of defects in the crystals thus enhancing nucleation reaction. The increase in nucleation thus enhances hydrogen generation. In this study, the effect of ball milling on hydrogen yield was investigated by comparing hydrogen generation from unmilled and ball-milled MgH_2 via hydrolysis technique.

2 Experimental Design and Method

Figure 1 is a schematic representation of the hydrogen generation experiment. The major components of the experimental set up include a three-neck round bottom flask within which the hydrolysis reaction takes place. Other components include temperature regulated water bath (Julabo TW20, Julabo GmbH), the moisture-absorbent unit, the flowmeter (T1000, Fujikin) and the data logger.

Hydrolysis method was used for hydrogen generation in the study. The substrate (MgH_2) was poured inside the round bottom flask, subsequently, the acetic acid-catalyzed solution in the funnel connected to the reaction vessel was released by opening the valve on the funnel for the commencement of the hydrolysis reaction. Other notable components are the thermometer fixed to one of the outer mouths of the

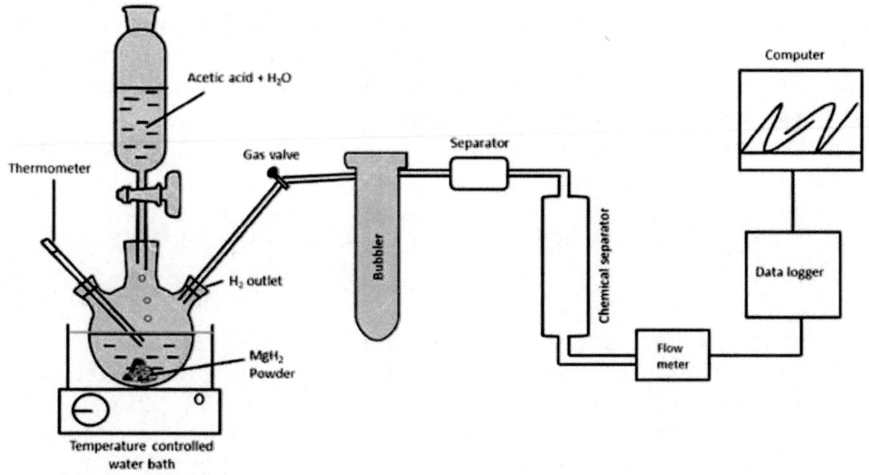

Fig. 1 Schematic representation of hydrogen generation process [12]

three-neck flask while the tube conveying the generated hydrogen from the reaction vessel is connected to the second outer neck. The line carrying hydrogen from the reaction vessel was joined to the bubbler for condensation of the gas. The second opening on the bubbler had another connected to it, through which the condensed hydrogen gas flows into the flow regulator. It is essential that only hydrogen gas gets to the flowmeter thus two major components (flow regulator and filter/chemical separator) were installed on the reactor. The water/moisture in the hydrogen gas is trapped in the regulator while the gas flows into the filter/chemical separator. The moisture-absorbent in the design trapped the moisture in hydrogen thus ensuring the only hydrogen is recorded by the flowmeter. The hydrogen generation was recorded using the datalogger connected to the flowmeter [13, 14].

The experiment was accelerated with the aid of acetic acid (99.8%, Labchem, SA (South Africa). The reason acetic acid was chosen as the reaction accelerator is that in reaction mixture the hydrolysis of the reaction is favour by the large excess of water employed in the reaction, here water is also serving as the solvent rather than the acid. This will drive the acetic metal hydrides equilibrium toward the acetic acid, thus favouring the product formation. The equilibrium will favour the thermodynamically more stable side and acetic acid is playing the role effectively couple with its pka value which makes it weaker than other carboxylic acids that could be used for the same reaction. Furthermore, acetic acid is environmentally friendly as well as the products of its reaction with MgH_2 for hydrogen generation through hydrolysis. Two sets of substrates were utilized in the study: commercial magnesium hydride powder 99.8% purity (Rockwood Lithium, Germany) without further processing (unmilled) hereafter referred to as as-received magnesium hydride (AR MgH_2) and one-hour ball-milled magnesium hydride (1 h BM MgH_2). The manufacturer reported average particle size of 50 μm for the AR MgH_2. Ball milling of the magnesium hydride powder was done using Fritsch F6 Planetary mill using tungsten carbide vial and balls with ball to powder ratio of 30:1. The sample was loaded into the vial inside glove box operated with argon gas and the sample was ball-milled under argon. The rotational speed of the ball mill was 300 rpm.

The equation for reaction of magnesium hydride acetic acid-accelerated hydrolysis is represented in Eq. (1):

$$2MgH_2 + 2H_2O + 2CH_3COOH \rightarrow Mg(OH)_2 + Mg(CH_3COO)_2 + 4H_2 + Hrxn \quad (1)$$

The heat of reaction (ΔHrxn) is approximately −277 kJ/mole.

Three different substrate weights (0.2, 0.4 and 0.6 g) were used in the study. To ensure uniformity of the substrate's weight, BM-200 analytical balance (United Kingdom) with 0.0001 g repeatability was used to measure the substrates. The study was conducted at 30 °C ambient temperature. Comparison of effect of ball milling and substrate weight on hydrogen yield was done by conducting the hydrolysis experiment with both AR and one-hour ball-milled MgH_2 at the three weight compositions. Thereafter, the best performing sample was selected for the investigation of catalyst concentration on hydrogen yield. To ascertain the catalyst concentration effect on hydrogen generation, the study was carried out using 30 wt%, 40 wt%, 50 wt%, 60 wt% and 70% acetic acid concentrations respectively.

2.1 *Material Characterization*

The X-ray diffraction (XRD) analysis was accomplished with the aid of the Bruker AXS D8 Advance instrument (USA). Using Cu-Kα, $\lambda_1 = 1.5406$ Å, $\lambda_2 = 1.5444$ Å, $2\theta = 10$–$90°$ with a step size of 0.035°. Furthermore, α-Al_2O_3 was employed as the standard sample for establishing the impact of the equipment on the peak profile of parameters. Thereafter, the Rietveld full profile analysis of the data was achieved using the General Structure Analysis System (GSAS) software [15, 16].

3 Results and Discussion

3.1 *Role of Ball Milling/Crystalline Size on Hydrogen Generation*

To ascertain the effect of crystalline size on hydrogen yield, the AR MgH_2 and 1 h BM MgH_2 were evaluated for hydrogen generation through three sample weights namely: 0.2 g, 0.4 g and 0.6 g, respectively. Figure 2 shows the result of the comparison of hydrogen yield from the two materials in three different weight compositions.

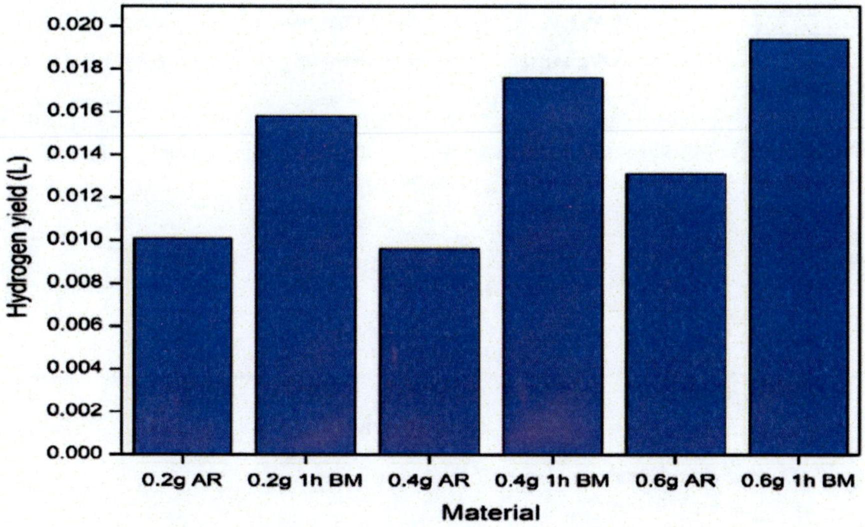

Fig. 2 Hydrogen generation from ball-milled and as-received MgH_2

Hydrogen generation from the BM samples was higher than AR samples in all the categories. The development is due to the improvement in the kinetics of the 1 h BM MgH_2 samples due to the increase in surface area relative to the AR MgH_2.

3.2 *Sample Weight Effect on Hydrogen Yield*

Figure 3 and Table 1 show the combined results of hydrogen yield from the six substrates used: 0.2 g (AR and 1 h BM MgH_2), 0.4 g (AR and 1 h BM MgH_2), and 0.6 g (AR and 1 h BM MgH_2), 0.4 g (AR and 1 h BM MgH_2). From the two forms of result presentations (Fig. 3 and Table 1), the weight of substrates influenced hydrogen yield with hydrogen generation proportional to weight of the substrates for each of the two categories of the six samples (i.e. AR MgH_2 and 1 h BM MgH_2). The observation agrees with some of the previously reported studies from our research team [13, 14]. In these sets of experiments, the highest hydrogen yield was 0.0194 L from 0.6 g 1 h BM MgH_2, however, 0.4 g AR MgH_2 produced the lowest hydrogen yield (0.0096 L) compared to approximately 0.0101 L from 0.2 g AR MgH_2. The reason for the development is not known now. Similarly, after observing the same trend of result from the sample when the experiment was repeated and for the fact that it is the only departure among the six samples, it may be considered an outlier.

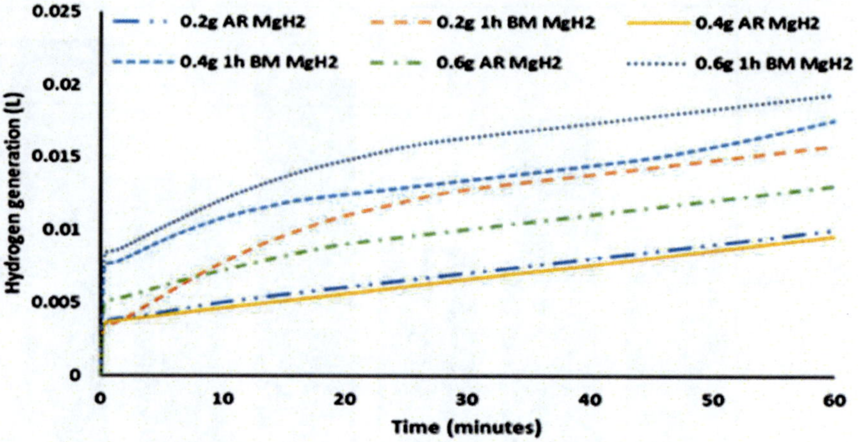

Fig. 3 Optimization of substrate quantity for hydrogen generation

Table 1 Hydrogen generation at 30 °C 40 wt% Acetic concentration

Material (M_gH_2)	Maximum H_2 yield (L)	Duration of experiment (min)	Ranking of H_2 yield
0.6 g 1 h BM	0.0194	60	1
0.6 g AR	0.0131	60	4
0.4 g 1 h BM	0.0176	60	2
0.4 g AR	0.0010	60	6
0.2 g 1 h BM	0.0158	60	3
0.2 g AR	0.0101	60	5

3.3 Optimization of Catalyst Concentration for Hydrogen Generation

Having observed that 0.6 g 1 h BM MgH_2 was the optimum sample for hydrogen generation, it was selected for optimization of acetic acid concentration on hydrogen yield. Figure 4 shows the hydrogen yield from the substrate at five different catalyst concentrations (30, 40, 50, 60 and 70 wt%). The 30 and 40 wt% recorded the best performance for maximum hydrogen yield at 60 min (approximately 0.0194 L). However, 30 wt% experiment recorded better hydrogen yield compared to 40 wt% within 10 min of study suggesting faster kinetics at that point. Moreover, 70 wt% recorded the fastest kinetics in the first 10 min of experimental observation.

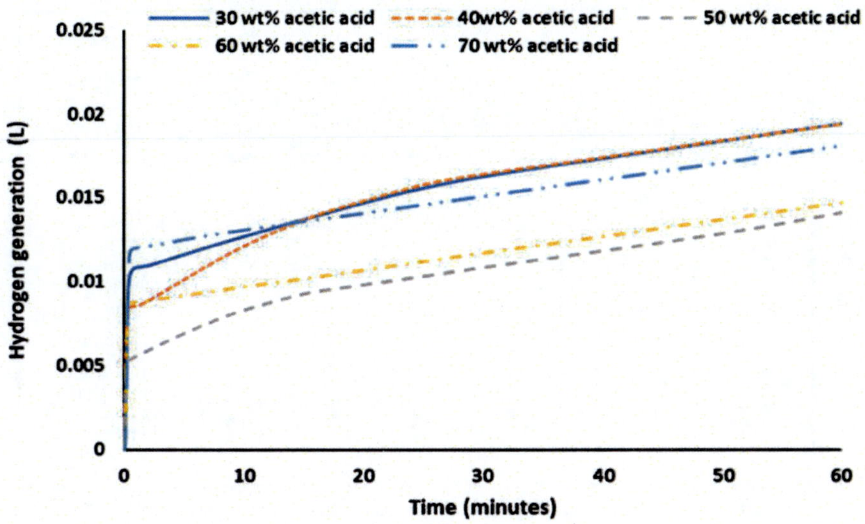

Fig. 4 Hydrogen generation at different catalyst concentrations

3.4 MgH_2 Material Characterization

In this section, the results from the XRD analyses of the two samples will be discussed.

X-ray diffraction (XRD) analyses

XRD characterization of the materials is important to provide useful information on their crystalline structures and phase information. Figure 5 shows the XRD plots of AR MgH_2 and 1 h BM MgH_2 stacked for easy comparison. From the plots, two major phases were observed namely: Mg and αMgH_2. The αMgH_2 is commonly observed in unmilled MgH_2 or samples that have been ball-milled for a short period of time [17]. Moreover, the crystallite size obtained using GSAS software for Rietveld analysis reported approximately 52.4 μm for the αMgH_2 (MgH_2) phase, a result that is close to the 50 reported by the manufacturer. After a 1 h ball milling, the crystallite size reduces to 92.25 nm. From the structural perspective, the XRD plots show broadening of peaks in 1 h BM sample and shifting of the peaks to the right compared to the AR MgH_2 an indication of reduction of crystallinity of the materials [18]. The impact of the crystalline size reduction on hydrogen yield was evident by the consistent higher hydrogen yield in the 1 h BM samples compared to the AR samples. The increased hydrogen yield was caused by fracturing of the crystals by ball milling leading to increased nucleation phenomenon which enhanced reaction kinetics.

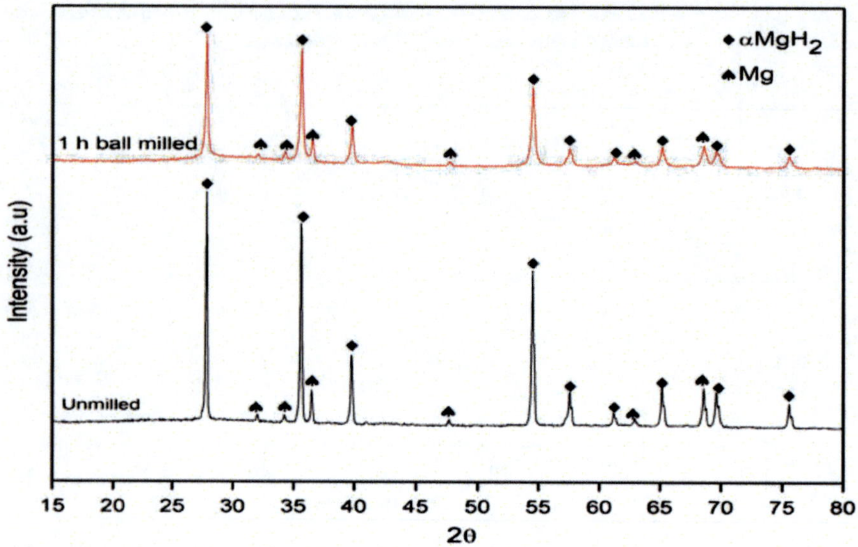

Fig. 5 X-ray powder diffraction pattern of AR MgH_2 and 1 h BM MgH_2

4 Conclusion

Ball milling of MgH_2 for 1 h reduced the crystallite size of the material 52.4 μm to 92.25 nm and improved hydrogen yield in the study. The result further underscores the importance of crystalline size reduction to improve the reaction kinetics and hydrogen generation through hydrolysis. The catalysis of the experiment with acetic acid being an environmentally friendly acid will also enhance the scale-up potential of the technology.

Acknowledgements Special thanks to the National Research Foundation (NRF) South Africa for funding the study of the first author. Authors also thank Dr. M Lototskyy for availing the ball milling facilities at the South African Institute for Advanced Materials Chemistry, University of the Western Cape, for this study and for assisting with the XRD and GSAS analyses of the MgH_2 powders. The support of Dr. SN Nyamsi during the ball milling process is also gratefully acknowledged. This project was funded by the TIA grant of the University of Johannesburg.

References

1. Song MY et al (2014) Hydrogen-storage properties of MgH_2–10Ni–2NaAlH_4–2Ti prepared by reactive mechanical grinding. J Ind Eng Chem 20(4):1591–1595
2. O'rourke JG, Stevenson DJ (2016) Powering Earth's dynamo with magnesium precipitation from the core. Nature 529(7586):387

3. Chao CH, Jen TC (2013) Maximized on-demand hydrogen generator design. Adv Mater Res. Trans Tech Publications
4. Sakintuna B, Lamrdi-Darkim F, Hirscher M (2007) Metal hydride materials for solid hydrogen storage: a review. Int J Hydrogen Energy 32(9):1121–1140
5. Huot J, Liang G, Schulz R (2003) Magnesium-based nanocomposites chemical hydrides. J Alloy Compd 353(1):L12–L15
6. Kojima Y, Kawai Y, Haga T (2006) Magnesium-based nano-composite materials for hydrogen storage. J Alloy Compd 424(1):294–298
7. Uan JY et al (2009) Evolution of hydrogen from magnesium alloy scraps in citric acid-added seawater without catalyst. Int J Hydrogen Energy 34(15):6137–6142
8. Akdim O, Demirci UB, Miele P (2009) Acetic acid, a relatively green single-use catalyst for hydrogen generation from sodium borohydride. Int J Hydrogen Energy 34(17):7231–7238
9. Akdim O, Demirci UB, Miele P (2009) Highly efficient acid-treated cobalt catalyst for hydrogen generation from $NaBH_4$ hydrolysis. Int J Hydrogen Energy 34(11):4780–4787
10. Gervasio D, Tasic S, Zenhausern F (2005) Room temperature micro-hydrogen-generator. J Power Sources 149:15–21
11. Webb CJ (2015) A review of catalyst-enhanced magnesium hydride as a hydrogen storage material. J Phys Chem Solids 84:96–106
12. Adeniran JA et al (2019) Temperature optimized hydrolysis of Acetic acid catalyzed magnesium Hydride for Hydrogen generation in a batch sytem Hydrogen reactor. In: Prabha D (ed)Transactions on engineering technologies, Springer Nature Singapore Pte: Singapore
13. Jen TC et al (2016) Hydrogen generation from Acetic acid catalyzed magnesium Hydride using an on-demand Hydrogen reactor. In: International mechanical engineering congress and exposition. Phoenix, Arizona, USA: American Society of Mechanical Engineers
14. Adeniran JA, et al (2017) Organic Acid-Catalyzed Hydrolysis of Magnesium Hydride for Generation of Hydrogen in a batch system Hydrogen reactor. In: Proceedings of the world congress on engineering and computer science
15. Larson AC, Von Dreele RB (1994) Gsas. Report lAUR, pp 86–748
16. Nyamsi SN, Yartysb V, Lototskyya M (2017) Synthesis of Mg 2 FeH 6 assisted by heat treatment of starting materials. In: 1st Africa energy materials conference. Pretoria: Elsevier
17. Huot J et al (1999) Structural study and hydrogen sorption kinetics of ball-milled magnesium hydride. J Alloy Compd 293:495–500
18. Grosjean MH et al (2006) Hydrogen production via hydrolysis reaction from ball-milled Mg-based materials. Int J Hydrogen Energy 31(1):109–119

A Review on the Synthesis of Activated Carbon from Natural Resources for Mechanical Applications

A. John Presin Kumar, S. Sivakumar, D. Prasanth, B. Guhanesh, and A. Ijas Ahamed

Abstract Activated carbon is a material which can be produced from various waste materials such as plant waste, animal waste and industrial waste resulting in reduction of material wastage, better utilization of natural materials and environmental friendliness. Activated carbon plays an important role in many applications in various forms like gas absorbent, composite fillers, etc. Activated carbon particles have many promising applications in various industries like water treatment, dye and sugar refining, degasification and composite fabrication as fillers. In composites, activated carbon can play a vital role as effective reinforcement materials for improving properties. During the last ten years, there had been a good work done in this field with regard to exploring the possibilities of activated carbon in both production and application. This work discusses contribution of several authors in the area of activated carbon in both synthesis and characterization toward various applications related to mechanical engineering.

Keywords Activated carbon · Crushing · Cleaning · Drying · Heating · Ball milling · Sieving · Furnace · Size · Morphology · Nano

A. John Presin Kumar (✉) · S. Sivakumar · D. Prasanth · B. Guhanesh · A. Ijas Ahamed
Department of Mechanical Engineering, Hindustan Institute of Technology and Science, Padur, Kelambakkam, Chennai, Tamil Nadu 603103, India
e-mail: johnpk@hindustanuniv.ac.in

S. Sivakumar
e-mail: ssivak@hindustanuniv.ac.in

D. Prasanth
e-mail: prasanthsekar22@gmail.com

B. Guhanesh
e-mail: bguhanesh@gmail.com

A. Ijas Ahamed
e-mail: ijas9794@gmail.com

S. Vijayan et al. (eds.), *Trends in Manufacturing and Engineering Management*, Lecture Notes in Mechanical Engineering,
https://doi.org/10.1007/978-981-15-4745-4_88

1 Activated Carbon

In India, it is always appreciable to develop methodologies based on natural resources for the production of useful materials. Natural resources when explored can provide raw materials at low cost for various purposes. Activated carbon is a form of carbon processed in various stages of heating to develop in it small, low-volume pores that can possibly increase the surface area available for adsorption as shown in Fig. 1. Due to the very high degree of microporosity that is contained in one gram of activated carbon, there could be a corresponding higher surface area available for better adsorption characteristics. This property of porosity provides lots of benefits during the crucial application areas of activated carbon. There are various levels of activation which are attainable through methods like chemical activation or heat activation. An activation level that is just sufficient for a useful application may be obtained by attaining high surface area as mentioned earlier. According to the requirements for application, the activation level is chosen and the appropriate method to attain it is also identified. Further chemical treatment if required often enhances adsorption properties.

As already mentioned, there are numerous applications for activated carbon. One such important industrial application involves use of activated carbon particles in metal finishing for the purpose of purification during electroplating solutions. One such example is that it is used as the main purification technique for removing organic impurities from bright nickel or chrome plating solutions. Activated carbon finds an astonishing usage in medical field also. It is used for treating poisonings and overdoses in humans and also for animals. Tablets or capsules of activated carbon are

Fig. 1 Activated carbon powder

popularly used in many countries as drug to treat diseases like diarrhea, indigestion and flatulence. The adsorption property of carbon or activated carbon is a much required property as it makes carbon an important material in the areas where adsorption is needed. For example, it finds an environment-friendly application of reducing pollution as mentioned in the following points. Activated carbon has numerous applications in removing pollutants from air or water streams in the field and in industrial processes such as: spill water cleanup, groundwater remediation, drinking water filtration, air purification, volatile organic compound capture from painting and other processes. Activated carbon is a readily allowed substance used by agricultural people and organic farmers in both livestock production and wine-making. In livestock production, it is normally used as a pesticide, animal feed additive, processing aid, nonagricultural ingredient and disinfectant. Filters along with activated carbon in the form of coatings are usually used in compressed air and gas purification to remove oil vapors, odor and other hydrocarbons from the air.

2 Review Summary

There had been a wide variety of synthesis methodologies and characterization techniques utilized by several authors based on their application scenarios. In all these works, the commonalities observed were like heat activation, chemical activation, popularly used chemicals, gas adsorption tests, ash content tests, etc. Synthesis methodologies can be chosen keeping in mind the budget, the planned application area of activated carbon, etc. Likewise, activation methods can be planned based on the properties desired for the activated carbon. As already pointed, more emphasis may be given for utilizing natural resources like plant materials or animal remaining. The advantages lying with choosing natural resources are in their plenty of availability, lower cost compared to artificial ones and most importantly the environment friendliness. The benefits provided to the society through environmental friendliness are getting much importance every day as the society is moving toward a much faster lifestyle through exploitation of fossil fuels.

Activated carbon can be produced from resources having both plant and animal origins. Initiated carbon can be created from assets having both plant and creature inceptions. Creature bones, for example, dairy animals, donkey, horse and chicken, can be utilized to deliver carbon singe for quality evaluation [1]. In this examination, tests broke down were carbonized in a shut pot at 400 °C and ground into powder subsequent to cooling. Foreordained amount of every one of the pounded tests was enacted utilizing hydrochloric corrosive, and another example was initiated by warm technique. To acquire the ideal weight reduction of the carbons, the carbonization time frame was shifted for couple of hours. From the outcome, the expansion in carbonization time caused a calculable increment in the rate weight reduction; this achieved a most extreme incentive at a specific carbonization time for both carbon tests at a comparing rate weight reduction of various creature bones separately. The level of carbon yield of different crude material assets, for example, cow, jackass,

Table 1 Variation of carbon yield

S. No.	Raw material	% of carbon yield	
		Acid activation	Heat activation
1	Resource 1	~48.9	~37.0
2	Resource 2	~48.6	~36.2
3	Resource 3	~46.3	~34.9
4	Resource 4	~44.8	~30.1

chicken and pony bones upon corrosive and warmth enactment, was in expanding request as appeared in Table 1. Among the bone examples, chicken bone has the least fiery remains content for corrosive, heat and non-actuated examples separately. The distinction in mass densities of corrosive-initiated and warmth-enacted carbons is minuscule.

The heat activated carbons show higher percentage of hardness than the acid activated and non-activated carbons. This study indicated that activated carbons from the animal residues were effective as adsorbents with those obtained from chicken having better features of adsorbents. Plant matters such as banana stem can also be a possible resource for producing activated carbon [2]. In this, the author performed the production of activated carbon electrodes from a banana stem for supercapacitor cell applications. The increase in pore properties was conducted using a potassium hydroxide KOH chemical activation agent at low concentrations, whereas physical activation was performed using CO_2 gas at a particular temperature and time period. The activated carbon electrode was fabricated without the addition of adhesive materials as can be noticeable in many works [1, 3, 4]. The density, degree of crystallinity, surface morphology, elemental content and surface area of the electrodes were analyzed. The study of electrochemical properties focused on the analysis of the specific capacitance of the supercapacitor cell using cyclic voltammeter. The physical properties of the activated carbon electrodes were correlated to generate the optimum conditions for the specific capacitance of the supercapacitor cells.

The ideal explicit capacitance got at a specific grouping of KOH achieved best an incentive with a particular surface territory. Initiated carbon can be a decent material for creation of composites for good quality and firmness. Lebedeva et al. [5] integrated composite materials, carbon/polyaniline, utilizing permeable actuated carbon delivered from carbonized rice husk [6]. This has a particular surface zone (Fig. 4). The arrangement of the composite materials was combined by oxidative aniline polymerization technique within the sight of the actuated carbon in acidic medium; polyaniline content was changed. Electrochemical properties of the composite materials were researched by cyclic voltammetry, galvano static charge/release and soundness tests. Results show that properties of the composites are administered by proportion of the parts: A type of cyclic bends and area of polyaniline crest esteems emphatically rely upon the polyaniline sum. Gravimetric capacitance develops with polymer content in the material and achieves greatest incentive, for example, with particular polyaniline content. In any case, toughness tests demonstrate that the steadiest is the example with the least polyaniline amount (Fig. 2).

Fig. 2 Rice husk

Coconut shells are often the promising natural material for many applications. The use of coconut shell (Fig. 5) activated carbon was investigated for a potential adsorbent for malachite green dye from aqueous solution as done by Haji Azaman et al. [7]. Abechi et al. studied the effect of various factors, such as initial dye concentration, contact time, pH and solution temperature [8]. The interaction between dye molecule and coconut shell activated carbon absorbent was strongly influenced by the pH of the solution [9]. Maximum adsorption of malachite green dye was obtained at particular pH, while the point of zero charge was obtained at another value of pH. The maximum monolayer adsorption capacity was increased with increment in temperature. The kinetics of adsorption followed a pseudo-second-order kinetic model. The intra-particle diffusion model was evaluated to determine the mechanism of the adsorption process (Vijayan et al. [10]). Based on the Boyd plots, the adsorption of malachite green dye on the coconut shell activated green absorbent was mainly governed by film diffusion (Fig. 3).

Thermodynamic parameters were determined, and it was found that MG dye adsorption on coconut shell activated carbon was spontaneous and endothermic in nature. The conducted reusability test disclosed the decreasing coconut shell activated carbon performance from higher value of malachite dye removal down to a lower value of the same after few adsorption/desorption cycles. Activated carbon has wide variety of applications have done a study in activated carbon [11]. Activated carbon is used in gas purification, decaffeination, gold purification, metal extraction, water

Fig. 3 Coconut shells

purification, medicine, sewage treatment, air filters in gas masks and respirators, filters in compressed air and many other applications. Due to its high degree of microporosity, just one gram of activated carbon has a particular surface area, as determined by gas adsorption. An activation level sufficient for useful application may be attained solely from high surface area; however, further chemical treatment often enhances adsorption properties.

One noteworthy mechanical application includes utilization of activated carbon in the metal finishing field. It is broadly utilized for refinement of electroplating arrangements. For instance, it is a fundamental sanitization system for expelling natural contaminations from brilliant nickel plating arrangements. Anticipating on the accessible crude materials, it is built up that modern interest for actuated carbon can be met locally. The portrayal of two sorts of manufacturing plant squanders demonstrated that paper squander was a superior choice over rice husk. The enacted carbon delivered from pyrolysis of paper factory squander was synthetically actuated with different initiating specialists like zinc chloride, potassium hydroxide and potassium chloride. A precise examination of the impact of impregnation proportion, actuation temperature and enactment time on the properties of the initiated carbon was finished. The ideal planning conditions together with the affected of these conditions on the yield and effectiveness of delivered actuated carbon have been examined. The artificially initiated carbons were described by estimating iodine and yield rate. The enacted carbon arranged from paper plant squander in this investigation had most extreme iodine with zinc chloride as the actuating operator.

A composite material is characterized as a blend of at least two materials that outcomes in preferable properties over the individual segments are utilized alone. The job of filler material in composites is essential as it advances improvement in properties. Actuated carbon can be a superior filler material in composites as in the work. Here, examination was practiced for the effect of started carbon as filler on the mechanical properties of wood composite. It on a very basic level revolved around the quality, stress and expulsion of the wood composites. The wood composite has certain positive conditions over wood, as they are moderate and have the potential for adaptable plans. The materials have incredible mechanical properties and have a long organization life. As a result of the significance of mechanical properties, wood composite materials have been commonly used all through the latest four decades. Essential and non-helper structured wood composites subject to squeezed wood, medium thickness fiber board, secured veneer timber, thermoplastic or wood fiber blends, and are right now used in both inside and outside applications.

With advances in existing development, especially in building field in which the usage of PC programming, for instance, SolidWorks reenactment is used to test show cautiously for huge specific information directly off the bat in the arrangement methodology. This sort of programming decreases weight and material cost, and improves sturdiness and gathering limit of the thing. Reenactment furthermore enables depiction of the exhibited real structure, and it is directly persistent by techniques for PC. From both proliferation and exploratory system used in this examination, medium thickness fiber composite models show higher-quality motivating force than packed wood composite precedents in view of the growing thickness of the sanctioned carbon filler. For the most part, the rate contrasts among generation and preliminary technique are lower than 10% which demonstrate that reenactment is sensible mechanical assembly use to anticipate the nature of wood composites.

3 Inferences

Activated carbon has wide utilization. On the one hand, it tends to be degasified; then again, it can likewise be filler for composites. Numerous creators have done work on enacted carbon under various objectives. Exploration and usage of plant and creature source have been done broadly by numerous creators. Figure 4 portrays roughly picture scatters of different waste asset usages over the most recent ten years by different creators in the field of enacted carbon amalgamation. This is just a rough picture showing a comparative analysis through review of selective works.

Various studies indicated that materials made of activated carbon are performing well compared to conventional materials. The adsorptive efficiency of each of the carbons can be enhanced through controlled and well-monitored activation. The study utilizing animal bones [1] indicated that activated carbons from various animals are effective as adsorbents with regard to their evaluated properties of moisture content, ash content, bulk density, fixed carbon, hardness and percentage yield. Activated carbon gotten from chicken having low ash content is a good indication of a better adsorbent. Plant materials are also good sources for activated carbon such as banana, rice and coconut. The preparation method for activated carbon electrodes from banana

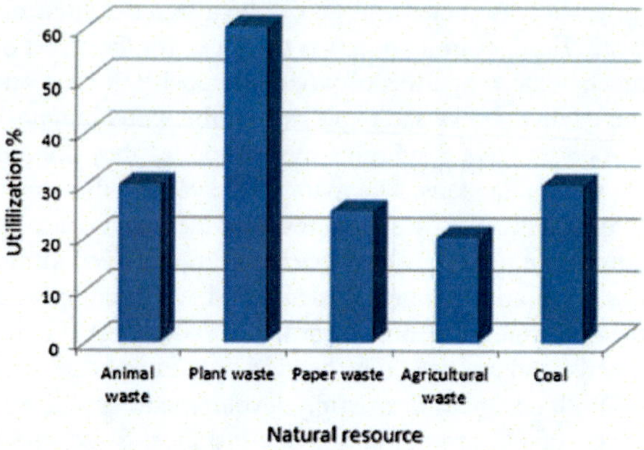

Fig. 4 Comparative utilization of various wastes

stems used herein is a very feasible method for the production of supercapacitor cells as in [2] which gave optimum results.

Figure 5 depicts the observed commonality in the procedural steps of synthesizing activated carbon from naturally available resources. The steps involved and the corresponding hardware requirements are also mentioned below. Different authors on different applications had synthesized activated carbon using activation processes involving pyrolysis. This is normally grouped as heat activation and chemical activation method. For most of the applications, activated carbon can be prepared economically and easily using the following seven-step procedure. The following diagram depicts this clearly followed by an explanation of what are the exact steps involved and the corresponding hardware requirements.

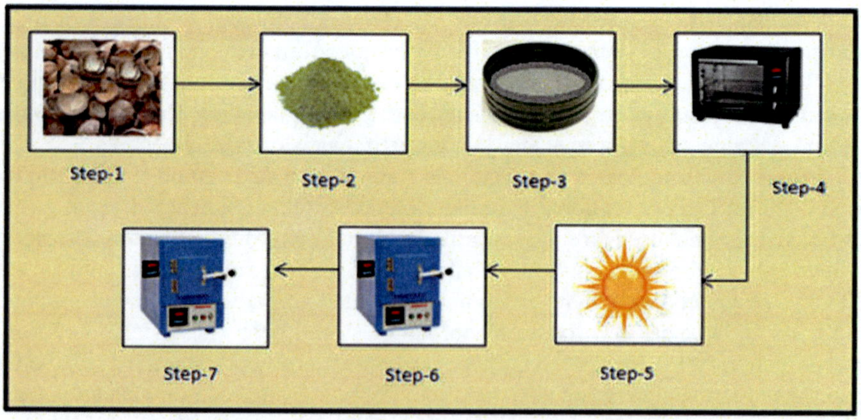

Fig. 5 Generalized steps for activated carbon synthesis from natural materials

Steps

1. Collecting naturally available raw material from locality.
2. Crushing into small pieces. (approx. 2–3 mm).
3. Cleaning the pieces through sieving.
4. Drying the pieces by heating in oven mildly. (approx. 100 °C).
5. Soaking it for particular time duration. (approx. 24 h).
6. Initial furnace heating. (approx. 460 °C and held const. for less than an hour).
7. Final furnace heating. (approx. 750 °C and held const. for less than an hour).

Hardware requirements:

1. Ball milling machine.
2. Sieving machine.
3. Oven.
4. Furnace.

Rice husk was also a promisingly utilized plant material like banana stem, which when used for activated carbon has provided many opportunities for making composite using polymerization [5]. It was shown that the specific capacitance of composite materials rises with the increase of the polymer amount, and in all cases, its value exceeds a simple sum of the capacitances of the individual components. Thus, the combination of the activated carbon material and the conductive polymer provides the synergistic effect leading to improved capacitive and operational characteristics and allows consideration of such composite materials as promising for the use as active component of supercapacitor electrodes. Rice husk has been successfully used by many authors as in [12–14] for polymer bio-composites in addition to other natural elements like walnut shell and coconut shell, whereas the same authors have used glass fibers in another work [15] for polymer composites for artificial limp applications [16, 17].

The coconut shell activated carbon has demonstrated good reusability properties, with the ability to sustain the removal efficiency of greater values of malachite green upon the completion of five consecutive adsorption/desorption cycles [7]. In addition to animal and plant materials, paper wastes are also possible materials for activated carbons [12] in which the results of the study showed that it is feasible to prepare activated carbons with relatively high surface areas and pore volumes from paper waste by direct chemical activation. Activation with zinc chloride produced activated carbons with better developed porosities than with potassium hydroxide. The iodine value of the activated carbon product increased with concentration of zinc chloride solution. As the impregnation time increased, the iodine value rose steeply, reaching a maximum value. The iodine value also increased with higher activation temperature, beyond which it gradually decreased, presumably due to excessive carbonization.

The porosity of the product was comparable with that of commercially achieved carbon. The waste-based activated carbon had proper mean pore diameters, pore volume and micro-details indicating its microporous and mesoporous characters.

Table 2 shows a comparative data of the natural resources used with respect to different application areas by many authors. Activated carbon not only is capable of improving strength but also has the ability to relieve stresses [18, 19]; it was shown that the stress value was increased with the increasing forces applied to the samples of activated carbon. Other than that, the stress value of plywood without activated carbon as filler was higher than the one without activated carbon filler. But, due to the addition of different thicknesses of activated carbon, the stress value of plywood became lower. This showed activated carbon limiting the energy inside the plywood sample [20] and thus lowering the stress value of the sample. SolidWorks simulation determined the sample that has high average displacement and average stress which indicate good performance of mechanical properties. The addition of activated carbon as filler increased the average displacement [6, 21, 22] and average stress which is higher than plywood. Thus, both the methods which were experimental and simulation showed that the mechanical properties were higher compared to plywood with the addition of activated filler. It was also proved that SolidWorks simulation was a suitable tool used to measure strength to get accurate result.

Activated carbon finds better applications as reinforcement fillers for composites [23–26]. Composites are materials nowadays getting more and more importance, especially for applications requiring less weight considerations. For instance, mechanical properties of carbon coir composites depend on many parameters mainly factors like stress–strain behavior of carbon and matrix phases, the phase volume fractions, the carbon concentration, the distribution and orientation of the carbon or reinforcement fillers relative to each other. In many works, it was found that the tensile strength of filler-based composites is higher compared to the other composites [27–31]. It could be easily predictable that there is always a trend toward carbon concentration in increasing order. At lower concentration of filler material, the composite specimens exhibit slightly linear behavior before or prior to a sharp failure.

Table 2 Materials used by authors and their applications

Author	Natural resource used	Application area
[1, 47]	Animal bone	Gas adsorption
[1–4]	Banana, rice, coconut, coal	Supercapacitor cell
[5, 6]	Rice husk and coffee residue	Polymer composites
[7–10]	Coconut shell and palm shell	Gas adsorption
[12–14]	Paper waste	Metal purification (electroplating)
[15–17]	Agricultural raw waste, wood, sawdust	Wood composite

This clearly means that the specimen deformed plastically immediately after elastic deformation. Normally, the impact performance of the fiber-reinforced composites depends on many factors including the nature of the individual constituents, interface between carbon and matrix, the construction and geometry of the composite material and the surrounding conditions of test. In addition, the impact strength of the composites depends and gets influenced by many factors like toughness properties of the fillers, the very nature of the interfacial region and frictional component [32–36]. Overall, the mechanical properties of composites filled with activated carbon particles as reinforcement fillers are influenced by many parameters as discussed. And in addition to that, the size and morphology of the activated carbon particles are having very much importance on the resulting mechanical properties of the composites. It is generally surprising to note that the size of a reinforcement particle is influencing the overall behavior of the composites [37–41]. Activated carbon particles should generally be of uniform shape and size as far as concerned in order to achieve betterness in properties and strength. Surface topography and morphology of activated carbon particles can be better studied by means of scanning electron microscopy. Recently, activated carbon particles are getting more importance and influence toward many environmental applications also like pollution control, agriculture field, etc., through latest developments in its synthesis procedures and conditions in achieving nano-size levels [42–46]. Nano-size of the particles can be better promising in achieving even better surface areas and thereby providing the possibilities of better gas adsorption and better mechanical properties.

4 Conclusions

- Thus, it was very evident from the work of several authors in the area of activated carbons that these materials can be better produced through proper selection and utilization of natural materials for various useful purposes such as natural materials like coconut shell, rice husk, walnut shell, banana stem and animal bones at an economical level suitable for our growing country like India.
- The application areas of activated carbon are vast, and especially like gas adsorption and in composites, the activated carbon plays an enhancing role of improving the performance characteristics.
- The effective utilization of waste materials which has plant or animal origin creates lots of benefits not only to the desired application but also to the society as a whole.
- Thus, it promises to be a good future material whose very nature and properties can be further explored for various better applications and the needs of the society.
- More mechanical properties were discussed with regard to the behavior of composites made of activated carbon particles in the form of reinforcement fillers.
- The main influencing factors were discussed for composites in relation to the filled activated carbon fillers.

- The relation between mechanical properties of composites made up of activated carbon particles as fillers with that of size, morphology and topography of particles was also discussed.
- Recent advancements in the field of activated carbon such as the nano-activated carbon were discussed with respect to its advantages and applications.
- Activated carbon could thus be a very promising material for many mechanical applications, especially as reinforcement filler for composites with regard to nano-size, morphology and recent developments.
- There is a good future scope for activated carbons. More emphasis should be given while choosing the raw materials based on natural resources. This gives the benefit of minimizing waste materials which is a very environmental-friendly act.

Acknowledgements We would like to express our sincere gratitude and thanks to Dr. Velayudham, Scientist 'G', Ministry of Defence, CVRDE, Avadi, Chennai, India; Dr. Ashish Selokar, Accendere, CL Educate Ltd., New Delhi, India, for their valuable suggestions; and R. Balaji, mechanical second-year student of Hindustan Institute of Technology and Science. We express our sincere gratitude and thanks for their helps in the whole duration of executing this research work.

References

1. Sunday NJ, Okechukwu NS, Elom N, Anthony O, Michael PT (2018) Quantitative characterization of activated Carbon from Cow, Donkey, Chicken and Horse Bones from Ezzangbo in Ebonyi State, Nigeria. Am J Appl Chem 6(5):169–174
2. Taer E, Taslim R, Mustika WS, Kurniasih B, Agustino A (2018) Afrianda and Apriwandi. Int J Electrochem Sci 13:8428–8439
3. Renugadevi N, Anisha G, Lalith P (2010) Adv App Sci Res 1(2):102
4. Mane PC, Bhosle AB, Deshmukh PD, Janagam CM (2010) Adv App Sci Res 1(3):212
5. Lebedeva MV, Ayupov AB, Yeletsky PM, Parmon VN (2018) Rice husk derived activated carbon/polyaniline composites as active materials for supercapacitors. Int J Electrochem Sci 13:3674–3690
6. Aloko DF, Adebayo GA (2007) J Eng Appl Sci 2(2):440–444
7. Haji Azaman SA, Afandi A, Hameed BH, Mohd Din AT (2018) J Appl Sci Eng 21(3):317–330
8. Abechi SE, Gimba CE, Uzairu A, Dallatu YA (2013) Res J Chem Sci 3(7):54–61
9. Cobb A, Warms M, Maurer Chiesa EPS (2012) Int J Ser Lear Eng 7(1):93–104
10. Nsi EW, Akpakpan AE, Ukpong EJ, Akpabio UD (2016) Int J Eng Sci 5(9):38–41
11. Subramani T, Revathi PK (2015) Production of activated Carbon from agricultural raw waste. IOSR J Eng 5(3):54–63
12. Chandramohan D, Kumar AJP (2017) Experimental data on the properties of natural fiber particle reinforced polymer composite material. Data in brief, vol 13, pp 460–468
13. Yusuf MI, Ariahu CC, Igbabul BD (2012) Afr J Pure Appl Chem 6(9):123–131
14. Darmawan S, Sofyan K, Pari G, Sugiyant K (2010) J Res 7(2):100–111
15. Chandramohan D, Presin Kumar AJ (2017) Data-enabled Discover. Appl CrossMark. Springer International Publication, pp 1–9
16. Irle M, Barbu MC (2010) Wood-based panel tech. In: Brunei University Press, London pp 190
17. Lam MK, Zakaria R (2008) (Icenv).Inter Conf Envir pp 1–12
18. Aziz NA, Mazlan M, Marinah M, Mohamad Amini MH, Abdul Aziz MS, Yusoff H, Ismael Rizman Z (2015) ARPN J Eng Appl Sci 10(1):376–386

19. Kumar A, Gupta A, Sharma KV, Gazali SB 2013 (1988) 8:6231–6241
20. Ambedkar G, Muniyan M (2011) Mater Adv App Sci Res 2(4):283
21. Roh KC, Park JB, Lee CT, Park CW (2008) J Ind Eng Chem 14:247–251
22. Kim T, Jin HJ, Park SS, Kim SJ, Cho SY (2008) J Ind Eng Chem 14(6):714–719
23. Fierro V, Fernandez VT, Celzard A (2007) Microp Mesop Mater 360–363
24. Rodenas MAL, Amoros DC, Solono AL (2003) Carbon 41:267–275
25. Rodenas MAL, Amoros DC, Solono AL (2005) Carbon 43:1758–1767
26. Rodenas MAL, Juan JJ, Amoros DC, Solono AL (2004) Carbon 42:1371–1375
27. Rodenas MAL, Castello DL, Amoros DC, Solono AL (2001) Carbon 39:751–759
28. Castello DL, Rodenas MAL, Amoros DC, Solono AL (2001) Carbon 39:741–749
29. Mastalerz M, Bustin RM (1995) Fuel 74:536–542
30. Otowa T, Nojiyama Y, Miyazaki T (1997) Carbon 35(9):1315–1319
31. Qadeer R, Akhtar S (2005) J Chem 29:95–99
32. Qadeer R, Rehan AH (2002) J Chem 26:357361
33. Sanchez AR, Elguézebal AA, Pliego JA (2005) Mater 85:331–339
34. Solano AL, Gullon IM, Lecea CSM, Talovera BS (2005) Fuel 79:635–643
35. Stavropulos GG, Zabaniotu AA (2005) Mater 82(1–2):79–85
36. Supaluknari S, Larkins FP, Redlich P, Jackson WR (1998) Fuel Proc Technol 19:123–140
37. Teker M, Imamoglu M (1999) J Chem 23:185–191
38. Guzel F, Uzun I (2000) Adsorption of some heavy metal ion from Aqueous solution by activated Carbon and comparison of percent adsorption result of activated Carbon with those of some other adsorbent. Turk J Chem 24:291–297
39. Yener J, Aksu Z (1999) Atıksulardaki fenol ve klorofenollerin aktif karbon ve kurutulmuş aktif çamura adsorpsiyonu. Tr J Eng Environ Sci 23:93–104
40. Zondlo JW, Velez MR (2007) Development of surface area and pore structure for activation of anthracite coal. Fuel Process Technol 88(4):369–374
41. Prauchner MJ, Reinoso FR (2008) Microp Mesop Mater 109:581–584
42. Najibi H, Chapoy A, Tohidi B (2008) Methane/natural gas storage and delivered capacity for activated carbons in dry and wet conditions. Fuel 87:7–13
43. Ridha FN, Yunus RM, Rashid M, Ismail AF (2007) Exper Ther Flu Sci 32:14–22
44. Celzard A, Perrin A, Albiniak A, Broniek E, Mareche JF (2007) The effect of wetting on pore texture and methane storage ability of NaOH activated anthracite. Fuel 86(1–2):287–293
45. Ouederni SBYA (2012) Hydrocarbons gas storage on activated Carbons. Int J Chem Eng Appl 3(3), 81:1777–1803 (2007)
46. Castello DL, Amoros DC, Solano AL, Quinn DF (2008) Carbon 40:989–1002
47. Abubakar M, Alechenu AA, Manase A, Mohammed J (2012) Adv Appl Sci Res 3(5):3089–3096
48. Aznar JS (2011) ARPN J Eng Appl Sci 10(1):376–386
49. Evans MJB, Halliop E, MacDonald JAF (1999) Carbon 37(2):269–274

Combined Casing Groove and Blade Tip Treatment for Wave Energy Harvesting Turbine

P. Madhan Kumar, Paresh Halder, and Abdus Samad

Abstract The Wells turbine is a self-rectifying air turbine, used in oscillating water column (OWC) to harvest wave energy. It produces unidirectional torque as the flow oscillates inside the OWC chamber. It has inherent disadvantage of narrow operating range due to stall at high airflow rate. Whereas, a wider operating range is essential to improve the turbine power output. A casing groove modifies the tip leakage flow pattern and improves the operating range. In addition, a radiused tip can alter the tip leakage flow and delay the stall. To enhance the performance further, this paper investigates the combined effect of tip groove and radiused tip (CG&RT) design modification. The flow was simulated by solving steady, incompressible Reynolds averaged Navier–Stokes equations in Ansys CFX 15.0. As expected, the CG&RT blade enhanced the relative operating range and the turbine power output by 44.4% and 23.8%, respectively.

Keywords Wells turbine · Casing groove · Radiused blade · Wave energy · OWC

1 Introduction

Wells turbine combined with oscillating water column (OWC) is used to extract power from ocean wave (Fig. 1). It comprises of symmetrical blades aligned perpendicular to the flow direction. It suffers narrow operating range due to flow separation at a

P. Madhan Kumar · P. Halder · A. Samad (✉)
Wave Energy and Fluids Engineering Laboratory, Department of Ocean Engineering, Indian Institute of Technology Madras, Chennai 600036, India
e-mail: samad@iitm.ac.in

P. Madhan Kumar
e-mail: madyypts@gmail.com

P. Halder
e-mail: paresh.halder@gmail.com

P. Halder
Quantum Wave Microscopy Unit, Okinawa Institute of Science and Technology Graduate University, 1919-1 Tancha, Onna-son, Okinawa 904-0495, Japan

S. Vijayan et al. (eds.), *Trends in Manufacturing and Engineering Management*, Lecture Notes in Mechanical Engineering,
https://doi.org/10.1007/978-981-15-4745-4_89

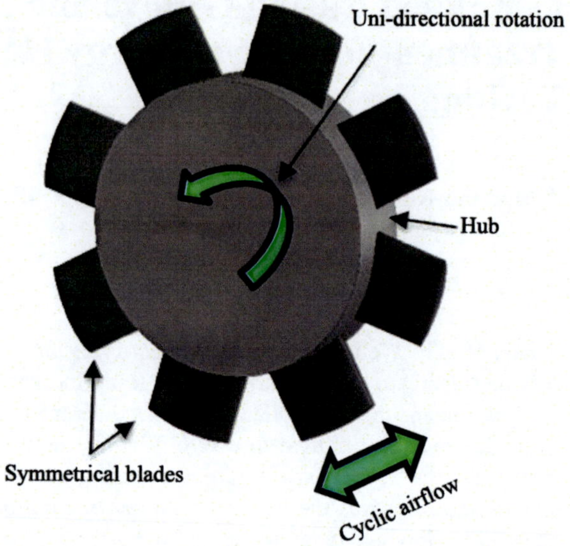

Fig. 1 Wells turbine

higher flow rate. The airflow inside the OWC is cyclic, and it varies from minimum velocity to maximum and vice versa. Hence, it is necessary to have a wider operating range to improve the energy harvest of the turbine.

Many researchers actively pursued different design modifications to enhance the operating range of Wells turbine. Raghunathan and Tan [1] introduced a biplane Wells turbine to improve the stall characteristics. The kinetic energy losses at downstream of Wells turbine can be reduced by implementing guide vanes. The Wells turbine with guide vanes also showed improved starting characteristics [2]. The blades with increased thickness improved starting characteristics [3]. Raghunathan and Beattie [4] used contra-rotating Wells turbine as an alternative to biplane configuration. They concluded that contra-rotating is superior to biplane turbine regarding operating range and efficiency. Webster and Gato [5] optimized blade profile to improve the stall characteristics. Gato and Webster [6] studied Wells turbine with blade sweep and guide vanes. They reported an enhanced operating range with reduced efficiency.

The blade with variable chord with guide vanes showed better starting characteristics [7]. The turbine performance can be improved by modifying blade setting angle [8]. Takao et al. [9] implemented end plates to improve the performance. Mohamed et al. [10] performed multi-objective optimization to improve efficiency and tangential force coefficient. The tangential force and efficiency improved by 8.8% and 1%, respectively. Shaaban and Hafiz [11] studied the Wells turbine with duct geometry. The optimized duct geometry enhanced the turbine power and efficiency by 14% and 9%, respectively. The downstream duct suppressed the interaction of tip leakage flow (TLF) and the suction side (SS). The nature of the interaction between TLF and primary flow affects the stall characteristic of Wells turbine. Halder et al. [12, 13]

Table 1 Turbine geometry [16]

Blade profile	NACA0015
Maximum thickness	15% chord length
Number of blades	8
Chord length (C)	125 mm
Tip radius (R_{tip})	300 mm
Hub radius	200 mm
Solidity	0.64
Tip clearance	1.25 mm (1%C)
Angular velocity	209.44 rad/s

introduced a casing groove treatment to modify the TLF behaviour, thereby delaying stall phenomenon. It improved the turbine power output by 26%. Nazeryan and Lakzian [14] used a variable thickness blade to weaken the interaction between TLF and blade SS. The variable thickness blade enhanced torque and operating range. Kumar et al. [15] used a radiused edge blade tip (REBT) to enhance the stall margin of Wells turbine. REBT showed weak TLF and primary flow (PF) interaction and delayed stall.

As explained above, various design modifications are proposed to alter the characteristics of the interaction between TLF and PF. In this study, a combined casing groove and blade tip (CG&RT) blade are investigated to study their effect on Wells turbine performance. The performance of the CG&RT blade is computed by solving steady, and incompressible Reynolds averaged Navier–Stokes (RANS) equations using commercial CFD solver ANSYS CFX 15.0. The performance parameters of the CG&RT blade are compared with the reference blade. Moreover, the flowfield analyses has been explained using the post-processed figures.

2 Reference Geometry

The reference geometry was taken from Torresi et al. [16]. Figure 1 shows the Wells turbine with eight blades. It has a unidirectional rotation for the cyclic airflow because of the symmetrical blades. Table 1 provides the details of the reference blade.

3 Casing Groove and Blade Tip Treatment

The reference geometry was CG&RT using CAD software to implement the casing groove and RETB (Fig. 2). The geometrical parameters for the casing groove were taken from the study of Halder et al. [12]. The width and depth of the casing groove were chosen as 40%C and 2%C, respectively. The tip diameter was maintained

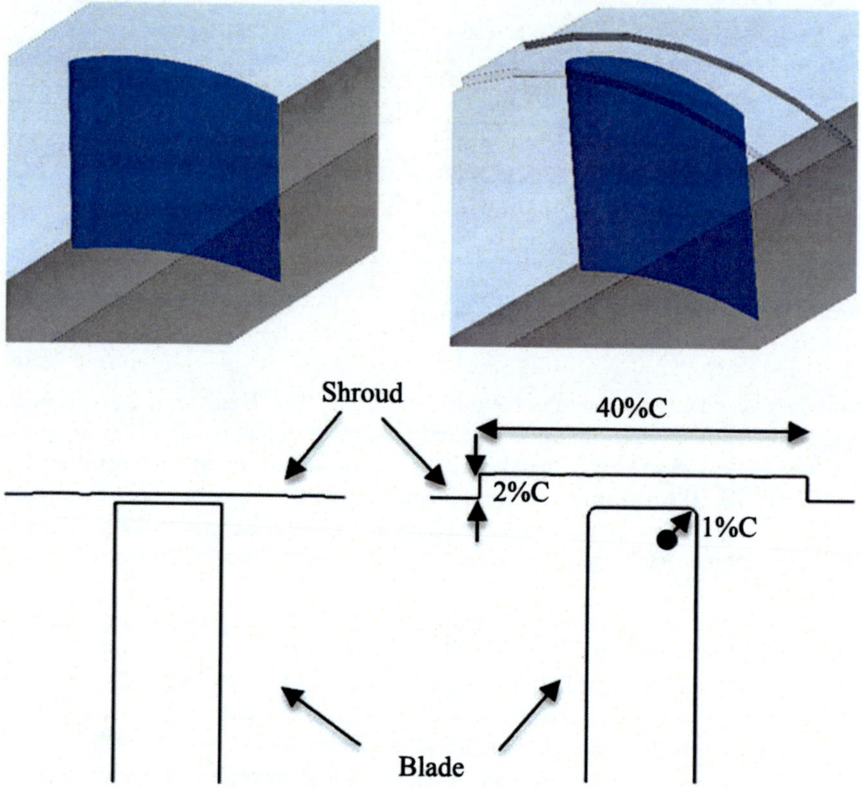

Fig. 2 Reference (left) and CG&RT (right) geometry

constant for both geometries, and the casing diameter was increased to accommodate the groove. Additionally, the radius of fillet used in blade is taken as 1%C as suggested in the work of Kumar et al. [15].

4 Numerical Methodology

The performance parameters of the CG&RT turbine was computed by solving 3D steady, incompressible RANS equations. A single blade passage was selected as computational domain (Fig. 3). The periodic boundary conditions are applied to the lateral faces to account for the full domain. ANSYS ICEM CFD 15.0 was used to discretize the computational domain with unstructured tetrahedral elements. The mesh closer to the blade vicinity was provided with higher refinement. In addition, to resolve the viscous sub-layer region, 20 prism layers with stretching ratio of 1.2

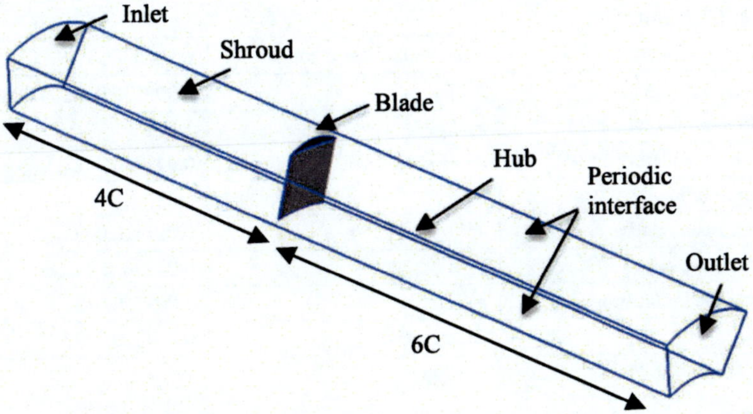

Fig. 3 Computational domain for flow analyses

were created around the blade. The initial cell height was fixed as 1E-5 m to retain the y + < 1.

The boundary conditions were imposed in ANSYS CFX-Pre. Table 2 provides the applied boundary conditions, and the simulations were run in ANSYS CFX 15.0. High-resolution advection scheme with second-order accuracy is used for spatial discretization. The *k-w SST* model was chosen as the turbulence model. It employs *k-w* model closer to wall and *k-ε* model far from the wall [17]. Double precision was enabled for all simulations to minimize the round-off error.

Table 2 Applied boundary conditions

Working fluid	Air
Nature of flow	Incompressible
Reference pressure	1 atm
Inlet	Velocity inlet (4–18 m/s)
Outlet	Pressure outlet
Blade and hub	No-slip wall
Shroud	Contra-rotating no-slip wall
Turbulence intensity	5%
Residual criteria	1e–5
Mass imbalance	0.001%

Table 3 GCI estimation

Number of elements (10^6)	Fine, medium and coarse	6.80, 3.07, 1.39
Average cell size (h)	h_1, h_2, h_3	0.0052, 0.0068, 0.0089
Grid refinement factor (r)	r_{21}, r_{32}	1.3, 1.3
Performance parameter (ϕ)	ϕ_1, ϕ_2, ϕ_3	0.0427, 0.0422, 0.0402
Apparent order	p	5.47
Approximate relative error	e_a^{21}, e_a^{32}	0.0114, 0.0495
Extrapolated values	$\phi_{\text{ext}}^{12}, \phi_{\text{ext}}^{32}$	0.04294, 0.04295
Extrapolated relative error	$e_{\text{ext}}^{21}, e_{\text{ext}}^{32}$	0.0035, 0.0152
GCI	$\text{GCI}_{\text{fine}}^{21}, \text{GCI}_{\text{medium}}^{32}$	0.0043, 0.0192

5 Results and Discussion

5.1 *Grid Convergence and Error Estimation*

The grid convergence index (GCI) was used to estimate discretization error and grid convergence. The methodology suggested by Celik et al. [18] is followed to determine GCI. The GCI can be expressed as

$$\text{GCI} = \frac{1.25 e_a^{21}}{r_{21}^p - 1} \tag{1}$$

where e_a^{21} is the approximate relative error given by $\frac{\phi_2 - \phi_1}{\phi_1}$, p is the apparent order of the mesh employed for the numerical study. In addition, r_{21} is the ratio between the average cell size of the fine and coarse meshes. Table 3 displays the calculation of GCI.

The GCI decreases with an increase in grid resolution, and it implies that the discretization error reduces with an increase in grid refinement. The GCI value of the medium grid is less than 2%. Hence, it can be assumed that the grid convergence was achieved and the medium grid was retained for all the computations performed in this study to reduce computational power.

5.2 *Numerical Validation*

The present numerical results were validation against the experimental [19] and numerical results [20, 16] available in open literature. The non-dimensional flow

Fig. 4 Numerical validation

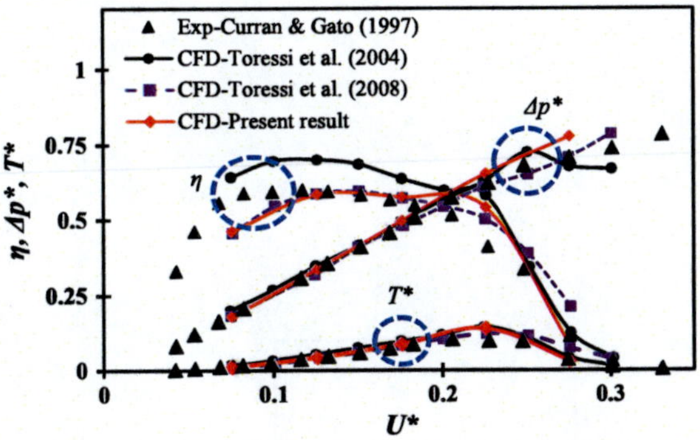

coefficient $\left(U^* = \frac{U_A}{U_{\text{tip}}}\right)$, torque $\left(T^* = \frac{T}{\rho\omega^2 R_{\text{tip}}^5}\right)$, pressure $\left(\Delta\ p^* = \frac{\Delta p}{\rho\omega^2 R_{\text{tip}}^2}\right)$ and efficiency $\left(\eta = \frac{T\omega}{Q\Delta p}\right)$ were chosen as the parameters to describe Wells turbine performance. The notation U_A is inlet axial velocity (m/s), U_{tip} is tip velocity (m/s), T is turbine torque (Nm), Q is the volumetric discharge (m^3/s), and Δp is static pressure drop (N/m^2). Figure 4 shows the validation of present numerical results with the existing results. It can be seen that the present CFD result matches well with the existing literature.

5.3 Aerodynamic Performance of CG&RT Blade

Figure 5 displays the performance parameters of the reference and the CG&RT blade. At low U^*, the reference blade shows higher torque compared to the CG&RT blade. However, the reference blade experiences stall at $U^* = 0.225$. On the contrary, the CG&RT blade stalls at $U^* = 0.325$ and produces higher torque than reference blade at $U^* > 0.225$. The reduced pressure drop in the CG&RT blade explains the decreased torque output. Similarly, the reference blade exhibits higher efficiency compared to CG&RT blade up to $U^* = 0.225$, and later it reduces due to stall condition. Furthermore, the CG&RT blade enhanced the relative stall margin and peak torque by 44.4% and 23.8%, respectively.

5.4 Flowfield Analyses

Figure 6 shows the pressure coefficient (Cp) distribution at blade midspan for different flow coefficients (U^*). The area enclosed by the Cp curve provides the lift force. At low $U^* = 0.075$, area of the Cp curve is small, which indicates lesser

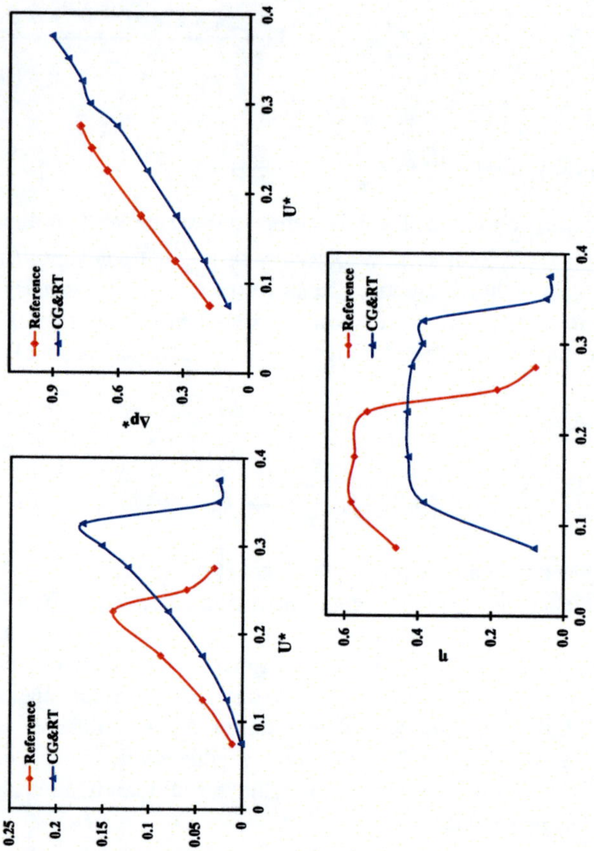

Fig. 5 Performance comparison of reference and modified blade

Fig. 6 Pressure coefficient distribution at blade midspan

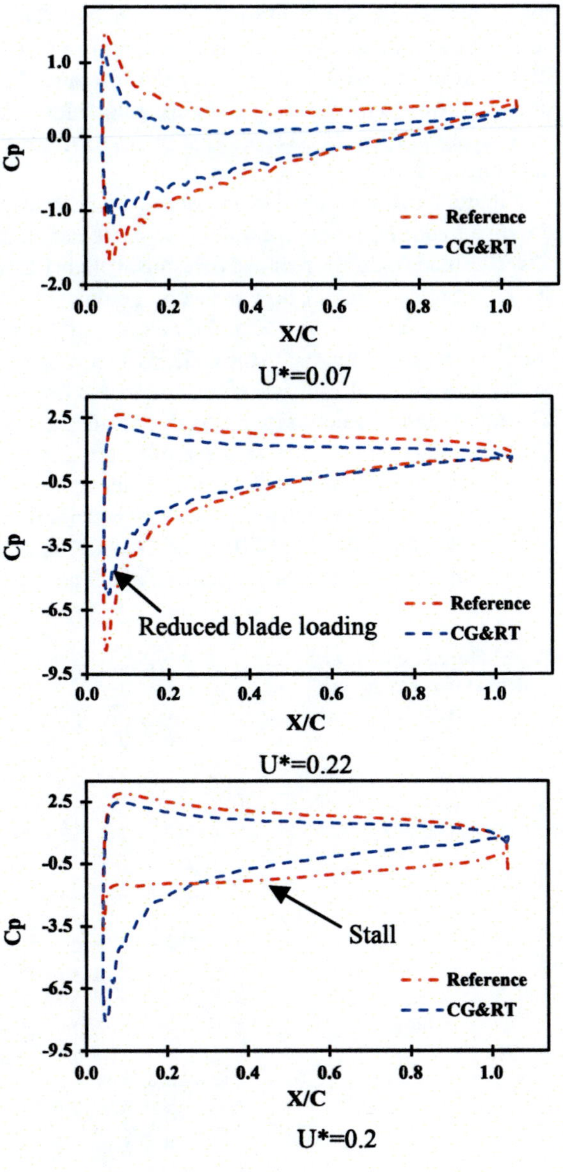

torque. As the U^* increases, the area enclosed by Cp curves increases which signifies lift increase. As revealed in Fig. 6, the reference blade shows higher pressure variation at blade leading edge (LE) compared to the CG&RT blade. The reduced suction at the CG&RT blade explains the decreased torque shown in Fig. 5. At U^* = 0.225, the suction pressure at LE increases for both blades. However, the CG&RT blade shows reduced blade loading near the LE and trailing edge (TE). To generate maximum torque, the enclosed Cp curve area should be higher, and the CG&RT blade shows reduced area and consequently reduced torque compared to reference blade. At increased U^* (0.275), for the reference blade, the suction peak vanishes and the pressure distribution is almost same along the chord. It implies stall condition because of the flow separation. In contrast, the CG&RT blade shows increased suction pressure and higher torque.

Figure 7 displays the Cp contour at blade midspan for different U^*. At $U^* = 0.075$, a maximum pressure region is visible near the blade LE for both blades. The CG&RT blade exhibits reduced pressure at blade LE compared to the reference blade. It can be noticed that the high pressure region at blade LE shifts towards the TE with increase in U^*. At $U^* = 0.225$, the pressure difference at blade LE is higher for the reference blade compared to the CG&RT blade, and it corroborates the results shown in Fig. 6. At $U^* = 0.275$, for reference blade, a low-pressure region engulfs the blade SS, and it signifies flow separation due to adverse pressure gradient. In contrast, in CG&RT blade the low-pressure region is confined near the blade LE and produces higher torque compared to reference blade.

Figure 8 illustrates the axial velocity distribution along the normalized span at blade LE location. At $U^* = 0.075$, the CG&RT blade shows reduced axial velocity at hub and tip in contrast to the reference blade. The decreased axial velocity reduces the

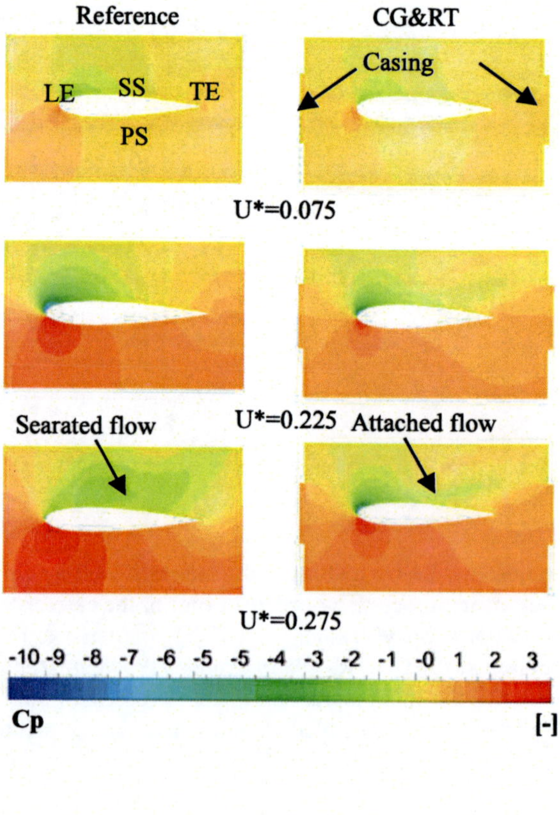

Fig. 7 Pressure coefficient contour at blade midspan

Fig. 8 Normalized span axial velocity distribution near blade leading edge (LE)

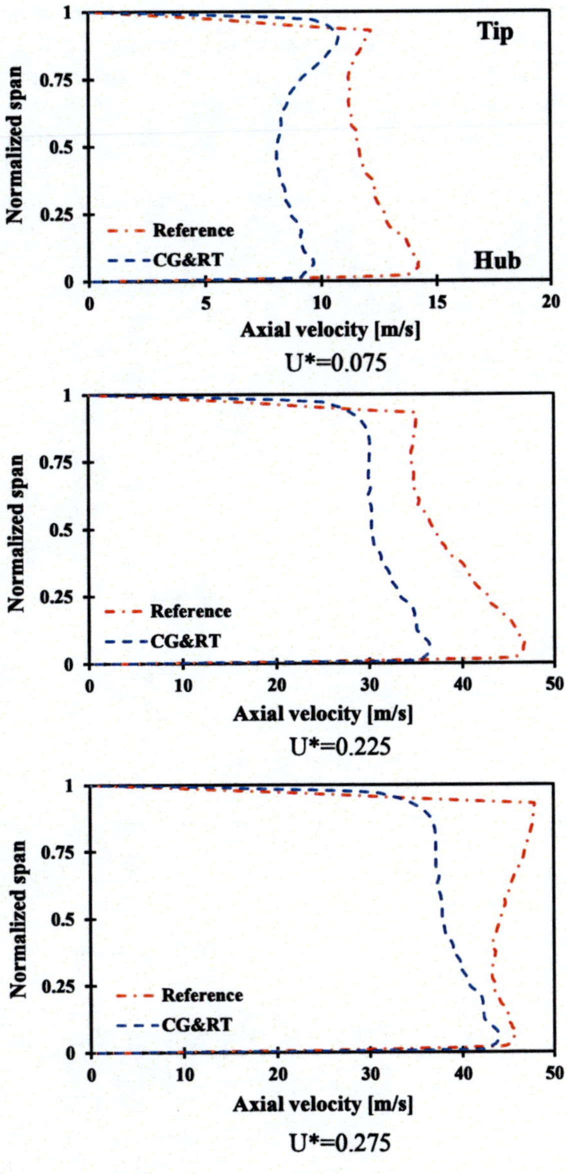

relative velocity and angle of attack which reduces the torque generated as shown in Fig. 5. The casing groove modification is the reason behind the reduced axial velocity near tip in the CG&RT blade. Moreover, the axial velocity increases with an increase in U^*. At $U^* = 0.225$, the axial velocity increases for both blades and the CG&RT blade shows reduced axial velocity compared to the reference case. Furthermore, at

$U^{*} = 0.275$, the reference blade shows increased axial velocity near the tip, and it indicates stall. Higher axial velocity produces a high angle of attack which signifies flow separation. In contrast, the CG&RT blade shows reduced axial velocity, and it is responsible for the delayed stall.

Figure 9 displays the tip leakage vortex at the blade mid-chord section for various U^{*}. The TLF occurs owing to the pressure variation between the pressure side (PS) and SS of the tip. Consequently, it rolls into a vortex on the blade SS known as

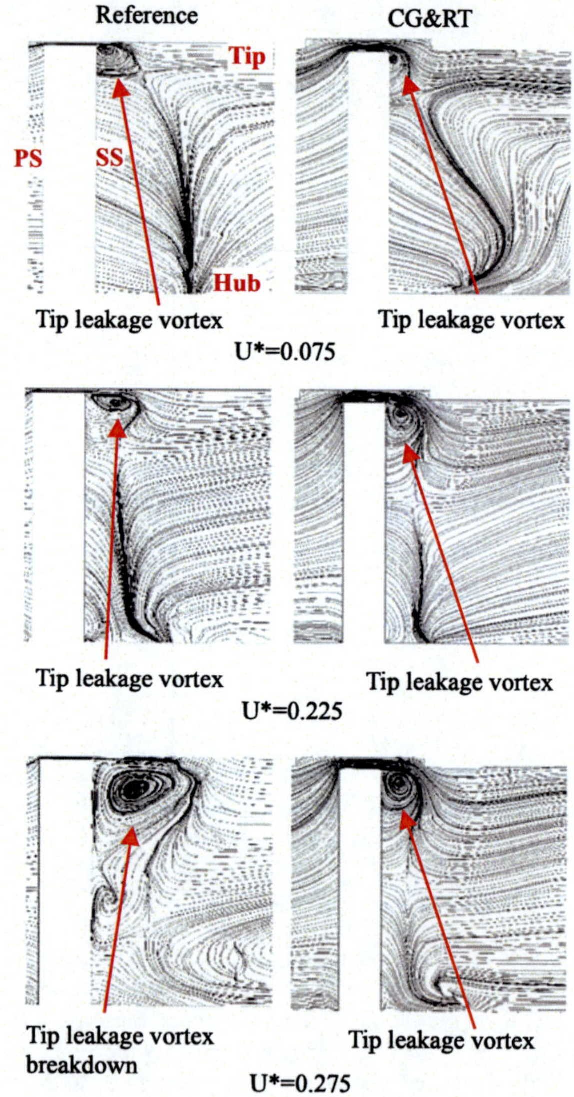

Fig. 9 Tip leakage vortex at blade mid chord section

tip leakage vortex. At $U^* = 0.075$, a clockwise tip leakage vortex is visible on the SS of blade tip for both blades. As revealed in figure, the strength of tip leakage vortex increases with U^*. The behavior of tip leakage vortex in the CG&RT blade is altered due to the casing groove and radiused edge tip. At $U^* = 0.275$, the tip leakage vortex breakdowns in reference blade and the increases the blockage causing stall. The presence of severe adverse pressure gradient causes the tip leakage vortex breakdown. In contrast, the tip leakage vortex is stable in the CG&RT blade. It is confined near the blade tip and offers less blockage to the primary flow. Moreover, the casing groove increases TLF, and it explains the reduced efficiency (Fig. 5) as the fluid passing through the tip clearance will not contribute to energy extraction. Furthermore, coupled with the casing groove and the radiused edge tip can be done to alleviate the decrement in efficiency.

6 Conclusion

The numerical investigation of Wells turbine with casing groove and radius edge tip was presented in this study. The problem was solved in ANSYS CFX 15.0, and the performance parameters obtained were compared with the reference blade. Moreover, the pressure coefficient, axial velocity distribution and tip leakage vortex were analysed. The prominent conclusions drawn were listed below.

- The blade with casing groove and radiused edge tip showed a wider operating range and enhance power output.
- The relative peak torque and stall margin were enhanced by 44.4% and 23.8%, respectively.
- The CG&RT blade displayed reduced axial velocity at the tip which decreased the relative velocity and angle of attack, thereby delaying stall.
- Moreover, optimization of the combined casing groove and radiused edge tip parameters should be performed for further performance enhancement.

References

1. Raghunathan S, Tan CP (1983) Performance of biplane wells turbine. J Energy 7(6):741–742
2. Inoue T, Kaneko M, Setoguchi K (1985) Studies on wells turbine for wave power generator: part 3; effect of guide vanes. Bul JSME 28(243):1986–1991
3. Raghunathan S, Tan CP (1985) Effect of blade profile on the performance of the Wells self-rectifying air turbine. Int J Heat Fluid Flow 6(1):17–22
4. Raghunathan S, Beattie WC (1996) Aerodynamic performance of contra-rotating wells turbine for wave energy conversion. Proc Inst Mech Eng Part A J Power Energy 210(6):431–447
5. Webster M, Gato LMC (2001) The effect of rotor blade shape on the performance of the Wells turbine. Int J Offshore Polar Eng 11(3):227–230

6. Gato LMC, Webster M (2001) An experimental investigation into the effect of rotor blade sweep on the performance of the variable-pitch Wells turbine. Proc Inst Mech Eng Part A J Power Energy 215(5):611–622
7. Govardhan M, Chauhan VS (2007) Numerical studies on performance improvement of self-rectifying air turbine for wave energy conversion. Eng Appl Comput Fluid Mech 1(1):57–70
8. Kaneko K, Setoguchi T, Santhakumar S, Takao M, Kim TH (2003) A modified Wells turbine for wave energy conversion. Renew Energy 28(1):79–91
9. Takao M, Setoguchi T, Kinoue Y, Kaneko K (2007) Wells turbine with end plates for wave energy conversion. Ocean Eng 34(11–12):1790–1795
10. Mohamed MH, Janiga G, Pap E, Thevenin D (2011) Multi-objective optimization of the airfoil shape of Wells turbine used for wave energy conversion. Energy 36(1):438–446
11. Shaaban S, Hafiz AA (2012) Effect of duct geometry on Wells turbine performance. Energy Convers Manag 61:51–58
12. Halder P, Samad A, Kim JH, Choi YS (2015) High performance ocean energy harvesting turbine design—a new casing treatment scheme. Energy 86:219–231
13. Halder P, Samad A (2015) Casing treatment of a wave energy extracting turbine. Aquatic Procedia 4:516–521
14. Nazeryan M, Lakzian E (2018) Detailed entropy generation analysis of a Wells turbine using the variation of the blade thickness. Energy 143:385–405
15. Kumar PM, Halder P, Samad A (2018) Performance analysis of Wells turbine with radiused blade tip. In: ASME 2018 37th international conference on ocean, offshore and arctic engineering. OMAE2018, pp 1–10
16. Torresi M, Camporeale SM, Strippoli PD, Pascazio G (2008) Accurate numerical simulation of a high solidity Wells turbine. Renew Energy 33(4):735–747
17. Menter FR (1994) Two-equation eddy-viscosity turbulence models for engineering applications. AIAA J 32(8):1598–1605
18. Celik IB, Ghia U, Roache PJ, Freitas CJ, Coleman H, Raad PE (2008) Procedure for estimation and reporting of uncertainty due to discretization in CFD applications. J Fluids Eng 130(7)
19. Curran R, Gato LMC (1997) The energy conversion performance of several types of Wells turbine designs. Proc Inst Mech Eng Part A J Power Energy 211(2):133–145
20. Torresi BFM, Camporeale SM, Pascazio G (2004) Fluid dynamic analysis low solidity Wells turbine, 59 Congresso ATI, Genvo, Italy, pp 277–288

Design and Development of Wind Tunnel to Study Smoldering Combustion

S. Sanjana, M. Rakshantha, Yeleti Bunny Venkat, and B. T. Kannan

Abstract Smoldering is flameless combustion that occurs on the surface of the condensed phase that is very difficult to detect and which causes serious fire risks in residential and industrial areas by the presence of turbulent winds. Statistics show that major fire accidents are due to smoldering for example cigarettes. Smoldering combustion is also a great concern in space facilities like ISS since the zero gravity environments can increase the rate of smoldering propagation. The main motivation of this paper is to design and develop the wind tunnel in order to study the effect of aerodynamic and combustible properties on the smoldering. A wind tunnel is an effective tool that is used to study the aerodynamic effects of air or smoke when it is allowed flowing past the solid bodies. The specific measurements that are required to construct the wind tunnel are mentioned in this paper, as well as the materials used in various sections of the wind tunnel are discussed. The power source and light source used for the visualization of the turbulent wind flow past the burning fuel are discussed. A low-speed tabletop wind tunnel is first designed using CATIA and then fabricated using necessary tools. The results from these experiments can be used to know how the combined effect of both laminar and turbulent flow on smoldering combustion is more harmful than the normal combustion.

Keywords Smoldering · Combustion · Wind tunnel · Design · Flow · Aerodynamics

S. Sanjana · M. Rakshantha · Y. B. Venkat · B. T. Kannan (✉)
Department of Aerospace Engineering, SRM Institute of Science and Technology, Kattankulathur, Kancheepuram District 603203, Tamil Nadu, India
e-mail: skyinventorbt@gmail.com

S. Sanjana
e-mail: sanjanasanjay1155@gmail.com

M. Rakshantha
e-mail: rakshanthamohan97@gmail.com

Y. B. Venkat
e-mail: bunnyyeleti@yahoo.com

S. Vijayan et al. (eds.), *Trends in Manufacturing and Engineering Management*,
Lecture Notes in Mechanical Engineering,
https://doi.org/10.1007/978-981-15-4745-4_90

1 Introduction

Kinbara et al. [1] did a study on smoldering combustion downward propagation through solid materials. The smoldering combustion of sold materials like the paper was conducted on the surface supported by the air to study the propagation mechanism. The sample ahead of the burning front when the smoldering combustion propagates upward is found to be ascending due to the combustion gases and the propagation velocity is found to be accelerating and resulting in a wide range of values. The samples used for studying the downward propagation of smoldering combustion were a circular rod, incense sticks, and cardboard. They concluded that the combustion gases found on the sample were the reason for the propagation velocity increase with respect to the depth of the layer. Ohlemiller et al. [2] worked on permeable fuel beds to conduct experiments on forward smoldering propagation and reverse smolder propagation. Wood fiber and polyisocyanurate were used as permeable fuels. The sample bed is tangentially insulated in the dispersion section where the uniform flow of oxygen or nitrogen mixture takes place. This study showed that the reverse smolder is 10 times faster than thse forward smolder because of conduction and radiation. And they also concluded that the high mass flow rate of oxygen increases the consumption of char in both the forward and reverse smoldering. Madara et al. [3] submitted a paper to a journal where the primary focus was on the design and fabrication process of a small-scale subsonic type wind tunnel which is an open circuit used to visualize the flow over an object and calculate the lift and drag coefficient using the experiments. CFD simulations give imperative results whereas the low circuit subsonic wind tunnel experiments provide a visual understanding of the impact of wind on different shapes. The test area was constructed based on the geometry of the object with a power source and instrumentation for weight estimation. The profile of the test area includes the cone area and square test area of different dimensions based on the geometry with the mean velocity of 15 m/s. The straight portion before the test section was made in order to stabilize the flow before it reaches the testing area. The estimation was made on the wind tunnel without the aerodynamic object which showed the uniform field which is the basic requirement for the study of streamlined bodies. The results calculated in a low-speed subsonic type wind tunnel were nearly accurate to the results calculated using NASA and MIT wind tunnel.

Mukunda et al. [4] did research on experiments and modeling of smoldering combustion on incense sticks. The experimental and modeling studies were done on smoldering as a function of air, velocity, and size. The results from both experimental and modeling cases were based on the surface smoldering combustion due to heat transfer of free and forced convection. Blower type wind tunnel was made and the mixed air of desired oxygen was supplied into the mixing chamber of the wind tunnel. The sample holder was used to hold the incense sticks and the anemometer was used to measure the flow inside the wind tunnel. Experimental data on the effects of smoldering combustion show that the forward smoldering increases with stream speed and for the reverse smoldering the spread rate are more than the forward smoldering. The reduction in the smoldering rate in the recirculation zone is due to the reduction of mass of the oxidizer. Bar-Ilan [5] did an experiment on small polyurethane foam to study the transition phenomenon of forward smoldering to flaming. The experiment was carried out on the polyurethane foam in order to find the velocity flow effect, oxygen concentration and the radiation flux on the transition at microgravity and low gravity environment. The samples of small size are selected because the smoldering combustion can self-propagate as the heat is lost to the environment. The experiments are carried out in a wind tunnel where the fuel is placed vertically. The lateral sides of the three samples were maintained at some peak temperature whereas the other three sides were maintained to the upward flow and radiation flux. From these experiments, it was found that there is a delay in the transition of smoldering to flaming by decreasing the flow velocity, and also by increasing the concentration of oxygen and radiation flux. They concluded that the transition can occur in the char region that was left behind by the smoldering reaction with the appropriate external conditions.

Buoyancy, Turbulence, and Combustion in Fire Field Modeling were discussed by Gao et al. [6]. The main purpose of this study was to known the effect of buoyancy, turbulence and combustion on the fire field model, for example, a test section in the form of the compartment with a window in the side and door in front. Simulations were carried out using Computational Fluid Dynamics (CFD). Two cases were studied under different operating conditions. Case A simulations were carried out with fire at the center of the room. Case B simulations were carried out at the corner of the room. Both cases were compared. The flame height is determined by knowing whether the fuel is burnt or not. The results from case B showed that because of the closed surface area the flame is height is more thereby increasing the upper ceiling temperature. In Case A, the flame height is less and doesn't reach the ceiling as the combustion propagates uniformly in all directions. They concluded that this is the reason for the combustion reactions to take place at the ceiling where the heat release rates are higher.

Schult et al. [7] did a study on porous material to know the propagation and extinction of smoldering waves caused by the opposed flow that was artificially generated. The main reason to use porous material is to allow the mixture of gas (oxygen and nitrogen) to flow through the pores which are present in the porous material. The cylindrical samples are considered where the gases are allowed to flow through one end. In the other end, smolder wave is initiated and it flows in the opposite direction to the flow of the oxidizer. In this paper, they have used graphical methods to learn about uniform propagation and smolder wave solutions. These results were compared with the experimental results which were taken by the same author in his other paper. The analytical results in this paper showed that the propagation velocity increases with an increase in mass flow rate which perfectly fits with the experimental results. Stocker et al. [8] did smoldering combustion experiments on flexible polyurethane foam in microgravity. The experiments were carried out in both microgravity and normal gravity and the results were compared. Four cases were conducted by varying the shapes and the environment both for micro and normal gravity. In this experiment temperature was measured along with the fuel sample at different locations, the video recording was done and post analyses were done on the char and gas composition. They found that heat loss was less in microgravity due to no natural convection in microgravity. This resulted in an increase in ignition temperature in the microgravity. They concluded that the burning of polyurethane foam gave rise to combustion products like carbon monoxide which is hazardous in space.

Hadden [9] did a study on the initiation of smoldering combustion and its spread in peat. The experimental study was done on the vertical samples of peatlands. The top surface of the sample was ignited by the radiation so that the smoldering combustion propagates downward in the direction opposite to the flow of oxidizer. The experiments were carried by varying the ignition conditions and the oxygen concentrations. The pyrolysis reaction and the oxidation reactions are the parameters focused in this paper. The char is formed in the pyrolysis oxidation and the char is converted into ash in the oxidation reactions. It was found that the smoldering front thickness depends mainly on the low mass flow rate and the mass of the leftover residues. The wider smoldering front is required to sustain the smoldering combustion. They concluded that the decreased oxygen concentration environment with less mass flow rate is required to avoid fire accidents. Sato et al. [10] conducted an experiment on cellulosic powder material to study the smoldering spread. The experimental set up involved a high bulk density cellulosic powder bed over which a stream of air was supplied in order to study the smolder spread in a horizontal layer. The results involved the 2D structure of a spread zone, spread rate, cross-sectional views of the burning region and temperature distributions. From this experiment, it was observed that the forward smoldering increases the velocity of the air and also increases the spread rate of the fuel. But above a certain value of the airstream velocity, the spread rate becomes independent and the smolder zone disappears. From experimental data, it is found that the balance of heat between the oxidation zone and unburnt part of the fuel bed is important to the behavior of smoldering spread. They also concluded that the aerodynamic effects and heat transfer control the behavior of spread rate over the smolder zone. The design and construction of wind tunnel were made to understand the combustion better and the behavior of smolder under the influence of forced flow effect by taking different parameters, i.e., shaped of fuel, different airspeeds, different sample holders and also the introduction of flow visualization to study the aerodynamic behavior of the different shapes under smoldering.

2 Wind Tunnel Design

2.1 Geometry and Meshing

The tabletop wind tunnel was initially designed using CATIA V5. The detailed design was carried out in order to give a clear picture of the fabrication process. The finalized CAD design of the tabletop wind tunnel is seen in the Figs. 1, 2 and 3.

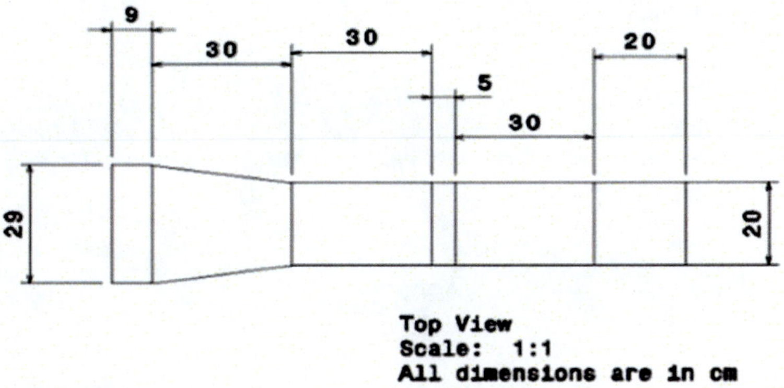

Fig. 1 Top View

Fig. 2 Side View

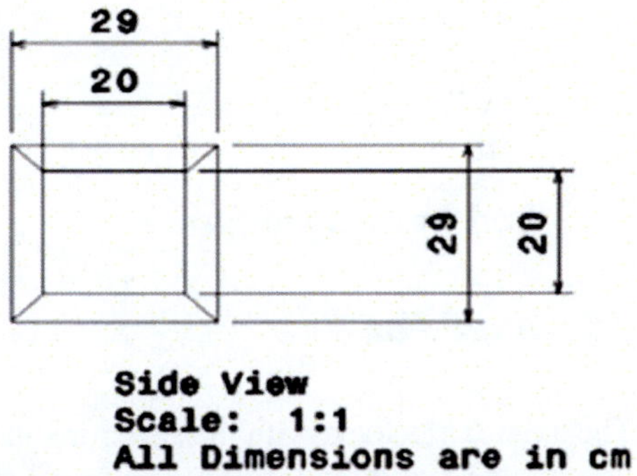

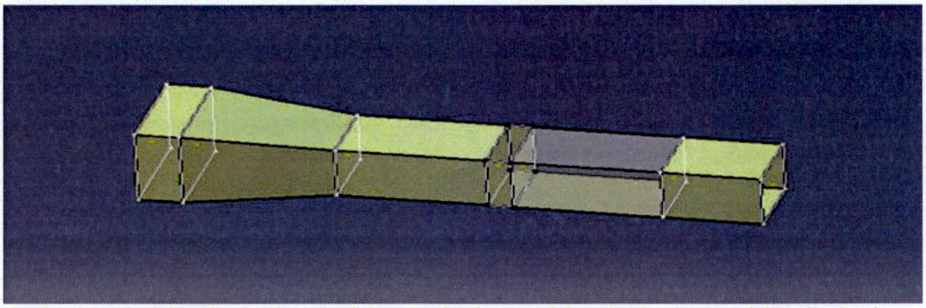

Fig. 3 The finalized CAD model of tabletop wind tunnel

The whole length of the wind tunnel is 129 cm and the maximum width is about 29 cm for the inlet to fix the axial fan and 20 cm uniform length for other sections. Inlet, contraction cone, flow conditioning unit, test section, and outlet are the different sections of the wind tunnel. Special attention was given on the turbulence intensity, uniformity, and flow control.

Fig. 4 Axial Flow Fan

3 Wind Tunnel Design

3.1 Axial Fan

The axial fan is fixed with the speed regulator of 5 speeds that allow the airflow to flow in an axial direction. The fan draws the air from the environment and blows into the wind tunnel. The speed reaches up to 4.7 m/s. The fan rotates to produce a pressure difference across it as the flow is axial in both entry and exit. The axial flow fan used in the present work helps in producing the flow inside the wind tunnel (Fig. 4).

This facilitates in studying the effect of both the laminar and turbulent flow on the smoldering combustion. The static rise is produced in order to compensate for the total pressure in the wind tunnel.

3.2 Contraction Cone

The contraction cone is the second part of the tabletop wind tunnel used to study the smoldering combustion. The contraction cone is used to accelerate the flow and allow the flow to pass into the test section. This section effectively reduces the velocity variation and helps in maintaining the flow uniformity across the length of the wind tunnel. The decrease in the area along the length of the contraction cone helps in increasing and decreasing velocity and pressure respectively.

Fig. 5 Honeycomb structure

3.3 Flow Conditioning Unit

Flow conditioning unit is used to straighten the air that was drawn from the surrounding environment and it is also used to channel the ambient air into the test section. It comprises of mesh and honeycomb structures. The flow conditioning unit reduces the turbulence of the air. The main purpose of the mesh is to avoid unnecessary items to enter the tunnel. There are different types of honeycomb structures used in the low-speed wind tunnel. For example, honeycomb structures with a circular cross-section, rectangular cross-section and hexagonal cross-section. The type of honeycomb structure used in the present work is of the circular cross-section. The cells are aligned in such a manner that it is in the streamwise direction. The honeycomb structures reduce the transverse velocity of the flow by reducing the pressure loss (Figs. 5 and 6).

3.4 Test Section

The test section is the basic and most important component of the wind tunnel. The incense materials of different shapes are placed inside the test section of the wind tunnel to study the effect of flow over them. The test section of the rectangular cross-section is used in this paper. The test section is provided with necessary facilities

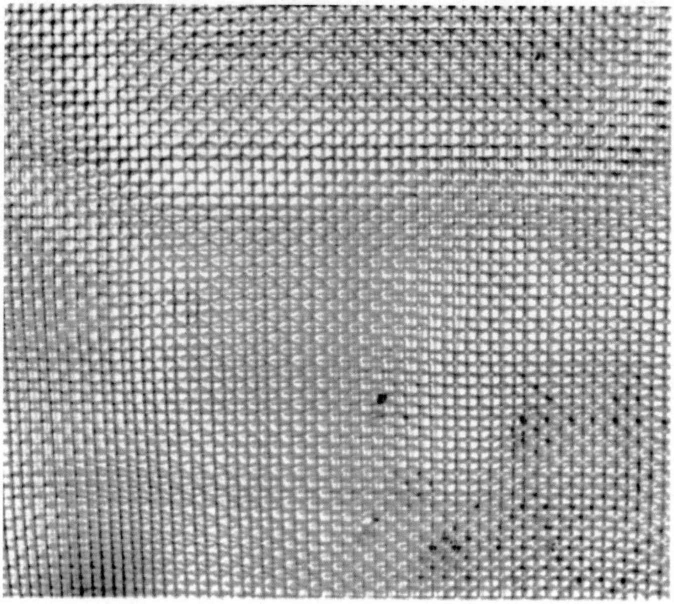

Fig. 6 Mesh

like glass for visual ability, laser light for flow illumination, sample holder, digital microscope as a camera, LED light and a black screen to capture clear images (Fig. 7).

Fig. 7 Test section

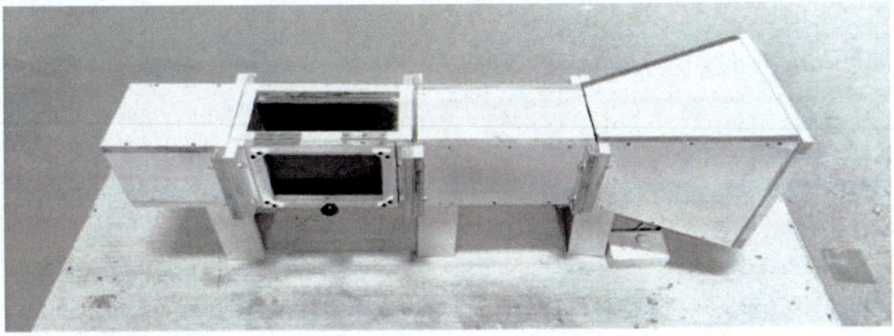

Fig. 8 Fabricated model

4 Fabrication of the Wind Tunnel

The fabrication of the tabletop wind tunnel was done using wood of thickness 180 mm, nuts, bolts, nails, glass in the test section, mesh and straws for honeycomb structures. Woodcutter, tape measures, hammer, chisel, nail gun, and clamps are some of the tools used in the fabrication process. The fan is the basic component with respect to which the design and fabrication were done. The contraction cone was the difficult part to fabricate and it was fabricated so that the fan is installed in such a manner that no vibration takes place. Similarly, the flow conditioning unit and the outlet were fabricated. Intense care was taken in fabricating the test section of the tabletop wind tunnel as it was fabricated using glass. The parts were fabricated in such a way that the parts can be dismantled for easier transportation, to upgrade the modification of tunnel for various experiments and installed for the experimentation (Fig. 8).

5 Schematic of Tabletop Wind Tunnel

The schematic diagram (see Fig. 9) gives a clear picture of experimental set up used to visualize the effect of flow on the smoldering combustion. The laser light of green color helps in flow visualization inside the wind tunnel test section. The camera used is a digital microscope that is connected to the computer to record the video.

6 Experimental Setup and Calculations

A sudden impact of the flow on the solid object, when placed in 90°, may not withstand the flow and can create a disturbance. In order to avoid such circumstances, the ramp (see Fig. 10) was placed inside the test section of the wind tunnel to develop a smooth

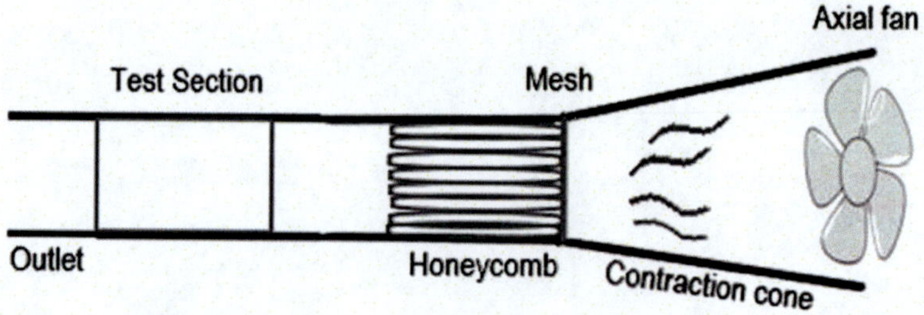

Fig. 9 Schematic diagram of tabletop wind tunnel

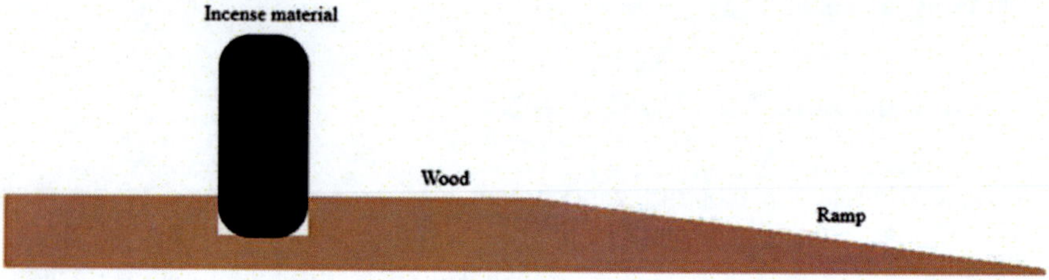

Fig. 10 Sample holder for boundary layer formation

boundary layer. The different types of sample holders were used for different sample shapes. Three possible cases were employed inside the test section to carry out the experimentations. They were along with the flow, transverse to the flow and against the flow. The angle can be varied by using the string as shown in (see Fig. 13). The velocity inside the test section was measured using the vent anemometer. The width of the test section was used, and the Reynolds number had been calculated for the test section (Figs. 11, 12, 13 and 14) (Table 1).

7 Results and Conclusion

The tabletop wind tunnel was initially designed using CATIA and then fabricated using different materials and tools. Axial fan, contraction cone, flow conditioning unit, test section, and outlet are the components of the wind tunnel used. The components of the tabletop wind tunnel were fabricated separately and then assembled to perform experiments. The incense materials with different shapes were used for the study. Sample holders based on different orientations and shapes were used. The experimentations were carried under normal gravity conditions (21% oxygen). The fabricated tabletop wind tunnel was used to study the effect of laminar and turbulent flow on the smoldering combustion using incense material. Five different fan speeds were employed over a burning incense material that was placed along with the flow, transverse to the flow and against the flow. The material is marked at regular intervals

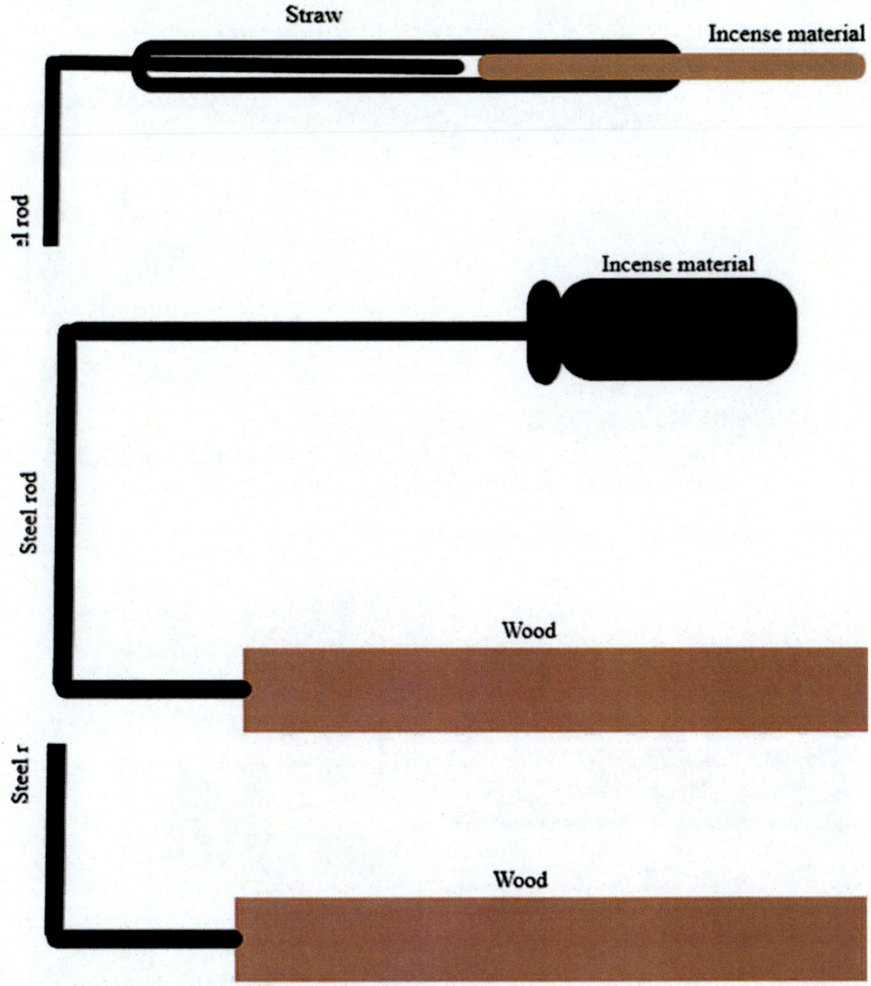

Fig. 11 Sample holders when placed against the flow

of fixed distance each along the whole length. The regression rates were calculated. The results obtained from the forced flow smoldering combustion experiments were compared with the results of natural smoldering combustion experiments. The natural smoldering combustion results were taken for three different orientations. They were vertical (90°), horizontal (180°), and inverted (270°). In conclusion, the tabletop wind tunnel was successfully designed and developed to study the effects of forced flow on the smoldering combustion. The burn rates of the fuel calculated were used to predict how smoldering combustion is more harmful than normal combustion. In addition, the tabletop wind tunnel can be used to conduct furthermore experiments

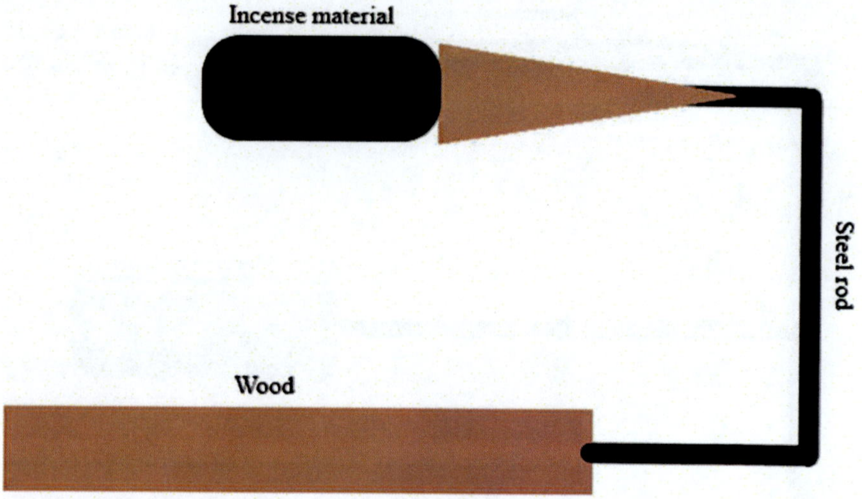

Fig. 12 Sample holder when placed along with the flow

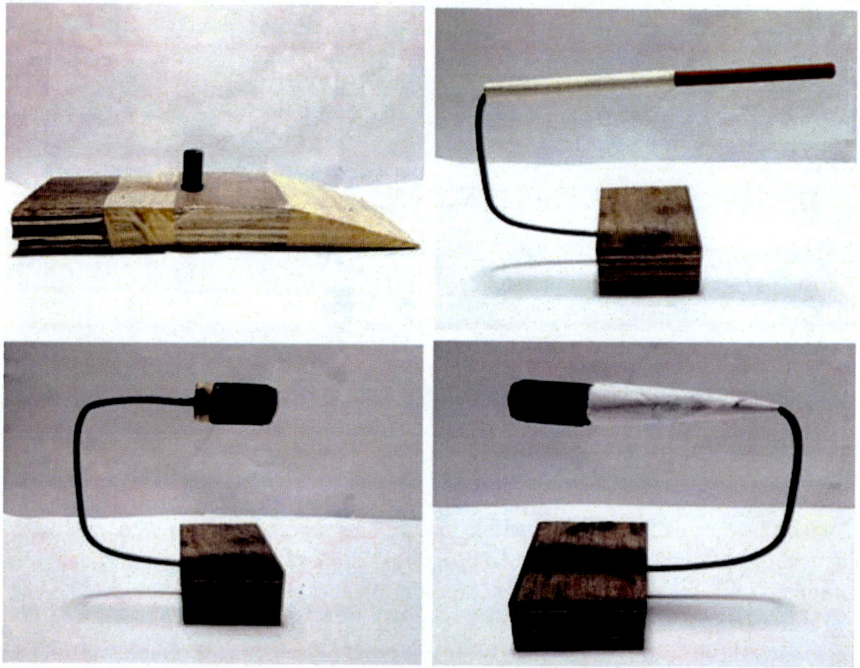

Fig. 13 Sample holders

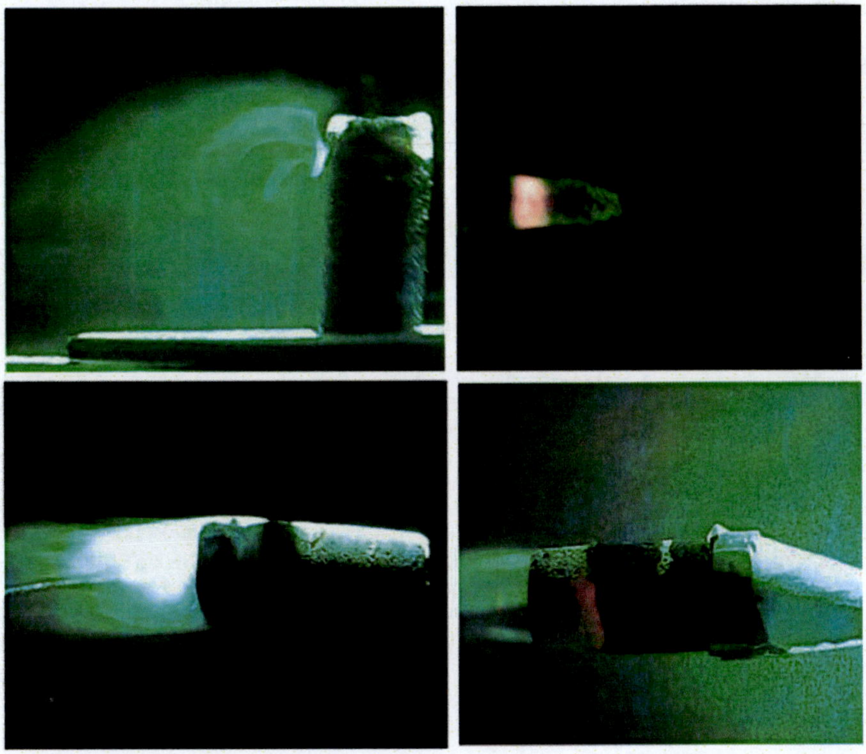

Fig. 14 Sample holders with burning incense materials inside the test section

Table 1 Different speeds inside the wind tunnel

Speed regulator	Speed (m/s)	Reynolds number inside the test section
1	0.8	10.256×10^3
2	1.8	23.077×10^3
3	2.2	28.205×10^3
4	2.3	29.487×10^3
5	2.4	30.769×10^3

to analyze and avoid the causes of residential fires, industrial fires, forest fires, rocket fires, and building fires.

Acknowledgements The authors acknowledge the usage of Flow Visualization Laboratory present in the Department of Aerospace Engineering, SRM Institute of Science and Technology (SRMIST) for the experiments performed towards the present research work.

References

1. Kinbara T, Endo H, Sega S (1967) Downward propagation of Smoldering combustion through solid materials. In: Symposium (International) on Combustion 11(1):525–531
2. Ohlemiller TJ, Lucca DA (1983) An experimental comparison of forward and reverse smolder propagation in permeable fuel beds. Combust Flame 54:131–147
3. SR Madara, Chithirai Pon Selvan M, JT Varghese (2017) Design and fabrication of low-cost open circuit subsonic wind tunnel. Int J Innov Technol Res 5(3):6154–6161
4. Mukunda HS, Basani J, Shravan HM, Philip B (2007) Smoldering combustion of incense sticks—experiments and modeling. J Combust Sci Technol 179(6):1113–1129
5. Bar-Ilan A, Putzeys OM, Rein G, Fernandez-Pello DL (2005) Urban, transition from forward smoldering to flaming in small polyurethane foam samples. In: Proceedings of the combustion institute 30:2295–2302
6. Gao Y, Chow WK (2005) Discussion on buoyancy, turbulence, and combustion in fire field modelling. Int J Architect Sci 6(3):93–105
7. Schult DA, Matkowsky BJ, Volpert VA, Fernandez-Pello AC (1995) Propagation and extinction of forced opposed flow smolder waves. Combust Flame 101:471–490
8. Stocker DP, Olson SL, Urban DL, Torero JL, Walther DC, Carlos Fernande-Pello A (1996) Small-scale smoldering combustion experiments in microgravity.In: Twenty-Sixth Symposium (International) on Combustion/The Combustion Institute, pp 1361–1368
9. Hadden RM, Rein G, Belcher CM (2013) Study of the competing chemical reactions in the initiation and spread of smoldering combustion in peat. Proc Combust Inst 34:2547–2553
10. Sato K, Sega S Smolder (1989) Spread in a Horizontal Layer of Cellulosic Powder. International Association for Fire Safety Science. Department of Physics, Faculty of science, Toho University

Spatial Location of Renewable Energy Plants: How Good Is Good Enough?

Paul A. Adedeji, Stephen Akinlabi, Nkosinathi Madushele, and Obafemi O. Olatunji

Abstract Land suitability analysis for renewable energy (RE) plants is gradually becoming multidisciplinary. The integration of Geographical Information System (GIS) with Multi-Criteria Decision Making (MCDM) tools has proven effective in such studies. From the trend in the literature, both experts in GIS and non-experts use these tools. There exist several factors that determine the suitability of a site for an intended purpose. In this study, factors specific to the reliability and accuracy of the results of GIS-MCDM approach to siting RE plants are reviewed. The reliability of the results from this integrated tool is hinged on the accuracy of principal milestones in the analysis. Four of these were identified in this article: data integrity, sensitivity analysis, correct use of tools, criteria weight assignment. It was concluded that apart from the identified dots in achieving high reliability and accuracy of the suitability analysis, ground verification is highly essential as this helps to verify the reality of the seemingly virtual analysis.

Keywords GIS · Land suitability analysis · Multi-criteria decision making · Renewable energy plant

1 Introduction

Spatial location of renewable energy (RE) plants largely depends on the availability of sources, which is a geospatial variant and as such vary from one geographical location to another. Asides climatic factors, RE facility location problem depends on other criteria, like economic, technological, social, and physiological factors [1–3].

P. A. Adedeji (✉) · N. Madushele · O. O. Olatunji
Department of Mechanical Engineering Science, University of Johannesburg, Johannesburg, South Africa
e-mail: pauladedeji2k5@gmail.com

S. Akinlabi
Department of Mechanical and Industrial Engineering, University of Johannesburg, Johannesburg, South Africa

Department of Mechanical Engineering, Covenant University Ota, Ota, Nigeria

S. Vijayan et al. (eds.), *Trends in Manufacturing and Engineering Management*, Lecture Notes in Mechanical Engineering,
https://doi.org/10.1007/978-981-15-4745-4_91

This results in a multi-criteria optimization problem. In recent times, siting RE plants have been established as a geospatial problem, hence the relevance of Geographical Information System (GIS) [4–7]. GIS technique fosters the analysis and synthesis of spatial data. It helps to integrate spatially referenced data for problem-solving [8]. The technique has found usefulness in several areas like crime area analysis, facility management and planning, resource monitoring, mineral exploration, water resource management, energy system facility location and planning, and so on [9–13]. Rather than the use of analytical models that offer a closed-form of solution, the GIS technique uses empirical modelling technique with georeferenced data such that the location component of the system been analyzed is not played down.

The literature is replete with solutions to the "how" and the "what" dimensions of RE resources, however, the "where" dimension (geospatial component) has lately attracted attention. GIS is highly effective in locating RE resources with multi-criteria considerations by means of what is called land suitability analysis. Land suitability analysis considers several criteria represented as layers in GIS software. These layers represent constraints to be satisfied, which varies from one energy resource to another. Presented in Table 1 are some criteria considered in the literature when carrying out land suitability analysis for specific RE resources.

While some criteria are common to almost all RE farms like distance from residential areas, the government reserved areas, road networks, some criteria are peculiar to each RE resource like distance from communication networks peculiar to locating wind farms [14] and gradient peculiar to solar farms [1, 29, 30], proximity to hot springs peculiar to geothermal resources [23] and so on. This results from peculiar interference with and dependence on identified physical features. The results of the suitability analysis give the best location of RE facilities based on stated criteria. However, the question of the degree of suitability of the geospatial solution domain is the focus of this study.

In a view to overcoming climate change and its associated effects on the biosphere, alternative sources of energy, which are clean and green are explored to replace the conventional sources. RE resources, which are energy sources that can be replenished within a short period [30] is fast gaining relevance both at national and global levels and are deployed as an off-grid, mini-grid, or on-grid system [31]. In any of these, proximity to energy sources or siting plants in energy hotspot among other factors is highly essential for the plant's efficiency, effectiveness, and sustainability. Two common approaches have been explored in the literature in solving this problem: the mathematical approach and the GIS-MCDM approach. This study, however, focuses on the GIS-MCDM approach.

Several techniques have been developed over the years to determine land suitability for RE resources. However, the most commonly used technique is the integration of MCDM techniques with GIS [32]. This technique considers the spatial component of criteria relative to identified feature raster files often in a projected coordinate system. This integrated method has recorded success over time in determining suitable locations for wind farms [1, 14, 33], solar farms [1, 29, 30, 34], biogas plants [26, 35, 36], geothermal plants [37], hydropower plant [38] and so on. This method is associated with uncertainties, which can be due to the nature of the data used, human

Table 1 Land suitability criteria for RE resources

Resource	Selection criteria	References
Wind Energy	Wind speed, Elevation, slope, Distance from 1. Energy transmission lines 2. Lakes and rivers 3. Roads 4. Protected Areas 5. Airports and Railway stations 6. Mining sites 7. Fault Lines 8. Radio and TV stations 9. Urban areas	[7, 14–17]
Solar Energy	Solar irradiation potential, average temperature, agrological capacity, land slope, land orientation, plot areas Distance from: 1. Power lines 2. Substations 3. Main roads 4. Protected areas	[16, 18]
Geothermal Energy	Heat degree Distance from: 1. Volcano craters 2. Faults 3. Volcanic rocks 4. Hot springs 5. Alteration zones 6. Fumarole	[19–23]
Hydropower	Head of hydropower, ground surface slope, conveyance length-head ratio, Distance to: 1. Road network 2. Power transmission line 3. Dam reservoir	[24]
Biomass/biogas	Biomass resource, slope, geomorphological factors, Distance from: 1. Industrial areas 2. Hydrological networks (e.g. rivers, streams, etc.) 3. Flora and fauna 4. Railways, airports, and heliports 5. Protected areas 6. Wetlands 7. Mining areas	[25–28]

judgement, weighting factors used, inherent variability in the model used [39, 40]. However, in reality, criteria for suitability analysis of a RE plant are inexhaustible. Attempts are often made by modellers to cover critical factors, that largely influence RE facility location.

MCDM-GIS approach to land suitability analysis is efficient, however, essential factors that predicate the reliability of its results most especially, when land suitability for RE in power generation is considered are discussed briefly. These are fundamental pitfalls that compromise the accuracy of MCDM-GIS approach to the location of RE plants. This study aims at carrying out a mini-review on four identified areas to take into consideration by energy researchers to ensure the reliability of land suitability analysis in energy research and RE plant location.

2 Reliability Factors for Geospatial Approach to RE Plant Location

2.1 Data Integrity

Data integrity emphasizes data accuracy and consistency in quality and recency. The recency of the geospatial data either raster or vector plays a significant role in the reliability of the suitable sites obtained for RE plant. The advent of GIS as a potent tool for spatial analysis has led to the proliferation of geographic databases globally. Individual data creator also makes efforts in populating these databases with digital maps of physical features and attributes; an example is raster files for road networks, places, locations, etc. The accuracy of these data, the expertise and integrity of the data creator is highly significant to the reliability of the study which uses the data. Government organizations also populate databases for public use. They often update these data with recent developments about the geospatial location of features. The analysis which makes use of data from trusted database (e.g. national database) however, without recency may likely falsify the results of land suitability analysis of RE plants. For example, protected areas in the database of a year may not be completely accurate for a study in the next year. Certain areas may have been considered as non-protected areas due to political or anthropogenic activities. The recency of the data reveals the present status of surfaces or features.

Data quality, on the other hand, depends on several factors among which is data projection. Incorrect projection of the data presents the wrong spatial location of the site. Geospatial data used for spatial analysis are obtained from metadata, which is either in the form of raster or vector [41]. These metadata contain information that describes the data like the projection (either spherical or projected coordinate system). The quality of the input data into a model determines the quality and reliability of its output [39].

2.2 Sensitivity Analysis

Aside tractability and feasibility, a good model is expected to be robust, which sensitivity analysis constitutes one of the tools in achieving this. As part of good modelling practice, it is expected that a modeller gives a level of confidence to which the model efficiently represents the system [39]. Sensitivity analysis helps to account for inaccuracy, ambiguity, and imprecision of GIS-multi-criteria evaluation process [42] and also enhances understanding of land suitability results for informed decision making [43]. It is a related practice to uncertainty analysis, and it is aimed at identifying principal inputs significantly responsible for output variations [44]. Many empirical facility location models follow this trend. However, few GIS integrated models, especially in RE plant consider this. Sensitivity analysis in spatial models differs from that of analytical and other empirical models due to the nature of the data. There are several methods, which have been developed over the years for performing sensitivity analysis on spatial models. Some of these are extensively discussed in [39, 45, 46]. However, in RE farm site suitability problems, a pixel by pixel validation to ensure satisfactory results is sometimes used. Here, a cadastral plot is selected at random and exported as a single layer, where a similar process of adding factors to its attribute table is performed [18]. The result is compared with the main analysis for consistency. The simple weighted model aggregate scoring method of essential factors of influence is also commonly used in RE plant suitability analysis. This method is efficient in RE plant land suitability analysis because sometimes, the suitability of a site for RE farm may be due to an aggregate of trivial factors rather than crucial and technical factors [47]. Sensitivity analysis in RE farm location also enables the modeller to identify principal factors, which aggregates to the ranking score. This analysis is highly essential before actual ground verification.

2.3 Choice of Appropriate Tool

GIS software like the ArcMap is multi-functional and so contains many tools, which can be harnessed for geospatially related problems in many fields. The choice of an appropriate tool for each analysis determines the validity and reliability of the results. For example, in determining land suitability for a RE plant using GIS-MCDM approach, several criteria are considered to ensure that the proposed site for the plant does not interfere with physical or environmental features. To ensure this, exclusion rules are developed with specific distances from physical features of significance to the study [7, 14]. Exclusion criteria use inequalities, which require it been carefully specified using appropriate tools. A commonly used distance function is the Euclidean distance. This calculates a straight-line distance from one point to another as a crow would fly. It has found usefulness in the RE land suitability analysis in the specification of distance from criteria like waterways, roads, seismic, volcanic faults [2, 18, 48], etc. The Euclidean distance tool contains several fields like maximum distance. The maximum distance displays raster files with subclasses to the maximum distance specified. Further exclusion criteria with distances greater than the already specified distance do become irresponsive to such criteria. Incorrect use of tools as such may lead to misleading results and wrong location for RE plant.

2.4 Subjectivity in Assigning Criteria Weights

Determining suitable land areas for RE plants requires a combination of critical, subjective, and objective inputs/criteria [49]. Criteria weight assignment contributes immensely to the controversy and uncertainty associated with MCDM-GIS approach to land suitability analysis in RE farm [46]. The impact of this is significant when several decision-makers are involved. Rather than one set of weights, a range of weights is derived. A slight perturbation in the weights can significantly affect the ranking of possible alternatives, which in some cases causes inaccuracy in the results and undesirable consequences [50].

While objective criteria are easily represented in quantitative forms, subjective criteria come in qualitative forms. This is often observed when maps are to be classified into fuzzy classes [23, 51]. On the other hand, critical inputs/criteria emanate from the successful integration of objective and subjective inputs/criteria. They form principal criteria in the model for further analysis. In spatial RE farm location problem, subjectivity in decision making is demonstrated when site choices are to be made based on criteria of influence [52]. Commonly used MCDM techniques for this include, analytic hierarchy process (AHP), analytic network process (ANP), technique for order processing by similarity to ideal solution (TOPSIS), elimination and choice expressing reality (ELECTRE), multi-criteria optimization and compromise solution (VIKOR) and their integrated fuzzy hybrids, as decision support system. These methods incorporate weight assignment techniques like rating, ranking, trade-off, pairwise comparison [23]. In RE studies, rating, ranking, and pairwise comparison techniques are commonly used for weight assignment and selection of alternatives. This process is subjective and may likely result in a disparity in the model result. Experts' knowledge is leveraged, as there is no technical documentation for this. These methods are chosen due to their simplicity in computation, though computation becomes complex for large criteria size.

One of the problems associated with subjectivity transformation is inconsistency in the formulation of the criteria matrix on the part of the modeller. To avoid this pitfall, Eigen method for criteria matrix is recommended [49], such that a square matrix of criteria has an aggregate of its eigenvalues equal to its trace. The trace is the sum of the diagonal elements of the criteria matrix.

3 Conclusions

It is observed in the recent decade that studies on suitability analysis for RE plants using GIS-MCDM as a tool is gradually becoming multidisciplinary. This study, therefore, focused on specific milestones in the analysis, which contributes to the reliability and credibility of results obtained. The study discussed the relevance of data integrity with a focus on obtaining recent data from credible sources. We also established the importance of sensitivity analysis and the use of appropriate tools in ensuring the reliability of the result. Effort is being made to reduce subjectivity in transforming qualitative variables to quantitative equivalence based on experts' knowledge. The study identified Eigen matrix method with singularity of the trace elements as a method of ensuring consistency in subjective to the objective transformation process. Asides these identified dots in achieving high reliability and accuracy of the results, ground verification is highly essential as this helps to verify the reality of the seemingly virtual analysis.

References

1. Janke JR (2010) Multicriteria GIS modeling of wind and solar farms in Colorado. Renew Energy 35:2228–2234. https://doi.org/10.1016/j.renene.2010.03.014
2. Villacreses G, Gaona G, Martínez-Gómez J, Jijón DJ (2017) Wind farms suitability location using geographical information system (GIS), based on multi-criteria decision making (MCDM) methods: the case of continental Ecuador. Renew Energy 109:275–286. https://doi.org/10.1016/j.renene.2017.03.041
3. Palmer D, Gottschalg R, Betts T (2018) The future scope of large-scale solar in the UK : site suitability and target analysis. Renew Energy 2018:1–11. https://doi.org/10.1016/j.renene.2018.08.109
4. Bonham-Carter GF (1994) Geographical information systems for geoscientists: modeling with GIS. Comput methods Geosci vol 13, New York: Pergamon
5. Noorollahi Y, Itoi R (2007) Geothermal power plant site selection with environmental consideration in Namafjall area, North Iceland. Trans Geotherm Resour Counc 31:193–198
6. Moghaddam MK, Meentemeyer RK, Samadzadegan F, Sharifi MA, Itoi R (2013) Spatial data analysis for exploration of regional scale geothermal resources. J Volcanol Geotherm Res 266
7. Noorollahi Y, Yousefi H, Mohammadi M (2016) Multi-criteria decision support system for wind farm site selection using GIS. Sustain Energy Technol Assess 13:38–50. https://doi.org/10.1016/j.seta.2015.11.007
8. Iyengar JV (1998) Application of geographical information systems. J Int Inf Manag 7
9. Case C, Hawthorne TL (2013) Served or unserved? A site suitability analysis of social services in Atlanta, Georgia using geographic information systems. Appl Geogr 38:96–106. https://doi.org/10.1016/j.apgeog.2012.11.012
10. Calvert K, Mabee W (2015) More solar farms or more bioenergy crops? Mapping and assessing potential land-use conflicts among renewable energy technologies in eastern Ontario, Canada. Appl Geogr 56:209–221. https://doi.org/10.1016/j.apgeog.2014.11.028
11. Castro-santos L, Prado G, Sim T (2019) Planning of the installation of offshore renewable energies : a GIS approach of the Portuguese roadmap. 132:1251–1262. https://doi.org/10.1016/j.renene.2018.09.031

12. Tien Bui D, Pradhan B, Lofman O, Revhaug I, Dick OB (2012) Landslide susceptibility mapping at Hoa Binh province (Vietnam) using an adaptive neuro-fuzzy inference system and GIS. Comput Geosci 45:199–211. https://doi.org/10.1016/j.cageo.2011.10.031
13. Boulos MNK (2004) Towards evidence-based, GIS-driven national spatial health Kingdom vol 50, pp 1–50
14. Atici KB, Simsek AB, Ulucan A, Tosun MU (2015) A GIS-based multiple criteria decision analysis approach for wind power plant site selection. Util Policy 37:86–96. https://doi.org/10.1016/j.jup.2015.06.001
15. Al-Yahyai S, Charabi Y, Gastli A, Al-Badi A (2012) Wind farm land suitability indexing using multi-criteria analysis. Renew Energy 44:80–87. https://doi.org/10.1016/j.renene.2012.01.004
16. Ali S, Taweekun J, Techato K, Waewsak J, Gyawali S (2019) GIS based site suitability assessment for wind and solar farms in Songkhla, Thailand. Renew Energy 132:1360–1372. https://doi.org/10.1016/j.renene.2018.09.035
17. Bili A, Vagiona DG (2018) Use of multicriteria analysis and GIS for selecting sites or onshore wind farms: the case of Andros Island (Greece). Eur J Environ Sci 8:5–13
18. Sánchez-Lozano JM, Teruel-Solano J, Soto-Elvira PL, Socorro García-Cascales M (2013) Geographical information systems (GIS) and multi-criteria decision making (MCDM) methods for the evaluation of solar farms locations: case study in south–eastern Spain. Renew Sustain Energy Rev 24:544–556. https://doi.org/10.1016/j.rser.2013.03.019
19. Kiavarz M, Jelokhani-Niaraki M (2017) Geothermal prospectivity mapping using GIS-based ordered weighted averaging approach: a case study in Japan's Akita and Iwate provinces. Geothermics 70:295–304. https://doi.org/10.1016/j.geothermics.2017.06.015
20. Kiavarz M, Samadzadegan F, Noorollahi Y (2014) Geothermics Spatial analysis and multi-criteria decision making for regional-scale geothermal favorability map. Geothermics 50:189–201. https://doi.org/10.1016/j.geothermics.2013.09.004
21. Alsuhaibani Z, Hepbasli A (2013) Environmental effects future aspects of geothermal energy in Saudi Arabia. Energy Sources Part A Recover Util Environ Eff 7036. https://doi.org/10.1080/15567036.2010.497805
22. Watts MD (1987) Geothermal exploration of Roccamonfina volcano, Italy. Geothermics 16:517–528
23. Yalcin M, Kilic GF (2017) A GIS-based multi criteria decision analysis approach for exploring geothermal resources: Akarcay basin (Afyonkarahisar). Geothermics 67:18–28. https://doi.org/10.1016/j.geothermics.2017.01.002
24. Ahmadi H, Shamsai A, Storage P, Selection S (2009) Preliminary site selection of pumped storage hydropower plants—a GIS-based approach. AUT J Model Simul 2009:25–32
25. Perpiña C, Martínez-Llario JC, Pérez-Navarro Á (2013) Multicriteria assessment in GIS environments for siting biomass plants. Land Use Policy 31:326–335. https://doi.org/10.1016/j.landusepol.2012.07.014
26. Sahoo K, Mani S, Das L, Bettinger P (2018) GIS-based assessment of sustainable crop residues for optimal siting of biogas plants. Biomass Bioenerg 110:63–74. https://doi.org/10.1016/j.biombioe.2018.01.006
27. Thompson E, Wang Q, Li M (2013) Anaerobic digester systems (ADS) for multiple dairy farms: a GIS analysis for optimal site selection. Energy Policy 61:114–124. https://doi.org/10.1016/j.enpol.2013.06.035
28. Brahma A, Saikia K, Hiloidhari M, Baruah DC (2016) GIS based planning of a biomethanation power plant in Assam, India. Renew Sustain Energy Rev 62:596–608. https://doi.org/10.1016/j.rser.2016.05.009
29. Uyan M (2013) GIS-based solar farms site selection using analytic hierarchy process (AHP) in Karapinar region Konya/Turkey. Renew Sustain Energy Rev 28:11–17. https://doi.org/10.1016/j.rser.2013.07.042
30. Asakereh A, Soleymani M, Sheikhdavoodi MJ (2017) A GIS-based Fuzzy-AHP method for the evaluation of solar farms locations: case study in Khuzestan province, Iran. Sol Energy 155:342–353. https://doi.org/10.1016/j.solener.2017.05.075

31. Bhandari B, Lee KT, Lee CS, Song CK, Maskey RK, Ahn SH (2014) A novel off-grid hybrid power system comprised of solar photovoltaic, wind, and hydro energy sources. Appl Energy 133:236–242. https://doi.org/10.1016/j.apenergy.2014.07.033
32. Cebi S, Ilbahar E, Atasoy A (2016) A fuzzy information axiom based method to determine the optimal location for a biomass power plant: a case study in Aegean region of Turkey. Energy 116:894–907. https://doi.org/10.1016/j.energy.2016.10.024
33. Kabu IK (2016) Wind farm site selection assessment in the greater Accra, Volta and Eastern Regions of Ghana : a GIS spatial multi-criteria assessment 2016
34. Chen CR, Huang CC, Tsuei HJ (2014) A hybrid MCDM model for improving GIS-based solar farms site selection. Int J Photoenergy 2014. https://doi.org/10.1155/2014/925370
35. Franco C, Bojesen M, Hougaard JL, Nielsen K (2015) A fuzzy approach to a multiple criteria and geographical information system for decision support on suitable locations for biogas plants. Appl Energy 140:304–315. https://doi.org/10.1016/j.apenergy.2014.11.060
36. Höhn J, Lehtonen E, Rasi S, Rintala J (2014) A geographical information system (GIS) based methodology for determination of potential biomasses and sites for biogas plants in southern Finland. Appl Energy 113:1–10. https://doi.org/10.1016/j.apenergy.2013.07.005
37. Noorollahi Y, Itoi R, Fujii H, Tanaka T (2008) GIS integration model for geothermal exploration and well siting. Geothermics 37:107–131. https://doi.org/10.1016/j.geothermics.2007.12.001
38. Setiawan D (2015) Potential sites screening for mini hydro power plant development in Kapuas Hulu, West Kalimantan: a GIS approach. Energy Procedia 65:76–82. https://doi.org/10.1016/j.egypro.2015.01.034
39. Crosetto M, Tarantola S, Saltelli A (2000) Sensitivity and uncertainty analysis in spatial modelling based on GIS. Agric Ecosyst Environ 81:71–79. https://doi.org/10.1016/S0167-8809(00)00169-9
40. Crosetto M, Tarantola S (2001) Uncertainty and sensitivity analysis: tools for GIS-based model implementation. Int J Geogr Inf Sci 15:415–437. https://doi.org/10.1080/13658810110053125
41. Gigović L, Pamučar D, Božanić D, Ljubojević S (2017) Application of the GIS-DANP-MABAC multi-criteria model for selecting the location of wind farms: a case study of Vojvodina. Serbia Renew Energy 103:501–521. https://doi.org/10.1016/j.renene.2016.11.057
42. Malczewski J (2004) GIS-based land-use suitability analysis: a critical overview. Prog Plann 62:3–65. https://doi.org/10.1016/j.progress.2003.09.002
43. Xu E, Zhang H (2013) Spatially-explicit sensitivity analysis for land suitability evaluation. Appl Geogr 45:1–9. https://doi.org/10.1016/j.apgeog.2013.08.005
44. Mavromatidis G, Orehounig K, Carmeliet J (2018) A review of uncertainty characterisation approaches for the optimal design of distributed energy systems. Renew Sustain Energy Rev 88:258–277. https://doi.org/10.1016/j.rser.2018.02.021
45. Bonin O (2006) Sensitivity analysis and uncertainty analysis for vector geographical applications. 7th Int Symp Spat Accuracy Assess Nat Resour Environ Sci 319–328
46. Feizizadeh B, Blaschke T (2014) An uncertainty and sensitivity analysis approach for GIS-based multicriteria landslide susceptibility mapping. Int J Geogr Inf Sci 28:610–638. https://doi.org/10.1080/13658816.2013.869821
47. Cevallos-Sierra J, Ramos-Martin J (2018) Spatial assessment of the potential of renewable energy: the case of ecuador. Renew Sustain Energy Rev 81:1154–1165. https://doi.org/10.1016/j.rser.2017.08.015
48. Al Garni HZ, Awasthi A (2017) Solar PV power plant site selection using a GIS-AHP based approach with application in Saudi Arabia. Appl Energy 206:1225–1240. https://doi.org/10.1016/j.apenergy.2017.10.024
49. Sule D (2001) Logistics of facility location and allocation. vol 20010505. https://doi.org/10.1201/9780203910405
50. Feizizadeh B, Blaschke T (2013) GIS-multicriteria decision analysis for landslide susceptibility mapping: comparing three methods for the Urmia lake basin. Iran Nat Hazards 65:2105–2128

51. Miller A, Li R (2014) A geospatial approach for prioritizing wind farm development in Northeast Nebraska, USA. ISPRS Int J Geo-Inf 3:968–979. https://doi.org/10.3390/ijgi3030968
52. Kazak J, van Hoof J, Szewranski S (2017) Challenges in the wind turbines location process in Central Europe—the use of spatial decision support systems. Renew Sustain Energy Rev 76:425–433. https://doi.org/10.1016/j.rser.2017.03.039

Analysis of Pendulum-Based Nonlinear Energy Sink for Energy Harvesting

Pradeep V. Malaji

Abstract Passive method is one of the best-suited methods for vibration reduction of the primary structure. Energy harvesting converts vibration energy into useful electrical energy which can be used to power up sensors used for monitoring. Nonlinear energy sinks (NES) are analyzed to passively reduce vibration of the structure as well as energy harvesting simultaneously in this article. Current work considers the applicability of common pendulum as the NES for mitigation linear primary structure and electromagnetic energy conversion. It is observed that the pendulum NES can overcome one of the main limitations of conventional tuned mass damper (TMD) designs, as it has the capability to mitigate primary structure excitation and harvest electrical energy over a wide range. This is because the pendulum has the capability to operate both in linear and nonlinear zones that can be utilized into a resonance of the primary oscillator. For small input excitation, the pendulum acts as a conventional tuned mass damper. However, for larger energy input, the pendulum acts in the nonlinear zone as NES to reduce primary mass vibration. Thus, a pendulum can dissipate energies in a relatively broad spectrum. This article presents a numerical analysis of pendulum NES for control and energy harvesting.

Keywords Nonlinear energy sink · Pendulum · Energy harvesting · Passive control

1 Introduction

Protection of structures, machines and vehicles from excessive vibration is vital as damage due to vibration can severely affect equipment's service life. With emerge of new sensing and control technologies can provide information about the health of the structure. These tools can help to detect the initial damage in the system [1].

P. V. Malaji (✉)
BLDEA's V P Dr. P G Halakatti College of Engineering and Technology, Vijayapur, Karnataka 586103, India
e-mail: pradeepmalaji@gmail.com

S. Vijayan et al. (eds.), *Trends in Manufacturing and Engineering Management*, Lecture Notes in Mechanical Engineering,
https://doi.org/10.1007/978-981-15-4745-4_92

Vibration isolation by passive methods has been widely studied by many researchers. The main limitation of conventional linear isolators is that they are effective only if the natural frequencies of these isolators are nearer or below the excitation frequency [2, 3]. This requires a low spring stiffness which makes the primary structure to undergo large static deflection. To address this issue, researchers analyzed anti-resonant vibration which leads to the generation of anti-resonance frequencies beyond the natural frequency [4]. This might have an advantage over conventional isolator but still has a limitation on the isolation bandwidth with a single narrow anti-resonance frequency.

To overcome the above limitations, nonlinearity has been used in designs in order to improve the bandwidth of the vibration isolation system. Various nonlinear isolators have been reviewed by Ibrahim [5] which indicates active research in this area. Nonlinear damper and nonlinear stiffness are the two kinds of nonlinear elements that have been studied [6, 7].

The nonlinear energy sink (NES) refers to an essentially nonlinear system consisting of a small mass attached to a primary mass to attenuate the vibration under various excitations [8]. NES provides the targeted energy transfer (TET) feature which refers to its ability to irreversibly transfer energy from the primary system and dissipating it within itself [9]. However, rather than dissipation, the vibration energy can be converted into useful electric energy for minor energy needs. The process of converting vibration energy into electric energy is called as energy harvesting.

Utilization of vibration energy for energy harvesting can be very useful for small portable equipment or for equipment present in isolated regions [10, 11]. Conventional energy harvesters have a limitation of harvesting energy at resonance [13]. Other options consisted of multimodal techniques [13, 15] or a switch to nonlinear techniques [13, 14, 16]. While tuned mass dampers (TMD) and linear vibration absorbers have been utilized for control of primary structure [17], limitations such as negligible frequency robustness and poor performance for random vibration render their utilization an infeasible task [17]. Attachment of auxiliary systems as energy harvesters for vibration suppression has an advantage of vibration control and useful energy harvesting [12]. Further investigations into nonlinear alternatives have led to the development of nonlinear energy sink (NES). While it is successful in suppressing primary systems' responses, it can also perform simultaneous energy harvesting. Such features allow vibration control in a broadband manner with simultaneous energy harvesting, making NES a lucrative and more advantageous system when compared to its counterparts. Also, while it has yet not been fully investigated, NES's ability to increase operational bandwidth also allows it to effectively suppress primary vibrations unlike the linear systems which falter when operated upon by frequency varying excitation [17].

This paper analyzes a two-degree vibration isolation system, consisting of primary mass coupled with a pendulum NES with electromagnetic conversion system. Amplitude of primary structure and voltage output from the NES are presented when the primary system is under harmonic base excitation. Parametric study is conducted to explore the effect of these on NES performance. Comparison of NES with TMD is also presented to enumerate the advantages of NES.

This paper has been divided into four sections. Section 2 describes the NES harvester model with mathematical equations used for the simulations. Section 3 deals with the results and discussion followed by summarizing and providing the relevant and observed conclusions in Sect. 4.

2 System Model

The system model is shown in Fig. 1. The primary system has mass M, damping c_1 and linear stiffness k. The pendulum is mounted on the primary mass with provision for electromagnetic conversion due to pendulum oscillation when primary system is subjected to support excitation x_g. It has mass m, mechanical and electrical damping c_{2m} and $c_{2e,}$ respectively, and length l.

Lagrangian of the system in terms of potential and kinetic energy is written as,

$$L = (M+m)\left(\dot{x}_g^2+\dot{x}^2\right) + M\dot{x}_g\dot{x} - \frac{1}{2}kx^2 - mgl(1-\cos\theta)$$
$$+\frac{1}{2}m\left(l^2\dot{\theta}^2 + 2\dot{x}_g\dot{x} + 2\dot{x}l\dot{\theta}\cos\theta + 2\dot{x}_g l\dot{\theta}\cos\theta\right) \quad (1)$$

Damping as through Rayleigh's dissipation function is taken as,

$$D = \frac{1}{2}\left(C_1x^2 + (C_{2m}+C_{2e})\theta^2\right) \quad (2)$$

From the Lagrangian equation of motion with the voltage generated is written as: where B is magnetic flux density, L is coil inductance and d is a gap between coil and magnet.

$$(M+m)\ddot{x} + ml\cos\theta\ddot{\theta} - ml^2\sin\theta\dot{\theta}^2 + c_1\dot{x} + Kx = -(M+m)\ddot{x}_g$$

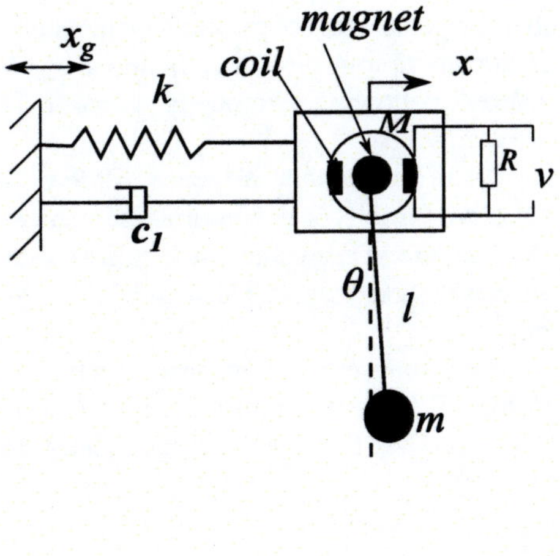

Fig. 1 System model with pendulum NES harvester

$$ml^2\ddot{\theta} + ml\ddot{x}\cos\theta + (c_{2m} + c_{2e})l^2\dot{\theta} + mgl\sin\theta = ml\ddot{x}_g\cos\theta \tag{3}$$

$$v = BLd\dot{\theta}$$

An equation of motion in non-dimensional form is written as,

$$\ddot{u} + \varepsilon\cos\theta\ddot{\theta} - \varepsilon\sin\theta\dot{\theta}^2 + \zeta_1\dot{u} + u = f\Omega^2\cos(\Omega\tau)$$
$$\ddot{\theta} + \cos\theta\ddot{u} + (\zeta_{2m} + \zeta_{2e})\dot{\theta} + r^2\sin\theta = f\Omega^2\cos(\Omega\tau)\cos\theta \tag{4}$$

$$v = \dot{\theta}$$

The non-dimensional parameters are,

$$\tau = \omega_1^2 t,\ x = ul,\ \omega_1 = \sqrt{\frac{k}{M+m}},\ \omega_2 = \sqrt{\frac{g}{l}}$$
$$x_g = X_g\cos(\omega t),\ \varepsilon = \frac{m}{(M+m)},\ \Omega = \frac{\omega}{\omega_1}$$
$$\varsigma_1 = \frac{c_1}{m\omega_1},\ \varsigma_{2m} = \frac{c_{2m}}{m\omega_1},\ \varsigma_{2e} = \frac{c_{2e}}{m\omega_1}$$
$$f = \frac{X_g}{l},\ r = \frac{\omega_2}{\omega_1} \tag{5}$$

Equation 4 will be solved numerically using ODE45 in MATLAB to obtain results which will be presented in the next section.

3 Results and Discussion

In order to study the effect of the pendulum, NES on the primary system and voltage output from NES parametric study on the system are considered. It is to be noted that the parameters of primary systems are not changed. Therefore, the influence of frequency ratio r, electrical damping ξ_{2e}, mass ratio ε and excitation level f is considered. Following parameters shown in Table 1 are used unless otherwise changes are mentioned.

Figure 2 shows the response of the primary structure without any absorber. The structure exhibits peak amplitude at resonance as expected. To safeguard the structure from failure at resonance, it is necessary to minimize the response of the primary structure at resonance. This can be done by adding secondary structure as shown in Fig. 1.

The response of the primary structure and voltage generated from NES is shown in Fig. 3. With the introduction of NES, the amplitude of the primary system at resonance has been reduced drastically. Ratio of primary and pendulum structure

Table 1 Parameters used for simulation

Parameter	Value
Mass ratio ε	0.2
Ratio of frequencies of secondary and primary system r	0.9
Damping factor of primary system ξ_1	0.025
Damping factor of secondary system ξ_{2m}	0.01
Electrical damping factor of secondary system ξ_{2e}	0.05
Ratio of excitation amplitude to length of pendulum f	0.03

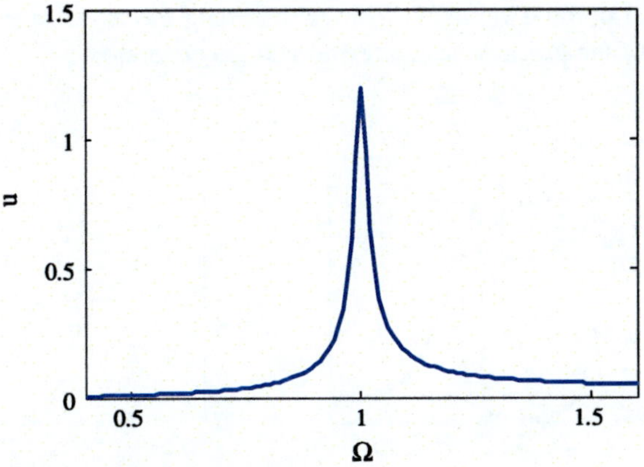

Fig. 2 Primary system response without absorber

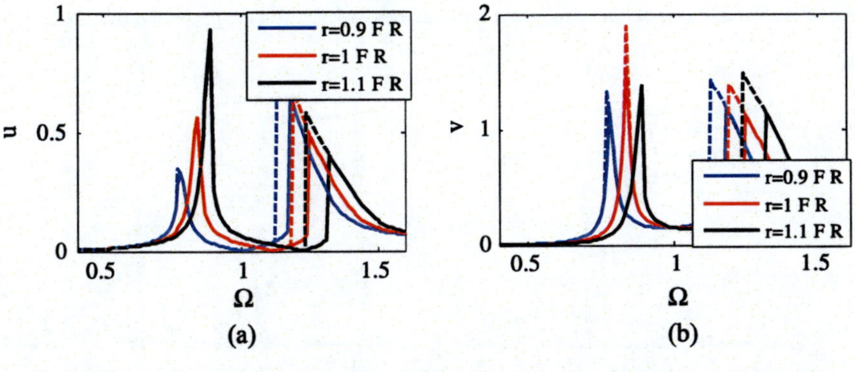

Fig. 3 Effect of frequency ratio of primary system and pendulum NES **a** on primary system, **b** voltage generated from NES. (Continuous line-F-forward sweep, dotted line-R-backward sweep)

frequency r is varied to study its effect. With an increase in r, the response curve of primary and NES structure shifts toward right. For $r < 1$, the amplitude u of the primary system is almost zero at resonance when $r > 1$ the amplitude of primary structure increases as shown in Fig. 3a. Multi-frequency voltage can be generated from NES indicating broadband harvesting. Spring softening effect can be observed at the second frequency which further increases frequency band as in Fig. 3b.

Effect of electrical damping on system performance is shown in Fig. 4. The electrical damping is varied between 0.005 and 0.02. With the increase in electrical damping, the amplitude of primary structure away from resonance decreases indicating vibration reduction. This also will affect the bandwidth of the voltage generated by NES. Hence, an optimal value of electrical damping has to be selected to optimize the amplitude of the primary structure and voltage generated.

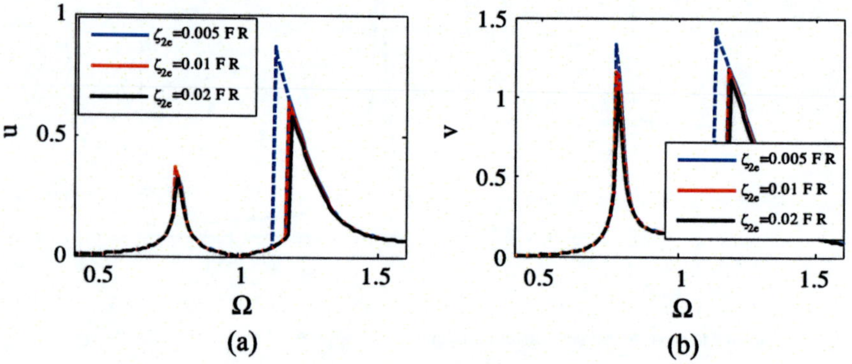

Fig. 4 Effect of electrical damping ξ_{2e} **a** on primary system, **b** voltage generated from NES

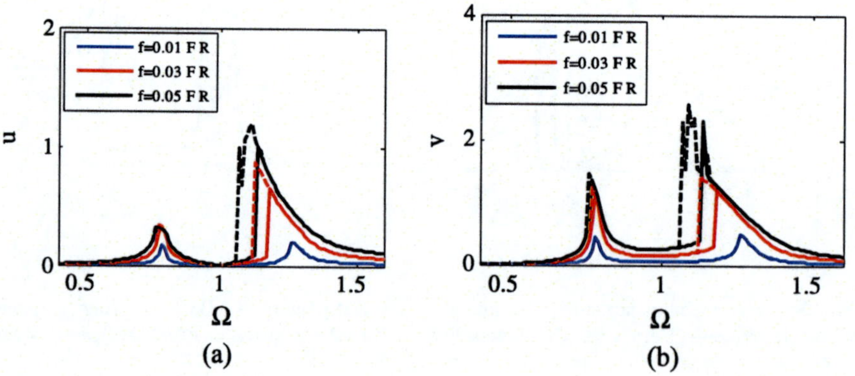

Fig. 5 Effect of excitation amplitude **a** on primary system, **b** voltage generated from NES, $\xi_{2e} = 0.01$

Figure 5 shows the effect of excitation amplitude f on the system which is varied from 0.01 to 0.05. With the increase in excitation level, amplitudes of primary structure away from resonance increase as expected whereas the amplitude at resonance remains the same as shown in Fig. 5a. Voltage amplitude and bandwidth from NES increase with the increase in excitation level as shown in Fig. 5b. Spring softening effect at the first frequency can also be observed at higher excitation.

Mass of the pendulum m has to be much smaller than the mass of primary system M. The effect of mass ratio on the system performance is shown in Fig. 6. Increases in mass ratio shift the response curve toward left. For the low value of mass ratio $\varepsilon = 0.1$, the amplitude of primary structure at resonance will be more compared to when $\varepsilon = 0.2$ as shown in Fig. 6a. Voltage generated by NES at a low mass ratio will have higher amplitude at both frequencies and peaks are nearer as shown in Fig. 6b.

To highlight the benefit of the NES with compared to the classic TMD, even at $r = 1$, a comparison is shown in Fig. 7. For the sake of comparison, all other parameters are kept the same for both NES and TMD. The amplitudes with both NES and TMD

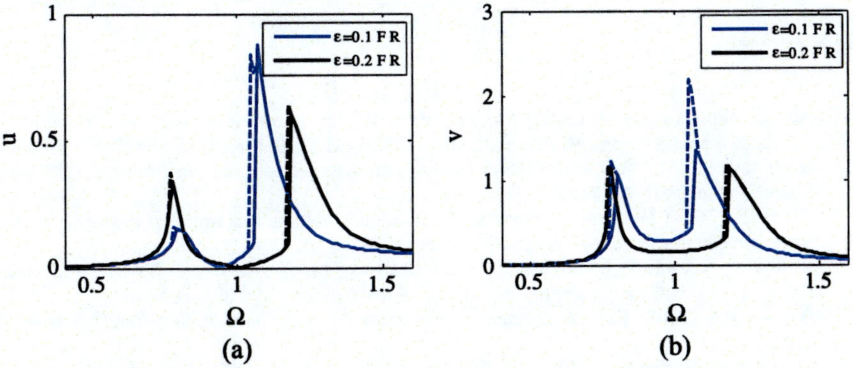

Fig. 6 Effect of mass ratio **a** on primary system, **b** voltage generated from NES. $\xi_{2e} = 0.01$

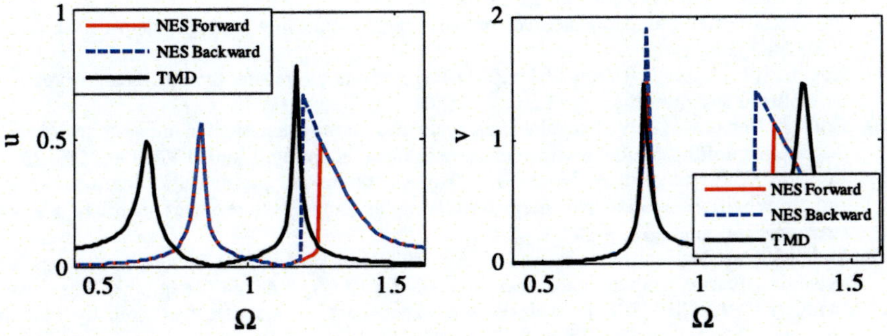

Fig. 7 Comparison of TMD and NES at $r = 1$ **a** primary system, **b** voltage generated from NES. $\xi_{2e} = 0.01$, $r = 1$

at resonance remain the same for the primary structure, whereas TMD produces comparatively low amplitudes away from the resonance. The voltage generated by NES has higher amplitude and better bandwidth compared to TMD.

4 Conclusion

From the analysis carried out in the manuscript, it can be mainly concluded that regular damped pendulum with electromagnetic conversion provision can be used as a nonlinear energy sink to mitigate vibration of primary structure and harvest useful energy from NES. Also, NES exhibits better performance at a broader range. Parametric study is carried out to understand the effect of parameters on system performance. Further, a theoretical and experimental study with an optimization study will be considered in the future.

References

1. Ali SF, Ramaswamy A (2009) Optimal dynamic inversion based semi-active control of benchmark bridge using MR dampers. Struct control Health Monitor 16:564–585
2. DeBra DB (1992) Vibration isolation of precision machine tools and instruments. CIRP Ann Manuf Technol 41:711–718
3. Rivin EI (2003) Dynamic properties of vibration isolation systems. In: Passive vibration isolation, Chapter 1, ASME press, New York, USA
4. Yilmaz C, Kikuchi N (2006) Analysis and design of passive band-stop filter-type vibration isolators for low frequency applications. J Sound Vibration 291:1004–1028
5. Ibrahim RA (2008) Recent advances in nonlinear passive vibration isolators. J Sound Vib 314:371–452
6. Peng Z, Lang Z, Zhao L, Billings SA, Tomlinson GR, Guo P (2011) The force transmissibility of MDOF structures with a non-linear viscous damping device. Int J Non-Linear Mech 46:1305–1314
7. Ahn HJ (2008) Performance limit of a passive vertical isolator using a negative stiffness mechanism. J Mech Sci Technol 22:2357–2365
8. Vakakis AF (2001) Inducing passive nonlinear energy sinks in vibrating systems. J Vib Acous 123:324–332
9. Kopidakis G, Aubry S, Tsironis GP (2001) Targeted energy transfer through discrete breathers in nonlinear systems. Phys Rev Lett 87:165501
10. Ertuk A, Inman DJ (2011) Broadband piezoelectric power generation on high-energy orbits of the bistable duffing oscillator with electromechanical coupling. J Sound Vib 330:2339–2353
11. Liuyang X, Tang L, Liu K, Mace BR (2018) On the use of piezoelectric nonlinear energy sink for vibration isolation and energy harvesting. In: Conference on smart materials, adaptive structures and intelligent systems, pp 1–6
12. Malaji PV, Rajarathinam M, Jaiswal V, Ali SF, Howard IM (2019) Energy harvesting from dynamic vibration pendulum absorber. Rec Adv Struct Eng 2:467–478 (Springer Singapore)
13. Malaji PV, Ali SF (2018) Analysis and experiment of magneto-mechanically coupled harvesters. Mech Syst Signal Process 108:304–316
14. Malaji PV, Ali SF (2017) Magneto-mechanically coupled multiple energy harvesters. In: 1st International Conference on Power Electronics, Intelligent Control and Energy Systems (ICPEICES 2016), IEEE, pp 1–5

15. Rajarathinam M, Ali S (2018) Energy generation in a hybrid harvester under harmonic excitation. Energy Convers Manage 155:10–19
16. Kumar KA, Ali SF, Arockiarajan A (2017) Magneto-elastic oscillator: modeling and analysis with nonlinear magnetic interaction. J Sound Vib 393:265–284
17. Xiong L, Tang L, Liu K, Mace BR (2018) Broadband piezoelectric vibration energy harvesting using a nonlinear energy sink. J Phys D Appl Phys 51:1–2

Effect of Input Velocity on the Output of Vertical Axis Wind Turbine (VAWT)

Sankgeeth Vennila Sigamani, Shami Jose Jose Robin, Arun Prakash Chandran, and B. Anand Ronald

Abstract Vertical axis wind turbines (VAWT) are finding applications in the advanced scientific fields and seem to be a promising and rapidly growing innovation. A rarely researched field regarding VAWT is on analyzing the effects of various input velocities for bidirectional wind flow on the vertical axis wind turbines due to the complex and unsteady wind flow characteristics. In the current study, three-dimensional simulations are carried out, and the static and dynamic results are then reported for different values of rotational speed. The outcomes like output velocities and pressure are analyzed for a range of input velocities in a Savonius wind turbine. Different input velocities in the range of 10 m/s, 15 m/s, 20 m/s and 25 m/s were studied. The model for the rotor was designed on Solidworks 2018, and numerical analysis was carried out for a bidirectional wind input on commercial CFD software, ANSYS Fluent v16.0. From the analysis, the output velocity and pressure graphs were plotted with respect to the input air velocity and compared.

Keywords Renewable energy · VAWT · Velocity · Wind turbine

1 Introduction

Over the past few years, development and modernization in the form of natural alterations and over-exploitation of available resources have led to global concern over pollution, global warming and possible depletion of non-renewable sources of energy, such as oil. This has led us to the exploration of renewable and sustainable energy resources. Conventionally, wind energy is harnessed on a large scale using horizontal axis wind turbines (HAWTs) spread over a large area. But, HAWT has several disadvantages like the requirement of large wind speeds, large area, high initial cost and more maintenance. Vertical axis wind turbine (VAWT), unlike a HAWT, is one whose axis of rotation is vertical with respect to ground. This arrangement

S. Vennila Sigamani · S. J. Jose Robin · A. P. Chandran (✉) · B. Anand Ronald
Department of Mechanical Engineering, Sri Sivasubramaniya Nadar College of Engineering, Kalavakkam, Chennai, Tamilnadu 603110, India
e-mail: arunprakashc@ssn.edu.in

S. Vijayan et al. (eds.), *Trends in Manufacturing and Engineering Management*, Lecture Notes in Mechanical Engineering,
https://doi.org/10.1007/978-981-15-4745-4_93

allows the generator and gearbox to be located close to the ground facilitating easy maintenance and repair. They have gained more interest for their excellent adaptability to the urban environment in terms of visual impact and noise pollution. Being axis symmetric, they are omnidirectional turbines, which respond well to changes in wind direction. A typical wind turbine, which utilizes natural wind power, faces problems regarding lack or excess of wind speed. But in this case, a considerable amount of wind energy is produced due to the pressure difference created by the moving vehicles on the highways. We need a system to recycle otherwise wasted energy of highway traffic. Moreover in remote locations extending power supply would itself be a tiring task which can be solved by having this type of turbine which could generate power enough to illuminate the road cautions or indicators needed for the road users. By setting VAWTs in highways, artificial wind can also be used to produce electricity. The other major advantage of this system is that it is dependent only on the number of vehicles on the road and not on other resources such as sunshine or wind, which are seasonal or location specific. Out of the two types of VAWT, called Darrieus VAWT and Savonius VAWT, the latter is known as quite a wind power source because of its lowest tip speed ratio and compact size. Savonius rotor was firstly developed and introduced by Finnish inventor S.J. Savonius. Though the Savonius rotor has low aerodynamic efficiency, it is preferably used because it runs at wind speeds as low as 2 m/s.

2 Literature Review

Although the concept of vertical axis wind turbines (VAWTs) was proposed by Darrieus as early as 1931, the research and development in this area are still of interest and in progress nowadays. In order to determine the performance of the VAWT, analytical and numerical approaches, as well as experimental tests, are generally used among others [1]. The Sutherland report on the stream tube model and vortex models allows one to analyze the aerodynamic response of VAWTs [2]. The first simplest approach developed by Templin is the single stream tube numerical model. Templin calculates the performance for a single blade whose chord equals the sum of the chords of the actual rotor's blade through the application of the blade element theory [3]. Strickland extends Templin's approach into a multiple single stream tube model by considering a number of adjacent smaller steam tubes, applying the conservation of momentum and blade element theory to each steam tube. Further improvement in the multiple steam tube model has been reported by Paraschivoiu, namely the double-multiple steamtube model (DSMT) [4]. A second class of models is those based on vortex representation of the blades and their wakes. Strickland et al.'s model is based on the replacement of the rotor blade by a bound vortex filament, called a lifting line, whose strength is a function of the azimuthal position [5]. Wang presents a potential flow 2D vortex panel model. Experimental tests can be performed by means of wind tunnel or open-field measurement in order to evaluate the real performance of wind turbines [6]. Bedon describes the results of an experimental campaign of measurements conducted on a VAWT with a helical blade having a radius of 1 m with a NACA-0018 airfoil section, tested in an open field. With the advent of computers, many authors have investigated the feasibility and accuracy of different computational fluid dynamics (CFD) approaches [7]. Wolfe uses CFD to investigate the two-dimensional aerodynamic performance of typical wind turbine airfoils with the standard turbulence model. Carrigan et al., fixing the tip speed ratio (TSR) of the wind turbine, developed an iterative design system to maximize the torque for different airfoil cross sections and solidities in two-dimensional CFD simulations. A CFD performance model based on blade element momentum theory was presented by Raciti Castelli et al. for evaluating the aerodynamic forces of straight blade VAWT [8]. Most of the authors have the adopted Reynolds averaged Navier–Stokes equations (RANS) approach with different turbulence models. However, Iida et al. used a model with large-eddy simulations (LES), proving that the as-obtained results are in good agreement at high TSR with that of momentum theory. Only recently have some researchers carried out three-dimensional CFD simulations. Howell et al. provided a three-dimensional CFD model based on the Re-Normalization Group (RNG) κ-turbulence model, and the effects of turbine solidity, roughness and tip vortices were considered.

3 Experimental Setup

Savonius VAWT is a drag-type device, consisting of two scoops. Looking down on the rotor from above, a two-scoop machine would look like an "S" shape in cross section. Because of the curvature, the scoops experience less drag when moving against the wind than when moving with the wind. The differential drag causes the Savonius turbine to spin. Effect of various parameters on the performance of Savonius rotor was studied in the previous literature such as aspect ratio, use of end plates, use of deflecting plates, overlap ratio and number of blades. The current experimental model consists of a mild steel rod of length 700 mm and a diameter of 20 mm supporting the blades. The blades are made of aluminum with an "S-"shaped profile and thickness of 2 mm. Each blade is curved approximately to a diameter of 260 mm and a length of 335 mm. The entire setup is supported on a mild steel base. A model was made in Solidworks 2018 using the appropriate dimensions. The Navier stokes equation was used as the governing equation and a 6 Degree of Freedom (6DOF) solver was used. Subsequently, the analysis was carried out in ANSYS Fluent v16.0.

4 3D Modeling

Using Solidworks 2018, the 3D part is generated in the fabricated dimensions by comprehensively considering structural features, processing requirements and solution calculation amount. For the purpose of reducing the number of elements and

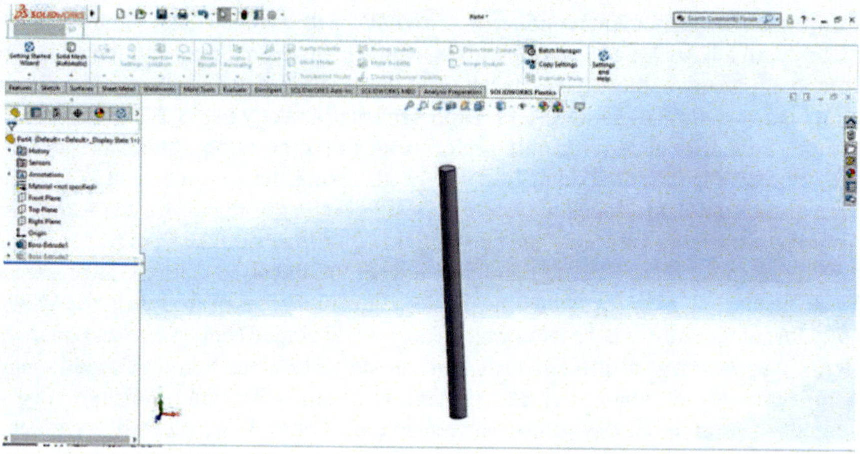

Fig. 1 Shaft rod profile

workload, when the finite model is established various reasonable simplifications were carried on, various petty features like fillet and chamfering were given up. An optimum aspect ratio (Ç 0.5) and power coefficient [2, 5] were considered.

4.1 Create the Shaft Rod Profile

The shaft rod has been modeled using Solidworks 2018 software and is shown in Fig. 1.

4.2 Generate the Blade Design (Fig. 2)

The CAD model of the blade design is shown in Fig. 2.

5 CFD Model

The sequence of operations involved in the modeling and analysis is shown in the flowchart (Fig. 3 and 4).

The profile of the rotor is symmetric and is extruded off the 2D geometry. The analysis was done in ANSYS Fluent v16.0. The finite element method (FEM) is used to perform the analysis. The finite element method is a powerful tool to obtain the numerical solution of a wide range of engineering problem. The method is general

Fig. 2 Design of the blade

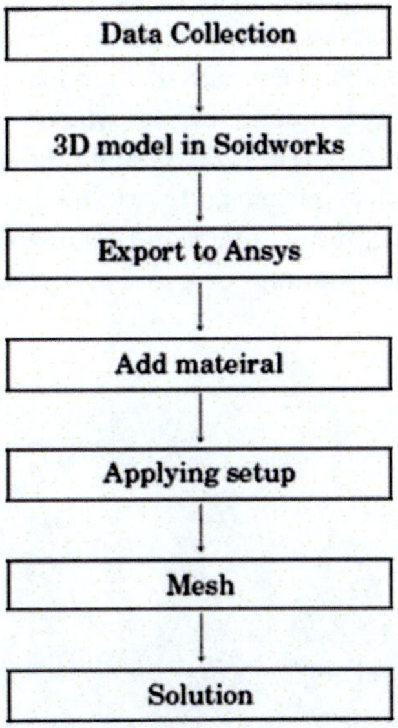

Fig. 3 Steps involved in analysis of various blade velocities for VAWT

enough to handle any complex shape or geometry, for any material under different boundary and loading conditions. The generality of the finite element method fits the analysis requirement in the current application. It is a method of piecewise approximation in which the structure or body is divided into small elements of finite dimensions called finite elements, and then, the original body or the structure is considered as an assemblage of these elements connected at the finite number of

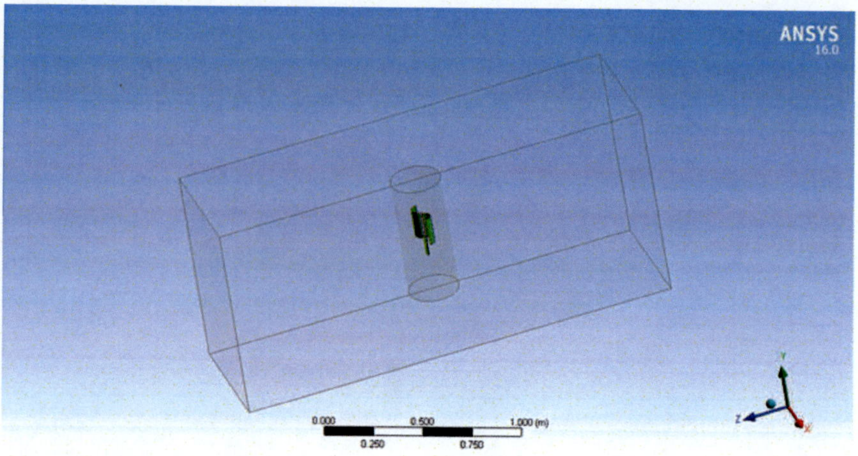

Fig. 4 Importing the file geometry to the design module

joints called nodal points or nodes. The constraints of the body are given through a C program. The enclosure has inlets and outlets diagonally on opposite sides of a box enclosure. The adjacent inlet and outlet are separated by a small region of nil effect considering practical conditions of bidirectional fluid flow. Re-meshing (as shown in Fig. 5) is carried out for each time step and smoothing is defined. The time steps are taken as 0.0025 s considering the accuracy and calculation time. A dynamic mesh is defined for the fluid domain. The procedures are repeated for each input velocities (Fig. 6 and 7).

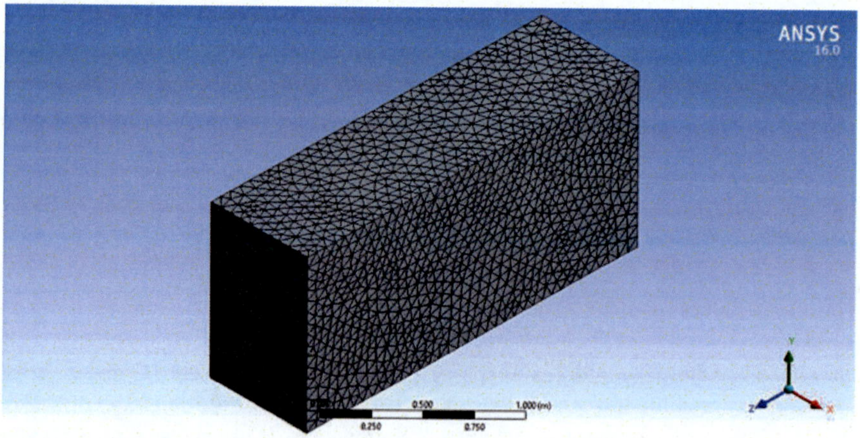

Fig. 5 Meshed view

Fig. 6 Setting up of boundary conditions

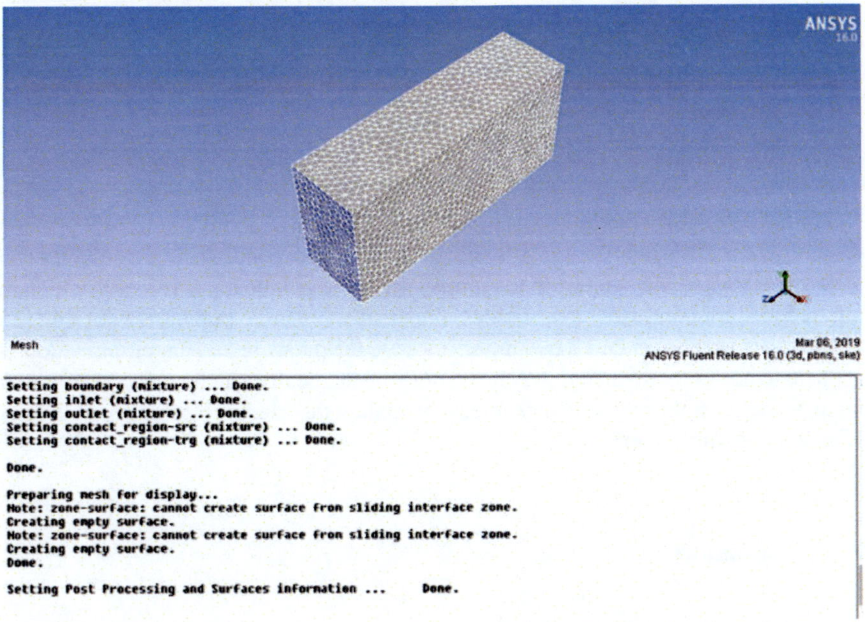

Fig. 7 Importing in CFD Fluent

6 Results

The results are calculated and plotted (Fig. 8).

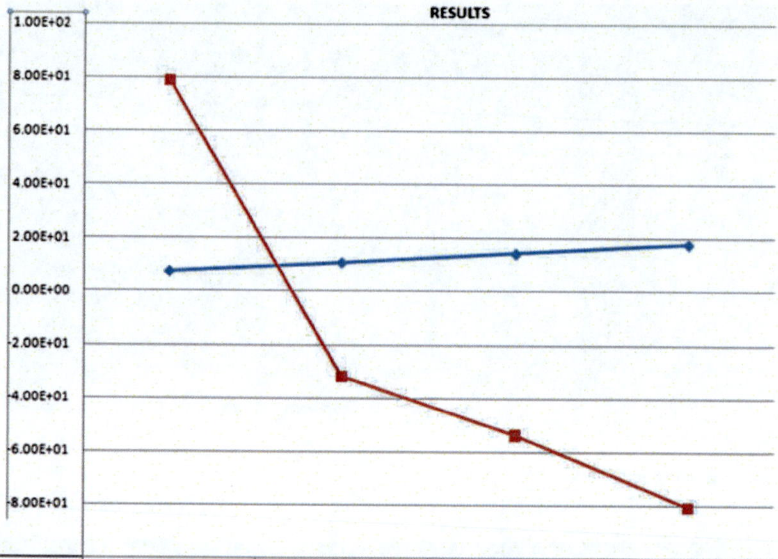

INPUT VELOCITY	10	15	20	25
OUTPUT VELOCITY	7.13E+00	1.07E+01	1.43E+01	1.78E+01
PRESSURE	7.86E+01	-3.19E+01	-5.37E+01	-8.06E+01

Fig. 8 Input velocity versus output velocity and pressure

For the Savonius wind turbine with a blade thickness of about 2 mm and by fixing the temperature, the input velocity is changed. The corresponding output velocity and pressure are noted. As seen from the graph, we increased the input velocity, which in turn reduced the pressure (on the blades) drastically and increased the output velocity linearly. We found that, at 15 m/s input velocity, the pressure and the output velocity are near to each other which make the turbine mechanically efficient. For a particular input velocity, both low and high pressure makes the blade unstable. And thus at 15 m/s, the turbine is efficient.

7 Conclusion

In this project, we investigated different velocities which exert on the vertical axis wind turbine. Here, the analysis is taken for the different velocity values such as 10, 15, 20, 25 m/s. The Savonius wind turbine is designed in Solidworks and which is imported in ANSYS Fluent launcher for the flow analysis, and the variations in the pressure and velocity values are obtained for each condition. The wind speed is taken into account, and it is given as angular velocity to the fan. An analysis was performed on ANSYS Fluent simulating the process, and it was concluded that, in case of complex bidirectional wind flow, a Savonius rotor with a curved blade of

2 mm thickness is efficient at 15 m/s of input velocity with maximum mechanical efficiency and minimum pressure difference, thus attaining stable speed faster and has a higher rate of acceleration.

References

1. Darrieus G (1931) Turbine having its rotating shaft transverse to the flow of the current. U.S. Patent1, 835, 018, 8 Dec 1931
2. Sutherland HJ, Berg DE, Ashwill TD (2012) A retrospective of VAWT technology; Sandia Report, SAND 2012-0304; Sandia National Laboratories: Albuquerque, NM, USA, Jan 2012
3. Islam M, Ting DSK, Fartaj A (2008) Aerodynamic models for Darrieus-type straight-bladed vertical axis wind turbines. Renew Sustain Energy Rev 12:1087–1109
4. Templin RJ (1974) Aerodynamic performance theory for the NRC Vertical-Axis Wind Turbine; Technical Report N-76-16618; LTR-LA-160; National Aeronautical Establishment: Ottawa, ON, Canada, 1 June 1974
5. Paraschivoiu I (1988) Double-multiple stream tube model for studying vertical-axis wind turbines. J Propuls Power 4:370–377
6. Brusca S, Lanzafame R, Messina M (2014) Design of a vertical-axis wind turbine: how the aspect ratio affects the turbine's performance. Int J Energy Environ Eng 5:333–340
7. Strickland J, Webster B, Nguyen T (1979) A vortex model of the Darrieus turbine: an analytical and experimental study. J Fluids Eng 101:500–505
8. Wang L, Zhang L, Zeng N (2007) A potential flow 2-D vortex panel model: applications to vertical axis straight blade tidal turbine. Energy Convers Manag 48:454–461

Radiative Heat Transfer of Magnetic Nanofluid Flow Past a Porous Inclined Plate: A Mathematical Model

M. Shanmugapriya

Abstract This paper numerically investigates radiative magnetohydrodynamic boundary layer flow of nanofluids over a porous inclined plate in the presence of viscous dissipation. The governing partial differential equations are converted into nonlinear differential equations by using a suitable similarity transformation, which are solved numerically using Runge–Kutta–Fehlberg fourth-fifth-order method along with shooting technique. Two types of nanoparticles are discussed, namely copper (Cu) and alumina (Al_2O_3) in the base fluid of water. Maxwell Garnetts and Brinkman models are used for the effective thermal conductivity and dynamic viscosity of the nanofluids, respectively. The velocity and temperature distributions are obtained for various governing physical parameters and are presented graphically. In addition, the evolution of skin friction coefficient and Nusselt number values with selected parameters is presented. Comparison is made with previously published work and found to be in good agreement.

Keywords Nanofluid · Thermal radiation · Magnetohydrodynamic flow · Heat transfer · Inclined plate

1 Introduction

A nanofluid are fluids that contain small volumetric quantities of nanometer-sized (diameter of about 1–100 nm) particles called nanoparticles. These fluids are engineered colloidal suspensions of nanoparticles in a base fluid like oil, water and ethylene glycol. Nanoparticles are made from various materials, such as Cu, Ag, Au, Fe, Hg, Ti metals and non-metallic, Al_2O_3, CuO, TiO_2, SiO_2, etc. The convectional heat transfer base fluids have poor heat transfer abilities due to their low thermal conductivity. To upgrade the thermal conductivity of these fluids, a lot of theoretical and experiential analysis has been made. Choi [1] first used the

M. Shanmugapriya (✉)
Department of Mathematics, Sri Sivasubramaniya Nadar College of Engineering, Chennai 603110, India
e-mail: shanmugapriyam@ssn.edu.in

S. Vijayan et al. (eds.), *Trends in Manufacturing and Engineering Management*, Lecture Notes in Mechanical Engineering,
https://doi.org/10.1007/978-981-15-4745-4_94

word nanoparticle to improve the thermal conductivity of base fluids. Eastman et al. [2] have studied that the thermal conductivity of the base fluid (water) was increased up to 60% when CuO nanoparticles of volume fraction 5% were added to the water. Furthermore, Eastman et al. [3] have also noticed that the effective thermal conductivity has shown to be increased 40% when copper nanoparticles of volume fraction less than 1% are added to the ethylene glycol or oil. Xuan et al. [4] reported that with a small volumetric fraction of nanoparticles less than 5% the thermal conductivity of the base fluid is enhanced by 10–20%. Xie et al. [5] have studied that Al_2O_3-ethylene glycol-based nanofluid thermal conductivity is increased in the range 25–30% when alumina nanoparticles are added. Therefore, these nanofluids are capable of promoting energy efficiency, improving thermal conductivity and thermal properties of heat transport of base fluid. Nanofluids have therefore found significant applications in major process industries, including material and chemicals, food and drink, oil and gas, paper and printing and textiles.

The word magnetohydrodynamics has several industrial applications such as MHD generators, pumps, petroleum technologies and liquid metal cooling blanket for fusion reactor. The MHD flow in electrically conducting fluid can control the rate of heat transfer at the surface, and hence, desired cooling effect can be achieved. Magneto-nanofluid has specific applications in biomedicine, optical modulators, magnetic cell separation, magneto-optical wavelength filters, silk flood separation, nonlinear optical materials, drug delivery, etc. A magnetic nanofluid has both liquid and magnetic properties. Hamad et al. [6] introduced a one-parameter group to represent similarity reduction for the problem of MHD-free convection flow of a nanofluid past a vertical semi-infinite flat plate. Sudarsana Reddy et al. [7, 8] have discussed the effect of chemical reaction on MHD natural convection boundary layer flow, heat and mass transfer characteristics of nanofluids over a rotating disk through a porous medium and vertical cone, respectively.

The radiative effects have important applications in physics and engineering. The radiative heat transfer effects on different flows are very important in space technology and high-temperature processes, and very little is known about the effects of radiation on the boundary layer of a radiative MHD fluid past a body. Shanmugapriya [9] has studied the effect of thermal radiation Cu-water nanofluid past a moving wedge using Gyarmati's variational principle. Shanmugapriya et al. [10] investigated the effect of entropy generation and thermal radiation of Cu-water nanofluid over a moving wedge using the shooting technique together with Runge–Kutta fourth-order integration scheme. Thiripathi Thumma et al. [11] have presented heat source/sink effects in addition to viscous dissipation for two different magnetic nanofluid flows from a nonlinear inclined stretching/shrinking sheet. Sudarsana Reddy et al. [12] have discussed the influence of thermal radiation and chemical reaction on MHD boundary layer flow over a nanofluid saturated inclined porous plate under boundary conditions. Sudarsana Reddy et al. [13] have reported MHD boundary layer flow, heat and mass transfer analysis of nanofluid over an inclined vertical plate saturated by the porous medium with thermal radiation and heat generation/absorption by using versatile, extensively validated, variational finite element method.

The main objective of this work is to address the impact of thermal radiation and viscous dissipation on MHD boundary layer flow and heat transfer over a porous inclined plate saturated by Cu-water and Al_2O_3-water-based nanofluid. The effects of the magnetic parameter, mixed convection parameter, radiation parameter, porous medium parameter, Eckert number, suction parameter, Prandtl number and angle of inclination on velocity and temperature fields are presented via figures and tables.

2 Mathematical Formulation of the Model

Consider the steady two-dimensional laminar MHD flow and heat transfer of an incompressible nanofluid past a porous inclined plate with a linear stretching velocity $u_w(x) = ax$ where $a(> 0)$ being the stretching constant. The stream function ψ can be defined as $u = \frac{\partial \psi}{\partial y}$ and $v = -\frac{\partial \psi}{\partial x}$, and similarly, the temperature at the surface and free stream are denoted as T_w and T_∞, respectively. The viscous dissipation and radiation effects are considered in the energy equation. The Rosseland approximation is used to describe the radiative heat flux in the energy equation. The uniform magnetic field B_0 is applied to the flow direction. The basic governing equations for such type of flow can be written as (Tiwari and Das [14]).

$$\frac{\partial^2 \psi}{\partial y \partial x} - \frac{\partial^2 \psi}{\partial x \partial y} = 0 \tag{1}$$

$$\begin{aligned}\left(\frac{\partial \psi}{\partial y}\frac{\partial^2 \psi}{\partial x \partial y} - \frac{\partial \psi}{\partial x}\frac{\partial^2 \psi}{\partial y^2}\right) &= \gamma_{nf}\frac{\partial^3 \psi}{\partial y^3} + \frac{g(\rho\beta)_{nf}(T - T_\infty)\cos\alpha}{\rho_{nf}} \\ &\quad - \frac{\sigma B_0^2}{\rho_{nf}}\left(\frac{\partial \psi}{\partial y}\right) - \frac{\gamma_{nf}}{K}\left(\frac{\partial \psi}{\partial y}\right)\end{aligned} \tag{2}$$

$$\begin{aligned}\left(\frac{\partial \psi}{\partial y}\frac{\partial T}{\partial x} - \frac{\partial \psi}{\partial x}\frac{\partial T}{\partial y}\right) &= \alpha_{nf}\frac{\partial^2 T}{\partial y^2} + \frac{\mu_{nf}}{(\rho c_p)_{nf}}\left(\frac{\partial^2 \psi}{\partial y^2}\right)^2 \\ &\quad - \frac{1}{(\rho c_p)_{nf}}\frac{\partial q_{\text{rad}}}{\partial y}\end{aligned} \tag{3}$$

subject to the boundary conditions

$$\frac{\partial \psi}{\partial y} = U_w(x) = ax, \quad \frac{\partial \psi}{\partial x} = V_w, \quad T = T_w \text{ at } y = 0 \tag{4}$$

$$\frac{\partial \psi}{\partial y} \to 0, T \to T_\infty \text{ as } y \to \infty \tag{5}$$

Here, u, v stand for components of velocity in x, y directions, respectively, T is the temperature of the nanofluid in the boundary layers, μ_{nf} is the viscosity of the nanofluid, ρ_{nf} is the density of the nanofluid, α_{nf} is the thermal diffusivity of the nanofluid and $(\rho c_p)_{nf}$ is the heat capacitance of the nanofluid, $\gamma_{nf} = \frac{\mu_{nf}}{\rho_{nf}}$ symbolizes the kinematic diffusivity of the nanofluid, β_{nf} is the thermal expansion coefficient of the nanofluid, α is the angle of inclination, B_0 is the strength of the magnetic field and K is the constant permeability of the porous medium. By the use of the Rosseland estimation for radiation in Eq. (3), we get the following form of the equation.

$$\left(\frac{\partial\psi}{\partial y}\frac{\partial T}{\partial x} - \frac{\partial\psi}{\partial x}\frac{\partial T}{\partial y}\right) = \alpha_{nf}\frac{\partial^2 T}{\partial y^2} + \frac{\mu_{nf}}{(\rho c_p)_{nf}}\left(\frac{\partial^2\psi}{\partial y^2}\right)^2 + \frac{1}{(\rho c_p)_{nf}}\frac{16\sigma^* T_\infty^3}{3k^*}\frac{\partial^2 T}{\partial y^2} \quad (6)$$

The relation between the physical quantities of nanofluid, water, copper and aluminum oxide is written as follows:

The thermophysical characteristics of the nanofluid with nanoparticle concentration φ are defined as (Oztop et al. [15])

$$\mu_{nf} = \frac{\mu_f}{(1-\varphi)^{2.5}},\ \rho_{nf} = (1-\varphi)\rho_f + \varphi\rho_s,$$

$$\alpha_{nf} = \frac{\kappa_{nf}}{(\rho c_p)_{nf}},\ (\rho c_p)_{nf} = (1-\varphi)(\rho c_p)_f + \varphi(\rho c_p)_s,$$

$$(\rho\beta)_{nf} = (1-\varphi)(\rho\beta)_f + \varphi(\rho\beta)_s,$$

$$\frac{\kappa_{nf}}{\kappa_f} = \frac{(k_s + 2k_f) - 2\varphi(k_f - k_s)}{(k_s + 2k_f) + 2\varphi(k_f - k_s)} \quad (7)$$

The following similarity transformations are invoked:

$$\Psi = x\sqrt{a\gamma_f}\,f(\eta),\ \eta = y\sqrt{\frac{a}{\gamma_f}},\ (\eta) = \frac{T - T_\infty}{T_w - T_\infty} \quad (8)$$

Using these variables in Eqs. (2) and (6), we get the following nonlinear ordinary differential equations:

$$f''' + (1-\varphi)^{2.5}\left[\left(1-\varphi+\varphi\frac{\rho_s}{\rho_f}\right)\left(ff'' - f'^2\right) + \lambda\left(1-\varphi+\varphi\frac{\rho\beta_s}{\rho\beta_f}\right)\theta\cos\alpha - Mf'\right] - K_1 = 0 \quad (9)$$

$$\left[1+\frac{4}{3}R\right]\theta'' + \Pr\left(\frac{k_f}{k_{nf}}\right)\left(1-\varphi+\varphi\frac{\rho\beta_s}{\rho\beta_f}\right)$$

$$\left[f\theta' - f'\theta + \frac{Ec}{(1-\varphi)^{2.5}\left(1-\varphi+\varphi\frac{\rho c_{ps}}{\rho c_{pf}}\right)}(f'')^2 \right] = 0 \quad (10)$$

subject to the boundary conditions

$$f(\eta) = S, \quad f'(\eta) = 1, \quad \theta(\eta) = 1 \quad \text{at} \quad \eta = 0 \quad (11)$$

$$f' = 0, \quad \theta(\eta) = 0 \quad \text{at} \quad \eta \to \infty. \quad (12)$$

where the prime denotes differentiation with respect to η,$M = \frac{\sigma B_0^2}{a\rho_f}$ is the magnetic parameter, $(Gr_x)_f = \frac{g\beta_f(T_w - T_\infty)x^3}{\vartheta_f^2}$ is the local Grashof number, $(\mathrm{Re}_x)_f = \frac{U_w(x)x}{\vartheta_f}$ is the local Reynolds number, $\lambda = \frac{(Gr_x)_f}{(Re_x)_f^2} = \frac{g\beta(T_w - T_\infty)}{a^2}$ is the mixed convection parameter, $R = \frac{4\sigma^* T_\infty^3}{k^* k_{nf}}$ is the radiation parameter, $K_1 = \frac{\vartheta_f}{aK}$ is the porous medium parameter,$\mathrm{Ec} = \frac{U_w(x)^2}{c_{pf}(T_w - T_\infty)}$ is the Eckert number, $S = -v_w(x)\left(a\vartheta_f\right)^{\frac{-1}{2}}$ is the suction parameter,$\mathrm{Pr} = \frac{\gamma_f}{\alpha_f}$ is the Prandtl number and α is the angle of inclination.

3 Skin Friction Coefficient and Nusselt Number

The important physical quantities in this study are skin friction coefficient C_f and the Nusselt number Nu_x, which are defined by the following equations:

$$C_f = \frac{\mu_{nf}}{\rho_f U_w(x)^2}\left(\frac{\partial u}{\partial y}\right)_{y=0},$$

$$Nu_x = \frac{-k_{nf}\left(\frac{\partial T}{\partial y}\right)_{y=0} - \frac{4\sigma^*}{3k^*}\left(\frac{\partial T^4}{\partial y}\right)_{y=0}}{k_f(T_w - T_\infty)/x} \quad (13)$$

Using Eqs. (8) and (13), the skin friction coefficient and the Nusselt number can be expressed as

$$(\mathrm{Re}_x)_f^{\frac{1}{2}} C_f = \frac{1}{(1-\varphi)^{2.5}} f''(0), \; (\mathrm{Re}_x)_f^{\frac{-1}{2}} Nu_x = \frac{-k_{nf}}{k_f}\left[1 + \frac{4R}{3}\right]\theta'(0). \quad (14)$$

Table 1 Thermal properties

Properties	Base fluid (water)	Cu	Al_2O_3
c_p (J/kg K)	4179	385	765
ρ (kg/m^3)	997.1	8933	3970
κ (W/m K)	0.613	400	40
$\beta \times 10^5$(k^{-1})	21	1.67	0.85

Table 2 Comparison of $-f''(0)$ and $-\theta'(0)$ for various values of M

M	Ishak et al. [16]		Anjali Devi et al. [17]		Present result	
	$-f''(0)$	$-\theta'(0)$	$-f''(0)$	$-\theta'(0)$	$-f''(0)$	$-\theta'(0)$
0.0	0.5607	1.0873	0.5608	1.0873	0.5606	1.0873
0.1	0.5658	1.0863	0.5659	1.0863	0.5659	1.0863
0.2	0.5810	1.0833	0.5810	1.0833	0.5810	1.0833
0.5	0.6830	1.0630	0.6830	1.0630	0.6830	1.0630
1.0	1.0000	1.0000	1.0000	1.0000	1.0000	1.0000
2.0	1.8968	0.8311	1.8968	0.8311	1.8968	0.8311
5.0	4.9155	0.4702	4.9155	0.4703	4.9155	0.4703

4 Numerical Method

Equations (9)–(12) are solved numerically using Runge–Kutta–Fehlberg fourth-fifth-order method along with shooting technique. A comparison of the obtained numerical results with previously published data is performed. The values of skin friction coefficient $f''(0)$ for various values of M are compared with those obtained by Ishak et al. [16] and Anjali Devi et al. [17], which is shown in Table 2 and observed that both the results are in good agreement.

5 Results and Discussion

Graphical distributions and table representation of the influence of various physical parameters on velocity, temperature profiles, local skin friction and local Nusselt number are presented in Figs. 1, 2, 3, 4, 5, 6, 7, 8, 9, 10, 11, and 12 and Table 3. In this paper, we computed the above results by fixing the value $Pr = 6.2$, $K_1 = 1$ and $S = 0.5$. Figures 1 and 2 represent the dimensionless velocity and temperature for different values of the magnetic parameter for both types of *Cu*-water and Al_2O_3-water nanofluids. It is clearly shown in Fig. 1, the velocity of the nanofluid decreases with increasing strength of the magnetic field for both the types of nanofluid. The

Table 3 Values of $-f''(0)$ and $-\theta'(0)$ for various values of M, φ, α, λ, R, Ec, S and Pr.

M	φ	α	λ	R	Ec	S	Pr	$-f''(0)$		$-\theta'(0)$	
								Cu	Al_2O_3	Cu	Al_2O_3
0.0	0.01	45°	1.0	1.0	0.1	0.5	6.2	1.1782	1.1922	2.4666	2.4579
0.5								1.2884	1.3017	2.4326	2.4241
1.0								1.5716	1.5834	2.3452	2.3371
0.5	0.02	45°	1.0	1.0	0.1	0.5	6.2	1.3230	1.3461	2.3638	2.3479
	0.03							1.3549	1.3872	2.2977	2.2748
	0.04							1.3843	1.4251	2.2339	2.2044
0.5	0.01	0°	1.0	1.0	0.1	0.5	6.2	1.1668	1.1770	2.4591	2.4514
		30°						1.3901	1.4012	2.4091	2.4011
		60°						1.6986	1.7111	2.3300	2.3212
0.5	0.01	45°	1.5	1.0	0.1	0.5	6.2	1.2220	1.2324	2.4472	2.4395
			2.0					1.1537	1.1638	2.4618	2.4542
			2.5					1.2185	1.0960	2.4680	2.4682
0.5	0.01	45°	1.0	2.0	0.1	0.5	6.2	1.2481	1.2619	1.7621	1.7554
				3.0				1.2156	1.2298	1.4167	1.4108
				4.0				1.1883	1.2029	1.2018	1.1964
0.5	0.01	45°	1.0	1.0	0.2	0.5	6.2	1.2829	1.2963	2.3031	2.2930
					0.3			1.2774	1.2909	2.1752	2.1636
					0.4			1.2720	1.2855	2.0490	2.0357
0.5	0.01	45°	1.0	1.0	0.1	1.0	6.2	1.6612	1.6799	3.3482	3.3365
						3.0		3.4628	3.5093	7.7218	7.695
						5.0		5.4588	5.5355	12.4381	12.3964
0.5	0.01	45°	1.0	1.0	0.1	0.5	1	1.0854	1.1011	0.6889	0.6852
							10	1.3224	1.3352	3.4448	3.4337
							100	1.5891	1.6008	23.4968	23.4393

Fig. 1 Velocity profiles for different values of *M*

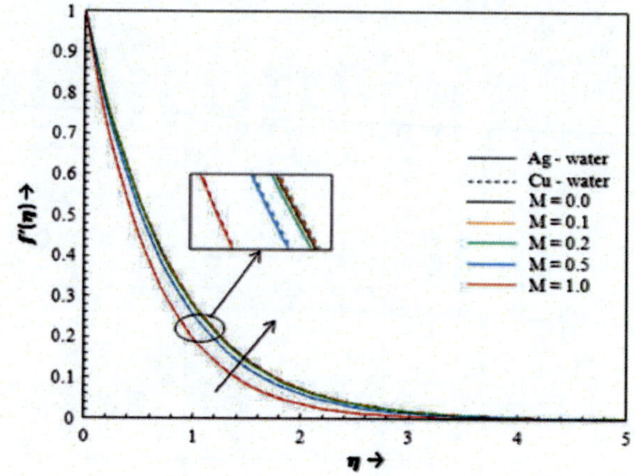

Fig. 2 Temperature profiles for different values of *M*

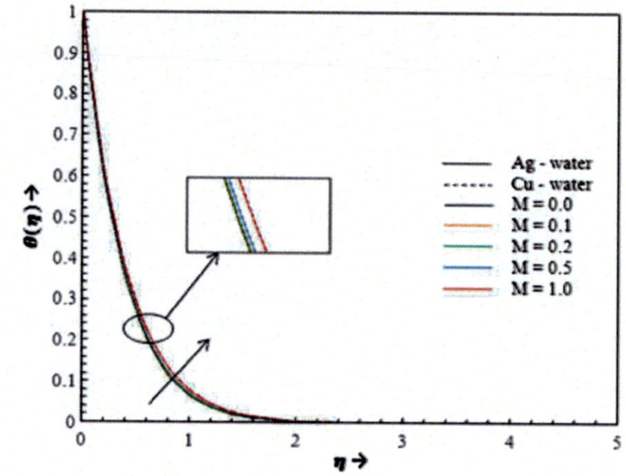

Fig. 3 Velocity profiles for different values of *φ*

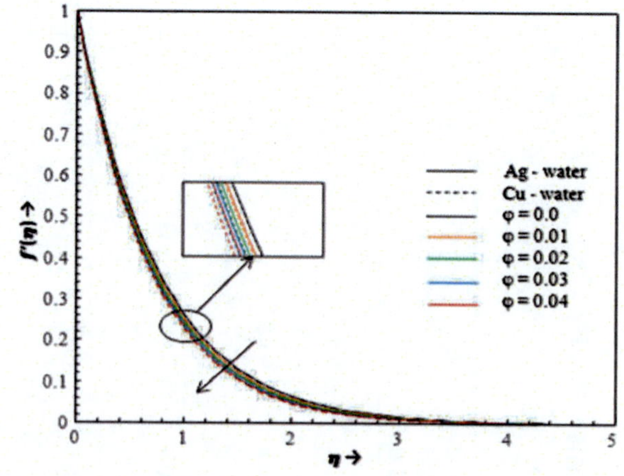

Fig. 4 Temperature profiles for different values of φ

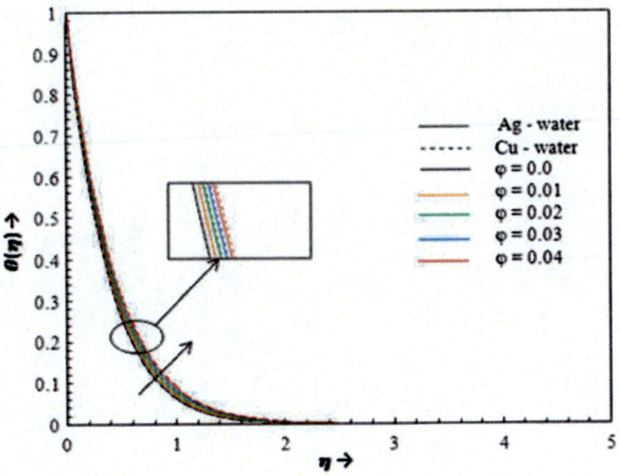

Fig. 5 Velocity profiles for different values of λ

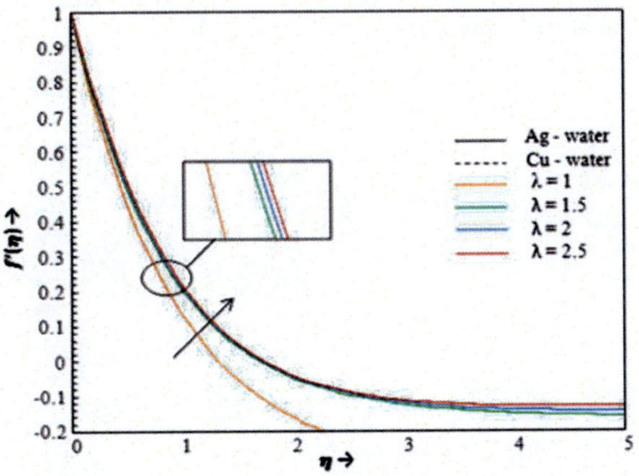

Fig. 6 Temperature profiles for different values of λ

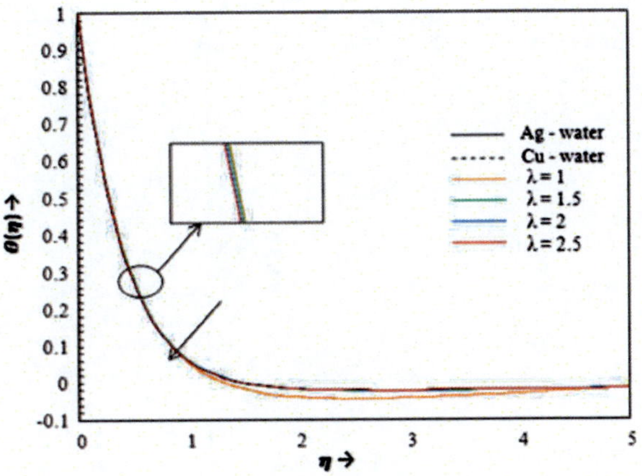

Fig. 7 Velocity profiles for different values of *R*

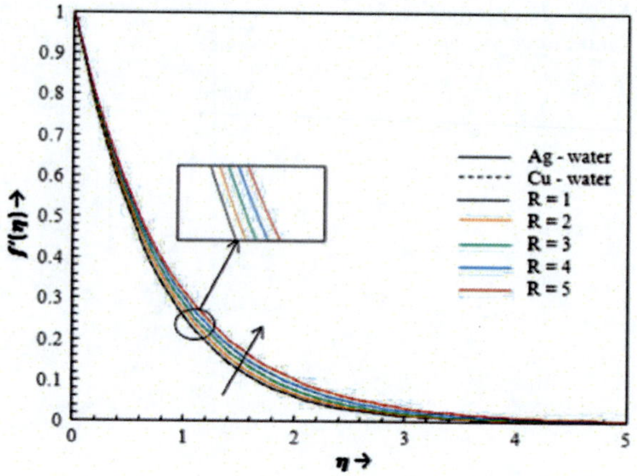

Fig. 8 Temperature profiles for different values of *R*

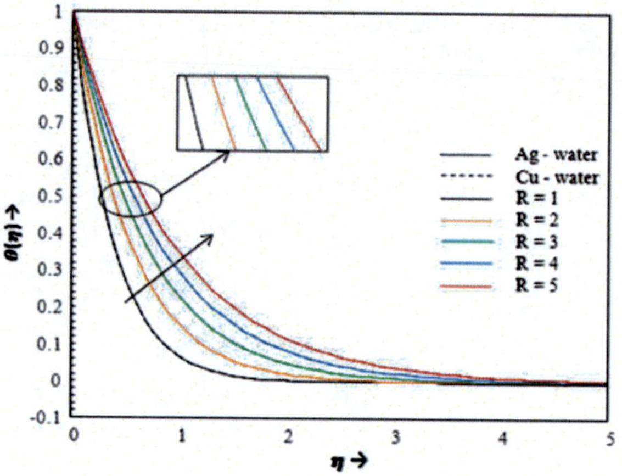

Fig. 9 Velocity profiles for different values of *Ec*

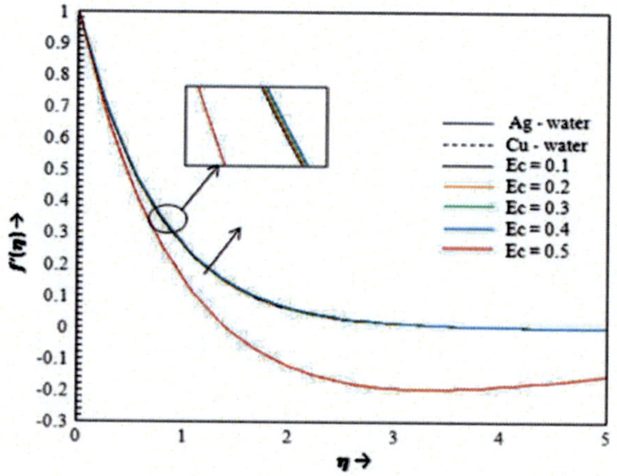

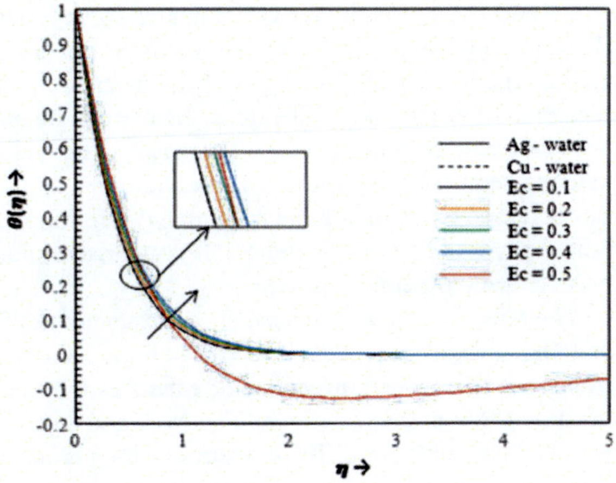

Fig. 10 Temperature profiles for different values of *Ec*

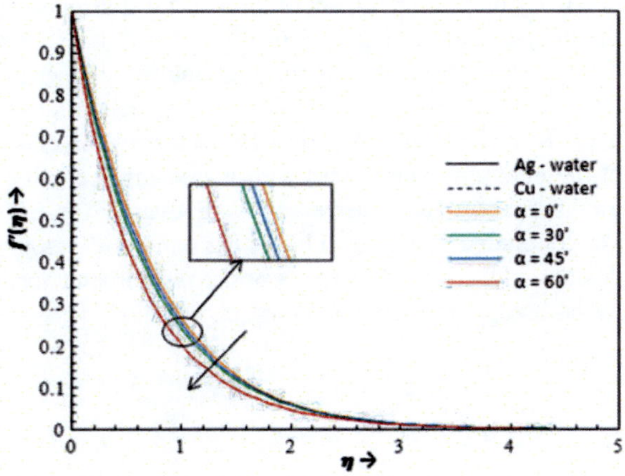

Fig. 11 Velocity profiles for different values of α

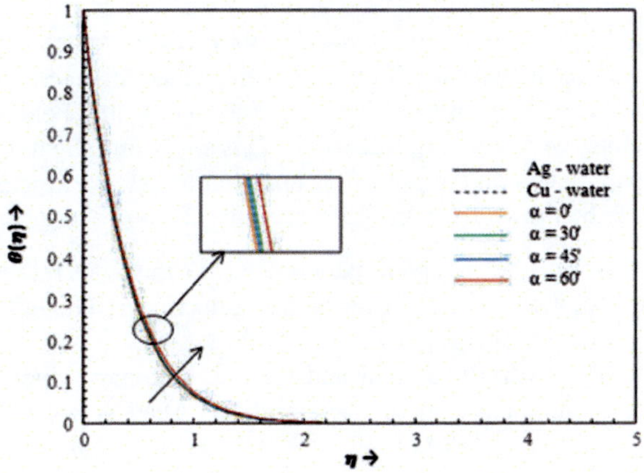

Fig. 12 Temperature profiles for different values of α

reductions in the velocity are accompanied by simultaneous reductions in the hydrodynamic boundary layer thickness for both the copper–water and alumina–water nanofluids. Increasing the values of the magnetic parameter (M) is to increase the temperature distribution. Therefore, the thermal boundary layer thickness increases with the increasing values of M. The effect of the volume fraction parameter (φ) on velocity, temperature profiles is illustrated in Figs. 3 and 4. It is shown in Fig. 3, for the increased range of volume fraction (φ) is to suppress the velocity distribution for both the types of Cu-water and Al_2O_3-water nanofluids and the opposite phenomena occur in the temperature profile.

The effect of the mixed convection parameter ($\boldsymbol{\lambda}$) for both the types of Cu-water and Al_2O_3-water nanofluids is plotted in Figs. 5 and 6. It is observed that the increase in the mixed convection parameter enhances the velocity but diminishes the temperature for both the nanofluids. Figures 7 and 8 show the deviation of velocity and temperature fields for different values of the radiation parameter (R). It is noticed that increasing the radiation parameter improves the hydrodynamical and thermal boundary layer. The impact of the viscous dissipation parameter on the velocity and the temperature is depicted in Figs. 9 and 10. It is found that the hydrodynamical boundary layer and the thermal boundary layer fall for the increasing values of the Eckert number (Ec). Figures 11 and 12 are plotted to investigate the effect of angle of inclination α for both Cu-water and Al_2O_3-water nanofluids. The velocity of the nanofluid suppresses with increasing the values of α. Physically increase in the inclination angle decreases the buoyancy forces, so velocity profile decreases. We concluded from Fig. 11 that the nanofluid velocity is maximum at the vertical position of the plate, and the opposite phenomena occur in the nanofluid temperature for both copper–water and alumina–water.

6 Conclusion

The boundary layer flow and heat transfer of a nanofluid over a porous inclined plate with the effect of viscous dissipation, radiation and magnetic field have been studied numerically. The nonlinear ordinary differential equations with given boundary conditions were solved using Runge–Kutta–Fehlberg fourth-fifth-order method along with shooting technique. This study investigates the effect of various physical parameters on the fluid flow and heat transfer characteristics. The following results can be drawn from the present study.

- When the magnetic parameter (M) raises, it reduces the velocity, skin friction coefficient and enhances the temperature, Nusselt number due to the effect of Lorentz force.
- The hydrodynamical and thermal boundary layer was increased by increasing radiation parameter (R), and this trend is reversed in the viscous dissipation parameter (Ec).

- Skin friction coefficient decreases for increasing the values of the suction parameter (S) and Prandtl number (Pr).
- Suction parameter (S) and Prandtl number (Pr) increases the thermal boundary layer thickness.
- It is found that, in general, the non-dimensionless heat transfer rate for the copper–water nanofluid is more compared with the alumina–water nanofluid.

Acknowledgements The author wishes to thank the Management of Sri Sivasubramaniya Nadar College of Engineering for providing the necessary facility to carry out the present work.

References

1. Choi SUS (1995) Enhancing thermal conductivity of fluids with nanoparticles, developments and applications of non-Newtonian flows. In: Siginer DA, Wang HP (eds) FED-vol 231/MD, vol 66, pp 99–105. The American Society of Mechanical Engineers
2. Eastman JA, Choi SUS, Li S, Thompson LJ, Lee S (1997) Enhanced thermal conductivity through the development of nanofluid. In: Komarneni S, Parker JC, Wollenberger HJ (eds) Nanophase and Nanocomposite Materials II. Pittsburg, PA, MRS, pp 3–11
3. Eastman JA, Choi SUS, Li S, Yu W, Thompson LJ (2001) Anomalously increased effective thermal conductivities of ethylene glycol-based nano-fluids containing copper nano-particles. Appl Phys Lett 78:718–720
4. Xuan Y, Lin Q (2003) Investigation on convective heat transfer and flow features of nanofluids. J Heat Trans 125:151–155
5. Xie H, Wang J, Xi T, Liu Y, Ai F, Wu Q (2002) Thermal conductivity enhancement of suspensions containing nanosized alumina particles. J Appl Phys 91:4568–4572
6. Hamad MA, Pop I, Ismail AI (2011) Magnetic field effects on free convection flow of a nanofluid past a vertical semi-infinite flat plate. Nonlinear Anal Real World Appl 12:1338–1346
7. Sudarsana Reddy P, Sreedevi P, Chamkha AJ (2017) MHD boundary layer flow, heat and mass transfer analysis over a rotating disk through porous medium saturated by Cu–water and Ag–water nanofluid with chemical reaction. Powder Technol 307:46–55
8. Sudarsana Reddy P, Suryanarayana Rao KV (2015) MHD natural convection heat and mass transfer of Al_2O_3—water and Ag–water nanofluids over a vertical cone with chemical reaction. Proc Eng 127:476–484
9. Shanmugapriya M (2018) Analysis of heat transfer of Cu-water nanofluid flow past a moving wedge. J Inf Math Sci 10(1&2):287–296
10. Shanmugapriya M, Sangeetha P (2018) Entropy generation analysis of copper–water nanofluid flow over a moving wedge. Desalination Water Treatment 121:14–21
11. Thumma T, Anwar Beg O, Kadir A (2017) Numerical study of heat source/sink effects on dissipative magnetic nanofluid flow from a non-linear inclined stretching/shrinking sheet. J Mol Liquids 232:159–173
12. Sudarsana Reddy P, Sreedevi P, Chamkha AJ (2016) Heat and mass transfer flow of a nanofluid over an inclined plate under enhanced boundary conditions with magnetic field and thermal radiation. Heat Transfer-Asian Res 46(7):815–839
13. Sudarsana Reddy P, Chamkha AJ, Al-Mudhaf A (2017) MHD heat and mass transfer flow of a nanofluid over an inclined vertical porous plate with radiation and heat generation/absorption. Adv Powder Technol 28(3):1008–1017
14. Tiwari RK, Das MK (2007) Heat transfer augmentation in a two-sided lid-driven differentially heated square cavity utilizing nanofluids. Int J Heat Mass Trans 50(9–10):2002–2018

15. Oztop H, Abu-Nada E (2008) Numerical study of natural convection in partially heated rectangular enclosures filled with nanofluids. Int J Heat Fluid Flow 29(5):1326–1336
16. Ishak A, Nazar R, Pop I (2008) Hydromagnetic flow and heat transfer adjacent to stretching vertical sheet. Heat Mass Trans 44:921–927
17. Anjali Devi SP, Suriyakumar P (2017) Effect of magnetic field on Blasius and Sakiadis flow of nanofluids past an inclined plate. J Taibah Univ Sci 11(6):1275–1288

A GA-ANFIS Model for the Prediction of Biomass Elemental Properties

Obafemi O. Olatunji, Stephen Akinlabi, Nkosinathi Madushele, and Paul A. Adedeji

Abstract The elemental composition of biomass is a significant property, which determines the energy content of biomass feedstock. This article develops a prediction model based on a hybrid adaptive neuro-fuzzy inference system (ANFIS) optimized with genetic algorithm (GA). The model inputs were the proximate constituents of biomass which are ash, fixed carbon, and volatile matter. These were used to predict the hydrogen (H), oxygen (O) and carbon (C) content of biomass fuels. The proposed algorithm was evaluated based on some known performance metrics. The root mean squared error (RMSE), mean absolute deviation (MAD), mean absolute percentage error (MAPE), coefficient of correlation (CC), mean absolute error (MAE) are 3.673, 2.4609, 5.1757, 0.9464, 0.309 at computation time (CT) of 33.65 secs for carbon (C); 0.6293, 0.4168, 8.3011, 0.75581, 0.0716 at CT of 40.21 secs for hydrogen (H); 4.4538, 3.1042, 13.3983,0.9167, 0.9899 at CT of 33.57 secs for oxygen (O), respectively. Regression analysis was also carried out to determine the level of dependence among the correlated variables. The model performance shows that GA-ANFIS can be applied in the computation of the elemental composition of biomass for strategic decision-making.

Keywords GA-ANFIS · Biomass feedstock · Efficient utilization · Elemental composition

1 Introduction

As of 2017 the United Nation, UN projected population growth from 7.6 billion to 9.8 billion by 2050 [1]. The clean and sustainable energy is the viable solution

O. O. Olatunji (✉) · N. Madushele · P. A. Adedeji
Department of Mechanical Engineering Science, University of Johannesburg, Johannesburg, South Africa
e-mail: tunjifemi@gmail.com

S. Akinlabi
Department of Mechanical and Industrial Engineering, University of Johannesburg, Johannesburg, South Africa

S. Vijayan et al. (eds.), *Trends in Manufacturing and Engineering Management*, Lecture Notes in Mechanical Engineering,
https://doi.org/10.1007/978-981-15-4745-4_95

to achieve safer and energy-efficient world since the rising population is leading to increasing concentration of CO_2, which was reported at 1.4% as at 2017 as the highest ever recorded. Considering the yardsticks such as availability, sustainability, energy efficiency, applicability, environmental impact, and flexibility, biomass is promising option to address over-dependence on fossil fuel and provide attractive feedstocks for energy generation [2]. Biomass elemental composition is an indispensable parameter required for the design and operation of biomass-to-fuel conversion technologies [3]. The carbon, hydrogen, oxygen content, and other constituents in negligible quantity have an impact on the gross heating value, the emission, and corrosion of conversion equipment. The knowledge of elemental composition helps in: calculating the material balance in bioprocess, development of biomass handling equipment such as boilers, gasifiers, combustors, estimating the volume and constituents of combustion gases, and fixing the amount of oxidant required for biomass gasification [4]. However, experimental determination of elemental composition using the essential apparatus is becoming very expensive and time-consuming compared to proximate analysis which can be conducted much easily. Therefore, the rapid and economical approach would be to develop a model which can be used to estimate the elemental constituent from the proximate value since it has been established that there is a relationship between these two experimental methods. Against all these odds, artificial intelligence, AI, has emerged with a potential to assist in fulfilling the objectives of biomass-to-energy exploration which centres around the estimation of the properties of biomass, such as the elemental composition and enthalpy of combustion [3, 5–10]. AI has the capability for modelling of complex systems with high degree of accuracy. AI can be grouped into artificial neural network (ANN), adaptive neuro-fuzzy inference systems (ANFIS), evolutionary algorithms (EA), and support vector machines (SVMs) [11–16]. Several optimization methods have been proposed by researchers to improve the accuracy of ANFIS [17–19], one of these methods is the genetic algorithm, GA. GA is an optimization technique which can be applied to solve problems built on natural selection process among the population such that a population of candidate solutions is advanced towards improved solutions [20–22]. It was developed in the 1970s by Holland with the basic idea of abstract modelling of natural evolution [23]. The GA has been extensively used in different fields [24–26] and most closely in the prediction of the heating value of biomass [13], but there is no any know GA application for the prediction of the elemental composition of biomass which covers different sources. There are quite a few published models which predict the elemental composition of solid biomass based on proximate analysis [3, 5, 8–10, 27–30]. Nhuchhen [30] developed a correlation which estimated the carbon, oxygen, and hydrogen constituent of untreated and torrefied biomass on the basis of proximate analysis. Similarly, Mahmudul et al. [9] developed an empirical correlation towards the prediction of the elemental composition of torrefied woody biomass. Also, the correlation developed by Vakkilainen [29] is only limited to black liquor. Parikh et al. [10] and Shen et al. [5] considered a data set for proximate values which are volatile matter, fixed carbon, ash for an equivalent elemental composition of carbon, hydrogen, and oxygen. The range of the data considered is as shown in Table 1. It can be observed that the present study has a wider datapoint coverage

Table 1 Datapoint coverage for the proximate and ultimate properties of biomass

Reference	Constituents	Datapoint range %
[10]	C	36.2–53.1
	H	4.4–8.3
	O	31.4–49.5
	VM	57.2–90.6
	FC	4.7–38.4
	ASH	0.12–77.7
[5]	C	36.2–53.1
	H	4.7–6.6
	O	31.4–48.0
	VM	57.2–90.6
	FC	9.2–32.8
	ASH	0.1–24.6
Present study	C	22.35–92.86
	H	0.52–11.42
	O	2.32–52.06
	VM	5.04–89.60
	FC	7.40–94.73
	ASH	0.10–56.10

compared to previous studies in the same domain. The performance of these correlations was evaluated as shown in Table 2. The contribution of ash content in the prediction of elemental composition was considered by Shen [5] which made the correlation to be more robust with better prediction than Parikh et al. [10].

All the mentioned correlations are linear and have not satisfactorily predicted the properties of the biomass since the elemental constituent of biomass and other dependent variables do not constantly follow a linear pattern, particularly when tested with various samples. Ghugare et al. [3] has proposed a hybrid genetic programming algorithm, support vector regression (SVR), and artificial intelligence in order to

Table 2 Correlations for the estimation of elemental composition

Reference	Equation*	R^2	RMSE	MAPE
[10]	$C = 0455VM + 0.637FC$	0.810	6.763	4.990
	$H = 0.062VM + 0.052FC$	0.500	1.082	19.734
	$O = 0.476VM + 0.304FC$	0.890	1.530	26.374
[5]	$C = 0.46VM - 0.095ASH + 0.635FC$	0.745	7.025	5.149
	$H = 0.06VM + 0.01ASH + 0.059FC$	0.353	1.169	13.11
	$O = 0.469VM - 0.023ASH + 0.34FC$	0.828	6.839	28.304

predict the elemental composition biomass. The result of their investigation showed that the model developed was more accurate than the existing linear and polynomial models.

Since there is no known GA-ANFIS model which has been used to predict the elemental composition of solid biomass with such a large data coverage, this article proposed a prediction model based on a hybrid adaptive neuro-fuzzy inference system (ANFIS) optimized with GA. The model inputs were the proximate properties of biomass which are ash content (ASH), fixed carbon (FC), and volatile matter (VM). These were used to predict the hydrogen (H), oxygen (O), and carbon (C) content of biomass fuels.

2 Materials and Methods

2.1 Collection of Data

In predicting the elemental constituents of biomass-based on the proximate analysis results, the component of proximate which are VM, FC, and ash in weight percentages (dry basis) were used as input variables to estimate the C, H, and O as the output variables. The ranges of the proximate value are shown as the present study in Table 1. A cumulative of 830 datapoints were extracted from the database compiled by Ghugare et al. [3]. After randomization, the data were divided into training and testing at ratio 7:3, respectively, for the development of the model.

2.2 Theory of ANFIS

ANFIS combines the dual benefits of fuzzy logic and the neural network in a single framework, which enables it to solve complex problems in a way that optimal parameters of membership function are obtained from input to output mapping [15, 31]. It is a variant of the Takagi–Sugeno fuzzy inference system [32, 33]. It consists of input–output parameters and fuzzy rules of Takagi–Sugeno class [15]. It is a class of adaptive multilayer feed-forward network which assumes a framework of two inputs H_1, H_2 and output G such that the entries are evaluated by linguistic variable (A_1, B_1). The first-order Takagi–Sugeno fuzzy model can be expressed as:
Principle 1: If H_1 is A_1 AND H_2 is B_1 then $G_1 = b_1J_1 + c_1J_2 + z_1$.
Principle 2: If H_1 is A_2 AND H_2 is B_2 then $G_2 = b_2J_1 + c_2J_2 + z_2$
where $b_1, c_1, z_1, b_2, c_2, z_2$ are output function parameters. In this article, the two rules above were adapted with input $J_i(i = 1\ldots3)$, output $G_i(i = 1\ldots3)$ and cluster $C_j(j = 1\ldots10)$ with the same weights using a fuzzy c-means clustering method.

Typically, the ANFIS network has five layers [16] in the following sequence: fuzzification layer, multiplication layer, rationalization layer, defuzzification layer, and summation layer. It should be noted that the first and fourth layer include the adaptive parameters capable of modification and update based on a learning technique used. The estimated output ultimately depends on the five layers in the ANFIS after the completion of the training phase. From previous studies, the model performs optimally when the fuzzy rules based on the Gaussian membership function are applied. This study also applied the Gaussian membership function on the input variables. The readers can refer to [15, 16, 22, 25, 32, 34, 35] for further information on ANFIS models. In this study, elemental constituents comprising of three inputs which are VM, FC, and ash were mapped to the output which is the principal elemental constituents (C, H, O) of biomass.

2.3 *Model Optimization Based on Genetic Algorithm (GA)*

GA is an evolutionary heuristic optimization algorithm which is based on population [20, 31, 36–38]. The process which leads to the generation of new solution in GA has a semblance of the natural selection process of living organisms. Usually, the recurrent application of three evolutionist operators occurs in three phases which are: selection, crossover, and mutation which all assist in the attainment of new solutions. Mutation can be defined as a mechanism through which a population is diversified in order to avoid stagnation [39, 40]. In GA, a population of candidate solutions is advanced to improve the solutions. Each potential solution has a set of characteristics termed genotype, which can be mutated. The objective function governs the selection of individuals for a new generation. During the selection process of the elites for the next generation, a fitness value is obtained for each member of the population. This value determines which member proceeds to the next generation. Consequentially, the optimized generation of the potential solutions is used in the following iteration [41, 42]. Apart from population density and the optimal number of iterations, the selection, crossover, and mutation techniques must be considered in GA decision-making process. Various selection methods may adopt different probability of assignment in order to ensure population diversity and obtain an optimal solution. The two most popular selection criteria are: tournament and roulette wheel. This study adopts the roulette wheel for the selection criteria during parameter optimization in the hybrid model. Crossover method should be selected carefully such that a simple method which may lead to unusable chromosome for the complex optimization problem is avoided. The concept of mutation has been comprehensively discussed by Gen et al. [39] Holland [23], Pal et al. [40]. The selection of mutation technique and probability will ensure that the diversity of the population is maintained.

The antecedent and consequence of a typical ANFIS model are optimized by the training function. This optimization is performed on the adaptive layers which are the first and the fourth layers [36]. In this investigation, the GA algorithm was written in MATLAB (R2015a) and used to train the ANFIS model. An iterative training process aimed at achieving optimal parameters for the antecedent was performed.

The Gaussian membership function (Eq. 1) was used due to its less disruptive effect and robustness when applied to the system [43, 44].

$$\emptyset(z) = e^{-\frac{1}{2}\left(\frac{z-\mu}{\sigma}\right)^2} \tag{1}$$

where $\emptyset$ = Gaussian function, σ and μ = the membership function parameters.

2.4 Implementation of ANFIS-GA Model

Shown in Fig. 1 is the flow diagram of GA-ANFIS model developed for the prediction of C, H, O components of biomass-based on VM, FC, and ash contents as the input

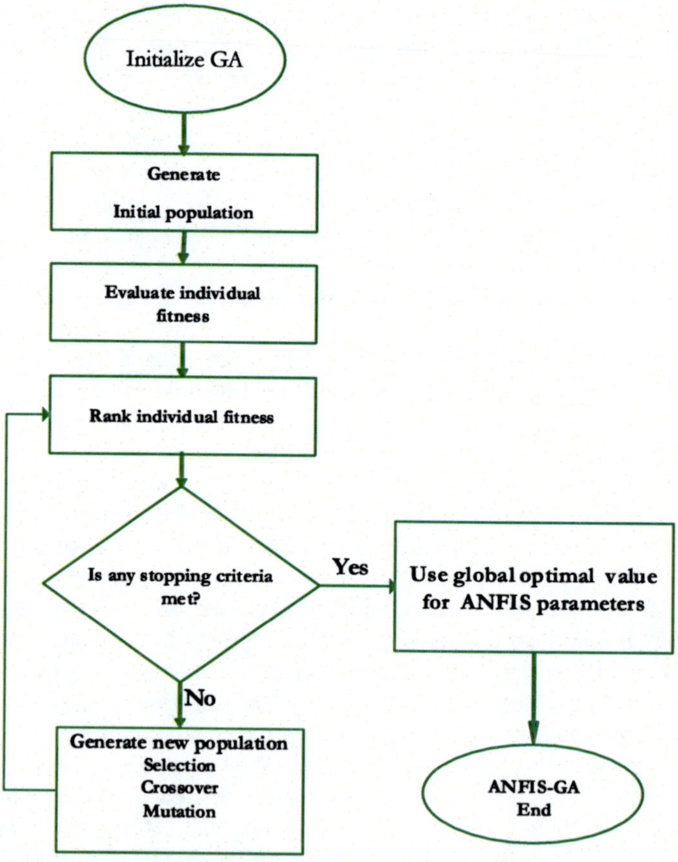

Fig. 1 Flow diagram for ANFIS-GA implementation

Table 3 Learning and network optimization parameters

User-defined parameters	Value
Number of rules	20
FIS type	Takagi–Sugeno
Initial population size	20
Population type	Double vector
Selection function	Roulette wheel
Mutation rate	0.15
Crossover function	Heuristic
Mutation function	Gaussian
Crossover percentage	0.4
Number of iterations	600
Stopping criteria	Minimum training error

variables. Genetic operators are applied to facilitate breeding in GA. Using GA, the ANFIS model was trained until the optimal solution was obtained.

The initial conditions used for the computation is presented in Table 3. The convergence values, which minimize the training error objective function, were made a stopping criterion. The computation was done on a desktop computer workstation with configuration 64 bits, 32 GB RAM Intel (R) Core (TM) i7 5960X. In general terms, the principle guiding the GA-ANFIS is as highlighted in the steps below:

Step 1: Generate the first-string population
Step 2: Assign and determine the fitness of each member of the population.
Step 3: Choose a membership pair from the population string for breeding
Step 4: Rank the fitness of the members of the population
Step 5: Allocate the pair into passing population membership
Step 6: Repeat step 5, until temporary population membership is filled up
Step 7: Replace the existing population with members of passing population and some members of existing population to complete the membership
Step 8: Use the optimal population value for the ANFIS parameters then end GA-ANFIS
Step 9: If the termination criteria are not met, return to step 2.

2.5 Model Performance Evaluation

Statistical estimation of errors was carried out to evaluate the performance of GA-ANFIS model which was designed in this study as follows: computation time (CT) was also estimated as a measure of the overall prediction time as to evaluate the time cost of the forecasting and to further underline the economic significance of this model.

Correlation Coefficient (CC):

$$CC = 1 - \frac{\sum_{i=1}^{n}\left(y_k - \widehat{y_k}\right)^2}{\sum_{i=1}^{n}\left(y_k - \overline{y_k}\right)^2} \quad (2)$$

Root Mean square Error (RMSE):

$$RMSE = \sqrt{\frac{\sum_{k=1}^{N}\left[y_k - \widehat{y_k}\right]^2}{N}} \quad (3)$$

Mean Absolute Deviation (MAD):

$$MAD = \frac{1}{N}\sum_{k=1}^{N}|y_k - \bar{y}_k| \quad (4)$$

Mean Absolute Percentage Error (MAPE)

$$MAPE = \left[\frac{1}{N}\sum_{k=1}^{N}\left|\frac{y_k - \bar{y}_k}{y_k}\right|\right] \times 100 \quad (5)$$

3 Results

3.1 Model Validation

To satisfactorily predict the elemental component of biomass, the model was trained, and the prediction capacity of the models was evaluated at the testing phase. Each elemental composition was considered independently during the training and testing phase. Each predicted elemental constituents and corresponding experimental values were plotted against the data index. Figures 2, 3 and 4 were plotted for carbon, hydrogen, and oxygen, respectively, at both testing and training phase to show the prediction progression from training to testing. The results show a satisfactory concurrence between the actual and predicted elemental constituents with a slight variation at training and testing phases. The concurrency between the actual and predicted carbon and oxygen revealed a closer pattern compared to hydrogen. In case of hydrogen, the lower accuracy could be as a result of experimental errors and narrower data range when compared to other experimental constituents [3].

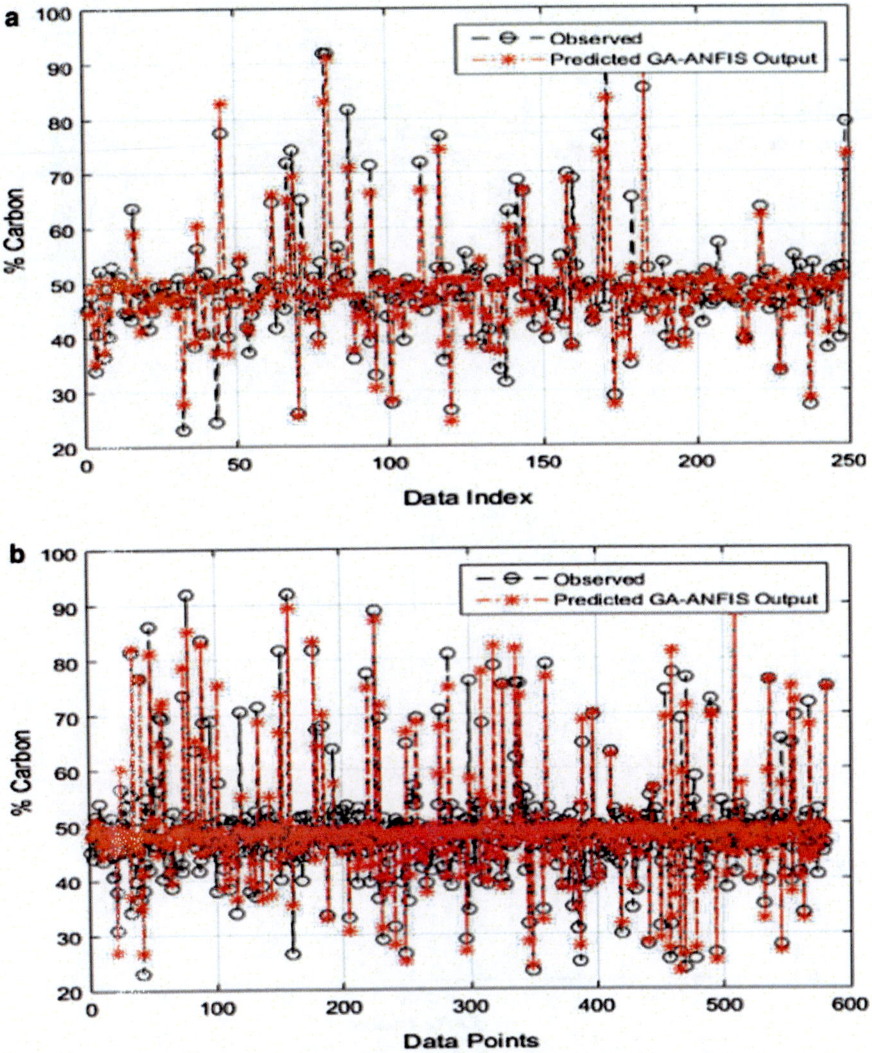

Fig. 2 a Testing phase for carbon **b** Training phase for carbon

3.2 *Model Evaluation*

For each model, the error was statistically calculated for the predicted and actual elemental constituents at training and testing stages as presented in Table 4. The RMSE, MAD, and STD are consistently lower for hydrogen at the training and testing phase when compared to carbon and oxygen. However, despite the lowest RMSE, MAD, and STD reported for hydrogen, the magnitude of MAPE and CC

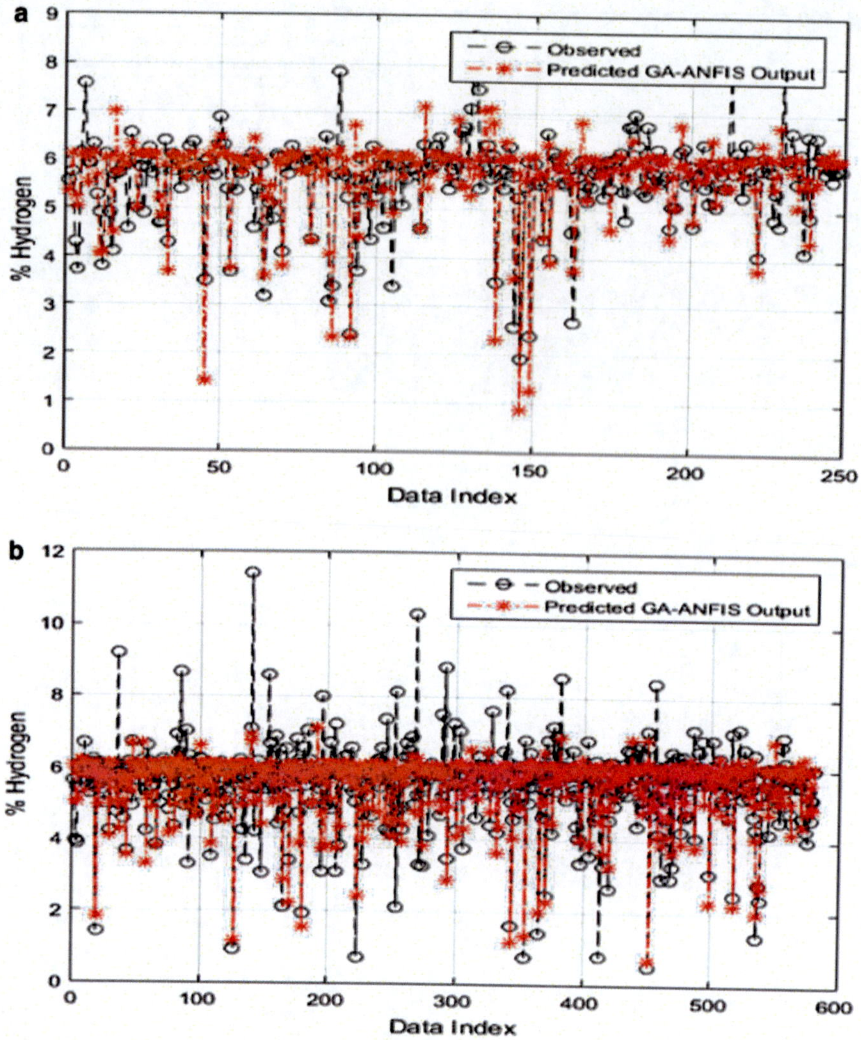

Fig. 3 **a** Testing phase for hydrogen **b** Training phase for hydrogen

shows that the model does not predict as much data correctly as the models for carbon, though it does better than the oxygen. The computation time of the carbon and oxygen were approximately the same, but significantly different from hydrogen. This further validates the initial observation that the hydrogen model is not as accurate and economical as the others. The MAPE revealed that around 80% of the datapoint were correctly predicted for hydrogen. Of all the three constituents, carbon was the most accurately predicted with 95% of the data set followed by hydrogen (>80%) and oxygen (>70%). This is at variance with various models (GP-MLP, GP-SVR,

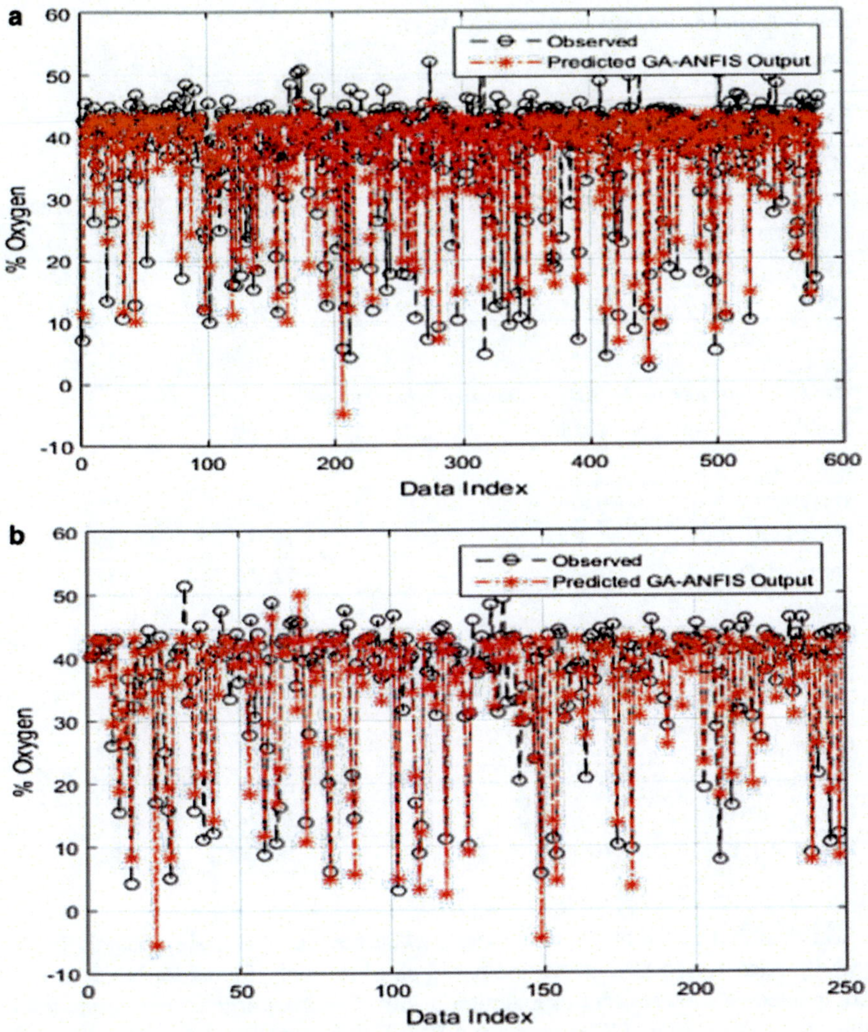

Fig. 4 **a** Training phase for Oxygen **b** Testing phase for Oxygen

MLP-SVR) developed by Ghugare [3] which proposed that the MAPE of carbon and oxygen constituents is generally and constantly better than hydrogen for all testing cases. The CT results revealed that the model can be tested in less than 1 min, though closely related to the computing power of the computer used. The result is important in the conversion of biomass to energy since the properties of biomass feedstock determine the energy quality.

The comparative analysis of the GA-ANFIS model with the models developed by other authors is indicated in Table 5. The RMSE of hydrogen is significantly

Table 4 Predicted Elemental components (C, H, O)

Evaluation parameters	Hydrogen		Carbon		Oxygen	
	Training	Testing	Training	Testing	Training	Testing
RMSE	0.696	0.6293	3.673	3.3923	4.2235	4.4538
MAD	0.4309	0.4168	2.4609	2.383	2.85	3.1042
MAPE	9.4125	8.3011	5.1757	4.9733	11.5911	13.3983
Error STD	0.6966	0.6265	3.6725	3.385	4.1843	4.3512
CT (s)		40.205		33.65		33.57
CC	0.7861	0.7559	0.9159	0.9464	0.9024	0.9167

Table 5 Performance of GA-ANFIS compared to other models

Models	RMSE	MAD	MAPE	CC
Hydrogen				
This study	0.629	0.417	8.301	0.756
Parikh et al. [10]	1.082	–	19.734	0.500
Shen et al. [5]	1.169	–	22.276	0.353
Carbon				
This study	3.392	2.383	4.973	0.946
Parikh et al. [10]	6.763	–	4.990	0.810
Shen et al. [5]	7.025	–	5.149	0.745
Oxygen				
This study	4.454	3.104	13.398	0.917
Parikh et al. [10]	6.301	–	26.374	0.890
Shen et al. [5]	6.839	–	28.304	0.829

lesser than that of the other two models, while the CC is also significantly higher [5, 10]. Also, the MAPE calculated based on Shen et al. [5] and Parikh et al. [10] are pointed of a lower order of magnitude with about half of the value estimated in this study. For carbon, CC is lower and RMSE is also significantly lower by more than 50% for the model developed in this study. However, the MAPE was close as against a significant disparity in case of hydrogen constituent prediction model. Also, for oxygen model developed in this study, RMSE value shows a significant order of difference and the CC, though not as much as RMSE. But MAPE showed that the other linear correlation could only accurately predict around 50% of the data. This further underlines the significance of the model developed in this study. The overall observation revealed that the GA-ANFIS model developed in this study provides more satisfactory performance for all the elemental components which were predicted in compared to the existing linear models.

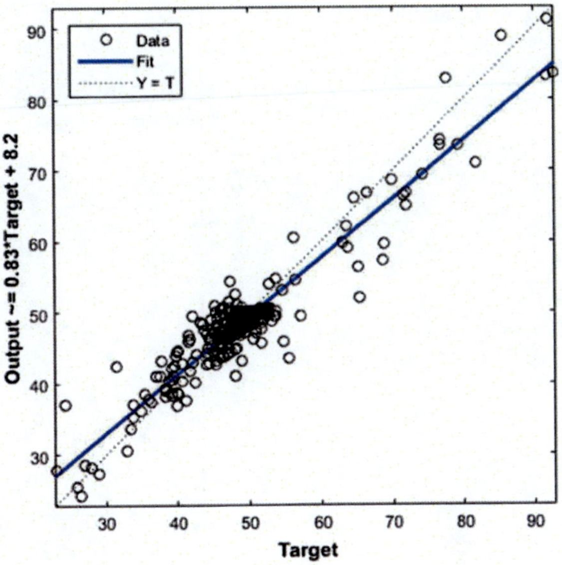

Fig. 5 Regression for C constituent

3.3 Post-regression Analysis

The regression analysis was based on the test results for all the predicted elements. The objective is to determine the model of greatest fitness based on minimum error criterion and the data dispersion along the line of the best fit (Figs. 5, 6 and 7). It can be inferred from the plots that carbon in Fig. 5 shows the lowest dispersion compared to other constituents with regression equation given as: $F(C)_{predicted} = 0.83F(C)_{exp} + 8.2$ in %weight.

4 Conclusion

The elemental constituents of biomass were predicted based on the proximate values using ANFIS model which was optimized by GA. The accuracy of the model developed in the study was more satisfactory compared to the other linear models since it resolved nonlinear dependencies inherent in biomass properties. Regression analysis was also carried out to determine the level of dependence among the correlated variables and the data scattering along the line of best fit. The report shows that GA-ANFIS can be used in the computation of the elemental composition of biomass-based on proximate properties. This result is significant for strategic decision-making regarding biomass-to-energy conversion operation.

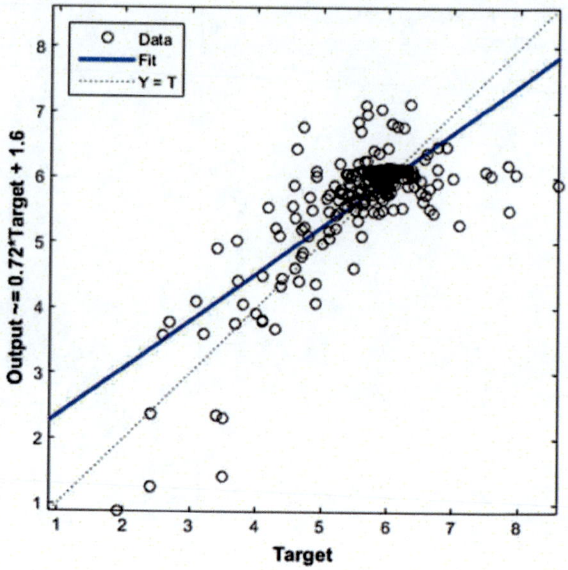

Fig. 6 Regression for H constituent

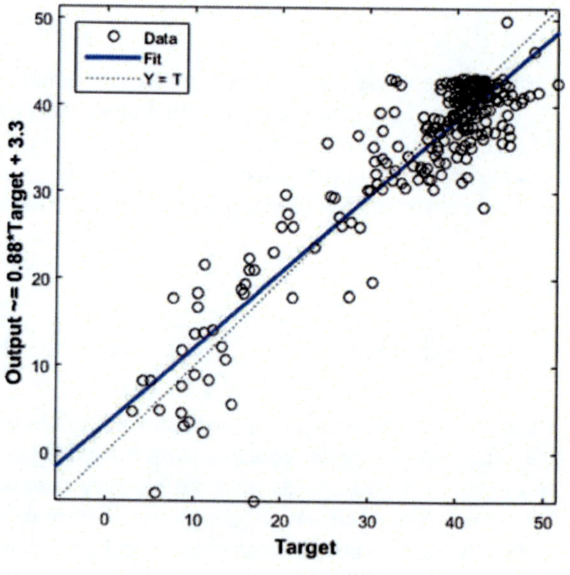

Fig. 7 Regression plot for oxygen constituent

References

1. WorldNuclearAssociation (2018) World energy needs and nuclear power. http://www.world-nuclear.org/information-library/current-and-future-generation/world-energy-needs-and-nuclear-power.aspx. Accessed on 8 Dec 2018

2. Callejón-Ferre A, Velázquez-Martí B, López-Martínez J, Manzano-Agugliaro F (2011) Greenhouse crop residues: energy potential and models for the prediction of their higher heating value. Renew Sustain Energy Rev 15(2), 948–955
3. Ghugare SB, Tiwary S, Tambe SS (2017) Computational intelligence based models for prediction of elemental composition of solid biomass fuels from proximate analysis. Int J Syst Assur Eng Manage 8(4):2083–2096
4. Tumuluru JS, Wright CT, Boardman RD, Yancey NA, Sokhansanj S (2011) A review on biomass classification and composition, co-firing issues and pretreatment methods. In: 2011 Louisville, Kentucky, American Society of Agricultural and Biological Engineers, p 1, 7–10 Aug 2011
5. Shen J, Zhu S, Liu X, Zhang H, Tan J (2010) The prediction of elemental composition of biomass based on proximate analysis. Energy Convers Manage 51(5):983–987
6. Uzun H, Yildiz Z, Goldfarb JL, and S. Ceylan, "Improved prediction of higher heating value of biomass using an artificial neural network model based on proximate analysis. Bioresour Technol vol 234, pp 122–130, Jun 2017
7. ObafemiO, Stephen A, Ajayi O, Mashinini P, Nkosinathi M (2018) Experimental investigation of thermal properties of Lignocellulosic biomass: a review. IOP Conference Series: Materials Science and Engineering, vol 413, no. 1, p 012054
8. Posom J, Sirisomboon P (2017) Evaluation of lower heating value and elemental composition of bamboo using near infrared spectroscopy. Energy 121, 147–158
9. Hasan M, Haseli Y, Karadogan E (2018) Correlations to predict elemental compositions and heating value of torrefied biomass. Energies 11(9):2443
10. Parikh J, Channiwala SA, Ghosal GK (2007) A correlation for calculating elemental composition from proximate analysis of biomass materials. Fuel 86(12–13):1710–1719
11. Vikhar PA (2016) Evolutionary algorithms: a critical review and its future prospects. In: 2016 International conference on Global Trends in Signal Processing, Information Computing and Communication (ICGTSPICC), IEEE, pp 261–265
12. Gong S, Sasanipour J, Shayesteh MR, Eslami M, Baghban A (2017) Radial basis function artificial neural network model to estimate higher heating value of solid wastes. Energy Sources Part A Recovery Utilization Environ Effects 39(16):1778–1784
13. Baghban A, Ebadi T (2019) GA-ANFIS modeling of higher heating value of wastes: application to fuel upgrading. Energy Sourc Part A Recovery Utilization Environ Effects 41(1):7–13
14. Baghban A, Kashiwao T, Bahadori M, Ahmad Z, Bahadori A (2016) Estimation of natural gases water content using adaptive neuro-fuzzy inference system. Petroleum Sci Technol 34(10):891–897
15. Jang JS (1993) ANFIS: adaptive-network-based fuzzy inference system. IEEE Trans Syst Man Cybernetics 23(3):665–685
16. Akkaya E (2016) ANFIS based prediction model for biomass heating value using proximate analysis components. Fuel 180:687–693
17. Bäck T, Fogel DB, Michalewicz Z (2018) Evolutionary computation 1: basic algorithms and operators. CRC press
18. Wang GG, Deb S, Gao XZ, Coelho LDS (2016) A new metaheuristic optimisation algorithm motivated by elephant herding behaviour. Int J Bio-Inspired Comput 8(6):394–409
19. Kachitvichyanukul V (2012) Comparison of three evolutionary algorithms: GA, PSO, and DE. Indus Eng Manage Syst 11(3):215–223
20. Malhotra R, Singh N, Singh Y (2011) Genetic algorithms: concepts, design for optimization of process controllers. Comput Inf Sci 4(2):39
21. Sarkheyli A, Zain AM, Sharif S (2015) Robust optimization of ANFIS based on a new modified GA. Neurocomputing 166:357–366
22. Karaboga D, Kaya E (2018) Adaptive network based fuzzy inference system (ANFIS) training approaches: a comprehensive survey. Artificial Intelligence Rev 1–31
23. Holland JH (1992) Adaptation in natural and artificial systems: an introductory analysis with applications to biology, control, and Artificial Intelligence. MIT press

24. Rouhibakhsh K, Darvish H, Sabzgholami H, Goodarzi MS (2018) Application of ANFIS-GA as a novel and accurate tool for estimation of interfacial tension of carbon dioxide and hydrocarbon. Petrol Sci Technol 36(15):1143–1149
25. Esfandyari M, Esfandyari M, Jafari D (2018) Prediction of thiophene removal from diesel using [BMIM][AlCl4] in EDS Process: GA-ANFIS and PSO-ANFIS modeling. Petroleum Sci Technol 1–7
26. Armaghani DJ, Mohamad ET, Hajihassani M, Yagiz S, Motaghedi H (2016) Application of several non-linear prediction tools for estimating uniaxial compressive strength of granitic rocks and comparison of their performances. Eng Comput 32(2):189–206
27. Fried A, Padouvas E, Rotter H, Varmuza K (2005) Prediction of heating values of biomass fuel from elemental composition. Analytica Chimica Acta 544(1–2):191–198
28. Yi L, Feng J, Qin YH, Li WY (2017) Prediction of elemental composition of coal using proximate analysis. Fuel 193:315–321
29. Vakkilainen E (2000) Estimation of elemental composition from proximate analysis of black liquor. Paperi ja puu 82(7):450–455
30. Nhuchhen DR (2016) Prediction of carbon, hydrogen, and oxygen compositions of raw and torrefied biomass using proximate analysis. Fuel 180:348–356
31. Yaghoobi A, Bakhshi-Jooybari M, Gorji A, Baseri H (2016) Application of adaptive neuro fuzzy inference system and genetic algorithm for pressure path optimization in sheet hydroforming process. Int J Adv Manuf Technol 86(9–12):2667–2677
32. Takagi T, Sugeno M (1985) Fuzzy identification of systems and its applications to modeling and control. IEEE Trans Syst Man Cybern 1:116–132
33. Sugeno M, Kang G (1988) Structure identification of fuzzy model. Fuzzy Sets Syst 28(1):15–33
34. Rezakazemi M, Dashti A, Asghari M, Shirazian S (2017) H2-selective mixed matrix membranes modeling using ANFIS, PSO-ANFIS, GA-ANFIS. Int J Hydrogen Energy 42(22):15211–15225
35. Suparta W, Alhasa KM (2016) Modeling of tropospheric delays using ANFIS. Springer
36. Haznedar B, Kalinli A (2016) Training ANFIS using genetic algorithm for dynamic systems identification. Int J Intell Syst Appl Eng 4(1):44–47
37. Qdais HA, Hani KB, Shatnawi N (2010) Modeling and optimization of biogas production from a waste digester using artificial neural network and genetic algorithm. Res Conserv Recycl 54(6):359–363
38. Jacob S, Banerjee R (2016) Modeling and optimization of anaerobic codigestion of potato waste and aquatic weed by response surface methodology and artificial neural network coupled genetic algorithm. Bioresour Technol 214:386–95, Aug 2016
39. Gen M, Cheng R, Lin L (2008) Network models and optimization: multiobjective genetic algorithm approach. Springer Science & Business Media
40. Pal SK, Wang PP (2017) Genetic algorithms for pattern recognition. CRC press
41. Mayilvaganan MK, Naidu K (2011) Comparison of membership functions in adaptive-network-based fuzzy inference system (ANFIS) for the prediction of groundwater level of a watershed. J Comput Appl Res Dev 1(1):35–42
42. Yagiz S, Ghasemi E, Adoko AC (2018) Prediction of rock brittleness using genetic algorithm and particle swarm optimization techniques. Geotech Geol Eng 36(6):3767–3777
43. Antonio LM, Coello CAC (2018) Coevolutionary multiobjective evolutionary algorithms: survey of the state-of-the-art. IEEE Trans Evol Comput 22(6):851–865
44. Corne D, Lones MA (2018) Evolutionary algorithms. In: Handbook of Heuristics, pp 1–22